D1473784

REVIEW OF PHYSIOLOGICAL CHEMISTRY

13
Edition

review of PHYSIOLOGICAL CHEMISTRY

HAROLD A. HARPER, PhD

Professor of Biochemistry
University of California School of Medicine
San Francisco
Biochemist Consultant to the Clinical Investigation Center, U.S. Naval Hospital, Oakland
Biochemist Consultant to St. Mary's Hospital
San Francisco

Lange Medical Publications

LOS ALTOS, CALIFORNIA

1971

Italian Edition: *Piccin Editore, Padua, Italy*
Spanish Edition: *El Manual Moderno, S.A., Mexico, D.F.*
Japanese Edition: *Maruzen Co., Ltd., Tokyo, Japan*
Portuguese Edition: *Atheneu Editora Sao Paulo, S.A., Sao Paulo, Brazil*
French Edition: *Les Presses de l'Université Laval, Quebec, Canada*

International Standard Book Number: *0–87041–031–8*
Library of Congress Catalogue Card Number: *61–10434*

A Concise Medical Library for Practitioner and Student

Current Diagnosis & Treatment 1971 M.A. Krupp, M.J. Chatton, S. Margen, Editors	$11.00
Current Pediatric Diagnosis & Treatment, 1970 C.H. Kempe, H.K. Silver, D. O'Brien, Editors	$11.00
Review of Physiological Chemistry, 13th Edition, 1971 H.A. Harper	$8.00
Review of Medical Physiology, 5th Edition, 1971 W.F. Ganong	$8.50
Review of Medical Microbiology, 9th Edition, 1970 E. Jawetz, J.L. Melnick, E.A. Adelberg	$7.50
Review of Medical Pharmacology, 2nd Edition, 1970 F.H. Meyers, E. Jawetz, A. Goldfien	$8.50
General Urology, 6th Edition, 1969 D.R. Smith	$8.00
General Ophthalmology, 6th Edition, 1971 D. Vaughan, T. Asbury, R. Cook	$8.00
Correlative Neuroanatomy & Functional Neurology, 14th Edition, 1970 J.G. Chusid	$7.50
Principles of Clinical Electrocardiography, 7th Edition, 1970 M.J. Goldman	$7.00
Handbook of Psychiatry, 2nd Edition, 1971 P. Solomon, V.D. Patch, Editors	$7.50
Handbook of Surgery, 4th Edition, 1969 J.L. Wilson, Editor	$6.00
Handbook of Obstetrics & Gynecology, 4th Edition, 1971 R.C. Benson	$6.50
Physician's Handbook, 16th Edition, 1970 M.A. Krupp, N.J. Sweet, E. Jawetz, E.G. Biglieri	$6.00
Handbook of Medical Treatment, 12th Edition, 1970 M.J. Chatton, S. Margen, H. Brainerd, Editors	$6.50
Handbook of Pediatrics, 9th Edition, 1971 H.K. Silver, C.H. Kempe, H.B. Bruyn	$6.50
Handbook of Poisoning: Diagnosis & Treatment, 7th Edition, 1971 R.H. Dreisbach	$6.00

Lithographed in USA

Preface

This *Review* is intended to serve as a compendium of those aspects of chemistry that are fundamental to the study of biology and medicine. A prominent feature of its success since the First Edition appeared in 1939 has been the opportunity afforded by frequent revisions to include recent contributions to the field. However, the accumulation of knowledge continues to pose the difficult problem of giving adequate attention to recent discoveries while still serving the original purpose of the *Review* as a more concise source of information than the standard textbooks. It is hoped that the current edition represents a reasonable solution to this problem and that those readers will continue to be served who seek to gain an introduction to the subject or who may desire a review such as is required in preparation for examinations of state and specialty boards.

Those who have been familiar with the *Review* in its previous editions will note that it has been completely reset in a new and more attractive format. It is for this reason that we have been able to add to the text in accordance with the necessity to consider important new findings in physiological chemistry; yet the 13th Edition contains slightly fewer pages than did the previous edition.

In the 13th Edition, the *Review* continues to reflect the substantial contributions of Drs. Gerold M. Grodsky, Peter A. Mayes, and Victor Rodwell as acknowledged in each of the chapters with which they are identified. Dr. Tawfik ElAttar collaborated in the writing of the section on steroid hormones in Chapter 20. The illustrations, including structural formulas and metabolic schemes, reflect the superb skill of Laurel V. Schaubert. Her contribution to the *Review* is one of its most precious assets.

I am grateful for the opportunity to serve students in many countries outside the United States, not only with the original text in English but also through the publication of the *Review* in French, Spanish, Italian, Portuguese, and Japanese. A translation into Polish is in preparation.

Harold A. Harper

San Francisco, California
August, 1971

Preface

Table of Contents

Notice

In keeping with the decision of several scientific societies to employ a uniform system of metric nomenclature, this edition of *Review of Physiological Chemistry* has been converted to such a uniform system. The following prefixes, some of which may be unfamiliar to the reader, have been used:

m	milli-	10^{-3}
μ	micro-	10^{-6}
n	nano-	10^{-9}
p	pico-	10^{-12}
f	femto-	10^{-15}
a	atto-	10^{-18}

Examples:

Former Usage		Present Usage
mμg (millimicrograms)	=	ng (nanograms)
mμ (millimicrons)	=	nm (nanometers)
μμg (micromicrograms)	=	pg (picograms)
cu mm (cubic millimeters)	=	μl (microliters)
cu μ (cubic microns)	=	fl (femtoliters)

Note:

The Ångstrom (Å) unit (10^{-10} m) is expressed as a ×10 multiple of the nanometer. Thus, 10 Å = 1 nm, etc.

Molecular weights are expressed as daltons (1 dalton = a unit of mass equal to 1/12 the mass of the ^{12}C atom, or approximately 1.65×10^{-24} g).

1...

Carbohydrates

The carbohydrates, often termed starches or sugars, are widely distributed both in animal and in plant tissues. In plants they are produced by photosynthesis and include the starches of the plant framework as well as of the plant cells. In animal cells carbohydrate serves as an important source of energy for vital activities. Some carbohydrates have highly specific functions in vital processes (eg, ribose in the nucleoprotein of the cells, galactose in certain fats, and the lactose of milk).

Carbohydrates are defined chemically as aldehyde or ketone derivatives of the higher polyhydric (more than one OH group) alcohols or as compounds which yield these derivatives on hydrolysis.

Classification

Carbohydrates are divided into 4 major groups as follows:

(1) Monosaccharides (often called "simple sugars") are those which cannot be hydrolyzed into a simpler form. The general formula is $C_nH_{2n}O_n$. The simple sugars may be subdivided as trioses, tetroses, pentoses, hexoses, or heptoses, depending upon the number of carbon atoms they possess; and as aldoses or ketoses, depending upon whether the aldehyde or ketone groups are present. Examples are:

		Aldo Sugars	Keto Sugars
Trioses	$(C_3H_6O_3)$	Glycerose	Dihydroxyacetone
Tetroses	$(C_4H_8O_4)$	Erythrose	Erythrulose
Pentoses	$(C_5H_{10}O_5)$	Ribose	Ribulose
Hexoses	$(C_6H_{12}O_6)$	Glucose	Fructose

(2) Disaccharides are carbohydrates which yield 2 molecules of the same or of different monosaccharides when hydrolyzed. The general formula is $C_n(H_2O)_{n-1}$. Examples are sucrose, lactose and maltose.

(3) Oligosaccharides are those which yield 2–10 monosaccharide units on hydrolysis.

(4) Polysaccharides yield more than 10 molecules of monosaccharides on hydrolysis. The general formula is $(C_6H_{10}O_5)_X$. Examples of polysaccharides are the starches and dextrins. These are sometimes designated as hexosans, pentosans, or mixed polysaccharides, depending upon the nature of the monosaccharides which they yield on hydrolysis.

Asymmetry

In the formulas for glucose shown at right it will be noted that a different group is attached to each of the 4 bonds of carbon atoms 2–5. For example, the 4 groups attached to carbon 2 are

O
//
C–H, H, OH,

and, in D-glucose,

|
HO--C–H.
|
R

A carbon atom to which 4 different atoms or groups of atoms are attached is said to be **asymmetric.**

Isomerism

The presence of asymmetric carbon atoms in a compound makes possible the formation of isomers of that compound. Such compounds, which are identical in composition and differ only in spatial configuration, are called **stereoisomers.** Two such isomers of glucose, one of which is the mirror image of the other, are shown below.

The number of possible isomers of any given compound depends upon the number of asymmetric carbon atoms in the molecule. According to the **rule of n** (where "n" represents the number of asymmetric carbon atoms in a compound), 2^n equals the number of possible isomers of that compound. Glucose, with 4 asymmetric carbon atoms, would thus be expected to have 2^4 or 16 isomers. Eight would belong to the D series, and the mirror images of each of these would comprise the L series.

Two sugars which differ from one another only in the configuration around a single carbon atom are

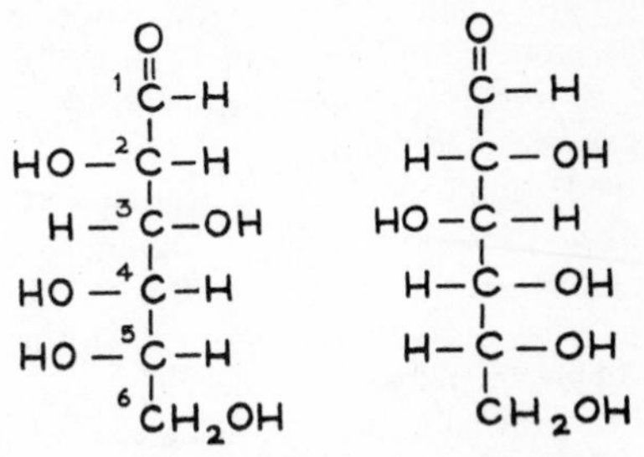

L-Glucose	D-Glucose
O ‖ ¹C–H	O ‖ C–H
HO–²C–H	H–C–OH
H–³C–OH	HO–C–H
HO–⁴C–H	H–C–OH
HO–⁵C–H	H–C–OH
⁶CH₂OH	CH₂OH

L-Glucose **D-Glucose**

termed **epimers.** Galactose and glucose are examples of an epimeric pair which differ with respect to carbon 4; mannose and glucose are epimers with respect to carbon 2. Epimerization or interconversion of epimers in the tissues is illustrated by the conversion in the liver of galactose to glucose. The conversion is catalyzed by a specific enzyme, designated as an epimerase.

The designation of an isomer as a D form or of its mirror image as an L form is determined by its spatial relationship to the parent substance of the carbohydrate family, the 3-carbon sugar, glycerose. The L and D forms of this sugar are shown below.

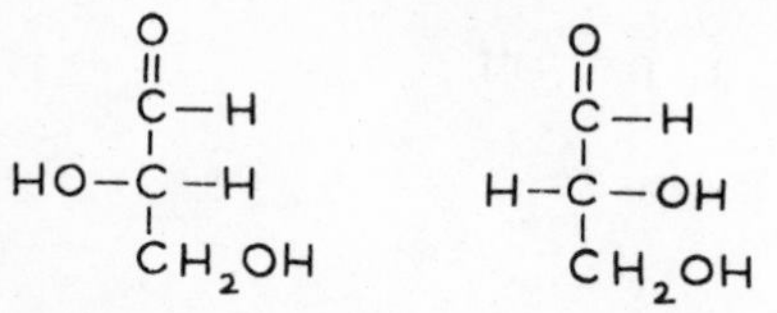

```
      O                O
      ||               ||
      C—H              C—H
      |                |
  HO—C—H           H—C—OH
      |                |
      CH2OH            CH2OH
```

L-Glycerose (L-glyceraldehyde) **D-Glycerose (D-glyceraldehyde)**

The orientation of the H and OH groups around the carbon atom just adjacent to the terminal primary alcohol carbon (eg, carbon atom 5 in glucose) determines the family to which the sugar belongs. When the OH group on this carbon is on the right, the sugar is a member of the D series; when it is on the left, it is a member of the L series. The distribution of the H and OH groups on the other carbon atoms in the molecule is of no importance in this connection.

The majority of the monosaccharides occurring in the body are of the D configuration.

THE KILIANI SYNTHESIS

The structural relationships of the monosaccharides may be readily visualized by the formulas shown in Fig 1–3. A method for the synthesis of these sugars was first proposed by Kiliani. It is based upon the addition of HCN to the carbonyl group of aldehydes or ketones. The application of the Kiliani synthesis to the production of the 2 tetroses, **erythrose** and **threose,** from glycerose is shown in Fig 1–1.

This process can be repeated and the 4 isomeric D-pentoses would be formed; from these, the 8 isomeric hexoses would result.

If the D- and L-aldoses be conceived as thus evolving from the trioses, the aldehyde carbon becomes the one most recently added. It is for this reason that the carbon adjacent to the terminal primary alcohol group (on the opposite end of the molecule) is the true indicator of the original family of origin, ie, of the D series or of the L series, since this carbon is the original and the only asymmetric carbon of the parent sugars.

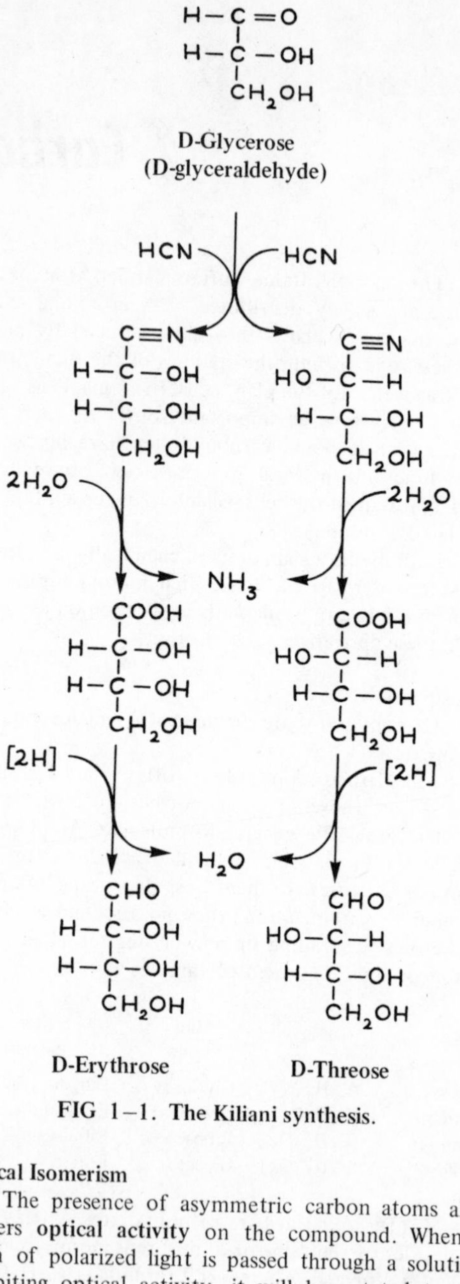

```
                 H—C=O
                   |
                 H—C—OH
                   |
                   CH2OH
```

D-Glycerose (D-glyceraldehyde)

```
          HCN              HCN
     C≡N                     C≡N
     |                       |
   H—C—OH                 HO—C—H
     |                       |
   H—C—OH                  H—C—OH
     |                       |
     CH2OH                   CH2OH
2H2O                              2H2O
                 NH3
     COOH                    COOH
     |                       |
   H—C—OH                 HO—C—H
     |                       |
   H—C—OH                  H—C—OH
     |                       |
     CH2OH                   CH2OH
[2H]                              [2H]
                 H2O
     CHO                     CHO
     |                       |
   H—C—OH                 HO—C—H
     |                       |
   H—C—OH                  H—C—OH
     |                       |
     CH2OH                   CH2OH
```

D-Erythrose **D-Threose**

FIG 1–1. The Kiliani synthesis.

Optical Isomerism

The presence of asymmetric carbon atoms also confers **optical activity** on the compound. When a beam of polarized light is passed through a solution exhibiting optical activity, it will be rotated to the right or left in accordance with the type of compound, ie, the **optical isomer,** which is present. A compound which causes rotation of polarized light to the right is said to be dextrorotatory and a plus (+) sign is used to designate this fact. Rotation of the beam to the left (levorotatory action) is designated by a minus (−) sign.

When equal amounts of dextrorotatory and levorotatory isomers are present, the resulting mixture has no optical activity since the activities of each isomer cancel one another. Such a mixture is said to be a **racemic,** or a DL mixture. Synthetically produced compounds are necessarily racemic because the opportunities for the formation of each optical isomer are identical. The separation of optically active isomers from a racemic mixture is called resolution, ie, the

TABLE 1–1. Examples of pentoses.

Sugar	Source	Importance	Reactions
D-Ribose	Nucleic acids.	Structural elements of nucleic acids and coenzymes, eg, ATP, NAD, NADP (DPN, TPN), flavoproteins.	Reduce Benedict's, Fehling's, Barfoed's, and Haynes' solutions. Forms distinctive osazones with phenylhydrazine.
D-Ribulose	Formed in metabolic processes.	Intermediates in direct oxidative pathway of glucose breakdown (see p 241).	Those of keto sugars.
D-Arabinose	Gum arabic. Plum and cherry gums.	These sugars are used in studies of bacterial metabolism, as in fermentation tests for identification of bacteria. They have no known physiologic function in man.	With orcinol-HCl reagent gives colors: violet, blue, red, and green.
D-Xylose	Wood gums.		With phloroglucinol-HCl gives a red color.
D-Lyxose	Heart muscle.	A constituent of a lyxoflavin isolated from human heart muscle.	

racemic mixture is said to be "resolved" into its optically active components.

Geometric isomerism and optical isomerism are independent properties. Thus a compound might be designated D (−) or L (+), indicating structural relationship to D or L glycerose but exhibiting the opposite rotatory power. The naturally occurring form of lactic acid, the L (+) isomer, is an example.

MONOSACCHARIDES

The monosaccharides include trioses, tetroses, pentoses, hexoses, and heptoses (3, 4, 5, 6, 7 carbon atoms). The trioses are formed in the course of the metabolic breakdown of the hexoses; pentose sugars (Table 1–1) are important constituents of nucleic acids and many coenzymes; they are also formed in the breakdown of glucose by the direct oxidative pathway, and the hexoses glucose, galactose, and fructose are physiologically the most important of the monosaccharides (Table 1–2). A 7-carbon keto sugar, sedoheptulose, was first discovered in 1917 in the sedum plant. It also occurs in animal tissues, where it is formed as a phosphate ester in the metabolism of pentose phosphates by the direct oxidative pathway.

The structures of the aldo sugars are shown in Fig 1–3. Five keto sugars which are important in metabolism are shown in Fig 1–2.

HEXOSES

The hexoses are most important physiologically (Table 1–2). Examples are D-glucose, D-fructose, D-galactose, and D-mannose.

Ring Structures

The formulas in Figs 1–1 and 1–2 illustrate the so-called "straight chain" structure of hexose sugars, but this does not satisfactorily explain many carbohydrate reactions. On the basis of the 2 ring structures known to exist in the glycosides (see below), Haworth in 1929 proposed similar structures for the sugars

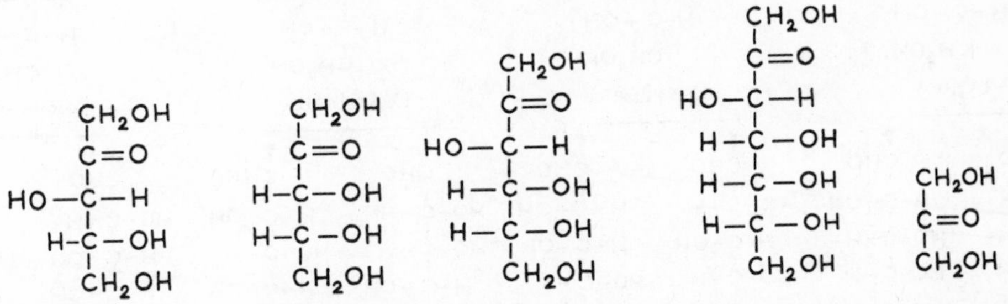

```
   CH2OH          CH2OH          CH2OH            CH2OH
    |              |              |                |
    C=O            C=O            C=O              C=O
    |              |              |                |
HO—C—H         H—C—OH        HO—C—H          HO—C—H
    |              |              |                |            CH2OH
 H—C—OH        H—C—OH         H—C—OH           H—C—OH          |
    |              |              |                |            C=O
   CH2OH          CH2OH       H—C—OH           H—C—OH           |
                                  |                |            CH2OH
                                 CH2OH          H—C—OH
                                                   |
                                                  CH2OH
```

D-Xylulose D-Ribulose D-Fructose D-Sedoheptulose Dihydroxyacetone

FIG 1–2. Examples of ketoses.

TABLE 1–2. Hexoses of physiologic importance.

Sugar	Source	Importance	Reactions
D-Glucose	Fruit juices. Hydrolysis of starch, cane sugar, maltose, and lactose.	The "sugar" of the body. The sugar carried by the blood, and the principal one used by the tissues. Glucose is usually the "sugar" of the urine when glycosuria occurs.	Reduces Benedict's, Haynes', Barfoed's reagents (a reducing sugar). Gives osazone with phenylhydrazine. Fermented by yeast. With HNO_3, forms soluble saccharic acid.
D-Fructose	Fruit juices. Honey. Hydrolysis of cane sugar and of inulin (from the Jerusalem artichoke).	Can be changed to glucose in the liver and intestine and so used in the body.	Reduces Benedict's, Haynes', Barfoed's reagents (a reducing sugar). Forms osazone identical with that of glucose. Fermented by yeast. Cherry-red color with Seliwanoff's resorcinol-HCl reagent.
D-Galactose	Hydrolysis of lactose.	Can be changed to glucose in the liver and metabolized. Synthesized in the mammary gland to make lactose of mother's milk. A constituent of glycolipids (see p 20).	Reduces Benedict's, Haynes', Barfoed's reagents (a reducing sugar). Forms osazone, distinct from above. Phloroglucinol-HCl reagent gives red color. With HNO_3, forms insoluble mucic acid. Not fermented by yeast.
D-Mannose	Hydrolysis of plant mannosans and gums.	A constituent of prosthetic polysaccharide of albumins, globulins, mucoids. Also the prosthetic polysaccharide of tuberculoprotein. Convertible to glucose in the body.	Reduces Benedict's, Haynes', Barfoed's reagents (a reducing sugar). Forms same osazone as glucose.

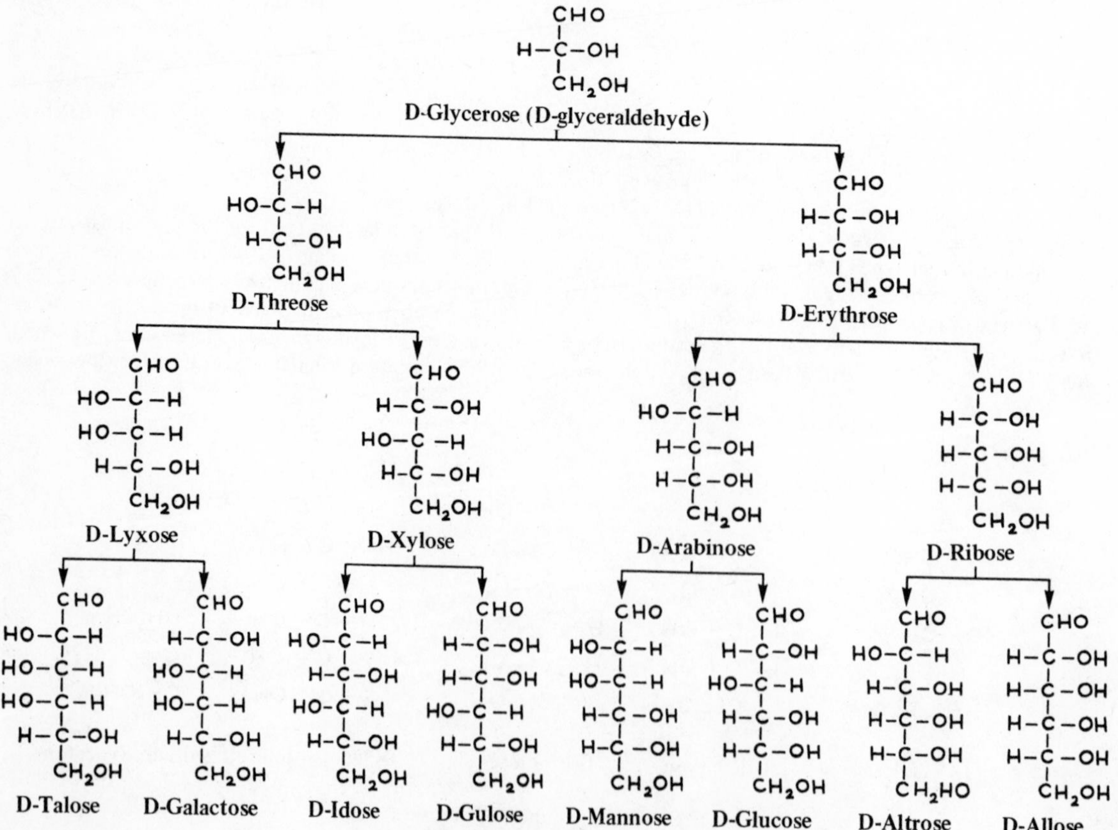

FIG 1–3. The structural relations of the aldoses, D series.

themselves. The terminology for such structures was based on the simplest organic compounds exhibiting a similar ring structure (ie, pyran and furan).

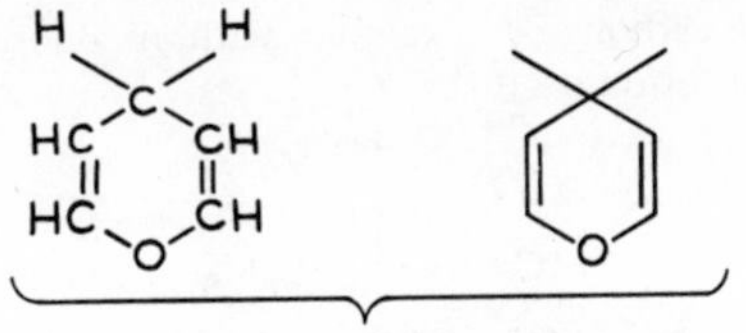

```
  H   H
   \ /
    C
 HC   CH
 ||   ||
 HC   CH
   \ /
    O
```

Pyran

```
 HC —— CH
 ||    ||
 HC    CH
   \  /
    O
```

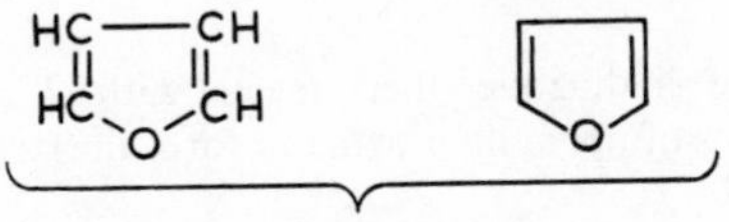

Furan

The ring structures shown below represent **hemiacetal** formation, ie, a condensation between the aldehyde group and a hydroxy group of the same compound. Because of the asymmetry present in the terminal carbon, 2 forms for each ring structure can exist. These are designated as *α* and *β*. For D-glucose, for example:

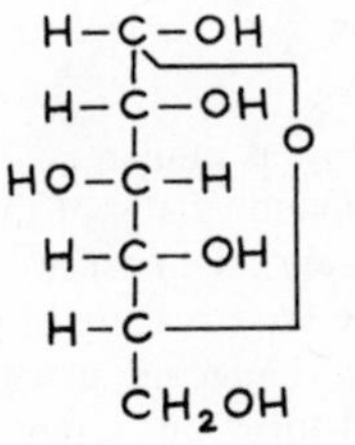

```
  H-C-OH          HO-C-H
  H-C-OH  |        H-C-OH  |
 HO-C-H   O       HO-C-H   O
  H-C-OH  |        H-C-OH  |
  H-C ----         H-C ----
   CH2OH            CH2OH
```

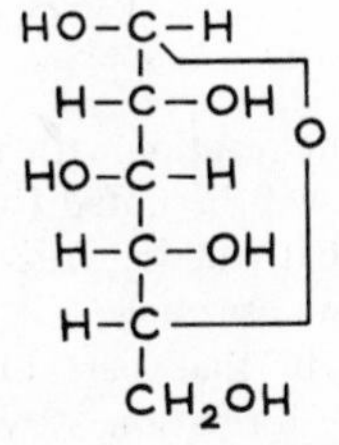

α-D-Glucopyranose **β-D-Glucopyranose**

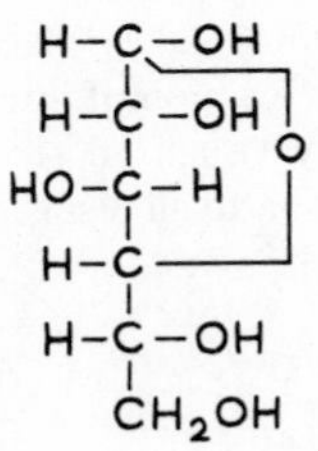

```
  H-C-OH          HO-C-H
  H-C-OH  |        H-C-OH  |
 HO-C-H   O       HO-C-H   O
  H-C ----         H-C ----
  H-C-OH           H-C-OH
   CH2OH            CH2OH
```

α-D-Glucofuranose **β-D-Glucofuranose**

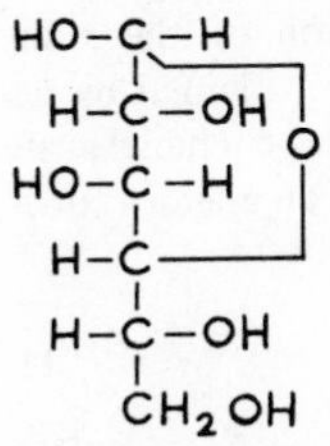

Ketoses may also show ring formation. Thus one may find α-D-fructofuranose, β-D-fructofuranose, or the corresponding pyranoses. Other ring forms (eg, between C atoms 1 and 2 or 1 and 3) may exist but are so unstable that when liberated from their glycosides as sugars they tend to mutate readily to the pyranose form, as do the freed furanoses also.

GLYCOSIDES

Glycosides are compounds containing a carbohydrate and a noncarbohydrate residue in the same molecule. In these compounds the carbohydrate residue is attached by an acetal linkage at carbon atom 1 to a noncarbohydrate residue or **aglycone.** If the carbohydrate portion is glucose, the resulting compound is a **glucoside**; if galactose, a **galactoside**, etc.

A simple example is the methyl glucoside formed when a solution of glucose in boiling methyl alcohol is treated with 0.5% hydrogen chloride as a catalyst. The reaction proceeds as follows:

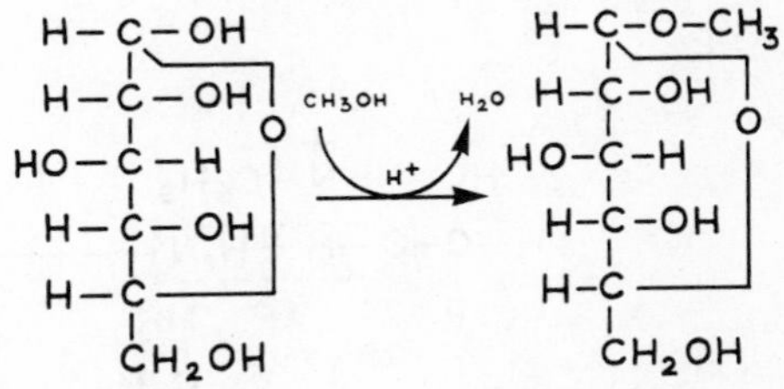

```
  H-C-OH                         H-C-O-CH3
  H-C-OH  |    CH3OH     H2O     H-C-OH  |
 HO-C-H   O   -------------->   HO-C-H   O
  H-C-OH  |         H+           H-C-OH  |
  H-C ----                       H-C ----
   CH2OH                          CH2OH
```

α-D-Glucose (pyranoside) **α-Methyl-D-glucoside (pyranose form)**

From β-D-glucose, β-methyl-D-glucoside would be formed.

Glycosides are found in many drugs, spices, and in the constituents of animal tissues. The aglycone may be methyl alcohol, glycerol, a sterol, a phenol, etc. The glycosides which are important in medicine because of their action on the heart all contain steroids as the aglycone component. These include derivatives of digitalis and strophanthus.

Mutarotation

When an aldohexose is first dissolved in water, its optical rotation gradually changes until a constant rotation characteristic of the sugar is reached. This phenomenon, known as mutarotation, appears to be due to changes of alpha to beta forms and vice versa, probably via the straight chain aldo or keto form. When equilibrium is reached, the characteristic constant rotation is observed.

IMPORTANT CHEMICAL REACTIONS OF MONOSACCHARIDES

Several reactions are of importance as proof of the structure of a typical monosaccharide such as glucose. These include the following:

Iodo Compounds

An aldose heated with concentrated hydriodic acid (HI) loses all of its oxygen and is converted into an iodo compound (glucose to iodohexane, $C_6H_{13}I$). Since the resulting derivative is a straight chain compound related to normal hexane, this is evidence of the lack of any branched chains in the structure of the sugar.

ture of phenylhydrazine hydrochloride and sodium acetate to the sugar solution and heating in a boiling water bath. The reaction involves only the carbonyl carbon (ie, aldehyde or ketone group) and the next adjacent carbon. For example, with an aldose the following reaction occurs:

$$\begin{matrix} HC{=}O \\ | \\ HC{-}OH \\ | \\ R \end{matrix} + H_2N{-}\overset{H}{N}{-}C_6H_5 \xrightarrow{-H_2O} \begin{matrix} HC{=}N{-}\overset{H}{N}{-}C_6H_5 \\ | \\ HC{-}OH \\ | \\ R \end{matrix}$$

Aldose + Phenylhydrazine → **Phenylhydrazone**

The hydrazone then reacts with 2 additional molecules of phenylhydrazine to form the osazone:

$$\begin{matrix} HC{=}N{-}\overset{H}{N}{-}C_6H_5 \\ | \\ H{-}C{-}O{-}H \\ | \\ R \end{matrix} + H_2N{-}NH{-}C_6H_5 \longrightarrow \begin{matrix} HC{=}N{-}\overset{H}{N}{-}C_6H_5 \\ | \\ C{=}O \\ | \\ R \end{matrix} + NH_3 + C_6H_5NH_2$$

Aniline

$$\begin{matrix} HC{=}N{-}\overset{H}{N}{-}C_6H_5 \\ | \\ C{=}O \\ | \\ R \end{matrix} + H_2N{-}\overset{H}{N}{-}C_6H_5 \xrightarrow{-H_2O} \begin{matrix} HC{=}N{-}\overset{H}{N}{-}C_6H_5 \\ | \\ C{=}N{-}\overset{H}{N}{-}C_6H_5 \\ | \\ R \end{matrix}$$

Osazone

Acetylation

The ability to form sugar esters, eg, acetylation with acetylchloride ($CH_3CO.Cl$), indicates the presence of alcohol groups. The total number of acyl groups which can thus be taken up by a molecule of the sugar is a measure of the number of such alcohol groups. Because of its 5 OH groups, the acetylation of glucose, for example, results in a penta acetate.

Other Reactions

Various reactions dependent upon the presence of aldehyde or ketone groups are particularly important because they form the basis for most analytical tests for the sugars. The best-known tests involve **reduction** of metallic hydroxides together with oxidation of the sugar. The alkaline metal is kept in solution with sodium potassium tartrate (Fehling's solution) or sodium citrate (Benedict's solution). Various modifications permit quantitative detection of the copper reduced as a measurement of the sugar content. Other metallic hydroxides may be used (bismuth, Nylander's test; ammoniacal silver, Tollens' test). Barfoed's test distinguishes between monosaccharides and disaccharides, since copper acetate in dilute acid is reduced by the former in 30 seconds but only after several minutes' boiling (to produce hydrolysis) of the disaccharides.

Osazone formation is a useful means of preparing crystalline derivatives of the sugars. These compounds have characteristic crystal structures, melting points, and precipitation times, and are valuable in the identification of sugars. They are obtained by adding a mix-

The reaction with a ketose is similar.

It will be noted from a comparison of their structures that glucose, fructose, and mannose would form the same osazones; but since the structure of galactose differs in that part of the molecule unaffected in osazone formation, it would form a different osazone.

Interconversion. Glucose, fructose, and mannose are interconvertible in solutions of weak alkalinity such as $Ba(OH)_2$ or $Ca(OH)_2$. Presumably, the interconversion of these sugars which takes place in the body is explained by a similar reaction of interconversion. These changes are easily visualized structurally through an enediol form common to all 3 sugars:

$$\begin{matrix} H{-}C{=}O \\ | \\ HO{-}C{-}H \\ | \\ R \end{matrix}$$

Mannose

$$\Updownarrow$$

$$\begin{matrix} H{-}C{=}O \\ | \\ H{-}C{-}OH \\ | \\ R \end{matrix} \rightleftharpoons \begin{matrix} H{-}C{-}OH \\ \| \\ C{-}OH \\ | \\ R \end{matrix} \rightleftharpoons \begin{matrix} H{-}C{-}OH \\ | \\ C{=}O \\ | \\ R \end{matrix}$$

Glucose — **Common enediol form of glucose, mannose, and fructose** — **Fructose**

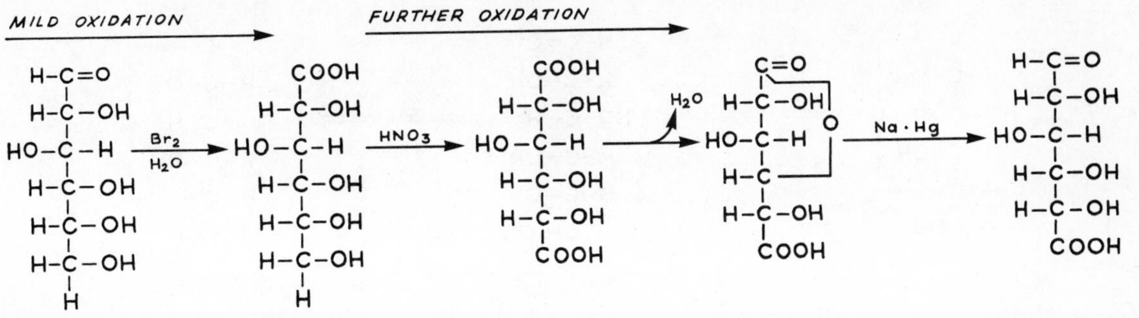

MILD OXIDATION → FURTHER OXIDATION →

Glucose —Br_2, H_2O→ Gluconic acid —HNO_3→ Saccharic acid —(H_2O)→ Saccharolactone —Na · Hg→ Glucuronic acid

FIG 1–4. Oxidation of glucose.

Oxidation of aldoses may form acids as end products (Fig 1–4). Oxidation of the aldehyde group forms "aldonic acids." However, if the aldehyde group remains intact and the primary alcohol group at the opposite end of the molecule is oxidized, "uronic acids" are formed instead (Fig 1–4).

Note that glucuronic acid exerts "reducing" activity because of the free aldehyde group. These so-called hexuronic acids are important in connection with conjugation reactions.

Oxidation of galactose with concentrated HNO_3 yields the dicarboxylic mucic acid. This compound crystallizes readily, and this is useful as an identifying test. Galacturonic acid is found in natural products (eg, pectins).

Reduction. The monosaccharides may be reduced to their corresponding alcohols by reducing agents such as sodium amalgam.

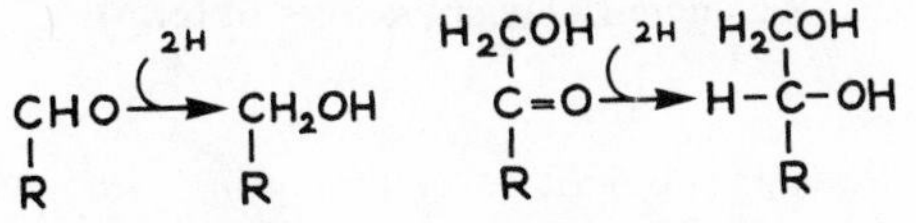

CHO–R —2H→ CH_2OH–R
H_2COH–C=O–R —2H→ H_2COH–H–C–OH–R

Aldo sugar **Keto sugar**

Thus, glucose yields sorbitol; galactose yields dulcitol; mannose yields mannitol; and fructose yields mannitol and sorbitol.

With **strong mineral acids** there is a shift of hydroxyl groups toward and of hydrogen away from the aldehyde end of the chain.

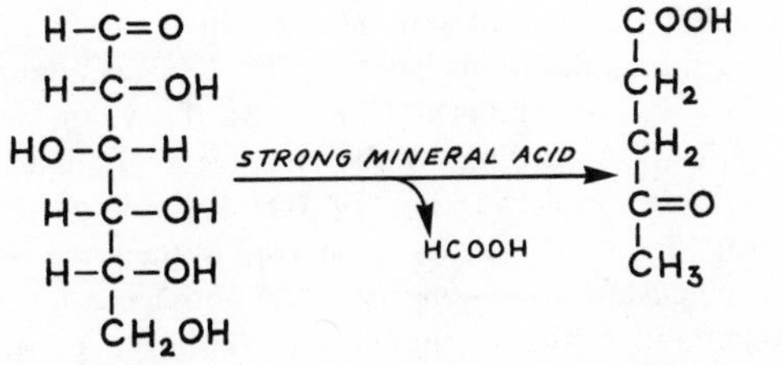

D-Glucose —STRONG MINERAL ACID (HCOOH)→ Levulinic acid

D-Glucose **Levulinic acid**

Reaction products with acid (furfural or one of its derivatives) will condense with certain organic phenols to form compounds of characteristic colors. Color tests for the various sugars are based on such reactions.

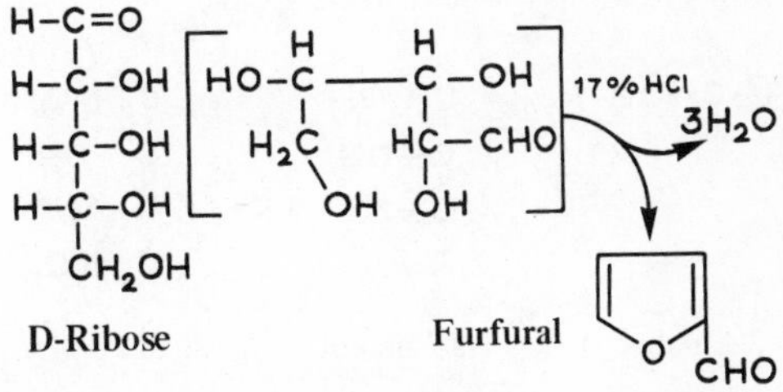

D-Ribose —17% HCl, $3H_2O$→ Furfural

D-Ribose **Furfural**

Heating of gluconic acid produces lactones. These are cyclic structures resembling the pyranoses and furanoses described on p 5.

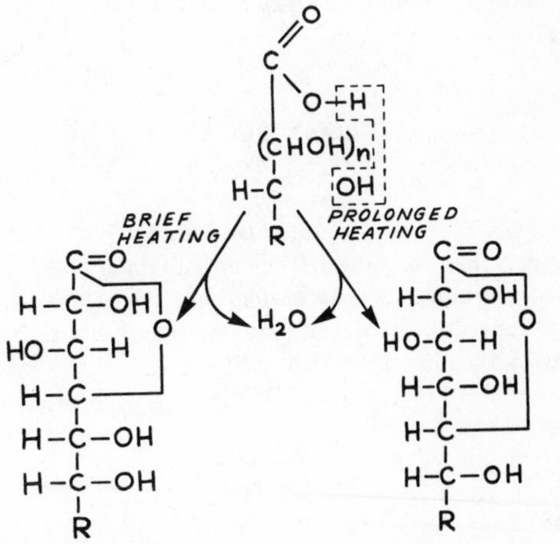

BRIEF HEATING ← (H_2O) → PROLONGED HEATING

γ-Gluconolactone **δ-Gluconolactone**

With **alkali** monosaccharides react in various ways:

(1) In dilute alkali the sugar will change to the cyclic alpha and beta structures, with an equilibrium

```
 H   OH                                                  HO   H
  \ /                        H-C=O                         \ /
   C-----                      |                            C-----
   |    |                    H-C-OH                         |    |
 H-C-OH |                      |                          H-C-OH |
   |    |                   HO-C-H                          |    |
HO-C-H  O  <============>      |      <============>     HO-C-H  O
   |    |                    H-C-OH                         |    |
 H-C-OH |                      |                          H-C-OH |
   |    |                    H-C-OH                         |    |
 H-C-----                      |                          H-C-----
   |                           CH2OH                        |
 H2COH                                                    H2COH

α-D-Glucose                  D-Glucose                  β-D-Glucose
```

FIG 1–5. Reaction of glucose in dilute alkali.

between the 2 isomeric forms. (See Mutarotation and Fig 1–5.)

On standing, a rearrangement will occur which produces an equilibrated mixture of glucose, fructose, and mannose through the enediol form (see p 6).

(2) If the mixture is heated to 37° C, the acidity increases and a series of enols are formed in which double bonds shift from the oxygen-carbon link to positions between various carbon atoms.

```
H-C-OH         CH2OH          C2H5O2
  ||             |              |
  C-OH           C-OH           C-OH
  |              ||             ||
  R              C-OH           C-OH
                 |              |
                 R              C2H5O2

1,2-Enediol    2,3-Enediol    3,4-Enediol, etc
```

(3) In concentrated alkali, sugar caramelizes and produces a series of decomposition products. Yellow and brown pigments develop, salts may form, many double bonds between carbon atoms are formed, and carbon-to-carbon bonds may rupture.

DEOXY SUGARS

Deoxy sugars are those containing fewer atoms of oxygen than of carbon. They are obtained on hydrolysis of certain substances which are important in biologic processes. An example is the deoxyribose occurring in nucleic acids (DNA, p 51).

```
H-C-OH
  |\-----
H-C-H   |
  |     O
H-C-OH  |
  |     |
H-C------
  |
  CH2OH
```

2-Deoxy-D-ribofuranose (α form)

AMINO SUGARS (HEXOSAMINES)

Sugars containing an amino group are called **amino sugars.** Examples are D-glucosamine, D-galactosamine, and D-mannosamine, all of which have been identified in nature. Glucosamine is a constituent of hyaluronic acid. Galactosamine (chondrosamine) is a constituent of chondroitin. Mannosamine is an important constituent of mucoprotein.

```
 H-C-OH
   |\-----
 H-C-NH2 |
   |     |
HO-C-H   O
   |     |
 H-C-OH  |
   |     |
 H-C------
   |
   CH2OH
```

Glucosamine
(2-amino-D-glucopyranose) (α form)

Several antibiotics (erythromycin, carbomycin) contain amino sugars. Erythromycin contains a dimethylamino sugar. Carbomycin contains the first known 3-amino sugar, 3-amino-D-ribose. The amino sugars are believed to be related to the antibiotic activity of these drugs.

DISACCHARIDES

The disaccharides are sugars which can be hydrolyzed into 2 monosaccharides. The 2 sugars are united by a glycosidic linkage (Fig 1–6). They are named chemically according to the structures of their component monosaccharides. In one system of nomenclature the disaccharides are named in the same way as the glycosides, with one of the monosaccharide components identified as though it were the aglycone. The hydroxyl group of the sugar to which this aglycone-like monosaccharide is attached is designated by a

MALTOSE (α FORM) (HAWORTH FORMULAS)

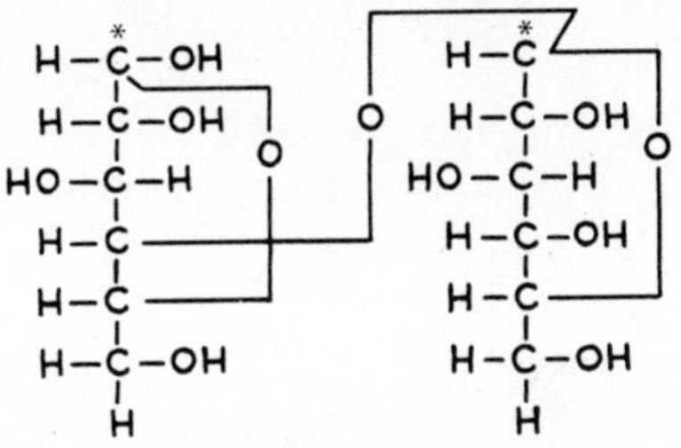

Two α-D-glucopyranose components

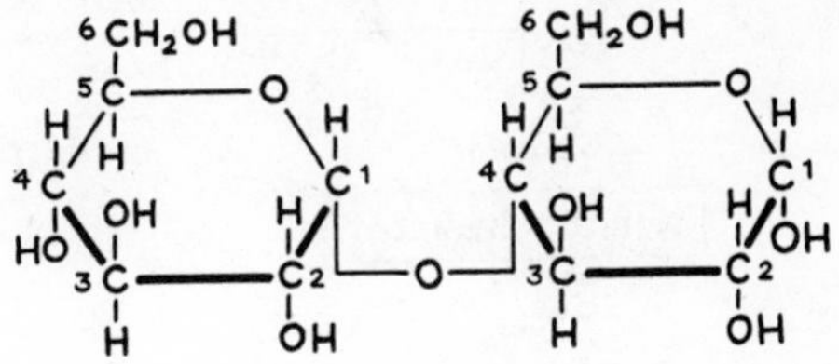

4-α-D-Glucopyranosido-α-D-glucopyranoside

SUCROSE

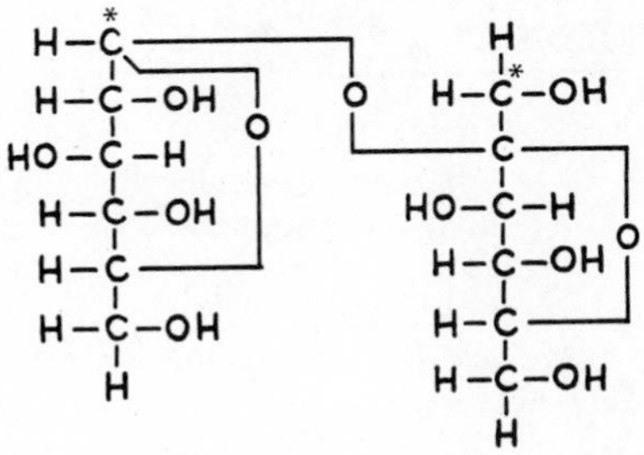

α-D-Glucopyranose component

β-D-Fructofuranose component

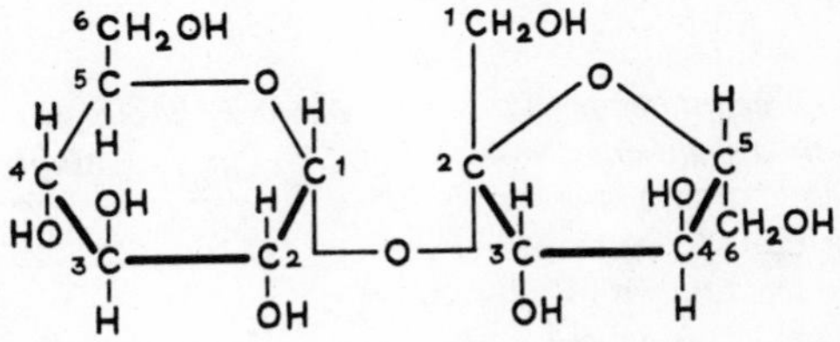

1-α-D-Glucopyranosido-β-D-fructofuranoside

LACTOSE (β FORM)

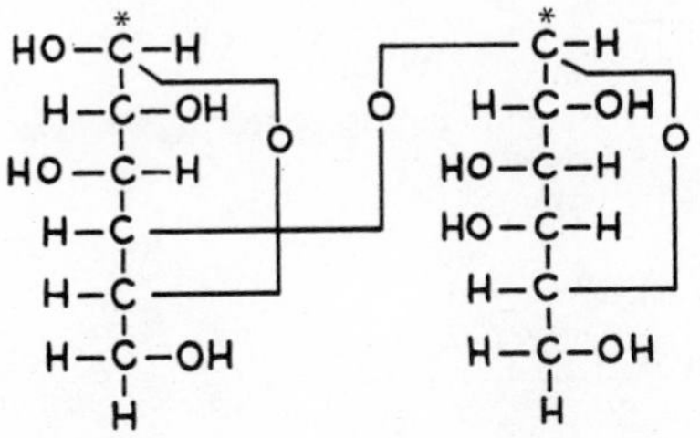

β-D-Glucopyranose component

β-D-Galactopyranose component

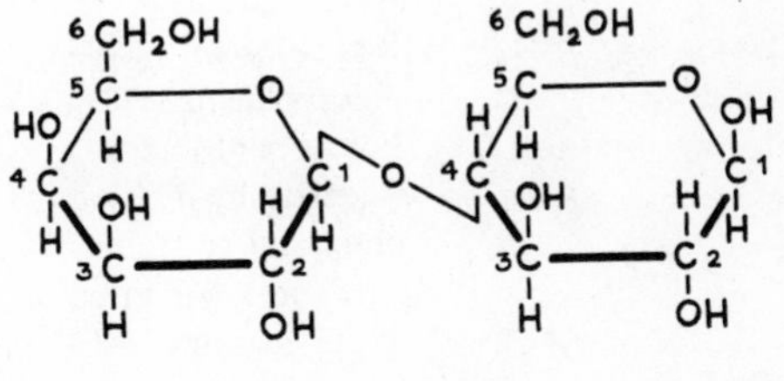

4-β-D-Glucopyranosido-β-D-galactopyranoside

TREHALOSE (α FORM)

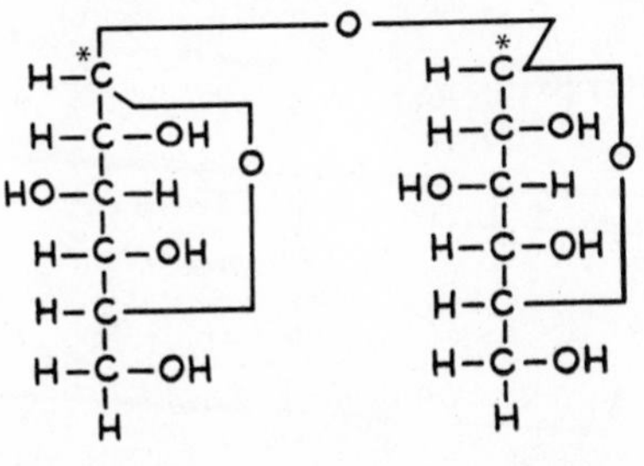

Two α-D-glucopyranose components

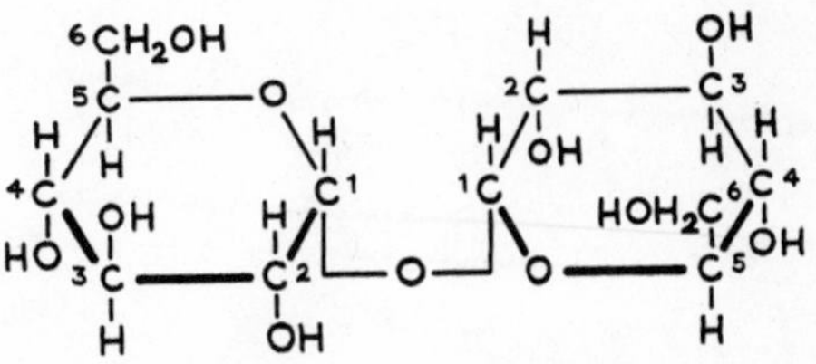

1-α-D-Glucopyranoside

FIG 1–6. Structures of representative disaccharides.

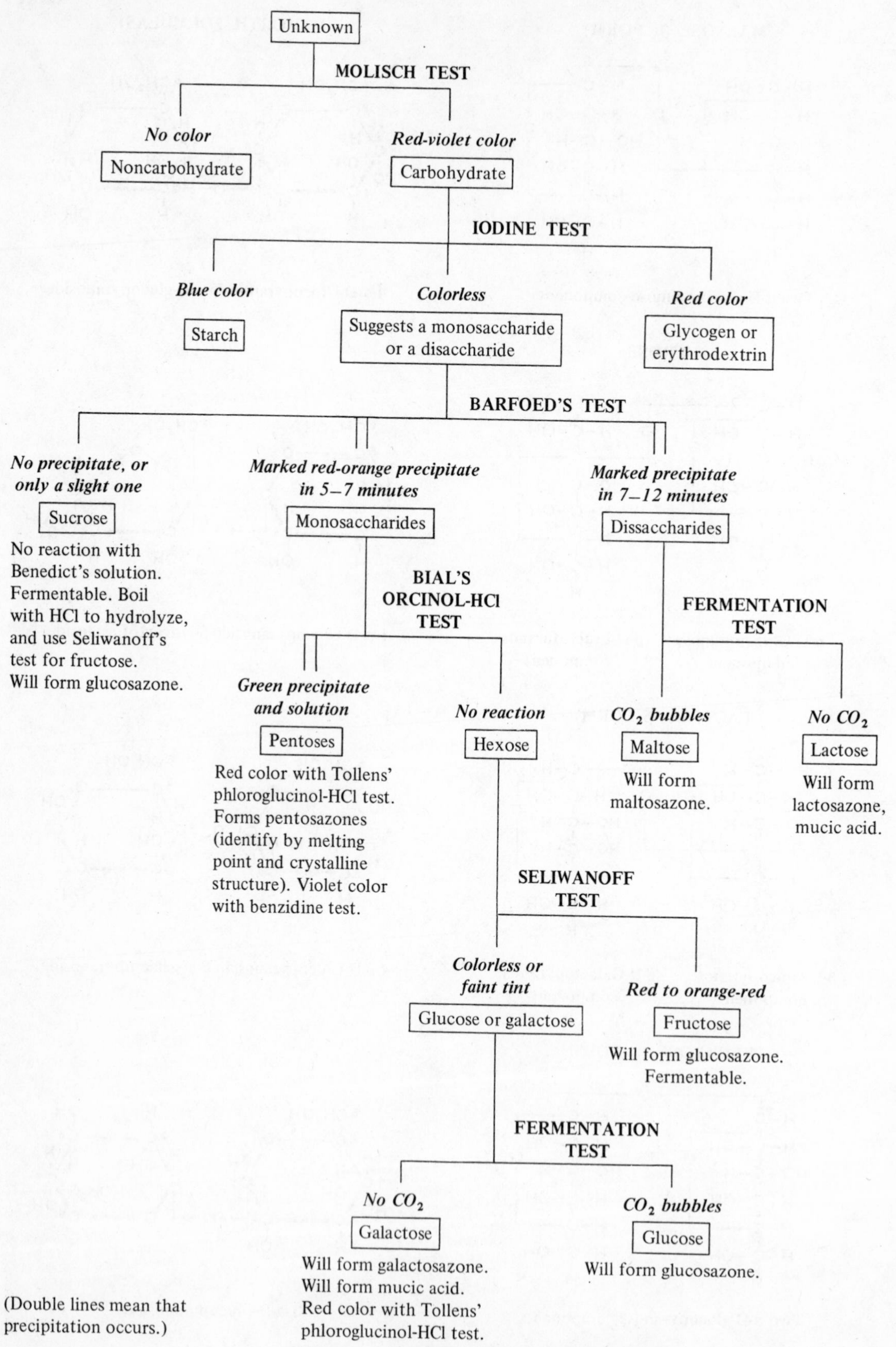

FIG 1–7. Qualitative tests for identification of carbohydrates.

TABLE 1–3. Disaccharides.

Sugar	Source	Reactions
Maltose	Diastatic digestion or hydrolysis of starch. Germinating cereals and malt.	Reducing sugar. Forms osazone with phenylhydrazine. Fermentable. Hydrolyzed to D-glucose.
Lactose	Milk. May occur in urine during pregnancy. Formed in the body from glucose.	Reducing sugar. Forms osazone with phenylhydrazine. Not fermentable by yeasts. Hydrolyzed to glucose and galactose.
Sucrose	Cane and beet sugar. Sorghum. Pineapple. Carrot roots.	Nonreducing sugar. Does not form osazone. Fermentable. Hydrolyzed to fructose and glucose.
Trehalose	Fungi and yeasts. The major sugar of insect hemolymph.	Nonreducing sugar. Does not form an osazone. Hydrolyzed to glucose.

number. The suffix **-furan** or **-pyran** refers to the structural resemblances to these compounds. The alpha and beta refer to the configuration at the starred (*) carbon atom, as indicated in examples shown in connection with the formulas in Fig 1–6. The physiologically important disaccharides are maltose, sucrose, lactose, and trehalose (Table 1–3).

Since sucrose has no free carbonyl group, it gives none of the reactions characteristic of "reducing" sugars. Thus it fails to reduce alkaline copper solutions, form an osazone, or exhibit mutarotation. Hydrolysis of sucrose yields a crude mixture which is often called "invert sugar" because the strongly levorotatory fructose thus produced changes (inverts) the previous dextrorotatory action of the sucrose.

Lactose gives rise to mucic acid on prolonged boiling with HNO_3. This is derived from the galactose produced as a result of hydrolysis of lactose by the boiling with acid.

TECHNIC OF TESTS FOR CARBOHYDRATES*
(See Fig 1–7.)

Anthrone Test: To 2 ml of anthrone test solution (0.2% in concentrated H_2SO_4) add 0.2 ml of unknown. A green or blue-green color indicates the presence of carbohydrate. The test is very sensitive; it will give a positive reaction with filter paper (cellulose). The anthrone reaction has been adapted to the quantitative colorimetric determination of glycogen, inulin, and sugar of blood.

Barfoed's Test (copper acetate and acetic acid): To 5 ml of reagent add 1 ml of unknown. Place in boiling water bath. (See Fig 1–7 for interpretation.)

Benedict's Test (copper sulfate, sodium citrate, sodium carbonate): To 5 ml of reagent in test tube add 8 drops of unknown. Place in a boiling water bath for 5 minutes. A green, yellow, or orange-red precipitate gives a semiquantitative estimate of the amounts of reducing sugar present.

Bial's Orcinol-HCl Test: To 5 ml of reagent add 2–3 ml of unknown and heat until bubbles of gas rise to the surface. Green solution and precipitate indicate pentose.

Fermentation Test: To 5 ml of a 20% suspension of ordinary baker's yeast add about 5 ml of unknown solution and 5 ml of phosphate buffer (pH 6.4–6.8). Place in a fermentation tube or test tube and let stand 1 hour. Bubbles of CO_2 indicate fermentation.

Haynes' Test (Rochelle salt, or potassium sodium tartrate, glycerol, copper sulfate): Performed similarly to Benedict's.

Iodine Test: Acidify the unknown solution with HCl and add 1 drop of the mixture to a solution of iodine in KI. The formation of a blue color indicates the presence of starch; a red color indicates the presence of glycogen or erythrodextrin.

Molisch Test: To 2 ml of unknown add 2 drops of fresh 10% α-naphthol reagent and mix. Pour 2 ml of concentrated H_2SO_4 so as to form a layer below the mixture. A red-violet ring indicates the presence of carbohydrate.

Pavy's Test (Rochelle salt, ammonium hydroxide, copper sulfate): Similar to Benedict's test.

Phenylhydrazine Reaction (osazone formation): Heat phenylhydrazine reagent with 2 ml of a solution of the sugar in a test tube in a boiling water bath for 30 minutes; cool, and examine crystals with a microscope. Compare with diagrams in laboratory manuals or with crystals prepared from known solutions.

Seliwanoff's Resorcinol Test: To 1 ml of unknown add 5 ml of freshly-prepared reagent. This is made by adding 3.5 ml of 0.5% resorcinol to 12 ml of concentrated HCl and diluting to 35 ml with distilled water. Place in boiling water bath for 10 minutes. Cherry-red color indicates fructose.

Tauber's Benzidine Test: To 1 ml of benzidine solution add 2 drops of unknown sugar; boil and cool quickly. A violet color indicates the presence of pentose.

Tollens' Naphthoresorcinol Reaction: To 5 ml of unknown in a test tube add 1 ml of 1% alcoholic solution of naphthoresorcinol. Heat gradually to

*Chromatographic technics (see p 35) are now more frequently used for separation and identification of carbohydrates.

boiling; boil for 1 minute with shaking; let stand for 4 minutes, then cool under tap. Then prepare ether extract. A violet-red color in the ether extract indicates presence of hexuronic acids and rules out pentoses.

Tollens' Phloroglucinol-HCl Test: To equal volumes of the unknown solution and HCl add phloroglucinol. Glucuronates may be distinguished from pentoses or galactose by the naphthoresorcinol test.

POLYSACCHARIDES

Polysaccharides include the following physiologically important substances:

Starch $(C_6H_{10}O_5)_x$ is formed of an α-glucosidic chain. Such a compound, yielding only glucose on hydrolysis, is called a **glucosan**. It is the most important food source of carbohydrate and is found in cereals, potatoes, legumes, and other vegetables. Natural starch is insoluble in water and gives a blue color with iodine solution. The microscopic form of the granules is characteristic of the source of the starch. The 2 chief constituents are amylose (15–20%), which is nonbranching in structure, and amylopectin (80–85%), which consists of highly branched chains. Each is composed of a number of α-glucosidic chains having 24–30 molecules of glucose apiece. The glucose residues are united by 1:4 or 1:6 linkages.

Glycogen is the polysaccharide of the animal body. It is often called animal starch. It is a branched structure with straight chain units of 11-18-α-D-glucopyranose (in α[1-4]-glucosidic linkage) with branching by means of α(1-6)-glucosidic bonds. It is nonreducing and gives a red color with iodine.

Inulin is a starch found in tubers and roots of dahlias, artichokes, and dandelions. It is hydrolyzable to fructose and hence it is a fructosan. No color is given when iodine is added to inulin solutions. This starch is easily soluble in warm water. It is used in physiologic investigation for determination of the rate of glomerular filtration.

Dextrins are substances which are formed in the course of the hydrolytic breakdown of starch. The partially digested starches are amorphous. Dextrins which give a red color when tested with iodine are first formed. These are called **erythrodextrins**. As hydrolysis proceeds the iodine color is no longer produced. These are the so-called **achroodextrins**. Finally, reducing sugars will appear.

Cellulose is the chief constituent of the framework of plants. It gives no color with iodine and is not soluble in ordinary solvents. Since it is not subject to attack by the digestive enzymes of man, it is an important source of "bulk" in the diet.

Chitin is an important structural polysaccharide of invertebrates. It is found, for example, in the shells of crustaceans. Structurally, chitin apparently consists of N-acetyl-D-glucosamine units joined by β(1-4)-glucosidic linkages.

Polysaccharides which are associated with the structure of animal tissues are analogous to the cellulose of the plant cells. Examples are **hyaluronic acid** and the **chondroitin sulfates**. These substances are members of a group of carbohydrates, the **mucopolysaccharides**, which are characterized by their content of amino sugars and uronic acids. Heparin, which occurs in several animal tissues, is also a mucopolysaccharide because on hydrolysis heparin yields glucuronic acid and glucosamine, as well as acetic and sulfuric acids.

Many of the mucopolysaccharides occur in the tissues as prosthetic groups of conjugated proteins, the **mucoproteins** or **glycoproteins**. Examples are found among the $alpha_1$ and $alpha_2$ globulins of plasma. The mucoproteins of the plasma are characterized by the presence of an acetyl hexosamine (N-acetyl glucosamine?) and a hexose (mannose or galactose) in their polysaccharide portion. In addition, a methyl pentose (**L-fucose**) and the **sialic acids** commonly occur in these conjugated proteins.

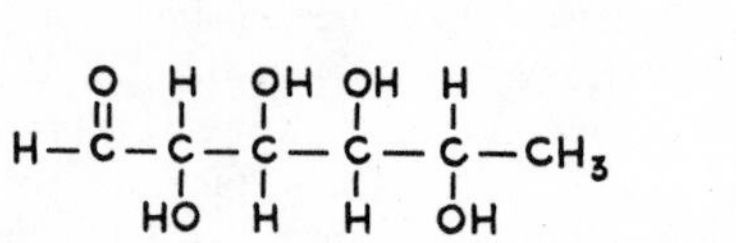

```
     O   H   OH  OH  H
     ‖   |   |   |   |
H —  C — C — C — C — C — CH3
         |   |   |   |
         HO  H   H   OH
```

L-Fucose (6-deoxy-L-galactose)

The sialic acids are actually a family of compounds derived from neuraminic acid. They are widely distributed in vertebrate tissues and have also been isolated from certain strains of bacteria. **N-Acetylneuraminic acid**, the structure of which is shown below, is an example of a sialic acid. Enzymes have been identified in the liver of the rat and in bovine submaxillary glands which can accomplish the biosynthesis of N-acetylneuraminic acid.

N-Acetylneuraminic acid undergoes cleavage to pyruvic acid and N-acetyl-D-mannosamine in a reverse

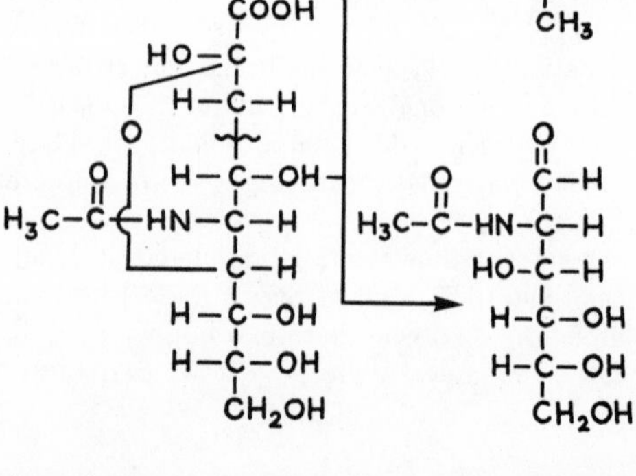

```
                                            COOH
                                            |
                         ┌────────────────► C=O
               COOH      │                  |
               |         │                  CH3
          HO — C         │
          ┌─   |         │
          │  H—C—H       │
          O    ┼         │                  O
     O    │    |         │                  ‖
     ‖    │  H—C—OH ─────┤       O          C—H
H3C— C ─┼─ HN—C—H        │       ‖          |
          │    |         │  H3C— C — HN —   C—H
          └─── C—H       │                  |
               |         │               HO—C—H
           H — C—OH      └──────────────► H—C—OH
               |                            |
           H — C—OH                       H—C—OH
               |                            |
               CH2OH                        CH2OH
```

N-Acetylneuraminic acid   N-Acetyl-D-mannosamine

reaction catalyzed by the enzyme **N-acetylneuraminic acid aldolase.**

The "blood group" substances of the erythrocytes (isoagglutinogens) which are responsible for the major immunologic reactions of blood (blood types) are also mucoproteins. L-Fucose is an important constituent of human blood group substances (19% in blood group B substance). Other examples of mucopolysaccharides which produce specific immune reactions are found among the bacteria. The capsular polysaccharides (haptenes) of pneumococci have been the most extensively studied in this connection. Preparations of capsular polysaccharide from type I pneumococci yield, on hydrolysis, glucosamine and glucuronic acid.

Some of the pituitary hormones, although mainly proteins (eg, the gonadotropins and thyrotropic hormone), also contain carbohydrate. This suggests that they may also be mucoproteins or glycoproteins.

• • •

General Bibliography

Advances in Carbohydrate Chemistry. Academic Press, 1945–current.

Florkin M, Stotz E: *Comprehensive Biochemistry; Carbohydrates.* Section II, vol 5. Elsevier, 1963.

Kabat EA: *Blood Group Substances.* Academic Press, 1956.

Meyer KH: *Natural and Synthetic High Polymers,* 2nd ed. Interscience, 1950.

Percival EGV, Percival E: *Structural Carbohydrate Chemistry.* Prentice-Hall, 1962.

Pigman WW (editor): *The Carbohydrates: Chemistry, Biochemistry, Physiology.* Academic Press, 1957.

West ES & others: *Textbook of Biochemistry,* 4th ed. Macmillan, 1966.

2...
Lipids

With Peter Mayes, PhD, DSc*

The lipids are a heterogeneous group of compounds related, either actually or potentially, to the fatty acids. They have the common property of being (1) relatively insoluble in water and (2) soluble in lipid solvents such as ether, chloroform, and benzene. Thus, the lipids include fats, oils, waxes, and related compounds.

A lipoid is a "fat-like" substance which may not actually be related to the fatty acids although occasionally the terms "lipid" and "lipoid" are used synonymously.

Lipids are important dietary constituents not only because of their high energy value but also because of the fat-soluble vitamins and the essential fatty acids which are found with the fat of natural foods. In the body, fat serves as an efficient source of energy—both directly and potentially, when stored in adipose tissue. It serves as an insulating material in the subcutaneous tissues and around certain organs. The fat content of nerve tissue is particularly high. Combinations of fat and protein (lipoproteins) are important cellular constituents, occurring both in the cell membrane and in the mitochondria within the cytoplasm, and serving also as the means of transporting lipids in the blood.

Classification

The following classification of lipids has been proposed by Bloor:

A. Simple Lipids: Esters of fatty acids with various alcohols.

1. **Fats**—Esters of fatty acids with glycerol. A fat which is in the liquid state is known as an oil.

2. **Waxes**—Esters of fatty acids with higher alcohols than glycerol.

B. Compound Lipids: Esters of fatty acids containing groups in addition to an alcohol and a fatty acid.

1. **Phospholipids**—Lipids containing, in addition to fatty acids and glycerol, a phosphoric acid residue, nitrogen-containing bases, and other substituents. These lipids include phosphatidyl choline, phosphatidyl ethanolamine, phosphatidyl inositol, phosphatidyl serine, plasmalogens, and sphingomyelins.

2. **Cerebrosides (glycolipids)**—Compounds of the fatty acids with carbohydrate, containing nitrogen but no phosphoric acid.

3. **Other compound lipids**, such as sulfolipids and aminolipids. Lipoproteins may also be placed in this category.

C. Derived Lipids: Substances derived from the above groups by hydrolysis. These include fatty acids (both saturated and unsaturated), glycerol, steroids, alcohols in addition to glycerol and sterols, fatty aldehydes, and ketone bodies (see Ketosis, Chapter 14).

Because they are uncharged, glycerides, cholesterol, and cholesterol esters are termed neutral lipids.

FATTY ACIDS

Fatty acids are obtained from the hydrolysis of fats. Fatty acids which occur in natural fats usually contain an even number of carbon atoms (because they are synthesized from 2-carbon units) and are straight-chain derivatives. The chain may be saturated (containing no double bonds) or unsaturated (containing one or more double bonds).

Nomenclature

The most frequently used systematic nomenclature is based on naming the fatty acid after the hydrocarbon with the same number of carbon atoms, **-oic** being substituted for the final **e** in the name of the hydrocarbon (Genevan system). Thus, saturated acids end in **-anoic**, eg, octanoic acid, and unsaturated acids with double bonds end in **-enoic**, eg, octadecenoic acid (oleic acid). Carbon atoms are numbered from the carboxyl carbon (carbon No. 1). The carbon atom adjacent to the carboxyl carbon (No. 2) is also known as the α-carbon. Carbon atom No. 3 is the β-carbon, and the end methyl carbon is known as the ω-carbon. Various conventions are in use for indicating the number and position of the double bonds, eg, Δ^9 indicates a double bond between carbon atoms 9 and 10 of the fatty acid. A widely used convention is to indicate the number of carbon atoms, number of double bonds, and the positions of the double bonds as in the following examples:

*Lecturer in Biochemistry, Royal Veterinary College, University of London.

18 : 1 ; 9

$$\overset{}{CH_3}(CH_2)_7\overset{10}{C}H=\overset{9}{C}H(CH_2)_7COOH$$

Oleic acid

18 : 2 ; 9, 12

$$CH_3(CH_2)_4\overset{13}{C}H=\overset{12}{C}H-CH_2-\overset{10}{C}H=\overset{9}{C}H(CH_2)_7COOH$$

Linoleic acid

A closer examination of the position of the double bonds in naturally occurring fatty acids reveals that they are related to the $-CH_3$ or ω-end of the fatty acid rather than the carboxyl group. This is because the biosynthetic elongation of fatty acid chains takes place at the carboxyl end of the molecule. Thus, a series of monounsaturated fatty acids of increasing chain length based on oleic acid may be described as ω-9 acids.

Saturated Fatty Acids

Saturated fatty acids may be envisaged as based on acetic acid as the first member of the series (general formula: $C_nH_{2n+1}COOH$). Examples of the acids in this series are shown in Table 2–1.

Other higher members of the series are known to occur, particularly in waxes. A few branched-chain fatty acids have also been isolated from both plant and animal sources.

Unsaturated Fatty Acids

These may be further subdivided according to degree of unsaturation.

A. Monounsaturated (Monoethenoid) Acids: General formula: $C_nH_{2n-1}COOH$. *Example:* Oleic acid, found in nearly all fats.

B. Polyunsaturated (Polyethenoid) Acids:

1. Two double bonds. General formula: $C_nH_{2n-3}COOH$. *Example:* Linoleic acid* (18:2; 9, 12). Occurs in many seed oils, eg, corn, peanut, cottonseed, soybean oils.

2. Three double bonds. General formula: $C_nH_{2n-5}COOH$. *Example:* Linolenic acid* (18:3; 9, 12, 15). Found frequently with linoleic acid but particularly in linseed oil.

3. Four double bonds. General formula: $C_nH_{2n-7}COOH$. *Example:* Arachidonic acid* (20:4; 5, 8, 11, 14). Found in small quantities with linoleic and linolenic acids but particularly in peanut oil.

A group of compounds known as prostaglandins (found in seminal plasma and other tissues) are of interest because of their pharmacologic and biochemical activity on smooth muscle, blood vessels, and adipose tissue. In vivo they are synthesized from arachidonic acid. *Example:* Prostaglandin E_2 (PGE_2):

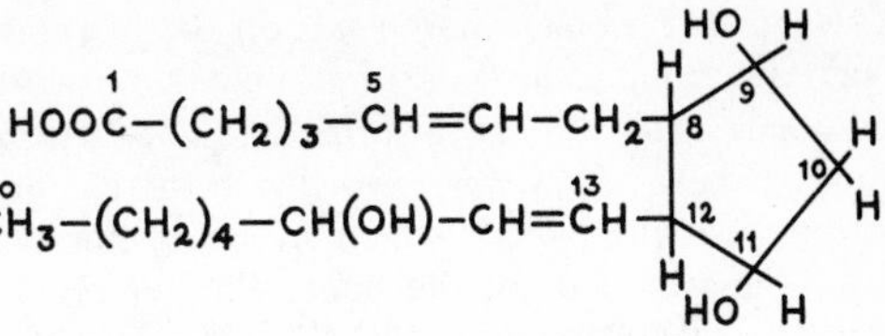

$$\overset{1}{HOOC}-(CH_2)_3-\overset{5}{C}H=CH-CH_2-[\text{ring C-8 to C-12; C-9: HO, H; C-10: H, H; C-11: HO, H}]$$
$$\overset{20}{CH_3}-(CH_2)_4-CH(OH)-CH=\overset{13}{C}H-$$

Prostaglandin E_2 (PGE_2)

C. Many other fatty acids have been detected in biologic material. For example, fish oil contains 22:5 and 22:6 unsaturated fatty acids. Various other structures, such as hydroxy groups (ricinoleic acid) or cyclic groups, have been found in nature. An example of the latter is chaulmoogric acid.

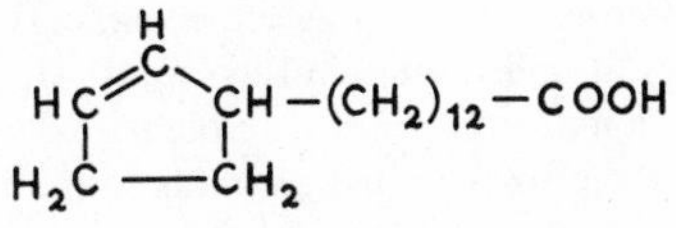

$$\begin{array}{l} \quad\ H \\ \quad\ C \\ HC\!\!=\!\!\quad CH-(CH_2)_{12}-COOH \\ H_2C\,-\!-\!-\,CH_2 \end{array}$$

Chaulmoogric acid

Isomerism in Unsaturated Fatty Acids

Variations in the location of the double bonds in unsaturated fatty acid chains produce isomers. Thus, oleic acid could have 15 different positional isomers.

TABLE 2–1. Saturated fatty acids.

Acetic	CH_3COOH	Major end product of carbohydrate fermentation by rumen organisms
Propionic	C_2H_5COOH	An end product of carbohydrate fermentation by rumen organisms
Butyric	C_3H_7COOH	In certain fats in small amounts (especially butter). An end product of carbohydrate fermentation by rumen organisms.
Caproic	$C_5H_{11}COOH$	
Caprylic (octanoic)	$C_7H_{15}COOH$	In small amounts in many fats (including butter), especially those of plant origin
Decanoic (capric)	$C_9H_{19}COOH$	
Lauric	$C_{11}H_{23}COOH$	Spermaceti, cinnamon, palm kernel, coconut oils, laurels
Myristic	$C_{13}H_{27}COOH$	Nutmeg, palm kernel, coconut oils, myrtles
Palmitic	$C_{15}H_{31}COOH$	Common in all animal and plant fats
Stearic	$C_{17}H_{35}COOH$	
Arachidic	$C_{19}H_{39}COOH$	Peanut (arachis) oil
Lignoceric	$C_{23}H_{47}COOH$	Cerebrosides, peanut oil

*Linoleic, linolenic, and arachidonic are the so-called "essential" fatty acids (EFA). See p 276.

Geometric isomerism depends on the orientation of radicals around the axis of double bonds. Some compounds differ only in the orientation of their parts around this axis. This is noteworthy in the chemistry of steroids. If the radicals which are being considered are on the same side of the bond, the compound is called "cis"; if on opposite sides, "trans." This can be illustrated with oleic and elaidic acids or with dicarboxylic acids such as fumaric and maleic acids (Fig 2–1).

In acids with a greater degree of unsaturation there are, of course, more geometric isomers. Naturally occurring unsaturated long chain fatty acids are nearly all of the "cis" configuration, the molecule being "bent" at the position of the double bond.

Alcohols

Alcohols found in lipid molecules include glycerol, cholesterol, and higher alcohols (eg, cetyl alcohol, $C_{16}H_{33}OH$), usually found in the waxes. The presence of glycerol is indicated by the acrolein test (Fig 2–2).

Among the unsaturated alcohols found in fats are a number of important pigments. These include phytol (phytyl alcohol), which is also a constituent of chlorophyll, and lycophyll ($C_{40}H_{56}O_2$), a polyunsaturated dihydroxy alcohol which occurs in tomatoes as a purple pigment.

H_2C–OH / HC–OH / C–H_2 / OH → ($2H_2O$; DEHYDRATED WITH HEAT PLUS $KHSO_4$) → CH_2 = CH – CHO

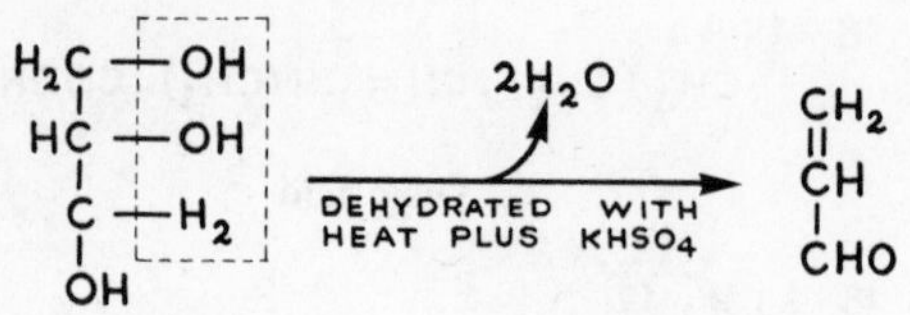

Glycerol **Acrolein (recognized by its pungent odor)**

FIG 2–2. Conversion of glycerol to acrolein.

Fatty Aldehydes

The fatty acids may be reduced to fatty aldehydes. These compounds are found either combined or free in natural fats.

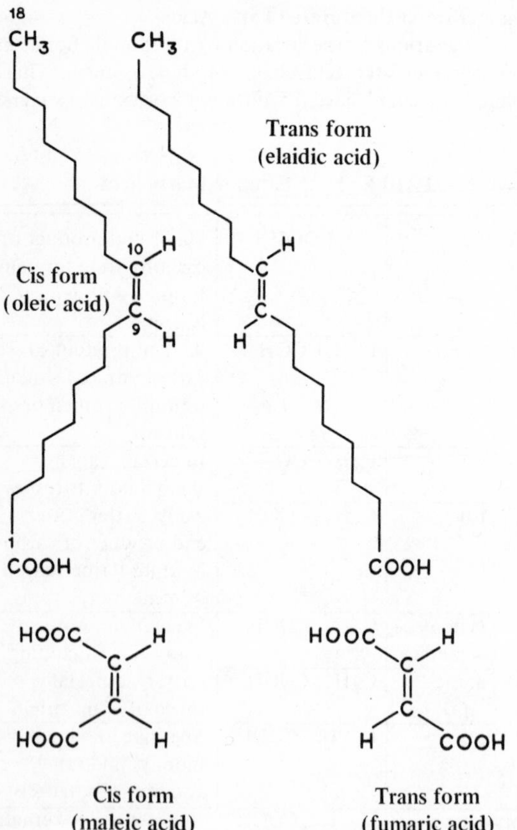

FIG 2–1. Geometric isomerism of oleic and elaidic acids and of maleic and fumaric acids.

STEROIDS

The steroids are often found in association with fat. They may be separated from the fat after the fat is saponified, since they occur in the "unsaponifiable residue." All of the steroids have a similar cyclic nucleus resembling phenanthrene (rings A, B, and C) to which a cyclopentane ring (D) is attached. However, the rings are not uniformly unsaturated, so the parent (completely saturated) substance is better designated as cyclopentanoperhydrophenanthrene. The positions on the steroid nucleus are numbered as shown below.

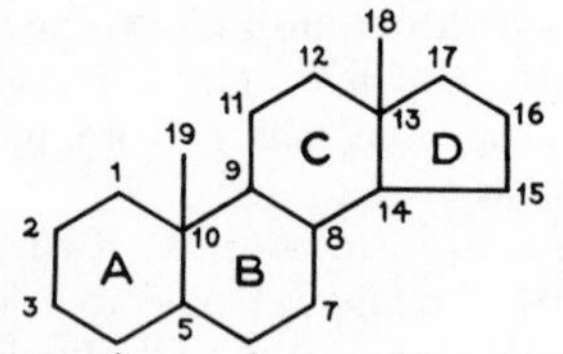

Cyclopentanoperhydrophenanthrene nucleus

It is important to realize that in considering structural formulas of steroids a simple hexagonal ring denotes a completely saturated 6-carbon ring with all valences satisfied by hydrogen bonds unless shown otherwise, ie, it is not a benzene ring. All double bonds are shown as such. Methyl side chains are shown as single bonds unattached at the farther (methyl) end. These occur typically at positions 10 and 13 (constituting C atoms 19 and 18). A side chain at position 17 is usual (as in cholesterol). If the compound has one or more hydroxyl groups and no carbonyl or carboxyl groups, it is a **sterol**, and the name terminates in -ol.

Stereochemical Aspects

Because of their complexity and the possibilities of asymmetry in the molecule, steroids have many potential stereoisomers. Each of the 6-carbon rings of the steroid nucleus is capable of existing in the 3-dimensional conformation either of a "chair" or of a "boat."

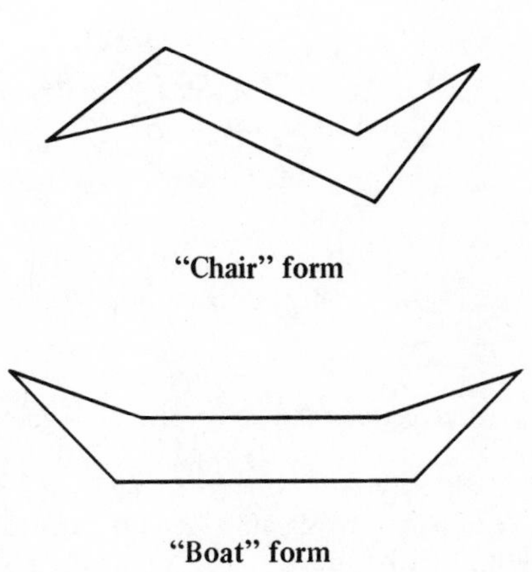

"Chair" form

"Boat" form

In naturally occurring steroids, virtually all the rings are in the "chair" form, which is the more stable conformation. With respect to each other, the rings can be either -cis or -trans. (See Fig 2–3.)

The junction between the A and B rings is -cis or -trans in naturally occurring steroids. That between B and C is -trans and the C/D junction is -trans except in cardiac glycosides and toad poisons (Klyne, 1965). Bonds attaching substituent groups above the plane of the rings are shown with bold solid lines (β), whereas those bonds attaching groups below are indicated with broken lines (α). The A ring of a 5α steroid is always -trans to the B ring, whereas it is -cis in a 5β steroid. The methyl groups attached to C_{10} and C_{13} are invariably in the β configuration.

Cholesterol

Cholesterol is widely distributed in all cells of the body, but particularly in nervous tissue. It occurs in animal fats but not in plant fats. The metabolism of cholesterol is discussed on p 294. Cholesterol is designated as 3-hydroxy-5,6-cholestene.

HO

Cholesterol

HO

Ergosterol

Ergosterol

Ergosterol occurs in ergot and yeast. It is important as a precursor of vitamin D. When irradiated with ultraviolet light, it acquires antirachitic properties consequent to the opening of ring B.

Coprosterol

Coprosterol (coprostanol) occurs in feces as a result of the reduction by bacteria in the intestine of the double bond of cholesterol between C_5 and C_6.

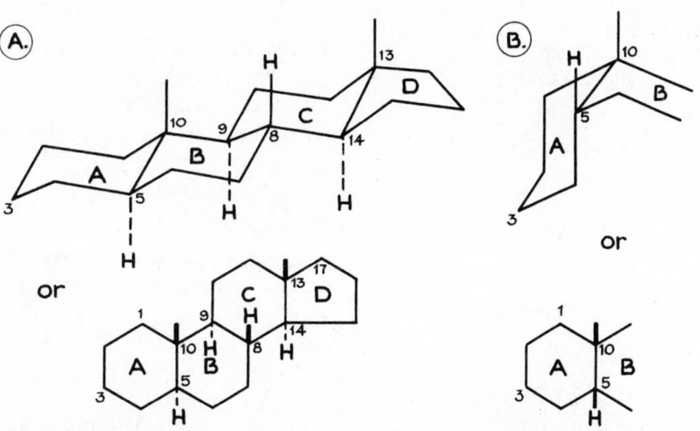

FIG 2–3. Generalized steroid nucleus, showing (a) an all-trans configuration between adjacent rings and (b) a cis configuration between rings A and B.

The orientation of rings A and B (between carbon atoms 5 and 10), which is **trans** in cholesterol, is **cis** in coprosterol.

Other Important Steroids

These include the bile acids, adrenocortical hormones, sex hormones, D vitamins, cardiac glycosides, the sitosterols of the plant kingdom, and some alkaloids.

Color Reactions to Detect Sterols

Saturated sterols (eg, coprosterol) do not give these color tests.

Liebermann-Burchard reaction. A chloroform solution of a sterol, when treated with acetic anhydride and sulfuric acid, gives a green color. The usefulness of this reaction is limited by the fact that various sterols give the same or a similar color. This reaction is the basis of a colorimetric estimation of blood cholesterol.

Salkowski test. A red to purple color appears when a chloroform solution of the sterol is treated with an equal volume of concentrated sulfuric acid.

Digitonin, $C_{56}H_{92}O_{29}$ (a glycoside occurring in digitalis leaves and seeds), precipitates cholesterol as the digitonide if the hydroxyl group in position 3 is free. This reaction serves as a method for the separation of free cholesterol and cholesterol esters.

For further information on the chemical properties and separation of steroids, see Klyne, 1965; Bush, 1954.

TRIGLYCERIDES

The triglycerides, or so-called neutral fats, are esters of the alcohol, glycerol, and fatty acids. In naturally occurring fats, the proportion of triglyceride molecules containing the same fatty acid residue in all 3 ester positions is very small. They are nearly all mixed glycerides.

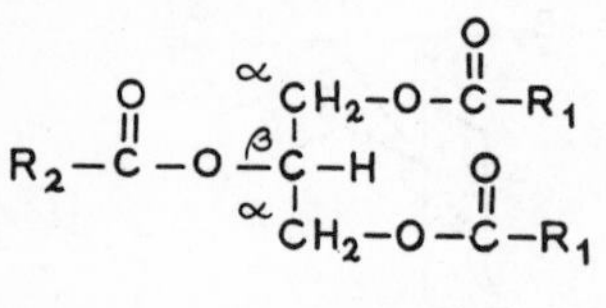

$$\begin{array}{l} \quad\quad\quad\quad\quad\quad\ ^{\alpha}CH_2-O-\overset{O}{\overset{\|}{C}}-R_1 \\ R_2-\overset{O}{\overset{\|}{C}}-O-\overset{\beta}{C}-H \\ \quad\quad\quad\quad\quad\quad\ ^{\alpha}CH_2-O-\overset{O}{\overset{\|}{C}}-R_1 \end{array}$$

or

$$\begin{array}{l} \quad\quad\quad\quad\quad\quad\ ^{1}CH_2-O-\overset{O}{\overset{\|}{C}}-R_1 \\ R_2-\overset{O}{\overset{\|}{C}}-O-\overset{2}{C}-H \\ \quad\quad\quad\quad\quad\quad\ ^{3}CH_2-O-\overset{O}{\overset{\|}{C}}-R_2 \end{array}$$

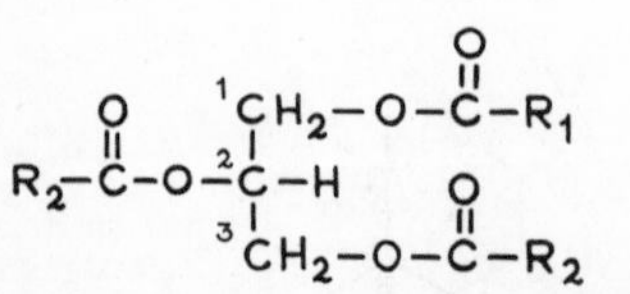

In the above example, if all 3 fatty acids were the same and if R were $C_{17}H_{35}$, the fat would be known as tristearin, since it consists of 3 stearic acid residues esterified with glycerol. In a mixed glyceride, more than one fatty acid is involved, eg:

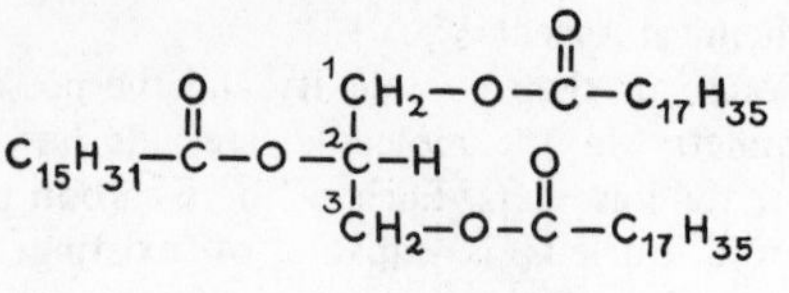

$$\begin{array}{l} \quad\quad\quad\quad\quad\quad\quad\ ^{1}CH_2-O-\overset{O}{\overset{\|}{C}}-C_{17}H_{35} \\ C_{15}H_{31}-\overset{O}{\overset{\|}{C}}-O-\overset{2}{C}-H \\ \quad\quad\quad\quad\quad\quad\quad\ ^{3}CH_2-O-\overset{O}{\overset{\|}{C}}-C_{17}H_{35} \end{array}$$

1,3-Distearopalmitin
(or α,α'-distearopalmitin)

$$\begin{array}{l} \quad\quad\quad\quad\quad\quad\quad\ ^{1}CH_2-O-\overset{O}{\overset{\|}{C}}-C_{17}H_{35} \\ C_{17}H_{35}-\overset{O}{\overset{\|}{C}}-O-\overset{2}{C}-H \\ \quad\quad\quad\quad\quad\quad\quad\ ^{3}CH_2-O-\overset{O}{\overset{\|}{C}}-C_{15}H_{31} \end{array}$$

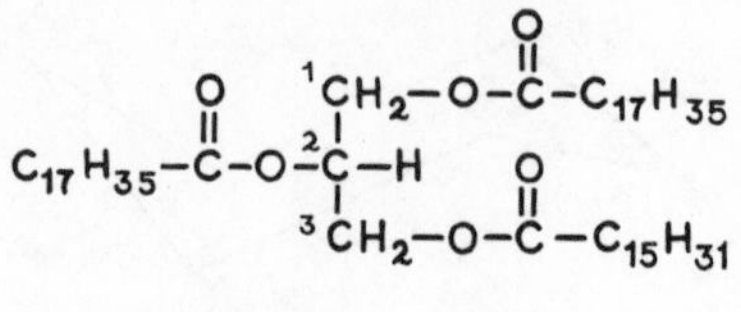

1,2-Distearopalmitin
(or α,β-distearopalmitin)

Partial glycerides consisting of mono- and diglycerides wherein a single fatty acid or 2 fatty acids are esterified with glycerol are also found in the tissues. These are of particular significance in the synthesis and hydrolysis of triglycerides.

Waxes

If the fatty acid is esterified with an alcohol of high molecular weight instead of with glycerol, the resulting compound is called a wax.

PHOSPHOLIPIDS

The phospholipids include the following groups: (1) phosphatidic acid and phosphatidyl glycerols, (2) phosphatidyl choline, (3) phosphatidyl ethanolamine, (4) phosphatidyl inositol, (5) phosphatidyl serine, (6) lysophospholipids, (7) plasmalogens, and (8) sphingomyelins.

Phosphatidic Acid & Phosphatidyl Glycerols

Phosphatidic acid is important as an intermediate in the synthesis of triglycerides and phospholipids but is not found in any great quantity in tissues.

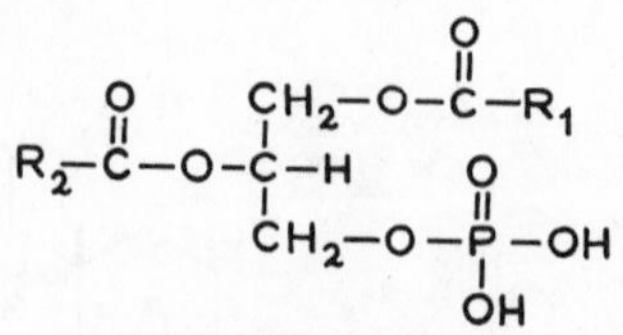

$$\begin{array}{l} \quad\quad\quad\quad\quad\quad\ CH_2-O-\overset{O}{\overset{\|}{C}}-R_1 \\ R_2-\overset{O}{\overset{\|}{C}}-O-C-H \\ \quad\quad\quad\quad\quad\quad\ CH_2-O-\overset{O}{\overset{\|}{P}}-OH \\ \quad\quad\quad\quad\quad\quad\quad\quad\quad\quad\quad\ \ OH \end{array}$$

Phosphatidic acid

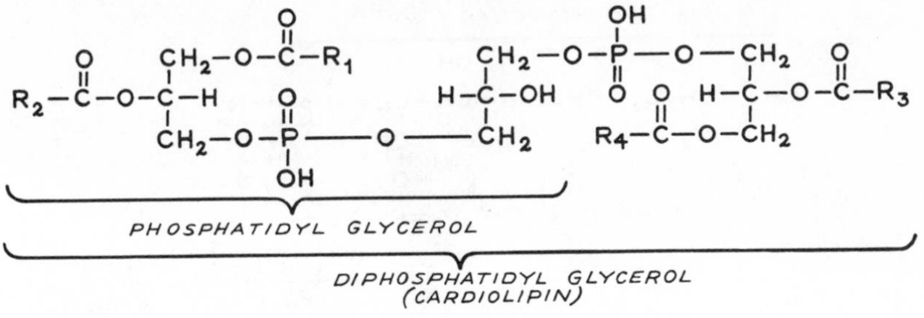

PHOSPHATIDYL GLYCEROL

DIPHOSPHATIDYL GLYCEROL (CARDIOLIPIN)

FIG 2–4. Diphosphatidyl glycerol (cardiolipin).

Cardiolipin is a phospholipid which is found in mitochondria. It is formed from **phosphatidyl glycerol** (Fig 2–4).

Phosphatidyl Choline (Lecithins)

The lecithins contain glycerol and fatty acids, as do the simple fats, but they also contain phosphoric acid and choline. The lecithins are widely distributed in the cells of the body, having both metabolic and structural functions. Dipalmityl lecithin is a very effective surface active agent, preventing adherence, due to surface tension, of the inner surfaces of the lungs. However, most phospholipids have a saturated acyl radical in the C_1 position and an unsaturated radical in the C_2 position.

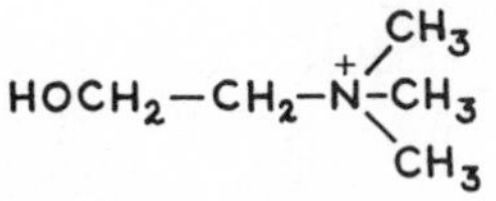

$HOCH_2{-}CH_2{-}\overset{+}{N}(CH_3)_3$ **Choline**

The β-Lecithin contains phosphoric acid-choline complex on center, or β carbon of glycerol

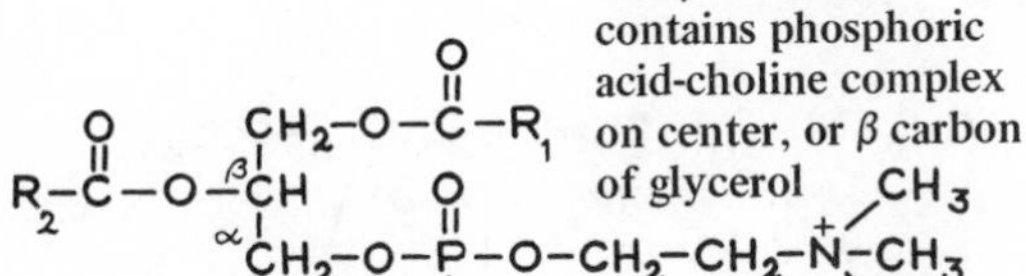

Example of an alpha lecithin (3-phosphatidyl choline)

Phosphatidyl Ethanolamine (Cephalins)

The cephalins differ from lecithins only in that ethanolamine replaces choline. Both alpha and beta cephalins are known.

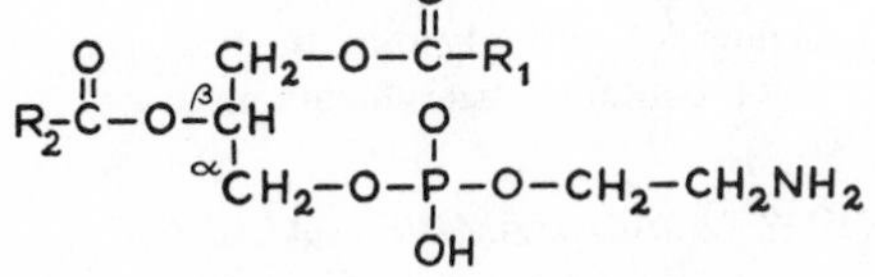

Example of an alpha cephalin (3-phosphatidyl ethanolamine)

Phosphatidyl Inositol (Lipositols)

Inositol (see p 114) as a constituent of lipids was first discovered in acid-fast bacteria. Later it was found to occur in phospholipids of brain tissue and of soybeans as well as in other plant phospholipids. The inositol is present as the stereoisomer, myo-inositol.

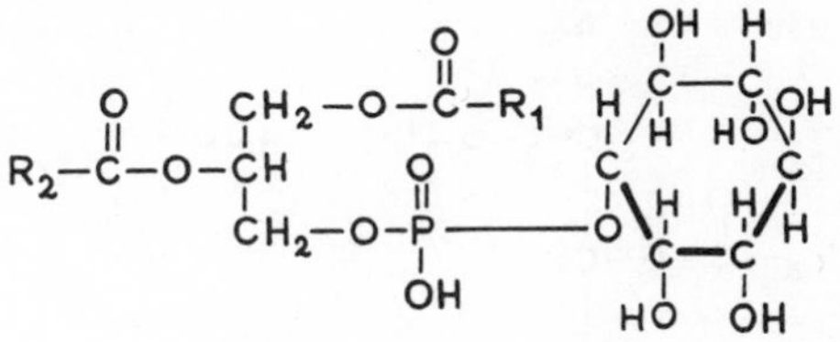

3-Phosphatidyl inositol

Phosphatidyl Serine

A cephalin-like phospholipid, phosphatidyl serine, which contains the amino acid serine rather than ethanolamine, has been found in tissues. In addition, phospholipids containing threonine have been isolated from natural sources.

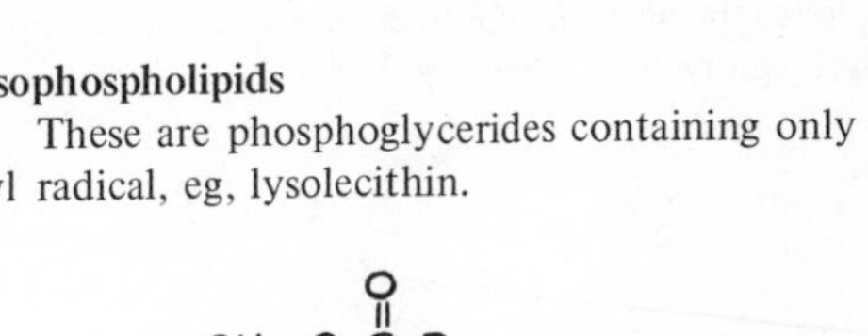

SERINE

3-Phosphatidyl serine

Lysophospholipids

These are phosphoglycerides containing only one acyl radical, eg, lysolecithin.

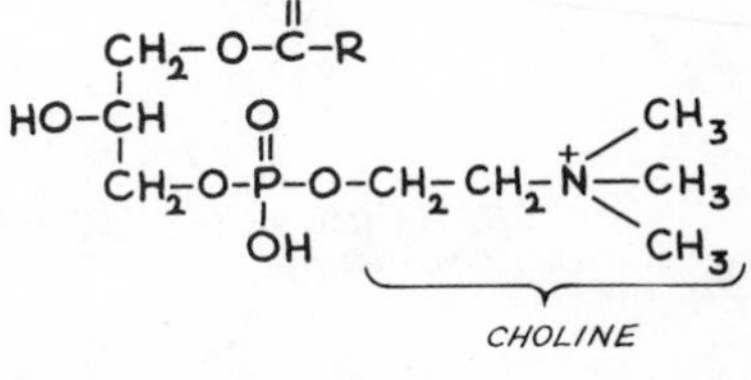

CHOLINE

Lysolecithin

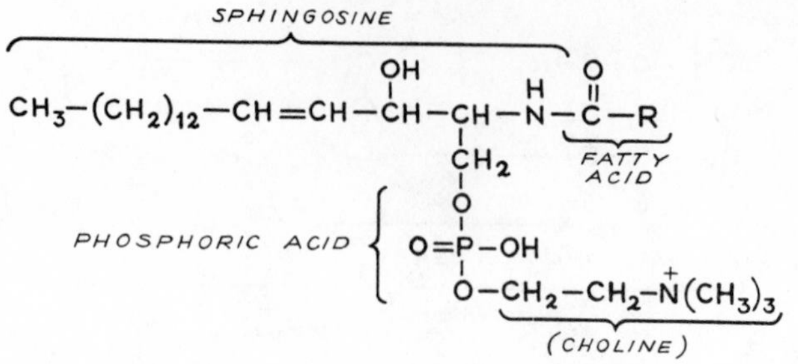

$$\underbrace{CH_3-(CH_2)_{12}-CH=CH-\underset{|}{\overset{OH}{CH}}-CH}_{SPHINGOSINE}(-CH_2-O-\underbrace{P(=O)(-OH)}_{PHOSPHORIC\ ACID}-O-\underbrace{CH_2-CH_2-\overset{+}{N}(CH_3)_3}_{(CHOLINE)})-\overset{H}{N}-\underbrace{\overset{O}{\overset{\|}{C}}-R}_{FATTY\ ACID}$$

FIG 2–5. Structure of a sphingomyelin.

Plasmalogens

These compounds constitute as much as 10% of the phospholipids of the brain and muscle. Structurally, the plasmalogens resemble lecithins and cephalins but give a positive reaction when tested for aldehydes with Schiff's reagent (fuchsin-sulfurous acid) after pretreatment of the phospholipid with mercuric chloride.

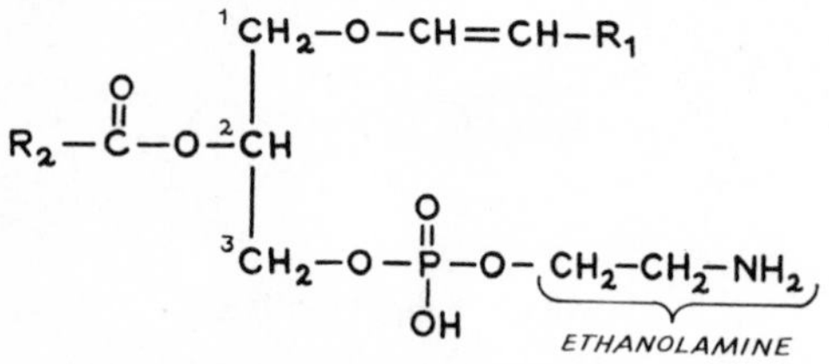

$$\begin{array}{l} {}^{1}CH_2-O-CH=CH-R_1 \\ R_2-\overset{O}{\overset{\|}{C}}-O-{}^{2}CH \\ {}^{3}CH_2-O-\underset{OH}{\overset{O}{\overset{\|}{P}}}-O-\underbrace{CH_2-CH_2-NH_2}_{ETHANOLAMINE} \end{array}$$

Structure of plasmalogen (phosphatidal ethanolamine)

Plasmalogens possess an ether link on the C_1 carbon instead of the normal ester link found in most glycerides. Typically, the alkyl radical is an unsaturated alcohol; it is this group that gives rise to a positive aldehyde test after the above treatment.

In some instances, choline, serine, or inositol may be substituted for ethanolamine.

Sphingomyelins

Sphingomyelins are found in large quantities in brain and nerve tissue (see Chapter 22). No glycerol is present. On hydrolysis the sphingomyelins yield a fatty acid, phosphoric acid, choline, and a complex amino alcohol, sphingosine. (See Fig 2–5.)

CEREBROSIDES (Glycolipids)

Cerebrosides contain galactose, a high molecular weight fatty acid, and sphingosine. Therefore, they may also be classified with the sphingomyelins as sphingolipids. Individual cerebrosides are differentiated by the type of fatty acid in the molecule. These are **kerasin**, containing lignoceric acid; **cerebron**, with a hydroxy lignoceric acid (cerebronic acid); **nervon**, containing an unsaturated homologue of lignoceric acid called nervonic acid; and **oxynervon**, having apparently the hydroxy derivative of nervonic acid as its constituent fatty acid. Stearic acid is a major component of the fatty acids of rat brain cerebrosides.

$$CH_3-(CH_2)_{22}-COOH$$

LIGNOCERIC ACID

$$CH_3-(CH_2)_{21}-CH(OH)-COOH$$

CEREBRONIC ACID

$$CH_3-(CH_2)_7-CH=CH-(CH_2)_{13}-COOH$$

NERVONIC ACID

$$CH_3-(CH_2)_7-CH=CH-(CH_2)_{12}-CH(OH)-COOH$$

OXYNERVONIC ACID

The cerebrosides are found in many tissues besides brain. In Gaucher's disease, the cerebroside content of the reticuloendothelial cells (eg, the spleen) is very high and the kerasin is characterized by glucose replacing galactose in the cerebroside molecule. The cerebrosides are in much higher concentration in medullated than in nonmedullated nerve fibers.

Sulfatides are sulfate derivatives of the galactosyl residue in cerebrosides.

Gangliosides are glycolipids, occurring in the brain, which contain **N-acetylneuraminic acid** (NANA)

CERAMIDE – GLUCOSE – GALACTOSE – N – ACETYLGALACTOSAMINE – GALACTOSE
(ACYLSPHINGOSINE)
NANA

A ganglioside

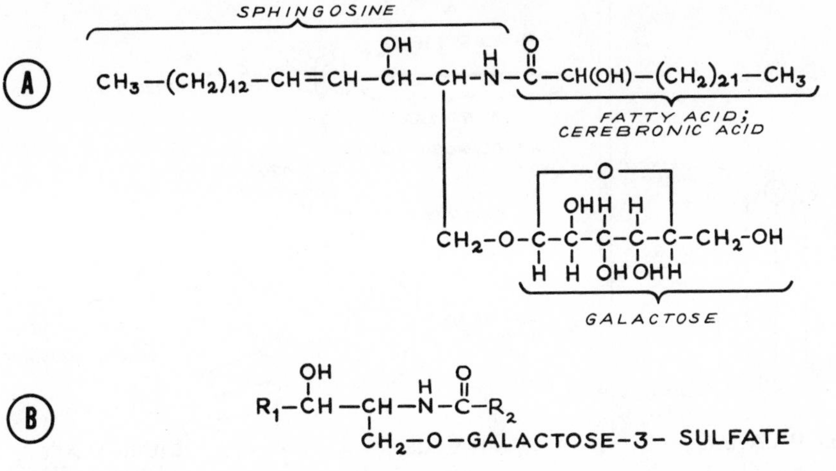

SPHINGOSINE

(A) $CH_3-(CH_2)_{12}-CH=CH-CH(OH)-CH-NH-C(=O)-CH(OH)-(CH_2)_{21}-CH_3$

FATTY ACID; CEREBRONIC ACID

CH_2-O- GALACTOSE

(B) $R_1-CH(OH)-CH-NH-C(=O)-R_2$

CH_2-O-GALACTOSE-3- SULFATE

FIG 2–6. Structure of (A) a cerebroside and (B) a sulfatide (cerebroside sulfate).

in addition to C_{24} or C_{22} fatty acids, sphingosine, and 3 molecules of hexose (glucose and galactose). A hexosamine may be substituted for neuraminic acid in some gangliosides, or both may be present.

CHARACTERISTIC CHEMICAL REACTIONS & PROPERTIES OF THE LIPIDS

Hydrolysis

Hydrolysis of a lipid such as a triglyceride may be accomplished enzymatically through the action of lipases, yielding fatty acids and glycerol. Use may be made of the property of pancreatic lipase to attack the ester bonds in positions 1 and 3 preferentially to position 2 of triglycerides. **Phospholipases** attack the various ester linkages in phospholipids. Their specificity may be used to analyze the components of phospholipids. The sites of action of the various phospholipases are shown in Fig 2–7.

Saponification

Hydrolysis of a fat by alkali is called **saponification.** The resultant products are glycerol and the alkali salts of the fatty acids, which are called **soaps.** Acid hydrolysis of a fat yields the free fatty acids and glycerol. Soaps are cleansing agents because of their emulsifying action. Some soaps of high molecular weight and a considerable degree of unsaturation are selective germicides. Others, such as sodium ricinoleate, have detoxifying activity against diphtheria and tetanus toxins.

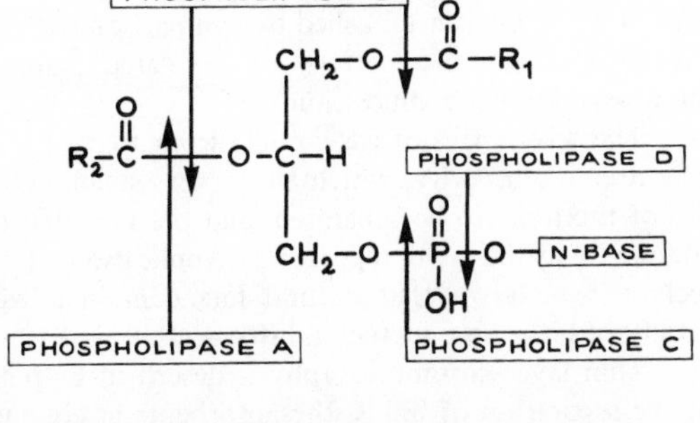

PHOSPHOLIPASE B

PHOSPHOLIPASE A

PHOSPHOLIPASE D

PHOSPHOLIPASE C

$CH_2-O-C(=O)-R_1$

$R_2-C(=O)-O-C-H$

$CH_2-O-P(=O)(OH)-O-$ N-BASE

FIG 2–7. Sites of hydrolysis of phospholipases.

Formation of Membranes, Micelles, & Emulsions

In general, lipids are insoluble in water since they contain a predominance of nonpolar (hydrocarbon) groups. However, fatty acids, some phospholipids, and sphingolipids (the polar lipids) contain a greater proportion of polar groups and are therefore partly soluble in water and partly soluble in nonpolar solvents. The molecules thus become oriented at oil-water interfaces with the polar group in the water phase and the nonpolar group in the oil phase. A bilayer of such polar lipids has been regarded as a basic structure in biologic membranes (Davson-Danielli model). When a critical concentration of polar lipids is present in an aqueous medium, they form micelles. Aggregations of bile salts into micelles and the formation of mixed micelles with the products of fat digestion may be important in facilitating absorption of lipids from the intestine. Emulsions are much larger particles, formed usually by nonpolar lipids in an aqueous medium. These are stabilized by emulsifying agents such as polar lipids (eg, lecithin), which form a surface layer separating the main bulk of the nonpolar material from the aqueous phase (Fig 2–8).

Analytic Methods for the Characterization of Fats

These include a determination of melting point, solidification temperature, and refractive index, as well as certain **chemical determinations,** as follows:

(1) Saponification number: The number of milligrams of KOH required to saponify 1 g of fat or oil. It varies inversely with the molecular weight of the fat or oil.

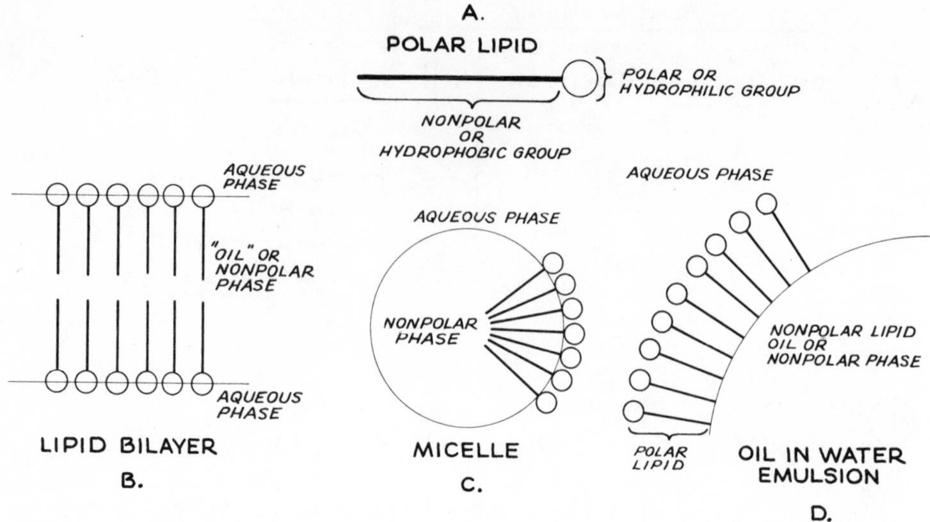

FIG 2–8. Formation of lipid membranes, micelles, and emulsions.

(2) Acid number: The number of milligrams of KOH required to neutralize the free fatty acid of 1 g of fat.

(3) Polenske number: The number of milliliters of 0.1 normal KOH required to neutralize the insoluble fatty acids (those not volatile with steam distillation) from 5 g of fat.

(4) Reichert-Meissl number: This is the same as the Polenske number except that, after a 5 g sample of the fat has been saponified, the **soluble** fatty acids are measured by titration of the distillate obtained by steam distillation of the saponification mixture.

(5) Iodine number: In the presence of iodine monobromide (Hanus method) or of iodine monochloride (Wijs method), unsaturated lipids will take up iodine. The iodine number is the amount (in grams) of iodine absorbed by 100 g of fat. This is a measure of the degree of unsaturation of a fat. Oils such as linseed oil or cottonseed oil have higher iodine numbers than solid fats such as tallow or beef fat because the former contain more unsaturated fatty acids in the fat molecule.

(6) Acetyl number: The number of milligrams of KOH required to neutralize the acetic acid obtained by saponification of 1 g of fat after it has been acetylated. This is a measure of the number of hydroxy-acid groups in the fat. Castor oil, because of its high content of ricinoleic acid, a fatty acid containing one OH group, has a high acetyl number (about 146).

The older, classical methods of separation and identification of fatty acids have now been largely supplanted by chromatographic procedures (see p 35). Particularly useful for analysis of fatty acids are the technics of gas-liquid chromatography (James, 1956, 1957, Farquhar, 1959) and thin layer chromatography. Before these technics are applied to wet tissues, the lipids are extracted by a solvent system based usually on a mixture of chloroform-methanol (2:1) introduced by Folch & others (1957).

Gas-liquid chromatography involves the physical separation of a moving gas phase by adsorption onto a stationary phase consisting of an inert solid such as silica gel or inert granules of ground firebrick coated with a nonvolatile liquid (eg, lubricating grease or silicone oils). In practice, a glass or metal column is packed with the inert solid and a mixture of the methyl esters of fatty acids is evaporated at one end of the column, the entire length of which is kept at temperatures of 170–225° C. A constantly flowing stream of an inert gas such as argon or helium keeps the volatilized esters moving through the column. As with other types of chromatography, separation of the vaporized fatty acid esters is dependent upon the differing affinities of the components of the gas mixture for the stationary phase. Gases which are strongly attracted to the stationary phase move through the column at a slower rate and therefore emerge at the end of the column later than those that are relatively less attracted. As the individual fatty acid esters emerge from the column, they are detected by physical or chemical means and recorded automatically as a series of peaks which appear at different times according to the tendency of each fatty acid ester to be retained by the stationary phase. The area under each peak is proportionate to the concentration of a particular component of the mixture. The identity of each component is established by comparison with the gas chromatographic pattern of a related standard mixture of known composition.

The advantages of gas-liquid chromatography are its extreme sensitivity, which allows very small quantities of mixtures to be separated, and the fact that the columns may be used repeatedly. Application of the technic has shown that natural fats contain a wide variety of hitherto undetected fatty acids.

Thin layer chromatography is described on p 35. In the separation of lipids, the absorbents in common use include silica gel, alumina, and infusorial earth (kieselguhr). As well as being used for analytical purposes, thin layer chromatography may be used in purifying lipids when milligram quantities of lipid may be applied as a band to one plate. For a recent review of general analytical methods applicable to lipids, see Lowenstein (1969).

Unsaponifiable Matter

Unsaponifiable matter includes substances in natural fats that cannot be saponified by alkali but are soluble in ether or petroleum ether. Since soaps are not ether-soluble, they may be separated from lipid mixtures by extraction with these solvents following saponification of the fat. Ketones, hydrocarbons, high molecular weight alcohols, and the steroids are examples of unsaponifiable residues of natural fats.

Hydrogenation

Hydrogenation of unsaturated fats in the presence of a catalyst (nickel) is known as "hardening." It is commercially valuable as a method of converting these fats, usually of plant origin, into solid fats as lard substitutes or margarines.

Rancidity

Rancidity is a chemical change that results in unpleasant odors and taste in a fat. The oxygen of the air is believed to attack the double bond to form a peroxide linkage. The iodine number is thus reduced, although little free fatty acid and glycerol are released. Lead or copper catalyzes rancidity; exclusion of oxygen or the addition of an antioxidant delays the process.

Spontaneous Oxidation

Oils that contain highly unsaturated fatty acids (eg, linseed oil) are spontaneously oxidized by atmospheric oxygen at ordinary temperatures and form a hard, waterproof material. Such oils are added for this purpose to paints and shellacs. They are then known as "drying oils."

• • •

References

Bush IE: Brit M Bull 10:232, 1954.

Farquhar JW, Insull W Jr, Rosen P, Stoffel W, Ahrens EH Jr: Nutr Rev, vol 17 (suppl), Aug 1959.

Folch J, Lees M, Sloane-Stanley GH: J Biol Chem 226:497, 1957.

James AT, Martin AJP: Biochem J 63:144, 1956.

James AT, Webb J: Biochem J 66:515, 1957.

General Bibliography

Ansell GB, Hawthorne JN: *Phospholipids: Chemistry, Metabolism and Function.* Elsevier, 1964.

Deuel HJ Jr: *The Lipids: Their Chemistry and Biochemistry.* 3 vols. Wiley, 1951–1957.

Florkin M, Mason HS (editors): *Comparative Biochemistry.* Vol 3, part A, Chapters 1–5 & 10. Vol 6, part B, Chapter 14. Academic Press, 1962.

Gunstone FD: *An Introduction to the Chemistry and Biochemistry of Fatty Acids and Their Glycerides,* 2nd ed. Chapman & Hall, 1967.

Hanahan DJ: *Lipide Chemistry.* Wiley, 1960.

Hilditch TP, Williams PN: *The Chemical Constitution of Natural Fats,* 4th ed. Chapman & Hall, 1964.

Klyne W: *The Chemistry of the Steroids.* Methuen, 1965.

Lowenstein JM (editor): *Methods in Enzymology.* Vol 14. Academic Press, 1969.

Pecsok RL (editor): *Principles and Practice of Gas Chromatography.* Wiley, 1959.

Ralston AW: *Fatty Acids and Their Derivatives.* Wiley, 1948.

3...

Amino Acids & Proteins

With Victor Rodwell, PhD *

Living cells produce an impressive variety of **macromolecules**, chiefly **proteins** and **nucleic acids**, which serve as structural components, as biocatalysts, as hormones, and as repositories for the genetic information characteristic of a species. These macromolecules are **biopolymers** constructed of distinct **monomer units** or **building blocks**. For nucleic acids, the monomer units are **nucleotides** (see Chapter 4); for proteins, the **L-α-amino acids**. While many proteins contain substances in addition to amino acids, the 3-dimensional structure and many of the biologic properties of proteins are determined largely by the **kinds of amino acids present**, the **order in which they are linked together** in a polypeptide chain, and the **spatial relationship of one amino acid to another**.

AMINO ACIDS

STRUCTURES OF AMINO ACIDS

Proteins may be broken down to their monomer units by acid-, base-, or enzyme-catalyzed **hydrolysis**.† Complete hydrolysis of most proteins produces approximately 20 different α-amino acids.

Alpha-amino acids have both an amino and a carboxylic acid function attached to the same (α) carbon atom.

Amino acids are classified into 7 groups in accordance with the structure of the side chain (R) (see Table 3–1).

$$R-\overset{\overset{H}{|\alpha}}{\underset{\underset{COOH}{|}}{C}}-NH_2$$

An α-amino acid

With a single exception (glycine), each amino acid has at least one **asymmetric carbon atom** and hence is **optically active**, ie, it can rotate the plane of plane-polarized light. Although some of the amino acids found in proteins are dextrorotatory and some are levorotatory at pH 7.0, all have **absolute configurations** comparable to that of L-glyceraldehyde (see p 2) and hence are **L-α-amino acids**. Although D-amino acids do occur in cells and even in polypeptides (eg, in polypeptide antibiotics elaborated by certain microorganisms), they are not present in proteins.

Threonine, isoleucine, 4-hydroxyproline, and hydroxylysine each have 2 asymmetric carbon atoms and therefore exist in 4 isomeric forms. Of these, 2 are forms of allothreonine or of alloisoleucine, etc (Fig 3–1). Although a sheep liver enzyme (allothreonine aldolase) acts on allothreonine, neither allothreonine nor alloisoleucine appears to occur in nature.

Note (in Fig 3–1) that I–II and III–IV form **enantiomeric pairs** (see p 502) and hence have similar chemical properties. I–III and II–IV are **diastereoisomeric pairs** with different chemical properties. The configurational structure for the 4-hydroxy-L-proline found in proteins is as shown in Fig 3–2.

Various other amino acids (see Table 3–2) in free or combined states fulfill important roles in metabolic processes other than as constituents of proteins. Many additional amino acids occur in plants or in antibiotics. Over 20 D-amino acids occur naturally. These include the D-alanine and D-glutamic acid of certain bacterial cell walls and a variety of D-amino acids in antibiotics.

Amino acids have both **trivial names** (eg, glycine, tryptophan) and **systematic** chemical names. Two methods of systematic nomenclature are used. In one, **the carbon atom bearing the carboxyl and amino groups is termed α, the next β**, etc. In the other, the familiar numbering system of organic chemistry is used. Note, however, that the **α-C is C-2, the β is C-3**, etc.

PROTONIC EQUILIBRIA OF AMINO ACIDS

The term "protonic equilibria" refers to association and dissociation of protons. Acids are **proton donors**; bases are **proton acceptors**. A distinction is

*Associate Professor of Biochemistry, Purdue University, Lafayette, Indiana.
†Hydrolysis = rupture of a covalent bond with addition of the elements of water.

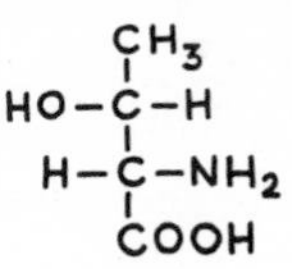

CH_3 HO–C–H H–C–NH_2 COOH	CH_3 H–C–OH H_2N–C–H COOH	CH_3 H–C–OH H–C–NH_2 COOH	CH_3 HO–C–H H_2N–C–H COOH
L-Threonine **(I)**	**D-Threonine** **(II)**	**L-Allothreonine** **(III)**	**D-Allothreonine** **(IV)**

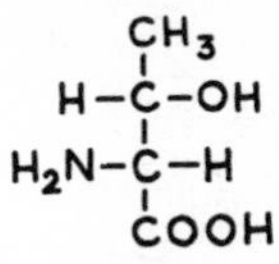

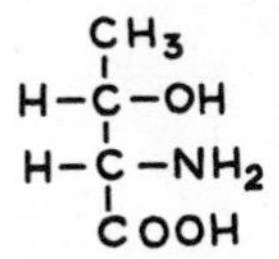

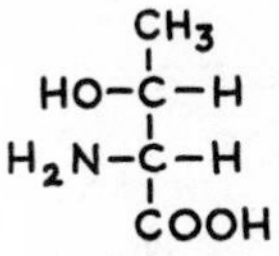

FIG 3–1. Isomers of threonine.

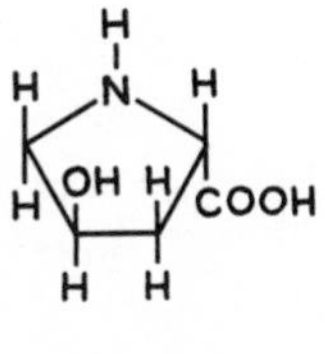

FIG 3–2. 4-Hydroxy-L-proline.

made between **strong acids** (eg, HCl, H_2SO_4), which are completely dissociated even in strongly acidic solutions (ie, at low pH); and **weak acids**, which dissociate only partially in acidic solution. A similar distinction is made between **strong bases** (eg, KOH, NaOH) and **weak bases** (eg, $Ca[OH]_2$). Only strong bases are dissociated at very high pH. Most charged biochemical intermediates, including amino acids, are **weak acids.** Exceptions include phosphorylated intermediates (eg, sugar phosphates), which also possess the strongly acidic phosphoric acid group.

Amino acids bear at least 2 ionizable weak acid groups, a –COOH and an $-NH_3^+$. In solution, 2 forms of these groups, one charged and one neutral, exist in protonic equilibrium with each other:

$$R\text{–}COOH \rightleftharpoons R\text{–}COO^- + H^+$$

$$R\text{–}NH_3^+ \rightleftharpoons H^+ + R\text{–}NH_2$$

R–COOH and $R\text{–}NH_3^+$ represent the **protonated** or **acid** partners in these equilibria. $R\text{–}COO^-$ and $R\text{–}NH_2$ are the **conjugate bases** (ie, proton acceptors) of the corresponding acids. Although both RCOOH and $R\text{–}NH_3^+$ are weak acids, RCOOH is a several thousand times stronger acid than is $R\text{–}NH_3^+$. At physiologic pH (7.4), carboxyl groups exist almost entirely as the conjugate base, ie, the **carboxylate ion**, $R\text{–}COO^-$. At the same pH, most amino groups are predominantly in the associated (protonated) form, $R\text{–}NH_3^+$. In terms of the prevalent ionic species present in blood and most tissues, we should draw amino acid structures as shown in A:

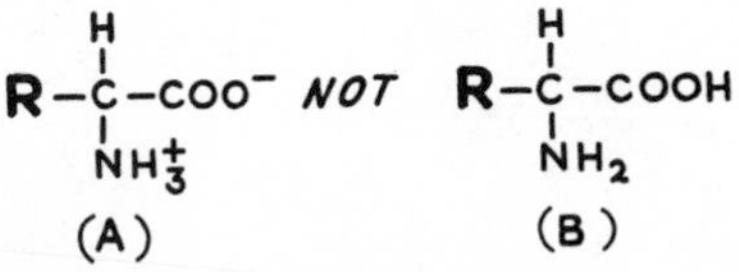

H R–C–COO^- NH_3^+	*NOT*	H R–C–COOH NH_2
(A)		(B)

Structure B cannot exist at **any** pH. At any pH sufficiently low to repress ionization of the comparatively strong carboxyl group, the more weakly acidic amino group would be protonated. If the pH is raised, the proton from the carboxyl will be lost long before that from the $R\text{–}NH_3^+$. At any pH sufficiently high to cause $R\text{–}NH_2$ to be the predominant species, the carboxylate ion ($R\text{–}COO^-$) must also be present. Convenience dictates, however, that the B representation be used for many equations involving reactions other than protonic equilibria.

The relative acid strengths of weak acids may be expressed in terms of their **dissociation constants** (see p 494). The larger the dissociation constant, the stronger the acid. Biochemists more often refer to the **pK** of an acid. This is merely the negative log of the dissociation constant, ie,

$$R\text{–}COOH \rightleftharpoons R\text{–}COO^- + H^+$$

$$K = \frac{[R\text{–}COO^-]\,[H^+]}{[R\text{–}COOH]}$$

$$pK = -\log K$$

$$R\text{–}NH_3^+ \rightleftharpoons R\text{–}NH_2 + H^+$$

$$K = \frac{[R\text{–}NH_2]\,[H^+]}{[R\text{–}NH_3]}$$

$$pK = -\log K$$

The form [x] means "concentration of x." Inspection of the above equation reveals that when

$$[R\text{–}COO^-] = [R\text{–}COOH]$$

or when

$$[RNH_2] = [R\text{–}NH_3^+]$$

then

$$K = [H^+]$$

In words, **when the associated (protonated) and dissociated (conjugate base) species are present in equal concentration, the prevailing hydrogen ion con-**

TABLE 3–1. L-α-Amino acids found in proteins.

Group	Trivial Name	Abbreviation	Chemical Name	Structural Formula
	With Aliphatic Side Chains			
I	Glycine*	Gly	Aminoacetic acid	$H-CH(NH_2)-COOH$
	Alanine	Ala	α-Aminopropionic acid	$CH_3-CH(NH_2)-COOH$
	Valine	Val	α-Aminoisovaleric acid	$(H_3C)_2CH-CH(NH_2)-COOH$
	Leucine	Leu	α-Aminoisocaproic acid	$(H_3C)_2CH-CH_2-CH(NH_2)-COOH$
	Isoleucine	Ile	α-Amino-β-methylvaleric acid	$CH_3-CH_2-CH(CH_3)-CH(NH_2)-COOH$
	With Side Chains Containing Hydroxylic (OH) Groups			
II	Serine	Ser	α-Amino-β-hydroxypropionic acid	$CH_2(OH)-CH(NH_2)-COOH$
	Threonine	Thr	α-Amino-β-hydroxy-n-butyric acid	$CH_3-CH(OH)-CH(NH_2)-COOH$
	With Side Chains Containing Sulfur Atoms			
III	Cysteine†	Cys	α-Amino-β-mercaptopropionic acid	$CH_2(SH)-CH(NH_2)-COOH$
	Methionine	Met	α-Amino-γ-methylthio-n-butyric acid	$CH_2(S-CH_3)-CH_2-CH(NH_2)-COOH$
	With Side Chains Containing Acidic Groups or Their Amides			
IV	Aspartic acid	Asp	α-Aminosuccinic acid	$HOOC-CH_2-CH(NH_2)-COOH$
	Asparagine	Asn	γ-Amide of α-aminosuccinic acid	$H_2N-C(=O)-CH_2-CH(NH_2)-COOH$

*Since glycine has no asymmetric carbon atom, there can be no D or L form.

†The amino acid cystine, β,β-dithio-(α-aminopropionic acid), consists of 2 cysteine residues linked by a disulfide bond:

$$HOOC-CH(NH_2)-CH_2-S-S-CH_2-CH(NH_2)-COOH$$

TABLE 3–1. L-α-Amino acids found in proteins (cont'd).

Group	Trivial Name	Abbre- viation	Chemical Name	Structural Formula
IV	Glutamic acid	Glu	α-Aminoglutaric acid	$HOOC-CH_2-CH_2-CH(NH_2)-COOH$
	Glutamine	Gln	δ-Amide of α-aminoglutaric acid	$H_2N-C(=O)-CH_2-CH_2-CH(NH_2)-COOH$
	With Side Chains Containing Basic Groups			
V	Arginine	Arg	α-Amino-δ-guanidino-n-valeric acid	$H-N(C(=NH)NH_2)-CH_2-CH_2-CH_2-CH(NH_2)-COOH$
	Lysine	Lys	α,ε-Diaminocaproic acid	$CH_2(NH_2)-CH_2-CH_2-CH_2-CH(NH_2)-COOH$
	Hydroxylysine*	Hyl	α,ε-Diamino-δ-hydroxy-n-caproic acid	$CH_2(NH_2)-CH(OH)-CH_2-CH_2-CH(NH_2)-COOH$
	Histidine	His	α-Amino-β-imidazolepropionic acid	(HN, N ring)$-CH_2-CH(NH_2)-COOH$
	Containing Aromatic Rings			
VI	Histidine (see above)			
	Phenylalanine	Phe	α-Amino-β-phenylpropionic acid	(ring)$-CH_2-CH(NH_2)-COOH$
	Tyrosine	Tyr	α-Amino-β-(*p*-hydroxyphenyl) propionic acid	$HO-$(ring)$-CH_2-CH(NH_2)-COOH$
	Tryptophan	Trp	α-Amino-β-3-indolepropionic acid	(ring, N–H)$-CH_2-CH(NH_2)-COOH$
	Imino Acids			
VII	Proline	Pro	Pyrrolidine-2-carboxylic acid	(ring, N–H)$-COOH$
	4-Hydroxyproline	Hyp	4-Hydroxypyrrolidine-2-carboxylic acid	$HO-$(ring, N–H)$-COOH$

*Thus far, found only in collagen and in gelatin.

TABLE 3–2. Naturally occurring amino acids which do not occur in proteins.

Trivial Name	Formula	Occurrence or Significance
β-Alanine	$CH_2(NH_2){-}CH_2{-}COOH$	Part of pantetheine and of coenzyme A
Taurine	$CH_2(NH_2){-}CH_2{-}SO_3H$	Free in cells; combined with bile acids (eg, taurocholate)
α-Aminobutyric acid	$CH_3{-}CH_2{-}CH(NH_2){-}COOH$	Animal and plant tissues
γ-Aminobutyric acid	$CH_2(NH_2){-}CH_2{-}CH_2{-}COOH$	Brain tissue
β-Aminoisobutyric acid	$H_2N{-}CH_2{-}CH(CH_3){-}COOH$	End product in pyrimidine metabolism; found in urine of patients with an inherited metabolic disease (see p 358)
Homocysteine	$CH_2(SH){-}CH_2{-}CH(NH_2){-}COOH$	Methionine biosynthesis
Homoserine	$CH_2(OH){-}CH_2{-}CH(NH_2){-}COOH$	Threonine, aspartate, and methionine metabolism
Cysteinesulfinic acid	$CH_2(SO_2H){-}CH(NH_2){-}COOH$	Rat brain tissue
Cysteic acid	$CH_2(SO_3H){-}CH(NH_2){-}COOH$	Wool
Felinine	$HO{-}CH_2{-}CH_2{-}C(CH_3)_2{-}S{-}CH_2{-}CH(NH_2){-}COOH$	Cat urine
Isovalthine	$CH_3{-}CH(CH_3){-}CH(COOH){-}S{-}CH_2{-}CH(NH_2){-}COOH$	Urine of cats and of certain hypothyroid patients
2,3-Diaminosuccinic acid	$HOOC{-}CH(NH_2){-}CH(NH_2){-}COOH$	Excreted by *Streptomyces rimosus,* which produces oxytetracycline
γ-Hydroxyglutamic acid	$HOOC{-}CH(OH){-}CH_2{-}CH(NH_2){-}COOH$	Catabolism of 4-hydroxyproline
α-Aminoadipic acid	$HOOC{-}CH_2{-}CH_2{-}CH_2{-}CH(NH_2){-}COOH$	Intermediate of lysine biosynthesis by yeast
α,ε-Diaminopimelic acid	$HOOC{-}CH(NH_2){-}CH_2{-}CH_2{-}CH_2{-}CH(NH_2){-}COOH$	Bacterial cell walls; intermediate of lysine biosynthesis by bacteria
α,β-Diaminopropionic acid	$CH_2(NH_2){-}CH(NH_2){-}COOH$	In the antibiotic viomycin

TABLE 3–2. Naturally occurring amino acids which do not occur in proteins (cont'd).

Trivial Name	Formula	Occurrence or Significance
α,γ-Diaminobutyric acid	$CH_2(NH_2)-CH_2-CH(NH_2)-COOH$	In polymyxin antibiotics
Ornithine	$CH_2(NH_2)-CH_2-CH_2-CH(NH_2)-COOH$	Urea cycle intermediate
Citrulline	$CH_2(NH-C(=O)-NH_2)-CH_2-CH_2-CH(NH_2)-COOH$	Urea cycle intermediate
Homocitrulline	$CH_2(NH-C(=O)-NH_2)-CH_2-CH_2-CH_2-CH(NH_2)-COOH$	Urine of normal children
Saccharopine	$HOOC-CHNH_2-CH_2-CH_2-CH_2-H_2C-N(H)-CH(COOH)-CH_2-CH_2-COOH$	Intermediate of lysine biosynthesis by yeast and Neurospora
Azetidine-2-carboxylic acid	four-membered ring: CH_2, H_2C, $N-H$, $CH-COOH$	Lilies
3-Hydroxyproline	pyrrolidine ring ($N-H$) bearing OH and $COOH$	Achilles tendon of cattle; in the antibiotic Telomycin
Pipecolic acid	piperidine ring ($N-H$) bearing $COOH$	In certain antibiotics; metabolic product of D-lysine breakdown in mammals.
5-Hydroxytryptophan	HO-indole ($N-H$) $-CH_2-CH(NH_2)-COOH$	Precursor of serotonin

TABLE 3–2. Naturally occurring amino acids which do not occur in proteins (cont'd).

Trivial Name	Formula	Occurrence or Significance
3,4-Dihydroxyphenylalanine (DOPA)	$(HO)_2C_6H_3-CH_2-CH(NH_2)-COOH$	Precursor of melanin
Monoiodotyrosine	$HO(I)C_6H_3-CH_2-CH(NH_2)-COOH$	Thyroid tissue and blood serum
3,5-Diiodotyrosine	$HO(I)_2C_6H_2-CH_2-CH(NH_2)-COOH$	In association with thyroid globulin
3,5,3′-Triiodothyronine	$HO(I)C_6H_3-O-C_6H_2(I)_2-CH_2-CH(NH_2)-COOH$	Thyroid tissue
Thyroxine (3,5,3′,5′-tetraiodothyronine)	$HO(I)_2C_6H_2-O-C_6H_2(I)_2-CH_2-CH(NH_2)-COOH$	In association with thyroid globulin
Azaserine*	$N_2CH-C(=O)-O-CH_2-CH(NH_2)-COOH$	Potent inhibitor of tumor growth

*Not a naturally occurring amino acid, but listed here because of its importance as an inhibitor at various steps of purine biosynthesis (see p 351).

TABLE 3–3. Weak acid groups of amino acids.

	Conjugate Acid	Conjugate Base	Approximate pK_a
α-Carboxyl	$R-COOH$	$R-COO^-$	2.1 ± 0.5
Non-α-carboxyl	$R-COOH$	$R-COO^-$	4.0 ± 0.3
Imidazolinium (histidine)	imidazolium ring (HN, NH^+), R	imidazole ring (H–N, N), R	6.0
α-Amino	$R-NH_3^+$	$R-NH_2$	9.8 ± 1.0
ε-Amino (lysine)	$R-NH_3^+$	$R-NH_2$	10.5
Phenolic OH (tyrosine)	$R-C_6H_4-OH$	$R-C_6H_4-O^-$	10.1
Guanidinium (arginine)	$R-NH-C(=NH_2^+)-NH_2$	$R-NH-C(=NH)-NH_2$	12.5
Sulfhydryl (cysteine)	$R-SH$	$R-S^-$	8.3

centration $[H^+]$ is numerically equal to the dissociation constant, K. If the logarithms of both sides of the above equation are taken and both sides are multiplied by −1, the expressions would be as follows:

$$K = [H^+]$$

$$-\log K = -\log [H^+]$$

−log K is defined as pK, and −log $[H^+]$ is the definition of pH. Consequently, the equation may be rewritten as:

$$pK = pH$$

ie, **the pK of an acid group is that pH at which the protonated and unprotonated species are present at equal concentrations.** From this, many interesting facts may be inferred. For example, the pK for an acid may be determined experimentally by adding 0.5 equivalent of alkali per equivalent of acid. The resulting pH will be equal to the pK of the acid.

A weak acid, HA, ionizes as follows:

$$HA \rightleftharpoons H^+ + A^-$$

The equilibrium constant for this dissociation is written:

$$K = \frac{[H^+]\,[A^-]}{[HA]}$$

cross-multiply,

$$[H^+]\,[A^-] = K\,[HA]$$

divide both sides by $[A^-]$,

$$[H^+] = K\,\frac{[HA]}{[A^-]}$$

take the log of both sides,

$$\log [H^+] = \log \left(K\,\frac{[HA]}{[A^-]}\right) = \log K + \log \frac{[HA]}{[A^-]}$$

multiply through by −1,

$$-\log [H^+] = -\log K - \log \frac{[HA]}{[A^-]}$$

substitute pH and pK for −log $[H^+]$ and −log K, respectively; then

$$pH = pK - \log \frac{[HA]}{[A^-]}$$

Then, to remove the minus sign, invert the last term.

$$\boxed{pH = pK + \log \frac{[A^-]}{[HA]}}$$

In its final form as shown above, this is the **Henderson-Hasselbalch equation.** It is an expression of great predictive value in protonic equilibria, as illustrated below.

(1) When $[A^-] = [HA]$: As mentioned above, this would occur when an acid is exactly ½ neutralized. Under these conditions:

$$pH = pK + \log \frac{[A^-]}{[HA]} = pK + \log \frac{1}{1} = pK + 0$$

Therefore, at ½ neutralization, pH = pK.

(2) When the ratio $[A^-]/[HA]$ = 100 to 1:

$$pH = pK + \log \frac{[A^-]}{[HA]}$$

$$pH = pK + \log 100/1 = pK + 2$$

(3) When the ratio $[HA]/[A^-]$ = 10 to 1:

$$pH = pK + \log 1/10 = pK - 1$$

If the equation is evaluated at several ratios of $[A^-]/[HA]$ between the limits 10^3 and 10^{-3}, and the calculated pH values plotted, the result obtained describes the titration curve for a weak acid (Fig 3–3).

pK values for α-amino groups of free amino acids average about 9.8. They are thus much weaker acid functions than are carboxyl groups. The weak acid groups of amino acids are shown in Table 3–3.

The **isoelectric pH (pI)** of an amino acid is that pH at which it bears no net charge and hence does not

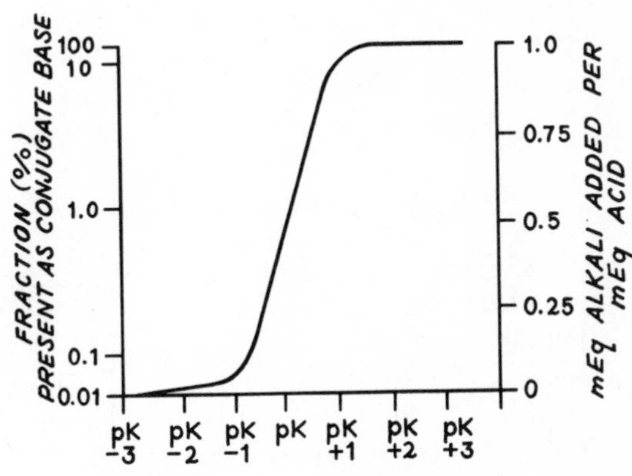

FIG 3–3. General form of a titration curve calculated from the Henderson-Hasselbalch equation.

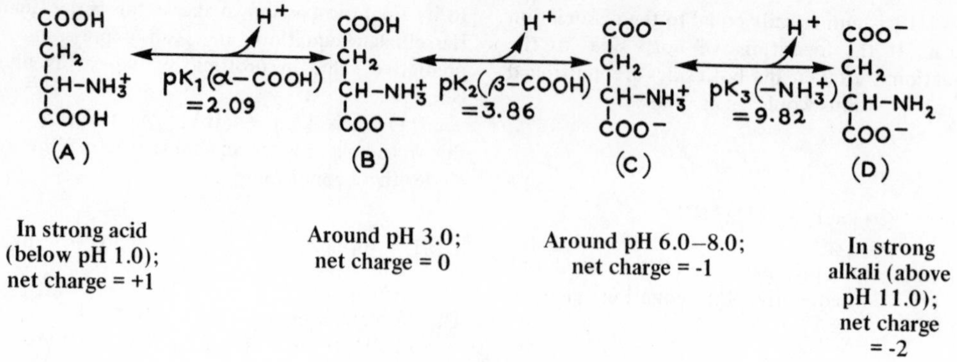

$$\underset{\text{(A)}}{\begin{matrix}COOH\\ CH_2\\ CH\text{-}NH_3^+\\ COOH\end{matrix}} \overset{H^+}{\underset{pK_1(\alpha\text{-COOH}) = 2.09}{\rightleftarrows}} \underset{\text{(B)}}{\begin{matrix}COOH\\ CH_2\\ CH\text{-}NH_3^+\\ COO^-\end{matrix}} \overset{H^+}{\underset{pK_2(\beta\text{-COOH}) = 3.86}{\rightleftarrows}} \underset{\text{(C)}}{\begin{matrix}COO^-\\ CH_2\\ CH\text{-}NH_3^+\\ COO^-\end{matrix}} \overset{H^+}{\underset{pK_3(\text{-}NH_3^+) = 9.82}{\rightleftarrows}} \underset{\text{(D)}}{\begin{matrix}COO^-\\ CH_2\\ CH\text{-}NH_2\\ COO^-\end{matrix}}$$

In strong acid (below pH 1.0); net charge = +1

Around pH 3.0; net charge = 0

Around pH 6.0–8.0; net charge = -1

In strong alkali (above pH 11.0); net charge = -2

FIG 3–4. Protonic equilibria of aspartic acid.

move in an electrical field. For an aliphatic amino acid such as alanine, the isoelectric species is

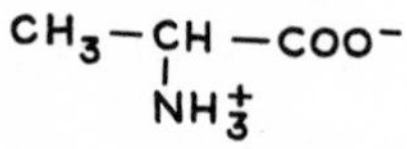

$$CH_3-\underset{\underset{NH_3^+}{|}}{CH}-COO^-$$

Since pK_1 (RCOOH) = 2.35 and pK_2 (RNH_3^+) = 9.69, the isoelectric pH (pI) is

$$pI = \frac{pK_1 + pK_2}{2} = \frac{2.35 + 9.69}{2} = 6.02$$

Dicarboxylic amino acids can exist in 4 different charged forms depending upon the pH, as shown for aspartic acid (Fig 3–4).

By writing the formulas for all possible charged species of the basic amino acids lysine and arginine, it will be observed that

$$pI = \frac{pK_2 + pK_3}{2}$$

For lysine, pI is 9.7, and for arginine 10.8. The isoelectric species of an amino acid is often referred to as a **zwitterion.**

The water solubility and high melting points of most amino acids reflect the presence of charged groups. They are readily solvated by and hence soluble in polar solvents such as water and ethanol, but they are insoluble in nonpolar solvents such as benzene, hexane, or ether. Their high melting points (above 250° F) reflect the energy needed to disrupt the ionic forces maintaining the crystal lattice.

CHEMICAL REACTIONS OF AMINO ACIDS

The carboxyl and amino groups of amino acids exhibit all the expected reactions of these functions, eg, salt formation, esterification, and acylation.

A variety of color reactions specific for particular functional groups in amino acids are known. These are useful in both the qualitative and quantitative identifi-

TABLE 3–4. Color reactions for specific amino acids.

Amino Acid Detected	Name	Reagents	Color
Arginine	Sakaguchi reaction	α-Naphthol and sodium hypochlorite	Red
Cysteine	Nitroprusside reaction	Sodium nitroprusside in dilute NH_4OH	Red
Cysteine	Sullivan reaction	Sodium 1,2-naphthoquinone-4-sulfonate and sodium hydrosulfite	Red
Histidine, tyrosine	Pauly reaction	Diazotized sulfanilic acid in alkaline solution	Red
Tryptophan	Glyoxylic acid reaction (Hopkins-Cole reaction)	Glyoxylic acid in 36 N H_2SO_4	Purple
Tryptophan	Ehrlich reaction	*p*-Dimethylaminobenzaldehyde in 12 N HCl	Blue
Tyrosine	Millon reaction	$HgNO_3$ in HNO_2; heat	Red
Tyrosine	Folin-Ciocalteu reaction	Phosphomolybdotungstic acid	Red
Tyrosine, tryptophan, phenylalanine	Xanthoproteic reaction	Boiling concentrated HNO_3	Yellow

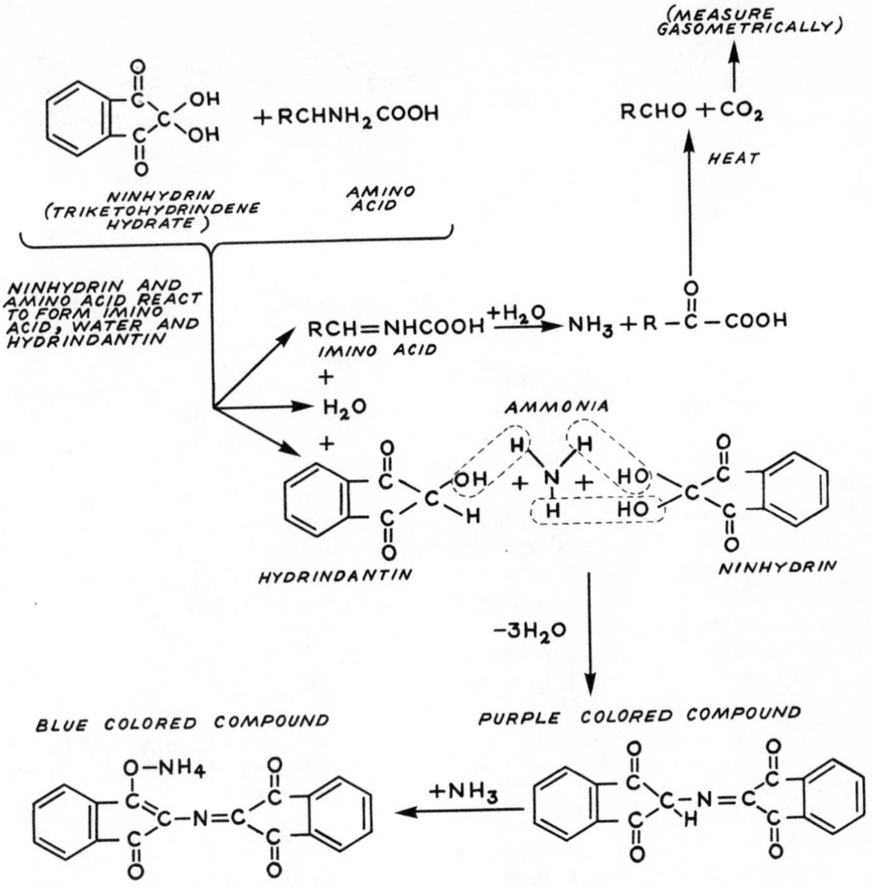

(MEASURE GASOMETRICALLY)

$RCHO + CO_2$

HEAT

NINHYDRIN (TRIKETOHYDRINDENE HYDRATE) + $RCHNH_2COOH$ AMINO ACID

NINHYDRIN AND AMINO ACID REACT TO FORM IMINO ACID, WATER AND HYDRINDANTIN

$RCH{=}NHCOOH \xrightarrow{+H_2O} NH_3 + R-\overset{O}{\overset{\|}{C}}-COOH$

IMINO ACID + H_2O + HYDRINDANTIN

AMMONIA

NINHYDRIN

$-3H_2O$

PURPLE COLORED COMPOUND

$+NH_3$

BLUE COLORED COMPOUND

FIG 3–5. Ninhydrin reaction with amino acids.

cation of particular amino acids. In many cases these color reactions may be used for amino acids combined in peptide or proteins. (See Table 3–4.)

Ninhydrin, a powerful oxidizing agent, causes oxidative decarboxylation of α-amino acids, producing CO_2, NH_3, and an aldehyde with one less carbon atom than the parent amino acid (Fig 3–5). The reduced ninhydrin then reacts with the liberated ammonia, forming a blue complex which maximally absorbs light of wavelength 570 nm. The intensity of the blue color produced under standard conditions is the basis of an extremely useful quantitative test for α-amino acids. Amines other than α-amino acids also react with ninhydrin, forming a blue color but without evolving CO_2. The evolution of CO_2 is thus indicative of an α-amino acid. Even NH_3 and peptides react, although more slowly than do α-amino acids. Proline and 4-hydroxyproline produce a yellow rather than a purple color with ninhydrin.

Ultraviolet Absorption Spectrum of Aromatic Amino Acids

The aromatic amino acids tryptophan, tyrosine, histidine, and phenylalanine absorb ultraviolet light. As shown in Fig 3–6, most of the ultraviolet absorption of proteins is due to their tryptophan content.

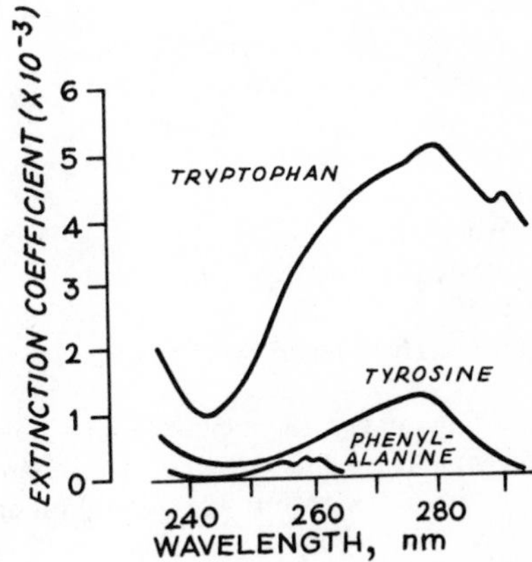

FIG 3–6. The ultraviolet absorption spectra of tryptophan, tyrosine, and phenylalanine.

PEPTIDES

Proteins consist of long chains of amino acids linked together by **peptide bonds** formed between the

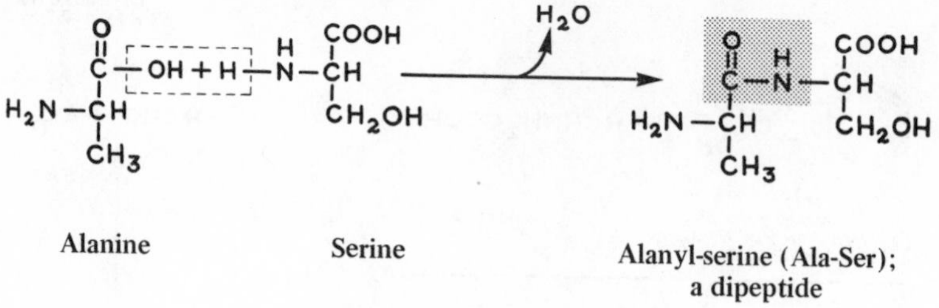

Alanine Serine Alanyl-serine (Ala-Ser); a dipeptide

FIG 3–7. Amino acids united by a peptide bond (shaded portion).

carboxyl group of one amino acid and the amino group of another. While a mol of water is removed during formation of a peptide bond (Fig 3–7), the synthesis of peptide bonds in cells involves a far more complex sequence of reactions (see Chapter 5).

A. Acid-Base Properties of Peptides: The peptide bond is an amide bond, and as such is neither basic nor acidic. In the above example, while the reactants on the left have 4 ionizable groups, that on the right has but 2. Condensation of additional amino acids with a dipeptide produces tripeptides, tetrapeptides, etc. (*Note:* a pentapeptide is one formed from 5 amino acids, not one with 5 peptide bonds.) A pentapeptide formed solely from neutral amino acids (eg, pentaglycine) has only 2 ionizable groups: the C-terminal carboxyl and the N-terminal amino. Peptide formed from acidic or basic amino acids contain, in addition, ionizable groups due to the presence of ionizable functional groups other than the α-carboxyl and α-amino groups involved in the peptide bond.

B. The Peptide Bond as a Structural Element in Proteins: The principal evidence for peptide bonds as the basic structural bond of proteins comprises the following observations:

(1) Proteases, enzymes which hydrolyze proteins, produce polypeptides as products. These enzymes also hydrolyze the peptide bonds of proteins.

(2) The infrared spectra of proteins suggest many peptide bonds.

(3) Two proteins, insulin and ribonuclease, have been synthesized solely by linking amino acids by peptide bonds.

(4) Proteins have few titratable carboxyl or amino groups.

(5) Proteins and synthetic polypeptides react with **biuret reagent** (an alkaline 0.02% cupric sulfate solution) to form a purple color. This reaction is specific for 2 or more peptide bonds.

(6) X-ray diffraction studies at the 0.2 nm level of resolution have conclusively identified the peptide bonds in the proteins, myoglobin and hemoglobin.

Polypeptides are long peptide chains containing large numbers of peptide bonds. Although the term polypeptide thus should include proteins, polymers consisting of less than 100 amino acid residues are arbitrarily termed polypeptides and those with more than 100 are generally termed proteins.

Even small changes in protein structure may produce profound physiologic effects. Substitution of a single amino acid for another in a linear sequence of possibly 100 or more amino acids may reduce or abolish biologic activity with potentially serious consequences (eg, sickle cell disease; see p 200). Indeed, many inherited metabolic errors may involve no more than a subtle change of this type. The introduction of new chemical and physical methods to determine protein structure has markedly increased knowledge of the biochemical bases for many inherited metabolic diseases.

DETERMINATION OF THE PRIMARY STRUCTURE OF PEPTIDES

The linear sequence of amino acid residues in a polypeptide is referred to as its **primary structure.** The determination of primary structure employs chemical methods whereas the higher orders of protein structure require physical technics such as x-ray crystallography. When the **number, kind,** and **linear order** of the amino acids are known, an accurate **primary structure** of a peptide or a protein may be obtained. In the conventional abbreviated form, a polypeptide might be shown as, for example:

Glu-Lys-Ala-Gly-Tyr-His-Ala

The **N-terminal** (amino terminal) amino acid is always shown at the left and the **C-terminal** (carboxyl terminal) amino acid at the right of the polypeptide chain. When the individual **amino acid residues** are linked by **straight lines**, as above, a definite and characteristic sequence is implied. The above structure (glutamyl-lysyl-alanyl-glycyl-tyrosyl-histidyl-alanine) is **named as a derivative of the C-terminal amino acid,** alanine. Uncertainty as to the exact order of specific residues is represented by enclosing that portion of the sequence, separated by **commas,** in **parentheses.**

Glu-Lys-(Ala,Gly,Tyr)-His-Ala

Determination of primary structure may conveniently be considered under 2 categories: (1) qualitative identification and quantitative estimation of the amino acid residues present; and (2) determination of sequence.

Determination of the Number & Kinds of Amino Acids Present

The peptide bonds linking the amino acids are first broken by hydrolysis. The free amino acids are then separated from one another and identified by chromatography and electrophoresis. The quantity of each amino acid present is then determined by quantitative chemical and physical technics (ninhydrin, specific color reactions, spectral properties of aromatic residues). Since peptide bonds are stable in water at pH 7.0, to accomplish hydrolytic cleavage of these bonds catalysis by acid, alkali, or enzymes must be employed. Each has advantages and disadvantages.

A. Hydrolysis of Proteins by Acid: Most proteins are completely hydrolyzed to their constituent amino acids by heating at 110° C for 20–70 hours in 6 N HCl. Hydrolysis is carried out in a sealed, evacuated tube to exclude oxygen and prevent oxidative side reactions.

Undesirable side-effects of acid-catalyzed hydrolysis include the following:

(1) All the tryptophan and variable amounts of serine and threonine are destroyed. Formation of **humin**, a black polymer of breakdown products of tryptophan, accompanies acid hydrolysis.

(2) Glutamine and asparagine are deamidated to glutamate and aspartate.

(3) Glutamic acid undergoes intramolecular dehydration to pyrollidone 5-carboxylic acid (Fig 3–8).

(4) Other amino acids may undergo intermolecular dehydration forming cyclic anhydrides or **diketopiperazines** (Fig 3–9).

B. Hydrolysis of Proteins by Alkali: This is used to recover tryptophan, which is not destroyed by alkaline hydrolysis. However, serine, threonine, arginine, and cysteine are lost, and all amino acids are racemized.

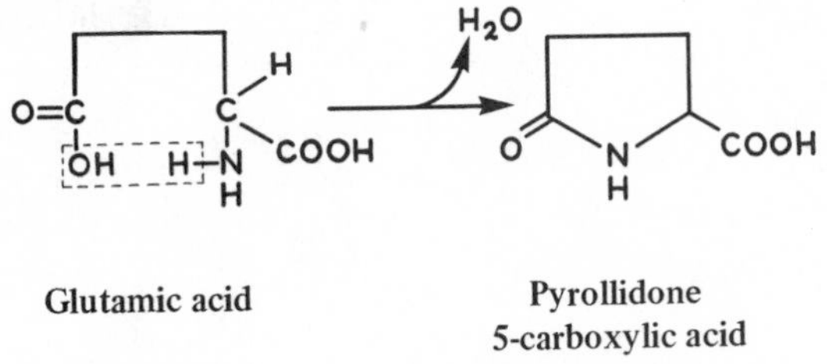

Glutamic acid → H_2O → **Pyrollidone 5-carboxylic acid**

FIG 3–8. Intramolecular dehydration of glutamic acid.

C. Enzyme-Catalyzed Hydrolysis: The bacterial peptidases **subtilisin** and **Pronase** catalyze hydrolysis of all peptide bonds, but the reaction is slow compared to acid-catalyzed hydrolysis. Other proteolytic enzymes (trypsin, chymotrypsin) catalyze hydrolysis of certain peptide bonds quite rapidly. Other bonds are hydrolyzed slowly or not at all, and the polypeptide is incompletely hydrolyzed to free amino acids. This specificity is used to advantage in sequence studies.

D. Separation and Identification of the Amino Acid Residues Present: The free amino acids of the protein hydrolysate are simultaneously separated one from another, identified, and prepared for quantitative analysis by partition chromatography or by electrophoresis. In paper, thin layer, column, or gas-liquid partition chromatography, amino acids are partitioned or divided between a stationary and a mobile phase. Separation depends on the relative tendencies of specific amino acids to reside in one or the other phase. Since the partitioning process is repeated hundreds or thousands of times, small differences in partition ratio permit excellent separations.

E. Paper Chromatography: A small volume (about 0.005 ml) of an amino acid solution containing about 0.01 mg of amino acids is applied at a marked point 5 cm from the end of a filter paper strip. This is suspended in a sealed cylindrical jar or cabinet (Fig 3–10). One end of the paper dips into an aqueous solution of a solvent which typically consists of water, an acid or base, and an organic solvent such as n-butyl alcohol. The solvent may be placed in a trough from which the paper strip hangs ("descending paper chromatography"), or the strip may be suspended from the top of the jar and dip into a trough at the bottom of the jar ("ascending paper chromatography").

Strips are removed when the solvent has migrated over most of the available distance. When dry, the strips are sprayed with 0.5% ninhydrin in acetone and heated for a few minutes at 80–100° C. Purple spots appear where amino acids are present (Fig 3–11). In paper partition chromatography, the stationary phase, hydrated cellulose, is more polar than the mobile organic phase. Fig 3–11, left, shows that amino acids with large nonpolar side-chains (Leu, Ile, Phe, Trp, Val, Met, Tyr) migrate farther in n-butanol:acetic acid: water than those with shorter nonpolar side-chains (Pro, Ala, Gly) or with polar side-chains (Thr, Glu, Ser, Arg, Asp, His, Lys, Cys). This reflects the greater relative solubility of polar molecules in the hydrophilic

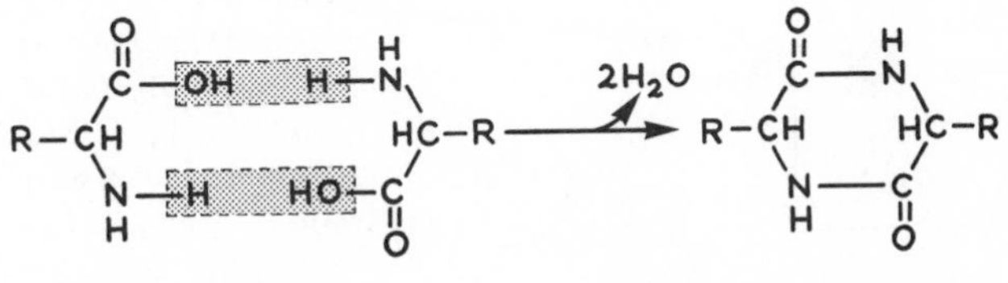

Two amino acids → $2H_2O$ → **A diketopiperazine**

FIG 3–9. Intermolecular dehydration of amino acids.

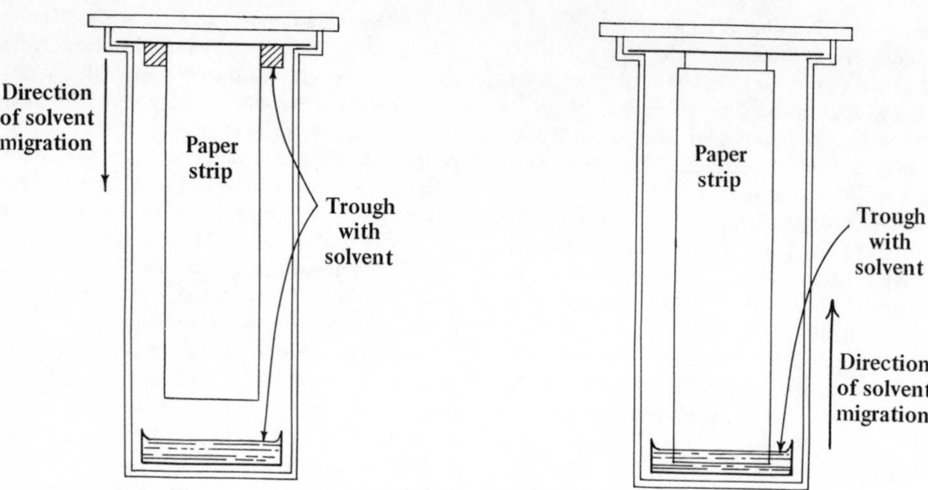

FIG 3–10. Cross-section of apparatus for descending *(left)* and ascending *(right)* paper chromatography.

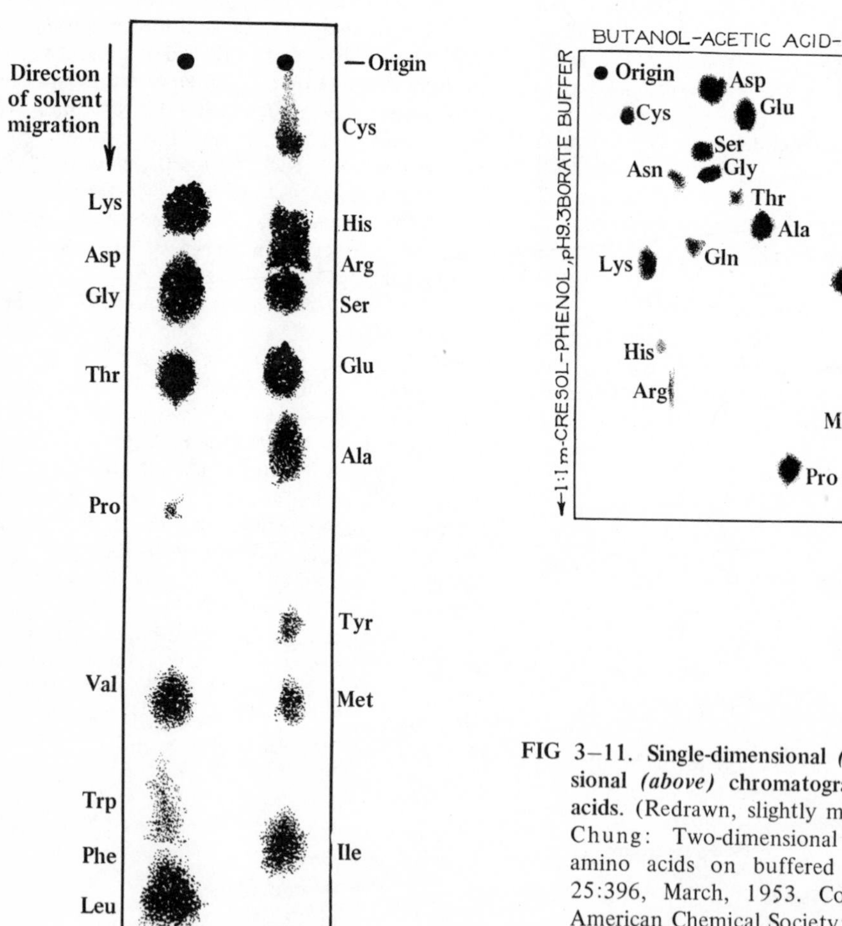

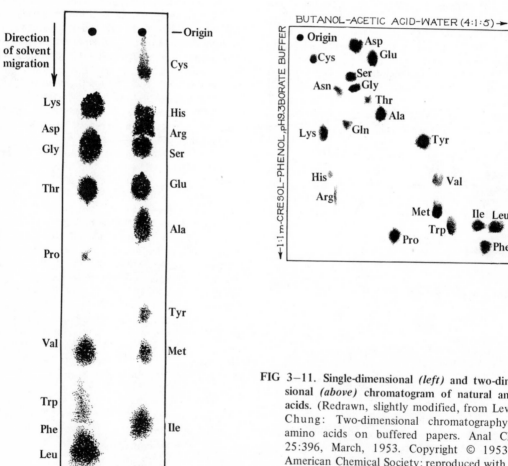

FIG 3–11. Single-dimensional *(left)* and two-dimensional *(above)* chromatogram of natural amino acids. (Redrawn, slightly modified, from Levy & Chung: Two-dimensional chromatography of amino acids on buffered papers. Anal Chem 25:396, March, 1953. Copyright © 1953 by American Chemical Society; reproduced with permission.)

stationary phase and of nonpolar molecules in organic solvents. Note further that for a nonpolar series (Gly, Ala, Val, Leu, Ile) increasing length of the nonpolar side-chain results in increased mobility.

The ratio of the distance traveled by an amino acid to that traveled by the solvent front, both measured from the marked point of application of the amino acid mixture, is called the **R_f value** for that amino acid. R_f values for a given amino acid vary with experimental conditions, eg, the solvent used. Although it is possible to tentatively identify an amino acid by its R_f value alone, it is preferable to chromatograph known amino acid standards simultaneously with the unknown mixture. The spots on the test strip may then be readily identified by comparison with the standards.

Quantitation of the amino acids may be accomplished by cutting out each spot, removing (eluting) the compound with a suitable solvent, and performing a colorimetric (ninhydrin) or chemical (nitrogen) analysis. Alternatively, the paper may be sprayed with ninhydrin and the color densities of the spots measured with a recording transmittance or reflectance photometer device (see also p 160).

Modifications introduced to obtain better separation of components and to improve their quantitation include **2-dimensional paper chromatography**. In this technic, a square sheet rather than a strip of filter paper is used. The sample is applied to the upper left corner and chromatographed for several hours with one solvent mixture (eg, n-butanol:acetic acid:water). After drying to remove this solvent, the paper is turned through 90° and chromatographed in a second solvent mixture (eg, collidine-water) (Fig 3–11, right).

F. Thin Layer Chromatography (TLC): One- or 2-dimensional thin layer chromatography (TLC) may be employed in place of paper chromatography. Thin layers of a chromatographic support or adsorbent (eg, cellulose powder, alumina, a cellulose ion exchange resin, Sephadex) are spread as a slurry on 8 × 8 inch glass plates. The plates are dried and used like paper sheets in paper chromatography. The advantage lies both in the choice of adsorbents, which permit separations not possible on paper, and in the rapidity of separation. Lipids, including sterols, may be rapidly and cleanly separated by adsorption chromatography on alumina. Amino acid mixtures which require 18 hours for separation on paper require as little as 3 hours using cellulose TLC.

G. Ion Exchange Chromatography: Moore and Stein utilized an ion exchange resin (Dowex) and various buffers as well as alterations in temperature to accomplish separation on a chromatographic column of all of the amino acids in a protein hydrolysate (Fig 3–12). Modifications in this procedure introduced by Moore, Spackman, & Stein permit a complete amino acid analysis of a peptide or of a protein hydrolysate in 24–48 hours. The modified system can be used either with a fraction collector or with automatic recording equipment. In this method, columns of finely pulverized 8% cross-linked sulfonated polystyrene resin (Amberlite IR-120) are used. The method has been used for analysis of proteins such as histones, hemoglobin, and ribonuclease. With modifications it

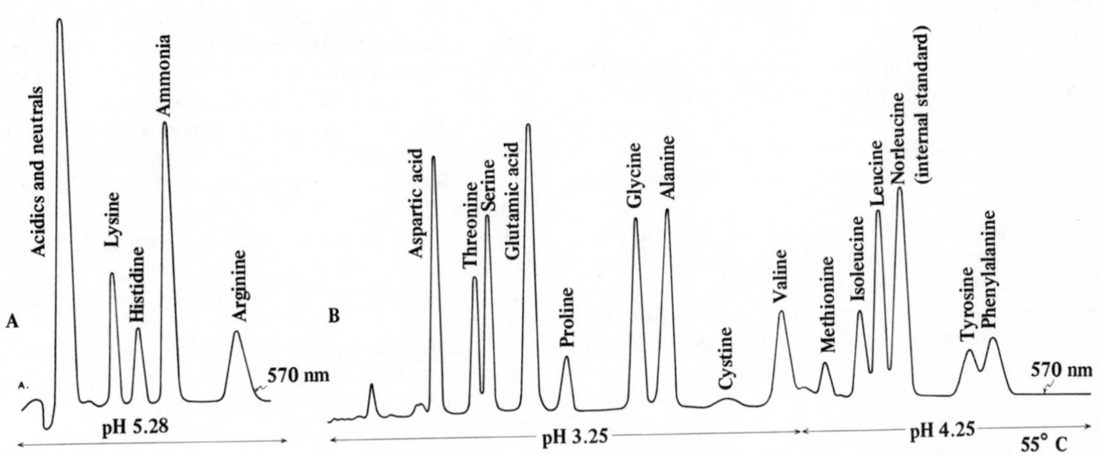

FIG 3–12. Automated analysis of an acid hydrolysate of corn endosperm on Moore-Stein Dowex 50 columns (at 55° C). *A:* A short (5.0 × 0.9 cm) column used to resolve basic amino acids by elution at pH 5.28. Time required = 60 minutes. *B:* A longer (55 × 0.9 cm) column used to resolve neutral and acidic amino acids by elution first with pH 3.25 and then with pH 4.25 buffer. An internal standard of norleucine is included for reference. Basic amino acids remain bound to the column. Time required = 180 minutes. Emerging samples are automatically reacted with ninhydrin and the optical density of samples recorded at 570 nm and 440 nm. The latter wavelength is used solely to detect proline and hydroxyproline (absent from corn endosperm). Ordinate = optical density plotted on a log scale. Abscissa = time in minutes. (Courtesy of Professor E.T. Mertz, Purdue University.)

can be used for determination of amino acids and related compounds in plasma, urine, plant, and animal tissues.

Determination of Amino acid Sequence in a Peptide Chain

Useful technics for sequencing amino acids include the following:

A. Hydrazinolysis: Peptides are reacted with hydrazine, converting all peptide and amide bonds to acid hydrazides. The C-terminal carboxyl group does not react. The unchanged amino acid is thus that originally at the C-terminal end.

B. Treatment With 1-Fluoro-2,4-dinitrobenzene DNB, Sanger's Reagent): This quantitatively arylates all free amino groups, producing intensely yellow 2,4-dinitrophenyl amino acids. These derivatives are readily quantitated by spectrophotometry.

In a peptide, only the N-terminal residue can form an α-DNP derivative. Following hydrolysis, this may be separated and identified (Fig 3–13).

Fluorodinitrobenzene also reacts with the ε-amino groups of lysine, the imidazole of histidine, the OH of tyrosine, and the SH of cysteine. Since the dinitrophenyl group is resistant to removal by acid hydrolysis, it is useful in structural analysis of the N-terminal amino acid of polypeptides and proteins.

C. Reactions With Phenylisothiocyanate (Edman Reagent) and With Phosgene: Phenylisothiocyanate reacts with the amino groups of amino acids, yielding phenylthiohydantoic acids. On treatment with acids in nonhydroxylic solvents, these cyclize to phenylthiohydantoins (Fig 3–14).

Phosgene reacts with amino groups to form N-carboxyanhydrides. These are useful in peptide biosynthesis (Fig 3–15).

D. Digestion With Aminopeptidases or Carboxypeptidases: These enzymes catalyze successive removal of N- or C-terminal residues, respectively. The order in which residues appear in the nonprotein fraction as a function of time tells much of their order in a polypeptide.

E. Digestion With Residue-Specific Endopeptidases: To facilitate sequencing large peptides, these are broken down to smaller peptides using residue-specific peptidases such as trypsin and chymotrypsin (Fig 3–16). This forms the basis of the peptide-mapping technic.

SYNTHESIS OF PEPTIDES

Peptides, like amides, may be synthesized by a reaction between an activated carboxyl group such as an acid chloride, an acid anhydride, or a thioester of one amino acid and the amino group of another, as for example between cysteine acid chloride and lysine (Fig 3–17). When this reaction is carried out, however, the activated carboxyl group also reacts with the ε-amino group of lysine, producing 2 isomeric Cys-Lys dipeptides. In addition, it may react with the amino group of another cysteine acid chloride producing Cys-Cys-Cl, and this process may continue, producing Cys-Cys-Cys-Cl, etc. To avoid these undesirable side products, all amino groups to be excluded from the reaction must be blocked. After formation of the peptide bond, the blocking group is then removed, leaving the desired peptide.

Synthesis of large polypeptides is difficult and extremely time-consuming. While classical chemical technics were adequate for synthesis of the octapeptides, vasopressin and oxytocin, and later of bradykinin (see below), the yields of final product are too low to permit synthesis of long polypeptides or proteins. This has recently been achieved by the automated, solid-phase synthesis technic developed by R.B. Merrifield. In this process, an automated synthesis is carried out in a single vessel by a machine programmed to add reagents, remove products, etc at timed intervals. The steps involved are as follows:

(1) The amino acid which ultimately will form the C-terminal end of the polypeptide is attached to an insoluble resin particle.

(2) The second amino acid bearing an appropriately blocked amino group is introduced and the peptide bond formed in the presence of the strong dehydrating agent, dicyclohexylcarbodiimide.

(3) The blocking group is removed with acid, forming gaseous products which are removed.

(4) Steps 2 and 3 are repeated with the next amino acid in sequence, then the next, until the entire polypeptide attached to the resin particle has been synthesized.

(5) The polypeptide is cleaved from the resin particle.

The process proceeds rapidly and with excellent yields of final product. About 3 hours are required per peptide bond synthesized. Using this technic, the A

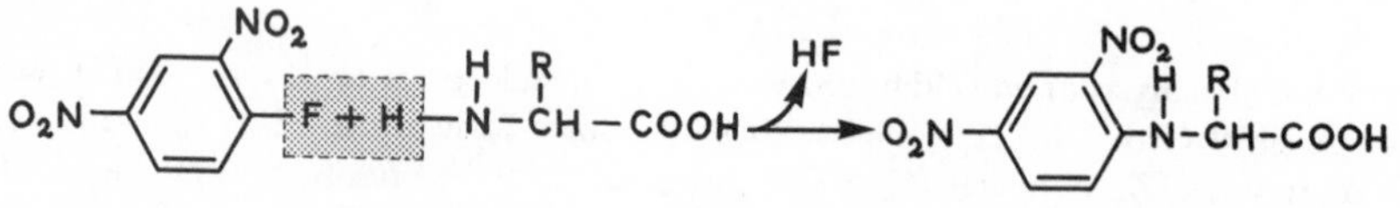

FIG 3–13. Reaction of Sanger's reagent with an amino acid or peptide.

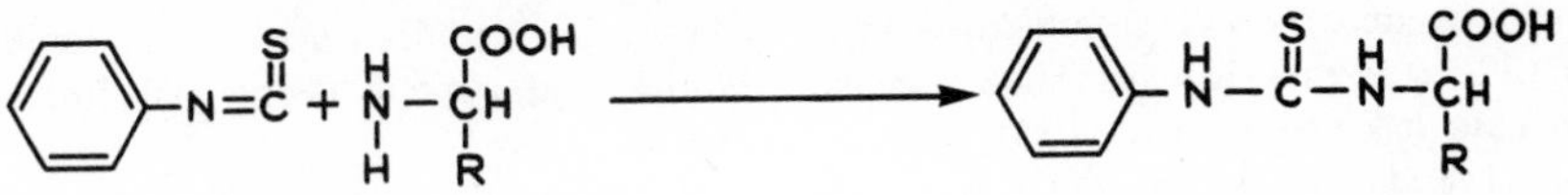

Phenylisothiocyanate (Edman reagent) and amino acid → **A phenylthiohydantoic acid**

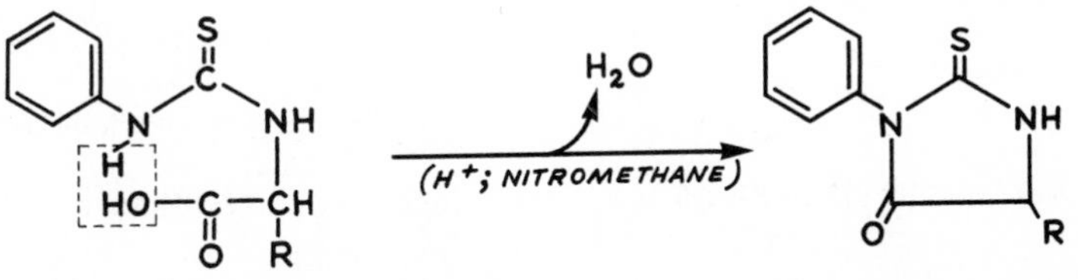

Phenylthiohydantoic acid → (H_2O; H^+; NITROMETHANE) → **A phenylthiohydantoin**

FIG 3–14. Conversion of an amino acid (or of the N-terminal residue of a polypeptide) to a phenylthiohydantoin derivative.

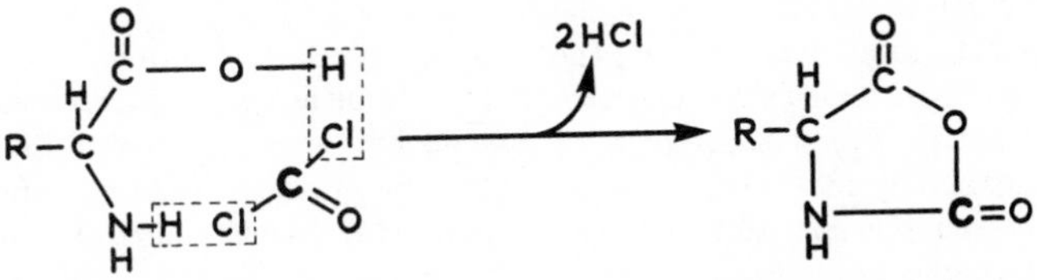

Amino acid and phosgene → 2HCl → An N-carboxyanhydride

FIG 3–15. Reaction with phosgene.

C C C
-Gly-Lys↓Val-Phe↓Arg↓Leu-Cys-Tyr↓Ile-Arg↓Trp↓Gln
T T T

FIG 3–16. Hydrolysis of a polypeptide at the indicated peptide bonds catalyzed by trypsin (T) and chymotrypsin (C).

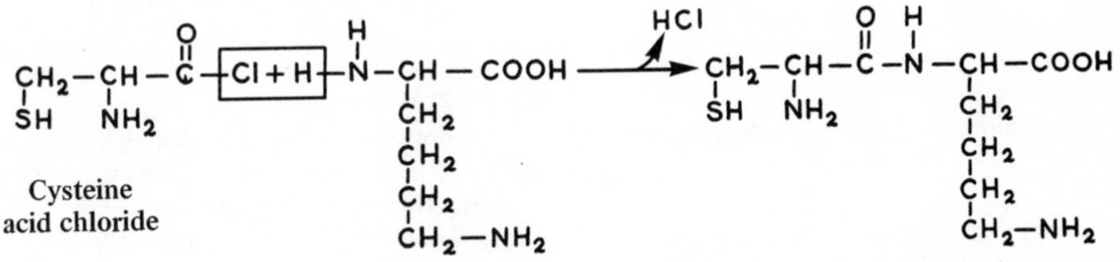

Cysteine acid chloride + Lysine (Lys) → HCl → Cysteinyl-lysine (Cys-Lys)

FIG 3–17. Reaction of an activated amino acid with the α-N of lysine.

chain of insulin (21 residues) was synthesized in 8 days and the B chain (30 residues) in 11 days. The crowning achievement to date has been the total synthesis of pancreatic ribonuclease (124 residues; see p 47) in 18% overall yield. This constitutes the first total synthesis of an enzyme. It foreshadows a new era not only in confirmation of protein structures but in related areas such as immunology and perhaps in the treatment of inborn errors of metabolism.

PHYSIOLOGICALLY ACTIVE PEPTIDES

A variety of physiologically important peptides, prepared synthetically from L-amino acids by routes which involve no racemization, have full physiologic activity. These include the octapeptides oxytocin and vasopressin, ACTH, and melanocyte-stimulating hormone (see Chapter 20).

Although a large number of peptides exist in the free state in cells of animals, plants, and bacteria, the physiologic function of most of them is not clear. In some cases, they are thought to represent products of protein turnover; in others, they may be hormones, antibiotics, precursors of bacterial cell walls, or even potent poisons. Many bacterial and fungal peptides, including most antibiotics, are cyclic peptides containing both unusual amino acids and the D-isomers of the familiar protein amino acids. Frequently the peptide bonds involve the non-*α* carboxyl of glutamate or aspartate. D-Amino acids, despite their wide occurrence in microbial polypeptides, have not been found in plant or animal cells.

An example of one of the many known naturally occurring polypeptides is shown in Fig 3–18. This is the widely distributed tripeptide, glutathione (GSH).

```
              O   H          O   H
              ‖   |          ‖   |
CH2 — C — N — CH — C — N — CH2 — COOH
 |                 |
CH2               CH2
 |                 |
HC — NH2           SH
 |
COOH
```

FIG 3–18. Glutathione (γ-glutamyl-cysteinyl-glycine).

Other important naturally occurring peptides include bradykinin and kallidin (lysyl-bradykinin), potent smooth muscle hypotensive agents liberated from specific plasma proteins by treatment with snake venom or the proteolytic enzyme, trypsin.

Arg-Pro-Pro-Gly-Phe-Ser-Pro-Phe-Arg

Bradykinin

Lys-Arg-Pro-Pro-Gly-Phe-Ser-Pro-Phe-Arg

Kallidin

The cyclic peptides tyrocidin and gramicidin are antibiotics, and contain D-phenylalanine. (See structures below.)

```
Val-Orn-Leu-D-Phe-Pro
|                   |  =  (Val-Orn-Leu-D-Phe-Pro)2
Pro-D-Phe-Leu-Orn-Val
```

Gramicidin S

```
Val-Orn-Leu-D-Phe-Pro
|                   |
Tyr-Gln-Asn-D-Phe-Phe
```

Tyrocidin

PROTEINS

Knowledge of the molecular architecture of proteins has increased enormously in the past decade. The total synthesis in 1969 of ribonuclease from its constituent amino acids ushered in a new era in protein chemistry wherein the effects of specific amino acid substitutions on structure and on biologic activity of proteins may be studied by the direct synthesis of such modified proteins. One consequence of this rapid increase in knowledge has been a sudden outdating of older terms and concepts. As an example, the system of protein classification given below has now fallen into disuse by protein chemists. Its inclusion here is warranted by the continued use of terms such as albumins or globulins in the medically related field of clinical chemistry. The classification which follows is based solely on solubility characteristics of proteins.

CLASSIFICATION

Simple Proteins

These contain only L-alpha-amino acids or their derivatives and include the following groups:

A. Albumins: Soluble in water, coagulated by heat, precipitated by saturated salt solutions. ***Examples:*** lactalbumin, serum albumin.

B. Globulins: Soluble in dilute salt solutions of the strong acids and bases; insoluble in pure water or in moderately concentrated salt solutions; coagulated by heat. ***Examples:*** serum globulin, ovoglobulin.

C. Glutelins: Soluble in dilute acids and alkalies; insoluble in neutral solvents; coagulated by heat. ***Example:*** glutenin from wheat.

D. Prolamines: Soluble in 70–80% alcohol; insoluble in absolute alcohol, water, and other neutral solvents. ***Examples:*** zein (corn) and gliadin (wheat).

E. Albuminoids (Scleroproteins): Insoluble in all neutral solvents and in dilute acids and alkalies. These are the proteins of supportive tissue. ***Examples:*** keratin, collagen.

F. Histones: Soluble in water and very dilute acids; insoluble in very dilute NH_4OH; not coagulated by heat. Basic amino acids predominate. ***Example:*** nucleohistones of nuclei.

G. Protamines: Basic polypeptides, soluble in water or in NH_4OH, not coagulated by heat; basic amino acids predominate in their structure; precipitate other proteins. Found principally in egg cells. ***Examples:*** salmine (salmon) and sturine (sturgeon).

Conjugated Proteins

Those which contain some nonprotein substance (the prosthetic group) linked by forces other than salt linkages.

A. Nucleoproteins: Compounds of one or of several molecules of proteins with nucleic acid. ***Examples:*** nuclein, nucleohistone from nuclei-rich material (eg, glandular tissues).

B. Glycoproteins and Mucoproteins: Compounds with carbohydrate prosthetic groups (mucopolysaccharides) which on hydrolysis yield amino sugars (hexosamines) and uronic acids. ***Examples:*** the protein, mucin, and proteins in the plasma which migrate electrophoretically with the $alpha_1$ and $alpha_2$ fractions.

C. Phosphoproteins: Compounds with a phosphorus-containing radical other than a phospholipid or nucleic acid. ***Example:*** casein.

D. Chromoproteins: Compounds with a chromophoric group in conjugation. ***Examples:*** hemoglobin, hemocyanin, cytochrome, flavoproteins.

E. Lipoproteins: Conjugated to neutral fats (triglycerides) or other lipids such as phospholipid and cholesterol.

F. Metalloproteins: Binding a metal such as copper (ceruloplasmin) or iron (siderophilin).

STRUCTURE OF PROTEINS

The primary structure of proteins derives ultimately from linkage of L-α-amino acids by α-peptide bonds. The principal evidence for the peptide bond as the primary structural bond of proteins is summarized on p 34).

Technics used to study protein structure include the following:

Electron Photomicrography

An actual picture of very small objects can be obtained with the electron microscope. Magnifications as high as 100,000 diameters can be obtained with this instrument. This permits the visualization of proteins of high molecular weight, such as virus particles.

X-Ray Diffraction

A single crystal of a protein or layers of protein or protein fibers will deflect x-rays, and the resultant image on a photographic plate can be analyzed to yield information on the crystal or on the structure of the fiber. This technic has been the single most important factor in determination of complex, higher orders of protein structure.

Streaming Birefringence or Double Refraction of Flow

A beam of light is passed through a polarizing lens. The polarized light is then passed through a solution of a protein and, finally, through a second polarizing lens which is oriented at right angles to the first lens. No light will emerge from the second lens if the protein solution does not affect the polarized light. This is the case with spherical protein molecules. It is also true of fibrous molecules when the solution is at rest because the protein molecules are then randomly oriented. However, when a solution of a fibrous protein is put in motion, the elongated molecules arrange themselves lengthwise in the axis of the stream and thus act as if another polarizer were added to the system. As a result the polarized light does pass through the second lens. This phenomenon is referred to as "streaming birefringence" or "double refraction of flow." It has been used to calculate the "axial ratio" of a fibrous protein, ie, the ratio of the length of the long axis to that of the short axis. An example is fibrinogen, which has a calculated axial ratio of 20:1.

Optical Rotatory Dispersion

The ability of solutions of proteins to rotate the plane of plane-polarized light is examined at various wavelengths. Since proteins are composed of L-α-amino acids which are themselves optically active, proteins are highly optically active. In certain instances the optical rotation of a protein far exceeds that due solely to the sum of the individual rotations of its constituent amino acids. This suggests that the protein possesses asymmetry in addition to that of the α-carbon atoms. Helical structures (see below) can exist in right- or left-handed forms and hence are optically active. The presence of a high fraction of helical structure therefore contributes to the optical rotation. Other asymmetrical structures can also contribute. The change in optical rotation accompanying the transition from helix to random coil or the reverse is used to assess the fraction of α-helix structure in proteins.

Overall Shape of Proteins

One may distinguish 2 broad classes of proteins on the basis of their overall dimensions: (1) globular proteins, which have an axial ratio (length:width) of less than 10; and (2) fibrous proteins, with axial ratios greater than 10. **Keratin**, the protein of hair, wool, and skin, is a typical fibrous protein. It consists of a long peptide chain or groups of such chains. The peptide chains may be coiled in a spiral or helix formation (see Fig 3–21) and cross-linked by S–S bonds as well as by

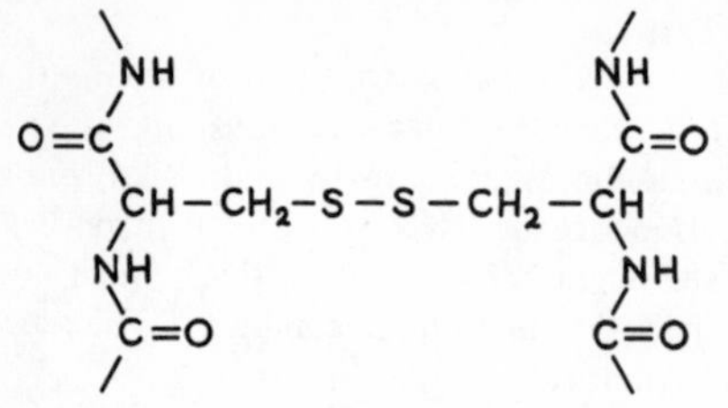

```
   \                         /
    NH                      NH
   /                          \
O=C                            C=O
   \                          /
    CH-CH2-S-S-CH2-CH
   /                          \
  NH                           NH
   \                          /
    C=O                    C=O
   /                          \
```

FIG 3–19. Two peptide chains united by a disulfide linkage.

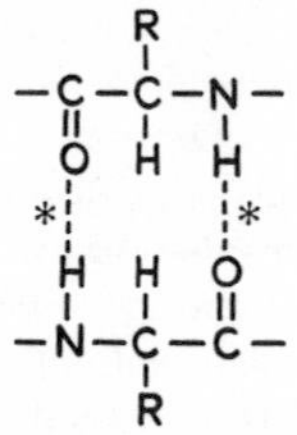

```
        R
        |
  -C  - C - N-
   ||   |   |
   O    H   H
  *:        :*
   H    H   O
   |    |   ||
  -N  - C - C-
        |
        R
```

FIG 3–20. *Hydrogen bonds.

hydrogen bonds (Figs 3–19 and 3–20). The condensed form is referred to as alpha-keratin. This changes to beta-keratin by unfolding.

Myosin, the major protein of muscle, is also a fibrous protein which undergoes a change in its structure during muscle contraction and relaxation. (See p 483.)

Globular proteins are characterized by the presence of peptide chains which are folded or coiled in a very compact manner. Axial ratios are usually not over 3:1 or 4:1. Examples of globular proteins are found among the fractions of the albumins and globulins in the plasma. **Insulin** is another globular protein.

MOLECULAR WEIGHT

Molecular weights of proteins have been most successfully investigated by physical methods such as osmotic pressure measurements or by freezing point (cryoscopic) determinations. In general, the results lack precision because of variables due to pH, electrolytes, and the degree of hydration of the protein molecule. The method developed by Svedberg is the best yet advanced for determination of the molecular weight of proteins. This method depends upon measurement of sedimentation rates as determined in the ultracentrifuge.

Many proteins are of extremely high molecular weight (eg, infectious causative agents of psittacosis, lymphogranuloma venereum, and trachoma: about 8.5 billion daltons). Proteins of such high molecular weight must be considered to contain units of about 400,000 daltons. These units are thought to combine (polymerize) in a reversible manner in accordance with concentration, pH, temperature, etc.

Molecular weights (in daltons) of representative proteins (ultracentrifuge measurements) are as follows: egg albumin, 44,000; bovine insulin, 12,000; serum albumin, 69,000; hemoglobin (horse), 68,000; serum globulin, 180,000; fibrinogen, 450,000; thyroglobulin, 630,000.

Bonds Responsible for Protein Structure

A. Peptide Bonds: Long chains of amino acids linked by the peptide bond are called **polypeptides.** The number and order of the amino acids in the polypeptide chains is referred to as the **"primary structure"** of the protein.

B. Disulfide Bonds: The disulfide bond may interconnect 2 parallel chains through cysteine residues within each polypeptide (see structure of ribonuclease, Fig 3–27). This bond is relatively stable and thus is not readily broken under the usual conditions of denaturation. Performic acid treatment oxidizes the S–S bonds. This reagent is used, for example, to oxidize insulin in order to separate the protein molecule into its constituent polypeptide chains without affecting the other parts of the molecule (see p 45). The union of 2 parallel peptide chains by an S–S linkage is illustrated in Fig 3–19.

C. Hydrogen Bonds: In addition to S–S bonds, one other major force is involved in the preservation of the structure of a protein molecule. This is the **hydrogen bond,** which is produced by the sharing of hydrogen atoms between the nitrogen and the carbonyl oxygen of the same or of different peptide chains (Fig 3–20).

The concept that peptide chains are folded in the form of a helix assumes that the coiled structure is maintained by the hydrogen bonds between the $-\overset{\overset{O}{\|}}{C}-$ and $-\overset{\overset{H}{|}}{N}-$ groups of a single peptide chain and that it is these bonds which are broken in the unfolding process which occurs during denaturation of proteins. Individual hydrogen bonds are very weak, but the reinforcing action of a large number of such bonds in the protein molecule produces a stable structure. The helical structure of a protein maintained by hydrogen bonds is illustrated in Figs 3–21 and 3–24.

Hydrophobic Interactions

The nonpolar side-chains of neutral amino acids tend to be closely associated with one another in proteins. The relationship is nonstoichiometric; hence no true bond may be said to exist. Nonetheless, these interactions play a significant role in maintaining protein structure.

Orders of Protein Structure

The folding of the polypeptide chains into a specific coiled structure held together by disulfide bonds and by hydrogen bonds is referred to as the **"secondary structure" of the protein,** and the arrangement and interrelationship of the twisted chains of protein into specific layers or fibers is called the **"ter-**

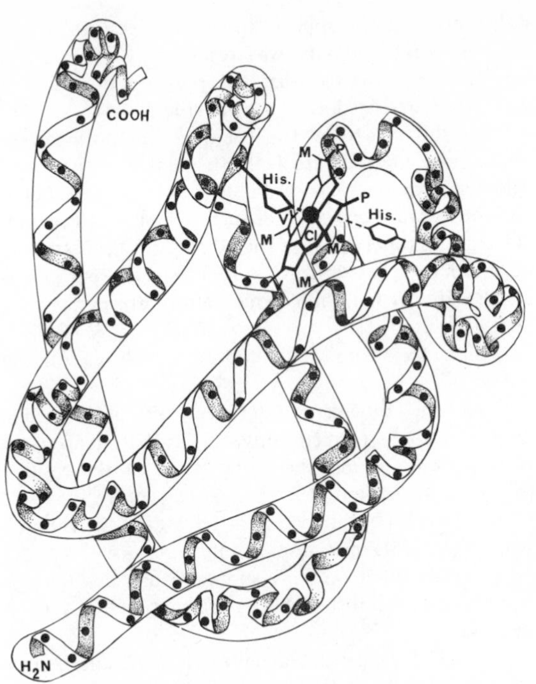

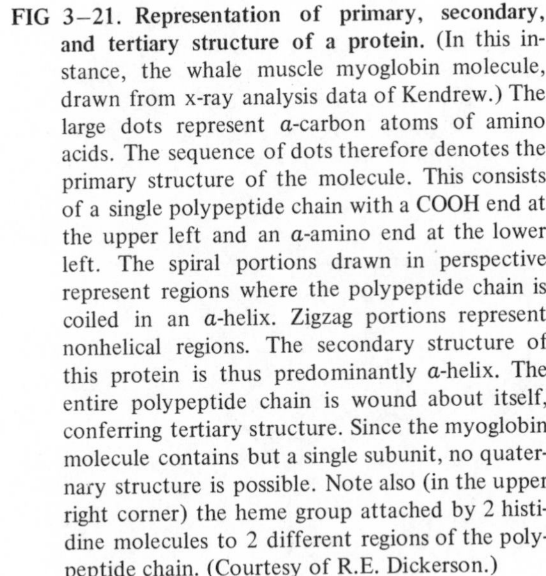

FIG 3–21. Representation of primary, secondary, and tertiary structure of a protein. (In this instance, the whale muscle myoglobin molecule, drawn from x-ray analysis data of Kendrew.) The large dots represent α-carbon atoms of amino acids. The sequence of dots therefore denotes the primary structure of the molecule. This consists of a single polypeptide chain with a COOH end at the upper left and an α-amino end at the lower left. The spiral portions drawn in perspective represent regions where the polypeptide chain is coiled in an α-helix. Zigzag portions represent nonhelical regions. The secondary structure of this protein is thus predominantly α-helix. The entire polypeptide chain is wound about itself, conferring tertiary structure. Since the myoglobin molecule contains but a single subunit, no quaternary structure is possible. Note also (in the upper right corner) the heme group attached by 2 histidine molecules to 2 different regions of the polypeptide chain. (Courtesy of R.E. Dickerson.)

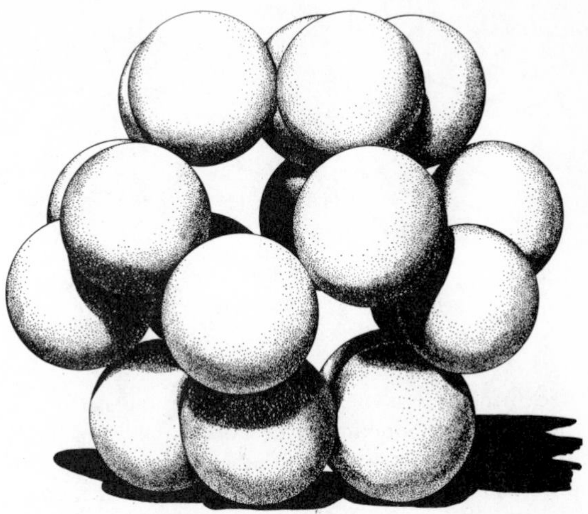

FIG 3–22. Representation of quaternary structure of a protein. "Ping-pong ball" model of the apoferritin molecule. This consists of 20 subunits, each with a molecular weight of about 20,000 daltons. The subunits are arranged to form a hollow sphere which may become packed with iron salts forming the iron storage protein ferritin (see p 406). (Courtesy of R.A. Fineberg.)

tiary structure" of the protein. This tertiary structure is maintained by weak interatomic forces such as hydrogen bonds or by what are termed Van der Waals' forces. For example, the protein of tobacco mosaic virus has a tertiary structure resembling a kernel of corn. These "kernels" line up along the "cob" of nucleic acid to produce the elongated nucleoprotein rods which have been seen under the electron microscope.

In addition to primary, secondary, and tertiary structures, some proteins may display a fourth level of organization wherein several monomeric units, each with appropriate primary, secondary, and tertiary structures, may combine. The association of similar or dissimilar subunits confers on the protein a **quaternary structure.** Such a higher level of organization of monomeric units may be essential to the activity of a protein such as an enzyme protein (see Table 8–4).

Representations of the various levels of organization of protein molecules are shown in Figs 3–21 and 3–22.

The secondary and tertiary structures of a protein are themselves determined by the amino acid structure of the primary polypeptide chain. Once the chain has been formed, the chemical groups which extend from the amino acids direct the specific coiling (secondary structure) and aggregation of the coiled chains (tertiary structure). Treatment of ribonuclease with a mild reducing agent inactivates it, but when it is gently reoxidized almost complete reactivation occurs. As illustrated in Figs 3–26 and 3–27, the disulfide bonds serve to maintain the specific coiled arrangement of the protein. When these are broken by reduction (to SH groups), the characteristic coiling is lost; but gentle reoxidation (to S–S linkages) will reestablish it exactly as before.

Protein Conformation

The term **conformation** refers to the relative positions in space of each of the constituent atoms of a molecule. For a simple molecule like ethane (C_2H_6), 2 conformations are possible—"staggered" and

Staggered Eclipsed

"eclipsed" (see above). Since the carbon atoms can rotate freely about the single bond connecting them, the conformational forms of ethane are interconvertible. The staggered form is, however, favored thermodynamically over the sterially hindered eclipsed form. Free rotation about the bond connecting the carbon atom would not be possible if it were a double bond.

While the "backbone" of polypeptide chains consists of single bonds, free rotation does not occur about all these bonds. The peptide bond itself is planar and has some double bond character. Rotation about this bond does not occur (Fig 3–23).

The α-Helix

X-ray data obtained in the early 1930s indicated that hair and wool α-keratins possessed repeating units spaced 0.5–0.55 nm along their longitudinal axis. As shown in Fig 3–23, no dimension of the extended polypeptide chain appears to measure 0.5–0.55 nm. This apparent anomaly was resolved by Pauling and Corey, who proposed that the polypeptide chain of α-keratin is arranged as an α-helix (Fig 3–24).

In this structure, the R-groups on the α-carbon atoms protrude outward from the center of the helix. There are 3.6 amino acid residues per turn of the helix, and the distance traveled per turn is 0.54 nm—a reasonable approximation of the 0.5–0.55 nm spacing observed by x-ray diffraction. The spacing per amino acid residue is 0.15 nm, which also corresponds with x-ray data.

The following are important features of the α-helix:

(1) The α-helix is stabilized by inter-residue hydrogen bonds formed between the H atom attached to a peptide N and the carbonyl O of the residue fourth in line behind in the primary structure.

(2) Each peptide bond participates in the H-bonding. This confers maximum stability.

(3) An α-helix forms spontaneously as it is the lowest energy, most stable conformation for a polypeptide chain.

(4) The right-handed helix which occurs in proteins is significantly more stable than the left-handed helix when the residues are L-amino acids.

Certain amino acids tend to disrupt the α-helix. Among these are proline (the N-atom is part of a rigid ring and no rotation of the N-C bond can occur) and amino acids with charged or bulky R-groups which either electrostatically or physically interfere with helix formation (see Table 3–5).

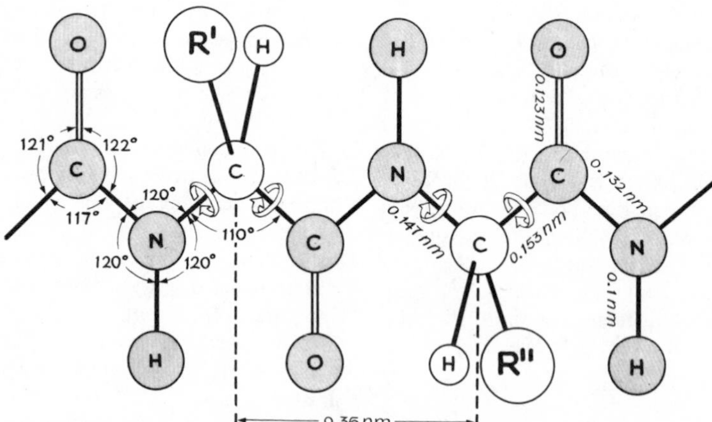

FIG 3–23. Dimensions of a fully extended polypeptide chain. Each group of 4 shaded atoms are **coplanar**, ie, they lie in the same plane. These same 4 atoms comprise the polypeptide bond. The unshaded atoms are the α-carbon atom, the α-hydrogen atom, and the α-R group of the particular amino acid. Free rotation can occur about the bonds connecting the α-carbon with the α-nitrogen and α-carbonyl functions (curved arrows). The extended polypeptide chain is thus a semirigid structure with 2/3 of the atoms of the backbone held in a fixed planar relationship one to another. The distance between adjacent α-carbon atoms is 0.36 nm. The interatomic distances and bond angles, which are not equivalent, are also shown. (Redrawn and reproduced, with permission, from Corey LP, Branson, HR: Proc Nat Acad Sc 37:205, 1951.)

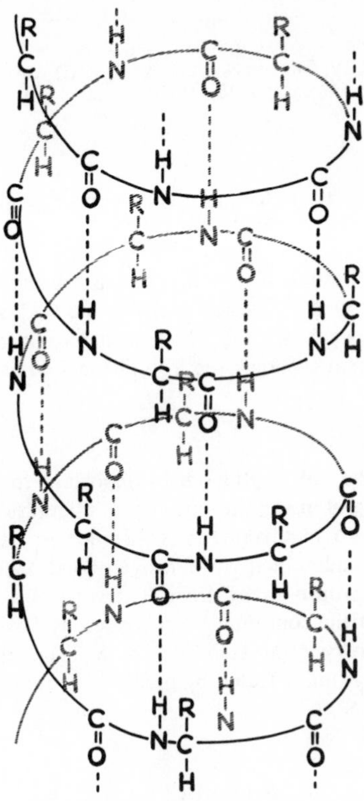

FIG 3–24. Alpha helix structure of a protein.

STRUCTURES OF SPECIFIC PROTEINS

INSULIN

Insulin contains 4 peptide chains per unit weight of 12,000 daltons. Two have glycine and 2 have phenylalanine at the free amino end of the chains. The chains themselves are connected by S–S linkages. Oxidation with performic acid breaks these linkages. By such a reaction, Sanger obtained 2 major polypeptide chains from the insulin molecule which he termed A and B. The A chain has a molecular weight of approximately 2750 daltons and the terminal amino acid is glycine; the B chain, with a molecular weight of approximately 3700 daltons, has phenylalanine as the terminal amino acid. The complete sequence of all of the amino acids in both A and B chains of beef insulin is shown in Fig 3–25. It is now known that the A and B chains of insulin are synthesized as a single sequence of amino acids (proinsulin). One or more peptides are then cleaved between residues 30 and 1, forming active insulin (see Chapter 20).

The positions at which the chains are interconnected in an insulin residue of molecular weight 6000 daltons (derived from beef pancreas) were also reported by Sanger. In the A chain, the cysteine residues at 6 and 11 (counting from the N-terminal acid, glycine) are connected by an S–S linkage. The A and B chains are interconnected by S–S linkages between 7A and 7B and between 20A and 19B respectively. This may be represented diagrammatically as follows:

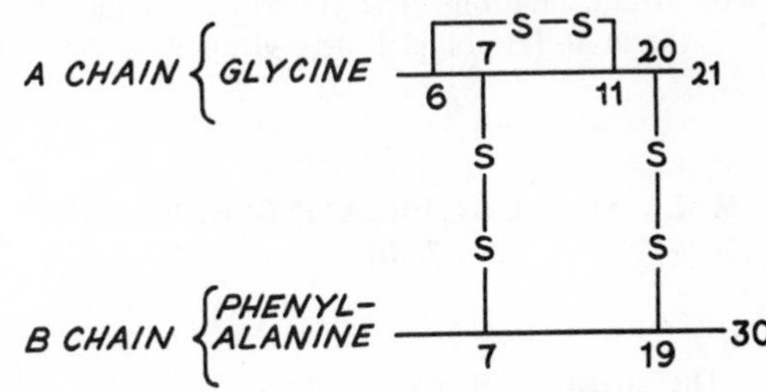

The insulins of various mammalian species (pig, sheep, horse, whale) exhibit differences in the amino acid sequences at positions 8–10 of the A chain. The structure of human insulin is shown in Fig 20–3.

TABLE 3–5. Effect of various amino acid residues on helix formation.

Permit Stable α-Helix	Destabilize α-Helix	Break α-Helix
Ala	Arg	Pro
Asn	Asp	Hyp
Cys	Glu	
Gln	Gly	
His	Lys	
Leu	Ile	
Met	Ser	
Phe	Thr	
Trp		
Tyr		
Val		

GLUCAGON

Glucagon is a polypeptide (minimal molecular weight 3485 daltons) containing 29 amino residues (Fig 20–9). It differs from insulin in several ways; only a few dipeptide sequences are similar, and there are no disulfide bridges in glucagon.

OXYTOCIN & VASOPRESSIN

Du Vigneaud and his colleagues have established the structure of **oxytocin** and of **vasopressin**, the 2 hormones of the posterior pituitary gland. These hormones are polypeptides with a molecular weight of about 1000 daltons and consist of 8 different amino acids arranged in a cyclic structure through S–S linkages. The exact structures of these hormones are shown in Fig 20–18.

A Chain

Gly-Ile-Val-Glu-Gln-CySO_3H-CySO_3H-Ala-Ser-Val-CySO_3H-Ser-Leu-Tyr-Gln-Leu-Glu-Asn-Tyr-CySO_3H-Asn
1 2 3 4 5 6 7 8 9 10 11 12 13 14 15 16 17 18 19 20 21

B Chain

Phe-Val-Asn-Gln-His-Leu-CySO_3H-Gly-Ser-His-Leu-Val-Glu-Ala-Leu-Tyr-Leu-Val-CySO_3H-Gly-Glu-Arg-Gly-Phe-Phe-Tyr-Thr-Pro-Lys-Ala
1 2 3 4 5 6 7 8 9 10 11 12 13 14 15 16 17 18 19 20 21 22 23 24 25 26 27 28 29 30

FIG 3–25. Amino acid sequence of the A and B chains of beef insulin. The abbreviations are those used to designate the amino acids in a peptide chain (Table 3–1). The acid at the left is the amino acid with a free amino group (the N-terminal amino acid). CySO_3H is cysteic acid (see Fig 15–43), the oxidized form of cysteine, which would be obtained after performic acid oxidation to break the S–S linkages which connect the chains in the intact molecule of insulin.

CORTICOTROPIN (ACTH)

An ACTH molecule with all of the biologic properties of the naturally occurring hormone has been synthesized. This synthetic compound is a polypeptide which contains 23 amino acids and has a molecular weight of 3200 daltons. The amino acid sequence of the synthetic ACTH polypeptide is given on p 463.

MELANOCYTE-STIMULATING HORMONE (MSH)

The structure of MSH is shown on p 465. The amino acid sequence from the N-terminus between 7 and 13 inclusive is identical with that between 4 and 10 in ACTH. A hexapeptide which is comprised of the amino acid chain from 8–13 in MSH has been synthesized and found to contain MSH activity. Removing the first amino acid of this hexapeptide (glutamic) results in complete inactivation; but substitution of the glutamyl residue with a glycyl residue restores activity, which suggests that the length of the polypeptide chain is important to MSH activity. The MSH obtained from beef and pig differ only in one amino acid in the number 2 position.

ANGIOTENSIN

Angiotensin I (formerly called hypertensin or angiotonin) is a polypeptide composed of 10 amino acids. This decapeptide is only slightly active, but when it is converted to angiotensin II, an octapeptide, by removal of the 2 amino acids at the C-terminal end of the molecule, it assumes its characteristic pressor activity. The structure of angiotensin is shown in Fig 18–8. However, not only is the sequence of amino acids in the polypeptide chain important to the activity of angiotensin; the specific configuration of the chain is also important. It is assumed to be a helical structure maintained by hydrogen bonds. When these bonds are ruptured by treatment with 10% urea, the activity of the compound is decreased by 50%. By contrast, a polypeptide stabilized by S–S bonds, such as oxytocin, is not affected by urea.

RIBONUCLEASE

The sequence of the amino acid residues in performic acid-oxidized ribonuclease has been established by Hirs, Moore, & Stein (1960). This enzyme protein was found to be a single chain of 124 amino acid residues beginning with lysine as the N-terminal amino acid and ending with valine as the C-terminal amino acid. As was described above for insulin, there are cysteine residues in the amino acid chain of ribonuclease which are joined to one another by disulfide bonds, thus accomplishing cross-linkages within the single polypeptide chain. The position of these disulfide bonds has been established as shown in the schematic diagram of the structure of bovine ribonuclease (Fig 3–26). It will be noted that there are 8 cysteine residues in the chain, thus providing 4 disulfide or S–S linkages.

The results of x-ray diffraction studies of bovine ribonuclease indicate that the molecule of the enzyme has dimensions of about 3.2 × 2.8 × 2.2 nm. There is very little alpha helix structure, such structures being limited to 2 turns of the helix at the region of amino acid residues 5–12 and 2 more turns near the residues 28–35. The 3-dimensional structure of myoglobin is shown in Fig 3–21. In contrast to myoglobin, ribonuclease has much of its structure exposed, no part being shielded by more than one layer of the main chain. A phosphate ion is associated directly with the active site of the enzyme. The amino acid residues nearest the phosphate are numbers 119 and 12. Both of these amino acid residues are histidine. Lysine resi-

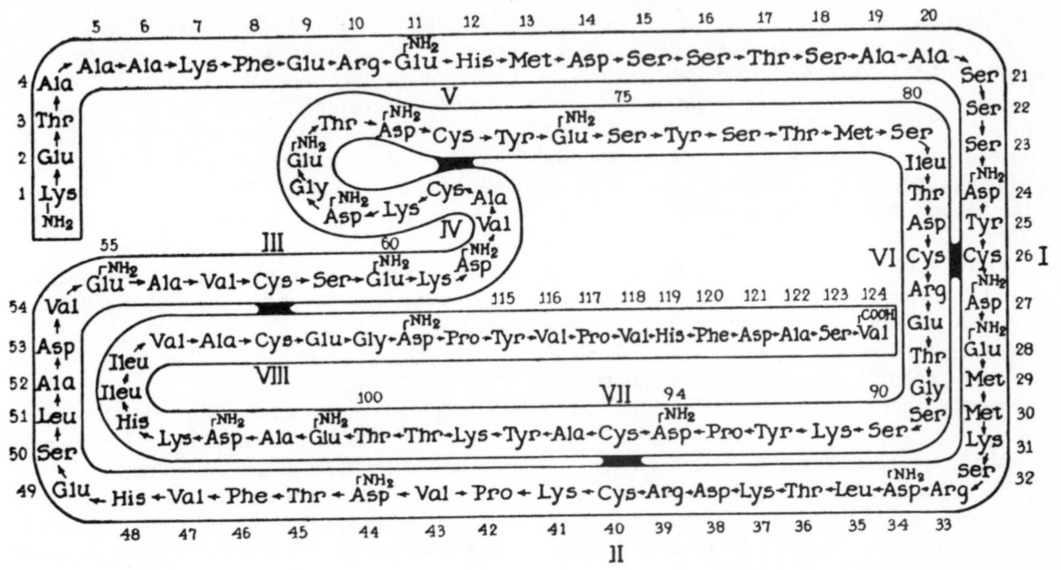

FIG 3–26. Structure of bovine ribonuclease. Two-dimensional schematic diagram showing the arrangement of the disulfide bonds and the sequence of the amino acid residues. Arrows indicate the direction of the peptide chain starting from the amino end. (Reproduced, with permission, from Smyth, Stein, & Moore: The sequence of amino acid residues in bovine pancreatic ribonuclease: Revisions and confirmations. J Biol Chem 238:227, 1963.)

dues at positions 7 and 41 and histidine at 48 have also been implicated as at the active sites, and all are also near the phosphate group. If ribonuclease is split by the enzyme subtilisin, 2 inactive peptides are produced. The shorter one (the so-called S-peptide) consists of the first 21 amino acids from the N-terminal. It can be reassociated with the longer peptide fragment (S-protein) and full activity of the enzyme returns.

A representation of the structure of ribonuclease as determined from x-ray diffraction data is shown in Fig 3–27 (see also Chapter 8).

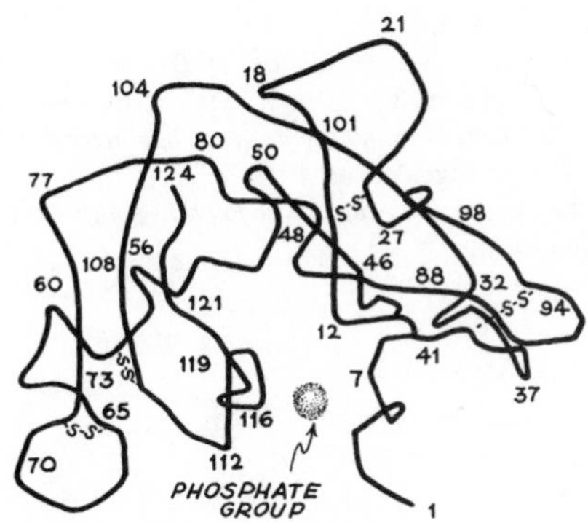

FIG 3–27. Structure of ribonuclease as determined from x-ray diffraction studies. (Numbers refer to specific amino acid residues.)

GENERAL REACTIONS OF PROTEINS

Precipitation Reactions

(1) Concentrated mineral acids. Heavy precipitates occur with small amounts of the acid, but further addition of acid redissolves the protein and hydrolysis may occur later.

(2) Alkalies do not precipitate, but hydrolysis and oxidative decomposition occur.

(3) Heavy metals act as protein precipitants, depending upon the hydrogen ion concentration, temperature, and the presence of other electrolytes. Mercuric chloride and silver nitrate produce a heavy precipitate which cannot be redissolved, whereas copper sulfate and ferric chloride cause precipitation with resolution in an excess of reagent.

(4) So-called "alkaloidal" reagents (trichloroacetic acid, tannic acid, phosphotungstic acid, phosphomolybdic acid) act as protein precipitants when the pH of the solution is on the acid side of the isoelectric point of the protein.

(5) Alcohol and other organic solvents are protein precipitants. They are most effective when the protein is at its isoelectric point.

(6) Heat will coagulate many proteins, although the effective temperature may range from 38–75° C. Various factors influence coagulation, but the protein is most easily coagulated at its isoelectric point. The resulting coagulum is insoluble unless the solvent causes hydrolysis or other decomposition.

Denaturation (See also p 138.)

Many of the agents listed above, as well as x-ray and ultraviolet irradiation, cause denaturation, a change in the physical and physiologic properties of the protein, as well as other changes which are not well understood. Denaturation results in an unfolding of the protein molecule. Hydrogen and nonpolar bonds are the linkages destroyed during denaturation. Urea acts in this way as a denaturing agent. In some cases denaturation without coagulation may be accomplished by heating, as occurs when the protein is heated in an acid or alkaline solution. The protein is transformed into a metaprotein which is insoluble at its isoelectric point. It will flocculate when returned to the isoelectric point but will redissolve in either acid or base.

Other changes resulting from denaturation include alteration in surface tension, loss of enzymatic activity, and loss or alteration of antigenicity. However, gross changes in the protein molecule can occur without permanent inactivation of what may be thought of as the biologically active "core." For example, 2/3 of the molecule of the protein-splitting enzyme, papain, can be removed by a proteolytic hydrolysis without altering the biologic activity of the enzyme. Similarly, ribonuclease, when treated with subtilisin, a protein-splitting endopeptidase, loses a fragment containing 20 amino acids with a resultant loss in enzymatic activity as well. However, if the purified fragment is returned to the denatured ribonuclease, enzymatic activity is restored although peptide bond formation does not occur. This seems to suggest that proximity of certain portions of the original peptide chain may be more essential to biologic activity than preservation of the intact chain itself.

Color Reactions

Certain color reactions previously mentioned (Table 3–4) as specific for various amino acids are useful also to detect the presence of protein containing the amino acids for which the color test is indicative. These include the Millon, xanthoproteic, and ninhydrin reactions as well as the Hopkins-Cole (glyoxylic acid test. The biuret reaction (see p 34), specific to peptide linkages, is also used as a test for the presence of proteins.

• • •

References

Edman P: Acta chem scandinav 4:283, 1950.
Gutte B, Merrifield RB: J Am Chem Soc 91:501, 1969.
Harte RA, Rupley JA: J Biol Chem 243:1663, 1968.
Hirs CHW, Moore S, Stein WH: J Biol Chem 235:633, 1960.
Klotz IM: Science 155:697, 1967.
Merrifield RB: Science 150:178, 1965.
Sanger F, Tuppy H: Biochem J 49:463, 1951.
Sanger F, Thompson EOP: Biochem J 53:353, 1963.
Witkop B: Science 162:318, 1968.

Bibliography

Bailey JL: *Techniques in Protein Chemistry.* Elsevier, 1962.
Dayhoff MO: *Atlas of Protein Structure.* Vol 4. National Biomedical Research Foundation, 1969.
Greenstein JP, Winitz M: *Chemistry of the Amino Acids.* 3 vols. Wiley, 1961.
Meister A: *Biochemistry of the Amino Acids,* 2nd ed. Academic Press, 1965.
Morris CJOR, Morris P: *Separation Methods in Biochemistry.* Interscience, 1964.
Ribeiro LP, Mitidieri E, Affonso OR: *Paper Electrophoresis: A Review of Methods and Results.* Elsevier, 1961.
Schachman HK: *Ultracentrifugation in Biochemistry.* Academic Press, 1959.
Scott RM: *Clinical Analysis by Thin-Layer Chromatography Techniques.* Ann Arbor, 1969.
Sober HA, Harte RA: *Handbook of Biochemistry: Selected Data for Molecular Biology.* Chemical Rubber Co., 1968.
Stahl E: *Thin-Layer Chromatography: A Laboratory Handbook,* 2nd ed. Springer-Verlag, 1969.
Stewart JM, Young JD: *Solid Phase Peptide Synthesis.* Freeman, 1969.

4...

Nucleoproteins & Nucleic Acids

Nucleoproteins are one of the groups of conjugated proteins. They are characterized by the presence of a nonprotein prosthetic group (nucleic acid) which is attached to one or more molecules of a simple protein. The simple protein is usually a basic protein such as a protamine or histone. These conjugated proteins are found in all animal and plant tissues, but are most easily isolated from yeast or from tissues with large nuclei where the cells are densely packed, such as in the thymus gland.

The nucleoproteins were so named because they constitute a large part of the nuclear material of the cell. However, these proteins are also found in the cytoplasm, associated particularly with the ribosomes, where the ribonucleic acid is intimately concerned with synthesis of proteins (see Chapter 5).

All living cells contain nucleoprotein, and some of the simplest systems, such as the viruses, seem to be almost entirely nucleoprotein. Furthermore, such an important cellular constituent as chromatin is largely composed of nucleoproteins, which indicates that these compounds are involved in cell division and the transmission of hereditary factors. For this reason any abnormality in the mechanism of nucleoprotein formation is followed by alterations in cell growth and reproduction. This may be exemplified by the effects of deficiencies of folic acid and vitamin B_{12}, radiations which induce mutations, or radiomimetic chemical agents such as the nitrogen and sulfur mustards, which not only affect chromosome structure and gene activity but also suppress mitosis.

Nucleoproteins have been extracted from a variety of plant and animal tissues. Extracting agents used include water, dilute alkali, sodium chloride solutions, and buffers ranging in pH from 4.0–11.0. In each case extraction is followed by precipitation with acid, saturated ammonium sulfate, or dilute calcium chloride. When the purified nucleoprotein is hydrolyzed with acid or by the use of enzymes, it is broken down into various components as shown in Fig 4–1.

By the use of careful acid hydrolysis, the purine or pyrimidine base may be removed from a nucleotide, leaving the sugar attached to the phosphate.

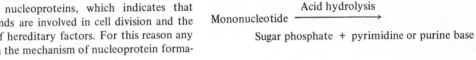

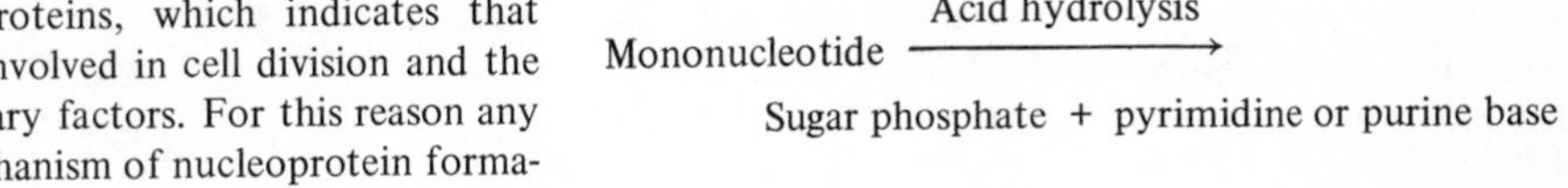

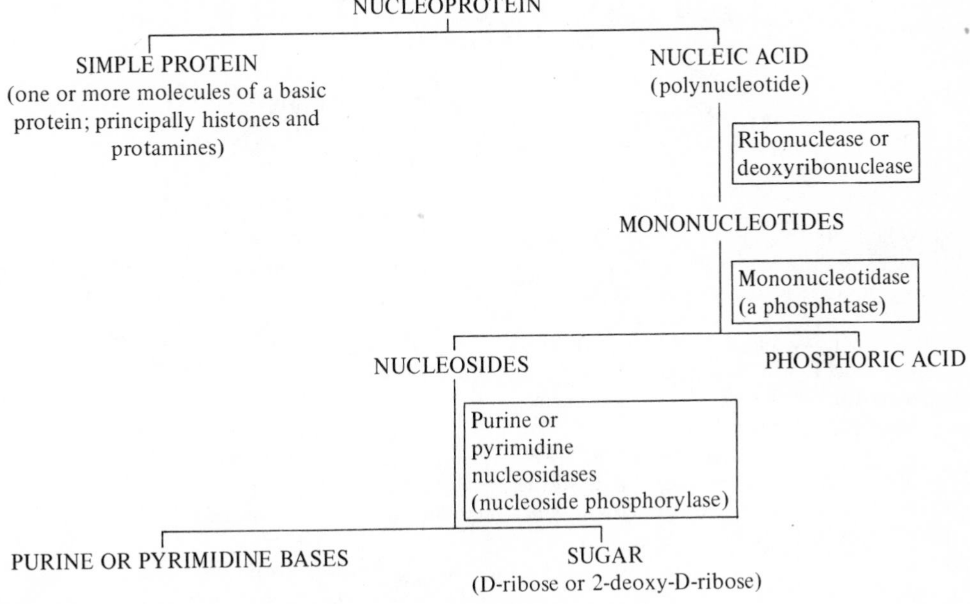

FIG 4–1. Hydrolysis of nucleoprotein.

THE PYRIMIDINE & PURINE BASES

The various purine and pyrimidine bases which occur in the nucleotides of nucleic acids are derived by appropriate substitution on the ring structures of the parent substances, purine or pyrimidine. The structures of these parent nitrogenous bases are as follows: (The positions on the rings are numbered according to the original Fischer system. The newer International System proposes to number the pyrimidine ring as shown in the brackets. The purine numbering system is not changed.)

Purine

Pyrimidine

The Pyrimidine Bases

There are 3 main pyrimidine bases which have been isolated from nucleic acids: cytosine, thymine, and uracil. **Cytosine** (2-oxy-6-aminopyrimidine) is found in all nucleic acids except the deoxyribonucleic acid (DNA) of certain bacterial viruses, viz, coli bacteriophage of the T-even series (T_2, T_4, T_6, etc); in its place there is **5-hydroxymethylcytosine** (HMC) (Wyatt, 1952).

Cytosine (2-oxy-6-aminopyrim- [2] [4] idine)

5-Hydroxymethyl-cytosine

Thymine (2,6-dioxy-5-methyl pyrimidine) occurs mainly in nucleic acids which contain deoxyribose as the characteristic carbohydrate, the so-called deoxyribonucleic acids (DNA) (see below). However, minor amounts have also been found in transfer RNA (tRNA). **Uracil** (2,6-dioxypyrimidine), on the other hand, is confined to the ribonucleic acids (RNA) which contain ribose rather than the deoxy sugar.

Thymine (2,6-dioxy-5- [2,4] methylpyrimidine)

Uracil (2,6-dioxypyrimidine) [2,4]

The Purine Bases

Adenine and guanine are the 2 purines found in all nucleic acids.

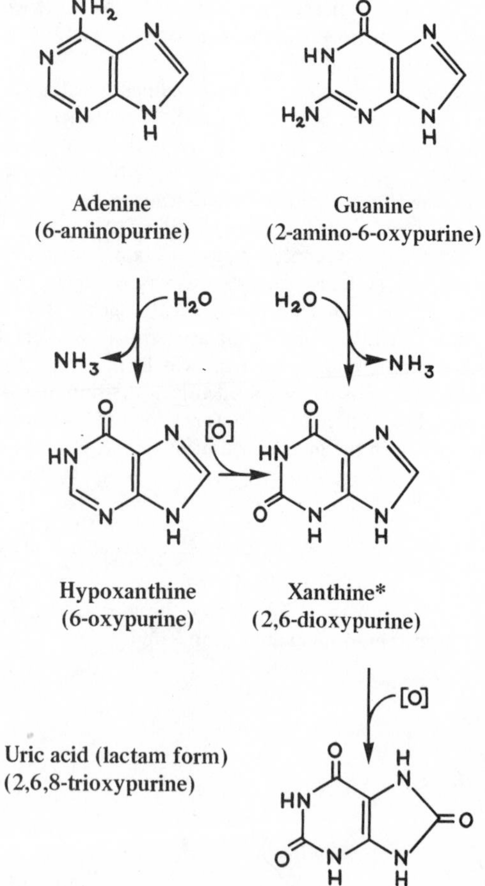

Adenine (6-aminopurine)

Guanine (2-amino-6-oxypurine)

Hypoxanthine (6-oxypurine)

Xanthine* (2,6-dioxypurine)

Uric acid (lactam form) (2,6,8-trioxypurine)

The oxypurines or oxypyrimidines may form enol derivatives by migration of hydrogen to the oxygen substituents. This is illustrated by the so-called lactim (hydroxy) structure of uric acid (shown below), which

*Three xanthine derivatives are important constituents of coffee, tea, and cocoa. Coffee: caffeine, 1,3,7-trimethyl-xanthine. Tea: theophylline, 1,3-dimethyl-xanthine. Cocoa: theobromine, 3,7-dimethyl-xanthine.

is formed by enolization of the lactam (oxy) form shown above. However, the lactam (oxy) form is the predominant tautomer.

The existence of these 2 forms is suggested by the fact that the oxypurines or oxypyrimidines form salts with alkali. The purines and pyrimidines also form salts with acids because of the nitrogen atoms which are weakly basic.

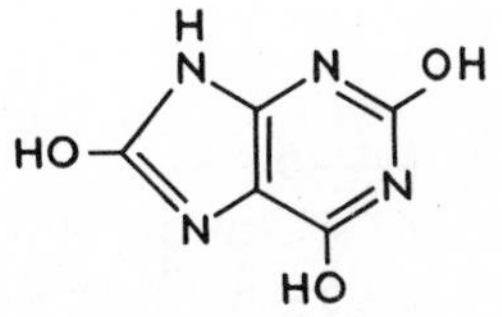

Lactim form of uric acid (2,6,8-trihydroxypurine)

Both ribonucleic acids and deoxyribonucleic acids may contain several minor base components in addition to the 4 main bases of their primary structure, adenine, guanine, cytosine, and thymine or uracil. In the DNA of bacteria, the minor component is **N-6-methyladenine,** and in that of plants and animals it is **5-methylcytosine.**

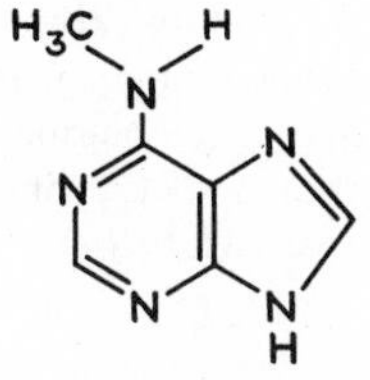

N-6-Methyladenine 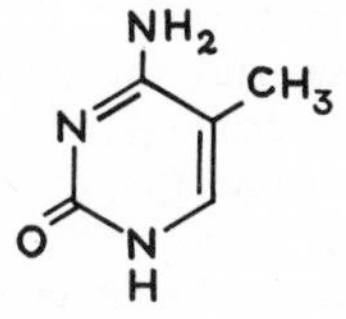 **5-Methylcytosine**

Ribosomal RNA also contains methylated purine or pyrimidine bases. Such bases are also present in transfer RNA, at least 10 such methylated bases having been demonstrated in these ribonucleic acids. According to Srinivasan & Borek (1964), these methylated bases are formed by the action of specific methylases which transfer methyl groups to the bases only after they have been incorporated into the nucleic acid polymers.

Transfer RNA also contains as minor components the peculiar uridine nucleoside, pseudouridine, and nucleotides containing thymine, hypoxanthine (6-oxypurine), 1-methylhypoxanthine, and 5,6-dihydrouracil.

THE NUCLEOTIDES & NUCLEOSIDES

A **nucleotide,** the structural unit of nucleic acids, is composed of a purine or a pyrimidine base attached to a sugar (ribose or 2-deoxyribose) by a glycosidic linkage; the sugar is then combined with phosphoric acid:

Purine or pyrimidine – Sugar – Phosphoric acid

Two general types of nucleotides are found in nucleic acids; one contains D-ribose (furanose form; see p 5); the other contains 2-deoxy-D-ribofuranose (see p 8). It is therefore customary to refer to these nucleic acids as ribonucleic acid (RNA) or deoxyribonucleic acid (DNA).

Removal of the phosphate moiety from a nucleotide produces a **nucleoside,** which is composed of a purine or a pyrimidine base and a sugar, either D-ribose or 2-deoxyribose.

Nomenclature & Structure of Nucleotides & Nucleosides

The name of each nucleotide may be derived from that of its constituent nitrogenous base. For example, the nucleotides present in ribonucleic acid are thus termed:

Adenylic acid: Adenine + ribose + phosphoric acid
Guanylic acid: Guanine + ribose + phosphoric acid
Cytidylic acid: Cytosine + ribose + phosphoric acid
Uridylic acid: Uracil + ribose + phosphoric acid

Deoxyribonucleic acid (DNA) contains thymidylic acid (thymine + deoxyribose + phosphoric acid) rather than uridylic acid.

The attachment of the sugar in a nucleotide (ribose or 2-deoxyribose, in the furanose form) is to the nitrogen of a purine base at position 9 or to the nitrogen at position 3 of a pyrimidine base. An exception occurs in the structure of nucleotides in which the sugar is attached to the pyrimidine, uracil, at position 5, thus establishing a carbon-to-carbon linkage instead of a nitrogen-to-carbon linkage, as is the usual structural arrangement in the nucleotides. An example is the 5-ribosyl uridine nucleotide (5-ribosyl uridylic acid) which has been termed **pseudouridylic acid.** 3-Ribosyl purines have also been described. These would represent compounds in which the attachment of the ribosyl moiety is to the nitrogen at position 3 of the purine rather than at position 9, as is the case with other naturally occurring purine ribonucleotides. Examples are 3-ribosyl uric acid-5′-phosphate and 3-ribosyl-xanthine-5′-phosphate, a precursor of the uric acid compound.

Nucleotides may also be designated by considering them to be nucleoside phosphates, the nucleoside being a compound composed of a purine or a pyrimidine base with either ribose or deoxyribose. If the sugar is deoxyribose, the prefix **deoxy** (abbreviated "d") is included in the naming of the compound. Thus, the nucleosides of DNA would be designated deoxyadenosine (dA), deoxyguanosine (dG), deoxycytidine (dC), and deoxythymidine (dT). If no prefix is used, the sugar in the nucleoside is assumed to be ribose as is the case with the nucleosides of RNA.

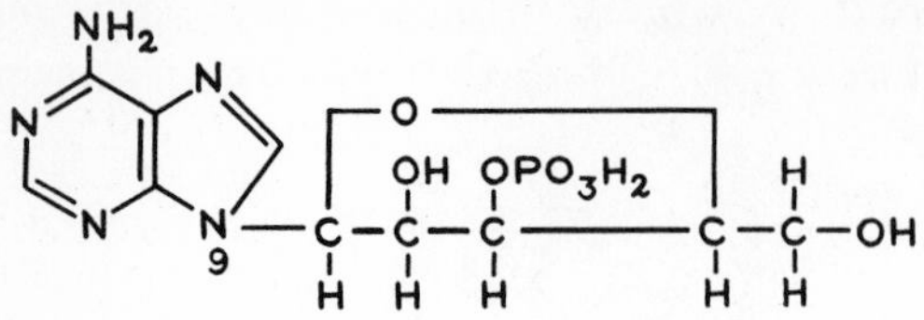

Adenylic acid (in RNA)
(adenosine-3′-phosphate)

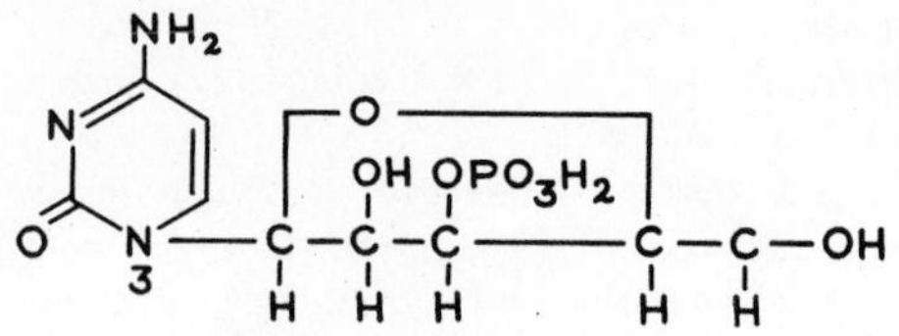

Cytidylic acid (in RNA)
(cytidine-3′-phosphate)

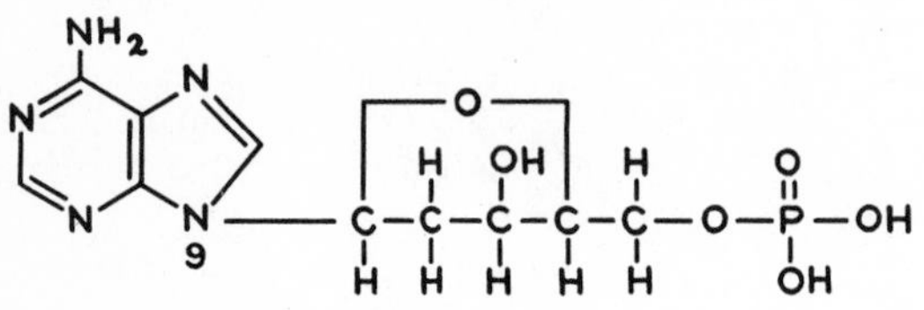

Adenylic acid (in DNA)
(5′-deoxyadenosine phosphate; 5′-dAMP)

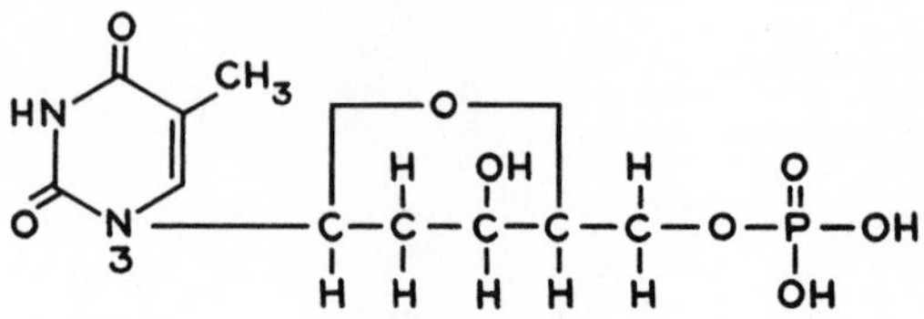

Thymidylic acid (in DNA)
(5′-deoxythymidine phosphate; 5′-dTMP)

Attachment of a phosphate to the pentose sugar of a nucleoside produces the corresponding nucleotide. The position of the phosphate is indicated by a numeral. For example, an adenine-containing nucleotide of RNA (adenylic acid) with the phosphate attached to carbon 3 of the sugar, ribose, would be designated adenosine-3′-phosphate. (The "prime" mark after the numeral is required to differentiate a numbered position on the sugar moiety from a numbered position on the purine or pyrimidine base which would be unmarked.) An adenine-containing nucleotide of DNA with the phosphate moiety attached to carbon 5 of the sugar (deoxyribose) would be designated 5′-deoxyadenosine phosphate. Examples of nucleotides illustrating both types of phosphate linkage are shown above.

The abbreviations A, G, C, T, or U may be used to designate a nucleoside in accordance with which purine or pyrimidine base it contains, with the addition of the prefix d if the sugar of the nucleoside is deoxyribose. When occurring in the free form as a nucleotide, MP (monophosphate) may be added. For example, a guanine-containing nucleoside in DNA would be designated dG (deoxyguanosine). The corresponding nucleotide with phosphate esterified to carbon 3 of the deoxyribose moiety is 3′-dGMP.

Nucleotides involved in metabolic reactions such as adenosine triphosphate (ATP) generally have the phosphate on carbon 5 of the sugar. In these instances, a number is not used; when thus unspecified, the phosphate in the nucleotide is assumed to be on carbon 5 of the sugar. Thus, ATP (adenosine triphosphate) designates the nucleoside adenosine (adenine-ribose) with 3 phosphates esterified at carbon 5 of the ribose component. The structure of ATP is shown below.

Biologically Important Nucleotides

Nucleotides which are not combined in nucleic acids are also found in the tissues. They have important special functions. Some of these compounds will be listed below.

Adenine derivatives. Adenosine diphosphate (ADP) and adenosine triphosphate (ATP) are important compounds in view of their participation in oxidative phosphorylation and, in the case of ATP, as the source of high-energy phosphate for energy-requiring reactions in the cells.

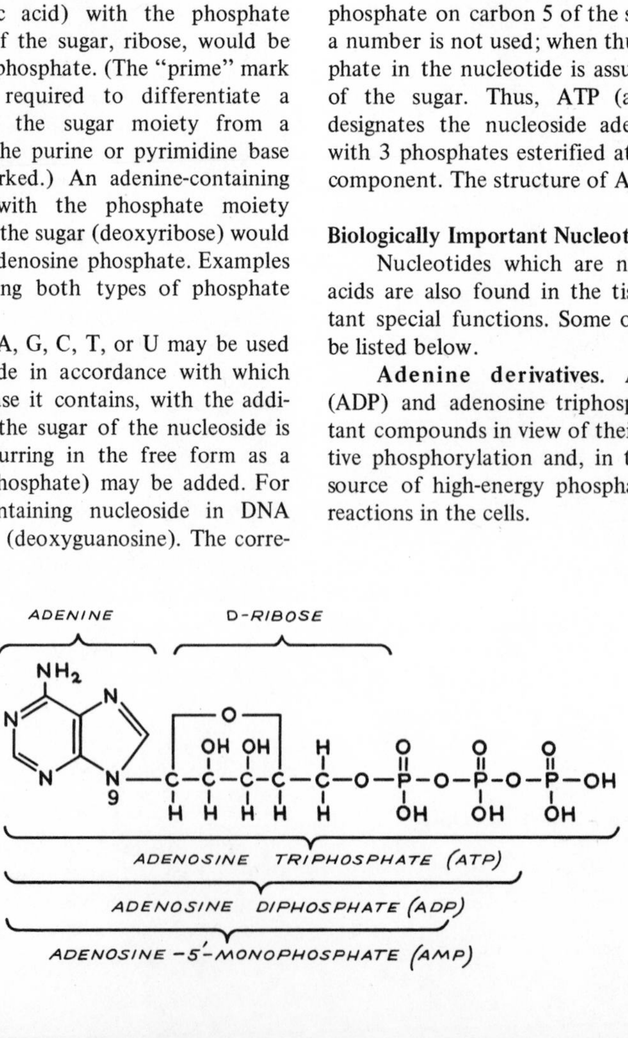

ADENINE *D-RIBOSE*

ADENOSINE TRIPHOSPHATE (ATP)

ADENOSINE DIPHOSPHATE (ADP)

ADENOSINE-5′-MONOPHOSPHATE (AMP)

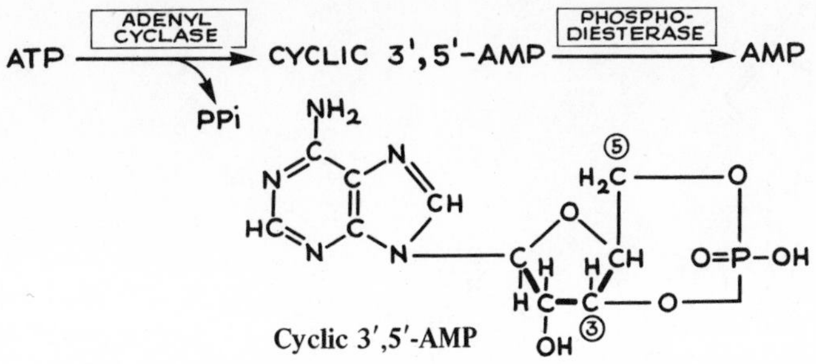

ATP —ADENYL CYCLASE→ CYCLIC 3′,5′-AMP —PHOSPHODIESTERASE→ AMP (PPi released)

Cyclic 3′,5′-AMP

Formation of cyclic AMP from ATP and destruction of cAMP by phosphodiesterase

Cyclic AMP (3′,5′-adenosine monophosphate) is derived from ATP in a reaction catalyzed by the enzyme **adenyl cyclase**. Cyclic AMP is destroyed in the tissues by conversion to AMP (adenosine monophosphate; adenylic acid) in a reaction catalyzed by **phosphodiesterase.**

Adenyl cyclase is bound to the plasma membranes of cells. Thus, many cells of animal tissues as well as those of bacteria and other microorganisms can synthesize cyclic AMP. It is of interest that cyclic AMP is not found in plants.

Cyclic AMP has been found to influence a number of metabolic processes, including those in muscle (eg, phosphorylase activity; see p 234), gastric secretion, and CNS function. As is described in Chapter 20, cyclic AMP plays a role both in the release of some hormones and the action of others. In the latter instance, a hormone may activate adenyl cyclase in the cells of its target tissue, which leads to production of cyclic AMP within these cells, and it is the cyclic AMP which actually brings about the effect attributed to the hormones. In this way cyclic AMP is acting as a so-called "second messenger," the hormone being the first messenger to the metabolic system affected. In other instances, cyclic AMP may bring about the synthesis of a hormone.

A relationship between hormones and protein synthesis may be explained by the finding that cyclic AMP activates a protein kinase that catalyzes phosphorylation and thus activation of an RNA polymerase. The activated polymerase, in turn, stimulates RNA production, which will function as the messenger for protein synthesis as described in Chapter 5. A hormone, by initiating production of cyclic AMP through activation of adenyl cyclase, would thus influence protein synthesis.

The incorporation of sulfate into ester linkages in such compounds as the sulfated mucopolysaccharides (eg, chondroitin sulfuric acid and the similar mucoitin sulfuric acid in mucosal tissues) requires the preliminary "activation" of the sulfate moiety. This is accomplished by the formation of "active sulfate" with ATP. The reaction may be depicted as shown below.

"Active" sulfate is also required in conjugation reactions involving sulfate.

Hypoxanthine derivatives. Deamination of adenosine monophosphate produces a hypoxanthine nucleotide, usually called inosinic acid; removal of the phosphate group of inosinic acid forms the nucleoside, inosine (hypoxanthine riboside).

Analogues of ADP and ATP in which the purine derivative is hypoxanthine rather than adenine have also been found to participate in phosphorylation reactions. These compounds are inosine diphosphate (IDP) and inosine triphosphate (ITP).

Guanine derivatives. Guanine analogues of ATP are also involved in metabolism. The oxidation of ketoglutaric acid to succinyl-Co A involves oxidative phosphorylation with transfer of phosphate to guanosine diphosphate (GDP) to form guanosine triphosphate (GTP). This phosphorylation reaction is identical with similar reactions involving ADP and ATP.

Uracil derivatives. Uridine (uracil-ribose) derivatives are important coenzymes in reactions involving epimerization of galactose and glucose (uridine diphosphate glucose, UDPG; and uridine diphosphate galactose, UDPgal). UDPG is also a precursor in glycogenesis. A uridine coenzyme, uridine diphosphate glucuronic acid (UDPgluc), serves as a source of "active" glucuronide in conjugation reactions requiring glucuronic acid, eg, formation of menthol glucuronide or bilirubin glucuronide.

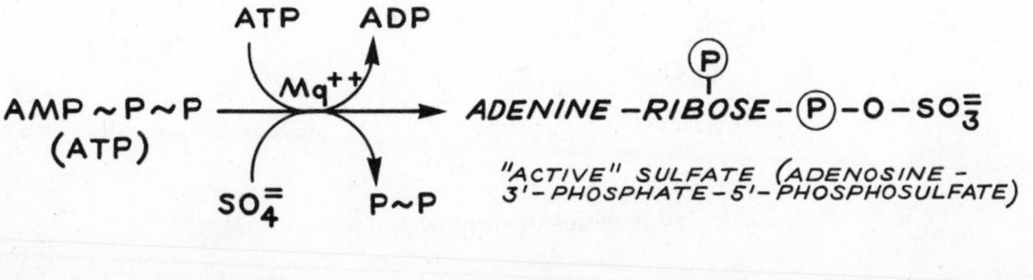

AMP~P~P (ATP) + $SO_4^{=}$ —(Mg^{++}; ATP → ADP; P~P released)→ ADENINE–RIBOSE(–Ⓟ)–Ⓟ–O–$SO_3^{=}$

"ACTIVE" SULFATE (ADENOSINE-3′-PHOSPHATE-5′-PHOSPHOSULFATE)

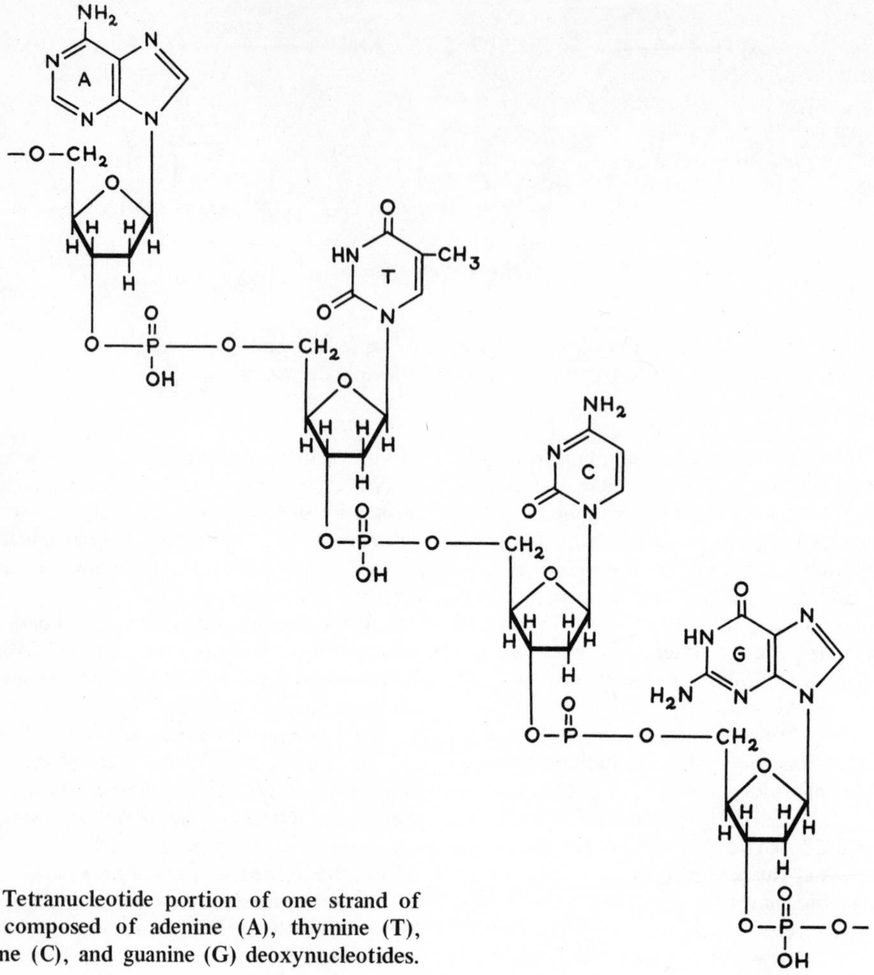

FIG 4–2. Tetranucleotide portion of one strand of DNA composed of adenine (A), thymine (T), cytosine (C), and guanine (G) deoxynucleotides.

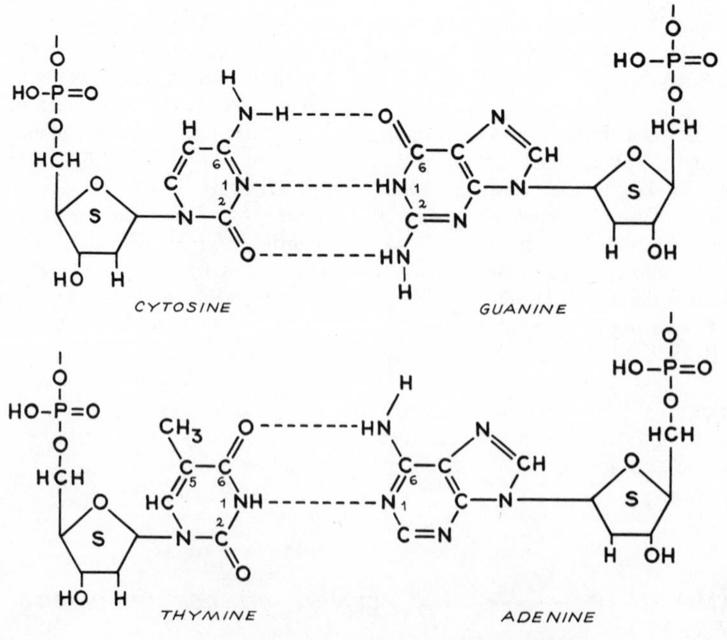

CYTOSINE GUANINE

THYMINE ADENINE

FIG 4–3. Hydrogen bonding of nucleotide pairs.

Uracil also participates in the formation of high-energy phosphate compounds analogous to ATP, ITP, or GTP, mentioned above. Uridine triphosphate (UTP) is formed, for example, in the reactions involving conversion of galactose to glucose. It may also be produced by the transfer of phosphate from ATP to UDP. Berg & Joklik (1954) described the synthesis of UTP (as well as ITP) by phosphorylation of the diphosphate with ATP as phosphate donor. The enzyme which catalyzes this phosphate transfer is called a **nucleoside diphosphokinase.** It has been partially purified from dried (brewer's) yeast and from rabbit muscle. Krebs has also detected it in pigeon breast muscle and in the intestinal mucosa of the rat.

Cytosine derivatives. Cytidine (cytosine-ribose) may form the high-energy phosphate compound, **cytidine triphosphate (CTP).** CTP is the only nucleotide which was found to be effective in an in vitro mitochondrial system for the biosynthesis of lecithin. CTP reacts with phosphoryl choline to form cytidine diphosphate choline (CDP-choline), which in turn combines with a diglyceride to form the phospholipid. CTP is also required in a similar series of reactions involving phosphoryl ethanolamine which leads to the biosynthesis of the cephalins.

Nucleotide Structures in B Vitamins

In their active form, several vitamins of the B complex are combined in nucleotide linkages. These include riboflavin, as in flavin adenine dinucleotide (FAD) or as riboflavin phosphate; niacin, in coenzymes I and II (ie, NAD [DPN] and NADP [TPN], respectively); and pantothenic acid in coenzyme A (coacetylase).

Thiamine (cocarboxylase) and pyridoxal (cotransaminase) are both phosphorylated when functioning as prosthetic groups of enzymes involved in intermediary metabolism. A nucleotide is also a component of the structure of vitamin B_{12}.

Biologically Important Nucleosides

The importance of the amino acid methionine as a source of labile methyl groups is discussed in Chapters 14 and 15. The participation of methionine in transmethylation reactions requires first the formation of "active" methionine. This compound has been found to be an adenine nucleoside, S-adenosyl methionine (Cantoni, 1953). A free nucleoside containing a thiopentose, adenine-5′-thiomethylriboside, has been isolated from yeast. (See structures below.)

NUCLEIC ACIDS

Structure of Nucleic Acids

Nucleic acids consist of long chains of nucleotides **(polynucleotides)** combined one with another through phosphate diester linkages. This is illustrated by the structure depicted in Fig 4–2 which represents a tetranucleotide portion of a strand of DNA. It will be noted that the nitrogenous purine or pyrimidine bases are bonded to a terminal carbon atom (No. 1) of the sugar whereas the individual nucleotides are united by means of a phosphate diester linkage between carbon atom 3 of one sugar molecule and a terminal carbon (No. 5) of the next. This 3′,5′-linkage is to be expected in DNA since the only sugar hydroxy groups available in deoxyribose for the formation of a phosphate ester are those in the 3 and 5 positions. In RNA, 3′,5′-linkages also predominate, although 2′,3′-linkages are possible.

The structure of a nucleic acid chain may be indicated in abbreviated form. The letters A, G, C, U, or T represent the nucleosides of the purine or pyrimidine bases (adenine, guanine, cytosine, uracil, or thymine); the letter p designates phosphate. When p is placed to the left of the nucleoside abbreviation, a 5′-sugar-phosphate linkage is indicated; to the right, a 3′-sugar-

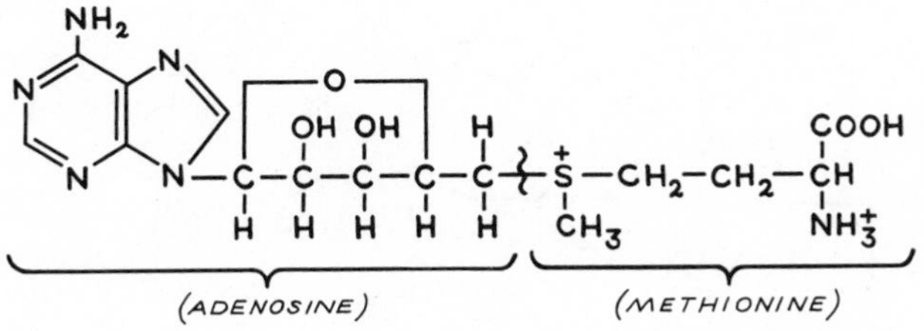

(ADENOSINE) (METHIONINE)

"Active" methionine

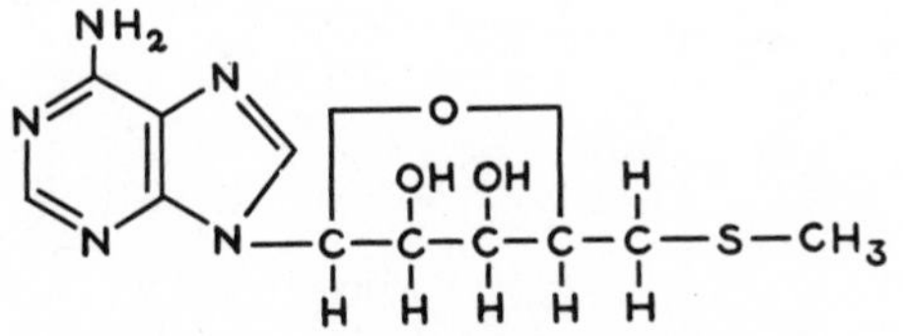

Adenine-5′-thiomethylriboside

phosphate linkage. *Example:* dpG is deoxyguanosine-5′-phosphate; dGp is deoxyguanosine-3′-phosphate. The tetranucleotide structure depicted in Fig 4–2 would thus be represented as d-ApTpCpGp, indicating a tetranucleotide fragment comprised of adenine, thymine, cytosine, and guanine deoxynucleosides united by phosphate diester linkages between carbon 3 of one sugar molecule and carbon 5 of the next. Each of the nucleotides is assumed to be a 3′-phosphate; thus, the linkage between each nucleoside phosphate must involve carbon 5 of the sugar on the adjacent nucleoside since this is the only sugar hydroxyl group available for the formation of the phosphate ester.

The molecular weight of DNA may be as high as 2 billion daltons, and this compound may thus comprise as many as 1 million or more purine and pyrimidine bases arranged in a continuous line. The DNA molecules of mammalian cells appear to be smaller than those from most organisms, containing in a typical molecule only about 30,000 nucleotides. X-ray diffraction patterns of fibrous DNA suggest that its form is that of a double helix. The presence of the double helix can also be shown by studies of ultraviolet light absorption of solutions of DNA. Nucleotides absorb ultraviolet light with wavelengths near 260 nm by virtue of the presence of purine and pyrimidine rings. When these rings are hydrogen-bonded, as they are in the helix, their ability to absorb light is decreased. If a solution of DNA is heated, it will be found that the absorption of ultraviolet light is increased when the hydrogen bonds are broken.

The 2 strands of the double helix of DNA are conceived of as running in opposite directions with respect to the 3,5-deoxyribose linkage (Fig 4–4). Cytochemical studies of DNA obtained from a variety of cell types indicate that the amount of adenine equals the amount of thymine and the amount of guanine equals the amount of cytosine. This "pairing" of the nitrogenous bases and the x-ray diffraction data constitute the essential foundation for the Watson and Crick model of DNA structure. These authors (Watson, 1953) have proposed a double helical structure in which the 2 polynucleotide strands are coiled in such a manner that an adenine of one strand is bonded by hydrogen bonds to a thymine of the complementary strand, and a guanine of one strand is similarly bonded to a cytosine of the complementary strand. This arrangement has come to be called the "pairing rule," namely, A bonded to T and G bonded to C. The hydrogen bonding is believed to involve the keto and amino groups of the nitrogenous bases. Therefore, as shown in Fig 4–3, guanine and cytosine are bonded by 3 hydrogen bonds and adenine and thymine by 2.

There are 3 characteristic forms of **ribonucleic acid (RNA).**

A. Messenger RNA (mRNA): mRNA is a single-stranded molecule. It is made in the nucleus of the cell as a complement to one strand of DNA, so that, according to the pairing rule, thymine, cytosine, guanine, and adenine in a DNA strand direct the incorporation, respectively, of adenine, guanine, cytosine, and uracil (the RNA analogue of thymine in DNA) in the complementary RNA strand. The reaction is catalyzed by **RNA polymerase.** Only one strand of DNA serves as a template for synthesis of mRNA, and a message may be derived from discrete sections of the DNA strand; that is to say, an active DNA strand may carry information for the synthesis of several different protein molecules. The unit of information for one protein peptide chain is termed the **cistron.** A DNA strand may therefore contain information for synthesis of several protein peptide chains, ie, several cistrons

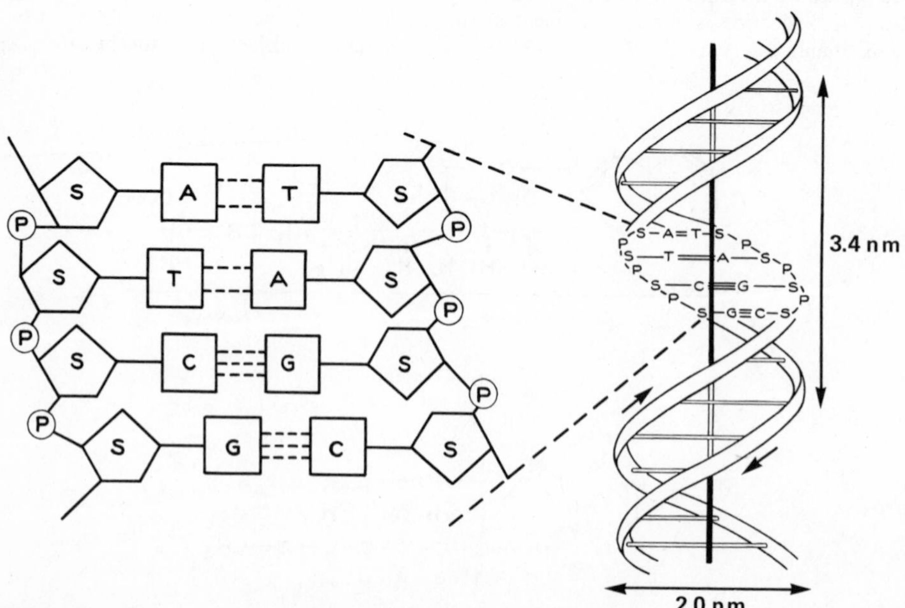

FIG 4–4. The Watson and Crick model of the double helical structure of DNA (modified). A = adenine, C = cytosine, G = guanine, T = thymine, P = phosphate, S = sugar (deoxyribose).

often working together as an operon to provide information necessary for the synthesis of a sequence of enzyme proteins involved in a metabolic pathway.

As just noted, an active DNA strand contains information for synthesis of several different proteins. Messenger RNA serves as the means of transcribing the information stored in DNA to determine the primary structure (amino acid sequence) of proteins being synthesized within the endoplasmic reticulum of the cell. In order to regulate the special metabolic activity of the cell, only certain of the protein peptides the cell is capable of synthesizing should be produced at a given time. Thus, only a small amount of the total information in DNA may be used by the cell at any one time. Chromosomes contain DNA associated with specialized proteins, the histones, a combination referred to as **nucleohistone.** Messenger RNA cannot be formed with a DNA template unless the DNA is exposed by dissociation from the histone. The particular type of mRNA made in the cell is thus controlled by the extent of dissociation of the DNA template from the protein (histone) with which it is associated.

B. Ribosomal RNA (rRNA): A second form of RNA is that associated with the ribosomes of the endoplasmic reticulum in the cytoplasm. Like mRNA, ribosomal RNA is formed so as to be complementary to some region of DNA. It is thus apparent that not all of the genetic information in DNA is associated with coding for synthesis of cellular protein through messenger RNA. Ribosomal RNA is required to bind messenger RNA and the specific enzymes utilized for peptide bond synthesis. The ribosomes are composed of 2 parts which can be separated by varying concentrations of divalent cations (Mg^{++} is often used). The subunits of the ribosomes are classified in terms of their sedimentation constants (S); the commonly studied ribosomes of *E coli,* for example, are comprised of 30 S and 50 S components; mammalian ribosomes are comprised of 40 S and 60 S components. Each part contains RNA along with several different proteins. Ribosomal RNA is made in the nucleolus as a single chain. After hydrolysis into several pieces, some of the pieces associate with the necessary proteins to form ribosomes. As will be further described in connection with the utilization of amino acids for protein synthesis, the 2 pieces of the ribosome are separated at the beginning of peptide synthesis. The 30 S component (in *E coli*) possesses an initiator site which binds the first codon on messenger RNA and the corresponding transfer RNA (described below); the 50 S component binds the growing polypeptide chain by means of a specific transfer RNA molecule.

C. Transfer RNA (tRNA), Also Called Soluble RNA (sRNA): This type of RNA accomplishes the transfer of amino acids to the proper site on the RNA template of mRNA. For each amino acid to be incorporated into the primary structure of a protein molecule, there is therefore a specific tRNA molecule to which the amino acid is attached. Although transfer RNAs are a mixture of different molecular species of similar size, they can be differentiated by the ability of each of them to accept only one amino acid and to transport that amino acid for use in protein synthesis.

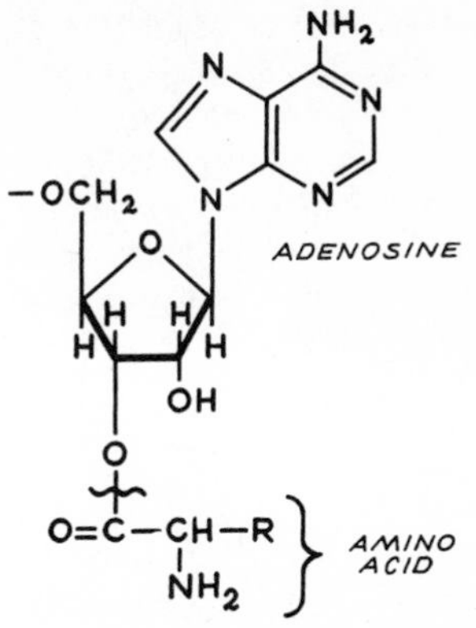

FIG 4–5. Attachment of amino acid to adenosine (AOH) terminus of tRNA.

There are certain similarities in the structures of all of the tRNAs so far studied. The amino acid carrying end of all of the tRNA molecules contains the same trinucleotide, having the base sequence A–C–C, ie, an adenylic acid and 2 cytidylic acid nucleotide residues, attached to a terminal adenosine (AOH) residue. On the hydroxyl group of carbon 3 of the ribose of the terminal AOH, the amino acid carried on the tRNA molecule is attached by an amino acyl linkage as shown in Fig 4–5. A guanine nucleotide (designated pG) frequently (although not invariably) occurs at the 5′-phosphate terminus of the tRNA molecule. Although thymine is generally stated to occur only in DNA, a thymine-containing nucleotide does occur at the 23rd position in the tRNA molecule. Finally, the sequences A–G–DiHU, G–T–ψ–C, and G–C–DiMeG also appear in each tRNA molecule.

Reference was made above to the so-called minor base components of nucleic acid. The nucleotide composition of tRNA is characterized by the presence of several such minor base-containing nucleotides, the term referring to the fact that these minor nucleotides occur much less frequently in the molecule than the major 4 nucleotides of adenine, guanine, cytosine, and uracil. The minor nucleotides include pseudouridylic acid (ψ), ribothymidylic acid (Tr), methylated derivatives of guanine (DiMeG and MeG), 5,6-dihydrouracil (DiHU), inosine (I) and 1-methylinosine (MeI), 6,6-dimethyladenine (DiMeA) and 6-methyladenine (MeA), 5-methylcytosine (MeC), and 2′-O-methylguanine (MeOG).

As already indicated above, specific tRNA molecules are required for each of the amino acids to be incorporated into proteins. Since the amino acid-carrying portion of the tRNA molecules are all alike (ie, –A–C–C–AOH), specificity for a given amino acid must reside elsewhere in the molecule. This is believed to be somewhere about the middle, where there would be a sequence of 3 bases specific for the amino acid to be carried by the tRNA molecule.

The complete analysis of a tRNA for alanine was accomplished by Holley & others (1965). The nucleotide sequence of a yeast tRNA for tyrosine was later reported by Madison & others (1966). The structures proposed for these 2 tRNA molecules are shown in Fig 4–6 (from Madison, 1966). It will be noted that the tRNA molecules are presumed to possess a secondary structure with base pairing by means of hydrogen bonds (shown in Fig 4–6 as dotted lines), as was described for DNA. The "cloverleaf" structure shown would provide for a maximum amount of base pairing. The proposed secondary structure for alanine tRNA has 17 GC, 2 AU, and one GU base pairs. The tyrosine tRNA has 15 GC pairs, 4 AU pairs, and one GU pair. That section of the molecule which is specific for the amino acid to be carried should contain a sequence of 3 bases, called the **anticodon**, which is complementary to the 3 bases on mRNA (the **codon**) which code for the amino acid carried. The amino acid code is described in Chapter 5. There it will be noted that a codon for alanine is GCC. This would pair with CGG anticodon. Examination of the proposed structure for the alanine tRNA shows at the lower loop a sequence reading (from right to left) CGI. I is inosinic acid (hypoxanthine nucleotide), which reads as a guanine nucleotide. The CGI sequence is evidently the anticodon for the alanine tRNA. For tyrosine tRNA, the sequence at the lower loop, A–ψ–G, is the anticodon. This is assumed from a codon for tyrosine, UAC, which would pair with AψG when ψ (pseudouridine) is read as uridine.

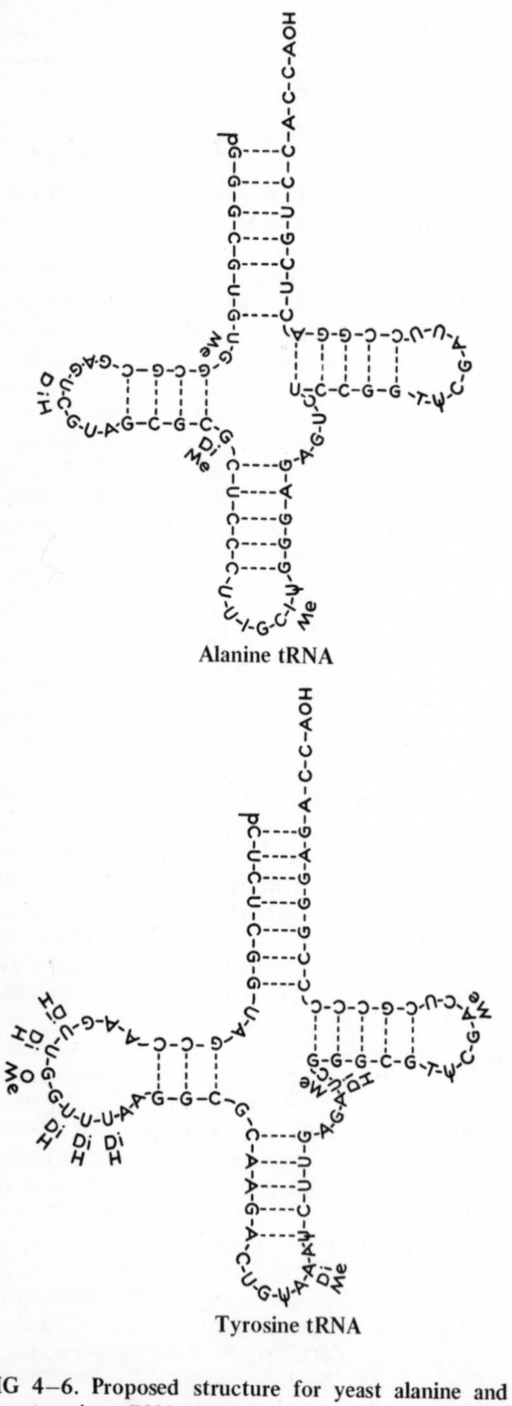

As was mentioned above, it is believed that a molecule of tRNA may assume a double-stranded form in certain regions where base pairing can occur through hydrogen bonds, as illustrated in Fig 4–6. Consideration has also been given to the possibility of a tertiary structure wherein the molecule may fold to constitute a more globular form. X-ray diffraction analyses as used in the study of the higher orders of protein structure are required. Purified species of tRNA have been crystallized and subjected to single-crystal x-ray diffraction investigations by Kim & Rich (1968) as well as by Hampel & others (1968).

tRNA molecules are the first pure nucleic acid molecules to be obtained as single crystals. The fact that these molecules can be crystallized indicates that they must have a definite regular size and shape which makes it possible for them to be organized into a lattice array. Further knowledge of the 3-dimensional structure of tRNA molecules may prove essential to an understanding of the mechanisms whereby these molecules bind to ribosomes and are "recognized" by specific enzymes such as the aminoacyl tRNA synthetases. These enzymes catalyze the attachment of an activated amino acid to its specific tRNA molecule as shown in Fig 5–2 to form the tRNA–amino acid complexes which are then aligned on the messenger RNA template as dictated by the anticodon found on another part of the tRNA molecule.

Alanine tRNA

Tyrosine tRNA

FIG 4–6. Proposed structure for yeast alanine and tyrosine tRNAs. (Reproduced, with permission, from Madison JT, Everett GA, King H: Science 153:531, 1966.)

Biologic Significance of Nucleic Acids

Viruses are notably rich in nucleoprotein, but their nucleic acid content varies both in amount and in composition. Tobacco mosaic virus, for example, contains about 6% as contrasted with as much as 40–60% in certain *Escherichia coli* bacteriophages. The plant viruses are mainly composed of RNA. Such viruses as, for example, tobacco mosaic and bushy stunt (tomato) viruses are comprised of a single strand of RNA rather than the double strand previously described for DNA. There are, however, other viruses, containing DNA, which are double stranded, eg, several *E coli* phages, although there are also RNA coliphage viruses which are single stranded.

The function of DNA in connection with the genetic function of the nucleus of the cell is indicated by several pieces of evidence. DNA is found in all nuclei and it is confined to the chromosomes. In studies of the DNA content of cell nuclei, the relationship between DNA molecules and the estimated number of genes present is such as to suggest a proportionality of nucleic acid molecules to genes. The amount of DNA in each somatic cell is constant for a given species, but the amount in the germ cell, which has only ½ the number of chromosomes, is ½ that in the somatic cell. In cells containing multiple sets of chromosomes (polyploidy), the amount of DNA is correspondingly increased. It is also reported that there is virtually no variation in the composition of DNA found in the sperm cell and that of other cells of the same organism, although the cell proteins are quite different. This would be expected if DNA is an integral part of the chromosomes, which are reduplicated in every other cell from the parent cells.

The manner in which DNA functions as the fundamental genetic substance has been elucidated by experimental studies of its role in directing the synthesis of cellular proteins. It is by means of the information stored in nuclear DNA that specific proteins are formed in accordance with the so-called genetic code. The code itself consists of a sequence of 3 nucleotides which directs the incorporation of a given amino acid into its proper position in the particular sequence of amino acids that is characteristic of the primary structure of each protein. (See Chapter 5.)

In cell division, the nucleic acid chain must be duplicated in such a manner as to preserve in each daughter cell the information contained in the parent cell. For this purpose, the double helix unravels and each of the 2 original strands then serves as a template for the synthesis of another, complementary chain. A direct demonstration of this phenomenon has been obtained by adding tritium-labeled purine and pyrimidine intermediates to a synthesizing system. Autoradiograms indicate that one strand of newly synthesized DNA is labeled by the presence of the labeled intermediates taken up from the medium whereas the complementary strand is unlabeled, representing the original or template strand. (See Fig 4–7.)

According to the "pairing rule" (see above), in DNA, adenine can combine only with thymine and guanine only with cytosine. The newly synthesized strand will thus be exactly constituted in its nucleotide sequence as was the original complementary strand of the parent (template) strand. The result is the synthesis of 2 pairs of strands (seen in cell preparations as "reduplication" of chromosomes) in which each pair is identical in nucleotide sequence—and hence in coding information—to the original parent pair.

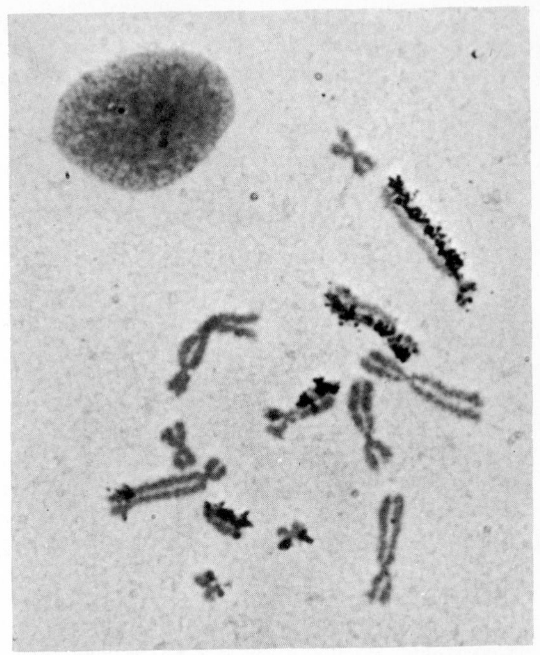

FIG 4–7. Autoradiogram of a tissue culture cell of *Potorous tridactylis* labeled with tritiated thymidine, third metaphase after labeling. (Reproduced, with permission, from Heddle JA, Wolff S, Whissel D, Cleaver JE: Science 158:929–931, 1967.)

Kornberg (1960) supported the above view of nucleic acid replication by accomplishing the synthesis of DNA with the aid of a DNA synthesizing enzyme (a polymerase) purified from *Escherichia coli.* It was shown that the enzyme did indeed catalyze the synthesis of a new DNA chain in response to directions from a DNA template. These directions are dictated by the hydrogen bonding relations of adenine to thymine and of guanine to cytosine (see Fig 4–3).

Experimental support for the role of hydrogen bonding during replication of DNA catalyzed by a DNA-polymerase has been obtained by observations on the effects of providing chemical analogues for the 4 naturally occurring purine or pyrimidine bases in a system capable of synthesizing DNA. In such experiments it was found that the analogues are incorporated into the newly synthesized nucleic acid molecule in place of the natural bases provided that the alterations in the chemical structure of the bases to produce the analogues do not affect the 6-keto or amino groups that Watson and Crick had postulated were involved in hydrogen bonding (see above),

Ribonucleic acid (RNA) functions primarily in the cytoplasm of the cell as a template in connection with synthesis of specific cellular proteins, as well as in the ribosomes. The formation of the RNA template is directed by nuclear DNA. This is possible because a strand of DNA and one of RNA can wrap around each other to form a hybrid helix. In this way the synthesis of a specific RNA molecule can be accomplished in the nucleus followed by migration of the RNA (messenger RNA) out of the nucleus via the endoplasmic reticulum into the cytoplasm where protein synthesis takes place (see Chapter 5).

Other types of RNA were described above.

It is established that RNA can be synthesized from ribonucleoside triphosphate precursors by an enzyme system (RNA polymerase) which is dependent on the presence of DNA, presumably acting as a template for the RNA to be synthesized since the base composition of the RNA produced is predictable from that of the DNA added, in accordance with the pairing rule. As an example, a reaction primed by polydeoxyribothymidylate as the DNA template produced only polyriboadenylate as the RNA product. A source of DNA which contained more adenine than thymine and more guanine than cytosine directed the synthesis of RNA containing more uracil than adenine and more cytosine than guanine.

Reference was made above to the results of experiments with purine and pyrimidine base analogues tending to support the role of hydrogen bonding and the application of the pairing rule in DNA replication. Similar experiments have been performed in a system in which RNA, with the aid of an RNA polymerase, was being synthesized with a DNA primer template. In these experiments, ribonucleoside triphosphates containing base analogues were found to be incorporated into the RNA formed, provided that the analogues were not so modified as to affect the chemical groups on the bases which are involved in hydrogen bonding. It was concluded that a fundamental similarity exists between the mechanisms by which DNA primes the formation of either complementary DNA or complementary RNA (Kahan & Hurwitz, 1962).

In Vitro Synthesis of RNA & DNA

In 1965, Haruna & Spiegelman accomplished the synthesis in vitro of a ribonucleic acid which was self-propagating and infectious in the manner of naturally occurring viruses. In 1967, Goulian, Kornberg, & Sinsheimer accomplished the in vitro synthesis of biologically active deoxyribonucleic acid. In these experiments, the template copied with the aid of the DNA polymerase was the DNA from a coliphage designated φX174, originally discovered by Sinsheimer. This phage virus is in many ways unique. It contains 5500 nucleotide residues normally occurring only as a single strand of DNA, although it has been obtained also in a double helical replicative form. The single strand of φX-DNA is not linear but occurs rather as a covalently bonded closed circle. When the circular DNA molecule of the virus (arbitrarily called the + strand) enters the susceptible coli cell, within the cell a complementary strand of DNA (the – strand) is synthesized, using the original infective (+) strand as a template, according to the pairing rule with respect to nucleotide pairs as described above. The resulting structure is a double-stranded, base-paired double circle rather than a helix. These covalent duplex circles have been isolated from infected *E coli* cells and each strand separated by physical methods. The newly synthesized (–) circles are themselves infective, ie, they can enter a susceptible cell, serving as a template for the synthesis of a complementary strand which would be identical to the original (+) strand.

The effort to synthesize biologically active DNA in vitro began with the preparation in Kornberg's laboratory of tritium-labeled (+) DNA of φX174. This was accomplished by addition of tritium-labeled thymidine to the solution in which φX174 was grown in *E coli*. Under these circumstances labeled thymidine was incorporated into the viral DNA. The labeled DNA was then purified by repeated washing and ultracentrifugation until it was ready for use as a template for the in vitro synthesis of a complementary strand of DNA. For this purpose the labeled DNA was incubated at 25° C for 3 hours with a mixture of the 4 deoxyribonucleotides required to synthesize DNA except that a bromouracil-containing nucleotide was used rather than one containing thymine, and radioactive phosphate (^{32}P) was also added to tag the newly synthesized nucleotides. Two enzymes were also required to complete the synthesis. The first is the DNA polymerase which catalyzed the joining together of the nucleotides to form the DNA polynucleotide

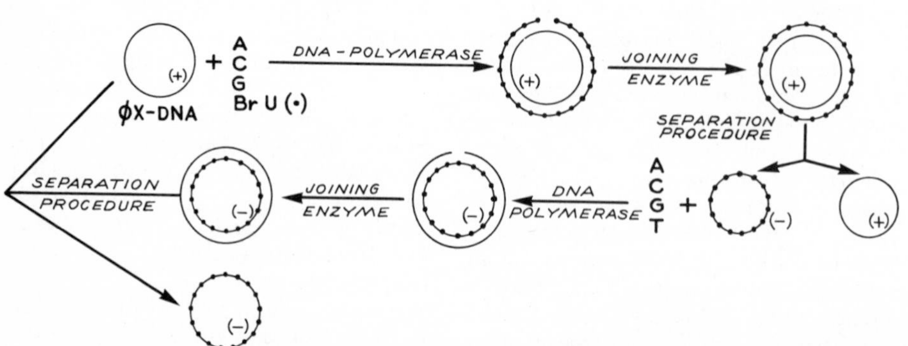

FIG 4–8. Synthesis of infective viral DNA. (Reproduced, with permission, from Goulian, Kornberg, & Sinsheimer: Science 158:1551, 1967.)

chain; the second is an enzyme called polynucleotide joining enzyme (DNA ligase) which catalyzed the covalent joining of the 2 opposite ends of the newly synthesized (–) chain of DNA to form the circular DNA characteristic of this viral strain. There was thus produced a partially synthetic double-stranded circular virus, one strand (+) having acted as a template for the synthesis of a complementary circular (–) strand which contained bromouridine nucleotides and was labeled with ^{32}P.

Next, with the aid of deoxyribonuclease and gradient centrifugation at 45,000 rpm for a period of 50 hours, the duplex circles were separated from one another. The newly synthesized (–) DNA containing the heavy bromouracil and ^{32}P accumulated at the lowest layer of the centrifugate, whereas the original (+) strand, being lighter, remained at the top layer. Now the synthetic (–) DNA was used as a template in an in vitro system containing polymerase and joining enzyme as well as the necessary constituents of the nucleotides to bring about synthesis of a new synthetic (+) strand and thus the formation of a completely synthetic duplex circular virus. When synthetic (+) or (–) DNA entered an *E coli* cell, it behaved exactly as the naturally occurring virus, giving rise to a new generation of intact, normal virus, thus proving that the completely synthetic DNA remained biologically active.

The diagram in Fig 4–8 summarizes the procedures described above for the preparation of a biologically active, completely synthetic DNA.

Synthesis of DNA from RNA; RNA-Dependent DNA Polymerase

In the preceding text, the synthesis of DNA or of RNA from a template of DNA or RNA, respectively, has been described, as well as the replication of nuclear DNA in the course of cell division. Until recently it was not generally believed that DNA could be synthesized using information derived from an RNA template. However, such "reverse transcription," wherein the usual flow of genetic information from DNA to RNA was in the opposite direction, was suggested by the discovery of an **RNA-dependent DNA polymerase.** Such an enzyme exists, for example, in the core of the Rous sarcoma virus. With the aid of this enzyme, this virus, with a genetic core of RNA acting as a template, can accomplish the synthesis within an infected cell of a DNA copy of itself. This newly synthesized DNA can then act as the template for synthesis of genetic information which can be incorporated into the chromosome of the original host cell infected with the virus.

Finding RNA-dependent DNA polymerase in the tumor virus suggested that the associated phenomenon of genetic information transfer could explain the transformation of normal to malignant cells as being due to the formation of a DNA virus derived from an RNA viral template. Several viruses associated with production of tumors, in addition to exhibiting RNA-dependent DNA polymerase activity, also have **DNA-dependent DNA polymerase** activity, indicating that the original DNA formed against the RNA viral template may then be replicated as DNA by the DNA-dependent polymerase enzyme.

There is evidence that an RNA-dependent DNA polymerase may also be present in normal cells. This supports the idea that the enzyme may function in normal cell growth and differentiation, particularly of embryonic cells.

Action of Reagents on DNA

Acids and alkalies cause irreversible reduction in the viscosity of DNA solutions, and certain chemical groups which were not available prior to the acid or alkaline treatment now become titratable. It is suggested that this is caused by rupturing of the hydrogen bonds between amino and hydroxyl groups of adjacent bases, and these groups are thus freed to react in the titrations. Guanidine, urea, and phenol in relatively high concentrations also irreversibly decrease the viscosity of DNA solutions. These observations all point to the loss of viscosity as due mainly to a destruction of the rigidity of the molecule, which is normally maintained by the inter- and intramolecular hydrogen bonds.

The nitrogen and the sulfur mustards are known to cause abnormalities in mitosis as well as a high incidence of mutations. In higher concentrations they may inhibit growth. These compounds, even in low concentrations, also reduce the viscosity of DNA solutions. It is believed that one chemical explanation for the biologic activity of the mustards is their ability to combine with (alkylate) the free amino groups of the purine or pyrimidine bases or of the proteins. In so doing, in the case of the nucleic acids, the bonding described above as maintaining the stiffness of the molecule is destroyed. This is evidenced by the reduction in viscosity.

It is noteworthy that those radiations in the ultraviolet range which induce mutations most readily are of the same wavelength as those which characterize the absorption spectra of nucleic acids. This further supports the idea that alterations in nucleic acid structure are involved in the causation of mutations.

Action of Ribonucleases & Deoxyribonucleases on Nucleic Acids

The highly polymerized nucleic acids or polynucleotides as they exist in nature are broken down into smaller components (oligonucleotides) by the action of specific enzymes. Examples are the enzymes ribonuclease and deoxyribonuclease, which split RNA or DNA, respectively. **Streptodornase** is a deoxyribonuclease-containing enzyme found in certain streptococci. It is used clinically, together with **streptokinase,*** which is also derived from certain hemolytic streptococci. The mixture attacks fibrin blood clots and digests and liquefies viscous accumulated pyogenic

*Streptokinase serves as an activator of plasminogen (profibrinolysin).

material, eg, in the chest cavity. The liquefied material may then be more easily removed by aspiration.

Streptodornase is actually a mixture of "nucleolytic" enzymes rather than a single one, so that it carries the breakdown of DNA to completion. A highly purified DNAase has been isolated and crystallized. It acts only on the initial depolymerization of DNA; further breakdown must be brought about by nucleotidases or nucleosidases as shown in Fig 4–1.

Ribonuclease has also been obtained in purified form, its chemical structure determined (see p 47), and synthesized—the first enzyme to be prepared synthetically.

• • •

References

Berg P, Joklik WK: J Biol Chem 210:657, 1954.
Cantoni GL: J Biol Chem 204:403, 1953.
Cohn WE: J Biol Chem 235:1488, 1960.
Goulian M, Kornberg A, Sinsheimer RL: Proc Nat Acad Sc 58:2321, 1967.
Hampel A, Labanauskas M, Connors PG, Kirkegard L, RajBhandary UL, Sigler PB, Bock RM: Science 162:1384, 1968.
Haruna I, Spiegelman S: Proc Nat Acad Sc 54:579, 1965.
Holley RW & others: Science 147:1462, 1965.
Kahan FM, Hurwitz J: J Biol Chem 237:3778, 1962.
Kim S-H, Rich A: Science 162:1381, 1968.
Kissane J, Robins E: J Biol Chem 233:184, 1958.
Kornberg A: Science 131:1503, 1960.
Madison JT, Everett GA, Kung H: Science 153:531, 1966.
Roberts J, Friedkin M: J Biol Chem 233:483, 1958.
Srinivasan PR, Borek E: Science 145:548, 1964.
Watson JD, Crick FHC: Nature 171:737, 1953.
Wyatt GR, Cohen SS: Nature 170:1072, 1952.

Bibliography

Chargaff E, Davidson JN (editors): *The Nucleic Acids.* 3 vols. Academic Press, 1955–1960.
Davidson JN: *The Biochemistry of Nucleic Acids,* 4th ed. Wiley, 1960.
Michelson AM: *The Chemistry of Nucleosides and Nucleotides.* Academic Press, 1963.

5...

Mechanism of Protein Synthesis & Genetic Regulation of Metabolism

Electron microscopy has revealed the existence within the cytoplasm of the cell of a network of what appear to be membranous tubes and saccular vesicles. This network has been designated the **endoplasmic reticulum.** Attached to its membranous tubules are numerous dense spherical granules with diameters of 10–15 nm which contain 80% of the ribonucleic acid (RNA) within the cell. The reticular granules have therefore been called **ribosomes.** The ribosomes are the sites of protein synthesis within the cell.

Elsewhere it is pointed out that deoxyribonucleic acid (DNA) is the fundamental substance involved in the genetic control of the synthesis of specific cellular proteins, including the enzymes which control metabolic events within the cell. Through the catalytic action of DNA polymerases, replication of nuclear DNA occurs as required in the formation of new cells during the process of cell division. However, DNA must also serve to direct the formation of the various types of RNA. Particularly important is the role of DNA in directing the synthesis of messenger RNA (mRNA), referred to as "**transcription,**" whereby the genetic information stored in DNA required to provide for the primary structure (ie, the amino acid sequence) of specific proteins can be transmitted to the protein synthesizing apparatus in the cytoplasm. This is accomplished by the synthesis within the nucleus of the "hybrid" helix whereby one strand of DNA directs the synthesis of a complementary strand of RNA so that T, C, G, and A in a DNA strand direct the incorporation, respectively, of A, G, C, and U into single strands of RNA. While both strands of DNA must be present, only one strand is actually "copied." The reaction is catalyzed by **RNA polymerase** (Weiss, 1960). The newly synthesized RNA molecule, **messenger RNA (mRNA),** may then migrate out of the nucleus by way of the endoplasmic reticulum where, in association with ribosomal RNA, it acts as a template to direct the synthesis of specific cellular proteins. This role of mRNA is referred to as "translation" of the genetic message.

The Genetic Code

The sequence of purine and pyrimidine bases on the mRNA strand, originally determined by information transcribed from the nuclear DNA, provides directions for the arrangement of the sequence of amino acids, ie, the primary structure, in the synthesis of proteins. This information carried on mRNA resides not in a single nucleotide but rather in the sequence of 3 nucleotides. Thus the so-called "**genetic code**" is comprised of a triplet of nonoverlapping bases. Considering that there are 4 different nucleotides in mRNA, ie, those containing adenine, guanine, cytosine, or uracil, if a nonoverlapping sequence of any 3 of these is required to code for an amino acid there are 64 (4^3) possible combinations that could occur. It has been found that 61 of the 64 triplets do indeed code for the 20 amino acids required in the synthesis of proteins. Consequently, there may be more than one **code word** or **codon** for a given amino acid (see Table 5–1). This is referred to as "degeneracy" of the code. Three of the 64 triplets do not code for any amino acid. They function rather as a signal (so-called nonsense triplets) to terminate a polypeptide chain at that point. As a result of this function, these triplets are now termed **chain-terminating triplets.**

Table 5–1. The genetic code (codon assignments in messenger RNA)*.

First Nucleotide	Second Nucleotide				Third Nucleotide
	U	C	A	G	
U	Phe	Ser	Tyr	Cys	U
	Phe	Ser	Tyr	Cys	C
	Leu	Ser	CT	CT	A
	Leu	Ser	CT	Trp	G
C	Leu	Pro	His	Arg	U
	Leu	Pro	His	Arg	C
	Leu	Pro	Gln	Arg	A
	Leu	Pro	Gln	Arg	G
A	Ile	Thr	Asn	Ser	U
	Ile	Thr	Asn	Ser	C
	Ile	Thr	Lys	Arg	A
	Met (CI)	Thr	Lys	Arg	G
G	Val	Ala	Asp	Gly	U
	Val	Ala	Asp	Gly	C
	Val	Ala	Glu	Gly	A
	Val (CI)	Ala	Glu	Gly	G

*The terms first, second, and third nucleotide refer to the individual nucleotides of a triplet codon. U = uridine nucleotide; C = cytosine nucleotide; A = adenine nucleotide; G = guanine nucleotide; CI = chain initiator codon; CT = chain terminator codon. (Abbreviations of amino acids are expanded in Table 3–1.)

Nirenberg & Matthaei (1961) were the first to discover information about the nature of the code when they found that a synthetic polynucleotide composed only of uridylic acid (polyuridylic acid) could serve as a template for the synthesis in a biologic system of a polypeptide containing only phenylalanine. These experimental studies on the nature of the amino acid code were carried out by combining ribosomes (prepared from the bacterium *Escherichia coli*) with a given synthetic trinucleotide such as the polyuridylic acid (poly U) mentioned above. This simulates the attachment of mRNA (poly U in the above experiment) to ribosomes as occurs in an in vivo system. To the ribosomal preparations having the synthetic template attached, there were added transfer RNA (tRNA) molecules (also prepared from *E coli*), each bearing their own specific amino acid, itself labeled with ^{14}C. The result of this experiment was to show that a poly U preparation bound only phenylalanine and thus served to direct the synthesis of a polypeptide composed entirely of phenylalanine (polyphenylalanine). It was concluded that a codon for phenylalanine is UUU, ie, a sequence of 3 uridylic acid nucleotides.

In similar experiments, synthetic preparations of the trinucleotides containing guanine or uracil as GUU directed the synthesis of entirely valine-containing polypeptides. In later studies, UGU was found to code for cysteine and UUG for leucine.

It also soon became clear that the first 2 bases of a triplet were more specific in coding than was the third. Thus AUU and AUC both coded for isoleucine, and poly AC preparations were found to stimulate the uptake of proline, threonine, and histidine, suggesting that 2 of the 3 bases in a triplet code carry most of the information.

Other studies to elucidate the nature of the triplet code were carried out by the use of what are termed "block copolymers" of ribonucleotides. These are synthetic polynucleotides consisting of repeating sequences of 2 or 3 bases, eg, AGAGAG or AAGAAGAAG. When used as mRNA in a polypeptide-synthesizing system, the copolymer AGAGAG directed the synthesis of a polypeptide consisting of alternate arginyl and glutamyl residues. This result is interpreted as evidence that the synthetic messenger was "read" as AGA (codon for arginine) and GAG (codon for glutamic acid) alternating triplets, ie, AGA.GAG.-AGA.GAG etc. In contrast, the AAGAAGAAG synthetic messenger directed the synthesis of peptides containing only glutamyl, or arginyl, or lysyl residues, ie, polyglutamic acid, polyarginine, or polylysine. The result is explained by assuming that the messenger was read at different starting points. For example, if the reading started as AAG, only lysine would be incorporated; if started at the next base, the triplet sequence would then be read as AGA.AGA etc, in which case arginine would be incorporated; if started at the third nucleotide, the triplet sequence could be read as GAA.GAA etc, which codes for glutamic acid. These results with the AAG copolymer are illustrated below.

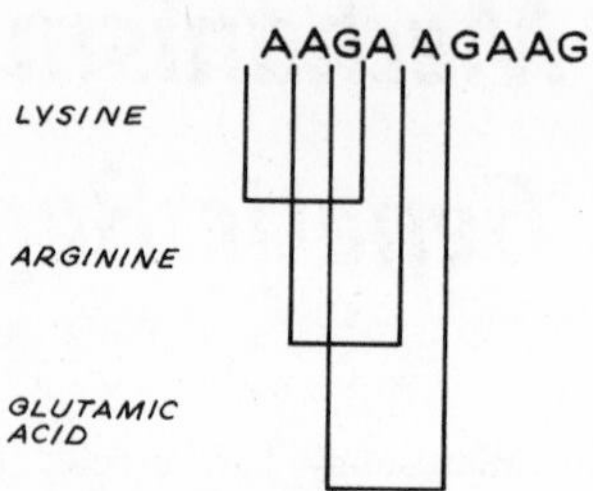

A copolymer consisting of repeated AAA triplets and ending with AAC was found to code for a polypeptide which consisted of repeated lysyl residues terminating with asparagine at the carboxyl end of the chain. Knowing that AAA is a codon for lysine and that AAC codes for asparagine, it was concluded that the mRNA message is read from left to right; ie, in aligning the amino acids, the codon on the left of the messenger codes for the N-terminal amino acid and that on the far right for the C-terminal amino acid.

As a result of many additional studies both with copolymers and with the trinucleotide-tRNA binding system used by Nirenberg's group, RNA codon assignments (the genetic code) for all the amino acids to be incorporated into a peptide have now been at least provisionally assigned as shown in Table 5–1. Three amino acids have 6 codons; 5 have 4 codons; and 10 have 2 codons. UAA, UAG, and UGA are **chain-terminating triplets** which signal that a peptide chain is to end at that point.

In addition to codons for peptide chain termination, there appear to be codons which are required for initiation of a peptide chain. Using extracts of *E coli*, it has been found that methionine attached to a methionine transfer RNA can be readily formylated with the aid of an enzyme catalyzing the transfer of a formyl group from f^{10}-tetrahydrofolate ($f^{10}.FH_4$), which serves as the formyl donor, to the amino group on methionine to form N-formylmethioninyl-tRNA. Other amino acid-tRNA compounds are not formylated. The significance of this observation was clarified when it was observed that all of the proteins synthesized by *E coli* have N-formylmethionine as the N-terminal amino acid residue. After synthesis of the protein chain is complete, depending on the specific protein observed, the formyl group, if not the methionine as well, is removed by hydrolysis. On the basis of the above findings, which are at present confined to *E coli*, it has been suggested that initiation of a peptide chain depends upon binding of N-formylmethionyl-tRNA to the ribosomes. Subsequently, the other amino acids are added in accordance with the relationship of the codons on the mRNA template to the anticodons of the amino acid-carrying tRNA. The chain is then terminated by the message transmitted by one of the chain-terminating codons mentioned above.

The only codon identified for methionine is AUG. This trinucleotide will act to bind both tRNA-methionine and tRNA-formylmethionine to ribosomes. Consequently, in Table 5–1, AUG is designated not only as a codon for methionine but also as a chain-initiating (CI) codon. The trinucleotide GUG which codes for valine appears to function in animals also as a chain-initiating codon; efforts to show that the mechanism involving N-formylmethionine which functions in *E coli* for chain initiation is also operative in animals have not been successful. It has also been suggested that, in animals, initiator codons may bind particular types of tRNA that do not carry amino acids, being used only to ensure that the ribosomes are assembled at the beginning of the mRNA molecule.

The release of newly formed peptide chains from ribosomes depends on the presence of a "release factor" which has been partially purified, as well as the chain-terminator codon. The release factor was discovered by Capecchi (1967). Caskey & others (1968) have shown that initiator and terminator codons sequentially stimulate N-formylmethionyl-tRNA binding to ribosomes and, in the presence of release factor, the release of free N-formylmethionine from the ribosome intermediate.

The formation of an abnormal protein molecule by the substitution of a single amino acid was first described in connection with the elucidation of the abnormality in sickle cell globin as compared to normal adult globin. There have now been described a number of different genetic mutations of this sort which result in the formation of an abnormal protein simply by the replacement of one amino acid with another. With present knowledge of the genetic code, most of these mutations can be interpreted as resulting from only a single base change in a coding triplet. As a result, a codon for one amino acid is changed into that for another. This is referred to as a **"missense" mutant.**

Terzaghi & others (1966) have reported the following illuminating experiment which supports the concept of the triplet code and also supplies further information about the effects of changes in base sequence on the production of mutations by amino acid substitutions. By treating a bacteriophage with an acridine dye, a purine or pyrimidine base may be either deleted from or inserted into the internal sequence of a DNA strand. As a result, it would be expected that the genetic message carried on the DNA would be shifted out of phase, as it were, beginning at the right side of the deleted or inserted base—remembering that the mRNA code is related to the sequence of 3 consecutive bases and is read from left to right. A mutation is thus produced because that portion of the protein synthesized in accordance with information from the altered region of the messenger will have an abnormal amino acid sequence.

Terzaghi & others found that an enzyme, lysozyme, synthesized by cells after being injected with a normal strain of bacteriophage T-4, could no longer be made if injected with one or another of 2 mutant phages. One mutant produced after acridine treatment was a so-called deletion mutant, the other an insertion mutant, ie, a base deleted from or inserted into a DNA strand. When the mutants were crossed, a crossing in the DNA took place between the deletion in the first mutant and the insertion in the second, producing a new hybrid phage which could produce a lysozyme except that the newly produced enzyme protein was only ½ as effective enzymatically as the normal enzyme. It was then of interest to compare the structure of the 2 enzymes. The normal (wild) type of lysozyme had a peptide with an amino acid sequence as follows:

- - - -Thr-Lys-[-Ser-Pro-Ser-Leu-Asn-] -Ala-Ala-Lys-

The mutated lysozyme was identical except in the corresponding region (shown in brackets above) of the peptide chain where it was as follows:

- - - -Thr-Lys-[-Val-His-His-Leu-Met-] -Ala-Ala-Lys-

Translating the bracketed amino acid sequences into mRNA codons gives, for the normal enzyme:

.AGU.CCA.UCA.CUU.AAU.

If the first base (A) is deleted and a G is inserted at the end, a new codon sequence results, as follows:

.GUC.CAU.CAC.UUA.AUG.

which translates as:

Val-His-His-Leu-Met-

exactly as was found to have occurred in the mutated lysozyme.

Utilization of Amino Acids for Protein Synthesis

The first step in the utilization of amino acids for protein synthesis requires their activation. This takes place by a reaction between the amino acid and ATP which requires also an **amino acid activating enzyme (E)** (aminoacyl-RNA synthetase). There results an activated complex of enzyme with an adenosine-monophosphate (AMP)-amino acid compound (aminoacyl-AMP-enzyme complex) in which the 5′-phosphate group of AMP is linked as a mixed anhydride to the carboxyl group of the amino acid and pyrophosphate is split off. The activation reaction is shown in Fig 5–1.

Specific activating enzymes are required for each of the naturally occurring amino acids which are to be incorporated into the peptide chain. The first highly purified amino acid activating enzyme to be described was a tryptophan-activating enzyme obtained from beef pancreas (Davie, 1956). A tyrosine-activating enzyme from hog pancreas has also been found (Schweet, 1958). **Arginyl-RNA synthetase,** an amino acid activating enzyme for arginine, has been prepared from rat liver (Allende & Allende, 1964); and the

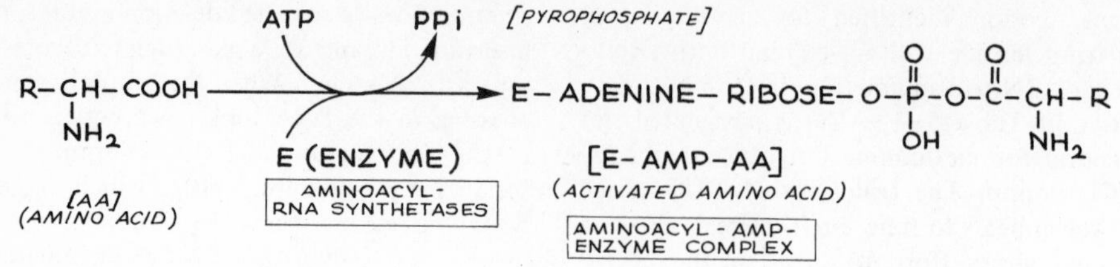

R–CH(NH₂)–COOH [AA] (AMINO ACID) + ATP + E (ENZYME) [AMINOACYL-RNA SYNTHETASES] → PPi [PYROPHOSPHATE] + E–ADENINE–RIBOSE–O–P(=O)(OH)–O–C(=O)–CH(NH₂)–R [E-AMP-AA] (ACTIVATED AMINO ACID) [AMINOACYL-AMP-ENZYME COMPLEX]

FIG 5–1. Activation reaction in the utilization of amino acids for protein synthesis.

properties of the enzyme complex, **threonyl-ribonucleic acid synthetase**, the activating enzyme for threonine, have been studied by Allende & others (1966).

The second stage in protein synthesis also requires catalysis by specific enzymes which are termed **aminoacyl-soluble (or transfer) ribonucleic acid synthetases.** During this stage the activated amino acid molecules are transferred to ribonucleic acids of relatively low molecular weight. These relatively short chain RNA molecules, which occur free in the cytoplasmic fluid, may be termed **soluble RNA (sRNA)** or, because of their function, **transfer RNA (tRNA).** There is a specific tRNA molecule for each amino acid which is provided for by the presence in the tRNA molecule of a triplet **anticodon** which relates to the **codon** for the amino acid to be carried. This is described on p 58, where the structures of alanine tRNA and tyrosine tRNA are illustrated. After transfer of the activated amino acid to a specific tRNA molecule, the amino acid activating enzyme and AMP are liberated, as shown in Fig 5–2.

The third step in protein synthesis involves the ribosomal RNA template, **messenger ribonucleic acid (mRNA),** previously described. It is here that the tRNA complex (aminoacyl-tRNA) is attached in accordance with the triplet nucleotide code, a codon for amino acid on mRNA "recognizing" the anticodon on the amino acid-carrying tRNA molecule. The result is an alignment of amino acids in a particular sequence as dictated by the mRNA template.

The essential steps whereby alignment of amino acids may occur to form the primary structure of a protein are shown diagrammatically in Fig 5–3 (Watson, 1963). The polypeptide chain is initiated at the signal from a chain initiator codon such as, in *E coli,* that which binds N-formylmethionyl-tRNA (see below), or, in animals, a special chain initiator codon binding a type of tRNA that does not carry an amino acid. The chain then elongates by the addition of individual amino acids in sequence beginning with the N-terminal amino acid. The other end of the growing chain attaches to a tRNA-amino acid molecule, the amino acid-carrying end of which is an adenosine moiety with a structure as shown in Fig 4–2. The chain elongation process may be indicated schematically as shown in Fig 5–4.

During protein synthesis, tRNA is specifically bound to ribosomes, but each ribosome binds only one tRNA molecule. The ribosome itself is composed of subunits which can be separated by varying concentrations of divalent cations (Mg^{++} is usually used). The ribosomal subunits are classified in terms of their sedimentation constants; in *E coli,* for example, the ribosomes may be separated into 30 S and 50 S subunits. An aggregate of a 30 S subunit with a 50 S subunit forms the 70 S ribosome, active in protein synthesis. It is the 30 S subunit of the 70 S ribosome that binds tRNA molecule over a distance of about 30 nucleotides, and the 50 S subunit to which the growing polypeptide chain is attached by means of a molecule of tRNA.

In the schematic representation shown in Fig 5–4, the N-terminal amino acid designated H_2N-AA_1 would, in *E coli,* be N-formylmethionine acting as the chain initiator; in animal cells, some other chain initiator–possibly valine or other special tRNA initiator. The C-terminal AA_4 attached to $tRNA_4$ would be associated with the 50 S subunit of the ribosome, and the newly arriving amino acid, designated AA_5, would be attached to the 30 S subunit.

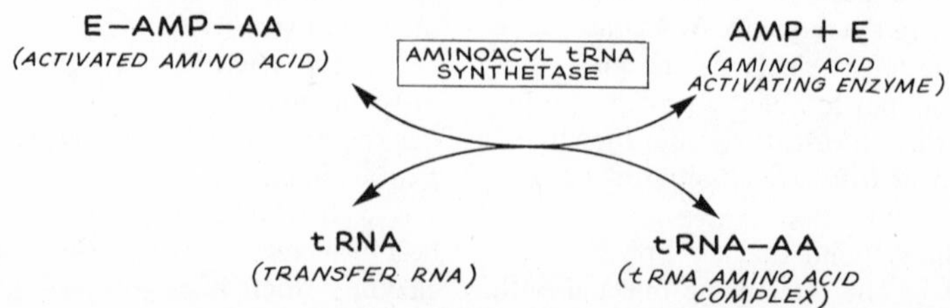

FIG 5–2. Transfer of activated amino acids to form tRNA-amino acid complexes.

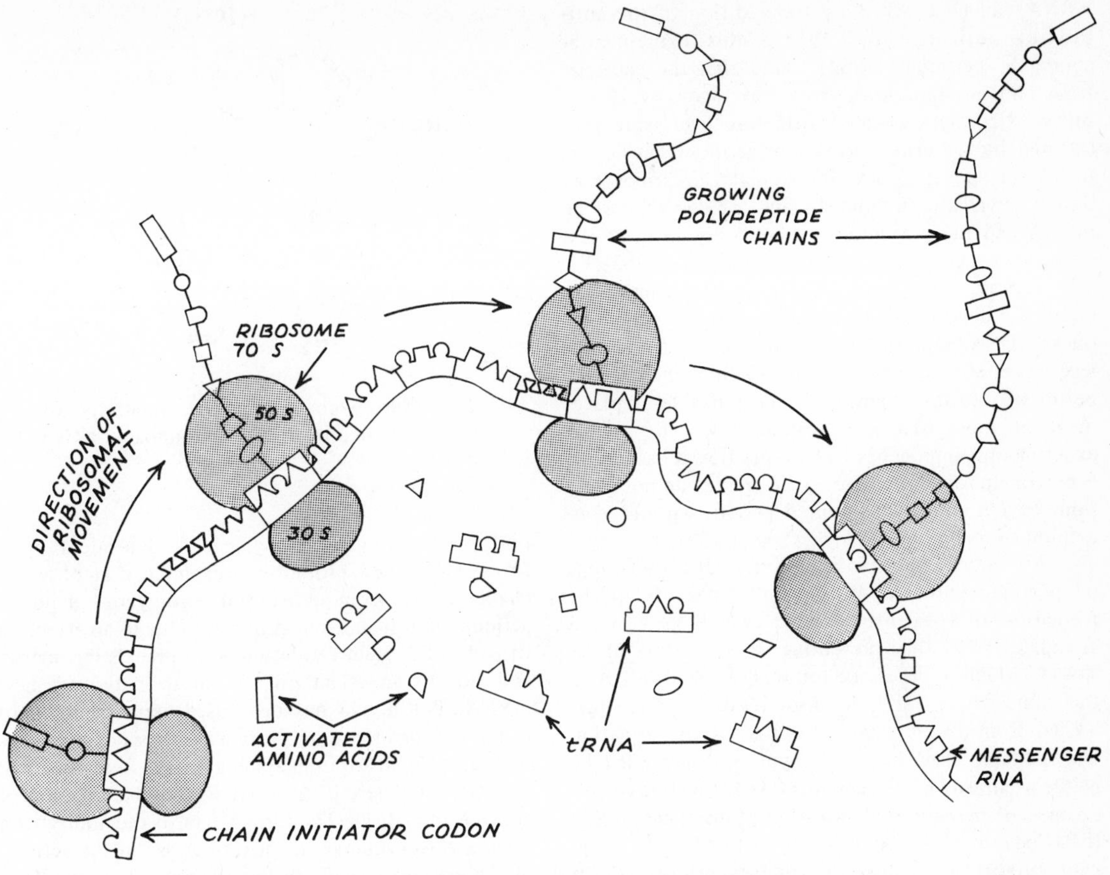

FIG 5–3. Relationship of messenger RNA to ribosomes in protein synthesis.

At a given interval, each functioning ribosome holds but one growing polypeptide chain. As the chain grows, the amino terminal end moves farther away from the point of addition of new amino acids which, as noted above, is at the attachment of the tRNA molecule. The tRNA molecule is itself bound to the 30 S subunit of the ribosome. Because there is only one specific site on the ribosome where peptide bond formation can occur, it is assumed that the ribosomes move along the RNA template to bring the next coded trinucleotide site on the template into that position on the ribosome where it can receive the tRNA complex of the next amino acid to be added to the growing peptide chain. In this way a single mRNA molecule can serve several ribosomes—as many as 6–8 70 S ribosomes for each mRNA molecule constituting what are termed **polysomes** or **ergosomes.** It would also follow that the polypeptide chains should be increasingly longer from one successive ribosome to the next in accordance with the length of RNA template over which the ribosomes have already passed. These ideas are diagrammatically expressed in Fig 5–3.

Messenger RNA molecules do not function indefinitely. In a bacterial system, synthesis of new

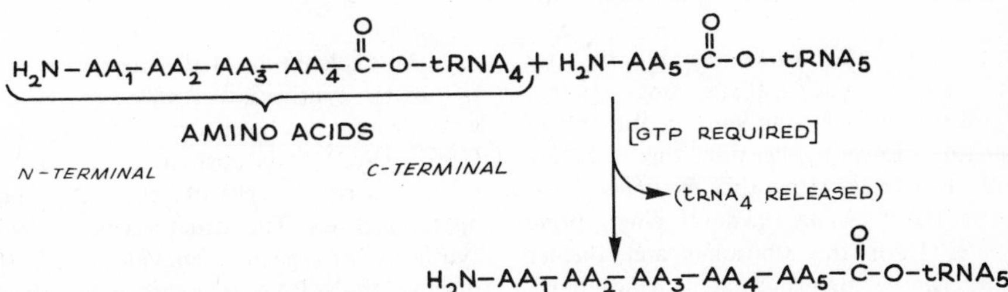

FIG 5–4. Schematic indication of process of elongation of polypeptide chain by successive addition of tRNA-amino acid complexes.

mRNA can be blocked by the addition of the antibiotic, actinomycin D. Using this experimental approach, Levinthal (1962) found that the bacterial RNA template functioned on the average only 10–20 times. Ribosomes isolated from rat liver were also affected by the prior injection of actinomycin D into the intact animal. It was found that these ribosomes broke down into subunits because of degradation of mRNA and the prevention by actinomycin D of the synthesis of new mRNA (Staehelin & others, 1963).

Because formation of the RNA template is originally dependent upon information supplied by nuclear DNA, alteration or destruction of DNA should lead to cessation of protein synthesis since there would be no way to form new RNA templates for replacement of those that are breaking down. By several experimental approaches to this question it has indeed been concluded that ribosomal templates do not function, except for a very limited period, without concomitant functioning DNA.

A number of antibiotics interfere at various steps in protein synthesis. **Mitomycin** brings about the formation of covalently bonded cross-links between strands of DNA, thus preventing the separation of the strands which is necessary for replication of DNA in the course of cell division. As a result, cells are prevented from dividing, but messenger RNA formation and the later stages in protein synthesis may still take place. As already noted, **actinomycin D** (dactinomycin; Cosmegen) prevents the formation of messenger RNA. It has been used for control of growth of some malignant tumors. It is also of value experimentally to determine the extent to which changes in protein formation are due to changes in the rate of mRNA formation, since actinomycin prevents the formation of more mRNA but has little effect on the production of peptides using existing mRNA. **Tetracycline, streptomycin,** and **chloramphenicol** are examples of antibiotics which inhibit ribosomal activity. Tetracycline inhibits the binding of the aminoacyl tRNA to mRNA by preventing the combination of the aminoacyl tRNA with the initiator site on the 30 S subunit of the ribosome. Streptomycin binds to 50 S subunits of ribosomes, resulting in a decreased rate of synthesis of proteins as well as the production of faulty proteins incident to misreading of the codons of mRNA. Chloramphenicol can compete with a messenger RNA for sites on bacterial ribosomes. Neomycin B alters the binding of aminoacyl tRNA onto mRNA-ribosome complexes.

Puromycin is a compound that resembles the sites on tRNA that bind amino acids. It also contains a free ammonium group to which a growing peptide chain can be attached. However, because this complex cannot become associated with other enzymes necessary to bind it to the ribosome, the developing peptide chain dissociates from the ribosome and further growth of the chain is thus prevented. Since puromycin acts to prevent peptide synthesis only in the final stages of this phase of synthesis of proteins, it is a useful compound experimentally to test whether or not a given physiologic change is dependent upon formation of new protein. For example, if an observed change results from addition of puromycin but not of actinomycin, it is assumed that it is due to an effect on the rate of protein formation from preexisting mRNA and not to an effect on rate of formation of new mRNA. Puromycin has been used experimentally to test the dependence on protein synthesis of a process such as an increase in enzyme activity. Reference will be made elsewhere in this text to the phenomenon of enzyme induction. The term refers to the ability of a substrate to evoke an adaptive increase in the activity of the enzyme specific to the substrate. An example in animal tissues is the induction of tryptophan pyrrolase activity by tryptophan. It may be demonstrated in the newborn animal as well as the adult. Puromycin completely blocks the normal developmental increase in the enzyme in the newborn and inhibits approximately 70% of the adaptive increase otherwise observed in the adult. It is thus concluded that the increase in pyrrolase activity which occurs in the newborn is entirely due to the formation of new protein, whereas that observed in the adult animal is partly attributable to synthesis of new protein and partly to activation of a preexisting protein precursor (Nemeth, 1962).

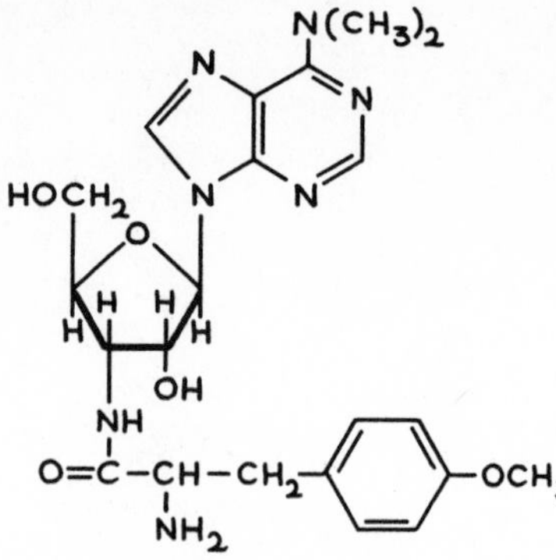

FIG 5–5. Puromycin. Note the similarity of the structure of puromycin to aminoacyl tRNA, Fig 4–5.

THE OPERON CONCEPT

An explanation of the mechanism of genetic control of the synthesis of enzyme proteins or of that of virus protein has been formulated by Monod & Jacob (1961; 1962). It is postulated that the chromosomes actually carry 2 types of genes: **structural genes** and **operator genes.** The structural gene DNA directs the synthesis of specific enzyme and other proteins through the mRNA template as described above. The operator gene, however, exerts control over the adjacent structural genes located on the same chromosome since formation of mRNA can be effected only at a point on DNA where the operator occurs. Thus, structural genes, even though present in a given cell, cannot initiate protein synthesis except when appropriate conditions prevail within the cell. One or more structural genes together with their operator gene constitute the genetic structure which is designated an **operon** (Fig 5–6).

Repression & Derepression

A third factor in the genetic control of protein synthesis relates the operon to metabolic events within the cell. This factor is the **regulator gene** which controls the activity of the operator gene. The regulator gene does so by its ability to induce the synthesis of protein macromolecules (probably RNA proteins) called **repressors.** The operator gene, when combined with a repressor, is unable to induce activity in the structural gene, in which circumstance the operon is said to be "repressed." When the operon is active because the repressor system is itself inactivated, the operon is said to be "derepressed" (Fig 5–6).

The concentration within the cell of a given metabolite is known to affect the activity of enzyme systems which are involved in the synthesis or metabolism of the metabolite itself. This phenomenon is involved in **enzyme induction** (see Chapter 8) and is therefore a very important factor in regulation of growth and metabolism within the cell. An example is the ability of galactose to induce in *E coli* preparations synthesis of enzymes necessary for metabolism of galactose itself. The operon concept applied to enzyme induction by a substrate (in this instance, galactose) would be illustrated as follows. In the presence of galactose, the repressor is inactivated so that the operator gene is now free to activate the adjacent structural gene, which, in turn, can bring about synthesis of enzyme protein for galactose metabolism. In this example, galactose is acting to effect derepression of the operon, and accumulation of the substrate (galactose) is thus acting as a control mechanism to cause enhancement of the enzymatic processes by which the substrate itself can be metabolized. As metabolism proceeds, the concentration of the substrate falls and repression once more sets in as the regulator is subjected to the diminishing influence of the substrate.

In another situation, a substrate may itself seem to act as a repressor. For example, *E coli* is able to synthesize an enzyme required in the metabolism of tryptophan (tryptophan synthetase) only in the absence of tryptophan. This is explained in terms of the operon concept by assuming that tryptophan reacts with the repressor to form a modified repressor which blocks the operator and thus prevents it from affecting the adjacent structural gene for tryptophan synthetase. It would follow that, in the absence of tryptophan, synthesis of tryptophan synthetase could proceed.

Another example of the genetic control of metabolism through the operon mechanism is supplied by a consideration of the hypothesis of Granick (1966)

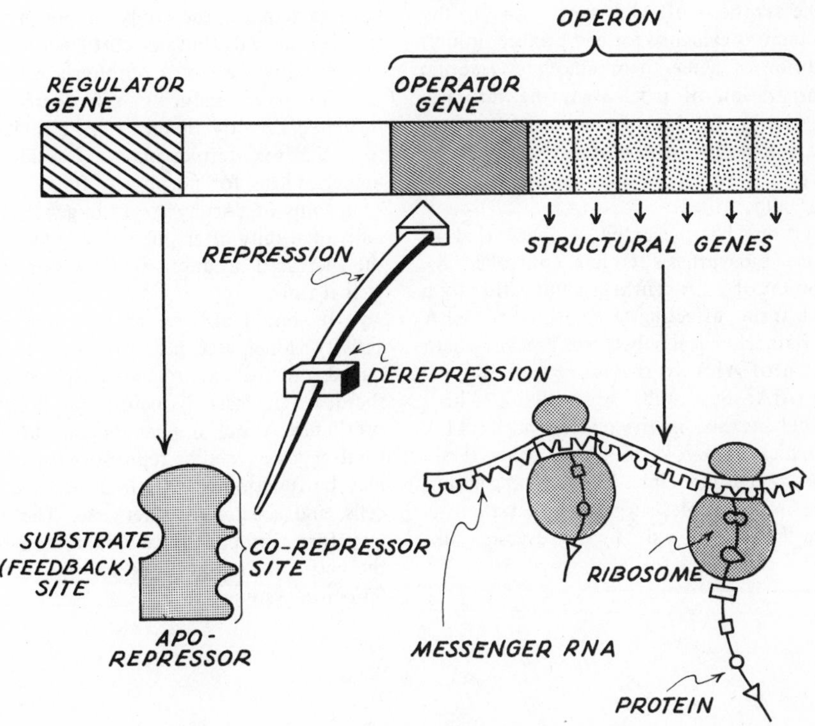

FIG 5–6. The operon concept.

with respect to the etiology of hepatic porphyria, an inherited defect in porphyrin metabolism further described in Chapter 6. It has been shown that hepatic porphyria is characterized biochemically by an increase in the activity of one enzyme required at an early stage in the biosynthetic pathway for porphyrin formation. This enzyme, **aminolevulinic acid synthetase (ALA synthetase)**, catalyzes the synthesis of δ-aminolevulinic acid from glycine and succinate by way of ketoadipic acid. ALA synthetase appears to be the limiting enzyme in porphyrin (including heme) biosynthesis since all other enzymes of the pathway are present in nonlimiting and adequate amounts. From studies on bacterial preparations, Jacob, Monod, & Wollman (see Lwoff, 1962) have postulated the existence of 2 general mechanisms for the control of the first enzyme of a biosynthetic pathway, such as is ALA synthetase. One mechanism is **end product inhibition** (see Chapter 8, concerning allosteric sites of inhibition), in which the end product of a biosynthetic pathway combines with enzyme protein at its allosteric site and thus effects immediate inhibition of the action of the enzyme. The other mechanism is **end product ("feedback") repression**, which involves the operon mechanism described above. The photosynthetic bacterium, *Rhodopseudomonas spheroides*, exhibits both mechanisms in regulation of its biosynthetic pathway to heme. In this instance the end product of the biosynthetic pathway, heme, inhibits the first enzyme in the pathway (ALA synthetase) both allosterically and by repression, as evidenced by the fact that heme, when added to cultures of these bacteria, will repress 3- to 4-fold the synthesis of ALA synthetase. In the red blood cell there is evidence for end product inhibition since addition of heme to reticulocytes (rabbit) decreases incorporation of labeled glycine into the heme of hemoglobin. In the differentiating erythroblast, however, control of heme synthesis seems to be due to end product repression although the repressor itself has not been identified.

Granick (1966) has presented evidence that in liver cells heme biosynthesis is not controlled by allosteric inhibition of ALA synthetase but rather by a repressor mechanism affecting formation of ALA synthetase. Certain chemical substances are known to induce formation of ALA synthetase, eg, allylisopropylacetylurea (AIA; Sedormid, Apronalide), which induces an 8-fold increase in the mitochondrial ALA synthetase of chick embryo liver. In no other tissue tested was this enzyme inducible. The inducing action of the chemical is reversible, suggesting that there is competition for the induction site by the reversing substance.

The following hypothesis has been advanced by Granick (1966) to fit the above described experimental observations. It is considered that the repressor controlling heme biosynthesis is composed of a protein, the apo-repressor, and a co-repressor which is the end product, heme. Competition for the active site on the apo-repressor normally occupied by heme could extend to the ALA synthetase-inducing chemical substances such as Sedormid. If such a chemical displaced the heme (which normally activates the repressor), the effect would be to prevent the repressor from acting; ie, the ALA synthetase system would be derepressed and more porphyrins would be formed, as is observed when such chemicals are introduced. It follows that the inducing action of an inducing chemical should be reversed or prevented by heme, and this seemed to be the case in the experiments reported by Granick.

Hepatic prophyria is known to be inherited as a mendelian-dominant trait. An explanation for the genetic background for this disease is the assumption that the mutation which has occurred has produced a defective operator gene which is poorly repressed by the repressor. This would explain the clinical observation that amounts of barbiturates which produce no detectable porphyria in normal individuals do cause a significant porphyria in a patient with hepatic porphyria. Symptoms of hepatic porphyria characteristically occur only after puberty. Furthermore, several investigators have noted that sex steroids may occasionally precipitate an acute attack of the disease in susceptible individuals. Granick found, in the chick embryo liver cell preparation which he used as an in vitro system for the study of porphyrin metabolism, that sex steroids (but not corticosteroids) could indeed induce a low degree of porphyria, which suggests that the porphyria-inducing effect of sex steroids is mediated directly in the liver cells. It appears, therefore, that sex steroids may be considered to be inducing chemicals for porphyria. If so, the emergence of symptoms of porphyria in the genetically predisposed individual only after puberty would be attributable to the increased production of sex steroids which occurs at that time.

It should also be recalled that ALA synthetase activity is increased not only in the hepatic porphyrias but also in the erythrogenic porphyrias. It may be that there are at least 3 operons for production of ALA synthetase, which implies separate operator and structural genes as well as repressors for each. One operon may be functioning only in liver, a second in erythroid cells, and a third in other cells. The different hereditary forms of porphyria would then be explained on the basis of a mutation in any one of the hypothetical 3 operon systems.

• • •

References

Allende CC, Allende JE: J Biol Chem 239:1102, 1964.

Allende CC, Allende JE, Gatica M, Celis J, Mora G, Matamala M: J Biol Chem 241:2245, 1966.

Capecchi MR: Proc Nat Acad Sc 58:1144, 1967.

Caskey CT, Tompkins R, Scolnick E, Caryk T, Nirenberg M: Science 162:135, 1968.

Davie EW, Koningsberger VV, Lipmann F: Arch Biochem 65:21, 1956.

Granick S: J Biol Chem 241:1359, 1966.

Jacob F, Monod J: J Molec Biol 3:318, 1961.

Levinthal C, Keyman A, Higa A: Proc Nat Acad Sc 48:1631, 1962.

Lwoff A: *Biological Order.* MIT Press, 1962.

Monod J, Jacob F: Cold Spring Harbor Symposia 26:193, 1961.

Monod J, Jacob F, Gros F: Biochem Soc Symposia 21:104, 1962.

Nemeth AM, de la Haba G: J Biol Chem 237:1190, 1962.

Nirenberg MW, Matthaei JH: Proc Nat Acad Sc 47:1588, 1961.

Schwett RS, Allen EH: J Biol Chem 233:1104, 1958.

Staehelin T, Wettstein FO, Noll H: Science 140:180, 1963.

Terzaghi E, Okada Y, Streisinger G, Emrich J, Inouye M, Tsugita A: Proc Nat Acad Sc 56:500, 1966.

Watson JD: Science 140:17, 1963.

Weiss SB: Proc Nat Acad Sc 46:1020, 1960.

Bibliography

Green DE, Goldberger RF: *Molecular insights into the Living Process.* Academic Press, 1967.

Hsia DY (editor): *Lectures in Medical Genetics.* Year Book, 1966.

Watson JD: *Molecular Biology of the Gene.* Benjamin, 1965.

6 . . .
Porphyrins & Bile Pigments

Porphyrins are cyclic compounds formed by the linkage of 4 pyrrole rings through methylene bridges (Fig 6–1). A characteristic property of the porphyrins is the formation of complexes with metal ions. The metal ion is bound to the nitrogen atom of the pyrrole rings. Examples are the iron porphyrins such as **heme** of hemoglobin and the magnesium-containing porphyrin **chlorophyll,** the photosynthetic pigment of plants.

In nature, the metalloporphyrins are conjugated to proteins to form a number of compounds of importance in biologic processes. These include the following:

A. Hemoglobins: Iron porphyrins attached to the protein, globin. These conjugated proteins possess the ability to combine reversibly with oxygen. They serve as the transport mechanism for oxygen within the blood. Hemoglobin has a molecular weight of 64,450 daltons; it contains 4 gram atoms of iron per mol in the ferrous (Fe^{++}) state.

B. Erythrocruorins: Iron porphyrinoproteins which occur in the blood and tissue fluids of some invertebrates. They correspond in function to hemoglobin.

C. Myoglobins: Respiratory pigments which occur in the muscle cells of vertebrates and invertebrates. An example is the myoglobin obtained from the heart muscle of the horse and crystallized by Theorell in 1934. The purified porphyrinoprotein has a molecular weight of about 17,000 daltons and contains only 1 gram atom of iron per mol.

D. Cytochromes: Compounds which act as electron transfer agents in oxidation-reduction reactions. An important example is **cytochrome c,** which has a molecular weight of about 13,000 daltons and contains 1 gram atom of iron per mol.

E. Catalases: Iron porphyrin enzymes, several of which have been obtained in crystalline form. They are assumed to have a molecular weight of about 225,000 daltons and to contain 4 gram atoms of iron per mol. In plants, catalase activity is minimal, but the iron porphyrin enzyme peroxidase performs similar functions. A peroxidase from horseradish has been crystallized; it has a molecular weight of 44,000 daltons and contains 1 gram atom of iron per mol.

F. The Enzyme, Tryptophan Pyrrolase: This enzyme catalyzes the oxidation of tryptophan to formyl kynurenine. It is an iron porphyrin protein.

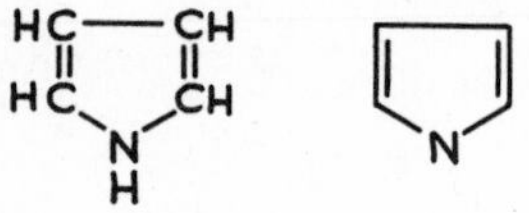

Pyrrole

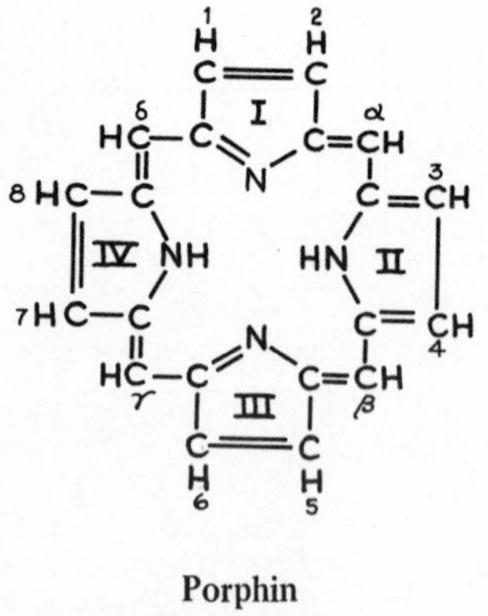

Porphin
($C_{20}H_{14}N_4$)

FIG 6–1. The porphin molecule. Rings are labeled I, II, III, IV. Substituent positions on rings are labeled 1, 2, 3, 4, 5, 6, 7, 8. Methylene bridges are labeled α, β, γ, δ.

Structure of Porphyrins

The porphyrins found in nature are compounds in which various side-chains are substituted for the 8 hydrogen atoms numbered in the porphin nucleus shown at left. As a simple means of showing these substitutions, Fischer proposed a shorthand formula in which the methylene bridges are omitted and each pyrrole ring is shown as a bracket with the 8 substituent positions numbered as shown below. Uroporphyrin, whose detailed structure is shown in Fig 6–11, would be represented as shown in Fig 6–2. (A = $-CH_2.COOH$; P = $-CH_2.CH_2.COOH$; M = $-CH_3$.)

The formation and occurrence of other porphyrin derivatives may be depicted as shown in Figs 6–3, 6–4, and 6–5.

It will be noted that the arrangement of the A and P substituents in the uroporphyrin shown in Fig

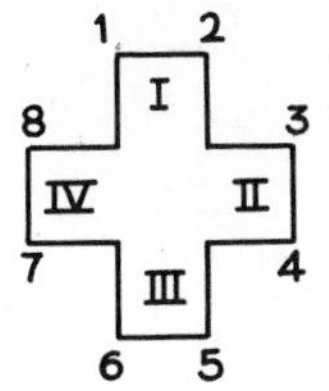

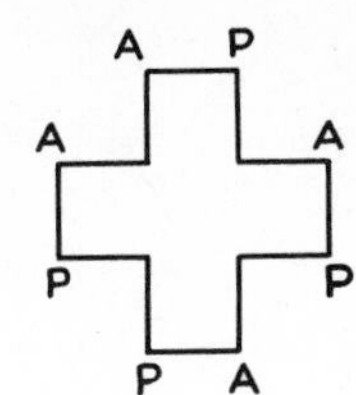

FIG 6–2. Uroporphyrin III.

6–2 is asymmetric (in ring IV, the expected order of the acetate and propionate substituents is reversed). This type of asymmetric substitution is classified as a type III porphyrin. A porphyrin with a completely symmetrical arrangement of the substituents is classified as a type I porphyrin. Only types I and III are found in nature, and the type III series is by far the more abundant.

The compounds shown in Figs 6–4 and 6–5 are all type III porphyrins (ie, the methyl groups are in the same substituent position as in type III coproporphyrin). However, they are sometimes identified as belonging to series 9 because they were designated ninth in a series of isomers isolated by Hans Fischer, the pioneer worker in the field of porphyrin chemistry. These derivatives have one of 3 types of substituents: ethyl ($-CH_2CH_3$), E; hydroxyethyl ($-CH_2CH_2OH$), EOH; or vinyl ($-CH{=}CH_2$), V.

Deuteroporphyrins (Fig 6–6) and mesoporphyrins (Fig 6–7) may be formed in the feces by bacterial activity on protoporphyrin III.

Biosynthesis of Porphyrins

Both chlorophyll, the photosynthetic pigment of plants, and heme, the iron protoporphyrin of hemoglobin in animals, are synthesized in living cells by a common pathway. The 2 starting materials are "active succinate," the coenzyme A derivative of succinic acid, derived from the citric acid cycle, and the amino acid glycine. Pyridoxal phosphate is also necessary in this reaction to "activate" glycine. It is probable that pyridoxal reacts with glycine to form a Schiff base, whereby the alpha carbon of glycine can be combined with the carbonyl carbon of succinate. The product of the condensation reaction between succinate and glycine is α-amino-β-ketoadipic acid, which is rapidly decarboxylated to form δ-aminolevulinic acid (ALA). This step is catalyzed by the enzyme **ALA synthetase.** This appears to be the rate-controlling enzyme in porphyrin biosynthesis in mammalian liver (see also p 70). Synthesis of ALA occurs in the mitochondria, where succinyl-Co A is being produced in the reactions of the citric acid cycle.

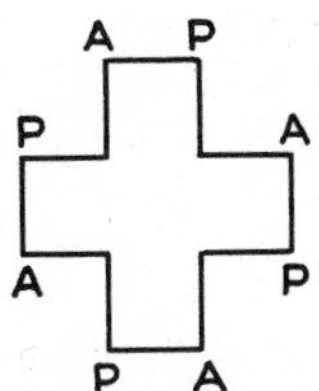

Uroporphyrin I

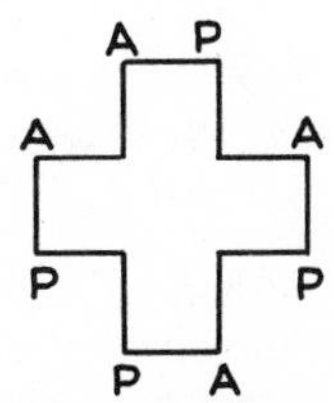

Uroporphyrin III

Uroporphyrins were first found in the urine, but they are not restricted to urine.

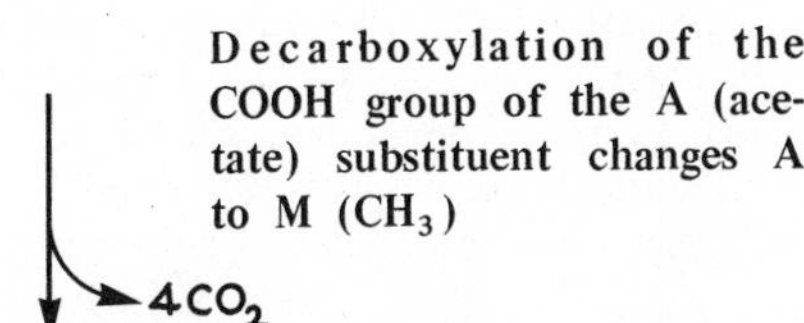

Decarboxylation of the COOH group of the A (acetate) substituent changes A to M (CH_3)
$4CO_2$

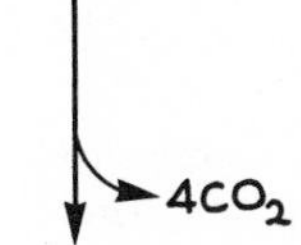

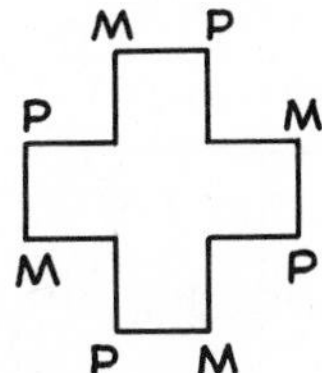

Coproporphyrin I

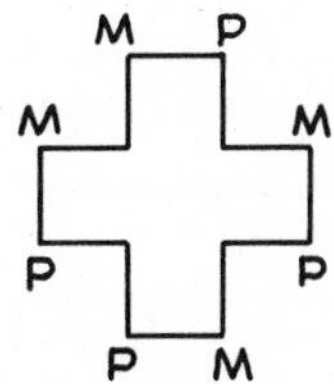

Coproporphyrin III

Coproporphyrins were first isolated from feces but are also found in the urine.

FIG 6–3. Conversion of uroporphyrins to coproporphyrins.

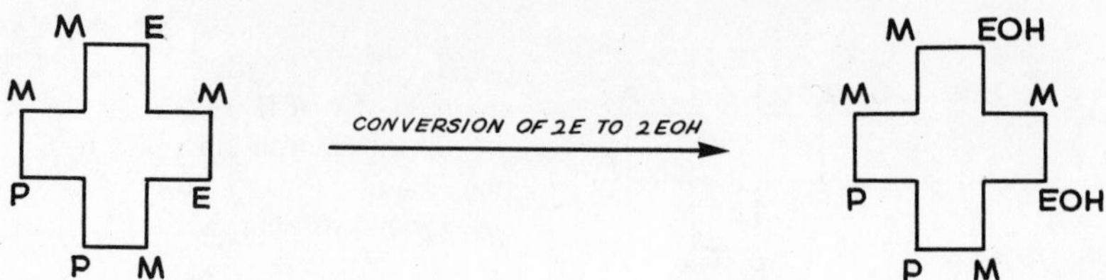

Mesoporphyrin III (9) (from coproporphyrin III by decarboxylation of propionates on positions 2 and 4)

Hematoporphyrin III (9) (may also be derived from coproporphyrin III by changing propionates on positions 2 and 4 to hydroxyethyl groups)

FIG 6–4. Conversion of mesoporphyrin to hematoporphyrin.

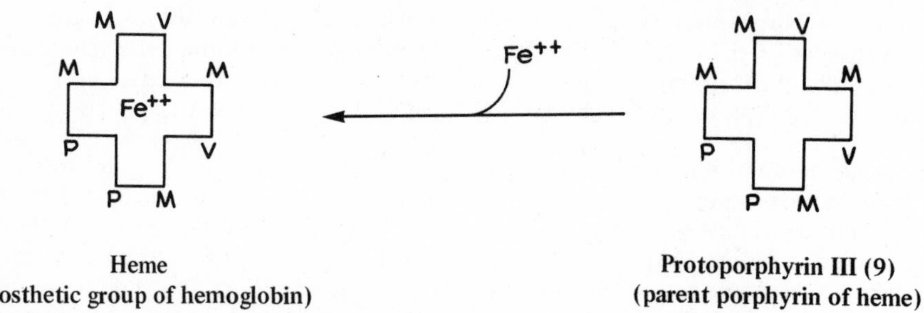

Heme
(prosthetic group of hemoglobin)

Protoporphyrin III (9)
(parent porphyrin of heme)

FIG 6–5. Addition of iron to protoporphyrin to form heme.

Schulman & Richert (1957) studied the incorporation of glycine and succinate into heme by avian red blood cells. It was found that blood samples from vitamin B_6- and from pantothenic acid-deficient ducklings utilized glycine and succinate for heme synthesis at a reduced rate, although δ-aminolevulinic acid incorporation into heme was essentially normal. The addition in vitro of pyridoxal-5-phosphate restored glycine and succinate incorporation without affecting aminolevulinic acid incorporation. Added coenzyme A was without effect in vitro, but the injection of calcium pantothenate 1 hour before the blood specimens were drawn restored to normal the rate of glycine incorporation into heme. These experiments suggest that the block in heme synthesis in pantothenic acid or in vitamin B_6 deficiency occurs at a very early step in heme synthesis, presumably in the formation of coenzyme A succinate and in the pyridoxal-dependent decarboxylation of α-amino-β-ketoadipic acid. The anemia which has been found to accompany vitamin B_6 or pantothenic acid deficiency in several species of experimental animals may be explained on a biochemical basis by these observations.

The next step in porphyrin biosynthesis following formation of ALA is characterized by the condensation of 2 mols of ALA to form **porphobilinogen,** the monopyrrole precursor of the porphyrins. The reaction is catalyzed by **δ-aminolevulinase (ALA dehydrase).** All of these reactions are shown in Fig 6–8.

The pathway for the synthesis of porphobilinogen as described above has received support by the finding of both δ-aminolevulinic acid and porphobilinogen in the urine of normal subjects (Mauzerall, 1956) as well as in the urine of patients with acute porphyria (Granick, 1955). In this group of patients these compounds, which are involved in the synthesis of porphyrins, occur in greatly increased amounts. The conversion of δ-aminolevulinic acid to porphobilinogen has

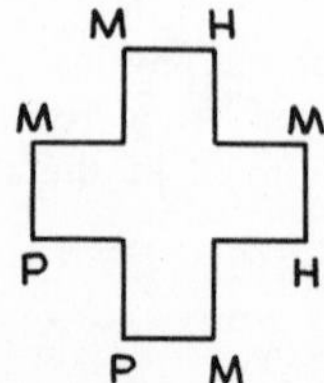

FIG 6–6. Deuteroporphyrin (type III).

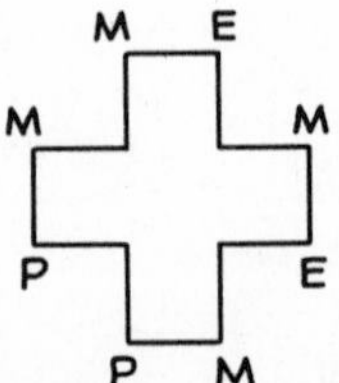

FIG 6–7. Mesoporphyrin (type III).

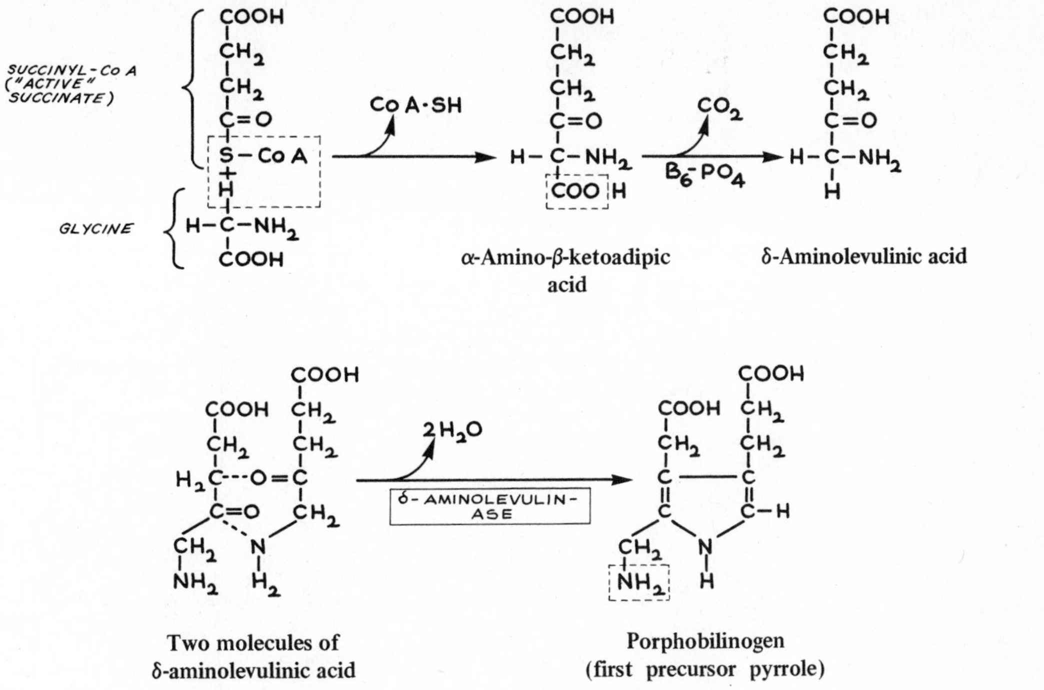

SUCCINYL-CoA ("ACTIVE" SUCCINATE)

GLYCINE

CoA·SH

CO_2

B_6-PO_4

α-Amino-β-ketoadipic acid

δ-Aminolevulinic acid

$2H_2O$

δ-AMINOLEVULINASE

Two molecules of δ-aminolevulinic acid

Porphobilinogen (first precursor pyrrole)

FIG 6–8. Biosynthesis of porphobilinogen.

been shown to occur by in vitro experiments using enzyme preparation from aqueous extracts of hemolysates of chick red blood cells (Granick, 1954) as well as from liver tissue (Gibson, 1954).

The formation of a tetrapyrrole, ie, a porphyrin, occurs by condensation of 4 monopyrroles derived from porphobilinogen. In each instance, the amino carbon (originally derived from the alpha carbon of glycine) serves as the source of the methylene (alpha, beta, gamma, delta) carbons which connect each pyrrole in the tetrapyrrole structure. The enzymes involved in the conversion of ALA to porphyrinogens occur in the soluble portion of the cell in contrast to the mitochondrial location of the ALA-forming enzymes. Although the conversion of porphobilinogen to a porphyrin can be accomplished simply by heating under acid conditions, in the tissues this conversion is catalyzed by specific enzymes. Such an enzyme isolated from avian red blood cells by Lockwood and Rimington transforms porphobilinogen to uroporphyrinogen, the first tetrapyrrole. This enzyme was termed **porphobilinogenase.**

It has been pointed out that only types I and III porphyrins occur in nature, and it may be assumed that the type III isomers are the more abundant since the biologically important porphyrins such as heme and the cytochromes are type III isomers. Shemin, Russell, & Abramsky (1955) have suggested that both type I and type III porphyrinogens may be formed as diagrammed in Fig 6–9. In this scheme, 3 mols of porphobilinogen condense first to form a tripyrrylmethane, which then breaks down into a dipyrrylmethane and a monopyrrole. The dipyrryl compounds are of 2 types depending upon where the split occurs on the tripyrryl precursor: at the point marked (A) or at (B). The formation of the tetrapyrrole occurs by condensation of 2 dipyrrylmethanes. If 2 of the (A) components condense, a type I porphyrin results; if one (A) and one (B) condense, a type III results. Because of the structure of the side-chains on porphobilinogen (acetate and propionate), it is clear that uroporphyrinogens types I and III would result as the first tetrapyrroles to be formed.

The uroporphyrinogens I and III are converted to coproporphyrinogens I and III by decarboxylation of all of the acetate (A) groups, which changes these to methyl (M) substituents. The reaction is catalyzed by **uroporphyrinogen decarboxylase.** The coproporphyrinogen III then enters the mitochondria, where it is converted to protoporphyrinogen and then to protoporphyrin. Several steps seem to be involved in this conversion. An enzyme, **coproporphyrinogen oxidase,** is believed to catalyze the decarboxylation and oxidation of 2 propionic side-chains to form protoporphyrinogen. This enzyme is able to act only on type III coproporphyrinogen, which would explain why a type I protoporphyrin has not been identified in natural materials. The oxidation of protoporphyrinogen to protoporphyrin is believed to be catalyzed by an enzyme, **protoporphyrinogen oxidase.** In mammalian liver the reaction of conversion of coproporphyrinogen to protoporphyrin requires molecular oxygen.

The final step in heme synthesis involves the incorporation of ferrous iron into protoporphyrin in a

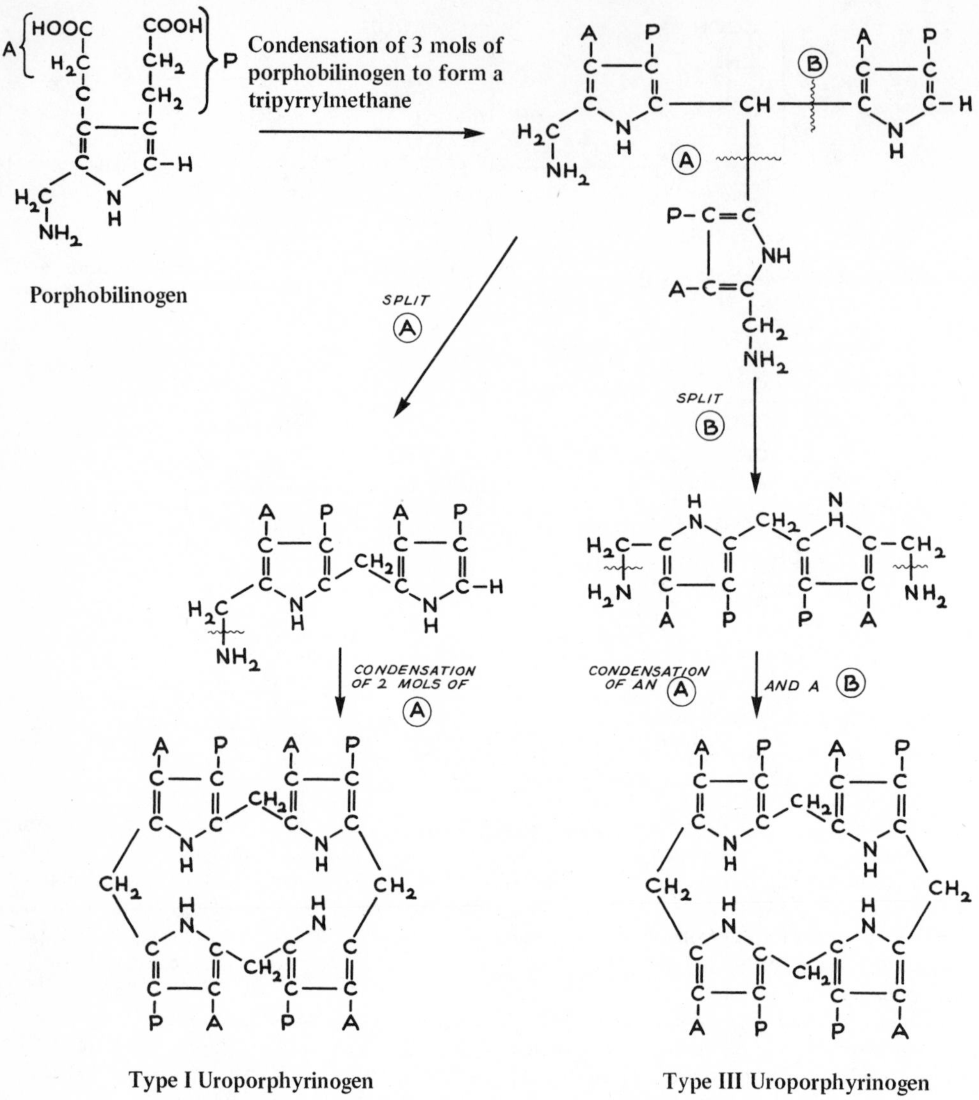

Porphobilinogen

Condensation of 3 mols of porphobilinogen to form a tripyrrylmethane

SPLIT (A)

SPLIT (B)

CONDENSATION OF 2 MOLS OF (A)

CONDENSATION OF AN (A) AND A (B)

Type I Uroporphyrinogen

Type III Uroporphyrinogen

FIG 6–9. Conversion of porphobilinogen to uroporphyrinogens.

reaction catalyzed by **heme synthetase** or **ferrochelatase.** This reaction occurs readily in the absence of enzymes; but it is noted to be much more rapid in the presence of tissue preparations, presumably because of the tissue contribution of enzymes active in catalyzing iron incorporation.

A summary of the steps in the biosynthesis of the porphyrin derivatives from porphobilinogen is given in Fig 6–10.

The porphyrinogens which have been described above are colorless reduced porphyrins containing 6 extra hydrogen atoms as compared to the corresponding porphyrins. They have been reported to occur in biologic material by Fischer and by Watson, Schwartz, and their associates. It is now apparent that these reduced porphyrins (the porphyrinogens) and not the corresponding porphyrins are the actual intermediates in the biosynthesis of protoporphyrin and of heme. This idea is supported by the observation that the oxidized porphyrins cannot be used for heme or for chlorophyll synthesis either by intact or disrupted cells. Furthermore, the condensation of 4 mols of porphobilinogen (Fig 6–9) would directly give rise to uroporphyrinogens rather than to uroporphyrins.

The porphyrinogens are readily auto-oxidized to the respective porphyrins as shown in Fig 6–11 for uroporphyrinogen. These oxidations are catalyzed in the presence of light and by the porphyrins that are formed. The amounts of these porphyrin byproducts that are produced depend not only on the activities of the various enzymes involved but also on the presence of catalysts (light) or inhibitors (reduced glutathione) of their auto-oxidation.

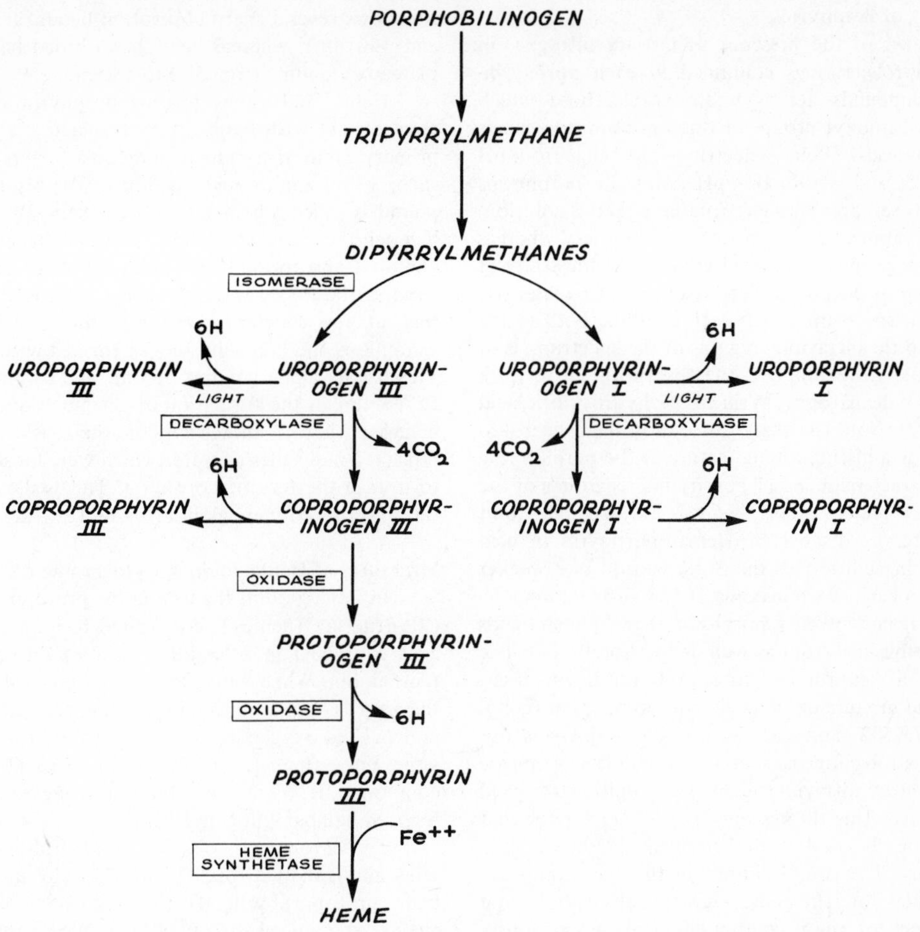

FIG 6–10. Steps in the biosynthesis of the porphyrin derivatives from porphobilinogen.

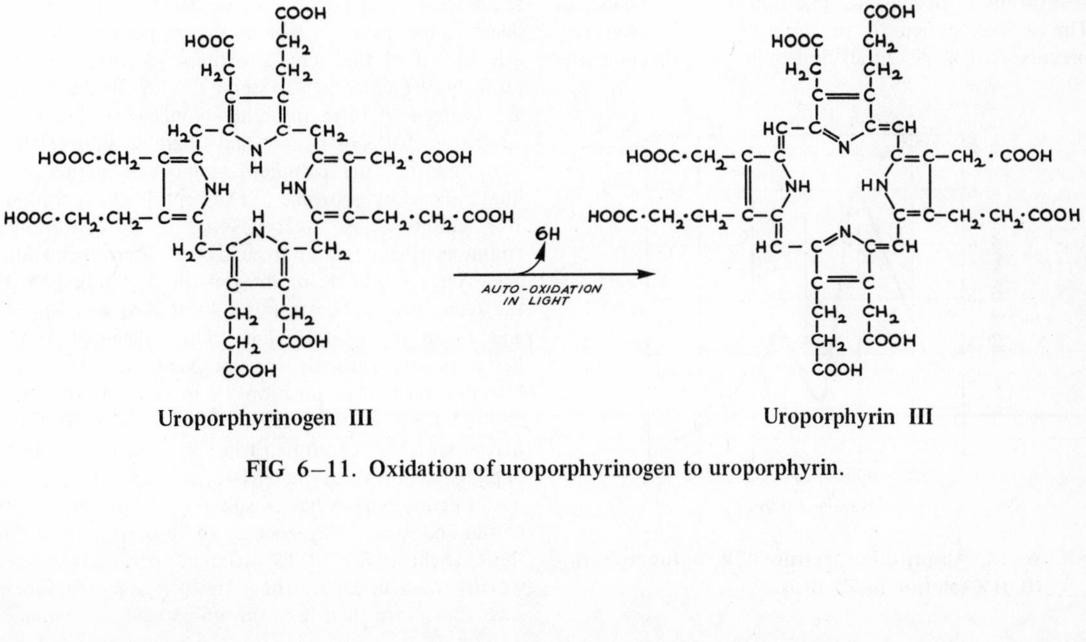

6H
AUTO-OXIDATION IN LIGHT

Uroporphyrinogen III

Uroporphyrin III

FIG 6–11. Oxidation of uroporphyrinogen to uroporphyrin.

Chemistry of Porphyrins

Because of the presence of tertiary nitrogens in the 2 pyrrolene rings contained in each porphyrin, these compounds act as weak bases. Those which possess a carboxyl group on one or more side-chains act also as acids. Their isoelectric points range from pH 3.0 to 4.5, and within this pH range the porphyrins may easily be precipitated from an aqueous solution.

The various porphyrinogens are colorless, whereas the various porphyrins are all colored. In the study of porphyrins or porphyrin derivatives, the characteristic absorption spectrum which each exhibits, both in the visible and the ultraviolet regions of the spectrum, is of great value. An example is the absorption curve for a solution of hematoporphyrin in 5% hydrochloric acid (Fig 6–12). Note the sharp absorption band near 400 nm. This is a distinguishing feature of the porphin ring and is characteristic of all porphyrins regardless of the side-chains present. This band is termed the **Soret band**, after its discoverer. Hematoporphyrin in acid solution, in addition to the Soret band, has 2 weaker absorption bands with maxima at 550 and 592 nm.

In organic solvents, porphyrins have 4 main bands in the visible spectrum as well as the Soret band. For example, a solution of protoporphyrin in an ether-acetic acid mixture exhibits absorption bands at 632.5, 576, 537, 502, and 395 nm. When porphyrins dissolved in strong mineral acids or in organic solvents are illuminated by ultraviolet light, they emit a strong red fluorescence. This fluorescence is so characteristic that it is frequently used to detect small amounts of free porphyrins. The double bonds in the porphyrins are responsible for the characteristic absorption and fluorescence of these compounds, and, as previously noted, the reduction (by addition of hydrogen) of the methylene (C=H) bridges to CH_2 leads to the formation of colorless compounds termed **porphyrinogens.**

When a porphyrin combines with a metal, its absorption in the visible spectrum becomes changed. This is exemplified by protoporphyrin, the iron-free precursor of heme. In alkaline solution, protoporphyrin shows several sharp absorption bands (at 645, 591, and 540 nm), whereas heme has a broad band with a plateau extending from 540 to 580 nm.

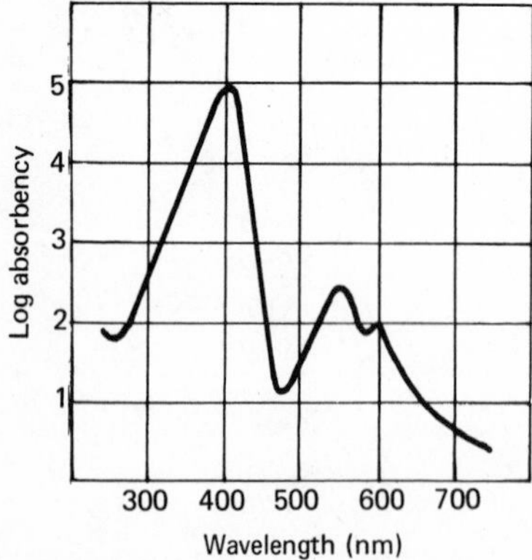

FIG 6–12. Absorption spectrum of hematoporphyrin (0.01% solution in 5% HCl).

Heme and other ferrous porphyrin complexes react readily with basic substances such as hydrazines, primary amines, pyridines, ammonia, or an imidazole such as the amino acid, histidine. The resulting compound is called a hemochromogen (hemochrome). The hemochromogens still show a Soret band but also exhibit 2 absorption bands in the visible spectrum. The band at the longer wavelength is called the alpha band; that at the shorter wavelength, the beta band. An example is the hemochromogen formed with pyridine. This has its alpha band at 559 nm and its beta band at 527.5 nm. In the formation of a hemochromogen, it is believed that 2 molecules of the basic substance replace 2 molecules of water which were loosely bound to iron in the ferrous porphyrin. This is illustrated by the structure of hemoglobin (Fig 6–13).

Structures of Hemoglobin & Cytochrome c

In hemoglobin, the iron in the protoporphyrin III (9) structure (heme) is conjugated to imidazole nitrogens contained in 2 histidine residues within the protein, globin. When hemoglobin combines with oxygen, the iron is displaced from one imidazole group and 1 molecule of oxygen is bound by 1 atom of iron. At the same time, there is an increase in the acidity of the compound, ie, oxyhemoglobin is a stronger acid than is reduced hemoglobin (Fig 6–13).

Cytochrome c is also an iron porphyrin; it contains about 0.43% iron. The iron porphyrin is derived from protoporphyrin III (9), but the 2 vinyl side-chains are reduced and linked by a thioether bond to 2 cysteine residues of the protein; in addition, the iron is linked to a histidine residue in the protein, as in the case of hemoglobin.

Theorell (1956) described the amino acid sequence of that portion of the heme protein in horse heart cytochrome c which was responsible for the attachment of the heme prosthetic group. To accomplish this attachment, 2 cysteine residues in the protein are utilized to form thioether bonds with 2 reduced vinyl side-chains of hemin. Matsuhara & Smith (1963) have reported the complete amino acid sequence of human heart cytochrome c. The peptide chain contains 104 amino acids. Acetylglycine is the N-terminal amino acid and glutamic acid the C-terminal amino acid. The 2 cysteine residues involved in linkage with the heme prosthetic group are located at positions 14 and 17 in the peptide chain. The linkage of iron in heme occurs through the imidazole nitrogen of a histidine residue at position 18 in the peptide chain. On the basis of the data of Matsuhara & Smith, a partial structure of human heart cytochrome c can be depicted as shown in Fig 6–14.

Further studies have resulted in the determination of the amino acid sequence of the protein moieties in the cytochromes c of 13 different species. It is noteworthy that in all of them the primary structure is such that more than ½ of the amino acids are arranged

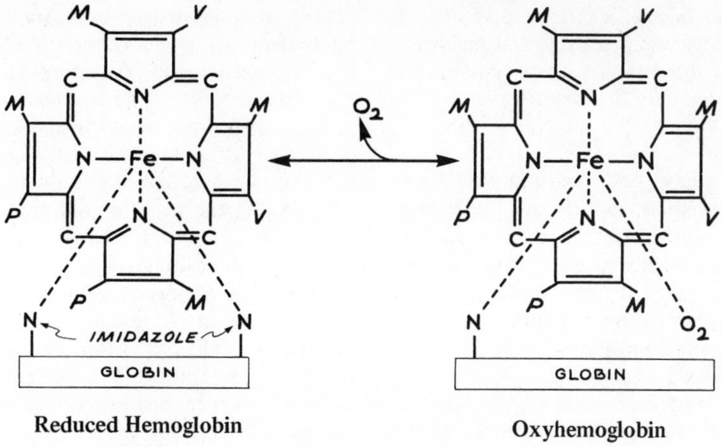

Reduced Hemoglobin

Oxyhemoglobin

FIG 6–13. Imidazole conjugation in hemoglobin.

in an identical sequence. It has been suggested that the degree of difference in primary structure among the 13 cytochromes c might be related to the degree of phylogenetic relationship between the species. For example, the cytochrome c of man as compared to that of a rhesus monkey differs by only one amino acid of the 104 amino acids comprising the whole chain. Human cytochrome c differs from that of the dog in 11 amino acid residues; from that of the horse, in 12. Between a species of fish (the tuna) and baker's yeast there are 48 differences in the amino acid sequences of the cytochromes c.

Tests for Porphyrins

The presence of coproporphyrins or of uroporphyrins is of clinical interest since these 2 types of compounds are excreted in increased amounts in the porphyrias. Coproporphyrins I and III are soluble in glacial acetic acid-ether mixtures, from which they may then be extracted by hydrochloric acid. Uroporphyrins, on the other hand, are not soluble in acetic acid-ether mixtures but are partially soluble in ethyl acetate, from which they may be extracted by hydrochloric acid. In the HCl solution, ultraviolet illumination gives a characteristic red fluorescence. A spectrophotometer may then be used to demonstrate the characteristic absorption bands. The melting point of the methyl esters of the various porphyrins may also be used to differentiate them (Dobriner, 1940; Watson, 1947). Paper chromatography (Petryka, 1968) has more recently been employed as a means of separating and identifying the porphyrins.

Porphyrinuria

The excretion of coproporphyrins may be increased under many circumstances, eg, acute febrile states; after the ingestion of certain poisons, particularly heavy metals such as lead or arsenic; blood

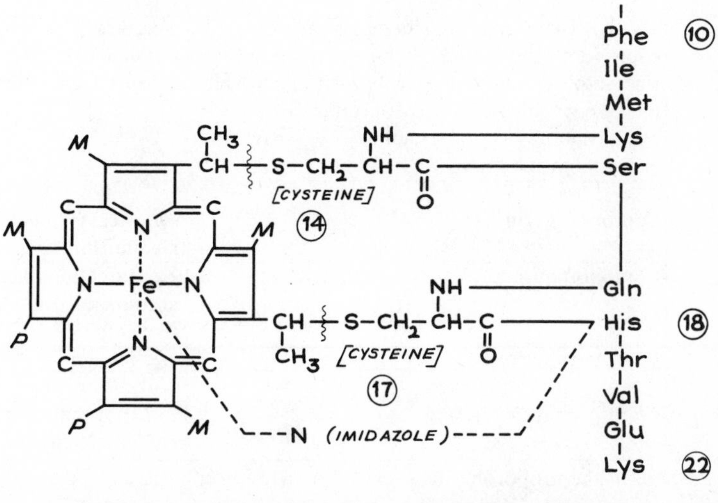

FIG 6–14. Partial structure of human heart cytochrome c.

dyscrasias, hemolytic anemia, or pernicious anemia; and sprue, cirrhosis, acute pancreatitis, and malignancies (eg, Hodgkin's disease). In these conditions, uroporphyrin is not present in the urine.

The Porphyrias

When the excretion of both coproporphyrin and uroporphyrin is increased because of their presence in the blood, the condition is referred to as **porphyria.** Under this term are included a number of syndromes; some are hereditary and familial and some are acquired, but all are characterized by increased excretion of uroporphyrin and coproporphyrin in the urine or feces or both. Reduced catalase activity of the liver has also been reported in these cases.

A summary of these observations on the excretion of porphyrins is given in Table 6–1.

Several different classifications of the porphyrias have been proposed. These are summarized by Goldberg & Rimington (see Bibliography). However, it is convenient to divide the porphyrias into 2 general groups based upon the porphyrin and porphyrin precursor content of the bone marrow or liver. By this system, the porphyrias are classified as **erythropoietic** or **hepatic**. Various subdivisions within each general type have been described as shown in Table 6–2.

Among the hepatic porphyrias there are 2 types (shown in the table as A and B) with focal centers in Sweden or in South Africa. In the Swedish type, early intermediates of porphyrin biosynthesis such as aminolevulinic acid (ALA) and porphobilinogen are excreted in the urine, whereas in the South African type coproporphyrin and protoporphyrin are excreted. It now seems established that the chemical manifestations of acute porphyria represent overproduction of porphyrin precursors as a result of increased activity of the hepatic enzyme **ALA synthetase**. This enzyme, which catalyzes the production of aminolevulinic acid, is the

TABLE 6–1. Excretion of porphyrins.

Type	Porphyrins Excreted	Remarks
Normal	Urine: Coproporphyrin I, Coproporphyrin III } 60–280 (avg 160) μg/day; about 30% is type III.	The excretion of uroporphyrins is negligible in normal individuals, averaging 15–30 μg/day, mostly type I.
	Feces: Coproporphyrin I, Coproporphyrin III } 300–1100 μg/day; 70–90% is type I.	
Hereditary		
Acute (hepatogenic) porphyria (increased porphyrins in the liver)	Mainly type III porphyrins. Coproporphyrin III: 144–2582 μg/day. Coproporphyrin I: Small amounts. Uroporphyrin I and III*: 61,000–147,000 μg/day. Porphobilinogen, δ-Aminolevulinic acid } in urine.	Hereditary as an autosomal dominant. Relatively common; metabolic defect is in the liver. Catalase activity of liver is markedly reduced. Patients are not light-sensitive.
Congenital (erythrogenic) porphyria (increased porphyrins in the marrow)	Mainly type I porphyrins. Type III coproporphyrin was reported in 2 cases. The fecal content of porphyrin is high.	Hereditary as an autosomal recessive. Rare. Patient shows sensitivity to light. Marrow is site of metabolic error.
Chronic porphyria (mixed)	Varies. There may be increased amounts of coproporphyrin I and III and no uroporphyrins; in other cases, uroporphyrin I and III are increased; still others have mixtures of copro- and uroporphyrins. Porphobilinogen may also be found in the urine.	Hereditary or acquired? Patient may be light-sensitive. Frequently associated with enlargement of the liver.
Acquired		
Toxic agents	Coproporphyrin III.	Eg, heavy metals, chemicals, acute alcoholism, and cirrhosis in alcoholics.
Liver disease	Coproporphyrin I.	Eg, infectious hepatitis, cirrhosis not accompanied by alcoholism, obstructive jaundice.
Blood dyscrasias	Coproporphyrin I.	Eg, leukemia, pernicious anemia, hemolytic anemias.
Miscellaneous	Coproporphyrin III.	Eg, poliomyelitis, aplastic anemias, Hodgkin's disease.

*According to Watson (1960), the uroporphyrins excreted in acute porphyria are a complex mixture of uroporphyrin I and a peculiar uroporphyrin III which has 7 rather than 8 carboxyl groups. The mixture is referred to as Waldenström porphyrin. It is extractable from urine by acetate but not by ether.

TABLE 6–2. Classification of porphyrias.*

I. Erythropoietic porphyrias
   A. Congenital erythropoietic porphyria (recessive)
   B. Erythropoietic protoporphyria (dominant?)
II. Hepatic porphyrias
   A. Acute intermittent porphyria, Swedish genetic porphyria, and pyrrolporphyria (dominant)
      1. Manifest
      2. Latent
   B. Porphyria variegata, mixed porphyria, South African genetic porphyria, porphyria cutanea tarda hereditaria, and protocoproporphyria (dominant)
      1. Cutaneous with little or no acute manifestations
      2. Acute intermittent without cutaneous symptoms
      3. Various combinations
      4. Latent
   C. Symptomatic porphyria, porphyria cutanea tarda symptomatica, urocoproporphyria, constitutional porphyria
      1. Idiosyncratic–Associated with alcoholism, liver disease, systemic disease, drugs, etc
      2. Acquired–Hexachlorobenzene-induced porphyria and hepatoma

*Reproduced, with permission, from Tschudy DP: Biochemical lesions in porphyria. JAMA 191:718, 1965.

rate-controlling enzyme in hepatic porphyrin synthesis (Granick, 1963). Normally it is under almost complete repression (see Chapter 5) in the liver, whereas in patients with hepatic porphyria ALA synthetase activity has been found to be more than 7 times that of nonporphyric controls (Tschudy, 1965). Granick (1966) has developed a hypothesis with respect to the etiology of hepatic porphyria which suggests that the increase in ALA synthetase activity in this disease is due to a genetic mutation affecting the operator gene-repressor mechanism normally controlling ALA synthetase production. This hypothesis extends also to an explanation of chemically induced porphyrias such as those following administration of certain drugs (eg, barbiturates) to sensitive individuals. A discussion of Granick's hypothesis is to be found on p 70.

Acute intermittent porphyria is the most common type. It is inherited as an autosomal dominant genetic trait, although it may not manifest itself until the third decade of life. As indicated above, the biochemical defect in this disease is located in the liver—hence the term "hepatogenic porphyria" sometimes applied to this disease. The presenting symptoms occur most often in the gastrointestinal tract and nervous system. In the urine, which is characteristically pigmented and darkens on standing, there is an increased excretion of porphyrin precursors (aminolevulinic acid and porphobilinogen) as well as of type III coproporphyrin and uroporphyrin.

Congenital erythropoietic porphyria is a rare disease inherited as a recessive trait. It occurs more often in males than in females and usually manifests itself early in life. Chemically, the disease is characterized by the production of large quantities of uroporphyrin I, which is excreted in the urine together with increased amounts of coproporphyrin I. Increased amounts of uroporphyrin and coproporphyrin III may also be detected in the urine. That the bone marrow is the site of the increased production of porphyrin is supported by the finding of large concentrations of porphyrin in many of the normoblasts in the marrow, most of the porphyrin being located either in the nucleus of the cell or on its surface. In addition, the concentration of uro- and coproporphyrin in the marrow itself is greatly increased, whereas that of the liver is only slightly so—and that only secondarily as a result of uptake by the liver of the excess porphyrins originally produced in the marrow.

A significant clinical finding in erythropoietic porphyrias is marked sensitivity to light. Areas of the body exposed to light may become necrotic, scarred, and deformed. This photosensitivity is explained by the excessive porphyrin production followed by hemolysis which results in liberation of increased amounts of porphyrin and its accumulation under the skin. Porphyrins are photosensitizing agents because of their ability to concentrate radiant energy by absorption particularly at the wavelength of the Soret band (405 nm) as well as at infrared wavelengths (2600 nm). The presence of porphyrins near the surface of the body, as occurs in erythropoietic porphyria, permits light sensitization to occur; this is in contrast to the circumstances which prevail in hepatic porphyria, wherein accumulation of porphyrin takes place in areas of the body not exposed to light.

From the red blood cells of chickens, Granick & Mauzerall (1958) obtained 3 soluble enzyme fractions which are involved in porphyrin biosynthesis. One fraction, **δ-aminolevulinase**, condenses δ-aminolevulinic acid to form porphobilinogen (Fig 6–8). The steps involving condensation of porphobilinogens to uroporphyrinogens (Fig 6–10) seem to require 2 kinds of enzymic reactions. By the action of an **isomerase,** uroporphyrinogen III is formed. In the absence of this enzyme, uroporphyrinogen I is produced. A marked increase of uroporphyrin I, such as occurs in congenital erythropoietic porphyria, could be explained as due to an imbalance between the action of the isomerase leading to the type III isomer and that of the enzyme catalysts for the production of the type I porphyrins. An absence or even a reduction in activity of the isomerase is not a likely explanation, since these patients can increase production of hemoglobin to compensate for hemolysis. This would not be possible unless type III porphyrins (from which heme is derived) can be produced.

Erythropoietic protoporphyria is believed to be inherited by a dominant mode of transmission. The disease may occur in an active or in a latent form, depending perhaps on the plasma protoporphyrin level.

In the so-called "complete syndrome" there is increased free erythrocyte and plasma protoporphyrin as well as increased fecal protoporphyrin. However, an increase in free erythrocyte protoporphyrin may occur without increase in plasma or fecal protoporphyrin, or there may occur only increased fecal protoporphyrin without other detectable abnormalities of porphyrin content in erythrocytes or plasma. When increased protoporphyrin is found, there is an increase in coproporphyrin as well. It is of interest that the urine does not contain increased amounts of porphyrins or their precursors.

The metabolic defect in erythropoietic protoporphyria is not yet known. It is apparently not due to inability to incorporate iron into protoporphyrin since studies of iron metabolism in vitro and in vivo have revealed no abnormalities. Photosensitivity in protoporphyria is related to the plasma content of protoporphyrin.

Porphyria cutanea tarda is a type of hepatic porphyria which occurs predominantly in older males. The patient characteristically develops blisters on the exposed surfaces of the skin following trauma or exposure to the sun. Other manifestations are hyperpigmentation, hypertrichosis, and abnormally high excretion of both uroporphyrin and coproporphyrin. In many of the reported cases there is an associated history of hepatic disease, so that porphyria cutanea tarda seems to develop secondary to a primary disease of the liver.

Schmid (1960) has described an outbreak of cutaneous porphyria which occurred in Turkey. All of the affected individuals excreted a dark or red urine which fluoresced readily under ultraviolet light and which contained large quantities of both ether-soluble and ether-insoluble porphyrins. Qualitative tests for urine porphobilinogen were negative. All of the patients were found to have consumed wheat intended for planting which had been treated with hexachlorobenzene to inhibit growth of a fungal parasite. This may be the first direct evidence for the occurrence of an acquired toxic form of porphyria in man.

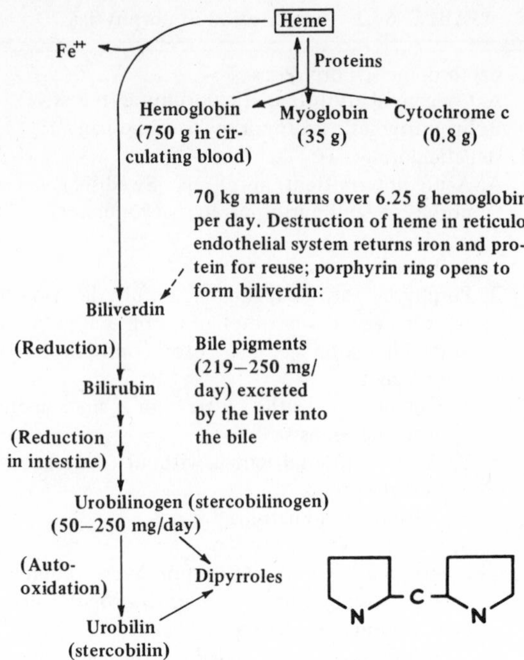

FIG 6–15. Catabolism of heme.

Catabolism of Heme; Formation of the Bile Pigments (See Fig 6–15.)

When hemoglobin is destroyed in the body the protein portion, globin, may be reutilized, either as such or in the form of its constituent amino acids, and the iron enters the iron "pool"—also for reuse. However, the porphyrin portion, heme, is broken down in all likelihood mainly in the reticuloendothelial cells of the liver, spleen, and bone marrow. The initial step in the metabolic degradation of heme involves opening of the porphyrin ring between pyrrole residues I and II and elimination of the alpha methylene carbon as carbon monoxide (Fig 6–16). It is possible that the porphyrin ring may be opened and that the iron is still present before the protoporphyrin is released from the globin. Such a green, conjugated protein has been produced by oxidation of hemoglobin by oxygen in the presence of ascorbic acid. The prosthetic group of the protein is the iron complex of a bile pigment resembling biliverdin. It has been named **choleglobin** by Lemberg.

After removal of iron and cleavage of the porphyrin ring of heme, as described above, **biliverdin**, the first of the bile pigments, is formed. Biliverdin is easily reduced to **bilirubin**, which is the major pigment in human bile. Biliverdin is the chief pigment of the bile in birds. Normally, there are only slight traces of biliverdin in human bile, but the color of biliverdin is so much more intense than that of bilirubin that a relatively small amount, if present in the serum, is detectable even in the presence of a much larger amount of bilirubin. There is a report of a patient with so-called biliverdin jaundice in whom a definite green color was detected in the skin and in the serum when the serum biliverdin concentration was 3 mg/100 ml while that of bilirubin was 25 mg/100 ml. Watson (1969) states that biliverdin jaundice is largely limited to 2 conditions of which biliary obstruction as a result of carcinoma is the more common, the other being severe parenchymal liver disease, as in subacute atrophy or advanced cirrhosis of the liver. The amounts of biliverdin in the serum are very small in obstructive jaundice caused by the presence of a stone in the common bile duct; no biliverdin occurs in the serum in hemolytic jaundice.

A specific enzyme catalyzing reduction of biliverdin to bilirubin has been studied by Singleton & Lasker (1965). The enzyme, **bilirubin reductase**, which utilizes either NADH or NADPH as hydrogen donor, was detected in preparations from guinea pig liver and spleen, but it has also been found in human liver.

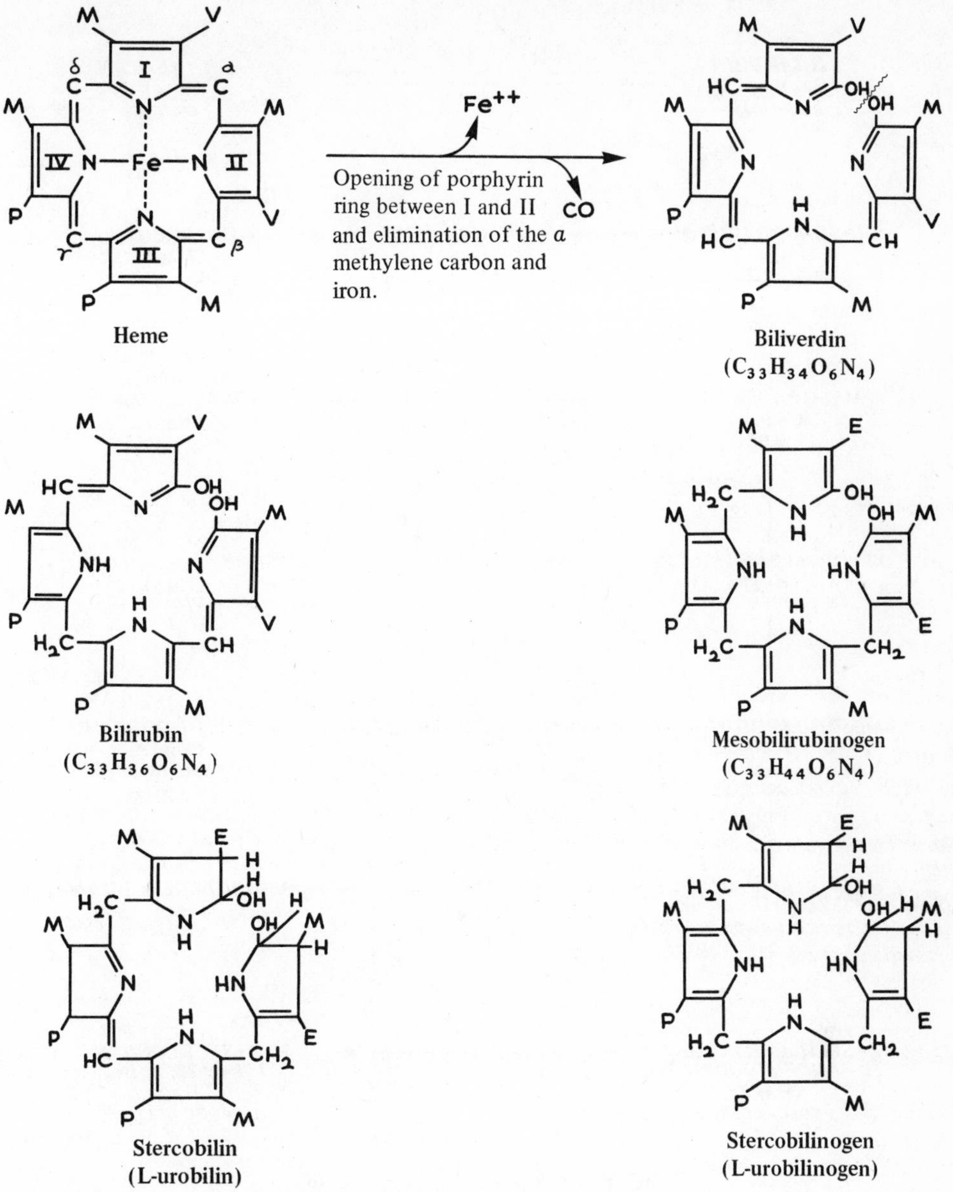

Fe⁺⁺

Opening of porphyrin ring between I and II and elimination of the *a* methylene carbon and iron.

CO

Heme

Biliverdin
($C_{33}H_{34}O_6N_4$)

Bilirubin
($C_{33}H_{36}O_6N_4$)

Mesobilirubinogen
($C_{33}H_{44}O_6N_4$)

Stercobilin
(L-urobilin)

Stercobilinogen
(L-urobilinogen)

FIG 6–16. Structure of some bile pigments.

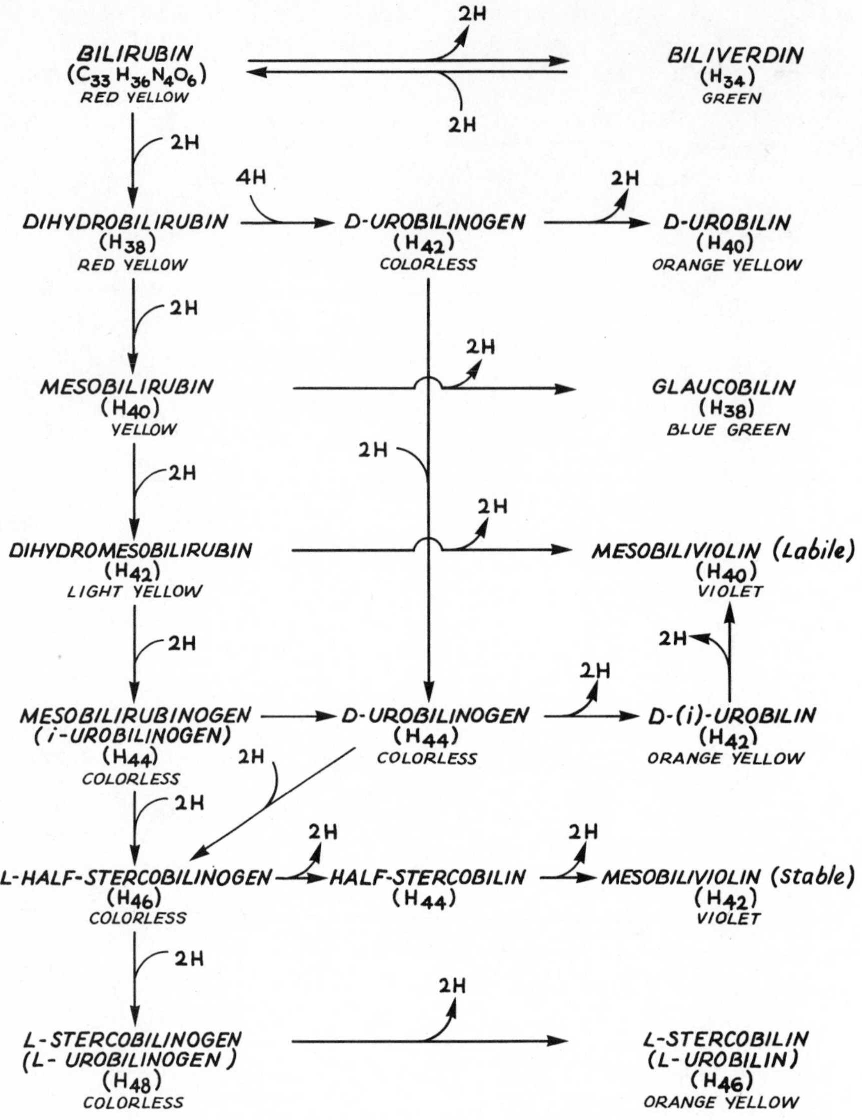

FIG 6–17. Formation of urobilinoids.

Bilirubin associated with serum albumin is transported in the plasma from the tissues in which it is formed to the liver, where it is conjugated with glucuronic acid to form bilirubin diglucuronide (Schmid, 1956). The conjugated bilirubin, being now water-soluble, is excreted by way of the bile into the intestines. Bile pigments constitute 15–20% of the dry weight of human bile.

In the lower portions of the intestinal tract, especially the cecum and the colon, the bilirubin is released from the glucuronide conjugate and then subjected to the reductive action of enzyme systems present in the intestinal tract, mainly derived from anaerobic bacteria in the cecum. Fecal flora as well as a pure strain of a clostridium derived from the rat colon have been demonstrated (in vitro) to be able to complete the reduction of bilirubin to L-stercobilinogen, the normal end product of bilirubin metabolism in the colon. If the intestinal flora is modified or diminished, as by the administration of orally effective antibiotic agents which are capable of producing partial sterilization of the intestinal tract, bilirubin may not be further reduced and may later be auto-oxidized, in contact with air, to biliverdin. Thus, the feces acquire a green tinge under these circumstances.

The metabolism of biliverdin and of bilirubin within the intestine is summarized in Fig 6–17 (Watson, 1969; Lightner & others, 1969). It will be noted that progressive hydrogenation occurs to produce a series of intermediary compounds which, beginning with mesobilirubinogen, comprise a number of colorless urobilinoids which may be oxidized, with loss of hydrogen, to colored compounds. The end product is colorless L-stercobilinogen (L-urobilinogen). Auto-oxidation in the presence of air produces stercobilin (L-urobilin), an orange-yellow pigment which contributes to the normal color of the feces. Stercobilin is strongly levorotatory ($[\alpha_D] = 3600°$).

Urobilin IX (D-[i]-urobilin, or inactive [i] urobilin), is an optically inactive urobilinoid that has been identified in the feces. It is less stable than stercobilin, becoming oxidized in air to form violet and blue-green pigments. In the feces of patients whose intestinal flora has been altered by oral administration of oxytetracycline or chlortetracycline, a dextrorotatory urobilinoid, D-urobilin ($[\alpha_D] = +5000°$) has been identified. It is believed to be derived from dihydrobilirubin by way of D-urobilinogen (Fig 6–17).

The various products derived from the progressive reduction of bilirubin may in part be absorbed from the intestine and returned to the liver for reexcretion or they may be excreted in the urine, which normally contains traces of urobilinogen and urobilin as well as mesobilirubinogen and perhaps other intermediary products. The great majority of the metabolites of bilirubin are, however, excreted with the feces.

The presence of dipyrroles in association with other bilirubin metabolites has also been reported. Such compounds are designated **mesobilifuscins** and are said to be identical with the **copronigrin** isolated from feces by Watson. It is not known whether the dipyrroles are precursors or breakdown products of the tetrapyrroles, or both. It will be recalled that dipyrroles may be intermediates in the formation of tetrapyrroles.

Summary of Porphyrin Metabolism

Types I and III uroporphyrinogens are synthesized from glycine and succinate precursors. By decarboxylation, these are converted to coproporphyrinogens. Both uroporphyrinogens and coproporphyrinogens are readily auto-oxidized to the corresponding uroporphyrins or coproporphyrins. The type I coproporphyrin is apparently a byproduct of the synthesis of heme protoporphyrin because the amount produced is proportionate to the production of hemoglobin. However, this pigment is useless and is excreted in the urine in amounts of 40–190 μg/day except under conditions of rapid hematopoiesis, as in hemolytic disease, where the daily urinary excretion of coproporphyrin I may exceed 200–400 μg/day. A somewhat larger amount of coproporphyrin is normally excreted in the feces: 300–1100 μg/day, 70–90% of which is coproporphyrin I.

Type III coproporphyrinogen is largely converted to protoporphyrin and then to heme, which conjugates with protein to form the hemoproteins: hemoglobin, myoglobin, cytochrome, etc. A small amount (20–90 μg/day) of type III coproporphyrin is normally excreted as such in the urine.

The excretion of uroporphyrins by normal subjects is negligible.

The breakdown of heme leads to the production of the bile pigments, so that the amount of bile pigment formed each day is closely related to the amount of hemoglobin created and destroyed. In hemoglobin, the porphyrin portion, exclusive of the iron, makes up 3.5% by weight of the hemoglobin molecule. Thus, 35 mg of bilirubin could be expected to appear for each gram of hemoglobin destroyed. It is estimated that in a 70 kg man about 6.25 g of hemoglobin are produced and destroyed each day (normal is 90 mg/kg/day). This means that about 219 mg of bilirubin (6.25 × 35) should be produced per day in this same individual. The bilirubin is excreted by the liver into the intestine by way of the bile and may be measured as fecal urobilinogen (stercobilinogen), to which it is converted by the action of the intestinal bacteria. However, 10–20% more than this estimated quantity actually appears each day as urobilinogen. This is most likely due to some porphyrin which is synthesized probably in the liver but never actually incorporated into red cells (early urobilinogen), as well as porphyrin derived from the catabolism of hemoproteins other than hemoglobin. Total daily bile pigment production is therefore close to 250 mg, but not all of this is recovered as urobilinogens because some of the urobilinogen is broken down by bacterial action to dipyrroles (Fig 6–9), which do not yield a color with the Ehrlich reagent (see Chapter 17) used to measure urobilinogen. Fig 6–15 summarizes the catabolism of heme and the formation of bile pigments.

• • •

References

Dobriner K, Rhoads CP: Physiol Rev 20:416, 1940.

Gibson KD, Neuberger A, Scott JJ: Biochem J 58:XLI, 1954.

Granick S: Science 120:1105, 1954.

Granick S, Mauzerall D: J Biol Chem 232:1119, 1958.

Granick S, Vanden Schrieck HG: Proc Soc Exper Biol Med 88:270, 1955.

Granick S: J Biol Chem 238:2247, 1963.

Granick S: J Biol Chem 241:1359, 1966.

Labbe RF, Talman EL, Aldrich RA: Biochim et biophys acta 15:590, 1954.

Lightner DA, Moscowitz A, Petryka ZJ, Jones S, Weimer M, Davis E, Beach NA, Watson CJ: Arch Biochem 131:566, 1969.

Matsuhara H, Smith EL: J Biol Chem 238:2732, 1963.

Mauzerall D, Granick S: J Biol Chem 219:435, 1956.

Petryka ZJ, Watson CJ: J Chromatogr 37:76, 1968.

Schmid R: New England J Med 263:397, 1960.

Schmid R: Science 124:76, 1956.

Schulman MP, Richert DA: J Biol Chem 226:181, 1957.

Shemin D, Russell CS: J Am Chem Soc 74:4873, 1953.

Shemin D, Russell CS, Abramsky T: J Biol Chem 215:613, 1955.

Singleton JW, Laster L: J Biol Chem 240:4780, 1965.

Theorell H: Science 124:467, 1956.

Tschudy DP, Perlroth MG, Marver HS, Collins A, Hunter G Jr: Proc Nat Acad Sc 53:841, 1965.

Watson CJ: Ann Int Med 70:839, 1969.

Watson CJ, Berg MH, Hawkinson VE, Bossenmaier I: Clin Chem 6:71, 1960.

Watson CJ, Larson EA: Physiol Rev 27:478, 1947.

Bibliography

Goldberg A, Rimington C: *Diseases of Porphyrin Metabolism.* Thomas, 1962.

Lemberg R, Legge JW: *Hematin Compounds and Bile Pigments.* Interscience, 1949.

Rimington C: Haem pigments and porphyrins. Ann Rev Biochem 26:561, 1957.

Schmid R: The porphyrias. In: *The Metabolic Basis of Inherited Disease.* Stanbury JB, Wyngaarden JB, Fredrickson DS (editors). McGraw-Hill, 1966.

Symposium: Porphyrin Biosynthesis and Metabolism. Ciba Foundation. Little, Brown, 1954.

7. . .
Vitamins

When animals are maintained on a chemically defined diet containing only purified proteins, carbohydrates, and fats, and the necessary minerals, it is not possible to sustain life. Additional factors present in natural foods are required, although often only minute amounts are necessary. These "accessory food factors" are called vitamins. The vitamins have no chemical resemblance to each other, but because of a similar general function in metabolism they are considered together.

Early studies of the vitamins emphasized the more obvious pathologic changes which occurred when animals were maintained on vitamin-deficient diets. Increased knowledge of the physiologic role of each vitamin has enabled attention to be concentrated on the metabolic defects which occur when these substances are lacking, and we may therefore refer to the biochemical changes as well as the anatomic lesions which are characteristic of the various vitamin deficiency states.

Before the chemical structures of the vitamins were known it was customary to identify these substances by letters of the alphabet. This system is gradually being replaced by a nomenclature based on the chemical nature of the compound or a description of its source or function.

The vitamins are generally divided into 2 major groups: fat-soluble and water-soluble. The fat-soluble vitamins, which are usually found associated with the lipids of natural foods, include vitamins A, D, E, and K. The vitamins of the B complex and vitamin C comprise the water-soluble group.

FAT-SOLUBLE VITAMINS

VITAMIN A

Chemistry

The structure of vitamin A_1 aldehyde (retinal) is shown in Fig 7–1. It may be derived from β-carotene by cleavage at the mid point of the carotene in the polyene chain connecting the 2 β-ionone rings. Goodman, Huang, & Shiratori (1966), as a result of studies on the mechanism of biosynthesis of vitamin A, have shown that the 2 hydrogen atoms attached to the 2 central carbon atoms of β-carotene are retained during the conversion of the carotene to vitamin A. They suggest that the biosynthesis of vitamin A from β-carotene is most likely a dioxygenase reaction in which molecular oxygen reacts with the 2 central carbon atoms of β-carotene followed by cleavage of the central double bond of β-carotene to yield 2 mols of vitamin A aldehyde (retinal). These reactions are shown in Fig 7–1. Vitamin A alcohol (retinol) may then be produced by reduction of the aldehyde in an NADH-dependent reaction catalyzed by retinene reductase.

Vitamin A_2 (3-dehydroretinol) has also been described. Its potency is 40% that of vitamin A_1; its structure differs from A_1 by the presence of an additional double bond (between carbons 3 and 4 of the β-ionone ring).

Vitamin A alcohol occurs only in the animal kingdom, mainly as an ester with higher fatty acids, in the liver, kidney, lung, and fat depots. The sources of all of the vitamin A in animals are probably certain plant pigments known as carotenes or carotenoid pigments, the provitamins A, which are synthesized by all plants except parasites and saprophytes. These provitamins A are transformed into vitamin A in the animal body. β-Carotene yields 2 mols of vitamin A, whereas the α- and γ-carotenes yield only 1 mol since they are not symmetric. The conversion of the carotenes to vitamin A occurs in the intestinal wall in rats, pigs, goats, rabbits, sheep, and chickens, although the liver may also participate. In man, the liver is believed to be the only organ which is capable of accomplishing this transformation.

These provitamins of the diet are not as well absorbed from the intestine as is vitamin A. Many factors affect the efficiency of absorption and utilization of preformed vitamin A and carotene (Roels, 1966). Small amounts of mineral oil added to the diet of experimental animals were found to interfere with the utilization of both vitamin A and carotene, although the inhibition of carotene utilization was much greater than that of vitamin A.

In general it is assumed that carotene has ½ or less the value of an equal quantity of vitamin A in the diet of man. This is due to the differences in intestinal absorption and the necessity for conversion to vitamin A by the liver.

β-Carotene

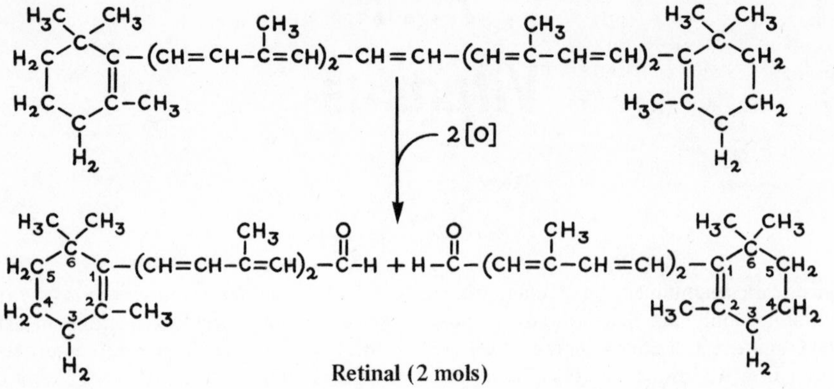

2[O]

Retinal (2 mols)

FIG 7–1. Conversion of β-carotene to retinal (vitamin A_1 aldehyde).

Physiologic Role

The maintenance of the integrity of epithelial tissue is an important function of vitamin A. In its absence, normal secretory epithelium is replaced by a dry, keratinized epithelium which is more susceptible to invasion by infectious organisms. Xerophthalmia, ie, keratinization of ocular tissue, which may progress to blindness, is a late result of vitamin A deficiency. Xerophthalmia is a major cause of blindness in childhood. It is still a major health problem in many parts of the world, especially in rapidly growing urban areas of the Far East such as Hong Kong, Djakarta, Manila, Saigon, and Dacca.

The specific role of vitamin A in the physiologic mechanisms of vision has been elucidated largely by Wald and by Morton. The retinal pigment **rhodopsin,** or visual purple, which has long been recognized in the rod cells of the retina, is a conjugated protein with a molecular weight of approximately 40,000 daltons. When light strikes the retina, rhodopsin is split into its protein component, **opsin,** and the associated nonprotein carotenoid, retinene. This latter compound has been identified as vitamin A_1 aldehyde and therefore is now termed **retinal.** In the light-bleached retina, vitamin A_1 itself, the alcohol, **retinol,** appears later, and it is therefore assumed that the aldehyde, retinal, is slowly converted by reduction to the alcohol, retinol. It follows that the regeneration of retinal from retinol requires oxidation of the terminal alcohol group to the aldehyde group. This is accomplished through the catalytic action of the enzyme **retinene reductase,** involving also NAD as coenzyme (Fig 7–3). These reactions are summarized below.

Retinene reductase appears to be very similar to alcohol dehydrogenase of liver; indeed, a crystalline preparation of alcohol dehydrogenase from horse liver was found to catalyze the retinol-retinal reaction in vitro.

It should be pointed out that at least 2 colored intermediate compounds are formed in the course of the reactions whereby retinal is liberated from rhodopsin. One is a red or orange-red compound, **lumirhodopsin,** stable only at temperatures below −50° C. Beginning at a temperature of about −20° C, lumirhodopsin is converted to **metarhodopsin,** also orange-red in color. At temperatures above −15° C and in the presence of water, this compound hydrolyzes to retinal and opsin.

Regeneration of rhodopsin takes place in the dark. Under normal circumstances equilibrium is maintained in the retina of the eye such that the rate of breakdown of rhodopsin is equaled by the rate of regeneration. If, however, a deficiency of vitamin A exists, the rate of regeneration of rhodopsin is retarded, probably because of a shortage of precursor substances. This concept is supported by the observation that the retinas of rats maintained on a vitamin A-deficient diet contain less rhodopsin than do those of animals on an adequate diet.

The biochemical mechanism of cone vision is analogous to that of rod vision, described above. The

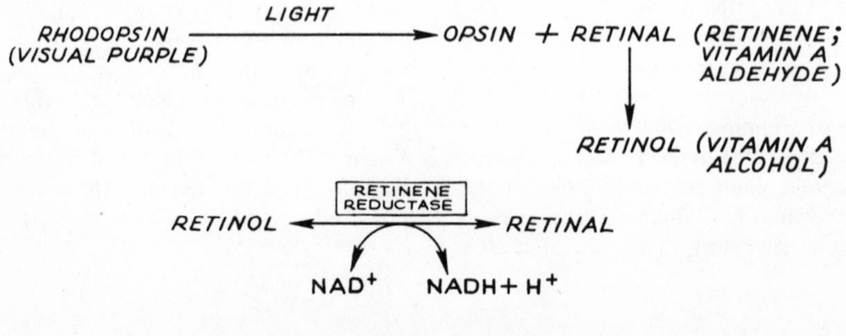

RHODOPSIN (VISUAL PURPLE) —LIGHT→ OPSIN + RETINAL (RETINENE; VITAMIN A ALDEHYDE)
↓
RETINOL (VITAMIN A ALCOHOL)

RETINOL ⇌ RETINAL (RETINENE REDUCTASE)
NAD⁺ → NADH + H⁺

photoreceptors of both rods and cones contain essentially the same chromophore (retinal), although the protein moiety (opsin) differs.

Night blindness (nyctalopia), which is a disturbance of rod vision, is one manifestation of vitamin A deficiency. Measurements of the rate of regeneration of a normal response to light have therefore been used to detect early vitamin A deficiency states because regeneration rates are considerably decreased by even moderate lack of the vitamin.

Although the role of vitamin A in the visual apparatus is now well established, the vitamin must also participate in the metabolism of the body in a much more generalized way. Animals on a diet free of vitamin A do not merely suffer visual impairment and ocular lesions but will eventually die unless the vitamin is supplied.

Reference has already been made to a function of vitamin A in connection with epithelial tissue. It has also been observed that in the absence of vitamin A the growth of experimental animals does not progress normally. The skeleton is affected first, and then the soft tissues. Mechanical damage to the brain and cord occurs when these structures attempt to grow within the arrested limits of the bony framework of the cranium and vertebral column. In the growing animal collagenous tissues are particularly affected by a deficiency of vitamin A. The mucopolysaccharides (see pp 259 and 480) which form the ground substance are an important constituent of such tissues. Consequently it is of considerable interest that the rate of mucopolysaccharide formation was found to be inhibited in the tissues of vitamin A-deficient animals and restored to normal when the vitamin was provided (Wolf & Varandani, 1960). That vitamin A may play a role in protein synthesis has also been indicated by recent work (Roels, 1964).

Vitamin A may also participate in reactions which affect the stability of cell membranes and of the membranes of subcellular particles. Lucy, Luscombe, & Dingle (1963) have shown that vitamin A alcohol causes swelling of mitochondria in vitro. Vitamin A aldehyde or acid has considerably less effect. Mitochondria from the liver were the most readily swollen by vitamin A; those from spleen and brain were least affected. Heart mitochondria exhibited about ¼ of the effect shown by liver mitochondria.

When rats were either made deficient in vitamin A or given an excess of the vitamin, oxidative phosphorylation was impaired as evidenced by studies with mitochondria obtained from the livers of the experimental animals. Supplementation with retinyl acetate (but not with retinoic acid) either in vivo or in vitro restored the phosphorus to oxygen (P:O) ratios to normal, indicating that oxidative phosphorylation was now proceeding normally (Seward, 1966). The results of these experiments suggest that vitamin A is required in mitochondrial membranes at an optimal concentration. Variations above or below this optimum cause these membranes to become unstable, and functional changes in enzymes associated with oxidative phosphorylation may also be induced.

Janoff & McCluskey (1962) demonstrated that vitamin A, when administered in excess to guinea pigs, caused a significant decrease in the activity of acid phosphatase, extractable from their peritoneal phagocytes. It was suggested that large doses of vitamin A reduced the stability or increased the permeability of the lysosomes within the peritoneal phagocytes. This would permit release of acid phosphatase (and presumably other enzymes) from the cells so that in subsequent measurements their acid phosphatase content would be lower than normal. The question thus arises whether a major function of vitamin A in normal metabolism may not be the preservation of the structural integrity and the normal permeability of the cell membrane as well as that of the membranes of subcellular particles such as lysosomes and mitochondria.

Sources & Daily Allowance

All yellow vegetables and fruits (eg, sweet potatoes, apricots, and yellow peaches) and the leafy green vegetables supply provitamin A in the diet. Preformed vitamin A is supplied by milk, fat, liver, and, to a lesser extent, by kidney and the fat of muscle meats.

The recommended daily allowance is 5000 IU/day for an adult (increase to 6000 to 8000 units during pregnancy and lactation). Allowances for infants and children are given in Table 21–3. An international unit is equivalent to the activity of 0.6 μg of pure β-carotene; 0.3 μg of vitamin A alcohol; 0.344 μg of vitamin A acetate.

A number of reports have emphasized the possibility of toxic effects as a result of the ingestion of excess amounts of vitamin A. Hypervitaminosis A may occur as a consequence of the administration of large doses (in the form of vitamin A concentrates) to infants and small children. The principal symptoms are painful joints, periosteal thickening of long bones, and loss of hair.

Various congenital defects have been produced experimentally in the offspring of rats given a single large dose (75,000–150,000 IU) of vitamin A on the ninth, tenth, or eleventh day of pregnancy. No correlation has yet been established between the results of these experiments and a similar occurrence in humans, but it seems prudent to exercise caution in giving repeated large doses of vitamin A to pregnant women.

Vitamin A deficiency can occur not only from inadequate intake but also because of poor intestinal absorption or inadequate conversion of provitamin A, as occurs in diseases of the liver. In such cases, a high plasma carotene content may coincide with a low vitamin A level.

It is important to point out that the fat-soluble group of vitamins are poorly absorbed from the intestine in the absence of bile. For this reason any defect in fat absorption is likely to foster deficiencies of fat-soluble vitamins as well.

Chemical and physical methods of determination of vitamin A are based on spectrophotometric measurements. Vitamin A_1 absorbs maximally at

610–620 nm and A_2 at 692–696 nm. A colorimetric determination of vitamin A utilizes the Carr-Price reaction, in which a blue color is obtained when a solution of antimony trichloride in chloroform is added to the vitamin-containing mixture. This reaction may be used to determine the vitamin A content of blood plasma. However, plasma vitamin A levels alone are not satisfactory as a means of detecting early deficiency of the vitamin because the levels are maintained at or near normal until there is advanced depletion of vitamin A. When lesions of the eyes are apparent, plasma levels of vitamin A are very low (5 μg/100 ml). Where it has been possible to measure the levels of vitamin A in the liver, it has been found that in malnourished children the normally large amounts of vitamin A stored in the liver are virtually exhausted, levels of less than 15 μg/g fresh liver tissue being reported.

THE VITAMINS D

The vitamins D are actually a group of compounds. All are sterols which occur in nature, chiefly in the animal organism. Certain of these sterols (known as provitamins D), when subjected to long-wave ultraviolet light (about 265 nm), acquire the physiologic property of curing or preventing rickets, a disease characterized by skeletal abnormalities, including failure of calcification.

Although all of the vitamins D possess antirachitic properties, there is a considerable difference in their potency when tested in various species. For example, irradiated ergosterol (vitamin D_2) is a powerful antirachitic vitamin for man and for the rat but not for the chicken. Vitamin D_3, on the other hand, is much more potent for the chicken than for the rat or the human organism.

Chemistry

For nutritional purposes the 2 most important D vitamins are D_2 (activated ergosterol; also known as ergocalciferol or viosterol) and D_3 (activated 7-dehydrocholesterol, cholecalciferol), the form which occurs in nature in the fish liver oils. Provitamin D_2 (ergosterol) occurs in the plant kingdom (eg, in ergot and in yeast).

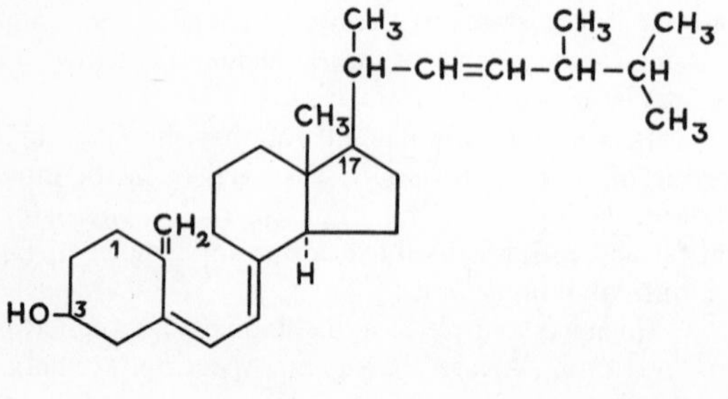

Vitamin D_2 (ergocalciferol)

The structure of vitamin D_3 is the same as that of D_2 given above except that the side-chain on position 17 is that of cholesterol:

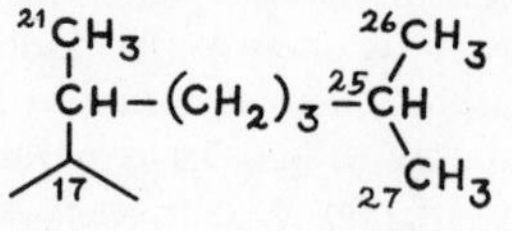

Vitamin D_3 (cholecalciferol)
(structure of side-chain)

Man and other mammals can synthesize provitamin D_3 in the body. It is believed that the vitamin is then activated in the skin by exposure to ultraviolet rays and carried to various organs in the body for utilization or storage (in liver).

Physiologic Role

The principal action of vitamin D is to increase the absorption of calcium and phosphorus from the intestine. The vitamin also has a direct effect on the calcification process. Evidence for this has been obtained by isotopic tracer studies, which indicate that the administration of vitamin D to animals deficient in this vitamin increases the rate of accretion and resorption of minerals in bone.

Vitamin D also influences the handling of phosphate by the kidney. In animals deficient in vitamin D, the excretion of phosphate and its renal clearance are decreased; in parathyroidectomized animals, vitamin D increases the clearance of phosphate and promotes lowering of the serum phosphate concentration.

As was noted above, cholecalciferol (vitamin D_3, activated 7-dehydrocholesterol) is the naturally occurring form of vitamin D which is present in dietary sources of the vitamin. However, this is not the active form of the vitamin in the tissues. According to Lund & DeLuca (1966), 25-hydroxycholecalciferol (OH-D_3) is the active metabolite. This compound is formed by a mitochondrial enzyme in the liver. More recently (Haussler & others, 1971), a vitamin D metabolite has been identified in the intestinal tissue which on a weight basis is at least 5 times more effective than cholecalciferol. This metabolite appears to be the active form of vitamin D operating on the calcium transport system in the intestine. This form of vitamin D, which functions in connection with intestinal transport of calcium, is 1,25-dihydroxycholecalciferol (1,25-OHCC), which is produced in the kidney from 25-hydroxycholecalciferol originally formed in the liver from cholecalciferol. The 1,25-dihydroxycholecalciferol is then transported to the intestinal mucosal cells, where it is active in calcium transport. It is suggested that in chronic renal disease 1,25-OHCC may not be formed. This would account for signs of vitamin D resistance in such patients and a resultant lowering of serum calcium together with increased production of parathyroid hormone. A 21,25-dihydroxycholecalciferol has also been detected; this form of vitamin D acts preferentially on bone and only slightly on the intestine.

Administration of vitamin D to chicks with a vitamin D deficiency produces an increase in biosynthesis of proteins necessary for mobilization and transport of calcium ion at the brush border side of intestinal epithelium. It is hypothesized that the active form of vitamin D induces formation of the messenger RNA (mRNA) for the synthesis of calcium-binding proteins that comprise the calcium ion transport system in the intestinal mucosal cells. The effect of vitamin D on the intestine is blocked by the administration of dactinomycin, an observation which supports the concept that the vitamin is acting to stimulate mRNA synthesis. Vitamin D is thus pictured as functioning in the regulation of Ca^{++} absorption from the intestine by controlling the expression of genetic information.

It has been suggested that the action of vitamin D on calcium metabolism is not confined to certain organs but rather is generalized, and further that the vitamin controls translocation of divalent cations in a number of tissues that are particularly concerned with the turnover of these cations. The distribution of radioactively labeled vitamins D_2 and D_3 supports the concept of a generalized or systemic action for the vitamin. It is found in liver, small intestine mucosal cells, the membranes of the heart and of striated muscle, proliferating chondrocytes, and the epiphyseal plates of long bones. In homogenates of liver, kidney, and small intestine, the vitamin was found in the microsomal fraction.

One unit of vitamin D (1 USP unit or IU) is defined as the biologic activity of 0.025 μg of ergocalciferol. Substances intended for human nutrition are biologically assayed for vitamin D_2 by the "line test." Twenty-eight-day-old rats are put on a rachitogenic diet, characterized by a high cereal content and a high (4:1) ratio of calcium to phosphorus, until depleted of vitamin D (18–25 days). For the following 8 days one group, the **reference group**, is given daily test doses of cod liver oil of known potency (the USP Reference Oil) in a quantity sufficient to produce a narrow continuous line of calcification across the metaphysis of the tibia. This degree of healing is designated in the experimental protocol as "unit" or 2-plus healing. Other groups of rachitic test animals are given the test substance in varying amounts for a similar 8-day test period. All animals are sacrificed 10 days after the reference or test samples were first administered and the tibial bones examined for the degree of healing. The potency of the test material is calculated from the quantity required to produce healing equivalent to that of the reference sample. Substances intended for the chick, which responds better to vitamin D_3, are assayed by a method which determines the amount of ash in the bones of growing chicks which have been fed with the test materials.

Physical (spectrophotometric) methods are also used in assay of the vitamins D.

Sources & Daily Allowance

In its active form vitamin D is not well distributed in nature, the only rich sources being the liver and viscera of fish and the liver of animals which feed on fish. However, certain foods, such as milk, may have their vitamin D content increased by irradiation with ultraviolet light.

For infants and children, a requirement of 400 IU/day has been suggested. Although the adult requirement is not known, 400 IU have been proposed for women during pregnancy and lactation as well as for other individuals of both sexes up to age 22 (see Table 21–3). The ingestion of large amounts of vitamin D has been shown to cause toxic reactions and widespread calcification of the soft tissues including lungs and kidneys. The quantities of the vitamin required to induce a state of hypervitaminosis are not obtainable from natural sources, and so this fact is of importance only when massive doses of vitamin D are given for therapeutic purposes.

THE VITAMINS E

Chemistry

Compounds possessing vitamin E activity are known chemically as tocopherols. There are 3 such substances, designated as α-, β-, and γ-tocopherols.

The structure of α-tocopherol is shown below:

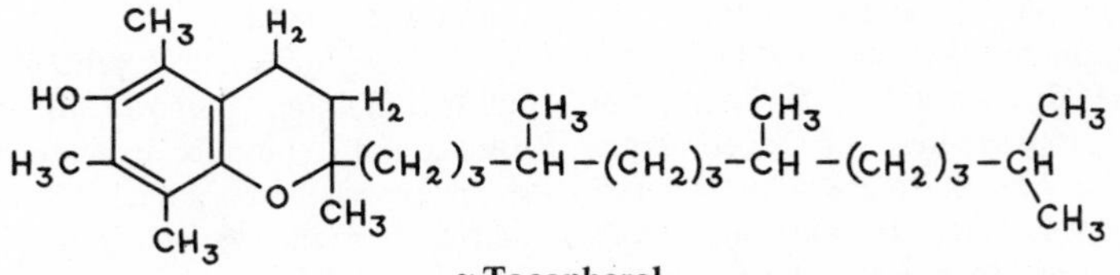

$$-(CH_2)_3-CH(CH_3)-(CH_2)_3-CH(CH_3)-(CH_2)_3-CH(CH_3)_2$$

α-Tocopherol

A deficiency of vitamin E in rats and some other animals causes resorption of the fetus in the female and, in the male, atrophy of spermatogenic tissue and permanent sterility. The susceptibility to hemolysis of erythrocytes treated in vitro with dilute solutions of hydrogen peroxide has been used as a test of vitamin E deficiency in humans. By this test it was shown that erythrocytes of full-term and, notably, of premature infants often had an increased susceptibility to hemolysis which could be reversed by administering vitamin E. It was later found that plasma levels of tocopherol may be low in newborn infants. In studies with adult male subjects on diets containing about 3 mg of α-tocopherol, deficiencies of vitamin E developed slowly as evidenced by the erythrocyte hemolysis test and direct measurements of plasma tocopherol levels. Depletion appeared to be hastened when the daily intake of polyunsaturated fatty acids, particularly linoleic acid, was increased. As a result of these and other studies, it has been concluded that a deficiency of vitamin E can occur in otherwise normal humans, and, further, that the intake of polyunsaturated fatty acids is the single most important factor in the determination of the requirement for vitamin E under normal circumstances.

In several studies of malnourished children in whom plasma tocopherol levels were low, it was found also that macrocytic anemia and a decreased erythrocyte survival time were associated abnormalities. After the administration of vitamin E, there were increases in plasma tocopherol, reticulocytosis, and disappearance of the anemia. Similar observations have been made in premature infants with hemolytic anemia.

Although vitamin E occurs in many foods, absorption of the vitamin from the intestine may be impaired in abnormal states characterized by malabsorption, particularly of lipids, as is the case with the other fat-soluble vitamins.

Physiologic Role

The most striking chemical characteristic of the vitamins E is their antioxidant property. Polyunsaturated fatty acids are easily attacked by molecular oxygen, resulting in formation of peroxides. The tocopherols prevent this. Indeed, some believe that the deleterious effects of vitamin E deficiency are related to the accumulation of fatty acid peroxides in the tissues. The relationship of vitamin E requirements to the unsaturated fatty acid dietary intake is thus also explainable.

It has been suggested that vitamin E and other antioxidants obtained from the diet, such as vitamin C, may be important in inhibiting damage to lung tissue from oxidants in the air such as may be present in smog-contaminated atmospheres. In experimental studies, rats deficient in vitamin E were more damaged by ozone and nitrogen dioxide, which are among the oxidants in polluted air, than were those animals supplemented with vitamin E. Lipid peroxidation appears to be a damaging mechanism in ozone toxicity.

In some animal species, a lack of vitamin E produces muscular dystrophy. Such dystrophic muscles exhibit increased respiration (oxygen uptake). Treatment with tocopherol reduces the oxygen uptake of such tissue. However, vitamin E has not been shown to benefit any type of muscular dystrophy seen in man. In fact, despite the occurrence of a number of disorders in experimental animals maintained on vitamin E-deficient diets, no comparable clinical signs have been conclusively shown to result from a deficiency of vitamin E in man although there are some reports which provide suggestive evidence that this may be the case.

The placental transfer of vitamin E is limited; mammary transfer is much more extensive. Thus the serum α-tocopherol level of breast-fed infants is increased more rapidly than that of bottle-fed infants. Furthermore, intake of vitamin E by the mother during her pregnancy is variable. As a result, the vitamin E nutriture of very young children could be inadequate. A relationship between vitamin E requirement and the quantity of unsaturated fats taken in the diet has been proposed. If such a relationship exists, it might be supposed that unsaturated fats in infant diets would further increase the need for vitamin E. Therefore, it is of interest that a group of infants fed a diet wherein the content of unsaturated fats was increased did develop anemia, edema, and certain changes in the skin. These changes, as well as a hemolytic anemia reported to have occurred in premature infants, responded to the administration of vitamin E.

Pathologic states characterized by malabsorption—eg, steatorrhea, cystic fibrosis, biliary atresia, nontropical sprue, and chronic pancreatitis—are reported to be associated with evidence of vitamin E deficiency as indicated by creatinuria, low levels of serum tocopherol, and increased hemolysis of erythrocytes. As a result of dietary surveys and studies of plasma tocopherol levels as well as resistance of erythrocytes to hemolysis, it is reported that there is widespread evidence for suboptimal vitamin E nutriture among the poorly nourished peoples of the world.

The level of tocopherol in the plasma after oral administration of DL-α-tocopheryl acetate has been measured. Single doses of 200, 400, 500, and occasionally of 100 mg were effective in increasing the free tocopherol levels of the plasma to a significant degree after 6 hours. Repeated daily oral doses produced somewhat greater maximum increases than did single doses. However, the parenteral administration of vitamin E failed to increase the free tocopherol level of the plasma regardless of the type of compound (free alcohol or monosodium phosphate) and the type of vehicle (oil or water) used.

Certain diets low in protein and especially in the sulfur-containing amino acids (particularly cystine) were found to produce an acute massive hepatic necrosis in experimental animals (Schwartz, 1954). A vitamin E deficiency enhances the effects of such diets, whereas added vitamin E exerts a preventive action upon the necrosis. Rats which have been kept on the deficient ration develop the fatal hepatic lesion suddenly (within a few hours or days) after a symptom-free latent period which averages 45 days. However, the occurrence of a metabolic defect in the livers of these animals can be demonstrated several weeks before the development of the necrotic lesion itself. Liver slices from these animals which are still histologically normal are able to respire in the Warburg apparatus for only 30–60 minutes; subsequently, oxygen consumption declines as incubation continues. The 3 major metabolic pathways for the utilization of acetate by the liver, viz, ketogenesis, lipogenesis, and oxidation to CO_2, are also deficient in the prenecrotic liver slice, probably as a result of the respiratory defect. If the diet is supplemented with cystine, vitamin E, or preparations of "factor 3" (see p 285), both the metabolic and the histologic lesions are prevented. A reversal of the respiratory decline in the necrotic liver can also be prevented by direct infusion of tocopherols into the portal vein.

Factor 3 has been identified as a selenium compound (Schwartz, 1957), and it is true that selenite gives complete protection against dietary liver necrosis in rats. However, the respiratory decline observed in the liver slices from rats kept on the deficient diet, as described above, is only partially prevented by supplementation with factor 3, whereas the liver slices from animals whose diets have been supplemented with vitamin E show no decline whatever (Corwin, 1959). It now seems clear that the apparent potency of cystine in preventing dietary liver necrosis was in reality due to contamination with traces of factor 3-active selenium. It has therefore been concluded that dietary liver necrosis is the result of a simultaneous lack of factor 3-selenium and of vitamin E. A lack of one or the other alone produces relatively mild chronic diseases, but a simultaneous deficiency of both leads to severe tissue damage and death (Schwartz, 1960). A number of lesions hitherto attributed solely to vitamin E deficiency may actually be of dual origin, and the presence or absence of factor 3-active selenium may determine the fate of an animal on a vitamin E-deficient diet. However, some diseases are apparently caused entirely by a deficiency of vitamin E (eg, resorption sterility in rats and encephalomalacia in chicks), and there seem also to be other diseases which are little affected by administration of vitamin E but are cured by small supplements of factor 3-active selenium (see also p 411).

Sources & Requirements

The α-tocopherol content of foods has been measured, using modern analytical technics, by Bunnell & others (1965). Good sources of vitamin E include eggs, muscle meats, liver, fish, chicken, oatmeal, the oils of corn, soya, and cottonseed, and products made with such oils such as margarine and mayonnaise. It is of interest, however, that foods which were fried in vegetable oils and then frozen were low in tocopherol, indicating substantial losses of tocopherol during freezer storage. Concentrates of natural tocopherols are prepared by molecular distillation of wheat germ oil, which is particularly rich in vitamin E.

Bunnell & others calculated that in the "average" American diet the daily intake of α-tocopherol ranged from 2.6–15.4 mg, with an average of 7.4 mg. This contrasts with a recommended daily allowance (Table 21–3) of 25–30 mg in adults.

Vitamin E was originally measured by a biologic assay based on the ability of the test material to support gestation when the pregnant rat is maintained on a vitamin E-deficient diet. A chemical method that permits the estimation of 2–5 μg of vitamin E has also been described (Nair, 1956). By this method, the vitamin E content of human blood was found to range from 361–412 μg/100 ml. At present, analyses for vitamin E may utilize paper, column, and gas-liquid chromatography. These technics permit more accurate and specific assays of α-tocopherol than were possible by the older methods. In particular, these more modern assay procedures differentiate between α-tocopherol and other less active or completely inactive forms of tocopherol.

Recommended daily allowances for vitamin E are listed in Table 21–3. One mg of DL-α-tocopheryl acetate is equal to one International Unit (IU) of vitamin E.

Herting (1966) has reviewed the nutritional status of vitamin E and concluded that deficiencies of this vitamin may be much more common than has previously been believed.

THE VITAMINS K

Chemistry

A large number of chemical compounds which are related to 2-methyl-1,4-naphthoquinone possess some degree of vitamin K activity.

The naturally occurring vitamins K possess a phytyl radical on position 3 (vitamin K_1) or a difarnesyl radical (K_2).

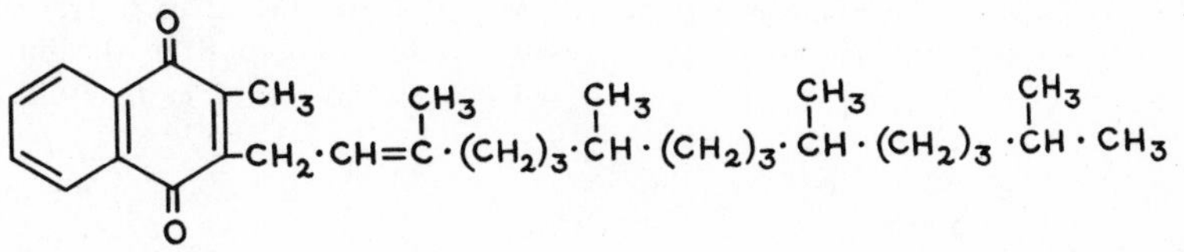

Vitamin K_1; phytonadione (phylloquinone; Mephyton (2-methyl-3-phytyl-1,4-naphthoquinone)

Several synthetic compounds containing the 2-methyl-1,4-naphthoquinone structure, such as menadione (shown below), exhibit vitamin K activity. This suggests that this portion of the molecule is essential for the formation of a second substance which actually exerts the biologic effects of the vitamin.

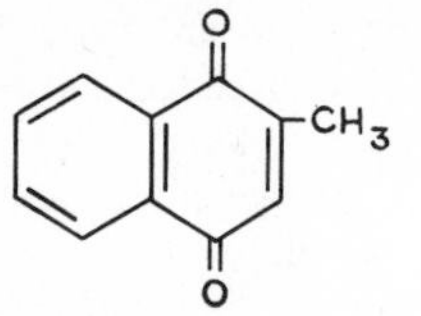

Menadione (2-methyl-1,4-naphthoquinone)

Physiologic Role

The best known function of vitamin K is to catalyze the synthesis of prothrombin by the liver. In the absence of vitamin K a hypoprothrombinemia occurs in which blood clotting time may be greatly prolonged. It must be emphasized that the effect of vitamin K in alleviation of hypoprothrombinemia is dependent upon the ability of the hepatic parenchyma to produce prothrombin. Advanced hepatic damage, as in carcinoma or cirrhosis, may be accompanied by a prothrombin deficiency which cannot be relieved by vitamin K.

The activities of several plasma thromboplastic factors are reduced in states of vitamin K deficiency or after administration of vitamin K antagonists such as bishydroxycoumarin (Dicumarol). The cause of delayed clotting in vitamin K deficiency states is therefore not confined to a prothrombin deficiency, although this is perhaps the most important factor.

Although, as noted above, vitamin K is required for the synthesis of prothrombin, it has never been proved whether or not this is a specific effect. Possibly it is only a manifestation of a more fundamental role for the vitamin. It is known that vitamin K_1 is an essential component of the phosphorylation processes involved in photosynthesis in green plants, and it may have a similar function in animal tissues, ie, that of a cofactor necessary in oxidative phosphorylation.

Vitamin K_1 is altered by the action of ultraviolet radiation. Rats fed beef sterilized by irradiation have developed vitamin K deficiency. Anderson & Dallam (1959) have reported that an impairment in the oxidative phosphorylative activity of mitochondria occurred when these cytoplasmic structures were irradiated with ultraviolet light at a wavelength of 2357 nm. A similar effect on oxidative phosphorylation was obtained by Beyer (1959) when he treated rat liver mitochondria with ultraviolet light. After addition of vitamin K_1 to the irradiated mitochondria, oxidative phosphorylation was restored almost to normal. The results of these experiments suggest that vitamin K, or a substance very closely related to it, does indeed play an important role in oxidative phosphorylation in the mitochondria.

The experiments with vitamin K_1 mentioned above are of considerable interest in the light of similar observations on the role of coenzyme Q. This biologically active quinone is so widely distributed in natural materials that it might be called "ubiquinone." The structure of coenzyme Q is shown below. It is that of a 2,3-dimethoxy-5-methylbenzoquinone with a polyisoprenoid side-chain at carbon 6. Thus far, 5 crystalline homologues that differ from one another in the number of isoprenoid units (formula: $-CH_2CH{=}C.CH_3{-}CH_2-$) in the side-chain have been obtained from various sources. For example, coenzyme Q from beef heart has 10 isoprenoid units in the side-chain (Q_{10}). From a yeast, *Saccharomyces cerevisiae,* a Q_6 was isolated; from *Torula utilis,* Q_9, etc. The function of coenzyme Q as an electron carrier in terminal electron transport and in oxidative phosphorylation has been demonstrated (Crane, 1957). (See Chapter 9.)

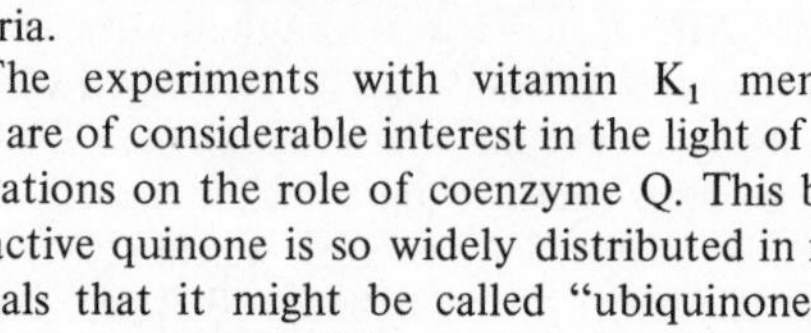

Coenzyme Q (ubiquinone)

A dietary deficiency of vitamin K is not likely to occur since the vitamin is fairly well distributed in foods and the intestinal microorganisms synthesize considerable vitamin K in the intestine. However, a deficiency may occur as a result of prolonged oral

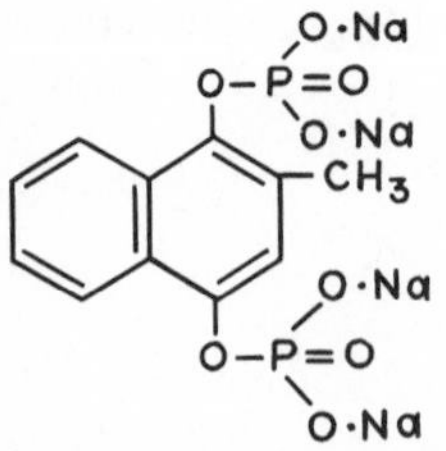

Sodium menadiol diphosphate (Synkayvite)

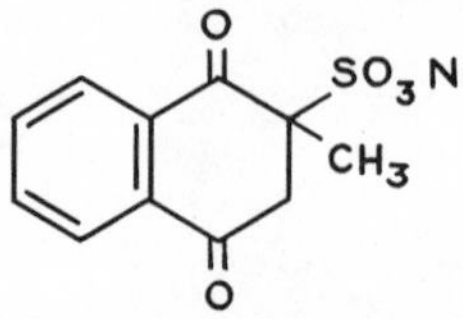

Menadione sodium bisulfite (Hykinone)

therapy with antibiotic drugs capable of suppressing vitamin K producing bacteria. Furthermore, as has already been noted for the other fat-soluble vitamins, the absorption of vitamin K from the intestine depends on the presence of bile. A deficiency state will therefore result in biliary tract obstruction or if there is a defect in fat absorption, such as in sprue and celiac disease. Short-circuiting of the bowel as a result of surgery may also foster a deficiency which will not respond even to large oral doses of vitamin K. For such situations water-soluble forms of vitamin K are available which may be absorbed even in the absence of bile. However, these derivatives are relatively ineffective in correcting the hypoprothrombinemia induced by oral anticoagulants, and they may produce some toxic manifestations in infants.

Synkayvite (sodium menadiol diphosphate) and Hykinone (menadione sodium bisulfite) are 2 water-soluble compounds with vitamin K activity. The chemical structures of these compounds are shown above.

In the immediate postnatal period the intestinal flora produces insufficient vitamin K since the intestine is sterile at birth. The quantity of the vitamin supplied by the mother during gestation is apparently not large. Thus during the first few days of life a hypoprothrombinemia may appear which will persist until the intestinal flora becomes active in the manufacture of the vitamin. This can be prevented by administering vitamin K to the mother before parturition or by giving the infant a small dose of the vitamin.

The parenteral administration to infants of too large doses of vitamin K (eg, 30 mg/day for 3 days) has been shown to produce hyperbilirubinemia in some cases. Three mg of sodium menadiol diphosphate, which is equivalent to 1 mg of vitamin K_1, is adequate to prevent hypoprothrombinemia in the newborn, and there is no danger of provoking jaundice with this dosage. The oral administration of vitamin K has not been found to produce jaundice.

coagulant drug, will usually return to normal in 12–36 hours after the administration of the vitamin provided liver function is adequate to manufacture prothrombin.

Sources

The green leafy tissues of plants are good sources of the vitamin. Fruits and cereals are poor sources. Molds, yeasts, and fungi contain very little; but since it occurs in many bacteria, most putrefied animal and plant materials contain considerable quantities of vitamin K.

WATER-SOLUBLE VITAMINS

VITAMIN C (Ascorbic Acid)

Chemistry

The chemical structure of ascorbic acid (vitamin C) resembles that of a monosaccharide.

Vitamin C is readily oxidized to the dehydro form. Both forms are physiologically active, and both are found in the body fluids. The enediol group of ascorbic acid (from which removal of hydrogen occurs to produce the dehydro form, as shown in the formulas below) may be involved in the physiologic function of this vitamin. It is conceivable that this chemical grouping functions in a hydrogen transfer system; a role of the vitamin in such a system, ie, the oxidation of tyrosine, is described below.

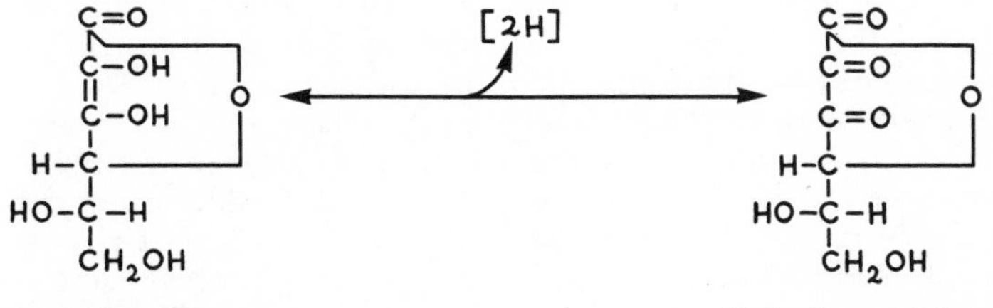

```
      C=O                                  C=O
      C-OH  ┐                              C=O  ┐
      ||    O   <-------[2H]------->       |    O
      C-OH  ┘                              C=O  ┘
   H-C─────┘                            H-C────┘
  HO-C-H                               HO-C-H
      CH2OH                                CH2OH
```

Ascorbic acid (reduced form) **Dehydroascorbic acid**

Uncontrollable hemorrhage is a symptom of vitamin K deficiency. The newborn child may bleed into the adrenal, brain, and gastrointestinal tract, and from the umbilical cord. In the adult, hemorrhage may also occur, most commonly after an operation on the biliary tract.

An important therapeutic use of vitamin K is as an antidote to the anticoagulant drugs such as bishydroxycoumarin (Dicumarol; see Fig 10–3). For this purpose large doses of vitamin K_1 may be used, either orally or, as an emulsion, intravenously. The prothrombin time, which is lengthened by the use of the anti-

The reducing action of ascorbic acid is the basis of the chemical determination of the compound. In most plant and animal tissues this is the only substance which exhibits this reducing action in acid solution. One of the most widely used analytic reactions for vitamin C is the quantitative reduction of the dye, 2,6-dichlorophenolindophenol, to the colorless leuco base by the reduced form of ascorbic acid. The method has been adapted to the microdetermination of blood ascorbic acid, so that only 0.01 ml of serum is needed for assay. Dehydroascorbic acid can be determined colorimetrically by the formation of a hydrazone with

2,4-dinitrophenylhydrazine. This method may also be used for the assay of total vitamin C after conversion of the reduced form.

Physiologic Role

Although ascorbic acid is undoubtedly widely required in metabolism, it can be synthesized in a variety of plants and in all animals studied except man and other primates and the guinea pig. The pathway of biosynthesis in animals (which is not the same as that of plants) is shown in Fig 13–15. Those animals which are unable to synthesize the vitamin presumably lack the enzyme system necessary to convert L-gulonic acid to ascorbic acid. In this sense scurvy may be considered to be the result of an inherited defect in carbohydrate metabolism (Burns, 1959).

Studies with L-ascorbic acid labeled in the various positions with isotopic carbon 14 have shown that the vitamin is extensively oxidized to respiratory CO_2 in rats and guinea pigs but that this is not the case in man. Correspondingly, ascorbic acid disappears slowly in man: it has a half-life of about 16 days in man compared to a half-life of about 4 days in the guinea pig. This correlates well with the fact that it takes 3–4 months for scurvy to develop in man on a diet containing no vitamin C, while the guinea pig becomes scorbutic in about 3 weeks.

L-Ascorbic acid-1-^{14}C is converted to labeled urinary oxalate in man, guinea pigs, and rats. In man, conversion of ascorbic acid to oxalate may account for the major part of the endogenous urinary oxalate (Hellman, 1958).

Severe ascorbic acid deficiency produces scurvy. The pathologic signs of this deficiency are almost entirely confined to supporting tissues of mesenchymal origin (bone, dentine, cartilage, and connective tissue). Scurvy is characterized by failure in the formation and maintenance of intercellular materials, which in turn causes typical symptoms, such as hemorrhages, loosening of the teeth, poor wound healing, and the easy fracturability of the bones.

The biochemical function of ascorbic acid is still not known. Probably the most clearly established functional role of the vitamin is in maintaining the normal intercellular material of cartilage, dentine, and bone, as mentioned above. There is increasing experimental evidence for a specific role of ascorbic acid in collagen synthesis, with special reference to the synthesis of hydroxyproline from a proline precursor. There are also a number of reports of a possible function of ascorbic acid in oxidation-reduction systems, coupled with glutathione, cytochrome c, pyridine nucleotides, or flavin nucleotides. The vitamin has been reported to be involved in the oxidation of tyrosine and in the metabolism of adrenal steroids and of various drugs. However, its role in these reactions does not seem to be specific because it can usually be replaced by other compounds having similar redox properties.

The adrenal cortex contains a large quantity of vitamin C, and this is rapidly depleted when the gland is stimulated by adrenocorticotropic hormone. A similar depletion of adrenocortical vitamin C is noted when experimental animals (guinea pigs) are injected with large quantities of diphtheria toxin. Increased losses of the vitamin accompany infection and fever. These losses are particularly notable when bacterial toxins are present. All of these observations suggest that the vitamin may play an important role in the reaction of the body to stress.

Sources & Daily Allowances

The infant is usually well supplied with vitamin C at birth. However, infants 6–12 months of age who are fed processed milk formulas not supplemented with fruits and vegetables are very susceptible to the development of infantile scurvy. Adult cases appear from time to time, particularly in patients studied in municipal hospitals who may be living in depressed areas of cities. Elderly bachelors and widowers who may prepare their own foods are particularly prone to the development of vitamin C deficiency, a syndrome termed "bachelor scurvy." Food faddists may also develop vitamin C deficiencies if their diet avoids raw foods, particularly fruits and vegetables.

The best food sources of vitamin C are citrus fruits, berries, melons, tomatoes, green peppers, raw cabbage, and leafy green vegetables, particularly salad greens. Fresh (but not dehydrated) potatoes, while only a fair source of vitamin C on a per gram basis, constitute an excellent source in the average diet because of the quantities which are commonly consumed.

The vitamin is easily destroyed by cooking, since it is readily oxidized. There may also be a considerable loss in mincing of fresh vegetables such as cabbage, or in the mashing of potatoes. Losses of vitamin C during the storage and processing of foods are also extensive, particularly where heat is involved. Traces of copper and other metals accelerate this destruction.

The tissues and body fluids contain varying amounts of vitamin C. With the exception of muscle, the tissues of the highest metabolic activity have the highest concentration. Fasting individuals who are given liberal quantities of vitamin C (75–100 mg/day) have serum ascorbic acid levels of 1–1.4 mg/100 ml. Those on diets which provide only 15–25 mg/day will have serum levels correspondingly lower: 0.1 to 0.3 mg/100 ml. When the blood levels of ascorbic acid exceed 1–1.2 mg/100 ml, excretion of the vitamin occurs readily. For this reason the intravenous administration of vitamin C is usually attended by a considerable urinary loss.

The recommended daily intakes of ascorbic acid for infants, children, and adults are listed in Table 21–3.

An ascorbic acid tolerance test which is useful for clinical diagnosis of vitamin C deficiency has been described (Dutra de Oliveira, 1959).

THE VITAMINS OF THE B COMPLEX

(1) **Thiamine**: Vitamin B_1, antiberiberi substance, antineuritic vitamin, aneurine.

(2) **Riboflavin**: Vitamin B_2, lactoflavin.

(3) **Niacin**: P-P factor of Goldberger, nicotinic acid.

(4) **Pyridoxine**: Vitamin B_6, rat antidermatitis factor.

(5) **Pantothenic acid**: Filtrate factor, chick antidermatitis factor.

(6) **Lipoic acid**: Thioctic acid, protogen, acetate replacement factor.

(7) **Biotin**: Vitamin H, anti-egg white injury factor.

(8) **Folic acid group**: Liver *Lactobacillus casei* factor, vitamin M, *Streptococcus lactis* R (SLR) factor, vitamin B_C, fermentation residue factor, pteroylglutamic acid.

(9) **Inositol**: Mouse anti-alopecia factor.

(10) ***p*-Aminobenzoic acid (PABA).**

(11) **Vitamin B_{12}**: Cyanocobalamin, cobamide, anti-pernicious anemia factor, extrinsic factor of Castle.

THIAMINE

Chemistry

The crystalline vitamin, thiamine hydrochloride ($C_{12}H_{17}ClN_4OS.HCl$), is a 2,5-dimethyl-6-amino pyrimidine combined with 4-methyl-5-hydroxyethyl thiazole. It is shown below as the chloride hydrochloride.

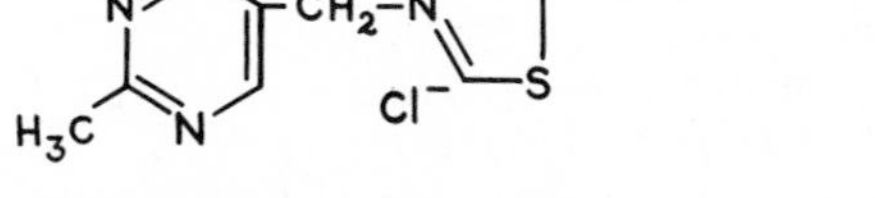

NH$_2$·HCl CH$_3$ CH$_2$–CH$_2$–OH N CH$_2$–N$^+$ Cl$^-$ S H$_3$C N

Physiologic Role

Thiamine, in the form of thiamine diphosphate (thiamine pyrophosphate), is the coenzyme for the decarboxylation of α-keto acids such as pyruvic acid or α-ketoglutaric acid. As such it is often referred to as cocarboxylase. The decarboxylation reaction with pyruvic acid is as follows:

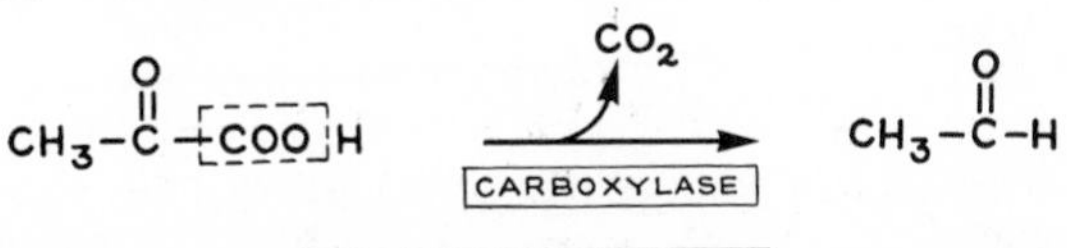

$$CH_3-\overset{O}{\overset{\|}{C}}-COOH \xrightarrow[\text{CARBOXYLASE}]{CO_2} CH_3-\overset{O}{\overset{\|}{C}}-H$$

Pyruvic acid | **Thiamine diphosphate (cocarboxylase)** | Acetaldehyde

This reaction as it occurs in yeasts is one of "straight" decarboxylation or simple removal of CO_2. It results in the production of acetaldehyde, which is reduced subsequently to ethyl alcohol. In animal tissues the decarboxylation of pyruvic acid results in the formation of acetyl-Co A (active acetate), which is an oxidation product of acetaldehyde. This reaction is therefore referred to as "oxidative" decarboxylation. It involves not only thiamine but also participation of other coenzymes, specifically lipoic acid, coenzyme A, flavin adenine dinucleotide (FAD), and nicotinamide adenine dinucleotide (NAD) (Fig 7–3).

Thiamine diphosphate is also a coenzyme in the reactions of transketolation which occur in the direct oxidative pathway for glucose metabolism (see p 241). The operation of this pathway in erythrocytes from thiamine-deficient rats is markedly retarded at the transketolase step so that pentose sugars accumulate to levels 3 times normal (Brin, 1958). The biochemical defect appears before growth ceases in the thiamine-deficient animal, and the defect can be significantly alleviated by addition of thiamine or of cocarboxylase to the cells in vitro or by the intraperitoneal injection of thiamine in vivo.

Thiamine deficiency affects predominantly the peripheral nervous system, the gastrointestinal tract, and the cardiovascular system. Thiamine has been shown to be of value in the treatment of beriberi, alcoholic neuritis, and the neuritis of pregnancy or of pellagra. Beriberi occurs in endemic form where polished milled rice is a staple food. The disease is still an important public health problem in South and East Asia, especially in the Philippines, Viet Nam, Thailand, and Burma.

In certain fish there is a heat-labile enzyme which destroys thiamine. Attention was drawn to this "thiaminase" by the appearance of "Chastek paralysis" in foxes fed a diet containing 10% or more of uncooked fish. The disease is characterized by anorexia, weakness, progressive ataxia, spastic paraplegia, and hyperesthesia. The similarity between the focal lesions of the nervous system in this paralysis in the fox and the lesions seen in Wernicke's syndrome in man have lent support to the concept that the latter is in part attributable to thiamine deficiency.

Chemical or microbiologic methods are used to determine thiamine in foods or in body fluids. The chemical procedures are based on conversion of thiamine to a compound which fluoresces under ultraviolet illumination. Quantitative measurement of the vitamin may then be accomplished with a photofluorometer. The older methods were based on the conversion of thiamine by oxidation to the fluorescent compound, thiochrome. In a procedure suggested by Teeri (1952), the vitamin reacts with cyanogen bromide to form a highly fluorescent compound.

For detection of thiamine deficiency in man, a determination of the amount of thiamine excreted in 4 hours may be used. This is sometimes modified to include the prior administration of a test dose of thiamine, and the percentage of the test dose which is excreted in the urine is observed (thiamine load test). Such studies may differentiate between persons with very high or moderate to low thiamine intakes, but their principal value in individual cases is to rule out thiamine deficiency. Another diagnostic test for thiamine deficiency is based on the measurement of the ratio of lactic to pyruvic acids in the blood after administration of glucose. Blood and urinary pyruvic acid levels are characteristically elevated in thiamine deficiency, as would be expected from the role of thiamine in pyruvic acid metabolism; but abnormal blood lactic acid-pyruvic acid ratios are said to be more specific indicators of vitamin B_1 deficiency than the levels of pyruvic acid alone.

Sources & Daily Allowances

Thiamine is present in practically all of the plant and animal tissues commonly used as food, but the content is usually small. Among the more abundant sources are unrefined cereal grains, liver, heart, kidney, and lean cuts of pork. With improper cooking the thiamine contained in these foods may be destroyed. Since the vitamin is water-soluble and somewhat heat-labile, particularly in alkaline solutions, it may be lost in the cooking water. The enrichment of flour, bread, corn, and macaroni products with thiamine has increased considerably the availability of this vitamin in the diet. On the basis of the average per capita consumption of flour and bread in the USA, as much as 40% of the daily thiamine requirement is now supplied by these foods.

It is difficult to fix a single requirement for vitamin B_1. The requirement is increased when metabolism is heightened, as in fever, hyperthyroidism, increased muscular activity, pregnancy, and lactation. There is also a relationship to the composition of the diet. Fat and protein reduce–while carbohydrate increases–the quantity of the vitamin required in the daily diet. It is also possible that some of the thiamine synthesized by the bacteria in the intestine may be available to the organism. Deficiencies of thiamine are likely not only in persons with poor dietary habits, or in the indigent, but also in many patients suffering from organic disease.

A detailed statement of thiamine requirements will be found in Table 21–3.

In a population subsisting on a high-carbohydrate (rice) diet low in fat and protein, an average daily intake of about 0.2 mg/1000 Calories is associated with widespread beriberi. Mild polyneuritis was produced in 2 women maintained on a daily thiamine intake of 0.175 mg/1000 Calories for a period of about 4 months.

RIBOFLAVIN

The existence of a water-soluble, yellow-green, fluorescent pigment in milk whey was noted as early as 1879; but this substance, riboflavin, was not isolated in pure form until 1932. At that time it was shown to be a constituent of oxidative tissue-enzyme systems and an essential growth factor for laboratory animals.

Riboflavin is relatively heat-stable but sensitive to light. On irradiation with ultraviolet rays or visible light it undergoes irreversible decomposition.

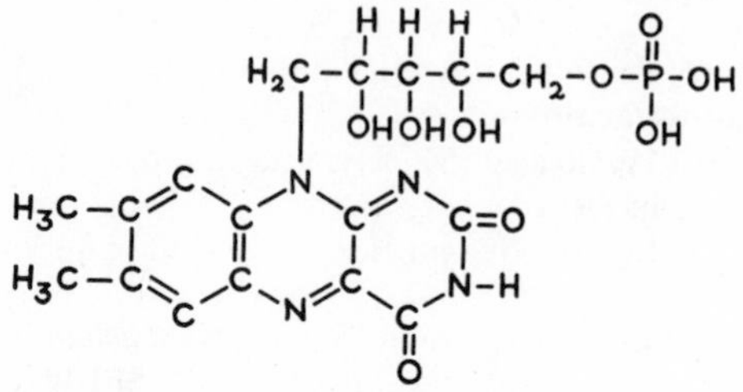

6,7-Dimethyl-9 (D-ribityl-5-phosphate)-isoalloxazine

Riboflavin phosphate (riboflavin mononucleotide)

Physiologic Role

Riboflavin is a constituent of several enzyme systems which are involved in intermediary metabolism. These enzymes are called flavoproteins. Riboflavin acts as a coenzyme for hydrogen transfer in the reactions catalyzed by these enzymes. In its active form riboflavin is combined with phosphate. This phosphorylation of riboflavin occurs in the intestinal mucosa as a condition for its absorption.

Two forms of riboflavin are known to exist in various enzyme systems. The first, riboflavin phosphate (riboflavin mononucleotide), is a constituent of the Warburg yellow enzyme, cytochrome c reductase, and the amino acid dehydrogenase for the naturally occurring L-amino acids. The other form is flavin adenine dinucleotide (FAD; see Fig 7–2), which contains 2 phosphate groups and adenine as well as ribose and ribitol. FAD is the prosthetic group of diaphorase, the D-amino acid dehydrogenase, glycine oxidase, and xanthine oxidase, which contains also iron and molybdenum. It is also an integral part of the prosthetic group of acyl-Co A dehydrogenase, the enzyme which mediates the first oxidative step in the oxidation of fatty acids.

Characteristic lesions of the lips, fissures at the angles of the mouth (cheilosis), localized seborrheic dermatitis of the face, a particular type of glossitis (magenta tongue), and certain functional and organic disorders of the eyes may result from riboflavin deficiency. However, these are not due to riboflavin deficiency alone and may result from various other conditions.

It has been suggested that a determination of the riboflavin content of the serum is of value in the diag-

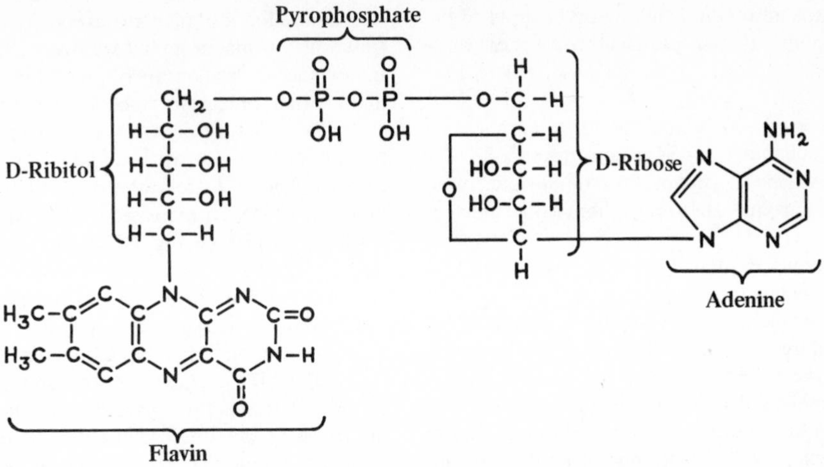

FIG 7–2. Flavin adenine dinucleotide (FAD).

nosis of riboflavin deficiencies. Methods for such determinations are given by Suvarnakich & others (1952). These authors report that the normal concentration of riboflavin in the serum is 3.16 μg/100 ml. Most of this is present as flavin adenine dinucleotide (FAD) (2.32 μg/100 ml); the remainder exists as free riboflavin (0.84 μg/100 ml). However, the relationship of blood levels of riboflavin to the amounts of the vitamin stored in the body remains to be elucidated. Urinary excretion of less than 50 μg riboflavin in 24 hours is usually associated with clinical signs of a deficiency of the vitamin.

From evidence gathered in dietary surveys, riboflavin deficiency should be among the most prevalent of the nutritional diseases attributable to lack of a vitamin. However, despite the fundamental role of riboflavin in metabolism, clinical signs of a riboflavin deficiency are relatively mild and rather nonspecific because of the frequent association of other nutritional deficiencies occurring simultaneously, such as pellagra and iron deficiency.

Sources & Daily Allowances

Riboflavin is widely distributed throughout the plant and animal kingdoms, with very rich sources in anaerobic fermenting bacteria. Milk, liver, kidney, and heart are excellent sources. Many vegetables are also good sources, but the cereals are rather low in riboflavin content. The riboflavin concentration in oats, wheat, barley, and corn is increased strikingly during germination.

Ordinary cooking procedures do not affect the riboflavin content of foods. Roasted, braised, or boiled meats retain 70–85% of the vitamin; an additional 15% is recovered in the drippings.

Unless proper precautions are taken, extensive losses of riboflavin in milk may occur during pasteurization, exposure to light in the course of bottling, or as a result of the irradiation of milk to increase its vitamin D content. Flour and bread, as a result of enrichment with crystalline riboflavin, may provide as much as 16% of the daily per capita requirement for this vitamin in the USA.

The riboflavin requirements are listed in Table 21–3.

NIACIN & NIACINAMIDE

Niacin and niacinamide are specific for the treatment of acute pellagra. It is important to remember that vitamin deficiencies seldom occur singly, as is well illustrated by patients with pellagra. Very often these patients exhibit symptoms caused by a lack of vitamins other than niacin, particularly a polyneuritis amenable to thiamine administration. Nevertheless, the dermatitis, diarrhea, dementia, stomatitis, and glossitis observed respond, often spectacularly, to niacin. Niacin is the P-P (pellagra-preventive) factor originally named by Goldberger.

Although the incidence of pellagra has declined as a result of greater diversification of the components of the diet, it still occurs in parts of the Near East, Africa, southeastern Europe, and in the USA, usually in populations subsisting on diets high in corn. Alcoholism is an important precipitating factor in some areas.

It has been shown that the amino acid tryptophan normally contributes to the niacin supply of the body. Many of the diets causing pellagra are low in good quality protein as well as in vitamins. For this reason, pellagra is usually due to a combined deficiency of tryptophan and niacin.

Niacin is not excreted to any extent as the free nicotinic acid. A small amount may occur in the urine as niacinamide or as nicotinuric acid, the glycine conjugate. By far the largest portion is excreted as methyl derivatives, viz, N-methylnicotinamide and the 6-pyridone of N-methylnicotinamide, and N-methylnicotinic acid and the glycine conjugates of these methyl derivatives. This methylation is accomplished in the liver at the expense of the labile methyl supply of the body. Methionine is the principal source of these methyl groups.

Physiologic Role

Niacinamide functions as a constituent of 2 coenzymes: coenzyme I, diphosphopyridine nucleotide (DPN); and coenzyme II, triphosphopyridine nucleotide (TPN).

The Enzyme Commission of the International Union of Biochemistry has recommended that the 2 niacinamide-containing coenzymes be redesignated by names that more accurately describe their chemical structure. In the new terminology, to be used henceforth in this book, DPN (coenzyme I) is called nicotinamide adenine dinucleotide (abbreviated form, **NAD**); TPN (coenzyme II) is called nicotinamide adenine dinucleotide phosphate **(NADP)**. The reduced form of either coenzyme is designated by the prefix **dihydro-**, eg, dihydronicotinamide adenine dinucleotide **(NADH)** for DPN.H; dihydronicotinamide adenine dinucleotide phosphate **(NADPH)** for TPN.H. The reasons for the changes are discussed by Dixon (1960).

These coenzymes, which operate as hydrogen and electron transfer agents by virtue of reversible oxidation and reduction, play a vital role in metabolism. The function of niacin in metabolism explains its great importance in human nutrition and its requirement by many other organisms, including bacteria and yeasts.

The structure of NAD is known to be a combination of niacinamide with 2 molecules of the pentose sugar, D-ribose; 2 molecules of phosphoric acid; and a molecule of the purine base, adenine. It is shown in the oxidized form (see Fig 7–3). In changing to the reduced form it accepts hydrogen and electrons.

The mechanism of the transfer of hydrogen from a metabolite to oxidized NAD, thus completing the oxidation of the metabolite and the formation of reduced NAD, is shown in the abbreviated formula below. These reactions have been studied by observing the transfer of deuterium (heavy hydrogen) from labeled ethanol, CH_3CD_2OH, as catalyzed by alcohol dehydrogenase.

$$CH_3CD_2OH + NAD^+ \rightleftharpoons CH_3CDO + NADD + H^+$$

Both the reduced NAD and the aldehyde formed have one atom of deuterium per molecule. Hence one atom of deuterium is transferred to NAD; the other

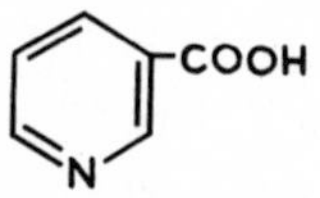

Niacin (nicotinic acid)

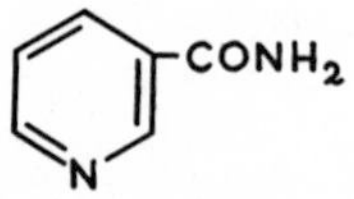

Niacinamide (nicotinamide)

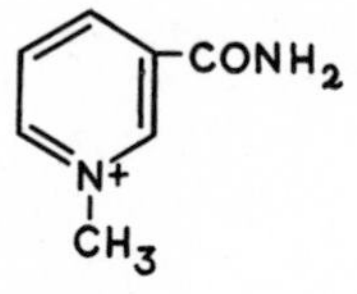

N-Methylnicotinamide

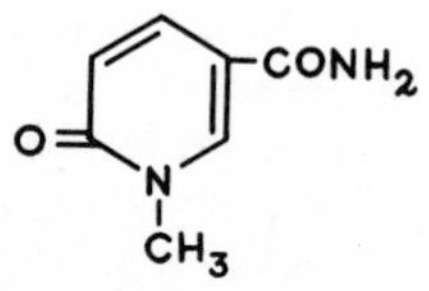

6-Pyridone-N-methylnicotinamide

Nicotinamide — Adenine — D-Ribose — D-Ribose — Pyrophosphate

*Phosphate attached here in NADP

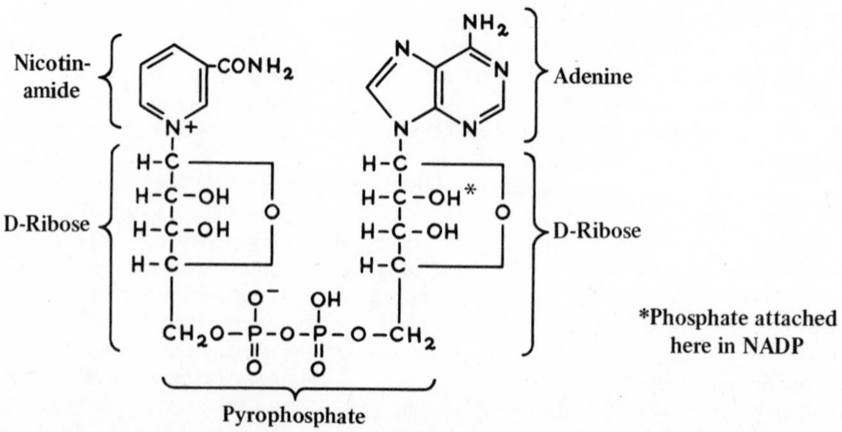

FIG 7–3. Nicotinamide adenine dinucleotide (NAD). (Oxidized form.)

remains attached to the aldehyde carbon. The H atom originally a part of the OH group of the alcohol loses an electron and enters the medium as H^+.

Reduction of NAD occurs in the **para**-position, as shown below.

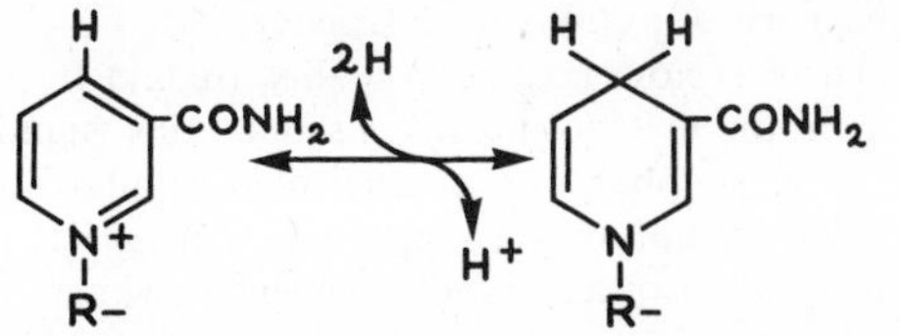

NAD^+ (oxidized) ⇌ **NADH (reduced)**

NADP differs only in the presence of one more phosphate moiety, esterified to the OH group on the second carbon of the ribose attached to the adenine (Fig 7–3). Its function is similar to that of NAD in hydrogen and electron transport. The 2 coenzymes are interconvertible.

Sources & Daily Allowances

Niacin is found most abundantly in yeast. Lean meats, liver, and poultry are good sources. Milk, tomatoes, canned salmon, and several leafy green vegetables contribute sufficient amounts of the vitamin to prevent disease, although they are not in themselves excellent sources. On the basis of the average per capita consumption, enriched bread and other enriched flour products may provide as much as 32% of the daily niacin requirement. Most fruits and vegetables are poor sources of niacin.

The recommended daily allowances for niacin are listed in Table 21–3.

The niacin requirements are influenced by the protein content of the diet because of the ability of the amino acid tryptophan to supply much of the niacin required by the body. Sixty mg of tryptophan are considered to give rise to 1 mg of niacin. There is also evidence that niacin may be synthesized by bacterial activity in the intestine and that some of this may be absorbed and utilized by the tissues. The niacin requirements specified in Table 21–3 are therefore to be considered "niacin equivalents," ie, to include both preformed niacin in the diet and that derived from tryptophan.

PYRIDOXINE

Pyridoxine was first discovered as essential for rats and named the rat antidermatitis factor or the rat antipellagra factor. Later work has shown that rats and man convert pyridoxine to other substances which far surpass pyridoxine in potency when tested with the lactobacilli or yeasts, which are used to assay foodstuffs for this vitamin. This suggested that pyridoxine is not the most active form of the vitamin in nature but that it is convertible to other derivatives which function as described below. These more active derivatives are pyridoxal and pyridoxamine phosphates. Vitamin B_6 as it occurs in nature is probably a mixture of all 3.

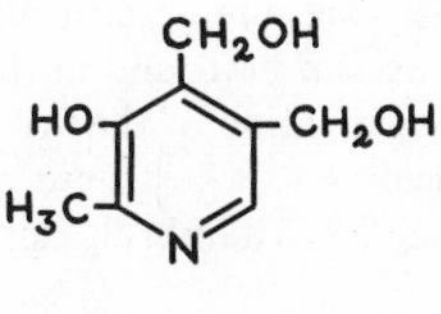

Pyridoxine 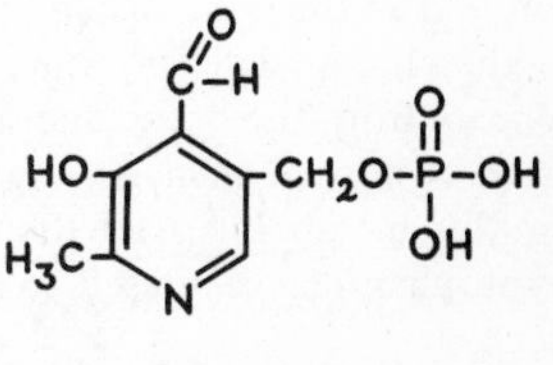 **Pyridoxal phosphate**

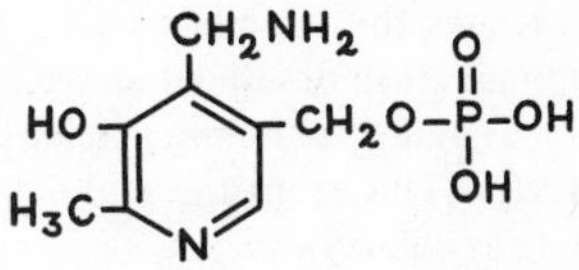

Pyridoxamine phosphate

The predominant metabolite of vitamin B_6, which is excreted in the urine either from dietary B_6 or after ingestion of any of the 3 B_6 derivatives, is 4-pyridoxic acid (2-methyl-3-hydroxy-4-carboxy-5-hydroxymethyl pyridine). This metabolite can be measured by a fluorometric method (Reddy & others, 1958).

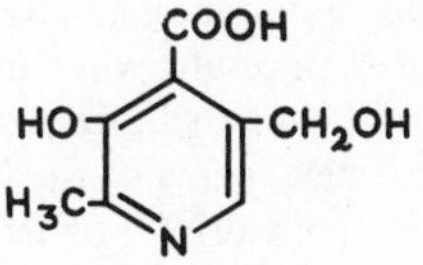

4-Pyridoxic acid

Physiologic Role

Pyridoxal phosphate is the prosthetic group of enzymes which decarboxylate tyrosine, arginine, glutamic acid, and certain other amino acids. In this way it functions as a **codecarboxylase**. The deaminases (dehydrases) for serine and threonine are also catalyzed by pyridoxal phosphate acting as coenzyme. A third and very important function of the vitamin is as a coenzyme for enzymes involved in transamination, ie, a **cotransaminase**. This function of pyridoxal phosphate probably is carried out by conversion to pyridoxamine phosphate. The reaction is reversible, so that the vitamin is actually functioning in an amino transfer system analogous to the hydrogen transfer systems described above in connection with niacinamide and riboflavin.

There is a specific relationship between vitamin B_6 and the metabolism of tryptophan because of the requirement for pyridoxal phosphate as a coenzyme for kynureninase. Failure to convert kynurenine to anthranilic acid results in the production of xanthurenic acid from kynurenine. In pyridoxine-deficient rats, dogs, swine, monkeys, and man, xanthurenic acid is found in the urine. When the vitamin is administered to the vitamin-deficient animals, xanthurenic acid disappears from the urine, and none can be found in the urine of normal animals. The examination of the urine for this metabolite after the feeding of a test dose of tryptophan has been used to diagnose vitamin B_6 deficiency.

The metabolism of cysteine is described on p 328. In these reactions, vitamin B_6 is concerned with the transfer of sulfur from methionine to serine to form cysteine. This relates the vitamin to **transulfuration** as well as to transamination described above. The removal of sulfur from cysteine or homocysteine is catalyzed by desulfhydrases. These enzymes also require pyridoxal phosphate as coenzymes.

In studies on the factors which affect the transport of amino acids into the cells, Christensen (1955) has concluded that pyridoxal participates in the mechanisms which influence intracellular accumulation of these metabolites. This may include a direct involvement of pyridoxal phosphate in the process of absorption of amino acids from the intestine (Jacobs, 1960).

It is thus apparent that vitamin B_6 is essential to amino acid metabolism in several roles: as a coenzyme for decarboxylation, deamination of serine and threonine, transamination, transulfuration, desulfuration of cysteine and homocysteine, the activity of kynureninase, and the transfer of amino acids into cells.

Vitamin B_6 is required by all animals investigated so far. Impaired growth results when immature animals are maintained on a vitamin B_6-free diet. Specific defects include acrodynia, edema of the connective tissue layer of the skin, convulsive seizures and muscular weakness in rats, and severe microcytic hypochromic anemia in dogs, swine, and monkeys accompanied by a 2- to 4-fold increase in the level of plasma iron and by hemosiderosis in the liver, spleen, and bone marrow. The anemia is not hemolytic in character, since there is no rise in icterus index or serum bilirubin. It will be recalled that pyridoxal is a coenzyme in the reaction by which α-amino-β-ketoadipic acid is decarboxylated to δ-aminolevulinic acid. The anemia of pyridoxine-deficient animals may be attributed to a defect at this point in the synthesis of heme.

Harris & others (1956) have established the fact that in humans a reversible hypochromic microcytic anemia may occur with a high serum iron similar to that observed in pyridoxine-deficient animals. An example of this type of anemia was recently described by Erslev, Lear, & Castle (1960). The patient suffered from a chronic mild hematologic disorder characterized by hypochromasia and microcytosis. He was observed intermittently for 7 years, during which time he had 8 episodes of severe hypochromic anemia associated with high serum iron levels. The anemic state was alleviated by administration of pyridoxine, and metabolic studies with tryptophan load tests supported the diagnosis of pyridoxine deficiency.

There is now no question that vitamin B_6 is required in the diet of humans, although this vitamin is adequately supplied in the usual diets of adults, children, and all but very young infants. However, deficiency states in infants and in pregnant women have been described. In the first instance, epileptiform seizures were reported in a small percentage (3–5/1000) of very young infants maintained on an unsupplemented diet of a liquid infant food preparation which had been autoclaved at a very high temperature. The method of preparation presumably destroyed most of the vitamin B_6 content of the product. Supplementation of this material with pyridoxine promptly alleviated the symptoms. In the second instance, pregnant women given 10 g of DL-tryptophan excreted various intermediary metabolites of tryptophan breakdown, including xanthurenic acid. The administration of vitamin B_6 to these women suppressed to a large degree the excretion of these metabolites. This suggests that in pregnancy there may exist a B_6 deficiency which is brought about by the increased demand of the fetus for this vitamin.

There is increasing evidence that vitamin B_6 is intimately concerned with the metabolism of the CNS. Swine, after 9–10 weeks on a vitamin B_6-free diet, exhibit demyelinization of the peripheral nerves and degeneration of the axon. In humans, the effects of pyridoxine deficiency have been best demonstrated in infants and children. The epileptiform seizures in infants which were described above are examples. The abnormal CNS activity that accompanies low vitamin B_6 intake during infancy is characterized by a syndrome of increasing hyperirritability, gastrointestinal distress, and increased startle responses as well as convulsive seizures. During the actual periods of seizure EEG changes may be noted. The clinical and EEG changes both respond quickly to pyridoxine therapy.

A syndrome resembling vitamin B_6 deficiency as observed in animals has also been noted in man during the treatment of tuberculosis with high doses of the tuberculostatic drug, isoniazid (isonicotinic acid hydrazide, INH). Two to 3% of patients receiving conventional doses of INH (2–3 mg/kg) developed neuritis; 40% of patients receiving 20 mg/kg developed neuropathy. Tryptophan metabolism (as indicated by xanthurenic acid excretion) was also altered. The signs and symptoms were alleviated by the administration of pyridoxine. Fifty mg of pyridoxine per day completely prevented the development of the neuritis. It is believed that INH forms a hydrazone complex with pyridoxal, resulting in incomplete activation of the vitamin.

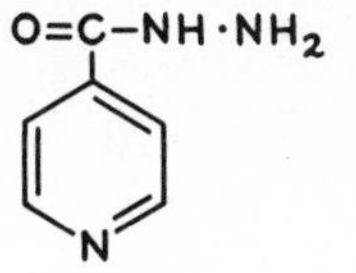

Isonicotinic acid hydrazide (isoniazid, INH)

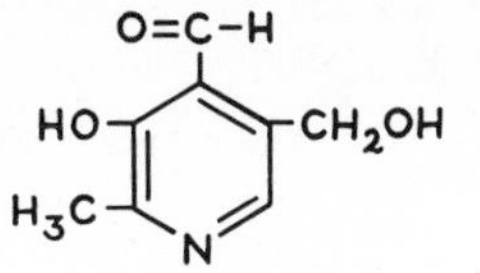

Pyridoxal

The role of pyridoxal in the metabolism of brain has recently been elucidated. In connection with the function of this vitamin as a codecarboxylase for amino acids, there is one pyridoxal-dependent reaction which is specific to the CNS. This reaction is the decarboxylation of glutamic acid to γ-aminobutyric acid, which is further metabolized to succinic acid by way of an NAD-dependent soluble dehydrogenase in brain. The glutamic decarboxylase and the product of the decarboxylation, γ-aminobutyric acid, are found in the CNS, principally in the gray matter.

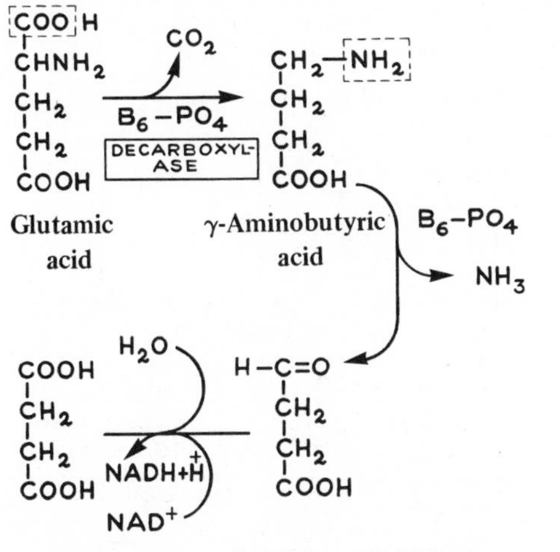

Glutamic acid — CO_2 — B_6-PO_4 DECARBOXYLASE — γ-Aminobutyric acid — B_6-PO_4 — NH_3 — Succinic semialdehyde — H_2O — NADH+H⁺ — NAD^+ — Succinic acid

The effects of γ-aminobutyric acid on peripheral as well as central synaptic activity suggest that this compound may function as a regulator of neuronal activity. It is now believed that the epileptiform seizures in animals produced by a deficiency of B_6, the action of INH, or the administration of pyridoxine antimetabolites, eg, deoxypyridoxine, may be related to a decrease in the activity of the glutamic acid decarboxylase with a resultant decrease in the amounts of γ-aminobutyric acid necessary to regulate neuronal activity in a normal manner. This idea is supported by the fact that the seizures can be controlled not only by the administration of vitamin B_6 but also by the administration of γ-aminobutyric acid.

Sources & Daily Allowances

It has been difficult to establish definitely the human requirement for vitamin B_6, probably because the quantity needed is not large and because bacterial synthesis in the intestine provides a portion of that requirement. There is some evidence that the requirement for vitamin B_6 is related to the dietary protein intake.

For an adult, 2 mg/day has been recommended. In a significant percentage of infants, intakes of less than 0.1 mg/day were associated with clinical manifestations of deficiency, as described above; with 0.3 mg/day, no symptoms developed, and in most cases there was no increase in the excretion of xanthurenic acid after a tryptophan load. However, in the case of some infants studied, as much as 2–5 mg/day were required to prevent convulsions, which suggests the existence of an abnormality in the metabolism of vitamin B_6 in these cases (vitamin B_6 dependency).

The currently recommended allowances for vitamin B_6 are listed in Table 21–3.

Good sources of the vitamin include yeast and certain seeds, such as wheat and corn, liver, and, to a limited extent, milk, eggs, and leafy green vegetables. There is little evidence that diets containing a reasonable balance of naturally occurring foodstuffs are ever seriously deficient in vitamin B_6. However, occasional cases of B_6 deficiency do arise as a result of malabsorption, alcoholism, antagonism to drugs, or the dependency mentioned above, which appears to be an inherited metabolic abnormality.

PANTOTHENIC ACID

Pantothenic acid is essential to the nutrition of many species of animals, plants, bacteria, and yeasts as well as for man. In experimental animals, symptoms due to pantothenic acid deficiency occur in such a wide variety of tissues that the basic function of this vitamin in cellular metabolism is amply confirmed. Gastrointestinal symptoms (gastritis and enteritis with diarrhea) are common to several species when a deficiency of this vitamin occurs. Skin symptoms, including cornification, depigmentation, desquamation, and alopecia also occur frequently. Lack of this vitamin also affects the adrenals. Animals deficient in pantothenic acid exhibit hemorrhage and necrosis of the adrenal cortex and an increased appetite for salt. If this condition persists the gland becomes exhausted, as shown by disappearance of lipoid material from the cortex and an acute state of adrenal cortical insufficiency, with sudden prostration and terminal dehydration.

Chemistry

In its active form, pantothenic acid is a constituent of **coenzyme A,** also known as coacetylase, the coenzyme for acetylation reactions. The coenzyme has a nucleotide structure, as shown in Fig 7–4.

The biosynthesis of coenzyme A in many forms of life, including man, appears to proceed as shown in

Pantothenic acid

Pantoic acid β-Alanine Thioethanolamine

Pantetheine

Pyro-phosphate

Adenine

D-Ribose-3-phosphate

FIG 7–4. Structure of coenzyme A.

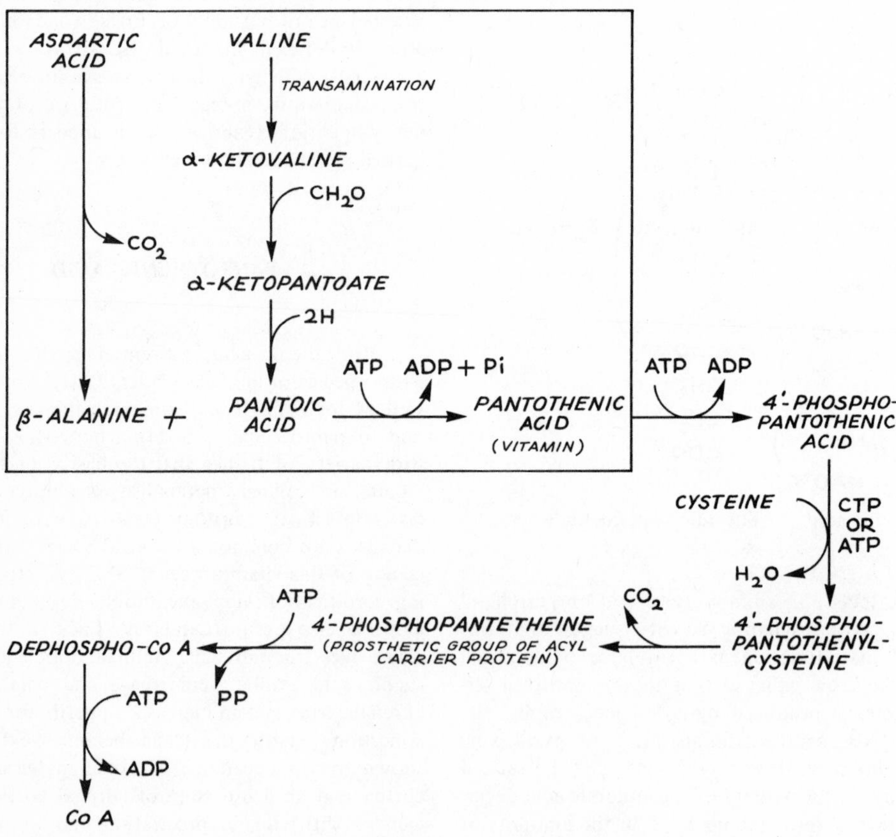

FIG 7–5. Biosynthesis of coenzyme A.

Fig 7–5. The reactions shown in the box are responsible for the biosynthesis of the vitamin pantothenic acid in plants and bacteria; these reactions do not occur in man.

In reactions involving coenzyme A, combination of the metabolite activated by the coenzyme occurs at the sulfhydryl (SH) group of the pantetheine moiety through a high-energy sulfur bond. It is therefore customary to abbreviate the structure of the free (reduced) coenzyme as Co A.SH, in which only the reactive SH group of the coenzyme is indicated.

Physiologic Role

As a constituent of coenzyme A, pantothenic acid is essential to several fundamental reactions in metabolism. An example is the combination of coenzyme A with acetate to form "active acetate." In the form of acetyl-coenzyme A (active acetate), acetic acid participates in a number of important metabolic processes. For example, it is utilized directly by combination with oxaloacetic acid to form citric acid, which initiates the citric acid cycle. Thus acetic acid derived from carbohydrates, fats, or many of the amino acids undergoes further metabolic breakdown via this "final common pathway" in metabolism. In the form of active acetate, acetic acid also combines with choline to form acetylcholine, or with the sulfonamide drugs which are acetylated prior to excretion.

The product of decarboxylation of α-ketoglutarate in the citric acid cycle is a coenzyme A derivative called "active" succinate (succinyl-Co A). Active succinate and glycine are involved in the first step leading to the biosynthesis of heme. Anemia frequently occurs in animals deficient in pantothenic acid. It may be assumed that this is referable to difficulty in formation of succinyl-Co A.

Coenzyme A has also an essential function in lipid metabolism. The first step in the oxidation of fatty acids catalyzed by thiokinases involves the "activation" of the acid by formation of the coenzyme A derivative, and the removal of a 2-carbon fragment in beta oxidation is accomplished by a "thiolytic" reaction, utilizing another mol of coenzyme A. The 2-carbon fragments thus produced are actually in the form of acetyl-Co A which may directly enter the citric acid cycle for degradation to carbon dioxide and water or combine to form ketone bodies.

Although a major function of pantothenic acid is in conjunction with its role as a constituent of coenzyme A, the total pantothenic acid content of the cell cannot be accounted for as Co A. Furthermore, a significant amount of the cellular pantothenic acid is protein-bound. This latter form of pantothenic acid is that contained in a compound known as **acyl carrier protein (ACP)** (Pugh & Wakil, 1965). Acyl carrier protein is a coenzyme required in the biosynthesis of fatty acids, whereas Co A is involved primarily in catabolism of fatty acids except for its function in connection with cholesterol biosynthesis.

Acyl carrier protein (molecular weight: 9100 daltons) possesses one SH group to which acetyl, malonyl, and intermediate chain-length acyl groups are attached in covalent linkages. The various changes that acyl groups undergo in fatty acid biosynthesis occur while they are in a thioester linkage with ACP. This thiol-containing residue of ACP is 2-mercaptoethanolamine (thioethanolamine), the same SH residue as on Co A. Indeed, the active SH-containing residue of ACP is the phosphopantetheine portion of Co A (4′-phosphopantetheine), the remaining (adenine nucleotide) portion of the Co A molecule not being involved. The amino acid composition of the phosphopantetheine-containing peptide of ACP has also been determined by Pugh & Wakil (1965). It appears that the pantetheine residue is linked covalently to a hydroxy group of a serine residue in the peptide. The amino acids which are immediately adjacent to this serine residue on the amino and carboxyl ends of the peptide are aspartic acid (or asparagine) and leucine, respectively.

Acetyl-Co A is a precursor of cholesterol and thus of the steroid hormones. As was noted above, a pantothenic acid deficiency inevitably produces profound effects on the adrenal gland. The anatomic changes are accompanied by evidence of functional insufficiency as well. This is due to poor synthesis of cholesterol by the pantothenic acid-deficient gland.

Activation of some amino acids may also involve Co A.

All of these facts point to an extremely important function for this vitamin in metabolism, involving as it does the utilization of carbohydrate, fat, and protein and the synthesis of cholesterol and steroid hormones as well as various acetylation reactions.

The combination of acetic acid or of fatty acids with Co A occurs at the terminal sulfhydryl group (SH) of the pantetheine residue. When an acyl group is transferred, the SH group is liberated to participate in the activation of another acyl radical. In bacteria, acetate may also be converted to acetyl phosphate, utilizing ATP as a phosphate donor. This high-energy compound can then be transferred directly to the SH site on Co A. The sulfur bond of acetyl-Co A (Co A.SH) is a high-energy bond equivalent to that of the high-energy phosphate bonds of ATP and other high-energy phosphorylated compounds. A similar high-energy sulfur bond is found in the derivatives of lipoic acid (thioctic acid). The formation of these high-energy bonds requires, therefore, a source of energy, either from a coupled exergonic reaction which yields the energy for incorporation into the bond, or from the transfer of energy from a high-energy phosphate bond or from another high-energy sulfur bond.

Examples of the function of coenzyme A in the formation of acetyl-Co A from pyruvate and in the formation of succinyl-Co A from ketoglutarate will be shown in the next section.

Although pantothenic acid contributes in a significant degree to many important biologic processes, there is no well substantiated evidence for the existence of a deficiency in human subjects. Consequently, no quantitative data with respect to the daily requirements for this vitamin have yet been gathered.

Sources

Excellent food sources (100–200 μg/g of dry material) include egg yolk, kidney, liver, and yeast. Broccoli, lean beef, skimmed milk, sweet potatoes, and molasses are fair sources (35–100 μg/g).

A 57% loss of pantothenic acid in wheat may occur during the manufacture of patent flour, and up to 33% is lost during the cooking of meat. Only a slight loss occurs in the preparation of vegetables.

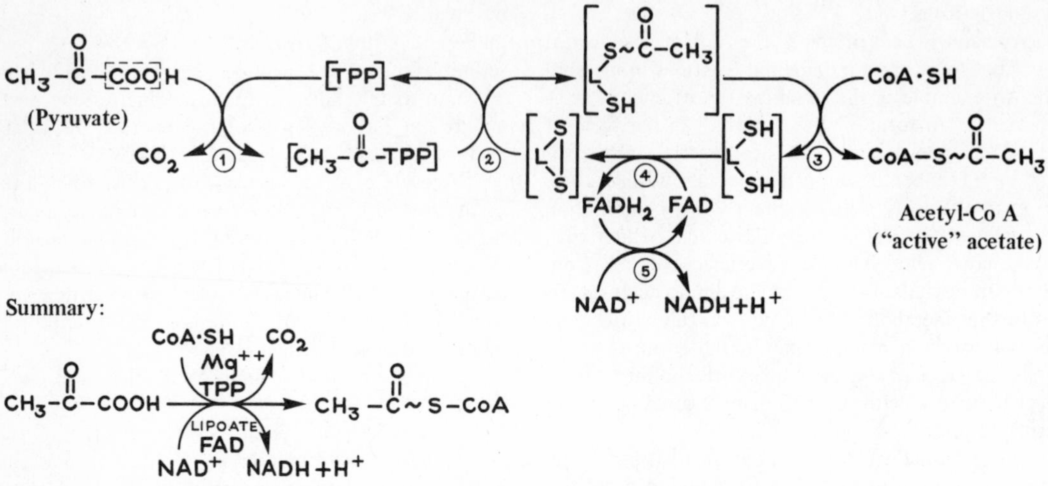

(Pyruvate) CoA·SH [TPP] CO₂ ① [$CH_3-\overset{O}{\overset{\|}{C}}-TPP$] ② ③ ④ ⑤ $FADH_2$ FAD NAD^+ $NADH+H^+$

Acetyl-Co A
("active" acetate)

Summary:

$$CH_3-\overset{O}{\overset{\|}{C}}-COOH \xrightarrow[\text{LIPOATE, FAD};\ NAD^+ \to NADH+H^+]{CoA\cdot SH \to CO_2;\ Mg^{++},\ TPP} CH_3-\overset{O}{\overset{\|}{C}}\sim S-CoA$$

FIG 7–6. Oxidative decarboxylation of pyruvate.

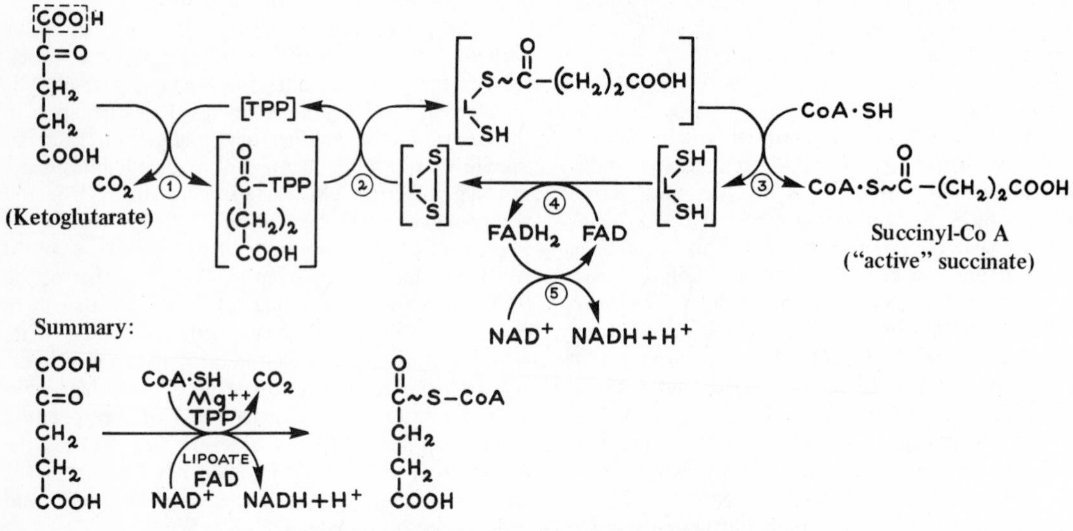

(Ketoglutarate) CoA·SH [TPP] CO_2 ① ② ③ ④ ⑤ $FADH_2$ FAD NAD^+ $NADH+H^+$

Succinyl-Co A
("active" succinate)

Summary:

$$HOOC-CH_2-CH_2-C(=O)-COOH \xrightarrow[\text{LIPOATE, FAD};\ NAD^+ \to NADH+H^+]{CoA\cdot SH \to CO_2;\ Mg^{++},\ TPP} HOOC-CH_2-CH_2-\overset{O}{\overset{\|}{C}}\sim S-CoA$$

FIG 7–7. Oxidative decarboxylation of ketoglutarate.

Abbreviations:

Square brackets mean that the intermediate is bound tightly to the enzyme interface.

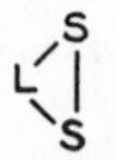

- $L\langle S-S \rangle$ = Oxidized lipoic acid 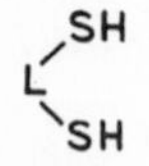
- $L\langle SH, SH \rangle$ = Reduced lipoic acid 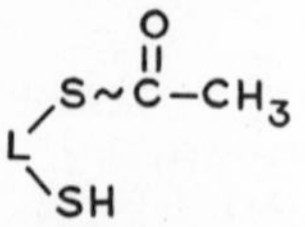
- $L\langle S\sim\overset{O}{\overset{\|}{C}}-CH_3,\ SH \rangle$ = S-Acetyl lipoic acid
- [TPP] = Thiamine pyrophosphate
- [$CH_3-CHO-TPP$] = "Active" acetaldehyde
- CoA·SH = Coenzyme A
- FAD, $FADH_2$ = Flavin adenine dinucleotide, oxidized and reduced forms, respectively
- NAD^+, NADH = Nicotinamide adenine dinucleotide, oxidized and reduced forms, respectively

The enzymes catalyzing the numbered reactions are as follows:

- Reactions ① & ② = Pyruvate (ketoglutarate) dehydrogenase
- Reaction ③ = Dihydrolipoyl transacetylase
- Reaction ④ = Dihydrolipoyl dehydrogenase

LIPOIC ACID

In connection with the oxidative decarboxylation of pyruvate to acetate it has been noted that in addition to thiamine, certain other coenzymes are required. Closely associated with thiamine in the initial decarboxylation of a α-keto acids is **lipoic acid** (thioctic acid). This factor was first detected in studies of the nutrition of lactic acid bacteria, where it was shown to replace the growth-stimulating effect of acetate. For this reason, the designation "acetate replacement factor" was assigned to it. A vitamin required for the nutrition of the protozoon *Tetrahymena geleii,* to which the term "protogen" was applied, and a factor from yeast (pyruvate oxidation factor) necessary for the oxidation of pyruvate to acetate by *Streptococcus faecalis,* which were studied at about the same time, were both later shown to be identical with the original acetate replacement factor of Guirard, Snell, and Williams. The active compound extracted from natural materials has been identified as a sulfur-containing fatty acid, **6,8-dithio-octanoic acid** (lipoic acid; thioctic acid).

$$\underset{\text{SH}}{\text{CH}_2}-\text{CH}_2-\underset{\text{SH}}{\text{CH}}-(\text{CH}_2)_4-\text{COOH} \xrightarrow{\ 2\text{H}\ }$$

α-Lipoic acid (reduced form)
(6,8-dithio-octanoic acid)

$$\underset{\text{S}}{\text{CH}_2}-\text{CH}_2-\underset{\text{S}}{\text{CH}}-(\text{CH}_2)_4-\text{COOH} \quad (\text{S}-\text{S bond})$$

α-Lipoic acid (oxidized form)

Physiologic Role

Lipoic acid occurs in a wide variety of natural materials. It is recognized as an essential component in metabolism, although it is active in extremely minute amounts. It has not yet been demonstrated to be required in the diet of higher animals, and attempts to induce a lipoic acid deficiency in animals have so far been unsuccessful.

The complete system for oxidative decarboxylation of pyruvic acid and of α-ketoglutaric acid (Fig 13–7) involves both thiamine and lipoic acid as well as pantothenic acid, riboflavin, and possibly nicotinamide derivatives (as coenzyme A, flavin adenine dinucleotide [FAD], and NAD, respectively).

In a bacterial system (*E coli*), the series of reactions for the decarboxylation of pyruvate, shown in Fig 7–6, has been proposed (Koike, 1960).

In reactions 1 and 2 in Fig 7–6 the energy of the decarboxylation step is utilized to form a high-energy bond at the linkage between the acetyl group and a sulfur of the protein-bound lipoic acid moiety. In reaction 3 the high-energy bond is maintained when the acetyl group is transferred to Co A to form acetyl-Co A ("active acetate"). The reoxidation of lipoic acid is accomplished in reaction 4 with the aid of FAD, so that lipoic acid is again ready to function as in reaction 2. In vivo it may be that $FAD.H_2$ is directly oxidized through the cytochrome system. In that case, reaction 5 would not be involved.

A similar series of reactions (Fig 7–7) accomplishes the decarboxylation of α-ketoglutaric acid to form succinyl-Co A (reaction of the citric acid cycle; see Fig 13–7). The high-energy sulfur bond is again formed at the expense of the decarboxylation reaction.

Succinyl-Co A ("active succinate") is a precursor of porphobilinogen, the first pyrrole to be synthesized in the production of the porphyrin nucleus. It may also serve as a source of coenzyme A for activation of acetoacetate in a reaction which seems specific to heart and skeletal muscle as well as kidney.

The reactions by which succinyl-Co A may be converted to free succinic acid are shown below. In the process, a high-energy phosphate bond is formed by transfer of the energy of the sulfur bond to inosine diphosphate (IDP) or to guanosine diphosphate (GDP). These transfer reactions are catalyzed by a specific magnesium-activated enzyme, **succinic thiokinase.** The enzyme which has been identified in heart and kidney tissues catalyzes specifically the formation of inosine triphosphate (ITP) or guanosine triphosphate (GTP) from succinyl-Co A. Subsequently, ITP or GTP may transfer a high-energy phosphate to ADP to form ATP. The formation of high-energy phosphate from succinyl-Co A is an example of "phosphorylation at the substrate level."

It was mentioned in the preceding paragraph that guanosine triphosphate (GTP) formed in the succinic thiokinase reaction transfers its high-energy phosphate to ADP to form ATP. There are, however, reactions in which GTP may function directly as a high-energy phosphate donor. A recently discovered example is in the activation of fatty acids catalyzed by a GTP-specific thiokinase isolated by Rossi & Gibson (1964) from beef liver mitochondria. This role of GTP may

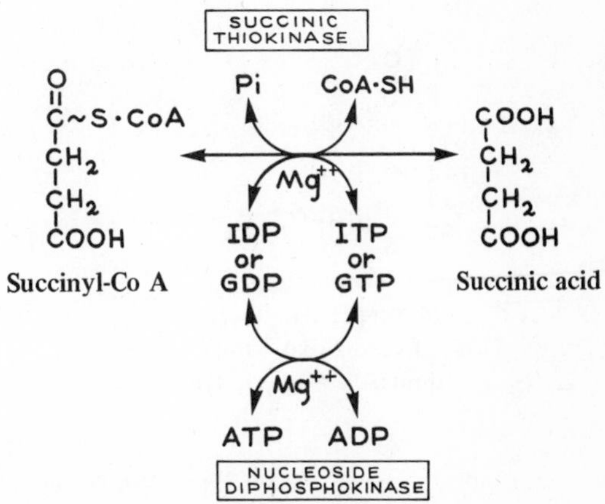

not be confined to activation of fatty acids, and the GTP thiokinase may exist in organs other than liver (although it has not yet been identified anywhere else). Further study may show that the GTP-specific system is of greater importance than is now evident.

BIOTIN

Biotin was first shown to be an extremely potent growth factor for microorganisms. As little as 0.005 μg permits the growth of test bacteria. In experiments with rats, dermatitis, retarded growth, loss of hair, and loss of muscular control occurred when egg white was fed as the sole source of protein. Certain foods were then found to contain a protective factor against these injurious effects of egg white protein, and this protective factor was named biotin, or the anti-egg white injury factor. The antagonistic substance in raw egg white is a protein **(avidin)** which combines with biotin, even in vitro, to prevent its absorption from the intestine.

It is difficult to arrive at a quantitative requirement for biotin. A large proportion of the biotin requirement probably is supplied by the action of the intestinal bacteria, since a biotin deficiency can be induced more readily in animals which have been fed with those sulfonamide drugs which reduce intestinal bacteria to a minimum. Careful balance studies in man showed that in many instances urinary excretion of biotin exceeded the dietary intake, and that in all cases fecal excretion was as much as 3–6 times greater than the dietary intake. It is therefore difficult to conceive of a dietary biotin deficiency under the usual circumstances. Highly purified diets have produced the deficiency in chickens and monkeys, but it is possible that the observed symptoms were really due to the effect of the diets on the intestinal flora.

Chemistry

In the free state biotin has the structure shown below.

```
        O
        ‖
        C
      /   \
H–N         N–H
  |         |
H–C ——————— C–H
  |         |
 H₂C        CH–(CH₂)₄COOH
     \    /
        S
```

Biotin ($C_{10}H_{16}O_3N_2S$)
(hexahydro-2-oxo-1-thieno-3,4-imidazole-4-valeric acid)

In biologic systems, biotin functions as the coenzyme for **carboxylases**, enzymes which catalyze carbon dioxide "fixation," or carboxylation. The biotin coenzyme is tightly bound to the enzyme protein (the apoenzyme), probably by an amide linkage between the biotin carboxyl group and the terminal (epsilon) nitrogen of a lysine residue in the enzyme protein, as indicated below. This is suggested by the discovery in natural materials of a combined form of biotin, **biocytin,** identified as a lysine-biotin conjugate (ϵ-N-biotinyl lysine). Evidently, in the isolation of biotin from the naturally occurring biotin-enzyme complex, the vitamin may be split off from the enzyme protein together with the amino acid, lysine, to which it is attached in the protein.

```
      O
      ‖
      C
    /   \
  N       N
  |       |
  C ————— C              O
  |       |              ‖
  C       C–(CH₂)₄–C
    \   /              |
      S            [   NH      ]
                   [   |       ]
                   [ LYSINE    ]
                   [   |       ]
                   [ PROTEIN   ]
```

Attachment of biotin coenzyme to apoenzyme through epsilon nitrogen of lysine

Physiologic Role

Human subjects placed on a refined diet containing a large amount of dehydrated egg white developed symptoms resembling those caused by biotin deficiency in animals. Urinary biotin decreased from a normal level of 30–60 μg to 4–7 μg. Prompt relief of symptoms occurred when a biotin concentrate was administered.

The best demonstrated biochemical role of biotin is in connection with the reactions of carboxylation (CO_2 "fixation"). For this purpose, the biotin coenzyme-apoenzyme complex attaches CO_2, which can later be transferred to other substances as described below. The mechanism for formation of the CO_2-biotin enzyme complex may proceed as shown in the reactions on p 109.

The CO_2 on the biotin enzyme complex may be transferred in a reaction catalyzed by **acetyl-Co A carboxylase** to acetyl-Co A to form **malonyl-Co A,** an important step in the extramitochondrial biosynthetic pathway for fatty acids.

In the Krebs-Henseleit cycle for formation of urea (see p 307), by a reaction of **"transcarbamylation,"** citrulline is formed by addition of CO_2 and ammonia to ornithine. The metabolically active source of the CO_2 and ammonia is **carbamyl phosphate.** Its formation is described on p 308, where it will be noted that the CO_2-biotin complex described above is the source of the "active CO_2" required in the formation of carbamyl phosphate.

The initial step in biosynthesis of pyrimidines involves a reaction between carbamyl phosphate and

CARBONIC PHOSPHORIC ANHYDRIDE

ATP + HCO_3^- —(Mn^{++})→ ADP + carbonic phosphoric anhydride + BIOTIN-ENZYME (ENZYME-NH) → Pi + CO_2 BIOTIN-ENZYME (ENZYME-NH)

aspartate. As indicated, biotin is involved in the formation of carbamyl phosphate, which is also required in pyrimidine biosynthesis.

Formation of carbamyl phosphate is effected by the catalytic activity of the enzyme, **carbamyl phosphate synthetase.** As might be expected, biotin is an integral component of the enzyme-protein system. This was indicated by the finding that the activity of the purified enzyme was inhibited by avidin (see above) and that a partially purified but inactive synthetase fraction was restored to activity by addition of biotin. Direct analysis of a purified preparation of the active enzyme also revealed the presence of biotin.

It is thus apparent that biotin contributes to the formation of an essential coenzyme for operation of the urea cycle, and thus the synthesis of arginine, as well as for the functioning of the biosynthetic pathway for pyrimidines and the extramitochondrial pathway of fatty acid biosynthesis.

Another example of a biotin-dependent reaction of carboxylation is the conversion of pyruvate to oxaloacetate, as shown below (see also Fig 13–7):

CH_3–C=O–COOH (**Pyruvate**) + CO_2* → COOH*–CH_2–C=O–COOH (**Oxaloacetate**) ⇌ (NH_3) COOH–CH_2–$CHNH_2$–COOH (**Aspartate**)

Transamination, as by transfer of an amino group from glutamate, forms aspartate. That biotin may be involved in this series of reactions is supported by the observation that both aspartic acid and oxaloacetic acid, when added to a nutrient solution for bacterial culture, reduce the requirement of the bacteria for the vitamin. The "sparing" action of the 2 compounds is presumably achieved by reducing the need for the vitamin in the synthesis of aspartic and oxaloacetic acids. Fixation of CO_2 in formation of carbon 6 in purine synthesis is impaired in biotin-deficient yeast, which suggests that the vitamin plays a role in purine synthesis in connection with this step in formation of the purine structure.

A number of other enzyme systems are reportedly influenced by biotin. These include succinic acid dehydrogenase and decarboxylase, and the deaminases of the amino acids aspartic acid, serine, and threonine.

Sources

It seems doubtful whether any but the most severely deficient diet would result in a biotin deficiency in man. The vitamin is widely distributed in natural foods. Egg yolk, kidney, liver, tomatoes, and yeast are excellent sources.

THE FOLIC ACID GROUP

Folic acid (folacin, pteroylglutamic acid, PGA) is a combination of the pteridine nucleus, *p*-aminobenzoic acid (PABA), and glutamic acid.

There are at least 3 chemically related compounds of nutritional importance which occur in natural products. All may be termed pteroyl glutamates. These 3 compounds differ only in the number of glutamic acid residues attached to the pteridine-PABA complex. The formula for folic acid shown below is the monoglutamate. This is synonymous with vitamin B_c. The substance once designated as the fermentation factor is a triglutamate, and vitamin B_c conjugate of yeast is a heptaglutamate. The deficiency syndrome which is now thought to be due to a lack of these substances has been recognized in the monkey since 1919. Since many investigators discovered the same deficiency state by varying technics, several names for these factors, in addition to those already mentioned, have appeared. These include vitamin M, factor U, factor R, norit eluate factor, and *Streptococcus lactis* R (SLR) factor.

Folic acid, when added to liver slices, is converted to a formyl derivative. Ascorbic acid enhances the activity of the liver in this reaction. This form of folic

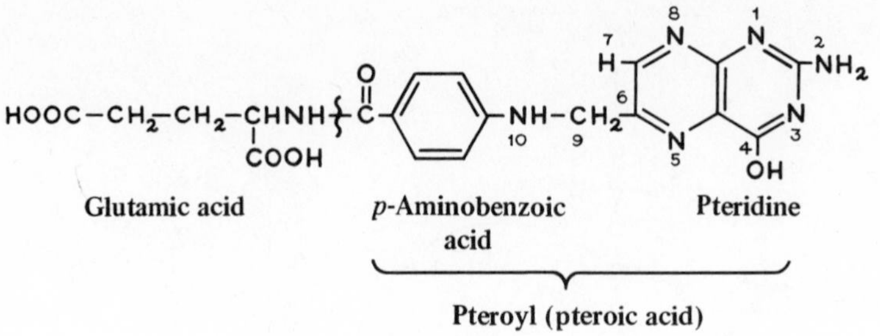

Glutamic acid — *p*-Aminobenzoic acid — Pteridine

Pteroyl (pteroic acid)

Folic acid
(folacin)

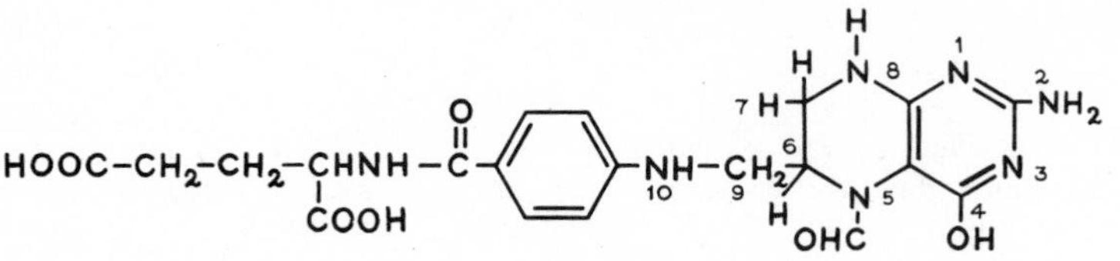

Folinic acid (leucovorin)

acid was first discovered in liver extracts when it was found to supply an essential growth factor for a lactobacillus, *Leuconostoc citrovorum.* Thus it was termed the **citrovorum factor (CF).** When its chemical structure was determined, the name **folinic acid** was applied.

The structure of folinic acid (leucovorin, folinic acid-SF [synthetic factor]) is shown above. It is the reduced (tetrahydro) form of folic acid with a formyl group on position 5.

A similar compound but with the formyl group on position 10 has been recovered from liver. Rhizopterin, or the *Streptococcus lactis* R (SLR) factor, is a naturally occurring compound which is a 10-formyl derivative of pteroic acid.

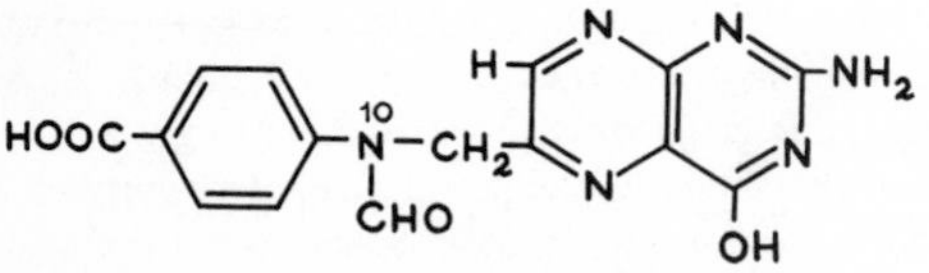

Rhizopterin
(10-formylpteroic acid)

Physiologic Role

The folic acid coenzymes are specifically concerned with biochemical reactions involving the transfer and utilization of the single carbon (C_1) moiety (Fig 7–8). Before functioning as a C_1 carrier, folic acid must be reduced, first to dihydrofolic acid, FH_2, and then to the tetrahydro compound, FH_4, catalyzed by a **folic acid reductase** and using NADPH as hydrogen donor. The reactions are as follows:

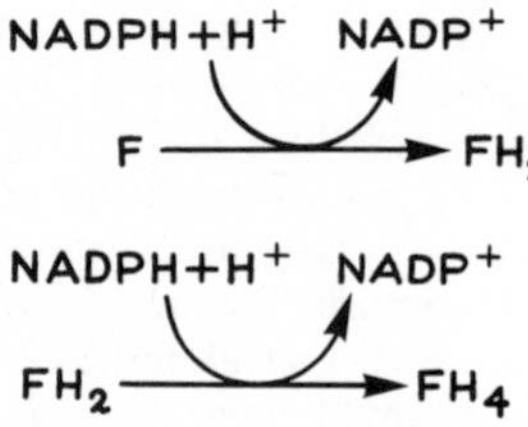

$$F \xrightarrow{NADPH+H^+ \;\to\; NADP^+} FH_2$$

$$FH_2 \xrightarrow{NADPH+H^+ \;\to\; NADP^+} FH_4$$

The folic acid antagonists such as methotrexate (amethopterin) are extremely potent competitive inhibitors of the reductase reaction ($FH_2 \rightarrow FH_4$).

The "one-carbon" (C_1) moiety may be either formyl (–CHO), formate (H.COOH), or hydroxymethyl ($–CH_2.OH$). These are metabolically interconvertible in a reaction catalyzed by a NADP-dependent hydroxymethyl dehydrogenase.

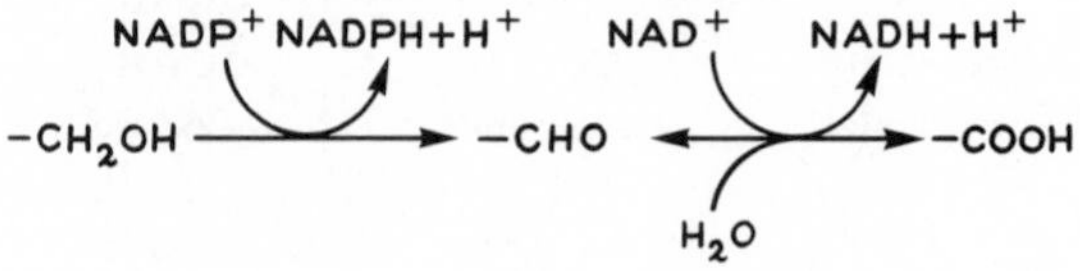

$$-CH_2OH \xrightarrow{NADP^+ \;\to\; NADPH+H^+} -CHO \xleftrightarrow[H_2O]{NAD^+ \;\to\; NADH+H^+} -COOH$$

As has been noted above, folinic acid is a 5-formyl (f^5) tetrahydrofolic acid (FH_4); in abbreviated form, $f^5.FH_4$. However, except for the formylation of glutamic acid in the course of the metabolic degradation of histidine (see p 331), the f^5 compound (folinic acid) is metabolically inert. Instead, the f^{10} ($f^{10}.FH_4$) or f^{5-10} ($f^{5-10}.FH_4$) tetrahydro derivatives, in which the single carbon is bound between positions 5 and 10 on tetrahydrofolic acid, are the active forms of the folic acid coenzymes in metabolism.

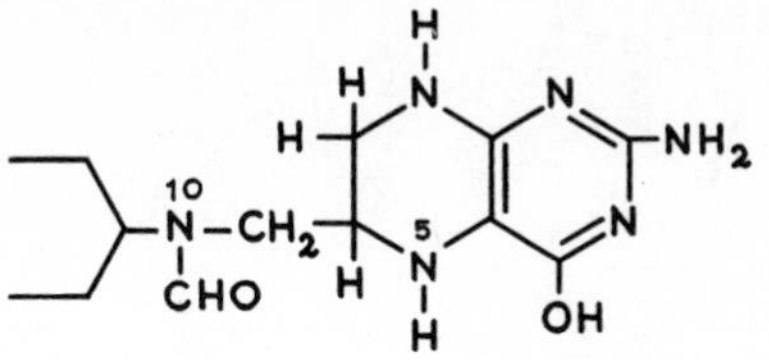

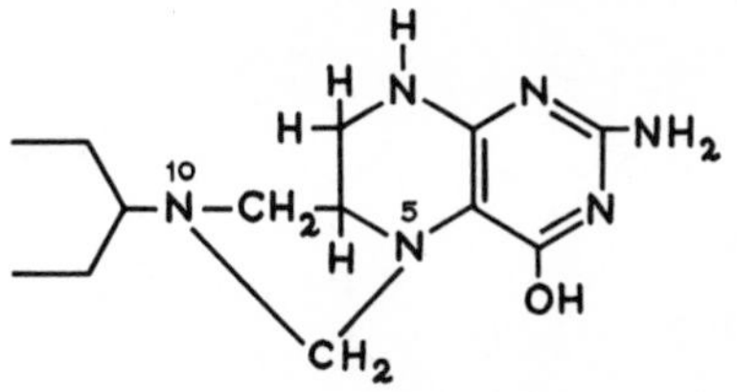

The f^5 can be converted to f^{10} by the action of an enzyme system, formyl tetrahydrofolic acid isomerase, as follows:

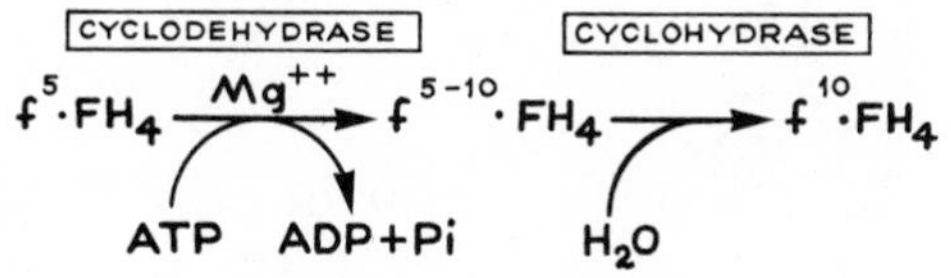

$$f^{5}\cdot FH_4 \xrightarrow[\text{ATP} \rightarrow \text{ADP+Pi}]{\text{CYCLODEHYDRASE},\ Mg^{++}} f^{5-10}\cdot FH_4 \xrightarrow[H_2O]{\text{CYCLOHYDRASE}} f^{10}\cdot FH_4$$

There is also present in a variety of natural materials (eg, pigeon liver, many bacteria) an enzyme, **formyltetrahydrofolate synthetase**, which catalyzes the direct addition of formate (H.COOH) to tetrahydrofolic acid. The enzyme is specific for formate. The reaction is as follows:

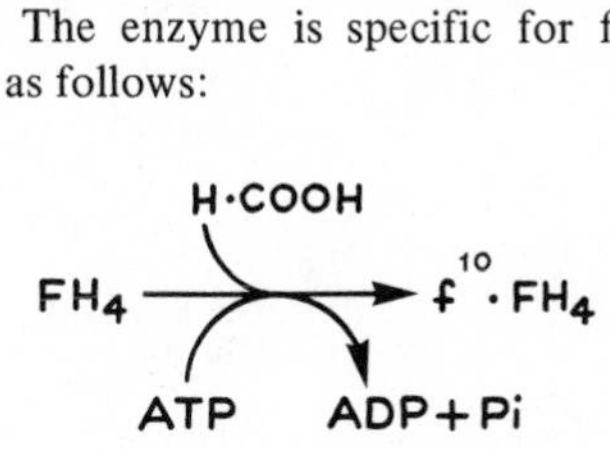

$$FH_4 \xrightarrow[\text{ATP} \rightarrow \text{ADP+Pi}]{H\cdot COOH} f^{10}\cdot FH_4$$

Sources & Utilization of the One-Carbon Moiety

The one-carbon moiety on tetrahydrofolic acid can be transferred to amino or to SH groups. An example of the first is the formimino (HC=NH) group on glutamic acid, a product of histidine breakdown (Fig 15–40); the second is exemplified by the formation of thiazolidine carboxylic acid with cysteine.

The formimino group (fi) on glutamic acid can serve as a source of the one-carbon moiety as follows:

$$\text{fi-glutamic acid} + FH_4 \longrightarrow fi^5 \cdot FH_4 + \text{glutamic acid}$$

$$fi^5 \cdot FH_4 \longrightarrow f^{5-10} \cdot FH_4 + NH_3$$

Other sources of the one-carbon moiety are the methyl groups of (a) methionine, (b) choline, by way of betaine (Fig 7–12), and (c) of thymine–all of which are oxidized to hydroxy methyl ($-CH_2OH$) groups and carried as such on $f^{5-10}.FH_4$. The hydroxy methyl group (h) is then oxidized in an NADP-dependent reaction to a formyl (f) group:

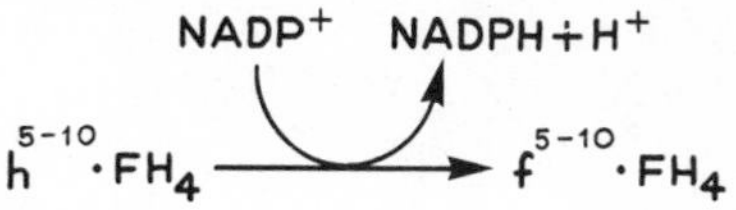

$$h^{5-10}\cdot FH_4 \xrightarrow{NADP^+ \rightarrow NADPH + H^+} f^{5-10}\cdot FH_4$$

The beta carbon of serine as a hydroxy methyl group may also contribute to the formation of a single carbon moiety.

The single formyl carbon which is present on the tetrahydrofolic acids is utilized in several important reactions (Fig 7–8). The first is as a source of carbons 2 and 8 in the purine nucleus as described on p 352. A second is the role of f^{10}-tetrahydrofolate ($f^{10}.FH_4$) as a source of the formyl group on N-formylmethionine-tRNA, which in microorganisms initiates synthesis of peptide chains on ribosomes (see p 65). A third reaction involving the formyl carbon on $f^{10}.FH_4$ is formation of the beta carbon of serine in conversion of glycine to serine (see p 328). A fourth reaction is in the synthesis of methyl groups for (a) methylation of homocysteine to form methionine or (b) of uracil to form thymine,* and (c) for the synthesis of choline by way of methyl groups for methionine. In the methylation reactions, the formyl group is first reduced to hydroxyl methyl before methylation will occur. Vitamin B_{12} may also be required in these methylation reactions.

The participation of the folic acid coenzymes in reactions leading to synthesis of purines and to thymine, the methylated pyrimidine of DNA, emphasizes the fundamental role of folic acid in growth and reproduction of cells. Because the blood cells are subject to a relatively rapid rate of synthesis and destruction, it is not surprising that interference with red blood cell formation would be an early sign of a deficiency of folic acid, or that the folic acid antagonists would readily inhibit the formation of leukocytes. But it must be remembered that the requirement for the folic acid coenzymes is undoubtedly generalized throughout the body and not confined to the hematopoietic system. This is supported by the observation that in folic acid-deficient monkeys there was a considerable decrease in the rate of synthesis of nucleoprotein, which rose to normal after administration of the vitamin. The function of the folic acid coenzymes in synthesis and utilization of methyl groups relates these vitamins also to phospholipid metabolism (choline synthesis; see Fig 7–12) and to amino acid metabolism.

A deficiency of folic acid has been produced experimentally in man by feeding a diet containing 5 μg of folate/day for a period of 4½ months (Herbert, 1962). During this period hemoglobin fell from 15.5 g to 13.9 g/100 ml of blood. At the termination of the

*Probably from β-carbon of serine only.

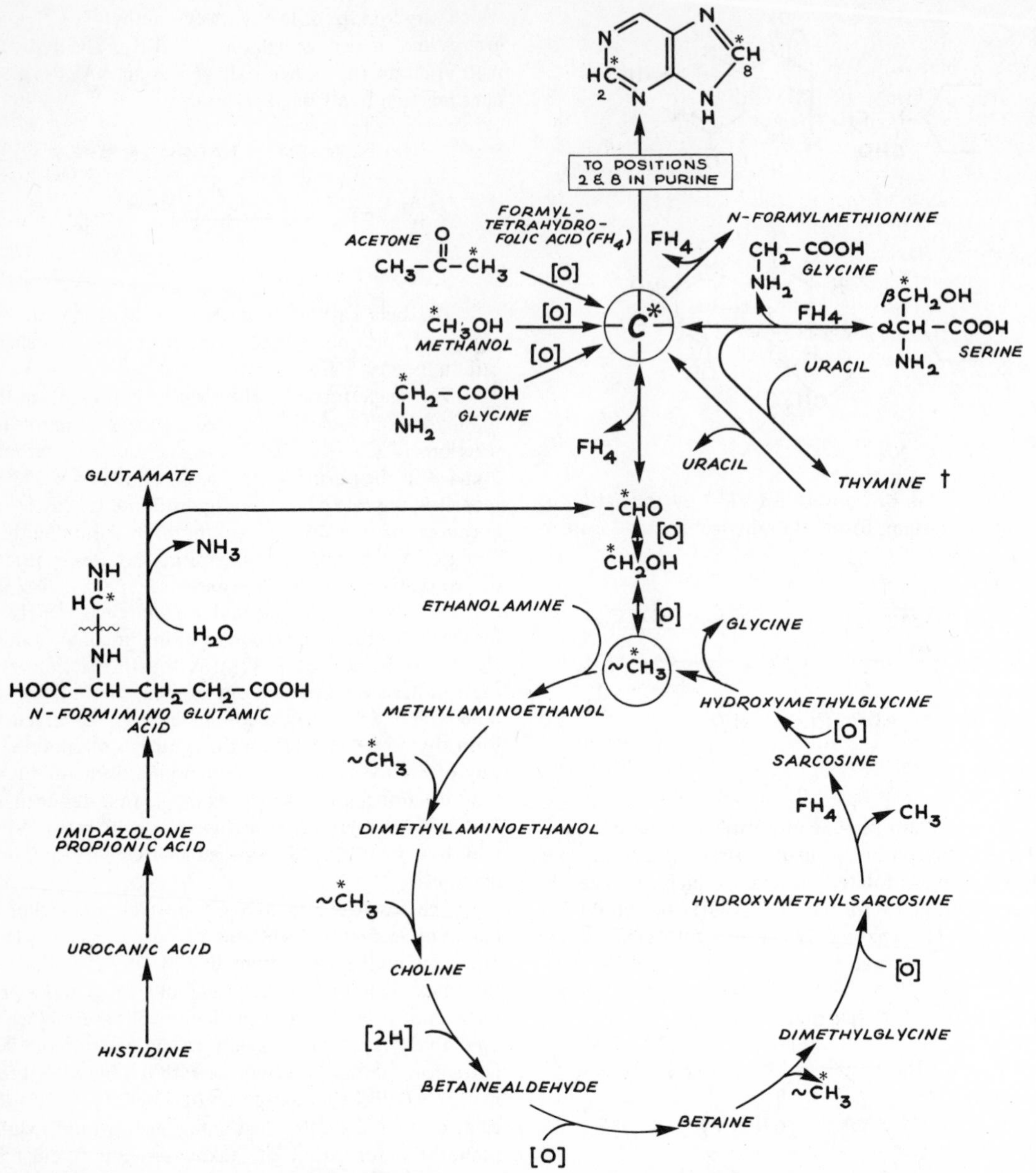

*Methyl groups from methionine (see Fig 7–12).
†C* for methylation of uracil to form thymine probably only from serine β-carbon via FH_4.

FIG 7–8. Sources and utilization of the one-carbon moiety.

experiment, examination of bone marrow aspirates clearly revealed abnormalities in megaloblast cells.

When folic acid was first made available in crystalline form, it excited considerable interest because of its therapeutic effect in nutritional macrocytic anemia, pernicious anemia, and the related macrocytic anemias. At that time, vitamin B_{12} was not known. It was soon discovered that in those anemias which were due to other than simple folic acid deficiencies, or to deficiencies of other factors (as is the case in uncomplicated pernicious anemia), the hematologic response to folic acid was not permanent and the neurologic symptoms in pernicious anemia (combined system disease) remained unchanged. It is now apparent that while folic acid derivatives do have an effect on hematopoiesis, other factors are also necessary for the complete development of the blood cells. In pernicious anemia the most important factor is vitamin B_{12}; uncomplicated cases respond to this vitamin alone. Furthermore, vitamin B_{12} controls both the hematologic and the neurologic defect. It has therefore been concluded that folic acid has no place in the treatment of uncomplicated pernicious anemia. Although folic acid is harmless when administered to patients adequately treated with vitamin B_{12}, some have recommended that folic acid should not be included in multivitamin preparations because of the danger that it may mask pernicious anemia in susceptible patients and thus permit the disease to progress to the much more serious stage involving neurologic damage.

However, there are well documented reports of megaloblastic anemias which were due apparently to simple nutritional deficiencies of folic acid. These reports, together with the results of the experimental production of folic acid deficiency mentioned above, would seem to be ample proof that a megaloblastic anemia due entirely to dietary folic acid deficiency can occur. The clinical reports of the occurrence of megaloblastic anemia as a result of folic acid deficiency were presented by Gough & others (1963). In their 7 cases, in addition to megaloblastic changes, there were also low levels of folic acid in the serum (less than 5 ng/ml) when the vitamin B_{12} content of the serum was normal. Treatment with 5 mg of folic acid 2 or 3 times a day produced a rapid reticulocyte response followed by substantial rises in hemoglobin to normal levels.

In the metabolism of the amino acid histidine (Fig 15–46), there is a folic acid-dependent step at the point where formiminoglutamic acid is converted to glutamic acid. When folic acid-deficient patients are given a loading dose of histidine, the increased excretion of **formiminoglutamic acid ("figlu")** into the urine which occurs can be used as a chemical test of a lack of folic acid. Evidently the folic acid deficiency results in a metabolic block which is reflected in an accumulation and consequent excretion of formiminoglutamic acid. The use of this **"figlu" excretion test**, as well as direct assays (by microbiologic methods) of the serum content of folic acid and of vitamin B_{12}, should improve the accuracy of diagnosis of the causes of megaloblastic anemias.

In sprue, the administration of synthetic folic acid (5–15 mg/day) has been followed by rapid and impressive remissions, both clinically and hematologically. The glossitis and diarrhea subside in a few days, and a reticulocytosis occurs which is followed by regeneration of the erythrocytes and hemoglobin. Roentgenologic evidence of improved gastrointestinal function, improved fat absorption, and a return of the glucose tolerance curve to normal are also observed. The vitamin seems therefore to correct both the hematopoietic and gastrointestinal abnormalities in sprue.

Sources & Daily Allowances

According to some clinical nutritionists, folic acid deficiency is possibly the most common vitamin deficiency in North America and western Europe. This is especially true in pregnancy, wherein folic acid deficiency is said to be the most common cause of megaloblastic anemia. A suggested reason for this is that folic acid may not be contained in some vitamin preparations prescribed for pregnant women because, as noted above, folic acid can correct the anemia of vitamin B_{12} deficiency and thus mask the lack of B_{12}, thereby permitting neurologic damage to progress. It should be emphasized that combined deficiencies of folic acid, vitamin B_{12}, ascorbic acid, and iron—as well as other nutrients—are more common than isolated deficiencies of any one of these nutritional components. Folic acid deficiency should be considered in connection with alcoholism, hemolytic anemias, tropical and nontropical sprue, and the anemias occurring in infancy, pregnancy, or malignancies.

Fresh leafy green vegetables, cauliflower, kidney, and liver are rich sources of folic acid. The lability of the vitamin to cooking processes is said to be similar to that of thiamine.

In a truly folic acid-deficient patient, the administration of 300–500 μg/day of folic acid will produce a hematologic response. However, this small dose will not effect a response in a patient with pernicious anemia. The use of this conservative but adequate dose of folic acid will thus serve as a method of differentiating between vitamin B_{12} and folic acid deficiency (Davidson, 1959).

Recommended daily allowances of folic acid (folacin) are listed in Table 21–3.

Folic Acid Antagonists

The concept of competitive inhibition or metabolic antagonism is discussed on p 146. Antagonists to folic acid have found some clinical application in the treatment of malignant disease, and confirmation of the action of folic acid in cell growth has been obtained in studies of the effect of these antagonists on cells maintained in tissue culture.

Maximal inhibitory action is obtained when an amino group is substituted for the hydroxy group on position 4 of the pteridine nucleus. Thus **aminopterin** (4-amino folic acid) is the most potent folic acid inhibitor yet discovered. Another antagonist is methotrexate (**amethopterin**, 4-amino-10-methylfolic acid). In animals the inhibitory effect of aminopterin cannot be reversed by folic acid but only by folinic acid. This suggests that aminopterin interferes with the formation of folinic acid from folic acid or with the utilization of the formyl group. Recent work suggests that the interference of the antimetabolites occurs in the reduction of folic acid to the tetrahydro compound. Reduction is a necessary preliminary to the carriage of the one-carbon moiety.

In tissue cultures it has been found that aminopterin blocks the synthesis of nucleic acids, presumably by preventing the reduction of folic acid to the tetrahydro derivative and thus transport of the formyl carbon into the purine ring. Such inhibited cells fail to complete their mitoses; they do not progress from metaphase to anaphase because of a failure in the synthesis of nucleoprotein, a synthesis which is essential to chromosome reduplication.

Aminopterin has been used in the treatment of leukemia, particularly in children. A remission is induced temporarily in some patients, but after a time the leukemic cells apparently acquire the power to overcome the effects of the antagonist.

INOSITOL

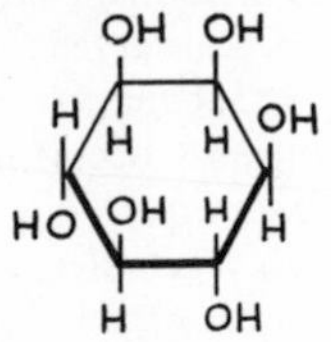

Meso-inositol
($C_6H_{12}O_6$,hexahydroxycyclohexane)

There are 9 isomers of inositol. Meso-inositol, also called myo-inositol (shown above), is the most important one in nature and the only isomer which is biologically effective.

The significance of this compound in human nutrition has not been established. However, in studies on the nutrient requirements of cells in tissue culture, Eagle (1957) found that 18 different human cell strains maintained on a semisynthetic medium failed to grow without the addition of meso-inositol. None of the other isomers were effective, a finding which is in agreement with the results of similar experiments in animals.

Together with choline, inositol has a lipotropic action in experimental animals. This lipotropic activity may be associated with the formation of inositol-containing lipids (lipositols; see Chapter 2).

Deficiency symptoms in mice include so-called spectacled eye, alopecia, and failure of lactation and growth. In inositol-deficient chicks an encephalomalacia and an exudative diathesis have been reported.

Sources

Inositol is found in fruits, meat, milk, nuts, vegetables, whole grains, and yeast.

PARA-AMINOBENZOIC ACID (PABA)

Para-aminobenzoic acid is a growth factor for certain microorganisms and an antagonist to the bacteriostatic action of sulfonamide drugs. It forms a portion of the folic acid molecule, and it is suggested that its actual role is to provide this component for the synthesis of folic acid by those organisms which do not require a preformed source of folic acid.

The successful use of para-aminobenzoic acid in the treatment of certain rickettsial diseases has been reported. It has been found that para-oxybenzoic acid is an essential metabolite for these organisms and that para-aminobenzoic acid acts as an antagonist to that substance.

VITAMIN B_{12}

Vitamin B_{12}, the anti-pernicious anemia factor (extrinsic factor of Castle) was first isolated in 1948 from liver as a red crystalline compound containing cobalt and phosphorus. The vitamin can be obtained as a product of fermentation by *Streptomyces griseus.* Its concentration, either in liver or in the fermentation liquor, is only about one part per million.

Chemistry

The structure of vitamin B_{12} is shown in Fig 7–9. The central portion of the molecule consists of 4 reduced and extensively substituted pyrrole rings surrounding a single cobalt atom. This central structure is referred to as a "corrin" ring system. It is very similar to that of the porphyrins but differs from the porphyrins in that 2 of the pyrrole rings (rings I and IV) are joined directly rather than through a single, methene carbon. It is of interest, however, that studies with an actinomyces organism that synthesizes vitamin B_{12} have revealed that the basic corrin structure is synthesized from the known precursors of porphyrins such as δ-aminolevulinic acid. The 6 "extra" methyl groups on B_{12} are derived from methionine (Bray, 1963).

Below the corrin ring system there is a 5,6-dimethylbenzimidazole riboside which is connected at one end to the central cobalt atom and at the other end from the ribose moiety through phosphate and aminopropanol to a side chain on ring IV of the tetrapyrrole nucleus. A cyanide group which is coordinately bound to the cobalt atom may be removed; the resulting compound is called **"cobalamin."** Addition of cyanide forms **"cyanocobalamin,"** identical with the originally isolated vitamin B_{12}. Substitution of the cyanide group with a hydroxy group forms **"hydroxocobalamin" (vitamin B_{12a})**; with a nitro group, **"nitrocobalamin."** The biologic action of these derivatives appears to be similar to that of cobalamin, although B_{12a} is more active in enzyme systems requiring B_{12} and therefore is used more often than B_{12} in experimental studies in vitro. Furthermore, although hydroxocobalamin given orally in large doses is absorbed as well as cyanocobalamin in similar doses, hydroxocobalamin is retained longer in the body; this suggests that hydroxocobalamin may be more useful for therapeutic administration of vitamin B_{12} by mouth.

Crystalline vitamin B_{12} is stable to heating at 100° C for long periods, and aqueous solutions at pH 4.0–7.0 can be autoclaved with very little loss. However, destruction is rapid when the vitamin is heated at pH 9.0 or above.

Although it had been suspected that vitamin B_{12} functions as a coenzyme in metabolism, such a function was not actually proved until Barker & others (1958) isolated 3 B_{12}-containing coenzymes from microbial sources. These coenzymes, the **cobamides,** do not contain the cyano group attached to cobalt as does vitamin B_{12} (cyanocobalamin). Instead there is

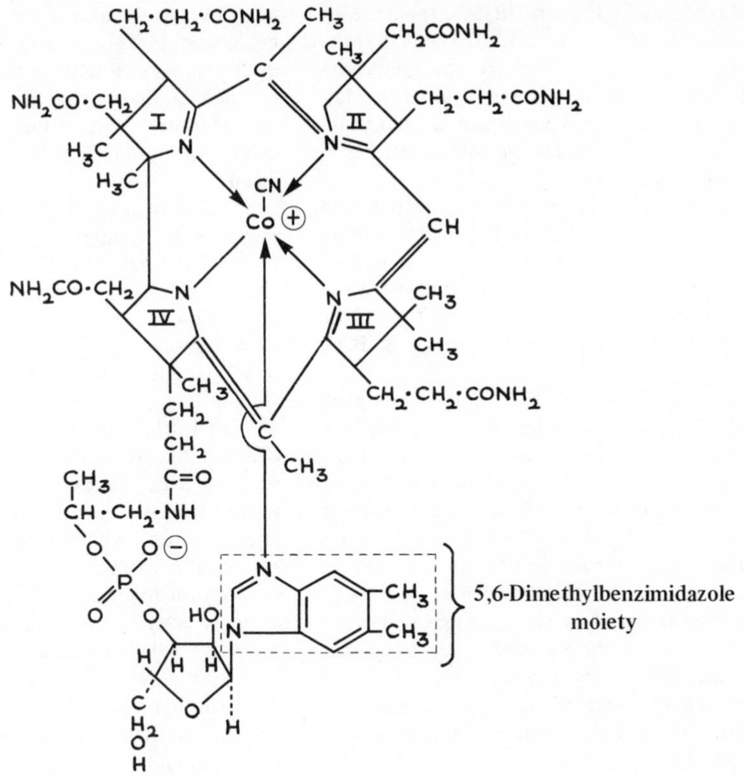

FIG 7–9. Cyanocobalamin; vitamin B_{12} ($C_{63}H_{88}O_{14}N_{14}PCo$).

an adenine deoxy nucleoside (5′-deoxyadenosine) which is linked to the cobalt by a carbon-to-cobalt bond, as shown in Fig 7–10.

The 5′-deoxyadenosyl moiety that replaces the cobalt-bound cyanide of cyanocobalamin when coenzyme B_{12} is formed is derived from adenosine triphosphate, which, after donating the adenosyl group, releases all 3 phosphate groups as inorganic tripolyphosphate (Peterkofsky, 1963). In the formation of

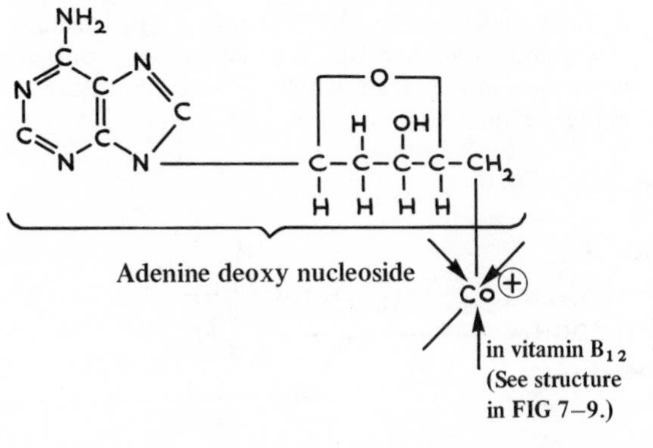

FIG 7–10. Attachment of deoxyadenosyl moiety to vitamin B_{12} through C′-5 to cobalt in the B_{12} coenzymes.

the adenosyl coenzyme, cobalt undergoes successive reduction in a series of steps catalyzed by B_{12a} reductase and requiring NADH and FAD. Thus, in B_{12a}, which is red-colored, cobalt is present as Co^{+++}. This progresses to B_{12r} (orange) with cobalt as Co^{++}, and to B_{12s} (gray-green) Co^{+}, the latter form reacting with ATP to form the adenosyl coenzyme as described at left.

All 3 of the coenzymes possess the adenine deoxynucleoside described above, but they differ from one another in the benzimidazole portion of the B_{12} molecule. Here there may be found the 5,6-dimethylbenzimidazole group, as occurs in vitamin B_{12}; an unsubstituted (methyl-free) benzimidazole; or an adenyl group. These 3 cobamide coenzymes are therefore designated 5,6-dimethylbenzimidazole cobamide (**DBC**), benzimidazole cobamide (**BC**), or adenyl cobamide (**AC**). DBC is the B_{12} coenzyme found in the largest quantities in natural materials. Its complete chemical designation is 5,6-dimethylbenzimidazole cobamide 5′-deoxyadenosine. Alternatively, it could also be called 5′-deoxyadenosyl cobalamin.

There has also been detected another B_{12} coenzyme, a cobamide in which a methyl group is attached to the cobalt atom rather than the adenosyl moiety.

The activity of the cobamide coenzymes has been determined by means of their function in catalyzing the reaction shown in Fig 7–11, whereby glutamic acid is converted to β-methylaspartic acid (and thence to mesaconic acid in a reaction catalyzed by the enzyme β-methylaspartase, which is not cobamide-dependent). The apoenzyme system required in addition to the cobamide coenzyme is obtained from strain Hl cultures of *Clostridium tetanomorphum* (Barker & others: J Biol Chem 235:181, 1960).

DBC also plays an essential role as a hydrogen transferring agent in the cobamide-dependent ribonucleotide reductase reaction whereby the ribose moiety in a ribonucleotide is converted to deoxyribose when DNA is to be formed. The mechanism of action of the cobamide coenzyme is discussed by Abeles & Beck (1967).

The B_{12} coenzymes have been isolated not only from several bacterial cultures but also from the liver of various animals (mainly dimethylbenzimidazole cobamide). The best source yet found is a culture of *Propionibacterium shermanii* (ATCC 9614) (Barker & others: J Biol Chem 235:480, 1960). The coenzymes are inactivated and converted to the vitamin form by visible light or by cyanide ion, the adenine nucleoside being removed or replaced by the cyano group (Weissbach, 1960). The methods previously used to extract the vitamin included heating in weak acid, addition of cyanide ion, and exposure to light. As a result it is likely that the coenzyme was converted to the vitamin and thus overlooked.

The coenzyme and the vitamin forms seem to function equally well in bacteria or animals, although further study is required to establish this point in the human.

Physiologic Role

Vitamin B_{12} is absorbed from the ileum. The efficient absorption of the vitamin from the intestine is dependent upon the presence of hydrochloric acid and a constituent of normal gastric juice which has been designated as **"intrinsic factor"** by Castle. The intrinsic factor is a constituent of gastric mucoprotein. It is found in the cardia and fundus of the stomach but not in the pylorus. Atrophy of the fundus and a lack of free hydrochloric acid (achlorhydria) are usually associated with pernicious anemia. Patients who have sustained total removal of the stomach will also develop a vitamin B_{12} deficiency and anemia because complete absence of intrinsic factor prevents absorption of vitamin B_{12} from their intestines, although as long as 3 years may elapse after the operation before anemia will be apparent. This is because the vitamin B_{12} stores disappear very slowly. In the liver the biologic half-life for the vitamin is estimated to be about 400 days.

Intrinsic factor concentrates prepared from one animal species are not able in all cases to increase the intestinal absorption of vitamin B_{12} in other species of animals or in man. With intrinsic factor concentrates prepared from hog mucosa, a refractory state may eventually develop in some pernicious anemia patients given also B_{12} by mouth, which suggests that a block in the absorptive mechanism has developed. However, human gastric juice remains effective in these "refractory" patients as a means of facilitating absorption of vitamin B_{12}.

At least 2 separate antigen-antibody systems specific to the stomach can be demonstrated in most patients with adult pernicious anemia. One involves gastric intrinsic factor; the other is a complement-fixing system wherein the antibodies are directed against the parietal cells of the stomach. This latter observation together with the common occurrence of atrophic gastric mucosa in pernicious anemia patients has suggested that pernicious anemia may be an autoimmune disease.

Another variety of pernicious anemia with vitamin B_{12} deficiency is that which occurs very early in the postnatal period–usually before age 2½. This appears to be due specifically to a lack of intrinsic factor, but it is not accompanied by absence of gastric acid secretion or by abnormalities in the histologic structure of the gastric mucosa. This syndrome does not appear to be related to adult pernicious anemia. It has been termed "congenital pernicious anemia." Later in childhood there may develop in other children a form of pernicious anemia which is characterized by a failure to secrete intrinsic factor and an associated achlorhydria and atrophic gastritis. Antibodies to intrinsic factor or to parietal cells have been detected in the serum of such patients. This latter form of pernicious anemia has been designated "juvenile pernicious anemia."

The serum from some patients with pernicious anemia will inhibit the intrinsic factor-mediated absorption of vitamin B_{12}. The factor responsible for this action in the serum is that of an antibody carried in the γ-globulin fraction.

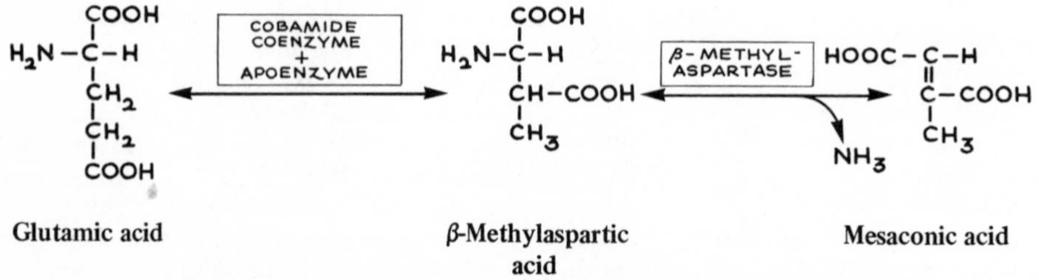

Glutamic acid — β-Methylaspartic acid — Mesaconic acid

FIG 7–11. Conversion of glutamic acid to β-methylaspartic acid.

The chemical nature of the intrinsic factor is not yet known, although it is believed to be a mucoprotein. It is destroyed by heating 30 minutes at 70–80° C and inactivated by prolonged digestion with pepsin or trypsin.

If very large doses of vitamin B_{12} (3000 μg) are given by mouth to a patient with pernicious anemia, an increase in the concentration of the vitamin in the plasma is observed. If the vitamin is given in a dose of 10–25 μg IM, the rise in its concentration in the serum is similar, and the vitamin is entirely effective without intrinsic factor. Apparently the only function of intrinsic factor is to provide for the absorption of the vitamin from the intestine, and then only when it is present in very small amounts, as it is in natural foods. Thus vitamin B_{12} itself is both the **extrinsic factor** and the **anti-pernicious anemia factor (APA)** as originally described by Castle.

The normal partial intestinal mucosal block to absorption of B_{12} is complete or almost complete in sprue and in pernicious anemia when tested by the oral administration of radioactive cobalt-labeled B_{12} followed by measurements of hepatic uptake. The defect in sprue is not corrected by the administration of intrinsic factor because it is due to a generalized defect inherent in the absorptive mechanisms in the intestinal wall. In pernicious anemia caused by a lack of intrinsic factor, the administration of a test dose of labeled B_{12} together with 75–100 ml of normal human gastric juice, or with a potent source of intrinsic factor, results in a satisfactory hepatic uptake.

The vitamin B_{12} content of the serum can be measured by microbiologic methods. According to Rosenthal & Sarett (1952) the content of vitamin B_{12} in human serum is between 0.008 and 0.042 μg/100 ml, with an average in 24 normal individuals of 0.02 μg/100 ml. In the pernicious anemia patient in relapse, the vitamin is reported to be absent or present only in very small amounts (less than 0.004 μg/100 ml).

It has been found that restoration of serum B_{12} levels to normal and complete remission of pernicious anemia can be accomplished by the oral administration of 300 μg of vitamin B_{12} per day without the supplementary use of intrinsic factor (Waife, 1963). However, oral therapy has given unpredictable responses. Consequently, administration of vitamin B_{12} by intramuscular injection is now the preferred route for treatment of pernicious anemia. There appears to be no objection to giving folic acid so long as adequate vitamin B_{12} therapy is also being provided.

Functions in Metabolism

The vitamin B_{12} (cobamide) coenzyme has been shown to catalyze the enzymatic conversion in bacterial systems of glutamate to β-methylaspartate, as shown in Fig 7–11, and in animal tissue (eg, rat or ox liver) to catalyze the isomerase reaction whereby methylmalonyl-coenzyme A is converted to succinyl-coenzyme A (see below).

Methylmalonic acid can scarcely be detected in the urine of healthy humans (less than 2 mg/day), but it is excreted in significant amounts by patients with vitamin B_{12} deficiency such as occurs in untreated patients with pernicious anemia (Cox & White, 1962). Excretion of methylmalonic acid appears to be a sensitive index of the adequacy of body stores of vitamin B_{12}. The increased levels of methylmalonic acid in the urine begin to decrease as soon as treatment is started, but normal levels do not occur until hematologic abnormalities have been corrected and the content of vitamin B_{12} in the serum has reached normal. The abnormality in methylmalonic acid metabolism may be ascribed to a lack of adequate amounts of the vitamin B_{12} cobamide coenzyme, which, as indicated in the reaction shown below, is an essential cofactor with the enzyme methylmalonyl isomerase in catalyzing conversion of methylmalonyl-Co A to succinyl-Co A.

$$H_3C{-}CH(COOH){-}C(=O){\sim}S{\cdot}CoA \xrightleftharpoons[\text{METHYL MALONYL ISOMERASE}]{\text{COBAMIDE COENZYME}} HOOC{\cdot}CH_2{-}CH_2{-}C(=O){\sim}S{\cdot}CoA$$

Methylmalonyl-Co A **Succinyl-Co A**

Methylmalonic aciduria has also been observed in infants and young children with severe metabolic acidosis who were not, however, deficient in vitamin B_{12}. The metabolic abnormality is ascribed to an inherited inability to convert methylmalonyl-Co A to succinyl-Co A because of a mutation which results in formation of an abnormal methylmalonyl isomerase enzyme protein. Rosenberg, Lilljequist, & Hsia (1968) have described the syndrome of methylmalonic aciduria in a 1-year-old male infant who during episodes of keto-acidosis excreted 800–1200 mg of methylmalonic acid per day as well as increased amounts of long-chain ketones. When the patient was given 1 mg of B_{12} IM each day, excretion of methylmalonic acid decreased to 220–280 mg/day. The authors suggest that this situation illustrates the existence in man of a vitamin B_{12} dependency state as distinguished from a B_{12} deficiency. The dependency is explained by the assumption that the mutant isomerase apoenzyme, in distinction to the normal apoenzyme, has a very low affinity for its coenzyme (cobamide). Thus it is only after relatively high concentrations of coenzyme, as would be present following daily administration of 1 mg of B_{12}, that effective binding of coenzyme to apoenzyme occurs, with resultant partial restoration of catalytic activity.

The most characteristic sign of a deficiency of vitamin B_{12} in man is the development of a macrocytic anemia or characteristic lesions of the nervous system (or both–so-called "combined system disease"). Neurologic symptoms may supervene in B_{12}-deficiency states without the prior development of anemia. In general it may be concluded that when the intake of B_{12} is low, the demand for this vitamin in hemopoiesis exceeds that for any other clinically recognizable physiologic function. Macrocytosis is therefore a sensitive indicator of a vitamin B_{12} deficiency.

The coenzyme functions of B_{12} described above cannot as yet be interpreted in relation to the role of the vitamin in hemopoiesis, but these findings certainly support the view that this vitamin functions in a fundamental manner in metabolic processes which are not limited to the hematopoietic system. The most consistent evidence of such a metabolic role for vitamin B_{12} is in connection with the neogenesis of methyl groups or as a cofactor in transmethylation reactions, as in the biosynthesis of methionine. Dinning & Young (1959) believe that vitamin B_{12} is concerned with reduction of the one-carbon moiety (formate) (Fig 7–8) to the methyl group of thymine, specifically reduction of formic acid to formaldehyde preparatory to methyl synthesis. Other experiments have been interpreted as suggesting that the vitamin is involved in nucleic acid synthesis (possibly in methylation of uridine to thymidine), and also in the synthesis of proteins in the ribosomes (see Chapter 5). Reference has also been made above to the role of B_{12} as a cofactor in the conversion of ribonucleotides to deoxyribonucleotides.

Sources & Daily Allowances

The exact amount of vitamin B_{12} required by a normal human subject is not precisely known. However, pernicious anemia patients have been satisfactorily maintained by an intramuscular injection of 45 μg of B_{12} every 6 weeks. The recommended daily requirements for vitamin B_{12} are listed in Table 21–3.

As noted below, foods of animal origin are the only important dietary sources of B_{12}. The ingestion of 1 cup of milk, 4 ounces of meat, and 1 egg per day provides 2–4 μg of vitamin B_{12}. The use of beef liver or kidney would increase the intake to 15–20 μg/day.

The amounts of B_{12} in foods are very low. The dietary sources of the vitamin are predominantly foods of animal origin, the richest being liver and kidney, which may contain as much as 40–50 μg/100 g. Muscle meats, milk, cheese, and eggs contain 1–5 μg/100 g. The vitamin is almost if not entirely absent from the products of higher plants. Symptoms including sore tongue, paresthesia, amenorrhea, and nonspecific "nervous symptoms" have been reported from Great Britain and the Netherlands in groups living exclusively on vegetable foods. These are the only instances in which a dietary deficiency of the vitamin has been discovered. A true dietary deficiency of vitamin B_{12} must be very rare. In most cases of B_{12} deficiency an intestinal absorptive defect is responsible.

It is of great interest that probably the only original source of vitamin B_{12} is microbial synthesis. There is no evidence for its synthesis by the tissues of higher plants or animals. The activity of microorganisms in synthesizing B_{12} extends to the bacteria of the intestine. This is best illustrated by the microbial flora of the rumen in ruminant animals. A vitamin B_{12} concentration of 50 μg/100 g of dried rumen contents has been reported. Presumably this accounts for the superior B_{12} content of livers from ruminant animals as compared to other animals such as the pig or rat. It is probable that the synthetic activity of the intestinal bacteria also provides B_{12} for herbivorous animals other than ruminants.

Vitamin B_{12} is excreted in the feces of human beings, and the amounts found in the feces of pernicious anemia patients may be even larger than that in normal subjects. This is due to an absorptive defect incident to a lack of intrinsic factor in these patients. It is thus apparent that B_{12} is synthesized by the bacteria present in the human intestine as well as by those in the intestine of other animals as described above. The contribution of this source of B_{12} in normal subjects is not known, but the restricted amounts available in the diet suggest that this may be an important auxiliary source of the vitamin.

CHOLINE

Choline is an important metabolite, although probably it cannot be classified as a vitamin since it is synthesized by the body. Furthermore, the quantities of choline which are required by the organism are considerably larger than most substances considered as vitamins. However, in many animal species, a deficiency of choline, or of choline precursors, leads to certain well defined symptoms which are suggestive of vitamin deficiency diseases. Disturbances in fat metabolism are most prominently evidenced by the development of fatty livers. In the young growing rat there is also hemorrhagic degeneration of the kidneys and hemorrhage into the eyeballs and other organs. Older rats and the young animals which survive the acute stage develop cirrhosis. In chicks and young turkeys, choline deficiency causes perosis, or slipped tendon disease, a condition in which there is a defect at the tibiotarsal joint of the bird. Many other animals, such as rabbits and dogs, are also susceptible to choline deficiency.

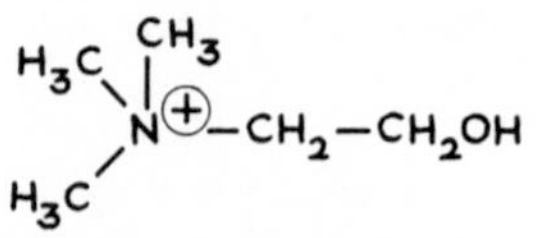

$$(H_3C)_2(CH_3)N^{\oplus}-CH_2-CH_2OH$$

Choline

Physiologic Role

Acetylcholine is well known as the chemical mediator of nerve activity. It is produced from choline and acetic acid. The reaction is preceded by the synthesis of "active acetate," ie, acetyl-coenzyme A which is formed from acetate and Co A. A source of high-energy phosphate, adenosine triphosphate, is also required. An enzyme (**acetyl thiokinase**) which catalyzes the formation of acetyl-Co A has been found in pigeon liver. After the active acetate has been formed, acetylation of choline by acetyl-Co A occurs in the presence of a second enzyme, **choline acetylase.**

This latter compound is then progressively methylated to choline, as shown in Fig 7–12. Experiments with rat liver (Wilson, 1960) show that the methyl group of methionine (by way of S-adenosyl methionine) is

ACETYL THIOKINASE

$$\underset{\text{Acetate}}{CH_3-\overset{O}{\overset{\|}{C}}-OH} \xrightarrow[\text{CoA·SH} \quad H_2O]{\text{ATP} \quad \text{AMP + PPi}} \underset{\text{Acetyl-Co A}}{CH_3-\overset{O}{\overset{\|}{C}}\sim S-CoA}$$

CHOLINE ACETYLASE

$$\underset{\text{Choline}}{HO-CH_2-CH_2-\overset{\oplus}{N}(CH_3)_3} \xrightarrow[CH_3-\overset{O}{\overset{\|}{C}}\sim S-CoA \quad \text{CoA·SH}]{} \underset{\text{Acetylcholine}}{CH_3-\overset{O}{\overset{\|}{C}}-O-CH_2-CH_2-\overset{\oplus}{N}(CH_3)_3}$$

Acetylcholine esterase (ACh-esterase) is an enzyme, present in many tissues, which hydrolyzes acetylcholine to choline and acetic acid. Its importance in nerve activity is discussed on p 486. It has been found that red blood cells can synthesize acetylcholine and that both choline acetylase and acetylcholine esterase are present in red cells. Choline acetylase has also been detected not only in brain but in skeletal muscle, spleen, and placental tissue as well. The presence of this enzyme in tissues like placenta or erythrocytes which have no nerve supply suggests a more general function for acetylcholine than that in nerve alone. According to Holland & Greig (1952), the formation and breakdown of acetylcholine may be related to cell permeability. With respect to red blood cells, they have noted that when the enzyme, choline acetylase, is inactive either because of drug inhibition or because of lack of substrate, the cell loses its selective permeability and undergoes hemolysis.

A chemical method for the determination of choline in the plasma has been developed by Appleton & others (1953). The free choline level in the plasma of normal male adults averages about 4.4 μg/ml. Analyses made over a period of several months indicate that each individual maintains a relatively constant plasma level and that this level is not increased after meals or by the oral administration of large amounts of choline. This suggests that there is a mechanism in the body to maintain the plasma choline at a constant level. Excretion into the urine is a minor factor in this regulatory process, which must therefore be metabolic in origin.

Metabolism of Choline

The biosynthesis of choline has been established in the intact animal as well as with liver slices in vitro. In this process, the amino acid serine is decarboxylated in a pyridoxal-dependent reaction to ethanolamine.

apparently the sole precursor of the choline methyl groups. Although conversion of one-carbon fragments to the methyl groups of choline does occur (Fig 7–8), this is preceded by the incorporation of the one-carbon fragment into a methyl group of methionine. Because of the contributions of the amino acids to the biosynthesis of choline, the quantity of protein in the diet affects the choline requirement and a choline deficiency is usually coincident with some degree of protein deficiency.

The first reaction in the catabolism of choline is oxidation to betaine aldehyde, which is further oxidized to betaine. This latter compound is an excellent methyl donor and in fact, choline itself functions as a methyl donor only after oxidation to betaine. After loss of a methyl group by a direct methylation reaction in which a methyl group is transferred to homocysteine to form methionine, or to activated methyl donors, betaine is converted to dimethylglycine. Oxidation of one of the methyl groups on dimethylglycine produces N-hydroxymethylsarcosine. The hydroxymethyl group is then lost and sarcosine is formed by transfer of the hydroxymethyl to tetrahydrofolic acid. It is by this oxidative reaction and transfer to folic acid derivatives that methyl groups contribute to the one-carbon (formyl) pool (Fig 7–8). Sarcosine (N-methylglycine) is now converted to glycine by oxidation and transfer of its methyl group as described above. Glycine is readily converted to serine by addition of a hydroxymethyl group derived from the one-carbon pool of the formylated tetrahydrofolic acid derivatives; finally, the decarboxylation of serine to produce ethanolamine starts the cycle of choline synthesis once again. These reactions are diagrammed in Fig 7–12.

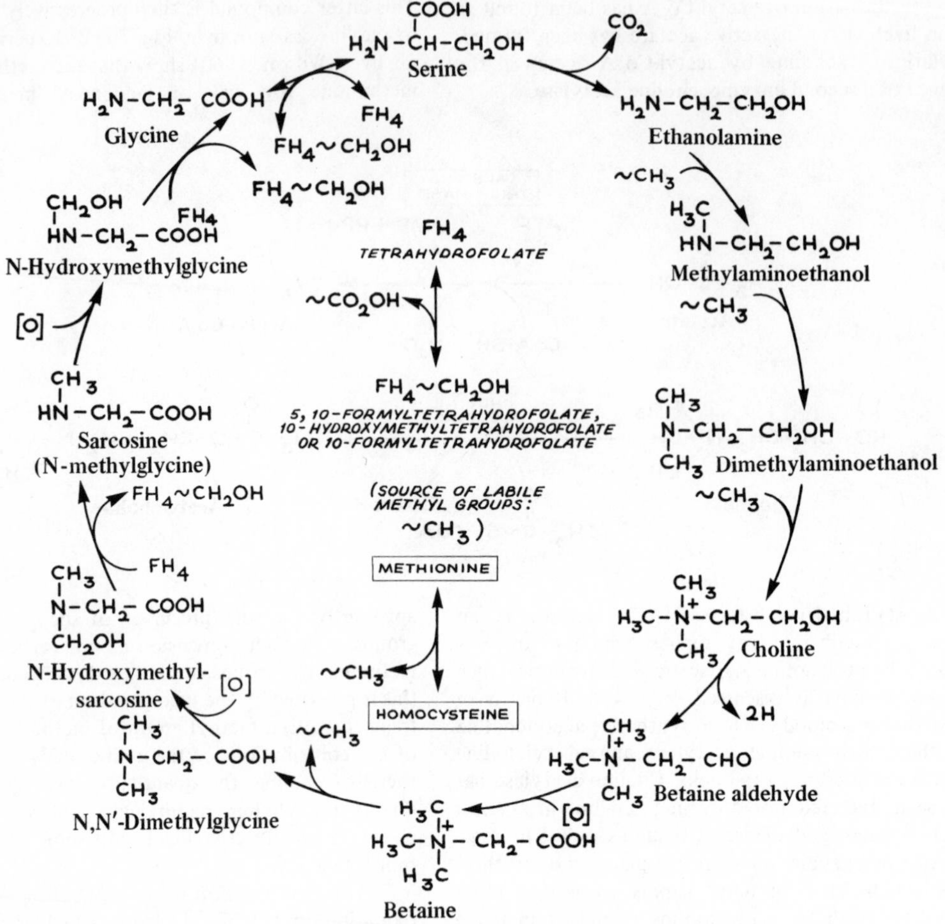

FIG 7–12. Metabolism of choline.

• • •

References

Abeles RH, Beck WS: J Biol Chem 242:3589, 1967.
Anderson WW, Dallam RD: J Biol Chem 234:409, 1959.
Appleton BN La Du Jr, Levy BB, Steele JM, Brodie BB: J Biol Chem 205:803, 1953.
Barker HA, Smyth RD, Wawszkiewicz EJ, Lee M, Wilson RM: Arch Biochem 78:468, 1958.
Barker HA, Smyth RD, Weissbach H, Munch-Petersen A, Toohey JI, Ladd JN, Volcani BE, Wilson RM: J Biol Chem 235:181, 1960.
Barker HA, Smyth RD, Weissbach H, Toohey JI, Ladd JN, Volcani BE: J Biol Chem 235:480, 1960.
Beyer RE: J Biol Chem 234:688, 1959.
Bray RC, Shemin D: J Biol Chem 238:1501, 1963.
Brin M, Shoshet SS, Davidson CS: J Biol Chem 230:319, 1958.
Bunnell RH, Keating J, Quaresimo A, Parman GK: Am J Clin Nutr 17:1, 1965.
Burns JJ: Am J Med 26:740, 1959.
Christensen HN, in: *Amino Acid Metabolism: Symposium.* McElroy WD, Glass HB (editors). Johns Hopkins Press, 1955.
Corwin LM, Schwartz K: J Biol Chem 234:191, 1959.
Cox EV, White AV: Lancet 2:853, 1962.
Crane FL, Hatef Y, Lester RL, Widmer C: Biochim biophys acta 25:220, 1957.
Davidson CS, Jandl JH: J Clin Nutr 7:711, 1959.
Dinning JS, Young RS: J Biol Chem 234:3241, 1959.
Dixon M: Science 132:1548, 1960.
Dutra de Oliviera JE, Pearson WN, Darby WJ: Am J Clin Nutr 7:630, 1959.
Eagle H, Oyama VI, Levy M, Freeman AE: J Biol Chem 226:191, 1957.
Erslev AJ, Lear AA, Castle WB: New England J Med 262:1209, 1960.
Goodman DS, Huang HS, Shiratori T: J Biol Chem 241:1929, 1966.
Gough KR, Read AE, McCarthy CF, Waters AH: Quart J Med 32:243, 1963.
Harris JW, Whittington RM, Weisman R Jr, Horrigan DL: Proc Soc Exper Biol Med 91:427, 1956.
Haussler MR, Boyce DW, Littledike ET, Rasmussen H: Proc Nat Acad Sc 68:177, 1971.
Hellman L, Burns JJ: J Biol Chem 230:923, 1958.
Herbert V: Tr A Am Physicians 75:307, 1962.
Herting DC: Am J Clin Nutr 19:210, 1966.
Holland WC, Greig ME: Arch Biochem 39:77, 1952.
Jacobs FA, Flaa RC, Belk WF: J Biol Chem 235:3224, 1960.
Janoff A, McCluskey RT: Proc Soc Exper Biol Med 110:586, 1962.
Koike M, Reed LJ: J Biol Chem 235:1931, 1960.
Lucy JA, Luscombe M, Dingle JT: Biochem J 89:419, 1963.
Lund J, DeLuca HF: J Lipid Res 7:739, 1966.
Nair P, Magar M: J Biol Chem 220:157, 1956.
Nason A, Lehman JR: J Biol Chem 222:511, 1956.
Peterkofsky A, Weissbach H: J Biol Chem 238:1491, 1963.
Pugh EL, Wakil SJ: J Biol Chem 240:4727, 1965.
Reddy SK, Reynolds MS, Price JM: J Biol Chem 233:691, 1958.
Roels OA: New York J Med 64:288, 1964.
Roels OA: Nutr Rev 24:129, 1966.
Rosenberg L, Lilljequist A-C, Hsia YE: New England J Med 278:1319, 1968.
Rosenberg L, Lilljequist A-C, Hsia YE: Science 162:805, 1968.
Rosenthal HL, Sarett HP: J Biol Chem 199:422, 1952.
Rossi CR, Gibson DM: J Biol Chem 239:1694, 1964.
Schwartz K (editor): Ann New York Acad Sc 57:378, 615, 1954.
Schwartz K: Nutr Rev 18:193, 1960.
Schwartz K, Foltz CM: J Am Chem Soc 79:3292, 1957.
Seward CR, Vaughan G, Hove EL: J Biol Chem 241:1229, 1966.
Suvarnakich K, Mann GV, Stare JF: J Nutr 47:105, 1952.
Teeri AE: J Biol Chem 196:547, 1952.
Waife SO, Jansen CJ Jr, Crabtree RE, Grinnan EL, Fouts PJ: Ann Int Med 58:810, 1963.
Weissbach H, Ladd JN, Volcani BE, Smyth RD, Barker HA: J Biol Chem 235:1462, 1960.
Wilson JD, Gibson KD, Udenfriend S: J Biol Chem 235:3213, 1960.
Wolf G, Varandani PT: Biochim biophys acta 43:501, 1960.

Bibliography

Annual Reviews of Biochemistry. Annual Reviews, Inc.
György P, Pearson WN (editors): *The Vitamins.* Vols 6 & 7. Academic Press, 1967.
Harris RS, Loraine TA, Wool IG (editors): *Vitamins and Hormones: Advances in Research and Applications.* (An annual publication.) Academic Press.
Johnson BC: *Methods of Vitamin Determination.* Burgess, 1948.
Recommended Dietary Allowances. Publication 1694. National Academy of Sciences-National Research Council, 1968.
Sebrell WH Jr, Harris RS (editors): *The Vitamins,* 2nd ed. Vols 1–5. Academic Press, 1967.

8 . . .
Enzymes

With Victor Rodwell, PhD *

Catalysts accelerate chemical reactions. Although a catalyst is a participant in a reaction and undergoes physical change during the reaction, it reverts to its original state when the reaction is complete. **Enzymes** are **protein catalysts** for chemical reactions in biologic systems. Most chemical reactions of living cells would occur very slowly were it not for catalysis by enzymes.

By contrast to nonprotein catalysts (H^+, OH^-, or metal ions), each enzyme catalyzes a small number of reactions, frequently only one. Enzymes are thus **reaction-specific** catalysts. Since **essentially all biochemical reactions are enzyme-catalyzed**, many different enzymes must exist. Indeed, for almost every organic compound in nature, and for many inorganic compounds as well, there is an enzyme in some living organism capable of reacting with it and catalyzing some chemical change.

Although the catalytic activity of enzymes was formerly thought to be expressed only in intact cells (hence the term *en-zyme,* ie, "in yeast"), most enzymes may be extracted from cells without loss of their biologic (catalytic) activity. They can therefore be studied outside the living cell. Enzyme-containing extracts are used in studies of metabolic reactions and their regulation, of structure and mechanism of action of enzymes, and even as catalysts in the industrial synthesis of biologically active compounds such as hormones and drugs. Since the enzyme content of human serum may change significantly in certain pathologic conditions, studies of serum enzyme levels provide an important diagnostic tool for the physician (see p 160).

Structure & Functions of Intracellular Components

In this and subsequent chapters reference will be made to various intracellular structures which are the sites of specific biochemical activities of the cell. The identity and the biochemical functions of many of these intracellular structures have been studied by electron microscopy, by histochemistry, and by technics involving separation by high-speed centrifugation.

Within multicellular organisms, diverse kinds of cells serve the specialized functions of the various tissues. Certain features are, however, common to all eukaryotic cells. All possess a cell membrane, a nucleus, and a cytoplasm containing cellular organelles and soluble proteins essential to the biochemical and physiologic functions of that cell. Although no "typical" cell exists, Fig 8–1 represents intracellular structures in general. In the electron micrograph of a section of an actual cell (rat pituitary gland), the cell membrane, mitochondria, and endoplasmic reticulum with associated ribosomes may be seen (Fig 8–2). In addition, the secretion granules, which in this instance are responsible for production of mammotropic (lactogenic) hormone, may also be noted.

Under a light microscope, the **cell membrane** appears only as a limiting boundary, since it is less than 10 nm thick. In electron micrographs, the membranes of many cells are seen to possess definite structures which often can be related to the specific functions of the tissue from which the cell is derived. That the membrane should possess such structures is not surprising, for it must maintain selective permeability in order to preserve within the cell the precise chemical environment necessary to the cell's function. Indeed, the mechanism of transport of metabolites in and out of cells is a highly significant aspect of cellular metabolic activity, and substances which affect transport across membranes thereby exert significant regulatory control of intracellular biochemical activities. Examples of substances which affect transport include hormones such as insulin (see p 426) and growth hormone (p 458).

If a solute of low molecular weight is present in extracellular fluid at a higher concentration than in the cell, it will tend to diffuse into the cell. The converse also holds true, for solutes tend to diffuse from a region of high to a region of low concentration (see p 495). **Passive transport** of solutes across cellular or intracellular membranes results from just such random molecular motion (see p 133). It is believed that water, for example, is passively transported across membranes. Certain solutes, however, migrate across membranes in a direction opposite to that predicted from their concentrations on either side of the membrane (eg, K^+ in cells, Na^+ in extracellular fluid). This establishes a **solute gradient** which can only be maintained by expending energy. Maintaining these gradients by **active transport** thus requires ATP, which must be supplied by the cell. Many small molecules appear to be actively transported across membranes.

In many cells, large molecules and other macromaterials enter by **pinocytosis** ("cell drinking"). An

*Associate Professor of Biochemistry, Purdue University, Lafayette, Indiana.

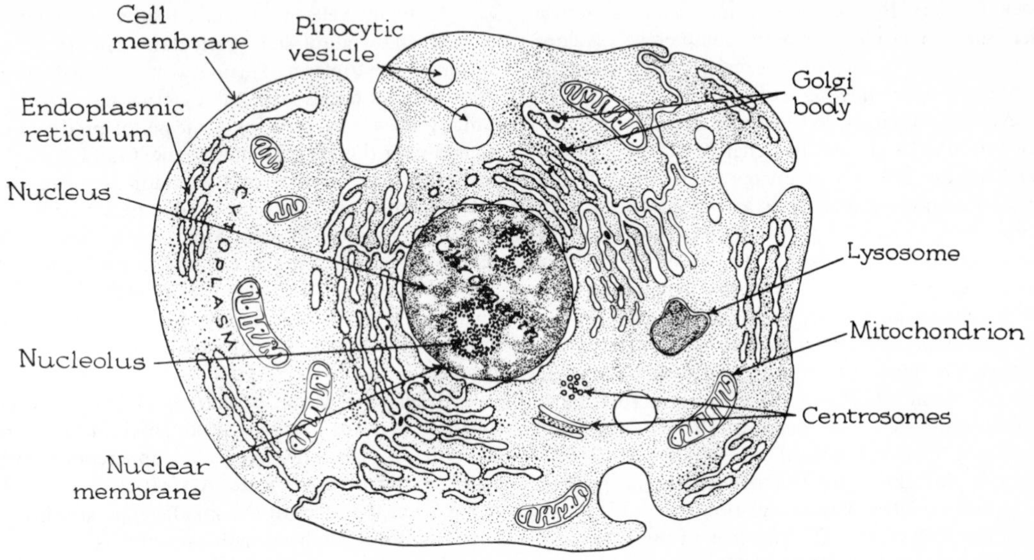

FIG 8–1. Structural components of an idealized "typical cell."

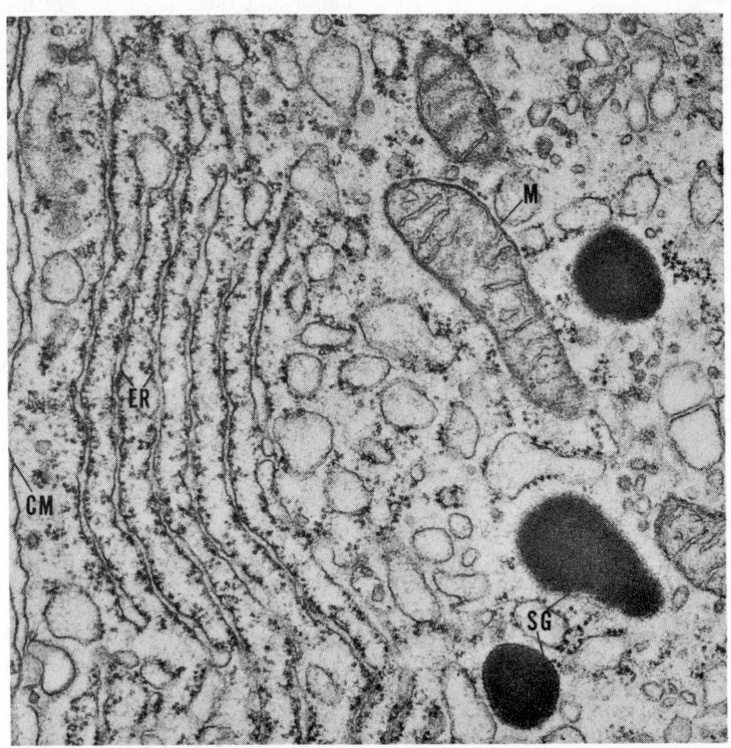

FIG 8–2. Portion of a mammotropic hormone-producing cell of rat anterior pituitary gland. (Reduced 30% from × 50,000.) Shown are several rows of endoplasmic reticulum with associated ribosomes (ER), mitochondria (M), secretion granules (SG), and a portion of cell membrane (CM). (Courtesy of R.E. Smith and M.G. Farquhar.)

inpocketing of the cell membrane forms a vesicle which surrounds and ultimately completely envelops the material to be ingested in a vacuole that enters the cytoplasm as a free-floating structure.

Among the largest cytoplasmic structures are the **mitochondria** (3–4 μm in length). These "power plants" of the cell extract energy from nutrients and trap the energy released by oxidative processes with simultaneous formation of the high-energy chemical bonds of adenosine triphosphate (ATP; see p 173). The fine structure of the mitochondrion as visualized by electron microscopy is shown diagrammatically in Fig 8–3. Note that the mitochondrion has 2 membranes, and that the inner one forms folded structures or **cristae** extending into the matrix of the structure. Each membrane is believed to consist of alternate layers of protein and lipid molecules. The respiratory chain (see p 171) is thought to be associated with the protein layer, while oxidative phosphorylation (see p 175) may involve the lipid layers. The enzymes of the citric acid cycle are located in the fluid matrix, the soluble part of the mitochondrial interior.

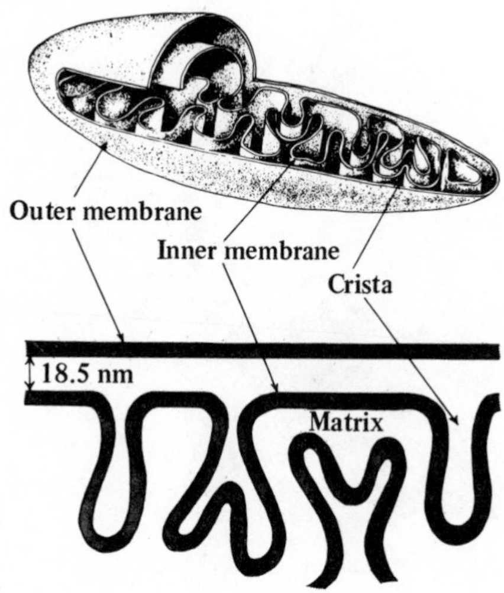

FIG 8–3. Representation of the structure of a mitochondrion.

Lysosomes, subcellular organelles approximately the same size as mitochondria, do not appear to possess internal structure. Lysosomes contain digestive enzymes which break down fats, proteins, nucleic acids, and other large molecules into smaller molecules capable of being metabolized by the enzyme systems of mitochondria. As long as the lipoprotein membrane of the lysosome remains intact, the enzymes within the lysosome are unable to act on substrates within the cytoplasm. Once the membrane is ruptured, release of lysosomal enzymes is quickly followed by dissolution (lysis) of the cell. This concept of lysosomal function as derived from studies of lysosomes from rat liver is illustrated in Fig 8–4.

In many cells a system of internal membranes termed the **endoplasmic reticulum** can be detected within the cytoplasm. This network of canaliculi may itself be continuous with the external membrane. The **Golgi bodies** may serve as a means of producing and maintaining this internal membrane. Closely associated with the inner surface of the endoplasmic reticulum are numerous granules rich in ribonucleic acid (RNA) termed **ribosomes.** The ribosomes are the sites of protein synthesis within the cell. As might be expected, the reticular system and the ribosomes are most highly developed in cells (such as those of the liver and pancreas) actively engaged in the production of proteins.

Other cytoplasmic structures include the **centrosomes** or centrioles which, although visible even under the ordinary light microscope, are apparent only when a cell is preparing to divide. At that time, the centrosomes form the poles of the spindle apparatus involved in chromosomal replication during mitosis.

The **nucleus** is characterized by its high content of chromatin, which contains most of the cellular deoxyribonucleic acid (DNA). When the cell is not in the process of dividing, the chromatin is distributed throughout the nucleus in a diffuse manner. Immediately before cell division, the chromatin assumes the organized structure of the chromosomes which will eventually be distributed equally to each daughter cell. The **nucleolus,** a discrete body within the nucleus, contains much ribonucleic acid (RNA), which under the electron microscope appears as extremely small granules resembling the ribosomes of the endoplasmic reticulum of the cytoplasm.

Intracellular Distribution of Enzymes

In addition to the morphologic details of the subcellular structures, there is great interest also in localization within the cell of various metabolic activities. The original concept of the cell as a "sack of enzymes" has given way to recognition of the cardinal significance of spatial arrangement and compartmentalization of enzymes, substrates, and cofactors within the cell. In rat liver cells, for example, the enzymes of the glycolytic pathway (see p 234) are located in the nonparticulate portion of the cytoplasm, whereas enzymes of the citric acid cycle (see p 239) are within the mitochondrion.

The metabolic functions of various cellular organelles may be studied following their separation by differential centrifugation. After rupture of the cell membrane, centrifugation of the cell contents in fields of 600–100,000 gravities (× *g**) separates the cell components into microscopically identifiable fractions: intact cells, cell debris and nuclei (600 × *g* for 5 minutes), mitochondria (10,000 × *g* for 30 minutes), microsomes (100,000 × *g* for 60 minutes), and the remaining soluble or nonsedimentable fraction. The enzyme content of each fraction may then be examined.

*× *g* refers to the amount by which the centrifugal field exceeds the force of gravity.

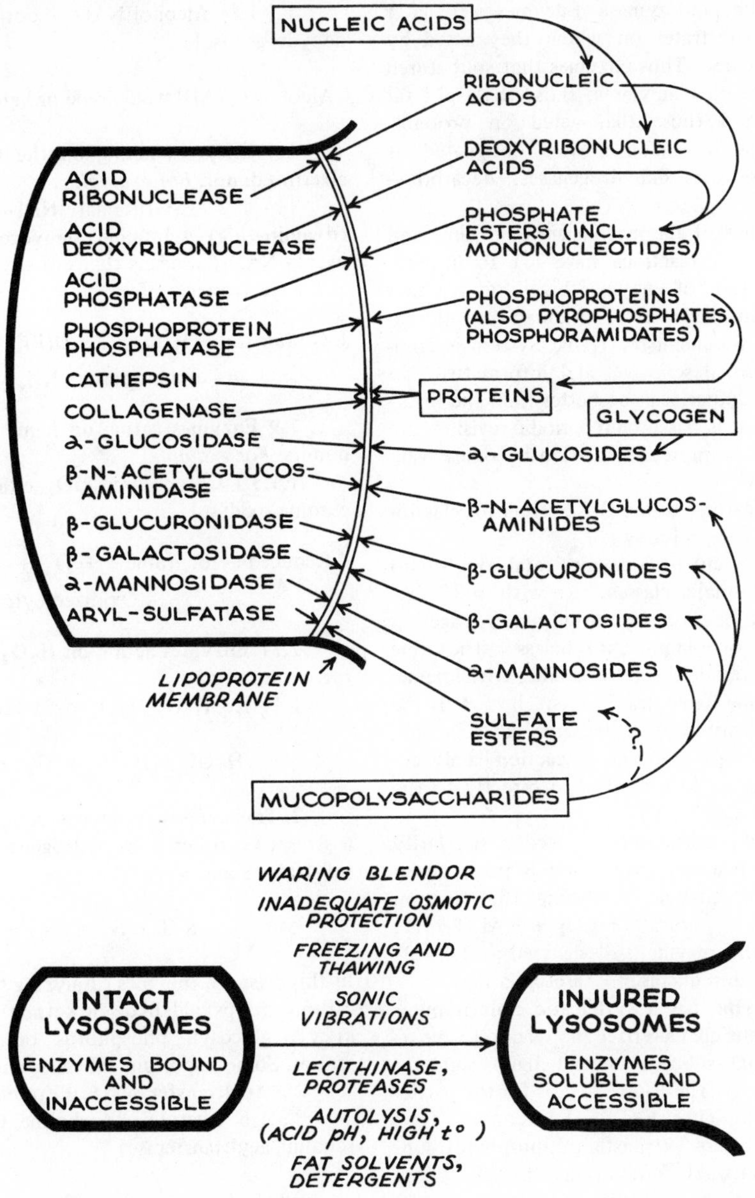

FIG 8–4. The lysosome. (Redrawn and reproduced, with permission, from De Duve C: The Lysosome Concept. Ciba Foundation Symposium: *Lysosomes.* Little, Brown, 1963.)

Localization of a particular enzyme activity in a tissue or cell in a relatively unaltered state may frequently be accomplished by histochemical procedures ("histoenzymology"). Thin (2–10 μm) frozen sections of tissue, prepared with a low-temperature microtome, are treated with a substrate for a particular enzyme. In regions where the enzyme is present, the product of the enzyme-catalyzed reaction is formed. If the product is colored and insoluble, it remains at the site of formation and serves as a marker for the localization of the enzyme. Although quantitatively reliable technics have yet to be perfected, "histoenzymology" provides a graphic and relatively physiologic picture of patterns of enzyme distribution. Enzymes for which satisfactory histochemical technics are available include the acid and alkaline phosphatases, monoamine oxidase, and a variety of dehydrogenase activities.

Enzyme Classification & Nomenclature

The function of classification is to emphasize relationships and similarities in a precise and concise way. In view of the limited information available, it was perhaps inevitable that early attempts at nomenclature and classification of enzymes produced a confused tangle of ambiguous and uninformative names such as amygdalin, ptyalin, and zymase. Enzymes were later named for the substrates on which they acted by adding the suffix **-ase**. Thus enzymes that split starch (amylon) were termed amylases; those that split fat (lipos), lipase; and those that acted on proteins, proteases. Groups of enzymes were designated as oxidases, glucosidases, dehydrogenases, decarboxylases, etc.

Recent studies of the mechanism of organic and of enzyme-catalyzed reactions have led to a more rational classification of enzymes based on reaction types and reaction mechanisms. Although this International Union of Biochemistry (IUB) System is complex, it is precise, descriptive, and informative. No classification, of course, can be better than the information on which it is based, and periodic revisions are to be anticipated as more information becomes available.

The major features of the **IUB System for classification of enzymes** are as follows:

A. Reactions (and the enzymes catalyzing them) are divided into 6 major classes, each with 4–13 subclasses. The 6 major classes are listed below, together with examples of some important subclasses. The name appearing in brackets is the more familiar trivial name.

B. The enzyme name has 2 parts. The first is the name of the substrate or substrates. The second, ending in **-ase**, indicates the **type of reaction catalyzed.** The suffix -ase is no longer attached directly to the name of the substrate.

C. Additional information, if needed to clarify the nature of the reaction, may follow in parentheses. For example, the enzyme catalyzing the reaction L-malate + NAD^+ = pyruvate + CO_2 + NADH + H^+, known as the malic enzyme, is designated as 1.1.1.37 L-malate:NAD oxidoreductase (decarboxylating).

D. Each enzyme has a systematic code number (E.C.). This number characterizes the reaction type as to class (first digit), subclass (second digit), and subsubclass (third digit). The fourth digit is for the particular enzyme named. Thus, E.C. 2.7.1.1 denotes class 2 (a transferase), subclass 7 (transfer of phosphate), subsubclass 1 (an alcohol function as the phosphate acceptor). The final digit denotes the enzyme, hexokinase, or ATP:D-hexose-6-phosphotransferase, an enzyme catalyzing phosphate transfer from ATP to the hydroxyl group on carbon 6 of glucose.

The 6 major classes of enzymes with some illustrative examples are given below.

1. Oxidoreductases. Enzymes catalyzing oxidoreductions between 2 substrates, S and S′.

$$S_{reduced} + S'_{oxidized} = S_{oxidized} + S'_{reduced}$$

This large and important class includes the enzymes formerly known either as dehydrogenases or as oxidases. Included are enzymes catalyzing oxidoreductions of CH–OH, CH–CH, C=O, CH–NH_2, and CH=NH groups. Representative subclasses include:

1.1 Enzymes acting on the CH–OH group as electron donor. *For example:*

1.1.1.1 Alcohol:NAD oxidoreductase [alcohol dehydrogenase].

$$\text{Alcohol} + NAD^+ = \text{aldehyde or ketone} + NADH + H^+$$

1.4 Enzymes acting on the CH–NH_2 group as electron donor. *For example:*

1.4.1.3 L-Glutamate:NAD(P) oxidoreductase (deaminating) [glutamic dehydrogenase of animal liver]. NAD(P) means that either NAD or NADP acts as the electron acceptor.

$$\text{L-Glutamate} + H_2O + NAD(P)^+ = \alpha\text{-ketoglutarate} + NH_4^+ + NAD(P)H + H^+$$

1.9 Enzymes acting on heme groups of electron donors. *For example:*

1.9.3.1 Cytochrome c:O_2 oxidoreductase [cytochrome oxidase].

$$4\ \text{Reduced cytochrome c} + O_2 + 4H^+ = 4\ \text{Oxidized cytochrome c} + 2H_2O$$

1.11 Enzymes acting on H_2O_2 as electron acceptor. *For example:*

1.11.1.6 H_2O_2:H_2O_2 oxidoreductase [catalase].

$$H_2O_2 + H_2O_2 = O_2 + 2H_2O$$

2. Transferases. Enzymes catalyzing a transfer of a group, G (other than hydrogen), between a pair of substrates S and S′.

$$S\text{-}G + S' = S'\text{-}G + S$$

In this class are enzymes catalyzing the transfer of one-carbon groups, aldehyde or ketone residues, and acyl, alkyl, glycosyl, phosphorus or sulfur containing groups. Some important subclasses include:

2.3 Acyltransferases. *For example:*

2.3.1.6 Acetyl-Co A:choline O-acetyltransferase [choline acyltransferase].

$$\text{Acetyl-Co A} + \text{choline} = \text{Co A} + \text{O-acetylcholine}$$

2.4 Glycosyltransferases. *For example:*

2.4.1.1 α-1,4-Glucan:orthophosphate glycosyl transferase [phosphorylase].

$$(\alpha\text{-1,4-Glucosyl})_n + \text{orthophosphate} = (\alpha\text{-1,4-Glucosyl})_{n-1} + \alpha\text{-D-glucose-1-phosphate}$$

2.7 Enzymes catalyzing transfer of phosphorus containing groups. *For example:*

2.7.1.1 ATP:D-hexose-6-phosphotransferase [hexokinase].

$$\text{ATP} + \text{D-hexose} = \text{ADP} + \text{D-hexose-6-phosphate}$$

3. Hydrolases. Enzymes catalyzing hydrolysis of ester, ether, peptide, glycosyl, acid-anhydride, C–C, C-halide, or P–N bonds. *For example:*

3.1 Enzymes acting on ester bonds. *For example:*

3.1.1.8 Acylcholine acyl-hydrolase [pseudocholinesterase].

An acylcholine + H_2O = choline + an acid

3.2 Enzymes acting on glycosyl compounds. *For example:*

3.2.1.23 β-D-Galactoside galactohydrolase [β-galactosidase].

A β-D-galactoside + H_2O = an alcohol + D-galactose

3.4 Enzymes acting on peptide bonds. The classical names (pepsin, plasmin, rennin, chymotrypsin) have been largely retained due to overlapping and dubious specificities which make systematic nomenclature impractical at this time.

4. Lyases. Enzymes that catalyze removal of groups from substrates by mechanisms other than hydrolysis, leaving double bonds.

```
X  Y
|  |
C–C = X–Y + C=C
```

Included are enzymes acting on C–C, C–O, C–N, C–S, and C-halide bonds. Representative subgroups include:

4.1.2 Aldehyde-lyases. *For example:*

4.1.2.7 Ketose-1-phosphate aldehyde-lyase [aldolase].

A ketose-1-phosphate = dihydroxyacetone phosphate + an aldehyde

4.2 Carbon-oxygen lyases. *For example:*

4.2.1.2 L-Malate hydro-lyase [fumarase].

L-Malate = fumarate + H_2O

5. Isomerases. This class includes all enzymes catalyzing interconversion of optical, geometric, or positional isomers. Some subclasses are:

5.1 Racemases and epimerases. *For example:*

5.1.1.1 Alanine racemase.

L-Alanine = D-alanine

5.2 Cis-trans isomerases. *For example:*

5.2.1.3 All trans-retinene 11-*cis-trans* isomerase [retinene isomerase].

All *trans*-retinene = 11-*cis*-retinine

5.3 Enzymes catalyzing interconversion of aldoses and ketoses. *For example:*

5.3.1.1 D-Glyceraldehyde-3-phosphate ketol-isomerase [triosephosphate isomerase].

D-Glyceraldehyde-3-phosphate = dihydroxyacetone phosphate

6. Ligases. (Ligare = "to bind.") Enzymes catalyzing the linking together of 2 compounds coupled to the breaking of a pyrophosphate bond in ATP or a similar compound. Included are enzymes catalyzing reactions forming C–O, C–S, C–N, and C–C bonds. Representative subclasses are:

6.2 Enzymes catalyzing formation of C–S bonds. *For example:*

6.2.1.4 Succinate:Co A ligase (GDP) [succinic thiokinase].

GTP + succinate + Co A = GDP + Pi + succinyl-Co A

6.3 Enzymes catalyzing formation of C–N bonds. *For example:*

6.3.1.2 L-Glutamate:ammonia ligase (ADP) [glutamine synthetase].

ATP + L-glutamate + NH_4^+ = ADP + orthophosphate + L-glutamine

6.4 Enzymes catalyzing formation of C–C bonds. *For example:*

6.4.1.2 Acetyl-Co A:CO_2 ligase (ADP) [acetyl-Co A carboxylase].

ATP + acetyl-Co A + CO_2 = ADP + Pi + malonyl-Co A

Enzyme Specificity

Nonprotein catalysts typically accelerate a wide variety of chemical reactions. By contrast, a given enzyme catalyzes only a very few reactions (frequently only one). The ability of an enzyme to catalyze one specific reaction and essentially no others is perhaps its most significant property. The rates of a multitude of metabolic processes may thus be minutely regulated by suitable changes in the catalytic efficiency of particular enzymes (see p 144). That such control be exerted via enzymes is essential if a cell, tissue, or whole organism is to function normally.

Close examination reveals that most enzymes can catalyze the same type of reaction (phosphate transfer, oxidation-reduction, etc) with several structurally related substrates. Frequently, reactions with alternate substrates take place if they are present in high concentration. Whether all of the possible reactions will occur in the living organism thus depends in part on the relative concentration of alternate substrates in the cell and their relative affinities for an enzyme. Some general aspects of enzyme specificity include:

A. Optical Specificity: With the exception of epimerases (racemases) which interconvert optical isomers, **enzymes generally show absolute optical specificity for at least a portion of a substrate molecule.** Thus, maltase catalyzes the hydrolysis of α- but not β-glucosides, while enzymes of the Embden-Meyerhof and direct oxidative pathways catalyze the interconversion of D- but not L-phosphosugars. With a few exceptions, such as the D-amino acid oxidase of kidney, the vast majority of mammalian enzymes act on the L-isomers of amino acids. Other life forms may have enzymes with equal specificity for D-amino acids.

Optical specificity may extend to a portion of the substrate molecule or to its entirety. The glycosidases provide examples of both extremes. These enzymes, which catalyze hydrolysis of glycosidic bonds between sugars and alcohols, are highly specific for the sugar portion and for the linkage (α or β), but relatively nonspecific for the alcohol portion or aglycone.

Many substrates apparently form 3 bonds with enzymes. This **"3-point attachment"** can thus confer asymmetry on an otherwise symmetric molecule. A substrate molecule, represented as a carbon atom having 3 different groups (Fig 8–5), is shown about to attach at 3 points to a planar enzyme site. If the site can be approached only from one side and only complementary atoms and sites can interact, the molecule can bind in only one way. The reaction itself–eg, dehydrogenation–may be confined to the atoms bound at sites 1 and 2 even though atoms 1 and 3 are identical. By mentally turning the substrate molecule in space, note that it can attach at 3 points to one side of the planar site with only one orientation. Consequently, atoms 1 and 3, although identical, become distinct when the substrate is attached to the enzyme. Extension of this line of reasoning can explain why the enzyme-catalyzed reduction of the optically inactive pyruvate molecule results in formation of L- and not D,L-lactate.

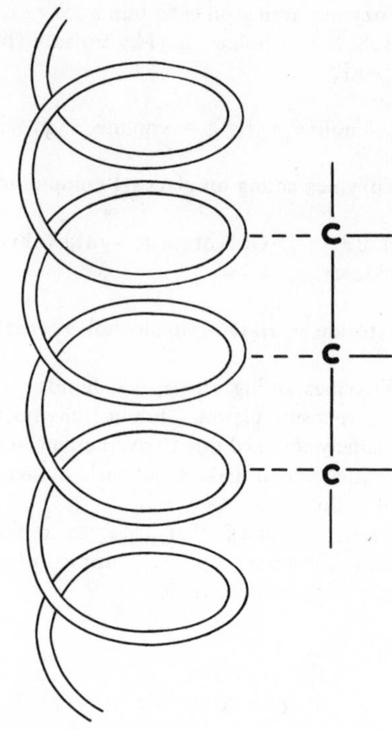

FIG 8–6. Representation of 3-point attachment of substrate to successive turns of a helical portion of an enzyme.

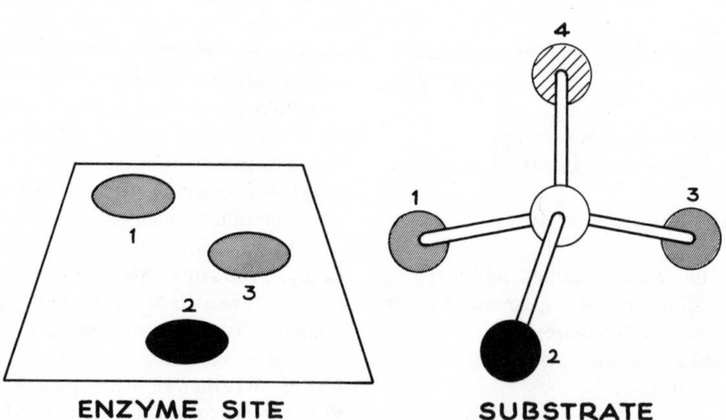

FIG 8–5. Representation of 3-point attachment of a substrate to a planar active site of an enzyme.

Representation of the substrate binding site (active site) of an enzyme as planar, while useful for illustrative purposes, is not strictly correct due to the 3-dimensional structure of the enzyme molecule. A somewhat more realistic representation might be to visualize the substrate bound to some organized portion of the enzyme molecule–perhaps a region of an α-helix, as shown in Fig 8–6.

B. Group Specificity: A particular enzyme acts only on particular chemical groupings, eg, glycosidases on glycosides, alcohol dehydrogenase on alcohols, pepsin and trypsin on peptide bonds, and esterases on ester linkages. Within these restrictions, however, a large number of substrates may be attacked, thus for example, lessening the number of digestive enzymes that might otherwise be required.

Certain enzymes exhibit a higher order of group specificity. Chymotrypsin preferentially hydrolyzes peptide bonds in which the carboxyl group is contributed by the aromatic amino acids phenylalanine, tyrosine, or tryptophan. Carboxypeptidases and aminopeptidases split off amino acids one at a time from the carboxyl- or amino-terminal end of polypeptide chains, respectively.

Although some oxidoreductases function equally well with either NAD or NADP as electron acceptor, most use one or the other preferentially. As a broad generalization, oxidoreductases functional in biosynthetic processes in mammalian systems (eg, fatty acid synthesis) tend to use NADPH as reductant, while those functional in degradative processes tend to use NAD as oxidant. Occasionally, a tissue may possess 2 oxidoreductases which differ only in their coenzyme specificity. One example is the NAD- and NADP-specific isocitrate dehydrogenases of rat mitochondria (Table 8–1).

TABLE 8–1. Distribution of NAD- and NADP-specific isocitrate dehydrogenases in mitochondria of rat tissue.*

Organ	Specific Activity (μmols/min/mg) of	
	NAD-Specific Enzyme	NADP-Specific Enzyme
Skeletal muscle	0.84	0.78
Heart	0.57	2.22
Kidney	0.28	1.20
Brain	0.25	0.054
Liver	0.16	0.33

*From Lowenstein JM: The tricarboxylic acid cycle. Page 168 in: *Metabolic Pathways.* Vol 1. Greenberg DM (editor). Academic Press, 1967.

In liver, about 90% of the NADP-specific enzyme occurs extramitochondrially. This may be concerned with biosynthetic processes, as the NAD-specific enzyme of mitochondria is specifically activated by ADP. Since ADP levels rise during depletion of ATP stores, this suggests a degradative role for the NAD-specific isocitrate dehydrogenase of mitochondria. High ADP (low ATP) levels would promote carbon flow through the citric acid cycle by activating the NAD-specific mitochondrial enzyme.

Quantitative Measurement of Enzyme Activity

The extremely small quantities present introduce problems in determining the amount of an enzyme in tissue extracts or fluids quite different from those of determining the concentration of more usual organic or inorganic substances. Fortunately, **the catalytic activity of an enzyme provides a sensitive and specific device for its own measurement.** To measure the amount of an enzyme in a sample of tissue extract or other biologic fluid, the **rate of the reaction** catalyzed by the enzyme in the sample is measured. Under appropriate conditions, **the measured rate is proportionate to the quantity of enzyme present.** Where possible, this rate is compared with the rate catalyzed by a known quantity of the highly purified enzyme. Provided that both are assayed under exactly comparable conditions and under conditions where the enzyme concentration is the rate-limiting factor (high substrate and low product concentration, and favorable pH and temperature), the micrograms (μg) of enzyme in the extract may be calculated.

Many enzymes of clinical interest are not available in a purified state. It thus is not generally possible to determine the μg of enzyme present. Results therefore are expressed in terms of arbitrarily defined enzyme units. The relative amounts of enzyme in different extracts may then be compared. Enzyme units are expressed in micromols (μmols) of substrate reacting or product produced per minute or per hour under specified assay conditions. Frequently other units are used for convenience. For example, 1 Bodansky unit of phosphatase activity is that amount of the enzyme which will catalyze the formation of 1 mg of phosphorus (as inorganic phosphate) per 100 ml of serum in 1 hour under standardized conditions of pH and temperature. In reactions involving NAD^+ (dehydrogenases), advantage is taken of the property of NADH or NADPH (but not NAD^+ or $NADP^+$) to absorb light at a wavelength of 330 nm (Fig 8–7). When NADH changes to NAD^+ (or vice versa), the absorbancy (optical density) at 340 nm changes. Under specified conditions, the rate of change in absorbancy depends directly on the enzyme activity. Hence, an enzyme unit of any dehydrogenase may be defined as a change in optical density at 340 nm of 0.001 per minute (Fig 8–7).

Isolation of Enzymes

Much knowledge about the pathways of metabolism and of regulatory mechanisms operating at the level of catalysis have come from studies of isolated, purified enzymes. Indeed, those areas of metabolism where the enzymes involved have not been purified are exactly those where information is fragmentary and controversial. In addition, reliable information concerning the kinetics, cofactors, active sites, structure, and mechanism of action also requires highly purified enzymes.

Enzyme purification involves the isolation of a specific enzyme protein from a crude extract of whole cells containing many other components. Small molecules may be removed by dialysis, nucleic acids by adsorption on charcoal, etc. The problem is to separate the desired enzyme from a mixture of hundreds of chemically and physically similar proteins. Useful methods include precipitation with varying salt concentrations (generally ammonium or sodium sulfate) or solvents (acetone or ethanol), differential heat or pH denaturation, differential centrifugation, gel filtration, and electrophoresis. Selective adsorption and elution of proteins from the cellulose anion exchange diethyl-

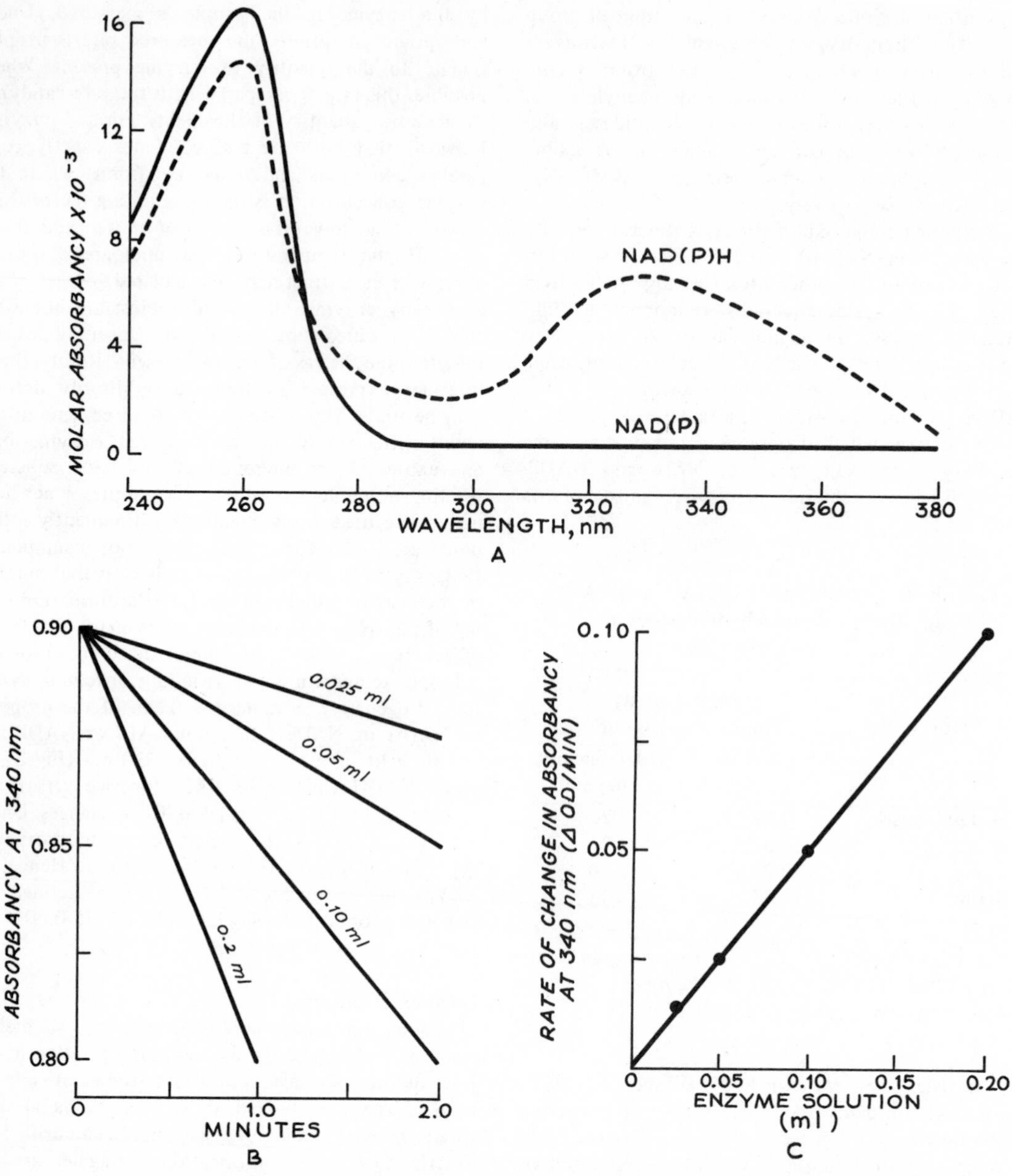

FIG 8–7. Assay of an NADH- or NADPH-dependent dehydrogenase. The rate of change in absorbancy at 340 nm due to conversion of reduced to oxidized coenzyme is observed. For an enzyme-catalyzed reaction: S + NADH + H^+ = SH_2 + NAD^+, the oxidized form of the substrate (S) and the reduced form of the coenzyme (NADH) plus buffer is added to a cuvette and light of 340 nm wavelength is passed through it. Initially *(A)* the absorbancy or optical density (OD) is high since NADH (or NADPH) absorbs at this wavelength. On addition of 0.025–0.2 ml of a standard enzyme solution, the OD decreases *(B)*. Plotting the rate of change in OD per minute versus ml of enzyme used produces a more or less linear calibration curve *(C)*. Using this calibration curve, a given rate of change in absorbency at 340 nm may be related to the activity of the standard enzyme solution used to prepare (C). If the enzyme concentration of this is known, results may be expressed in terms of μg of enzyme present per ml of unknown tissue fluid. More frequently, results are expressed in enzyme units.

aminoethylcellulose (DEAE cellulose) and the cation exchanger carboxymethylcellulose (CM cellulose) have also been extremely successful for extensive and rapid purification. Recently, **affinity column chromatography** has been used with great success. A small molecule–eg, a substrate analogue–is bonded chemically to an inert support. Proteins which interact strongly with this material (eg, an enzyme whose active site "recognizes" the analogue) may then be separated from other proteins.

Ultimately, crystallization of the enzyme may be achieved, generally from an ammonium sulfate solution. Crystallinity does not, however, imply homogeneity. Exhaustive physical, chemical, and biologic tests must still be applied as criteria of purity.

The progress of a typical enzyme purification for a liver enzyme with good recovery and 490-fold overall purification is shown in Table 8–2. Note how specific activity and recovery of initial activity are calculated. The aim is to achieve the maximum specific activity (enzyme units per mg protein) with the best possible recovery of initial activity.

TABLE 8–2. Summary of a typical enzyme purification scheme.

Enzyme Fraction	Total Activity (enzyme units)	Total Protein (mg)	Specific Activity (enzyme units/mg)	Overall Recovery (%)
Crude liver homogenate	100,000	10,000	10	(100)
100,000 × *g* supernatant liquid	98,000	8,000	12.2	98
40–50% $(NH_4)_2SO_4$ precipitate	90,000	1,500	60	90
20–35% acetone precipitate	60,000	250	240	60
DEAE column fractions 80–110	58,000	29	2,000	58
43–48% $(NH_4)_2SO_4$ precipitate	52,000	20	2,600	52
First crystals	50,000	12	4,160	50
Recrystallization	49,000	10	4,900	49

BIOENERGETICS

The energy changes associated with biochemical reactions constitute the study of **bioenergetics** or **biochemical thermodynamics**. A chemical change is said to be spontaneous if accompanied by a decrease in chemical potential (roughly the chemical equivalent of potential energy). Examples of spontaneous physical processes include the unwinding of a tightly-coiled spring or the flow of water from a raised reservoir to a lower level. Examples of spontaneous chemical changes include the combustion of gasoline, sugar, or glycogen to CO_2 and water, the combination of hydrogen with oxygen to form water, or the hydrolysis of a protein to its constituent amino acids. In each case, the products of the reaction are at a lower chemical potential (energy state) than were the reactants.

A spontaneous chemical or physical process does not necessarily take place rapidly. **The potential chemical energy of covalent bonds may be stored for future release and utilization.** Just as the energy of a coiled spring or the hydrostatic pressure of water may be stored, so also may the chemical potential of blood glucose, muscle or liver glycogen, or ATP be stored for future use. If all spontaneous processes took place rapidly, life would not be possible. **The release of stored chemical energy by accelerating the rates of spontaneous reactions is one vital function of enzymes.**

Chemical thermodynamics is the study of overall energy changes in chemical reactions. Both living and nonliving systems obey the same laws of thermodynamics. Spontaneous reactions are said to be **exergonic** or energy-yielding, and nonspontaneous reactions **endergonic** or energy-requiring. If the reaction A → B is exergonic, it follows that the reaction B → A is endergonic. To compare several exergonic (or endergonic) reactions one with another, their relative spontaneities must be expressed in quantitative terms. This is done in terms of the overall change in free energy (ΔG or ΔF*) for a reaction.

The reaction A → B will be spontaneous in the direction shown at high concentrations of A and low concentrations of B, but spontaneous in the reverse direction only at sufficiently high B and low A concentrations. A standard set of reactant and product concentrations is therefore chosen (1 M concentration of reactants and products†). A and B are mixed at 1 M concentration and the reaction allowed to reach equilibrium. The ΔG for this chemical change is referred to as the **standard change in free energy, ΔG^0**, for the reaction. If the standard state reaction is spontaneous as written (A → B), the ΔG^0 value is negative. If it is spontaneous in the reverse direction as written (B → A), the ΔG^0 is positive. A redefinition of **a spontaneous process is one with a negative value of ΔG.**

ΔG^0 for A → B may be positive (nonspontaneous). Nevertheless at sufficiently high concentrations of A and low concentrations of B (non-standard state conditions), ΔG may be negative. The reaction will be

*Some texts use ΔG and some ΔF. The terms are synonymous.
†The concentration of water is arbitrarily defined as 1 M. Strictly speaking, activity rather than concentration should be used. Here, the reference is to concentration, however.

spontaneous under the latter condition. **In biochemical systems, product concentrations frequently are kept low either by transfer across a semipermeable membrane or by further metabolism to other compounds. In biologic systems it is therefore commonplace to encounter reactions for which ΔG^0 is positive, but because the prevailing concentrations of reactants and products are so far from standard state conditions, ΔG for the actual reactant and product concentrations is negative.** The relative spontaneity of chemical reactions is always expressed in tables in terms of standard free energy changes, ΔG^0. In biochemical systems, $\Delta G^{0\prime}$ values are tabulated. These are ΔG^0 values at pH 7.0. Where H^+ is a reactant or product, it too must be at 1 M concentration if ΔG^0 is to be calculated. $\Delta G^{0\prime}$, ie, the ΔG^0 at a hydrogen ion activity of 10^{-7} M (close to the physiologic activity), is more useful for biochemical calculations.

ENZYMES & HOMEOSTASIS

The thermodynamic principles outlined above apply equally to enzyme-catalyzed and to noncatalyzed reactions. Although it is hoped ultimately to explain physiologic processes in terms of simple chemical events, our present knowledge of the chemical and physical factors responsible for enzyme specificity, catalytic efficiency, and regulation of enzyme activity is incomplete, although substantial progress has been made in recent years.

The concept of homeostatic regulation of the internal milieu advanced by Claude Bernard the latter half of the 19th century stressed the ability of animals to maintain the constancy of their intracellular environments. This implies not only that all the necessary enzyme-catalyzed reactions proceed, but that they proceed at rates responsive to changes in both the internal and external environment. A cell or organism might be defined as diseased when it fails to respond or responds inadequately or incorrectly to an internal or external stress. Knowledge of factors affecting the rates of enzyme-catalyzed reactions is essential not only to understand homeostasis in normal cells but also to comprehend the molecular basis of disease.

All chemical reactions, including enzyme-catalyzed reactions, are to some extent reversible.* Within living cells, however, reversibility may not in fact occur because reaction products are promptly removed by further reactions catalyzed by other enzymes. The flow of metabolites in a living cell is analogous to the flow of water in a water main. Although the main can transfer water in either direction, in practice the flow is unidirectional. Metabolite flow in the living cell also is largely unidirectional. True equilibrium, far from being characteristic of life, is approached only on the death of the cell.

The living cell is a steady-state system maintained by a unidirectional flow of metabolites (Fig 8–8). In the mature cell, the average concentration of a particular chemical compound remains relatively constant over considerable periods of time.† The flexibility of the steady-state system is well illustrated in the delicate shifts and balances by which an organism maintains the constancy of the internal environment in spite of wide variations in food, water, and mineral intake, work output, or external temperature.

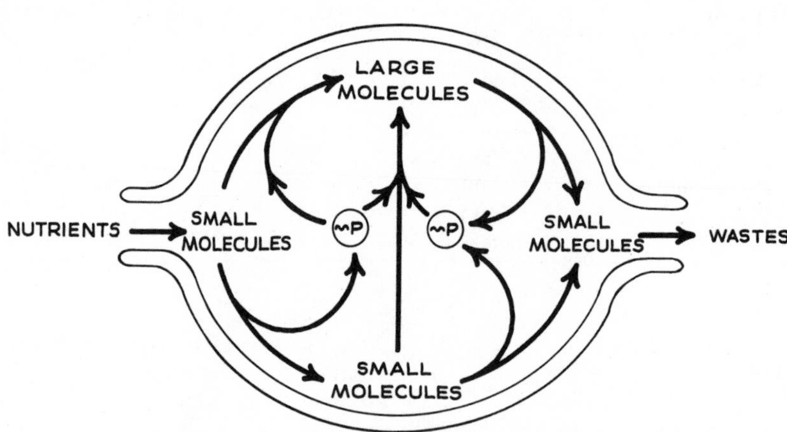

FIG 8–8. An idealized cell in steady state.

*A readily reversible reaction has a small numerical value of ΔG. One with a large negative value for ΔG (> −5000 cal) might be termed "effectively irreversible" in most biochemical situations.

†Short-term oscillations of metabolite concentrations and of enzyme levels do occur, however, and are of profound physiologic importance.

FACTORS AFFECTING THE VELOCITY OF ENZYME-CATALYZED REACTIONS

To understand how enzymes control the rates of individual reactions and of overall metabolic processes, we must review briefly how certain factors affect the rates of chemical reactions in general.

The **kinetic** or **collision theory** states that for molecules to react they must collide and must also possess sufficient energy to overcome the **energy barrier for reaction.** If the molecules have sufficient kinetic energy to react, anything that increases the frequency of collision between molecules will increase their rate of reaction. Factors that decrease either the frequency of collision or the kinetic energy will decrease the rate of reaction.

If some molecules have insufficient energy to react, factors such as increased temperature, which increases their kinetic energy, will increase the rate of the reaction. These concepts are illustrated diagrammatically in Fig 8–9. In A none, in B a portion, and in C all of the molecules have sufficient kinetic energy to overcome the energy barrier for reaction.

Molecules are in motion at all temperatures above absolute zero (–273° C), the temperature at which all molecular motion ceases. Concrete evidence of molecular motion is provided by the phenomenon of diffusion. This may be seen by use of a colored solute or gas which will in time become uniformly distributed throughout a solvent or container. With increasing temperature, the rate of diffusion (a result of increased molecular motion due to increased kinetic energy) increases. The pressure of a gas results from gas molecules colliding with the container walls. As the temperature of the gas increases, molecular motion, and hence the number of collisions with the vessel walls, increases. In a rigid container this causes increased pressure; in a flexible container it causes expansion. Lowering the temperature decreases the frequency of collisions with the container walls, causing a drop in pressure or a contraction in volume.

In the absence of enzymic catalysis, many chemical reactions proceed exceedingly slowly at the temperature of living cells. However, even at this temperature molecules are in active motion and are undergoing collision. **They fail to react rapidly because most possess insufficient kinetic energy to overcome the energy barrier for reaction.** At a considerably higher temperature (and higher kinetic energy), the reaction will occur more rapidly. That the reaction takes place at all shows that it is spontaneous (ΔG = negative). At the lower temperature it is spontaneous but slow; at the higher temperature, spontaneous and fast. What **enzymes** do is to **make spontaneous reactions proceed rapidly under the conditions prevailing in living cells.**

The mechanism by which enzymes accelerate reactions may be illustrated by a mechanical analogy. Consider a boulder on the side of a hill (Fig 8–10 [A]).

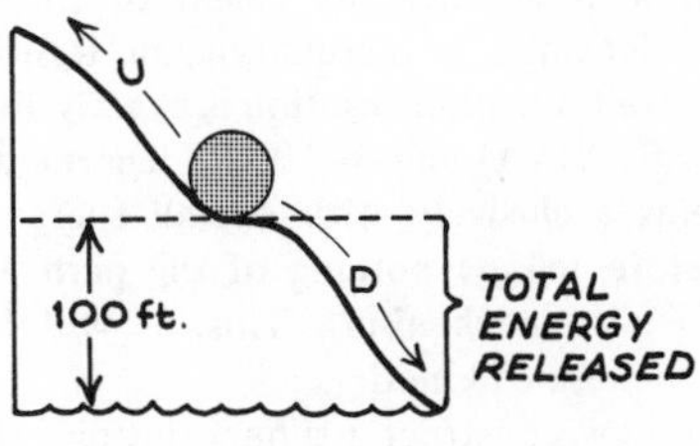

FIG 8–10 (A).

Although the boulder might move up or down the hill, neither reaction is probable or likely to proceed rapidly. It is necessary to supply a small amount of energy to send the boulder rolling downhill. This energy represents the **energy barrier** for reaction D. Similarly, the energy required to move the boulder uphill corresponds to the energy barrier for reaction U.

Consider now the same boulder on a different hill, but the same height above ground level (Fig 8–10 [B]).

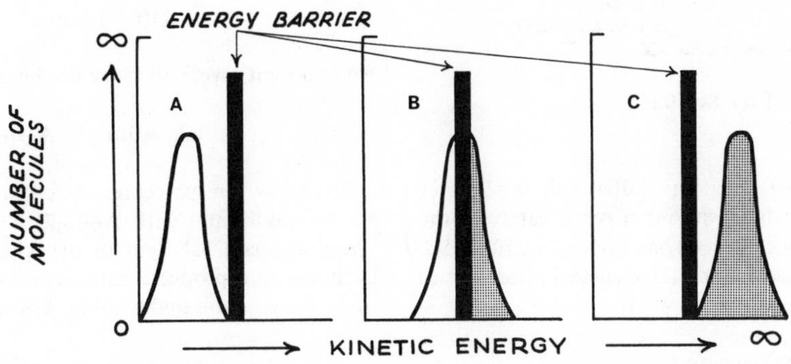

FIG 8–9. The energy barrier for chemical reactions.

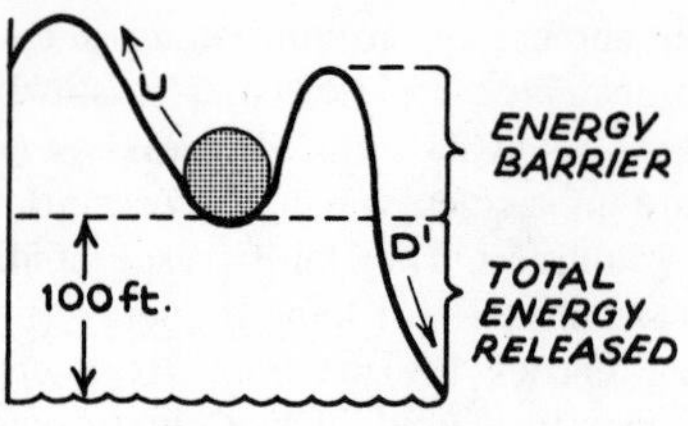

FIG 8–10 (B).

The energy barrier for reaction D is now far greater. Note, however, that since the energy supplied in moving the boulder up from the initial position to the hump is released in its fall to ground level, the total energy released is the same as in Fig 8–10(A). This illustrates the concept that the **overall energy changes in chemical reactions are independent of the path or mechanism of the reaction.** The mechanism of the reaction determines the height of the hump or energy barrier only. In thermodynamic terms, the ΔG for the overall downhill reaction is exactly the same in both Figs 8–10 (A) and 8–10 (B). **Thermodynamics, which deals exclusively with overall energy changes, can therefore tell us nothing of the path a reaction follows (ie, its mechanism).** This, as will be shown below, is the task of kinetics.

If we now construct a tunnel through the energy barrier (Fig 8–10 [C]), reaction D would become more probable (proceed faster). Although ΔG remains the same, the activation energy requirement is reduced. The tunnel removes or lowers the energy barrier for the reaction. Enzymes may be considered to lower energy barriers for chemical reactions in roughly this way—**by providing an alternate path with the same overall change in energy**, ie, by "tunneling through" the energy barrier.

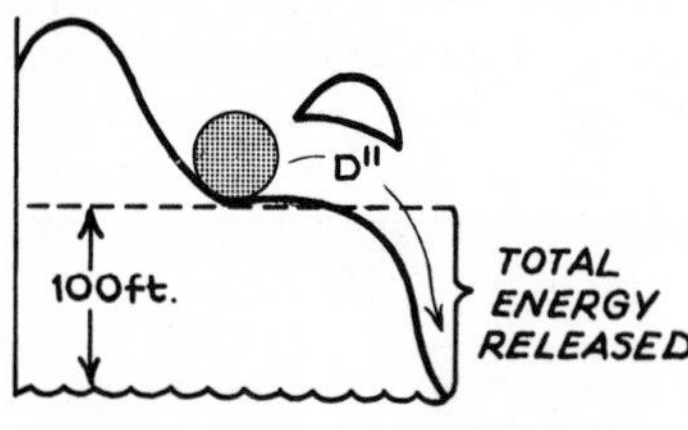

FIG 8–10 (C).

Note also that since the initial and final states remain the same with or without enzymic catalysis, the presence or absence of catalysts does not affect ΔG. This is determined solely by the chemical potentials of the initial and final states.

Effect of Reactant Concentration

At high reactant concentrations, both the number of molecules with sufficient energy to react and their frequency of collision is high. This is true whether all or only a fraction of the molecules have sufficient energy to react. For reactions involving 2 different molecules, A and B,

$$A + B \rightarrow AB$$

doubling the concentration either of A or of B will double the reaction rate. Doubling the concentration of both A and B will increase the probability of collision 4-fold. The reaction rate therefore increases 4-fold. **The reaction rate is proportionate to the concentrations of the reacting molecules.** Square brackets ([ ]) are used to denote molar concentrations;* α means "proportionate to." The rate expression is:

$$\text{Rate } \alpha \text{ [reacting molecules]}$$

or

$$\text{Rate } \alpha \, [A][B]$$

For the situation represented by

$$A + 2B \rightarrow AB_2$$

the rate expression is given by

$$\text{Rate } \alpha \, [A][B][B]$$

or

$$\text{Rate } \alpha \, [A][B]^2$$

For the general case where n molecules of A react with m molecules of B

$$nA + mB \rightarrow A_nB_m$$

the rate expression is

$$\text{Rate } \alpha \, [A]^n[B]^m$$

Since all chemical reactions are reversible, for the reverse reaction:

$$A_nB_m \rightarrow nA + mB$$

the appropriate rate expression is

$$\text{Rate } \alpha \, [A_nB_m]$$

We represent reversibility by double arrows:

$$nA + mB \rightleftharpoons A_nB_m$$

which reads: "n molecules of A and m molecules of B are in equilibrium with A_nB_m." We may replace the "proportionate to" symbol (α) with an equality sign by inserting a proportionality constant, k, characteristic of the reaction under study. For the general case

$$nA + mB \rightleftharpoons A_nB_m$$

*Strictly speaking, molar activities rather than concentrations should be used.

expressions for the rates of the forward ($Rate_1$) and back ($Rate_{-1}$) reactions are:

$$Rate_1 = k_1 [A]^n [B]^m$$

and

$$Rate_{-1} = k_{-1} [A_nB_m]$$

When the rates of the forward and back reactions are equal, the system is said to be **at equilibrium**, ie,

$$Rate_1 = Rate_{-1}$$

Then $$k_1 [A]^n [B]^m = k_{-1} [A_nB_m]$$

and $$\frac{k_1}{k_{-1}} = \frac{[A_nB_m]}{[A]^n [B]^m} = K_{eq}$$

The ratio of k_1 to k_{-1} is termed the **equilibrium constant, K_{eq}**. The following important properties of a system at equilibrium should be kept in mind.

1. The equilibrium constant is the ratio of the reaction rate constants k_1/k_{-1}.

2. At equilibrium the reaction rates (not the reaction rate constants) **of the forward and back reactions are equal.**

3. Equilibrium is a dynamic state. Although no **net** change in concentration of reactant or product molecules occurs at equilibrium, A and B are continually being converted to A_nB_m and vice versa.

4. The equilibrium constant may be given a numerical value if we know the concentrations of A, B, and A_nB_m at equilibrium. The equilibrium constant is related to ΔG^0 as follows:

$$\Delta G^0 = -RT \ln K_{eq}$$

R is the gas constant and T the absolute temperature. Since these are known, **knowledge of the numerical value of K_{eq} permits one to calculate a value for ΔG^0.** If the equilibrium constant is greater than 1, the reaction is spontaneous, ie, the reaction as written (from left to right) is favored. If it is less than 1, the opposite is true, ie, the reaction is more likely to proceed from right to left. In terms of the mechanical analogy (Fig 8–10), **if the equilibrium constant is greater than 1, the reaction from left to right is "downhill" and the reverse reaction "uphill."** Note, however, that although the equilibrium constant for a reaction indicates the **direction** in which a reaction is spontaneous, it does not indicate whether it will take place **rapidly**. That is, it does not tell us anything about the **magnitude of the energy barrier** for reaction. This follows from the fact that K_{eq} determines ΔG^0, which previously was shown to concern only initial and final states. **Reaction rates depend on the magnitude of the energy barrier, not that of ΔG^0.**

Most factors affecting the velocity of enzyme-catalyzed reactions do so by **changing reactant concentration.** These include:

A. Enzyme Concentration: The **initial velocity, v,** of an enzyme-catalyzed reaction is **directly proportionate to the enzyme concentration [E].** (See Fig 8–7 [C].) The initial velocity is that measured when almost no substrate has reacted. That the rate is not always proportionate to enzyme concentration may be seen by considering the situation at equilibrium. Although the reaction is proceeding, the rate of the reverse reaction equals it. In most practical situations it thus will appear that the reaction velocity is zero.

The enzyme is a reactant that combines with substrate forming an **enzyme-substrate complex, ES,** which decomposes to form a product, P, and free enzyme:

$$E + S \rightleftharpoons ES \rightleftharpoons E + P$$

Note that although the rate expressions for the forward, back and overall reactions include the term [E]:

$$E + S \rightleftharpoons E + P$$

$$Rate_1 = k_1 [E][S]$$

$$Rate_{-1} = k_{-1} [E][P]$$

in the expression for the overall equilibrium constant, [E] cancels out.

$$K_{eq} = \frac{k_1}{k_{-1}} = \frac{[E][P]}{[E][S]} = \frac{[P]}{[S]}$$

The enzyme concentration thus has no effect on the equilibrium constant. Stated another way, since enzymes affect rates, not rate constants, they cannot affect K_{eq}, which is a ratio of rate constants. **The K_{eq} of a reaction is the same regardless of whether equilibrium is approached with or without enzymatic catalysis** (recall ΔG^0). In terms of the mechanical analogy (Fig 8–10), enzymes "dig tunnels" and change the path of the reaction, but do not affect the initial and final positions of the boulder which determine K_{eq} and ΔG^0.

B. Substrate Concentration: If the concentration of the substrate [S] is increased while all other conditions are kept constant, the **measured initial velocity, v** (the velocity measured when very little substrate has reacted), increases to a maximum value, V, and no further (Fig 8–11). The velocity increases as the substrate concentration is increased up to a point where the enzyme is said to be "saturated" with substrate. The reason that the measured initial velocity reaches a maximal value and is unaffected by further increases in substrate concentration is that, even at very low substrate concentrations, the substrate is still present in excess of the enzyme by a large molar ratio. For example, if an enzyme with a molecular weight of 100,000 daltons acts on a substrate with a molecular weight of 100 daltons and both are present at a concentration of 1 mg/ml, there are 1000 mols of substrate for every mol of enzyme. More realistic figures might be

$$[E] = 0.1\ \mu g/ml = 10^{-9} \text{ molar}$$
$$[S] = 0.1\ mg/ml = 10^{-3} \text{ molar}$$

giving a 10^6 molar excess of substrate over enzyme. Even if [S] is decreased 100-fold, substrate is present in 10,000-fold molar excess over enzyme.

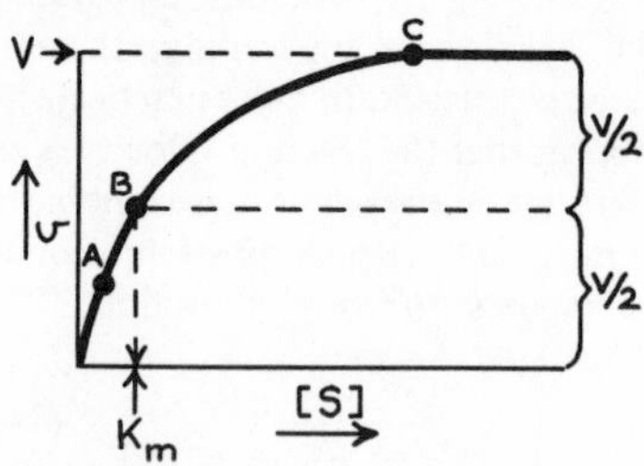

FIG 8–11. Effect of substrate concentration on the velocity of an enzyme-catalyzed reaction.

The situation at points A, B, and C in Fig 8–11 is illustrated in Fig 8–12. At points A and B not all the enzyme present is combined with substrate, even though there are many more molecules of substrate than of enzyme. This is because the equilibrium constant for the reaction E + S ⇌ ES is not infinitely large. **At points A or B, increasing or decreasing [S] will therefore increase or decrease the amount of E associated with S as ES, and v will thus depend on [S].** At C, essentially all the enzyme is combined with substrate, so that a further increase in [S], although it increases the frequency of collision between E and S, cannot result in increased rates of reaction since no free enzyme is available to react.

Case B depicts a situation where exactly ½ the enzyme molecules "hold" or are "saturated with" substrate. The velocity is accordingly **½ the maximal velocity** attainable at that particular enzyme concentration. **The substrate concentration that produces half-maximal velocity, termed the K_m value or Michaelis constant,** may be determined experimentally by graphing v as a function of [S] (Fig 8–11).

The Michaelis-Menten expression

$$v = \frac{V[S]}{K_m + [S]}$$

describes the behavior of many enzymes as substrate concentration is varied. The dependence of the initial velocity of an enzyme-catalyzed reaction on [S] and on K_m may be illustrated by evaluating the Michaelis-Menten equation as follows:

1. When [S] is very much less than K_m (point A in Figs 8–11 and 8–12). Adding [S] to K_m now changes its value very little, so the [S] term is dropped from the denominator. Since V and K_m are both constants, we can replace their ratio by a new constant, K. [≈ means "approximately equal to."]

$$v = \frac{V[S]}{K_m + [S]},\quad v \approx \frac{V[S]}{K_m} \approx \frac{V}{K_m}[S] \approx K[S]$$

In other words, **when the substrate concentration is considerably below that required to produce half-maximal velocity (the K_m value), the velocity, v, depends upon the substrate concentration [S].**

2. When [S] is very much greater than K_m (point C, Figs 8–11 and 8–12). Now adding K_m to [S] changes the value of [S] very little, so the term K_m is dropped from the denominator.

$$v = \frac{V[S]}{K_m + [S]},\quad v \approx \frac{V[S]}{[S]} \approx V$$

This states that when the substrate concentration [S] far exceeds the K_m value, the measured velocity, v, is maximal, V.

3. When [S] = K_m (point B, Figs 8–11 and 8–12),

$$v = \frac{V[S]}{K_m + [S]},\quad v = \frac{V[S]}{[S] + [S]} = \frac{V[S]}{2[S]} = \frac{V}{2}$$

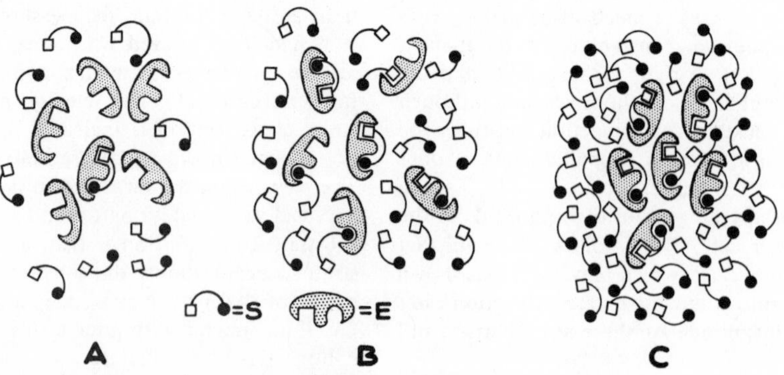

FIG 8–12. Representation of an enzyme at low *(A)*, high *(C)*, and at the K_m concentration of substrate *(B)*. (Points A, B, and C correspond to those of Fig 8–11.)

This states that **when the substrate concentration is equal to the K_m value, the observed velocity, v, is half-maximal.** It also tells how **to evaluate K_m,** namely, to **find the substrate concentration where the velocity is half-maximal.**

Since few enzymes give saturation curves which readily permit evaluation of V (and hence of K_m) when v is plotted versus S, it is convenient to rearrange the Michaelis-Menten expression to simplify evaluation of K_m and V. The Michaelis-Menten equation may be inverted and factored as follows:

$$v = \frac{V[S]}{K_m + [S]}$$

Invert: $$\frac{1}{v} = \frac{K_m + [S]}{V[S]}$$

Factor: $$\frac{1}{v} = \frac{K_m}{V} \times \frac{1}{[S]} + \frac{[S]}{V[S]}$$

$$\frac{1}{v} = \frac{K_m}{V} \times \frac{1}{[S]} + \frac{1}{V}$$

This is the equation for a **straight line**

$$y = ax + b$$

where if y, or 1/v, is plotted as a function of x, or 1/[S], the y intercept, b, is 1/V, and the slope, a, is K_m/V. The negative x intercept may be evaluated by setting y = o. Then

$$x = -\frac{b}{a} = -\frac{1}{K_m}$$

From the **double-reciprocal or Lineweaver-Burk plot,** K_m may be estimated (Fig 8–13) either from the slope and y intercept or from the negative x intercept. Since [S] is expressed in molarity, the dimension of K_m is molarity or mols per liter. Velocity, v, may be expressed in any units, since **K_m is independent of [E].** The double-reciprocal treatment requires relatively few points to define K_m, and is therefore the method most often used to determine K_m. K_m values, apart from their usefulness in interpretation of the mechanisms of enzyme-catalyzed reactions, are of considerable practical value. At a substrate concentration of 100 times the K_m value, the enzyme will act at essentially maximum rate. This is generally desirable in the assay of enzymes. The **K_m value tells how much substrate to use.** The double-reciprocal treatment finds extensive use in the evaluation of inhibitors (see p 146).

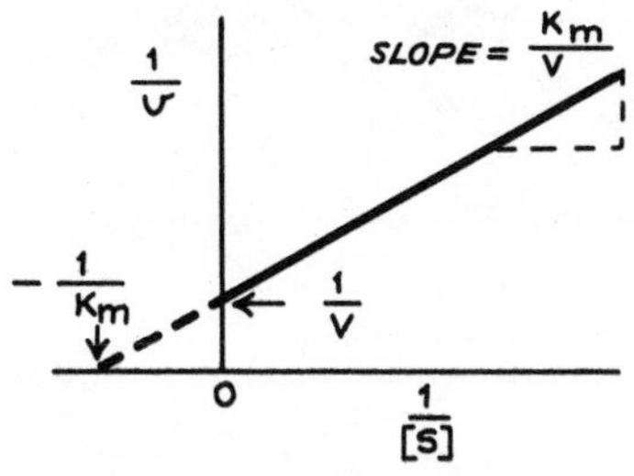

FIG 8–13. Double-reciprocal or Lineweaver-Burk plot of $\frac{1}{v}$ versus $\frac{1}{[S]}$ used for graphic evaluation of K_m and V.

The Michaelis-Menten expression may be rearranged in other ways to give the equation of a straight line. If, for example, the substrate concentration [S] is plotted on the x axis and [S]/v on the y axis, the negative x intercept gives $-K_m$ directly. V may then be evaluated from the y intercept, which is K_m/V. Other forms are possible.

C. Temperature: Over a limited range of temperatures, the velocity of enzyme-catalyzed reactions increases as temperature rises. The exact ratio by which the velocity changes for a 10° C temperature rise is the **Q_{10}, or temperature coefficient.** The velocity of many biologic reactions roughly doubles with a 10° C rise in temperature (Q_{10} = 2), and is halved if the temperature is decreased by 10° C. Many physiologic processes—eg, the rate of contraction of an excised heart—consequently exhibit Q_{10} of about 2.

When the rate of enzyme-catalyzed reactions is measured at several temperatures, the result shown in Fig 8–14 is typical. There is an optimal temperature at which the reaction is most rapid. Above this, the reaction rate decreases sharply, generally due to heat-denaturation of the enzyme.

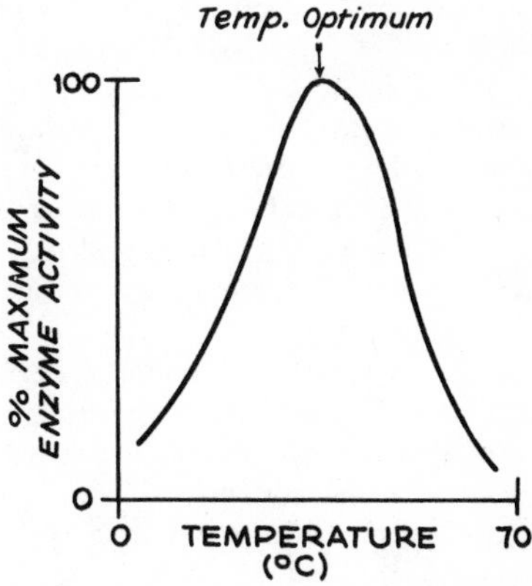

FIG 8–14. Effect of temperature on the velocity of a hypothetical enzyme-catalyzed reaction.

For most enzymes, optimal temperatures approximate those of the environment of the cell. For the homeothermic organism man, this is 37° C. Enzymes from microorganisms adapted to growth in natural hot springs may exhibit optimal temperatures close to the boiling point of water.

The increase in rate below optimal temperature results from the increased kinetic energy of the reacting molecules. As the temperature is raised still further, however, the kinetic energy of the enzyme molecule becomes so great that it exceeds the energy barrier for breaking the secondary bonds that hold the enzyme in its native or catalytically active state. There is consequently a loss of secondary and tertiary structure (see p 43) and a parallel loss of biologic activity.

Denaturation of a protein is best defined as loss of biologic activity. Denaturation does not involve covalent bond cleavage, but **results from a rearrangement of secondary and tertiary structure.** Proteins are denatured by a variety of reagents including heat, acid, high salt concentrations, or heavy metals. Frequently the denatured protein is less soluble and may be removed by filtration. Serum may thus be deproteinized by treatment with certain acids (phosphotungstate, trichloroacetate, or perchlorate) or with a heavy metal (Zn^{++}).

Denaturation by heat, acid, or high salt concentrations probably results from rupture of the relatively weak ionic, salt, and nonpolar bonds responsible for maintaining the enzyme in its "native" or active tertiary configuration (Fig 8–15).

2. Effects on the charged state of the substrate or enzyme. For the enzyme, charge changes may affect activity either by changing structure or by changing the charge on an amino acid residue functional in substrate-binding or catalysis. If a negatively charged enzyme (E^-) reacts with a positively charged substrate (SH^+):

$$E^- + SH^+ \rightarrow ESH$$

then at low pH values E^- will be protonated and lose its negative charge.

$$E^- + H^+ \rightarrow E\text{-}H$$

Similarly, at very high pH values, SH^+ will ionize and lose its positive charge:

$$SH^+ \rightarrow S + H^+$$

Since the only forms that will interact are SH^+ and E^-, extreme pH values will lower the effective concentrations of E^- and SH^+, thus lowering the reaction velocity, as follows (Fig 8–16):

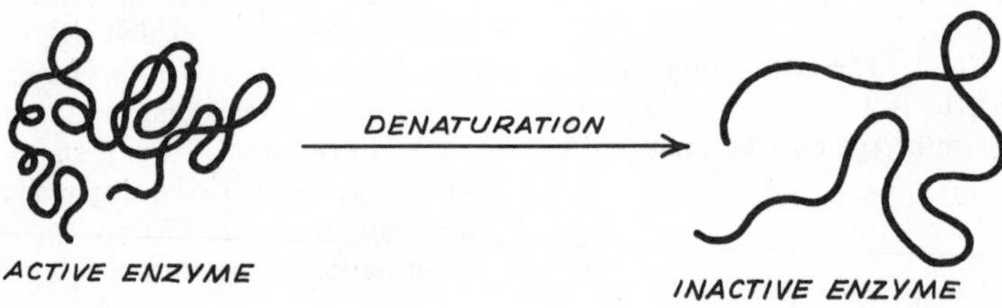

FIG 8–15. Representation of denaturation of an enzyme.

The heat and acid lability of most enzymes provides a simple test to decide whether a reaction is enzyme-catalyzed. If a cell extract having catalytic activity loses this activity when boiled or when acidified and reneutralized, the catalyst probably was an enzyme.

Frequently enzymes are either more or less readily denatured if their substrate is present. Either effect is attributed to a conformational change in the enzyme structure occurring when substrate is bound (see p 143). The new conformation may be either more or less stable than before.

D. pH: Moderate pH changes affect the **ionic state of the enzyme** and frequently that of the substrate also. When enzyme activity is measured at several pH values, optimal activity is generally observed between pH values of 5.0 and 9.0. However, a few enzymes, eg, pepsin, are active at pH values well outside this range.

The shape of pH-activity curves is determined by the following factors:

1. Enzyme denaturation at extremely high or low pH values.

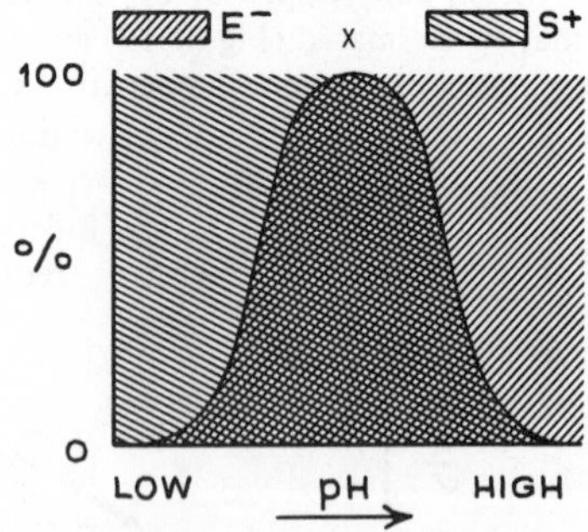

FIG 8–16. Effect of pH on enzyme activity.

Only in the cross-hatched area are both E and S in the appropriate ionic state, and the maximal concentrations of E and S are correctly charged at X. The result is a bell-shaped pH-activity curve.

Another important factor is a change in conformation of the enzyme when the pH is varied. A charged group far removed from the region where the substrate is bound may be necessary to maintain an active tertiary or quaternary structure. As the charge on this group is changed, the protein may unravel, or become more compact, or dissociate into subunits—all with a resulting loss of activity.

E. Oxidation: The sulfhydryl (SH) groups of many enzymes, notably the oxidoreductases (dehydrogenases), are essential for enzymatic activity. Oxidation (dehydrogenation) of these SH groups, forming disulfide linkages (S–S), brought about by many oxidizing agents including the O_2 of air, results in loss of activity. Frequently this also may cause a conformational change in the enzyme. Full activity may often be restored by reduced sulfhydryl compounds such as glutathione or cysteine (R–SH). These reduce the enzyme S–S to SH by disulfide exchange.

$$E\langle{}^{S}_{S}| + 2\,R{-}SH \rightleftharpoons E\langle{}^{SH}_{SH} + R{-}S{-}S{-}R$$

The reverse is true for ribonuclease. Here reduction of certain disulfide bridges results in loss of secondary structure and of catalytic activity.

F. Radiation: Enzymes are highly sensitive to short wavelength (high-energy) radiation such as ultraviolet light, x-, β-, or γ-rays. This is in part due to oxidation of the enzyme by peroxides formed by high-energy radiation. In the intact cell, loss of enzyme activity upon irradiation may also be due to indirect effects on the DNA of genes.

COENZYMES

Many enzymes catalyze reactions of their substrates only in the presence of a specific nonprotein organic molecule called the coenzyme. Only when both enzyme and coenzyme are present will catalysis occur. Where coenzymes are required, the complete system or **holoenzyme** consists of the protein part or **apoenzyme** plus a heat-stable, dialyzable nonprotein **coenzyme*** that is bound to the apoenzyme protein. Types of reactions that frequently require the participation of coenzymes are oxidoreductions, group transfer and isomerization reactions, and reactions resulting in the formation of covalent bonds (classes 1, 2, 5, and 6; see pp 126–127). By contrast, lytic reactions, including hydrolytic reactions such as those catalyzed by the enzymes of the digestive tract, are not known to require coenzymes (classes 3 and 4; see pp 126–127).

Coenzymes frequently contain B vitamins as part of their structure (see p 97). Thus many enzymes concerned with the metabolism of amino acids require enzymes containing vitamin B_6 (see p 101). The B vitamins **nicotinamide, thiamine, riboflavin, pantothenic acid**, and **lipoic acid** are important constituents of coenzymes for biologic oxidations and reductions (see p 126); and folic acid and cobamide coenzymes function in one-carbon metabolism (see pp 112 and 118).

It often is helpful to regard the coenzyme as a second substrate, ie, a **cosubstrate.** This is so for at least 2 reasons. In the first instance, **the chemical changes in the coenzyme exactly counterbalance those taking place in the substrate**; eg, in transphosphorylation reactions involved in the metabolism of sugars, for every molecule of sugar phosphorylated, one molecule of ATP is dephosphorylated and converted to ADP:

ATP → ADP; SUGAR → SUGAR–PHOSPHATE

Similarly, in oxidoreduction (dehydrogenase) reactions (E.C. class 1.1), one molecule of substrate is oxidized (dehydrogenated) and one molecule of coenzyme is reduced (hydrogenated):

$CH_3{-}CH(OH){-}COOH$ (L-LACTATE) → $CH_3{-}C(=O){-}COOH$ (PYRUVATE); NAD^+ → $NADH + H^+$

In transamination reactions (class 2.6), pyridoxal phosphate acts as a second substrate in 2 concerted reactions, and acts as carrier for transfer of an amino group between different α-keto acids (Fig 8–17).

For every molecule of alanine converted to pyruvate, one molecule of the aldehyde form of pyridoxal phosphate is aminated. The amino form of the coenzyme does not appear as a reaction product since the aldehyde form is regenerated by transfer of the amino group to α-ketoglutarate, forming glutamate.

A second reason for giving equal emphasis to the reactions of the coenzyme is that this aspect of the reaction actually may be of greater fundamental physiologic significance. For example, the importance of the ability of muscle working anaerobically to convert pyruvate to lactate resides not in pyruvate nor lactate themselves. The reaction serves merely to convert NADH to NAD^+. Without NAD^+, glycolysis cannot continue and anaerobic ATP synthesis (and hence muscular work) ceases. In summary, under anaerobic conditions, the conversion of pyruvate to lactate serves to reoxidize NADH and permit synthesis of ATP. Other reactions can serve the identical function equally well. In bacteria or yeast growing anaerobically a number of substances derived more or less directly from pyruvate are utilized as oxidants for NADH and are in the process themselves reduced (Table 8–3).

*The term "prosthetic group" was formerly employed to denote coenzymes covalently bonded to the apoenzyme. The term is now largely obsolete.

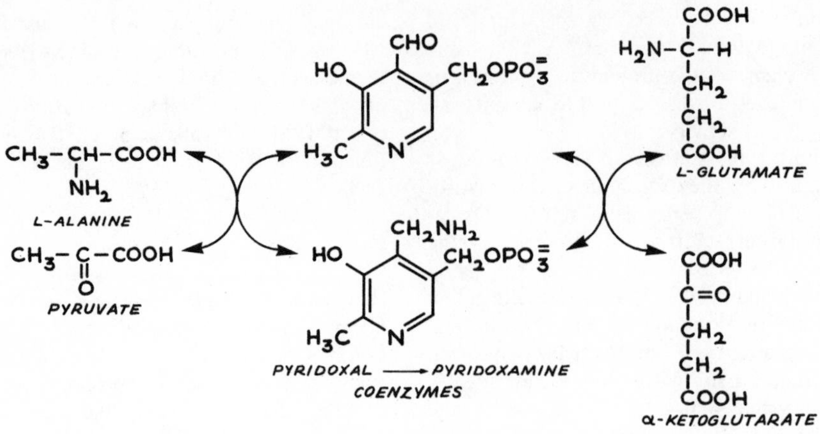

FIG 8–17. Participation of pyridoxal phosphate in transamination reactions.

TABLE 8–3. Representative mechanisms for anaerobic regeneration of NAD.

Oxidant	Reduced Product	Life Form
Pyruvate	Lactate	Muscle, homolactic bacteria
Acetaldehyde	Ethanol	Yeast
Dihydroxyacetone phosphate	α-Glycerophosphate	*E coli*
Fructose	Mannitol	Heterolactic bacteria

Classification of Coenzymes Based on Functional Characteristics

Coenzymes might be classified in many ways, each emphasizing various features of significance. The feature emphasized here is the reaction type in which a given coenzyme is functional.

A. Coenzymes for group transfer of groups other than H.
   1. ATP and its relatives.
   2. Sugar phosphates.
   3. Co A.
   4. Thiamine pyrophosphate.
   5. B_6 phosphate.
   6. Folate coenzymes.
   7. Biotin.
   8. Cobamide coenzymes (B_{12}).
   9. Lipoic acid.

B. Coenzymes for transfer of H.
   1. NAD, NADP.
   2. FMN, FAD.
   3. Lipoic acid.
   4. Coenzyme Q.

Note that all coenzymes function in one or another type of group transfer reaction. Particularly striking is the frequent occurrence in the structure of coenzymes of the adenine ring joined to D-ribose and phosphate. Many coenzymes may therefore be regarded as derivatives of adenosine monophosphate

TABLE 8–4. Coenzymes and related compounds which are derivatives of adenosine monophosphate.

Coenzyme	R	R′	R″	n
AMP	H	H	H	1
ADP	H	H	H	2
ATP	H	H	H	3
Active methionine	Methionine	H	H	0
Amino acid adenylates	H	Amino acid		1
Active sulfate	SO_3H_2	H	PO_3H	1
3′,5′-Cyclic AMP	H	H	PO_3H	
NAD	*	H	H	2
NADP	*	PO_3H	H	2
FAD	*	H	H	2
Co A	*	H	PO_3H	2

*See Chapter 9.

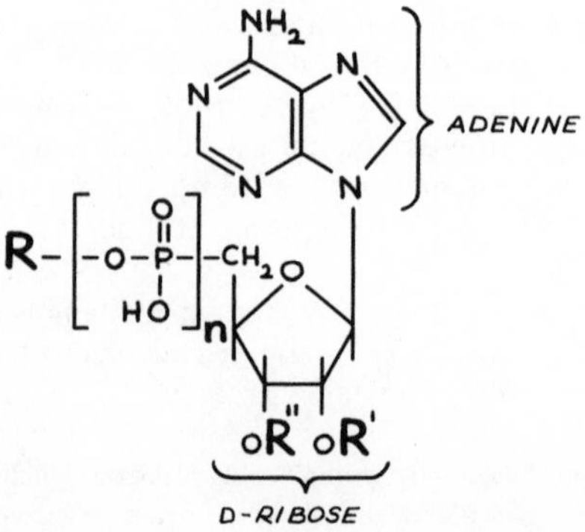

which differ only in the substituents at positions R,R′, and R″, and in the number of phosphate groups, n, attached to the 5′ position of the ribose (Fig 8–18). This is illustrated in Table 8–4. Also included are 3 forms of activated amino acids.

FIG 8–18. Adenosine monophosphate derivatives.

THREE-DIMENSIONAL STRUCTURE OF ENZYMES AS REVEALED BY X-RAY CRYSTALLOGRAPHY

X-ray crystallographic analysis (see p 41) has provided detailed 3-dimensional structures, not only of myoglobin and hemoglobin, but of the enzymes ribonuclease, lysozyme, chymotrypsin, and lactate dehydrogenase as well. Progress has also been made in the elucidation of the structures of glyceraldehyde-3-phosphate dehydrogenase, carboxypeptidase A, carbonic anhydrase, papain, and cytochome c.

Lysozyme

Lysozyme is an enzyme, present in tears, nasal mucus, sputum, tissues, gastric secretions, milk, and egg white, which catalyzes the hydrolysis of β-1,4-linkages of N-acetylneuramic acid (see p 12) in mucopolysaccharides or mucopeptides. It performs the function, in tears and nasal mucus, of destroying the cell walls of many airborne gram-positive bacteria. Lysozyme (molecular weight about 15,000 daltons) consists of a single polypeptide chain of 129 amino acid residues having no coenzyme or metal ion cofactors. Since the lysozyme molecule may readily be unfolded and refolded, not only catalysis and specificity but also the 3-dimensional

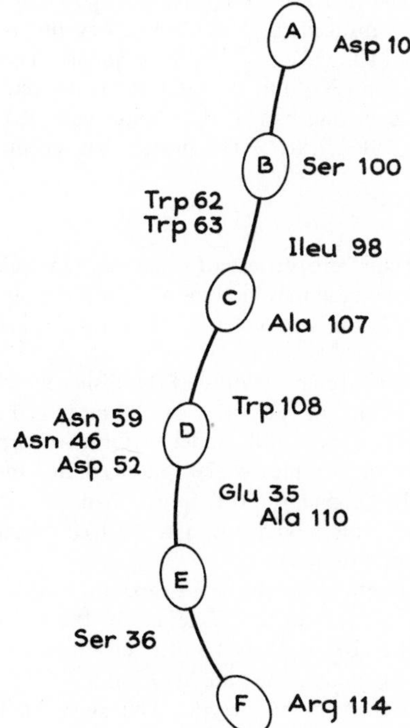

FIG 8–19. Schematic representation of the active site in the cleft region of lysozyme. A to F represent the glycosyl moieties of a hexasaccharide. Some of the amino acids in the cleft region near these subsites of the active site are shown together with their numbers in the lysozyme sequence. (Adapted from Koshland & Neet: Ann Rev Biochem 37:364, 1968.

3-dimensional structure are determined solely by these residues. There are small regions of pleated sheet structure, little α-helix, and large regions without ordered secondary structure (see p 00). A satisfactory representation of the structure of the enzyme consisting of a model of the molecule and its substrate, photographed in 3-dimensional color by the Parallax-Panoramagram technic, appears in J Biol Chem 243:1633, 1968. The molecule is seen to bear a deep central cleft which harbors an active site with 6 subsites (Fig 8–19) which bind various substrates or inhibitors. The residues responsible for bond cleavage are thought to lie between sites D and E close to the carboxyl groups of asp 52 and glu 35. It is thought that glu 35 protonates the acetal bond of the substrate while the negatively charged asp 52 stabilizes the resulting carbonium ion from the back side.

Ribonuclease

Unlike lysozyme, considerable information about the active site of ribonuclease (see p 47) was available prior to solution of the 3-dimensional structure by x-ray crystallography (Fig 8–20). The conclusions based on chemical investigations were largely confirmed by crystallography. The structure contains a cleft similar to that of lysozyme across which lie 2 amino acid residues, his 12 and his 119. These previously were implicated by chemical evidence as being at the active site. Both residues are near the binding site for uridylic acid.

Carboxypeptidase A

X-ray crystallographic data indicate that substrates are bound in the region of the essential zinc atom and that the enzyme undergoes a conformational change when substrate is bound. Two groups, a tyrosyl and an aspartyl residue (which might well perform a role in catalysis) undergo a spatial shift of about 1.5 and 0.2 nm respectively. Since removal of the zinc or its replacement by mercury destroys the activity without noticeable effect on the 3-dimensional structure, the zinc may be inferred to perform catalytic rather than structural functions.

Carbonic Anhydrase

X-ray data at the 0.55 nm level are available for human erythrocyte carbonic anhydrase C, a zinc metalloenzyme with a molecular weight of 30,000 daltons which contains less than 1/3 α-helix. The zinc is located near the center of the molecule at the bottom of a large cavity. One sulfonamide inhibitor, acetoxymercurisulfanilamide, binds so that the sulfonamide group is adjacent to the zinc atom.

Lactate Dehydrogenase

Although the primary structure has yet to be elucidated by chemical means, the complete tertiary structure of the LDH subunit has been described solely by interpretation of x-ray crystallographic data. This is the first and only example of the successful application of this technic to a subunit-containing enzyme; it is a

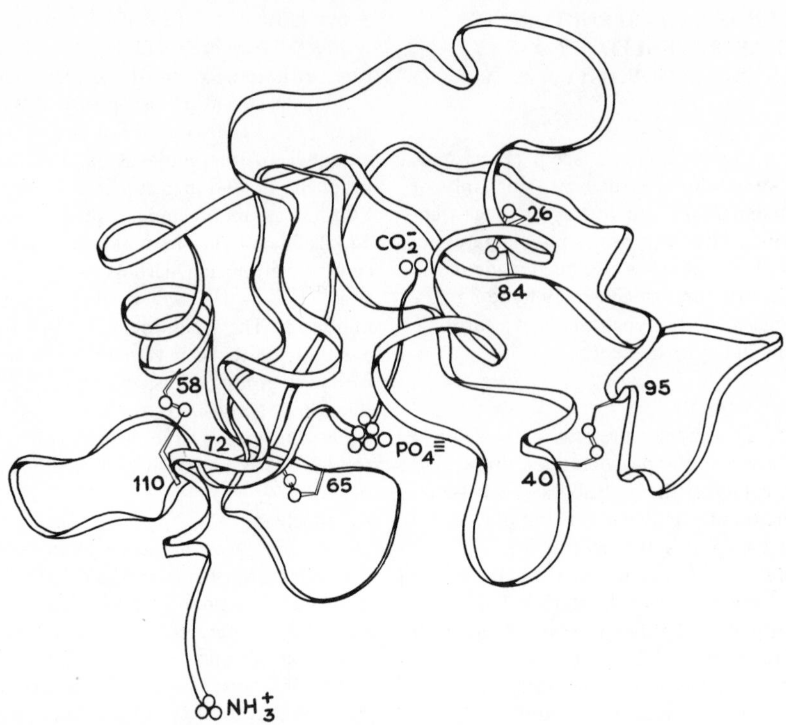

FIG 8–20. Schematic diagram of the main chain folding of bovine pancreatic ribonuclease. This protein is a single chain of 124 amino acid residues starting at the amino end (marked NH_3^+) and ending at the carboxy terminal (marked CO_2^-). The chain is crosslinked at 4 places by disulfide bridges from half cystine residues. The disulfide pairing for these bridges are 26–84, 40–95, 58–110, and 65–72 in the sequence. The region of the active site is indicated by the binding of the phosphate ion ($PO_4^{\equiv}$) in the cleft of the molecule. This model was obtained by x-ray diffraction studies of crystalline bovine pancreatic ribonuclease at 0.2 nm resolution. (Adapted from Kartha, Bello, & Harker: Nature 213:862, 1967.) The protein has recently been chemically synthesized in its entirety.

noteworthy achievement. The subunit contains regions both of helix and of pleated sheet and is relatively compact save for the N-terminal end of the polypeptide chain which projects out and is thought to interact with other subunits. The subunit contains a deep cleft which accommodates the coenzyme molecule (NAD). The nicotinamide portion of the NAD lies close to the essential thiol group in the region thought to contain the active site.

Summary of X-Ray Crystallography

From these data, x-ray crystallography has confirmed the major conclusions reached previously on chemical grounds. It appears also that this technic has greater promise as a means of localization of active sites than any other.

The Active Site

The concept that a specific region of the enzyme functions as an **"active site," "catalytic site,"** or **"substrate site"** arose soon after biochemists began to study enzymes in cell-free systems. Initially, interaction of substrate and enzyme was visualized in terms of the "template" or "lock and key" model of Fisher (Fig 8–21). The template model still is useful for explaining certain properties of enzymes—for example, ordered or sequential binding of 2 or more substrates (Fig 8–22) or the kinetics of a simple substrate-saturation curve (Fig 8–12).

An unfortunate feature of the Fisher model is the implied rigidity of the active site. A more refined and certainly a more useful model in terms of explaining properties of enzymes is the **"induced fit" model** of Koshland. Originally little more than an attractive hypothesis, this model now has received considerable experimental support.

An essential feature is the flexibility of the region of the active site. In the Fisher model the active site is presumed to be pre-shaped to fit the substrate. In the induced fit model, the substrate induces a conformational change* in the enzyme. This aligns amino acid residues or other groups on the enzyme in the correct

*Conformational change: A change in the average positions of atomic nuclei, but does not include bond changes. Unfolding or rotation about bonds of a protein is a conformational change, but the dissociation of a proton is not, even though it may accompany a conformational change. Polarizations of electrons without changes of the atomic nuclei are not conformational changes.

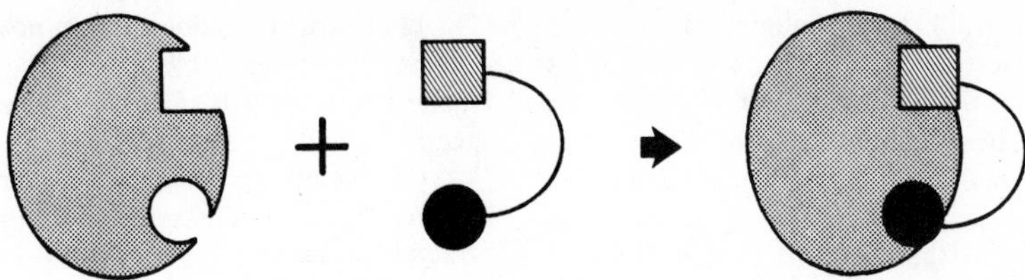

FIG 8–21. Representation of formation of an ES complex according to the Fisher template hypothesis.

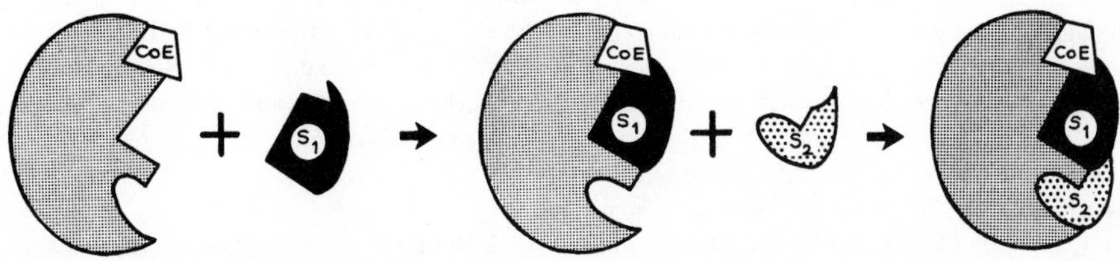

FIG 8–22. Representation of sequential adsorption of a coenzyme (CoE) and of 2 substrates (S_1 and S_2) to an enzyme in terms of the template hypothesis. The coenzyme is assumed to bear a group essential for binding the first substrate (S_1) which in turn facilitates binding of S_2.

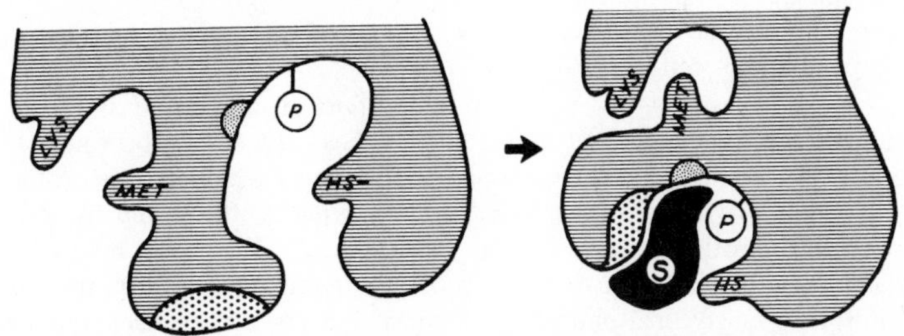

FIG 8–23. Representation of an induced fit by a conformational change in the protein structure. (After Koshland.)

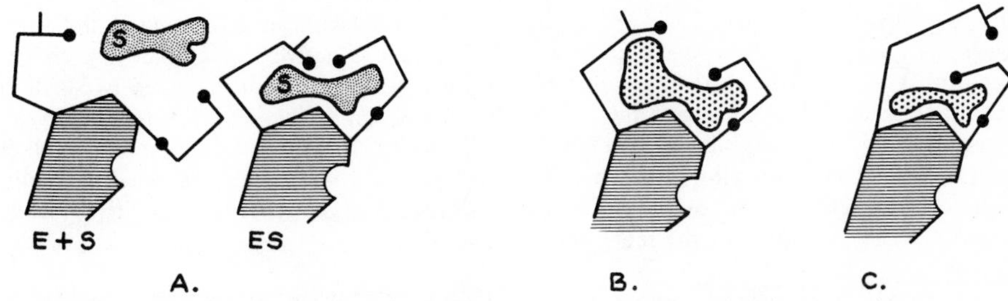

FIG 8–24. Representation of conformational changes in an enzyme protein when binding substrate (A) or inactive substrate analogues (B, C). (After Koshland.)

spatial orientation for substrate binding, catalysis, or for both. At the same time, other amino acid residues may become buried in the interior of the molecule.

In the hypothetical example (Fig 8–23), hydrophobic groups (hatched portion) and charged groups (dots) both are involved in substrate binding. A phosphoserine (–P) and the –SH of a cysteine residue are involved in catalysis. Other residues involved in neither process are represented by the side chains of 2 amino acids, lysine and methionine. In the absence of substrate, the catalytic and the substrate-binding groups are several bond distances removed from one another. Approach of the substrate induces a conformational change in the enzyme protein, aligning the groups correctly for substrate binding and for catalysis. At the same time, the spatial orientations of other regions are also altered–the lysine and methionine are now closer together (Fig 8–23). An alternate representation is shown in Fig 8–24. Substrate analogues may cause some, but not all, of the correct conformational changes. On attachment of the true substrate (A), all groups (shown as closed circles in the illustrations) are brought into correct alignment. Attachment of a substrate analogue that is too "bulky" (Fig 8–24,B) or too "slim" (Fig 8–24,C) induces incorrect alignment. One final feature is the site shown as a small notch on the right of the enzyme. One may perhaps visualize a regulatory molecule attaching at this point and "holding down" one of the polypeptide arms bearing a catalytic group. Substrate binding, but not catalysis, might then occur.

Experimental evidence for the induced fit model includes demonstration of conformational changes during substrate binding and catalysis with creatine kinase, phosphoglucomutase, and several other enzymes. With phosphoglucomutase, compounds similar to the substrate (eg, inorganic phosphate or glycerol phosphate) produced noticeable but less extensive conformational changes. With carboxypeptidase, substrate binding induces an appreciable change in the location of 2 amino acid residues which may also be involved in catalysis.

The exact sequence of events in a substrate-induced conformational change remains to be established. Several possibilities exist (Fig 8–25).

Even when the complete primary structure of an enzyme is known, it may still be difficult to decide exactly which amino acid residues constitute the active site. As so well illustrated by the induced fit model, these may be far distant one from another in terms of primary structure but spatially close in the sense of 3-dimensional or tertiary structure. In the representation of an active site shown in Fig 8–26, several regions of a polypeptide chain each contribute amino acid residues to the active site. Furthermore, the residues contributed generally are not all sequential within a polypeptide chain. One may ask: "How far does the active site extend?" To facilitate discussion, Koshland distinguishes 3 types of amino acid residues in enzymes.

(1) **Contact residue.** An amino acid residue within one bond distance (0.2 nm) of the substrate or other ligand* concerned. The term may include both specificity and catalytic residues.

(2) **Specificity residue.** An amino acid residue involved both in substrate binding and in subsequent catalytic process.

(3) **Catalytic residue.** An amino acid residue directly involved in covalent bond changes during enzyme action.

These are illustrated in Fig 8–26.

Progress in partial decoding of the primary structures at some active sites has led to the discovery that there are many similarities between the sequences of several hydrolytic enzymes. (See Table 8–5.) This may mean that the number of bond-breaking mechanisms operating in biologic systems is relatively small.

TABLE 8–5. Amino acid composition near the active site of some esterases and proteases.

Enzyme	Sequence Around Serine
Chymotrypsin, trypsin, thrombin, and elastase	Asp-Ser-Gly
Liver aliesterase, pseudocholinesterase	Glu-Ser-Ala

Modifiers of Enzyme Activity

The flow of carbon and of energy into various pathways of metabolism is profoundly influenced both by enzyme synthesis and by activation of pre-enzymes (see p 152). However, these processes are irreversible. Like all mammalian proteins, enzymes eventually are degraded to amino acids (protein turnover). In bacteria, the activity is diluted out among daughter cells on successive divisions.† Although both mechanisms effectively reduce enzyme concentration and hence catalytic activity, they are slow, wasteful of carbon and energy, and rather like turning out a light by smashing the bulb, then inserting a new one when light is needed. An "on-off" switch for enzymes clearly would be advantageous. It thus is not surprising that the **catalytic activity** of certain key enzymes can be reversibly decreased or increased by small molecules–in many cases themselves intermediary metabolites (Fig 8–27). Small molecule **modifiers** which decrease catalytic activity are termed **negative modifiers**; those which increase or stimulate activity are called **positive modifiers.**

Inorganic Modifiers (Metals)

Many metal ions act as positive modifiers. Apart from a requirement for a coenzyme, certain enzymes

*Ligand: Any small molecule bound to the enzyme by noncovalent forces. The term includes activators, substrates, and inhibitors.

†However, protein turnover does occur in bacteria at slow rates.

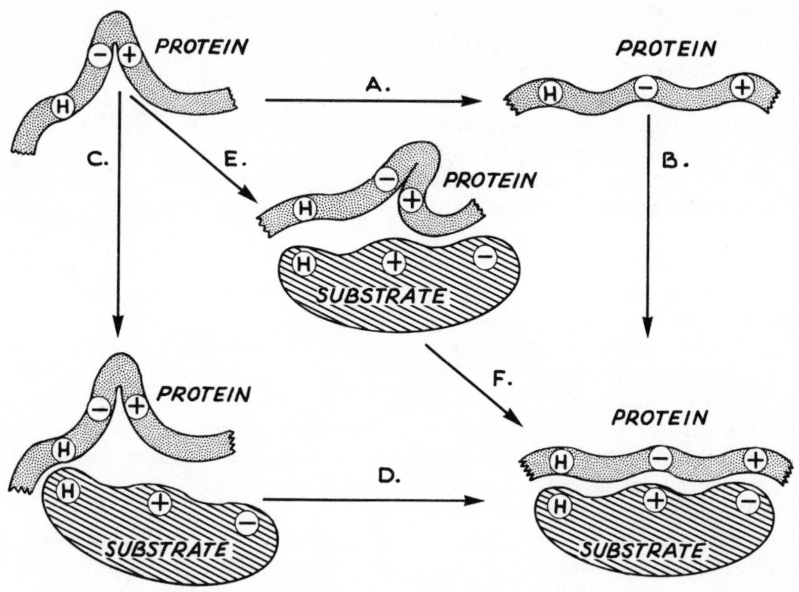

FIG 8–25. Representation of alternate reaction paths for a substrate-induced conformational change. The enzyme may first undergo a conformational change (A), then bind substrate (B). Alternatively, substrate may first be bound (C) whereupon a conformational change occurs (D). Finally, both processes may occur in a concerted manner (E) with further isomerization to the final conformation (F). (Adapted from Koshland & Neet: Ann Rev Biochem 37:387, 1968.)

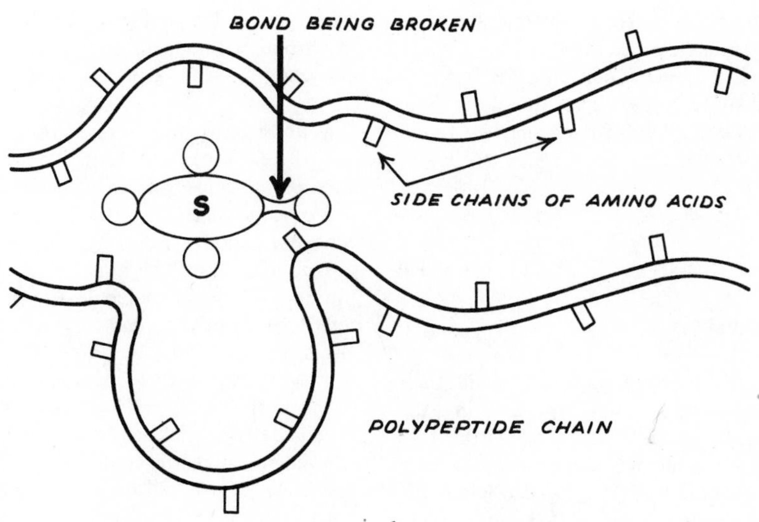

FIG 8–26. The active site. This may be composed of amino acid residues far removed from one another in terms of primary structure. (After Koshland.)

require a metal ion for full activity. Many pure enzymes (alcohol dehydrogenase, catalase, xanthine oxidase) contain a low, reproducible number of tightly bound metal ions per mol or per subunit of protein. Removal of the metal (by complexing with chelating agents such as EDTA) often results in partial or total loss of enzymatic activity. Activity frequently may be restored by replacing either the original or a similar metal ion. Cations present in cells and required for full activity of one or more enzymes include K^+, Cu^+, Fe^{++} and Fe^{+++}, Mo^{++++}, and divalent Mg, Mn, Ca, Zn, Cu, Co, and Mo. Fe, Mo, and Cu participate principally in oxidoreduction reactions (Fe in catalase, peroxidases, and cytochromes; Cu in ascorbic acid oxidase and tyrosinase; Mo in xanthine oxidase). Mg^{++} is required for all phosphate-transfer reactions. For isolated, purified phosphotransferases, one or more of divalent Mn, Ca, Zn, or Co may be equally effective. Thus, although the physiologic effects of these latter metals differ appreciably, they may be interchangeable in their ability to activate purified enzymes.

No single mechanism can explain the function of metals in enzymes. Some possible ways metals may accelerate enzyme-catalyzed reactions include:

(1) Direct participation in catalysis. Metal ions may undergo a valence change during an oxidation-reduction reaction and function in electron transport (eg, Fe in cytochromes or in catalase).

(2) Combination with a substrate with a negative charge or unshared electrons to form a metal-substrate (MS) complex which is the true substrate for the enzyme. For example, Mg^{++}-ATP^{4-} is the true substrate for many phosphotransferase reactions.

(3) Formation of a metalloenzyme (ME) which then binds the substrate in an enzyme-metal-substrate (EMS) complex. Possibilities 2 and 3 may be represented as:

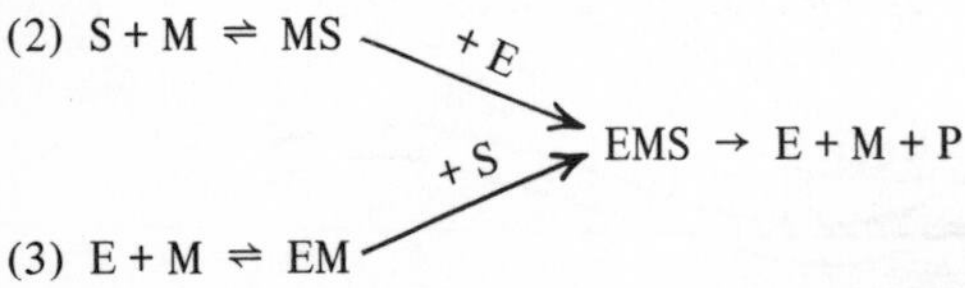

(4) Metals may alter the apparent equilibrium constant of the overall reaction by changing the nature of the reacting species. Where MS is the true substrate, an increase in metal ion concentration may increase the concentration of MS and accelerate the reaction. The same result is achieved if the metal complexes with the product of the reaction and thus effectively prevents its conversion back to substrate (eg, stimulation of lipase by Ca^{++}, which removes the fatty acid products as insoluble soaps).

(5) A metal ion may bring about a conformational change in the protein, converting it from an inactive to an active conformation. In this case, the metal may be bound at a point far removed from the substrate and may serve to maintain an active quaternary structure (eg, Zn^{++} in liver alcohol dehydrogenase) or tertiary structure (Fig 8–27).

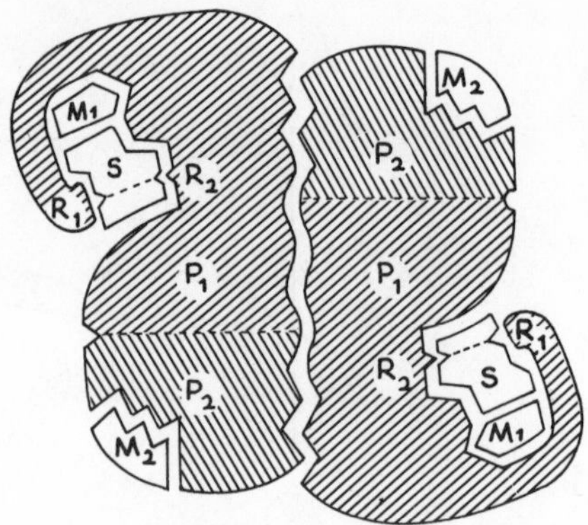

FIG 8–27. Schematic illustration of bond rupture by an enzyme consisting of 2 identical subunits. The substrates (S) are in contact with 2 catalytic residues (R_1 and R_2). Modifiers at the active site (M_1) and at an allosteric site (M_2) aid in maintaining the correct enzyme conformation. P_1 and P_2 may represent either 2 different polypeptide chains or parts of the same polypeptide chain. (Adapted from Koshland & Neet: Ann Rev Biochem 37:361, 1968.)

Metals may also act as negative modifiers. Addition of a metal may convert an active into an inactive conformation. Thus the kind and concentration of various metal ions near specific enzymes in cells can profoundly affect both the direction and rate of flow of carbon within cells. In this way, metal ions play important roles in regulation of cellular metabolism.

Organic Negative Modifiers: Inhibitors

In one sense, the term "modifier" should apply only to molecules occurring naturally in a particular cell or organism. This restricted usage preserves the concept of modifiers as physiologic regulators of metabolic processes. There are, however, other "nonphysiologic" molecules which affect enzyme activity both in vitro and in vivo by analogous mechanisms. These too may be termed modifiers.

It is customary to distinguish 2 broad classes of inhibitors—competitive and noncompetitive—depending on whether the inhibition is or is not relieved by increasing concentrations of substrate. In practice, many inhibitors do not exhibit the idealized properties of pure competitive or noncompetitive inhibition discussed below. An alternate way to classify inhibitors is by their site of action. Some bind to the enzyme at the same place or site as does the substrate (the catalytic or active site), while others bind at some region (the allosteric site) other than the substrate site.

Competitive or Substrate Analogue Inhibition

Classical competitive inhibition occurs at the substrate-binding or catalytic site. The chemical structure of a substrate analog inhibitor (I) closely resembles that of the substrate (S). It may therefore combine reversibly with the enzyme, forming an enzyme inhibitor (EI) complex rather than an ES complex. When both the substrate and this type of inhibitor are present, they compete for the same binding sites on the enzyme surface. A much studied case of competitive inhibition is that of malonate (I) with succinate (S) for succinate dehydrogenase.

COOH–CH₂–COOH	COOH–CH₂–CH₂–COOH
Malonate	**Succinate**

Succinate dehydrogenase catalyzes formation of fumarate by removal of one hydrogen atom from each of the 2 α-carbon atoms of succinate.

$$\mathrm{HOOC{-}CH_2{-}CH_2{-}COOH} \xrightarrow[\text{SUCCINATE DEHYDROGENASE}]{-2H} \mathrm{HOOC{-}CH{=}CH{-}COOH}$$

Succinate **Fumarate**

FIG 8–28. Reaction catalyzed by succinate dehydrogenase.

Malonate (I) can combine with the dehydrogenase, forming an EI complex. This, however, cannot be dehydrogenated since there is no way to remove even one H atom from the single α-carbon atom of malonate without forming a pentavalent carbon atom. The only reaction the EI complex can undergo is decomposition back to free enzyme plus inhibitor. For the reversible reaction,

$$\mathrm{EI} \underset{k_{-1}}{\overset{k_1}{\rightleftharpoons}} \mathrm{E} + \mathrm{I}$$

the equilibrium constant, K_i, is

$$K_i = \frac{[E][I]}{[EI]} = \frac{k_1}{k_{-1}}$$

The action of competitive inhibitors may be understood in terms of the following reactions:

$$\mathrm{E} \underset{\pm I}{\rightleftharpoons} \mathrm{EI\ (inactive)} \not\rightarrow \mathrm{E + P}$$

$$\mathrm{E} \underset{\pm S}{\rightleftharpoons} \mathrm{ES} \rightarrow \mathrm{E + P}$$

The rate of product formation, which is what is measured, depends solely on the concentration of ES. Suppose I binds very tightly to the enzyme (K_i = a small number). There now is little free enzyme (E) available to combine with S to form ES and to decompose to E + P. The measured reaction rate will thus be slow. For analogous reasons, an equal concentration of a less tightly bound inhibitor (K_i = a larger number) will not decrease the rate of the catalyzed reaction so markedly. Suppose that, at a fixed concentration of I, more S is added. This increases the probability that E will combine with S rather than with I. The ratio of ES/EI and the reaction rate also rises. At a sufficiently high concentration of S, the concentration of EI should be vanishingly small. If so, the rate of the catalyzed reaction will be the same as in the absence of I. This is shown in Fig 8–29.

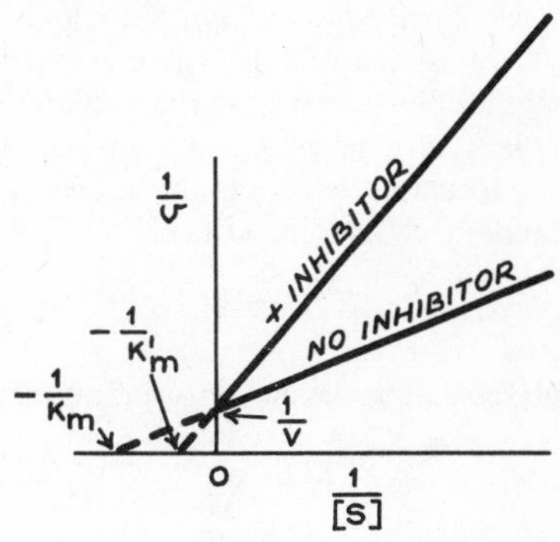

FIG 8–29. Classical competitive inhibition.

The reaction velocity (v) at a fixed concentration of inhibitor was measured at various concentrations of S. The lines drawn through the experimental points coincide at the y-axis. Since the y-intercept is 1/V, this states that **at an infinitely high concentration of S (1/S = O), v is the same as in the absence of inhibitor.** However, the intercept on the x-axis (which is related to K_m) varies with inhibitor concentration and becomes a larger number ($-1/K'_m$ is smaller than $-1/K_m$) in the presence of I. Thus, **a competitive inhibitor raises the apparent K_m** (K'_m) for the substrate. Since K_m is the substrate concentration where the concentration of free enzyme is equal to the concentration of enzyme as ES, substantial free enzyme is available to combine with inhibitor. For simple competitive inhibition, the intercept on the x-axis is

$$y = \frac{1}{K_m\left(1 + \frac{[I]}{K_i}\right)}$$

K_m may be evaluated in the absence of I, and K_i evaluated using the above equation. If the number of mols of I added is very much greater than the number of mols of enzyme present, [I] may generally be taken as the added (known) concentration of inhibitor. The K_i

values for a series of substrate analogue (competitive) inhibitors indicate which are most effective. **At a low concentration, those with the lowest K_i values will cause the greatest degree of inhibition.**

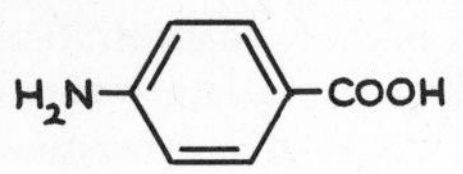

p-**Aminobenzoic acid** 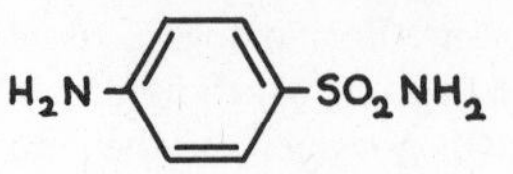 **Sulfanilamide**

Competitive inhibitors that block enzyme reactions in a parasite are potent **chemotherapeutic agents.** For example, many microorganisms form the vitamin folic acid (Fig 8–30) from *p*-aminobenzoic acid. Sulfanilamide, a structural analogue of *p*-aminobenzoate, will block folic acid synthesis, and the resulting deficiency of this essential vitamin is fatal to the microorganism. Since man lacks the enzymes necessary to form folic acid from *p*-aminobenzoate, folic acid is required as a vitamin in the diet. It follows that sulfonamides cannot act as competitive inhibitors of folic acid synthesis in man.

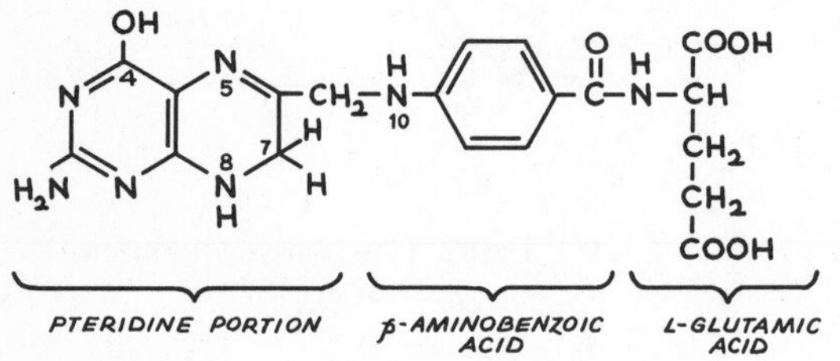

PTERIDINE PORTION *p*-AMINOBENZOIC ACID L-GLUTAMIC ACID

FIG 8–30. 7,8-Dihydrofolic acid.

Folic acid analogues studied as chemotherapeutic agents against tumors include aminopterin (4-aminofolic acid) and amethopterin (Fig 8–31) which inhibit growth of Ehrlich ascites tumor cells. Amethopterin is a competitive inhibitor for dihydrofolate in the dihydrofolate reductase reaction (see p 110). An aminopterin-resistant strain of Ehrlich cells has been shown to have as much as 14 times more dihydrofolate reductase than the nonresistant strain, although levels of other folate-utilizing enzymes are unchanged. This illustrates one mechanism of drug resistance, namely hyperproduction of the drug-sensitive enzyme.

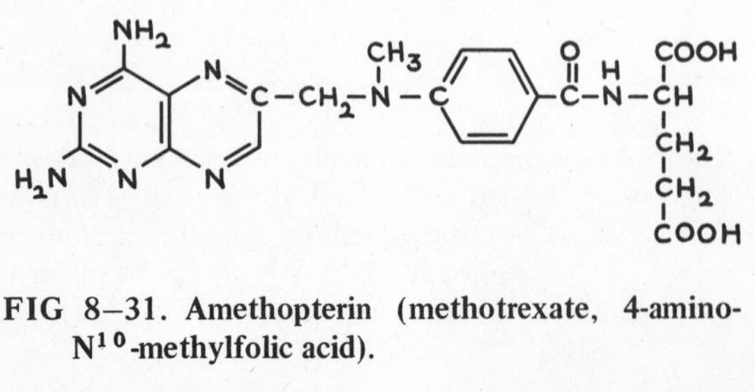

FIG 8–31. Amethopterin (methotrexate, 4-amino-N^{10}-methylfolic acid).

Other **antagonists to B vitamins** include pyrithiamine and oxythiamine (antagonists to thiamine), pyridine-3-sulfonic acid (to nicotinamide), pantoyl taurine and ω-methylpantothenic acid (to pantothenic acid), deoxypyridoxine (to pyridoxine), desthiobiotin (to biotin), and bishydroxycoumarin (Dicumarol; see p 000) to vitamin K. Purine and pyrimidine antimetabolites have also been prepared and studied as possible chemotherapeutic agents in the treatment of tumors. Examples are 6-mercaptopurine (Purinethol, 6-MP), which may be a hypoxanthine antagonist, 5-fluorouracil, 5-fluorouridylic acid, and 5-iodouridine (see p 356).

Many other drugs that inhibit enzyme action operate in a similar manner. D-Histidine competitively inhibits the action of histidase on L-histidine. **Physostigmine** competitively inhibits the hydrolysis of acetylcholine by cholinesterase, probably because it is structurally similar to acetylcholine. Even ATP and ADP are competitive inhibitors for many oxidoreductases where NAD and NADP are required as coenzymes. Recall that both the coenzymes and the inhibitors are derivatives of AMP (see Table 8–4).

The sulfonamide derivative **acetazolamide** (Diamox), although not bacteriostatic, is a potent inhibitor of carbonic anhydrase. Acetazolamide has been used to intensify renal excretion of water and electrolytes because of the importance of carbonic anhydrase in those functions of the renal tubule which affect reabsorption of electrolyte and thus of water (see p 381).

Reversible Noncompetitive Inhibition

As the name implies, in this case no competition occurs between S and I. I usually bears little or no structural resemblance to S and may be assumed to bind to a different region on the enzyme. **Reversible noncompetitive inhibitors lower the maximum velocity attainable with a given amount of enzyme (lower V) but do not affect K_m.** Since I and S may combine at different sites, formation of both EI and EIS complexes is possible. Since EIS may break down to form product at a slower rate than does ES, the reaction is slowed but not halted. The following competing reactions may occur:

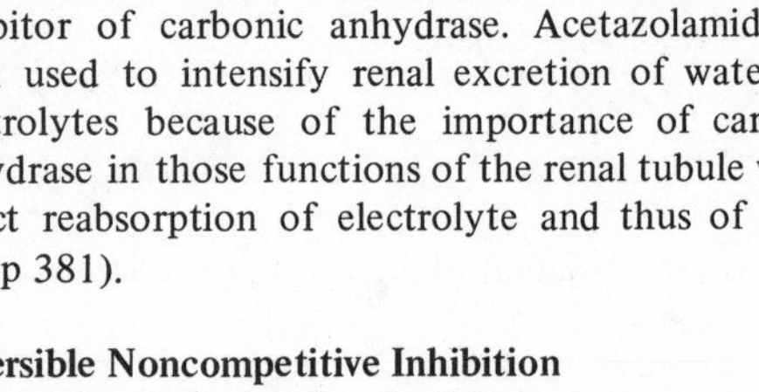

$$E \underset{}{\overset{\pm I}{\rightleftharpoons}} EI \overset{\pm S}{\rightleftharpoons} EIS \rightarrow EI + S$$
$$E \overset{\pm S}{\rightleftharpoons} ES \overset{\pm I}{\rightleftharpoons} EIS$$
$$ES \rightarrow E + P$$

If S has equal affinity both for E and for EI (I does not affect the affinity of E for S), the results shown in Fig 8–32 are obtained when 1/v is plotted against 1/S at various concentrations of inhibitor. It is assumed that there has been no significant alteration of the conformation of the active site when I is bound.

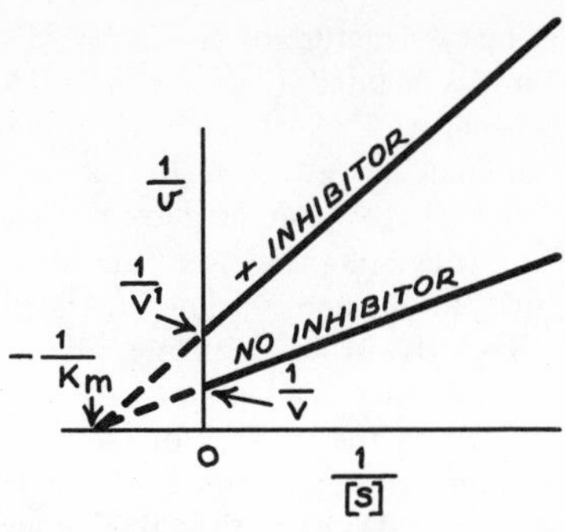

FIG 8–32. Noncompetitive inhibition.

Irreversible Noncompetitive Inhibition

A wide variety of enzyme "poisons" such as iodoacetamide, heavy metal ions (Ag^+, Hg^{++}), oxidizing agents, etc reduce enzyme activity. Since these inhibitors bear no structural resemblance to the substrate, an increase in the substrate concentrations generally is ineffective in relieving this inhibition. Simple kinetic analysis of the type discussed above may not distinguish between enzyme poisons and true reversible competitive inhibitors. Reversible noncompetitive inhibition is, in any case, rare. Unfortunately this is not always appreciated since both reversibly and irreversibly bound noncompetitive inhibition exhibit similar kinetics.

Extracts of the intestinal parasite Ascaris contain pepsin and trypsin inhibitors. The parasitic worm thus escapes digestion in the intestine. These protein inhibitors occur also in pancreas, soybeans, and raw egg white. Animals may also produce antibodies that irreversibly inactivate enzymes in response to the parenteral injection of the enzyme which functions as a foreign protein or antigen. This seriously limits the use of enzymes as chemotherapeutic agents.

Inhibition at an Allosteric Site

An enzyme may possess 2 separate sites which can bind ligands–the active site, where catalysis occurs, and the allosteric ("other") site where a modifier may bind (Figs 8–27 and 8–34). A modifier present at the allosteric site may affect the conformation at the active site. To do so, it is not necessary that the allosteric and substrate sites be physically close. The resulting conformational change at the active site may affect either the affinity of the active site for substrate or the relative spatial orientation of the catalytic groups and the substrate (or both), and hence affect the catalytic rate. Such a modifier can thus affect either the **binding constant** for the substrate, the **velocity** of the enzyme-catalyzed reaction, or both. Similarly, the binding of the modifier (M) will be affected if substrate at the active site causes a conformational change at the allosteric site. If both situations hold, the binding of both M and S will depend on the concentrations of both M and S, and a complex kinetic pattern will result. In the simplified case where a negative modifier (allosteric inhibitor) does not affect the binding constant for the substrate but only slows the catalytic rate, noncompetitive inhibition will be observed. If only substrate binding is decreased without effect on the catalytic rate, the inhibition will be of a competitive type. In most cases, neither extreme case will be true. The observed kinetics will therefore be mixed, tending more toward competitive or toward noncompetitive inhibition depending on whether the catalyzed reaction or substrate binding is more affected. In some cases, when v is plotted against S, instead of the simple hyperbolic saturation curve observed in the presence of substrate alone, a sigmoid saturation curve is observed (Fig 8–33).

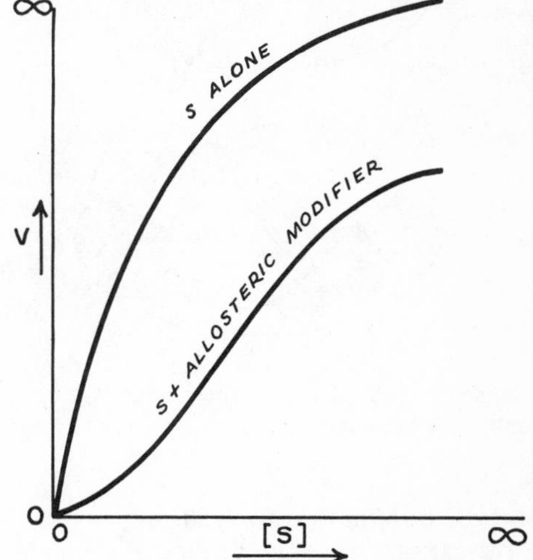

FIG 8–33. Sigmoid saturation curve for substrate in the presence of an allosteric inhibitor.

At low substrate concentrations, the reaction is slowed considerably (M is an effective inhibitor). As more and more substrate is added, the modifier has less and less effect, a circumstance reminiscent of simple competitive inhibition. Fig 8–34 attempts to represent one way this may be imagined to occur. Binding of I at the allosteric site deforms the active site so that S is less tightly bound. A typical example is inhibition of aspartate transcarbamylase (see pp 157 and 356) by cytidine triphosphate (CTP). As the concentration of aspartate is increased, CTP becomes a less effective negative modifier. At high concentrations of aspartate, the inhibition by CTP is more or less competitive with aspartate. However, since there is no discernible structural resemblance between CTP and aspartate, CTP is not a substrate-analogue inhibitor. The phenomenon is best understood in terms of conformational changes at both substrate and allosteric sites. At low concentrations of aspartate, the substrate sites are largely unoccupied. CTP binds to the allosteric sites and causes a conformational change at the active site which decreases its affinity for the substrate aspartate and thus slows the rate. As more and more aspartate is added, more binds to the active site. This causes a conformational change at the allosteric site, decreasing its affinity for CTP. CTP thus becomes a less effective inhibitor at high concentrations of aspartate. Regulation of enzyme activity via allosteric inhibition is now a well established mechanism for regulation of the flow of carbon into different metabolic pathways (see p 154), particularly in bacteria.

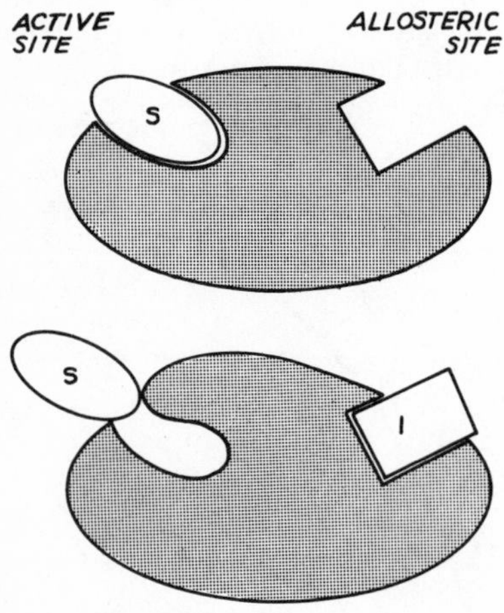

FIG 8–34. Modification of the active site by an allosteric inhibitor.

The Genetic Basis for Enzyme Synthesis

The extreme specificity of an enzyme is closely associated with a specific chemical and physical structure. This structure may be studied at several levels of organization. Simple proteins consist primarily of long chains of amino acids linked together by peptide bonds (see p 42). The sequence or order of the amino acids in a polypeptide chain is referred to as its **primary structure**. Portions of a polypeptide chain may be held together by one or more disulfide bonds, hydrogen bonds, hydrophobic bonds, or salt linkages (see p 42). These secondary bonds form an organized 3-dimensional structure such as an α-helix (see p 45), and confer secondary structure. Several helices, or portions of a single helix, may be wound about each so as to confer further 3-dimensional or tertiary structure. Finally, several of these coiled, intertwined subunits may combine to form the active enzyme protein (Table 8–6). The association of similar or dissimilar subunits is said to confer quaternary structure. Although all enzymes have primary, secondary, and some tertiary structure, the possession of quaternary structure is not universal.

Each level of structural organization has biologic significance. For example, an enzyme with correct primary structure but incorrect secondary, tertiary, or quaternary structure is not catalytically active. Similarly, subunits with correct primary, secondary, and tertiary structures but lacking quaternary structure may also be totally inactive.

The primary structure of an enzyme, like that of other proteins, is dictated by the trinucleotide (triplet) code of messenger RNA attached to polyribosomes (which in animals are attached to the endoplasmic reticulum) and by the matching bases of a transfer RNA-amino acid complex (see Chapter 5). The sequence of purine and pyrimidine bases of the messenger RNA is in turn dictated by a complementary base sequence that is part of a master DNA template or gene in the nucleus of the cell. Information for protein synthesis, stored in the form of a triplet code on the DNA, is thus the ultimate determinant of the ability of a cell to synthesize a particular enzyme. The enzyme content of the cell is therefore under strict genetic control. These concepts form the basis for what has been designated the **one gene, one enzyme theory**. This term is not strictly applicable to a protein which contains 2 or more different subunits. The synthesis of these subunits may be controlled by different genes and hence be under independent genetic control. It appears that **one gene, one subunit** may be the rule. It follows that **one or more** genes may actually control synthesis of one enzyme.

The secondary, tertiary, and quaternary structure of enzymes may be dictated by the primary structure. It appears that, although many conformations are possible, the biologically active form is that with the lowest energy level and greatest stability. Supporting evidence that the **lowest energy conformation** (3-dimensional structure) may be the biologically active form includes the ability of several enzymes whose active conformations have been destroyed by mild denaturation (see p 138) to regain activity on prolonged standing under conditions favoring breaking and reformation of secondary bonds. Assuming low energy conformations to be biologically active forms, computer programs have been designed which predict low-energy structures from the primary sequence and a limited amount of other data. These structures may then be projected on an oscilloscope screen and viewed from any angle. The ultimate aim of this approach is to predict correct, enzymically active conformations from primary structure data alone.

At present, it is not necessary to postulate independent genetic control of orders of protein structure above the primary level. A genetic **mutation** alters the DNA code and results in synthesis of a protein molecule with a modified primary structure. On occasion, this may result in altered structure at higher levels of organization also. Particularly if the new amino acid is significantly different from the old, changes in higher orders of structure may result. Depending on the nature of the structural change, a mutation may cause partial or complete loss of catalytic activity (see Chapter 5). Very rarely, a mutation may result in enhanced catalytic activity, as is sometimes seen in revertants of bacterial mutants to modified wild-types. Provided the mutation results in change which is not lethal, the modified genetic information is transmitted to the progeny of the cell. As a result, there frequently arises a **transmissible metabolic defect** which occurs at

TABLE 8–6. Quaternary structure of enzymes.*

Enzyme	Number of Monomer Units	Molecular Weight of Monomer Units (daltons)
E coli galactoside acetyltransferase (acetyl-Co A:galactoside 6-0-acetyltransferase, E.C. 2.3.1.18)	2	29,700
Rat liver malate dehydrogenase (L-malate:NAD oxidoreductase, E.C. 1.1.1.37)	2	37,500
Rabbit muscle glycerol-3-phosphate dehydrogenase (L-glycerol-3-phosphate:NAD oxido-reductase, E.C. 1.1.1.8)	2	39,000
E coli UDPglucose epimerase (UDPglucose 4-epimerase, E.C. 5.1.3.2)	2	39,000
E coli alkaline phosphatase (orthophosphoric monoester phosphohydrolase, E.C. 3.1.3.1)	2	40,000
Chicken or rabbit muscle creatine kinase (ATP:creatine phosphotransferase, E.C. 2.7.3.2)	2	40,000
Horse liver alcohol dehydrogenase (alcohol:NAD oxidoreductase, E.C. 1.1.1.1)	4	20,000
Yeast aldolase (fructose-1,6-diphosphate D-glyceraldehyde-3-phosphate-lyase, E.C. 4.1.2.13)	2	40,000
Rabbit muscle enolase (2-phospho-D-glycerate hydro-lyase, E.C. 4.2.1.11)	2	41,000
E coli methionyl-t-RNA synthetase (L-methionine:t-RNA ligase, [AMP], E.C. 6.1.1.10)	2	48,000
Chicken heart aspartate transaminase (L-aspartate:2-oxoglutarate aminotransferase, E.C. 2.6.1.1)	2	50,000
Yeast hexokinase (ATP:D-hexose 6-phosphotransferase, E.C. 2.7.1.1)	4	27,500
Rabbit liver fructose diphosphatase (D-fructose-1,6-diphosphate 1-phosphohydrolase, E.C. 3.1.3.11)	2†	29,000
	2†	37,000
Rat mammary gland glucose-6-phosphate dehydrogenase (D-glucose-6-phosphate:NADP oxidoreductase, E.C. 1.1.1.49)	2	63,000
Rat liver ornithine transaminase (L-ornithine:2-oxoacid aminotransferase, E.C. 2.6.1.13)	4	33,000
Aspartate transcarbamylase (carbamoylphosphate:L-aspartate carbamoyltransferase, E.C. 2.1.3.2)	2†	17,000
	3†	33,000
Rattlesnake venom L-amino acid oxidase (L-amino acid:O_2 oxidoreductase [deaminating], E.C. 1.4.3.2)	2	70,000
Beef heart, liver, or muscle LDH (L-lactate:NAD oxidoreductase, E.C. 1.1.1.27)	4†	35,000
Rabbit muscle glyceraldehyde-3-phosphate dehydrogenase (D-glyceraldehyde-3-phosphate:NAD oxidoreductase [phosphorylating], E.C. 1.2.1.12)	4†	37,000
Yeast alcohol dehydrogenase (alcohol:NAD oxidoreductase, E.C. 1.1.1.1)	4	37,000
Rabbit muscle aldolase (ketose-1-phosphate aldehyde-lyase, E.C. 4.1.2.7)	4	40,000
Pig heart fumarase (L-malate hydro-lyase, E.C. 4.2.1.2)	4	48,500
Rabbit muscle pyruvate kinase (ATP:pyruvate phosphotransferase, E.C. 2.7.1.40)	4	57,200
Beef liver catalase (H_2O_2:H_2O_2 oxidoreductase, E.C. 1.11.1.6)	4	57,500
Beef heart mitochondrial ATPase (ATP phosphohydrolase, E.C. 3.6.1.3)	10	26,000
Pigeon liver fatty acid synthetase	2	230,000
Jack bean meal urease (urea amidohydrolase, E.C. 3.5.1.5)	6	83,000
E coli glutamine synthetase (L-glutamate:NH_3 ligase [ADP], E.C. 6.3.1.2)	12	48,500
Pig heart propionyl-Co A carboxylase (propionyl-Co A:CO_2 ligase [ADP], E.C. 6.4.1.3)	4	175,000
E coli RNA polymerase (nucleosidetriphosphate:RNA nucleotidyltransferase, E.C. 2.7.7.6)	2	440,000
Chicken liver acetyl-Co A carboxylase (acetyl-Co A:CO_2 ligase [ADP], E.C. 6.4.1.2)	2†	4,100,000
	10†	409,000

*Adapted from Klotz IM, Langerman NR, Darnall DW: Quaternary structure of enzymes. Ann Rev Biochem 39:25, 1970.
†Nonidentical subunits.

that step formerly catalyzed by the now defective enzyme. Known examples of these inherited "molecular diseases" (Pauling) or "inborn errors of metabolism" (Garrod) include phenylketonuria, alkaptonuria, pentosuria, galactosemia, cystinuria, maple syrup urine disease, and the glycogen storage diseases. Many more heritable diseases are thought to be due to this phenomenon.

A much studied inherited defect is the formation in man of one or another of a variety of **abnormal hemoglobin molecules** (see p 201). Some abnormal hemoglobins are of no clinical significance. With others, such as hemoglobin S, homozygous inheritance of the defect may result in early death, while the heterozygote (sickle cell trait) may experience only a mild anemia, although this is rare.

Since mutations at various genetic loci can produce an enzyme with impaired activity, in theory an almost infinitely large number of molecular diseases can result. However, not all genetic loci appear to be equally susceptible to mutagenic agents, and code changes tend to cluster about particular regions of a gene. Some portions of a polypeptide chain are therefore more susceptible to change than are others.

Enzyme Formation From Precursors

Certain proteolytic enzymes concerned either with digestion or with the clotting of the blood are originally produced and secreted as inactive enzyme precursors. These **pre-enzymes** (formerly called **zymogens**) are subsequently converted to the active enzymes. The names of these pre-enzymes are formed by attaching the prefix **pro-** or **pre-** or the suffix **-ogen** to the name of the active enzyme. Pre-enzymes may be regarded as a way of protecting the tissues that secrete proteolytic enzymes from autodigestion.

Conversion of the pre-enzyme to the active enzyme is catalyzed either by proteolytic enzymes or by hydrogen ions, as in the following examples:

$$\text{Pepsinogen} \xrightarrow{H^+ \text{ or pepsin}} \text{Pepsin}$$

$$\text{Trypsinogen} \xrightarrow{\text{Trypsin or enterokinase}} \text{Trypsin}$$

$$\text{Chymotrypsinogen} \xrightarrow{\text{Trypsin}} \text{Chymotrypsin}$$

$$\text{Procarboxypeptidase} \xrightarrow{\text{Trypsin}} \text{Carboxypeptidase}$$

Since activation of the enzyme precursor is catalyzed by the active form of the enzyme itself, the activation of pepsinogen and of trypsinogen proceeds with ever increasing velocity and is said to be **autocatalytic**.

The activation process involves hydrolysis of peptide bonds, and results in "unmasking" the active or catalytic center (see p 144) of the enzyme protein. Frequently, large portions of the pre-enzyme are removed on activation. The conversion of pepsinogen (molecular weight of 42,500 daltons) to pepsin (molecular weight of 34,500 daltons) involves the loss of almost 1/5 of the molecule. Similarly, the conversion of procarboxypeptidase to carboxypeptidase is accompanied by a drop in molecular weight from 96,000 to 34,300 daltons, a decrease of 2/3. The conversion of trypsinogen to trypsin, however, involves the removal of only 6 amino acids.

Certain nondigestive enzymes, eg, a bacterial histidase, may under certain conditions exist in an inactive pre-enzyme form.

Regulation of Metabolism

For life to proceed in an orderly fashion, the flow of metabolites through anabolic and catabolic pathways must be regulated. Our concept of normal life incorporates the idea that not only all the requisite chemical events occur but also that they proceed at rates consistent with the activities and requirements of the intact organism in relation to its environment. Events such as ATP production, synthesis of macromolecular precursors, transport, secretion, and tubular reabsorption all must be responsive to subtle changes in the environment at the cellular, organ, and intact animal level. These processes must be coordinated and, in addition, must respond both to short-term changes in the external environment (such as the addition or removal of an essential nutrient) as well as to periodic intracellular events (such as DNA production prior to cell division). The mechanisms by which cells and intact organisms regulate and coordinate overall metabolism are of concern to biochemists with as seemingly diverse research interests as cancer, heart disease, aging, microbial physiology, differentiation and metamorphosis, or the mechanism of hormone action. At present, the molecular details of regulation are best understood in microorganisms (particularly bacteria) which appear to lack the complexities of rapid protein turnover or of hormonal or nervous control that exist in higher animal systems.

The study of cellular regulatory mechanisms encompasses sensory devices, trigger mechanisms, chemical messengers or effectors, and regulated enzyme activities. Regulation is nevertheless ultimately achieved via control of the **rate** of a specific enzyme-catalyzed reaction. Since chemical events in cells are enzyme-catalyzed, control of enzyme activity ultimately can control all aspects of metabolism.

The rate at which a particular reaction proceeds depends both on the **absolute amount of enzyme present** and the **catalytic efficiency** of the enzyme (ie, the number of molecules of substrate converted to product per molecule of enzyme per second). Both mechanisms are utilized by cells to achieve regulation.

Regulation of Bacterial Metabolism

An understanding of cellular regulatory processes is central to an understanding of metabolic disease as well as to rational therapy. The physician concerned with normal and abnormal metabolism in mammals must, however, recognize that the molecular details of regulation in mammalian cells are imperfectly understood, and clearly differ in detail from superficially similar phenomena in microorganisms.

Regulation of Enzyme Synthesis

Induction and derepression are terms applied to distinct but related phenomena studied principally in microorganisms. Both involve synthesis of new enzyme protein from amino acids (de novo synthesis of protein) in response to a "signal" transmitted by a small molecule. Induction refers to de novo synthesis of protein in response to the presence of a specific small molecule termed an inducer. Derepression refers to de novo synthesis of protein in response to the absence of a specific small molecule termed a co-repressor. Since both induction and derepression involve de novo synthesis of protein, they are inhibited by drugs (eg, chloramphenicol or actinomycin D) which block protein synthesis. Both processes thus are distinct from yet a third mechanism which results in increased enzyme levels, ie, activation of an inactive enzyme precursor (see p 152). Activation, which does not involve de novo synthesis of protein, is unaffected by chloramphenicol, actinomycin D, or other inhibitors of protein synthesis.

Induction

For a molecule to be metabolized or for an inducer to act, it first must enter the cell. In some cases, a specific transport system or **permease** is needed. The permease itself may also be inducible. Permeases share many properties in common with enzymes, and appear to perform functions analogous to the cytochromes in electron transport insofar as they appear to transport substrates without causing a net change in substrate structure.

The phenomenon of enzyme induction is illustrated by the following experiment: *Escherichia coli* grown on glucose will not ferment lactose. Its inability to do so is due to the absence both of a specific permease for a β-galactoside (lactose) and of the enzyme β-galactosidase, which hydrolyzes lactose to glucose and galactose. If lactose or certain other β-galactosides are added to the medium during growth, both the permease and the β-galactosidase are induced and the culture can now ferment lactose.

In the example given, the inducer (lactose) is a substrate for the induced proteins, the permease, and the β-galactosidase. Although in general inducers serve as substrates for the enzymes or permeases they induce, compounds structurally similar to the substrate may be inducers but not substrates. These are termed **gratuitous inducers.** Conversely, a compound may be a substrate but not an inducer.

Frequently a compound induces several enzymes which form part of a catabolic pathway. In the case cited above, β-galactoside permease and β-galactosidase were both induced by lactose.

Enzymes whose concentration in a cell is independent of an added inducer are termed **constitutive enzymes.** A particular enzyme may be constitutive in one strain of an organism, inducible in another, and neither constitutive nor inducible (ie, totally absent) in a third.

Cells capable of being induced for a particular enzyme always contain a small but measurable **basal level** of the inducible enzyme even when grown in the absence of added inducer. The extent to which a particular organism responds to the presence of an inducer is also genetically determined and varies greatly from strain to strain. Increases in enzyme content of from 2- to 1000-fold may be observed on induction in different strains. The genetic heritage of the cell thus determines not only the nature but also the magnitude of the response to an inducer. The terms "constitutive" and "inducible" are therefore relative terms, like "hot" and "cold," which represent the extremes of a spectrum of responses to added inducers.

Examples of inducible enzymes in animals are tryptophan pyrrolase, threonine dehydrase, tyrosine-α-ketoglutaric transaminase, invertase, and enzymes of the urea cycle (Table 8–8). An important example in bacteria is the inducible penicillinase that provides *Bacillus cereus* with a defense against penicillin.

Induction permits a microorganism to respond to the presence of a given nutrient in the surrounding medium by producing enzymes for its catabolism. In the absence of the inducer, little or no enzyme is produced. The ability to avoid synthesizing the enzyme in the absence of the nutrient permits the bacterium to use its available nutrients to maximum advantage, ie, it does not synthesize "unnecessary enzymes."

The genetic alternative to induction is **constitutivity**, the production of enzyme independently of the presence of small molecules acting as inducers. It appears that induction is restricted to enzymes catalyzing degradative or catabolic sequences of reactions.

Repression & Derepression (See also Chapter 5.)

In bacteria capable of synthesizing a particular amino acid, the presence of that amino acid in the culture medium curtails new synthesis of that amino acid via **repression**. The phenomenon is not restricted to amino acids and may operate in all biosynthetic pathways in microorganisms. A small molecule such as leucine, acting as a **co-repressor**, can inhibit ultimately the de novo synthesis of enzymes involved in its own synthesis.

In the presence of exogenous leucine, 3 enzymes of leucine biosynthesis are repressed in *Salmonella typhimurium.* Derepression can occur whenever the supply of leucine is removed. The mechanism by which repression and derepression are mediated is not known. It is thought that the actual repressor substance is a macromolecule such as a protein, a nucleic acid, or a nucleoprotein which can bind the co-repressor (see p 69). In the absence of the co-repressor, the macromolecule does not repress enzyme synthesis.

Whether repression occurs at the level of transcription or translation of genetic information is also not settled, although most results favor a transcriptional mechanism. The "holo-repressor" may bind to a region of DNA termed the operator region with the result that an adjacent region of DNA coding for synthesis of specific enzymes is repressed or rendered inoperable.

Examples are noted of the control the uptake of nutrients (mediated by permeases) may exert on the many reactions in intermediary metabolism and on the synthesis of large molecules. An organism capable of efficiently using substrates under a wide variety of conditions obviously has a biologic advantage. This may explain the ubiquity of these specific control mechanisms among living systems.

Allosteric Regulation of Enzyme Activity

The catalytic efficiency of an enzyme is affected by changes in the concentration of substrates, coenzymes, activators or inhibitors. Each can play a homeostatic role in regulation of catalytic efficiency. What follows relates solely to effects of activators or inhibitors on enzyme activity. The activity of certain key **regulatory enzymes** is modulated by low molecular weight **allosteric effectors** which generally have little or no structural similarity to the substrates or coenzymes for the regulatory enzyme.

Feedback regulation, which results in altered catalytic activity of a regulatory enzyme in response to binding an allosteric effector is discussed below. In the biosynthetic reaction sequence leading from A to D catalyzed by enzymes E_1 through E_3,

$$A \xrightarrow{E_1} B \xrightarrow{E_2} C \xrightarrow{E_3} D$$

a high concentration of D typically will inhibit conversion of A to B. This does not involve a simple "backing up" of intermediates, but reflects the ability of D specifically to bind to and inhibit E_1. D thus acts as a **negative allosteric effector** or **feedback inhibitor** of E_1. This negative feedback, or **feedback inhibition** on an **early enzyme*** by an end product of its own biosynthesis, achieves regulation of synthesis of D. Typically, D binds to an **allosteric site** on the inhibited enzyme which is remote from the catalytic site. An electronic analogy would be to regard the reaction sequence $A \rightarrow \rightarrow D$ as an amplification circuit with an input signal A and an output signal D. In an amplifier with negative feedback, the output signal, D, decreases the magnitude of the input signal, A, by changing the grid bias of a triode or the base current of a transistor. The result is a decreased signal output.

The kinetics of feedback inhibition may be noncompetitive, competitive, or mixed. The same enzyme, E_1, may also be subject to activation by other small molecules acting either at the catalytic site **(autosteric effectors)** or at an allosteric site **(positive allosteric effectors)**. Feedback inhibition is best illustrated within biosynthetic pathways. **Frequently the feedback inhibitor is the last small molecule before a macromolecule** as, eg, amino acids before proteins or nucleotides before nucleic acids. In general, **feedback regulation is exerted at the earliest functionally irreversible* step unique to a particular biosynthetic sequence**. Its relevance to metabolic regulation was emphasized by Umbarger, who in 1956 reported feedback inhibition by isoleucine of threonine dehydrase, an early enzyme of isoleucine biosynthesis.

Uncomplicated feedback inhibition of the type described above occurs in amino acid and in purine biosynthesis in microorganisms. Examples include inhibition by histidine of phosphoribosyl:ATP pyrophosphorylase, by tryptophan of anthranilate synthetase, and by CTP of aspartate transcarbamylase. In each case the regulated enzyme is involved in biosynthesis of a single end product—histidine, tryptophan, or CTP. Frequently a biosynthetic pathway may be branched, with the initial portion serving for synthesis of 2 or more essential metabolites. Further branching may occur, as in the biosynthetic pathways for the essential amino acids (see Chapter 15). In the hypothetical pathway of Fig 8–35, A, B, and C serve as precursors of G, I, and K; D and E as precursors of G and I; and F, H, and J solely as precursors of G, I, and K, respectively. The sequences

$$E \rightarrow F \rightarrow G$$
$$E \rightarrow H \rightarrow I$$
$$C \rightarrow J \rightarrow K$$

constitute linear reaction sequences and might be expected to be feedback-regulated by their end products at the earliest step unique to their biosynthesis. In general, this is what occurs. Frequently, however, additional feedback controls at reactions 1 and 2 of Fig 8–35 are superimposed upon this simple pattern. Any or all end products (G, I, or K) may inhibit reaction 1 and either G, I, or both may inhibit reaction 2 (Fig 8–36).

These **multiple feedback loops** provide additional fine control of metabolism. For example, if G is present in excess, the requirement for B decreases. The ability of G to decrease the rate of production of B thus confers a distinct biological advantage. However, the very existence of multiple feedback loops poses difficulties. If excess G can inhibit not only the portion of the pathway unique to its own synthesis but also portions common to that for synthesis of I or K, a large excess of G may cut off synthesis of all 3 end products. Clearly this is undesirable. Several mechanisms have evolved which circumvent this difficulty but retain the additional fine control conferred by multiple feedback loops. These include: (1) **cumulative feedback inhibition**, (2) **concerted, or multivalent feedback inhibition**, (3) **cooperative feedback inhibition**, and (4) **enzyme multiplicity**, or the existence of 2

*The term "early enzyme" means one catalyzing a remote or early step in a protracted reaction sequence.

*One strongly favored (in thermodynamic terms) in a single direction, ie, one with a large ΔG.

or more regulatory enzymes catalyzing the same biosynthetic reaction but having differing specificities with respect to their feedback inhibition by end products.

In **cumulative feedback inhibition** the inhibitory effect of 2 or more end products on a single regulatory enzyme is strictly additive. Cumulative feedback inhibition is encountered in regulation of glutamine utilization by *E coli* for synthesis of a spectrum of end products.

For **concerted** or **multivalent feedback inhibition,** no single end product alone greatly inhibits the regulatory enzyme. Marked inhibition occurs only when 2 or more end products are present in excess. Aspartokinase, which catalyzes conversion of aspartate to β-aspartyl phosphate (see reaction next column),

COO^-–CH_2–$H{-}C{-}NH_3^+$–COO^- —(ATP → ADP; ASPARTOKINASE)→ $C(=O){-}O{-}PO_3^=$–CH_2–$H{-}C{-}NH_3^+$–COO^-

L-Aspartate **β-Aspartyl phosphate**

is a regulatory enzyme of the so-called "aspartate family" of amino acids–lysine, threonine, methionine, isoleucine, and homoserine (Fig 8–37). The aspartokinase of *Bacillus polymyxa* is only slightly inhibited by excess lysine, threonine, isoleucine, or methionine alone, but is essentially inactive when both lysine and threonine are simultaneously present in excess. Since inhibition requires the concerted action of multiple inhibitors, it is therefore termed **concerted** or **multivalent feedback inhibition.**

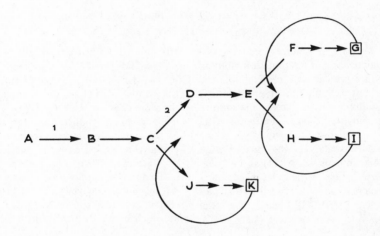

FIG 8–35. Sites of feedback inhibition in a branched biosynthetic pathway.

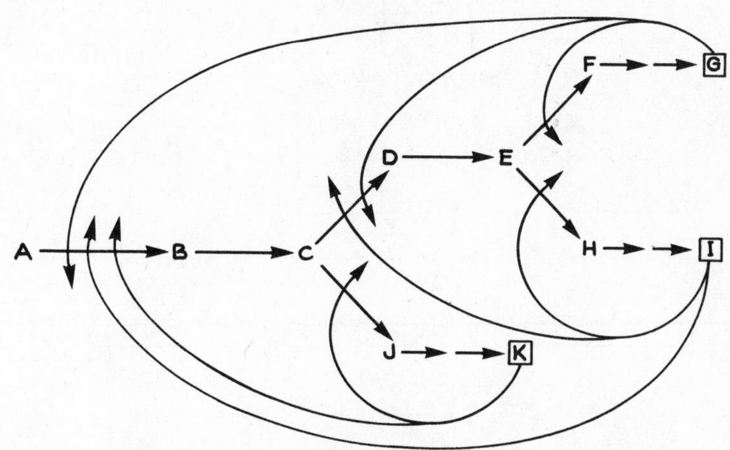

FIG 8–36. Representation of multiple feedback loops.

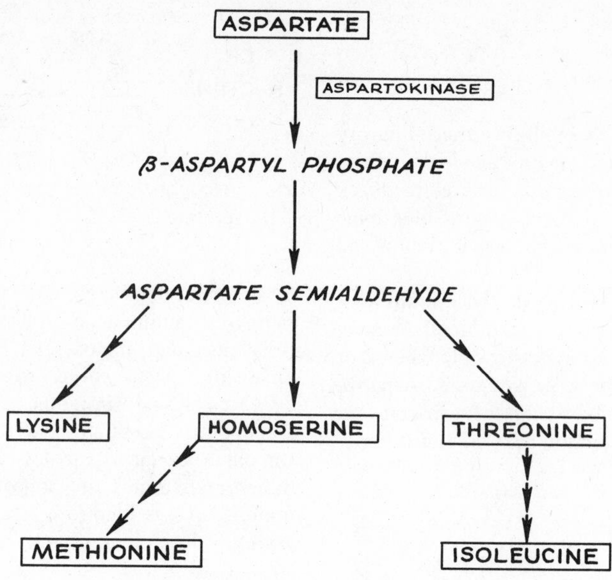

FIG 8–37. The aspartate family of amino acids.

Cooperative feedback inhibition embodies features both of cumulative and of multivalent inhibition. A single end product present in excess inhibits the regulatory enzyme, but **the inhibition observed when 2 or more inhibitors are present far exceeds the additive effects seen in cumulative feedback inhibition.** The regulatory enzyme phosphoribosylpyrophosphate amidotransferase, which catalyzes the first reaction unique to purine biosynthesis (see p 351), is feedback-inhibited by several purine nucleotides. The mammalian amidotransferase is controlled by 6-aminopurine ribonucleotides (AMP, ADP) and the bacterial enzyme by 6-oxypurine ribonucleotides (GMP, IMP). Mixtures of both types of ribonucleotides (GMP + AMP or IMP + ADP etc) are more effective than the sum of the inhibitory activities of either tested alone. Cooperative effects are not observed with purines of the same class (ie, AMP + ADP or GMP + IMP). Since the purine end products are interconvertible, the ability of only 2 purines to curtail production of all purines does not pose special problems.

The aspartate family provides yet another variant of feedback inhibition: the existence within a single cell of **multiple enzymes** each with distinct regulatory characteristics. *E coli* produces 3 aspartokinases, each of which catalyzes formation of β-aspartyl phosphate

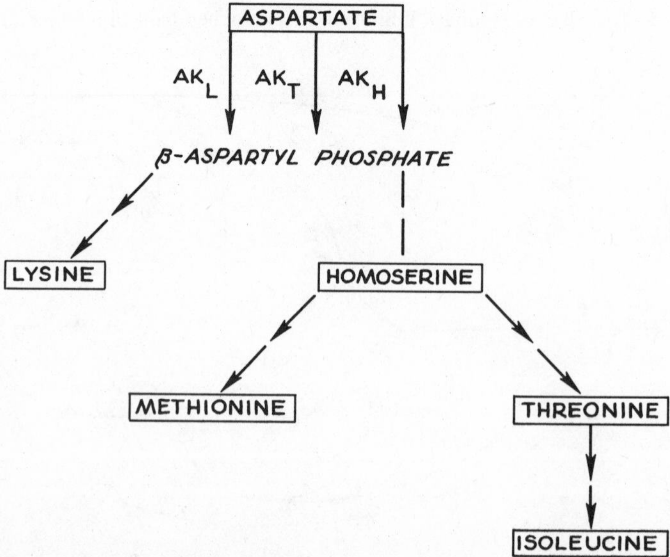

FIG 8–38. Regulation of aspartokinase (AK) activity in *E coli*. Multiple enzymes are subject to end product inhibition by lysine (AK_L), threonine (AK_T), or homoserine (AK_H).

from aspartate. Of these, one (AK_L) is specifically and completely inhibited by lysine, a second (AK_T) by threonine, and the third (AK_H) by homoserine, a precursor of methionine, threonine, and of isoleucine (Fig 8–38).

In the presence of excess lysine, AK_L is inhibited and β-aspartyl phosphate production decreases. This alone would not suffice to channel metabolites toward synthesis of homoserine and its products. Channeling is achieved by feedback inhibition at a secondary site or sites further along the pathway. Lysine thus also inhibits the first enzyme in the linear reaction sequence leading from β-aspartyl phosphate to lysine. This facilitates unrestricted synthesis of homoserine, and hence of threonine and isoleucine. Additional control points exist at the branch point where homoserine leads both to methionine and to threonine and isoleucine. Enzyme multiplicity in *E coli* occurs also in aromatic amino acid biosynthesis. Two distinct enzymes catalyze a reaction common both to tyrosine and phenylalanine synthesis. Each is separately and specifically inhibited either by tyrosine or by phenylalanine. As with the aspartate pathway, the individual aromatic amino acids also inhibit later steps in their own biosynthesis.

The above examples, chosen largely from microorganisms and from amino acid biosynthesis, illustrate the basic phenomenon of feedback inhibition and its major variants. Other examples are given throughout this book. That all the variations described are capable of exerting effective regulation of metabolism is suggested by the persistence in different strains of microorganisms of distinctive patterns of feedback inhibition of a single biosynthetic pathway (Table 8–7).

TABLE 8–7. Patterns of allosteric regulation of aspartokinase.

Organism	Feedback Inhibitor	Repressor
E coli (kinase I)	Homoser	. . .
E coli (kinase II)	Lys	Lys
E coli (kinase III)	Thr	. . .
R rubrium	Thr	. . .
B subtilis	Thr + lys	. . .
S cerevisiae	Thr, met, homoser	Thr, lys, homoser

The most studied allosteric enzyme is **aspartate transcarbamylase (ATC'ase)** which catalyzes the first reaction (see below) unique to purine biosynthesis (see p 351), condensation of carbamyl phosphate with aspartate forming carbamyl aspartate. **ATC'ase is feedback-inhibited by cytidine triphosphate (CTP).** Following treatment with mercurials, ATC'ase loses its sensitivity to inhibition by CTP but retains its full activity for carbamyl aspartate synthesis. This strongly suggests that CTP is bound at a different (allosteric) site from either substrate. ATC'ase apparently consists of 2 catalytic and 3 or 4 regulatory subunits. Each catalytic subunit contains 4 aspartate (substrate) sites and each regulatory subunit at least 2 CTP (regulatory) sites. Each type of subunit is subject to independent genetic control, as shown by the production of mutants lacking normal feedback control of CTP and, from these, of revertants with essentially normal regulatory properties.

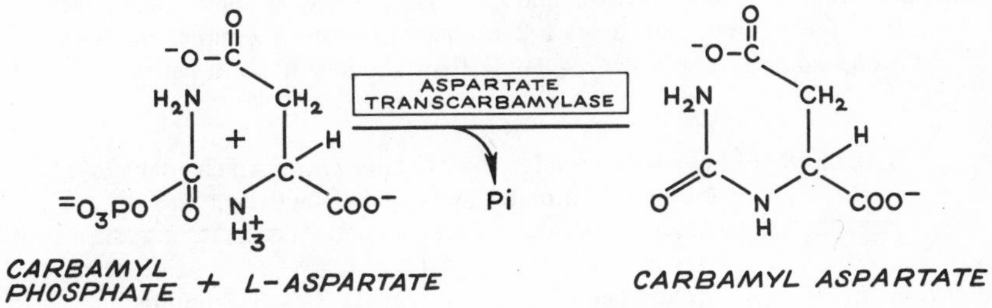

ASPARTATE TRANSCARBAMYLASE

Pi

CARBAMYL PHOSPHATE + L-ASPARTATE

CARBAMYL ASPARTATE

Some general properties of some, but not all, allosteric enzymes include:

1. **Desensitization to allosteric control**–Treatment with mercurials, urea, proteolytic enzymes, high or low pH, etc may produce loss of feedback control with retention of catalytic activity.

2. **Heat stability**–Most allosteric effectors, but not the substrates for the catalyzed reaction, confer enhanced resistance to heat denaturation of the allosteric enzyme.

3. **Cold sensitivity**–Unlike most enzymes, many regulatory enzymes undergo reversible inactivation (loss of catalytic activity) at 0°.

4. **Tertiary and quaternary structure**–All known allosteric enzymes possess tertiary, and in some cases also quaternary, structure (see p 33). ATC'ase, glycogen phosphorylase, pyruvate carboxylase, and acetyl-Co A carboxylase all undergo reversible association-dissociation reactions of subunits.

Models of Allosteric Enzymes

Two useful models for allosteric enzymes, both of which fit the known behavior of the best studied enzymes, have received wide attention. In what follows, it is well to remember that the information on which the models are based is fragmentary and in part specific to a particular enzyme rather than general. The essential features of both models are given below.

The Monod-Changeux-Wyman Model for Allosteric Enzymes

Four assumptions are made:

1. All allosteric enzymes are polymers of 2 or more identical subunits, each capable of existing in at least 2 conformational states in dynamic equilibrium with each other.

2. Each subunit bears both a catalytic (substrate) and an allosteric (inhibitor) site.

3. In each conformational state the sites have equal affinities for their respective ligands (substrate or allosteric effector).

4. The transition from one state to another involves simultaneous changes in all identical subunits.

The mathematical formulation of this model fits experimental observations on a number of allosteric enzymes.

The 2-Site Model for Allosteric Enzymes

This alternate hypothesis incorporates the concept of a flexible active site as proposed by Koshland (see p 143) and assumes the existence of at least 2 substrate-binding sites—one catalytic, the other regulatory. According to this model, substrate binding at the regulatory site induces a conformational change favoring substrate-binding at the catalytic site. Allosteric activators bind at the regulatory site producing similar effects. Allosteric inhibitors displace these activators, producing a conformational change favoring dissociation of substrate at the catalytic site. Mathematical formulations based on this model also fit the known facts.

Both hypotheses have yet to stand the test of time and of accumulated experimental evidence. Since both models fit the known facts it is clear that correspondence of mathematical predictions with experimental data is of itself no proof of validity. Probably both are at least partly correct, and possibly both may prove essentially correct in specific instances.

Regulation of Mammalian Metabolism

There is now a considerable body of evidence that enzyme levels in mammalian tissues may be altered by a wide range of physiologic, hormonal, or dietary manipulations. Examples are known for a variety of tissues and metabolic pathways (Table 8–8), but our

TABLE 8–8. Selected examples of rat liver enzymes which adapt to an environmental stimulus by changes in activity.*

Enzyme	E.C. Number	$t_{1/2}$	Stimulus	Fold Change
Amino acid metabolism				
Arginase	3.5.3.1	4–5 days	Starvation or glucocorticoids.	+2
			Change from high- to low-protein diet.	–2
Alanine transaminase	2.6.1.2	3.5 days	Glucocorticoids.	+5
Serine dehydratase	4.2.1.13	20 hours	Glucagon or dietary amino acids.	+100
Tyrosine transaminase	2.6.1.5	1.5 hours	Glucocorticoids, glucagon, or insulin.	+4
Ornithine transaminase	2.6.1.13	20 hours	Glucocorticoids, dietary amino acids, or high-protein diet.	+20
Histidase	4.3.1.3	2.5 days	Change from low- to high-protein diet.	+20
Carbohydrate metabolism				
Glucokinase	2.7.1.2		Starvation or alloxan diabetes.	–5
			Refed glucose or insulin to diabetic.	+5
PEP-pyruvate carboxykinase	4.1.1.32		Insulin to diabetic rat.	+4
Glucose-6-P dehydrogenase	1.1.1.49	15 hours	Thyroid hormone or fasted rats refed a high-carbohydrate diet.	+10
α-Glycerophosphate dehydrogenase	1.1.2.1	4 days	Thyroid hormone.	+10
Malate:NADP dehydrogenase	1.1.1.40	4 days	Thyroid hormone.	+10
Fructose-1,6-diphosphatase	3.1.3.11		Glucose.	+10
Lipid metabolism				
Citrate cleavage enzyme	4.1.3.8		Starved rats refed a high-carbohydrate, low-fat diet.	+30
Fatty acid synthetase			Starvation.	–10
			Starved animals refed a fat-free diet.	+30
HMG-Co A reductase	1.1.1.34	2–3 hours	Fasting or 5% cholesterol diet.	–10
			Twenty-four hour diurnal variation.	±10
Purine or pyrimidine metabolism				
Xanthine oxidase	1.2.3.2		Change to high-protein diet.	–10
Aspartate transcarbamylase	2.1.3.2	2.5 days	One percent orotic acid diet.	+2
Dihydroorotase	3.5.2.3	12 hours	One percent orotic acid diet.	+3

*Data, with the exception of that for HMG-Co A reductase, from Schimke RT, Doyle D: Ann Rev Biochem 39:929, 1970.

knowledge of the molecular details which account for these changes is fragmentary. The extent to which concepts, models, and mechanisms suitable for bacterial regulation are applicable to mammals is not presently clear. Complications include the possibility of hormonal or nervous control and regulation via control of the rate of degradation of an enzyme as well as by control of its rate of synthesis or of its catalytic efficiency. While both in mammals and in bacteria alterations may occur in the rate at which an enzyme is synthesized, only in mammals is regulation by control of the rate of enzyme degradation thought important. Protein turnover in logarithmically growing cultures of bacteria is relatively slow (2–5% per hour). In mammals it appears that stabilization of an enzyme, decreasing its rate of destruction, or acceleration of its removal by specific degradative enzymes may play major roles in determining the net flux of metabolites through both anabolic and catabolic pathways.

Examples of Control of the Synthesis & Degradation of Mammalian Enzymes

Glucocorticoids increase the concentration of tyrosine transaminase by stimulating its rate of synthesis. This was the first clear case of a hormone regulating the synthesis of a mammalian enzyme. Insulin and glucagon—despite their mutually antagonistic physiologic effects—both independently increase the rate of synthesis 4- to 5-fold. The effect of glucagon probably is mediated via cyclic AMP since this can mimic the effect of the hormone in organ cultures of rat liver.

Although both hydrocortisone and dietary tryptophan increase the activity of hepatic tryptophan oxygenase, the mechanism of the 2 effects differs. Tryptophan has no effect on the rate of enzyme synthesis but retards its degradation . Hydrocortisone, however, increases the rate of enzyme synthesis 5-fold. Glucocorticoids appear to act in a similar manner with respect to hepatic glutamate-pyruvate transaminase. The rate of synthesis is increased without apparent effects on the rate of degradation.

Although arginase levels increase via increased enzyme synthesis when the diet is rich in amino acids, transfer to a diet of lower protein content stimulates the rate of enzyme degradation.

The activity of δ-aminolevulinate synthetase, the first enzyme of heme biosynthesis (see p 73), is increased as much as 50-fold by drugs which produce experimental porphyria, and this effect is blocked by glucose. Although the exact site of regulation is not known, the extremely short half-life ($t_{1/2}$ = 1 hour) suggests control of enzyme degradation as a plausible site.

ISOZYMES

Isozymes are **physically distinct forms of the same catalytic activity.** They thus catalyze the same reaction. Isozymes of a particular enzyme are analogous to dimes minted at Philadelphia, Denver, and San Francisco. The currency value (reactions catalyzed = biologic value) is identical, and each dime (isozyme) is physically quite similar. By analogy to mint marked, however, subtle physical, chemical, and immunologic differences between isozymes become apparent on careful examination.

Medical interest in isozymes was stimulated by the discovery in 1957 that **human sera contained several lactate dehydrogenase (LDH) isozymes and that their relative proportions changed significantly in certain pathologic conditions.** Isozymes have also been reported in the sera and tissues not only of mammals but also of amphibians, birds, insects, plants, and unicellular organisms. Both the kind and number of enzymes involved are equally diverse. Isozymes of numerous dehydrogenases and of several oxidases, transaminases, phosphatases, transphosphorylases, and proteolytic enzymes have been reported.

Serum LDH isozymes may be visualized by subjecting a serum sample to electrophoresis, usually at pH 8.6, using a starch, agar, or polyacrylamide gel supporting medium. The isozymes have different charges at this pH and migrate to 5 regions of the electrophoretogram. Isozymes are then localized by means of their ability to catalyze reduction of a colorless dye to a colored form.

A typical dehydrogenase assay reagent contains the following:

(1) Reduced substrate (eg, lactate).

(2) Coenzyme (NAD).

(3) Oxidized dye (eg, nitro blue tetrazolium salt [NBT]).

(4) An intermediate electron carrier to transport electrons between NADH and the dye (eg, phenazine methosulfate [PMS]•).

(5) Buffer; activating ions if required.

Lactate dehydrogenase (LDH) catalyzes the transfer of 2 electrons and one hydrogen ion from lactate to NAD.

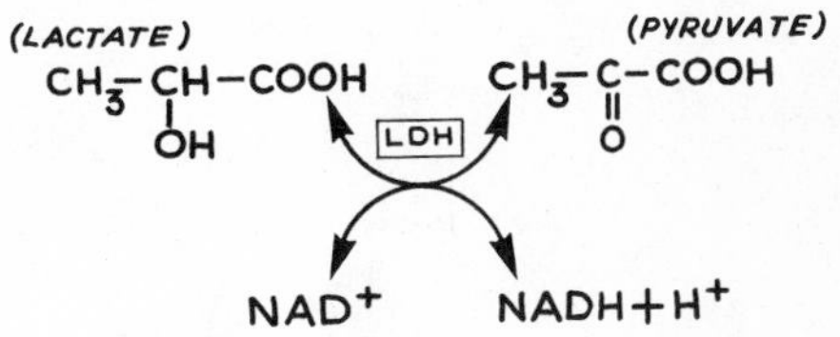

(LACTATE) $CH_3-CH(OH)-COOH$ ⇌ (PYRUVATE) $CH_3-C(=O)-COOH$

LDH

NAD^+ → $NADH + H^+$

The reaction proceeds at a measurable rate only in the presence of the enzyme catalyst, LDH. When the assay mixture is spread on the electrophoretogram and incubated at 37° C, concerted electron transfer reactions take place only in those regions where LDH is present (Fig 8–39).

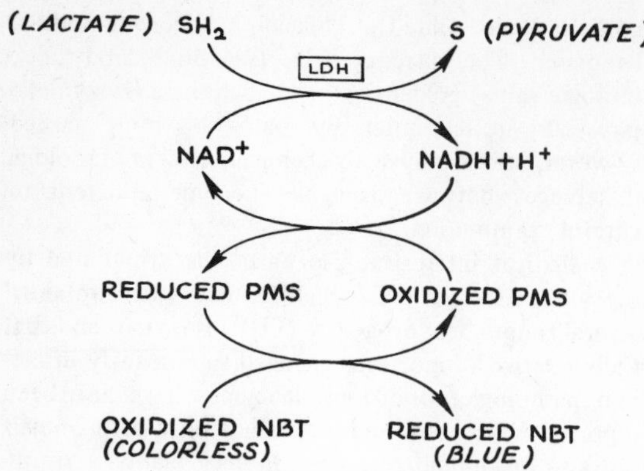

FIG 8–39. Coupled reactions in detection of LDH activity on an electrophoretogram.

The bands are visible to the naked eye and their relative intensities may be quantitated by a suitable scanning photometer (Fig 8–40). The most positive isoenzyme, as detected in an electrophoretogram, is I_2.

Lactate dehydrogenase (LDH) isozymes differ from one another at the level of the quaternary structure. The active lactate dehydrogenase molecule (molecular weight 130,000 daltons) consists of 4 subunits of 2 types, H and M (molecular weight about 34,000 daltons). Only the tetrameric molecule possesses catalytic activity. If order is unimportant, these subunits might be combined in the following 5 ways:

HHHH
HHHM
HHMM
HMMM
MMMM

CL Markert used conditions known to disrupt and reform quaternary structure to clarify the relationships between the LDH isozymes. Splitting and reconstitution of LDH-I_1 or LDH-I_5 produces no new isozymes. These therefore each consist of a single subunit type. When a mixture of purified LDH-I_1 and LDH-I_5 are subjected to the same treatment, LDH-I_2, -I_3, and -I_4 are also produced. The approximate proportions of the isozymes found are those that would result if the relationship were:

LDH Isozyme	Subunits
I_1	HHHH
I_2	HHHM
I_3	HHMM
I_4	HMMM
I_5	MMMM

Syntheses of H and M subunits have recently been shown to be controlled by distinct genetic loci.

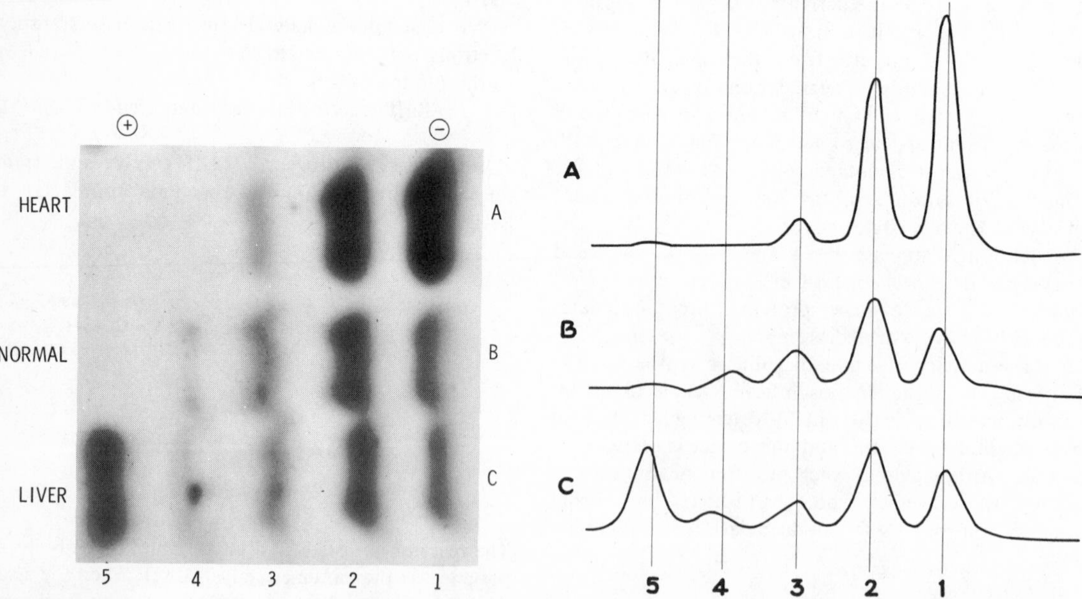

FIG 8–40. Normal and pathologic patterns of LDH isozymes in human serum. LDH isozymes of serum were separated on cellulose acetate at pH 8.6 and stained for enzyme (see p 154). The photometer scan shows the relative proportion of the isozymes. Pattern A is serum from a patient with a myocardial infarct, B is normal serum, and C is serum from a patient with liver disease. (Courtesy of Dr Melvin Black & Mr Hugh Miller, St Luke's Hospital, San Francisco.)

DIAGNOSTIC APPLICATIONS

The measurement of enzyme activity in body fluids such as plasma or serum is often of importance in clinical practice. **Functional plasma enzymes** are actively secreted into the circulation, frequently by the liver, and serve a physiologic function in plasma. Examples include pseudocholinesterase, lipases, and the enzymes involved in blood coagulation. The quantities of these functional enzymes in the plasma are normally high, and greater than those of the same enzymes occurring in the tissues. **Nonfunctional plasma enzymes,** whose substrates and cofactors may indeed be absent from plasma, apparently have no function there. Their normal levels in the plasma are low, up to a million times lower than found in the tissues.

The measurement of nonfunctional enzymes in the plasma is most frequently of value in diagnosis. In this group we may distinguish between enzymes of exocrine secretions and true intracellular enzymes. The exocrine enzymes–pancreatic amylase, lipase, bile alkaline phosphatase, and prostatic acid phosphatase–diffuse passively into the plasma. The true intracellular enzymes constitute the working machinery of the cell, and those which are tightly bound to particulate elements of the cell (see p 124) are generally absent from the circulation.

The low levels of nonfunctional enzymes found normally in plasma apparently arise from the routine normal destruction of erythrocytes, leukocytes, and other cells. With accelerated cell death, soluble enzymes enter the circulation. Although elevated plasma enzyme levels are generally interpreted as evidence of cellular necrosis, vigorous exercise also results in release of small quantities of muscle enzymes. Variable factors to consider include (1) the intracellular location of the enzyme and the permeability of nuclear, mitochondrial, and cellular membranes; (2) the solubility of the enzyme in extracellular fluid; (3) the rate of circulation of extracellular fluid and vascularity of the injured area and the presence or absence of an inflammatory barrier; and (4) the rate of destruction or excretion.

The diagnostic value of certain plasma enzymes is listed below:

(1) The plasma **lipase** level may be low in liver disease, vitamin A deficiency, some malignancies, and in diabetes mellitus. It may be elevated in acute pancreatitis and pancreatic carcinoma.

(2) The plasma **amylase** level may be low in liver disease, increased in high intestinal obstruction, parotitis, acute pancreatitis, and diabetes.

(3) Elevations of **trypsin** in the plasma occur during acute disease of the pancreas, with resultant changes in the coagulability of the blood reported as antithrombin titers. Direct measurement of the plasma trypsin in pancreatic disease may also be made. It is stated that elevation in concentration of plasma trypsin is a more sensitive and reliable indicator of pancreatic disease than either the plasma amylase or lipase.

(4) **Cholinesterase** has been measured in plasma in a number of disease states. In general, low levels are found in patients ill with liver disease, malnutrition, chronic debilitating and acute infectious diseases, and anemias. High levels occur in the nephrotic syndrome. A large number of drugs produce a temporary decrease in cholinesterase activity, but the alkyl fluorophosphates (see p 486) cause irreversible inhibition of the enzyme. Some insecticides in common use depress cholinesterase activity, and tests for the activity of this enzyme in the plasma may be useful in detecting overexposure to these agents.

The content of cholinesterase in young red blood cells is considerably higher than in the adult red blood cells; consequently the cholinesterase titer of erythrocytes in the peripheral blood may be used as an indicator of hematopoietic activity.

(5) The plasma **alkaline phosphatase** level may be increased in rickets, hyperparathyroidism, Paget's disease, osteoblastic sarcoma, obstructive jaundice, and metastatic carcinoma.

As is the case with several other enzymes, isozymes of alkaline phosphatase can be detected in body fluids. These include specific isozymes originating from bone, liver, placenta, and intestine. Measurement of specific alkaline phosphatase isozymes may therefore improve the diagnostic value of this test. Thus, serum alkaline phosphatase levels may increase in congestive heart failure as a result of injury to the liver. Of great value is the use of alkaline phosphatase isozyme measurements to distinguish liver lesions from bone lesions in cases of metastatic carcinoma.

The plasma **acid phosphate** level may be elevated in prostatic carcinoma with metastases.

(6) Two **transaminases** are of clinical interest. **Glutamic oxaloacetic transaminase (GOT)** catalyzes the transfer of the amino group of aspartic acid to α-ketoglutaric acid, forming glutamic and oxaloacetic acids; **glutamic pyruvic transaminase (GPT)** transfers the amino group of alanine to α-ketoglutaric acid, forming glutamic and pyruvic acids (see p 304). Normally the serum transaminase levels are low, but after extensive tissue destruction these enzymes are liberated into the serum. An example is heart muscle, which is rich in transaminase; consequently, myocardial infarcts are followed by rapid and striking increases in serum transaminase levels. Values decrease toward normal within a few days. The estimation of GOT is now widely used to confirm a diagnosis of myocardial infarction.

Liver tissue is rich in both transaminases, but it contains more of GPT than of GOT. Although both transaminases are elevated in sera of patients with acute hepatic disease, GPT, which is only slightly elevated by cardiac necrosis, is therefore a more specific indicator of liver damage.

Extensive skeletal muscle damage, as in severe trauma, also elevates serum transaminase levels.

(7) **Lactic dehydrogenase** (LDH) is an enzyme that can be detected by its ability to catalyze the reduction of pyruvate (see p 159) in the presence of

NADH. In myocardial infarction the concentration of serum LDH rises within 24 hours of the occurrence of the infarct and returns to the normal range within 5–6 days. High levels of LDH also occur in patients with acute and chronic leukemia in relapse, generalized carcinomatosis, and, occasionally, with acute hepatitis during its clinical peak, but not in patients with jaundice due to other causes. Serum LDH is normal in patients with acute febrile and chronic infectious diseases as well as those with anemia, pulmonary infarction, localized neoplastic disease, and chronic disease processes.

Reference was made above to **isozymes of lactic dehydrogenase.** Cardiac muscle contains a preponderance of one form of lactic dehydrogenase (designated LD-I_1). By measurement of the plasma isozyme pattern it has been discovered that the pattern found in the course of myocardial infarction appears to be a more sensitive and lasting indication of myocardial necrosis than is simple measurement of the total serum or plasma lactic dehydrogenase activity. (See also p 159.)

(8) Measurement of serum **isocitric dehydrogenase** activity has been found useful in the diagnosis of liver disease. The isocitric dehydrogenase levels of cerebrospinal fluid are also elevated in patients with cerebral tumors or meningitis of various types. With tumors, the values are about 10 times normal. With meningitis, the values may be as much as 50 times normal, but gradually decrease to normal as the patient recovers.

(9) **Ceruloplasmin**, a copper-containing serum globulin, shows oxidase activity **in vitro** toward several amines, including epinephrine, 5-hydroxytryptamine, and dihydroxyphenylalanine. Plasma ceruloplasmin levels, determined as oxidase activity, are elevated in several circumstances, such as cirrhosis, hepatitis, bacterial infections, pregnancy, etc. Decreased levels, however, provide a useful confirmatory test for Wilson's disease (hepatolenticular degeneration).

(10) **Creatine phosphokinase (CPK).** The measurement of serum creatine phosphokinase (creatine kinase, p 483) activity is of value in the diagnosis of disorders affecting skeletal and cardiac muscle as well as in studies of families affected with pseudohypertrophic muscular dystrophy. Nonmuscular tissues other than brain do not contain high levels of CPK, so that determinations of activity of this enzyme should be more specific to particular tissues than the transaminases or dehydrogenases which are more widely distributed. Crowley has evaluated CPK activity in myocardial infarction and heart failure and following various diagnostic and therapeutic procedures. He has concluded that, by the method used, CPK was somewhat less discriminatory than GOT in detecting myocardial infarction but that the CPK test was more specific in that it was not elevated in congestive failure. (GOT was elevated in 25% of the patients with heart failure.) The CPK test thus appeared to offer a distinct advantage when evaluating suspected infarction in the presence of cardiac failure. Furthermore, CPK may be elevated when GOT is normal. It was thus concluded that the CPK test is a valuable additional procedure to supplement the GOT test in evaluation of a suspected myocardial infarction, and it was suggested that the performance of both tests would improve the diagnostic value of enzyme measurements for the diagnosis of myocardial infarction.

A number of other enzymes have been studied in connection with the diagnosis of disease, particularly as an aid to the diagnosis and measurement of response to therapy in malignant disease (see bibliography).

• • •

Bibliography

General Enzymology

Boyer PD, Lardy H, Myrbäck K (editors): *The Enzymes.* 8 vols. Academic Press, 1959–1963.

Dawes EA: *Quantitative Problems in Biochemistry.* Williams & Wilkins, 1962.

Dixon M, Webb EC: *Enzymes,* 2nd ed. Academic Press, 1964.

Nord FF (editor): *Advances in Enzymology.* Interscience. Issued annually.

Biochemical Genetics

Aeibi HE: Inborn errors of metabolism. Ann Rev Biochem 36:271, 1967.

Epstein W, Beckwith JR: Regulation of gene expression. Ann Rev Biochem 37:411, 1968.

Hsia DY: *Inborn Errors of Metabolism,* 2nd ed. Year Book, 1966.

Stanbury JB, Wyngaarden JB, Frederickson DS (editors): *The Metabolic Basis of Inherited Disease.* McGraw-Hill, 1966.

Intracellular Distribution of Enzymes

Albers RW: Biochemical aspects of active transport. Ann Rev Biochem 36:727, 1967.

Greer DE, MacLennan DH: The mitochondrial system of enzymes. In: *Metabolic Pathways,* vol 1. Greenberg DM (editor). Academic Press, 1967.

Korn ED: Cell membranes: Structure and synthesis. Ann Rev Biochem 38:263, 1969.

Lehninger AL: *The Mitochondrion: Molecular Basis of Structure and Function.* Benjamin, 1964.

Osawa S: Ribosome formation and structure. Ann Rev Biochem 37:109, 1968.

Reed LJ, Cox DJ: Macromolecular organization of enzyme systems. Ann Rev Biochem 35:57, 1966.

Nomenclature

Enzyme Nomenclature. Recommendations of the International Union of Biochemistry on the Nomenclature and Classification of Enzymes, Together With Their Units and Symbols of Enzyme Kinetics. Elsevier, 1964.

Assay & Purification of Enzymes

Bergmeyer H-U (editor): *Methods of Enzymatic Analysis.* Academic Press, 1963.

Boyer PD, Lardy H, Myrbäck K (editors): *The Enzymes.* 8 vols. Academic Press, 1959–1963.

Colowick SP, Kaplan, NO (editors): *Methods in Enzymology.* 18 vols. Academic Press, 1955–1971.

Bioenergetics

Ingraham LL, Pardee AB: Free energy and entropy in metabolism. In: *Metabolic Pathways,* vol 1. Greenberg DM (editor). Academic Press, 1967.

Klotz IM: *Energetics in Biochemical Reactions.* Academic Press, 1957.

Racker E: *Mechanisms in Bioenergetics.* Academic Press, 1965.

Kinetics

Cleland WW: The statistical analysis of enzyme kinetic data. Advances Enzymol 29:1, 1967.

Cleland WW: Enzyme kinetics. Ann Rev Biochem 36:77, 1967.

Garfinkel D, Garfinkel L, Pring M, Green SB, Chance B: Computer applications to biochemical kinetics. Ann Rev Biochem 39:473, 1970.

Coenzymes

Chaikin S: Nicotinamide coenzymes. Ann Rev Biochem 37:149, 1968.

Hogenkamp HPC: Enzymatic reactions involving corrinoids. Ann Rev Biochem 37:225, 1968.

Krampitz LO: Catalytic functions of thiamine diphosphate. Ann Rev Biochem 38:213, 1969.

Neims AH, Hellerman L: Flavoenzyme catalysis. Ann Rev Biochem 39:867, 1970.

The Active Site

Koshland DE Jr, Neet KE: The catalytic and regulatory properties of enzymes. Ann Rev Biochem 37:359, 1968.

Vallee BL, Riordan JF: Chemical approaches to the active sites of enzymes. Ann Rev Biochem 38:733, 1969.

Inhibitors

Webb JL: *Enzyme and Metabolic Inhibitors.* 2 vols. Academic Press, 1966.

Regulation of Enzyme Activity

Exton JH, Park CR: The role of cyclic AMP in the control of liver metabolism. Advances Enzym Regulat 6:391, 1968.

Schimke RT, Doyle D: Control of enzyme levels in animal tissues. Ann Rev Biochem 39:929, 1970.

Stadtman ER: Allosteric regulation of enzyme activity. Advances Enzymol 28:41, 1966.

Umbarger HE: Regulation of amino acid metabolism. Ann Rev Biochem 38:323, 1969.

Weber G (editor): *Advances in Enzyme Regulation,* vols 1–6. Pergamon, 1963–1969.

Enzyme Structure

Blow DM, Steitz TA: X-ray diffraction studies of enzymes. Ann Rev Biochem 39:63, 1970.

Editorial: Evergreen enzyme. Nature 218:1202, 1968.

Ginsberg A, Stadtman ER: Multienzyme systems. Ann Rev Biochem 39:429, 1970.

Klotz IM, Langerman NR, Darnall DW: Quaternary structure of enzymes. Ann Rev Biochem 39:25, 1970.

Marglin A, Merrifield RB: Chemical synthesis of peptides and proteins. Ann Rev Biochem 39:841, 1970.

Isozymes

Katunuma N, Katsunuma T, Tomino I, Matsuda Y: Regulation of glutaminase activity and differentiation of the isozyme during development. Advances Enzym Regulat 6:227, 1968.

Moog F, Vire HR, Grey RD: The multiple forms of alkaline phosphatase in the small intestine of the young mouse. Biochem biophys acta 113:336, 1966.

Multiple molecular forms of enzymes. Ann New York Acad Sc 94:art 3, 1961.

Stambaugh R, Post D: Substrate and product inhibition of rabbit muscle lactic dehydrogenase heart (H_4) and muscle (M_4) isozymes. J Biol Chem 241:1462, 1966.

Clinical Enzymology

Brewer GJ, Sing CF: *An Introduction to Isozyme Techniques.* Academic Press, 1970.

Crowley LV: Creatine phosphokinase activity in myocardial infarction, heart failure, and following various diagnostic and therapeutic procedures. Clin Chem 14:1185, 1968.

Esnouf MP, McFarlane RG: Enzymology and the blood clotting mechanism. Advances Enzymol 30:255, 1968.

Latner AL, Skiller AW: *Isozymes in Biology and Medicine.* Academic Press, 1968.

Wilkinson JH: Clinical applications of isozymes. Clin Chem 16:733, 1970.

Wilkinson JH: Clinical significance of enzyme activity measurements. Clin Chem 16:882, 1970.

Wacker WEC, Coombs TL: Clinical biochemistry: Enzymatic methods: Automation and atomic adsorption spectroscopy. Ann Rev Biochem 38:539, 1969.

Weil-Malherbe H: The biochemistry of the functional psychoses. Advances Enzymol 29:479, 1967.

Wenner CE: Progress in tumor enzymology. Advances Enzymol 29:321, 1967.

9 . . .

Biologic Oxidation

With Peter Mayes, PhD, DSc*

Oxidation reactions are accompanied by liberation of energy as the reacting chemical system moves from a higher to a lower energy level. Most frequently the energy is liberated in the form of heat. In nonbiologic systems heat energy may be transformed into mechanical or electrical energy. Since biologic systems are essentially isothermic, no direct use can be made of heat liberated in biologic oxidations to drive the vital processes that require energy. These processes–eg, synthetic reactions, muscular contraction, nerve conduction, and active transport–obtain energy by chemical linkage or **coupling** to oxidative reactions. In its simplest form this type of coupling may be represented as follows:

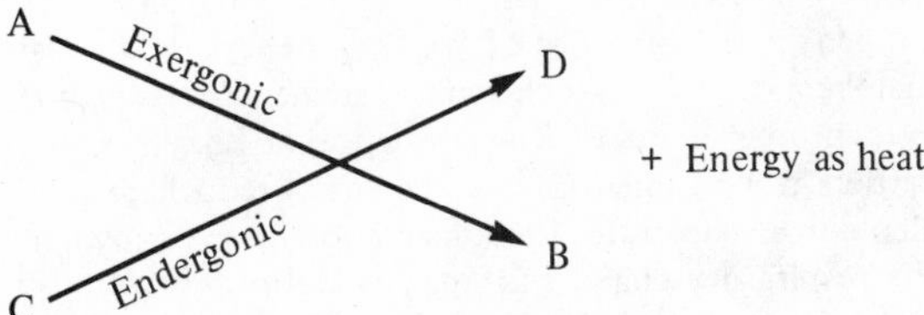

or A + C → B + D + calories (heat)

FIG 9–1. Coupling of an exergonic to an endergonic reaction.

The oxidation of metabolite A to metabolite B occurs with release of energy. It is coupled to another reaction, in which energy is required to convert metabolite C to metabolite D. As some of the energy liberated in the degradative reaction is transferred to the synthetic reaction in a form other than heat, the normal chemical terms exothermic and endothermic cannot be applied to these reactions. Rather, the terms **exergonic** and **endergonic** are used to indicate that a process is accompanied by loss of free energy or gain of free energy respectively, regardless of the form of energy involved.

If the above reaction is to go from left to right, then the overall process must be accompanied by loss of free energy as heat. One possible mechanism of coupling could be envisaged if a common obligatory intermediate (I) took part in both reactions, ie,

Some exergonic and endergonic reactions in biologic systems are coupled in this way. It should be appreciated that this type of system has a built-in mechanism for biologic control of the rate at which oxidative processes are allowed to occur since the existence of a common obligatory intermediate for both the exergonic and endergonic reactions allows the rate of utilization of the product of the synthetic path (D) to determine by mass action the rate at which A is oxidized. Indeed, these relationships supply a basis for the concept of **respiratory control.**

An alternative method of coupling an exergonic to an endergonic process is to synthesize a compound of high-energy potential in the exergonic reaction and to incorporate this new compound into the endergonic reaction, thus effecting a transference of free energy from the exergonic to the endergonic pathway.

Example:

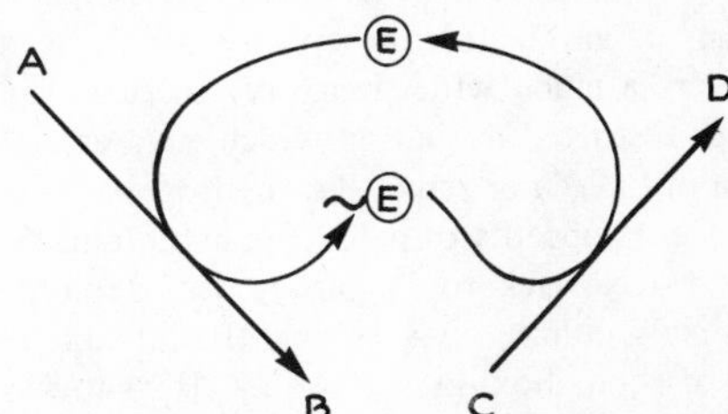

FIG 9–2. Transference of free energy from an exergonic to an endergonic pathway through formation of a high-energy intermediate compound.

where ~Ⓔ is a compound of high potential energy and Ⓔ is the corresponding compound of low potential energy. The biologic advantage of this mechanism is that Ⓔ, unlike I in the previous system, need not be structurally related to A, B, C, or D. This would allow Ⓔ to serve as a transducer of energy from a wide range of exergonic reactions to an equally wide range of endergonic reactions or processes.

Example: See Fig 9–3.

*Lecturer in Biochemistry, Royal Veterinary College, University of London.

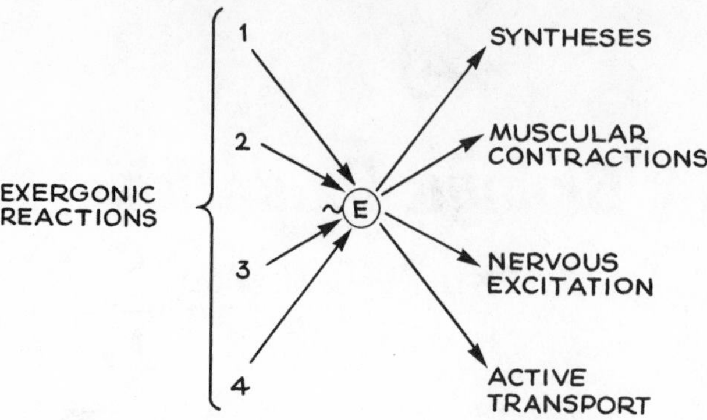

FIG 9–3. Transduction of energy through a common high-energy compound to energy-requiring (endergonic) biological processes.

In the living cell the high-energy intermediate or carrier compound (designated ~Ⓔ) is **adenosine triphosphate** or **ATP**.

Historical Review

Chemically, oxidation is defined as the removal of electrons and reduction is the gain of electrons, as illustrated by the conversion of ferrous to ferric ion.

$$Fe^{2+} \xrightarrow{e^- \text{(ELECTRON)}} Fe^{3+}$$

It follows that oxidation is always accompanied by reduction of an electron acceptor. The above definition covers a much wider range of reactions than did the older restricted definition which covered only the addition of oxygen or removal of hydrogen.

Modern concepts of oxidation in biologic systems may be traced back to Lavoisier, who demonstrated that animals utilize oxygen from the air and replace this by carbon dioxide and water. He showed that respiration was similar in this respect to the burning of a candle. However, Pasteur, in his studies of the fermentation of glucose by yeast, firmly established that living organisms could respire in the absence of oxygen, ie, under anaerobic conditions. In the period around 1930, 2 diametrically opposed concepts of biologic oxidation prevailed. Warburg advocated the view that a widely distributed enzyme (**Atmungsferment**) catalyzed the activation of oxygen and its combination with substrate molecules. He believed that heavy metals, particularly iron, played a part in the catalysis, a conclusion that was supported by the fact that carbon monoxide and cyanide were potent inhibitors not only of respiration but also of heavy metal catalysis of the oxidation of organic molecules. Opposed to this concept was the thesis of Wieland that substrate molecules were activated and oxidized by removal of hydrogen in reactions catalyzed by specific enzymes called **dehydrogenases**. It was shown that the reduced dehydrogenases that resulted could in turn become reoxidized not only by oxygen but also by other acceptors such as methylene blue. With the discovery by Keilin of a group of respiratory catalysts designated the **cytochrome system**, the 2 concepts were reconciled, as it became clear that most substrates were in fact oxidized by a combination of both processes. Dehydrogenation initiated oxidation, and the reducing equivalents were transported via the cytochrome system to react ultimately with molecular oxygen in the presence of Warburg's enzyme, the last member of the cytochrome system, now renamed **cytochrome oxidase**. The sequence of enzymes and carriers responsible for the transport of reducing equivalents from substrates to molecular oxygen is known as the **respiratory chain**. Further elucidation by Warburg and others of the role of **nicotinamide nucleotides** and **flavoproteins** made it possible by 1940 to construct the following sequence of components of the respiratory chain. The arrows indicate the direction of flow of reducing equivalents (H or electrons).

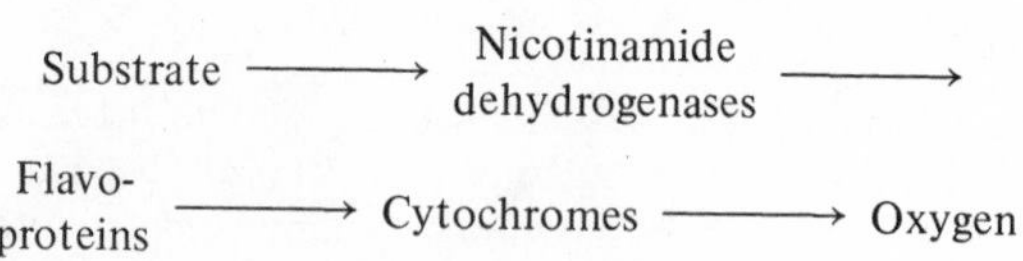

The respiratory chain is now known to be localized within mitochondria. By using mitochondria as a source, much purer preparations of the respiratory catalysts can now be studied than was possible with older methods of isolation from whole organs or tissue. At the present time the mechanism of coupling of oxidation in the respiratory chain to the production of the high-energy carrier, ATP, is a particularly active area of research.

Oxidation-Reduction Equilibria; Redox Potential

The change in free energy resulting from a chemical reaction, ΔG (ΔF) is a measure of the useful or available energy which is absorbed or which results from that reaction. If ΔG is negative in sign, the reaction proceeds spontaneously with loss of free energy, ie, it is **exergonic**. If, in addition, ΔG is large in magnitude, the reaction goes virtually to completion and is essentially irreversible. On the other hand, if ΔG is positive, the reaction proceeds only if free energy can be gained, ie, it is **endergonic**. If, in addition, the magnitude of ΔG is large, the system is stable with little or no tendency for a reaction to occur. In reactions involving oxidation and reduction, the free energy change is proportionate to the tendency of reactants to donate or accept electrons. This is expressed numerically as an **oxidation-reduction** or **redox potential.** It is usual to compare the redox potential of a system (E_O) against the potential of the hydrogen electrode, which at pH 0 is designated as 0.0 volts. However, for biologic systems it is normal to express the redox potential (E_O') at pH 7.0, at which pH the electrode potential of the hydrogen electrode is −0.42 volts. The redox potentials of some redox systems of special interest in mammalian physiology are shown in Table 9–1.

TABLE 9–1. Some redox potentials of special interest in mammalian oxidation systems.

System	E_O' volts
Oxygen/water	+0.82
Cytochrome a; Fe^{+++}/Fe^{++}	+0.29
Cytochrome c; Fe^{+++}/Fe^{++}	+0.22
Ubiquinone; ox/red	+0.10
Cytochrome b; Fe^{+++}/Fe^{++}	+0.08
Fumarate/succinate	+0.03
Flavoprotein-old yellow enzyme; ox/red	−0.12
Oxaloacetate/malate	−0.17
Pyruvate/lactate	−0.19
Acetoacetate/β-hydroxybutyrate	−0.27
Lipoate; ox/red	−0.29
$NAD^+/NADH$	−0.32
H^+/H_2	−0.42
Succinate/α-ketoglutarate	−0.67

ENZYMES & COENZYMES INVOLVED IN OXIDATION & REDUCTION

In the Report of the International Union of Biochemistry, 1961, all enzymes concerned in oxidative processes are designated oxidoreductases. In the following account, oxidoreductases are classified into 5 groups.

(1) **Oxidases**: Enzymes that catalyze the removal of hydrogen from a substrate but use only oxygen as a hydrogen acceptor. They invariably contain copper and form water as a reaction product (with the exception of uricase and monoamine oxidase, which form H_2O_2).

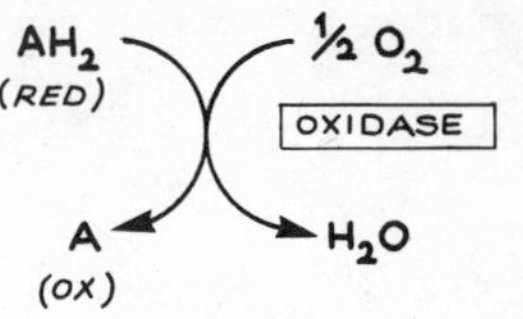

FIG 9–4. Oxidation of a metabolite catalyzed by an oxidase.

(2) Aerobic dehydrogenases: Enzymes catalyzing the removal of hydrogen from a substrate but which, as distinct from oxidases, can use either oxygen or artificial substances such as methylene blue as a hydrogen acceptor. Characteristically, these dehydrogenases are flavoproteins. Hydrogen peroxide rather than water is formed as a product.

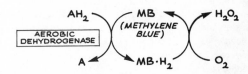

FIG 9–5. Oxidation of a metabolite catalyzed by an aerobic dehydrogenase.

(3) Anaerobic dehydrogenases: Enzymes catalyzing the removal of hydrogen from a substrate but not able to use oxygen as hydrogen acceptor. There are a large number of enzymes in this class. They perform 2 main functions:

(a) Transfer of hydrogen from one substrate to another in a coupled oxidation-reduction reaction not involving a respiratory chain.

FIG 9–6. Oxidation of a metabolite catalyzed by anaerobic dehydrogenases, not involving a respiratory chain.

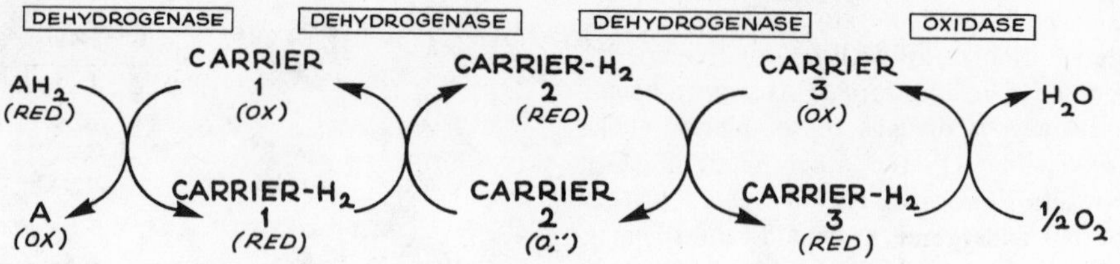

FIG 9–7. Oxidation of a metabolite by anaerobic dehydrogenases utilizing several components of a respiratory chain.

These dehydrogenases are specific for their substrates but often utilize the same coenzyme or hydrogen carrier as other dehydrogenases. As the reactions are reversible, these properties enable reducing equivalents to be freely transferred within the cell. This type of reaction, which enables a substrate to be oxidized at the expense of another, is particularly useful in enabling oxidative processes to occur in the absence of oxygen.

(b) As components in a respiratory chain of electron transport from substrate to oxygen (Fig 9–7).

(4) Hydroperoxidases: Enzymes utilizing hydrogen peroxide as a substrate. Two enzymes fall into this category: **peroxidase,** found in milk and in plants; and **catalase**, found in animals and plants.

(5) Oxygenases: Enzymes that catalyze the direct transfer and incorporation of oxygen into a substrate molecule.

Oxidases

Oxidases are conjugated proteins whose prosthetic groups contain copper.

Cytochrome oxidase is a hemoprotein widely distributed in many plant and animal tissues. It is the terminal component of the chain of respiratory carriers found in mitochondria and is therefore responsible for the reaction whereby electrons resulting from the oxidation of substrate molecules by dehydrogenases are transferred to their final acceptor, oxygen. The enzyme is poisoned by carbon monoxide (only in the dark), cyanide, and hydrogen sulfide. It is considered to be identical with Warburg's respiratory enzyme and with what has also been termed Cytochrome a_3. Two heme groups are present per molecule of enzyme, and the iron atoms oscillate between the Fe^{+++} and Fe^{++} valence during oxidation and reduction. The mechanisms of reaction of cytochrome oxidase and the possible role of copper in the interaction with oxygen are discussed by Gibson (1968).

Phenolase (tyrosinase, polyphenol oxidase, catechol oxidase) is a copper-containing enzyme that is specific for more than one type of reaction (see Malmström & Ryden, 1968). It is able to convert monophenols or *o*-diphenols to *o*-quinones. Other enzymes containing copper are **laccase**, which is widely distributed in plants and animals (converts *p*-hydroquinones to *p*-quinones), and **ascorbic oxidase**, found only in plants. Copper has been claimed to be present in a number of other enzymes such as **uricase**, which catalyzes the oxidation of uric acid to allantoin, and **monoamine oxidase**, found in the mitochondria of several tissues, an enzyme that oxidizes epinephrine (Fig 20–12) and tyramine, for example.

Aerobic Dehydrogenases

Aerobic dehydrogenases are flavoprotein enzymes having **flavin mononucleotide (FMN)** or **flavin adenine dinucleotide (FAD)** as prosthetic groups. The flavin groups vary in their affinity for their respective apoenzyme protein, some being detached easily and others not detached without destroying the enzyme. Many of these flavoprotein enzymes contain, in addition, a metal which is essential for the functioning of the enzyme; these are known as **metalloflavoproteins.**

Enzymes belonging to this group of aerobic dehydrogenases include **D-amino acid dehydrogenase** (D-amino acid oxidase), an FAD-linked enzyme, found particularly in liver and kidney, that catalyzes the oxidative deamination of the unnatural (D-) forms of amino acids. Other substrates include glycine, D-lactate, and L-proline, demonstrating that the enzyme is not completely specific for D-amino acids. **L-Amino acid dehydrogenase** (L-amino acid oxidase) is an FMN-linked enzyme found in kidney with general specificity for the oxidative deamination of the naturally occurring L-amino acids. **Xanthine dehydrogenase** (xanthine oxidase) has a wide distribution, occurring in milk and in liver. In the liver, it plays an important role in the conversion of purine bases to uric acid. It is of particular significance in the liver and kidneys of birds, which excrete uric acid as the main nitrogenous end product not only of purine metabolism but also of protein and amino acid catabolism. Xanthine dehydrogenase contains FAD as the prosthetic group. It is a metalloflavoprotein containing both nonheme iron and molybdenum, and it has a dual specificity in that it also oxidizes all aldehydes.

Aldehyde dehydrogenase (aldehyde oxidase) is an FAD-linked enzyme present in pig and other mammalian livers. It is similar to xanthine dehydrogenase in being a metalloflavoprotein containing molybdenum and nonheme iron and in acting upon aldehydes and N-heterocyclic substrates. However, it is distinguished from xanthine dehydrogenase by virtue of its inability to oxidize xanthine.

Of interest because of its use in estimating glucose is **glucose oxidase**, an FAD-specific enzyme prepared from fungi.

All of the above-mentioned aerobic dehydrogenases contain 2 molecules of the flavin nucleotide per mol. The metalloflavoproteins also have a fixed stoichiometry with regard to the number of atoms of metal per molecule, usually Mo:Fe as 2:8. The mechanisms of oxidation and reduction of these enzymes are complex. There seem to be different detailed mechanisms for each enzyme with the possible involvement of free radicals (Rajagopalan & Handler, 1968). However, evidence points to reduction of the isoalloxazine ring taking place in 2 steps via a semiquinone (free radical) intermediate, as indicated below:

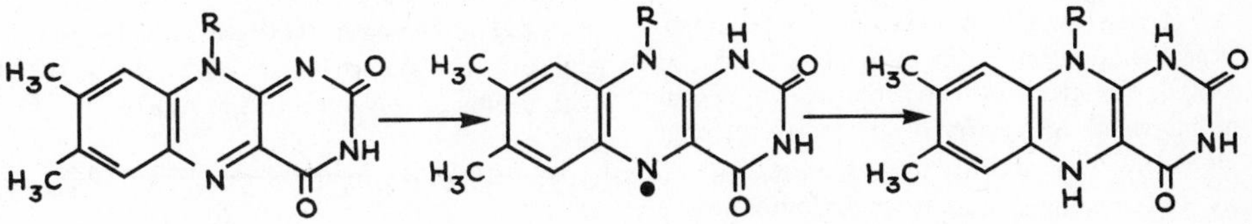

There is stereospecificity about position 4 of nicotinamide when it is reduced by a substrate AH_2. One of the hydrogen atoms is removed from the substrate as a hydrogen nucleus with 2 electrons (hydride ion) and is transferred to the 4 position where it may be attached in either the A- or B- position according to the specificity determined by the particular dehydrogenase catalyzing the reaction. The remaining hydrogen of the hydrogen pair removed from the substrate remains free as a hydrogen ion. Deuterium-labeled substrates have been used in elucidating these mechanisms. For a further discussion, see Sund (1968).

Generally, NAD-linked dehydrogenases catalyze oxidoreduction reactions in the oxidative pathways of metabolism, particularly in glycolysis, the citric acid cycle, and in the respiratory chain of mitochondria. NADPH-linked dehydrogenases are found characteristically in reductive syntheses, as in the extramitochondrial pathway of fatty acid synthesis and steroid synthesis. They are also to be found as coenzymes to the dehydrogenases of the hexose monophosphate shunt. Some nicotinamide coenzyme-dependent dehydrogenases have been found to contain zinc, notably alcohol dehydrogenase from liver and glyceraldehyde-3-phosphate dehydrogenase from skeletal muscle. The zinc ions are not considered to take part in the oxidation and reduction.

Anaerobic Dehydrogenases

A. Dehydrogenases Dependent on Nicotinamide Coenzymes: A large number of dehydrogenase enzymes fall into this category. They are linked as coenzymes either to **nicotinamide adenine dinucleotide (NAD)** or to **nicotinamide adenine dinucleotide phosphate (NADP)**. The coenzymes are reduced by the specific substrate of the dehydrogenase and reoxidized by a suitable electron acceptor. They may freely and reversibly dissociate from their respective apoenzymes. The nicotinamide nucleotides are synthesized from the vitamin niacin (nicotinic acid and nicotinamide). The mechanism of oxidation of the coenzymes is as shown in Fig 9–8.

B. Dehydrogenases Dependent on Riboflavin Prosthetic Groups: The prosthetic groups associated

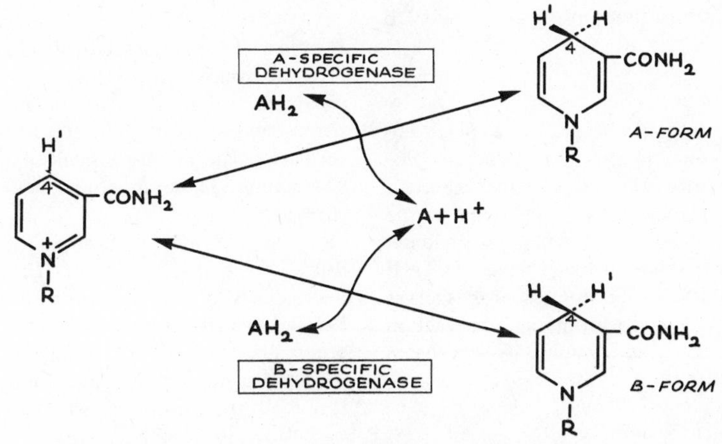

A-SPECIFIC DEHYDROGENASE

B-SPECIFIC DEHYDROGENASE

A-FORM

B-FORM

$$NAD^+ + AH_2 \longleftrightarrow NADH + H^+ + A$$

FIG 9–8. Mechanism of oxidation of nicotinamide coenzymes.

with these flavoprotein dehydrogenases are similar to those of the aerobic dehydrogenase group, namely FMN and FAD. They are in the main more tightly bound to their apoenzymes than the nicotinamide coenzymes. Most of the riboflavin-linked anaerobic dehydrogenases are concerned with electron transport in (or to) the respiratory chain. **NADH dehydrogenase** is a member of the respiratory chain acting as a carrier of electrons between NADH and the more electropositive components. Other dehydrogenases such as **succinate dehydrogenase**, **acyl-Co A dehydrogenase**, and **mitochondrial α-glycerophosphate dehydrogenase** transfer electrons directly from the substrate to the respiratory chain. Another role of the flavin-dependent dehydrogenases is in the dehydrogenation of reduced lipoate, an intermediate in the oxidative decarboxylation of pyruvate and α-ketoglutarate. In this particular instance, due to the low redox potential, the flavoprotein (FAD) acts as a carrier of electrons from reduced lipoate to NAD^+. The **electron-transferring flavoprotein (ETF)** is an intermediary carrier of electrons between acyl-Co A dehydrogenase and the respiratory chain.

C. The Cytochromes: Except for cytochrome oxidase (previously described), the cytochromes are classified as anaerobic dehydrogenases. Their identification and study are facilitated by the presence in the reduced state of characteristic absorption bands which disappear on oxidation. In the respiratory chain they are involved as carriers of electrons from flavoproteins on the one hand to cytochrome oxidase on the other. The cytochromes are iron-containing hemoproteins, in which the iron atom oscillates between Fe^{+++} and Fe^{++} during oxidation and reduction. Several identifiable cytochromes occur in the respiratory chain, viz, cytochromes b, c_1, c, a, and a_3 (cytochrome oxidase). Of these, only cytochrome c is soluble. Study of its structure has revealed that the iron porphyrin group is attached to the apoprotein by 2 thioether bridges derived from cystine residues of the protein (see p 79). Many other cytochromes are known from widely different types of tissues, eg, microsomes, plant cells, bacteria, and yeasts. For further information, refer to Gibson (1968).

Hydroperoxidases

A. Peroxidase: Although typically a plant enzyme, peroxidase is found in milk and leukocytes. The prosthetic group is protoheme, which, unlike the situation in most hemoproteins, is loosely bound to the apoprotein. In the reaction catalyzed by peroxidase, hydrogen peroxide is reduced at the expense of several substances that will act as electron acceptors such as ascorbate, quinones, and cytochrome c. The reaction catalyzed by peroxidase is complex (see Gibson, 1968), but the overall reaction is as follows:

$$H_2O_2 + AH_2 \xrightarrow{\text{PEROXIDASE}} 2H_2O + A$$

B. Catalase: Catalase is a hemoprotein containing 4 heme groups. In addition to possessing peroxidase activity, it is able to use one molecule of H_2O_2 as a substrate electron donor and another molecule of H_2O_2 as oxidant or electron acceptor.

$$2H_2O_2 \xrightarrow{\text{CATALASE}} 2H_2O + O_2$$

Catalase is found in blood and liver. Its function is assumed to be the destruction of hydrogen peroxide formed by the action of aerobic dehydrogenases.

Oxygenases

Enzymes in this group catalyze the incorporation of oxygen into a substrate molecule. They may be divided into 2 subgroups:

A. Dioxygenases (Oxygen Transferases, True Oxygenases): These enzymes catalyze the incorporation of both atoms of oxygen into the substrate:

$$A + O_2 \longrightarrow AO_2$$

Examples of this type include enzymes that contain iron as a prosthetic group such as **homogentisate dioxygenase** and **3-hydroxyanthranilate dioxygenase** from the supernatant fraction of the liver, and enzymes utilizing heme as a prosthetic group such as **L-tryptophan dioxygenase** (tryptophan pyrrolase) from the liver.

B. Mono-oxygenase (Mixed Function Oxidases, Hydroxylases): These enzymes catalyze the incorporation of only one atom of the oxygen molecule into a substrate. The other oxygen atom is reduced to water, an additional electron donor or cosubstrate being necessary for this purpose.

Example:

$$A{-}H + O_2 + ZH_2 \longrightarrow A{-}OH + H_2O + Z$$

Hayaishi (1968) has subdivided the mono-oxygenases into subgroups according to the nature of the cosubstrate electron donor involved. For example, many of the enzymes involved in steroid syntheses or transformations are mono-oxygenases utilizing NADPH as a cosubstrate. These are found mainly in the liver and adrenal glands.

The enzymes involved in the hydroxylation of drugs belong to this group. They are found in the microsomes of the liver together with cytochrome P-450 and cytochrome b_5. According to Estabrook & others (1970), both NADH and NADPH donate reducing equivalents for the reduction of these cytochromes (Fig 9–9), which in turn are oxidized by substrates in a reaction catalyzed by a hydroxylase enzyme.

Example: See below.

$$\underset{}{DRUG{-}H} + O_2 + \underset{(P\text{-}450)}{2Fe^{++}} + 2H^+ \xrightarrow{\text{HYDROXYLASE}} DRUG{-}OH + H_2O + \underset{(P\text{-}450)}{2Fe^{+++}}$$

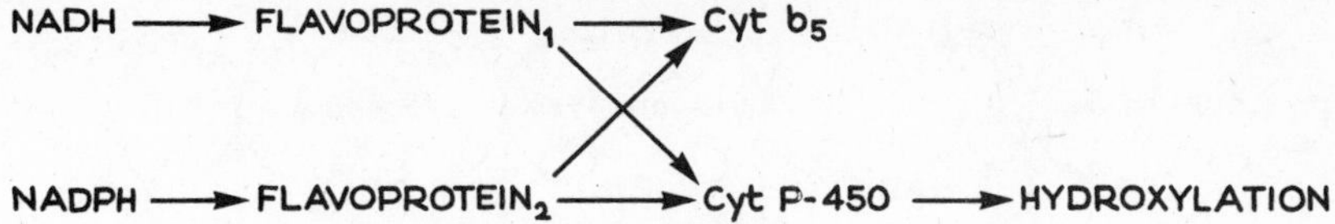

FIG 9–9. Electron transport chain in microsomes.

Many drugs such as phenobarbital have the ability to induce the formation of microsomal enzymes and cytochrome P-450.

Oxygenases do not take part in reactions that have as their purpose the provision of energy to the cell; rather, they are concerned with the synthesis or degradation of many different types of metabolites.

For a recent review on enzymic hydroxylation, see Hayaishi (1969).

THE RESPIRATORY CHAIN

It is not without justification that the mitochondrion has been called the "power house" of the cell since it is within the mitochondria that most of the energy derived from tissue oxidation is captured in the form of the high-energy intermediate, ATP. All the useful energy formed during the oxidation of fatty acids and amino acids and virtually all of the energy from the oxidation of carbohydrate is liberated within the mitochondria. To accomplish this, the mitochondria contain the series of catalysts known as the respiratory chain, which are concerned with the transport of reducing equivalents (hydrogen and electrons) and their final reaction with oxygen to form water. Mitochondria also contain the enzyme systems responsible for producing those reducing equivalents in the first place, ie, the enzymes of β-oxidation and of the citric acid cycle. The latter is the final common metabolic pathway for the oxidation of all the major foodstuffs. These relationships are shown in Fig 9–10.

Organization of the Respiratory Chain in Mitochondria

The major components of the respiratory chain (Fig 9–10) are arranged sequentially in order of increasing redox potential (Table 9–1). Electrons flow through the chain in a stepwise manner from the more electronegative components to the more electropositive, oxygen. Thus, the redox potential of a component of the respiratory chain contributes to the information necessary to assign it a tentative position in the chain. Several other approaches have been used to identify components and their relative positions. Chance and his associates have developed sophisticated technics for following the absorption spectra of the individual components in intact mitochondria. Other investigators, including Green, have broken the chain down into separate components or complexes and attempted to reconstruct it from the separate parts. Slater and others have used inhibitors which block specific reactions in the chain; these are frequently employed with artificial electron acceptors and donors. Chance & Williams (1956) introduced the concept of "crossover" to locate the site of action of inhibitors. The concept is based upon the assumption that when an inhibitor is introduced into an active series of redox components of the respiratory chain in the steady state, those components on the electronegative side of the block become more reduced while those on the electropositive side become more oxidized.

The main respiratory chain in mitochondria may be considered as proceeding from the NAD-linked dehydrogenase systems on the one hand, through flavoproteins and cytochromes, to molecular oxygen on the other. The reducing equivalents are transported either as H^+ or as covalent hydrogen (Fig 9–11). Not

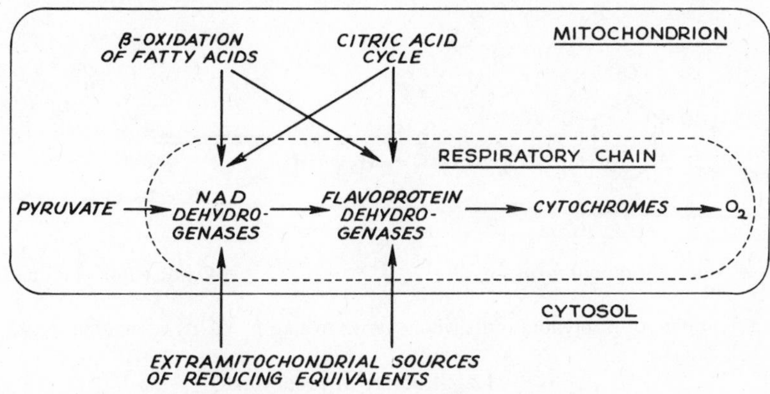

FIG 9–10. Relationship of electron transport in the respiratory chain to the β-oxidation of fatty acids, to the citric acid cycle, and to extramitochondrial sources of reducing equivalents.

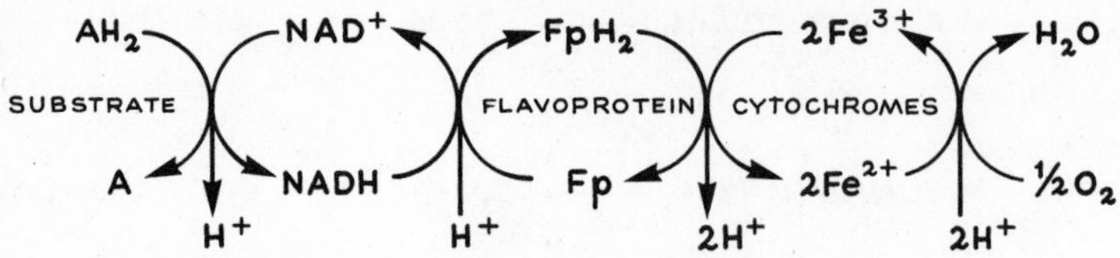

FIG 9–11. Transport of reducing equivalents through the respiratory chain.

all substrates are linked to the respiratory chain through NAD-specific dehydrogenases; some, which, because their redox potentials are more positive (eg, succinate), are linked directly to flavoprotein dehydrogenases which in turn are linked to the cytochromes of the respiratory chain.

In recent years it has become clear that an additional carrier is present in the respiratory chain linking the flavoproteins to cytochrome b, the member of the cytochrome chain of lowest redox potential. This substance, which has been named **ubiquinone** or **coenzyme Q** (Co Q; see Fig 9–12), exists in mitochondria in the oxidized quinone form under aerobic conditions and in the reduced quinol form under anaerobic conditions. Co Q is a constituent of the mitochondrial lipids, the other lipids being predominantly phospholipids that constitute part of the mitochondrial membrane. Co Q has a structure that is very similar to vitamin K and vitamin E. It is also similar to plastoquinone found in chloroplasts. All of these substances are characterized by the possession of a polyisoprenoid side chain. In mitochondria there is a large stoichiometric excess of Co Q to other members of the respiratory chain. It is possible that there is more than one pool of Co Q and that some of it is not in the direct pathway of oxidation. For further discussion of the role of quinones in electron transport, see Crane (1968).

An additional component found in respiratory chain preparations is **nonheme iron (NHI)** which is combined with protein and is associated with the flavoproteins (metalloflavoproteins) and with cytochrome b. The presence of a paramagnetic metal allows such preparations to be examined using the technic of electron paramagnetic resonance spectroscopy (EPR). The nonheme iron-protein complex is similar to the ferredoxins of bacteria and to iron proteins present in plants. On denaturation with acid or heat, H_2S is liberated in an amount which is stoichiometrically related to the iron present. The H_2S is derived from what is called "labile sulfur" attached to the iron. Both the sulfur and iron are thought to take part in the oxidoreduction mechanisms.

A current view of the principal components of the respiratory chain is shown in Fig 9–13. At the electronegative end of the chain, dehydrogenase enzymes catalyze the transfer of electrons from substrates to NAD of the chain. Several differences exist in the manner in which this is carried out. The α-keto acids pyruvate and α-ketoglutarate have complex dehydrogenase systems involving lipoate and FAD prior to the passage of electrons to NAD of the respiratory chain. Electron transfers from other dehydrogenases such as L(+)-β-hydroxyacyl-Co A, D(−)-β-hydroxybutyrate, glutamate, malate, and isocitrate dehydrogenases appear to couple directly with NAD of the respiratory chain, although it is possible, as Lehninger (1964) has pointed out, that more than one pool of NAD may exist. Of these dehydrogenases, β-hydroxybutyrate dehydrogenase is much more firmly bound to the mitochondrion.

The reduced NAD of the respiratory chain is in turn oxidized by a metalloflavoprotein enzyme—**NADH dehydrogenase**. This enzyme contains nonheme iron and the prosthetic group FMN and is tightly bound to the respiratory chain. Lately, Chance & others (1967) have postulated 2 flavoproteins in the respiratory chain between NAD and Co Q. The first of these has a redox potential similar to that of the NAD/NADH system, but the second has a considerably higher redox potential. According to Klingenberg &

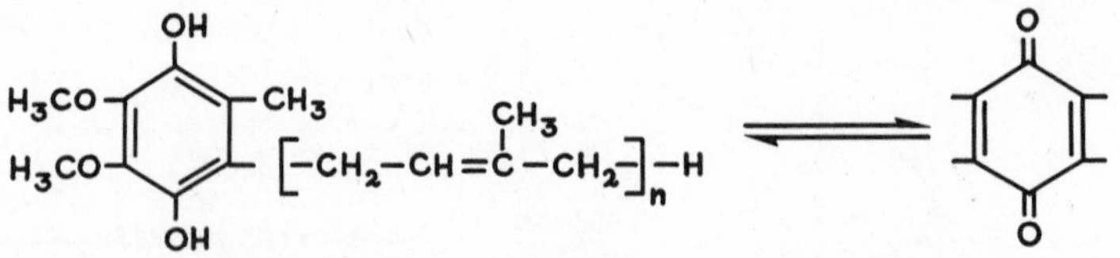

Reduced or quinol form **Oxidized or quinone form**

n = number of isoprenoid units which varies from 6 to 10, ie, coenzyme Q_{6-10}

FIG 9–12. Structure of coenzyme Q.

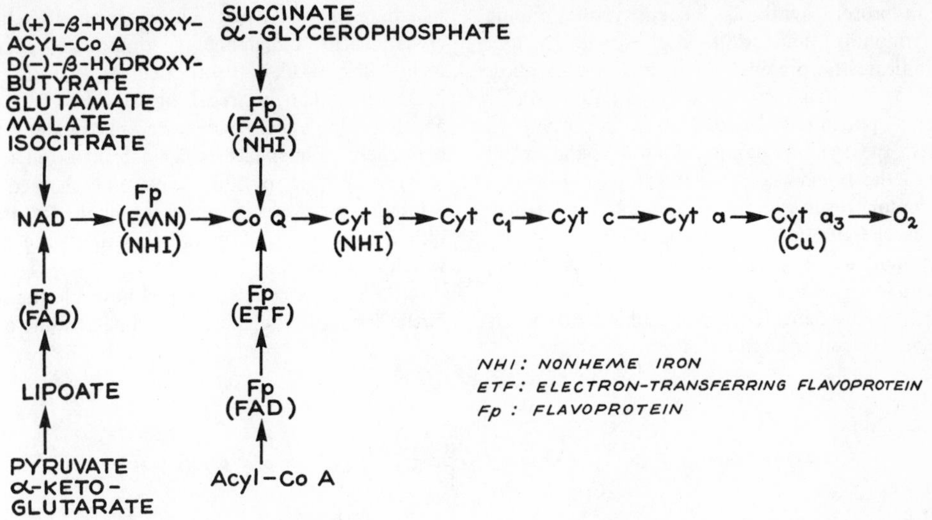

L(+)-β-HYDROXY-ACYL-Co A
D(−)-β-HYDROXY-BUTYRATE
GLUTAMATE
MALATE
ISOCITRATE
SUCCINATE
α-GLYCEROPHOSPHATE
Fp (FAD) (NHI)
NAD → Fp (FMN) (NHI) → Co Q → Cyt b (NHI) → Cyt c_1 → Cyt c → Cyt a → Cyt a_3 (Cu) → O_2
Fp (FAD)
LIPOATE
PYRUVATE
α-KETO-GLUTARATE
Fp (ETF)
Fp (FAD)
Acyl-Co A
NHI: NONHEME IRON
ETF: ELECTRON-TRANSFERRING FLAVOPROTEIN
Fp: FLAVOPROTEIN

FIG 9–13. Components of the respiratory chain in mitochondria.

Kröger (1967), Co Q is also the collecting point in the respiratory chain for reducing equivalents derived from other substrates that are linked directly to the respiratory chain through flavoprotein dehydrogenases. These substrates include succinate, α-glycerophosphate, and acyl-Co A. The flavin moiety of all these dehydrogenases appears to be FAD, and those catalyzing the dehydrogenation of succinate and α-glycerophosphate contain nonheme iron. In the dehydrogenation of acyl-Co A, an additional flavoprotein, the **electron-transporting flavoprotein (ETF)** is necessary to effect transference of electrons to the respiratory chain.

Electrons flow from Co Q, through the series of cytochromes shown in Fig 9–13, to molecular oxygen. The cytochromes are arranged in order of increasing redox potential. The terminal cytochrome a_3 (cytochrome oxidase) is responsible for the final combination of reducing equivalents with molecular oxygen. It has been noted that this enzyme system contains copper, an essential component of true oxidase enzymes. Cytochrome oxidase has a very high affinity for oxygen, which allows the respiratory chain to function at the maximum rate until the tissue has become virtually anoxic.

The structural organization of the respiratory chain has been the subject of considerable speculation. Of significance is the finding of nearly constant molar proportions between the components. The cytochromes are present in the approximate molar proportions, one with another, of 1:1. These findings, together with the fact that many of the components appear to be structurally integrated with the mitochondrial membranes, have suggested that these components have a definite spatial orientation in the membranes.

THE ROLE OF HIGH-ENERGY PHOSPHATES IN BIOLOGIC OXIDATION & ENERGY CAPTURE

The importance of phosphates in intermediary metabolism became evident in the period between 1930 and 1940 with the discovery of the chemical details of glycolysis and the role of **adenosine triphosphate (ATP), adenosine diphosphate (ADP),** and inorganic phosphate (Pi) in this process. ATP was considered to be a means of transferring phosphate radicals in the process of phosphorylation. The role of ATP in biochemical energetics was indicated in experiments demonstrating that ATP and creatine phosphate were broken down in muscular contraction and that their resynthesis depended on supplying energy from oxidative processes in the muscle. It was not until 1941, when Lipman introduced the concept of "high-energy phosphates" and the "high-energy phosphate bond," that the role of these compounds in bioenergetics was clearly appreciated.

Measurement of the free energy of hydrolysis of biologically occurring organophosphates indicates that they fall approximately into 2 groups. In one group, designated "low-energy phosphates," exemplified by the ester phosphates found in the glycolytic intermediates, the free energy of hydrolysis is small (about −4000 cal/mol). In another group, designated "high-energy phosphates," the free energy of hydrolysis is larger (more negative than −7000 cal/mol). Compounds of this type are usually anhydrides (ATP, ADP, the 1-phosphate of 1,3-diphosphoglycerate), enol phosphates (phosphoenolpyruvate), and phosphoguanidines (phosphocreatine, phosphoarginine). Other biologically important compounds that are classed as "high-energy compounds" are thiol esters involving coenzyme A or acyl carrier protein (eg, acetyl-Co A), amino acid esters involved in protein synthesis, S-adenosyl methionine (active methionine), and uridine diphosphate glucose.

To indicate the presence of the high-energy phosphate group, Lipman introduced the symbol ~P, or **"high-energy phosphate bond."** It is important to realize that the symbol denotes that it is the group attached to the bond which, on transfer to an appropriate acceptor, results in transfer of the larger quantity of free energy. For this reason, the term **"group transfer potential"** is preferred by some to "high-energy bond."

ATP contains 2 high-energy phosphate groups. Its structure may be represented as follows:

$$\text{ADENOSINE} - \text{O}-\overset{\overset{\displaystyle \text{O}}{\|}}{\underset{\underset{\displaystyle \text{OH}}{|}}{\text{P}}}-\text{O}\sim\overset{\overset{\displaystyle \text{O}}{\|}}{\underset{\underset{\displaystyle \text{OH}}{|}}{\text{P}}}-\text{O}\sim\overset{\overset{\displaystyle \text{O}}{\|}}{\underset{\underset{\displaystyle \text{OH}}{|}}{\text{P}}}-\text{OH} \quad \text{or} \quad \text{A}-Ⓟ\sim Ⓟ\sim Ⓟ$$

Within the "high-energy phosphate" class of compounds, high-energy phosphate groups are generally transferable by the process of "transphosphorylation" without great loss of free energy, resulting in reversible transference of free energy, eg,

1. A–Ⓟ~Ⓟ~Ⓟ + A–Ⓟ ⇌ 2A–Ⓟ~Ⓟ
   (ATP) (AMP) (ADP)
2. CREATINE + A–Ⓟ~Ⓟ~Ⓟ ⇌ CREATINE ~Ⓟ + A–Ⓟ~Ⓟ

The latter reaction is important in maintaining a store of high-energy phosphate in muscle ready to meet the energy needs of sudden muscular activity (Fig 22–4). However, when ATP is used as a phosphate donor in its role as an "activator" of substrate molecules, there is loss of a high-energy phosphate group. Part of the loss of free energy is used to raise the energy level of the substrate; but a considerable portion is lost from the system as heat, ensuring that these reactions go virtually to completion under physiologic conditions, eg,

GLUCOSE + A–Ⓟ~Ⓟ~Ⓟ ⟶ GLUCOSE –Ⓟ + A–Ⓟ~Ⓟ

During the course of catabolism of glucose to lactic acid in the series of reactions known as the Embden-Meyerhof pathway of glycolysis, there is net formation of 2 high-energy phosphate groups, resulting in new formation of 2 molecules of ATP from ADP. Thus, part of the free energy liberated in the catabolic process is conserved or captured as high-energy phosphate. The process is known as **energy conservation** or **energy capture.** The chemical processes resulting in this net formation of ATP involve the incorporation of inorganic phosphate into 3-phosphoglyceraldehyde, which, after dehydrogenation, forms 1,3-diphosphoglyceric acid, a high-energy compound which in turn reacts with ADP to form ATP. As a result of further molecular changes resulting in the formation of another high-energy substrate, phosphoenolpyruvate, high-energy phosphate is again transferred to ADP to form ATP (see p 230). The ATP formed in these catabolic processes becomes available to drive those other processes (enumerated earlier; see Fig 9–3) which require energy.

ADP is thus envisaged as a molecule that captures, in the form of high-energy phosphate, some of the free energy resulting from catabolic processes and which as ATP passes on this free energy to drive those processes requiring energy. ATP has thus been called the energy "currency" of the cell.

THE ROLE OF THE RESPIRATORY CHAIN IN ENERGY CAPTURE

As indicated above, there is a net capture of 2 high-energy phosphate groups in the glycolytic reactions under anaerobic conditions, equivalent to approximately 15,000 calories per mol of glucose. Since 1 mol of glucose yields approximately 686,000 calories on complete combustion, the energy captured by phosphorylation in glycolysis is negligible. The reactions of the citric acid cycle, the final pathway for the complete oxidation of glucose, include only one phosphorylation step, the conversion of succinyl-Co A to succinate, which allows the capture of 2 more high-energy phosphates per mol of glucose. All of the phosphorylation described so far occur **at the substrate level.** Examination of intact respiring mitochondria reveals that when substrates are oxidized via an NAD-linked dehydrogenase, 3 mols of inorganic phosphate are incorporated into 3 mols of ADP to form 3 mols of ATP per atom of oxygen consumed, ie, the P:O ratio =

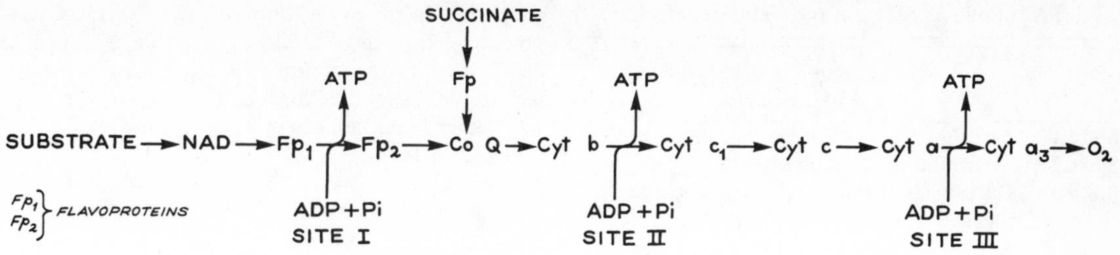

SUCCINATE

Fp

ATP ATP ATP

SUBSTRATE → NAD → Fp_1 → Fp_2 → Co Q → Cyt b → Cyt c_1 → Cyt c → Cyt a → Cyt a_3 → O_2

ADP+Pi SITE I ADP+Pi SITE II ADP+Pi SITE III

Fp_1, Fp_2 } FLAVOPROTEINS

FIG 9–14. Probable sites of phosphorylation in the respiratory chain.

3. On the other hand, when a substrate is oxidized via a flavoprotein-linked dehydrogenase, only 2 mols of ATP are formed, ie, P:O = 2. These reactions are known as **oxidative phosphorylation at the respiratory chain level.** Taking into account dehydrogenations in the pathway of catabolism of glucose in both glycolysis and the citric acid cycle, plus phosphorylations at substrate level, it is now possible to account for at least 42% of the free energy resulting from the combustion of glucose, captured in the form of high-energy phosphate.

Assuming that phosphorylation is coupled directly to certain reactions in the respiratory chain in a manner analogous to phosphorylation in the glycolytic sequence of reactions, it is pertinent to inquire at what sites this could occur. Lehninger (1964) suggests that there must be a redox potential of approximately 0.2 volts or a free energy change of approximately 9000 calories between components of the respiratory chain if that particular site is to support the coupled formation of 1 mol of ATP. Four sites in the respiratory chain fulfill these requirements, one between NAD and a flavoprotein, one between flavoprotein and cytochrome b, one between cytochrome b and cytochrome c, and one between cytochrome a and oxygen. Location of the phosphorylation sites has been elucidated by experiments in which the P:O ratio is measured in the presence of inhibitors of known reactions in the chain and in the presence of artificial electron acceptors. Chance & Williams (1956) showed that the rate of respiration of mitochondria can be controlled by the concentration of ADP. This is because oxidation and phosphorylation are **tightly coupled** and ADP is an essential component of the phosphorylation process. When ADP was deficient in the presence of excess substrate, 3 crossover points could be identified since the component at the substrate side of the crossover point became more reduced and that on the oxygen side became more oxidized. These crossover points coincided with 3 of the possible sites previously identified on thermodynamic grounds. The span flavoprotein–cytochrome b did not appear to have a crossover point. The 3 possible sites of phosphorylation have been designated as sites I, II, and III, respectively. Chance, Lee, & Mela (1967) regard site I as occurring between 2 flavoproteins, one of lower and the other of higher redox potential. The above findings explain why oxidation of succinate via the respiratory chain produces a P:O ratio of only 2, as site I would be bypassed by the flavoprotein-linked succinate dehydrogenase (Fig 9–14).

Respiratory Control

As stated above, oxidation and phosphorylation are tightly coupled in mitochondria. Thus, respiration cannot occur via the respiratory chain without concomitant phosphorylation of ADP. Chance & Williams (1956) have defined 5 conditions or states that can control the rate of respiration in mitochondria (Table 9–2).

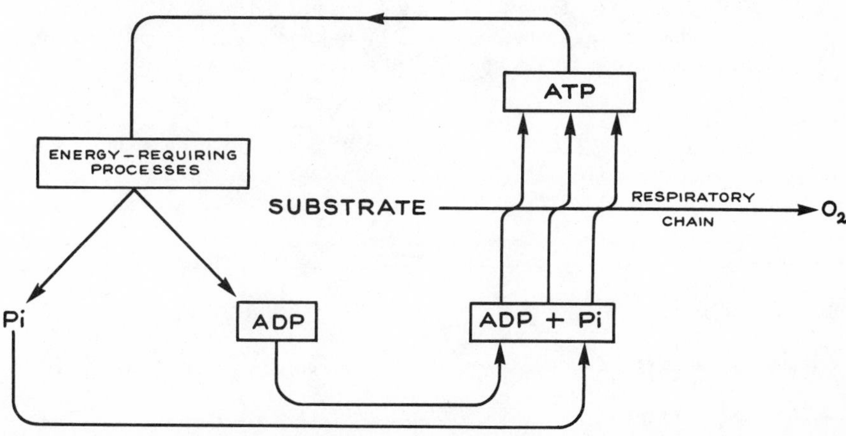

FIG 9–15. The role of ADP in respiratory control.

TABLE 9–2. States of respiratory control.

	Conditions Limiting the Rate of Respiration
State 1	Availability of ADP and substrate only
State 2	Availability of substrate only
State 3	The capacity of the respiratory chain itself, when all substrates and components are present in adequate amounts
State 4	Availability of ADP only
State 5	Availability of oxygen only

Generally most cells in the resting state seem to be in state 4, respiration being controlled by the availability of ADP. When work is performed, ATP is converted to ADP, allowing more respiration to occur, which in turn replenishes the store of ATP (Fig 9–15). It would appear that under certain conditions the concentration of inorganic phosphate and ATP could also affect the rate of functioning of the respiratory chain.

Thus, the manner in which biologic oxidative processes allow the free energy resulting from the oxidation of foodstuffs to become available and to be captured is stepwise, efficient (40–50%), and controlled rather than explosive, inefficient, and uncontrolled. The remaining free energy which is not captured is liberated as heat. This need not be considered as "wasted," since in the warm-blooded animal it contributes to maintenance of body temperature.

MECHANISMS OF OXIDATIVE PHOSPHORYLATION

Two principal hypotheses have been advanced to account for the coupling of oxidation and phosphorylation. The **chemical hypothesis** postulates direct chemical coupling at all stages of the process, as in the reactions that generate ATP in glycolysis. The **chemiosmotic hypothesis** postulates that oxidation of components in the respiratory chain generates hydrogen ions which are ejected to the outside of a coupling membrane in the mitochondrion. The electrochemical potential difference resulting from the asymmetric distribution of the hydrogen ions is used to drive the mechanism responsible for the formation of ATP.

Other hypotheses have been advanced in which it is envisaged that energy from oxidation is conserved in conformational changes of molecules which in turn lead to the generation of high-energy phosphate bonds. The subject of oxidative phosphorylation has been recently reviewed by Lardy & Ferguson (1969).

1. THE CHEMICAL HYPOTHESIS

Oxidative phosphorylation occurs in certain reactions of the Embden-Meyerhof system of glycolysis, in the citric acid cycle, and in the respiratory chain. However, it is only in those phosphorylations occurring at the substrate level in glycolysis and the citric acid cycle that the chemical mechanisms involved are known with any certainty. As these reactions have been used as models for investigating oxidative phosphorylation in the mitochondrial respiratory chain, the equations for these phosphorylations are shown below.

Several fundamental differences are evident in these equations. In equation 1, phosphate is incorporated into the product of the reaction **after** the redox reaction. In equation 2, phosphate is incorporated into the substrate **before** the internal rearrangement (redox change). In equation 3, the redox reaction leads to the generation of a high-energy compound other than a phosphate, which in a subsequent reaction leads to the formation of high-energy phosphate.

It is generally considered that oxidative phosphorylations in the respiratory chain follow the pattern shown in reactions 1 and 3, the latter being in effect an extension of 1 in which an extra nonphosphorylated high-energy intermediate stage is introduced. Of the possible mechanisms shown in Fig 9–16, mechanism C is favored since, in the presence of agents that uncouple phosphorylation from electron transport (eg, dinitrophenol), oxidation-reduction in the respiratory chain is independent of Pi. At present there is no definite knowledge of the identities of the hypothetical high-energy carrier (Car ~ I) at any of the coupling

1. 3-Phosphoglyceraldehyde + NAD⁺ + Pi ⟶ 1~,3-Diphosphoglycerate + NADH + H⁺
   1~,3-Diphosphoglycerate + ADP ⟶ 3-Phosphoglycerate + ATP

2. 2-Phosphoglycerate ⟶ 2~Phosphoenolpyruvate
   2~Phosphoenolpyruvate + ADP ⟶ Pyruvate + ATP

3. α-Ketoglutarate + NAD⁺ + Co A ⟶ Succinyl~Co A + NADH + H⁺
   Succinyl~Co A + GDP* + Pi ⟶ Succinate + GTP

*GDP is an analogous compound to ADP where adenine is replaced by guanine.

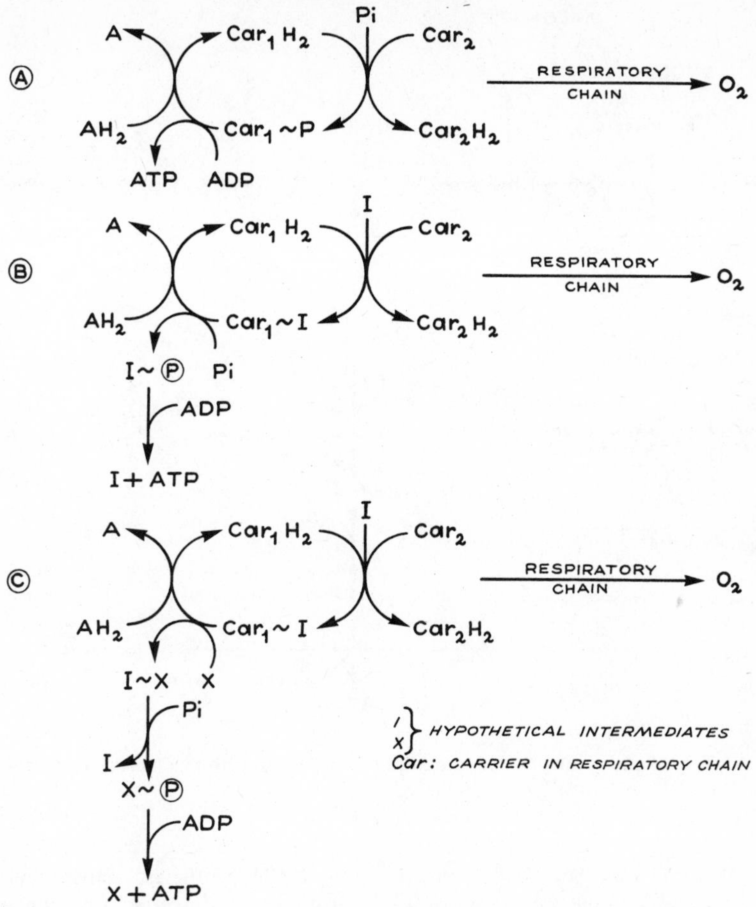

FIG 9–16. Possible mechanisms for the chemical coupling of oxidation and phosphorylation in the respiratory chain.

sites of the respiratory chain. Likewise, the identities of the postulated intermediates I and X are not known. In recent years, several so-called "coupling factors" have been isolated that restore phosphorylation when added to disrupted mitochondria. These factors are mainly protein in nature and usually exhibit "ATPase" activity. In addition, attempts have been made to isolate high-energy intermediates of the respiratory chain and of its branches. A current view of the chemical coupling of phosphorylation to oxidation in the respiratory chain is shown in Fig 9–17 (see Ernster & others, 1967).

Inhibitors of the Respiratory Chain & of Oxidative Phosphorylation

Much of the information shown in Fig 9–17 has been obtained by the use of inhibitors, and their proposed loci of action are shown in this figure. For descriptive purposes they may be divided into inhibitors of the respiratory chain, inhibitors of oxidative phosphorylation, and uncouplers of oxidative phosphorylation.

Inhibitors that arrest respiration by blocking the respiratory chain appear to act at 3 loci that may be identical to the energy transfer sites I, II, and III. Site I is inhibited by barbiturates such as amobarbital (Amytal), by the antibiotic piericidin A, and by the fish poison, rotenone. Some steroids and mercurials also affect this site. These inhibitors prevent the oxidation of substrates that communicate directly with the respiratory chain via an NAD-linked dehydrogenase, eg, β-hydroxybutyrate.

Dimercaprol (BAL) and **antimycin A** inhibit the respiratory chain at or around site II, between cytochrome b and cytochrome c; and the inhibitors of cytochrome oxidase that have been known for many years are considered to act at or near site III.

The antibiotic **oligomycin** completely blocks oxidation and phosphorylation in intact mitochondria. However, in the presence of the uncoupler, dinitrophenol, oxidation proceeds without phosphorylation, indicating that oligomycin does not act directly on the respiratory chain but subsequently on a step in phosphorylation (Fig 9–17). As energy generated at one

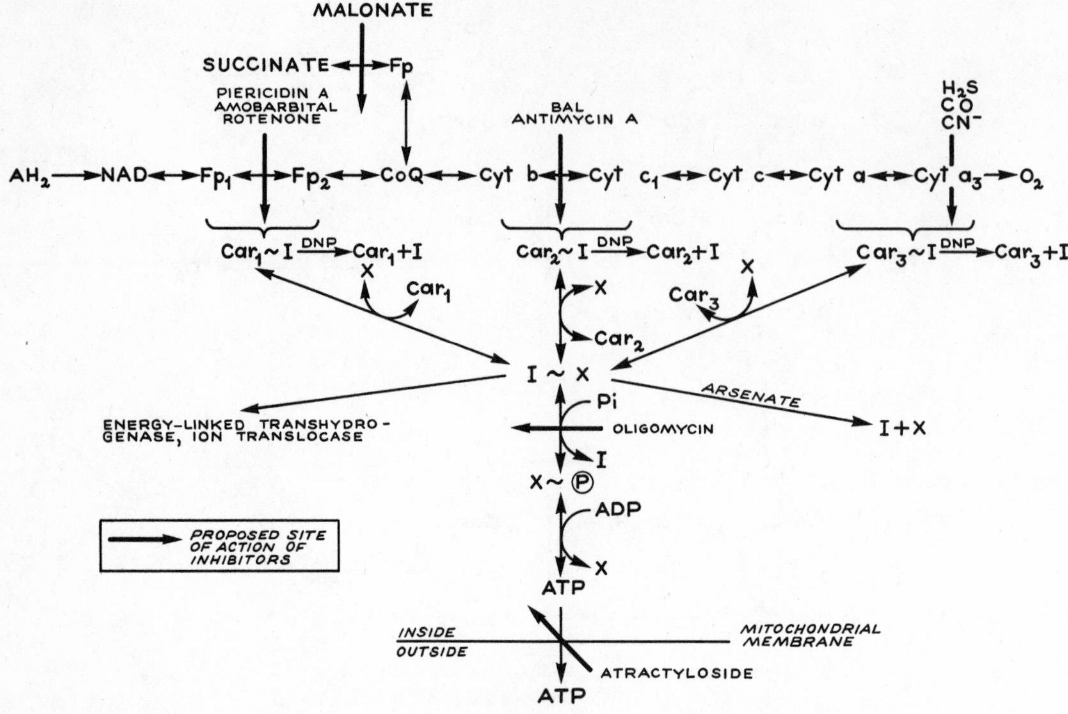

FIG 9–17. The coupling of phosphorylation to oxidation in the respiratory chain as interpreted by the chemical hypothesis.

site of energy conservation may be used to reverse oxidation at another site, even in the presence of oligomycin and in the absence of Pi, this inhibitor is considered to act after the stage represented by I ~ X.

Atractyloside inhibits oxidative phosphorylation which is dependent on the transport of adenine nucleotides across the inner mitochondrial membrane. It is considered to inhibit a "permease" or "translocase" of ADP into the mitochondrion and of ATP out of the mitochondrion. Thus, atractyloside acts only on intact mitochondria but does not inhibit phosphorylation in submitochondrial particles which have no intact membrane (see Pullman & Schatz, 1967).

The action of **uncouplers** is to dissociate oxidation in the respiratory chain from phosphorylation. This results in respiration becoming uncontrolled, the concentration of ADP or Pi no longer limiting the rate of respiration. The uncoupler that has been used most frequently is 2,4-dinitrophenol (DNP), but other compounds act in a similar manner, including dinitrocresol, pentadichlorophenol, bishydroxycoumarin (Dicumarol), and CCCP (*m*-chlorocarbonyl cyanide phenylhydrazone). The latter, compared with DNP, is about 100 times as active. DNP is considered to cause the hydrolysis of one of the high-energy intermediates (eg, Car ~ I), resulting in the release of the constituents Car + I and energy as heat. Because Car ~ I reacts to form, ultimately, ATP and because these reactions are reversible, DNP allows a reversal of the phosphorylation reactions, resulting in ATP hydrolysis, ie, release of latent ATP synthetase activity. As mentioned earlier, oligomycin has no effect on respiration in the presence of DNP, but it does inhibit the latent ATP synthetase resulting from the action of DNP because it inhibits the reactions involving phosphorylation.

Arsenate is another uncoupler that acts at a site more proximal to phosphorylation than DNP. Arsenate may uncouple respiration by reaction with I ~ X to form an unstable intermediate which is rapidly hydrolyzed. Pi may compete with arsenate preventing uncoupling. However, another interpretation of the action of arsenate is that it forms an unstable arsenyl-ADP that is easily hydrolyzed (see Ernster & others, 1967).

Reversal of Electron Transport

Mitochondria catalyze the energy-dependent reversal of electron transport through the respiratory chain. The energy is provided either by ATP, where the effect is mediated by a complete reversal of oxidative phosphorylation; or the high-energy intermediate I ~ X can serve as a source of energy so that phosphorylation is not involved. A system demonstrating the latter effect may consist of mitochondria plus the addition of succinate as electron donor and acetoacetate as electron acceptor. As succinate becomes oxidized, acetoacetate is reduced to β-hydroxybutyrate. It is considered that I ~ X is generated from sites II and III and used to drive reversed electron flow at site I, which enables NADH to be formed, which in turn reduces

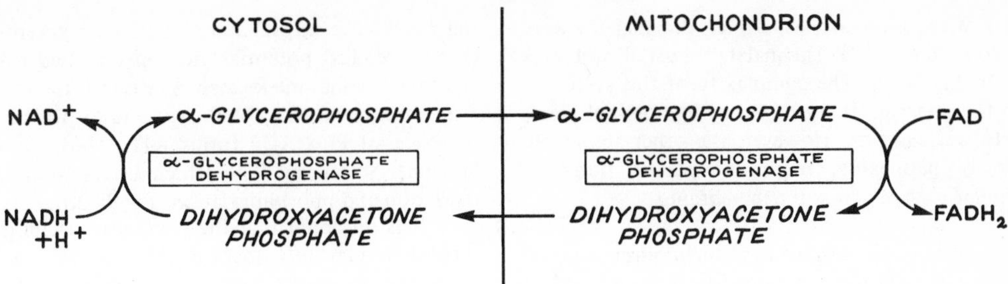

FIG 9–18. α-Glycerophosphate shuttle for transfer of reducing equivalents from the cytosol into the mitochondrion.

acetoacetate to β-hydroxybutyrate (Fig 9–17). Although these experiments demonstrate the reversibility of electron transport, the physiologic significance of the process is unknown.

Energy-Linked Transhydrogenase

There is evidence for an energy-linked transhydrogenase that can catalyze the transfer of hydrogen from NADH to NADP. Ernster & Lee (1964) consider that the nonphosphorylated intermediate I ~ X supplies the energy for the transfer as the process is only inhibited by oligomycin if ATP is used as the source of energy.

$$NADH + NADP^+ + I \sim X \longrightarrow NAD^+ + NADPH + I + X$$

Oxidation of Extramitochondrial NADH

Although NADH cannot penetrate the mitochondrial membrane, it is produced continuously in the cytosol by 3-phosphoglyceraldehyde dehydrogenase, an enzyme in the Embden-Meyerhof glycolysis sequence. However, under aerobic conditions, extramitochondrial NADH does not accumulate and is presumed to be oxidized by the respiratory chain in mitochondria. Several possible mechanisms have been considered to permit this process (Krebs, 1967). These involve transfer of reducing equivalents through the mitochondrial membrane via substrate pairs, linked by suitable dehydrogenases. Substrate pairs that have been considered include acetoacetate/β-hydroxybutyrate, lactate/pyruvate, dihydroxyacetone phosphate/α-glycerophosphate, and malate/oxaloacetate. It is necessary that the specific dehydrogenase be present on both sides of the mitochondrial membrane. However, β-hydroxybutyrate dehydrogenase is found only in mitochondria and lactate dehydrogenase only in the cytosol, ruling out these substrate pairs. α-Glycerophosphate dehydrogenase is NAD-linked in the cytosol, whereas the enzyme found in the mitochondria is a flavoprotein enzyme. The mechanism of transfer using this system is shown in Fig 9–18. Although it might be important in liver, in other tissues (eg, heart muscle) the mitochondrial α-glycerophosphate dehydrogenase is deficient. It is therefore believed that a transport system involving malate and malate dehydrogenase is of more universal occurrence. This system has been studied by Chappell (1968), who found that rapid oxidation of NADH occurred only when aspartate-α-ketoglutarate transaminase and malate dehydrogenase,

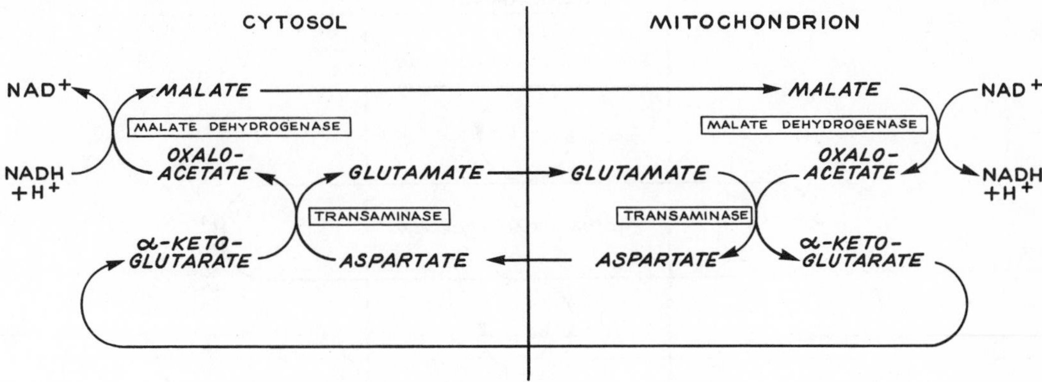

FIG 9–19. Malate shuttle for transfer of reducing equivalents from the cytosol into the mitochondria.

together with glutamate, aspartate, and malate, were added to mitochondria. The malate "shuttle" system is shown in Fig 9–19. The complexity of this system is due to the impermeability of the mitochondrial membrane to oxaloacetate. However, the other anions are not freely permeable, requiring specific transport systems for passage across the membrane.

Energy-Linked Ion Transport in Mitochondria

Actively respiring mitochondria in which oxidative phosphorylation is taking place maintain or accumulate cations such as K^+, Na^+, Ca^{++}, and Mg^{++}, and Pi. Uncoupling with DNP leads to loss of ions from the mitochondria but the ion uptake is not inhibited by oligomycin, suggesting that the energy is supplied by a nonphosphorylated high-energy intermediate, eg, I ~ X (Fig 9–17).

2. THE CHEMIOSMOTIC HYPOTHESIS

Mitchell (1961, 1966) has provided a hypothesis that is able to explain the coupling of oxidation and phosphorylation without having to postulate an "energy-rich" intermediate common to both the respiratory chain and to the phosphorylation pathway (ie, Car ~ I in Fig 9–17).

According to Mitchell, the primary event in oxidative phosphorylation is the translocation of protons (H^+) to the exterior of a coupling membrane driven by oxidation in the respiratory chain. It is also postulated that the membrane is impermeable to ions in general but particularly to protons which accumulate outside the membrane, creating an electrochemical potential difference. This consists of a chemical potential (difference in pH) and a membrane potential. The electrochemical potential difference is used to drive a vectorial, membrane-located ATPase (or the reversal of a membrane-located ATP synthetase) which in the presence of Pi + ADP forms ATP. Thus, there is no high-energy intermediate which is common to both oxidation and phosphorylation (Fig 9–20).

It is proposed that the respiratory chain is folded into 3 oxidation/reduction (**o/r**) loops, each loop corresponding functionally to coupling site I, site II, and site III of the chemical hypothesis, respectively. An idealized single loop consisting of a hydrogen carrier and an electron carrier is shown in Fig 9–21. A possible configuration of the respiratory chain folded into 3 **o/r** loops is shown in Fig 9–22.

The mechanism of coupling of proton translocation to the anisotropic (vectorial) ATP synthetase system is the most conjectural aspect of the hypothesis. Mitchell has postulated an anhydride intermediate X ~ I in a system depicted in Fig 9–23.

The existence of a membrane potential required to synthesize ATP would cause ions of a charge opposite to the internal phase to leak in through the coupling membrane. To prevent swelling and lysis, the ion leakage would have to be balanced by extrusion of ions against the electric gradient. It was therefore necessary to postulate that the coupling membrane contains exchange diffusion systems for exchange of anions against OH^- ions and of cations against H^+ ions. Such systems would be necessary for uptake of ionized metabolites through the membrane.

The chemiosmotic hypothesis can account for the phenomenon of respiratory control. The electrochemical potential difference across the membrane, once built up as a result of proton translocation, would inhibit further transport of reducing equivalents through the **o/r** loops unless it is discharged by back-

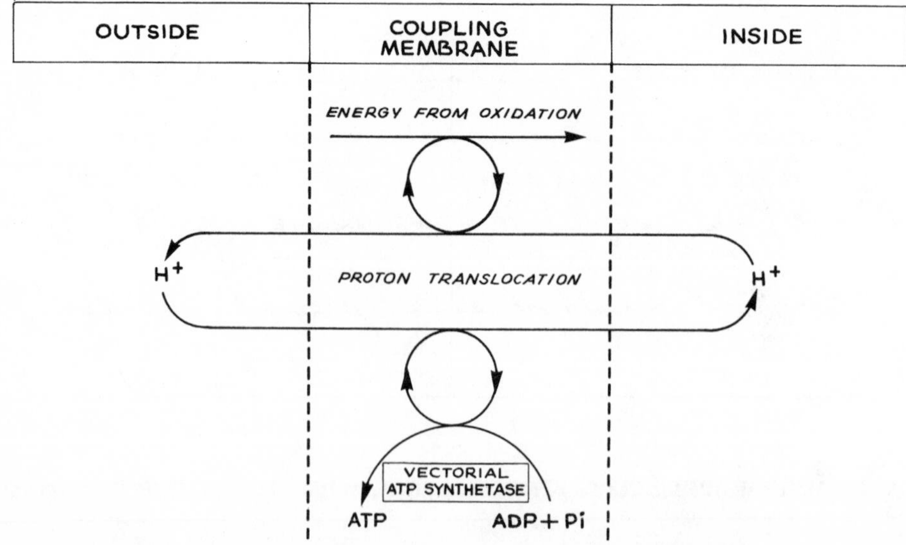

FIG 9–20. Principles of the chemiosmotic hypothesis of oxidative phosphorylation.

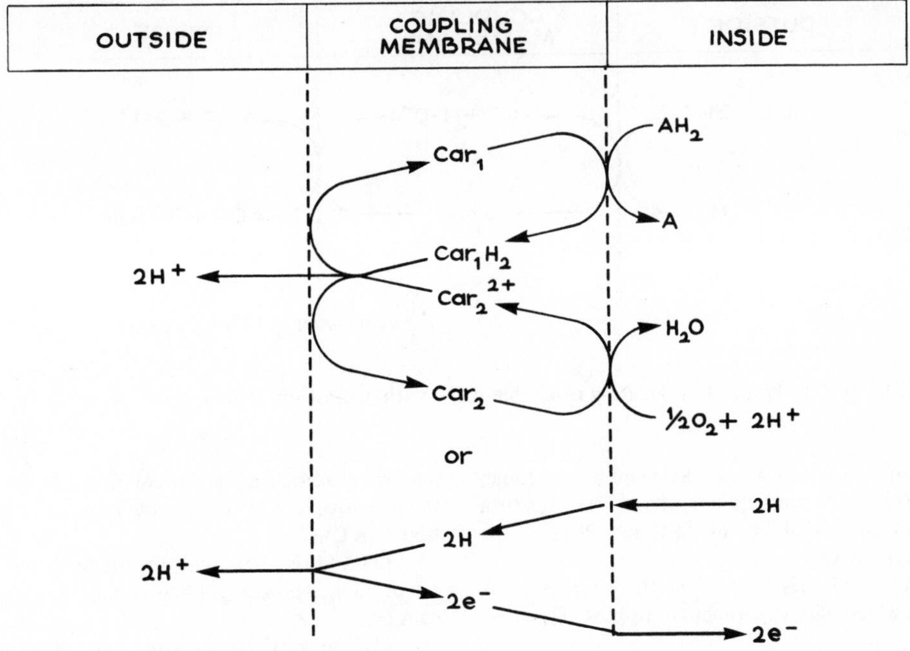

FIG 9–21. Proton-translocating oxidation/reduction (o/r) loop (chemiosmotic hypothesis).

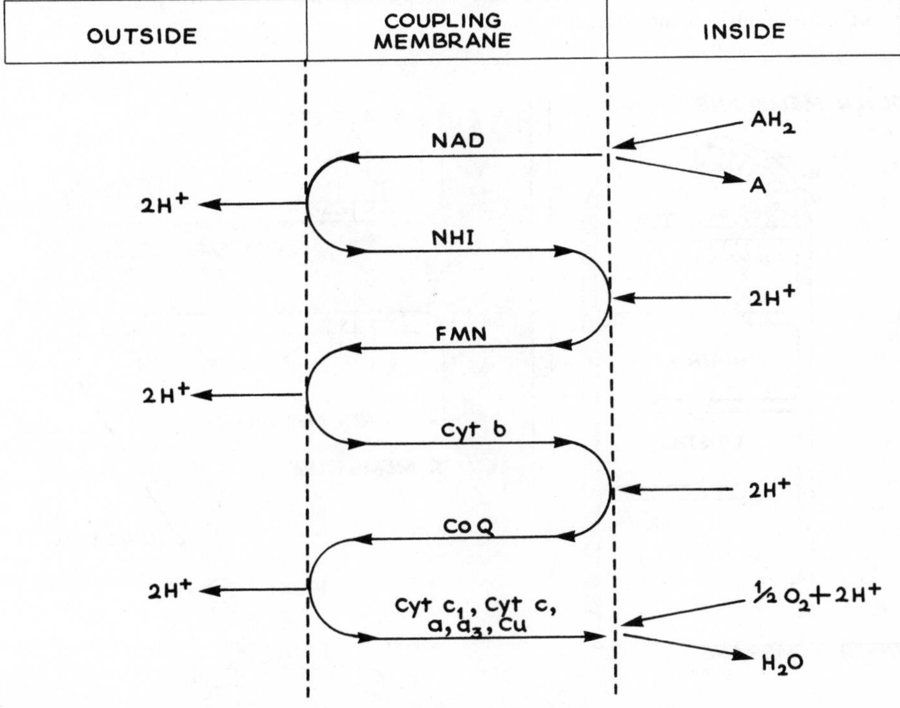

FIG 9–22. Possible configuration of o/r loops in the respiratory chain (chemiosmotic hypothesis). There is uncertainty with respect to the relative positions of cyt b and Co Q in the respiratory chain. In this scheme cyt b is shown on the substrate side of Co Q as it fits the hypothesis and the redox potentials better in this position.

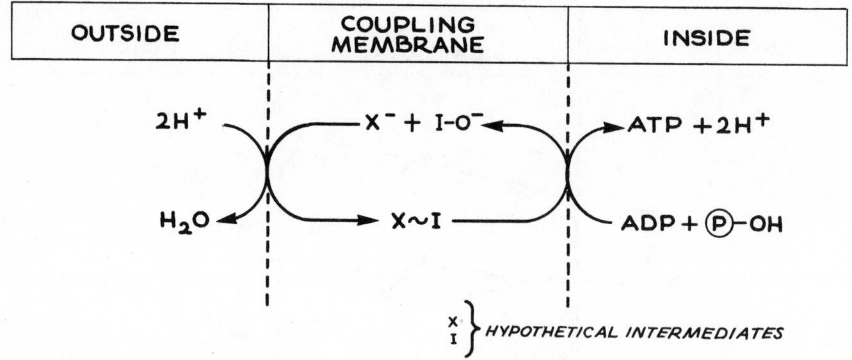

FIG 9–23. Proton-translocating reversible ATP synthetase of the chemiosmotic hypothesis.

translocation of protons across the membrane through the vectorial ATP synthetase system. This in turn depends on the availability of ADP and Pi as in the chemical hypothesis.

Several corollaries arise from the chemiosmotic hypothesis which have experimental support. These are as follows:

(1) Mitochondria are generally impermeable to protons and other ions. Chappell (1968) has reviewed the evidence for the existence of specific transport systems which enable ions to penetrate the inner mitochondrial membrane.

(2) Uncouplers such as DNP increase the permeability of mitochondria to protons, thus reducing the electrochemical potential and short-circuiting the anisotropic ATP synthetase system for the generation of ATP.

(3) Addition of acid to the external medium, establishing a proton gradient, leads to the generation of ATP.

(4) The P/H^+ (transported out) quotient of the ATP synthetase is ½ and the H^+ (transported out)/0 quotients for succinate and β-hydroxybutyrate oxidation are 4 and 6 respectively, conforming with the expected P/O ratios of 2 and 3, respectively. These ratios are compatible with the postulated existence of 3 o/r loops in the respiratory chain.

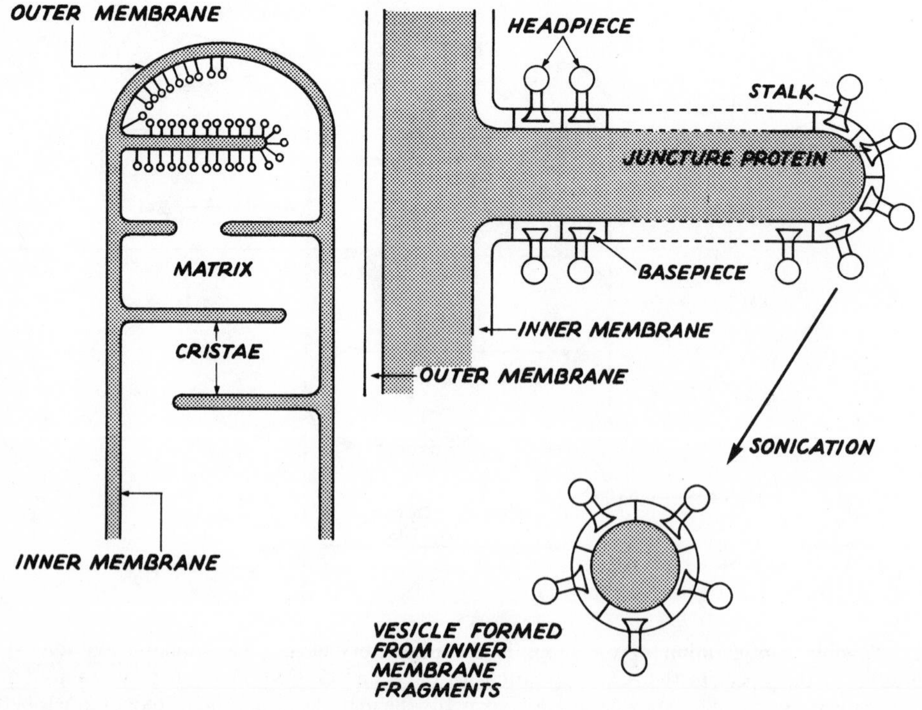

FIG 9–24. Structure of the mitochondrial membranes.

(5) Oxidative phosphorylation does not occur in soluble systems where there is no possibility of a vectorial ATP synthetase. Some structural element involving a closed membrane must be present in the system to obtain oxidative phosphorylation.

Tyler (1970) has demonstrated that the respiratory chain contains components organized in a sided manner as required by the chemiosmotic hypothesis.

Summary

The chemiosmotic hypothesis can explain oxidative phosphorylation, respiratory control, ion transport, the action of uncouplers, and the action of inhibitors as satisfactorily as the chemical hypothesis; but it has the virtue of being simpler, as it does not necessitate the postulation of a common high-energy intermediate of oxidation and phosphorylation. However, this subject is still surrounded by considerable controversy.

Anatomy & Function of the Mitochondrial Membranes

An account of the general structure of mitochondria is to be found in Chapter 8. A more detailed account integrating structure and function has been given by MacLennan (1970). Mitochondria have an outer membrane which is permeable to most metabolites, an inner membrane which is selectively permeable and which is thrown into folds or cristae, and a matrix within the inner membrane. The outer membrane may be removed by treatment with digitonin and is characterized by the presence of monoamine oxidase. Cardiolipin is concentrated in the inner membrane where most of the lipid is phospholipid. The exact relationship of the lipid to the protein of the membranes is not understood. Delipidation of the inner membrane does not lead to its disruption.

The inner membrane consists of repeating units, each composed of a headpiece which projects into the matrix, joined by a stalk to a basepiece. Sonication leads to vesicle formation by the inner membrane with the headpieces facing the external medium (Fig 9–24). The whole of the electron transfer chain is found in the basepieces, and it is suggested that each basepiece is a complex containing a portion of the enzymes of the respiratory chain. Four of these complexes, together with mobile components (NADH, Co Q, and cytochrome c), constitute a complete respiratory chain. Cytochrome c is located in the outer side of the basepiece, and Racker & others (1965) showed that the ATPase (ATP synthetase) system resides in the headpiece. To be sensitive to oligomycin, the headpiece, stalk, and a juncture protein in the basepiece must be present. It has been postulated that each headpiece, stalk, and basepiece, together with one of the respiratory chain complexes, constitute a phosphorylating unit. Thus, energy generated by electron transport in the basepiece could be conserved in the juncture protein of the basepiece and be transmitted through the stalk protein to the headpiece, where it is transduced into the phosphoryl bond of ATP. The soluble enzymes of the citric acid cycle and the enzymes of β-oxidation are found in the matrix, necessitating mechanisms for transporting ions, fatty and other organic acids, and nucleotides across the inner membrane.

• • •

References

Chance B, Ernster L, Garland PB, Lee CP, Light T, Ohnishi CI, Ragan CI, Wond D: Proc Nat Acad Sc 57:1498, 1967.

Chance B, Lee CP, Mela L: Fed Proc 26:1341, 1967.

Chance B, Williams GR: Advances Enzymol 17:65, 1956.

Crane FL: P 533 in: *Biological Oxidations.* Singer TP (editor). Interscience, 1968.

Chappell JB: Brit M Bull 24:150, 1968.

Ernster L, Lee CP: Ann Rev Biochem 33:729, 1964.

Ernster L, Lee CP, Janda S: P 29 in: *Biochemistry of Mitochondria.* Slater EC, Kaniuga Z, Wojtczak L (editors). Academic Press, 1967.

Estabrook RW, Shigematsu A, Schenkman JB: Advances Enzym Regulat 8:121, 1970.

Gibson QH: P 379 in: *Biological Oxidations.* Singer TP (editor). Interscience, 1968.

Hayaishi O: P 581 in: *Biological Oxidations.* Singer TP (editor). Interscience, 1968.

Hayaishi O: Ann Rev Biochem 38:21, 1969.

Klingenberg M, Kröger A: P 11 in: *Biochemistry of Mitochondria.* Slater EC, Kaniuga Z, Wojtczak L (editors). Academic Press, 1967.

Krebs HA: P 105 in: *Biochemistry of Mitochondria.* Slater EC, Kaniuga Z, Wojtczak L (editors). Academic Press, 1967.

Lardy HA, Ferguson SM: Ann Rev Biochem 38:991, 1969.

MacLennan DH: P 177 in: *Current Topics in Membranes and Transport.* Vol 1. Bronner F, Kleinzeller A (editors). Academic Press, 1970.

Malmström BG, Ryden L: P 415 in: *Biological Oxidations.* Singer TP (editor). Interscience, 1968.

Mitchell P: Biol Rev 41:445, 1966.

Mitchell P: Nature 191:144, 1961.

Palmer G, Massey V: P 263 in: *Biological Oxidations.* Singer TP (editor). Interscience, 1968.

Pullman ME, Schatz G: Ann Rev Biochem 36:539, 1967.

Racker E, Tyler DD, Estabrook RW, Conover TE, Parsons DF, Chance B: P 1077 in: *Oxidases and Related Redox Systems.* Vol 2. King TE, Mason HS, Morrison M (editors). Wiley, 1965.

Rajagopalan KV, Handler P: P 301 in: *Biological Oxidations.* Singer TP (editor). Interscience, 1968.

Slater EC: P 166 in: *Regulation of Metabolic Processes in Mitochondria.* Vol 7. Tager JM & others (editors). B.B.A. Library, 1966.

Strittmatter P: Ann Rev Biochem 35:125, 1966.

Sund H: Pp 603, 641 in: *Biological Oxidations.* Singer TP (editor). Interscience, 1968.

Tyler DD: Biochem J 116:30P, 1970.

Wellner D: Ann Rev Biochem 36:699, 1967.

Bibliography

Boyer PD, Lardy H, Myrbäck K (editors): *The Enzymes.* Vols 7 & 8. *Oxidation* and *Reduction.* Academic Press, 1963.

Chance B (editor): *Energy-Linked Functions of Mitochondria.* Academic Press, 1963.

Kaplan NO, Kennedy EP (editors): *Current Aspects of Biochemical Energetics.* Academic Press, 1966.

Lehninger AL: *The Mitochondrion.* Benjamin, 1964.

Slater EC, Kaniuga Z, Wojtczak L (editors): *Biochemistry of Mitochondria.* Academic Press, 1967.

10 . . .

The Blood, Lymph, & Cerebrospinal Fluid

BLOOD

Blood is a tissue which circulates in what is virtually a closed system of blood vessels. It consists of solid elements—the red and white blood cells and the platelets—suspended in a liquid medium, the plasma.

The Functions of the Blood

(1) Respiration: Transport of oxygen from the lungs to the tissues and of CO_2 from the tissues to the lungs.

(2) Nutrition: Transport of absorbed food materials.

(3) Excretion: Transport of metabolic wastes to the kidneys, lungs, skin, and intestines for removal.

(4) Maintenance of normal acid-base balance in the body.

(5) Regulation of water balance through the effects of blood on the exchange of water between the circulating fluid and the tissue fluid.

(6) Regulation of body temperature by the distribution of body heat.

(7) Defense against infection in the white cells and the circulating antibodies.

(8) Transport of hormones; regulation of metabolism.

(9) Transport of metabolites.

Packed Cell Volume

When blood which has been prevented from clotting by the use of a suitable anticoagulant is centrifuged, the cells will settle to the bottom of the tube while the plasma, a straw-colored liquid, will rise to the top. Normally the cells comprise about 45% of the total volume. This reading (45%) is a normal **hematocrit**, or **packed cell volume**, for males; for females, the normal packed cell volume is about 41%.

The specific gravity of whole blood varies between 1.054 and 1.060; the specific gravity of plasma is about 1.024–1.028.

The viscosity of blood is about 4.5 times that of water. Viscosity of blood varies in accordance with the number of cells present and with the temperature and degree of hydration of the body. Because these 3 factors are relatively constant under normal conditions, the viscosity of the blood does not ordinarily influence the physiology of the circulation.

Blood Volume

The volume of the blood can be measured by several procedures, although the results obtained will vary somewhat with the method used. The method of Gregersen, which is employed frequently in clinical practice, utilizes T-1824, a blue dye (Evans blue). The dye is injected intravenously, and its concentration in the plasma is determined after sufficient time has elapsed to allow for adequate mixing (usually 10 minutes). The plasma volume thus obtained is then converted to total blood volume by the following formula:

$$\text{Total blood volume (ml)} = \frac{\text{Plasma volume (ml)} \times 100}{100 - \text{hematocrit}}$$

Presumably the dye is bound to the plasma protein; for this reason it remains in the circulation for some time. Actually some of the dye does escape into the extracellular fluid; therefore the results obtained by this method are somewhat high. Normal volumes in males as found by the T-1824 method are given as follows: plasma volume, 45 ml/kg body weight; blood volume, 85 ml/kg. The corresponding values in the female are somewhat lower.

^{131}I-labeled human serum albumin is also used to determine plasma volume. A carefully measured dose is injected, and the dilution of the label is then obtained.

There are objections to measurements of plasma volume as a means of determining whole blood volume. The blood volume is calculated from the plasma volume by the use of the hematocrit, which, since peripheral blood is used, may not represent the ratio of cells to plasma throughout the circulation. Consequently, a direct measurement of the whole blood volume by the use of labeled red cells is preferred when greater accuracy is required, particularly in states where there is an impairment of circulatory efficiency (eg, shock or cardiac failure). ^{32}P and radiochromium have both been used to label the red cells for the purpose of measuring whole blood volume.

Blood Osmotic Pressure

The osmotic pressure of the blood is kept relatively constant mainly by the kidney. Osmotic pressure can be determined by measurement of the freezing point depression. The average freezing point for whole blood has been established as −0.537° C. This corresponds to an osmotic pressure of 7–8 atmospheres at body temperature. A solution of sodium chloride containing 0.9 g/100 ml has an osmotic pressure equal to that of whole blood. Such a saline solution is termed "isotonic" or "physiologic" saline. Actually these sodium chloride solutions are not physiologic, since additional ions are lacking which are necessary for the function of the tissues. Other solutions, which are not only isotonic but which contain these ions in proper proportions, are more appropriate from a physiologic standpoint. Examples of these balanced ionic solutions are Ringer's, Ringer-Locke, and Tyrode's solutions. The formulas are given in Table 10–1.

TABLE 10–1. Composition of some saline solutions isotonic with blood.

	Saline (%)	Mammalian Ringer (%)	Ringer-Locke (%)	Tyrode (%)
NaCl	0.9	0.86	0.9	0.8
$CaCl_2$	...	0.033	0.024	0.02
KCl	...	0.03	0.042	0.02
$NaHCO_3$	...	...	0.01–0.03	0.1
Glucose	...	...	0.10–0.2	0.01
$MgCl_2$	...	...	...	0.01
NaH_2PO_4	...	...	...	0.005

THE CLOTTING OF BLOOD

When blood is drawn and allowed to clot, a clear liquid (serum) exudes from the clotted blood. Plasma, on the other hand, separates from the cells only when blood is prevented from clotting.

The blood clot is formed by a protein **(fibrinogen)** which is present in soluble form in the plasma and which is transformed to an insoluble network of fibrous material **(fibrin,** the substance of the blood clot) by the clotting mechanism.

According to the original **Howell theory** of blood coagulation, the change of fibrinogen into fibrin is caused by **thrombin,** which in fluid blood exists as **prothrombin**. The conversion of prothrombin to thrombin depends on the action of **thromboplastin** and calcium. These stages in the clotting process may be diagrammed as follows:

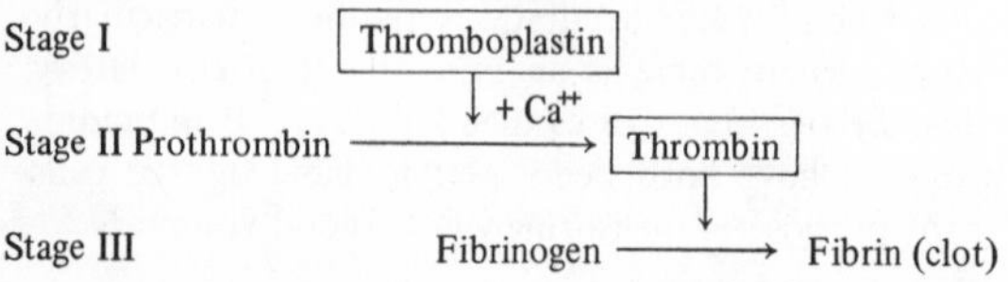

Stage I: Thromboplastin

↓ + Ca^{++}

Stage II: Prothrombin ⟶ Thrombin

↓

Stage III: Fibrinogen ⟶ Fibrin (clot)

The many continuing studies on the details of the coagulation process have served to indicate how complex the system actually is. Lack of knowledge of the chemical nature of many of the factors involved has also resulted in much confusion in terminology. In an effort to introduce uniformity in nomenclature, the International Committee for the Standardization of the Nomenclature of Blood Clotting Factors has recommended a numerical system to designate the various factors presently accepted as involved in the clotting process. This system is given in Table 10–2 (Wright, 1962), and each of the 12 factors will be described, together with some comments on function, in the text which follows.

Although a factor VI has been described, it is currently believed that no such separate factor exists; consequently it has been deleted.

TABLE 10–2. Numerical system for nomenclature of blood clotting factors.

Factor	Name
I	Fibrinogen
II	Prothrombin
III	Thromboplastin
IV	Calcium
V	Labile factor, proaccelerin, accelerator (Ac-) globulin
VII	Proconvertin, serum prothrombin conversion accelerator (SPCA), cothromboplastin, autoprothrombin I
VIII	Antihemophilic factor, antihemophilic globulin (AHG)
IX	Plasma thromboplastin component (PTC) (Christmas factor)
X	Stuart-Prower factor
XI	Plasma thromboplastin antecedent (PTA)
XII	Hageman factor
XIII	Laki-Lorand factor (LLF)

Stage I: Origin of Thromboplastin

The term "thromboplastin" should not be interpreted as referring to a single substance. Instead it is meant to describe a function—namely, that of activating or catalyzing the conversion of prothrombin to thrombin.

Substances with thromboplastic activity are contributed by the plasma, the platelets, and the tissues.

A. From the *Plasma:*

1. Antihemophilic globulin (AHG)—This factor is relatively heat-stable but labile on storage. The antihemophilic globulins are principally β-2 globulins which occur in Cohn fraction I (see p 193). A deficiency of AHG is the cause of the classical type of hemophilia, sometimes designated **hemophilia A.**

2. Plasma thromboplastin component (White, 1953) **(PTC) (Christmas factor)**—This factor is found in both serum and plasma. It occurs in the β-2 globulins in a concentration of less than 1 mg/100 ml of plasma. A deficiency of **PTC** is the cause of a hemophilioid disease sometimes referred to as Christmas disease after the surname of the first patient diagnosed as having this inherited coagulation defect.

3. **Plasma thromboplastin antecedent (PTA)**—A factor described by Rosenthal & others (1953). Patients with PTA deficiency are usually rather mild bleeders.

4. **Hageman factor**—The absence of this factor is evident only from appropriate laboratory studies of the clotting system. If whole normal blood is collected in glass and in silicone-treated tubes, the blood in the silicone tube takes longer to clot than that in the glass tube. It is hypothesized that some reaction occurs on exposure of the blood to glass, although many other substances such as kaolin, barium carbonate, Super-Cel Celite, bentonite, asbestos, and silicic acid will act like glass in this respect. This involves the Hageman factor and also PTA. The Hageman factor is thought to be activated by contact. The activated factor then reacts with a PTA factor to produce another activated product which is thought to be concerned with activation of PTA.

As noted above, Hageman factor deficiency is assumed to exist in patients with no clinically obvious clotting defect but who manifest abnormalities in the clotting system when tested in vitro. In a laboratory test, blood from either PTA-deficient or Hageman factor-deficient patients exhibits a prolonged clotting time and an abnormal consumption of prothrombin (ie, excess prothrombin remains in the serum after clotting has taken place). Consequently, from laboratory tests alone the results of either defect are similar. However, the in vivo conditions are readily differentiated because of the bleeding tendency of the PTA-deficient patient which is not evident in the Hageman factor-deficient person.

5. **Stuart factor (Stuart-Prower factor)**—This plasma factor is required not only for the formation of thromboplastin but also for the conversion of prothrombin to thrombin. There appear to be no specific distinguishing clinical features corresponding to this defect, but it is detectable in a laboratory test by the presence of a prolonged clotting time or a long one-stage prothrombin time.

6. **Factor V (labile factor, proaccelerin, accelerator [Ac] globulin)**—A plasma (or serum) factor which disappears on heating or storing of oxalated plasma. The factor occurs in a more active form in serum than in plasma.

B. From the *Platelets*: The **platelet thromboplastic factor** (thromboplastinogenase) is obtained upon disintegration of the platelets, as by contact with a rough surface. Thrombin catalyzes the formation of the platelet factor.

C. From the *Tissues*: Thromboplastic precursors are also supplied by the tissues. These may be very important in initiating coagulation reactions because they are supplied from outside the circulation itself.

Stage II: Conversion of Prothrombin to Thrombin

Prothrombin is a globulin circulating in plasma. In the second stage of the clotting process it is activated by conversion to thrombin. The conversion requires a number of reactions involving the interaction of thromboplastic factors and including Stuart factor, factor V, and calcium. **Factor VII** (called "stable factor" because of its stability on storage) is also required. Factor VII is not consumed in coagulation. Synonyms applied to factor VII include the terms proconvertin, serum prothrombin conversion accelerator (SPCA), cothromboplastin, and autoprothrombin I.

Although factor VII is found among the coagulation factors circulating in plasma, it is believed to be required as an accessory only for the activity of tissue thromboplastin and not for plasma thromboplastin.

A prolonged one-stage plasma prothrombin time is characteristic of defects in any of the components of the second stage of clotting because each factor mentioned above is necessary for the conversion of prothrombin to thrombin. A lack of any one factor results in incomplete conversion of prothrombin to thrombin. Consequently, an excess of prothrombin remains in the serum after the clot has formed (abnormal prothrombin consumption).

Stage III: Formation of Fibrin From Fibrinogen

In the third stage of clotting the protein, fibrinogen, loses one or more peptides under the influence of thrombin, which is actually a proteolytic enzyme. A factor from the platelets (platelet accelerator factor II) also catalyzes the conversion of fibrinogen to fibrin. The result of these reactions is the formation, first, of activated fibrinogen (F′), which then undergoes a spontaneous but reversible polymerization to produce fibrin, a protein of much larger molecular weight than the original fibrinogen.

Reversible polymerization is indeed the case in vitro, wherein a system composed of purified fibrinogen and thrombin and added calcium produces a clot which is soluble in 0.03% HCl, so-called **fibrin S** (soluble fibrin). However, in vivo an acid-soluble clot of fibrin I (insoluble fibrin) is formed, but, if small amounts of serum are added to the in vitro system, fibrin I is then also formed. The serum evidently contains a factor responsible for inhibition of reversible polymerization of fibrin. This stabilizing factor in serum which enters the clotting sequence after fibrin has formed is called the **Laki-Lorand factor (LLF)** or **fibrin stabilizing factor (FSF)**. Using the numerical system for nomenclature of blood clotting factors shown in Table 10–2, LLF would be designated **factor XIII**.

To understand the role of the many factors involved in the clotting process, it is helpful to note that the protein clotting factors generally interact in pairs. As a result of this interaction, each clotting factor is in turn converted from an inactive to an active form. The process thus resembles conversion of enzymes from an inactive to an active form by specific activators, although the resemblance is not entirely accurate since some of the clotting factors are probably not enzymes.

In Fig 10–1, the sequence of action of factors necessary for initiation of clotting is outlined. The intrinsic system is that which is entirely present in the

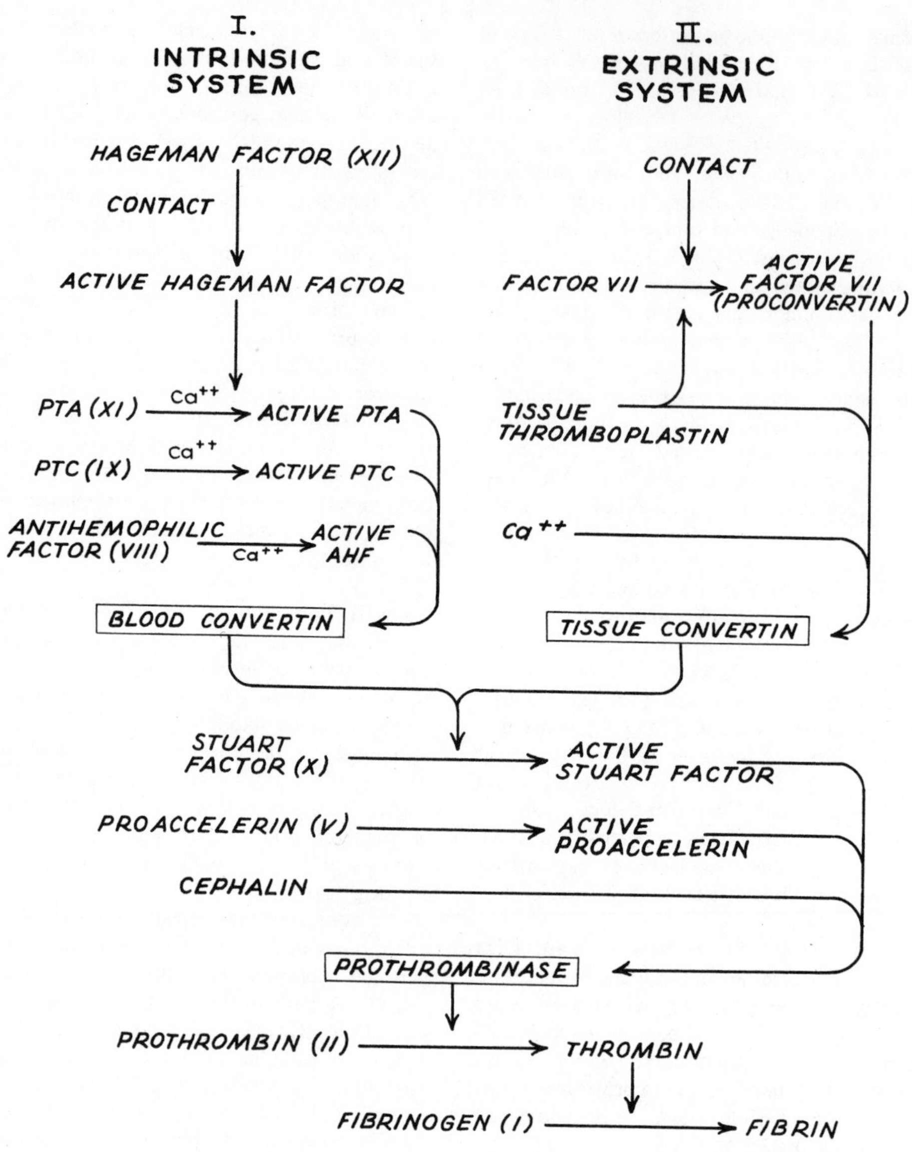

FIG 10–1. Intrinsic and extrinsic systems for initiation of clotting.

plasma and thus permits formation of fibrin without the need for contribution of clotting factors from the tissues. The extrinsic system differs from the intrinsic system by the presence of factors derived from outside the plasma. These extrinsic factors interact to produce active factor X (Stuart factor), and at that point the extrinsic and the intrinsic systems follow the same pathway to fibrin. The presence of both systems results in the formation of a greater amount of fibrin than is the case if the intrinsic system is operating alone.

Other Aspects of the Clotting Mechanism

A. Autocatalysis: In the first few seconds following an injury, little or nothing observable happens with respect to the clotting of shed blood. Clotting then begins suddenly, and the reactions become accelerated with the passage of time. This acceleration phenomenon is caused by **autocatalysis,** whereby certain products formed in the coagulation process actually catalyze the reactions by which they themselves were formed. The principal autocatalyst is thrombin.

B. Vasoconstrictor Action: In addition to the coagulation reactions at the site of the injury other factors may aid in securing hemostasis. These include:

1. A prompt reflex vasoconstriction in the region of the injury.

2. Compression of the vessels in the area of injury by the mass of clotted blood in the tissues (capillary adhesion). The capillary blood vessels become so compressed that their endothelial linings may actually adhere to one another.

3. Liberation of a vasoconstrictor principle upon lysis of the platelets. This may be **serotonin** (hydroxytryptamine; see p 342), which is known to be adsorbed and concentrated in the platelets although it is not manufactured there.

C. **Inhibitors of Prothrombin Activation and Conversion:** The clotting of blood may be prevented by the action of substances which interfere with the conversion of prothrombin to thrombin. The best known inhibitor of this reaction is **heparin,** a water-soluble, thermostable compound. It is extremely potent; as little as 1 mg will prevent the clotting of 100 ml or more of blood. Heparin is formed and stored in the metachromatic granules of the mast cells. These cells are located in the connective tissue surrounding capillaries and the walls of blood vessels. Liver and lung tissue are notably rich in heparin.

Heparin can act both in vivo and in vitro to prevent clotting of blood. In part, its anticoagulant effect is due to the direct and immediate combination of the highly sulfated heparin molecule with various coagulation factors.

Another inhibitor of clotting is referred to as "antithromboplastin," although it has not been proved that this substance acts specifically against thromboplastin. All that is known is that it inhibits the first phase of the process by which prothrombin is activated.

Plasma contains **antithrombic activity,** since it will cause the destruction of large quantities of thrombin by irreversible conversion to **metathrombin.** The antithrombin activity of the plasma is not influenced by heparin.

Calcium is also necessary for the conversion of prothrombin to thrombin. Citrates and oxalates which are commonly used as anticoagulants are effective because they remove calcium from the blood by the formation of insoluble citrate or oxalate salts of calcium. If calcium is added in excess, the clotting power of the blood is restored.

Blood may also be prevented from clotting by defibrination, ie, removal of fibrin by allowing it to form around a glass rod with which the blood has been stirred or on glass beads shaken in the flask with the blood.

The Fibrinolytic System

In addition to the mechanisms for the formation of a clot in the blood, there is also a mechanism which is concerned with the lysis of the clot. In plasma and serum there is a substance called **profibrinolysin (plasminogen)** which becomes activated to **fibrinolysin (plasmin),** the enzyme which lyses the clot. As was the case with prothrombin, profibrinolysin must be activated before it is converted to fibrinolysin. Plasminogen can be activated in various ways to yield the proteolytic enzyme, plasmin. One activator which has been found in many tissues and in the plasma is called **fibrinolysokinase** (fibrinokinase). An activator is also present in the urine; it is termed **urokinase.** Simply shaking a concentrate of plasminogen with chloroform will also activate plasminogen.

Enzymes found in certain bacteria are effective as plasminogen activators. The ability of some bacteria to lyse fibrin clots is undoubtedly due to the presence of such activating enzymes. Examples of bacterial activating enzymes for plasminogen are **staphylokinase** and **streptokinase.** According to Markus & Werkheiser (1964), activation by streptokinase (SK) appears to proceed in 2 stages. First, there is an interaction of SK with a substance called "proactivator I." This reaction then yields "activator I" which in turn catalytically converts plasminogen to plasmin. The conversion of plasminogen to plasmin by SK results in a decrease in molecular weight from 143,000 to 120,000 daltons. It is this lower molecular weight product that exerts fibrinolytic activity.

There are also **inhibitors of plasminogen activation.** One such inhibitor is **ε-amino caproic acid,** which competitively inhibits the activation of human or bovine plasminogen by streptokinase, urokinase, and probably fibrinokinase.

The fibrinolytic system is influenced by the neuroendocrine (pituitary) system. Thyroid stimulating hormone (TSH) and growth hormone (PGH) promote fibrinolysis, whereas adrenocorticotropic hormone (ACTH) inhibits it.

The essentials of the fibrinolytic system are diagrammed in Fig 10–2. Note the analogies to the clotting system as it pertains to prothrombin activation and conversion, and to the action of thrombin and antithrombin.

Prothrombin Production

Prothrombin, factor V, and factor VII are manufactured in the liver and vitamin K is necessary for their production and activity. Deficiencies of any of these substances or of Stuart factor are all characterized by a single laboratory finding–that of a prolonged one-stage prothrombin time. Most abnormalities associated with thrombin formation, the second stage of clotting, and formerly attributed to a defi-

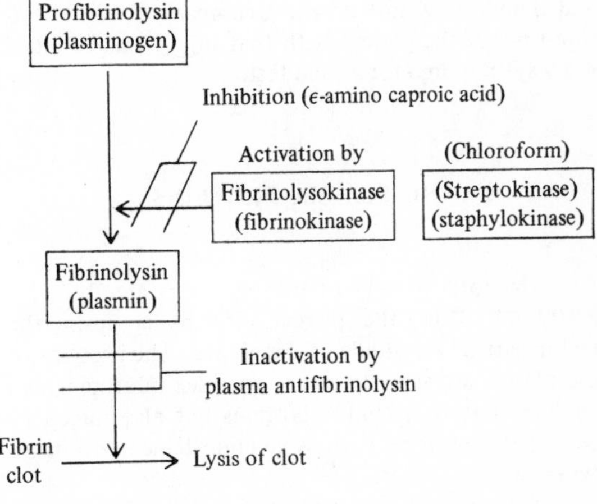

FIG 10–2. Fibrinolytic system.

ciency of prothrombin itself, are now known to be related to deficiencies of factors V and VII. Stuart factor deficiency produces a hemorrhagic disease clinically indistinguishable from factor VII deficiency.

Although deficiencies of factor V (parahemophilia), factor VII, or of Stuart factor are among the inherited second-stage clotting diseases, these deficiencies may also occur in severe hepatic disease or as a result of long-term anticoagulant therapy.

A lack of prothrombin (hypoprothrombinemia) may be acquired or inherited. The latter is rare. Acquired hypoprothrombinemia results when liver damage is so extensive as to interfere with prothrombin synthesis or when the absorption of vitamin K from the intestine is impaired—particularly when the flow of bile to the intestine is prevented, as in obstructive jaundice.

Bishydroxycoumarin (Dicumarol) (Fig 10–3), acting as an antagonist to vitamin K, produces hypoprothrombinemia. This was the first drug to be used clinically to prolong the clotting time of blood, but a number of other antagonists (anticoagulant drugs) are also available to produce hypoprothrombinemia.

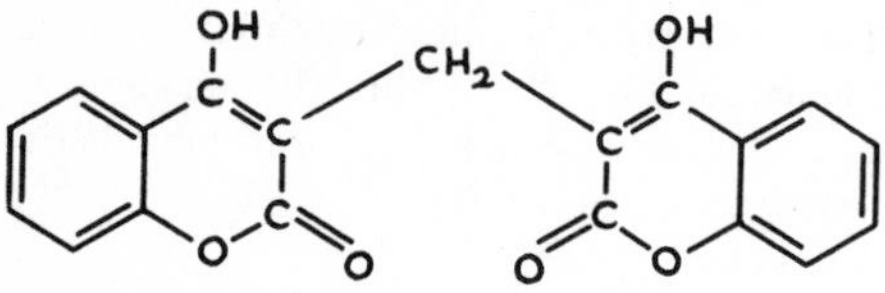

FIG 10–3. Bishydroxycoumarin (Dicumarol, 3,3′-methylene-bishydroxycoumarin).

Dicumarol is used in the treatment of thrombosis, or intravascular clotting. To control the dosage, frequent determinations of the **prothrombin level** of the blood are necessary. One technic for making these measurements is the method of Quick, in which an excess of thromboplastic substance (obtained from rabbit brain) and calcium are added to diluted plasma and the clotting time noted. A comparison of the clotting time of the patient with that of a normal control is always required for a valid test.

THE PLASMA PROTEINS

The total protein of the plasma is about 7–7.5 g/100 ml. Thus, the plasma proteins comprise the major part of the solids of the plasma. The proteins of the plasma are actually a very complex mixture which includes not only simple proteins but also mixed or conjugated proteins such as glycoproteins and various types of lipoproteins.

The separation of individual proteins from a complex mixture is frequently accomplished by the use of various solvents or electrolytes (or both) to remove different protein fractions in accordance with their solubility characteristics. This is the basis of the so-called "salting-out" methods commonly utilized in the determination of protein fractions in the clinical laboratory. Thus it is customary to separate the proteins of the plasma into 3 major groups (fibrinogen, albumin, and globulin) by the use of varying concentrations of sodium or ammonium sulfate. Since it is likely that the subsequent analysis of the protein fractions will require a nitrogen analysis, sodium sulfate is preferred to ammonium sulfate.

Fibrinogen is the precursor of fibrin, the substance of the blood clot. It resembles the globulins in being precipitated by half-saturation with ammonium sulfate; it differs from them in being precipitated in a 0.75 molar solution of Na_2SO_4 or by half-saturation with NaCl. In a quantitative determination of fibrinogen, these reactions are used to separate this protein from other closely related globulins.

Fibrinogen is a large, asymmetric molecule (see diagrams in Fig 10–4) which is highly elongated, having an axial ratio of about 20:1. The molecular weight is between 350,000 and 450,000 daltons. It normally constitutes 4–6% of the total proteins of the plasma. This protein is manufactured in the liver; in any situation where excessive destruction of liver tissue has occurred, a sharp fall in blood fibrinogen results.

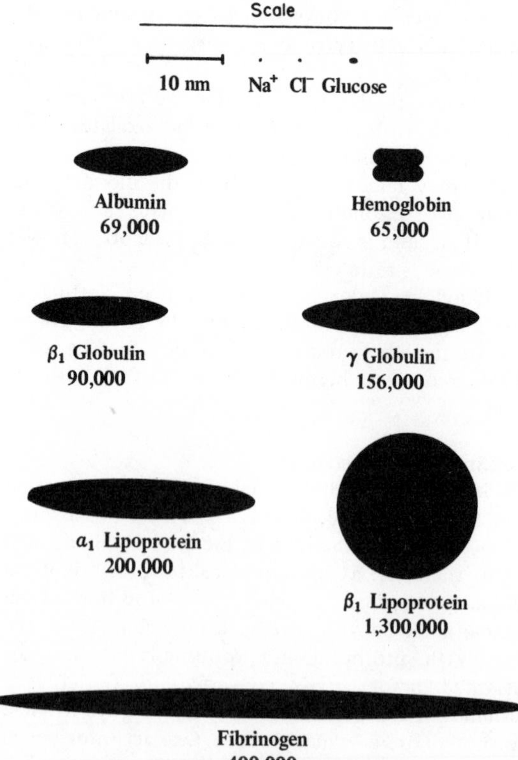

FIG 10–4. Relative dimensions and molecular weights (in daltons) of protein molecules in the blood (Oncley).

The **serum proteins** include mainly the **albumin and globulin** fractions of the plasma since most of the fibrinogen is removed in the clotting process, which is incident to the preparation of the serum. These 2 fractions may be separated by the use of a 27% solution of sodium sulfate, which precipitates the globulins and leaves the albumins in solution. By analysis of the nitrogen* in the filtrate following such a separation, a measure of the serum albumin concentration is obtained.

Direct colorimetric methods for determination of albumin in the serum have also been proposed. These methods depend upon the ability of albumin to bind various ions, including certain dyes such as methyl orange or phenol red. Bromcresol green is a dye which appears to be more specific for albumin than the other dyes which have been employed for the colorimetric measurement of albumin. Analytical methods using bromcresol green for analysis of serum albumin are now in use in some clinical laboratories.

Analysis of the total protein of the serum, when corrected for the nonprotein nitrogen, may be used to estimate the total of the albumin and globulin. The globulin concentration is obtained by subtracting the albumin concentration (determined by direct analysis) from the total. The concentration of these 2 major protein fractions is often expressed as the ratio of albumin to globulin (A/G ratio). If proper separation of albumin and globulin fractions is accomplished, the normal value for this ratio is about 1.2:1. In many clinical situations, this ratio is reversed or "inverted."

Electrophoretic Determination of Serum Proteins

Electrophoresis is the migration of charged particles in an electrolyte solution which occurs when an electric current is passed through the solution. Various protein components of a mixture, such as plasma, at pH values above and below their isoelectric points will migrate at varying rates in such a solution because they possess different surface charges. The proteins will thus tend to separate into distinct layers. Tiselius has applied this principle to the analysis of plasma proteins. The sample for analysis is dissolved in a suitable buffer (for plasma, usually 0.1 N sodium diethylbarbiturate at pH 8.6). This mixture is then placed in the U-shaped glass cell of the Tiselius electrophoresis apparatus, and positive and negative electrodes are connected to each limb of the cell. When the current is applied, migration of the protein components begins. The albumin molecules, which are smaller and more highly charged, exhibit the most rapid rate of migration, followed by various globulins. After a time, boundaries between the separate fractions can be detected because of differences in the index of refraction due to variations in concentrations of protein. A photographic record of these variations constitutes what is termed an electrophoretic pattern. Fig 10–5 illustrates typical patterns as seen in each limb of the cell; these are called descending or ascending patterns in accordance with the direction of protein migration.

*It is customary to convert the nitrogen into protein by the use of the factor 6.25 (N × 6.25 = protein). However, the average nitrogen factor found by analyses of dried proteins from pooled human plasma was reported as 6.73 by Armstrong & others (1947).

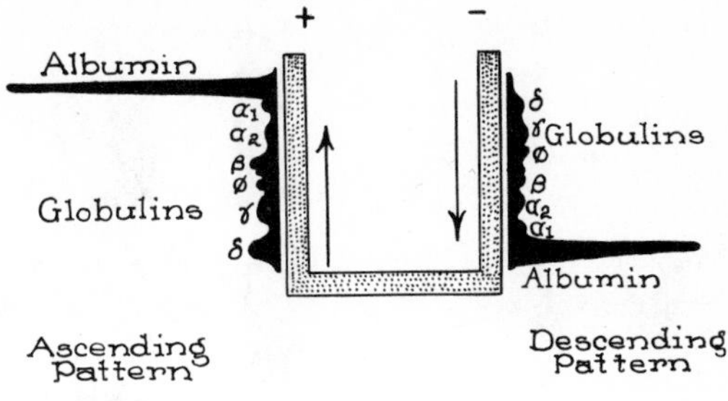

FIG 10–5. Diagrammatic representation of electrophoresis cell and electrophoretic patterns obtained from normal human plasma.

In normal human plasma, 6 distinct moving boundaries have been identified. These are designated in order of decreasing mobility as albumin, alpha$_1$ and alpha$_2$ globulins, beta globulin, fibrinogen, and gamma globulin. The distribution of electrophoretic components of normal human plasma is reported by Armstrong & others as follows (1947):

Albumin		55.2%	(of total plasma protein)
Globulin		44.8%	
α_1	5.3%		
α_2	8.6%		
β	13.4%		
γ	11.0%		
Fibrinogen	6.5%		
A/G ratio		1.2:1	

The electrophoresis of proteins on filter paper is also possible (Fig 10–6). This method, which requires relatively simplified and inexpensive equipment, is now widely used in clinical medicine. A detailed review of the clinical significance of analyses of serum proteins by paper electrophoresis has been prepared by Jencks, Smith, & Durrum (1956). Other reviews are those of Owen (1958) and of Ogryzlo & others (1959).

Other Methods of Separation of Proteins in Plasma

The separation of proteins by electrophoretic analysis depends on a single property, the mobility of the proteins in an electric field. It is known that some plasma proteins which differ in size, shape, composition, and physiologic functions may nonetheless have identical or nearly identical mobilities under the usual conditions of electrophoretic analysis. Thus the conventional electrophoretic fractions are by no means single protein components. Other methods of analysis, such as ultracentrifugation, alcohol precipitation, or immunologic analysis, reveal a considerable number of individual entities within each electrophoretic component.

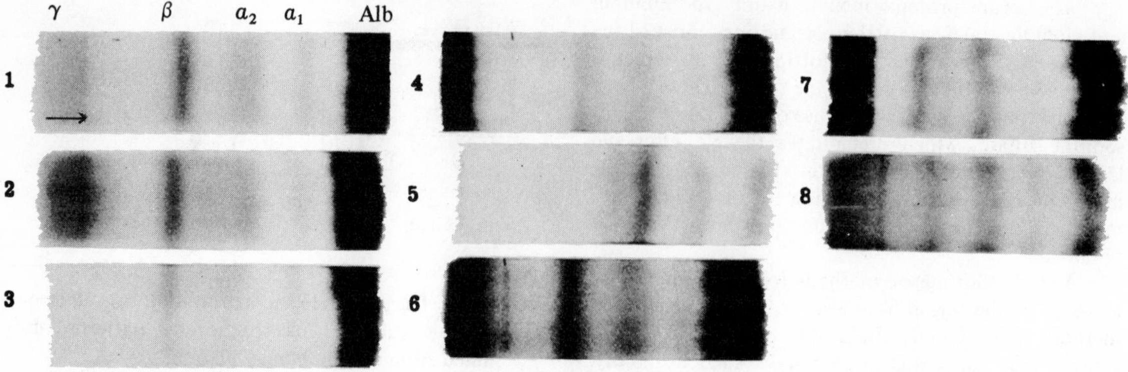

FIG 10–6. Examples of abnormalities in serum protein distribution which are evident on inspection of electrophoretic patterns obtained by electrophoresis on filter paper at pH 8.6. The direction of migration is indicated by the arrow. (1) Normal. (2) Infectious mononucleosis. (3) Hypogammaglobulinemia. (4) Leukemia (type undetermined). (5) Nephrotic syndrome. (6) Infectious hepatitis. (7) Multiple myeloma. (8) Sarcoidosis. (Reproduced, with permission, from Jencks & others: Am J Med 21:387, 1956.)

E.J. Cohn and his collaborators developed methods for the fractionation of plasma proteins which are particularly useful for the isolation in quantity of individual components. Their method is carried out at low temperatures and with low salt concentrations. Differential precipitation of the proteins is accomplished by variation of the pH of the solution and the use of different concentrations of ethyl alcohol.

The results of fractionation of pooled normal human plasma by the method of Cohn are shown in Fig 10–7. Five major fractions are obtained. These account for the vast majority of the total proteins of the plasma. The supernatant, after removal of fractions I–V (ie, fraction VI), contains less than 2% of the total protein. Fractions II and V are relatively homogeneous; the other fractions are very complex mixtures; subfractionation has revealed more than 30 protein components. By reprecipitation, albumin which is electrophoretically 97–99% homogeneous can be prepared from fraction V. This is the salt-poor human serum albumin which is used clinically. Fraction II is almost pure gamma globulin. It is rich in antibodies and has thus found application in prophylaxis and modification of measles and infectious (viral) hepatitis (epidemic jaundice).

A. Albumin: This fraction of the serum proteins, the most abundant of the proteins in the serum, is synthesized in the liver. It has a molecular weight of approximately 69,000 daltons. The primary structure of serum albumin consists of 610 amino acids arranged in a single peptide chain. The secondary structure of albumin appears to be one in which the chain is folded back upon itself to form layers which can be unfolded by lowering of the pH and refolded by raising the pH again.

B. Globulins: The globulin fraction of the serum proteins is a very complex mixture. Certain components of particular interest will be described:

1. Mucoproteins and glycoproteins–These are combinations of carbohydrate (hexosamine) moieties with globulin, found principally in the $alpha_1$ and $alpha_2$ globulin fractions. Meyer defines mucoproteins (mucoids) as those containing more than 4% hexosamine and glycoproteins as those containing less.

2. The lipoproteins–About 3% of the plasma protein consists of combinations of lipid and protein migrating with the alpha globulins and about 5% of similar mixtures migrating with the beta globulins. Using the ultracentrifuge, Hillyard & others (1955) separated human serum into various fractions to account for the total serum lipoproteins. Each fraction was analyzed for protein, phospholipid, free and esterified cholesterol, and triglycerides. The results of these analyses are shown in Table 10–3. Fraction A contains the beta lipoproteins, with densities less than 1.063; fraction B the $alpha_2$ lipoproteins, with densities of 1.063–1.107; and fraction C the $alpha_1$ lipoproteins, with densities of 1.107–1.220.

It will be noted that the beta lipoproteins (fraction A, above) are rich in fat and, consequently, low in protein. They are very large molecules, having molecu-

TABLE 10–3. Percentage composition of lipoproteins in man.

	Fraction		
	A	**B**	**C**
Density	**< 1.063**	**1.063–1.107**	**1.107–1.220**
Lipids			
Phospholipid	21%	29%	20%
Cholesterol			
Free	8%	7%	2%
Esterified	29%	23%	13%
Triglyceride	25%	8%	6%
Total Lipid	83%	67%	41%
Protein	17%	33%	59%

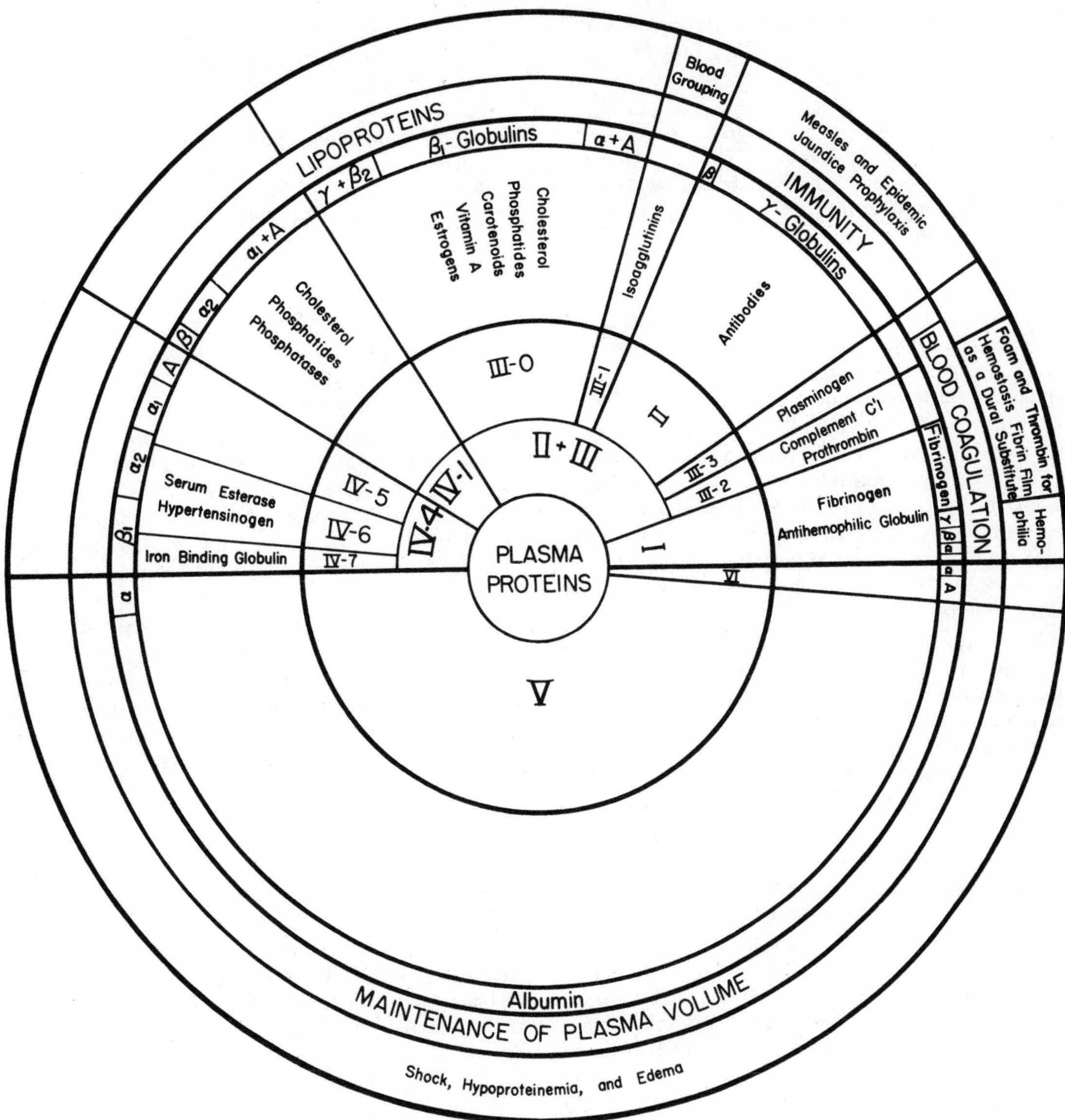

FIG 10–7. Plasma proteins: Their natural functions and clinical uses and separation into fractions. (Revised by LE Strong from Figure 1 *in:* Cohn EJ: Blood proteins and their therapeutic value. Science 101:54, 1945.)

lar weights in the range of 1,300,000 daltons. In contrast, the lipoproteins migrating with the alpha globulins have smaller amounts of fat and more protein, and their molecular weights tend to be much lower (in the range of 200,000 daltons). It is also apparent that the higher the fat and the lower the protein content of a lipoprotein, the lower is its specific gravity. As the fat content declines and the protein rises, there is a concomitant rise in the specific gravity of the lipoprotein so that the so-called high-density lipoproteins (sp gr > 1.220) contain relatively small amounts of lipid.

The lipoproteins probably function as major carriers of the lipids of the plasma since most of the plasma fat is associated with them. Such combinations of lipid with protein provide a vehicle for the transport of fat in a predominantly aqueous medium such as plasma.

3. Metal-binding proteins—Globulins which combine stoichiometrically with iron and copper comprise about 3% of the plasma protein. **Siderophilin (transferrin),** found in Cohn fraction IV-7, is an example of a protein in the plasma which binds iron. The main function of this protein is to transport iron in the plasma. In states of iron deficiency or in pregnancy, there is a significant increase in the concentration of this metal-binding protein in the plasma. In disease, such as in pernicious anemia, chronic infections, or liver disease, there is a reduction in the amounts of this protein.

A blue-green copper-binding protein has been isolated from normal plasma. This protein, ceruloplasmin (molecular weight, 150,000 daltons), contains about 0.34% copper.

4. Gamma globulins–This fraction of the serum proteins is the principal site of the circulating antibodies, the so-called **immunoglobulins**, which constitute a family of closely related proteins possessing all of the known antibody activity of the serum. On the basis of electrophoretic, immunologic, and ultracentrifugal studies, the immunoglobulins (Ig) have been divided into 3 groups. In order of increasing electrophoretic mobility, these are as follows:

a. IgG (γ_2 globulin, γG), the major antibody-containing fraction, comprising approximately 80% of the gamma globulins. By ultracentrifugal analysis it would be designated a 7 S gamma globulin in accordance with its sedimentation constant. Molecular weights are between 150,000 and 160,000 daltons.

b. IgA (γA, β_2A), differentiated from the above (IgG) by a higher content of carbohydrate.

c. IgM (γM, β_2M, 19 S gamma globulin, or β_2 macroglobulin), the high molecular weight (about 750,000 daltons) antibody-containing fraction.

The technic of **immunoelectrophoresis** (Fig 10–8) permits analysis of the various immunoglobulin fractions in the serum. This procedure combines electrophoresis with specific antigen-antibody precipitin reactions. Migration of the protein fractions is generally carried out in an agar gel medium. Immunologic identification of the electrophoretically separated proteins is accomplished by the addition of immune serum to a trough in the agar block adjacent to the separated protein fractions, the specific precipitin antibodies in the immune serum reacting with the protein fractions, acting as antigens, to form visible lines of precipitation within a few hours. This is illustrated in Fig 10–8, which is a drawing of an immunoelectrophoretic pattern of normal human serum reacting with horse antiserum prepared against normal human serum as antigen. The broad arrow marks the point at which electrophoretic migration was started.

Quantitation of the separated protein fractions is related to the extent and intensity of the observed precipitin reaction. Normal values are reported as follows:

IgG = 700–1450 mg/100 ml
IgA = 140–260 mg/100 ml
IgM = 70–130 mg/100 ml

Structure of Immunoglobulins

The most extensive studies of antibody structure have been carried out on IgG. Proteolytic enzymes such as papain or trypsin cleave IgG in the presence of cystine into 3 fragments: 2 are called Fab fragments; the third is called an Fc fragment. Each Fab fragment has a molecular weight of about 52,000 daltons; that of Fc is about 48,000 daltons. The site for combination of the antibody with the antigen is entirely contained within an Fab fragment.

A molecule of IgG consists of a pair of symmetrical halves joined by S–S bonds (Fig 10–9). Two polypeptide chains comprise each ½ of the molecule. One chain is designated a light (L) chain; the other, a heavy (H) chain. Each L and H chain is itself joined by S–S bonds. A heavy chain contains 420–440 amino acid residues; a light chain, 214 residues. Each ½ molecule possesses the binding site for an antigen, which site is present in a globular-ordered region comprising approximately ½ an H chain and an entire L chain. It is this portion of the ½ molecule that is easily separated by proteolytic enzymes such as trypsin or papain to form the Fab fragments mentioned above as the combining sites on the antibody for the antigen. The lower halves of the H chains combine to form another globular region, the Fc fragment, not involved in antibody specificity.

Several classes of L and H chains have been described. There are, however, only 2 major types of L chains in man: the κ (kappa) and λ (lambda) chains, distinguishable serologically as well as by their specific amino acid sequences. Although all of the 3 major classes of immunoglobulins may contain either κ or λ chains or both, the H chain is unique to the class. In IgG the H chain is termed a gamma (γ) chain; in IgA, an alpha (α) chain; and in IgM, a mu (μ) chain.

Carbohydrate residues are found within the immunoglobulin molecule attached to a polypeptide chain. There appear to be 2 carbohydrate units per molecule of IgG and 3 per molecule of IgM. Carbohydrate residues identified in the immunoglobulins include D-mannose, D-galactose, L-fucose, and D-N-acetylneuraminic acid.

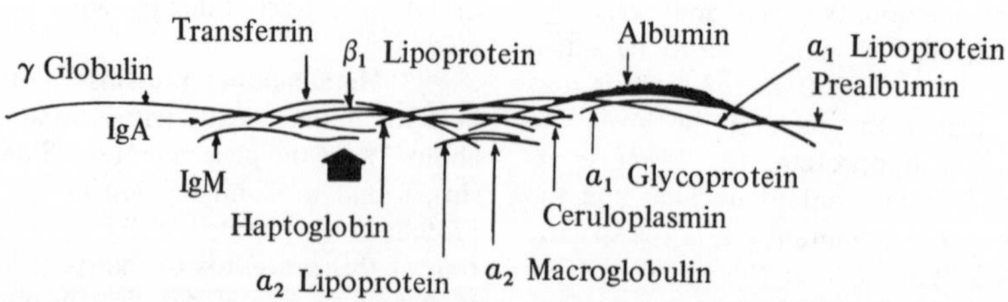

FIG 10–8. Immunoelectrophoresis of normal human serum developed with a horse antiserum to whole normal serum. The broad vertical arrow indicates the starting point. (Reproduced, with permission, from: *Principles of Biochemistry*, 4th ed. White, Handler, & Smith. Blakiston, 1968.)

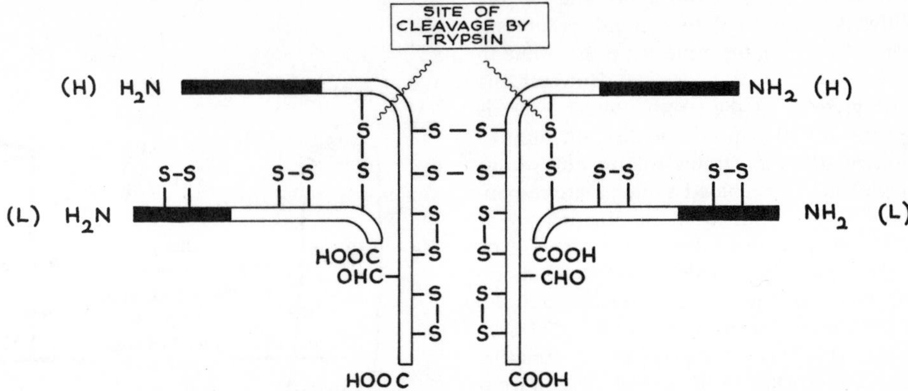

FIG 10–9. Diagrammatic representation of the structure of IgG showing heavy (H) and light (L) polypeptide chains joined by S–S bonds. The amino (NH_2) and carboxy (COOH) terminal portions of the peptide chains are also indicated. Shaded areas at the amino end of each chain outline the sites of variable amino acid composition; the unshaded areas, those which appear to be constant in amino acid sequence when the structures of different antibody molecules are compared. The 2 carbohydrate (CHO) units which are present in IgG are indicated. The sites of cleavage by proteolytic enzymes, such as trypsin, to form 2 Fab fragments and one Fc fragment, are also shown on the diagram.

Differences between IgG molecules are attributable to differences in the amino acid sequences, ie, the primary structures of the H chain, L chain, or both. As is characteristic of proteins in general, these differences in primary structure bring about changes in the conformation, ie, the secondary and higher orders of structure of the protein.

Direct studies of amino acid sequence on a single species of an IgG protein are possible with an immunoglobulin obtained from the serum of a patient with **multiple myeloma.** In this form of hypergammaglobulinemia, proliferation of a single type of an IgG-producing cell leads to the production of a large quantity of a single protein species. This is an example of a so-called monoclonal gammopathy in which one immunoglobulin with homologous composition predominates as a single peak on the electrophoretic pattern. L chains from a number of individual IgG molecules from myeloma patients have been partially or totally sequenced. None was found to be identical with another. Data on H chains are not complete, but the evidence so far obtained is such as to indicate that the same degree of individuality in amino acid sequence prevails among H chains as among L chains. The variability in amino acid sequence inherent in L chains is confined entirely to the amino terminal ½ of the chain. For a given animal species, the other ½ of the chain has an amino acid sequence which is constant. It is probable that the variant portion of the chain is the binding site on the antibody for a specific antigen, although maximal activity toward an antigen is present only when the specific L chain is combined with a specific H chain.

Origin of the Plasma Proteins

The liver is the sole source of fibrinogen, prothrombin, and albumin. Most of the alpha and beta globulins are also of hepatic origin, but the gamma globulins originate from plasma cells and lymphoid tissue. Indeed, gamma globulins are the only proteins secreted by isolated lymph node cells. In plasmapheretic studies,* Whipple and co-workers found that dogs on which an Eck fistula had been performed (portal blood diverted to the vena cava) were able to regenerate plasma proteins at a rate only 10% of that of control dogs. The Eck operation results in progressive impairment of liver function. A similar decline in liver function occurs in chronic liver disease (cirrhosis), and here also low plasma protein levels (particularly albumin levels) are very characteristic.

There is good evidence that the reticuloendothelial system participates in the formation of antibodies. This relates this system to the production of gamma globulin.

Dietary protein serves as a precursor of plasma protein. Many experiments have demonstrated a direct relationship between the quantity and quality of ingested protein and the formation of plasma protein, including also antibody formation. All dietary proteins are not equally effective in supplying materials for the regeneration of plasma protein. In studies on the dog after plasmapheresis, fresh and dried beef serum and lactalbumin, a protein of milk, were most effective. Egg white, beef muscle, liver, casein, and gelatin follow in that order.

Functions of the Serum Proteins

A. Water Exchange: An important function of the serum proteins is the maintenance of osmotic relations between the circulating blood and the tissue spaces.

*Plasmapheresis is a technic for depleting the plasma proteins by withdrawal of blood and reinjection of washed cells suspended in Ringer's solution.

The concentration of the electrolytes and of the organic solutes in plasma and tissue fluids is substantially the same; therefore, the osmotic pressures due to these substances are practically identical. However, the total osmotic pressure of the plasma, which exceeds 6.5 atmospheres (4940 mm Hg), is due not only to inorganic electrolytes and organic solutes but also to the plasma protein. These blood proteins are responsible for about 25 mm of the total osmotic pressure of the plasma. Because there is also a small amount of protein in the tissue fluids which exerts an osmotic pressure of about 10 mm Hg, the effective osmotic pressure of the blood over that of the tissue fluid is 15 (25 − 10) mm Hg. This has the effect of attracting fluid and dissolved substances into the circulation from the tissue spaces. Opposing this force is the hydrostatic pressure of the blood, which tends to force fluids out of the circulation and into the tissue spaces. On the arterial side of the capillary loop, the hydrostatic pressure may be considered to be about 30 mm Hg, and as the blood flows farther from the heart this pressure gradually decreases until it has fallen as low as 15 mm Hg in the venous capillaries and even lower in the lymphatics. The hydrostatic pressures of the capillary are opposed by approximately 8 mm Hg hydrostatic pressure in the tissue spaces. The effective hydrostatic pressure in the arterial capillary is therefore 22 mm Hg (30 − 8); in the venous capillary, 7 mm Hg (15 − 8). On the arterial side, the net result of these opposing pressures is a 7 mm Hg (22 − 15) excess of hydrostatic pressure over osmotic pressure; this favors filtration of materials outward from the capillary to the tissue spaces. On the venous side, the 8 mm Hg difference (15 − 7) is in favor of reabsorption of materials, because the intravascular osmotic pressure now predominates. Reabsorption is also aided by the lymphatics. This explanation of the mechanism of exchange of fluids and dissolved materials between the blood and tissue spaces is called the "Starling hypothesis." It is diagrammed in Fig 10–10.

The accumulation of excess fluid in the tissue spaces is termed "edema." Any alteration of the balance described above may result in edema. Decreases in serum protein concentration or increases in venous pressure, as in heart disease, are examples of pathologic processes which by altering the balance between osmotic and hydrostatic pressure would foster edema.

Each g% of serum albumin exerts an osmotic pressure of 5.54 mm Hg, whereas the same quantity of serum globulin exerts a pressure of only 1.43 mm Hg. This is due to the fact that the albumin fractions consist of proteins of considerably lower molecular weight than those of the globulin fractions. Albumin is of major importance in maintaining serum osmotic pressure. One g of albumin will hold 18 ml of fluid in the blood stream. Concentrated albumin infusions (25 g in 100 ml of diluent) are equivalent in osmotic effect to 500 ml of citrated plasma. This effect may be beneficial in treatment of shock or in any situation where it is desired to remove fluid from the tissues or to increase the blood volume.

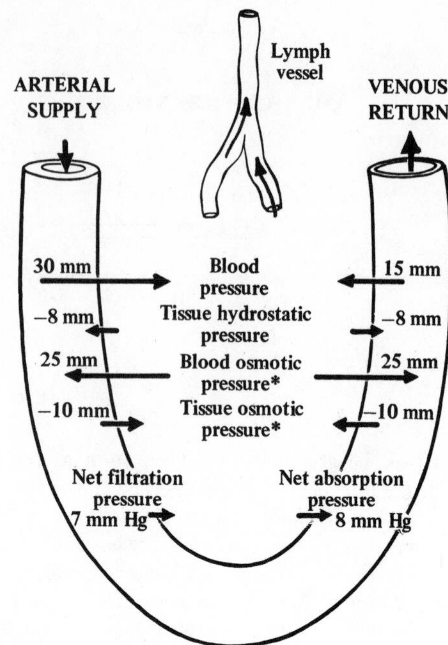

FIG 10–10. Capillary filtration and reabsorption ("Starling hypothesis"). The starred osmotic pressures are actually due only to the protein content of the respective fluids. They do not represent the total osmotic pressure.

B. Blood Buffers: The serum proteins, like other proteins, are amphoteric and can combine with acids or bases. At the normal pH of the blood, the proteins act as an acid and combine with cations (mainly sodium). The buffer pair which is formed constitutes a relatively small fraction of the total blood buffers, since only about 16 mEq/liter of sodium are combined with protein anions.

C. A Reserve of Body Protein: Serum albumin, when administered parenterally, is effective as a source of protein in hypoproteinemic patients. It is well assimilated and is not excreted unless a proteinuria already exists. Plasma is also effective as a source of nutrient protein.

The circulating plasma protein is not static; it constantly interchanges with a labile tissue reserve equal in quantity to the circulating protein. The term "dynamic equilibrium" has been applied to this interchange. In protein starvation, the body draws upon this tissue reserve as well as upon plasma protein for its metabolic needs.

D. Other Functions of Serum Proteins: These include transport of substances otherwise insoluble in plasma. Examples are the transport of bilirubin, steroid hormones, and the so-called free fatty acids by albumin. Many drugs are also transported in the plasma bound to albumin, such as certain antibiotics, coumarins, salicylates, and barbiturates. Lipids, including fat-soluble vitamins, are also transported by various fractions of the serum proteins, particularly the globulins, as illustrated by the lipoproteins which are lipid-globulin complexes. The gamma globulins are the sites of the circulating antibodies.

Plasma Protein Changes in Disease

The albumin component of the serum is either unchanged or, more usually, lowered in pathologic states. Albumin levels do not rise above normal except in the presence of hemoconcentration or dehydration. A decline in serum albumin levels (hypoalbuminemia) may follow prolonged malnutrition due to inadequate dietary intake of protein, impaired digestion of protein (as in pancreatic insufficiency), or inadequate absorption from the intestine. Hypoalbuminemia may also follow the chronic loss of protein, either in the urine (as in the nephrotic syndrome) or by extravasation (as in burns). Inability to synthesize albumin is a prominent feature of chronic liver disease (cirrhosis); indeed, hypoalbuminemia is a characteristic diagnostic and prognostic sign in this type of liver disease. Some types of hypoproteinemia are apparently due to an inherited inability to synthesize plasma protein fractions (so-called familial dysproteinemia), including albumin or the globulins. However, there are also patients with reduced levels of protein in the serum for which none of the above explanations suffice. Because these cases of "idiopathic hypoproteinemia" are characterized by excessively rapid disappearance of administered proteins from the plasma, the term "hypercatabolic hypoproteinemia" has been used to describe them. The rapid disappearance of protein from the plasma is now believed to be due to an excessive loss of protein into the gastrointestinal tract (Holman, 1959). Thus, the term "exudative enteropathy" has been suggested as more descriptive of the etiology of this syndrome (Gordon, 1959).

The alterations which occur in the alpha and beta globulins are of considerable interest in disease. An increase in alpha globulins, particularly in the glycoproteins and mucoproteins is a noteworthy feature of acute febrile disease. This seems to be related to inflammation or tissue destruction. It is also noted in moderate to advanced tuberculosis and in advanced carcinoma where tissue wasting is also occurring. In many diseases there is a constant association between decreased albumin and increased alpha globulin, eg, nephrosis, cirrhosis, and acute infections such as pneumonia, acute rheumatic fever, and typhus fever.

An abnormal constituent in the alpha globulin fraction of human serum is the so-called **C-reactive protein** (CRP) (Abernethy, 1941). This protein is formed by the body in response to an inflammatory reaction. It is called C-reactive protein because it forms a precipitate with the somatic C-polysaccharide of the pneumococcus. Small amounts of this protein may be detected in human serum by a precipitin test using a specific antiserum from rabbits hyperimmunized with purified C-reactive protein.

Increases in beta globulins are often associated with accumulations of lipids. The relation of the lipoproteins to the genesis of atherosclerosis is under study at the present time. Changes in the lipoproteins of the beta globulin component of the serum are responsible for positive thymol turbidity tests. This test is positive not only in hepatitis and cirrhosis but also in many other diseases in which increased serum lipids are a feature.

Elevated levels of serum gamma globulin are characteristic of several diseases. As already mentioned, in the monoclonal gammopathies, one immunoglobulin predominates; it is detected as a single peak on the electrophoretic pattern. Upon analysis, the predominant globulin will be found to be a single protein species whose composition may be unique to a single patient. Examples of diseases in which hypergammaglobulinemias occur are multiple myeloma, Waldenström's macroglobulinemia, carcinoma, and lymphoma. In macroglobulinemia, large quantities of a protein of the IgM class are synthesized. In myeloma, the electrophoretic pattern is normal except for a superimposed protein boundary which generally possesses the mobility of a gamma or a beta globulin, although there have been a few reports of excessive alpha globulins.

The proteins described above as occurring in large amounts in the serum of myeloma patients have molecular weights in excess of 160,000 daltons. However, in about 30% of cases of this disease, an additional group of peculiar proteins of lower molecular weight may be detected in the serum and, because of their low molecular weights, may be excreted into the urine. These globulins, first described by Bence Jones in 1848, occur only in multiple myeloma and in a few cases of myeloid leukemia or other diseases which extensively involve the bone marrow. **Bence Jones proteins** have very unusual properties of solubility, precipitating at 45–60° C but redissolving on boiling. Recent chemical studies of the structure of Bence Jones proteins have revealed that these proteins are identical in composition to the light chains of normal immunoglobulins either of the κ or λ type. They appear as a result of synthesis by plasma cells of light chains in excess of heavy chains. The excess light chains, identical in composition to those of the myeloma proteins in the serum, are then excreted into the urine.

The Bence Jones proteins formed in any given interval are almost entirely excreted within 12 hours. As a result, a patient exhibiting Bence Jones proteinuria may excrete as much as ½ his daily nitrogen intake as Bence Jones protein.

Inherited Deficiencies of Plasma Protein Fractions (Gitlin, 1958)

A. Thromboplastic Factors: The best example of a disease entity which is caused by an inherited lack of an essential factor in the clotting process is classical hemophilia. This hemorrhagic disease is due to a deficiency in the plasma content of antihemophilic globulin, a component of Cohn fraction I (see p 193). The defect is believed to be inherited as an X-linked recessive trait, transmitted exclusively through the females of an affected family although the females do not exhibit the bleeding tendency. The males, on the other hand, exhibit the disease but do not transmit the defect. The abnormality is characterized by a marked prolongation of the coagulation time of the blood with no abnormality in the prothrombin time. When antihemophilic globulin is added to hemophilic blood, clotting of such blood becomes normal. In the presence of hemorrhage, the hemophilic patient may be treated by injections of Cohn fraction I or by transfusions of fresh blood or plasma from normal donors.

Hemophilioid states have also been discovered, although these are less common than classical hemophilia. Each is due to a specific deficiency of a plasma thromboplastic factor.

Inherited as well as acquired deficiencies of the accelerator factors involved in conversion of prothrombin to thrombin (see p 187) have also been described.

B. Afibrinogenemia and Fibrinogenopenia: Afibrinogenemia is another inherited hemorrhagic disease which in its clinical manifestations superficially resembles hemophilia. It is characterized by the absence or near absence of fibrinogen, and transmitted as a non-X-linked recessive trait although it occurs slightly more frequently in males. In a typical case the clotting time and the prothrombin time are prolonged indefinitely. All of the clotting factors of the blood other than fibrinogen are present in normal amounts. In case of injury, death will occur from uncontrollable hemorrhage unless fibrinogen is supplied. Administered fibrinogen is lost from the body by normal decay in 12–21 days; therefore, the protein must be replaced every 10–14 days in quantities sufficient to maintain the fibrinogen level above 50 mg/100 ml of plasma.

Fibrinogenopenia and afibrinogenemia may also be acquired, most commonly as a complication of pregnancy where a long-standing intrauterine death of the fetus has occurred or where there has been premature separation of the placenta or the occurrence of amniotic fluid embolism following administration of oxytocin. It is thought that acute depletion of fibrinogen is brought about by release of thromboplastin-like substances from placenta and amniotic fluid, which results in extensive intravascular clotting and defibrination of the blood.

Acquired fibrinogen deficits may also occur as a consequence of surgical trauma, notably after thoracic or prostatic surgery. The cause may be an increase in fibrinolytic activity.

Totally incoagulable blood occurs only as a result of afibrinogenemia or in the presence of an excess of heparin. In the latter instance, correction of the clotting defect can be accomplished with either thrombin or thromboplastin; in the former situation, only fibrinogen will suffice to restore clotting.

C. Agammaglobulinemia and Hypogammaglobulinemia: Another apparently X-linked recessive factor is involved in the transmission of a defect in plasma protein production which is characterized by the complete or near-complete absence of gamma globulin from the serum. The cases reported so far have all been among males. Acquired forms of this disease have also been found; these occur in both sexes.

Patients afflicted with this inherited disorder of protein formation exhibit a greatly increased susceptibility to bacterial infection and an absence of gamma globulin from the serum and of circulating antibodies in the blood and tissues. Furthermore, there is a complete failure to produce antibodies in response to antigenic stimulation. Either the zinc turbidity test or a direct turbidimetric measurement of gamma globulin (see p 371) may be used to detect these cases. Injections of gamma globulin may be used to aid in controlling bacterial infections in these patients. Peculiarly, their resistance to viral diseases seems normal.

In a study of several patients with agammaglobulinemia, some acquired and others congenital in origin, no abnormalities in the plasma clotting factors were found (Frick, 1956). This indicates that none of the factors involved in clotting mechanisms are gamma globulins. It is probable that congenital agammaglobulinemia is the result of an isolated deficiency of protein synthesis resulting from a lack of a single enzyme system. Furthermore, the deficiency of gamma globulin synthesis in these patients does not involve the liver but depends rather upon an anomaly of protein metabolism existing elsewhere in the reticuloendothelial system.

It is reported that in hypogammaglobulinemia there is a disturbance in the architecture of the lymphoid follicles, a lack of plasma cells, and a failure to form plasma cells after antigenic stimulation. There may also be a deficiency of at least 2 beta globulins which are immunochemically unrelated to gamma globulin.

HEMOGLOBIN

The red coloring matter of the blood is a conjugated protein, hemoglobin. The normal concentration of hemoglobin is 14–16 g/100 ml of blood, all confined to the red cell. It is estimated that there are about 750 g of hemoglobin in the total circulating blood of a 70 kg man and that about 6.25 g (90 mg/kg) are produced and destroyed each day.

Dilute acid will readily hydrolyze hemoglobin into the protein, globin, which belongs to the class of proteins known as histones, and its prosthetic group, **heme** (hematin). Crystals of the hydrochloride of heme, **hemin**, can be easily prepared and its chemical structure then determined. Heme is an iron porphyrin. The formation of the porphyrins and of heme is described in Chapter 6.

Studies of globin, the protein moiety of hemoglobin, have revealed that it is composed of 4 polypeptide chains arranged in the configuration of a tetrahedron. Two of the chains with identical amino acid composition and having valine-leucine as the N-terminal sequence are designated the **alpha chains**; the other

2, also identical with one another and having an amino (N) terminal sequence valine-histidine-leucine, are called the **beta chains.** The chains can be dissociated from each other at either a low or a high pH. The separation of the alpha and beta chains of human hemoglobin has been accomplished by countercurrent distribution (Hill & Craig, 1959) and their amino acid composition determined (Hill, 1962; Guidotti, 1962; Hill & Konigsberg, 1962; Konigsberg, 1963). An alpha chain has 141 amino acids and a molecular weight of 15,126 daltons; a beta chain, 146 amino acids and a molecular weight of 15,866 daltons. From the fact that there are 2 alpha and 2 beta chains in the entire globin molecule, it may be concluded that it contains a total of 574 amino acids, which, with the 4 heme prosthetic groups (one for each chain), gives the hemoglobin molecule a weight of 64,450 daltons. (Ultracentrifugal determinations of the molecular weight of hemoglobin had previously suggested a weight of 64,500 daltons.)

The average life of a red blood cell in the human body is about 120 days. When the red cells are destroyed the porphyrin moiety of hemoglobin is broken down to form the bile pigments, biliverdin and bilirubin, which are carried to the liver for excretion into the intestine by way of the bile. The details of this process are described in the Summary of Chapter 6.

The most characteristic property of hemoglobin is its ability to combine with oxygen to form **oxyhemoglobin.** This combination is reversed merely by exposing the oxyhemoglobin to a low oxygen tension. The absorption spectra which are obtained when white light is passed through hemoglobin solutions or closely related derivatives are useful in distinguishing these compounds from one another. Oxyhemoglobin or diluted arterial blood shows 3 absorption bands: a narrow band of light absorption at a wavelength of λ = 578 nm, a wider band at λ = 542 nm, and a third with its center at λ = 415 nm in the extreme violet end of the spectrum. Reduced (ie, deoxygenated) hemoglobin, on the other hand, shows only one broad band with its center at λ = 559 nm.

When blood is treated with ozone, potassium permanganate, potassium ferricyanide, chlorates, nitrites, nitrobenzene, pyrogallic acid, acetanilid, or certain other substances, **methemoglobin** is formed. In this compound the iron, which is in the ferrous (Fe^{++}) state in hemoglobin, is oxidized to the ferric (Fe^{+++}) state. In acid solution, methemoglobin has one absorption band with its center at λ = 634 nm.

Carbon monoxide combines with hemoglobin even more readily than does oxygen. The carbon monoxide hemoglobin shows 2 absorption bands, the middle of the first at λ = 570 nm and the second at λ = 542 nm. Combinations of hemoglobin with hydrogen sulfide or hydrocyanic acid also give characteristic absorption spectra. This is a valuable means of detecting these compounds in the blood of individuals suspected of having been poisoned with H_2S or HCN.

Abnormal Hemoglobins

The red blood cell of an individual may contain 2 or 3 different molecular species of hemoglobin. The differences are in the protein (globin) portion of the hemoglobin molecule. The fetus of a given animal species produces a hemoglobin which is different from that of the adult of the same species; in certain anemic states it has been observed that the anemic patient may still be producing this fetal (hemoglobin F) type of hemoglobin at an age when it has disappeared from normal adult individuals.

The mechanism for the synthesis of hemoglobin protein is inherited from each parent in a manner similar to that of other genetically controlled protein synthesizing systems. Most individuals have inherited a normal mechanism from each parent, and their red blood cells contain only normal adult (A) hemoglobin, except for the first few months of postnatal life, when some fetal hemoglobin is still present.

For purposes of clinical diagnosis, some of the abnormal hemoglobins may be readily identified by the use of paper electrophoresis because differences in amino acid composition of the globins may produce a variation in the mobilities of the hemoglobins on the paper strip. The electrophoretic identification of a number of hemoglobins is illustrated in Fig 10–11.

The presence of abnormal hemoglobins in the blood is often associated with abnormalities in red cell morphology as well as definite clinical manifestations. Each abnormality appears to be transmitted as a mendelian-recessive characteristic. If the abnormality is heterozygous, ie, inherited from only one parent, and is associated with normal hemoglobin inherited from the other parent, the patient will have only a so-called "trait" (eg, sickle cell trait) and may be free of clinical findings, although the presence of the abnormality can still be detected electrophoretically (see Fig 10–11, showing pattern of S and A hemoglobin, as would be found in a patient with sickle cell trait).

The chemical differences between the hemoglobins are established by identification of the amino acid sequences (primary structures) of the globins. In the earliest studies of the composition of the globins, a purified preparation was split into a number of peptide fragments by treating the protein preparation with the proteolytic enzyme, trypsin. This enzyme splits the polypeptide chain at sites occupied by lysine or arginine residues. The peptide fragments were then separated by chromatography on paper, using electrophoresis to induce amino acid migration in one direction, and conventional solvent chromatography further to separate the peptides in another direction (see the discussion of 2-dimensional chromatography in Chapter 3).

The pattern of the separated peptides as seen on the developed chromatogram is referred to as the "fingerprint" of the protein. By such a fingerprinting technic, Ingram (1956, 1957) was able to study the amino acid composition of the abnormal hemoglobin (S) which occurs in so-called sickle cell anemia, as had been indicated by Pauling (1949) in his illustration of

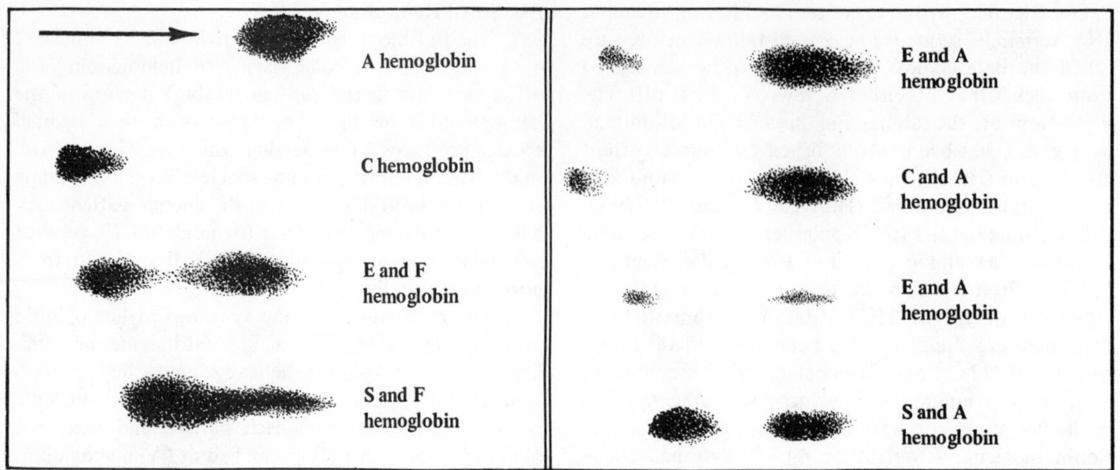

FIG 10–11. Paper electrophoresis of a number of hemoglobin specimens containing various types of hemoglobins. (Redrawn from Chernoff & Minnich: Hemoglobin E, a hereditary abnormality of human hemoglobin. Science 120:605, 1954.)

what he termed a "molecular disease." As a result of these studies of the amino acid composition of sickle cell globin (hemoglobin S), it was discovered that while the alpha chains have the same amino acid sequence as those of normal hemoglobin, in both of the beta chains a valine residue occurs at position 6 (counting from the N-terminal amino acid), whereas in normal hemoglobin A glutamic acid occurs at that position. This single amino acid substitution (valine for glutamic acid) is the only chemical difference between hemoglobins A and S among the 574 amino acids in the molecule. It is explained genetically as due to a mutation in the gene that controls synthesis of the beta chains. Such a mutation resulting in the replacement of one amino acid by another could result from only a single nucleotide base change in a coding triplet (see p 65).

The solubility of hemoglobin S in the oxygenated state is not very different from that of hemoglobin A, but in the reduced (deoxygenated) state hemoglobin S is only about 1/50 as soluble as reduced hemoglobin A and about 1/100 as soluble as its own oxy- form. It is likely that the loss of the glutamate residue with its 2 carboxylic residues alters the distribution of positive and negative charges on the protein surface and thus changes its solubility. As a result of this insolubility of hemoglobin S, a significant increase in viscosity occurs when concentrated solutions of this protein become deoxygenated. Crystals of this form of the protein are crescent-shaped, resembling the so-called "sickled erythrocytes" of the patient with sickle cell disease. Thus it is the marked insolubility of the reduced hemoglobin S that is responsible for the clinical tests used to diagnose sickle cell disease as well as all of the signs and symptoms of the disease itself.

The blood of normal human adults contains a small amount (2.5% of the total) of a slowly migrating (in paper electrophoresis) type of hemoglobin A. This type is called hemoglobin A_2 to distinguish it from the much more abundant hemoglobin A_1. In hemoglobin A_2, replacing the beta chains of A_1 are chains having an arginine residue at position 16 instead of glycine as would be found at that position in a normal beta chain. These variant chains are designated **delta chains.**

In the fetal hemoglobin (F) referred to above, the alpha chains of A are present; however, instead of the beta chains, there are 2 **gamma chains**, which differ in composition from that of the beta chains of A_1 by having a lysine residue at position 6 rather than glutamic acid.

Hemoglobin H is an abnormal hemoglobin which has no alpha chains; instead, the molecule is composed of 4 beta chains, thus designated as $\beta_4{}^A$. Hemoglobin Bart's contains only gamma chains. These 2 types of hemoglobin may be found in the thalassemias, further discussed below.

The oxygen affinity of hemoglobin M (Iwate) is much lower than that of normal hemoglobin A (Hayashi, 1966). Several variants of hemoglobin M other than Iwate have also been discovered (Osaka, Saskatoon, Milwaukee). In all of these hemoglobin M variants, the heme iron in the hemoglobin of the variant chains is spontaneously oxidized to the ferric (Fe^{+++}) state, thus producing methemoglobin (hence, the letter M used to designate these abnormal hemoglobins). It is evident that, in the hemoglobins M, 2 of the peptides (those in which the amino acid sequence is normal) will have ferrous iron heme molecules; the other 2, the variant peptides, will have ferric iron hemin groups. The hemin ferric iron peptides cannot bind oxygen and therefore will not serve as oxygen carriers. The heme ferrous iron peptides function normally, but the fact that 2 of the 4 peptides of the hemoglobin molecule are not functioning in oxygen transport results in only ½ the normal capacity for oxygen binding. Despite this circumstance, individuals with heterozygous inheritance of hemoglobin M do not exhibit signs of an oxygen deficit, although the increased concentration of methemoglobin in their blood produces a dusky appearance of the skin. It will be noted that 3 of the M variants involve replacement of histidine residues, resulting in disruption of the peptide-iron complex associated with oxygen transport (Fig 6–6). Under these circumstances, oxygen can react with the ferrous porphyrin, oxidizing it to the ferric state.

Hemoglobin **Gun Hill** is another variant of adult hemoglobin (Bradley & others, 1967). This abnormal hemoglobin has only ½ the expected number of heme groups. It appears that 5 amino acid residues are missing from the beta chains of the globin. The missing residues normally occur in a linear sequence in the beta chain in the region which is involved in binding of heme to globin. It would appear that the reason for the absence of ½ of the normal number of heme groups is related to the deletion of the 5 amino acids which normally would be involved in binding of heme to globin beta chains. The origin of the Gun Hill mutant may be genetically explained by unequal crossing over during meiosis.

Nomenclature of the Abnormal Hemoglobins

In the years intervening since the discovery of the abnormality of the composition of sickle cell hemoglobin, many additional hemoglobins have been described which may be considered abnormal in the sense that the amino acid composition of certain of the globin chains differs from that of hemoglobin A. The amino acid substitution may or may not change notably the properties of the hemoglobin molecule, but it often leads to diminished rate of production of messenger RNA, resulting in a reduced amount of the altered hemoglobin and an increased amount of hemoglobin F in the red blood cells of an individual heterozygous for the defect. If the defect is inherited from both parents, ie, the inheritance is homozygous, anemia will occur because of the decreased rate of hemoglobin synthesis.

In certain areas of the world (West Africa, the Mediterranean area, Southeast Asia), characteristic hemoglobin variants occur in a large proportion of the population, usually inherited as a heterozygous character and thus occurring as a "trait." Examples are hemoglobins S, C, D (Punjab), and E.

Originally, newly found hemoglobins were designated alphabetically in the order of their discovery. For example, the next abnormal hemoglobin to be reported after S was called C (B having at one time been used to designate sickle cell hemoglobin). This lettering system is no longer entirely suitable because, as the actual differences in composition of the hemoglobins are established, it becomes clear that in several instances different letters have been assigned by different investigators to the same protein. Furthermore, some hemoglobins have differences in amino acid composition which are not revealed electrophoretically; as a result, the same letter may be used when in fact the hemoglobins so designated are not identical. This has necessitated the addition of the name of the original geographic origin of the variant. Examples are found among hemoglobins M, C, D, G, and J.

In the present system of nomenclature for the hemoglobins (Gerald & Ingram, 1961), the peptide chains of the major components of normal adult (A_1) and fetal (F) hemoglobin are designated as alpha, beta, or gamma. By this system, adult hemoglobin (A_1) is written as $\alpha_2{}^A\beta_2{}^A$ and fetal hemoglobin as $\alpha_2{}^A\gamma_2{}^F$. The superscripts refer to the fact that the particular chain is that found in human adult or fetal hemoglobin. Hemoglobin S would be written as $\alpha_2{}^A\beta_2{}^S$, indicating that the chains corresponding to the beta chains of A are of a different type—namely, that of hemoglobin S. When the actual site and the nature of the amino acid substitution become known, the nomenclature will so designate. For example, the defect in hemoglobin S consists in the substitution of a valine residue for a glutamic acid residue at position 6 (counting from the N-terminal of the peptide chain) within the beta chains. Hemoglobin S is therefore designated $\alpha_2{}^A\beta_2{}^{6\,Glu \rightarrow Val}$, or $\alpha_2{}^A\beta_2{}^{6\,Val}$. The symbol $\beta_2{}^{6\,Glu \rightarrow Val}$ is read as, "in both of the beta chains, at positions 6 a glutamic residue is replaced by a valine."

Table 10–4 summarizes the current information on the nature of the amino acid substitutions in a number of abnormal hemoglobins exhibiting variations in composition of either alpha or beta chains.

Thalassemia

The abnormal hemoglobins discussed above result from mutations affecting coding for amino acid sequences in globin peptide chains. Other mutations may occur which affect the *rate* of synthesis of these chains, the amino acid sequences remaining unaffected. These latter mutations are the causes of the **thalassemias**. They appear to result from mutations affecting regulator genes rather than structural genes, as would be the case in the abnormal hemoglobins exhibiting amino acid substitutions. The thalassemia gene may affect the synthesis of either the alpha or the beta chain of normal hemoglobin. In so-called alpha chain thalassemia, wherein synthesis of α chains is repressed, there will be a compensatory increase in synthesis of other chains of which the cell is capable. An example is the pure β chain hemoglobin H mentioned above or the pure γ chain of hemoglobin Bart's. Likewise, in β chain thalassemia, when the thalassemia gene represses β chain synthesis, an excess of α chains results which can combine with δ chains, producing an increase in A_2 hemoglobin, or with γ chains, producing an increase in fetal hemoglobin (F). Beta thalassemia is allelic with hemoglobins S or C, so that by interaction with the S or C gene there results a mixture of hemoglobins wherein S or C may comprise as much as 80–90% of the total hemoglobin. Alpha thalassemia is not allelic with S or C. Consequently, there is no interaction; hemoglobin S or C then constitutes but ½ of the total hemoglobin.

TABLE 10–4. Composition of some abnormal hemoglobins.

Hemoglobin	Substitution	Designation
I. Amino Acid Substitutions in the Alpha Chains		
J (Toronto)	Ala → Asp at 5	$\alpha_2{}^{5\,Asp}\beta_2{}^A$
J (Paris)	Ala → Asp at 12	$\alpha_2{}^{12\,Asp}\beta_2{}^A$
J (Oxford)	Gly → Asp at 15	$\alpha_2{}^{15\,Asp}\beta_2{}^A$
I	Lys → Glu at 16	$\alpha_2{}^{16\,Glu}\beta_2{}^A$
J (Medellin)	Gly → Asp at 22	$\alpha_2{}^{22\,Asp}\beta_2{}^A$
(Memphis)	Glu → Gln at 23	$\alpha_2{}^{23\,Gln}\beta_2{}^A$
G (Honolulu; Singapore; Hong Kong)	Glu → Gln at 30	$\alpha_2{}^{30\,Gln}\beta_2{}^A$
L (Ferrara)	Asp → Gly at 47	$\alpha_2{}^{47\,Gly}\beta_2{}^A$
(Mexico)	Gln → Glu at 54	$\alpha_2{}^{54\,Glu}\beta_2{}^A$
(Shimonoseki)	Gln → Arg at 54	$\alpha_2{}^{54\,Arg}\beta_2{}^A$
(Norfolk)	Gly → Asp at 57	$\alpha_2{}^{57\,Asp}\beta_2{}^A$
M (Osaka; Boston)	His → Tyr at 58	$\alpha_2{}^{58\,Tyr}\beta_2{}^A$
G (Philadelphia)	Asn → Lys at 68	$\alpha_2{}^{68\,Lys}\beta_2{}^A$
M (Iwate; Kankakee)	His → Tyr at 87	$\alpha_2{}^{87\,Tyr}\beta_2{}^A$
(Chesapeake)	Arg → Leu at 92	$\alpha_2{}^{92\,Leu}\beta_2{}^A$
J (Capetown)	Arg → Gln at 92	$\alpha_2{}^{92\,Gln}\beta_2{}^A$
J (Tongariki)	Ala → Asp at 115	$\alpha_2{}^{115\,Asp}\beta_2{}^A$
O (Indonesia)	Glu → Lys at 116	$\alpha_2{}^{116\,Lys}\beta_2{}^A$
II. Amino Acid Substitutions in the Beta Chains		
C	Glu → Lys at 6	$\alpha_2{}^A\beta_2{}^{6\,Lys}$
S (Sickle cell)	Glu → Val at 6	$\alpha_2{}^A\beta_2{}^{6\,Val}$
C (Harlem)	Glu → Val at 6 Asp → Asn at 73	$\alpha_2{}^A\beta_2{}^{6\,Val\,+\,73Asn}$
G (San Jose, Calif.)	Glu → Gly at 7	$\alpha_2{}^A\beta_2{}^{7\,Gly}$
J (Baltimore)	Gly → Asp at 16	$\alpha_2{}^A\beta_2{}^{16\,Asp}$
E (Saskatoon)	Glu → Lys at 22	$\alpha_2{}^A\beta_2{}^{22\,Lys}$
G (Saskatoon)	Glu → Ala at 22	$\alpha_2{}^A\beta_2{}^{22\,Ala}$
E	Glu → Lys at 26	$\alpha_2{}^A\beta_2{}^{26\,Lys}$
E (Genoa)	Leu → Pro at 28	$\alpha_2{}^A\beta_2{}^{28\,Pro}$
G (Copenhagen)	Asp → Asn at 47	$\alpha_2{}^A\beta_2{}^{47\,Asn}$
(Hikari)	Lys → Asn at 61	$\alpha_2{}^A\beta_2{}^{61\,Asn}$
M (Saskatoon)	His → Tyr at 63	$\alpha_2{}^A\beta_2{}^{63\,Tyr}$
(Zurich)	His → Arg at 63	$\alpha_2{}^A\beta_2{}^{63\,Arg}$
(Sydney)	Val → Ala at 67	$\alpha_2{}^A\beta_2{}^{67\,Ala}$
M (Milwaukee)	Val → Glu at 67	$\alpha_2{}^A\beta_2{}^{67\,Glu}$
J (Cambridge)	Gly → Asp at 69	$\alpha_2{}^A\beta_2{}^{69\,Asp}$
(Seattle)	Ala → Glu at 70	$\alpha_2{}^A\beta_2{}^{70\,Glu}$
G (Accra)	Asp → Asn at 79	$\alpha_2{}^A\beta_2{}^{79\,Asn}$
N (Baltimore)	Lys → Glu at 95	$\alpha_2{}^A\beta_2{}^{95\,Glu}$
(Yakima)	Asp → His at 99	$\alpha_2{}^A\beta_2{}^{99\,His}$
(Kansas)	Asn → Thr at 102	$\alpha_2{}^A\beta_2{}^{102\,Thr}$
(New York)	Val → Glu at 113	$\alpha_2{}^A\beta_2{}^{113\,Glu}$
D (Punjab)	Glu → Gln at 121	$\alpha_2{}^A\beta_2{}^{121\,Gln}$
O (Arab)	Glu → Lys at 121	$\alpha_2{}^A\beta_2{}^{121\,Lys}$
K (Woolwich)	Lys → Glu at 132	$\alpha_2{}^A\beta_2{}^{132\,Glu}$
(Hope)	Gly → Asp at 136	$\alpha_2{}^A\beta_2{}^{136\,Asp}$
(Kenwood)	His → Asp at 143	$\alpha_2{}^A\beta_2{}^{143\,Asp}$
(Rainier)	Tyr → His at 145	$\alpha_2{}^A\beta_2{}^{145\,His}$

The abnormal hemoglobins consisting of only beta, gamma, or delta chains do not function as normal hemoglobin, having an abnormal oxygen dissociation curve including absence of a Bohr effect. **Thalassemia major** (Cooley's anemia; Mediterranean anemia) is the result of homozygous inheritance of thalassemia genes. **Thalassemia minor** is the heterozygous form.

A review of the subject of the abnormal hemoglobins and the diseases with which they are associated has been prepared by Lehman, Huntsman, & Ager (1966).

TABLE 10–5. Characteristic values in 3 types of anemia.*

	MCV (fl)	MCH (pg)	MCHC (Concen. %)
Normal blood†	84–95	28–32	33–38
Macrocytic	95–160	30–52	31–38
Microcytic	72–79	22–26	31–38
Hypochromic	50–71	14–21	21–29

*Modified and reproduced, with permission, from Krupp & others: *Physician's Handbook,* 16th ed. Lange, 1970.

†Or normocytic anemia.

THE ANEMIAS

The concentration of hemoglobin in the blood may be measured by a number of methods, including iron analysis and oxygen combining power and, most commonly, by the color of the acid heme formed when a measured quantity of blood is treated with acid or alkali. Anemia exists when the hemoglobin content of the blood falls below normal.

Anemia may result from a decreased rate of production or from an increased loss or destruction of red blood cells. This may occur in acute or chronic hemorrhages, or may be produced by toxic factors (poisons or infections) which cause hemolysis and increased erythrocyte destruction. Decreased production of blood may be due to destruction or loss of function of the blood-forming tissue, as in the leukemias, Hodgkin's disease, multiple myeloma, and aplastic anemia. Certain drugs (benzene, gold salts, arsphenamine), chronic infections, and radiations may also lead to severe anemias because of their destructive or suppressive effect on erythrogenic tissue. Anemias related to inherited defects in production of hemoglobin have been discussed above.

Failure of erythrocyte production may also be caused by a lack of iron and protein in the diet. These nutritional or hypochromic anemias are common in infancy and childhood as well as pregnancy and in chronic blood loss where the iron or protein intake (or both) is inadequate.

Pernicious anemia is due to a failure in red cell production occasioned by a lack of a factor (or factors) necessary to erythrocyte maturation. In uncomplicated pernicious anemia this deficiency is completely corrected by vitamin B_{12} (see p 114). According to Castle, 2 factors are necessary for the formation of red blood cells: (1) **extrinsic factor** found in meat, yeast, liver, rice-polishings, eggs, and milk; and (2) the **intrinsic factor**, produced by the gastric mucosa and possibly also by the duodenal mucosa. Vitamin B_{12} is the extrinsic factor. It is now apparent that the function of the intrinsic factor is to assure absorption of vitamin B_{12} from the intestine. It has no direct effect on erythrocyte production, since parenterally administered vitamin B_{12} is itself sufficient to correct pernicious anemia (see p 117).

Anemias are classified in one or both of 2 ways: (1) according to the predominating size of the erythrocytes, ie, macrocytic (large cell), microcytic (small cell), or normocytic (no significant alteration); and (2) according to the hemoglobin content of the red cell, ie, hyperchromic, hypochromic, or normochromic. There are 3 red cell indices which are useful in the differential diagnosis of the anemias:

A. Mean Corpuscular Volume (MCV): The range of normal MCV is 80–94 fl;* average, 87 fl. MCV is calculated from the packed cell volume (PCV) and red blood count. ***Example:*** PCV = 45%. RBC = 5,340,000/μl.

$$\frac{\text{Vol rbc (ml/liter blood)}}{\text{RBC} \times 10^6/\mu\text{l blood}} = \frac{450}{5.34} = 84.3 \text{ fl}$$

B. Mean Corpuscular Hemoglobin (MCH): This expresses the amount of hemoglobin per red blood cell. It is reported in picograms (pg).* The normal range is 27–32 pg; average, 29.5 pg. For children, the range is 20–27 pg. MCH is calculated from the red blood count and the hemoglobin concentration. ***Example:*** Hemoglobin = 15.6 g/100 ml. RBC = 5,340,000.

$$\frac{\text{Hgb (g/liter blood)}}{\text{RBC} \times 10^6/\mu\text{l blood}} = \frac{156}{5.34} = 29.2 \text{ pg}$$

C. Mean Corpuscular Hemoglobin Concentration (MCHC): This is the amount of hemoglobin expressed as percentage of the volume of a red blood cell. The normal range is 33–38%; average, 35%. It is calculated from the hemoglobin concentration and the packed cell volume. ***Example:*** Hemoglobin = 15.6 g/100 ml blood. PCV = 45%.

$$\frac{\text{Hgb (g/100 ml blood)}}{\text{PCV}} \times 100 =$$

$$\frac{15.6}{45} \times 100 = 34.7\%$$

*fl = femtoliter(s) (10^{-15} liter); pg = picogram(s) (10^{-12} g).

TABLE 10–6. Blood, plasma, or serum values.*

Determination	Material Analyzed	Amount Required† (F = fasting)	Normal Values (Values vary with procedure used.)
Acetone bodies	Plasma	2 ml	0.3–2 mg/100 ml
Aldosterone	Plasma		0.003–0.01 μg/100 ml
Amino acid nitrogen	Plasma	2 ml F	3–5.5 mg/100 ml
Ammonia‡	Blood	2 ml	40–70 μg/100 ml
Amylase	Serum	2 ml	80–180 Somogyi units/100 ml
Ascorbic acid	Plasma	1 ml F	0.4–1.5 mg/100 ml (fasting)
	White cells (blood)	10 ml F	25–40 mg/100 ml
Bilirubin	Serum	2 ml	Direct: 0.1–0.4 mg/100 ml Indirect: 0.2–0.7 mg/100 ml
Calcium	Serum	2 ml F	9–11 mg/100 ml; 4.5–5.5 mEq/liter (varies with protein concentration)
Carbon dioxide: Content	Serum or plasma	1 ml	24–29 mEq/liter 55–65 vol %
Combining power	Serum or plasma	1 ml	55–75 vol %
Carotenoids	Serum	2 ml F	100–300 IU/100 ml
Vitamin A	Serum	2 ml F	40–100 IU/100 ml; 24–60 μg/100 ml
Chloride	Serum	1 ml	100–106 mEq/liter; 350–375 mg/100 ml (as chloride)
Cholesterol	Serum	1 ml	150–280 mg/100 ml
Cholesterol esters	Serum	1 ml	65–75% of total cholesterol
Copper	Serum	5 ml	100–200 μg/100 ml
Cortisol (free)	Plasma		5–18 μg/100 ml
Creatinine	Blood or serum	1 ml	0.7–1.5 mg/100 ml
Glucose (Folin)	Blood	0.1–1 ml F	80–120 mg/100 ml (fasting)
Glucose (true)	Blood	0.1–1 ml F	60–100 mg/100 ml
Hemoglobin	Blood	0.05 ml	Women: 12–16 gm/100 ml Men: 14–18 gm/100 ml
Iodine (BEI)	Serum	2 ml	3–6.5 μg/100 ml
Iodine, protein-bound	Serum	5 ml	4–8 μg/100 ml
Iron	Serum	5 ml	50–175 μg/100 ml
Iron-binding capacity	Serum	5 ml	150–300 μg/100 ml
Lactic acid	Blood (in iodoacetate)	2 ml	0.44–1.8 mM/liter; 4–16 mg/100 ml
Lactic dehydrogenase	Serum		60/100 units/ml
Lipase	Serum	2 ml	0.2–1.5 units (ml of 0.1 N NaOH)
Lipids, total	Serum	5 ml	500–600 mg/100 ml
Magnesium	Serum	2 ml	1.5–2.5 mEq/liter; 1–3 mg/100 ml
Nonprotein nitrogen‡	Serum or blood	1 ml	15–35 mg/100 ml
Oxygen: Capacity	Blood	5 ml	16–24 vol % (varies with Hgb concentration)
Arterial content	Blood	5 ml	15–23 vol % (varies with Hgb content)
Arterial % sat.			94–100% of capacity
Venous content	Blood	5 ml	10–16 vol %
Venous % sat.			60–85% of capacity
Phosphatase, acid	Plasma	2 ml	1–5 units (King-Armstrong); 0.5–2 units (Bodansky); 0.5–2 units (Gutman); 0.1–1 unit (Shinowara); 0.1–0.63 unit (Bessey-Lowry)
Phosphatase, alkaline	Plasma	2 ml	5–13 units (King-Armstrong); 2–4.5 units (Bodansky); 3–10 units (Gutman); 2.2–8.6 units (Shinowara); 2.8–6.7 units (Bessey-Lowry)
Phospholipid	Serum	2 ml	145–200 mg/100 ml
Phosphorus, inorganic	Serum	1 ml F	3–4.5 mg/100 ml (children, 4–7 mg); 0.9–1.5 mM/liter
Potassium	Serum	1 ml	3.5–5 mEq/liter; 14–20 mg/100 ml

*Modified from Krupp & others: *Physician's Handbook*, 16th ed. Lange, 1970.
†Minimum amount required for any procedure.
‡Do not use anticoagulant containing ammonium oxalate.

BLOOD CHEMISTRY

Determination of the content of various compounds in the blood is of increasing importance in the diagnosis and treatment of disease. The blood not only reflects the overall metabolism of the tissues but affords the most accessible method for the sampling of body fluids.

Many of the methods of blood chemistry require the preparation of a protein-free filtrate which is then analyzed for those constituents which remain. The most common technic for the preparation of a protein-free filtrate is the method of Folin and Wu, which utilizes sodium tungstate and sulfuric acid to make tungstic acid for the precipitation of the plasma proteins. Other protein precipitants include trichloroacetic acid or picric acid and a mixture of sodium hydroxide and zinc sulfate (the Somogyi precipitants). Urea, nonprotein nitrogen, uric acid, sugar, and creatinine are examples of blood constituents commonly determined in protein-free filtrates. Many other substances are determined directly, using oxalated whole blood, blood serum, or blood plasma without prior removal of protein.

The normal ranges in concentration of many of the important constituents of the blood are listed in Table 10–6.

LYMPH

The lymphatic fluid is a transudate formed from the plasma by filtration through the wall of the capillary. It resembles plasma in its content of substances which can permeate the capillary wall, although there are some differences in the electrolyte concentrations. The distribution of the nonelectrolytes such as glucose and urea is about equal in plasma and lymph, but the protein concentration of the lymph is definitely lower than that of plasma.

In its broadest aspects, the term "lymph" includes not only the fluid in the lymph vessels but also the fluid which bathes the cells, the "tissue" or "interstitial" fluid. The chemical composition of lymph would therefore be expected to vary with the source of the sample investigated. Thus, the fluid from the leg contains 2–3% protein, whereas that from the intestines contains 4–6% and that from the liver 6–8%.

The lymphatic vessels of the abdominal viscera, the "lacteals," absorb the majority of fat from the intestine. After a meal, this chylous fluid, milky-white in appearance because of its high content of neutral fat, can be readily demonstrated. Except for this high fat content, the chyle is similar in chemical composition to the lymph in other parts of the body.

CEREBROSPINAL FLUID

The CSF is formed as an ultrafiltrate of the plasma by the choroid plexuses of the brain. The process is not one of simple filtration, since active secretory processes are involved. The normal fluid is water-clear, with a specific gravity of 1.003–1.008. Normally, the protein content is low, about 20–45 mg/100 ml, with an albumin-globulin ratio of 3:1, but in disease an increase in protein, particularly in globulin, is characteristic. In inflammatory meningitis, for example, the protein may rise as high as 125 mg to over 1 g/100 ml.

Various diseases of the brain (neurosyphilis, encephalitis, abscess, tumor) show protein elevations above normal of 20–300 mg/100 ml. The Pandy globulin test and the Lange colloidal gold test are diagnostic tests based on changes in the spinal fluid proteins.

The sugar in spinal fluid is somewhat less than in blood, 50–85 mg/100 ml in the fasting adult. It is raised in encephalitis, central nervous system syphilis, abscesses, and tumors. It is decreased in purulent meningitis.

The chloride concentration is normally 700–750 mg/100 ml (expressed as NaCl) or 120–130 mEq/liter. It is generally decreased in meningitis and unchanged in syphilis, encephalitis, poliomyelitis, and other diseases of the central nervous system. The chloride is especially low in tuberculous meningitis.

The concentration of calcium in normal human cerebrospinal fluid is 2.43 ± 0.05 mEq/liter, that of magnesium, 2.40 ± 0.14 mEq/liter. Thus the calcium content of the cerebrospinal fluid is considerably less than that of the serum, whereas that of magnesium is slightly higher; the ratio of calcium to magnesium was found to be 1.01 ± 0.06.

• • •

References

Abernethy TJ, Avery OT: J Exper Med 73:173, 1941.
Armstrong SH Jr, Budka MJE, Morrison KC: J Am Chem Soc 69:416, 1947.
Bradley TB Jr, Wohl RC, Rieder RF: Science 157:1581, 1967.
Frick PG, Good RA: Proc Soc Exper Biol Med 91:169, 1956.
Gerald PS, Ingram VM: J Biol Chem 236:2155, 1961.
Gitlin D, Janeway CA: Pediatrics 21:1034, 1958.
Gordon RS: Lancet 1:325, 1959.
Guidotti G, Hill RJ, Konigsberg W: J Biol Chem 237:2184, 1962.
Hayashi N, Motokawa Y, Kikucchi G: J Biol Chem 241:79, 1966.
Hill RJ, Craig LC: J Am Chem Soc 81:2272, 1959.
Hill RJ, Konigsberg W: J Biol Chem 237:3151, 1962.
Hill RJ, Konigsberg W, Guidotti G, Craig LC: J Biol Chem 237:1549, 1962.
Hillyard LA, Entenman C, Feinberg H, Chaikoff IL: J Biol Chem 214:79, 1955.
Holman H, Nickel WF Jr, Sleisenger MH: Am J Med 27:963, 1959.
Hughes WL, Dintzis HM: J Biol Chem 239:845, 1964.
Ingram VM: Nature 178:792, 1956.
Ingram VM: Nature 180:326, 1957.
Jencks WP, Smith ERB, Durrum EL: Am J Med 21:387, 1956.
Konigsberg W, Goldstein J, Hill RJ: J Biol Chem 238:2028, 1963.
Lehman H, Huntsman RS, Ager JAM: Page 1100 in: *The Metabolic Basis of Inherited Disease.* Stanbury JB, Wyngaarden JB, Fredrickson DS (editors). McGraw-Hill, 1966.
Markus G, Werkheiser WC: J Biol Chem 239:2637, 1964.
Ogryzlo MA, Maclachlan M, Dauphinee JA, Fletcher AA: Am J Med 27:596, 1959.
Owen JA: Advances Clin Chem 1:238, 1958.
Pauling L, Itano HA, Singer SJ, Wells IC: Science 110:543, 1949.
Rosenthal RL, Dreskin IH, Rosenthal N: Proc Soc Exper Biol Med 82:171, 1953.
White SG, Aggeler PM, Glendening MB: Blood 8:101, 1953.
Wright IS: Thromb diath haemorrh 7:381, 1962.

Bibliography

Albritton EC (editor): *Standard Values in Blood.* Saunders, 1952.
Best CH, Taylor NB: *Physiological Basis of Medical Practice,* 7th ed. Williams & Wilkins, 1960.
Biggs R, Macfarlane RG: *Human Blood Coagulation and Its Disorders.* Davis, 1962.
Cantarow A, Trumper M: *Clinical Biochemistry,* 6th ed. Saunders, 1962.
Christensen HN: *Body Fluids and Their Neutrality.* Oxford Univ Press, 1963.
Haurowitz F: *Immunochemistry and the Biosynthesis of Antibodies.* Wiley, 1968.
Henry RJ: *Clinical Chemistry: Principles and Technics.* Hoeber, 1964.
Hepler OE: *Manual of Clinical Laboratory Methods,* 4th ed. Thomas, 1949.
Ingram VM: *Hemoglobin and Its Abnormalities.* Thomas, 1961.
Janeway CA, Rosen FS, Merler E, Alper CA: *The Gamma Globulins.* Little, Brown, 1967.
Killander J (editor): *Nobel Symposium: Gamma Globulins.* Interscience, 1967.
Peters JP, Van Slyke DD: *Quantitative Clinical Chemistry.* Vol II: *Methods.* Williams & Wilkins, 1932. Reprinted 1956.
Putnam FW, (editor): *The Plasma Proteins,* 2 vols. Academic Press, 1960.
Roughton FJW, Kendrew JC (editors): *Haemoglobin.* Interscience, 1949.
Schultze HE, Heremans JF: *Molecular Biology of Human Proteins.* Elsevier, 1966.

11...

The Chemistry of Respiration

PHYSICAL EXCHANGE OF GASES

The term "respiration" is here applied to the interchange of the 2 gases, oxygen and CO_2, between the body and its environment.

Composition of Atmospheric Air

The atmospheric air which we inhale has the following composition: oxygen, 20.96%; CO_2, 0.04%; and nitrogen, 79%. Other gases are present in trace amounts but are not of physiologic importance.

Composition of Expired Air

The expired air contains the same amount of nitrogen as the inspired air, but the oxygen has been reduced to about 15% and the CO_2 increased to about 5%. About ¼ of the oxygen of the inspired air has passed into the blood and has been replaced in the expired air by an equal amount of CO_2 which has left the blood.

Partial Pressure of Gases

In the mixture of gases in air, each gas exerts its own partial pressure. For example, the partial pressure of oxygen at sea level would be 20% of the total pressure of 760 mm Hg; ie, the partial pressure of oxygen (P_{O_2}) = 760 × 0.20 = 152 mm Hg. In the alveoli of the lung the oxygen content is 15%. The total pressure after correction for the vapor pressure of water in the alveolar air (47 mm Hg at 37° C) is 760 − 47 = 713 mm Hg. The partial pressure of oxygen in the lung is therefore about 107 mm Hg (713 × 0.15 = 107); that of carbon dioxide, 36 mm Hg (713 × 0.05 = 36).

Diffusion of Gases in the Lungs

When the gases of the inspired air come in contact with the alveolar membrane of the lung, it is assumed that the exchange of gases takes place in accordance with the usual physical laws of diffusion. Thus, the gas passes through the membrane and into the blood, or in the reverse direction, in accordance with the difference in the pressure of that particular gas on either side of the membrane. The gas pressures in the blood are usually expressed as gas "tensions"; for example, the CO_2 "tension" (P_{CO_2}) is the pressure of the dry gas (mm Hg) with which the dissolved carbonic acid in the blood is in equilibrium; similarly P_{O_2} (oxygen tension) is the pressure of the dry gas with which the dissolved oxygen in the blood is in equilibrium.

The exchange of gases between the alveoli and the blood is illustrated by the following:

Oxygen tension in alveolar air: 107 mm Hg
Oxygen tension in venous blood: 40 mm Hg

A pressure difference of 67 mm Hg serves to drive oxygen from the alveoli of the lung into the blood.

CO_2 tension in alveolar air: 36 mm Hg
CO_2 tension in venous blood: 46 mm Hg

A relatively small difference of 10 mm Hg is sufficient to drive CO_2 from the blood into the lung. This small difference in pressure is adequate because of the rapidity of the diffusion of carbon dioxide through the alveolar membrane. In the resting state, a difference of as little as 0.12 mm Hg in CO_2 tension will still provide for the elimination of this gas.

The tension of nitrogen is essentially the same in both venous blood and lung alveoli (570 mm Hg). This gas is therefore physiologically inert.

After this exchange of gases has occurred, the blood becomes arterial (in a chemical sense). Arterial blood has an oxygen tension of about 100 mm Hg and a CO_2 tension of 40 mm Hg. The nitrogen tension is, of course, unchanged (570 mm Hg). These gases are dissolved in the blood in simple physical solution, and the quantity of each gas which might be carried in the blood in this manner can be calculated according to Henry's law from their absorption coefficients.

It is of interest to compare the quantities of each of these gases which could be dissolved (under physiologic conditions of temperature and pressure) with the actual quantities found in the blood.

TABLE 11–1. Comparison of the calculated content with the actual content of oxygen, CO_2, and nitrogen in the blood.

	ml/100 ml		
	O_2	CO_2	N_2
Calculated content in blood	0.393	2.96	1.04
Actually present:			
Arterial blood	20.0	50.0	1.70
Venous blood	14.0	56.0	1.70

It is apparent that considerable quantities of oxygen and CO_2 are carried in the blood in other than simple solution. The mechanisms by which these increased amounts of oxygen and CO_2 are transported will now be discussed.

THE TRANSPORT OF OXYGEN BY THE BLOOD

Function of Hemoglobin

The transport of oxygen by the blood from the lungs to the tissues is due mainly to the ability of hemoglobin to combine reversibly with oxygen. This may be represented by the equation:

$$Hb + O_2 \rightleftharpoons HbO_2$$

(Hb = reduced [ie, deoxygenated] hemoglobin; HbO_2 = oxyhemoglobin)

The combination of hemoglobin and oxygen is best thought of as a loose affinity rather than as a chemical combination such as an oxide. The degree of combination or of its reversal, ie, dissociation of oxyhemoglobin to release oxygen, is determined by the tension of the oxygen in the medium surrounding the hemoglobin. At a tension of 100 mm Hg or more, hemoglobin is completely saturated. Under these conditions, approximately 1.34 ml of oxygen are combined with each g of hemoglobin. Assuming a hemoglobin concentration of 14.5 g/100 ml of blood, the total oxygen which would be carried as oxyhemoglobin would be 14.5 × 1.34, or 19.43 ml/100 ml (19.43 Vol %). To this may be added the 0.393 ml physically dissolved; the total, approximately 20 Vol %, is the oxygen capacity of blood which contains 14.5 g/100 ml of hemoglobin. It is apparent that the oxygen carrying power of the blood (the oxygen content) is largely a function of the hemoglobin (red cell) concentration.

Dissociation of Oxyhemoglobin (See Fig 11–1.)

The important relationship between the saturation of hemoglobin and the oxygen tension may be perceived by an examination of the dissociation curve of oxyhemoglobin, in which the percent saturation is plotted against the oxygen tension. The shape of the dissociation curve varies with the tension of CO_2. The curve drawn with CO_2 at a tension of 40 mm Hg is to be considered as representative of the normal physiologic condition. It will be noted that at the oxygen tension which exists in arterial blood (100 mm Hg), the hemoglobin is 95–98% saturated; that is, almost complete formation of oxyhemoglobin has occurred. A further increase in oxygen tension has only a slight effect on the saturation of hemoglobin.

As the oxygen tension falls, the saturation of hemoglobin declines slowly until the oxygen tension drops to about 50 mm Hg, at which point a rapid evolution of oxygen occurs. This is the "unloading tension" of hemoglobin. This initial lag in dissociation of oxyhemoglobin provides a fairly wide margin of safety which permits the oxygen tension in the lung to fall as low as 80 mm Hg before any significant decrease in the oxygenation of hemoglobin occurs.

In the tissues, where the oxygen tension is about 40 mm Hg (approximately the unloading tension of hemoglobin), oxyhemoglobin dissociates and oxygen is readily made available to the cells. In the course of a single passage of the blood through the tissues, the

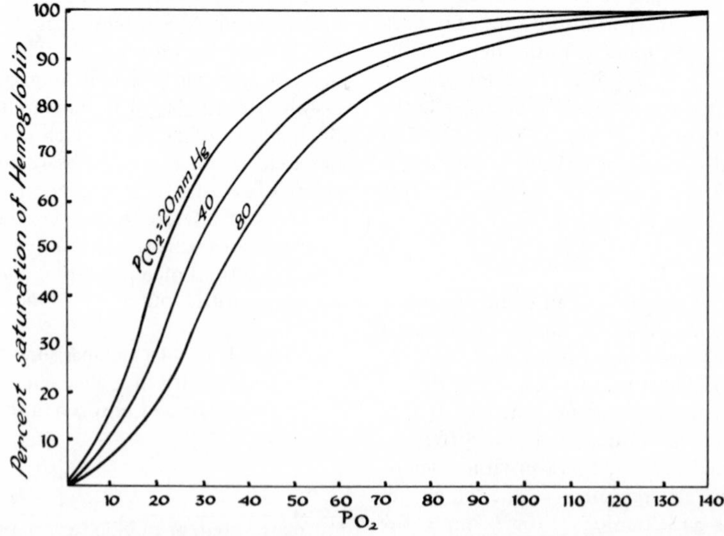

FIG 11–1. The dissociation curves of hemoglobin at 38° C and at partial pressures of carbon dioxide equal to 20, 40, and 80 mm Hg. (Redrawn from Davenport, *The ABC of Acid-Base Chemistry*, 4th ed. Univ of Chicago Press, 1958.)

oxygen content of the blood falls only from 20 to about 15 Vol %. This provides a considerable reserve of oxygenated blood in the event of inadequate oxygenation at the lung.

Clinical Signs of Variation in Hemoglobin Saturation

The red color of deoxygenated hemoglobin is darker than the bright red of oxyhemoglobin. For this reason arterial blood is always brighter than venous blood. A decrease in normal oxygenation of the blood, with a consequent increase in deoxygenated hemoglobin, gives a characteristic bluish appearance to the skin. This is spoken of as cyanosis. It is characteristic of cyanide poisoning, where respiration is also impaired. A cyanotic appearance is dependent on the presence of at least 5 g of **deoxygenated hemoglobin** per 100 ml of capillary blood (Lundsgaard). In severe anemia, the concentration of hemoglobin may be so low as to make cyanosis impossible even though the oxygen content of the blood is reduced. In carbon monoxide poisoning, the formation of the cherry-red carbon monoxide-hemoglobin produces a very characteristic ruddy appearance, particularly noticeable in the lips.

Factors Which Affect the Dissociation of Oxyhemoglobin

A. Temperature: A rise in temperature decreases hemoglobin saturation. For example, at an oxygen tension of 100 mm Hg, hemoglobin is 93% saturated at 38° C but 98% saturated at 25° C. If the saturation of hemoglobin is measured at 10 mm Hg oxygen tension, these differences are even greater: at 25° C, hemoglobin is still 88% saturated, whereas at 37° C it is only 56% saturated. This last observation is of physiologic interest since it indicates that in warm-blooded animals hemoglobin gives up oxygen more readily when passing from high to low oxygen tensions (as from lungs to tissues) than it does in cold-blooded animals.

B. Electrolytes: At low oxygen tensions, oxyhemoglobin gives up oxygen more readily in the presence of electrolytes than it does in pure solution.

C. Effect of CO_2: The effect of CO_2 is illustrated in the graph shown above, in which the curves for percentage saturation of hemoglobin at various tensions of oxygen are shown to vary with different tensions of CO_2. It is probable that the influence of CO_2 on the shape of the dissociation curve is actually the effect of carbonic acid formation, with consequent lowering of the pH of the environment. The increase in acidity, by altering the pH of the medium to the acid side of the isoelectric point of hemoglobin, apparently facilitates the dissociation of oxyhemoglobin. The ability of CO_2 to shift the slope of the oxyhemoglobin dissociation curve to the right is known as the **Bohr effect.**

Under physiologic circumstances, this action of the electrolytes and of CO_2 on the liberation of oxygen from oxyhemoglobin is important in the delivery of oxygen to the tissues.

Carboxyhemoglobin

As has been noted, hemoglobin combines with carbon monoxide even more readily than with oxygen (210 times as fast) to form cherry-red carboxyhemoglobin. This reduces the amount of hemoglobin available to carry oxygen. When the carbon monoxide in the inspired air is as low as 0.02%, headache and nausea occur. If the carbon monoxide concentration is only 1/210 that of oxygen in the air (approximately 0.1% carbon monoxide), unconsciousness will occur in 1 hour and death in 4 hours.

THE TRANSPORT OF CO_2 IN THE BLOOD

CO_2 is carried by the blood both in the cells and in the plasma. The CO_2 content of the arterial blood is 50–53 Vol %, and in venous blood it is 54–60 Vol % (ml CO_2 per 100 ml = Vol %). In accordance with the data on the solubility of CO_2, 100 ml of blood at 37° C, exposed to a CO_2 tension of 40 mm Hg (as it is, for example, in alveolar air or arterial blood), would dissolve only about 2.9 ml. It is obvious that, as was shown for oxygen, the large majority of the blood CO_2 is not physically dissolved in the plasma but must exist in other forms. These comprise 3 main fractions: (1) a small amount of carbonic acid; (2) the "carbamino-bound" CO_2, which is transported in combination with proteins (mainly hemoglobin); and (3) that carried as bicarbonate in combination with the cations, sodium or potassium.

The carbamino-bound CO_2, although it constitutes only about 20% of the total blood CO_2, is important in the exchange of this gas because of the relatively high rate of the reaction:

$$Hb{\cdot}NH_2 \xleftrightarrow{CO_2} Hb{\cdot}NH{\cdot}COOH$$

where $Hb{\cdot}NH_2$ represents a free amino group of hemoglobin (or other blood protein) which is capable of combination with CO_2 to form the carbamino compound.

The amount of CO_2 physically dissolved in the blood is not large, but it is important because any change in its concentration will cause the following equilibrium to shift:

$$\uparrow CO_2 + H_2O \rightleftharpoons H_2CO_3 \rightleftharpoons H^+ + HCO_3^-$$

Carbonic Anhydrase

The rate at which equilibrium in the above reaction is attained is almost 100 times too slow to account for the amount of CO_2 which is eliminated from the blood in the 1 second allowed for passage through the pulmonary capillaries. Nevertheless, about 70% of the CO_2 is derived from that fraction which is carried as bicarbonates in the blood. This apparent inconsistency is explained by the action of an enzyme,

carbonic anhydrase, which is associated with the hemoglobin in the red cells (never in the plasma). The enzyme has been isolated in highly purified form and shown to be a zinc protein complex. It specifically catalyzes the removal of CO_2 from H_2CO_3. The reaction is, however, reversible. At the tissues, the formation of H_2CO_3 from CO_2 and H_2O is also accelerated by carbonic anhydrase. Small amounts of carbonic anhydrase are also found in muscle tissue, in the pancreas, and in spermatozoa. Much larger quantities occur in the parietal cells of the stomach, where the enzyme is involved in the secretion of hydrochloric acid (Fig 12–2). Carbonic anhydrase also occurs in the tubules of the kidney; here its function is also involved in hydrogen ion secretion (Fig 18–4).

Effect of CO_2 on Blood pH

Although it is true that the CO_2 evolved from the tissues will form carbonic acid, as shown in the above reaction, very little CO_2 can actually be carried in this form because of the effect of carbonic acid on the pH of the blood. It is estimated that in 24 hours the lungs remove the equivalent of 20–40 liters of 1-normal acid as carbonic acid. This large acid load is successfully transported by the blood with hardly any variation in the blood pH, since most of the carbonic acid formed is promptly converted to bicarbonate, as shown in the equation below (B^+ represents cations in the blood, principally Na^+ or K^+).

$$H_2CO_3 \rightleftharpoons H^+ + HCO_3^- + B^+ \rightleftharpoons BHCO_3 + H^+$$

At the pH of blood (7.40), a ratio of 20:1 must exist between the bicarbonate and carbonic acid fractions. This ratio is calculated from the Henderson-Hasselbalch equation as follows:

$$7.4 = \text{pH of blood}$$

$$6.1 = \text{pKa, } H_2CO_3$$

$$\text{pH} = \text{pKa} + \log \frac{\text{salt}}{\text{acid}}$$

$$7.40 = 6.10 + \log \frac{S}{A}\left(\frac{BHCO_3}{H_2CO_3}\right)$$

$$1.30 = \log \frac{BHCO_3}{H_2CO_3}$$

$$\text{antilog } 1.3 = 20$$

$$\text{therefore} \quad \frac{20}{1} = \frac{BHCO_3}{H_2CO_3}$$

Any increase or decrease in H ion activity will be met by an adjustment in the above reaction. **As long as this ratio is maintained, the pH of the blood will be normal.** Any alteration in the ratio will disturb the acid-base balance of the blood in the direction of acidemia or alkalemia.

THE BUFFER SYSTEMS OF THE BLOOD

Although venous blood carries considerably more CO_2 than does arterial blood, the buffers of the blood are so efficient that the pH of venous blood is more acid than that of arterial blood by only 0.01–0.03 units, ie, pH 7.40 vs pH 7.43. These blood buffers consist of the plasma proteins, hemoglobin, and oxyhemoglobin, and bicarbonates and inorganic phosphates. The small decrease in pH which occurs when CO_2 enters the venous blood at the tissues has the effect of shifting the ratio of acid to salt in all of these buffer pairs. In the sense that less cation is required to balance anions of the salt component of each of these buffers, when the ratio is shifted to form more of the acid, cation becomes available to form additional bicarbonate derived from the incoming CO_2. In this respect, the plasma phosphates and bicarbonates play a minor role. The buffering effect of the plasma proteins is of greater importance since they release sufficient cation to account for the carriage of about 10% of the total CO_2. The phosphates within the red cell are responsible for about 25% of the total CO_2 carried. Most important, however, is the unique buffering role of hemoglobin and oxyhemoglobin, which accounts for 60% of the CO_2 carrying capacity of whole blood. This role, described below, is, of course, in addition to the part hemoglobin plays in the carriage of carbamino-bound CO_2.

The Hemoglobin Buffers

The remarkable buffering capacity of hemoglobin is due to the fact that this protein in the oxy form is a stronger acid than in the reduced (deoxygenated) form. This is shown by the respective dissociation constants of the 2 forms of hemoglobin:

$$\text{K, oxyhemoglobin} = 2.4 \times 10^{-7}$$
$$\text{K, reduced oxyhemoglobin} = 6.6 \times 10^{-9}$$

At the lungs the formation of oxyhemoglobin from reduced hemoglobin must therefore release hydrogen ions, which will react with bicarbonate to form H_2CO_3. Because of the low CO_2 tension in the lung, the equilibrium then shifts toward the production of CO_2, which is continually eliminated in the expired air:

$$H^+ + HCO_3^- \leftrightarrow H_2CO_3 \leftrightarrow H_2O + CO_2$$

However, in the tissues, where oxygen tension is reduced, oxyhemoglobin dissociates (aided by CO_2, Bohr effect; see p 209), delivering oxygen to the cells, and reduced hemoglobin is formed. At the same time, CO_2 produced in the course of metabolism enters the blood, where it is hydrated to form H_2CO_3, which ionizes to form H^+ and HCO_3^-. Reduced hemoglobin acting as an anion accepts the H^+ ions, forming so-called acid-reduced hemoglobin, HHb. Very little change in pH occurs because the newly arrived H^+ ions

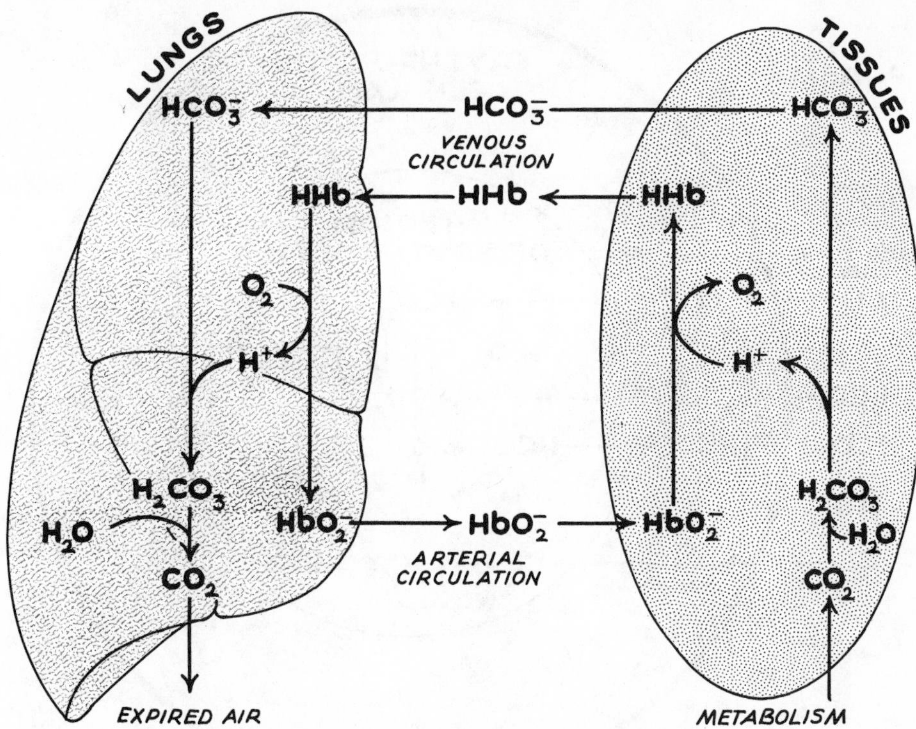

FIG 11–2. The buffering action of hemoglobin.

are buffered by formation of a very weak acid, ie, one in which ionization of hydrogen is suppressed. Then, as noted above, when the blood returns to the lungs these H^+ ions will be released as a result of the formation of a stronger acid; oxyhemoglobin, and the newly released H^+ will promptly be neutralized by HCO_3^-. Indeed, this reaction is essential to the liberation of CO_2 in the lungs.

These relationships of hemoglobin and oxyhemoglobin to the transport of CO_2 are illustrated in Fig 11–2.

At a pH of 7.25, 1 mol of oxyhemoglobin donates 1.88 mEq of H^+; 1 mol of reduced hemoglobin, on the other hand, because it is less ionized, donates only 1.28 mEq of H^+. It may therefore be calculated that at the tissues a change of 1 mol of oxyhemoglobin to reduced hemoglobin allows 0.6 mEq of H^+ to be bound (buffered) so that these newly formed H^+ ions do not bring about a change in pH. This circumstance as it relates to the role of the hemoglobin buffers is sometimes referred to as the **isohydric transport of CO_2**.

The Chloride Shift

It has been shown above that hemoglobin is responsible for about 60% of the buffering capacity of the blood. The red blood cell phosphates contribute another 25%. Thus some 85% of the CO_2 carrying power of the blood resides within the red cell. It is for this reason that the buffering power of whole blood greatly exceeds that of plasma or serum. However, it is also true that most of the buffered CO_2 is carried as bicarbonate in the plasma. These observations pose the question of how it is possible for the majority of the buffer capacity of the blood to reside in the red cells but to be exerted in the plasma.

CO_2 reacts with water to form carbonic acid, mainly inside the red cell since the catalyzing enzyme, carbonic anhydrase, is found only within the erythrocyte. The carbonic acid is then buffered by the intracellular buffers, phosphate and hemoglobin, combining in this case with potassium. Bicarbonate ion also returns to the plasma and exchanges with chloride, which shifts into the cell when the tension of CO_2 increases in the blood. The process is reversible, so that chloride leaves the cells and enters the plasma when the CO_2 tension is reduced. This fact is confirmed by the finding of a higher chloride content in arterial plasma than in venous plasma.

It is considered that under normal conditions the red cell is virtually impermeable to sodium or potassium. But since it is permeable to hydrogen, bicarbonate, and chloride ions, intracellular sources of cation (potassium) are indirectly made available to the plasma by chloride (anion) exchange. This permits the carriage of additional CO_2 (as sodium bicarbonate) by plasma.

The reactions of the chloride shift as they occur at the tissues between the plasma and red cells are summarized in Fig 11–3.

The CO_2 entering the blood from the tissues passes into the red cells, where it forms carbonic acid, a reaction catalyzed by carbonic anhydrase. Some of the carbonic acid returns to the plasma. The remainder reacts with the hemoglobin buffers to form bicarbon-

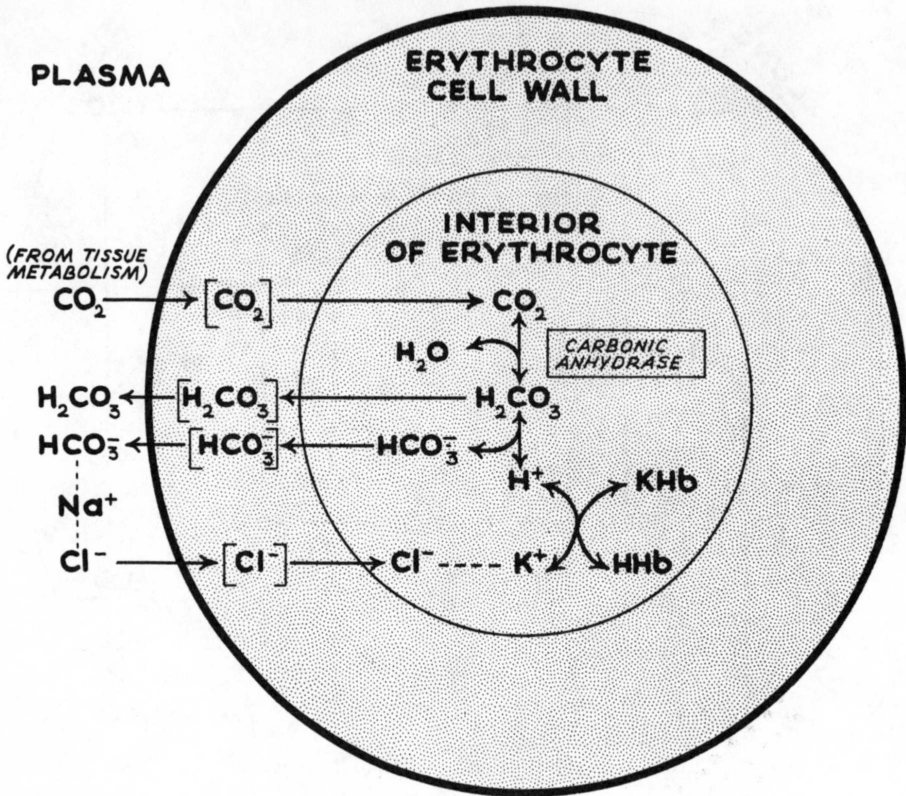

FIG 11–3. The chloride shift.

ate, which then returns to the plasma where it exchanges with chloride. Sodium bicarbonate is formed in the plasma; and the chloride which the bicarbonate has replaced enters the red cells, where it is neutralized by potassium.

All of these reactions are reversible. At the lung, when the blood becomes arterial, chloride shifts back into the plasma, thus liberating intracellular potassium to buffer the newly-formed oxyhemoglobin and, in the plasma, neutralizing the sodium liberated by the removal of CO_2 during respiration.

ACID-BASE BALANCE

It has been noted above that as long as the ratio of carbonic acid to bicarbonate in the blood is 1:20, the pH of the blood remains normal; and that any alteration in this ratio, which was calculated from the Henderson-Hasselbalch equation at the normal pH of the blood (see above), will disturb the acid-base balance of the blood and tissues in the direction of acidosis or alkalosis.

The content of H_2CO_3 in the blood is under the control of the respiratory system because of the dependence of carbonic acid on the P_{CO_2}, which in turn is controlled by the organs of respiration. In consequence, disturbances in acid-base balance which are due to alterations in content of H_2CO_3 of the blood are said to be respiratory in origin. Thus **respiratory acidosis** will occur when circumstances are such as to cause an accumulation of H_2CO_3 in the blood; and **respiratory alkalosis** will occur when the rate of elimination of CO_2 is excessive, so that a reduction of H_2CO_3 occurs in the blood. In either instance, the normal 1:20 ratio of H_2CO_3 to bicarbonate is disturbed, and the pH of the blood will fall or rise in accordance with the retention or the excessive elimination of CO_2. If, however, the bicarbonate content of the blood can be adjusted to restore the 1:20 ratio between carbonic acid and bicarbonate, the pH will once more return to normal. Such an adjustment can be accomplished by the kidneys (see p 381)–in respiratory acidosis by reabsorption of more bicarbonate in the renal tubules, and in respiratory alkalosis by permitting more bicarbonate to escape reabsorption and thus to be excreted into the urine. The respiratory acidosis or alkalosis is then said to be **compensated,** which means that even though the amounts of H_2CO_3 and of bicarbonate in the blood are abnormal, the pH is normal because the ratio of the two has been restored to the normal 1:20. It follows from the above discussion that the CO_2 content (see below) of the plasma, which is a measure of both carbonic acid and bicarbonate, will be higher than normal in compensated respiratory acidosis and lower than normal in compensated respiratory alkalosis.

Disturbances in acid-base balance which are due to alterations in the content of bicarbonate in the blood are said to be metabolic in origin. A deficit of bicarbonate without any change in H_2CO_3 will produce a **metabolic acidosis**; an excess of bicarbonate, a **metabolic alkalosis.** Compensation will occur by adjustments of the carbonic acid concentrations, in the first instance by elimination of more CO_2 (hyperventilation) and in the latter instance by retention of CO_2 (depressed respirations). The CO_2 content of the plasma will obviously be lower than normal in metabolic acidosis and higher than normal in metabolic alkalosis.

The biochemical changes which occur in the various types of acidosis and alkalosis, both uncompensated and compensated, are summarized in Fig 11–4.

Causes of Disturbances in Acid-Base Balance

A. Metabolic Acidosis: Caused by a decrease in the bicarbonate fraction, with either no change or a relatively smaller change in the carbonic acid fraction. This is the most common, classical type of acidosis. It occurs in uncontrolled diabetes with ketosis, in some cases of vomiting when the fluids lost are not acid, in renal disease, poisoning by an acid salt, excessive loss of intestinal fluids (particularly from the lower small intestine and colon, as in diarrheas or colitis), and whenever excessive losses of electrolyte have occurred. Increased respirations (hyperpnea) may be an important sign of an uncompensated acidosis.

B. Respiratory Acidosis: Caused by an increase in carbonic acid relative to bicarbonate. This may occur in any disease which impairs respiration, such as pneumonia, emphysema, congestive failure, asthma, or in depression of the respiratory center (as by morphine poisoning). A poorly functioning respirator may also contribute to respiratory acidosis.

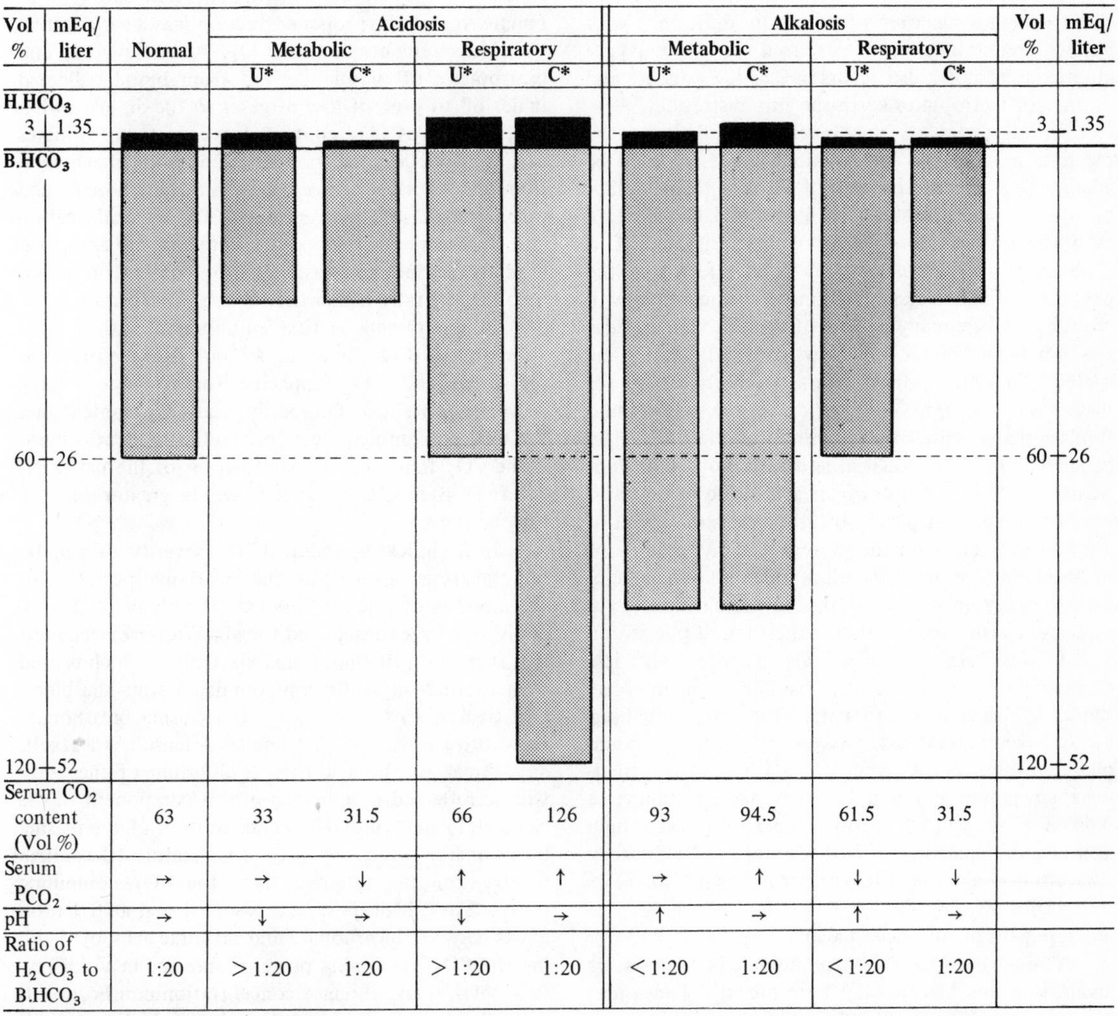

Vol % / mEq/liter	Normal	Acidosis				Alkalosis				Vol % / mEq/liter
		Metabolic		Respiratory		Metabolic		Respiratory		
		U*	C*	U*	C*	U*	C*	U*	C*	
H.HCO₃ 3 / 1.35; B.HCO₃										3 / 1.35
60 / 26										60 / 26
120 / 52										120 / 52
Serum CO_2 content (Vol %)	63	33	31.5	66	126	93	94.5	61.5	31.5	
Serum P_{CO_2}	→	→	↓	↑	↑	→	↑	↓	↓	
pH	→	↓	→	↓	→	↑	→	↑	→	
Ratio of H_2CO_3 to $B.HCO_3$	1:20	> 1:20	1:20	> 1:20	1:20	< 1:20	1:20	< 1:20	1:20	

*U = Uncompensated. C = Compensated.

FIG 11–4. Biochemical changes in acidosis and alkalosis.

C. Metabolic Alkalosis: Occurs when there is an increase in the bicarbonate fraction, with either no change or a relatively smaller change in the carbonic acid fraction. A simple alkali excess leading to alkalosis is produced by the ingestion of large quantities of alkali, such as might occur in patients under treatment for peptic ulcer. But this type of alkalosis occurs much more commonly as a consequence of high intestinal obstruction (as in pyloric stenosis), after prolonged vomiting, or after the excessive removal of gastric secretions containing hydrochloric acid (as in gastric suction). The elevated blood pH of an uncompensated alkalosis often leads to tetany, possibly by inducing a decrease in ionized serum calcium. This is sometimes referred to as gastric tetany, although its relation to the stomach is, of course, incidental. The common denominator in this form of alkalosis is a chloride deficit caused by the removal of gastric secretions which are low in sodium but high in chloride (ie, as hydrochloric acid). The chloride ions which are lost are then replaced by bicarbonate. This type of metabolic alkalosis is aptly termed "hypochloremic" alkalosis. The frequent association of potassium deficiency with hypochloremic alkalosis is discussed on p 403. Hypochloremic alkalosis also occurs in Cushing's disease and during corticotropin or cortisone administration.

In all types of uncompensated alkalosis, the respirations are slow and shallow; the urine may be alkaline, but usually, because of a concomitant deficit of sodium and potassium, will give an acid reaction even though the blood bicarbonate is elevated. This paradox is attributable in part to the fact that the excretion of the excess bicarbonate by the kidney will require an accompanying loss of sodium which under the conditions described (low sodium) cannot be spared. Thus the kidney defers to the necessity for maintaining sodium concentrations in the extracellular fluid at the expense of acid-base balance. However, an equal–if not, in the usual situations, a more important–cause of the excretion of an acid urine in the presence of an elevated plasma bicarbonate is the effect of a potassium deficit on the excretion of hydrogen ions by the kidney as described on p 380. Metabolic alkalosis as usually encountered clinically is almost always associated with a concomitant deficiency of potassium.

D. Respiratory Alkalosis: Occurs when there is a decrease in the carbonic acid fraction with no corresponding change in bicarbonate. This is brought about by hyperventilation, either voluntary or forced. Examples are hysterical hyperventilation, CNS disease affecting the respiratory system, the early stages of salicylate poisoning (see below), the hyperpnea observed at high altitude, or injudicious use of respirators. Respiratory alkalosis may also occur in patients in hepatic coma.

Measurement of Acid-Base Balance

The existence of uncompensated acidosis or alkalosis is most accurately determined by measurement of the pH of the blood. In respiratory acidosis or alkalosis, blood pH determination is essential to a satisfactory biochemical diagnosis. However, determination of the pH of the blood is often not feasible clinically. Furthermore, it is necessary to know to what extent the electrolyte pattern of the blood is disturbed in order to prescribe the proper corrective therapy. For these reasons, a determination of the CO_2 derived from a sample of blood plasma after treatment with acid (CO_2 capacity or CO_2 combining power) is used instead. This measures essentially the total quantity of H_2CO_3 and of bicarbonate in the plasma but gives no information as to the ratio of distribution of the 2 components of the bicarbonate buffer system (and hence of the blood pH). It should also be noted that such a single determination also fails to take into account the concentration of other buffer systems such as hemoglobin (both oxygenated and reduced), serum protein, and phosphates. In disease, these may be notably altered and thus exert important effects on acid-base balance. However, the total blood CO_2 determination is reasonably satisfactory when taken in association with clinical observations and the history of the case. In addition, as noted above, it yields information on the degree of depletion or excess of bicarbonate so that the proper correction may be instituted.

Measurement of the CO_2, carbonic acid, and bicarbonate of plasma derived from blood collected under oil to prevent loss of gases to the air gives what is designated as **CO_2 content.** It is reported as volumes of CO_2 per 100 ml, at standard conditions of temperature and pressure, since all of the bicarbonate and carbonic acid are converted to CO_2 by acidification and by the imposition of a vacuum in the apparatus used to measure the gas. The CO_2 content of venous blood is naturally higher than that of arterial blood.

If the plasma is first equilibrated with normal alveolar air (CO_2 tension, 40 mm Hg) before it is measured, the **CO_2 capacity** (or CO_2 combining power) is obtained. Ordinarily, the CO_2 content and the CO_2 combining power are practically identical; but if the CO_2 tension in the alveolar air of the patient is less than 40 mm, CO_2 capacity will be greater than the CO_2 content.

In a clinical appraisal of the severity of a metabolic acidosis or alkalosis, the bicarbonate fraction of the blood is of primary interest. The plasma bicarbonate is sometimes designated the **alkali reserve** because it is this fraction of the plasma electrolyte which is used to neutralize all acidic compounds entering the blood and tissues. In this capacity, the plasma bicarbonate constitutes a sort of first line of defense. As a result, any threat to the acid-base equilibrium of the body will be reflected in a change in this component of the electrolyte structure. The concentration of the plasma bicarbonate, which is used to measure the alkali reserve, can be obtained from the CO_2 combining power. For this purpose, it is assumed that a 20:1 ratio exists between bicarbonate and carbonic acid; by dividing the CO_2 combining power (expressed in Vol %) by 2.24, plasma bicarbonate concentration in mEq/liter is derived. A reduction in the plasma bicarbonate is usually sufficient to make a diagnosis of acidosis, although this may be erroneous since the ratio of carbonic acid to bicarbonate, which determines the blood pH, is not known.

An important illustration of the limitations of CO_2 content determinations is found in the chemical imbalance which prevails in salicylate poisoning. In the initial states, respiratory alkalosis occurs because of hyperventilation induced by the toxic effect of the drug on the respiratory center. Compensation produces a decrease in CO_2 content which by itself suggests the presence of metabolic acidosis; however, the blood pH will be found to be elevated above normal, which confirms the presence of an uncompensated respiratory alkalosis. Later, because of renal failure and other metabolic disturbances, a metabolic acidosis supervenes. This is of course aggravated by the lowered alkali reserve brought about by attempts at compensating for the pre-existing respiratory alkalosis. The time at which the metabolic acidosis occurs can be detected only by measurement of the blood pH, an observation which is of great importance in respect to electrolyte therapy. The sequence of events described above as characteristic of salicylate poisoning also occurs consequent to respiratory alkalosis from any other cause.

The Role of the Kidney in Acid-Base Balance

In addition to carbonic acid, which is eliminated by the respiratory organs as CO_2, other acids, which are not volatile, are produced by metabolic processes. These include lactic and pyruvic acids and the more important inorganic acids, hydrochloric, phosphoric, and sulfuric. About 50–150 mEq of these inorganic acids are eliminated by the kidneys in a 24-hour period. It is of course necessary that these acids be partially buffered with cation, largely sodium; but in the distal tubules of the kidney some of this cation is reabsorbed (actually exchanged for hydrogen ion), and the pH of the urine is allowed to fall. This acidification of the urine in the distal tubule is a valuable function of the kidney in conserving the reserves of cation in the body.

Another device used by the kidney to buffer acids and thus to conserve fixed base (cation) is the production of ammonia from amino acids. The ammonia is substituted for alkali cations, and the amounts of ammonia mobilized for this purpose may be markedly increased when the production of acid within the body is excessive (eg, as in metabolic acidosis such as occurs as a result of the ketosis of uncontrolled diabetes).

When alkali is in excess, the kidney excretes an alkaline urine to correct this imbalance. The details of the renal regulation of acid-base equilibrium are discussed in Chapter 18.

In kidney disease, glomerular and tubular damage results in considerable impairment of these important renal mechanisms for the regulation of acid-base balance. Tubular reabsorption of sodium in exchange for hydrogen is poor, and excessive retention of acid catabolites, such as phosphates and sulfates, occurs because of decreased glomerular filtration. In addition, the mechanism for ammonia production by the tubules is inoperative. As a result, acidosis is a common finding in nephritis.

• • •

Bibliography

Anderson OS: The acid-base status of the blood. Scandinav J Clin Lab Invest 15 (Suppl 70):1–134, 1963.

Astrup P: A new approach to acid-base metabolism. Clin Chem 7:1, 1961.

Best CH, Taylor NB: *Physiological Basis of Medical Practice,* 8th ed. Williams & Wilkins, 1966.

Blumentals AS: Symposium on acid-base balance. Arch Int Med 116:647–750, 1965.

Christensen, HN: *Body Fluids and the Acid-Base Balance.* Saunders, 1964.

Davenport HW: *The ABC of Acid-Base Chemistry,* 4th ed. Univ of Chicago Press, 1958.

Weisberg HF: *Water, Electrolyte and Acid-Base Balance,* 2nd ed. Williams & Wilkins, 1962.

Welt LG: *Clinical Disorders of Hydration and Acid-Base Equilibrium,* 2nd ed. Little, Brown, 1959.

12 . . .
Digestion & Absorption From the Gastrointestinal Tract

Most foodstuffs are ingested in forms which are unavailable to the organism, since they cannot be absorbed from the digestive tract until they have been reduced to smaller molecules. This breakdown of the naturally occurring foodstuffs into assimilable forms is the work of digestion.

The chemical changes incident to digestion are accomplished with the aid of the enzymes of the digestive tract. These enzymes catalyze the hydrolysis of native proteins to amino acids, of starches to monosaccharides, and of fats to glycerol and fatty acids. It is probable that in the course of these digestive reactions, the minerals and vitamins of the foodstuffs are also made more assimilable. This is certainly true of the fat-soluble vitamins, which are not absorbed unless fat digestion is proceeding normally.

DIGESTION IN THE MOUTH

Constituents of the Saliva

The oral cavity contains saliva secreted by 3 pairs of salivary glands: parotid, submaxillary, and sublingual. The saliva consists of about 99.5% water, although the content varies with the nature of the factors exciting its secretion. The saliva acts as a lubricant for the oral cavity, and by moistening the food as it is chewed it reduces the dry food to a semisolid mass which is easily swallowed. The saliva is also a vehicle for the excretion of certain drugs (eg, alcohol and morphine) and of certain inorganic ions such as K^+, Ca^{++}, HCO_3^-, iodine, and thiocyanate (SCN^-).

The pH of the saliva is usually slightly on the acid side, about 6.8, although it may vary on either side of neutrality.

Salivary Digestion

The saliva is relatively unimportant in digestion. It contains a starch-splitting enzyme, a salivary amylase known as **ptyalin**; but the enzyme is readily inactivated at pH 4.0 or less, so that digestive action on food in the mouth will soon cease in the acid environment of the stomach. Although saliva is capable of bringing about the hydrolysis of starch to maltose in the test tube, this actually is of little significance in the body because of the short time it can act on the food. Furthermore, other amylases of the intestine are capable of accomplishing complete starch digestion. In many animals, a salivary amylase is entirely absent.

DIGESTION IN THE STOMACH

Stimulation of Gastric Secretion

Gastric secretion is initiated by nervous or reflex mechanisms. The effective stimuli for these reflexes are similar to those which operate in salivary secretion. The continued secretion of gastric juice is, however, due to a hormonal stimulus, **gastrin** (gastric secretin). This chemical stimulant is produced by the gastric glands and absorbed into the blood, which carries it back to the stomach where it excites gastric secretion. Histamine, produced by decarboxylation of the amino acid, histidine, also acts as a potent gastric secretagogue.

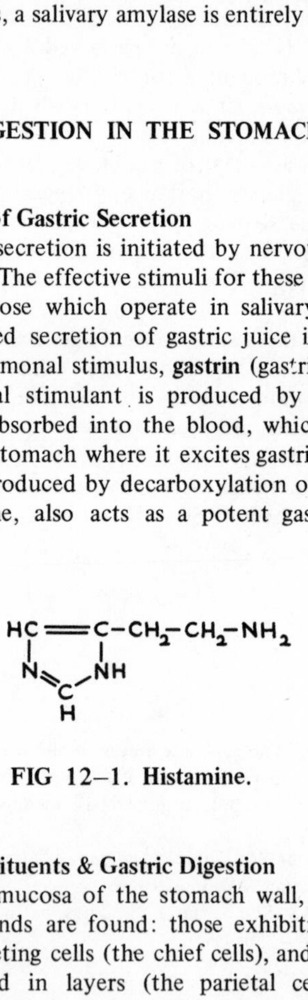

HC═C–CH₂–CH₂–NH₂ (imidazole ring: N, C–H, NH)

$$HC{=}C{-}CH_2{-}CH_2{-}NH_2 \quad (\text{ring: } HC{-}N{=}CH{-}NH{-}C)$$

FIG 12–1. Histamine.

Gastric Constituents & Gastric Digestion

In the mucosa of the stomach wall, 2 types of secretory glands are found: those exhibiting a single layer of secreting cells (the chief cells), and those with cells arranged in layers (the parietal cells), which secrete directly into the gastric glands. The mixed secretion is known as gastric juice. It is normally a clear, pale yellow fluid of high acidity, 0.2–0.5% HCl, with a pH of about 1.0.

A. Hydrochloric Acid: The parietal cells are the sole source of gastric hydrochloric acid. HCl is said to originate according to the reactions shown in Fig 12–2.

The process is essentially similar to that of the "chloride shift" described for the red blood cell on p 211. There is also a resemblance to the renal tubular mechanisms for secretion of H^+ as described on p 380, wherein the source of H^+ is also the carbonic anhydrase catalyzed formation of H_2CO_3 from H_2O and CO_2.

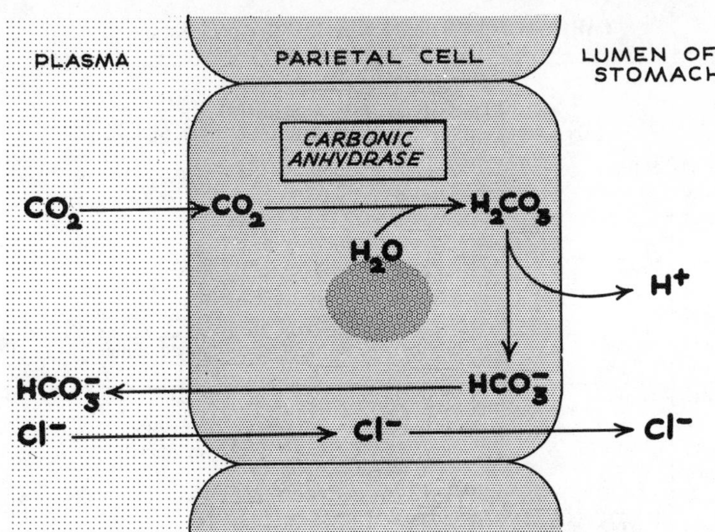

FIG 12–2. Production of gastric hydrochloric acid.

An alkaline urine often follows the ingestion of a meal ("alkaline tide"), presumably as a result of the formation of extra bicarbonate in the process of hydrochloric acid secretion by the stomach in accordance with the reaction shown in Fig 12–2.

The gastric juice is 97–99% water. The remainder consists of mucin and inorganic salts, the digestive enzymes (pepsin and rennin), and a lipase.

B. Pepsin: The chief digestive function of the stomach is the partial digestion of protein. Gastric pepsin is produced in the chief cells as the inactive zymogen, **pepsinogen,** which is activated to pepsin by the action of HCl and, autocatalytically, by itself, ie, a small amount of pepsin can cause the activation of the remaining pepsinogen. The enzyme transforms native protein into proteoses and peptones which are still reasonably large protein derivatives.

C. Rennin (Chymosin, Rennet): This enzyme causes the coagulation of milk. It is important in the digestive processes of infants because it prevents the rapid passage of milk from the stomach. In the presence of calcium, rennin changes irreversibly the casein of milk to a paracasein which is then acted on by pepsin. This enzyme is said to be absent from the stomachs of adults.

D. Lipase: The lipolytic action of gastric juice is not important, although a gastric lipase capable of mild fat-splitting action is found in gastric juice.

PANCREATIC & INTESTINAL DIGESTION

The stomach contents, or **chyme,** which are of a thick creamy consistency, are intermittently introduced during digestion into the duodenum through the pyloric valve. The pancreatic and bile ducts open into the duodenum at a point very close to the pylorus. The high alkaline content of pancreatic and biliary secretions neutralizes the acid of the chyme and changes the pH of this material to the alkaline side; this shift of pH is necessary for the activity of the enzymes contained in pancreatic and intestinal juice.

Stimulation of Pancreatic Secretion (Dreiling, 1956)

Like the stomach, the pancreas secretes its digestive juice almost entirely by means of hormonal stimulation. The hormones are secreted in the duodenum and upper jejunum as a result of stimulation by hydrochloric acid, fats, proteins, carbohydrates, and partially digested foodstuffs, and are carried by the blood to the pancreas, liver, and gallbladder after absorption from the small intestine through the portal blood. The active hormonal components of the duodenum (originally termed "secretin" by Bayliss & Starling) have now been separated into 5 separate factors: (1) secretin, which stimulates the production by the pancreas of a thin, watery fluid, high in bicarbonate but low in enzyme content; (2) pancreozymin, stimulating the production by the pancreas of a viscous fluid low in bicarbonate but high in enzyme content; (3) hepatocrinin, which causes the liver to secrete a thin, salt-poor type of bile; (4) cholecystokinin, which induces contraction and emptying of the gallbladder; and (5) enterocrinin, which induces the flow of succus entericus (intestinal juice).

Secretin, the first of the stimulatory factors described above, has recently been synthesized. The hormone is a polypeptide containing 27 amino acids.

Constituents of Pancreatic Secretion

Pancreatic juice is a nonviscid watery fluid which is similar to saliva in its content of water and contains some protein and other organic and inorganic compounds, mainly Na^+, K^+, HCO_3^-, and Cl^-. Ca^{++}, Zn^{++},

TABLE 12–1. Summary of digestive processes.

Source of Enzyme and Stimulus for Secretion	Enzyme	Method of Activation and Optimal Conditions for Activity	Substrate	End Products or Action
Salivary glands of mouth: Secrete saliva in reflex response to presence of food in mouth.	Salivary amylase	Chloride ion necessary. pH 6.6–6.8.	Starches	Maltose
Stomach glands: Chief cells and parietal cells secrete gastric juice in response to reflex stimulation and chemical action of gastrin.	Pepsin	Pepsinogen converted to active pepsin by HCl. pH 1.0–2.0.	Protein	Proteoses Peptones
	Rennin	Calcium necessary for activity. pH 4.0.	Casein of milk	Coagulates milk
Pancreas: Presence of acid chyme from the stomach activates duodenum to produce (1) secretin, which hormonally stimulates flow of pancreatic juice; (2) pancreozymin, which stimulates the production of enzymes.	Trypsin	Trypsinogen converted to active trypsin by enterokinase of intestine at pH 5.2–6.0. Autocatalytic at pH 7.9.	Protein Proteoses Peptones	Polypeptides Dipeptides
	Chymotrypsin	Secreted as chymotrypsinogen and converted to active form by trypsin. pH 8.0.	Protein Proteoses Peptones	Same as trypsin. More coagulating power for milk.
	Carboxypeptidase	Secreted as procarboxypeptidase activated by trypsin.	Polypeptides with free carboxyl groups	Lower peptides. Free amino acids.
	Amylase (amylopsin)	pH 7.1	Starch	Maltose
	Lipase (steapsin)	Activated by bile salts? pH 8.0.	Primary ester linkages of fats	Fatty acids, monoglycerides, diglycerides, glycerol
	Ribonuclease		Ribonucleic acid	Nucleotides
	Deoxyribonuclease		Deoxyribonucleic acids	Nucleotides
	Cholesterol esterase	Activated by bile salts.	Free cholesterol	Esters of cholesterol with fatty acids
Liver and gallbladder	Bile salts and alkali	Cholecystokinin and hepatocrinin–hormones from intestine; stimulate gallbladder and secretion of bile by the liver.	Fats–also neutralize acid chyme	Fatty acid-bile salt conjugates and finely emulsified neutral fat
Small intestine: Secretions of Brunner's glands of the duodenum and glands of Lieberkühn. Enterocrinin induces flow of succus entericus.	Aminopeptidase		Polypeptides with free amino groups	Lower peptides. Free amino acids.
	Dipeptidases		Dipeptides	Amino acids
	Sucrase	pH 5.0–7.0	Sucrose	Fructose, glucose
	Maltase	pH 5.8–6.2	Maltose	Glucose
	Lactase	pH 5.4–6.0	Lactose	Glucose, galactose
	Phosphatase	pH 8.6	Organic phosphates	Free phosphate
	Isomaltase or 1:6 glucosidase		1:6 glucosides	Glucose
	Polynucleotidase		Nucleic acid	Nucleotides
	Nucleosidases		Purine or pyrimidine nucleosides	Purine or pyrimidine bases, pentose-phosphate
	Lecithinase		Lecithin	Glycerol, fatty acids, phosphoric acid, choline

$HPO_4^=$, and $SO_4^=$ are present in small amounts. The pH of pancreatic juice is distinctly alkaline, 7.5–8.0 or higher.

The enzymes contained in pancreatic juice include trypsin, chymotrypsin, and carboxypeptidases, alpha-amylase, lipase, cholesterol esterase, ribonuclease, deoxyribonuclease, and collagenase. Some of these enzymes are secreted as inactive precursors (zymogens) such as trypsinogen or chymotrypsinogen, but are activated on contact with the intestinal mucosa. The activation of trypsinogen is attributed to **enterokinase**, which is produced by the intestinal glands. A small amount of active trypsin then autocatalytically activates additional trypsinogen and chymotrypsinogen.

A. Trypsin and Chymotrypsin: The protein-splitting action of pancreatic juice (proteolytic action) is due to trypsin and chymotrypsin, which attack native protein, proteoses, and peptones from the stomach to produce polypeptides. Chymotrypsin has more coagulative power for milk than trypsin, and, as previously noted, it is activated not by enterokinase but by active trypsin.

B. The "Peptidases": The further attack on protein breakdown products, ie, on the polypeptides, is accomplished by a mixture of **carboxypeptidase**, a zinc-containing enzyme of the pancreatic juice, and **aminopeptidase** and **dipeptidase** of the intestinal juices. Such a mixture was formerly termed "erepsin." The carboxypeptidase is an exopeptidase hydrolyzing the terminal peptide bond at the carboxyl end of the polypeptide chain. Likewise, the aminopeptidase attacks the terminal peptide bond at the free amino end of the chain. This system of intestinal proteases converts food proteins into their constituent amino acids for absorption by the intestinal mucosa and transfer to the circulation.

C. Amylase: The starch-splitting action of pancreatic juice is due to a pancreatic alpha-amylase (amylopsin). It is similar in action to salivary amylase, hydrolyzing starch to maltose at an optimum pH of 7.1.

D. Lipase: Fats are hydrolyzed by a pancreatic lipase (steapsin) to fatty acids, glycerol, monoglycerides, and diglycerides. This is an important enzyme in digestion. It is possibly activated by bile salts.

E. Cholesterol Esterase: This enzyme may catalyze either the esterification of free cholesterol with fatty acids or, depending upon the conditions of equilibrium, cholesterol esterase may catalyze the opposite reaction, ie, hydrolysis of cholesterol esters. According to Goodman (1965), under the conditions existing within the lumen of the intestine, cholesterol esterase catalyzes the hydrolysis of cholesterol esters which are thus mainly absorbed from the intestine in a non-esterified, free form.

F. Ribonuclease (RNase) and **deoxyribonuclease (DNase)** have been prepared from pancreatic tissue (see Chapter 4).

Constituents of Intestinal Secretions

The intestinal juice secreted, under the influence of enterocrinin (see p 217), by the glands of Brunner and of Lieberkühn also contains digestive enzymes. In addition to the proteolytic enzymes already mentioned, these include the following:

(1) The specific disaccharidases, ie, sucrase, maltase (including an isomaltase for splitting 1:6 glycosidase linkages), and lactase, which convert sucrose, maltose, or lactose into their constituent monosaccharides for absorption.

(2) A phosphatase, which removes phosphate from certain organic phosphates such as hexosephosphates, glycerophosphate, and the nucleotides (a nucleotidase) derived from the diet.

(3) Polynucleotidases (nucleinases, phosphodiesterases), which split nucleic acids into nucleotides.

(4) Nucleosidases (nucleoside phosphorylases), one of which attacks only purine-containing nucleosides, liberating adenine or guanine and the pentose sugar which is simultaneously phosphorylated. The pyrimidine nucleosides (uridine, cytidine, and thymidine) are broken down by another enzyme which differs from the purine nucleosidase.

(5) The intestinal juice is also said to contain a lecithinase, which attacks lecithins to produce glycerol, fatty acids, phosphoric acid, and choline.

The Results of Digestion

The final result of the action of the digestive enzymes already described is to reduce the foodstuffs of the diet to forms which can be absorbed and assimilated. These end products of digestion are, for carbohydrates, the monosaccharides (principally glucose); for proteins, the amino acids; and for fats, the fatty acids, glycerol, and the monoglycerides, although some unhydrolyzed fat is probably also absorbed.

THE BILE

In addition to many functions in intermediary metabolism, the liver, by producing bile, plays an important role in digestion. The gallbladder, a saccular organ attached to the hepatic duct, stores a certain amount of the bile produced by the liver between meals. During digestion, the gallbladder contracts and supplies bile rapidly to the small intestine by way of the common bile duct. The pancreatic secretions mix with the bile, since they empty into the common duct shortly before its entry into the duodenum.

Composition of Bile

The composition of hepatic bile differs from that of gallbladder bile. As shown in Table 12–2, the latter is more concentrated.

TABLE 12–2. The composition of hepatic and of gallbladder bile.

	Hepatic Bile (as secreted)		Bladder Bile
	Percent of Total Bile	Percent of Total Solids	Percent of Total Bile
Water	97.00	...	85.92
Solids	2.52	...	14.08
Bile acids	1.93	36.9	9.14
Mucin and pigments	0.53	21.3	2.98
Cholesterol	0.06	2.4	0.26
Fatty acids and fat	0.14	5.6	0.32
Inorganic salts	0.84	33.3	0.65
Specific gravity	1.01	...	1.04
pH	7.1–7.3	...	6.9–7.7

Stimulation of Gallbladder & Bile Formation

Contraction of the gallbladder and relaxation of its sphincter are initiated by a hormonal mechanism. This hormone, **cholecystokinin** (see Table 12–1), is secreted by the intestine in response to the presence of foods, mainly meats and fats. Bile salts and various other chemical substances, such as calomel, act as stimulants to bile flow. These biliary stimulants are known as **cholagogues.** A hormone of duodenal mucosal origin (hepatocrinin; see Table 12–1), is also active in stimulation of hepatic secretion of bile.

Bile Acids

Because of their detergent and emulsifying effects on fats, the conjugated bile acids are important in digestion. Four bile acids have been isolated from human bile. Cholic acid, whose structure is shown below, is the acid which is found in the largest amounts in the bile itself. Its structure is that of a completely saturated sterol, having 3 OH groups on the nucleus (at positions 3, 7, and 12). The other bile acids are deoxycholic acid, which lacks the OH group at position 7; chenodeoxycholic acid, lacking an OH at position 12; and lithocholic acid, which has only one OH group (at position 3). Deoxycholic acid is the main bile acid found in the feces of normal adult human beings.

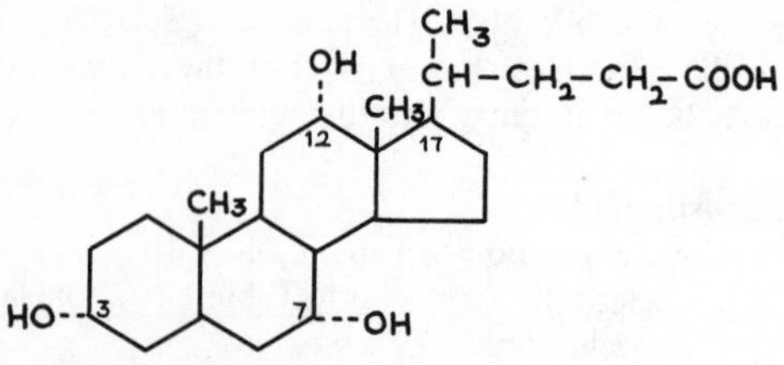

FIG 12–3. Cholic acid.

The bile acids are important end products in the metabolism of cholesterol which the liver removes from the blood. However, cholesterol itself is also present in the bile. This is a reflection of the synthesis of cholesterol by the liver. Measurement of the output of bile acids is the most accurate means of estimating the amount of cholesterol lost from the body by removal from the blood.

The mechanism of formation of the bile acids from cholesterol by rat liver mitochondria has been studied by Suld & others (1962).

The bile acids are not excreted in the bile in the free state but are conjugated by the liver with glycine or taurine, a cysteine derivative. In this conjugated form the bile acids are water-soluble. The conjugation with glycine or taurine occurs through the carboxyl group on the side-chain. Because of the alkalinity of the bile, the conjugated bile acids may also be largely neutralized with sodium or potassium to form the glycocholates or taurocholates. These are the so-called bile salts; they exert a powerful emulsifying and surface tension lowering effect within the intestine and thus aid in the digestion of fats.

The conjugation of cholic acid has been found to require preliminary activation with coenzyme A (Fig 7–5). The activating enzyme (Siperstein, 1956) which occurs only in the microsomes of the liver catalyzes the following reaction:

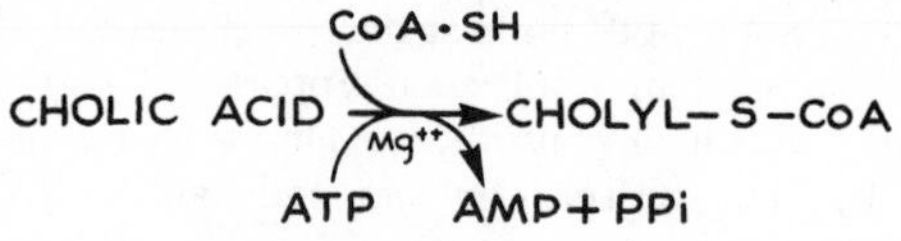

$$\text{CHOLIC ACID} + \text{CoA·SH} + \text{ATP} \xrightarrow{Mg^{++}} \text{CHOLYL–S–CoA} + \text{AMP} + \text{PPi}$$

A second enzyme in the liver catalyzes conjugation of cholyl-Co A with taurine.

Functions of the Bile System

A. Emulsification: The bile salts have considerable ability to lower the surface tension of water. This enables them to emulsify fats in the intestine and to dissolve fatty acids and water-insoluble soaps. The presence of bile in the intestine is necessary to accomplish the digestion and absorption of fats as well as the absorption of the fat-soluble vitamins A, D, E, and K. When fat digestion is impaired, other foodstuffs are also poorly digested, since the fat covers the food particles and prevents enzymes from attacking them. Under these conditions, the activity of the intestinal bacteria causes considerable putrefaction and production of gas.

B. Neutralization of Acid: In addition to its functions in digestion, the bile is a reservoir of alkali, which helps to neutralize the acid chyme from the stomach.

C. Excretion: Bile is also an important vehicle of excretion. It removes many drugs, toxins, bile pigments, and various inorganic substances such as copper, zinc, and mercury. Cholesterol, either derived from the diet or synthesized by the body, is eliminated almost entirely in the bile as cholesterol itself or as cholic acids. Free cholesterol is not soluble in water but is emulsified by the bile salts. Very often it precipitates from the bile and forms stones in the gallbladder or ducts. These stones, or calculi, may also contain a mixture of cholesterol and calcium, although pure cholesterol stones are the most common, probably because of the inability of the gallbladder to cope with an excess of cholesterol.

D. Bile Pigment Metabolism: The origin of the bile pigments from hemoglobin is discussed on p 85. Further consideration of bile pigment metabolism will be given in the discussion of liver function (see Chapter 17).

INTESTINAL PUTREFACTION & FERMENTATION

Most ingested food is absorbed from the small intestine. The residue passes into the large intestine. Here considerable absorption of water takes place, and the semiliquid intestinal contents gradually become more solid. During this period, considerable bacterial activity occurs. By fermentation and putrefaction, the bacteria produce various gases, such as CO_2, methane, hydrogen, nitrogen, and hydrogen sulfide, as well as acetic, lactic, and butyric acids. The bacterial decomposition of lecithin may produce choline and related toxic amines such as neurine.

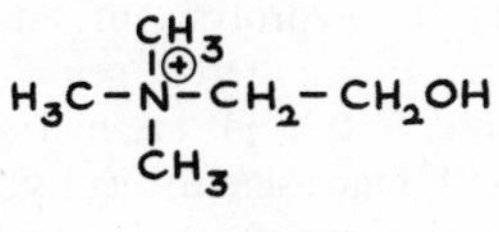

$$H_3C-\overset{CH_3}{\underset{CH_3}{\overset{|}{\underset{|}{N^{\oplus}}}}}-CH_2-CH_2OH$$

Choline

$$H_3C-\overset{CH_3}{\underset{CH_3}{\overset{|}{\underset{|}{N^{\oplus}}}}}-CH=CH_2$$

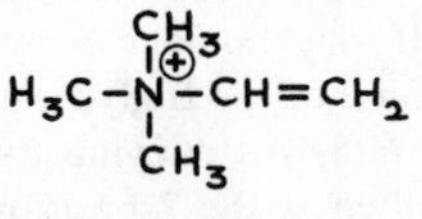

Neurine

Fate of Amino Acids

Many amino acids undergo decarboxylation as a result of the action of intestinal bacteria to produce toxic amines (ptomaines).

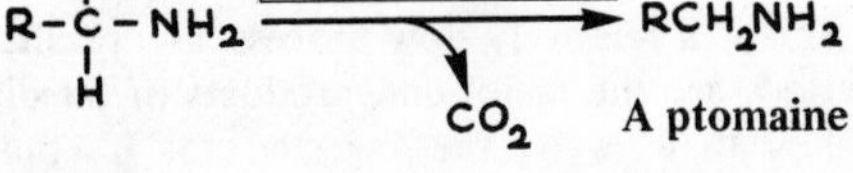

$$R-\underset{H}{\overset{COO\,H}{\overset{|}{\underset{|}{C}}}}-NH_2 \xrightarrow[\searrow CO_2]{\text{BACTERIAL DECARBOXYLASE}} RCH_2NH_2$$

A ptomaine

Such decarboxylation reactions produce cadaverine from lysine; agmatine from arginine; tyramine from tyrosine; putrescine from ornithine; and histamine from histidine. Many of these amines are powerful vasopressor substances.

The amino acid tryptophan undergoes a series of reactions to form indole and methylindole (skatole), the substances particularly responsible for the odor of the feces.

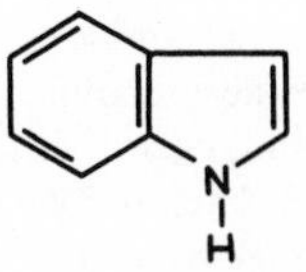

Indole 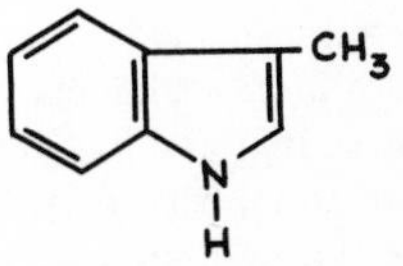 **Skatole**

The sulfur-containing amino acid cysteine undergoes a series of transformations to form mercaptans such as ethyl and methyl mercaptan as well as H_2S.

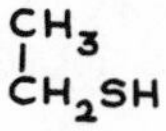

$$\underset{CH_2SH}{\overset{CH_3}{|}} \qquad CH_3SH$$

Ethyl mercaptan **Methyl mercaptan**

$$CH_3SH \xrightarrow{[2H]} CH_4 + H_2S$$

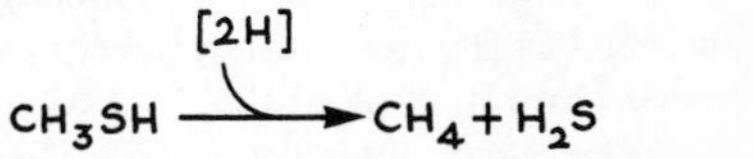

Methyl mercaptan **Methane and hydrogen sulfide**

The large intestine is a source of considerable quantities of ammonia, presumably as a product of the putrefactive activity on nitrogenous substrates by the intestinal bacteria. This ammonia is absorbed into the portal circulation, but under normal conditions it is rapidly removed from the blood by the liver. In liver disease this function of the liver may be impaired, in which case the concentration of ammonia in the peripheral blood will rise to toxic levels. It is believed that ammonia intoxication may play a role in the genesis of hepatic coma in some patients. In dogs on whom an Eck fistula has been performed (complete diversion of the portal blood to the vena cava), the feeding of large quantities of raw meat will induce symptoms of ammonia intoxication (meat intoxication) accompanied by elevated levels of ammonia in the blood. The oral administration of neomycin has been shown to reduce the quantity of ammonia delivered from the intestine to the blood (Silen, 1955). This is undoubtedly due to the antibacterial action of the drug. The feeding of high-protein diets to patients suffering from advanced liver disease, or the occurrence of gastrointestinal hemorrhage in such patients, may contribute to the development of ammonia intoxication. Under these circumstances neomycin is also beneficial.

Intestinal Bacteria

The intestinal flora may comprise as much as 25% of the dry weight of the feces. In herbivora, whose diet consists largely of cellulose, the intestinal or ruminal bacteria are essential to digestion, since they decompose this polysaccharide and make it available for absorption. In addition, these symbiotic bacteria may accomplish the synthesis of essential amino acids for these animals. In man, although the intestinal flora is not as important as in the herbivora, nevertheless some nutritional benefit is derived from bacterial activity in the synthesis of certain vitamins, particularly vitamin K, and possibly certain members of the B complex, which are made available to the body. Information gained from current experiments with animals raised under strictly aseptic conditions should help to define further the precise role of the intestinal bacteria.

ABSORPTION FROM THE GASTROINTESTINAL TRACT

There is little absorption from the stomach, even of foodstuffs like glucose which can be absorbed directly from the intestine. Although water is not absorbed to any extent from the stomach, considerable gastric absorption of alcohol is possible.

The small intestine is the main digestive and absorptive organ. About 90% of the ingested foodstuffs is absorbed in the course of passage through the approximately 25 feet of small intestine. Water is absorbed from the small intestine at the same time. Considerably more water is absorbed after the foodstuffs pass into the large intestine, so that the contents, which were fluid in the small intestine, gradually become more solid in the colon.

There are 2 general pathways for the transport of materials absorbed by the intestine: the veins of the portal system, which lead directly to the liver; and the lymphatic vessels of the intestinal area, which eventually lead to the blood by way of the lymphatic system and the thoracic duct.

Absorption of Carbohydrates

Carbohydrates are absorbed from the intestine into the blood of the portal venous system in the form of monosaccharides, chiefly the hexoses (glucose, fructose, mannose, and galactose), although the pentose sugars, if present in the food ingested, will also be absorbed. The passage of the hexose sugars across the intestinal barriers occurs at a fixed rate even against an osmotic gradient and independent of their concentration in the intestinal lumen. Furthermore, they are absorbed much faster than the pentose sugars. The constant and rapid absorption of the hexoses indicates that a mechanism other than simple diffusion is operating in the intestinal absorption of these sugars. The exact biochemical mechanism by which the hexose sugars are absorbed by the intestine against a concentration gradient has yet to be established. It has been assumed that the absorptive mechanism involves phosphorylation of the sugars as a necessary preliminary to absorption. Much of the experimental evidence on which this assumption was based was derived from studies in which monoiodoacetic acid or phlorhizin was used with a view to specific inhibition of the phosphorylating enzymes. Landau & Wilson (1959) point out, however, that iodoacetate is not a specific inhibitor but reacts with the sulfhydryl (SH) groups of many enzymes (see p 138) and that phlorhizin does not have as its primary site of action the phosphorylating enzyme, hexokinase, or that of a dephosphorylating enzyme such as a phosphatase. Instead, phlorhizin seems to act as an inhibitor at some point in an oxidative pathway. It is concluded, however, that although much of the original evidence for phosphorylation and dephosphorylation in the course of absorption of sugars from the intestine has proved inadequate, conclusive evidence against it is still lacking.

The rate of absorption of different hexoses is not the same. If the rate of absorption of glucose is taken as 100, the absorption rates of certain other sugars are as follows (data from Cori, using the rat as experimental animal):

Galactose:	110	Mannose:	19
Glucose:	100	Xylose:	15
Fructose:	43	Arabinose:	9

Absorption of Fats

The complete hydrolysis of fats (triglycerides) produces glycerol and fatty acids. However, the second and third fatty acids are hydrolyzed from the triglycerides with increasing difficulty, the removal of the last fatty acid requiring special conditions. Mattson & Beck (1956) have presented evidence that pancreatic lipase is virtually specific for the hydrolysis of primary ester linkages. If this lipase can hydrolyze the ester linkage at position 2 of the triglyceride at all, it does so at a very slow rate. An enzyme that can hydrolyze fatty acids esterified at the 2 position of a triglyceride was, however, found in rat pancreatic juice by Mattson & Volpenhein (1968). This enzyme is not pancreatic lipase; it may be sterol ester hydrolase.

Because of the difficulty of hydrolysis of the secondary ester linkage in the triglyceride, it is suggested that the digestion of a triglyceride proceeds first by removal of a terminal fatty acid to produce an α,β-diglyceride; the other terminal fatty acid is then removed to produce a β-monoglyceride. Since this last fatty acid is linked by a secondary ester group, its removal requires isomerization to a primary ester linkage. This is a relatively slow process; as a result, monoglycerides are the major end products of fat digestion and less than ¼ of the ingested fat is completely broken down to glycerol and fatty acids.

Within the intestinal wall, α-monoglycerides are further hydrolyzed to produce free glycerol and fatty acids, whereas β-monoglycerides may be reconverted to triglycerides. The utilization of fatty acids for resynthesis of triglycerides requires first their "activation." This is accomplished by formation of a coenzyme A (acyl) derivative of the fatty acid (see p 103). The reaction (which also requires ATP) is catalyzed by the enzyme **thiokinase.**

$$R-COOH \xrightarrow[Mg^{++}]{\text{THIOKINASE}} R-\overset{O}{\overset{\|}{C}}\sim S-CoA$$

$$ATP \quad AMP + PP$$

An ATP-dependent fatty acid thiokinase has been shown to be present in the mucosal cells of the intestine. It is likely, therefore, that the synthesis of triglycerides proceeds in the intestinal mucosa in a manner similar to that which takes place in other tissues, as described on p 271.

The free glycerol released in the intestinal lumen (from approximately 22% of the total amount of triglyceride originally present) is not reutilized but passes directly to the portal vein. However, the glycerol released within the intestinal wall cells can be reutilized for triglyceride synthesis by combining with free fatty acids present in the intestinal wall as a result of absorption from the intestine or produced by hydrolysis of monoglycerides within the intestinal wall. Thus, normally, all free fatty acids present in the intestinal wall are ultimately reincorporated into triglycerides which then are transported to the lymphatic vessels of the abdominal region (the so-called lacteals) for distribution to the rest of the body.

All of the above factors relating to digestion and absorption of lipid are shown in Fig 12–4 (Mattson & Volpenhein, 1964).

Triglycerides, having been synthesized in the intestinal mucosa after absorption as described above, are not transported to any extent in the portal venous blood. Instead, the great majority of absorbed fat appears in the form of **chylomicrons** which appear first in the lymphatic vessels of the abdominal region and later in the systemic blood. Chylomicrons are synthesized in the intestinal wall, and, although they contain largely triglyceride, cholesterol (both free and esterified), phospholipid, and a small (0.5%) but important amount of protein are also present. Indeed, administration of an inhibitor of protein synthesis (eg, puromycin) to rats prevents the formation of chylomicrons and results in accumulation of fat in the intestinal epithelial cells (Sabesin & Isselbacher, 1965).

The majority of absorbed fatty acids of more than 10 carbon atoms in length, irrespective of the form in which they are absorbed, are found as esterified fatty acids in the lymph of the thoracic duct. Fatty acids with carbon chains shorter than 10–12 carbons are transported in the portal venous blood as unesterified (free) fatty acids. This was demonstrated in experiments in which labeled fatty acids were fed either as free acids or in neutral fat. About 90% of labeled palmitic (C_{16}) acid was found in the lymph after it was fed to experimental animals. Stearic (C_{18}) acid was poorly absorbed. Myristic (C_{14}) acid was well absorbed, mostly into the lymph; however, while almost all of the lauric (C_{12}) acid or decanoic (C_{10}) acid was absorbed, these were not found to any extent in the lymph. Presumably they passed into the portal blood. These lower fatty acids are, however, not important constituents of fats ordinarily taken in the diet except in the fats of milk.

Chyluria is an abnormality in which the patient excretes milky urine because of the presence of an abnormal connection between the urinary tract and the lymphatic drainage system of the intestine, a so-called "chylous fistula." In a similar abnormality, **chylothorax,** there is an abnormal connection between the pleural space and the lymphatic drainage of the small intestine which results in the accumulation of milky pleural fluid. As noted above, fatty acids of chain lengths less than 10–12 carbon atoms are transported almost entirely in the portal blood rather than in the lymph. With this fact in mind, Hashim & others (1964) compared the effect on patients with chylous fistulas of feeding triglycerides in which the fatty acids were of medium chain length (less than 12 carbons) with that of feeding dietary fat. Substitution of the medium chain length fatty acids for other fat resulted in a disappearance of chyluria which reappeared when corn oil was given. In the patient with chylothorax, the use of triglyceride with short chain fatty acids resulted in the appearance of clear pleural fluid as well as a reduction in its accumulation so that less frequent removal of the fluid by thoracentesis was required. There was also a reduction in protein losses to the body which would otherwise occur because of the presence of protein in the chylous fluid whether it accumulates in the pleural cavity or is lost into the urine.

A lipase distinct from that of the pancreas is present in the intestinal mucosa. The properties of this enzyme have recently been studied by DiNella (1960). It is probable that the principal action of the intestinal lipase is not within the lumen of the intestine, as is the case with pancreatic lipase, but within the cells of the intestinal wall. Here it may continue hydrolysis of monoglycerides that are not readily split by pancreatic lipase and thus prepare fatty acids for resynthesis into neutral fat as described above. That resynthesis occurs from free fatty acids (after formation of the "active" Co A derivatives) seems evident from the observation of Borgstrom (1955) that ^{14}C-labeled palmitic acid given orally in olive oil to rats was later found to be randomly distributed in the 1, 2, and 3 positions of the glycerides recovered from the thoracic duct lymph.

There is evidence that unhydrolyzed fat can be absorbed if it is dispersed in very fine particles (not over 0.5 μm . diameter). A combination of bile salts,

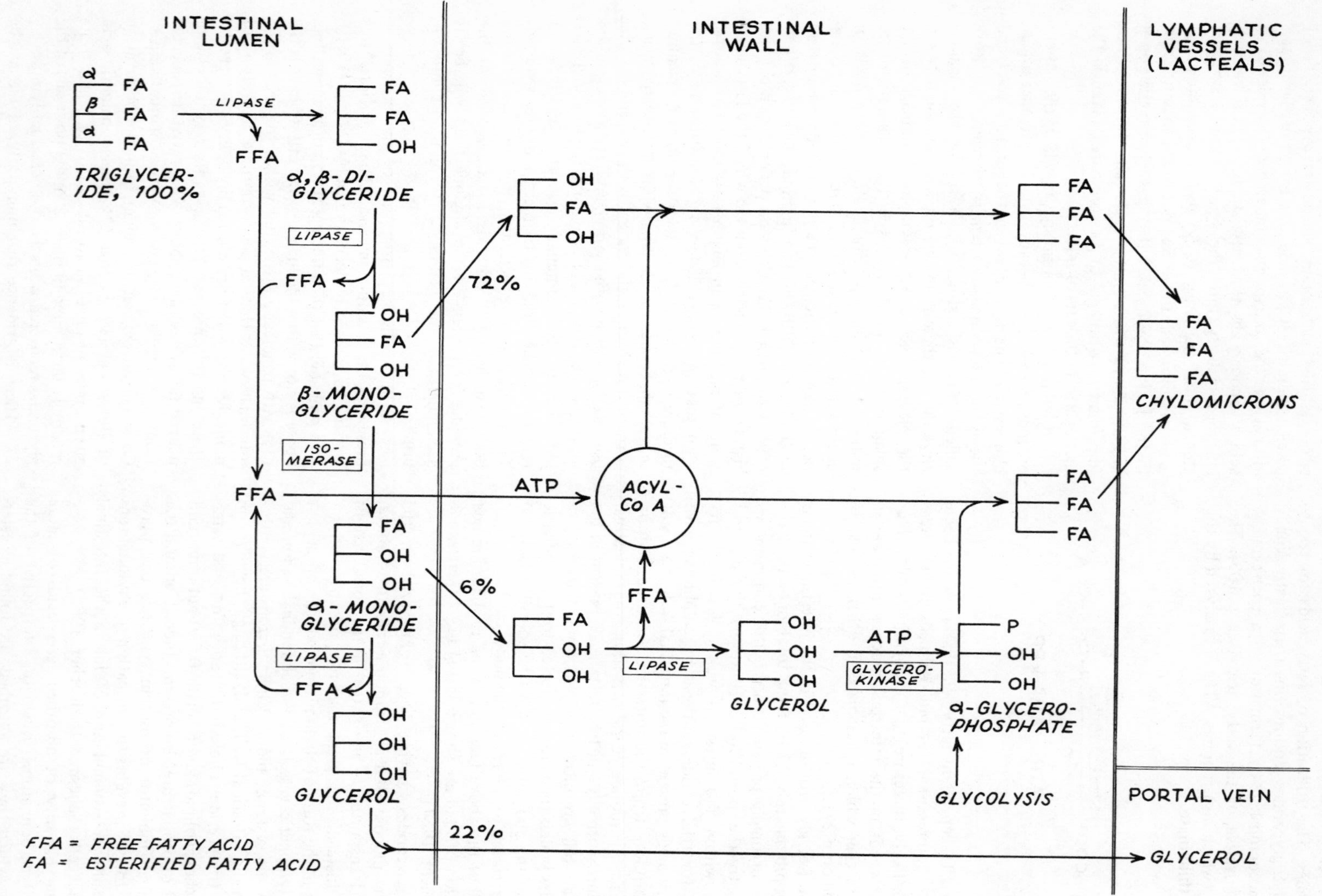

FIG 12–4. Chemical mechanisms of digestion and absorption of triglycerides.
(Modified from Mattson & Volpenhein, J Biol Chem 239:2772, 1964.)

fatty acids, and a monoglyceride will bring about this fine degree of dispersion of neutral fats. Certain synthetic "wetting agents" such as Tween 80 (sorbitan monooleate) have a similar effect and are used therapeutically to promote fat absorption. Since bile salts play such an important role in the absorption of fats, it is obvious that fat absorption is seriously hampered by a lack of bile in the intestine such as results when the bile duct is completely obstructed.

Cholesterol appears to be absorbed from the intestine almost entirely in the free (unesterified) form (Goodman, 1965). Nonetheless, 85–90% of the cholesterol in the lymph is in the esterified form, indicating that esterification of cholesterol, like that of fatty acids, must take place within the intestinal mucosal cells.

Summary of Fat Absorption (Senior, 1964.)

The dietary fat is digested, by the action of the pancreatic lipase present in the intestine, partially to glycerol and fatty acids and partially to split products such as monoglycerides and diglycerides. With the aid of the bile salts, these products of fat digestion enter the mucosal cells of the small intestine where digestion of fats may be completed through the action of the intestinal lipase, thus liberating the free fatty acids and glycerol. Resynthesis of triglycerides now occurs, utilizing the partial glycerides and the liberated free fatty acids. Surplus free fatty acids are esterified with α-glycerophosphate. The resynthesized fat then passes into the lymphatics (the lacteals) of the abdominal cavity and thence by way of the thoracic duct to the blood, where it may be detected as lipoprotein particles about 0.5 μm in diameter, the so-called chylomicrons. The bile salts are carried by the portal blood to the liver and excreted in the bile back to the intestine; this is the so-called enterohepatic circulation of the bile salts.

Phospholipid synthesis and turnover increase in the intestinal wall during the absorption of fat, although there is no evidence that they contribute substantially to lipid transport in the lymph or blood. Phospholipid obtained from the diet may be absorbed as such because of its hydrophilic nature; it may then travel by way of the portal blood directly to the liver.

Cholesterol is absorbed into the lymphatics and recovered therein mainly as cholesterol esters, although both free and esterified cholesterol are found in the blood plasma. Of the plant sterols (phytosterols), none is absorbed from the intestine except ergosterol, which is absorbed after it has been converted by irradiation to a vitamin D.

Absorption of Amino Acids & Protein

It is probable that under normal circumstances the dietary proteins are almost completely digested to their constituent amino acids and that these end products of protein digestion are then rapidly absorbed from the intestine into the portal blood. The amino acid content of the portal blood rises during the absorption of a protein meal, and the rise of individual amino acids in the blood a short time after their oral administration can be readily detected. It is possible that some hydrolysis, eg, of dipeptides, is completed in the intestinal wall. Animals may be successfully maintained in a satisfactory nutritional state with respect to protein when a complete amino acid mixture is fed to them. This indicates that intact protein is not necessary.

There is a difference in the rate of absorption from the intestine of the 2 isomers of an amino acid. The natural (L) isomer is actively transported across the intestine from the mucosa to the serosa; vitamin B_6 (pyridoxal phosphate) is involved in this transfer (see Chapter 7). The D-isomers, on the other hand, are transported only by free diffusion. This active transport of the L-amino acids is energy-dependent, as evidenced by the fact that, in studies of small pieces of segmented intestine, 2,4-dinitrophenol (DNP) inhibits the concentration of L-amino acids. There is no such effect with the D-amino acids. It will be recalled (see p 40) that DNP acts as an uncoupling agent in oxidative phosphorylation and thus interferes with production of ATP, which, in this case, is the energy source for active transport.

A valuable tool for the study of amino acid transport is the synthetic amino acid, α-aminoisobutyric acid. This compound is transported across cell membranes as are the natural amino acids; but once within the cells it cannot be metabolized, so that it remains for identification and analysis. Another amino acid model has also been used to study transport of amino acids across the intestine (Akedo & Christensen, 1962). This compound, 1-aminocyclopentane-1-carboxylic acid, was found to behave somewhat differently from α-aminoisobutyric acid in intestinal transport. The latter resembled glycine, whereas the former behaved like valine and methionine in respect to transport across the intestinal wall.

When groups of amino acids are fed, there is some evidence of competitive effects whereby one amino acid fed in excess can retard the absorption of another. These observations are similar to those made with respect to reabsorption of amino acids by the renal tubules.

The absorption of small peptide fragments from the intestine is undoubtedly possible, and it is very likely that this normally occurs. During the digestion and absorption of protein, an increase in the peptide nitrogen of the portal blood has in fact been found.

A puzzling feature of protein absorption is that in some individuals sensitivity to protein (in the immunologic sense) results when they eat certain proteins. It is known that a protein is antigenic, ie, able to stimulate an immunologic response, only if it is in the form of a relatively large molecule; the digestion of a protein even to the polypeptide stage destroys its antigenicity. Those individuals in which an immunologic response to ingested protein occurs must therefore be able to absorb some unhydrolyzed protein. This is not entirely undocumented, since the antibodies of the colostrum are known to be available to the infant.

There is increasing support for the hypothesis that the basic defect in nontropical sprue is located within the mucosal cells of the intestine and that it consists of an enzymatic defect which permits the polypeptides resulting from the peptic and tryptic digestion of gluten, the principal protein of wheat, not only to exert a local harmful effect within the intestine but also to be absorbed into the circulation and thus to elicit the production of antibodies. It has been definitely established that circulating antibodies to wheat gluten or its fractions are frequently present in patients with nontropical sprue (Alarcón-Segovia, 1964). Efforts to characterize chemically the peptic-tryptic digest of gluten which appears to be the harmful entity in the sprue patient have resulted in the finding that the digest is a mixture of peptides with a mean molecular weight between 820 and 928 daltons, suggesting that they are composed of 6 or 7 amino acids of which glutamine and proline must be present to ensure the harmful properties of the peptide. Mild acid hydrolysis which results in deamidation of glutamine renders the peptide harmless.

These observations on nontropical sprue, a disease entity which is undoubtedly the adult analogue of celiac disease in children, are further evidence for the possibility that protein fragments of larger molecular size than amino acids are indeed absorbed from the intestine under certain conditions.

• • •

References

Akedo H, Christensen HN: J Biol Chem 237:113, 1962.
Alarcón-Segovia D, Herskovic T, Wakim KG, Green PA, Scudamore HH: Am J Med 36:485, 1964.
Borgstrom B: J Biol Chem 214:671, 1955.
DiNella RR, Meng HC, Park CR: J Biol Chem 235:3076, 1960.
Dreiling DA, Janowitz, HD: Am J Med 21:98, 1956.
Goodman DS: Physiol Rev 45:747, 1965.
Hashim SA, Roholt HB, Babayan VK, Van Itallie TB: New England J Med 270:756, 1964.
Landau BR, Wilson TH: J Biol Chem 234:749, 1959.
Mattson FH, Beck LW: J Biol Chem 219:735, 1956.
Mattson FH, Volpenhein RA: J Biol Chem 239:2772, 1964.
Mattson FH, Volpenhein RA: J Lipid Res 9:79, 1968.
Reiser R, Williams MC: J Biol Chem 202:815, 1953.
Sabesin SM, Isselbacher KJ: Science 147:1149, 1965.
Senior JR: J Lipid Res 5:495, 1964.
Senior JR, Isselbacher KJ: J Biol Chem 237:1454, 1962.
Silen W, Harper HA, Mawdsley DL, Weirich WL: Proc Soc Exp Biol Med 88:138, 1955.
Siperstein MD, Murray AW: Science 123:377, 1956.
Suld HM, Staple E, Gurin S: J Biol Chem 237:338, 1962.

13...

Metabolism of Carbohydrate

With Peter Mayes, PhD, DSc*

Although the human diet is variable, in most instances carbohydrate accounts for a large proportion of the daily intake. However, much of the dietary carbohydrate is converted to fat and consequently is metabolized as fat. The extent of this process (lipogenesis) depends on whether or not the animal is a "meal eater" or a more continuous feeder. It is possible that in man the frequency of taking meals and the extent to which carbohydrates are converted to fat could have a bearing on disease states such as atherosclerosis, obesity, and diabetes mellitus (Cohn, 1960). In herbivores, especially ruminants, much of the intake of carbohydrate is fermented by microorganisms to lower fatty acids prior to absorption from the alimentary tract.

The major function of carbohydrate in metabolism is as a fuel to be oxidized and provide energy for other metabolic processes. In this role, carbohydrate is utilized by cells mainly in the form of glucose. The 3 principal monosaccharides resulting from the digestive processes are glucose, fructose, and galactose. Fructose may assume considerable quantitative importance if there is a large intake of sucrose. Galactose is of major quantitative significance only when lactose is the principal carbohydrate of the diet. Both fructose and galactose are readily converted to glucose by the liver.

Pentose sugars such as xylose, arabinose, and ribose may be present in the diet, but their fate after absorption is obscure. D-Ribose and D-2-deoxyribose are synthesized in the body for incorporation into nucleotides.

INTERMEDIARY METABOLISM OF CARBOHYDRATE

The metabolism of carbohydrate in the mammalian organism may be subdivided as follows:

(1) **Glycolysis**: The oxidation of glucose or glycogen to pyruvate and lactate by the Embden-Meyerhof pathway.

(2) **Glycogenesis**: The synthesis of glycogen from glucose.

(3) **Glycogenolysis**: The breakdown of glycogen: Glucose is the main end product in the liver, and pyruvate and lactate are the main products in muscle.

(4) **The citric acid cycle (Krebs cycle or tricarboxylic acid cycle)**: The final common pathway of oxidation of carbohydrate, fat, and protein through which acetyl-Co A is completely oxidized to carbon dioxide and, ultimately, water.

(5) **The hexose monophosphate shunt (HMS, direct oxidative pathway, phosphogluconate oxidative pathway, pentose phosphate cycle)**: An alternative pathway to the Embden-Meyerhof pathway and the citric acid cycle for the oxidation of glucose to carbon dioxide and water.

(6) **Gluconeogenesis**: The formation of glucose or glycogen from noncarbohydrate sources. The pathways involved in gluconeogenesis are mainly the citric acid cycle and glycolysis. The principal substrates for gluconeogenesis are glucogenic amino acids, lactate, and glycerol, and, in the ruminant, propionate (Fig 13–1).

GLYCOLYSIS

(See Fig 13–3.)

At an early period in the course of investigations on carbohydrate metabolism it was realized that the process of fermentation in yeast was similar to the breakdown of glycogen in muscle. Although many of the early investigations of the glycolytic pathway were carried out on these 2 systems, the process is now known to occur in virtually all tissues.

In many of the first studies on the biochemical changes which occur during muscular contraction it was noted that when a muscle contracts in an anaerobic medium, ie, one from which oxygen is excluded, glycogen disappears and pyruvate and lactate appear as the principal end products. When oxygen is admitted, aerobic recovery takes place and glycogen reappears, while pyruvate and lactate disappear. However, if contraction takes place under aerobic conditions, lactate does not accumulate and pyruvate is oxidized further to CO_2 and water. As a result of these observations, it has been customary to separate carbohydrate metabolism into anaerobic and aerobic phases. However, this

*Lecturer in Biochemistry, Royal Veterinary College, University of London.

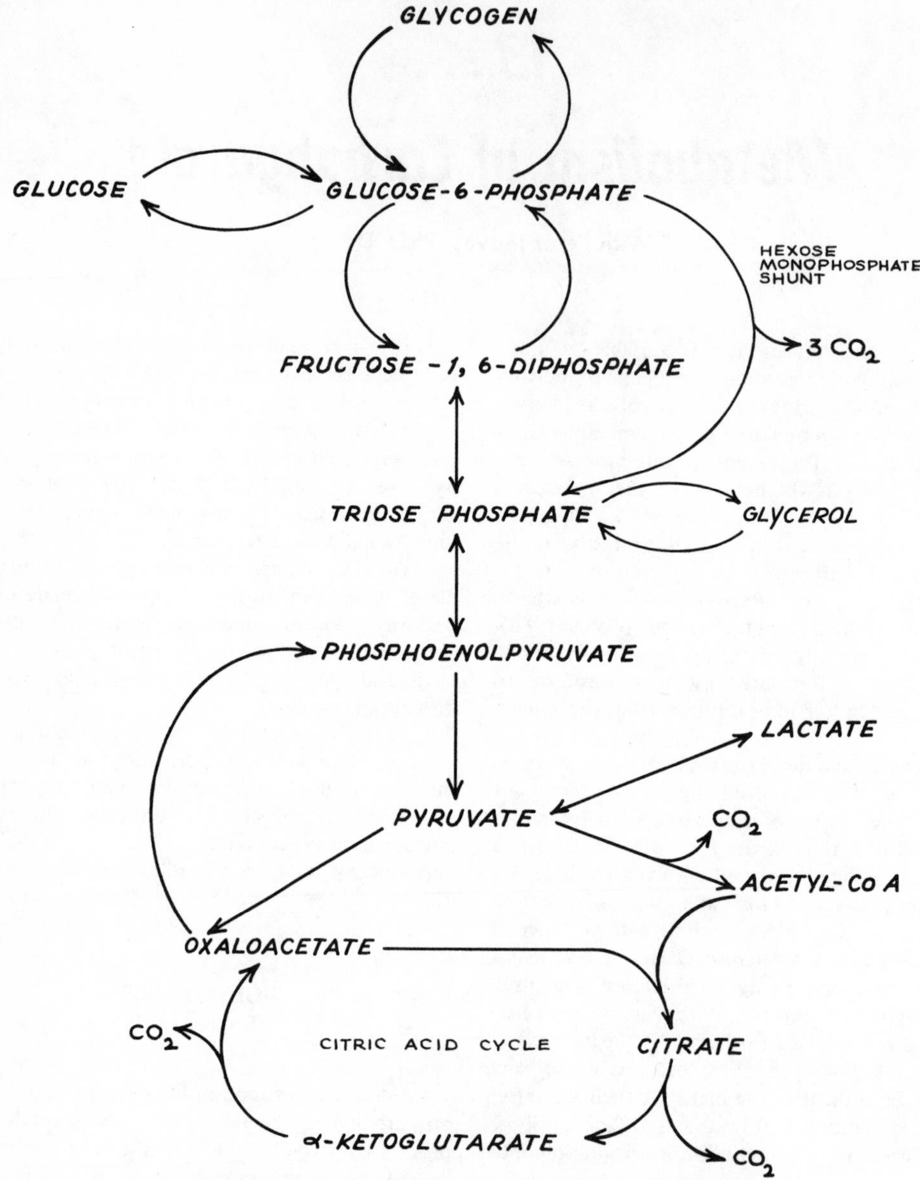

FIG 13–1. Major pathways of carbohydrate metabolism.

distinction is arbitrary since the reactions in glycolysis are the same in the presence of oxygen as in its absence except in extent and end products. When oxygen is in short supply, reoxidation of NADH formed during glycolysis is impaired. Under these circumstances NADH is reoxidized by being coupled to the reduction of pyruvate to lactate, the NAD so formed being used to allow further glycolysis to proceed (Fig 13–3). Glycolysis can thus take place under anaerobic conditions, but this limits the amount of energy produced per mol of glucose oxidized. Consequently, to provide a given amount of energy, more glucose must undergo glycolysis under anaerobic as compared with aerobic conditions.

All of the enzymes of the Embden-Meyerhof pathway (Fig 13–3) are found in the extramitochondrial soluble fraction of the cell. They catalyze the reactions involved in the glycolysis of glucose to lactate, which are as follows:

Glucose enters into the glycolytic pathway by phosphorylation to glucose-6-phosphate. This is accomplished by the enzyme **hexokinase** and by an additional enzyme in the livers of nonruminants, **glucokinase**, whose activity is inducible and affected by changes in the nutritional state. ATP is required as phosphate donor, and, as in many reactions involving phosphorylation, Mg^{++} must be present. One high-energy phosphate bond of ATP is utilized and ADP is

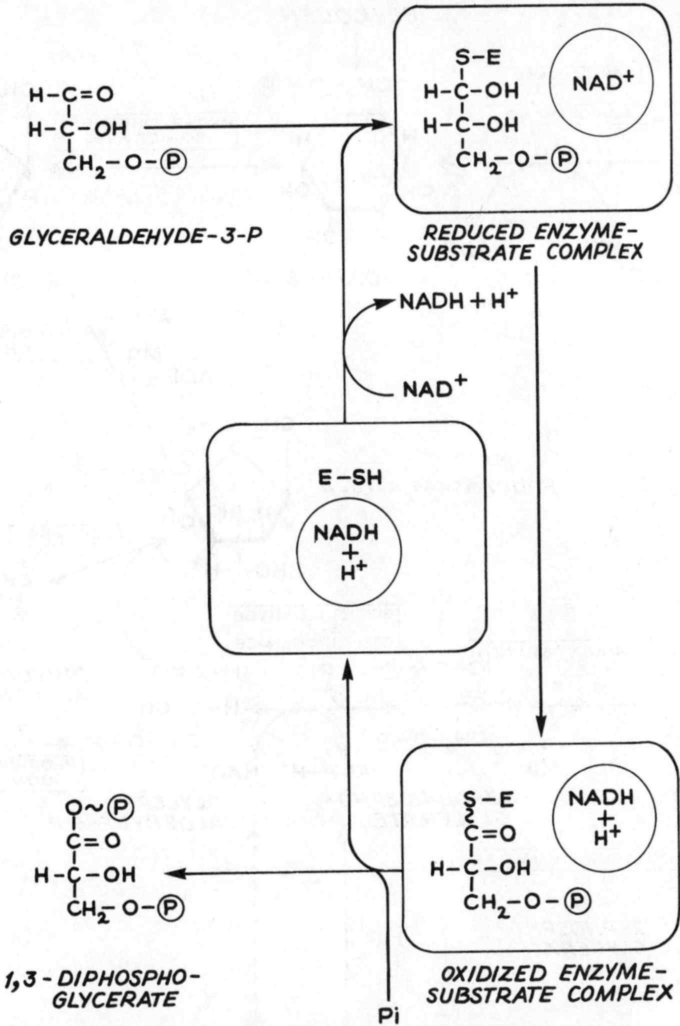

FIG 13–2. Oxidation of glyceraldehyde-3-phosphate.

produced. Hexokinase activity is present in rat liver in as many as 3 different enzyme proteins (isozymes). It is inhibited in an allosteric manner by the product, glucose-6-phosphate. The absence of glucokinase from the liver of ruminants is compatible with the negligible direct contribution of dietary carbohydrate to the blood glucose and therefore to a diminished requirement in these species for the liver to remove glucose from the blood.

Glucose-6-phosphate is an important compound, being at the junction of several metabolic pathways (glycolysis, gluconeogenesis, the HMS, glycogenesis, and glycogenolysis). In glycolysis it is converted to fructose-6-phosphate by **phosphohexose isomerase** followed by another phosphorylation with ATP catalyzed by the enzyme **phosphofructokinase** to produce fructose-1,6-diphosphate. Phosphofructokinase is another inducible enzyme whose activity is considered to play a major role in the regulation of the rate of glycolysis.

The hexose phosphate, fructose-1,6-diphosphate, is split by **aldolase** into 2 triose phosphates, glyceraldehyde-3-phosphate and dihydroxyacetone phosphate, which are interconverted by the enzyme **phosphotriose isomerase**. Several different aldolases have been detected; aldolase A occurs in most tissues, and aldolase B occurs in liver and kidney.

Glycolysis proceeds by the oxidation of glyceraldehyde-3-phosphate to 1,3-diphosphoglycerate, and, because of the activity of phosphotriose isomerase, the dihydroxyacetone phosphate is also oxidized to 1,3-diphosphoglycerate via glyceraldehyde-3-phosphate. The enzyme responsible for the oxidation, **glyceraldehyde-3-phosphate dehydrogenase**, is NAD-dependent. Structurally, it consists of 4 identical polypeptides (monomers) forming a tetramer. Four SH groups are present on each polypeptide, probably derived from cysteine residues within the polypeptide chain. One of the SH groups is found at the active site of the enzyme. It is believed that the SH group partici-

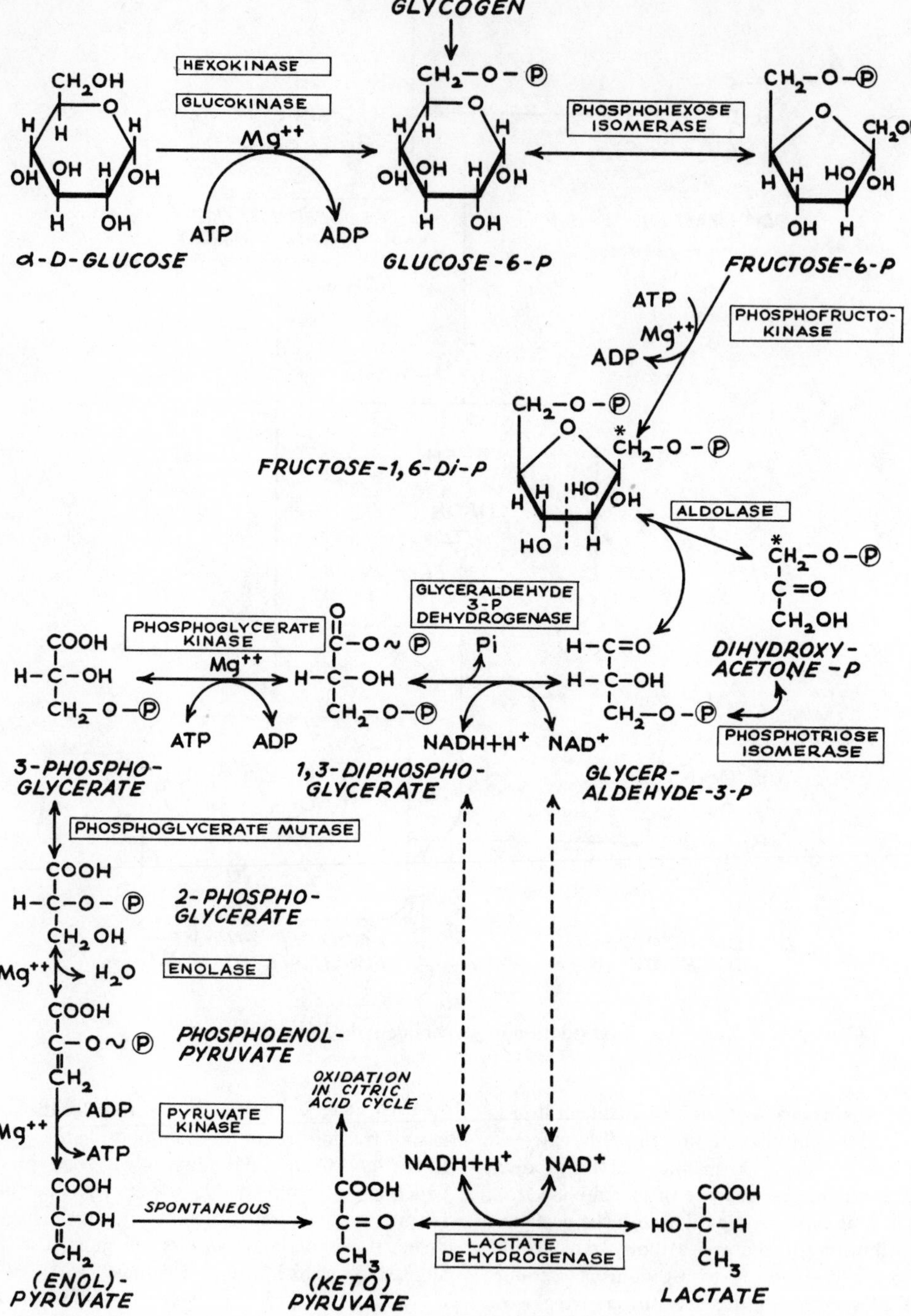

FIG 13–3. Embden-Meyerhof pathway of glycolysis.

pates in the reaction by which glyceraldehyde is oxidized—whereby the substrate initially combines with a cysteinyl moiety on the dehydrogenase, forming a thiohemiacetal* which is then converted to a thiol ester by oxidation, the hydrogens removed in this oxidation being transferred to NAD to form NADH.† Finally, by phosphorolysis, inorganic phosphate (Pi) is added, forming 1,3-diphosphoglycerate, and the free enzyme with a reconstituted SH group is liberated (Fig 13–2). Energy generated during the oxidation is retained by the formation of a high-energy sulfur bond which becomes a high-energy phosphate bond in position 1 of 1,3-diphosphoglycerate after phosphorolysis. This high-energy phosphate is captured as ATP in a further reaction with ADP catalyzed by **phosphoglycerate kinase**, leaving 3-phosphoglycerate. As 2 molecules of triose phosphate are formed per molecule of glucose undergoing glycolysis, 2 molecules of ATP are generated at this stage per molecule of glucose, an example of phosphorylation at the substrate level. If arsenate is present it will compete with inorganic phosphate (Pi) in the above reactions to give 1-arseno-3-phosphoglycerate, which hydrolyzes to give 3-phosphoglycerate without generating ATP. This property is similar to the uncoupling by arsenate of oxidative phosphorylation in mitochondria.

3-Phosphoglycerate arising from the above reactions is converted to 2-phosphoglycerate by the enzyme **phosphoglycerate mutase**. It is possible that 2,3-diphosphoglycerate is an intermediate in this reaction. The subsequent step is catalyzed by **enolase** and involves a dehydration and redistribution of energy within the molecule, raising the phosphate on position 2 to the high-energy state, thus forming phosphoenolpyruvate. Enolase is inhibited by fluoride.

The high-energy phosphate of phosphoenolpyruvate is transferred to ADP by the enzyme **pyruvate kinase** to generate, at this stage, 2 mols of ATP per mol of glucose oxidized. Enolpyruvate formed in this reaction is converted spontaneously to the keto form of pyruvate.

The redox state of the tissue now determines which of 2 pathways is followed. If anaerobic conditions prevail, the reoxidation of NADH by H transfer through the respiratory chain to oxygen is prevented. Pyruvate, which is the normal end product of glycolysis under aerobic conditions, is reduced under anaerobic conditions by the NADH to lactate, the reaction being catalyzed by **lactate dehydrogenase**. Several isozymes of this enzyme have been described in Chapter 8. The reoxidation of NADH via lactate formation allows glycolysis to proceed in the absence of oxygen by regenerating sufficient NAD^+ for the reaction catalyzed by glyceraldehyde-3-phosphate dehydrogenase. Thus, tissues which may function under hypoxic circumstances tend to produce lactate. This is particularly true of skeletal muscle, where the rate at which the organ performs work is not limited by its capacity for oxygenation. The additional quantities of lactate produced may be detected in the tissues and in the blood and urine. Glycolysis in erythrocytes, even under aerobic conditions, always terminates in lactate, because the enzymatic machinery for the oxidation of pyruvate is not present. The mammalian erythrocyte is unique in that about 90% of its total energy requirement is provided from glycolysis. In many mammalian erythrocytes, there is a bypass of the step catalyzed by phosphoglycerate kinase. Another enzyme, **diphosphoglyceromutase**, catalyzes the conversion of 1,3-diphosphoglycerate to 2,3-diphosphoglycerate. The latter is converted to 3-phosphoglycerate by **2,3-diphosphoglycerate phosphatase**, an activity which is also attributed to **phosphoglycerate mutase**. 2,3-Diphosphoglycerate combines with hemoglobin, causing a decrease in affinity for oxygen and a displacement of the oxyhemoglobin dissociation curve to the right. Thus, its presence in the red cell aids oxyhemoglobin to unload oxygen (Benesch, 1968).

Although most of the glycolytic reactions are reversible, 3 of them are markedly exergonic and must therefore be considered physiologically irreversible. These reactions are catalyzed by hexokinase (and glucokinase), phosphofructokinase, and pyruvate kinase. Cells which are capable of effecting a net movement of metabolites in the synthetic direction of the glycolytic pathway do so because of the presence of different enzyme systems which provide alternative routes to the irreversible reactions catalyzed by the above mentioned enzymes. These will be discussed under gluconeogenesis.

FORMATION & DEGRADATION OF GLYCOGEN; GLYCOGENESIS & GLYCOGENOLYSIS

The formation of glycogen occurs in practically every tissue of the body, but chiefly in liver and mus-

TABLE 13–1. Storage of carbohydrate in normal adult man (70 kg).

Liver glycogen	6.0%	=	108 g*
Muscle glycogen	0.7%	=	245 g†
Extracellular sugar	0.1%	=	10 g‡
Total: 363 g X 4		=	1452 Cal

*Liver weight, 1800 g.
†Muscle mass, 35 kg.
‡Total volume, 10 liters.

*The dehydrogenase enzyme may be inactivated by the –SH poison, iodoacetate, which is thus able to inhibit glycolysis at this point.

†The NADH produced on the enzyme is not as firmly bound to the enzyme as is NAD. Consequently, NADH is easily displaced by a molecule of NAD and the reconstituted enzyme thus carries NAD as would be necessary for the dehydrogenase to function again in the oxidative reaction.

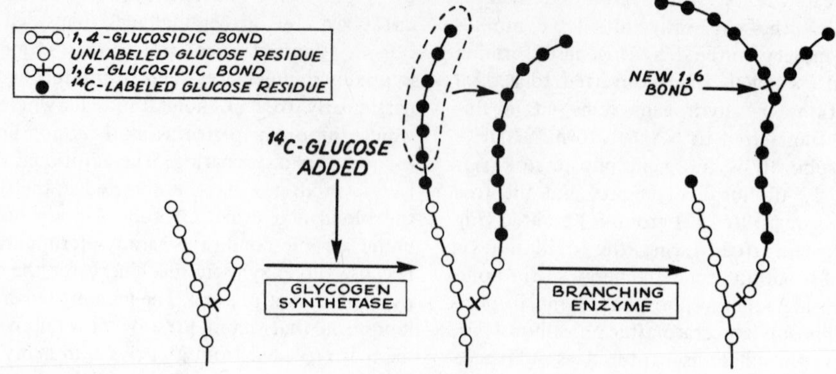

(a) Synthesis.

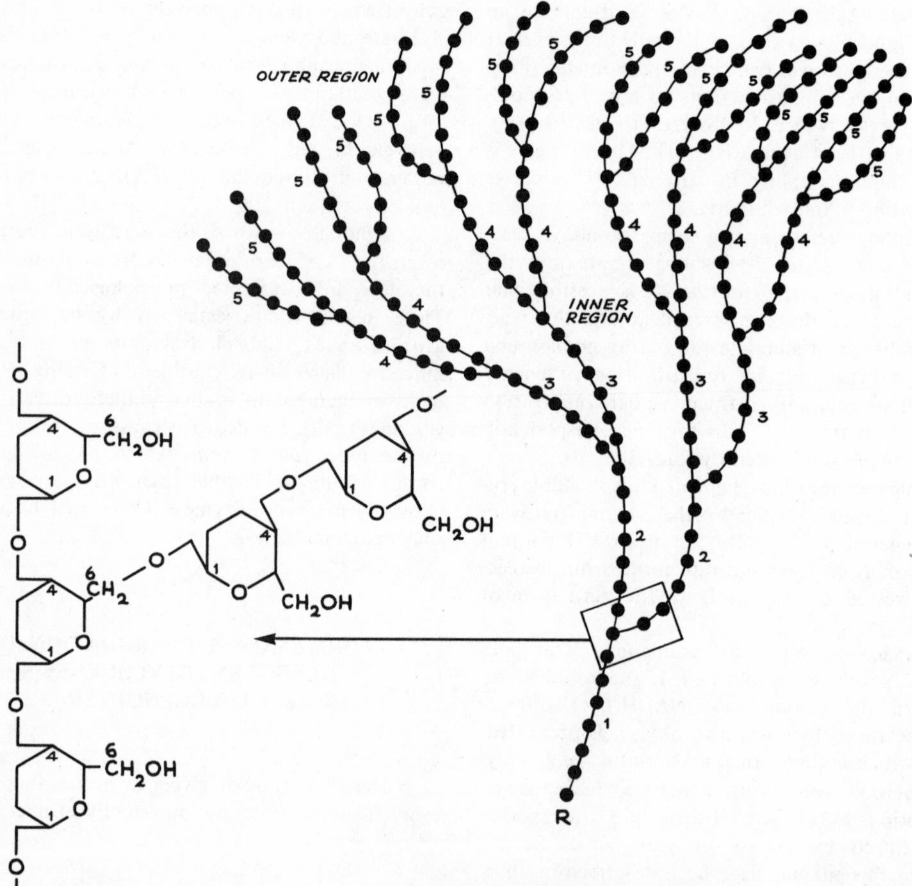

(c) Enlargement of structure at a branch point.

(b) Structure–The numbers refer to equivalent stages in the growth of the macromolecule. R, primary glucose residue with free reducing-CHO group (carbon No. 1). The branching is more variable than shown, the ratio of 1,4 to 1,6 bonds being from 12 to 18.

FIG 13–4. The glycogen molecule.

cle (Table 13–1). In man, the liver may contain as much as 5% glycogen when analyzed shortly after a meal high in carbohydrate. After 12–18 hours of fasting, the liver becomes almost totally depleted of glycogen.

Muscle glycogen is only rarely elevated above 1% of the wet weight of the tissue.

GLYCOGENESIS

Glucose is phosphorylated to glucose-6-phosphate, a reaction which is common to the first reaction in the pathway of glycolysis from glucose. Glucose-6-phosphate is then converted to glucose-1-phosphate in a reaction catalyzed by the enzyme **phosphoglucomutase.** This enzyme is phosphorylated and the phospho- group takes part in a reversible reaction in which glucose-1,6-diphosphate is an intermediate. Next, glucose-1-phosphate reacts with uridine triphosphate (UTP) to form the active nucleotide **uridine diphosphate glucose (UDPG).***

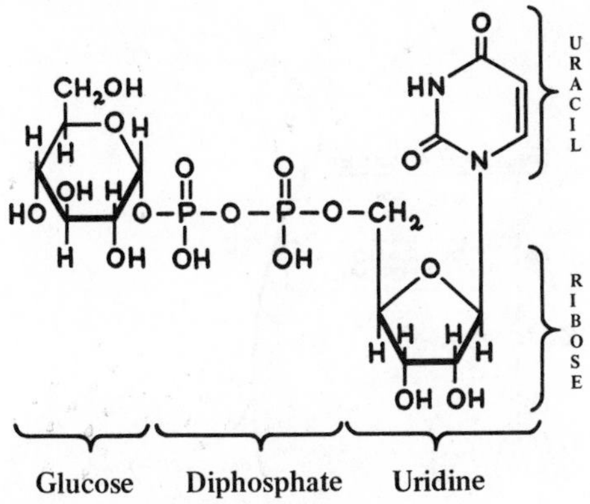

Glucose Diphosphate Uridine

Uridine diphosphate glucose (UDPG)

The reaction between glucose-1-phosphate and uridine triphosphate is catalyzed by the enzyme **UDPG pyrophosphorylase.**

By the action of the enzyme **UDPG-glycogen-transglucosylase** (also known as **glycogen synthetase** or **glucosyl transferase**), the C_1 of the activated glucose of UDPG forms a glycosidic bond with the C_4 of a terminal glucose residue of glycogen, liberating uridine diphosphate (UDP) (Fig 13–5).

The addition of a glucose residue to a preexisting glycogen chain occurs at the nonreducing, outer end of the molecule so that the "branches" of the glycogen "tree" become elongated as successive -1,4- linkages occur (Fig 13–4). When the chain has been lengthened to between 6 and 11 glucose residues, a second enzyme, the **branching enzyme (amylo-1,4→1,6-transglucosidase)** acts on the glycogen. This enzyme transfers a part of the -1,4- chain (minimum length 6 glucose residues) to a neighboring chain to form a -1,6- linkage, thus establishing a branch point in the molecule.

The action of the branching enzyme has been studied in the living animal by feeding ^{14}C-labeled glucose and examining the liver glycogen at intervals thereafter. At first only the outer branches of the chain are labeled, indicating that the new glucose residues are added at this point. Later, some of these outside chains are transferred to the inner portion of the molecule, appearing as labeled -1,6- linked branches (Fig 13–4). Thus, under the combined action of glycogen synthetase and branching enzyme, the glycogen molecule is assembled.

The structure of glycogen is shown in Fig 13–4. It will be seen to be a branched polysaccharide composed entirely of α-D-glucose units. These glucose units are connected to one another by glucosidic linkages between the first and fourth carbon atoms except at branch points, where the linkages are between carbon atoms 1 and 6. The molecular weight of glycogen may vary from 1 million daltons to 4 million or more. If glycogen is a regularly branched structure as shown in Fig 13–4, a maximum molecular weight of $10–20 \times 10^6$ daltons would be possible because of the fact that the molecule becomes more dense toward the periphery. However, if some of the glucose chains terminate in the interior, a larger molecule is theoretically possible.

In muscle (and possibly liver), **glycogen synthetase** is present in 2 interconvertible forms: **synthetase D** (dependent), which is totally dependent for its activity on the presence of glucose-6-phosphate; and **synthetase I** (independent), whose K_m for UDPG decreases in the presence of glucose-6-phosphate (Fig 13–7). However, only the latter effect occurs with physiologic concentrations of glucose-6-phosphate, implying that synthetase I is the active form of the enzyme. Synthetase D is converted to synthetase I by **synthetase phosphatase,** a reaction involving dephosphorylation of a serine residue within the enzyme protein. The reaction is inhibited by increasing concentrations of glycogen, which may thus act as a feedback control on glycogen synthesis. It is possible that muscular contraction and increased concentrations of Ca^{++} may cause direct inactivation of synthetase I to synthetase D. Synthetase I is phosphorylated to form synthetase D, with ATP acting as a phosphate donor, by the enzyme **synthetase kinase,** which itself exists in 2 forms: **synthetase kinase D,** active only in the presence of **3′,5′-cyclic adenylic acid (cyclic AMP)** and **synthetase kinase I,** which, while not dependent on the presence of cyclic AMP, exhibits a lowered K_m for ATP when cyclic AMP is present.

Cyclic AMP (see below) is the intracellular compound through which many hormones appear to act. It

*There are other nucleoside diphosphate sugar compounds which contain nucleotides. In addition, the same sugar may be linked to different nucleotides. For example, glucose may be linked to uridine (as shown above) as well as to guanine, thymine, adenine, or cytosine nucleotides.

is formed from ATP by an enzyme, **adenyl cyclase**, occurring in cell membranes. Adenyl cyclase is activated by hormones such as epinephrine, norepinephrine, and glucagon, all of which lead to an increase in cyclic AMP. Cyclic AMP is destroyed by a **phosphodiesterase** (Fig 13–7), and it is this activity that maintains the level of cyclic AMP at its normally low level. Insulin has been reported to increase its activity in liver. Thyroid hormones may increase the synthesis of adenyl cyclase, thus potentiating the effects of epinephrine in stimulating the formation of cyclic AMP. Insulin appears also to increase the activity of synthetase I, but independently of any action on the level of cyclic AMP. In muscle, this could be due to conversion of active synthetase kinase to the inactive form. However, the effect of insulin in elevating hepatic glycogen synthetase activity seems to depend on its ability to bring about an increase in synthetase phosphatase. It is not generally considered that insulin causes new synthesis of synthetase enzyme protein. Hepatic synthetase activity has also been reported to be increased by glucocorticoids.

The reactions of glycogenesis in liver are summarized in Fig 13–5. Those in muscle are shown in greater detail in Fig 13–7.

GLYCOGENOLYSIS

The breakdown of glycogen is initiated by the action of the enzyme **phosphorylase**, which is specific for the phosphorylytic breaking of the -1,4- linkages of glycogen to yield glucose-1-phosphate (Fig 13–5).

Phosphorylase Activation & Inactivation

(See Figs 13–6 and 13–7.)

Sutherland & Wosilait (1956) prepared phosphorylase from liver. The enzyme protein was found to exist in both an active and an inactive form. The active phosphorylase (**phosphophosphorylase**) has one of its serine residues phosphorylated in an ester linkage with

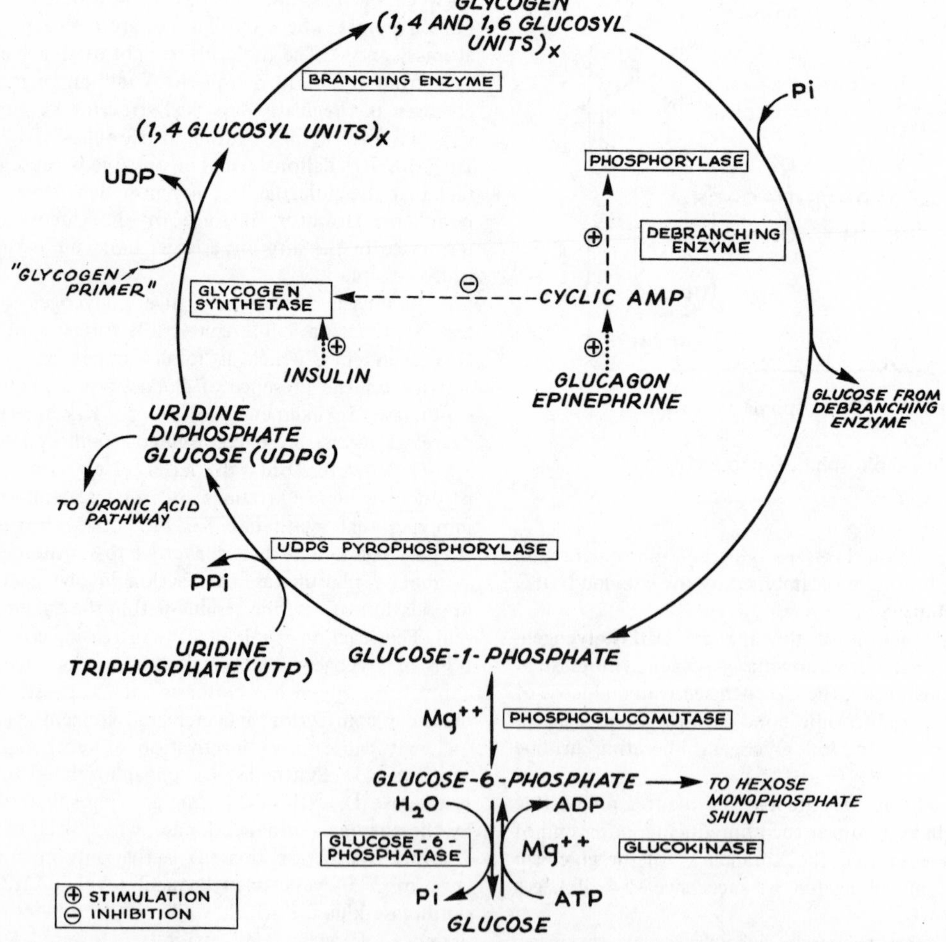

FIG 13–5. Pathway of glycogenesis and of glycogenolysis in the liver.

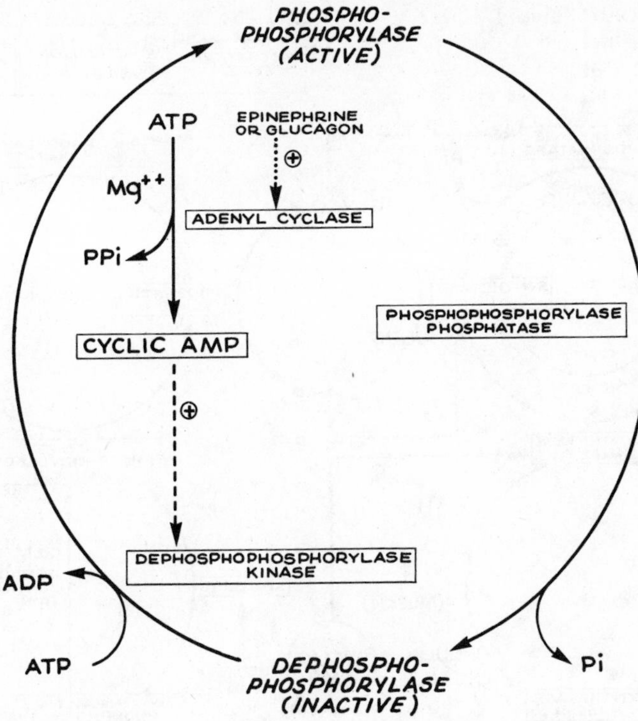

FIG 13–6. Summary of the reactions involved in activation and inactivation of liver phosphorylase.

the hydroxyl group of the serine. By the action of a specific phosphatase **(phosphophosphorylase phosphatase),** the enzyme can be inactivated to **dephosphophosphorylase** in a reaction which involves hydrolytic removal of the phosphate from the serine residue. Reactivation requires rephosphorylation with ATP and a specific enzyme, **dephosphophosphorylase kinase.** This kinase is stimulated by cyclic AMP.

A summary of the reactions involved in activation and inactivation of liver phosphorylase, the enzyme catalyzing glycogenolysis in liver, is shown in Fig 13–6.

Muscle phosphorylase is immunologically distinct from that of liver. It is present in 2 forms: **phosphorylase a,** which is active in the absence of 5′-AMP, and **phosphorylase b,** which is active only in the presence of 5′-AMP. Phosphorylase a is the physiologically active form of the enzyme. It is a tetramer containing 4 mols of pyridoxal phosphate. When it is hydrolytically converted to a dimer by **phosphorylase phosphatase,** which removes phosphate from phosphoserine residues, phosphorylase b is formed. This contains 2 mols of pyridoxal phosphate. The conversion of phosphorylase a (active enzyme) to phosphorylase b (the inactive form of the enzyme) is illustrated below.

3′,5′-Adenylic acid (cyclic AMP)

Conversion of phosphorylase b to phosphorylase a is considered to be the mechanism for increasing glycogenolysis.

$$\text{PHOSPHORYLASE a} + 4H_2O \xrightarrow{\text{PHOSPHORYLASE PHOSPHATASE}} \text{2 PHOSPHORYLASE b} + 4Pi$$

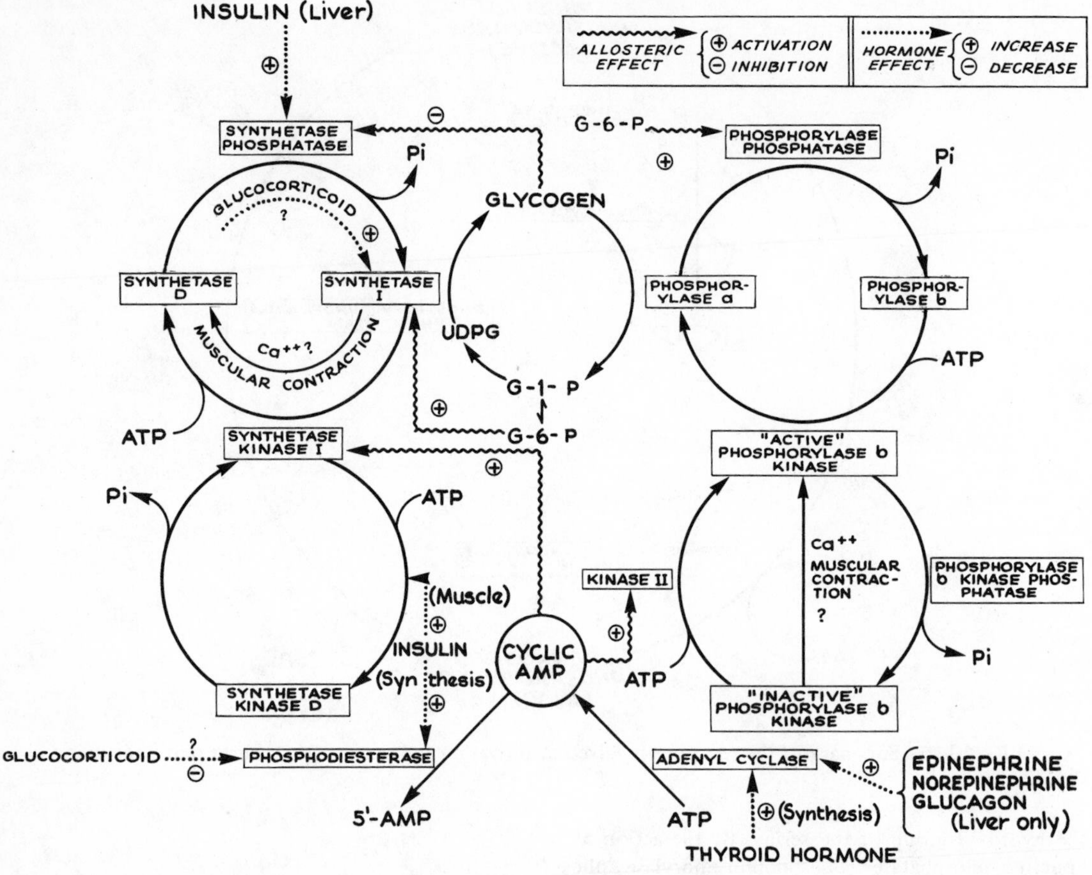

FIG 13–7 Glycogenesis and glycogenolysis in muscle, including the enzymes catalyzing each step and the mechanisms of activation and inactivation of these enzymes. The control mechanisms for these pathways, with special reference to the roles of cyclic AMP and hormones, are also shown. (Modified from Villar-Palasi & Larner [1968].)

Two dimers of phosphorylase b may recondense to an active phosphorylase a tetramer in the presence of a specific enzyme, **phosphorylase b kinase,** which rephosphorylates the serine residues at the expense of ATP.

$$\text{2 PHOSPHORYLASE b} + \text{4ATP} \xrightarrow[\text{Mg}^{++}]{\text{PHOSPHORYLASE b KINASE}} \text{PHOSPHORYLASE a} + \text{4 ADP}$$

Phosphorylase in muscle is activated by epinephrine. However, it is probable that this occurs not as a direct effect but rather by way of the effect of epinephrine on adenyl cyclase to form cyclic AMP, which then serves to produce activation of phosphorylase b kinase by way of a second kinase system (Fig 13–7). Activation of phosphorylase b kinase is also caused by muscular contraction, an effect which appears to be due to Ca^{++} rather than cyclic AMP.

These reactions for the inactivation of muscle phosphorylase are in some respects similar to those of the liver enzyme except that no cleavage of the protein molecular structure is involved in the case of liver phosphorylase. Furthermore, skeletal muscle phosphorylase is not affected by glucagon although heart muscle is.

It is the phosphorylase step which is rate-limiting in glycogenolysis. However, in the presence of inorganic phosphate, which favors the action of the enzyme in breaking the glucosyl-1,4 linkages, this enzyme removes the 1,4-glucosyl residues from the outermost chains of the glycogen molecule until approximately 4 glucose residues remain on either side of a -1,6- branch (Fig 13–4). Another enzyme **(α-1,4→α-1,4 glucan transferase)** transfers a trisaccharide unit from one side to the other, thus exposing the -1,6- branch points (Illingworth & Brown, 1962). The hydrolytic splitting of the -1,6- linkages requires the action of a specific **debranching enzyme (amylo-1,6-glucosidase).*** The combined action of phosphorylase and these other enzymes converts glycogen to glucose-1-phosphate. The action of phosphoglucomutase is reversible, so that glucose-6-phosphate can be formed from glucose-1-phosphate. In liver and kidney (but not in muscle), there is a specific enzyme, **glucose-6-phosphatase**, that removes phosphate from glucose-6-phosphate, enabling the free glucose to diffuse from the cell into the extracellular spaces, including the blood. This is the final step in hepatic glycogenolysis which is reflected by a rise in the blood glucose.

The reactions of glycogenolysis in liver are summarized in Fig 13–5. The reactions of glycogenesis and glycogenolysis are summarized in Fig 13–7, which includes also the control mechanisms involved.

Control of glycogen metabolism is due to the balance in activities of the enzymes of glycogen synthesis and breakdown which are seen to be under both substrate (through allosteric activity) and hormonal control. Not only is phosphorylase activated by a rise in concentration of cyclic AMP, but glycogen synthetase is at the same time converted to the inactive form. Thus, breakdown and synthesis of glycogen do not occur simultaneously, an important aspect of metabolic control of glycogen metabolism (Hers & others, 1970). It is apparent that mechanisms are present which explain both the glycogenolytic effects of epinephrine in muscle and of glucagon in liver and, to a less satisfactory degree, the glycogenic properties of insulin.

Further details of the metabolism of glycogen may be found in Villar-Palasi & Larner (1968, 1970), and Smith & others (1968).

The Diseases of Glycogen Storage

Abnormal metabolism of glycogen is a feature of the so-called diseases of glycogen storage or glycogen deposition. The term "glycogen storage disease" is a generic one intended to describe a group of inherited disorders characterized by deposition of an abnormal type or quantity of glycogen in the tissues. The subject has been reviewed by Brown & Brown (1968).

*Because the -1,6- linkage is hydrolytically split, 1 mol of free glucose is produced rather than 1 mol of glucose-1-phosphate. In this way it is possible for some rise in the blood glucose to take place even in the absence of glucose-6-phosphatase, as occurs in type I glycogen storage disease (Von Gierke's disease; see below) after glucagon or epinephrine is administered.

In **type I glycogenosis (Von Gierke's disease),** both the liver cells and the cells of the renal convoluted tubules are characteristically loaded with glycogen. However, these glycogen stores seem to be metabolically unavailable, as evidenced by the occurrence of hypoglycemia and a lack of glycogenolysis under stimulus by epinephrine or glucagon. Ketosis and hyperlipemia are also present in these patients, as would be characteristic of an organism deprived of carbohydrate. In liver, kidney, and intestinal tissue obtained from these patients, the activity of glucose-6-phosphatase is either extremely low or entirely absent.

Other types of glycogen storage disease include the following: **type II (Pompe's disease),** which is characterized by a deficiency of lysosomal α-1,4-glucosidase (acid maltase) whose function is to degrade glycogen, which otherwise accumulates in the lysosomes; **type III (limit dextrinosis),** characterized by the absence of debranching enzyme, which causes the accumulation of a polysaccharide of the limit dextrin type; and **type IV (amylopectinosis),** characterized by the absence of branching enzyme, with the result that a polysaccharide having few branch points accumulates.

An absence of muscle phosphorylase (myophosphorylase) is the cause of **type V glycogenosis (myophosphorylase deficiency glycogenosis; McArdle's syndrome).** Patients with this disease exhibit a markedly diminished tolerance to exercise. Although their skeletal muscles have an abnormally high content of glycogen (2.5–4.1%), little or no lactate is detectable in their blood after exercise. A rise in blood sugar does occur, however, after administration of glucagon or epinephrine, which indicates that hepatic phosphorylase activity is normal. In some of the reported cases, myoglobinuria has been an associated finding.

Also described among the glycogen storage diseases are **type VI glycogenosis,** involving phosphoglucomutase deficiency in the liver, and **type VII,** characterized by a deficiency of phosphofructokinase in the muscles.

THE CITRIC ACID CYCLE

In contrast to the enzymes responsible for glycolysis, the complete set of enzymes responsible for the reactions of the citric acid cycle is found in the mitochondrial fraction of the cell, mostly in the matrix, in proximity to the enzymes of the respiratory chain. As the citric acid cycle generates many reducing equivalents in the form of reduced coenzymes, the functional significance of the location of the 2 sets of enzymes may be appreciated.

Before pyruvate can enter the citric acid cycle, it must be oxidatively decarboxylated to acetyl-Co A ("active acetate"). This reaction is catalyzed by several different enzymes working sequentially in a multienzyme complex. They are collectively designated as

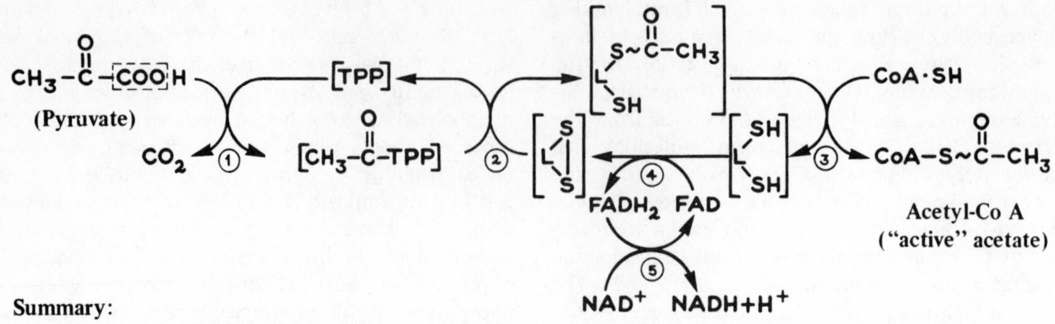

$CH_3-\overset{O}{\overset{\|}{C}}-COOH$ (Pyruvate) → ① CO_2; [TPP] → $[CH_3-\overset{O}{\overset{\|}{C}}-TPP]$ → ② → $\left[L\begin{smallmatrix}S{\sim}\overset{O}{\overset{\|}{C}}-CH_3\\SH\end{smallmatrix}\right]$ → ③ CoA·SH → $CoA-S{\sim}\overset{O}{\overset{\|}{C}}-CH_3$ Acetyl-Co A ("active" acetate); $\left[L\begin{smallmatrix}SH\\SH\end{smallmatrix}\right]$ → ④ $FADH_2$ FAD → $\left[L\begin{smallmatrix}S\\S\end{smallmatrix}\right]$; ⑤ NAD^+ → $NADH+H^+$

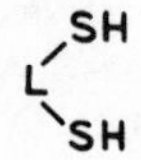

Summary:

$$CH_3-\overset{O}{\overset{\|}{C}}-COOH \xrightarrow[\text{LIPOATE, FAD; } NAD^+ \to NADH+H^+]{CoA\cdot SH \to CO_2;\ Mg^{++},\ TPP} CH_3-\overset{O}{\overset{\|}{C}}{\sim}S-CoA$$

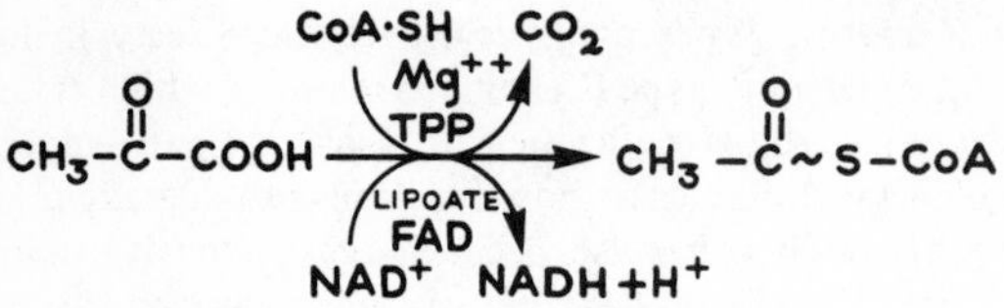

Abbreviations:

Square brackets mean that the intermediate is bound tightly to the enzyme interface.

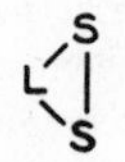

$L\begin{smallmatrix}S\\S\end{smallmatrix}$	= **Oxidized lipoic acid**
$L\begin{smallmatrix}SH\\SH\end{smallmatrix}$	= **Reduced lipoic acid**
$L\begin{smallmatrix}S{\sim}\overset{O}{\overset{\|}{C}}-CH_3\\SH\end{smallmatrix}$	= **S-Acetyl lipoic acid**
[TPP]	= **Thiamine pyrophosphate**
$[CH_3-CHO-TPP]$	= **"Active" acetaldehyde**
CoA·SH	= **Coenzyme A**
FAD, $FADH_2$	= **Flavin adenine dinucleotide, oxidized and reduced forms, respectively**
NAD^+, NADH	= **Nicotinamide adenine dinucleotide, oxidized and reduced forms, respectively**

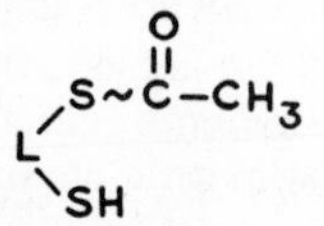

The enzymes catalyzing the numbered reactions are as follows:

Reactions ① & ②	= Pyruvate dehydrogenase
Reaction ③	= Dihydrolipoyl transacetylase
Reaction ④	= Dihydrolipoyl dehydrogenase

FIG 13–8. Oxidative decarboxylation of pyruvate.

the **pyruvate dehydrogenase** complex, shown above in Fig 13–8. Pyruvate is decarboxylated in the presence of thiamine pyrophosphate to a derivative of acetaldehyde, which in turn reacts with oxidized lipoate to form acetyl-S-lipoate, all catalyzed by pyruvate dehydrogenase. In the presence of **dihydrolipoyl transacetylase**, acetyl-S-lipoate reacts with coenzyme A to form acetyl-Co A and reduced lipoate. The cycle of reactions is completed when the latter is reoxidized by a flavoprotein in the presence of **dihydrolipoyl dehydrogenase**. Finally, the reduced flavoprotein is oxidized by NAD, which in turn transfers reducing equivalents to the respiratory chain. The pyruvate dehydrogenase complex consists of about 29 mols of pyruvate dehydrogenase and about 8 mols of flavoprotein (dihydrolipoyl dehydrogenase) distributed around 1 mol of transacetylase. Movement of the individual enzymes appears to be restricted, and the metabolic intermediates do not dissociate freely but remain bound to the enzymes (Reed, 1969).

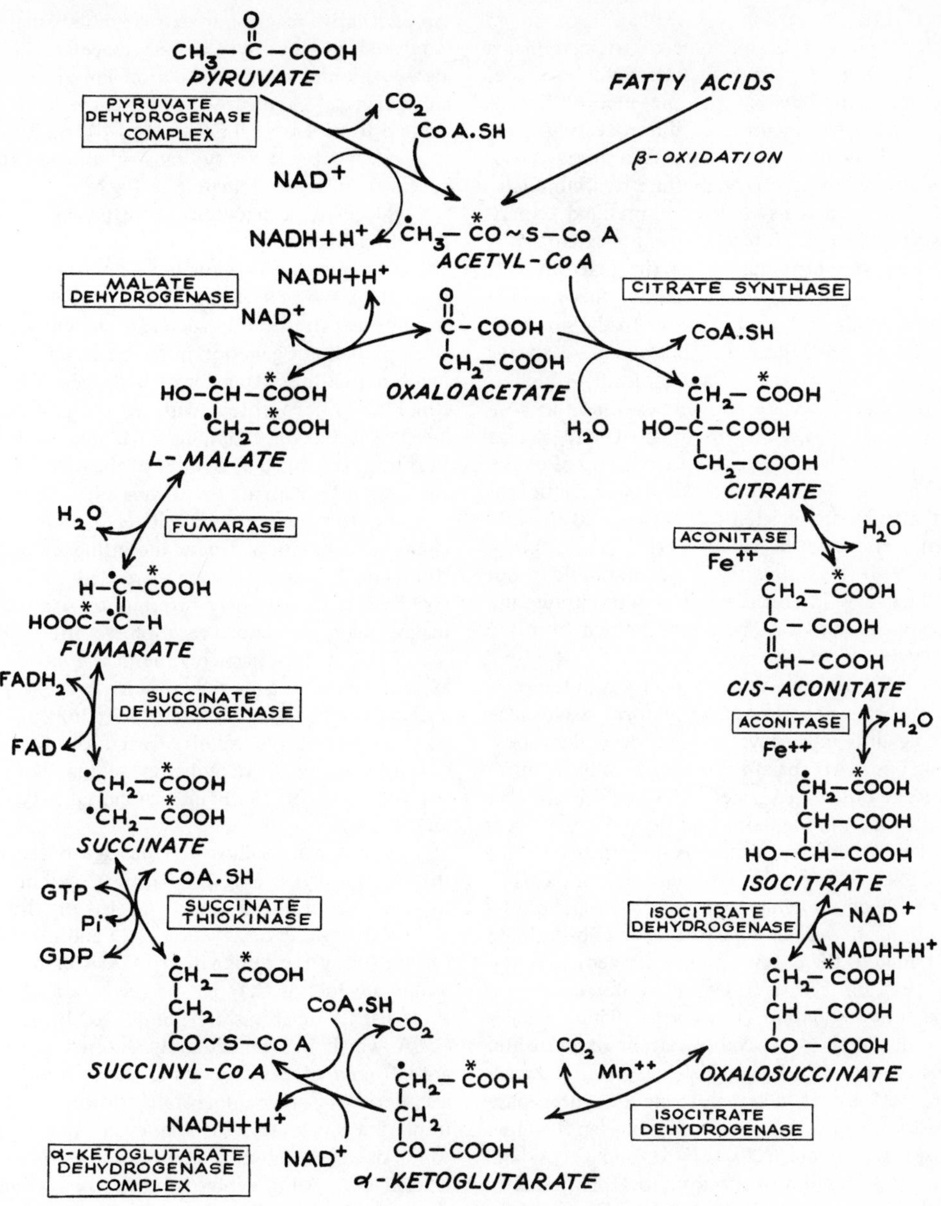

FIG 13–9. The citric acid (Krebs) cycle.

It is to be noted that the pyruvate dehydrogenase system is sufficiently electronegative with respect to the respiratory chain that, in addition to generating a reduced coenzyme (NADH), it also generates a high-energy thio ester bond in acetyl-Co A. For practical purposes, this reaction is irreversible, which prevents a net synthesis of carbohydrate from fatty acids having an even number of carbon atoms in their molecules. The presence of arsenite inhibits pyruvate dehydrogenase, allowing pyruvate to accumulate.

Acetyl-Co A, derived mainly from the oxidation of either carbohydrate or fatty acids, combines with oxaloacetate to form citrate. This is the first reaction of the citric acid cycle proper. Subsequently, citrate is oxidized in a series of reactions which liberate CO_2 and NADH and which finally regenerate oxaloacetate. The oxaloacetate functions therefore in a catalytic manner in the oxidation of acetyl-Co A to 2 molecules of CO_2.

Reactions of the Citric Acid Cycle (See Fig 13–9.)

The condensation of acetyl-Co A with oxaloacetate to form citrate is catalyzed by a condensing enzyme, **citrate synthase**, which effects a carbon-to-carbon bond between the methyl carbon of acetyl-Co A and the carbonyl carbon of oxaloacetate. The reaction may be regarded as irreversible. Citrate is converted to isocitrate by the enzyme **aconitase** (aconitate hydratase). This conversion takes place in 2 steps: dehydration to cis-aconitate, some of which remains bound to the enzyme, and rehydration to isocitrate. The reaction is inhibited in the presence of fluoroacetate, which, in the form of fluoroacetyl-Co A, condenses with oxaloacetate to form fluorocitrate. The latter inhibits aconitase, causing citrate to accumulate. Although citric acid is a symmetric compound, experiments using ^{14}C-labeled intermediates indicated that it reacts in an asymmetric manner in the reaction with aconitase. Ogston (1948) suggested that this was due to a 3-point attachment of the enzyme to the substrate (see Chapter 8), with the result that aconitase always acts on that part of the citrate molecule which is derived from oxaloacetate. It follows that the condensation of acetyl-Co A with oxaloacetate is also stereospecific with respect to the keto group of oxaloacetate. The consequences of the asymmetric action of aconitase may be appreciated by reference to the fate of labeled acetyl-Co A in the citric acid cycle as shown in Fig 13–9. It is possible that cis-aconitate may not be an obligatory intermediate between citrate and isocitrate but may in fact be a side branch from the main pathway.

Isocitrate undergoes dehydrogenation in the presence of **isocitrate dehydrogenase** to form oxalosuccinate. Three different enzymes have been described. One, which is NAD-specific, is found only in mitochondria. The other 2 enzymes are NADP-specific and are found in the mitochondria and the cytosol, respectively. There follows a decarboxylation to α-ketoglutarate, also catalyzed by isocitrate dehydrogenase. Mn^{++} is an important component of the decarboxylation reaction. It would appear that oxalosuccinate remains bound to the enzyme as an intermediate in the overall reaction. Next, α-ketoglutarate undergoes oxidative decarboxylation in a manner which is analogous to the oxidative decarboxylation of pyruvate, both substrates being α-keto acids (Fig 7–7). The reaction catalyzed by an **α-ketoglutarate dehydrogenase** complex also requires identical cofactors–eg, thiamine pyrophosphate, lipoate, NAD^+, FAD, and Co A–and results in the formation of succinyl-Co A. The equilibrium of this reaction is so much in favor of succinyl-Co A formation that the reaction must be considered as physiologically unidirectional. Again, as in the case of pyruvate oxidation, arsenite inhibits the reaction, causing the substrate, **α-ketoglutarate**, to accumulate.

To continue the cycle, succinyl-Co A is converted to succinate by the enzyme **succinate thiokinase (succinyl-Co A synthetase)**. This reaction requires GDP or IDP, which is converted in the presence of inorganic phosphate to either GTP or ITP. This is the only example in the citric acid cycle of the generation of a high-energy phosphate at the substrate level. By means of a phosphokinase, ATP may be formed from either GTP or ITP,

$$\text{eg, GTP + ADP} \rightleftharpoons \text{GDP + ATP}$$

An alternative reaction in extrahepatic tissues, which is catalyzed by **succinyl-Co A-acetoacetate Co A transferase (thiophorase)**, is the conversion of succinyl-Co A to succinate coupled with the conversion of acetoacetate to acetoacetyl-Co A (Fig 14–16). In liver there is also deacylase activity, causing some hydrolysis of succinyl-Co A to succinate plus Co A.

Succinate is metabolized further by undergoing a dehydrogenation followed by the addition of water, and subsequently by a further dehydrogenation which regenerates oxaloacetate. The first dehydrogenation reaction is catalyzed by **succinate dehydrogenase**. It is the only dehydrogenation in the citric acid cycle which involves the direct transfer of hydrogen from the substrate to a flavoprotein without the participation of NAD. The enzyme contains FAD and nonheme iron. Fumarate is formed as a result of the dehydrogenation. Addition of malonate or oxaloacetate inhibits succinate dehydrogenase competitively, resulting in succinate accumulation. Under the influence of **fumarase (fumarate hydratase)**, water is added to fumarate to give malate. Malate is converted to oxaloacetate by **malate dehydrogenase**, a reaction requiring NAD^+.

The enzymes participating in the citric acid cycle are also found outside the mitochondria except for the α-ketoglutarate and succinate dehydrogenases. While they may catalyze similar reactions, some of the enzymes, eg, malate dehydrogenase, may not in fact be the same proteins as the mitochondrial enzymes of the same name.

In order to follow the passage of acetyl-Co A through the cycle (Fig 13–9), the 2 carbon atoms of the acetyl radical are shown labeled on the carboxyl carbon (using the designation [*]) and on the methyl carbon (using the designation [•]). Although 2 carbon atoms are lost as CO_2 in one revolution of the cycle, these particular atoms are not derived from the acetyl-Co A which has immediately entered the cycle but arise from that portion of the citrate molecule which was derived from oxaloacetate. However, on completion of a single turn of the cycle, the oxaloacetate which is regenerated is now labeled, which leads to labeled CO_2 being evolved during the second turn of the cycle. It is to be noted that because succinate is a symmetric compound and because succinate dehydrogenase does not differentiate between its 2 carboxyl groups, "randomization" of label occurs at this step such that all 4 carbon atoms of oxaloacetate appear to be labeled after one turn of the cycle. When gluconeogenesis takes place, some of the label in oxaloacetate makes its way into glucose and glycogen. In this process, oxaloacetate is decarboxylated in the carboxyl group adjacent to the $-CH_2-$ group. As a result of recombination of the resulting 3-carbon residues in a process which is essentially a reversal of glycolysis, the eventual location of label from acetate in glucose (or glycogen) is distributed in a characteristic manner. Thus, if oxaloacetate leaves the citric acid cycle after only one turn from the entry of labeled acetyl-Co A ("acetate"), label in the carboxyl carbon of acetate is

TABLE 13–2. Generation of high-energy bonds in the catabolism of glucose.

Pathway	Reaction Catalyzed By	Method of ~ P Production	Number of ~ P Formed per Mol Glucose
Glycolysis	Glyceraldehyde-3-phosphate dehydrogenase	Respiratory chain oxidation of 2 NADH	6
	Phosphoglycerate kinase	Oxidation at substrate level	2
	Pyruvate kinase	Oxidation at substrate level	2
			10
Allow for consumption of ATP by reactions catalyzed by hexokinase and phosphofructokinase			−2
			Net 8
Citric acid cycle	Pyruvate dehydrogenase	Respiratory chain oxidation of 2 NADH	6
	Isocitrate dehydrogenase	Respiratory chain oxidation of 2 NADH	6
	α-Ketoglutarate dehydrogenase	Respiratory chain oxidation of 2 NADH	6
	Succinate thiokinase	Oxidation at substrate level	2
	Succinate dehydrogenase	Respiratory chain oxidation of 2 $FADH_2$	4
	Malate dehydrogenase	Respiratory chain oxidation of 2 NADH	6
			Net 30
Total per mol of glucose under aerobic conditions			38
Total per mol of glucose under anaerobic conditions			2

found in carbon atoms 3 and 4 of glucose, whereas label from the methyl carbon of acetate is found in carbon atoms 1, 2, 5, and 6. For a discussion of the stereochemical aspects of the citric acid cycle, see Greville (1968).

There cannot be a net conversion of acetyl-Co A to oxaloacetate via the cycle, as one molecule of oxaloacetate is needed to condense with acetyl-Co A and only one molecule of oxaloacetate is regenerated. For similar reasons there cannot be a net conversion of fatty acids having an even number of carbon atoms (which form acetyl-Co A) to glucose or glycogen.

Energetics of Carbohydrate Oxidation

When 1 mol of glucose is combusted in a calorimeter to CO_2 and water, approximately 686,000 calories are liberated as heat. When oxidation occurs in the tissues, some of this energy is not lost immediately as heat but is "captured" in high-energy phosphate bonds. At least 38 high-energy phosphate bonds are generated per molecule of glucose oxidized to CO_2 and water. Assuming each high-energy bond to be equivalent to 7600 calories, the total energy captured in ATP per mol of glucose oxidized is 288,800 calories, or approximately 42% of the energy of combustion. Most of the ATP is formed as a consequence of oxidative phosphorylation resulting from the reoxidation of reduced coenzymes by the respiratory chain. The remainder is generated by phosphorylation at the "substrate level." Table 13–2 indicates the reactions responsible for the generation of the new high-energy bonds.

THE HEXOSE MONOPHOSPHATE (HMP) SHUNT OR PENTOSE PHOSPHATE PATHWAY

This pathway for the oxidation of glucose occurs in certain tissues, notably liver, lactating mammary gland, and adipose tissue, in addition to the Embden-Meyerhof (E-M) pathway of glycolysis. It is in effect a multicyclic process whereby 3 molecules of glucose-6-phosphate give rise to 3 molecules of CO_2 and 3 5-carbon residues. The latter are rearranged to regenerate 2 molecules of glucose-6-phosphate and one molecule of glyceraldehyde-3-phosphate. Since 2 molecules of glyceraldehyde-3-phosphate can regenerate a molecule of glucose-6-phosphate by reactions which are essentially a reversal of glycolysis, the pathway can account for the complete oxidation of glucose. As in the E-M glycolysis pathway, oxidation is achieved by dehydrogenation; but in the case of the shunt pathway, NADP and not NAD is used as a hydrogen acceptor. The enzymes of the shunt pathway are found in the extramitochondrial soluble portion of the cell.

A summary of the reactions of the hexose monophosphate shunt is shown below.

$$\begin{matrix} 3\ \textit{GLUCOSE}\text{-6-P} \\ + \\ 6\ NADP^+ \end{matrix} \longrightarrow \begin{matrix} 3\ CO_2 + 2\ \textit{GLUCOSE}\text{-6-P} \\ + \\ \textit{GLYCERALDEHYDE}\text{-3-P} + 6\ NADPH + 6\ H^+ \end{matrix}$$

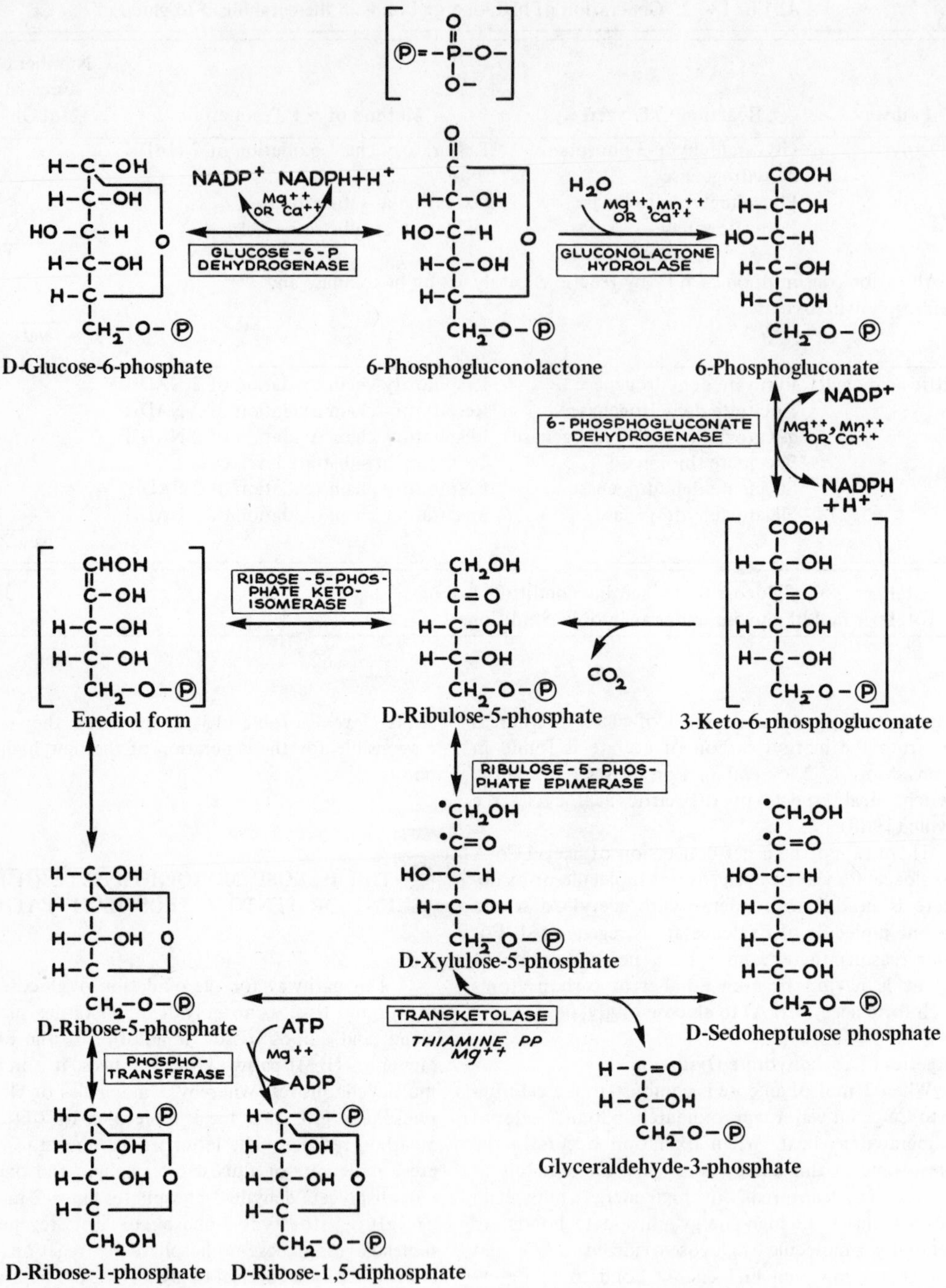

NADP⁺ NADPH+H⁺
Mg^{++} OR Ca^{++}
GLUCOSE-6-P DEHYDROGENASE

H_2O
Mg^{++}, Mn^{++} OR Ca^{++}
GLUCONOLACTONE HYDROLASE

D-Glucose-6-phosphate

6-Phosphogluconolactone

6-Phosphogluconate

6-PHOSPHOGLUCONATE DEHYDROGENASE
$NADP^+$
Mg^{++}, Mn^{++} OR Ca^{++}
NADPH $+H^+$

RIBOSE-5-PHOSPHATE KETO-ISOMERASE

CO_2

Enediol form

D-Ribulose-5-phosphate

3-Keto-6-phosphogluconate

RIBULOSE-5-PHOSPHATE EPIMERASE

D-Xylulose-5-phosphate

D-Ribose-5-phosphate

TRANSKETOLASE
THIAMINE PP
Mg^{++}

D-Sedoheptulose phosphate

ATP
Mg^{++}
ADP
PHOSPHO-TRANSFERASE

Glyceraldehyde-3-phosphate

D-Ribose-1-phosphate

D-Ribose-1,5-diphosphate

FIG 13–10. The hexose monophosphate (HMP) shunt (pentose phosphate pathway).

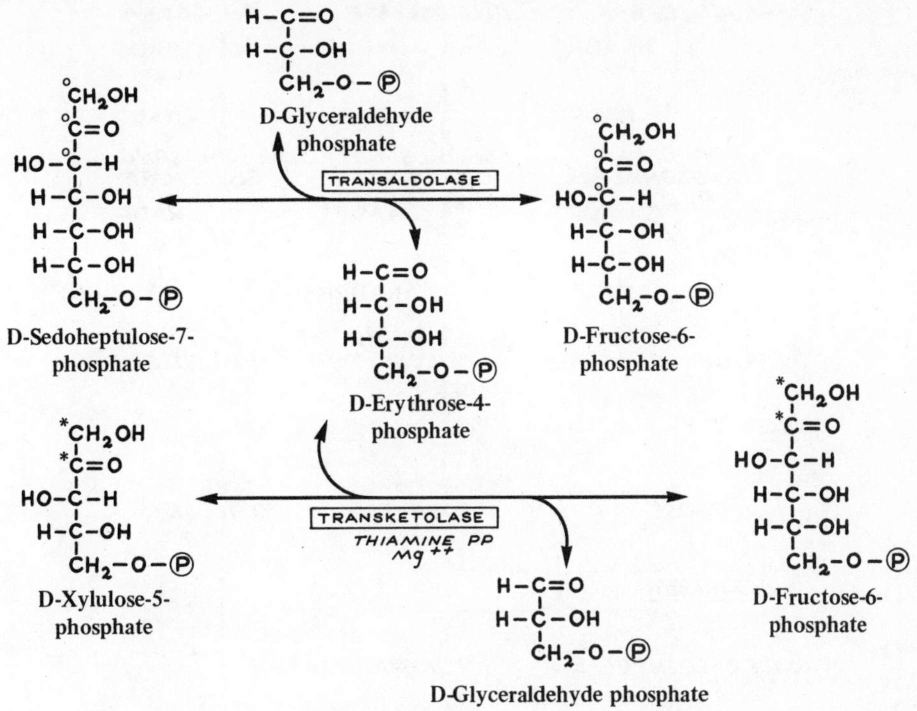

D-Glyceraldehyde phosphate

TRANSALDOLASE

D-Sedoheptulose-7-phosphate

D-Erythrose-4-phosphate

D-Fructose-6-phosphate

TRANSKETOLASE

THIAMINE PP

Mg^{++}

D-Xylulose-5-phosphate

D-Glyceraldehyde phosphate

D-Fructose-6-phosphate

FIG 13–10. The hexose monophosphate (HMP) shunt. (Cont'd.)

The sequence of reactions of the shunt pathway may be divided into 2 phases. In the first, glucose-6-phosphate undergoes dehydrogenation and decarboxylation to give the pentose ribulose-5-phosphate. In the second phase, ribulose-5-phosphate is converted back to glucose-6-phosphate by a series of reactions involving mainly 2 enzymes: **transketolase** and **transaldolase.**

Dehydrogenation of glucose-6-phosphate to 6-phosphogluconate occurs via the formation of 6-phosphogluconolactone catalyzed by **glucose-6-phosphate dehydrogenase**, an NADP-dependent enzyme. This reaction is inhibited by certain drugs such as sulfonamides and quinacrine (Atabrine). The hydrolysis of 6-phosphogluconolactone is accomplished by the enzyme **gluconolactone hydrolase.** A second oxidative step is catalyzed by **6-phosphogluconate dehydrogenase**, which also requires $NADP^+$ as hydrogen acceptor. Decarboxylation follows with the formation of the ketopentose, ribulose-5-phosphate. The reaction probably takes place in 2 steps through the intermediate 3-keto-6-phosphogluconate.

Ribulose-5-phosphate now serves as substrate for 2 different enzymes. **Ribulose-5-phosphate epimerase** alters the configuration about carbon 3, forming the epimer, xylulose-5-phosphate, another ketopentose. **Ribose-5-phosphate ketoisomerase** converts ribulose-5-phosphate to the corresponding aldopentose, ribose-5-phosphate. This reaction is analogous to the interconversion of fructose-6-phosphate and glucose-6-phosphate in the Embden-Meyerhof pathway.

Transketolase transfers the 2-carbon unit comprising carbons 1 and 2 of a ketose to the aldehyde carbon of an aldose sugar. It therefore effects the conversion of a ketose sugar into an aldose with 2 carbons less, and simultaneously converts an aldose sugar into a ketose with 2 carbons more. In addition to the enzyme transketolase, the reaction requires thiamine pyrophosphate as coenzyme and Mg^{++} ions. The 2-carbon moiety transferred is probably glycolaldehyde bound to thiamine pyrophosphate, ie, "active glycolaldehyde," resembling the "active acetaldehyde" formed during the oxidation of pyruvate (Fig 13–8). In the HMP shunt, transketolase catalyzes the transfer of the 2-carbon unit from xylulose-5-phosphate to ribose-5-phosphate, producing the 7-carbon ketose sedoheptulose-7-phosphate and the aldose glyceraldehyde-3-phosphate. These 2 products then enter another reaction known as transaldolation. Transaldolase allows the transfer of a 3-carbon moiety, "active dihydroxyacetone" (carbons 1–3), from the ketose sedoheptulose-7-phosphate to the aldose glyceraldehyde-3-phosphate to form the ketose fructose-6-phosphate and the 4-carbon aldose erythrose-4-phosphate.

A further reaction takes place, again involving transketolase, in which xylulose-5-phosphate serves as a donor of "active glycolaldehyde." In this case the erythrose-4-phosphate, formed above, acts as acceptor and the products of the reaction are fructose-6-phosphate and glyceraldehyde-3-phosphate.

In order to oxidize glucose completely to CO_2 via the shunt pathway, it is necessary that the enzymes are

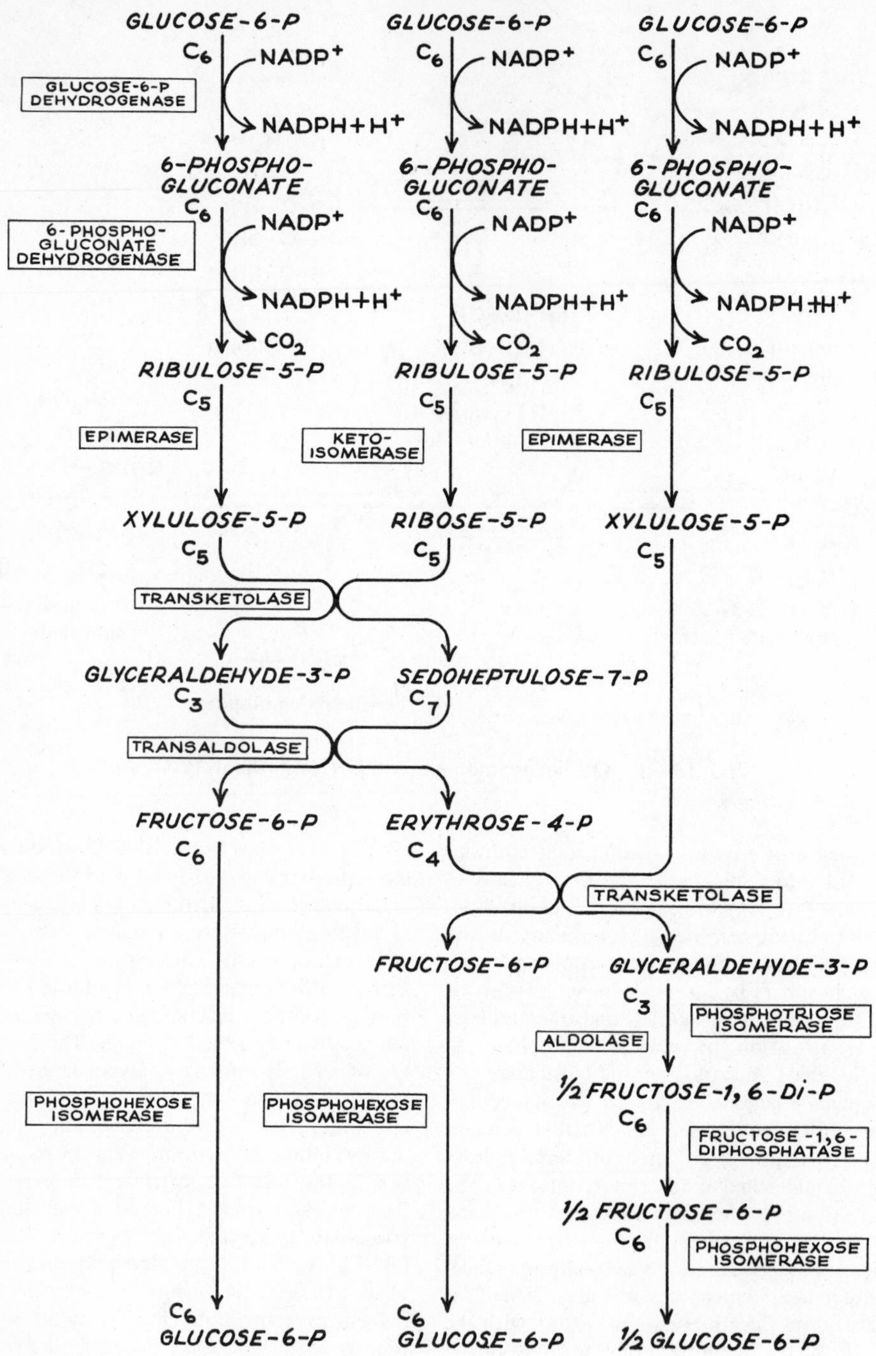

FIG 13–11. Flow chart of hexose monophosphate shunt and its connections with the Embden-Meyerhof pathway of glycolysis.

present in the tissue to convert glyceraldehyde-3-phosphate to glucose-6-phosphate. This involves the enzymes of the E-M pathway working in a reverse direction and, in addition, the enzyme **fructose-1,6-diphosphatase** (see p 247). A summary of the reactions of the direct oxidative pathway is shown in Figs 13–10 and 13–11. Most of the reactions are reversible, but the complete pathway is probably irreversible at the gluconolactone hydrolase step.

Metabolic Significance of the Hexose Monophosphate (HMP) Shunt

It is clear that the direct oxidative pathway is markedly different from the Embden-Meyerhof pathway of glycolysis. Oxidation occurs in the first reactions, and CO_2, which is not produced at all in the E-M pathway, is a characteristic product. This fact has been utilized in experiments designed to evaluate the relative proportions of glucose metabolized by the E-M pathway compared with the shunt pathway. Most studies have been based on measurement of differences in the rate of liberation of $^{14}CO_2$ from glucose-1-^{14}C and from glucose-6-^{14}C. In the glycolytic pathway, carbons 1 and 6 of glucose are both converted to the methyl carbon of pyruvic acid and therefore metabolized in the same manner. In the direct oxidative pathway, carbons 1 and 6 of glucose are treated differently, carbon 1 being removed early by decarboxylation with the formation of labeled CO_2. Interpretation of experimental results, however, is difficult because of the many assumptions which must be made, eg, whether or not fructose-6-phosphate formed by the direct oxidative pathway is recycled via glucose-6-phosphate. These matters are discussed in detail by Wood & others (1963).

Estimates of the activity of the shunt pathway in various tissues give an indication of its metabolic significance. It is active in liver, adipose tissue, adrenal cortex, thyroid, erythrocytes, testis, and lactating mammary gland. It is not active in nonlactating mammary gland, and its activity is low in skeletal muscle. Most of the tissues in which the pathway is active specialize in using NADPH in the synthesis of fatty acids or steroids and in the synthesis of amino acids via glutamate dehydrogenase. It is probable that the presence of active lipogenesis or of a system which utilizes NADPH stimulates an active degradation of glucose via the shunt pathway. The synthesis of glucose-6-phosphate dehydrogenase and 6-phosphogluconate dehydrogenase may also be induced during conditions associated with the "fed state." One of the major functions of the HMP shunt would appear to be, therefore, the provision of reduced NADP required by anabolic processes outside the mitochondria.

Another important function of the HMP shunt is to provide pentoses for nucleotide and nucleic acid synthesis. The source of the ribose is the ribose-5-phosphate intermediate. This compound may be isomerized to the 1-phosphate (cf glucose-6-phosphate ⟷ glucose-1-phosphate interconversion, p 233), or it can react with ATP to give ribose-1,5-diphosphate (cf fructose-6-phosphate → fructose-1,6-diphosphate). Muscle tissue contains very small amounts of glucose-6-phosphate dehydrogenase and 6-phosphogluconate dehydrogenase. Nevertheless, skeletal muscle is capable of synthesizing ribose. This is probably accomplished by a reversal of the shunt pathway utilizing fructose-6-phosphate, glyceraldehyde-3-phosphate, and the enzymes transketolase and transaldolase. Thus, it is not necessary to have a completely functioning shunt pathway in order that a tissue might synthesize ribose. Hiatt & Lareau (1960) concluded from studies in vitro that each of several tissues examined was capable of synthesizing ribose. In human tissues, ribose seems to be derived primarily by way of the oxidative reactions of the shunt pathway, whereas in the rat and mouse—except in muscle—the nonoxidative reactions appear to play a larger role than the oxidative.

In studying the metabolism of rabbit lens tissue, Kinoshita & Wachtl (1958) found that, while most of the glucose utilized was converted to lactate, at least 10% was metabolized via the shunt pathway. A relation between galactose accumulation and the experimental production of cataracts in rats has been recognized. Development of cataracts sometimes occurs as a complication of galactosemia, an inherited metabolic disease associated with the inability to convert galactose to glucose. In the light of the existence of a shunt pathway in the metabolism of the lens, it is of interest that galactose has been found to inhibit the activity of glucose-6-phosphate dehydrogenase of the lens in vivo when the sugar is fed to experimental animals and in vitro when galactose-1-phosphate is added to a homogenate of lens tissue (Lerman, 1959).

Formation of NADPH seems to be an important function of the operation of the shunt pathway in red blood cells, and a direct correlation has been found between the activity of enzymes of the direct oxidative pathway, particularly of glucose-6-phosphate dehydrogenase, and the fragility of red cells (susceptibility to hemolysis), especially when the cells are subjected to the toxic effects of certain drugs (primaquine, acetylphenylhydrazine) or the susceptible individual has ingested fava beans (*Vicia fava*—favism). The majority of patients whose red cells are hemolyzed by these toxic agents have been found to possess a hereditary deficiency in the oxidative enzymes of the shunt pathway of the red blood cell.

L-Phenylalanine and phenylpyruvate, metabolites which accumulate in the blood in phenylketonuria, inhibit hexokinase, pyruvate kinase, and 6-phosphogluconate dehydrogenase of differentiating brain (Weber & others, 1970).

GLUCONEOGENESIS

Gluconeogenesis meets the needs of the body for glucose when carbohydrate is not available in sufficient amounts from the diet. A continual supply of glucose is necessary as a source of energy especially for the

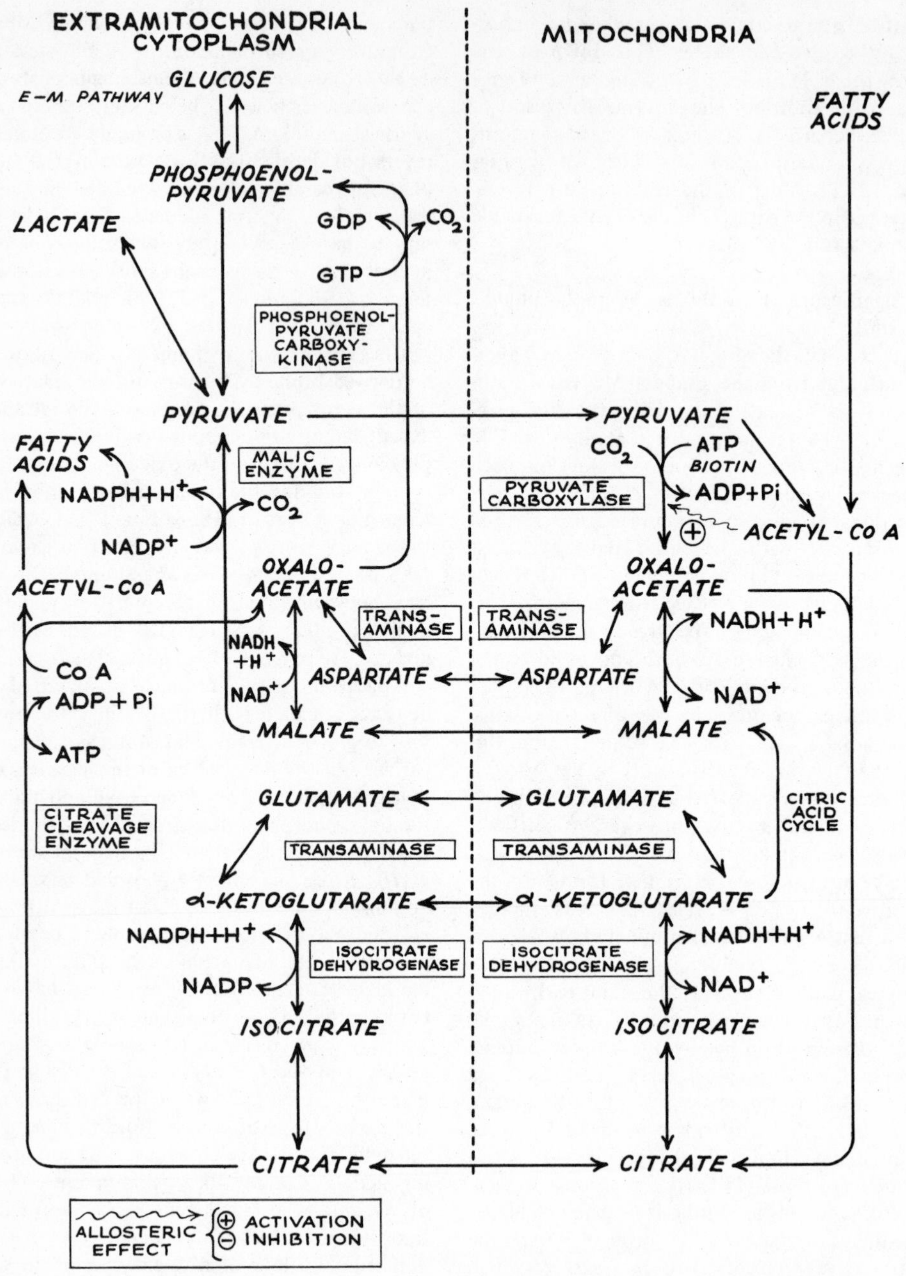

FIG 13–12. Some key enzymes of gluconeogenesis and lipogenesis showing distribution within the liver cell.

nervous system and the erythrocytes. Glucose is also required in adipose tissue as a source of glyceride-glycerol, and it probably plays a role in maintaining the level of intermediates of the citric acid cycle in many tissues. It is clear that even under conditions where fat may be supplying most of the caloric requirement of the organism, there is always a certain basal requirement for glucose. In addition, glucose is the only fuel which will supply energy to skeletal muscle under anaerobic conditions. It is the precursor of milk sugar (lactose) in the mammary gland and it is taken up actively by the fetus. It is not surprising, therefore, to find that enzymatic pathways have been developed in certain specialized tissues for the conversion of noncarbohydrates to glucose. In addition, these gluconeogenic mechanisms are used to clear the products of the metabolism of other tissues from the blood, eg, lactate, produced by muscle and erythrocytes, and glycerol, which is continuously produced by adipose tissue.

In mammals, the liver and the kidney are the principal organs responsible for gluconeogenesis. As the main pathway of gluconeogenesis is essentially a reversal of glycolysis, this can explain why the glycolytic activity of liver and kidney is low when there is active gluconeogenesis.

Metabolic Pathways Involved in Gluconeogenesis

These pathways are modifications and adaptations of the Embden-Meyerhof pathways and the citric acid cycle. They are concerned with the conversion of glucogenic amino acids, lactate, glycerol, and, in ruminants, propionate, to glucose or glycogen. It has been pointed out by Krebs (1963) that energy barriers obstruct a simple reversal of glycolysis (1) between pyruvate and phosphoenolpyruvate, (2) between fructose-1,6-diphosphate and fructose-6-phosphate, (3) between glucose-6-phosphate and glucose, and (4) between glucose-1-phosphate and glycogen. These barriers are circumvented by special reactions described below:

(1) Present in mitochondria is an enzyme, **pyruvate carboxylase**, which in the presence of ATP, biotin, and CO_2 converts pyruvate to oxaloacetate. The function of the biotin is to bind CO_2 from bicarbonate onto the enzyme prior to the addition of the CO_2 to pyruvate. In the extramitochondrial part of the cell is found a second enzyme, **phosphoenolpyruvate carboxykinase**, which catalyzes the conversion of oxaloacetate to phosphoenolpyruvate. High-energy phosphate in the form of GTP or ITP is required in this reaction, and CO_2 is liberated. Thus, with the help of these 2 enzymes, and lactic dehydrogenase, lactate can be converted to phosphoenolpyruvate. However, Lardy & others (1965) have pointed out that oxaloacetate does not diffuse readily from mitochondria. They have demonstrated that alternative means are available to achieve the same end by converting oxaloacetate into compounds which can diffuse from the mitochondria, followed by their reconversion to oxaloacetate in the extramitochondrial portion of the cell. Such compounds are malate, aspartate, α-ketoglutarate, glutamate, and citrate. Their formation from oxaloacetate within mitochondria and their conversion back to oxaloacetate in the extramitochondrial compartment involve citric acid cycle reactions and transaminations (Fig 13–12). There are species differences with regard to the distribution of phosphoenolpyruvate carboxykinase. The extramitochondrial location is true for the rat and mouse; but in the rabbit and chicken the enzyme is located in the mitochondria, and in the guinea pig it is found in both the mitochondria and cytosol (Scrutton & Utter, 1968).

(2) The conversion of fructose-1,6-diphosphate to fructose-6-phosphate, necessary to achieve a reversal of glycolysis, is catalyzed by a specific enzyme, **fructose-1,6-diphosphatase**. This is a key enzyme in the sense that its presence determines whether or not a tissue is capable of resynthesizing glycogen from pyruvate and triosephosphates. It is present in liver and kidney and has been demonstrated in striated muscle by Krebs & Woodford (1965). It is held to be absent from adipose tissue (Vaughan, 1961).

(3) The conversion of glucose-6-phosphate to glucose is catalyzed by another specific phosphatase, **glucose-6-phosphatase**. It is present in intestine, liver, and kidney where it allows these particular tissues to add glucose to the blood. The enzyme, which is microsomal, also possesses pyrophosphatase activity. It is absent from muscle and adipose tissue.

(4) The breakdown of glycogen to glucose-1-phosphate is carried out by phosphorylase. The synthesis of glycogen involves an entirely different pathway through the formation of uridine diphosphate glucose and the activity of **glycogen synthetase**.

The relationships between these key enzymes of gluconeogenesis and the E-M glycolytic pathway are shown in Fig 13–13. After transamination or deamination, glucogenic amino acids form either pyruvate or members of the citric acid cycle. Therefore, the reactions described above can account for the conversion of both glucogenic amino acids and lactate to glucose or glycogen. Similarly, propionate, which is a major source of glucose in ruminants, enters the main gluconeogenic pathway via the citric acid cycle after conversion to succinyl-Co A. Propionate, as with other fatty acids, is first activated with ATP and Co A by an appropriate **thiokinase**. Propionyl-Co A, the product of this reaction, undergoes a CO_2 fixation reaction to form D-methylmalonyl-Co A, catalyzed by **propionyl-Co A carboxylase** (Fig 13–14). This reaction is analogous to the fixation of CO_2 in acetyl-Co A by acetyl-Co A carboxylase (see Chapter 14) in that it forms a malonyl derivative and requires biotin as a coenzyme. D-Methylmalonyl-Co A must be converted to its stereoisomer, L-methylmalonyl-Co A, by **methylmalonyl-Co A racemase** before its final isomerization to succinyl-Co A by the enzyme **methylmalonyl-Co A isomerase**, which requires vitamin B_{12} as a coenzyme.

Although the pathway to succinate is its main route of metabolism, propionate may also be used as the priming molecule for the synthesis–in adipose tissue and mammary gland–of fatty acids which have an odd number of carbon atoms in the molecule. C-15 and C-17 fatty acids are found particularly in the lipids of ruminants.

Glycerol is a product of the metabolism of adipose tissue, and only tissues that possess the activating enzyme, **glycerokinase**, can utilize it. This enzyme, which requires ATP, is found, among other tissues, in liver and kidney. Glycerokinase catalyzes the conversion of glycerol to α-glycerophosphate. This pathway connects with the triosephosphate stages of the E-M pathway because α-glycerophosphate may be oxidized to dihydroxyacetone phosphate by NAD^+ in the presence of another enzyme, **α-glycerophosphate dehydrogenase**, although the equilibrium constant is very much in favor of α-glycerophosphate formation. Liver and kidney are thus able to convert glycerol to blood glucose by making use of the above enzymes, some of the enzymes of the E-M pathway, and the specific enzymes of the gluconeogenic pathway, fructose-1,6-diphosphatase and glucose-6-phosphatase.

GLYCOGEN SYNTHETASE
GLYCOGEN
UDPG
CYCLIC AMP
Pi
PHOSPHORYLASE
GLUCOSE-1-P
ADP
GLUCOKINASE
ATP
HEXOKINASE
GLUCOSE
GLUCOSE-6-P
H_2O
Pi
GLUCOSE-6-P-ASE
AMP
FRUCTOSE-6-P
Pi
FRUCTOSE 1,6-Di-P-ASE
H_2O
ATP
AMP
PHOSPHOFRUCTOKINASE
ADP
FRUCTOSE-1,6-Di-P
CITRATE
NADH+H$^+$
NAD$^+$
GLYCERALDEHYDE-3-P
DIHYDROXY-ACETONE-P
α-GLYCERO-P DEHYDROGENASE
α-GLYCERO-P
ADP
GLYCEROKINASE
ATP
GLYCEROL
E.M. PATHWAY
PHOSPHOENOL-PYRUVATE
PYRUVATE KINASE
PHOSPHOENOLPYRUVATE CARBOXYKINASE
ADP
GDP
CO_2
ATP
GTP
PYRUVATE
OXALOACETATE
ALLOSTERIC EFFECT
⊕ ACTIVATION
⊖ INHIBITION

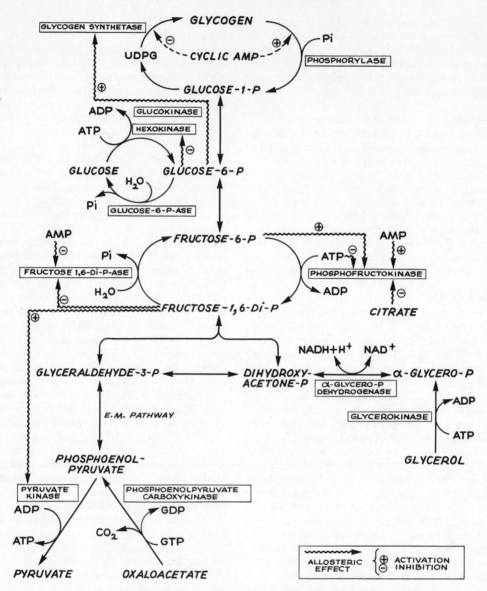

FIG 13–13. Key enzymes of glycolysis and gluconeogenesis in the liver.

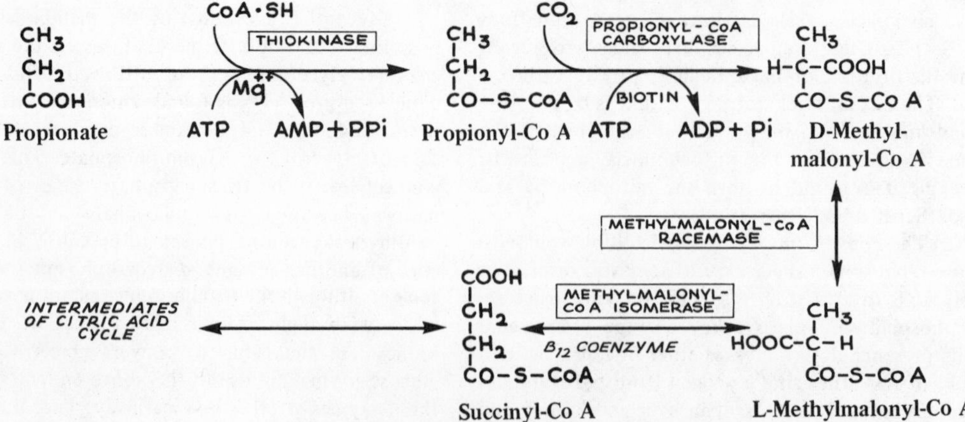

FIG 13–14. Metabolism of propionic acid.

Extramitochondrial **isocitrate dehydrogenase** is NADP-dependent. **"Malic enzyme,"** which catalyzes the oxidative decarboxylation of malate to pyruvate with the production of NADPH, is also present in the extramitochondrial compartment of the cell. Thus, extramitochondrial production of NADPH by the hexose monophosphate shunt is augmented by the action of these 2 enzymes. Probably they are concerned more with lipogenesis than with gluconeogenesis. Another extramitochondrial enzyme, which again is probably more concerned in supplying acetyl-Co A for lipogenesis, is the **citrate cleavage enzyme,** which catalyzes the conversion of citrate to oxaloacetate plus acetyl-Co A; ATP, Co A, and Mg^{++} are necessary for this reaction. The interrelationships of these various pathways with respect to gluconeogenesis are shown in Fig 13–12.

THE REGULATION OF CARBOHYDRATE METABOLISM

It is convenient to divide the regulation of carbohydrate metabolism into 2 parts: (1) factors affecting the blood glucose, and (2) the regulation of carbohydrate metabolism at the cellular and enzymatic level. However, this is an arbitrary division as the 2 parts are functionally related.

THE BLOOD GLUCOSE

Sources of Blood Glucose

A. From Carbohydrates of the Diet: Most carbohydrates in the diet form glucose or fructose upon digestion. These are absorbed into the portal vein. Fructose is readily converted to glucose in the liver.

B. From Various Glucogenic Compounds Which Undergo Gluconeogenesis: These compounds fall into 2 categories–those which involve a direct net conversion to glucose without significant recycling, such as amino acids and propionate; and those which are the products of the partial metabolism of glucose in certain tissues and which are conveyed to the liver and kidney, where they are resynthesized to glucose. Thus, lactate, formed by the oxidation of glucose in skeletal muscle and by erythrocytes, is transported to the liver and kidney where it re-forms glucose, which again becomes available via the circulation for oxidation in the tissues. This process is known as the Cori cycle or lactic acid cycle (Fig 13–15). Similarly, glyceride-glycerol of adipose tissue is derived initially from the blood glucose since free glycerol cannot be utilized readily for the synthesis of triglycerides in this tissue. Glycerides of adipose tissue are continually undergoing hydrolysis to form free glycerol, which diffuses out of the tissue into the blood. It is converted back to glucose by gluconeogenic mechanisms in the liver and kidney. Thus, a continuous cycle exists in which glucose is transported from the liver and kidney to adipose tissue and whence glycerol is returned to be synthesized into glucose by the liver and kidney.

C. From liver glycogen by glycogenolysis.

The Concentration of the Blood Glucose

In the postabsorptive state the blood glucose concentration of man varies between 80 and 100 mg/100 ml. After the ingestion of a carbohydrate meal it may rise to 120–130 mg/100 ml. During fasting, the level falls to around 60–70 mg/100 ml. Under normal circumstances, the level is controlled within these limits. The normal blood glucose level of ruminants is considerably lower, being approximately 40 mg/100 ml in sheep and 60 mg/100 ml in cattle. These lower normal levels appear to be associated with the fact that ruminants ferment virtually all dietary carbohydrate to lower (volatile) fatty acids, and these largely replace glucose as the main metabolic fuel of the tissues in the fed condition.

Regulation of the Blood Glucose

The maintenance of stable levels of glucose in the blood is one of the most finely regulated of all homeostatic mechanisms and one in which the liver, the extrahepatic tissues, and several hormones play a part. Liver cells appear to be freely permeable to glucose (Cahill & others, 1959) whereas cells of extrahepatic tissues are relatively impermeable. This results in the passage through the cell membrane being the rate-limiting step in the uptake of glucose in extrahepatic tissues, whereas it is probable that the activity of certain enzymes and the concentration of key intermediates exert a much more direct effect on the uptake or output of glucose from liver. Nevertheless, the concentration of glucose in the blood is an important parameter in determining the rate of uptake of glucose in both liver and extrahepatic tissues. It is to be noted that hexokinase is inhibited by glucose-6-phosphate, so that some feedback control may be exerted on glucose uptake in extrahepatic tissues dependent on hexokinase for glucose phosphorylation. The liver is not subject to this constraint since glucokinase is not affected by glucose-6-phosphate.

At normal blood glucose concentrations (80–100 mg/100 ml), the liver appears to be a net producer of glucose. However, as the glucose level rises, the output of glucose ceases, and at high levels there is a net uptake. In the rat, it has been estimated that the rate of uptake of glucose and the rate of output are equal at a blood glucose concentration of 150 mg/100 ml. Landau & others (1961) have shown that in dogs the blood glucose level at which there is net uptake by the liver varies with the type of diet. Thus, infusion of glucose into dogs maintained on a high-protein diet resulted in a rise in blood glucose, with a cessation of net hepatic glucose production only at hyperglycemic

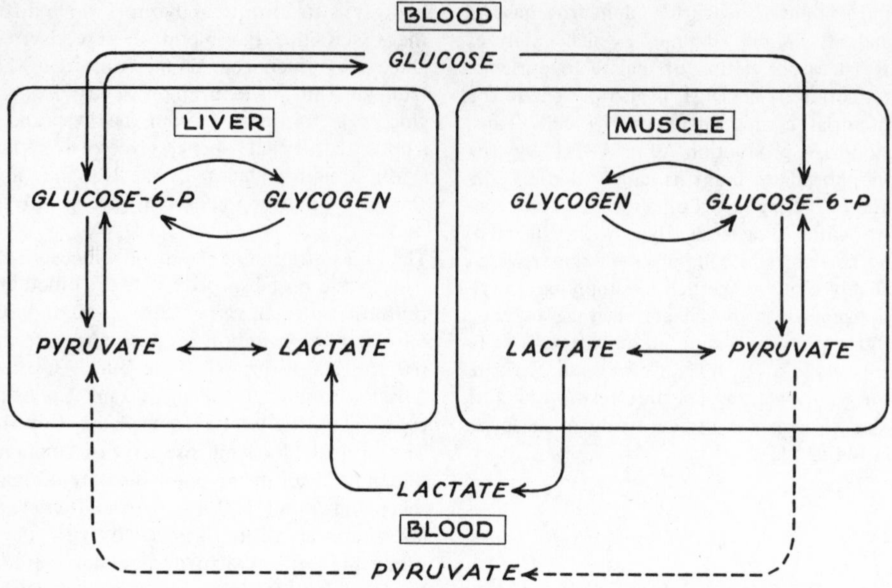

FIG 13–15. The lactic acid (Cori) cycle.

levels. In contrast, in carbohydrate fed dogs, the blood glucose increased in concentration very little upon glucose infusion, and there was an immediate net uptake of glucose by the liver. An explanation of these differences due to changes in diet is probably to be found in changes in activity of enzymes in the liver concerned with glycolysis and gluconeogenesis.

In addition to the direct effects of hyperglycemia in enhancing the uptake of glucose into both the liver and peripheral tissues, the hormone **insulin** plays a central role in the regulation of the blood glucose concentration. It is produced by the beta cells of the islets of Langerhans in the pancreas and is secreted into the blood as a direct response to hyperglycemia. Its concentration in the blood parallels that of the blood glucose, and its administration results in prompt hypoglycemia. Substances causing release of insulin include amino acids, free fatty acids, ketone bodies, glucagon, secretin, and tolbutamide. Epinephrine and norepinephrine block the release of insulin. In vitro (and probably in vivo), insulin has an immediate effect on tissues such as adipose tissue and muscle in increasing the rate of glucose uptake. It is considered that this action is due to an enhancement of glucose transport through the cell membrane. In contrast, it is not easy to demonstrate an immediate effect of insulin on glucose uptake by liver tissue. This agrees with other findings which show that glucose penetration of hepatic cells is not rate limited by their permeability. However, increased glucose uptake by liver slices can be demonstrated after administration of insulin in vivo, which indicates that other mechanisms are sensitive to the hormone. Perhaps the more unequivocal evidence that insulin does have a direct effect on glucose metabolism in the liver has been obtained in experiments with the isolated perfused liver by Mortimore (1963) and by Exton & others (1966), who showed that insulin suppresses glucose production; and by Miller (1965), who demonstrated reduced oxidation of glucose U-^{14}C to $^{14}CO_2$.

The **anterior pituitary gland** secretes hormones that tend to elevate the blood sugar and therefore antagonize the action of insulin. These are growth hormone, ACTH (corticotropin), and possibly other "diabetogenic" principles. Growth hormone secretion is stimulated by hypoglycemia. Growth hormone decreases glucose uptake in certain tissues, eg, muscle. Some of this effect may not be direct since it mobilizes free fatty acids from adipose tissue which themselves inhibit glucose utilization (see p 253).* Chronic administration of growth hormone leads to diabetes. By producing hyperglycemia it stimulates secretion of insulin, eventually causing beta cell exhaustion. Although ACTH could have an indirect effect upon glucose utilization, since it enhances the release of free fatty acids from adipose tissue, its major effect on carbohydrate metabolism is due to its stimulation of the secretion of hormones of the adrenal cortex.

The **adrenal cortex** secretes a number of steroid hormones of which the glucocorticoids (11-oxysteroids) are important in carbohydrate metabolism. Upon administration, the glucocorticoids lead to gluconeogenesis. This is as a result of increased protein catabolism in the tissues, increased hepatic uptake of amino acids, and increased activity of transaminases and other enzymes concerned with gluconeogenesis in the liver. In addition, glucocorticoids inhibit the utilization of glucose in extrahepatic tissues. In all these actions, glucocorticoids act in an antagonistic manner to insulin.

Epinephrine, secreted by the adrenal medulla, stimulates glycogen breakdown in both the liver and in muscle. Administration of epinephrine leads to an outpouring of glucose from the liver provided glycogen is present. In muscle, as a result of the absence of glucose-6-phosphatase, glycolysis ensues with the formation of lactate. The lactate which diffuses into the blood is converted by the gluconeogenic mechanisms back to glycogen in the liver (Cori cycle). The stimulation of glycogenolysis by epinephrine is due to its ability to activate the enzyme phosphorylase (see p 234). Hypoglycemia causes a sympathetic discharge. The increased epinephrine secretion stimulates glycogenolysis, which is followed by a rise in the level of blood glucose.

Glucagon is the hormone produced by the alpha cells of the islets of Langerhans of the pancreas. Its secretion is stimulated by hypoglycemia, and, when it reaches the liver (via the portal vein), it causes glycogenolysis by activating phosphorylase in a manner similar to epinephrine. Most of the endogenous glucagon is cleared from the circulation by the liver. Unlike epinephrine, glucagon does not have an action on muscle phosphorylase. Glucagon also enhances gluconeogenesis from amino acids and lactate.

Thyroid hormone should also be considered as affecting the blood sugar. There is experimental evidence that thyroxine has a diabetogenic action and that thyroidectomy inhibits the development of diabetes. It has also been noted that there is a complete absence of glycogen from the livers of thyrotoxic animals. In humans, the fasting blood sugar is elevated in hyperthyroid patients and decreased in hypothyroid patients. However, hyperthyroid patients apparently utilize glucose at a normal rate, whereas hypothyroid patients have a decreased ability to utilize glucose. In addition, hypothyroid patients are much less sensitive to insulin than normal or hyperthyroid individuals. All of these effects of thyroid hormone on carbohydrate metabolism may be related to differences in end-organ response, rates of destruction of insulin, or both.

The Renal Threshold for Glucose

When the blood sugar rises to relatively high levels, the kidney also exerts a regulatory effect. Glucose is continually filtered by the glomeruli but is ordinarily returned completely to the blood by the reabsorptive system of the renal tubules. The reabsorption of glucose is effected by phosphorylation in the tubular cells, a process which is similar to that responsible for the absorption of this sugar from the intestine. The phosphorylation reaction is enzymatically catalyzed, and the capacity of the tubular system to reabsorb glucose is limited by the concentration of the enzymatic components of the tubule cell to a rate of about 350 mg/minute. When the blood levels of glucose are elevated, the glomerular filtrate may contain more glucose than can be reabsorbed; the excess passes into the urine to produce **glycosuria.**

In normal individuals, glycosuria occurs when the venous blood sugar exceeds 170–180 mg/100 ml. This level of the venous blood sugar is termed the **renal threshold** for glucose. Since the maximal rate of reabsorption of glucose by the tubule (**Tm_G**–the tubular maximum for glucose) is a constant, it is a more accurate measurement than the renal threshold, which varies with changes in the glomerular filtration rate.

Glycosuria may be produced in experimental animals with phlorhizin, which inhibits the glucose reabsorptive system in the tubule. This is known as **renal glycosuria** since it is caused by a defect in the renal tubule and may occur even when blood glucose levels are normal. Glycosuria of renal origin is also found in human subjects. It may result from inherited defects in the kidney (see Chapter 18), or it may be acquired as a result of disease processes.

Carbohydrate Tolerance

The ability of the body to utilize carbohydrates may be ascertained by measuring its **carbohydrate tolerance.** It is indicated by the nature of the blood glucose curve following the administration of glucose. **Diabetes mellitus** ("sugar" diabetes) is characterized by decreased tolerance to carbohydrate due to decreased secretion of insulin. This is manifested by elevated blood glucose levels (hyperglycemia) and glycosuria and may be accompanied by changes in fat metabolism. Tolerance to carbohydrate is decreased not only in diabetes but also in conditions where the liver is damaged, in some infections, in obesity, and sometimes in atherosclerosis. It would also be expected to occur in the presence of hyperactivity of the pituitary or adrenal cortex because of the antagonism of the hormones of these endocrine glands to the action of insulin.

Insulin, the hormone of the islets of Langerhans of the pancreas, increases tolerance to carbohydrate. Injection of insulin lowers the content of the glucose in the blood and increases its utilization and its storage in the liver and muscle as glycogen. An excess of insulin may lower the blood glucose level to such an extent that severe hypoglycemia occurs which results in convulsions and even in death unless glucose is administered promptly. In man, hypoglycemic convulsions may occur when the blood glucose is lowered to about 20 mg/100 ml or less. Increased tolerance to carbohydrate is also observed in pituitary or adrenocortical insufficiency; presumably this is attributable to a decrease in the normal antagonism to insulin which results in a relative excess of that hormone.

Measurement of Glucose Tolerance

The glucose tolerance test is a valuable diagnostic aid. Glucose tolerance (ability to utilize carbohydrate) is decreased in diabetes and increased in hypopituitarism, hyperinsulinism, and adrenocortical hypofunction (such as in Addison's disease).

A. Standard Oral Glucose Tolerance Test: After an overnight fast of 12 hours, the patient is given 0.75–1.5 g of glucose/kg (or a standard dose of between 50 and 100 g of glucose may be used). Specimens of blood and urine are taken before the administration of glucose and at intervals of ½ or 1 hour thereafter for 3–4 hours. The concentration of glucose in the blood is measured and plotted against time. As a result of the administration of glucose as described, the blood glucose in normal individuals increases in 1 hour from about 80 mg/100 ml to about 130 mg/100 ml; at the end of 2–2½ hours, a return to normal levels occurs. In a diabetic patient, the increase in the blood glucose level is greater than in normal subjects and a much slower return to the pre-test level is observed; ie, the glucose tolerance curve is typically higher and more prolonged than normal.

B. Intravenous Glucose Tolerance Test: An intravenous test is preferred if abnormalities in absorption of glucose from the intestine, as might occur in hypothyroidism or in sprue, are suspected. A 20% solution of glucose (0.5 g/kg) is given intravenously at a uniform rate over a period of ½ hour. A control (fasting) blood specimen is taken, and additional blood samples are obtained ½, 1, 2, 3, and 4 hours after the glucose injection. In normal individuals, the control specimen of blood contains a normal amount of glucose; the concentration does not exceed 250 mg/100 ml after the infusion has been completed; by 2 hours the concentration of glucose in the blood has fallen below the control level, and between the third and fourth hours it has returned to the normal fasting level.

C. Corticosteroid Tests: Administration of corticosteroids (eg, prednisone) causes glucose intolerance. This property has been utilized to detect latent diabetes.

Regulation of Carbohydrate Metabolism at the Cellular & Enzymatic Level

Gross effects on metabolism of changes in nutritional state or in the endocrine balance of an animal may be studied by observing changes in the concentration of blood metabolites. By such technics as catheterization it is also possible to study effects on individual organs by measuring arteriovenous differences, etc. However, the changes which occur in the metabolic balance of the intact animal are due to shifts in the pattern of metabolism in individual tissues which are usually associated with changes in activity of key enzymes or changes in availability of metabolites. Three types of mechanisms can be described which are responsible for regulating the activity of enzymes concerned in carbohydrate metabolism. The first type of regulation is due to changes in the rate of enzyme synthesis (enzyme induction or repression) in which the hormones appear to play a leading role. Effects of changes of this type, which involve protein synthesis, usually take several hours to manifest themselves. The second type of mechanism, which is usually rapid, concerns the conversion of an inactive enzyme into an active one. This frequently involves phosphorylation of the enzyme by a protein kinase and ATP, and dephosphorylation by a phosphatase enzyme. The active form of the enzyme can be either the phosphorylated enzyme (eg, phosphorylase a) or the dephosphorylated enzyme (eg, glycogen synthetase I). The third type of regulation is due to changes in affinity of the enzyme for its substrate (ie, changes in K_m) usually brought about by another substance, perhaps a product, which attaches itself to the apoenzyme changing the protein conformation (allosteric effect). These effects are usually rapid and form the basis of several feedback control mechanisms.

Some of the better documented changes in enzyme activity that are considered to occur under various metabolic conditions are listed in Table 13–3. The information in this table applies mainly to the liver. The enzymes involved catalyze reactions which are not in equilibrium and which may be regarded physiologically as "one way" rather than balanced reactions. As a result, changes in their activity cause a net change in rate of flow of metabolites along the respective pathways they catalyze. This phenomenon is analogous to opening and closing a one-way valve. Also, the effect is reinforced because, invariably, the activity of the enzyme catalyzing the change in the opposite direction varies reciprocally. Thus, glucokinase catalyzes the conversion of glucose to glucose-6-phosphate. In the same compartment of the cell (the extramitochondrial region) is found glucose-6-phosphatase, the enzyme catalyzing the reaction in the reverse direction. Under conditions of a plentiful supply of carbohydrate, glucokinase activity is high whereas glucose-6-phosphatase activity is depressed. In starvation, glucokinase activity falls relative to glucose-6-phosphatase activity. In this way a so-called "futile cycle" whose net result would be hydrolysis of ATP is prevented. It is also of importance that the key enzymes involved in a metabolic pathway are all activated or depressed in a coordinated manner. Table 13–3 shows that this is clearly the case. The enzymes involved in the utilization of glucose are all activated under the circumstance of a superfluity of glucose, and under these conditions the enzymes responsible for producing glucose by the pathway of gluconeogenesis are all low in activity. It has been proposed by Weber (1968) that glucocorticoid hormones function as inducers and that insulin acts as a suppressor (repressor) of the biosynthesis of key hepatic gluconeogenic enzymes. Insulin is able to prevent induction of new enzymes by glucocorticoids; on the other hand, glucocorticoids have no effect on the activity of glycolytic enzymes which are sensitive to induction by insulin. Thus, according to this view, the secretion of insulin, which is responsive to the blood glucose concentration, controls the activity both of the enzymes responsible for glycolysis and those responsible for gluconeogenesis. All of these effects, which can be explained on the basis of new enzyme synthesis, can be prevented by agents which block the synthesis of new protein, such as actinomycin D, puromycin, and ethionine.

Both dehydrogenases of the HMP pathway can be classified as adaptive enzymes since they increase in activity in the well fed animal and when insulin is given to a diabetic animal. Activity is low in diabetes or

TABLE 13–3. Regulatory and adaptive enzymes of the rat (mainly liver).

	Activity In				
	Carbohydrate Feeding	Starvation	Diabetes	Activator or Inducer	Inhibitor or Repressor
Enzymes of glycolysis and glycogenesis, etc					
Glucokinase	↑	↓	↓	Insulin	
Glycogen synthetase	↑	↓	↓	Glucose-6-phosphate, insulin	Cyclic AMP
Phosphofructokinase	↑	↓	↓	AMP, insulin, fructose-6-P, P_i, fructose-1,6-diphosphate	Citrate (fatty acids, ketone bodies), ATP
Pyruvate kinase	↑	↓	↓	Insulin, fructose-1, 6-diphosphate	ATP, alanine
Pyruvate dehydrogenase	↑	↓	↓	Co A, NAD	Acetyl-Co A, NADH
Enzymes of gluconeogenesis					
Pyruvate carboxylase	↓	↑	↑	Acetyl-Co A, glucocorticoids, glucagon, epinephrine	ADP, insulin
Phosphoenolpyruvate carboxykinase	↑	↑	↑	Glucocorticoids	Insulin
Fructose-1,6-diphosphatase	↓	↑	↑	Glucocorticoids, glucagon, epinephrine	Fructose-1,6-diphosphate, AMP, insulin
Glucose-6-phosphatase	↓	↑	↑	Glucocorticoids, glucagon, epinephrine	Insulin
Enzymes of the hexose monophosphate shunt and lipogenesis					
Glucose-6-phosphate dehydrogenase	↑	↓	↓	Insulin	
6-Phosphogluconate dehydrogenase	↑	↓	↓	Insulin	
"Malic enzyme"	↑	↓	↓	Insulin	
Citrate cleavage enzyme	↑	↓	↓	Insulin	

fasting. "Malic enzyme" and citrate cleavage enzyme behave in a similar fashion, indicating that these 2 enzymes are probably involved in lipogenesis rather than in gluconeogenesis.

In addition to phosphorylase and glycogen synthetase previously cited as examples of the second method of controlling enzyme activity (ie, conversion of an inactive to an active form of an enzyme), it has recently been shown that pyruvate dehydrogenase may exist in a phosphorylated as well as a dephosphorylated form. Thus, this enzyme may be regulated by phosphorylation involving an ATP-specific kinase which causes a decrease in activity, and by dephosphorylation by a phosphatase which causes an increase in activity of the dehydrogenase. The concentration of ATP may determine whether pyruvate dehydrogenase is in the active or inactive form (Reed, 1969).

Several examples are available from carbohydrate metabolism to illustrate the third method of control of the effective activity of an enzyme. The synthesis of oxaloacetate from bicarbonate and pyruvate, catalyzed by the enzyme **pyruvate carboxylase,** requires the presence of acetyl-Co A. Cooper & Benedict (1966) have presented data which support the conclusion that the addition of acetyl-Co A results in a change in the tertiary structure of the protein, lowering the K_m value for bicarbonate. This effect has important implications for the self-regulation of intermediary metabolism, for, as acetyl-Co A is formed from pyruvate, it automatically ensures the provision of oxaloacetate and its further oxidation in the citric acid cycle by activating pyruvate carboxylase. The activation of pyruvate carboxylase by acetyl-Co A formed from the oxidation of fatty acids helps to explain the sparing action of fatty acid oxidation on the oxidation of pyruvate and the effect of FFA in promoting gluconeogenesis in the liver. It is possible that the gluconeogenic effect of glucagon may be mediated via increased formation of cyclic AMP, which in turn causes lipolysis and increased availability of fatty acids for oxidation to acetyl-Co A in the liver. The acetyl-Co A finally activates pyruvate carboxylase and gluconeogenesis. However, probably the main role of fatty acid oxidation in promoting gluconeogenesis is to supply energy in the form of ATP and reduced NAD (NADH). For a further discussion of the various metabolic and endocrine factors involved in gluconeogenesis, see Weber (1968) and Scrutton & Utter (1968).

Another enzyme which is subject to feedback control is **phosphofructokinase,** which occupies a key position in regulating glycolysis. Phosphofructokinase is inhibited by citrate and by ATP and is activated by

AMP. The presence of **adenylate kinase** in liver allows rapid equilibration of the reaction: ATP + AMP ⇋ 2 ADP. Thus, when ATP is used in energy-requiring processes resulting in formation of ADP, the concentration of AMP rises. This mechanism may control the quantity of carbohydrate undergoing glycolysis prior to its entry into the citric acid cycle. The inhibition by citrate and ATP could explain the sparing action of fatty acid oxidation on glucose oxidation and also the Pasteur effect whereby aerobic oxidation (via the citric acid cycle) inhibits the anaerobic degradation of glucose. The inhibition by ATP and citrate and activation by AMP would also explain how the total quantity of carbohydrate undergoing oxidation is adjusted to fit the requirements of the tissue. The inhibition of pyruvate dehydrogenase by NADH and acetyl-Co A also reinforces the control of oxidation of carbohydrate at the phosphofructokinase step. As both inhibitors are removed by addition of NAD and Co A, respectively, control would operate through the NAD/NADH and acetyl Co A/Co A ratios. There appears to be a reciprocal relationship between the regulation of pyruvate dehydrogenase and pyruvate carboxylase in both liver and kidney which alters the metabolic fate of pyruvate as the tissue changes from carbohydrate oxidation via glycolysis to gluconeogenesis. The reciprocal relationship in activities between phosphofructokinase and fructose-1,6-diphosphatase in liver and the various factors affecting their activity (Table 13–3) with respect to the control of glycolysis and gluconeogenesis have been discussed by Underwood & Newsholme (1965).

Regulation of the Citric Acid Cycle

The identification of regulatory enzymes of the citric acid cycle is difficult because of the many pathways with which the cycle interacts as well as its location within the mitochondrion wherein measurement of enzyme activity and substrate levels is relatively uncertain. In most tissues, where the primary function of the citric acid cycle is to provide energy, there is little doubt that respiratory control via the respiratory chain and oxidative phosphorylation is the overriding control on citric acid cycle activity. Thus, activity is immediately dependent on the supply of reduced dehydrogenase cofactors (eg, NADH), which in turn is dependent on the availability of ADP and Pi and ultimately, therefore, on the rate of utilization of ATP. In addition to this overall control, the properties of some of the enzymes of the cycle indicate that control might also be exerted at the level of the cycle itself. In a tissue such as brain, which is largely dependent on carbohydrate to supply acetyl-Co A, control of the citric acid cycle may occur at the pyruvate dehydrogenase step. In the cycle proper, control may be exercised by allosteric inhibition of citrate synthase by ATP or long chain fatty-acyl-Co A, and allosteric activation of mitochondrial NAD-dependent isocitrate dehydrogenase by ADP, which is counteracted by ATP and NADH. The α-ketoglutarate dehydrogenase complex appears to be under control analogous to that of pyruvate dehydrogenase. Succinate dehydrogenase is inhibited by oxaloacetate, and the formation of oxaloacetate, as controlled by malate dehydrogenase, depends on the NADH/NAD ratio. Which (if any) of these mechanisms operates in vivo has still to be resolved.

Interconversion of the Major Foodstuffs

(See Fig 13–16.)

That animals may be fattened on a predominantly carbohydrate diet demonstrates the ease of conversion of carbohydrate into fat. A most significant reaction in this respect is the conversion of pyruvate to acetyl-Co A, as acetyl-Co A is the starting material for the synthesis of long-chain fatty acids. However, the pyruvate dehydrogenase reaction is essentially nonreversible, which prevents the direct conversion of acetyl-Co A, formed from the oxidation of fatty acids to pyruvate. As a result there is no net conversion of long chain fatty acids to carbohydrate. Only the terminal 3-carbon portion of a fatty acid having an odd number of carbon atoms is glycogenic, as this portion of the molecule will form propionate upon oxidation. Nevertheless, it is possible for labeled carbon atoms of fatty acids to be found ultimately in glycogen after traversing the citric acid cycle; this is because oxaloacetate is an intermediate both in the citric acid cycle and in the pathway of gluconeogenesis. Many of the carbon skeletons of the nonessential amino acids can be produced from carbohydrate via the citric acid cycle and transamination. By reversal of these processes, glycogenic amino acids yield carbon skeletons which are either members or precursors of the members of the citric acid cycle. They are therefore readily converted by gluconeogenic pathways to glucose and glycogen. The ketogenic amino acids give rise to acetoacetate, which will in turn be metabolized as ketone bodies, forming acetyl-Co A in extrahepatic tissues.

For the same reasons that it is not possible for a net conversion of fatty acids to carbohydrate to occur, it is not possible for a net conversion of fatty acids to glucogenic amino acids to take place. Neither is it possible to reverse the pathways of breakdown of ketogenic amino acids, all of which fall into the category of "essential amino acids." Conversion of the carbon skeletons of glucogenic amino acids to fatty acids is possible, either by formation of pyruvate and acetyl-Co A or by reversal of nonmitochondrial reactions of the citric acid cycle from α-ketoglutarate to citrate followed by the action of the citrate cleavage enzyme to give acetyl-Co A (Fig 13–12). However, under most natural conditions, eg, starvation, a net breakdown of protein and amino acids is usually accompanied by a net breakdown of fat. The net conversion of amino acids to fat is therefore not a significant process except possibly in animals receiving a high-protein diet.

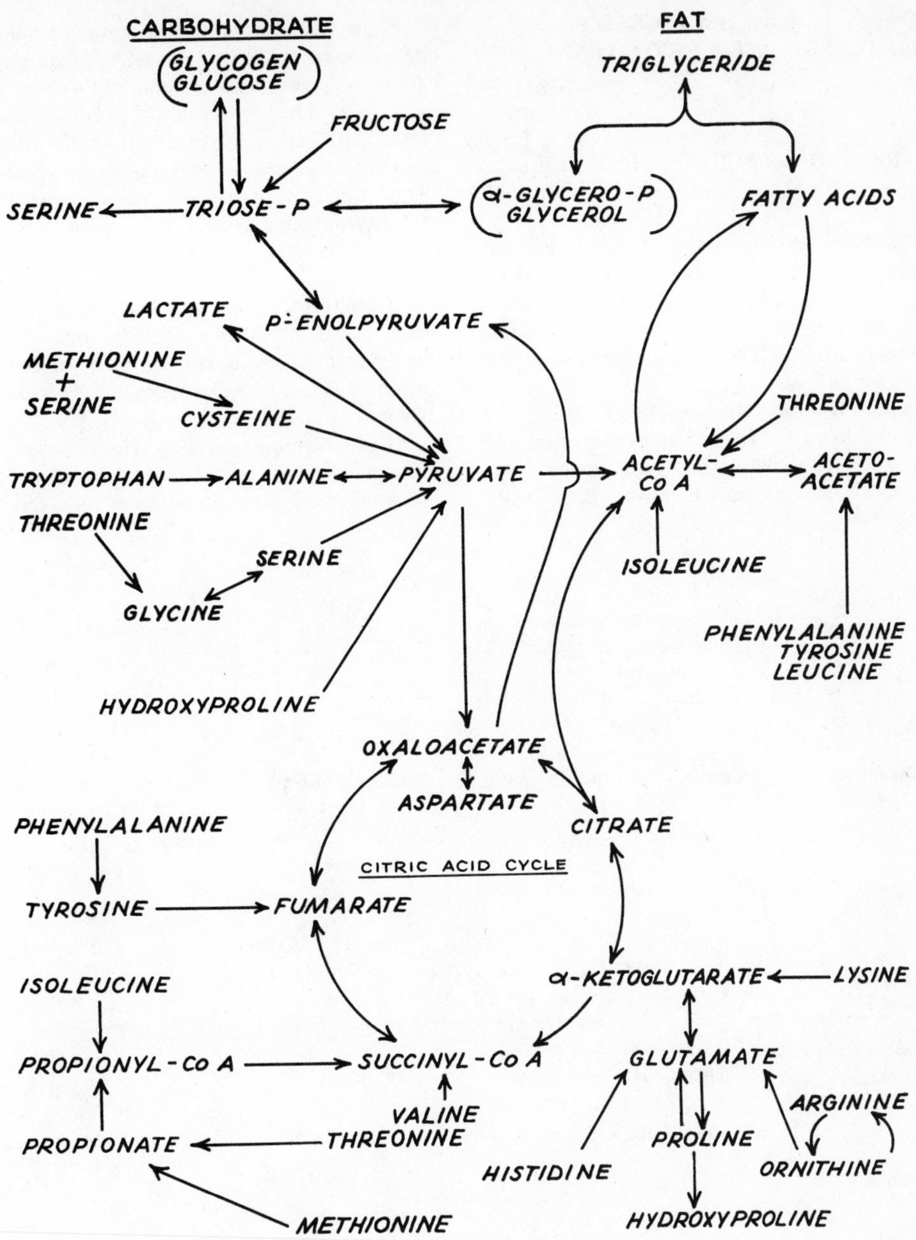

FIG 13–16. Interconversion of the major foodstuffs.

METABOLISM OF HEXOSES

Phosphorylation

The hexoses of metabolic importance–glucose, fructose, and galactose–enter most metabolic pathways, including glycolysis, after phosphorylation. As mentioned previously, glucose is phosphorylated by ATP in the presence of the enzyme **hexokinase**; but in liver there is in addition a more specific **glucokinase.** Hexokinase differs from glucokinase in that it is inhibited by glucose-6-phosphate (allosteric inhibition); it does not change in activity as a response to the nutritional or hormonal state of the animal; and it has a high affinity for glucose (low K_m). When glucose is the substrate, the product of the reaction with glucokinase or hexokinase is glucose-6-phosphate.

Fructose and galactose are not phosphorylated in the presence of glucokinase, but they have their own specific enzymes, **fructokinase** and **galactokinase,** which carry out phosphorylation in the liver. Unlike glucokinase or hexokinase, these enzymes always convert the hexose to the corresponding hexose-1-phosphate.

OTHER PATHWAYS OF GLUCOSE METABOLISM

THE URONIC ACID PATHWAY

Besides the major pathways of metabolism of glucose-6-phosphate that have been described, there exists a pathway for the conversion of glucose to glucuronic acid, ascorbic acid, and pentoses known as the **uronic acid pathway**. It is also an alternative oxidative pathway for glucose.

In the uronic acid pathway, glucuronic acid is formed from glucose by the reactions shown in Fig 13–17. Glucose-6-phosphate is converted to glucose-1-phosphate, which then reacts with uridine triphosphate (UTP) to form the active nucleotide, uridine diphosphate glucose (UDPG). This latter reaction is catalyzed by the enzyme **UDPG pyrophosphorylase.** All of the steps up to this point are those previously indicated as in the pathway of glycogenesis in the liver. Uridine diphosphate glucose is now oxidized at carbon 6 by a 2-step process to glucuronic acid. The product of the oxidation which is catalyzed by an NAD-dependent **UDPG dehydrogenase** is therefore, UDP-glucuronic acid.

Galacturonic acid is an important constituent of many natural products such as the pectins. It may be formed from UDP-glucuronic acid by inversion around carbon 4, as occurs when UDP-glucose is converted to UDP-galactose.

UDP-glucuronic acid is the "active" form of glucuronic acid for reactions involving incorporation of glucuronic acid into chondroitin sulfate or for reac-

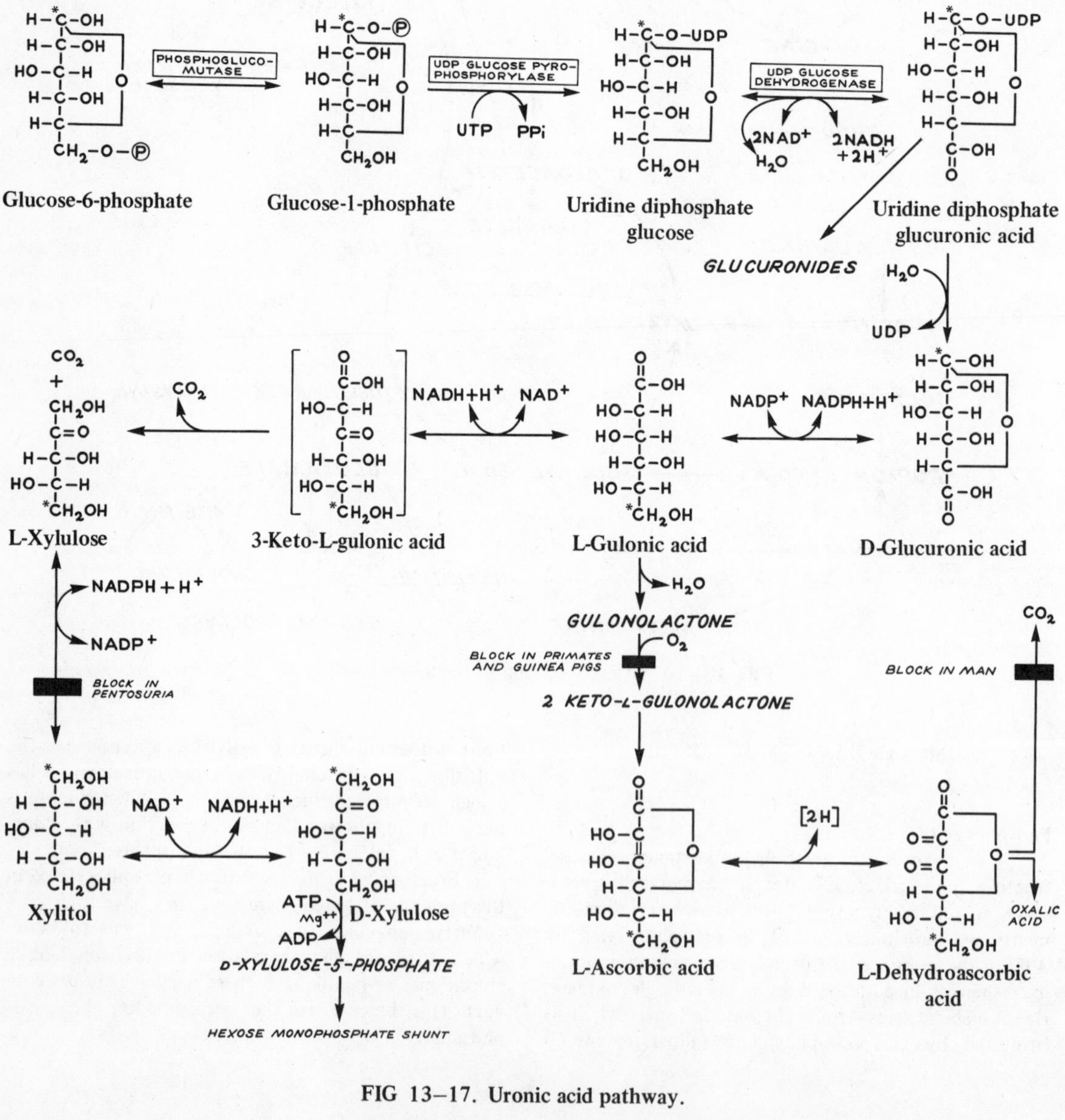

FIG 13–17. Uronic acid pathway.

tions in which glucuronic acid is conjugated to such substrates as steroid hormones, certain drugs, or bilirubin (formation of "direct" bilirubin).

The further metabolism of glucuronic acid is shown in Fig 13–17 (Burns, 1967). In an NADPH-dependent reaction, glucuronic acid is reduced to L-gulonic acid. This latter compound is the direct precursor of ascorbic acid in those animals which are capable of synthesizing this vitamin. In man and other primates as well as in guinea pigs, ascorbic acid cannot be synthesized and gulonic acid is oxidized to 3-keto-L-gulonic acid, which is then decarboxylated to the pentose, L-xylulose.

Xylulose is a constituent of the hexose monophosphate shunt pathway; but in the reactions shown in Fig 13–17, the L-isomer of xylulose is formed from ketogulonic acid. If the 2 pathways are to connect, it is therefore necessary to convert L-xylulose to the D-isomer. This is accomplished by an NADPH-dependent reduction to xylitol, which is then oxidized in an NAD-dependent reaction to D-xylulose; this latter compound, after conversion to D-xylulose-5-phosphate with ATP as phosphate donor, is further metabolized in the hexose monophosphate shunt.

In the rare hereditary disease termed **"essential pentosuria,"** considerable quantities of L-xylulose appear in the urine (Touster, 1959). It is now believed that this may be explained by the absence in pentosuric patients of the enzyme necessary to accomplish reduction of L-xylulose to xylitol, and hence inability to convert the L-form of the pentose to the D-form.

Various drugs markedly increase the rate at which glucose enters the uronic acid pathway. For example, administration of barbital or of chlorobutanol to rats results in a significant increase in the conversion of glucose to glucuronic acid, L-gulonic acid, and ascorbic acid. This effect on L-ascorbic acid biosynthesis is shown by many drugs, including various barbiturates, aminopyrine, and antipyrine. It is of interest that these last 2 drugs have also been reported to increase the excretion of L-xylulose in pentosuric subjects.

METABOLISM OF FRUCTOSE

Fructose may be phosphorylated to form fructose-6-phosphate, catalyzed by the same enzyme, hexokinase, that accomplishes the phosphorylation of glucose (or mannose). (See Fig 13–18.) However, the affinity of the enzyme for fructose is very small compared with its affinity for glucose. It is unlikely, therefore, that this is a major pathway for fructose utilization.

Another enzyme, **fructokinase,** is present in liver which effects the transfer of phosphate from ATP to fructose, forming fructose-1-phosphate. It has also been demonstrated in kidney and intestine. This enzyme will not phosphorylate glucose, and, unlike glucokinase, its activity is not affected by fasting or by insulin, which may explain why fructose disappears from the blood of diabetic patients at a normal rate. It is probable that this is the major route for the phosphorylation of fructose. The K_m for fructose of the enzyme in liver is very low, indicating a very high affinity of the enzyme for its substrate.

Fructose-1-phosphate is split into D-glyceraldehyde and dihydroxyacetone phosphate by **aldolase B,** an enzyme found in the liver. The enzyme also attacks fructose-1,6-diphosphate. Absence of this enzyme leads to a **hereditary fructose intolerance.** D-Glyceraldehyde may gain entry to the glycolysis sequence of reactions via 3 possible routes. One is by the action of **alcohol dehydrogenase** to form glycerol, which, in the presence of **glycerokinase,** forms α-glycerophosphate. A second alternative involves **aldehyde dehydrogenase** which forms D-glycerate from D-glyceraldehyde. In rat liver, **D-glycerate kinase** catalyzes the formation of 2-phosphoglycerate, but this enzyme is not active in human liver (Heinz & others, 1968). Another enzyme present in liver, **triokinase,** catalyzes the phosphorylation of D-glyceraldehyde to glyceraldehyde-3-phosphate. This appears to be the major pathway for the further metabolism of D-glyceraldehyde (Herman & Zakim, 1968). The 2 triose phosphates, dihydroxyacetone phosphate and glyceraldehyde-3-phosphate, may be degraded via the Embden-Meyerhof pathway or they may combine under the influence of aldolase and be converted to glucose. The latter is the fate of much of the fructose metabolized in the liver.

There is in addition the possibility that fructose-1-phosphate may be phosphorylated directly in position 6 to form fructose-1,6-diphosphate, an intermediate of glycolysis. The enzyme catalyzing this reaction, **1-phosphofructokinase,** has been found in muscle and liver. However, if this were a major pathway for fructose metabolism, hereditary fructose intolerance would probably not occur.

If the liver and intestines of an experimental animal are removed, the conversion of injected fructose to glucose does not take place and the animal succumbs to hypoglycemia unless glucose is administered. It appears that brain and muscle can utilize significant quantities of fructose only after its conversion to glucose in the liver. In man but not in the rat, a significant amount of the fructose resulting from the digestion of sucrose is converted to glucose in the intestinal wall prior to passage into the portal circulation. Fructose is more rapidly metabolized by the liver than glucose. This is due most probably to the fact that it bypasses the steps in glucose metabolism catalyzed by glucokinase and phosphofructokinase, at which points metabolic control is exerted on the rate of catabolism of glucose.

Studies have indicated that fructose is metabolized actively by adipose tissue and that it is metabolized independently of glucose. At low concentrations, fructose is utilized by adipose tissue (epididymal adipose tissue of the rat) more slowly than glucose; at high concentrations, fructose is metabolized at a faster rate than glucose.

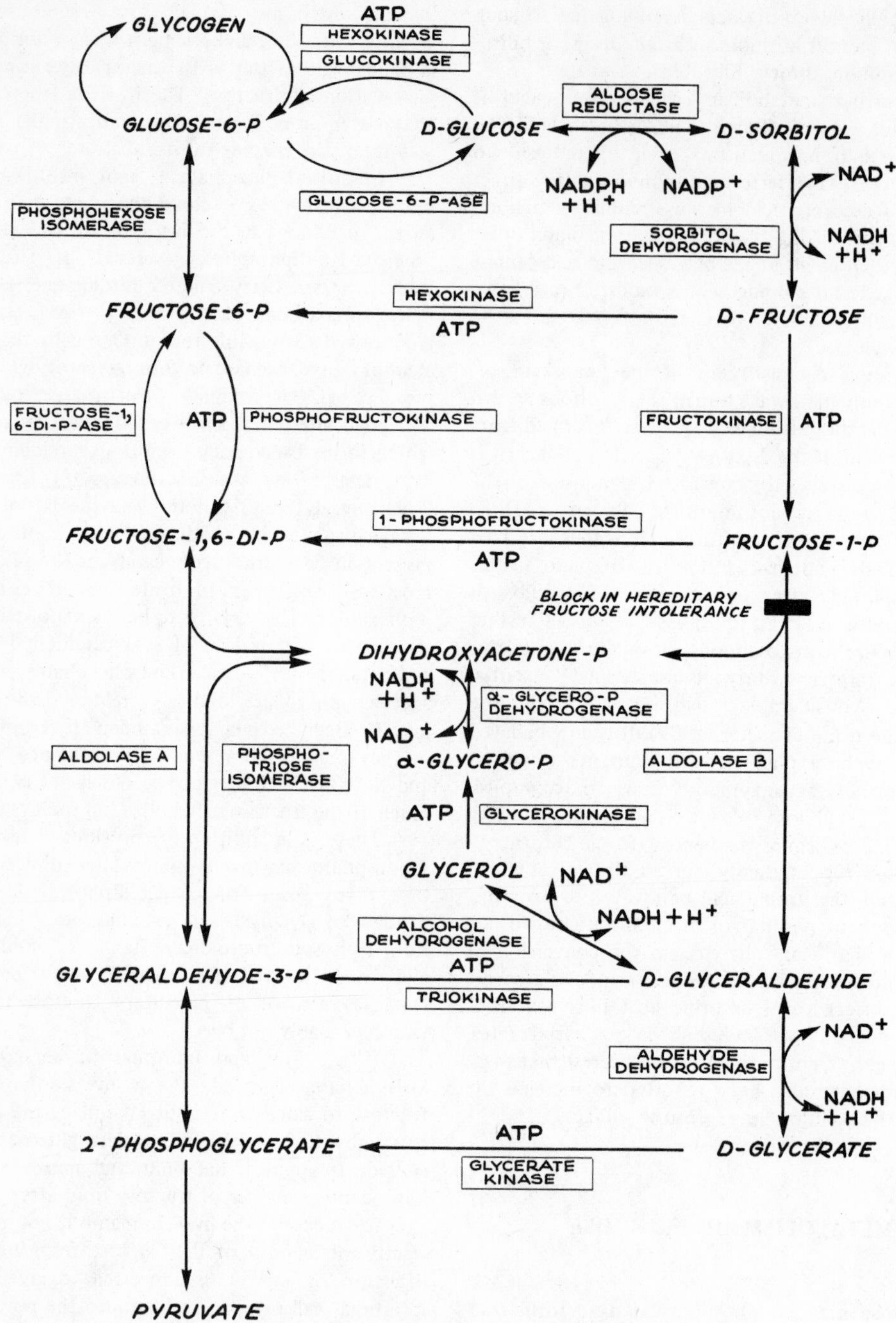

FIG 13–18. Metabolism of fructose.

Free fructose is found in seminal plasma and is secreted in quantity by the placenta into the fetal circulation of ungulates and whales (Huggett, 1961), where it accumulates in the amniotic and allantoic fluids. Experiments demonstrated that glucose was the precursor of fructose. One pathway proposed for this conversion is via sorbitol. Glucose undergoes reduction to sorbitol catalyzed by **aldose reductase (polyol dehydrogenase)** and NADPH. This is followed by oxidation of sorbitol to fructose in the presence of NAD and **ketose reductase** (sorbitol dehydrogenase).

METABOLISM OF GALACTOSE

Galactose is derived from the hydrolysis in the intestine of the disaccharide, lactose, the sugar of milk. It is readily converted in the liver to glucose. The ability of the liver to accomplish this conversion may be used as a test of hepatic function in the galactose tolerance test (see Chapter 17). The pathway by which galactose is converted to glucose is shown in Fig 13–19.

In reaction 1, galactose is phosphorylated with the aid of **galactokinase**, using ATP as phosphate donor. The product, galactose-1-phosphate, reacts with **uridine diphosphate glucose** (UDPG) to form **uridine diphosphate galactose** and glucose-1-phosphate. In this step (reaction 2), which is catalyzed by an enzyme called **galactose-1-phosphate uridyl transferase**, galactose is transferred to a position on UDPG, replacing glucose. The conversion of galactose to glucose takes place (reaction 3) in a reaction of the galactose-containing nucleotide which is catalyzed by an **epimerase**. The product is uridine diphosphate glucose, UDPG. Epimerization probably involves an oxidation and reduction at carbon 4 with NAD as coenzyme. Finally (reaction 4), glucose is liberated from UDPG as glucose-1-phosphate, probably after incorporation into glycogen followed by phosphorolysis.

Reaction 3 is freely reversible. In this manner glucose can be converted to galactose, so that preformed galactose is not essential in the diet. It will be recalled that galactose is required in the body not only in the formation of milk but also as a constituent of glycolipids (cerebrosides), chondromucoids, and mucoproteins.

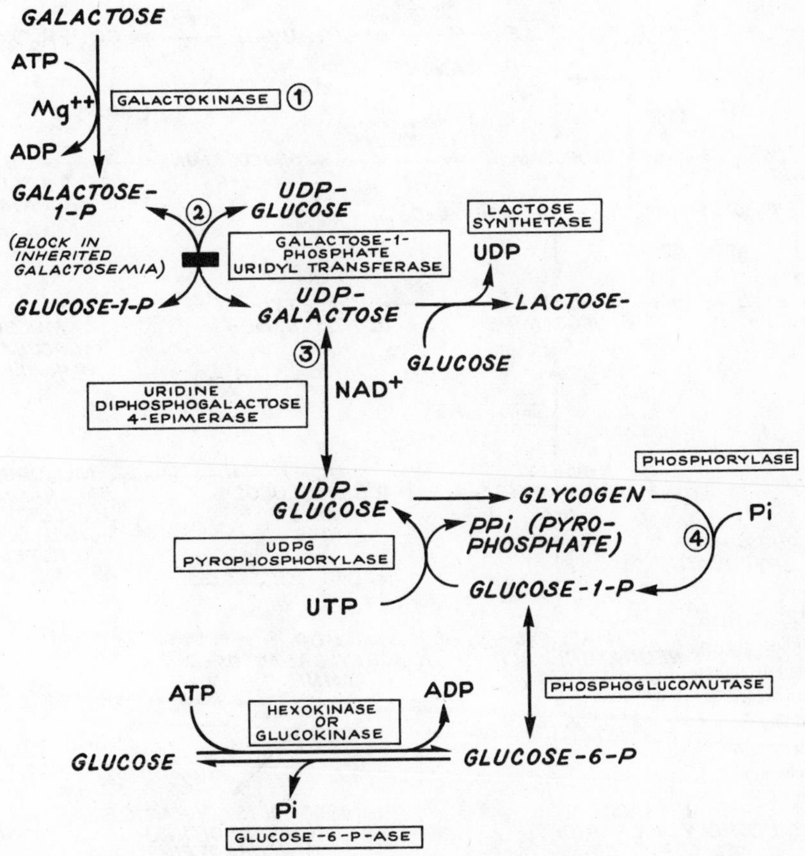

FIG 13–19. The pathway for conversion of galactose to glucose and for the synthesis of lactose.

Galactokinase is an adaptive enzyme, responding with an increased activity upon the feeding of galactose. Young animals particularly show higher activity than do adults.

In the synthesis of lactose in the mammary gland, glucose is converted to UDP-galactose by the enzymes described above. UDP-galactose condenses with glucose to yield lactose, catalyzed by **lactose synthetase** (Caputto & others, 1967).

Inability to metabolize dietary galactose occurs in **galactosemia**, an inherited metabolic disease in which galactose accumulates in the blood and spills over into the urine when this sugar or lactose are ingested. However, there is also marked accumulation of galactose-1-phosphate in the red blood cells of the galactosemic individual, which indicates that there is no deficit of galactokinase (reaction 1).

Recent studies suggest that an inherited lack of galactose-1-phosphate uridyl transferase in the liver and red blood cells is responsible for galactosemia. As a result reaction 2 (see above) is blocked. The epimerase (reaction 3) is, however, present in adequate amounts, so that the galactosemic individual can still form UDP-galactose from glucose. This explains how it is possible for normal growth and development of affected children to occur on the galactose-free diets which are used to control the symptoms of the disease.

Metabolism of Amino Sugars (Hexosamines)

(See Fig 13–20.)

Amino sugars are important components of the carbohydrate which is widely distributed throughout the body as a part of the structural elements of the tissues. The mucopolysaccharides are examples of

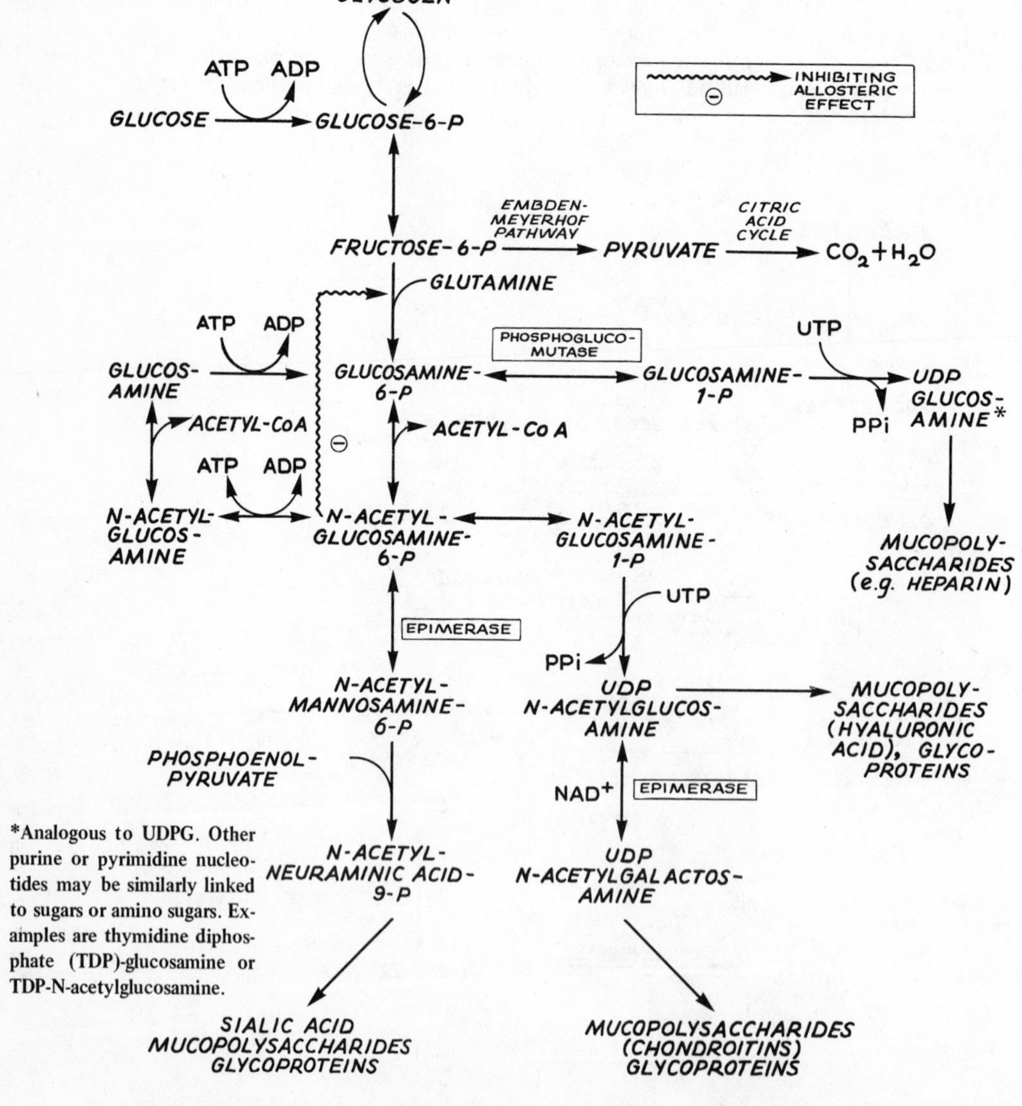

*Analogous to UDPG. Other purine or pyrimidine nucleotides may be similarly linked to sugars or amino sugars. Examples are thymidine diphosphate (TDP)-glucosamine or TDP-N-acetylglucosamine.

FIG 13–20. A summary of the interrelationships in metabolism of amino sugars.

these "structural" carbohydrates. In contrast to glycogen, in which each unit of the polysaccharide is identical (a glucosyl unit), the mucopolysaccharides appear to consist of 2 or more different units, one of which is an amino sugar. A familiar example of these mucopolysaccharides is hyaluronic acid, which occurs in synovial fluid, vitreous humor, umbilical cord, skin, and bone. This compound contains as a basic disaccharide unit, 1 mol of glucuronic acid linked to 1 mol of N-acetylglucosamine through a 1,3-glycosidic bond, as shown in Fig 13–21. UDP-glucuronic acid and UDP-N-acetylglucosamine are the precursors.

Another important group of mucopolysaccharides consists of the chondroitin sulfuric acids, the prosthetic compounds of the polysaccharide-protein complexes known as chondromucoids. The chondromucoids are the chief components of cartilage, but they also occur in the walls of the large blood vessels and in tendons, in the valves of the heart, and in the skin. Chondroitin sulfuric acids A and C contain glucuronic acid, but the amino sugar is an acetylated galactosamine, N-acetyl-D-galactosamine. This is sulfated by "active sulfate"–3′-phosphoadenosine-5′-phosphosulfate (PAPS, p 53).

The enzymatic synthesis of glucosamine has been studied by Ghosh & others (1960). Using a purified enzyme obtained from typical bacterial, fungal, or mammalian (rat liver) cells, they observed the conversion of fructose-6-phosphate to glucosamine-6-phosphate by a transamidation reaction from glutamine. The catalyzing enzyme was termed **L-glutamine-D-fructose-6-phosphate transamidase (amino-transferase).** The transamidase reaction may be unique in that the necessary energy is derived solely from the cleavage of the amide bond of glutamine in contrast to the usual reactions involving transfer of the amide nitrogen of glutamine to an acceptor other than water, all of which require an additional source of energy such as ATP.

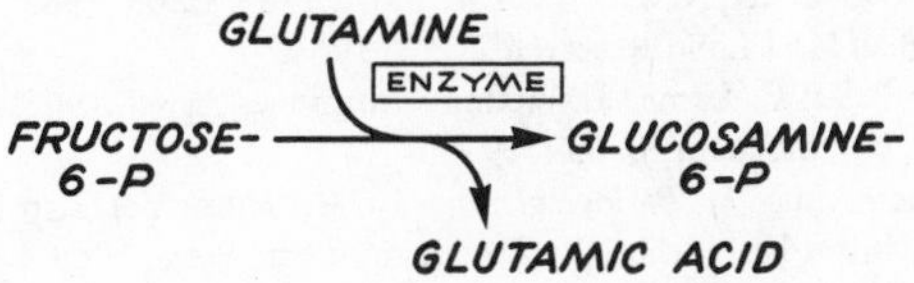

GLUTAMINE
ENZYME
FRUCTOSE-6-P → GLUCOSAMINE-6-P
GLUTAMIC ACID

The biosynthesis of N-acetylgalactosamine is accomplished through the catalytic action of an enzyme which has been identified in the liver. This enzyme acts on UDP-N-acetylglucosamine and epimerizes the glucosamine moiety to galactosamine, thus producing UDP-N-acetylgalactosamine. The enzyme is called **uridine diphosphate-N-acetylglucosamine epimerase** (cf uridine diphosphogalactose epimerase, Fig 13–19).

A summary of the interrelationships in the metabolism of the amino sugars is shown in Fig 13–20. Note the pathways for the synthesis from glucose of N-acetylglucosamine and N-acetylgalactosamine as their active uridine diphosphate derivatives. Also note the pathway of synthesis of N-acetylneuraminic acid, another important amino sugar in glycoproteins and mucopolysaccharides. For further details concerning glycoproteins consult Marshall & Neuberger (1968). For details concerning mucopolysaccharides, consult Dodgson & Lloyd (1968).

A
COOH
CH$_2$OH
NH–CO–CH$_3$
GLUCURONIC ACID
N-ACETYLGLUCOSAMINE

B
CH$_2$O–SO$_3$H
COOH
NH–SO$_3$H
O–SO$_3$H
SULFATED GLUCOSAMINE
SULFATED GLUCURONIC ACID

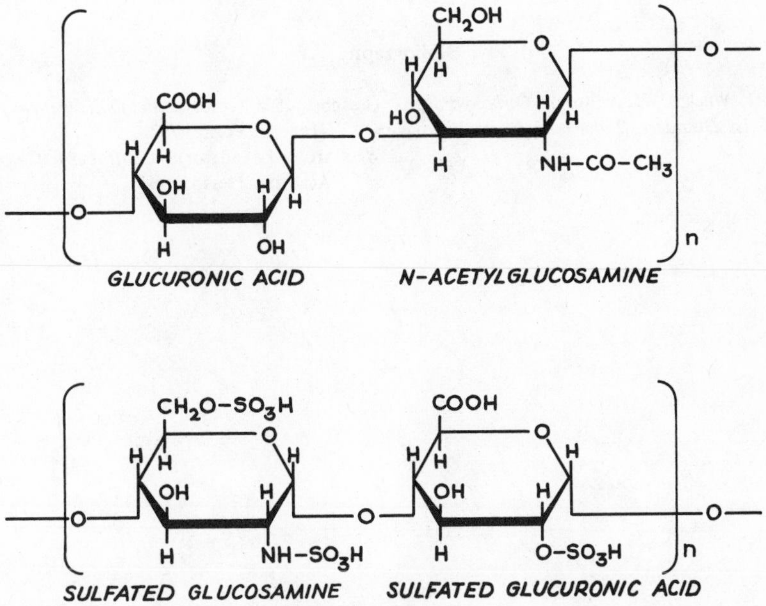

FIG 13–21. *A:* **The fundamental disaccharide unit of hyaluronic acid.** *B:* **Disaccharide unit of condroitin sulfuric acids.**

• • •

References

Benesch R, Benesch RE, Yu CI: Proc Nat Acad Sc 59:526, 1968.

Brown BI, Brown DH in: *Carbohydrate Metabolism and Its Disorders.* Vol 2, p 123. Dickens F, Randle PJ, Whelan WJ (editors). Academic Press, 1968.

Burns JJ in: *Metabolic Pathways,* 3rd ed. Vol 1, p 394. Greenberg DM (editor). Academic Press, 1967.

Cahill GF, Ashmore J, Renold AE, Hastings AB: Am J Med 26:264, 1959.

Caputto R, Barra HS, Cumar FA: Am Rev Biochem 36:211, 1967.

Cohn C, Joseph D: Am J Clin Nutr 8:682, 1960.

Cooper TG, Benedict CR: Biochem Biophys Res Comm 22:285, 1966.

Dodgson KS, Lloyd AG in: *Carbohydrate Metabolism and Its Disorders.* Vol 1, p 169. Dickens F, Randle PJ, Whelan WJ (editors). Academic Press, 1968.

Exton JH, Jefferson LS, Butcher RW, Park CR: Am J Med 40:709, 1966.

Ghosh S, Blumenthal HJ, Davidson E, Roseman S: J Biol Chem 235:1265, 1960.

Greville GD in: *Carbohydrate Metabolism and Its Disorders.* Vol 1, p 297. Dickens F, Randle PJ, Whelan WJ (editors). Academic Press, 1968.

Heinz F, Lamprecht W, Kirsch J: J Clin Invest 47:1826, 1968.

Herman RH, Zakim D: Am J Clin Nutr 21:245, 1968.

Hers HG, de Wulf H, Stalmaus W, van den Berghe G: Advances Enzym Reg 8:171, 1970.

Hers HG: Rev internat d'hépatol 9:35, 1959.

Hiatt HH, Lareau J: J Biol Chem 235:1241, 1960.

Huggett A StG: Brit M Bull 17:122, 1961.

Illingworth B, Brown DH: Proc Nat Acad Sc 48:1619, 1962.

Kinoshita JH, Wachtl C: J Biol Chem 233:5, 1958.

Krebs HA: Proc Roy Soc London s.B 159:545, 1963.

Krebs HA, Woodford M: Biochem J 94:463, 1965.

Landau BR, Leanards JR, Barry FM: Am J Physiol 201:41, 1961.

Lardy HA, Paetkau V, Walter P: Proc Nat Acad Sc 53:1410, 1965.

Lerman S: Science 130:1473, 1959.

Marshall RD, Neuberger A in: *Carbohydrate Metabolism and Its Disorders.* Vol 1, p 213. Dickens F, Randle PJ, Whelan WJ (editors). Academic Press, 1968.

Miller LL: Fed Proc 24:737, 1965.

Miller ON, Olson RE: Arch Biochem 50:257, 1954.

Mortimore GE: Am J Physiol 204:699, 1963.

Ogston AG: Nature, London 162:963, 1948.

Rapoport S in: *Essays in Biochemistry.* Vol 4, p 69. Campbell PN, Greville GD (editors). Academic Press, 1968.

Reed LJ in: *Current Topics in Cellular Regulation.* Vol 1, p 233. Academic Press, 1969.

Scrutton MC, Utter MF: Ann Rev Biochem 37:249, 1968.

Shull KH, Miller ON: J Biol Chem 235:551, 1960.

Smith EE, Taylor PM, Whelan WJ in: *Carbohydrate Metabolism and Its Disorders.* Vol 1, p 89. Dickens F, Randle PJ, Whelan WJ (editors). Academic Press, 1968.

Stetten D Jr: Am J Med 28:867, 1960.

Sutherland EW, Rall TW, Menon, T: J Biol Chem 237:1220, 1962.

Sutherland EW, Wosilait WD: J Biol Chem 218:459, 1956.

Touster O: Am J Med 26:724, 1959.

Traut RR, Lipmann F: J Biol Chem 238:1213, 1963.

Underwood AH, Newsholme EA: Biochem J 95:868, 1965.

Vaughan M: J Lipid Research 2:293, 1961.

Villar-Palasi C, Larner J: Vitamins Hormones 26:65, 1968; Ann Rev Biochem 39:639, 1970.

Weber G in: *The Biological Basis of Medicine.* Vol 2, p 263. Bittar EE, Bittar N (editors). Academic Press, 1968.

Weber G, Glazer RI, Ross RA: Advances Enzym Reg 8:13, 1970.

Wood HG, Katz J, Landau BR: Biochem Ztschr 338:809, 1963.

Bibliography

Dickens F, Randle PJ, Whelan WJ (editors): *Carbohydrate Metabolism and Its Disorders.* 2 vols. Academic Press, 1968.

Greenberg DM (editor): *Metabolic Pathways,* 3rd ed. Vol 1. Academic Press, 1967.

Whelan WJ (editor): *Control of Glycogen Metabolism.* Academic Press, 1968.

14 . . .

Metabolism of Lipids

With Peter Mayes, PhD, D Sc*

The lipids of metabolic significance in the mammalian organism include triglycerides (neutral fat), phospholipids, and steroids, together with products of their metabolism such as long chain fatty acids (free fatty acids), glycerol, and ketone bodies. For many years the tissue lipids were considered to be inactive storehouses of calorigenic material, called upon only in times of shortage of calories. However, Schoenheimer & Rittenberg (1935) showed by experiments in which deuterium-labeled fatty acids were fed to mice in caloric equilibrium that in only 4 days a considerable proportion of the depot fat had been formed from the dietary fat. Since the total mass of fat in the depots remained constant, a corresponding quantity of fat must have been mobilized during this period. These investigations demonstrated the dynamic state of body fat, a concept that forms the basis of present understanding of lipid metabolism.

Much of the carbohydrate of the diet is converted to fat before it is utilized for the purpose of providing energy. As a result, fat may be the major source of energy for many tissues; indeed, there is accumulating evidence that in certain organs fat may be used as a fuel in preference to carbohydrate.

As the principal form in which energy is stored in the body, fat has definite advantages over carbohydrate or protein. Its caloric value is over twice as great (9.3 Cal/g) and it is associated with less water in storage. Fat is, therefore, the most concentrated form in which potential energy can be stored.

A minimal amount of fat is essential in the diet to provide an adequate supply of certain polyunsaturated fatty acids (the essential fatty acids) and of fat-soluble vitamins which cannot be synthesized in adequate amounts for optimal body function. As well as acting as a carrier of these essential compounds, dietary fat is necessary for their efficient absorption from the gastrointestinal tract. Apart from these functions, it is not certain how essential fat is as a constituent of the diet. As a source of energy it can be replaced completely by either carbohydrate or protein, although the efficiency with which foodstuffs are utilized may suffer as a consequence.

THE BLOOD LIPIDS

Extraction of the plasma lipids with a suitable lipid solvent and subsequent separation of the extract into various classes of lipids shows the presence of triglyceride phospholipid, cholesterol and cholesterol ester, and in addition, the existence of a much smaller fraction of unesterified long chain fatty acids (free fatty acids, FFA) that accounts for less than 5% of the total fatty acid present in the plasma. This latter fraction, the FFA, is now known to be metabolically the most active of the plasma lipids. An analysis of blood plasma showing the major lipid classes is shown in Table 14–1. (See also Fig 14–1.)

Since lipids account for much of the energy expenditure of the body, the problem is presented of transporting a large quantity of hydrophobic material (lipid) in an aqueous environment. This is solved by associating the more insoluble lipids with more polar ones such as phospholipids and then combining them with protein to form a hydrophilic lipoprotein complex. It is in this way that triglycerides derived from intestinal absorption of fat or from the liver are transported in the blood as chylomicrons and very low density lipoproteins. Fat is released from adipose tissue in the form of FFA and carried in the unesterified state in the plasma as an albumin-FFA complex. Many classes of lipids are, therefore, transported in the blood as lipoproteins.

Pure fat is less dense than water; it follows that as the proportion of lipid to protein in lipoproteins increases, the density decreases. Use is made of this property in separating the various lipoproteins in plasma by ultracentrifugation. The rate at which each lipoprotein floats up through a solution of NaCl (specific gravity 1.063) may be expressed in Svedberg (Sf) units of flotation. One Sf unit is equal to 10^{-13} cm/second/dyne/g at 26° C. The composition of the various lipoprotein fractions obtained by centrifugation is shown in Table 14–2, which shows the increase in density of lipoproteins as the protein content rises and the lipid content falls. The various chemical classes of lipids are seen to occur in varying amounts in most of the lipoprotein fractions. Since the fractions represent approximately the lipid entities present in the plasma, mere chemical analysis of the plasma lipids

*Lecturer in Biochemistry, Royal Veterinary College, University of London.

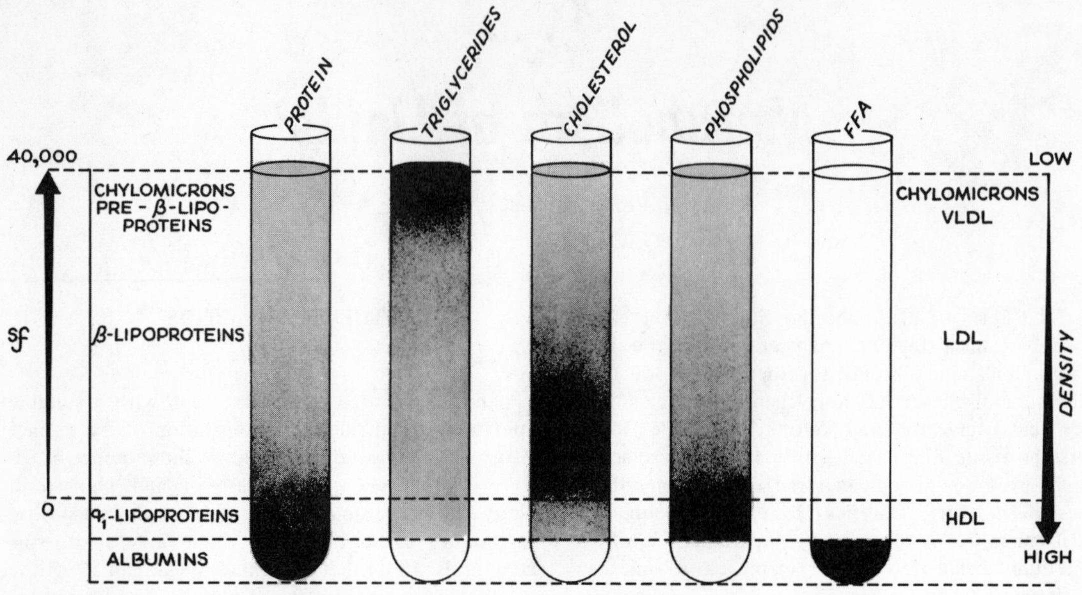

FIG 14–1. Distribution of lipids in serum. VLDL = very low density lipoproteins; LDL = low density lipoproteins; HDL = high density lipoproteins; FFA = free fatty acids.

(apart from FFA) yields little information on their physiology. Many of the fractions shown in the table are not necessarily distinct physiologic entities. Olson & Vester (1960) have pointed out with regard to the lipoproteins Sf 0–400 that as the protein, cholesterol, cholesterol ester, and phospholipid content decreases (with decreasing density), the triglyceride content increases progressively.

It is not possible (as Table 14–2 suggests) to separate completely the various fractions by centrifugation. For example, there is overlap in density between chylomicrons and very low density lipoproteins (VLDL) which prevents a clean separation. Chylomicrons (Sf > 400 lipoproteins) have a diameter of 0.5 μm, whereas VLDL (Sf 20–400) range from 0.025–0.075 μm.

In addition to the use of technics depending on their density, lipoproteins may be separated according to their electrophoretic properties and identified more accurately using immunoelectrophoresis. According to Fredrickson & others (1967), 4 groups of lipoproteins have been identified, and these are important in clinical diagnosis: high density or α-lipoproteins; low density or β-lipoproteins; very low density or pre-β-(α_2-) lipoproteins (VLDL); and chylomicrons (Fig 14–2). A characteristic HDL-apoprotein and an LDL-apoprotein, each consisting of possibly more than one polypeptide, are found in α- and β-lipoproteins, respectively. They differ in their amino acid content and terminal residues and in their immunochemical properties. Both apoproteins are found in VLDL (pre-β-) and chylomicrons, lending support to the concept that both α- and β-lipoproteins take part in triglyceride transport in the blood. Other proteins have been detected in the VLDL fraction. Chylomicrons and VLDL consist of a core of triglyceride and cholesterol ester surrounded by a more polar layer of protein, phospholipid, and cholesterol. Carbohydrates have been detected in most lipoproteins, indicating that they contain glycoprotein components.

The metabolism of the lipoproteins is described later in this chapter.

TABLE 14–1. Lipids of the blood plasma in man.

Lipid Fraction	mg/100 ml Mean	mg/100 ml Range
Total lipid	570	360–820
Triglyceride	142	80–180*
Total phospholipid†	215	123–390
Lecithin		50–200
Cephalin		50–130
Sphingomyelins		15–35
Total cholesterol	200	107–320
Free cholesterol (nonesterified)	55	26–106
Free fatty acids (nonesterified)	12	6–16*

Total fatty acids (as stearic) range from 200–800 mg/100 ml; 45% are triglycerides, 35% phospholipids, 15% cholesterol ester, and less than 5% free fatty acids.

*Varies with nutritional state.

†Analyzed as lipid phosphorus; mean lipid phosphorus = 9.2 mg/100 ml (range, 6.1–14.5). Lipid phosphorus X 25 = phospholipid as lecithin (4% phosphorus).

TABLE 14–2. Composition of the lipoproteins in plasma of man.
(Adapted from Olson & Vester, 1960.)

				Composition						
						Percentages of Total Lipid				
Fraction	**Source**	**Density**	**Sf**	**Protein (%)**	**Total Lipid (%)**	**Triglyceride**	**Phospholipid**	**Cholesterol Ester**	**Cholesterol (Free)**	**Free Fatty Acids**
Chylomicrons	Intestine	<0.96	>400	1	99	88	8	3	1	...
Very low density lipoproteins (VLDL)	Liver and intestine	0.96 –1.006	20–400	7	93	56	20	15	8	1
Low density lipoproteins LDL 1	Liver and intestine	1.006–1.019	12–20	11	89	29	26	34	9	1
LDL 2		1.019–1.063	2–12	21	79	13	28	48	10	1
High density lipoproteins HDL 1*	Liver; ? intestine	1.063	0–2							
HDL 2		1.063–1.125		33	67	16	43	31	10	...
HDL 3		1.125–1.210		57	43	13	46	29	6	6
Albumin-FFA	Adipose tissue	> 1.2810		99	1	0	0	0	0	100

LDH, low density fraction; HDL, high density fraction; VLDL, very low density fraction.
*This fraction is quantitatively insignificant.

OXIDATION OF FATS

Triglycerides must be hydrolyzed to their constituent fatty acids and glycerol before further catabolism can proceed. Much of this hydrolysis occurs in adipose tissue with release of FFA into the plasma, followed by FFA uptake into tissues and subsequent oxidation. Many tissues (including liver, heart, kidney, muscle, lung, testis, brain, and adipose tissue) have the ability to oxidize long chain fatty acids. The utilization of glycerol depends upon whether the tissue in question possesses the necessary activating enzyme, **glycerokinase**. The enzyme has been found in significant amounts in liver, kidney, intestine, brown adipose tissue, and lactating mammary gland.

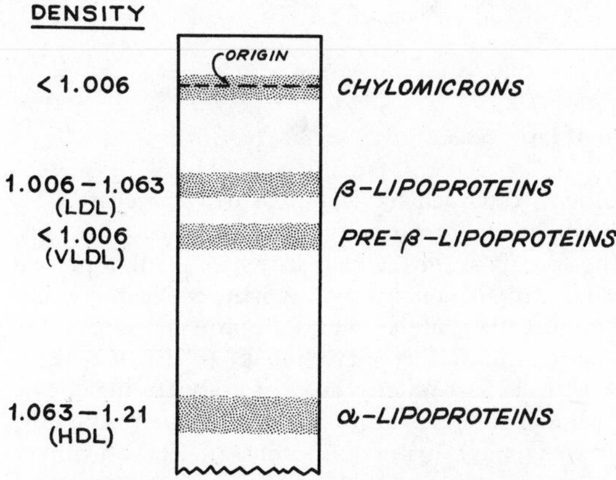

FIG 14–2. Separation of plasma lipoproteins by paper electrophoresis or by agarose gel electrophoresis.

Oxidation of Fatty Acids

Knoop (1905) proposed that fatty acids were oxidized physiologically by β-oxidation. In experiments which were the forerunners of the modern technic of labeling, he tagged the methyl end of fatty acids by substitution of a phenyl radical. This prevented the complete oxidation of the fatty acids and resulted in urinary excretion of phenyl derivatives as end products of their metabolism. On feeding to dogs fatty acids with an even number of carbon atoms labeled in this manner, he noticed that phenylacetic acid was always excreted into the urine (as the glycine conjugate, phenylaceturic acid). However, on feeding labeled fatty acids with an odd number of carbon atoms, benzoic acid was always excreted (as the glycine conjugate, hippuric acid). These results could only be explained if the fatty acids were metabolized by a pathway involving the removal of 2 carbon atoms at a time from the carboxyl end of the molecule, ie, β-oxidation.

β-Oxidation is recognized today as the principal method by which fatty acids are oxidized (Fig 14–3). Several enzymes, known collectively as the "fatty acid oxidase" complex, are found in the mitochondrial matrix adjacent to the respiratory chain which is found in the inner membrane. These catalyze the oxidation of fatty acids to acetyl-Co A, the system being coupled with the phosphorylation of ADP to ATP.

As in the metabolism of glucose, fatty acids must first be converted in a reaction with ATP to an active intermediate before they will react with the enzymes responsible for their further metabolism. This is the only step in the complete degradation of a fatty acid that requires energy from ATP. In the presence of ATP and coenzyme A, the enzyme **thiokinase** (acyl-Co A

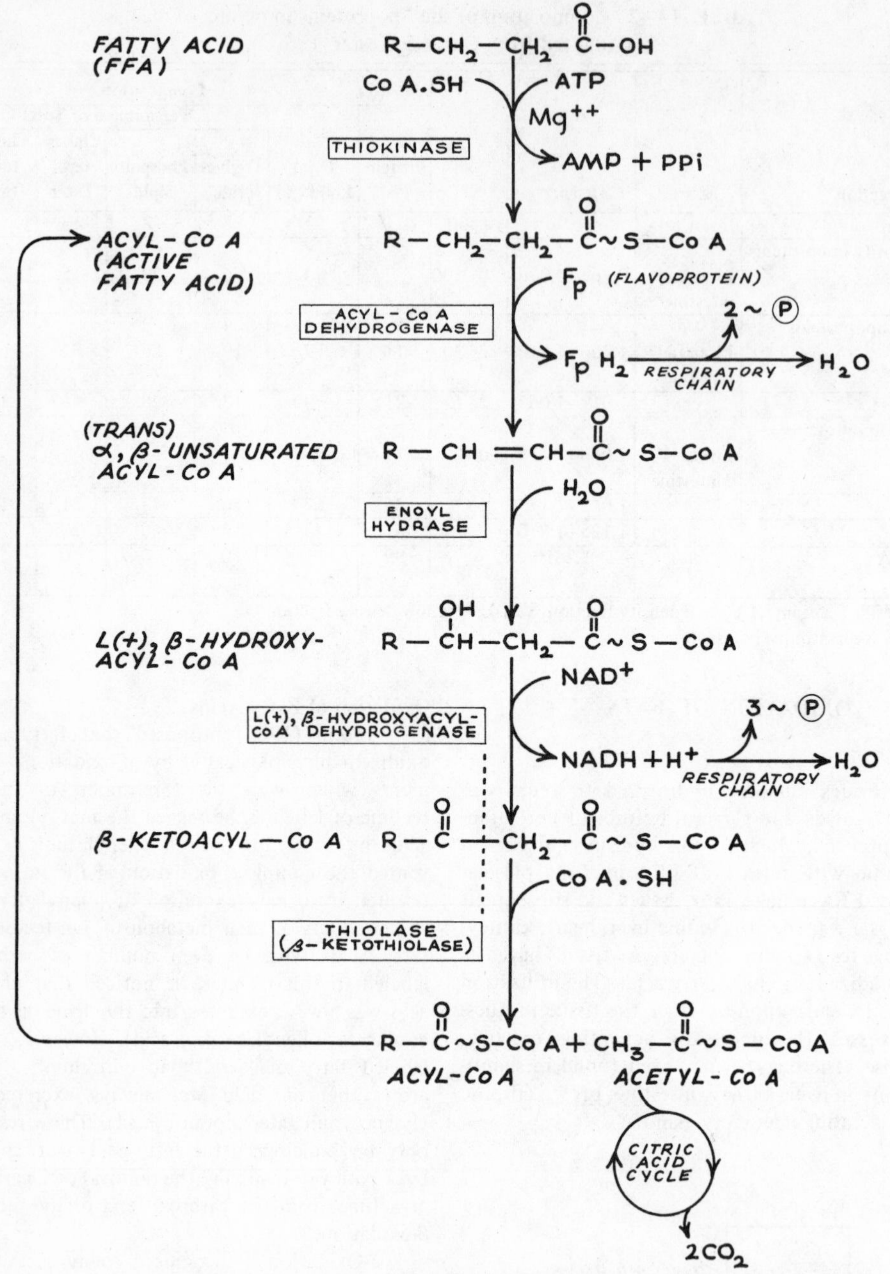

FIG 14–3. β-Oxidation of fatty acids.

synthetase) catalyzes the conversion of a fatty acid (or FFA) to an "active fatty acid" or acyl-Co A.

Thiokinases are found both inside and outside the mitochondria. Several thiokinases have been described, each specific for fatty acids of different chain length. In addition, there is a GTP-specific mitochondrial thiokinase which, unlike the ATP-specific enzyme, forms GDP + Pi as a product and not pyrophosphate. High-energy intermediates in oxidative phosphorylation may also serve as donors of free energy in fatty acid activation.

After the formation of acyl-Co A, there follows the removal of 2 hydrogen atoms from the α and β carbons, catalyzed by **acyl-Co A dehydrogenase.** This results in the formation of α,β-unsaturated acyl-Co A. The coenzyme for the dehydrogenase is a flavoprotein whose reoxidation by the respiratory chain requires the mediation of another flavoprotein, termed the electron-transferring flavoprotein (ETF) (Crane & Beinert, 1956). Water is added to saturate the double bond and form β-hydroxy-acyl-Co A, catalyzed by the enzyme **enoyl hydrase** (crotonase). The β-hydroxy derivative undergoes further dehydrogenation on the β carbon **(β-hydroxyl-acyl-Co A dehydrogenase)** to form the corresponding β-keto-acyl-Co A compound. In this case NAD is the coenzyme involved in the dehydrogenation. Finally, β-keto-acyl-Co A is split at the β position by **thiolase** (β-ketothiolase), which catalyzes a thiolytic cleavage involving another molecule of Co A. The products of this reaction are acetyl-Co A and an acyl-Co A derivative, containing 2 carbons less than the original acyl-Co A molecule which underwent oxidation. The acyl-Co A formed in the cleavage reaction reenters the oxidative pathway at reaction (2) (Fig 14–3). In this way, a long chain fatty acid may be degraded completely to acetyl-Co A (C_2-units). As acetyl-Co A can be oxidized to CO_2 and water via the citric acid cycle (which is also found within the mitochondria), the complete oxidation of fatty acids is achieved.

Energetics of Fatty Acid Oxidation

Transport in the respiratory chain of electrons from reduced flavoprotein and NAD will lead to the synthesis of at least 5 high-energy phosphate bonds (see Chapter 9) for each of the first 7 acetyl-Co A molecules formed by β-oxidation of palmitate ($7 \times 5 = 35$). The total of 8 mols of acetyl-Co A formed will each give rise to at least 12 high-energy bonds on oxidation in the citric acid cycle, making $8 \times 12 = 96$ high-energy bonds derived from the acetyl-Co A formed from palmitate, minus 2 for the initial activation of the fatty acid, yielding a net gain of 129 high-energy bonds/mol, or $129 \times 7.6 = 980$ kilocalories (Cal). As the caloric value of palmitic acid is 2340 kilocalories/mol, the process captures as high-energy phosphate at least 41% ($980/2340 \times 100$) of the total energy of combustion of the fatty acid.

BIOSYNTHESIS OF LIPIDS

Synthesis of Fatty Acids

Like many other degradative and synthetic processes (eg, glycogenolysis and glycogenesis), fatty acid synthesis was formerly considered to be merely the reversal of oxidation. However, it now seems clear that a **mitochondrial** system for fatty acid synthesis, involving some modification of the β-oxidation sequence, is responsible only for elongation of existing fatty acids of moderate chain length, whereas a radically different and highly active **extramitochondrial** system is responsible for the complete synthesis of palmitate from acetyl-Co A (Wakil, 1961). There is also an active system for chain elongation present in rat liver microsomes (Nugteren, 1965).

A. Mitochondrial System: (Fig 14–3.) Under anaerobic conditions, mitochondria will catalyze the incorporation of acetyl-Co A into long chain fatty acids (mainly stearate [C_{18}] and palmitate [C_{16}], with some C_{20} and C_{14} fatty acids). The system requires the addition of ATP, NADH, and NADPH. The enzymes are probably the same as those involved in β-oxidation except for the conversion of the α,β-unsaturated acyl-Co A to the corresponding saturated compound which is catalyzed by **α,β-unsaturated acyl-Co A reductase** (enoyl-Co A reductase), requiring NADPH. It is likely that the incorporation of acetyl-Co A into the long chain fatty acids is due to its addition to existing fatty acids rather than to the synthesis de novo of the long chain fatty acids from acetyl-Co A. The ATP is probably required for the formation of acyl-Co A from endogenous fatty acids. A role for pyridoxal phosphate has been suggested as a coenzyme for the enzyme condensing acetyl-Co A with acyl-Co A; thus, thiolase may not be used in this synthetic pathway. The physiologic significance of this pathway is uncertain since it will operate only under anaerobic conditions.

B. Extramitochondrial System for De Novo Synthesis: (Fig 14–4.) This system has been found in the soluble fraction of many tissues, including liver, kidney, brain, lung, mammary gland, and adipose tissue. Its cofactor requirements include NADPH, ATP, Mn^{++}, and HCO_3^- (as a source of CO_2). Free palmitate is the main end product. These characteristics contrast markedly with those of the mitochondrial system.

CO_2 is required in the initial reaction for the carboxylation of acetyl-Co A to malonyl-Co A in the presence of ATP and **acetyl-Co A carboxylase.** Acetyl-Co A carboxylase has a requirement for the vitamin biotin. Activity is inhibited when biotin is bound by the protein, avidin, from egg white. Lynen & Tada (1961) found that acyl-Co A derivatives were inactive in the system—unlike the situation in the mitochondria. They concluded that acyl derivatives of Co A were not intermediates in the extramitochondrial pathway during the synthesis of palmitate and proposed that the acyl moiety remained attached to the enzyme as an acyl-S-enzyme complex.

There appear to be 2 types of fatty acid synthetase systems found in the soluble portion of the cell (cytosol). In bacteria, plants, and lower forms like Euglena, the individual enzymes of the system may be separate and the acyl radicals are found in combination with a protein called the **acyl carrier protein** or ACP. However, in yeast, mammals, and birds, the synthetase system is a multienzyme complex which may not be subdivided without loss of activity (Majerus & Vagelos, 1967; Lynen & others, 1968). The following account is based principally on the yeast system described by Lynen (Fig 14–4).

The multienzyme complex contains 2 types of –SH groups, "central" and "peripheral." In the "priming reaction," acetyl-Co A reacts with the "peripheral" –SH group and malonyl-Co A reacts with the "central" –SH group to transfer acetyl and malonyl residues to the enzyme. The acetyl group attacks the methylene group of the malonyl residue to liberate CO_2 and form acetoacetyl enzyme attached to the central –SH group. This decarboxylation allows the reaction to go to completion and acts as a driving force for the whole system. While attached to the central –SH group, acetoacetyl enzyme is reduced, dehydrated, and reduced again to form the corresponding saturated acyl-enzyme compound. The main stages

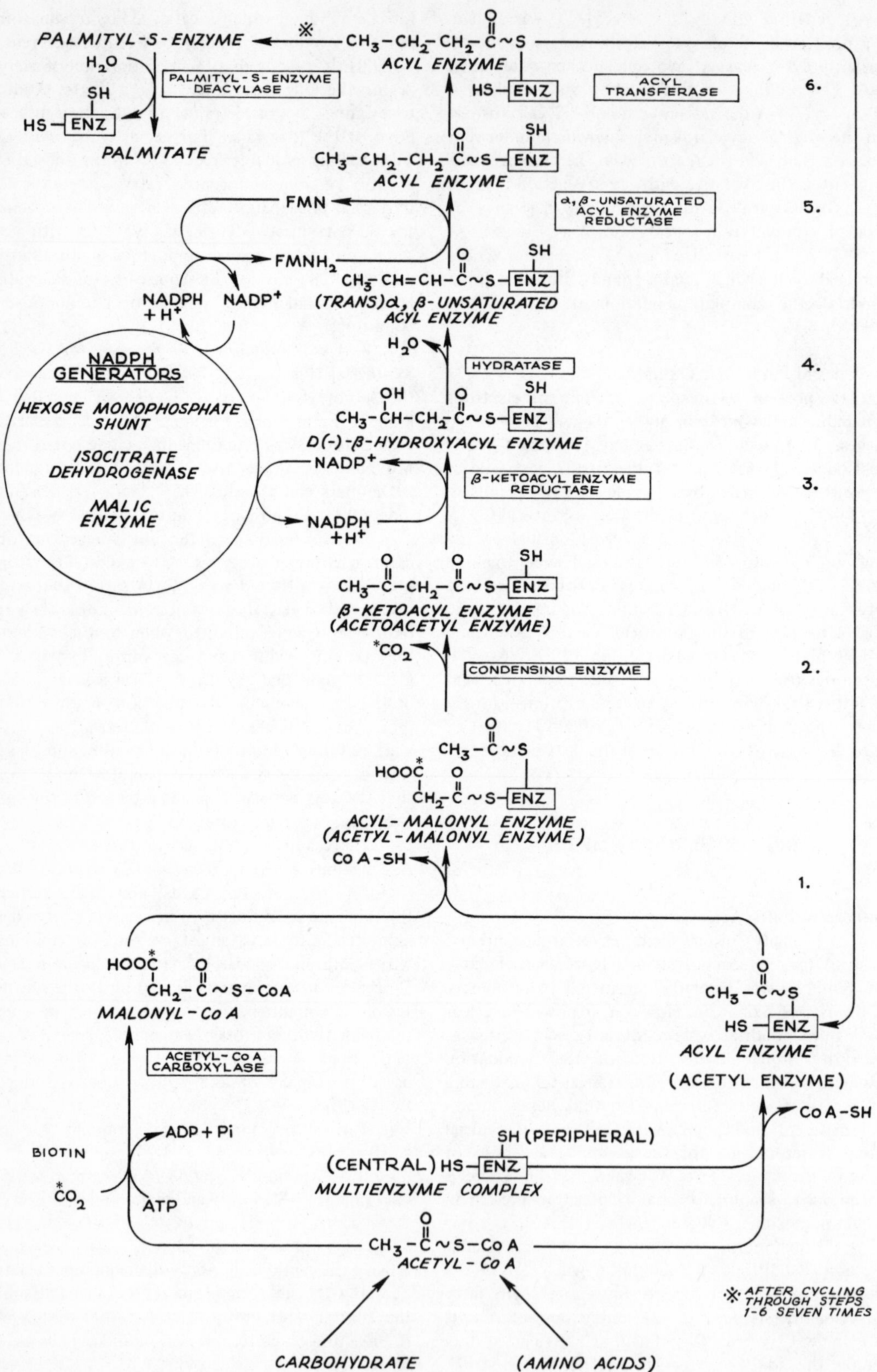

FIG 14–4. Extramitochondrial synthesis of palmitate.

of the reactions are analogous to those in β-oxidation except that the β-hydroxy acid is the D(−) isomer instead of the L(+) isomer. NADPH serves as the hydrogen donor in both reductions, with the mediation of FMN (flavin mononucleotide) in the reaction that saturates the double bond. Finally, the saturated acyl radical is transferred to the "peripheral" −SH group, a new malonyl residue takes its place on the "central" −SH group, and the process is repeated until a saturated acyl radical of 16 carbon atoms is formed.

In mammalian systems, free palmitate is liberated from the enzyme complex by hydrolysis. The free palmitate must be activated to acyl-Co A before it can proceed via any other metabolic pathway. Its usual fate is esterification into glycerides. Both ACP of bacteria and the multienzyme complex of yeast contain the vitamin, pantothenic acid, in the form of 4′-phosphopantetheine. This is the carrier of the "central" −SH group and is responsible for the binding of acyl groups as in coenzyme A.

The aggregation of all the enzymes of a particular pathway into one multienzyme functional unit offers great efficiency and freedom from interference by competing processes, thus achieving the effect of compartmentalization of the process without the erection of permeability barriers.

The equation for the overall synthesis of palmitate from acetyl-Co A and malonyl-Co A is as follows:

$$CH_3CO \cdot S \cdot CoA + 7HOOC \cdot CH_2CO \cdot S \cdot CoA + 14NADPH + 14H^+ \longrightarrow$$
$$CH_3(CH_2)_{14}COOH + 7CO_2 + 6H_2O + 8CoA \cdot SH + 14NADP^+$$

The 1 mol of acetyl-Co A used as a primer forms carbon atoms 15 and 16 of palmitate. The addition of the subsequent C_2 units is via malonyl-Co A formation. If propionyl-Co A acts as primer, long chain fatty acids having an odd number of carbon atoms result. These are found particularly in ruminants, where propionate is formed by microbial action in the rumen. NADPH is involved as coenzyme in both the reduction of the β-keto-acyl and of the α,β-unsaturated acyl derivatives. The oxidative reactions of the hexose monophosphate shunt are probably the chief source of the hydrogen required for the reductive synthesis of fatty acids. Tissues which possess an active hexose monophosphate shunt are also the tissues specializing in active lipogenesis, ie, liver, adipose tissue, and the lactating mammary gland. Moreover, both metabolic pathways are found in the extramitochondrial region of the cell, so that there are no membranes or permeability barriers for the transfer of NADPH/NADP from one pathway to the other. Other sources of NADPH include the extramitochondrial isocitrate dehydrogenase reaction and the reaction that converts malate to pyruvate catalyzed by the "malic enzyme" (see Chapter 13).

C. Microsomal System for Chain Elongation: This is probably the main site for the elongation of existing fatty acid molecules. The pathway converts acyl-Co A compounds of fatty acids to higher derivatives, using malonyl-Co A as acetyl donor and NADPH as reductant. Intermediates in the process are the Co A thioesters. The end product is the next higher homologue of the primer acyl-Co A molecule. The acyl groups that may act as a primer molecule include the saturated series from C_{10}–C_{16} and some unsaturated C_{18} fatty acids. Fasting largely abolishes chain elongation.

Acetyl-Co A, the main building block for fatty acids, is formed from carbohydrate via the oxidation of pyruvate within the mitochondria, but acetyl-Co A does not diffuse readily into the extramitochondrial compartment, the principal site of fatty acid synthesis. The rate of incorporation of citrate into the fatty acids of a supernatant preparation of lactating mammary gland is greater than that of acetate (Spencer & Lowenstein, 1962). In addition, the activity of the extramitochondrial citrate cleavage enzyme varies markedly with the nutritional state of the animal, closely paralleling the activity of the fatty acid synthesizing system. Utilization of pyruvate for lipogenesis by way of citrate involves the oxidative decarboxylation of pyruvate to acetyl-Co A and subsequent condensation with oxaloacetate to form citrate within the mitochondria, followed by the diffusion of citrate into the extramitochondrial compartment, where it undergoes cleavage to acetyl-Co A and oxaloacetate catalyzed by the citrate cleavage enzyme. The acetyl-Co A is then available for malonyl-Co A formation and synthesis to palmitate (Fig 13–10). The oxaloacetate can form malate via NADH-linked malate dehydrogenase, followed by the generation of NADPH via the malic enzyme. In turn, the NADPH becomes available for lipogenesis. This pathway is a means of transferring reducing equivalents from extramitochondrial NADH to NADP. There is little citrate cleavage enzyme in ruminants. This is probably because in these species acetate (derived from the rumen) is the main source of acetyl-Co A. As the acetate is activated to acetyl-Co A extramitochondrially, there is no necessity for it to enter mitochondria prior to incorporation into long chain fatty acids.

Regulation of Lipogenesis

Many animals, including man, take their food as spaced meals and therefore need to store much of the energy in their diet for use between meals. The process of lipogenesis is concerned with the conversion of glucose and intermediates such as pyruvate and acetyl-Co A to fat, which facilitates the anabolic phase of this cycle. The nutritional state of the organism and tissues is the main factor controlling the rate of lipogenesis. Thus the rate is high in the well fed animal whose diet contains a high proportion of carbohydrate. It is depressed under conditions of restricted caloric intake, on a high-fat diet, or when there is a deficiency of insulin, as in diabetes mellitus. As little as 2.5% of fat in the diet causes a measurable depression of lipogenesis in the liver as measured by the incorporation of acetate carbon into fatty acids (Hill & others, 1958). Lipogenesis from ^{14}C-acetate is higher in liver from rats consuming all their food in 2 hours. It is also higher when sucrose is fed instead of glucose (Fabry & others, 1968). Masoro & others (1950) showed that almost no ^{14}C-glucose was incorporated into fatty acids of liver slices from fasting rats, whereas its conversion to $^{14}CO_2$ was unaltered, demonstrating a metabolic block in the pathway of synthesis between acetyl-Co A and fatty acids. Because of the close association between the activities of the hexose monophosphate shunt on the one hand and of the lipogenic pathway on the other, it was considered that the block in lipogenesis was due to lack of NADPH generation from the shunt pathway. However, subsequent work in which an NADPH generating system was added to a liver homogenate from fasting rats failed to promote fatty acid synthesis (Sauer, 1960).

At present it is recognized that the rate-limiting reaction in the lipogenic pathway is at the acetyl-Co A carboxylase step, and more than one factor has been described which regulates the activity of this enzyme. Long chain acyl-Co A molecules competitively inhibit acetyl-Co A carboxylase (Bortz & Lynen, 1963), an example of metabolic negative feedback inhibition by a product of a reaction sequence inhibiting the initial reaction. Thus, if acyl-Co A accumulates because it is not esterified quickly enough, it will automatically damp down synthesis of new fatty acid. Likewise, if acyl-Co A accumulates as a result of an influx of FFA into the tissue, this will also inhibit synthesis of new fatty acid. This is the probable explanation of the depressed lipogenesis recorded under conditions of caloric deficiency, a high-fat diet, or diabetes mellitus—all of which are associated with increased levels of plasma FFA.

Acetyl-Co A carboxylase is activated in an allosteric manner by citrate. However, whether citrate plays such a role in vivo is not clear. Long chain acyl-Co A has also been reported to inhibit citrate formation at the citrate synthase step, essential in the pathway of fatty acid synthesis. It is also possible that oxidation of fatty acids—owing to increased levels of FFA or to lack of insulin, allowing increased lipolysis of triglycerides—may increase the concentrations of acetyl-Co A/Co A and NADH/NAD$^+$ in mitochondria, inhibiting pyruvate dehydrogenase and thus blocking the supply of acetyl-Co A from carbohydrate via pyruvate.

Flatt (1970) has suggested that lipogenesis from glucose in adipose tissue is an energy-producing process and may be self-limiting because of respiratory control and availability of ADP.

Various reports indicate that both the fatty acid synthetase complex and acetyl-Co A carboxylase may be adaptive enzymes, increasing in total amount in the fed state and decreasing in fasting, feeding of fat, and diabetes.

Microsomes have a stimulatory effect on fatty acid synthesis when added to the extramitochondrial system present in the supernatant fraction of the cell. Since microsomes catalyze the esterification of acyl-Co A to triglycerides and phospholipids, Lorch & others (1963) suggested that the mechanism of their stimulatory effect on fatty acid synthesis is by removing the feedback inhibition of acyl-Co A on acetyl-Co A carboxylase.

Role of Carnitine in Fatty Acid Metabolism

(−)Carnitine (β-hydroxy-γ-trimethylammonium butyrate), $(CH_3)_3N^+$-CH_2-CH(OH)-CH_2-COO^-, stimulates the oxidation of long chain fatty acids by mitochondria. It is widely distributed, being particularly abundant in muscle. Activation of long chain fatty acids to acyl-Co A occurs in microsomes and on the other membranes of mitochondria. Activation of lower fatty acids may occur within the mitochondria. This process seems to be independent of carnitine. Long

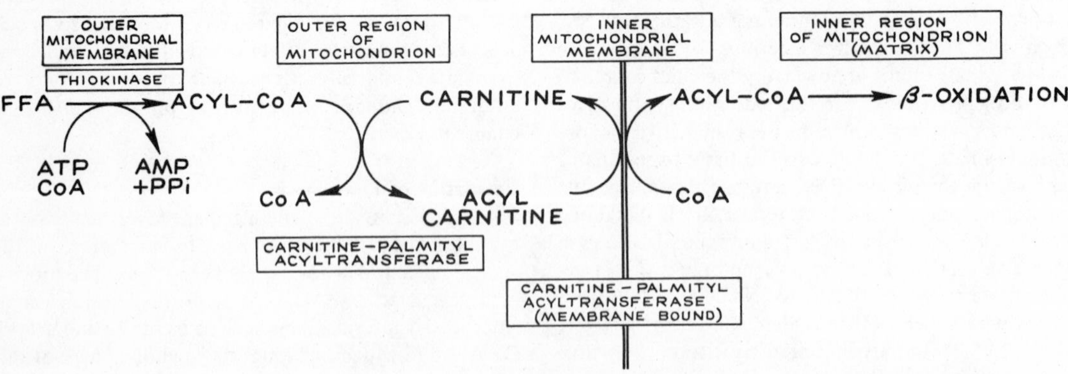

FIG 14–5. Proposed role of carnitine in the metabolism of long chain fatty acids.
(After Tubbs & Garland, 1968.)

chain acyl-Co A will not penetrate mitochondria and become oxidized unless carnitine is present, but carnitine itself will not penetrate mitochondria. An enzyme, **carnitine-palmityl acyltransferase**, is associated with the mitochondrial membranes and allows long chain acyl groups to penetrate the mitochondria and gain access to the β-oxidation system of enzymes (see Fritz, 1968; Bremer, 1968; and Tubbs & Garland, 1968). A possible mechanism to account for the action of carnitine in facilitating the oxidation of fatty acids by mitochondria is shown in Fig 14–5. In addition, another enzyme, **carnitine-acetyl acyltransferase**, is present in mitochondria which catalyzes the transfer of short chain acyl groups between Co A and carnitine. The function of this enzyme is somewhat obscure. It has been suggested by Fritz (1968) that 2 acetyl-Co A pools are present in mitochondria, one derived from fatty acid oxidation and the other from pyruvate oxidation, and that these 2 pools intercommunicate via the activity of carnitine-acetyl acyltransferase and reaction with carnitine.

ACYL–CoA + CARNITINE ⟷ ACYLCARNITINE + CoA
(CARNITINE-PALMITYL ACYLTRANSFERASE)

ACETYL–CoA + CARNITINE ⟷ ACETYL-CARNITINE + CoA
(CARNITINE-ACETYL ACYLTRANSFERASE)

BIOSYNTHESIS OF GLYCERIDES & METABOLISM OF PHOSPHOLIPIDS & SPHINGOLIPIDS

Although reactions involving the hydrolysis of triglycerides by lipase can be reversed, this does not seem to be the mechanism by which ester bonds of glycerides are synthesized in tissues. Tietz & Shapiro (1956) showed that ATP was required for the synthesis of neutral fat from fatty acids, and indeed both glycerol and fatty acids must be activated by ATP before they become incorporated into glycerides. If the tissue is liver, kidney, lactating mammary gland, or intestinal mucosa, the enzyme **glycerokinase** will catalyze the activation, by phosphorylation, of glycerol to α-glycerophosphate. If this enzyme is absent—or low in activity, as it is in muscle or adipose tissue—most of the α-glycerophosphate must be derived from an intermediate of the glycolytic system, dihydroxyacetone phosphate, which forms α-glycerophosphate by reduction with NADH catalyzed by **α-glycerophosphate dehydrogenase.**

Fatty acids are activated to acyl-Co A by the enzyme **thiokinase**, utilizing ATP and Co A. Two molecules of acyl-Co A combine with α-glycerophosphate to form α,β-diglyceride phosphate (phosphatidic acid), catalyzed by **glycerophosphate acyltransferase**. This may take place in 2 stages via lysophosphatidic acid. Phosphatidic acid is converted by a phosphatase (**phosphatidate phosphohydrolase**) to an α,β-diglyceride. In intestinal mucosa, a monoglyceride pathway exists whereby monoglyceride is converted to diglyceride as a result of the presence of **monoglyceride acyltransferase**. A further molecule of acyl-Co A is esterified with the diglyceride to form a triglyceride, catalyzed by **diglyceride acyltransferase** (Fig 14–6). Most of the activity of these enzymes resides in the microsomal fraction of the cell (Wilgram & Kennedy, 1963), but some is found also in mitochondria. It has been reported (Hajra & Agranoff, 1968) that dihydroxyacetone phosphate may be acylated and converted to lysophosphatidic acid after reduction by NADP. The quantitative significance of this pathway remains to be evaluated.

Phospholipids

Phospholipids are synthesized either from phosphatidic acid (phosphatidyl inositol) or from α,β-diglyceride (phosphatidyl choline or phosphatidyl ethanolamine). The synthesis of phosphatidyl inositol has been studied by Paulus & Kennedy (1960). They have proposed that cytidine triphosphate (CTP; see Chapter 4) reacts with phosphatidic acid to form a cytidine-diphosphate-diglyceride (CDP-diglyceride). Finally this compound reacts with inositol, catalyzed by the enzyme **CDP-diglyceride inositol transferase,** to form a phosphatidyl inositol (Fig 14–6).

In the biosynthesis of phosphatidyl choline and phosphatidyl ethanolamine (lecithins and cephalins) (Fig 14–6), choline or ethanolamine must first be converted to "active choline" or "active ethanolamine," respectively. This is a 2-stage process involving, first, a reaction with ATP to form the corresponding monophosphate, followed by a further reaction with CTP to form either cytidine diphosphocholine (CDP-choline) or cytidine diphosphoethanolamine (CDP-ethanolamine). In this form, choline or ethanolamine reacts with α,β-diglyceride so that a phosphorylated base (either phosphoryl choline or phosphoryl ethanolamine) is transferred to the diglyceride to form either phosphatidyl choline or phosphatidyl ethanolamine. The enzyme responsible for the formation of phosphatidyl ethanolamine, **phosphoryl-ethanolamine-glyceride transferase**, is not present in liver. Phospha-

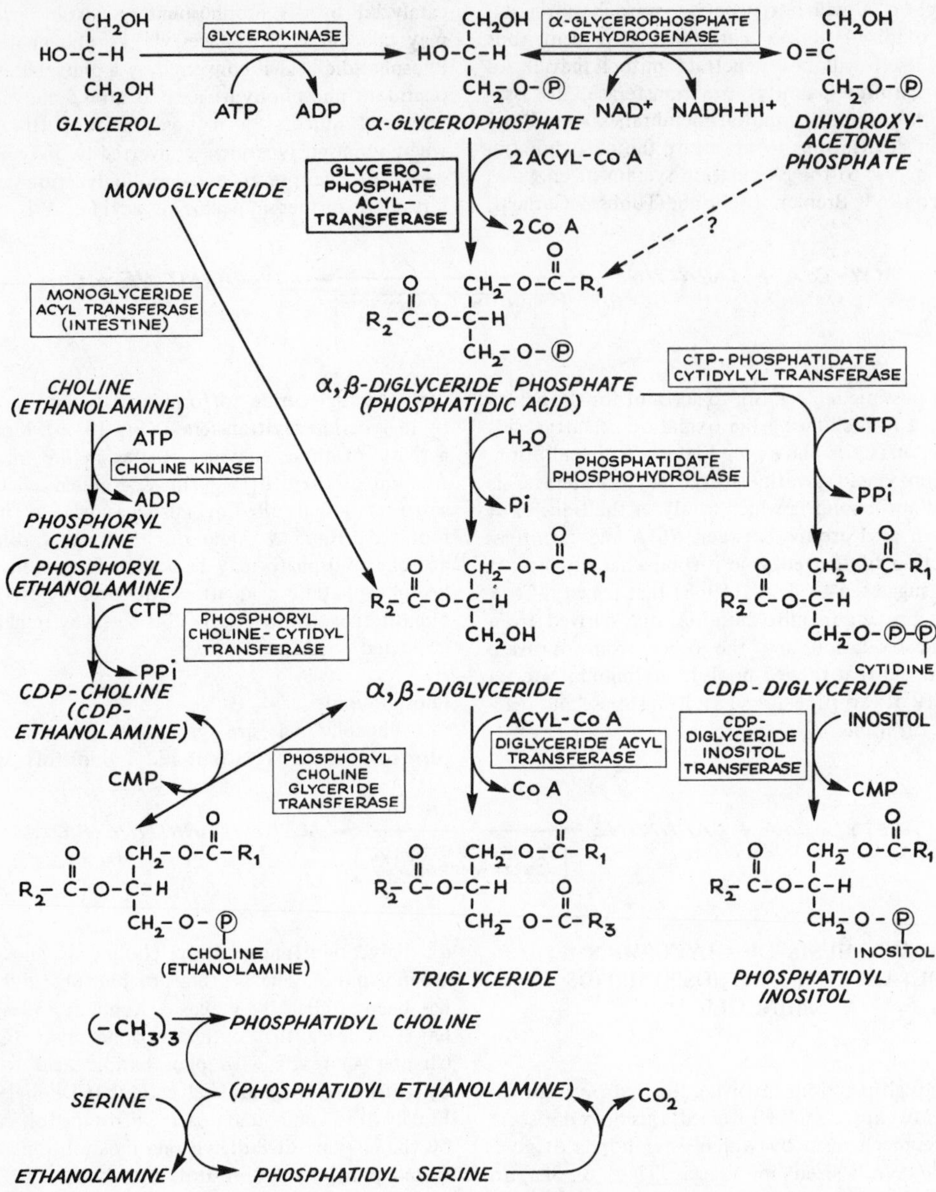

FIG 14–6. Biosynthesis of triglycerides and phospholipids.

tidyl serine is formed from phosphatidyl ethanolamine directly by reaction with serine. Phosphatidyl serine may re-form phosphatidyl ethanolamine by decarboxylation. In the liver, but not in brain, an alternative pathway enables phosphatidyl ethanolamine to give rise directly to phosphatidyl choline by progressive methylation of the ethanolamine residue.

Long chain saturated fatty acids are found predominantly in the α-position of phospholipids, whereas the unsaturated acids are incorporated more into the β-position. The incorporation of fatty acids into lecithins occurs by complete synthesis of the glyceride, by transacylation between cholesterol ester and lysolecithin and by direct acylation of lysolecithin by acyl-Co A. Thus, a continuous exchange of the fatty acids is possible, particularly with regard to introducing essential fatty acids into phospholipid molecules.

A phospholipid present in mitochondria is **cardiolipin** or diphosphatidyl glycerol (Fig 14–7). It is formed from phosphatidyl glycerol, which in turn is synthesized from CDP-diglyceride and α-glycerophosphate according to the scheme shown in Fig 14–6.

A pathway for the biosynthesis of plasmalogens in rat liver has been demonstrated by Kiyasu &

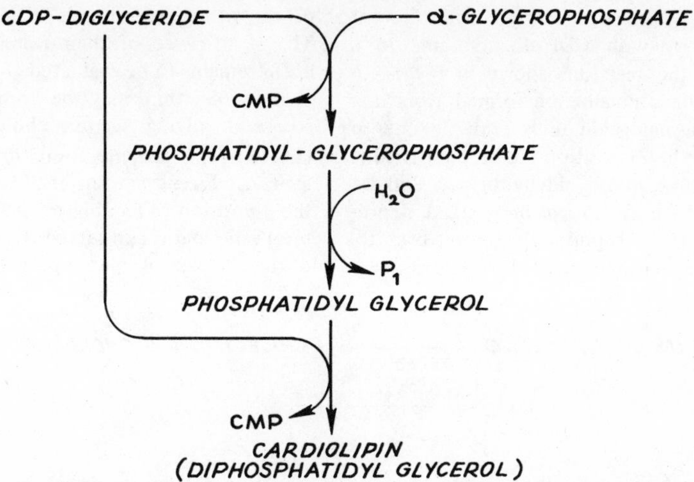

FIG 14–7. Biosynthesis of cardiolipin.

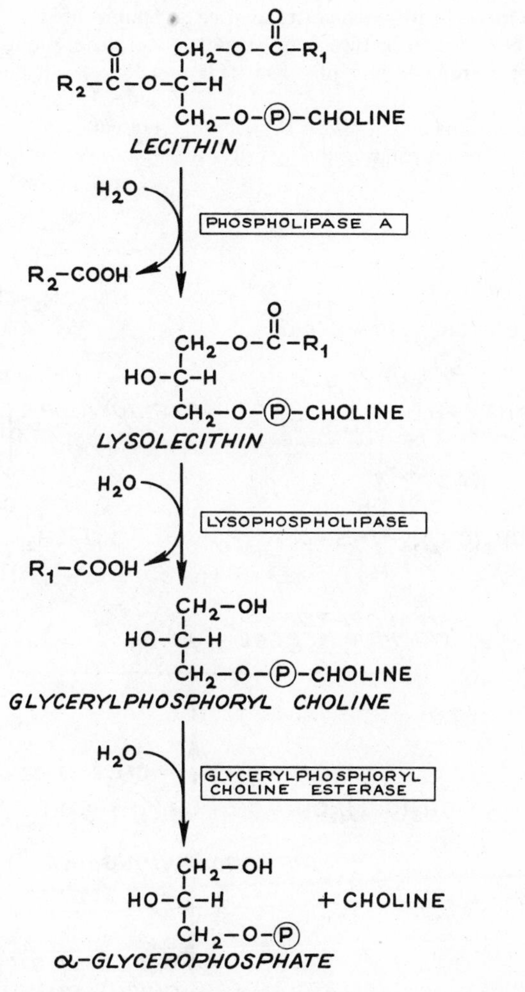

Fig 14–8. Catabolism of lecithin.

Kennedy (1960). "Plasmalogenic diglycerides" react with CDP-choline or with CDP-ethanolamine in a manner similar to the reactions shown in Fig 14–6 whereby lecithins or cephalins are formed from α,β-diglycerides. A plasmalogenic diglyceride is one in which the alpha (or beta) position has an acyl residue containing the vinyl ether aldehydogenic linkage ($-CH_2-O-CH=CH-R'$) as shown on p 20. A significant proportion of the phospholipids present in mitochondria are plasmalogens (Goldfine, 1968).

of the other ester bonds in phospholipid molecules. The significance of their role in mammalian metabolism remains to be evaluated.

Lysolecithin may be formed by an alternative route involving **lecithin:cholesterol acyltransferase (LCAT)**. This enzyme, found in plasma and possibly in liver, catalyzes the transfer of a fatty acid residue from the β-position of lecithin to cholesterol to form cholesterol ester and is considered to be responsible for much of the cholesterol ester in plasma lipoproteins.

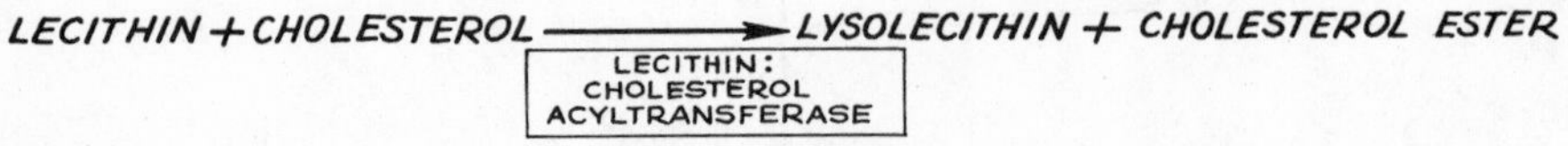

LECITHIN + CHOLESTEROL → *LYSOLECITHIN + CHOLESTEROL ESTER* (LECITHIN: CHOLESTEROL ACYLTRANSFERASE)

Degradation of Phospholipids by Phospholipases

Phospholipase enzymes are widely distributed and are responsible for catabolism of phospholipids. **Phospholipase A** hydrolyzes the fatty acyl ester bond in the β-position. The resulting lysophospholipid is hydrolyzed by **lysophospholipase** to form the corresponding glyceryl phosphoryl base, which in turn may be split by an esterase to α-glycerophosphate plus base (Fig 14–8).

Other phospholipases (named B-, C-, or D-) occurring both in animals and plants catalyze the hydrolysis

For further information on phospholipids and phospholipases, see Ansell & Hawthorne (1964), Van Deenen & de Haas (1966), Goldfine (1968), Rossiter (1968), and Thompson (1970).

Sphingolipids

The **sphingomyelins** are phospholipids containing a fatty acid, phosphoric acid, choline, and a complex amino alcohol, sphingol (sphingosine). No glycerol is present.

The synthesis of **sphingosine** (Fig 14–9) has been studied by Brady & Koval (1958) in brain tissue. The

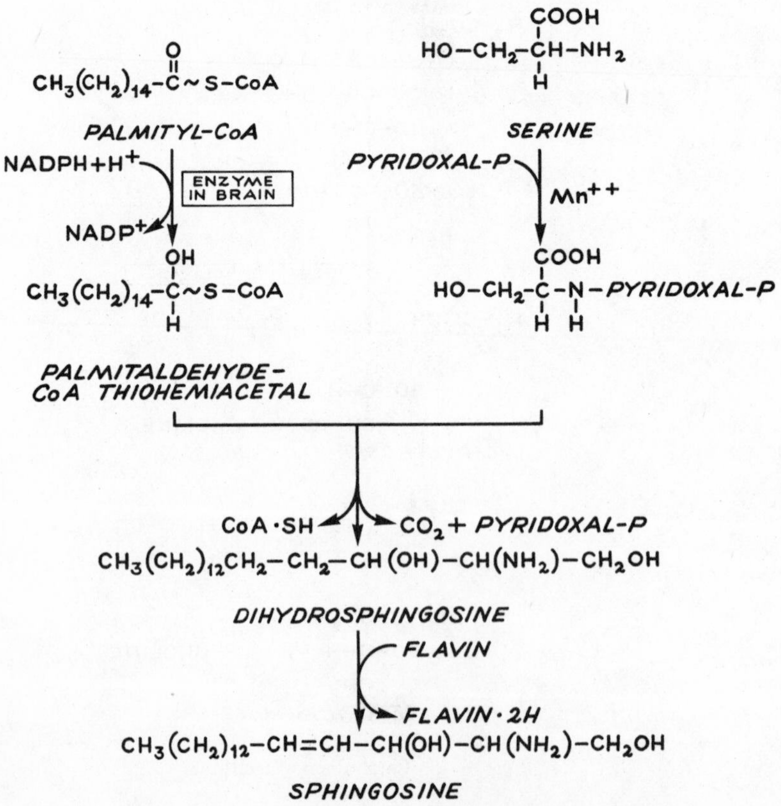

$CH_3(CH_2)_{14}-C(=O)\sim S-CoA$

PALMITYL-CoA

$NADPH+H^+$ → $NADP^+$ (ENZYME IN BRAIN)

$CH_3(CH_2)_{14}-CH(OH)\sim S-CoA$

PALMITALDEHYDE-CoA THIOHEMIACETAL

$HO-CH_2-CH(COOH)-NH_2$

SERINE

PYRIDOXAL-P, Mn^{++}

$HO-CH_2-CH(COOH)-NH-$*PYRIDOXAL-P*

$CoA\cdot SH$; CO_2 + *PYRIDOXAL-P*

$CH_3(CH_2)_{12}CH_2-CH_2-CH(OH)-CH(NH_2)-CH_2OH$

DIHYDROSPHINGOSINE

FLAVIN → *FLAVIN·2H*

$CH_3(CH_2)_{12}-CH=CH-CH(OH)-CH(NH_2)-CH_2OH$

SPHINGOSINE

FIG 14–9. Biosynthesis of sphingosine.

reduction of palmityl-Co A to the aldehyde is the first step in the synthetic pathway. The amino acid serine, after activation by combination with vitamin B_6 (pyridoxal phosphate) and after decarboxylation, condenses with palmityl aldehyde-Co A to form dihydrosphingosine, which in the presence of a flavoprotein dehydrogenase loses 2 H atoms to form sphingosine.

Kanfer & Gal (1966) showed that, in vivo, sphingomyelin is synthesized from sphingosine phosphoryl choline. This is formed by the reaction of sphingosine with CDP-choline. Sphingosine phosphoryl choline is acylated at the amino group by an acyl-Co A of a long chain fatty acid to form sphingomyelin. Alternatively, sphingomyelin may be synthesized from sphingosine via the formation of ceramide (N-acyl sphingosine), which in turn reacts with CDP-choline, giving CMP and sphingomyelin.

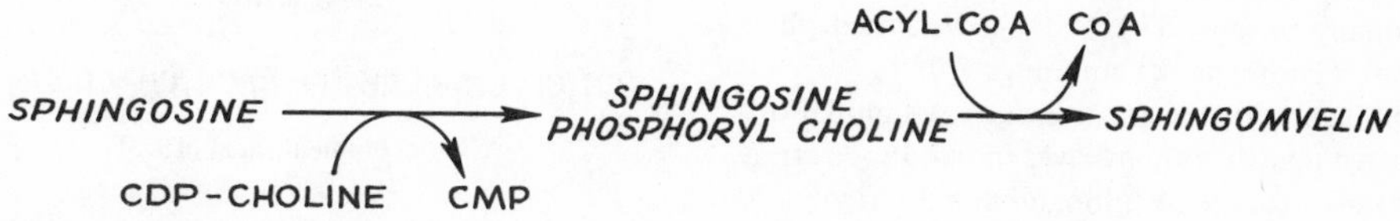

Cerebrosides, Sulfatides, & Gangliosides

The cerebrosides are glycolipids which contain the sphingosine-fatty acid combination (ceramide) found in the sphingomyelins, but a galactose moiety is attached to the ceramide in the place of the phosphoryl choline residue found in sphingomyelin. The biosynthesis of the characteristic C_{24} fatty acids which occur in cerebrosides (lignoceric, cerebronic, and nervonic acids) has been studied by Fulco & Mead (1961). Lignoceric acid ($C_{23}H_{47}COOH$) is completely synthesized from acetate. Cerebronic acid, the 2-hydroxy derivative of lignoceric acid, is formed from it. Nervonic acid ($C_{23}H_{45}COOH$), a mono-unsaturated acid, is formed by elongation of oleic acid.

The requirement for galactose in the formation of cerebrosides, chondromucoids, and mucoproteins is the only known physiologic role of this sugar other than in the formation of lactose in milk.

Brady (1962) has reported on his studies of the biosynthesis of the complete cerebroside molecule catalyzed by an enzyme preparation obtained from young rat brain (Fig 14–10). He found that **uridine diphosphogalactose epimerase** utilizes uridine diphosphate glucose (UDP-glucose) as substrate and accomplishes epimerization of the glucose moiety to galactose, thus forming uridine diphosphate galactose (UDP-galactose). The reaction in brain is similar to that described on p 259 for the liver and mammary gland.

The reaction sequence suggested for the biosynthesis of cerebrosides in brain tissue is shown in Fig 14–10.

In this reaction sequence, acyl-Co A represents the Co A derivative of a fatty acid which is to be incorporated into the cerebroside. Examples would be lignoceric, cerebronic, or nervonic acids, as noted above. In his experiments, Brady used labeled (1-^{14}C) stearic acid, a C_{18} saturated fatty acid which is in fact a major component among the fatty acids of the cerebrosides in rat brain. The cerebrosides are found in high concentration in the myelin sheaths of nerves. **Sulfatides** are formed from cerebrosides after reaction with 3′-phosphoadenosine-5′-phosphosulfate (PAPS; "active sulfate"). Gangliosides are synthesized from ceramide (acyl sphingosine) by the stepwise addition of the activated sugars (eg, UDPG and UDPGal) and N-acetylneuraminic acid (NANA) (Fig 14–11). A large number of gangliosides of increasing molecular weight may be formed. For further details of ganglioside metabolism, see Svennerholm (1970).

Phospholipids & Sphingolipids in Disease

Certain diseases are characterized by abnormal quantities of these lipids in the tissues, often in the nervous system. They may be classified into 3 groups: (1) true demyelinating diseases, (2) sphingolipidoses, and (3) leukodystrophies.

In **multiple sclerosis**, which is a demyelinating disease, there is loss of both phospholipids, particularly ethanolamine plasmalogen, and of sphingolipids from white matter, such that an analysis of it resembles

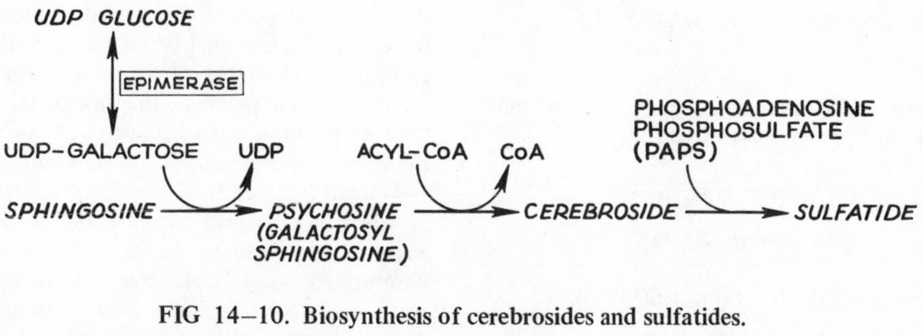

FIG 14–10. Biosynthesis of cerebrosides and sulfatides.

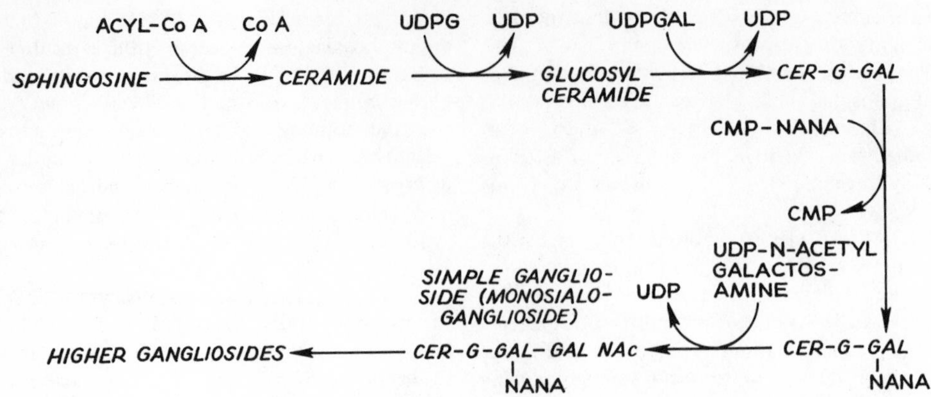

FIG 14–11. Biosynthesis of gangliosides.

more the composition of gray matter (Ansell & Hawthorne, 1964). Cholesterol esters are also found, though normally absent. The CSF shows raised phospholipid levels (Thomson & Cummings, 1964).

The **sphingolipidoses** are a group of familial diseases that often manifest themselves in childhood; they are due to excess storage of sphingolipids. In **Tay-Sachs disease**, ganglion cells, particularly in the cerebral cortex, become swollen with lipid, mostly ganglioside of a type that normally occurs only in trace amounts. In **Niemann-Pick disease**, cells of the liver and spleen become foamy in appearance and there is swelling of ganglion cells. In contrast to Tay-Sachs disease, the lipid deposited is sphingomyelin. In **Gaucher's disease**, the cerebroside content of cells of the enlarged spleen, liver, and lymph nodes (but not of the nervous system) is increased (Zöllner & Thannhauser, 1964). Only the cerebrosides containing glucose (but not galactose) are affected. It appears that diseases characterized by the accumulation of sphingolipids are due to malfunction of the catabolic pathways of these substances rather than of the mechanism of their syntheses. For a further summary, see Shapiro (1967).

In **metachromatic leukodystrophy**, there is general demyelination characterized by the accumulation of sulfatides containing galactose rather than glucose.

Lipidoses have been reviewed in Schettler (1967).

METABOLISM OF THE UNSATURATED & ESSENTIAL FATTY ACIDS (EFA)

The long chain unsaturated fatty acids of metabolic significance in mammals are as follows:

$$CH_3(CH_2)_5CH{=}CH(CH_2)_7COOH$$

Palmitoleic acid (16:1)

$$CH_3(CH_2)_7CH{=}CH(CH_2)_7COOH$$

Oleic acid (18:1)

$$CH_3(CH_2)_4CH{=}CHCH_2CH{=}CH(CH_2)_7COOH$$

Linoleic acid (18:2)

$$CH_3CH_2CH{=}CHCH_2CH{=}CHCH_2CH{=}CH(CH_2)_7COOH$$

Linolenic acid (18:3)

$$CH_3(CH_2)_4(CH{=}CHCH_2)_4(CH_2)_2COOH$$

Arachidonic acid (20:4)

Other C_{20} and C_{22} polyenoic fatty acids may be detected by gas-liquid chromatography. These are derived from linoleic and linolenic acids by chain elongation. It is to be noted that all double bonds present in naturally occurring unsaturated fatty acids of mammals are of the cis configuration.

Palmitoleic and oleic acids are not essential in the diet because the tissues are capable of introducing a single double bond into the corresponding saturated fatty acid. Experiments with labeled palmitate have demonstrated that the label enters freely into palmitoleic and oleic acids but is absent from linoleic, linolenic, and arachidonic acids. These latter fatty acids, as a group, cannot be synthesized in the mammalian organism and must be supplied in adequate amounts in the diet in order to maintain health. They are known as the **essential fatty acids**. Fasting abolishes desaturation of both stearic and palmitic acid (Elovson, 1965).

It is a common finding in the husbandry of animals that the degree of saturation of the fat laid down in the depots can be altered by dietary means. If, for example, an animal is fed a diet containing a large quantity of vegetable oil (ie, a high proportion of the unsaturated fatty acids), the animal lays down a soft type of depot fat. The converse situation is found in ruminants, where a characteristic hard, saturated fat is laid down as a result of the action of microorganisms in the rumen, which saturate the unsaturated fatty acids of the diet. As far as the nonessential monounsaturated fatty acids are concerned, the liver is considered to be the main organ responsible for their interconversion with the saturated fatty acids. Holloway & others (1962) have shown that an enzyme system in liver microsomes will catalyze the conversion of stearyl-Co A to oleyl-Co A. Oxygen, NADPH, or NADH is necessary for the reaction. The enzymes appear to be those of a typical mono-oxygenase system (hydroxylase; see p 170). The sequence of reactions is as follows:

$$\text{STEARYL-Co A} + \text{ENZYME} \longrightarrow \text{STEARYL-E} + \text{Co A}$$

$$\text{STEARYL-E} + O_2 + \text{NADPH} + H^+ \xrightarrow{\text{HYDROXYLASE}} \text{HYDROXYSTEARYL-E} + \text{NADP}^+ + H_2O$$

$$\text{HYDROXYSTEARYL-E} \xrightarrow{\text{HYDRATASE}} \text{OLEYL-E} + H_2O$$

$$\text{OLEYL-E} + \text{Co A} \longrightarrow \text{OLEYL-Co A} + \text{ENZYME}$$

The enzymatic reactions responsible for the oxidation of mono-unsaturated and poly-unsaturated fatty acids have been elucidated by Stoffel & Caesar (1965). The Co A esters of these acids are degraded by the enzymes normally responsible for β-oxidation until either a Δ^3-cis-acyl-Co A compound or a Δ^2-cis-acyl-Co A compound is formed, depending upon the position of the double bonds. The former compound is isomerized to the corresponding Δ^2-trans-Co A stage, which in turn is hydrated by **enoyl hydrase** to L(+)-β-hydroxy-acyl-Co A. The Δ^2-cis-acyl-Co A compound is first hydrated by enoyl hydrase to the D(−)-β-hydroxy-acyl-Co A derivative. This undergoes epimerization to give the normal L(+)-β-hydroxy-acyl-Co A stage in β-oxidation.

In 1928, Evans & Burr noticed that rats fed on a purified nonlipid diet to which vitamins A and D were added exhibited a reduced growth rate and a reproductive deficiency. Later work showed that the deficiency syndrome was cured by the addition of linoleic, linolenic, and arachidonic acids to the diet. Further diagnostic features of the syndrome include scaly skin, necrosis of the tail, and lesions in the urinary system, but the condition is not fatal.

Although the **essential fatty acids (EFA)** are not synthesized from other sources, linoleate may be converted to arachidonate (Mead, 1961). The pathway is first by dehydrogenation of the Co A ester through γ-linolenate followed by the addition of a 2-carbon unit (probably as acetyl-Co A in the mitochondrial system for chain elongation or as malonyl-Co A in the microsomal system, which appears to be the more active system) to give eicosatrienoate (homo γ-linolenate). The latter forms arachidonate by a further dehydrogenation. The dehydrogenating system is the same as that described for saturated fatty acids above.

The nutritional requirement for arachidonate may thus be spared provided there is adequate linoleate in the diet.

Corresponding derivatives to arachidonate can also be formed from palmitoleate, oleate, and linolenate by a similar sequence of reactions. The reaction responsible for elongation of the fatty acid molecule by the addition of a 2-carbon unit could involve, presumably, the mitochondrial system or the microsomal system described earlier. In the former pathway, pyridoxal phosphate may be required as a coenzyme for the condensing enzyme. It is interesting that Witten & Holman (1952), after experimenting with rats, concluded that pyridoxine was involved in the conversion of linoleate to arachidonate. Although the outward signs of essential fatty acid deficiency and vitamin B_6 deficiency are similar, the resemblances, according to Sinclair (1964), are only superficial.

The desaturation and chain elongation system is greatly diminished in the fasting state and in the absence of insulin.

The functions of the EFA appear to be various though not well defined. EFA are found in the structural lipids of the cell, are concerned with the structural integrity of the mitochondrial membrane, and occur in high concentration in the reproductive organs. In many of their structural functions, EFA are present in phospholipids, mainly in the 2-position. The roles of essential fatty acids in the genesis of fatty livers and in the metabolism of cholesterol are discussed later.

$CH_3(CH_2)_4CH{=}CHCH_2CH{=}CH(CH_2)_7COOH$

LINOLEATE

↓ (−2 H)

$CH_3(CH_2)_4CH{=}CHCH_2CH{=}CHCH_2CH{=}CH{-}(CH_2)_4COOH$

γ-LINOLENATE

↓ ($+C_2$)

$CH_3(CH_2)_4(CH{=}CHCH_2)_3(CH_2)_5COOH$

$\Delta^{5,8,11}$ EICOSATRIENOATE

↓ (−2 H)

$CH_3(CH_2)_4(CH{=}CHCH_2)_4(CH_2)_2COOH$

ARACHIDONATE

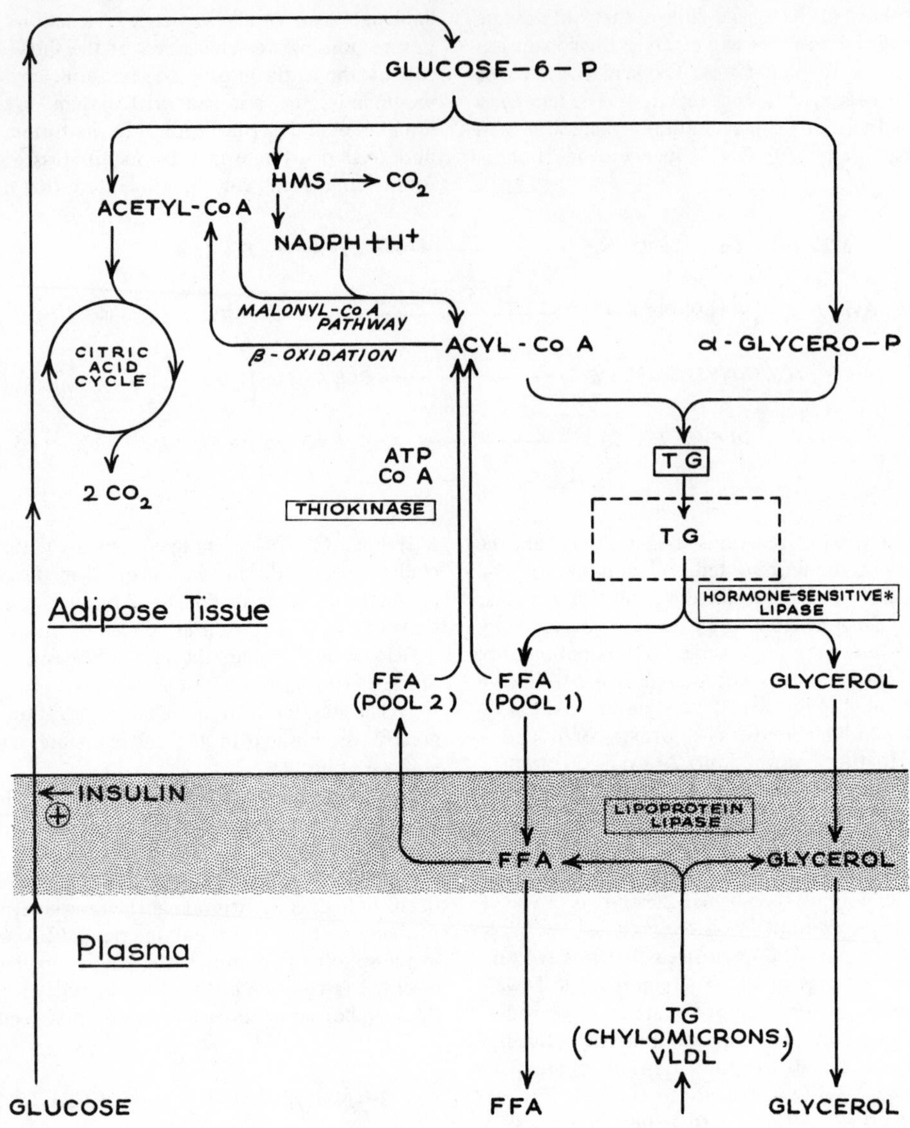

*Hormone-sensitive lipase activated by ACTH, TSH, glucagon, epinephrine, norepinephrine, and vasopressin. Inhibited by insulin, prostaglandin E_1, and nicotinic acid.

HMS = Hexose monophosphate shunt
TG = Triglyceride
FFA = Free fatty acids
VLDL = Very low density lipoproteins
[shaded] = Lipoprotein lipase region of the capillary wall

FIG 14–12. Metabolism of adipose tissue.

An essential fatty acid deficiency in man has not been demonstrated unequivocally. However, linoleate has been observed to cure skin lesions in infants receiving formula diets low in fat (Adam & others, 1958).

Isotopic experiments have indicated that arachidonate and some related C_{20} fatty acids with methylene-interrupted bonds give rise to the group of pharmacologically active compounds known as prostaglandins (see p 15). Prostaglandins do not relieve the symptoms of EFA deficiency, indicating that it is not via prostaglandin synthesis that EFA have their effect.

Trans-fatty acids. The presence of trans-unsaturated fatty acids in partially hydrogenated vegetable oils (eg, margarine) raises the question of their safety as food additives. Their long-term effects in man are not known, but up to 15% of tissue fatty acids have been found at autopsy to be in the trans configuration. They are metabolized more like saturated than like the cis-unsaturated fatty acids. This may be due to their similar straight chain conformation (see p 16). Trans-polyunsaturated fatty acids do not possess EFA activity. For reviews, see Sgoutas & Kummerow (1970) and Guarnieri & Johnson (1970), who have reviewed the entire field of EFA metabolism.

THE METABOLISM OF ADIPOSE TISSUE & THE REGULATION OF THE MOBILIZATION OF FAT

The triglyceride stores in adipose tissue are continually undergoing lipolysis (hydrolysis) and reesterification (Fig 14–12). These 2 processes are not the forward and reverse phases of the same reaction but are entirely different pathways involving different reactants and enzymes. Many of the nutritional, metabolic, and hormonal factors that regulate the metabolism of adipose tissue act either upon the process of esterification or on lipolysis. The resultant of these 2 processes determines the magnitude of the free fatty acid (FFA) pool in adipose tissue, which in turn is the source and determinant of the level of FFA circulating in the plasma. Since the level of plasma FFA has the most profound effects upon the metabolism of other tissues, particularly liver and muscle, the factors operating in adipose tissue which regulate the efflux of FFA exert an influence far beyond the tissue itself.

In adipose tissue, triglyceride is synthesized from acyl-Co A and α-glycerophosphate according to the mechanism shown in Fig 14–6. Because the enzyme **glycerokinase** is low in activity in adipose tissue, glycerol cannot be utilized to any great extent in the esterification of acyl-Co A. For the provision of α-glycerophosphate needed in this reaction the tissue is dependent on a supply of glucose. The triglyceride undergoes hydrolysis by a **hormone-sensitive lipase*** to form FFA and glycerol. Since glycerol cannot be utilized readily in the tissues, it diffuses out into the plasma, from which it is utilized by such tissues as liver and kidney which possess an active glycerokinase. The FFA formed by lipolysis can be resynthesized in the tissue to acyl-Co A by a **thiokinase** and reesterified with α-glycerophosphate to form triglyceride. Thus, there is a continual cycle within the tissue of lipolysis and reesterification. However, when the rate of reesterification is not sufficient to match the rate of lipolysis, FFA accumulates and diffuses into the plasma, where it raises the level of FFA.

Under conditions of adequate nutritional intake or when the utilization of glucose by adipose tissue is increased, the FFA outflow decreases and the level of plasma FFA falls. However, in vitro, the release of glycerol continues, demonstrating that the effect of glucose in reducing plasma FFA is not mediated by reducing the rate of lipolysis. It is believed that the effect is due to the provision of α-glycerophosphate from glucose, which enhances esterification of FFA via acyl-Co A. When the availability of glucose in adipose tissue is reduced, as in starvation or diabetes mellitus, less α-glycerophosphate is formed, allowing the rate of lipolysis to exceed the rate of esterification, with subsequent accumulation of FFA and their release into the plasma.

Glucose can take several pathways in adipose tissue, including oxidation to CO_2 via the citric acid cycle, oxidation in the hexose monophosphate shunt, conversion to long chain fatty acids, and formation of glyceride-glycerol via α-glycerophosphate. When glucose utilization is high, a larger proportion of the uptake is oxidized to CO_2 and converted to fatty acids. However, as total glucose utilization decreases, the greater proportion of the glucose is directed to the formation of α-glycerophosphate and glyceride-glycerol, which helps to minimize the efflux of FFA.

FFA liberated by adipose tissue may also be metabolized in that same tissue. In vitro studies (Shapiro & others, 1957) have demonstrated uptake of ^{14}C-stearate (FFA) into adipose tissue of both fed and fasting rats, but the uptake into the fed preparation was double that of the fasted one. However, in vivo the nutritional state has little effect on the deposition of 1-^{14}C-palmitate in adipose tissue, possibly because the FFA uptake is limited by the blood flow (Havel & Carlson, 1962). In the re-fed condition (Shapiro & others, 1957), most of the uptake was esterified and only a small percentage oxidized to CO_2, whereas in the fasting condition approximately equal amounts were oxidized and esterified.

*Adipose tissue contains at least 2 types of lipase, a so-called hormone-sensitive lipase or triglyceride lipase, responsible for fat mobilization, and lipoprotein lipase, concerned with the uptake of lipoprotein triglycerides into adipose tissue. The nomenclature of these lipases is confusing; lipoprotein lipase not only hydrolyzes triglycerides, but evidence is accumulating that it too is hormone-sensitive.

From all of the foregoing observations, it would appear that when carbohydrate is abundant adipose tissue tends to emphasize the utilization of glucose for energy production and to esterify FFA; when carbohydrate is in short supply, it conserves glucose for esterification via α-glycerophosphate formation and utilizes fatty acids for energy production.

Several laboratories have furnished evidence pointing to the existence of more than one FFA pool within adipose tissue. Dole (1961) has shown that the FFA pool (Fig 14–12, pool 1) formed by lipolysis of TG is the same pool that supplies fatty acids for reesterification; also, it releases them into the external medium (plasma). This latter process is not reversible, since labeled fatty acids taken up from the external medium do not label pool 1 before they are incorporated into triglyceride. It is necessary to postulate the existence of a second FFA pool (pool 2) through which FFA pass after uptake before they are incorporated into TG or oxidized to CO_2. The work of Dole indicates that this second pool would be small and have a high turnover rate, since FFA from the medium become esterified immediately upon entering the cell.

When unsaturated fatty acids are fed, they do not become incorporated very rapidly into all the depot fat but appear first of all to enter smaller and more active compartments, indicating that there are also several pools of triglyceride in adipose tissue.

Influence of Hormones on Adipose Tissue

The rate of release of FFA from adipose tissue is affected by many hormones that influence either the rate of esterification or the rate of lipolysis. Insulin administration is followed by a fall in circulating plasma FFA. In vitro, it inhibits the release of FFA from adipose tissue, enhances lipogenesis and the synthesis of glyceride glycerol, and increases the oxidation of glucose to CO_2 via the hexose monophosphate shunt. All of these effects are dependent on the presence of glucose in the medium and can be explained, therefore, on the basis of the ability of insulin to enhance the uptake of glucose into adipose tissue cells. A second action of insulin in adipose tissue is to inhibit the activity of the hormone-sensitive lipase (Ball & Jungas, 1963), reducing the release not only of FFA but of glycerol as well. Adipose tissue is much more sensitive to insulin than is diaphragm muscle, which points to adipose tissue as a major site of insulin action in vivo. Both glucose oxidation and lipogenesis are reduced to the extent of 80–90% in adipose tissue from alloxan-diabetic rats. These metabolic effects are reversed by the addition of insulin in vitro (Winegrad & Renold, 1958). Prolactin has an effect upon adipose tissue similar to that of insulin, but only if it is given in large doses.

Other hormones accelerate the release of FFA from adipose tissue and raise the plasma FFA level by increasing the rate of lipolysis of the triglyceride stores. These include adrenocorticotropic hormone (ACTH), α- and β-melanocyte stimulating hormones (MSH), thyroid stimulating hormone (TSH), growth hormone, vasopressin, epinephrine, norepinephrine, and glucagon. Many of these activate the hormone-sensitive lipase and increase glucose utilization as well. The latter process has been attributed to stimulation of esterification by the increased production of FFA. For an optimum effect, most of these lipolytic processes require the presence of glucocorticoids and thyroid hormones. On their own, these particular hormones do not increase lipolysis markedly, but act in a facilitatory or permissive capacity with respect to other lipolytic endocrine factors. These properties can be demonstrated in vivo using hypophysectomized or adrenalectomized animals. Levin & Farber (1952) showed the need for a minimal level of circulating glucocorticoid in order to evoke the adipokinetic properties of GH. An explanation of the mechanism of action of glucocorticoids has been given by Shafrir & Kerpel (1964). They suggest that cortisol, by depressing glucose uptake and depleting glycogen, causes a reduction in the capacity of adipose tissue to respond to the increased liberation of FFA by increasing esterification. Jeanrenaud & Renold (1966) showed that glucocorticoid increases the output of FFA from adipose tissue in vitro by decreasing the rate of esterification and not by stimulating lipolysis. To obtain this effect, prolonged incubation is necessary, suggesting that the effect may be due to a change in rate of protein synthesis. The effect is independent of the presence of glucose in the medium. The opposite effect may prevail after adrenalectomy, when, through enhanced esterification, the release of FFA is diminished in the presence of a lipolytic stimulus. These concepts aid in understanding the intensely adipokinetic properties of glucocorticoids when administered to completely depancreatized rats, where insulin is absent (Scow & Chernick, 1960).

A greater understanding of the properties of the adipose tissue has supplied a more unifying concept of the effects of the several hormones affecting lipolysis. Adipose tissue contains a number of lipases, one of which is a hormone-sensitive triglyceride lipase. In addition, there is present a diglyceride lipase and monoglyceride lipase, which are not hormone-sensitive, but they are considerably more active than the hormone-sensitive triglyceride lipase; therefore, the latter is considered to catalyze the rate-limiting step in lipolysis (Steinberg, 1966). It appears that cyclic AMP converts inactive hormone-sensitive triglyceride lipase into active lipase. The hormones that act rapidly in promoting lipolysis do so by stimulating the activity of adenyl cyclase, the enzyme that converts ATP to cyclic AMP. The mechanism is analogous to that responsible for hormonal stimulation of glycogenolysis. Lipolysis is controlled largely by the amount of cyclic AMP present in the tissue. It follows that processes that destroy or preserve cyclic AMP have an effect on lipolysis. Cyclic AMP is degraded to 5′-AMP by the enzyme cyclic **3′,5′-nucleotide phosphodiesterase.** This enzyme is inhibited by methyl xanthines such as caffeine and theophylline. Thus, at concentrations at which caffeine itself does not cause any increase in cyclic AMP in isolated fat cells, and in the presence of a lipolytic hormone such as epinephrine, caffeine acts synergistically to cause a considerable increase in cyclic AMP over that which would be caused by the epinephrine alone (Butcher & others, 1968). It is significant that the drinking of coffee or the administration of caffeine causes marked and prolonged elevation of plasma FFA in humans (Bellet & others, 1968). Insulin has a pronounced antilipolytic effect both in vivo and in vitro and antagonizes the effect of the lipolytic hormones. It is now considered that lipolysis may be more sensitive to changes in concentration of insulin than glucose utilization and esterification. Nicotinic acid and prostaglandin E_1 also suppress FFA mobilization. The antilipolytic effects of insulin, nicotinic acid, and prostaglandin E_1 may be accounted for by inhibition of the synthesis of cyclic AMP, possibly at the adenyl cyclase site (see Butcher & others, 1968, and Davies, 1968) or by stimulating phosphodiesterase. The site of action of thyroid hormones in facilitating lipolysis in adipose tissue is not clear. However, possible sites that have been reported include an augmentation of the level of cyclic AMP and an inhibition of phosphodiesterase activity. The effect of growth hormone (in the presence of glucocorticoids) in promoting lipolysis is a slow process. It is dependent on new formation of hormone-sensitive triglyceride lipase via RNA-dependent synthesis and is blocked by insulin (Fain & others, 1965). This finding also helps to explain the role of the pituitary gland and the adrenal cortex in enhancing fat mobilization. These relationships are summarized in Fig 14–13.

Besides the recognized hormones, certain other adipokinetic principles have been isolated from pituitary glands (Rudman, 1963). A "fat mobilizing substance" has been isolated from the urine of several fasting species, including man, provided the pituitary gland is intact (Chalmers & others, 1958). This substance is highly active both in vivo and in vitro.

The sympathetic nervous system, through liberation of norepinephrine in adipose tissue, plays a central role in the mobilization of FFA by exerting a tonic influence even in the absence of augmented nervous activity (Havel, 1964). Thus, the increased lipolysis caused by many of the factors described previously can be reduced or abolished by denervation of adipose tissue, ganglionic blockade with hexamethonium, or by depleting norepinephrine stores with reserpine.

Many of the facts reported above concern only the metabolism of adipose tissue in the young rat. However, data which have been reviewed by Rudman & Di Girolamo (1967) are now accumulating with

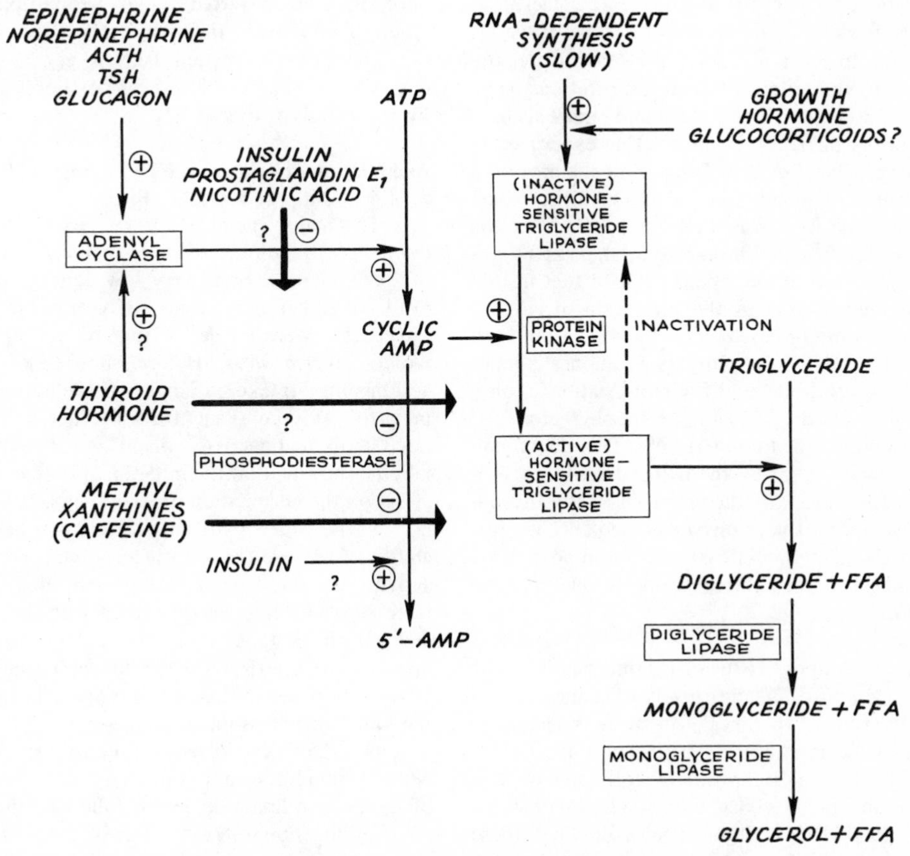

FIG 14–13. Control of adipose tissue lipolysis.

respect to older rats, humans, and other species. In older rats (> 350 g), a much greater proportion of the glucose metabolized is converted to glyceride glycerol and much less is synthesized into fatty acids, implying that in older rats there is a shift in lipogenesis from adipose to other tissues. The tissue is also less sensitive to insulin. These changes in adipose tissue of the older rat are related to adiposity rather than age, since weight reduction is followed by a return in the metabolism of adipose tissue to a pattern similar to that of the young rat. Human adipose tissue is unresponsive to most of the lipolytic hormones apart from the catecholamines. Of further interest is the lack of lipolytic response to epinephrine in the rabbit, guinea pig, pig, and chicken, the pronounced lipolytic effect of glucagon in birds, and the lack of glyceride glycerol synthesis from glucose in the pigeon. It would appear that in the various species studied a variety of mechanisms have been evolved for fine control of adipose tissue metabolism.

In spite of the mass of information which has been gathered on the action of administered hormones or hormone-like substances on the mobilization of FFA, little can be said about the physiologic role of hormones secreted endogenously. Their method of secretion, usually as a continuous infusion into the blood, is to be contrasted with the single injection approach using unphysiologic quantities, which is a characteristic of most investigations. The fact that adipose tissue varies markedly between various species in its ability to react to lipolytic hormonal preparations, even to the extent of some preparations being inactive against tissue from the homologous species, poses a further difficulty in interpreting experimental results. Nevertheless, on consideration of the profound derangement of metabolism in diabetes mellitus (which is due in the main to increased release of FFA from the depots) and the fact that insulin to a large extent corrects the condition, it must be concluded that insulin plays a prominent role in the regulation of adipose tissue metabolism. To reach as firm a conclusion with respect to the role of the pituitary hormones is more difficult, since the rate of FFA mobilization is only slightly depressed in fasting hypophysectomized animals (Goodman & Knobil, 1959). This depression could be accounted for by the reduced facilitatory or potentiating influence of the secretion of the thyroid and adrenal glands. Under physiologic conditions, it is likely that the main lipolytic stimulus in adipose tissue is due to liberation of norepinephrine through sympathetic activity.

Role of Brown Adipose Tissue in Thermogenesis

Brown adipose tissue is involved in metabolism particularly at times when heat generation is necessary. Thus, the tissue is extremely active in arousal from hibernation, in animals exposed to cold, and in heat production in the newborn animal (Rudman & Di Girolamo, 1967; Hull & Segall, 1965). Brown adipose tissue is characterized by a high content of mitochondria, cytochromes, and a well developed blood supply. Metabolic emphasis is placed on oxidative processes, O_2 consumption being high with a large conversion of both glucose and fatty acids to CO_2. Lipolysis is active, but reesterification with glycerol could occur as glycerokinase is present in significant amounts in this tissue. According to Himms-Hagen (1970), norepinephrine liberated from sympathetic nerve endings is important in increasing lipolysis in the tissue. Smith & others (1966) showed that mitochondria from brown adipose tissue of cold-acclimatized rats oxidized α-ketoglutarate rapidly with a P:O $\leqslant$ 1 and succinate and α-glycerophosphate with a P:O = 0. Addition of dinitrophenol had no effect, and there was no respiratory control by ADP. These experiments indicate that oxidation and phosphorylation are not coupled in mitochondria of this tissue. The phosphorylation that does occur appears to be at the substrate level. Thus, oxidation produces much heat, and little free energy is trapped in ATP. Results reported by Kornacker & Ball (1968) indicate that α-glycerophosphate is oxidized readily via the mitochondrial flavoprotein-linked α-glycerophosphate dehydrogenase. If substrate level phosphorylation is important in brown adipose tissue, this pathway would be a means of maintaining glycolysis by transporting reducing equivalents generated in glycolysis into the mitochondria for oxidation in the respiratory chain (see p 171). The presence of glycerokinase would enable free glycerol resulting from lipolysis to be converted to α-glycerophosphate and be oxidized directly in the tissue. The latter authors do not consider that much heat is generated by the energy-consuming cyclic process of lipolysis followed by resynthesis of triglyceride.

Assimilation of Triglyceride Fatty Acids (TGFA) by Adipose & Other Tissues

The major chemical form in which plasma lipid interacts with adipose tissue is triglyceride in the form of chylomicrons, or as very low density lipoproteins (VLDL). Experiments using triglycerides in which the fatty acids were labeled with ^{14}C and the glycerol moiety labeled with ^{3}H have shown that hydrolysis accompanies uptake by adipose tissue both in vivo and in vitro. Adipose tissue from re-fed rabbits incorporated more radioactivity than did tissue from fasted rabbits, showing that the nutritional state has a marked effect on the assimilation process (Felts, 1965).

There is a significant correlation between the ability of adipose tissue to incorporate TGFA and the activity of the enzyme **lipoprotein lipase** (clearing factor lipase), whose activity varies with the nutritional and hormonal state. Thus, the activity of lipoprotein lipase in adipose tissue is high in the fed state and low in starvation and in diabetes. Lipoprotein lipase is not the same enzyme as the hormone-sensitive lipase which is responsible for lipolysis within adipose tissue cells. Normal blood does not contain appreciable quantities of lipoprotein lipase; however, following the injection of heparin, lipoprotein lipase is released into the circulation from certain tissues and is accompanied by the clearing of any lipemia. It has been suggested that the enzyme is located in the walls of the blood capillaries (Robinson, 1963). It has been found in extracts of heart, adipose tissue, spleen, lung, renal medulla, aorta, diaphragm, and lactating mammary gland. It may be demonstrated in liver only in the presence of large quantities of added heparin. Evidence has been provided that it is not functional in this organ (Mayes & Felts, 1968). Present evidence supports the thesis of triglyceride hydrolysis by lipoprotein lipase occurring as a necessary preliminary to uptake of the TGFA into the extrahepatic tissues. The liberated FFA are considered to undergo the same fate in adipose tissue as has been described previously for the FFA taken up directly from plasma. The fate of the liberated glycerol will depend on whether or not the tissue in question contains glycerokinase. The low activity of this enzyme in adipose tissue means that the glycerol formed as a result of the uptake of TGFA is returned to the plasma (Fig 14–12). It has been suggested that the FFA released as a result of the action of lipoprotein lipase situated on the capillary wall produce a localized area of high concentration which facilitates their diffusion into the cells (Felts, 1964).

The decline in the lipoprotein lipase activity of adipose tissue which occurs on starvation may be due to a fall in enzyme content or conversion of the enzyme to a less active form. Resynthesis of the enzyme in vitro is stimulated by glucose and insulin and possibly antagonized by catecholamines, ACTH, growth hormone, and corticosteroids, ie, in a manner opposite to their effects on the adipolytic lipase. The effects of glucose and insulin are prevented by inhibitors of protein synthesis such as puromycin. As caffeine also inhibits the increase in activity of the enzyme, it is considered that at least some of these hormonal effects might be mediated by 3′,5′-cyclic AMP (Robinson, 1967).

As the activity of lipoprotein lipase falls in adipose tissue due to fasting, so it rises in heart and skeletal muscle. This may be a means of directing plasma triglyceride to the tissues where it is needed.

Lipoprotein lipase is activated by certain polypeptides present in HDL-apoprotein. It has also been shown that heparin is a specific ligand for the enzyme, allowing allosteric activation by the lipoprotein substrate (Whayne & Felts, 1970). For general reviews, see Renold & Cahill (1965) and Jeanrenaud & Hepp (1970).

METABOLISM OF THE LIPOPROTEINS*

Free Fatty Acids (FFA)

The free fatty acids (nonesterified fatty acids, NEFA; unesterified fatty acids, UFA) arise in the plasma from lipolysis of triglyceride in adipose tissue or as a result of the action of lipoprotein lipase during uptake of plasma triglycerides into tissues. They are found in combination with serum albumin in concentrations varying between 0.1 and 2 μEq/ml plasma and comprise the long chain fatty acids found in adipose tissue, ie, palmitic, stearic, oleic, palmitoleic, linoleic, other polyunsaturated acids, and smaller quantities of other long chain fatty acids. Binding sites on albumin of varying affinity for the fatty acids have been described. Low levels of FFA are recorded in the fully fed condition rising to about 0.5 μEq/ml postabsorptive and between 0.7 and 0.8 μEq/ml in the fully fasting state. In uncontrolled diabetes mellitus, the level may rise to as much as 2 μEq/ml. In meal eaters, the level falls just after eating and rises again prior to the next meal, whereas in continual feeders such as ruminants, where there is a continual influx of nutrient from the intestine, the FFA remain relatively constant and at a low level.

The rate of removal of FFA from the blood is extremely rapid. Estimates suggest that the FFA supply about 25–50% of the energy requirements in fasting. The remainder of the uptake is esterified and, according to evidence using radioactive FFA, eventually recycled. In starvation, the respiratory quotient (RQ) (see Chapter 21) would indicate that considerably more fat is being oxidized than can be traced to the oxidation of FFA. This difference may be accounted for by the oxidation of esterified lipids of the circulation or of those present in tissues. The latter are thought to occur particularly in heart and skeletal muscle, where considerable stores of lipid are to be found in the muscle cells. Armstrong & others (1961) showed that the FFA turnover was related directly to FFA concentration. Thus, the rate of FFA production in adipose tissue controls the FFA concentration in plasma, which in turn determines the FFA uptake by other tissues. The nutritional condition does not appear to have a great effect on the fractional uptake of FFA by tissues. It does, however, alter the proportion of the uptake which is oxidized to CO_2 compared to the fraction which is esterified.

Chylomicrons & Very Low Density Lipoproteins (VLDL)

Chylomicrons are the least dense of the plasma lipoproteins, forming particles sufficiently large (0.1–0.5 μm in diameter) to be seen in a darkfield microscope. By definition, chylomicrons are found in chyle, which is formed only in the lymphatic system draining the intestine. It is now realized that an appreciable quantity of VLDL is also found in chyle.

The intestinal cells esterify all long chain fatty acids absorbed through the intestinal wall and secrete the resulting chylomicrons and VLDL into the extracellular spaces between the intestinal cells, whence they enter lymph channels to be collected into the thoracic duct. The clearance of labeled chylomicrons from the blood is rapid, the half-time of disappearance being only a few minutes. The mechanism of removal becomes saturated with high chylomicron loads, causing a lower fractional rate of removal. Their rate of

*For reviews see Fredrickson & Gordon (1958), Fritz (1968), Nestel (1967), Nikkilä (1969), Robinson (1970), Schettler (1967).

removal by tissues parallels the activity of lipoprotein lipase in the tissue itself. The plasma FFA level has been reported to rise, remain the same, or fall upon the injection or administration of fat. Nevertheless, on high-fat diets, the plasma FFA is often elevated, and coincident with uptake of labeled chylomicrons there is a rapid labeling of the plasma FFA, indicating hydrolysis of the labeled triglyceride occurring concurrently with its uptake (Havel & Fredrickson, 1956). Available evidence points to the enzyme lipoprotein lipase as playing an integral part in the uptake of triglycerides into the extrahepatic tissues.

The liver is the main source of VLDL (very low density lipoproteins) that result from endogenous production of triglyceride to be secreted into the blood stream. Their metabolism by tissues appears in essence to be somewhat similar to that of chylomicrons. Disappearance of labeled VLDL from the circulation is considerably faster in re-fed than in fasted animals. Less label appears in CO_2, and much more label is incorporated by esterification into depot triglycerides.

Experiments using chylomicrons labeled in both the glycerol and fatty acid positions and then given as a single dose to intact rats have shown that 25% or so of the chylomicrons have been taken up by the liver without hydrolysis. However, negligible quantities of infused ^{14}C chylomicrons are oxidized by the perfused liver as compared with ^{14}C-FFA (Felts & Mayes, 1965), and similar results have been obtained with ^{14}C-VLDL. Subsequent studies have indicated that the chylomicrons taken up by the liver from a single dose can be washed out intact by retrograde perfusion from what must be an extracellular compartment confluent with the hepatic sinusoids (Felts, 1965). It appears from these experiments that, compared with FFA, the uptake of chylomicrons or VLDL into liver parenchymal cells is not a very significant physiologic process. The absence of active lipoprotein lipase in liver supports this view.

High & Low Density Lipoproteins (HDL, LDL)

Both chylomicrons and VLDL contain LDL and HDL apoproteins. It appears that apo-LDL is synthesized in the intestine as well as in the liver concurrently with the formation of the triglyceride of chylomicrons or VLDL. Apo-HDL is probably synthesized in the liver independently of triglyceride synthesis, forming the HDL which appears in the circulation. Upon secretion, the triglyceride-rich LDL particles take up HDL from the circulation to form typical chylomicrons and VLDL that can be isolated from the plasma. They also contain some protein in addition to apo-LDL and apo-HDL. In extrahepatic capillary beds they are attacked by lipoprotein lipase, liberating FFA from the triglyceride and ultimately releasing LDL and HDL into the circulation. Present evidence indicates that LDL is eventually broken down, probably in the liver, whereas the HDL is recycled (Fig 14–23).

THE ROLE OF THE LIVER IN LIPID METABOLISM

Much of the lipid metabolism of the body was formerly thought to be the prerogative of the liver. The discovery that most tissues have the ability to oxidize fatty acids completely and the knowledge that has accumulated showing that adipose tissue is extremely active metabolically have tended to modify the former emphasis on the role of the liver in lipid metabolism. Nonetheless, the concept of a central and unique role for the liver in lipid metabolism is still an important one. Apart from its role in facilitating the digestion and absorption of lipids by the production of bile, which contains cholesterol and bile salts synthesized within the liver, the liver has active enzyme systems for synthesizing and oxidizing fatty acids, for synthesizing triglycerides, phospholipids, cholesterol, and plasma lipoproteins, and for converting fatty acids to ketone bodies (ketogenesis). Some of these processes have already been described.

Triglyceride Synthesis & the Formation of VLDL

Experiments involving a comparison between hepatectomized and intact animals have shown that the liver is the main source of plasma lipoproteins derived from endogenous sources. Hepatic triglycerides are the immediate precursors of triglycerides contained in plasma VLDL (Havel & others, 1962). The fatty acids used in the synthesis of hepatic triglycerides are derived from 2 possible sources: (1) synthesis within the liver from acetyl-Co A derived in the main from carbohydrate and (2) uptake of FFA from the circulation. The first source would appear to be predominant in the well fed condition, when fatty acid synthesis is high and the level of circulating FFA is low. As triglyceride does not normally accumulate in the liver under this condition, it must be inferred that it is transported from the liver as rapidly as it is synthesized. On the other hand, during fasting, the feeding of high-fat diets, or in diabetes mellitus, the level of circulating FFA is raised and more is abstracted into the liver. Under these conditions, FFA are the main source of triglyceride fatty acids in the liver and in plasma lipoproteins because lipogenesis from acetyl-Co A is depressed. The enzyme mechanism responsible for the synthesis of triglycerides and phospholipids has been described on p 271. Factors which enhance both the synthesis of triglyceride and the secretion of VLDL by the liver include the feeding of diets high in carbohydrate (particularly if they contain sucrose or fructose), high levels of circulating FFA, ingestion of ethanol, and the presence of high levels of insulin (Topping & Mayes, 1970).

Lipoprotein apoproteins are probably synthesized in the rough endoplasmic reticulum and pass into the tubules of the smooth endoplasmic reticulum and Golgi bodies, from which they are ultimately secreted into the perisinusoidal spaces of Disse and thence into the circulation. The lipids appear to be formed into particles in the smooth endoplasmic reticulum. They are assembled into lipoproteins either in this location or in the Golgi apparatus. The Golgi bodies are known to be concerned in glycoprotein synthesis and secretion. It is of interest that the lipoproteins are also glycoproteins (Jones & others, 1967).

Fatty Livers & Lipotropic Factors

For a variety of reasons, lipid–mainly as triglyceride–can accumulate in the liver. Extensive accumulation is regarded as a pathologic condition. When accumulation of lipid in the liver becomes chronic, fibrotic changes occur in the cells which progress to cirrhosis and impaired liver function.

Fatty livers fall into 2 main categories. The first type is associated with raised levels of plasma FFA resulting from mobilization of fat from adipose tissue or from the hydrolysis of lipoprotein or chylomicron triglyceride by lipoprotein lipase in extrahepatic tissues. Increasing amounts of FFA are taken up by the liver and esterified. The production of plasma lipoprotein does not keep pace with the influx of FFA, allowing triglyceride to accumulate, causing a fatty liver. The quantity of triglyceride present in the liver is significantly increased during starvation and the feeding of high-fat diets. In many instances (eg, in starvation), the ability to secrete VLDL is also impaired. In uncontrolled diabetes mellitus, pregnancy toxemia of ewes, or ketosis in cattle, fatty infiltration is sufficiently severe to cause visible pallor or fatty appearance and enlargement of the liver.

The second type of fatty liver is usually due to a metabolic block in the production of plasma lipoproteins. Theoretically, the lesion may be due to a block in lipoprotein apoprotein synthesis, a block in the synthesis of the lipoprotein from lipid and apoprotein, a failure in provision of phospholipids which are found in lipoproteins, or to a failure in the secretory mechanism itself. It is often associated with deficiency of a substance known as a **lipotropic factor**. The deficiency causes triglyceride to accumulate even though only a normal rate of fatty acid synthesis and uptake of FFA may be occurring. The exact mechanism by which many fatty livers falling into this category arise is still far from clear. One type of fatty liver which has been studied extensively is due to a deficiency of choline. As choline may be synthesized using labile methyl groups donated by methionine in the process of **transmethylation** (see Chapter 15), the deficiency is basically due to a shortage of the type of methyl group donated by methionine. Thus, choline, methionine, and betaine can all act as lipotropic agents in curing fatty livers due to choline deficiency, and, conversely, processes which utilize methyl groups excessively or diets poor in protein (containing methionine) or lecithin (containing choline) will all tend to favor the production of fatty livers.

Several mechanisms have been suggested to explain the role of choline as a lipotropic agent. Olson & others (1958) demonstrated that the low density lipoproteins were virtually absent from the blood of choline deficient rats, indicating that the defect lay in the transport of triglyceride from the liver. Mishkel & Morris (1964), using the perfused rat liver, showed that the uptake of labeled FFA and their oxidation was not reduced in choline-deficient livers; however, more of the label was incorporated into liver triglyceride and non-choline-containing phospholipids, and significantly less was incorporated into the choline-containing phospholipids. Mookerjea (1971) has suggested that, in addition to causing an impairment in synthesis of lipoprotein phospholipids containing choline, a choline deficiency may impair availability of phosphoryl choline, which stimulates incorporation of glucosamine into glycolipoproteins. Lombardi (1971) has suggested that deficiency of phospholipids containing choline may impair synthesis of intracellular membranes concerned in lipoprotein synthesis. Corredor & others (1967) have shown that depression of long chain fatty acid oxidation, which they claim occurs in choline deficiency, may be due to depressed levels of carnitine (carnitine synthesis also being dependent on the provision of methyl groups). Reduced oxidation of fatty acids might be expected to enhance triglyceride formation. It is to be noted that the antibiotic puromycin, which inhibits protein synthesis, causes a fatty liver and a marked reduction in concentration of plasma esterified fatty acids in rats (Robinson & Seakins, 1962).

Other substances which cause fatty livers include ethionine (α-amino-γ-ethylmercaptobutyric acid), carbon tetrachloride, chloroform, phosphorus, lead, and arsenic. Choline will not protect the organism against these agents but appears to aid in recovery. The action of most of these substances is associated with inhibition of hepatic protein synthesis (Smuckler & Barker, 1966). However, while it appears that protein synthesis is impaired in most of these conditions, the rapidity of action of carbon tetrachloride (within minutes) compared with the several hours required to elicit an effect with ethionine, indicates some difference in mode of action. It is very likely that carbon tetrachloride also affects the secretory mechanism itself or the conjugation of the lipid with lipoprotein apoprotein. The action of ethionine is thought to be due to a decline in messenger RNA and protein synthesis caused by a reduction in availability of ATP. This results when ethionine, replacing methionine in S-adenosyl methionine, traps available adenine and prevents synthesis of ATP. This hypothesis is supported by the fact that the effect of ethionine may be reversed by administration of ATP or adenine. Administration of orotic acid also causes fatty livers due to a specific block in lipoprotein formation, either by inhibiting apo-LDL synthesis or by inhibiting synthesis of the lipoprotein particle.

A deficiency of vitamin E enhances the hepatic necrosis of the choline deficiency type of fatty liver. Added vitamin E or a dietary factor termed "factor 3" (an organic compound containing selenium; see Chapter 19) has a protective effect. In addition to protein deficiency, essential fatty acid and vitamin deficiencies

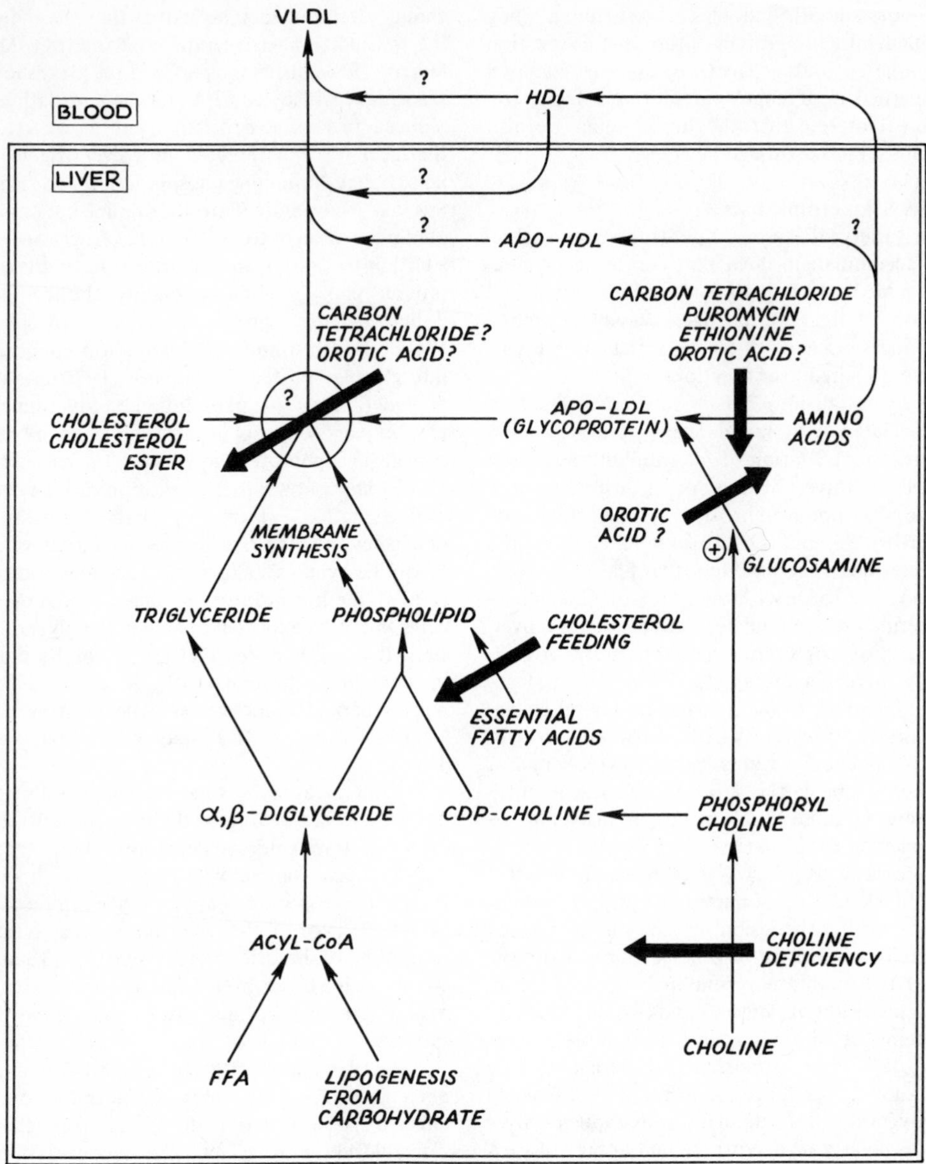

FIG 14–14. The secretion of very low density lipoprotein and the possible loci of action of factors causing a fatty liver.

(eg, pyridoxine and pantothenic acid) can cause fatty infiltration of the liver. A deficiency of essential fatty acids is thought to depress the synthesis of phospholipids; therefore, other substances such as cholesterol which compete for available essential fatty acids for esterification can also cause fatty livers. Alcoholism also leads to fat accumulation in the liver, hyperlipidemia, and ultimately cirrhosis. The exact mechanism of action of alcohol in this respect is still uncertain (see Nestel, 1967). Whether or not extra FFA mobilization plays some part in causing the accumulation of fat is not clear, but several studies have demonstrated elevated levels of FFA in the rat after administration of a single intoxicating dose of ethanol. There is good evidence of increased hepatic triglyceride synthesis and decreased fatty acid oxidation, caused possibly by increased concentrations of α-glycerophosphate (Nikkilä & Ojala, 1963). This may result from an increased NADH/NAD ratio generated by the oxidation of ethanol by alcohol dehydrogenase, which causes a shift to the right in the equilibrium dihydroxyacetone phosphate $\rightleftharpoons$ α-glycerophosphate and which would also inhibit the citric acid cycle. However, it has recently been demonstrated that, while the administration of pyrazole (which inhibits alcohol dehydrogenase) completely prevents the disappearance of ethanol from the blood and the changes in the NADH/NAD ratio, it did not prevent accumulation of triglyceride in the liver after administration of ethanol (Bustos & others, 1970). Other effects of alcohol may include increased lipogenesis and cholesterol synthesis from acetyl-Co A. Lieber (1966) has pointed out that alcohol consumption over a long period leads to the accumulation of fatty acids in the liver which are derived from endogenous synthesis rather than from adipose tissue. There is no impairment of hepatic synthesis of protein after ethanol ingestion (Ashworth & others, 1965). Some of the constituents of lipoprotein secretion and factors concerned in the production of fatty livers are shown in Fig 14–14.

KETOSIS

Under certain metabolic conditions associated with a high rate of fatty acid oxidation, the liver produces considerable quantities of acetoacetate and D(–)-β-hydroxybutyrate which pass by diffusion into the blood. Acetoacetate continually undergoes spontaneous decarboxylation to yield acetone. These 3 substances are collectively known as the **ketone bodies** (also known as acetone bodies or "ketones") (Fig 14–15).

The concentration of total ketone bodies in the blood of well fed mammals does not normally exceed 1 mg/100 ml (as acetone equivalents). It is somewhat higher than this in ruminants. Loss via the urine is usually less than 1 mg/24 hours in man. Higher than normal quantities present in the blood or urine constitute **ketonemia** (hyperketonemia) or **ketonuria**, respectively. The overall condition is called **ketosis.** Acetoacetic and β-hydroxybutyric acids are both moderately strong acids and are buffered when present in blood or the tissues. However, their continual excretion in quantity entails some loss of buffer cation (in spite of ammonia production by the kidney) which progressively depletes the alkali reserve, causing **ketoacidosis.** This may be fatal in uncontrolled diabetes mellitus.

The simplest form of ketosis occurs in starvation and involves depletion of available carbohydrate coupled with mobilization of FFA. No other condition in which ketosis occurs seems to differ qualitatively from this general pattern of metabolism, but quantitatively it may be exaggerated to produce the pathologic states found in diabetes mellitus, pregnancy toxemia in sheep, and ketosis in lactating cattle. Other nonpathologic forms of ketosis are found under conditions of high-fat feeding and after severe exercise in the postabsorptive state.

In vivo the liver appears to be the only organ in nonruminants to add ketone bodies to the blood. Extrahepatic tissues utilize them as respiratory substrates. In ruminants, the rumen wall converts butyric acid, formed as a result of ruminal fermentation, to β-hydroxybutyrate, which enters the blood stream. The ruminant lactating mammary gland is also reported to produce ketone bodies. It is believed that these other sources of ketone bodies do not contribute significantly to the occurrence of ketosis in these species.

Enzymatic Mechanism for Ketogenesis in the Liver & for the Utilization of Ketone Bodies in Extrahepatic Tissues

The net flow of ketone bodies from the liver to the extrahepatic tissues results from an active enzymatic mechanism in the liver for the production of

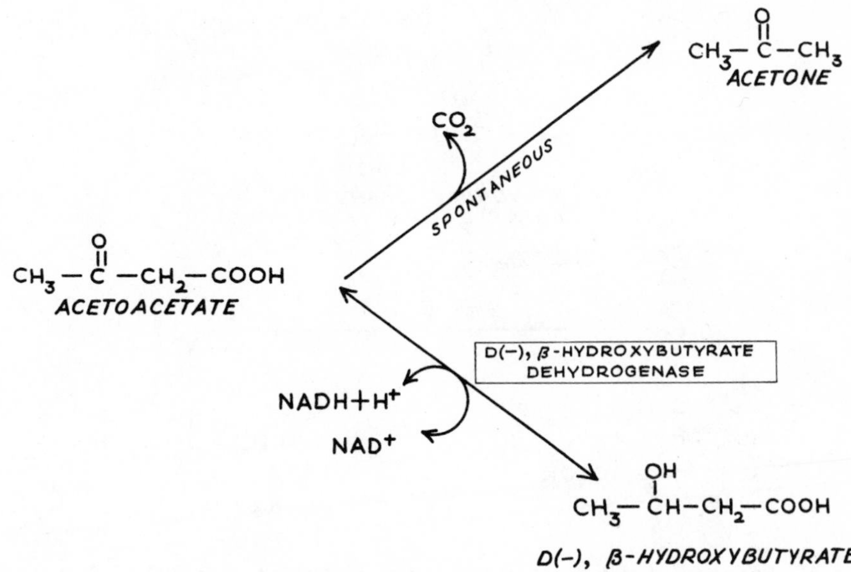

FIG 14–15. Interrelationships of the ketone bodies.

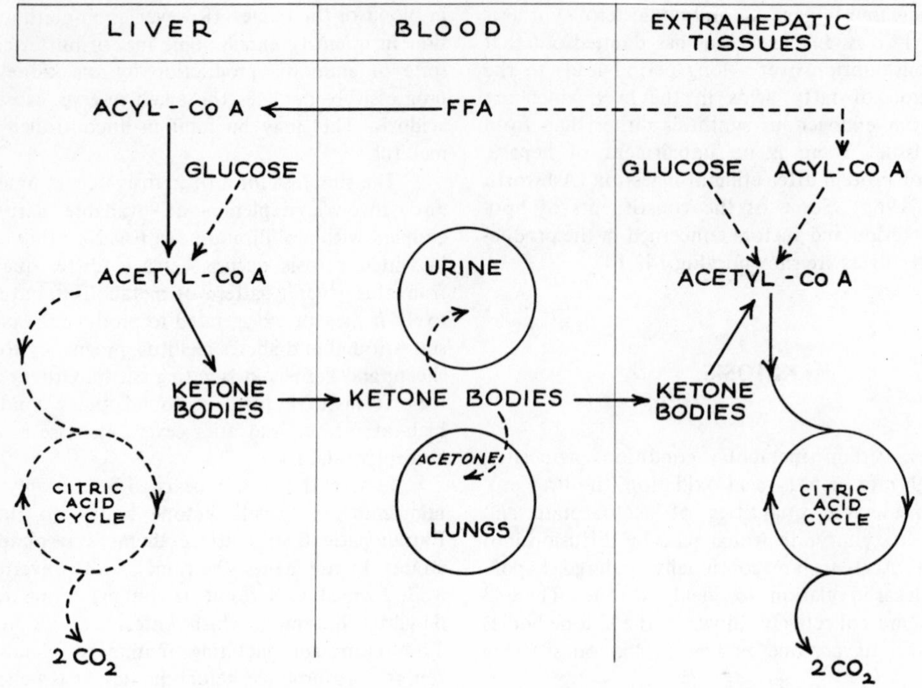

FIG 14–16. Formation, utilization, and excretion of ketone bodies.
(The main pathway is indicated by the heavier arrows.)

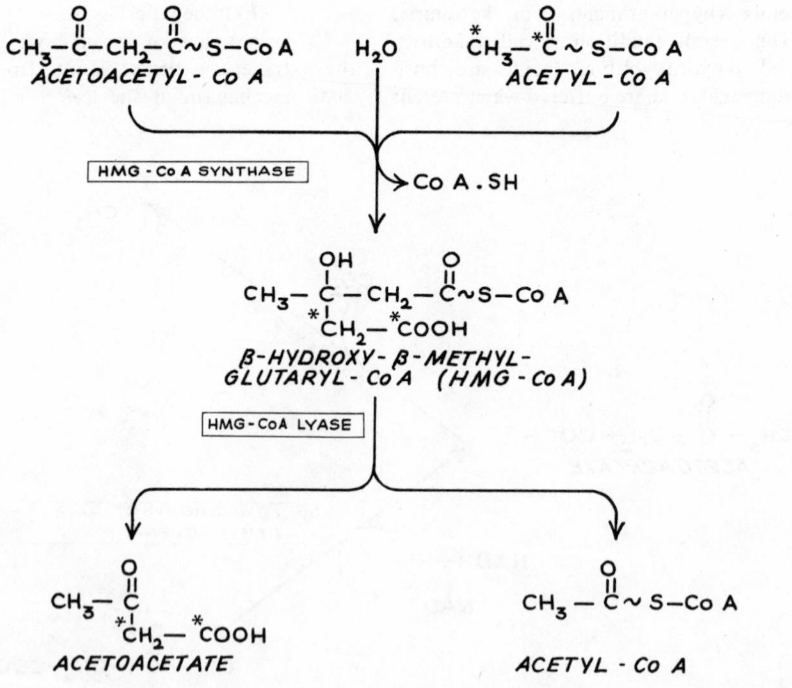

FIG 14–17. Formation of acetoacetate through intermediate production of HMG-Co A (Lynen).

$$CH_3COCH_2{\cdot}CO{\cdot}S{\cdot}CoA + H_2O \xrightarrow{\text{ACETOACETYL-Co A DEACYLASE}} CH_3COCH_2COOH + CoA{\cdot}SH$$

ketone bodies coupled with very low activity of enzymes responsible for their utilization. The reverse situation occurs in extrahepatic tissues (Fig 14–16).

Enzymes responsible for ketone body formation are associated mainly with the mitochondria. Originally it was thought that only 1 molecule of acetoacetate was formed from the terminal 4 carbons of a fatty acid upon oxidation. Later, to explain both the production of more than one equivalent of acetoacetate from a long chain fatty acid and the formation of ketone bodies from acetic acid, it was proposed that C_2 units formed in β-oxidation condensed with one another to form acetoacetate (MacKay & others, 1940). This may occur by a reversal of the **thiolase** reaction whereby 2 molecules of acetyl-Co A condense to form acetoacetyl-Co A. Thus, acetoacetyl-Co A, which is the starting material for ketogenesis, arises either directly during the course of β-oxidation or as a result of the condensation of acetyl-Co A. Acetoacetate is formed from acetoacetyl-Co A by 2 pathways (Fig 14–18). The first is by simple deacylation catalyzed by the enzyme **acetoacetyl-Co A deacylase** (Stern & Miller, 1959) (see above).

The second pathway (Lynen & others, 1958; Fig 14–17) involves the condensation of acetoacetyl-Co A with another molecule of acetyl-Co A to form butyryl-Co A. This compound can reform acetoacetyl-Co A (for further details, see Bressler, 1963), but in view of the active degradation of acetoacetyl-Co A to acetoacetate in the ketotic liver it is doubtful whether this pathway of ketone body utilization is of great quantitative significance.

Two reactions take place in extrahepatic tissues which will activate acetoacetate to acetoacetyl-Co A. The enzymes responsible are absent from liver. One mechanism involves succinyl-Co A and the enzyme **succinyl-Co A-acetoacetate-Co A transferase** (thiophorase). Acetoacetate reacts with succinyl-Co A, the Co A being transferred to form acetoacetyl-Co A and succinate.

CH_3COCH_2COOH *ACETOACETATE* + $CH_2COOH{-}CH_2CO{\cdot}S{\cdot}CoA$ *SUCCINYL-Co A* ⇌ (CoA TRANSFERASE) $CH_3COCH_2CO{\cdot}S{\cdot}CoA$ *ACETOACETYL-CoA* + $CH_2COOH{-}CH_2COOH$ *SUCCINATE*

The other reaction involves the activation of acetoacetate with ATP in the presence of Co A catalyzed by **acetoacetic thiokinase**.

$$\underset{\textit{ACETOACETATE}}{CH_3COCH_2COOH} + ATP + CoA{\cdot}SH \xrightarrow{\text{ACETOACETIC THIOKINASE}} \underset{\textit{ACETOACETYL-CoA}}{CH_3COCH_2CO{\cdot}S{\cdot}CoA} + AMP + PPi$$

β-hydroxy-β-methylglutaryl-Co A (HMG-Co A), catalyzed by **HMG-Co A synthase**. The presence of another enzyme in the mitochondria, **HMG-Co A lyase**, causes acetyl-Co A to split off from the HMG-Co A, leaving free acetoacetate. The carbon atoms of the acetyl-Co A molecule split off are derived from the original acetoacetyl-Co A molecule (Fig 14–17).

Present opinion favors the HMG-Co A pathway as the major route of ketone body formation. Wieland & others (1960) have noted that in fasting there is a marked increase in activity of HMG-Co A cleavage enzyme.

Acetoacetate may be converted to D(–)-β-hydroxybutyrate by **D(–)-β-hydroxybutyrate dehydrogenase**, which is present in many tissues, including the liver. D(–)-β-hydroxybutyrate is quantitatively the predominant ketone body present in the blood and urine in ketosis.

While the liver is equipped with an active enzymatic mechanism for the production of acetoacetate from acetoacetyl-Co A, acetoacetate once formed cannot be reactivated directly in the liver. This accounts for the net production of ketone bodies by the liver. A thiokinase is present, however, which can convert D(–)-β-hydroxybutyrate to D(–)-β-hydroxybutyryl-Co A ...

D(–)-β-hydroxybutyrate may be activated directly in extrahepatic tissues by a thiokinase similar to the enzyme reported to be present in the liver. Conversion to acetoacetate with D(–)-β-hydroxybutyrate dehydrogenase and NAD^+ followed by activation to acetoacetyl-Co A is an alternative route leading to its further metabolism. The acetoacetyl-Co A formed by these reactions is split to acetyl-Co A by thiolase and oxidized in the citric acid cycle (Fig 14–18).

Ketone bodies are oxidized in extrahepatic tissues proportionately to their concentration in the blood (Wick & Drury, 1941). They are also oxidized in preference to glucose (Williamson & Krebs, 1961) and to FFA (Olson, 1962). If the blood level is raised, oxidation of ketone bodies increases until, at a concentration of approximately 70 mg/100 ml, they saturate the oxidative machinery; any further increase in the rate of ketogenesis merely serves to raise the blood concentration and the rate of urinary excretion precipitously. At this point, approximately 90% of the oxygen consumption of the animal may be accounted for by the oxidation of ketone bodies (Wick & Drury, 1941; Nelson & others, 1941).

Most of the evidence suggests that ketonemia is due to increased production of ketone bodies by the

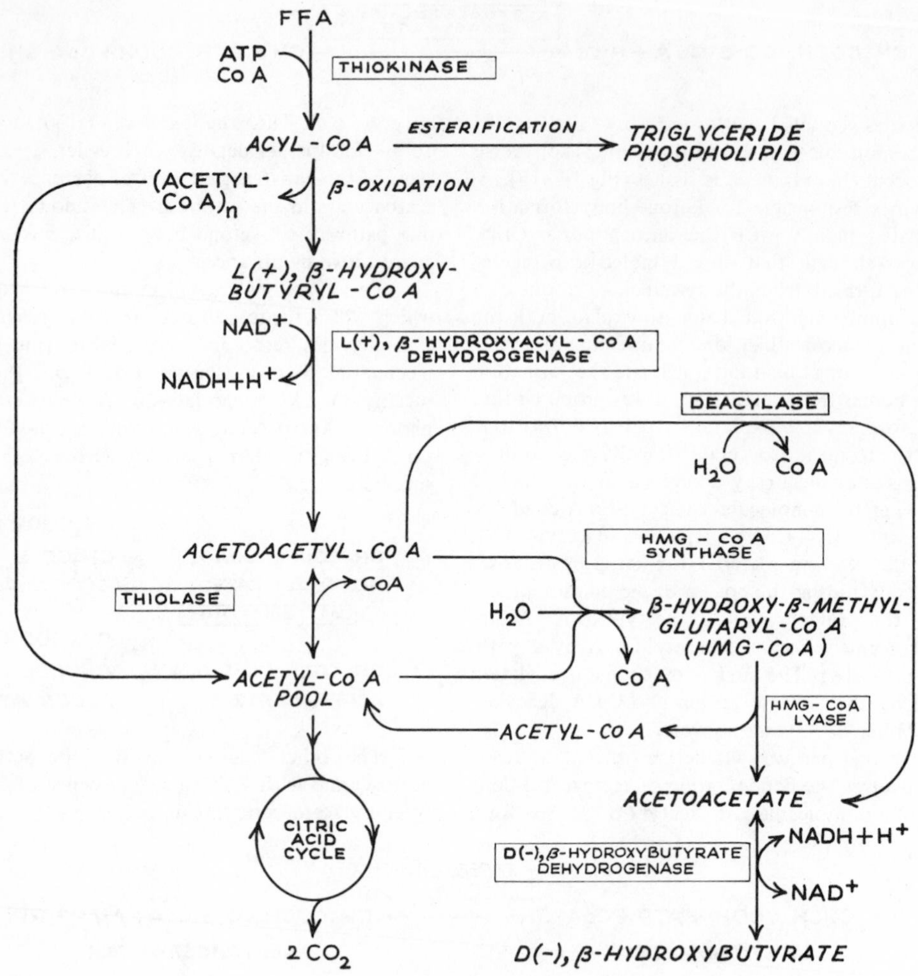

FIG 14–18. Pathways of ketogenesis in the liver.

liver rather than to a deficiency in their utilization by extrahepatic tissues. However, experiments on depancreatized rats (Beatty & others, 1960; Scow & Chernick, 1960) support the possibility that ketosis in the severe diabetic may be aggravated by a reduced ability to catabolize ketone bodies.

In moderate ketonemia, the loss of ketone bodies via the urine is only a few percent of the total ketone body production and utilization. As there are renal threshold-like effects (there is not a true threshold) which vary between species and individuals, measurement of the ketonemia, not the ketonuria, is the preferred method of assessing the severity of ketosis.

While acetoacetate and D(−)-β-hydroxybutyrate are readily oxidized by extrahepatic tissues, acetone is difficult to oxidize in vivo. Koehler & others (1941) injected acetone into human subjects and found that its concentration in the blood rose sharply and was maintained at a high level for several hours, indicating a very slow rate of utilization. However, experiments with ^{14}C-labeled acetone have shown it to be converted, to a limited extent at least, to $^{14}CO_2$. Several pathways for the utilization of acetone have been proposed. One is that acetone is converted to acetoacetate by a reversal of decarboxylation. Another pathway involves the formation of propanediol, which may provide, if it is a significant process, a route for the net conversion of fatty acids to carbohydrate. Alternatively, propanediol may form a 1-carbon (formate) unit plus a 2-carbon unit (acetate).

Factors Determining the Magnitude of Ketogenesis

In general, ketosis does not occur in vivo unless there is a concomitant rise in the level of circulating FFA, severe ketosis being accompanied invariably by very high concentrations of plasma FFA. In addition, numerous experiments in vitro have demonstrated that fatty acids are the precursors of ketone bodies. The liver, both in the fed and fasting condition, has the ability to extract about 30% or more of the FFA passing through it (Hillyard & others, 1959; Fine & Williams, 1960), so that at high concentrations of FFA the flux passing into the liver is substantial. One of 2 fates awaits the FFA upon uptake and after they are activated to acyl-Co A: they are esterified to triglyceride, phospholipid, or cholesterol ester; or they are β-oxidized to acetyl-Co A (Fig 14–18). In turn, acetyl-Co A is oxidized in the citric acid cycle or used to form ketone bodies. Experiments with fasting rats have demonstrated that the magnitude of ketonemia is more directly related to the quantity of fat present in the depots than to the quantity present in the liver (Mayes, 1962), indicating that plasma FFA (derived from the fat depots) are a more significant source of ketone bodies than fatty acids derived from lipolysis of liver triglyceride. Thus, esterification of FFA in the liver may be regarded as a significant antiketogenic mechanism.

Among several possible factors, the capacity for esterification depends on the availability of precursors in the liver to supply sufficient α-glycerophosphate. Bortz & Lynen (1963) have shown that the concentration of α-glycerophosphate in the livers of fasted rats is depressed when compared to that in fed animals. However, it was noted by Mayes and Felts (1967) that the availability of α-glycerophosphate did not appear to limit esterification in fasting perfused livers, where, irrespective of the mass of FFA taken up, a constant fraction was esterified. It has also been found in vivo that the antiketogenic effects of glycerol and dihydroxyacetone are not correlated with the levels of α-glycerophosphate in the liver (Williamson & others, 1969). Thus, whether the availability of α-glycerophosphate in the liver is ever rate-limiting on esterification is not clear; neither is there any information on whether the in vivo activities of the enzymes involved in esterification are rate-limiting.

Using the perfused liver, Mayes & Felts (1967) have shown that livers from fed rats esterify considerably more ^{14}C FFA than livers from fasted rats, the balance not esterified in the livers from fasted rats being oxidized to either $^{14}CO_2$ or ^{14}C ketone bodies. As the level of serum FFA was raised, proportionately more of the FFA was converted to ketone bodies and less was oxidized via the citric acid cycle to CO_2. The partition of acetyl-Co A between the ketogenic pathway and the pathway of oxidation to CO_2 was so regulated that the total energy production from FFA (as ATP) remained constant. It will be appreciated that complete oxidation of 1 mol of palmitate involves a net production of 129 mols of ATP via CO_2 production in the citric acid cycle, whereas only 33 mols of ATP are produced when acetoacetate is the end product. Thus, ketogenesis may be regarded as a mechanism that allows the liver to oxidize large quantities of fatty acids within an apparently tightly coupled system of oxidative phosphorylation without increasing its total energy production.

Several hypotheses have been advanced to account for the diversion of fatty acid oxidation from CO_2 formation to ketogenesis. Theoretically, a fall in concentration of oxaloacetate, particularly within the mitochondria, could cause impairment of the citric acid cycle to metabolize acetyl-Co A. This has been considered to occur because of a decrease in ratio of $NAD^+/NADH$ (Wieland & others (1964). Krebs (1966) has suggested that, since oxaloacetate is also on the main pathway of gluconeogenesis, enhanced gluconeogenesis leading to a fall in the level of oxaloacetate may be the cause of the severe forms of ketosis found in diabetes and the ketosis of cattle. Alternatively, it has been postulated that citrate synthase is inhibited, either by long chain acyl-Co A (Wieland & others, 1966) or by increased concentrations of ATP (Shepherd & Garland, 1966). However, there is considerable disagreement among the experimental findings and conclusions concerning all of these hypotheses. (See Greville & Tubbs [1968].)

Utter & Keech (1963) have shown that **pyruvate carboxylase**, which catalyzes the conversion of pyruvate to oxaloacetate, is activated by acetyl-Co A. Consequently, when there are significant amounts of acetyl-Co A, there should be sufficient oxaloacetate to initiate the condensing reaction of the citric acid cycle.

Evidence is accumulating which shows that lipolysis of liver triglycerides is under the control of a hormone-sensitive lipase, as in adipose tissue. This lipase is activated by increase in concentration of cyclic AMP, which in turn is raised in concentration by the action of glucagon on adenyl cyclase and depressed by the presence of insulin. The combined action of these hormones may regulate lipolysis in the liver and, therefore, net esterification of FFA. Thus, insulin and glucagon not only determine the rate of endogenous lipolysis but also affect the balance between esterification and oxidation of incoming plasma FFA.

In summary, ketosis arises as a result of a deficiency in available carbohydrate. This has 2 principal actions in fostering ketogenesis: (1) It causes an imbalance between esterification and lipolysis in adipose tissue, with consequent release of FFA into the circulation. FFA are the principal substrates for ketone body formation in the liver, and therefore all factors, metabolic or endocrine, affecting the release of FFA from adipose tissue influence ketogenesis. (2) Upon entry of FFA into the liver, the balance between esterification and oxidation of FFA is influenced by the hormonal state of the liver and possibly by the availability of α-glycerophosphate. That which remains unesterified is oxidized to CO_2 and ketone bodies. As the quantity of fatty acids presented for oxidation increases, more form ketone bodies and less form CO_2, regulated in such a manner that the total energy production remains constant. Ketone bodies are not oxidized significantly by the liver; they diffuse into the circulation whence they are extracted and oxidized by extrahepatic tissues preferentially to other fuels.

Ketosis in Vivo

The ketosis that occurs in starvation and fat feeding is relatively mild compared with the condition encountered in uncontrolled diabetes mellitus, pregnancy toxemia of ewes, ketosis of lactating cattle, or animals administered phlorhizin. The main reason appears to be that in the severe conditions carbohydrate is still less available to the tissues than in the mild conditions. Thus, in the milder forms of diabetes mellitus, in fat feeding, and in chronic starvation, glycogen is present in the liver in variable amounts, and this, together with lower levels of FFA, probably accounts for the less severe ketosis associated with these conditions. It is to be expected that the presence of glycogen in the liver is indicative of a greater capacity for esterification of fatty acids.

In ketosis of ruminants or in phlorhizin poisoning, there is a severe drain of glucose from the blood due to excessive fetal demands, the demands of heavy lactation, or impaired reabsorption by the kidney, respectively. Extreme hypoglycemia results, coupled with negligible amounts of glycogen in the liver. Ketosis in these conditions tends to be severe. As hypoglycemia develops, the secretion of insulin diminishes, allowing not only less glucose utilization in adipose tissue but also enhancement of lipolysis in adipose tissue and liver.

In diabetes mellitus, the lack (or relative lack) of insulin probably affects adipose tissue more than any other tissue because of its extreme sensitivity to this hormone. As a result, FFA are released in quantities that give rise to plasma FFA levels more than twice those in fasting normal subjects. Many changes also occur in the activity of enzymes within the liver which enhance the rate of gluconeogenesis and transfer of glucose to the blood despite high levels of circulating glucose. As a result, there is a tendency for the liver glycogen to be reduced, although this is variable.

THE ECONOMICS OF CARBOHYDRATE & LIPID METABOLISM IN THE WHOLE BODY

Many of the details of the interplay between carbohydrate and lipid metabolism in various tissues have been described. The conversion of glucose to fat is a process which occurs readily under conditions of optimal nutritional intake. With the exception of glycerol, fat (as fatty acids) cannot give rise to a net formation of glucose because of the irreversible nature of the oxidative decarboxylation of pyruvate to acetyl-Co A. Certain tissues, including the CNS and the erythrocytes, are much more dependent upon a continual supply of glucose than others. A minimal supply of glucose is probably necessary in extrahepatic tissues to maintain the integrity of the citric acid cycle. In addition, glucose appears to be the main source of α-glycerophosphate in tissues devoid of glycerokinase. There is in all probability a minimal and obligatory rate of glucose oxidation. Large quantities of glucose are also required for the nutrition of the fetus and for the synthesis of milk, particularly in ruminants. Certain mechanisms operate which safeguard essential supplies of glucose in times of shortage, allowing other substrates to spare its general oxidation.

Randle & others (1963) have demonstrated that ketone bodies and FFA spare the oxidation of glucose in muscle by impairing its entry into the cell, its phosphorylation to glucose-6-phosphate, the phosphofructokinase reaction, and the oxidative decarboxylation of pyruvate. Garland & others (1963) demonstrated that oxidation of FFA and ketone bodies caused an increase in the concentration of intracellular citrate which in turn inhibited phosphofructokinase. These observations, taken with those of Olson (1962), who demonstrated that acetoacetate was oxidized in the perfused heart preferentially to free fatty acids, justify the conclusion that under conditions of carbohydrate shortage available fuels are oxidized in the following order of preference: (1) ketone bodies (and probably other short chain fatty acids, eg, acetate), (2) FFA, and (3) glucose. This does not imply that any particular fuel is oxidized to the total exclusion of any other.

These facts help to explain the experiments of several investigators who have shown in vivo that under certain conditions fat mobilization can be reduced after the administration of noncarbohydrate substrates. Munkner (1959) showed that oral administration of fat reduced the level of circulating FFA, and Lindsay (1959) obtained a similar result in sheep after the administration of acetate. Fat mobilization and ketogenesis in rats on all-fat diets can be reduced substantially provided the quantity of fat ingested is increased to satisfy the caloric requirement of the animal. This has led to the hypothesis that in vivo the mobilization of FFA from adipose tissue is stimulated by a deficiency of calorigenic substrates (Mayes, 1962). If substrates such as FFA and ketone bodies spare the oxidation of glucose in muscle, more glucose will be available, causing a reduction in output of FFA from adipose tissue (either directly or via stimulation of insulin secretion) and allowing the plasma level of FFA to fall. As glucose is the fuel which is "burned last," it may be appreciated how adipose tissue is sensitive to a general deficiency in calorigenic substrates in the whole body through a mechanism based specifically on the availability of glucose. The combination of the effects of FFA in sparing glucose utilization in muscle and heart and the effect of the spared glucose in inhibiting FFA mobilization in adipose tissue has been called the "glucose-fatty acid cycle" (Randle & others, 1963).

On high-carbohydrate diets, FFA oxidation is spared; it is generally considered that this is due to the high capacity in the tissues for esterification (Fritz, 1961). As the animal passes from the fed to the fasting condition, glucose availability becomes less, liver glycogen being drawn upon in an attempt to maintain the

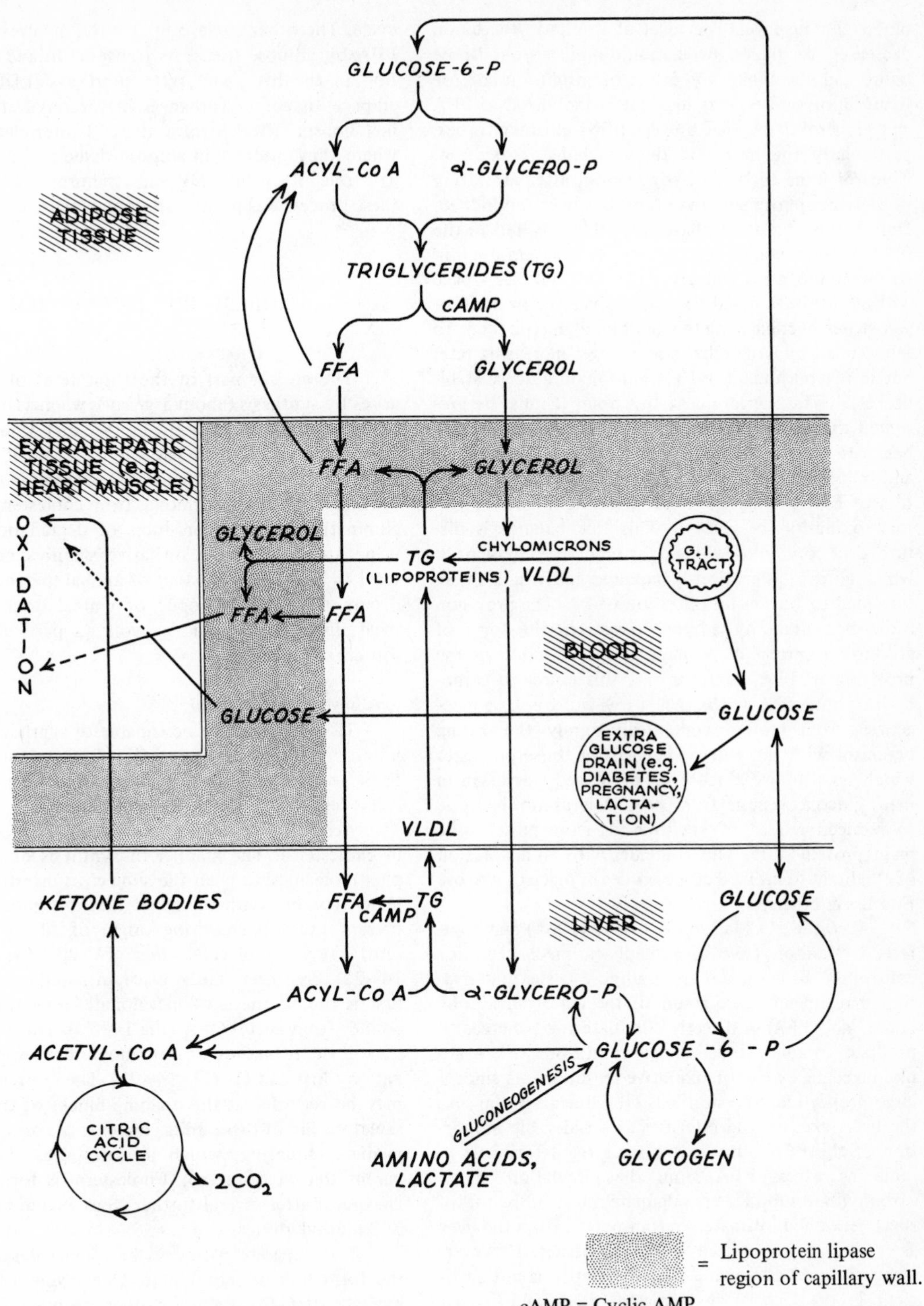

FIG 14–19. Metabolic interrelationships between adipose tissue, the liver, and extrahepatic tissues.

blood glucose level. The level of insulin in the blood decreases. As glucose utilization diminishes in adipose tissue and the inhibitory effect of insulin on adipose tissue lipolysis becomes less, fat is mobilized as FFA and glycerol. The FFA are esterified in other tissues, particularly the liver, and the remainder is oxidized. Glycerol joins the carbohydrate pool after activation to α-glycerophosphate, mainly in the liver and kidney. During this transition phase from the fully fed to the fully fasting state, endogenous glucose production (from amino acids and glycerol) does not keep pace with its utilization and oxidation since the liver glycogen stores become depleted and blood glucose tends to fall. Thus, fat is mobilized at an ever increasing rate, but in several hours the FFA and blood glucose stabilize at the fasting level. At this point it must be presumed that in the whole animal the supply of glucose balances the obligatory demands for glucose utilization and oxidation. This is achieved by the increased oxidation of FFA and ketone bodies, sparing the nonobligatory oxidation of glucose. This fine balance is disturbed in conditions which demand more glucose or in which glucose utilization is impaired and which therefore lead to further mobilization of fat. The provision of carbohydrate by adipose tissue, in the form of glycerol, is probably as important a function as the provision of FFA, for it is only this source of carbohydrate together with that provided by gluconeogenesis from protein which can supply the fasting organism with the glucose needed for those processes which must utilize glucose. In prolonged starvation in man, gluconeogenesis from protein is diminished due to reduced release of amino acids from muscle, the main protein store. This coincides with an adaptation of the brain to utilize ketone bodies in place of glucose (Marliss & others, 1970).

According to Madison & others (1964) and Seyffert & Madison (1967), a feedback mechanism for controlling FFA output from adipose tissue in starvation may operate as a result of the action of ketone bodies and FFA to directly stimulate the pancreas to produce insulin. Under most conditions, FFA are mobilized in excess of oxidative requirements since a large proportion are esterified, even during fasting. As the liver takes up and esterifies a considerable proportion of the FFA output, it plays a regulatory role in removing excess FFA from the circulation. When carbohydrate supplies are adequate, most of the influx is esterified and ultimately retransported from the liver as VLDL to be utilized by other tissues. However, when the capacity of the liver to esterify is not sufficient in the face of an increased influx of FFA, an alternative route, ketogenesis, is available which enables the liver to continue to retransport much of the influx of FFA in a form that is readily utilized by extrahepatic tissues under all nutritional conditions.

Most of these principles are depicted in Fig 14–19. It will be noted that there is a carbohydrate cycle involving release of glycerol from adipose tissue and its conversion in the liver to glucose, followed by its transport back to adipose tissue to complete the cycle. The other cycle, a lipid cycle, involves release of FFA by adipose tissue, its transport to and esterification in the liver, and retransport as VLDL back to adipose tissue. Disturbances in carbohydrate or lipid metabolism often involve these 2 interrelated cycles where they interact in adipose tissue and in the liver. The role of cyclic AMP and its hormonal control in these processes is particularly noteworthy.

CHOLESTEROL METABOLISM

The greater part of the cholesterol of the body arises by synthesis (about 1 g/day), whereas only about 0.3 g/day is provided by the average diet. Cholesterol is eliminated via 2 main pathways: conversion to bile acids and excretion of neutral sterols in the feces. The synthesis of steroid hormones from cholesterol and the elimination of their products of degradation in the urine are of minor quantitative significance. Cholesterol is typically a product of animal metabolism and occurs therefore in foods of animal origin such as meat, liver, brain, and egg yolk (a particularly rich source).

Synthesis of Cholesterol

Tissues known to be capable of synthesizing cholesterol include the liver, adrenal cortex, skin, intestines, testis, and aorta. The microsomal fraction of the cell is responsible for cholesterol synthesis.

Acetyl-Co A is the source of all the carbon atoms in cholesterol. The manner of synthesis of this complex molecule has been the subject of investigation by many workers, with the result that it is possible at the present time to chart the origin of all parts of the cholesterol molecule (Figs 14–20, 14–21, and 14–22). Synthesis takes place in several stages. The first is the synthesis of mevalonate, a 6-carbon compound, from acetyl-Co A (Fig 14–20). The next major stage is the formation of isoprenoid units from mevalonate by loss of CO_2 (Fig 14–21). The isoprenoid units may be regarded as the building blocks of the steroid skeleton. Six of these units condense to form the intermediate, squalene, which in turn gives rise to the parent steroid lanosterol. Cholesterol is formed from lanosterol after several further steps, including the loss of 3 methyl groups.

Two separate pathways have been described for the formation of mevalonate. One involves the intermediate HMG-Co A and the other, proposed by Brodie & others (1964), is through an HMG-S-enzyme complex. The pathway through HMG-Co A is considered to be quantitatively the more significant (Myant, 1968) and follows the same sequence of reactions described above for the synthesis of HMG-Co A from acetyl-Co A. It has also been proposed that the pathway may involve formation of malonyl-Co A, but this is unlikely because avidin, which inhibits biotin-linked enzymes such as acetyl-Co A carboxylase, does not inhibit the production of mevalonate from acetyl-Co A.

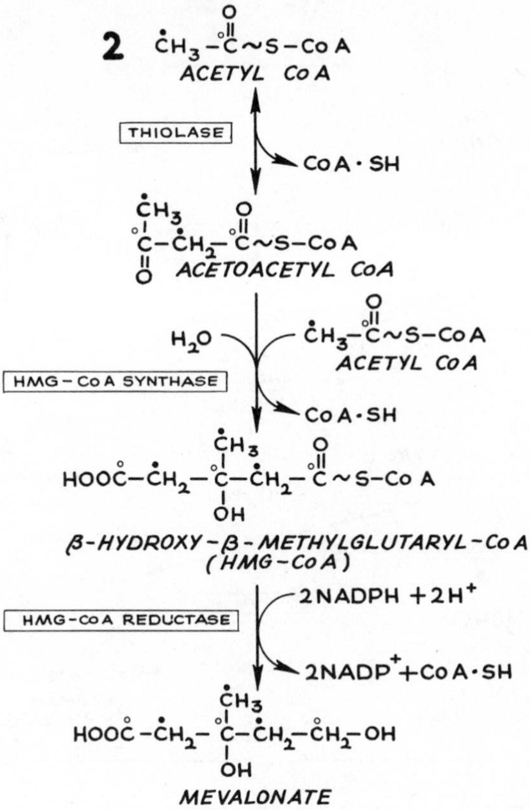

FIG 14–20. Biosynthesis of mevalonate.

HMG-Co A is converted to mevalonate in a 2-stage reduction by NADPH catalyzed by HMG-Co A reductase (Fig 14–20).

In the second stage, mevalonate is phosphorylated by ATP to form several active phosphorylated intermediates. By means of a decarboxylation, the active isoprenoid unit, isopentenylpyrophosphate, is formed. The next stage involves the condensation of molecules of isopentenylpyrophosphate to form farnesyl pyrophosphate. This occurs via an isomerization of isopentenylpyrophosphate involving a shift of the double bond to form dimethylallyl pyrophosphate, followed most probably by condensation with another molecule of isopentenylpyrophosphate to form the 10-carbon intermediate, geranyl pyrophosphate. A further condensation with isopentenylpyrophosphate forms farnesyl pyrophosphate. (For further details, consult Popják & Cornforth, 1966.) Two molecules of farnesyl pyrophosphate condense at the pyrophosphate end in a reaction involving a reduction with NADPH with elimination of the pyrophosphate radicals. The resulting compound is squalene.

Squalene has a structure which resembles the steroid nucleus very closely (Fig 14–22). It is converted to lanosterol by ring closures. Before closure occurs, the methyl group on C_{14} is transferred to C_{13} and that on C_8 to C_{14} and C_3 is hydroxylated. The latter reaction involves molecular oxygen, and the reaction is catalyzed by a microsomal hydroxylase system.

The last stage (Fig 14–22), the formation of cholesterol from lanosterol, involves changes to the steroid nucleus and side chain. The methyl group on C_{14} is oxidized to CO_2 to form 14-desmethyl lanosterol. Likewise, 2 more methyl groups on C_4 are removed to produce zymosterol. Δ7,24-Cholestadienol is formed from zymosterol by the double bond between C_8 and C_9, moving to a position between C_8 and C_7. Desmosterol is formed at this point by a further shift in the double bond in ring II to take up a position between C_5 and C_6, as in cholesterol. Finally, cholesterol is produced when the double bond of the side chain is reduced. The exact order in which the steps described actually take place is not known with certainty. Some investigators favor the view that the double bond at C_{24} is reduced early and that desmosterol is not the immediate precursor of cholesterol. These various possibilities are discussed by Frantz & Schroepfer (1967).

Control of cholesterol synthesis is exerted near the beginning of the pathway. Wieland & others (1960) noticed that there was a marked decrease in the activity of HMG-Co A reductase in fasting rats, which might explain the reduced synthesis of cholesterol during fasting. On the other hand, the activity of this

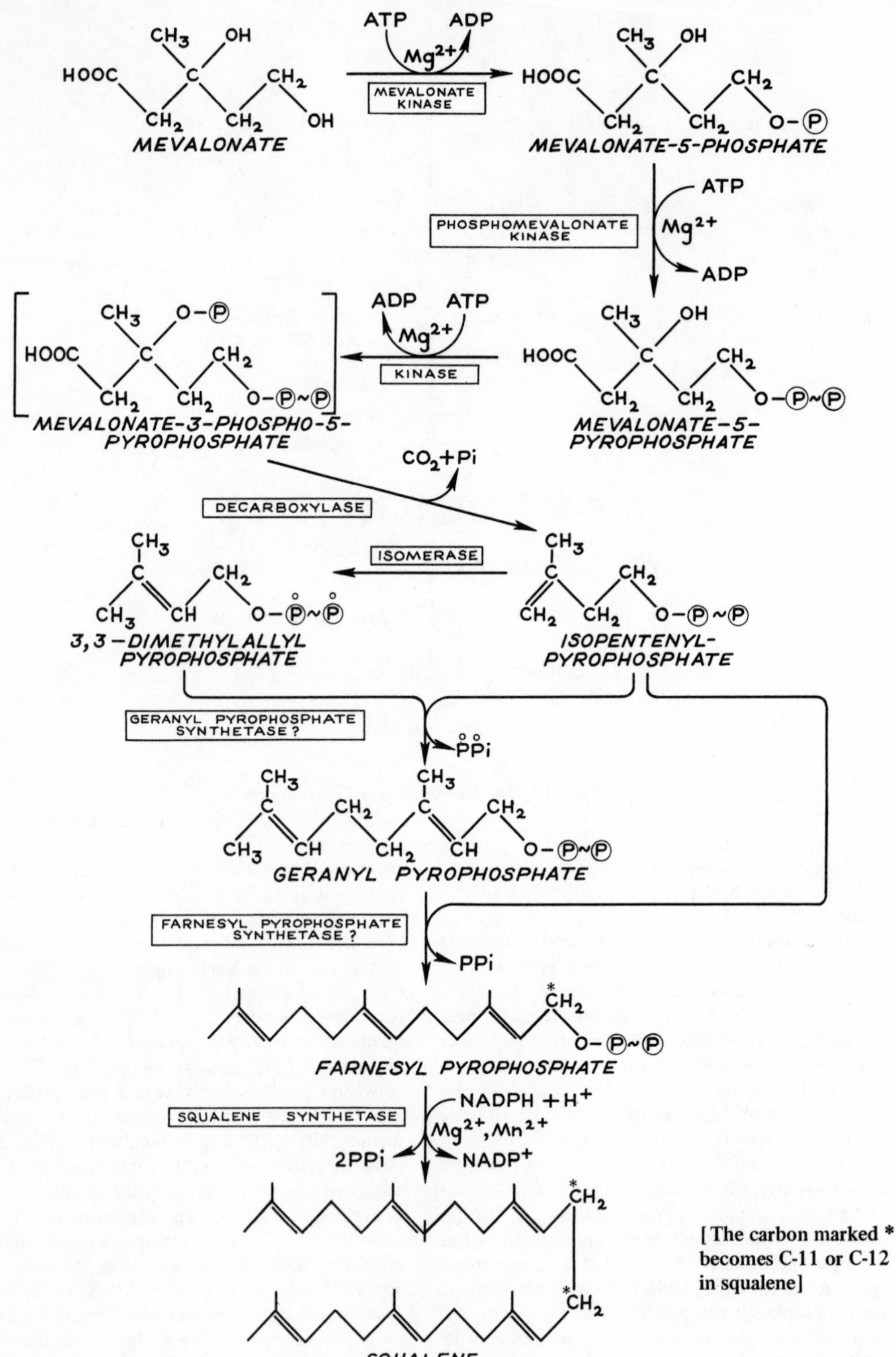

FIG 14–21. Biosynthesis of squalene.

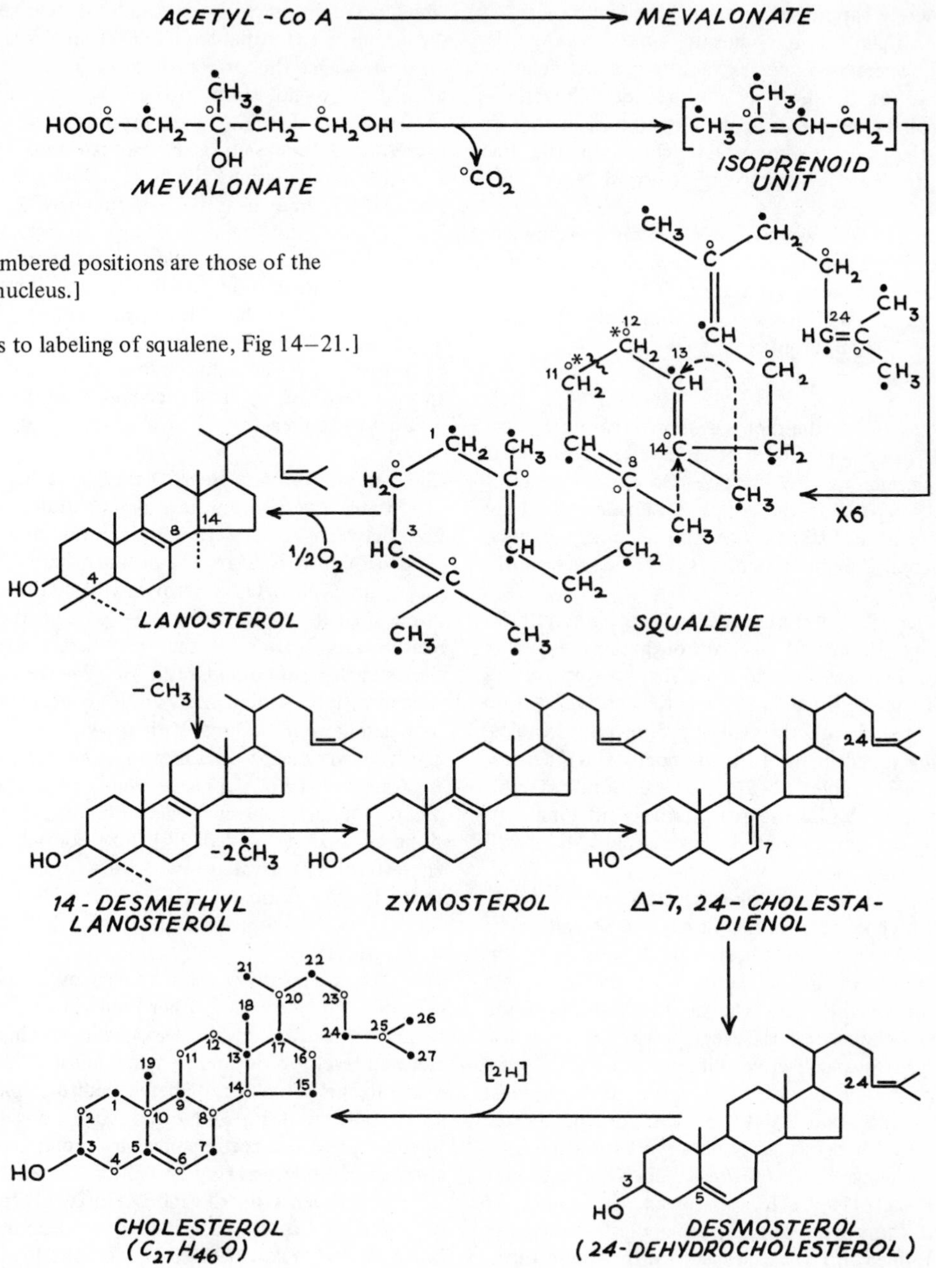

[The numbered positions are those of the steroid nucleus.]

[*Refers to labeling of squalene, Fig 14–21.]

FIG 14–22. Biosynthesis of cholesterol.

enzyme was not reduced in the livers of diabetic rats, which correlates well with the continued synthesis of cholesterol observed in the diabetic state. This enzyme is also involved in a feedback control mechanism proposed by Siperstein (1960), whereby cholesterol inhibits cholesterol synthesis in the liver by inhibition of its activity.

After the administration of ^{14}C-labeled acetate, the label can be detected in plasma cholesterol within a few minutes. Synthesis takes place in the liver, and the cholesterol is incorporated into low density lipoproteins. The effect of variations in the amount of cholesterol in the diet on the endogenous production of cholesterol in rats has been studied by Morris & Chaikoff (1959). When there was only 0.05% cholesterol in the diet, 70–80% of the cholesterol of the liver, small intestine, and adrenal gland was synthesized within the body, whereas on a diet containing 2% cholesterol, the endogenous production fell to 10–30%. However, endogenous production could not be completely suppressed by raising the dietary intake. It appears that it is only hepatic synthesis which is inhibited. There is a species variation in the relative importance of the liver as a source of endogenous cholesterol. In man, extrahepatic synthesis, mainly in the intestine, is more important, whereas in the dog and rat the liver is responsible for most cholesterol synthesis.

Attempts to lower plasma cholesterol in humans by reducing the amount of cholesterol in the diet may be effective. According to Stare (1966), an increase of 100 mg in dietary cholesterol causes a rise of 5 mg cholesterol/100 ml serum.

Transport

Cholesterol in the diet is absorbed from the intestine and, in company with other lipids, incorporated into chylomicrons and VLDL. Of the cholesterol absorbed, 80–90% in the lymph is esterified with long chain fatty acids. Esterification may occur in the intestinal mucosa. The plant sterols (sitosterols) are poorly absorbed.

In man, the total plasma cholesterol is about 200 mg/100 ml, rising with age, although there are wide variations between individuals. The greater part is found in the esterified form. It is transported as lipoprotein in the plasma, the highest proportion of cholesterol being found in the LDL (β-lipoproteins), density 1.019–1.063 (Table 14–2). However, under conditions where the VLDL are quantitatively predominant, an increased proportion of the plasma cholesterol will reside in this fraction.

Dietary cholesterol takes several days to equilibrate with cholesterol in the plasma and several weeks to equilibrate with cholesterol of the tissues. The turnover of cholesterol in the liver is relatively fast, with a half-life of several days, whereas the half-life of the total body cholesterol is several weeks. Free cholesterol in plasma and liver equilibrates in several hours. Equilibration of cholesterol ester with free cholesterol in plasma takes several days in man. In general, free cholesterol exchanges readily between tissues and lipoproteins, whereas cholesterol ester does not exchange freely. Nestel (1970) has discussed the nature of cholesterol turnover. Some plasma cholesterol ester may be formed in HDL as a result of the transesterification reaction in plasma between cholesterol and the fatty acid in position 2 of phosphatidyl choline, catalyzed by lecithin:cholesterol acyltransferase (LCAT). A familial deficiency of this enzyme has been reported (Norum & Gjione, 1967). In affected subjects, there are raised free cholesterol levels and lowered cholesterol ester levels in plasma. HDL and VLDL concentrations are also low.

Excretion of Cholesterol

Approximately ½ of the cholesterol eliminated from the body is excreted in the feces after conversion to bile salts. The remainder is excreted as neutral steroids. Much of the cholesterol secreted in the bile is reabsorbed, and it is believed that the cholesterol which serves as precursor for the fecal sterols is derived from the intestinal mucosa (Danielsson, 1960). Coprostanol is the principal sterol in the feces; it is formed from cholesterol in the lower intestine by the bacterial flora therein. A large proportion of the biliary excretion of bile salts is reabsorbed into the portal circulation, taken up by the liver, and reexcreted in the bile. This is known as the enterohepatic circulation. The bile salts not reabsorbed, or their derivatives, are excreted in the feces. Bile salts undergo changes brought about by intestinal bacteria. Bergström & Danielsson (1958) showed that the rate of production of bile acids from cholesterol in the liver was reduced by infusion of bile salts, indicating the existence of another feedback control mechanism initiated by the product of the reaction.

Cholesterol, Coronary Heart Disease, & Atherosclerosis

Many investigators have demonstrated a correlation between raised serum lipid levels and the incidence of coronary heart disease and atherosclerosis. Of the serum lipids, cholesterol has been the one most often singled out as being chiefly concerned in the relationship. However, other parameters such as the cholesterol:phospholipid ratio, Sf 12–400 lipoprotein concentration, serum triglyceride concentration, etc show similar correlations. Patients with arterial disease can have any one of the following abnormalities (Havel & Carlson, 1962): (1) elevated concentrations of VLDL (mainly triglycerides), with normal concentrations of LDL (D = 1.019–1.063) containing chiefly cholesterol; (2) elevated low density lipoproteins (cholesterol) with normal VLDL (triglycerides); (3) elevation of both lipoprotein fractions (cholesterol plus triglycerides).

Atherosclerosis is characterized by the deposition of cholesterol ester and other lipids in the connective tissue of the arterial walls. Diseases in which prolonged elevated levels of low and very low density lipoproteins occur in the blood (eg, diabetes mellitus, lipid nephrosis, hypothyroidism, and other conditions of hyperlipemia) are often accompanied by premature or more severe atherosclerosis (Fig 14–23).

Experiments on the induction of atherosclerosis in animals indicate a wide species variation in susceptibility (Kritchevsky, 1964). The rabbit, pig, monkey, and man are species in which atherosclerosis can be induced by feeding cholesterol. The rat, dog, and cat are resistant. Thyroidectomy or treatment with thiouracil drugs will allow induction of atherosclerosis in the dog and rat. Low blood cholesterol is a characteristic of hyperthyroidism. However, hyperthyroidism is associated with an increased rate of cholesterol synthesis. The fall in level of plasma cholesterol may be due to an increased rate of turnover and excretion.

Of the factors which lower blood cholesterol, the substitution in the diet of polyunsaturated fatty acids for some of the saturated fatty acids has been the most intensely studied (Kinsell & others, 1952; Groen & others, 1952). Naturally occurring oils which are beneficial in lowering plasma cholesterol include

TABLE 14–3. Typical fatty acid analyses of some fats of animal and plant origin.*
(All values in weight percentages of component fatty acids.)

	Saturated			Unsaturated		
	Palmitic	**Stearic**	**Other**	**Oleic**	**Linoleic**	**Other**
Animal Fats						
Lard	29.8	12.7	1.0	47.8	3.1	5.6
Chicken	25.6	7.0	0.3	39.4	21.8	5.9
Butterfat	25.2	9.2	25.6	29.5	3.6	7.2
Beef fat	29.2	21.0	3.4	41.1	1.8	3.5
Vegetable Oils						
Corn	8.1	2.5	0.1	30.1	56.3	2.9
Peanuts	6.3	4.9	5.9	61.1	21.8	. . .
Cottonseed	23.4	1.1	2.7	22.9	47.8	2.1
Soybean	9.8	2.4	1.2	28.9	50.7	7.0†
Olive	10.0	3.3	0.6	77.5	8.6	. . .
Coconut	10.5	2.3	78.4	7.5	trace	1.3

*Reproduced from NRC Publication No. 575: *The Role of Dietary Fat in Human Health: A Report.* Food and Nutrition Board, National Academy of Sciences.
†Mostly linolenic acid.

peanut, cottonseed, corn, and soybean oil, whereas butterfat and coconut oil raise the level. Reference to Table 14–3 indicates the high proportion of linoleic acid in the first group of oils and its relative deficiency or absence in butterfat or coconut oil, respectively. Gunning & others (1964) have shown that total plasma glycerides are reduced most significantly by increasing the proportion of fat in the diet. These investigators concluded that maximal reduction of total plasma lipids (mainly cholesterol, phospholipids, and glycerides) is achieved with diets which are relatively and absolutely high in fat, and in which the mean fatty acid composition has an iodine number in excess of 100. The effect of the low-fat (ie, high-carbohydrate) diet in raising total plasma lipids is presumably due to stimulation of the synthesis of VLDL in the liver and possibly also to a reduction in rate of removal. Sucrose (and fructose) have a greater effect in raising blood lipids than other carbohydrates. A correlation between the increased consumption of sucrose and atherosclerosis has been demonstrated by Yudkin & Roddy (1964).

The reason for the cholesterol lowering effect of polyunsaturated fatty acids is still not clear. However, several hypotheses have been advanced to explain the effect, including the stimulation of cholesterol excretion into the intestine and the stimulation of the oxidation of cholesterol to bile acids (Antonis & Bersohn, 1962). It is possible that cholesterol esters of polyunsaturated fatty acids are more rapidly metabolized by the liver and other tissues, which might enhance their rate of turnover and excretion. There is other evidence that the effect is largely due to a shift in distribution of cholesterol from the plasma into the tissues.

Additional factors considered to play a part in atherosclerosis include high blood pressure, obesity, lack of exercise, and soft as opposed to hard drinking water. Elevation of plasma FFA will also lead to increased VLDL secretion by the liver, involving extra triglyceride and cholesterol output into the circulation. Factors leading to higher or fluctuating levels of FFA include emotional stress, nicotine from cigarette smoking, coffee drinking, and partaking of few large meals rather than more continuous feeding. Premenopausal women appear to be protected against many of these deleterious factors.

When dietary measures fail to achieve reduced serum lipid levels, resort may be made to certain drugs. Several drugs are known which block the formation of cholesterol at various stages in the biosynthetic pathway. Many of these drugs have harmful effects, and it is now considered that direct interference with cholesterol synthesis is to be avoided (Connor, 1968). Sitosterol is a hypocholesterolemic agent which acts by blocking the esterification of cholesterol in the gastrointestinal tract, thereby reducing cholesterol absorption. Drugs which are considered to increase the fecal excretion of cholesterol and bile acids include dextrothyroxine (Choloxin), neomycin, and possibly clofibrate (Atromid S). On the other hand, cholestyramine (Cuemid, Questran) prevents the reabsorption of bile salts by combining with them, thereby increasing their fecal loss. Clofibrate may also act by inhibiting the secretion of VLDL by the liver or by inhibiting hepatic cholesterol synthesis. Other hypocholesterolemic drugs include nicotinic acid and estrogens.

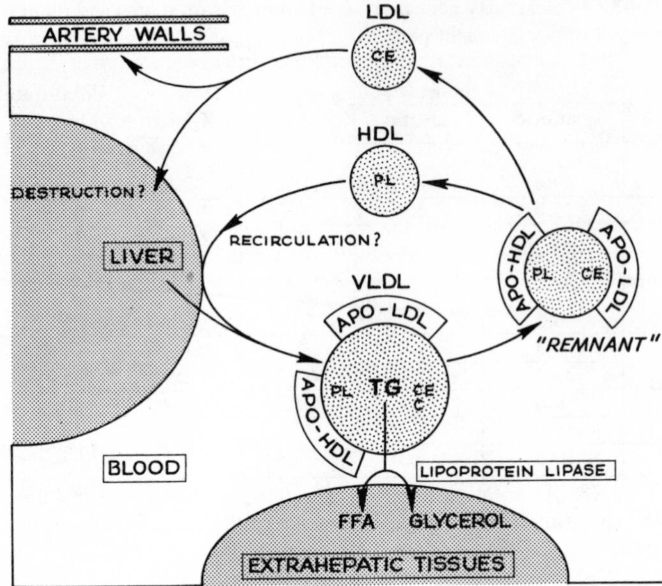

FIG 14–23. Postulated interrelationships between plasma lipoproteins and atherosclerosis. Chylomicrons are metabolized in a similar manner to VLDL. Only the prominent lipids are shown in LDL and HDL. VLDL = very low density lipoprotein; LDL = low density lipoprotein; HDL = high density lipoprotein; TG = triglyceride; PL = phospholipid; C = cholesterol; CE = cholesterol ester.

Disorders of the Plasma Lipoproteins

A few individuals in the population exhibit inherited defects in their lipoproteins leading to the primary condition of either hypo- or hyperlipoproteinemia. Many others having defects such as diabetes mellitus, hypothyroidism, and atherosclerosis show abnormal lipoprotein patterns which are very similar to one or the other of the primary inherited conditions. By making particular use of the technic of paper electrophoresis (Fig 14–2), Fredrickson & others (1967) have characterized the various types of lipoproteinemias.

A. Hypolipoproteinemia:

1. Abetalipoproteinemia—This is a rare inherited disease characterized by absence of β-lipoprotein (LDL) in plasma. Most of the blood lipids are present in low concentrations—especially glycerides, which are virtually absent since no chylomicrons or pre-β-lipoproteins (VLDL) are formed. Both the intestine and the liver accumulate glycerides.

2. Hypobetalipoproteinemia—In hypobetalipoproteinemia, LDL or β-lipoprotein concentration is between 10–50% of normal, but chylomicron formation occurs. It must be concluded that β-lipoprotein is essential for triglyceride transport.

3. Familial alphalipoprotein deficiency (Tangier disease)—In the homozygous individual there is near absence of plasma HDL or α-lipoproteins and accumulation of cholesterol esters in the tissues. There is no impairment of chylomicron formation or secretion of endogenous triglyceride by the liver. However, on electrophoresis, there is no pre-β-lipoprotein, but a broad β-band is found containing the endogenous triglyceride. This finding provides evidence that the normal pre-β-band contains α-lipoprotein. Although α-lipoprotein does not appear to be essential for glyceride transport, when it is absent, clearance from the plasma is slow, the patients tending to develop hyperglyceridemia.

B. Hyperlipoproteinemia:

1. Type I—Characterized by very slow clearing of chylomicrons from the circulation, leading to abnormally raised levels of chylomicrons. The condition is due to a deficiency of lipoprotein lipase. Pre-β-lipoproteins may be raised, but there is a decrease in α- and β-lipoproteins. Thus, the condition is fat induced. It may be corrected by reducing the quantity of fat in the diet, but high-carbohydrate diets lead to raised levels of pre-β-lipoproteins due to synthesis in the liver.

2. Type II—Characterized by hyperbetalipoproteinemia, which is associated with increased plasma total cholesterol. There may also be a tendency for the pre-β-lipoproteins to be elevated. Therefore, the patient may have somewhat elevated triglyceride levels but the plasma—as is not true in the other types of hyperlipoproteinemia—remains clear. Lipid deposition in the tissues (eg, xanthomas, atheromas) are common. A type II pattern may also arise as a secondary result of hypothyroidism. The disease appears to be associated with reduced rates of clearance of β-lipoprotein (LDL) from the circulation. Reduction of dietary cholesterol and saturated fats may be of use in treatment.

3. Type III—Characterized by an increase in both β- and pre-β-lipoproteins, causing hypercholesterolemia and hyperglyceridemia. Most of the β-lipoproteins tend to be in the D > 1.006 fraction after ultracentrifugation. Xanthomas and atherosclerosis are again present. Treatment by weight reduction and high fat diets containing unsaturated fats and little cholesterol is recommended.

4. Type IV—Characterized by hyperprebetalipoproteinemia with associated high levels of endogenously produced triglyceride. Cholesterol levels rise in proportion to the hypertriglyceridemia, and glucose intolerance is frequently present. Both α- and β-lipoproteins are subnormal in quantity. This lipoprotein pattern is also commonly associated with maturity onset diabetes, obesity, and many other conditions, including alcoholism and the taking of progestational hormones. Treatment of primary type IV is by weight reduction, replacement of much of the carbohydrate in the diet with unsaturated fat, low cholesterol diets, and with hypolipemic agents.

5. Type V—The lipoprotein pattern is complex since both chylomicrons and pre-β-lipoproteins are elevated, causing both triglyceridemia and cholesterolemia. Concentrations of α- and β-lipoproteins are low. Xanthomas are frequently present, but the incidence of atherosclerosis is apparently not striking. Glucose tolerance is abnormal and frequently associated with obesity and diabetes. The reason for the condition, which is familial, is not clear. Treatment has consisted of weight reduction followed by a diet not too high in either carbohydrate or fat.

• • •

References

Adam DJD, Hansen AE, Wiese HF: J Nutr 66:555, 1958.

Ansell GB, Hawthorne JN: *Phospholipids.* B.B.A. Library 3. Elsevier, 1964.

Antonis A, Bersohn I: Am J Clin Nutr 11:142, 1962.

Armstrong DT, Steele R, Altszuler N, Dunn A, Bishop JS, De Bodo RC: Am J Physiol 201:9, 1961.

Ashworth CT, Johnson CF, Wrightsman FJ: Am J Path 46:757, 1965.

Ball EG, Jungas RL: Biochemistry 2:586, 1963.

Beatty CH, Marcó A, Peterson RD, Book RM, West ES: J Biol Chem 235:2774, 1960.

Bellet S, Kershbaum A, Finck EM: Metabolism 17:702, 1968.

Bergström S, Danielsson H: Acta physiol scandinav 43:1, 1958.

Bortz WM, Lynen F: Biochem Ztschr 337:505, 1963.

Brady RO: J Biol Chem 237:PC 2416, 1962.

Brady RO, Koval GJ: J Biol Chem 233:26, 1958.

Bremer J: Page 65 in: *Cellular Compartmentalization and Control of Fatty Acid Metabolism.* Gran FC (editor). Universitetsforlaget, Oslo, 1968.

Bressler R: Ann New York Acad Sc 104:735, 1963.

Brodie JD, Wasson G, Porter JW: J Biol Chem 239:1346, 1964.

Bustos GO, Kalant H, Khanna JM, Loth J: Science 168:1598, 1970.

Butcher RW, Baird CE, Sutherland EW: J Biol Chem 243:1705, 1968.

Chalmers TM, Kekwick A, Pawan GLS, Smith I: Lancet 1:866, 1958.

Connor WE: M Clin North America 52:1249, 1968.

Corredor C, Mansbach C, Bressler R: Biochim Biophys Acta 144:366, 1967.

Crane FL, Beinert H: J Biol Chem 218:717, 1956.

Danielsson H: Acta physiol scandinav 48:364, 1960.

Davies JI: Nature 218:349, 1968.

Dole VP: J Biol Chem 236:3121, 1961.

Elovson J: Biochim Biophys Acta 106:291, 1965.

Evans HM, Burr GO: Proc Soc Exper Biol Med 25:390, 1928.

Fabry P, Poledne R, Kazdova L, Braun T: Nutr Dieta 10:81, 1968.

Fain JN, Kovacev VP, Scow RD: J Biol Chem 240:3522, 1965.

Felts JM: Page 95 in: *Fat as a Tissue.* Rodahl K, Issekutz B (editors). McGraw-Hill, 1964.

Felts JM: Ann New York Acad Sc 131:24, 1965.

Felts JM, Mayes PA: Nature 206:195, 1965.

Fine MB, Williams RH: Am J Physiol 199:403, 1960.

Flatt JP: J Lipid Research 11:131, 1970.

Frantz ID, Schroepfer GJ: Ann Rev Biochem 36:691, 1967.

Fredrickson DS, Gordon RS Jr: Physiol Rev 38:585, 1958.

Fredrickson DS, Levy RI, Lees RS: New England J Med 276:32–281, 1967.

Fritz IB: Page 39 in: *Cellular Compartmentàlization and Control of Fatty Acid Metabolism.* Gran FC (editor). Universitetsforlaget, Oslo, 1968.

Fulco AJ, Mead JF: J Biol Chem 236:2416, 1961.

Garland PB, Randle PJ, Newsholme EA: Nature 200:169, 1963.

Goldfine H: Ann Rev Biochem 37:303, 1968.

Goodman HM, Knobil E: Endocrinology 65:451, 1959.

Greville GD, Tubbs PK: Page 155 in: *Essays in Biochemistry.* Vol 4. Campbell PN, Greville GD (editors). Academic Press, 1968.

Groen J, Tjiong BK, Kamminga CE, Willebrands AF: Voeding 13:556, 1952.

Guarnieri M, Johnson RM: Advances Lipid Res 8:115, 1970.

Gunning BE, Imaichi K, Splitter SD, Kinsell LW: Lancet 2:336, 1964.

Hajra AK, Agranoff BW: J Biol Chem 243:3542, 1968.

Havel RJ: Page 357 in: *Lipid Pharmacology.* Paoletti R (editor). Academic Press, 1964.

Havel RJ, Carlson LA: Metabolism 11:195, 1962.

Havel RJ, Felts JM, Van Duyne CM: J Lipid Research 3:297, 1962.

Havel RJ, Fredrickson DS: J Clin Invest 35:1025, 1956.

Hill R, Linozasoro JM, Chevallier F, Chaikoff IL: J Biol Chem 233:305, 1958.

Hillyard LA, Cornelius CE, Chaikoff IL: J Biol Chem 234:2240, 1959.

Himms-Hagen J: Advances Enzyme Regulat 8:131, 1970.

Holloway PW, Peluffo R, Wakil SJ: Biochem Biophys Res Commun 6:270, 1962.

Hull D, Segall MM: J Physiol 181:449, 1965.

Jeanrenaud B, Hepp D (editors): *Adipose Tissue Regulation and Metabolic Functions.* Academic Press, 1970.

Jeanrenaud B, Renold AE: Excerpta Medica Internat Congr Series No. 132:769, 1966.

Jones AL, Ruderman NB, Herrera MG: J Lipid Research 8:429, 1967.

Kanfer JN, Gal AE: Biochem Biophys Res Commun 22:442, 1966.

Kinsell LW, Partridge J, Boling L, Margen S, Michaels GD: J Clin Endocrinol 12:909, 1952.

Kiyasu JY, Kennedy EP: J Biol Chem 235:2590, 1960.

Knoop F: Beitr chem Physiol u Path 6:160, 1905.

Koehler AE, Windsor E, Hill E: J Biol Chem 140:811, 1941.

Kornacker MS, Ball EG: J Biol Chem 243:1638, 1968.

Krebs HA: Vet Rec 78:187, 1966.

Kritchevsky D: Page 63 in: *Lipid Pharmacology*. Paoletti R (editor). Academic Press, 1964.

Levin L, Farber RK: Recent Progr Hormone Res 7:399, 1952.

Lieber CS: Gastroenterology 50:119, 1966.

Lindsay DB: Biochem J 73:10P, 1959.

Lombardi B: Fed Proc 30:139, 1971.

Lorch E, Abraham S, Chaikoff IL: Biochim et biophys acta 70:627, 1963.

Lynen F, Henning U, Bublitz C, Sorbo B, Kröplin-Rueff L: Biochem Ztschr 330:269, 1958.

Lynen F, Tada M: Angew Chem 73:513, 1961.

Lynen F, Oesterhelt D, Schweizer E, Willecke K: Page 1 in: *Cellular Compartmentilization and Control of Fatty Acid Metabolism*. Gran FC (editor). Universitetsforlaget, Oslo, 1968.

Madison LL, Mebane D, Unger RH, Lochner A: J Clin Invest 43:408, 1964.

Majerus PW, Vagelos PR: Advances Lipid Res 5:2, 1967.

Marliss E, Aoki TT, Felig P, Pozefsky T, Cahill GF: Advances Enzyme Reg 8:3, 1970.

Masoro EJ, Chaikoff IL, Chernick SS, Felts JM: J Biol Chem 185:845, 1950.

Mayes PA: Metabolism 11:781, 1962.

Mayes PA, Felts JM: Biochem J 108:483, 1968.

Mayes PA, Felts JM: Nature 215:716, 1967.

MacKay EM, Barnes RH, Carne HO, Wick AN: J Biol Chem 135:157, 1940.

Mead JF: Fed Proc 20:952, 1961.

Mishkel MA, Morris B: Quart J Exp Physiol 49:21, 1964.

Mookerjea S: Fed Proc 30:143, 1971.

Morris MD, Chaikoff IL: J Biol Chem 234:1095, 1959.

Munkner C: Scandinav J Clin Lab Invest 11:394, 1959.

Myant NB: Page 193 in: *The Biological Basis of Medicine*. Vol 2. Bittar ED, Bittar N (editors). Academic Press, 1968.

Nelson N, Grayman I, Mirsky IA: J Biol Chem 140:361, 1941.

Nestel PJ: Page 243 in: *Newer Methods of Nutritional Biochemistry*. Vol 3. Albanese AA (editor). Academic Press, 1967.

Nestel PJ: Advances Lipid Res 8:1, 1970.

Nikkilä EA, Ojala K: Proc Soc Exper Biol Med 113:814, 1963.

Nikkilä EA: Advances Lipid Res 7:63, 1969.

Norum KR, Gjione E: Scandinav J Clin Lab Invest 20:231, 1967.

Nugteren DH: Biochim Biophys Acta 106:280, 1965.

Olson RE: Nature 195:597, 1962.

Olson RE, Jablonski JR, Taylor, E: Am J Clin Nutr 6:111, 1958.

Olson RE, Vester JW: Physiol Rev 40:677, 1960.

Paulus H, Kennedy EP: J Biol Chem 235:1303, 1960.

Popják G, Cornforth JN: Biochem J 101:553, 1966.

Randle PJ, Garland PB, Hales CN, Newsholme EA: Lancet 1:785, 1963.

Renold AE, Cahill GF (editors): *Handbook of Physiology*. Section 5. American Physiological Society, 1965.

Robinson DS: Advances Lipid Res 1:134, 1963.

Robinson DS: Page 166 in: *Proceedings Deuel Conference*. Cowgill G, Kinsell LW (editors). US Department Health, Education & Welfare, 1967.

Robinson DS, Seakins A: Biochim et biophys acta 62:163, 1962.

Robinson DS: Page 51 in: *Comprehensive Biochemistry*. Vol 18. Florkin M, Stotz EH (editors). Elsevier, 1970.

Rossiter RJ: Page 69 in: *Metabolic Pathways*. Vol 2. Greenberg DM (editor). Academic Press, 1968.

Rudman D: J Lipid Research 4:119, 1963.

Rudman D, Di Girolamo M: Advances Lipid Res 5:35, 1967.

Sauer F: Canad J Biochem Physiol 38:635, 1960.

Schettler G (editor): *Lipids and Lipidoses*. Springer-Verlag, 1967.

Schoenheimer R, Rittenberg D: J Biol Chem 114:175, 1935.

Scow RD, Chernick SS: Recent Progr Hormone Res 16:497, 1960.

Seyffert WA, Madison LL: Diabetes 16:765, 1967.

Sgoutas D, Kummerow FA: Am J Clin Nutr 23:1111, 1970.

Shafrir E, Kerpel S: Arch Biochem 105:237, 1964.

Shapiro B: Ann Rev Biochem 36:247, 1967.

Shapiro B, Chowers I, Rose G: Biochim et biophys acta 23:115, 1957.

Shepherd D, Garland PB: Biochem Biophys Res Commun 22:89, 1966.

Sinclair HM: Page 237 in: *Lipid Pharmacology*. Paoletti R (editor). Academic Press, 1964.

Siperstein MD: Am J Clin Nutr 8:645, 1960.

Smith RE, Roberts JC, Hittelman KJ: Science 154:653, 1966.

Smuckler EA, Barker EA: Proc European Soc for the Study of Drug Toxicity 7:83, 1966.

Spencer AF, Lowenstein JM: J Biol Chem 237:3640, 1962.

Stare FJ: J Am Dietet A 48:88, 1966.

Steinberg D: Pharmacol Rev 18:217, 1966.

Stern JR, Miller GE: Biochim et biophys acta 35:576, 1959.

Stoffel W, Caesar H: Zeitschr Physiol Chem 341:76, 1965.

Svennerholm L: Page 201 in: *Comprehensive Biochemistry*. Vol 18. Florkin M, Stotz EH (editors). Elsevier, 1970.

Thompson GA: Page 157 in: *Comprehensive Biochemistry*. Vol 18. Florkin M, Stotz EH (editors). Elsevier, 1970.

Thompson RHS, Cumings JN: Page 548 in: *Biochemical Disorders in Human Disease*. Thompson RHS, King EJ (editors). Churchill, 1964.

Tietz A, Shapiro B: Biochim et biophys acta 19:374, 1956.

Topping DL, Mayes PA: Biochem J 119:48P, 1970.

Tubbs PK, Garland PB: Brit M Bull 24:158, 1968.

Utter MF, Keech DB: J Biol Chem 238:2603, 1963.

Van Deenen LLM, de Haas GH: Ann Rev Biochem 35:157, 1966.

Wakil SJ: J Lipid Research 2:1, 1961.

Whayne TF, Felts JM: Circulation Res 27:941, 1970.

Wick AN, Drury DR: J Biol Chem 138:129, 1941.

Wieland O, Löffler G, Neufeldt I: Biochem Ztschr 333:10, 1960.

Wieland O, Weiss L, Eger-Neufeldt I, Teinzer A, Westermann B: Klin Wschr 43:645, 1966.

Wieland O, Weiss L, Eger-Neufeldt I: Advances Enzyme Reg 2:85, 1964.

Wilgram GF, Kennedy EP: J Biol Chem 238:2615, 1963.

Williamson DH, Veloso D, Ellington EV, Krebs HA: Biochem J 114:575, 1969.

Williamson JR, Krebs HA: Biochem J 80:540, 1961.

Winegrad AI, Renold AE: J Biol Chem 233:267, 1958.

Witten PW, Holman RT: Arch Biochem 41:266, 1952.

Yudkin J, Roddy J: Lancet 2:6, 1964.

Zöllner N, Thannhauser SJ: Page 898 in: *Biochemical Disorders in Human Disease*. Thomson RHS, King EJ (editors). Churchill, 1964.

15 . . .

Protein & Amino Acid Metabolism

With Victor Rodwell, PhD*

Although the subject of amino acid metabolism includes the topics outlined in Fig 15–1, several of these are considered elsewhere–eg, synthesis and catabolism of proteins (Chapters 5 and 12), and conversion of certain amino acids to portions of nucleotides (Chapter 16) or porphyrins (Chapter 6). The latter portion of this chapter discusses the metabolic pathways for conversion of specific amino acids to certain specialized products of physiologic significance–eg, products such as histamine, serotonin, and melanin.

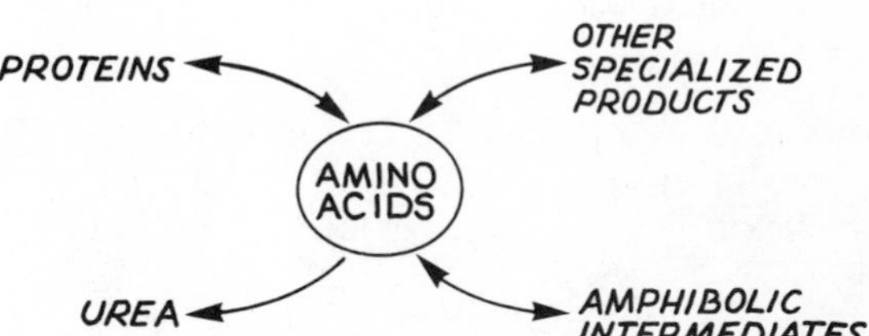

FIG 15–1. Outline of amino acid metabolism. All the indicated processes except urea formation proceed reversibly in intact cells. However, the catalysts and intermediates in biosynthetic and degradative processes frequently differ.

This chapter deals primarily with the catabolism of the nitrogen of amino acids to urea and of the carbon skeletons to amphibolic intermediates of the citric acid cycle, as well as with the biosynthesis of amino acids from amphibolic intermediates. The discussion is for the most part limited to mammalian metabolism. An important exception is the inclusion of pathways for biosynthesis of the nutritionally essential amino acids which are made by bacteria and plants but not by man or other animals. Inclusion of this material reflects the wealth of information and concepts concerning regulation of amino acid biosynthesis in procaryotic organisms–concepts that have implications in amino acid metabolism both in procaryotic and in eurcaryotic organisms, including man.

NITROGEN CATABOLISM OF AMINO ACIDS

In mammalian tissues the α-amino groups of amino acids, derived either from the diet or from breakdown of tissue proteins, ultimately are excreted in the urine as urea. The biosynthesis of urea involves the action of several enzymes. It may conveniently be divided for discussion into 4 processes: (1) transamination, (2) oxidative deamination, (3) ammonia transport, and (4) reactions of the urea cycle.

The relationship of these areas to the overall catabolism of amino acid nitrogen is shown in Fig 15–2. Vertebrates other than mammals share all features of this scheme except urea synthesis. Urea, the characteristic end product of amino acid nitrogen metabolism in man and other ureotelic organisms, is replaced by uric acid in uricotelic organisms (eg, reptiles and birds) or by ammonia in ammonotelic organisms (eg, bony fish).

Each of the above 4 processes will now be considered in detail. Although each also plays a role in amino acid biosynthesis (see p 326), what follows is first discussed from the viewpoint of amino acid catabolism.

Transamination

Transamination, catalyzed by enzymes termed **transaminases** or **aminotransferases**, involves interconversion of a pair of amino acids and a pair of keto acids. These generally are α-amino and α-keto acids (Fig 15–3). Pyridoxal phosphate, the coenzyme form of vitamin B_6, forms an essential part of the active site of transaminases and of many other enzymes with amino acid substrates. In all pyridoxal phosphate dependent reactions of amino acids, the initial step is formation of an enzyme-bound Schiff base intermediate (Fig 15–4). This intermediate, stabilized by interaction with a cationic region of the active site, can be rearranged in ways that include release of a keto acid with formation of enzyme-bound pyridoxamine phosphate. The bound, amino form of the coenzyme can then form an analogous Schiff base intermediate with a keto acid. During transamination, the bound coenzyme thus serves as an intermediate carrier of amino groups (Fig 15–5). Since the equilibrium constant for most transaminase reactions is close to unity, transamination is a freely reversible process. This reversibility permits transaminases to function both in amino acid catabolism and biosynthesis.

*Associate Professor of Biochemistry, Purdue University, Lafayette, Indiana.

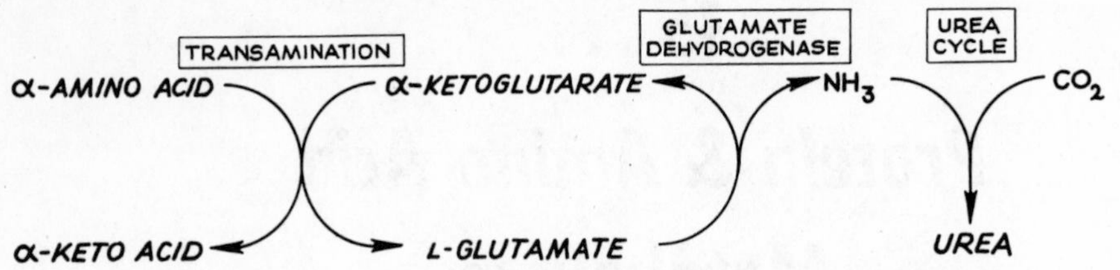

TRANSAMINATION — GLUTAMATE DEHYDROGENASE — UREA CYCLE

α-AMINO ACID → α-KETO ACID; α-KETOGLUTARATE ⇄ L-GLUTAMATE → NH_3; NH_3 + CO_2 → UREA

FIG 15–2. **Overall flow of nitrogen in amino acid catabolism.** Although the reactions shown are reversible, they are represented as being unidirectional to emphasize the direction of metabolic flow in mammalian amino acid catabolism.

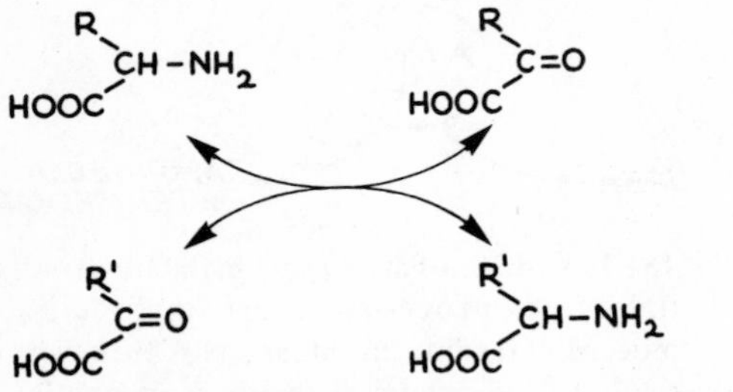

FIG 15–3. **Transamination.** The reaction is shown for 2 *a*-amino and 2 *a*-keto acids. Non-*a*-amino or carbonyl groups also participate in transamination, although this is relatively uncommon. The reaction is freely reversible with an equilibrium constant of about 1.

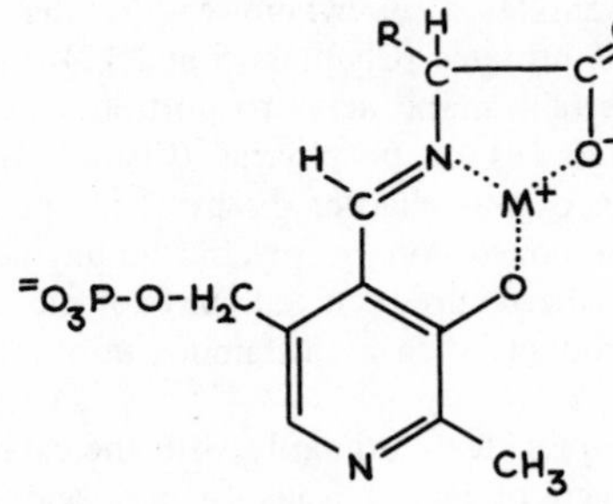

FIG 15–4. **Condensation product of an amino acid with enzyme-bound pyridoxal phosphate at the active site of an enzyme.** M^+ represents a cationic region of the active site. A mol of water is split out between the *a*-amino group of the amino acid and the carbonyl oxygen of enzyme-bound pyridoxal phosphate.

Two transaminases, alanine-pyruvate transaminase (**alanine transaminase**) and glutamate-*a*-ketoglutarate transaminase (**glutamate transaminase**), present in most mammalian tissues, catalyze transfer of amino groups from most amino acids to form alanine (from pyruvate) or glutamate (from *a*-ketoglutarate).

ALANINE TRANSAMINASE

α-AMINO ACID + PYRUVATE ⇌ α-KETO ACID + ALANINE

GLUTAMATE TRANSAMINASE

α-AMINO ACID + α-KETOGLUTARATE ⇌ α-KETO ACID + GLUTAMATE

Each transaminase is specific for the specified pair of amino and keto acids as one pair of substrates but nonspecific for the other pair, which may be any of a wide variety of amino acids and their corresponding keto acids. Since alanine is also a substrate for the glutamate transaminase reaction, all of the amino nitrogen from amino acids which can undergo transamination can be concentrated in glutamate. This is important because L-glutamate is the only amino acid in mammalian tissues which undergoes oxidative deamination at an appreciable rate. The formation of ammonia from *a*-amino groups thus requires their prior conversion to the *a*-amino nitrogen of L-glutamate.

Most (but not all) amino acids are substrates for transamination. Exceptions include lysine, threonine, and the cyclic imino acids, proline and hydroxyproline. Transamination is not restricted to *a*-amino groups. The δ-amino group of ornithine is, for example, readily transaminated, forming glutamate γ-semialdehyde (Fig 15–15). The serum levels of transaminases are markedly elevated in certain disease states (see p 161).

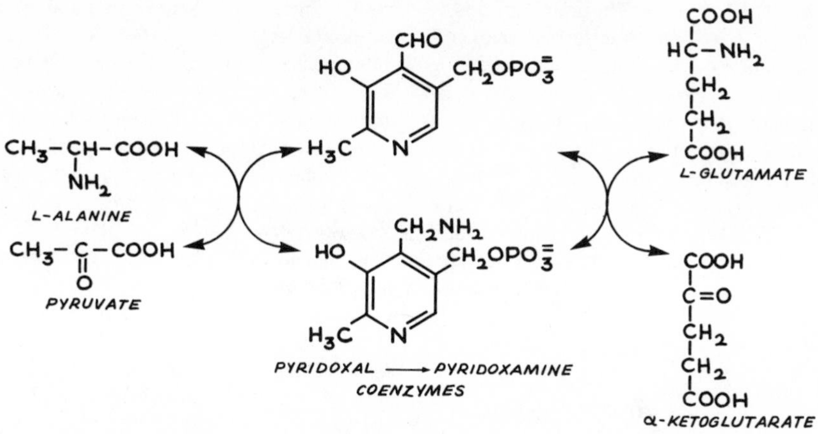

FIG 15–5. Participation of pyridoxal phosphate in transamination reactions.

Oxidative Deamination

Oxidative conversion of amino acids to their corresponding α-keto acids can be shown with homogenates of mammalian liver and kidney tissue. Although most of the activity of homogenates toward L-α-amino acids is due to the coupled action of transaminases plus L-glutamate dehydrogenase (see p 306), both L- and D-amino acid oxidase activities do occur in mammalian liver and kidney tissue and are widely distributed in other animals and microorganisms. It must be noted, however, that the physiologic function of L- and D-amino acid oxidase of mammalian tissue is not known.

The amino acid oxidases are **autoxidizable flavoproteins,** ie, the reduced FMN or FAD is reoxidized directly by molecular oxygen, forming hydrogen peroxide (H_2O_2) without participation of cytochromes or other electron carriers (see p 173). The toxic product H_2O_2 is then split to O_2 and H_2O by **catalase,** which occurs widely in tissues, especially liver (Fig 15–6). Although the amino acid oxidase reactions are reversible, if catalase is absent the α-keto acid product is nonenzymically decarboxylated by H_2O_2, forming a carboxylic acid with one less carbon atom.

Both L- and D-amino acid oxidase activities are present in renal tissue, although the function of the D-amino acid oxidase is obscure. In the amino acid oxidase reaction, the amino acid is first dehydrogenated by the flavoprotein of the oxidase, forming an α-imino acid. This spontaneously adds water, then decomposes to the corresponding α-keto acid with loss of the α-imino nitrogen as ammonia.

Mammalian L-amino acid oxidase, an FMN-flavoprotein, is restricted to kidney and liver tissue. Its activity is quite low, and it is essentially without effect on glycine or the L-isomers of the dicarboxylic or β-hydroxy α-amino acids. It thus is not likely that this enzyme fulfills a major role in mammalian amino acid catabolism. An active L-amino acid oxidase also occurs in snake venom.

Mammalian D-amino acid oxidase, an FAD-flavoprotein of broad substrate specificity, occurs in the liver and kidney tissue of most mammals. D-Asparagine and D-glutamine are not oxidized, and glycine and the D-isomers of the acidic and basic amino acids are poor substrates. The physiologic significance of this enzyme in mammals is not known.

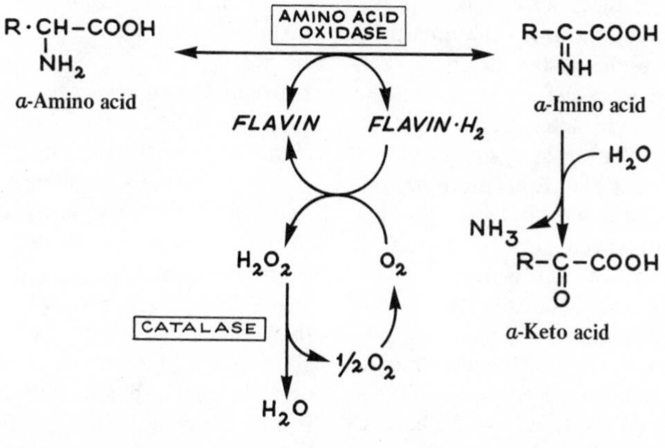

FIG 15–6. Oxidative deamination catalyzed by L-amino acid oxidase (L-α-amino acid:O_2 oxidoreductase).

L-Glutamate dehydrogenase. The amino groups of most amino acids ultimately are transferred to α-ketoglutarate by transamination, forming L-glutamate (Fig 15–2). Release of this nitrogen as ammonia is catalyzed by **L-glutamate dehydrogenase,** an enzyme of high activity widely distributed in mammalian tissues (Fig 15–7). Liver glutamate dehydrogenase is a regulated enzyme whose activity is affected by allosteric modifiers such as ATP, GTP, and NADP, which inhibit the enzyme; and ADP, which activates the enzyme. Certain hormones appear also to influence glutamate dehydrogenase activity.

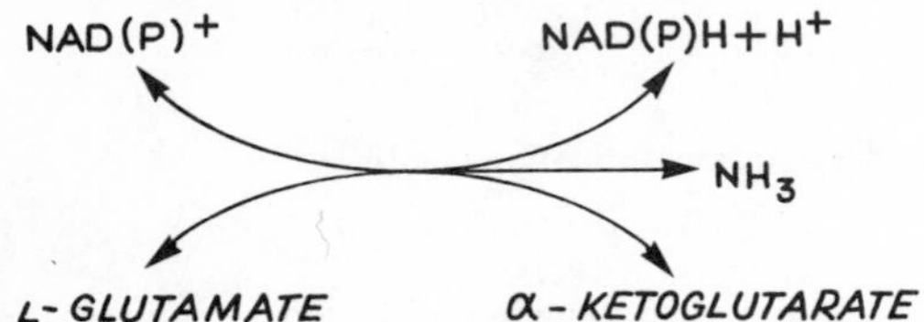

FIG 15–7. The L-glutamate dehydrogenase reaction. The designation $NAD(P)^+$ means that either NAD^+ or $NADP^+$ can serve as cosubstrate. The reaction is reversible, but the equilibrium constant favors glutamate formation.

Glutamate dehydrogenase uses either NAD^+ or $NADP^+$ as cosubstrate. The reaction is reversible and functions both in amino acid catabolism and biosynthesis. It therefore functions not only to funnel nitrogen from glutamate to urea (catabolism) but also to catalyze amination of α-ketoglutarate by free ammonia. This latter (biosynthetic) function is of particular importance in plants and bacteria, which can synthesize large quantities of amino acids from glucose plus ammonia (see p 326). When beef cattle are fed diets rich in carbohydrate plus nitrogen in the form of urea, the rumen bacteria first convert the urea to ammonia, then utilize the glutamate dehydrogenase reaction to provide the cattle with a diet rich in glutamate and other amino acids.

In addition to ammonia formed in the tissues, a considerable quantity is produced by intestinal bacteria both from dietary protein and from the urea present in fluids secreted into the gastrointestinal tract. This ammonia is absorbed from the intestine into the portal venous blood, which characteristically contains higher levels of ammonia than does systemic blood. Under normal circumstances the liver promptly removes the ammonia from the portal blood, so that blood leaving the liver (and indeed all of the peripheral blood) is virtually ammonia-free. This is essential since even minute quantities of ammonia are toxic to the CNS. The symptoms of **ammonia intoxication** include a peculiar flapping tremor, slurring of speech, blurring of vision, and, in severe cases, coma and death. These symptoms resemble those of the syndrome of hepatic coma which occurs when blood and, presumably, brain ammonia levels are elevated. Ammonia intoxication is assumed to be a factor in the etiology of hepatic coma. Therefore, treatment includes measures designed to reduce blood ammonia levels (see p 221).

With severely impaired hepatic function or the development of collateral communications between the portal and systemic veins (as may occur in cirrhosis), the portal blood may bypass the liver. Ammonia from the intestines may thus rise to toxic levels in the systemic blood. Surgically produced shunting procedures (so-called Eck fistula, or other forms of portacaval shunts) are also conducive to ammonia intoxication, particularly after ingestion of large quantities of protein or after hemorrhage into the gastrointestinal tract.

The ammonia content of the blood leaving the kidneys via the renal veins always exceeds that of the renal arteries, indicating that the kidneys produce ammonia and add it to the blood. However, the excretion into the urine of the ammonia produced by renal tubular cells constitutes a far more significant aspect of renal ammonia metabolism. Ammonia production forms part of the renal tubular mechanisms for regulation of acid-base balance as well as conservation of cations (see Chapter 18). Ammonia production by the kidneys is markedly increased in metabolic acidosis and depressed in alkalosis. It is derived not from urea but from intracellular amino acids, particularly glutamine. Ammonia release is catalyzed by renal **glutaminase** (Fig 15–8).

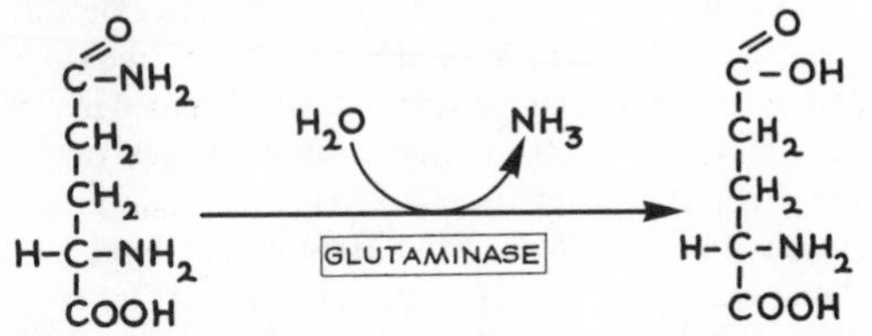

$$\underset{\text{L-Glutamine}}{HOOC{-}CH(NH_2){-}CH_2{-}CH_2{-}C(=O){-}NH_2} \xrightarrow[\text{GLUTAMINASE}]{H_2O \;\; NH_3} \underset{\text{L-Glutamate}}{HOOC{-}CH(NH_2){-}CH_2{-}CH_2{-}C(=O){-}OH}$$

FIG 15–8. The glutaminase reaction. The reaction proceeds essentially irreversibly in the direction of glutamate formation.

Transport of Ammonia

Although ammonia may be excreted as ammonium salts–particularly in states of metabolic acidosis (see p 381)–the vast majority is excreted as urea, the principal nitrogenous component of urine. Ammonia, constantly produced in the tissues by the processes described above, is present only in traces in blood (10–20 μg/100 ml) since it is rapidly removed from the circulation by the liver and converted either to glutamate, to glutamine, or to urea. These trace levels of ammonia in blood contrast sharply with the more considerable quantities of free amino acids, particularly glutamine, in the blood (Table 15–1).

TABLE 15–1. Plasma amino acid concentrations after a 12-hour fast.*

Amino Acid	Concentration in Plasma (Mean ± SD) (mg/100 ml)
Glutamine	7.5 ± 1.6
Alanine	4.0 ± 1.5
Lysine	3.7 ± 1.2
Valine	3.2 ± 0.5
Cysteine + cystine	3.0 ± 0.2
Glycine	2.9 ± 1.4
Proline	2.6 ± 1.3
Leucine	2.5 ± 0.9
Arginine	2.3 ± 0.5
Histidine	2.1 ± 0.6
Threonine	2.1 ± 0.7
Isoleucine	2.0 ± 0.8
Phenylalanine	2.0 ± 0.8
Tryptophan	1.7 ± 0.5
Serine	1.4 ± 0.4
Tyrosine	1.3 ± 0.4
Glutamate	0.9 ± 0.4
Methionine	0.6 ± 0.3
Aspartate	0.3 ± 0.3

*Harper, Hutchin, & Kimmel, 1952.

Removal of ammonia via the **glutamate dehydrogenase** reaction was mentioned above. Formation of glutamine is catalyzed by **glutamine synthetase** (Fig 15–9), a mitochondrial enzyme present in highest quantities in renal tissue. Synthesis of the amide bond of glutamine is accomplished at the expense of hydrolysis of one equivalent of ATP to ADP and Pi. The reaction is thus strongly favored in the direction of glutamine synthesis. As noted above (Table 15–1), glutamine is the principal amino acid found in blood.

NH_3 + L-Glutamate → L-Glutamine
ATP, Mg^{++} → ADP+Pi, H_2O
GLUTAMINE SYNTHETASE

FIG 15–9. The glutamine synthetase reaction. The reaction strongly favors glutamine synthesis.

Liberation of the amide nitrogen of glutamine as ammonia occurs not by reversal of the glutamine synthetase reaction but by hydrolytic removal of ammonia catalyzed by **glutaminase** (Fig 15–8). The glutaminase reaction, unlike the glutamine synthetase reaction, does not involve participation of adenine nucleotides, strongly favors glutamate formation, and does not function in glutamine synthesis. These 2 enzymes, glutamine synthetase and glutaminase (Fig 15–10), serve to catalyze interconversion of free ammonium ion and glutamine in a manner reminiscent of the interconversion of glucose and glucose-6-phosphate by hexokinase and glucose-6-phosphatase (Fig 13–6). An analogous reaction is catalyzed by **L-asparaginase** of animal, plant, and microbial tissue. Asparaginase and glutaminase have both been employed as antitumor agents since certain tumors exhibit abnormally high requirements for glutamine and asparagine.

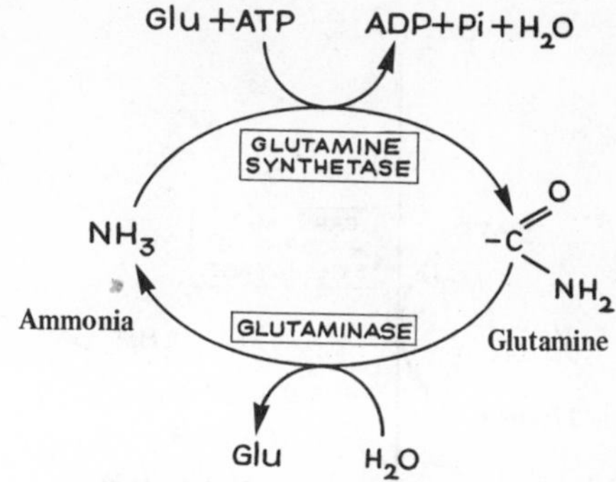

FIG 15–10. Interconversion of ammonia and of glutamine catalyzed by glutamine synthetase and glutaminase. Both reactions are strongly favored in the directions indicated by the arrows. Glutaminase thus serves solely for glutamine deamidation and glutamine synthetase solely for synthesis of glutamine from glutamate. (Glu = glutamate.)

Whereas in brain the major mechanism for removal of ammonia is glutamine formation, in the liver the most important pathway is urea formation. Brain tissue can form urea, although this does not play a significant role in ammonia removal. Formation of glutamine in the brain must be preceded by synthesis of glutamate in the brain itself because the supply of blood glutamate is inadequate to account for the increased amounts of glutamine formed in brain in the presence of high levels of blood ammonia. The immediate source of glutamate for this purpose is α-ketoglutarate. This would rapidly deplete the supply of citric acid cycle intermediates unless they could be replaced by CO_2 fixation with conversion of pyruvate to oxaloacetate (see p 247). A significant fixation of CO_2 into amino acids does indeed occur in the brain, presumably by way of the citric acid cycle, and after infusion of ammonia more oxaloacetate is diverted to the synthesis of glutamine (rather than to aspartate) via α-ketoglutarate.

Reaction of the Urea Cycle (Urea Synthesis)

The reactions and intermediates in biosynthesis of 1 mol of urea from 1 mol each of ammonia, carbon dioxide (activated with biotin and ATP), and of the α-amino nitrogen of aspartate are shown in Fig 15–11. The overall process requires 3 mols of ATP (2 of which are converted to ADP + Pi and 1 to AMP + PPi), and the successive participation of 5 enzymes catalyzing the numbered reactions of Fig 15–11. Of the 6 amino acids involved in urea synthesis, one, N-acetylglutam-

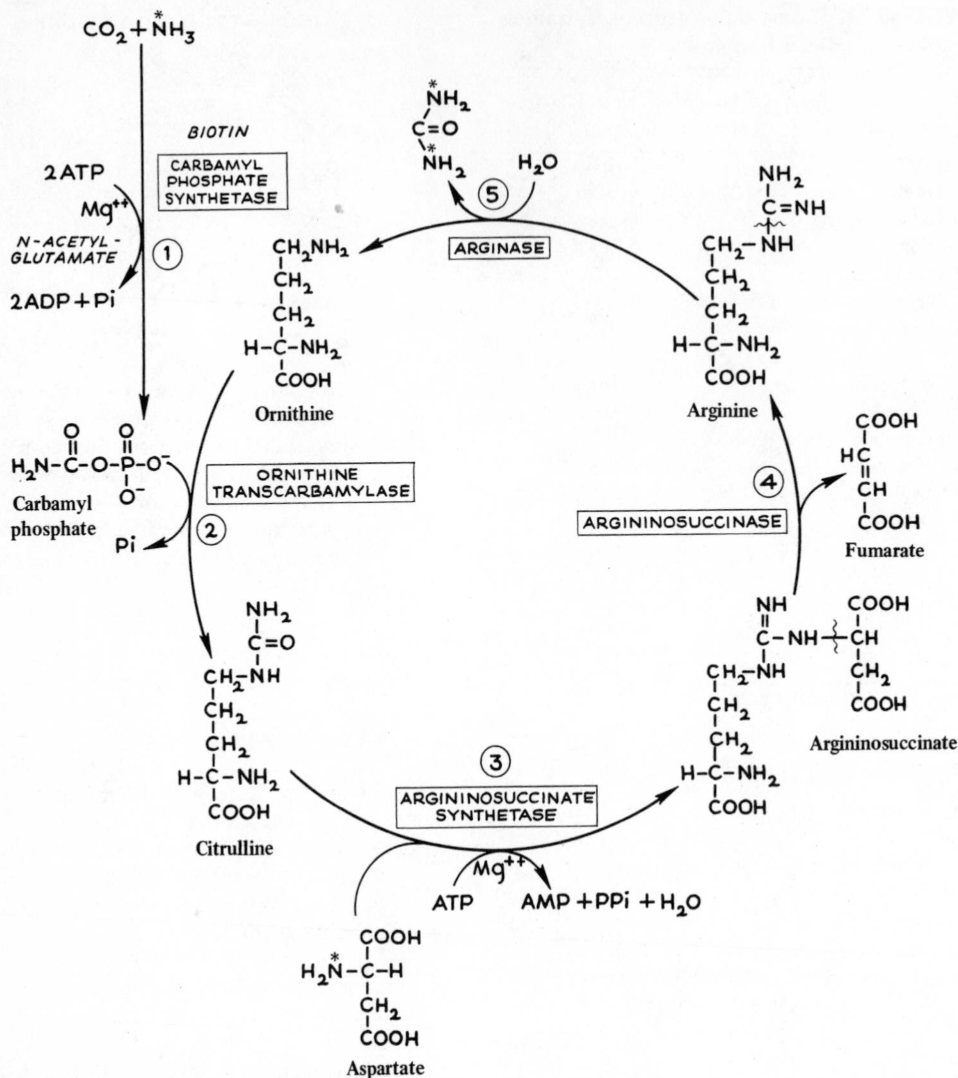

15–11. Reactions and intermediates of urea biosynthesis. The nitrogen atoms contributing to the formation of urea are starred.

ate, functions as an enzyme activator rather than as an intermediate. The remaining 5 amino acids—aspartate, arginine, ornithine, citrulline, and argininosuccinate—all function as carriers of atoms which ultimately become urea. Two (aspartate and arginine) occur in proteins, while the remaining 3 (ornithine, citrulline, and argininosuccinate) do not. The major metabolic role of these latter 3 amino acids in mammals is urea synthesis. Note that urea formation is in part a cyclical process. The ornithine used in reaction 2 is regenerated in reaction 5. There is thus no net loss or gain of ornithine, citrulline, argininosuccinate, or arginine during urea synthesis; however, ammonia, CO_2, ATP, and aspartate are consumed.

Reaction 1: Synthesis of carbamyl phosphate. Condensation of 1 mol each of ammonia, carbon dioxide (biotin-activated), and of phosphate (derived from ATP) to form carbamyl phosphate is catalyzed by **carbamyl phosphate synthetase**, an enzyme present in liver mitochondria of all ureotelic organisms, including man. The 2 mols of ATP hydrolyzed during this reaction provide the driving force for synthesis of 2 covalent bonds—the amide bond and the mixed carboxylic acid-phosphoric acid anhydride bond of carbamyl phosphate. In addition to Mg^{++}, a dicarboxylic acid, preferably N-acetylglutamate, is required. The exact role of N-acetylglutamate is not known with certainty. Its presence brings about a profound conformational change in the structure of carbamyl phosphate synthetase which exposes certain sulfhydryl groups, conceals others, and affects the affinity of the enzyme for ATP.

In bacteria, glutamine rather than ammonia serves as a substrate for carbamyl phosphate synthesis. A similar reaction catalyzed by carbamate kinase is also important in citrulline utilization by bacteria.

Reaction 2: Synthesis of citrulline. Transfer of a carbamyl moiety from carbamyl phosphate to ornithine, forming citrulline + Pi, is catalyzed by **L-ornithine transcarbamylase** of liver mitochondria. The reaction is highly specific for ornithine, and the equilibrium strongly favors citrulline synthesis.

Reaction 3: Synthesis of argininosuccinate. In the **argininosuccinate synthetase reaction,** aspartate and citrulline are linked together via the amino group of aspartate. The reaction requires ATP, and the equilibrium strongly favors argininosuccinate synthesis.

Reaction 4: Cleavage of argininosuccinate to arginine and fumarate. The reversible cleavage of argininosuccinate to arginine plus fumarate is catalyzed by **argininosuccinase**, a cold-labile enzyme of mammalian liver and kidney tissues. Loss of activity in the cold is associated with dissociation into 2 protein components. This dissociation is prevented by Pi, arginine, and argininosuccinate or by *p*-hydroxymercuribenzoate, which has no adverse effect on activity. The reaction proceeds via a **trans** elimination mechanism. The fumarate formed may be converted to oxaloacetate via the fumarase and malate dehydrogenase reactions (Fig 13–9) and then transaminated to regenerate aspartate.

Reaction 5: Cleavage of arginine to ornithine and urea. This reaction completes the urea cycle and regenerates ornithine, a substrate for reaction 2. Hydrolytic cleavage of the guanidino group of arginine is catalyzed by **arginase**, which is present in the livers of all ureotelic organisms (Buniatian & Davtian, 1966). Smaller quantities of arginase also occur in renal tissue, brain, mammary gland, testicular tissue, and skin. Highly purified arginase prepared from mammalian liver is activated by Co^{++} or Mn^{++}. Ornithine and lysine are potent inhibitors competitive with arginine.

Regulation of Urea Synthesis by Linkage of Glutamate Dehydrogenase with Carbamyl Phosphate Synthetase

Carbamyl phosphate synthetase is thought to act in conjunction with mitochondrial glutamate dehydrogenase to channel nitrogen from glutamate (and therefore from all amino acids; see Fig 15–2) into carbamyl phosphate and thus into urea. While the equilibrium constant of the glutamate dehydrogenase reaction favors glutamate rather than ammonia formation, removal of ammonia by the carbamyl phosphate synthetase reaction and oxidation of α-ketoglutarate by the citric acid cycle enzymes in the matrix of the mitochondrion serve to favor glutamate catabolism. This effect is enhanced by the presence of ATP, which, in addition to being substrate for carbamyl phosphate synthesis, stimulates glutamate dehydrogenase activity unidirectionally in the direction of ammonia formation.

Metabolic Fates of Carbamyl Phosphate

Carbamyl phosphate has 2 major metabolic fates in mammals. The first is urea synthesis. The second is synthesis of pyrimidines destined primarily for incorporation into nucleic acids (Fig 16–4). Taking carbamyl phosphate as a point of departure, the first enzyme of urea synthesis is **ornithine transcarbamylase** and the first enzyme of pyrimidine synthesis is **aspartate transcarbamylase.** Since carbamyl phosphate is a **branch point compound,** we might anticipate independent regulation of its metabolism at the levels of aspartate transcarbamylase and ornithine transcarbamylase (Fig 15–12).

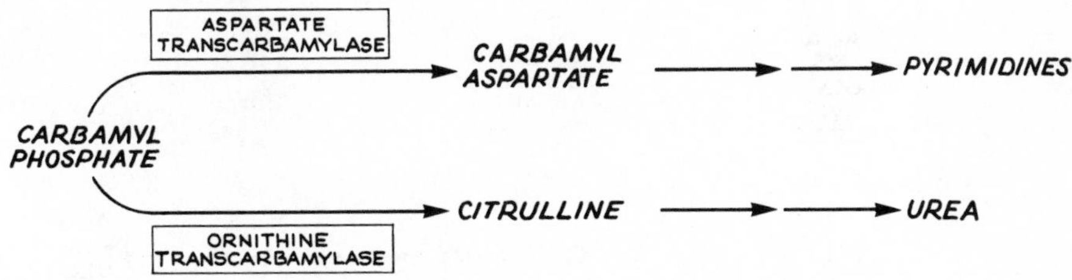

FIG 15–12. Metabolic fates of carbamyl phosphate.

Aspartate transcarbamylase is an allosteric, feedback-regulated enzyme in procaryotic cells (see p 356). Although similar evidence for regulation of ornithine transcarbamylase in eurcaryotic cells is lacking, independent regulation of these 2 enzymes is seen in regenerating mammalian liver tissue. Regeneration requires increased synthesis of nucleic acids, which would be facilitated by decreased use of carbamyl phosphate for urea synthesis. During regeneration, ornithine transcarbamylase levels decrease while aspartate transcarbamylase levels increase. The process may be viewed as an example of **biochemical dedifferentiation.** The urea cycle, a distinctive process of hepatic tissue, is an example of a biochemically differentiated system. Pyrimidine synthesis, which is necessary for all cell division, may be regarded as an undifferentiated process. When regeneration is complete, biochemical differentiation occurs, accompanied by a decrease in aspartate transcarbamylase levels and an increase in those of ornithine transcarbamylase. The physiologic factors regulating these processes are imperfectly understood at this time.

CONVERSION OF THE CARBON SKELETONS OF THE COMMON L-α-AMINO ACIDS TO AMPHIBOLIC INTERMEDIATES

This section deals with the conversion of the carbon skeletons of the common L-amino acids to amphibolic intermediates. A subsequent section will consider the conversion of the carbon skeletons or of the intact L-amino acids themselves to certain specialized products.

The conclusion that the carbon skeletons of each of the common amino acids are converted to amphibolic intermediates arose from the results of nutritional studies performed in the period 1920–1940. These data, reinforced and confirmed by studies using

TABLE 15–2. Amphibolic fates of the carbon skeletons of the common amino acids.

Carbon skeleton ultimately converted to amphibolic intermediates forming:			
Glycogen ("Glycogenic" Amino Acids)		Fat ("Ketogenic" Amino Acids)	Both Glycogen and Fat ("Glycogenic" and "Ketogenic" Amino Acids)
L-Alanine	L-Hydroxypoline	L-Leucine	L-Isoleucine
L-Arginine	L-Methionine		L-Lysine
L-Aspartate	L-Proline		L-Phenylalanine
L-Cystine	L-Serine		L-Tyrosine
L-Glutamate	L-Threonine		L-Tryptophan
L-Glycine	L-Valine		
L-Histidine			

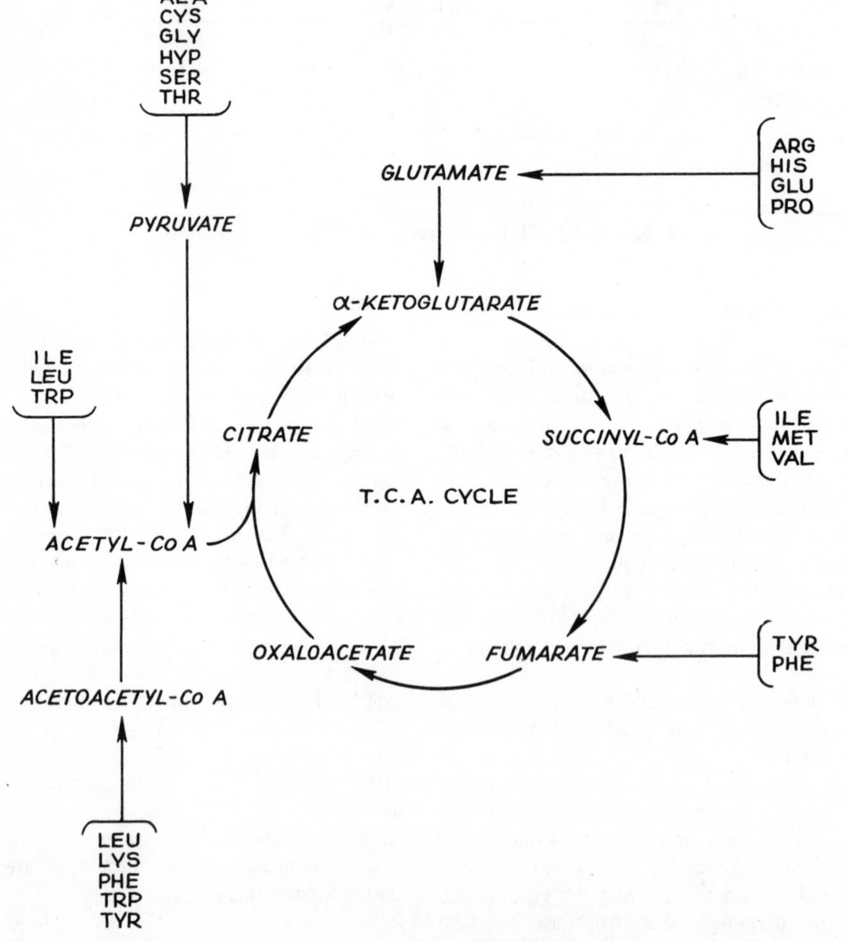

FIG 15–13. Amphibolic intermediates formed from the carbon skeleton of amino acids.

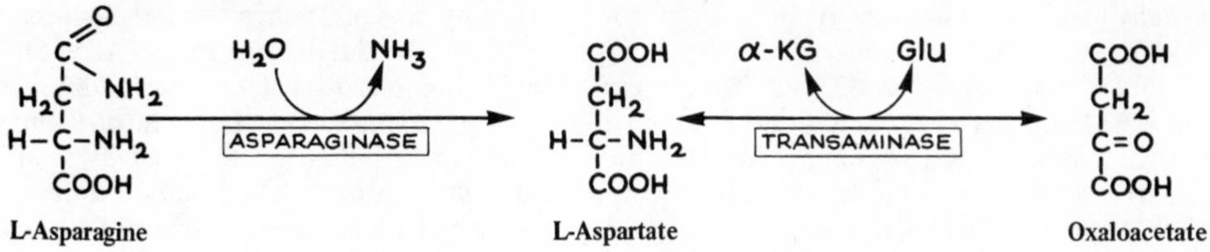

H_2O NH_3
ASPARAGINASE
α-KG Glu
TRANSAMINASE
L-Asparagine
L-Aspartate
Oxaloacetate

Fig 15–14. Catabolism of L-asparagine and of L-aspartate to oxaloacetate. (α-KG = α-ketoglutarate; Glu = glutamate.)

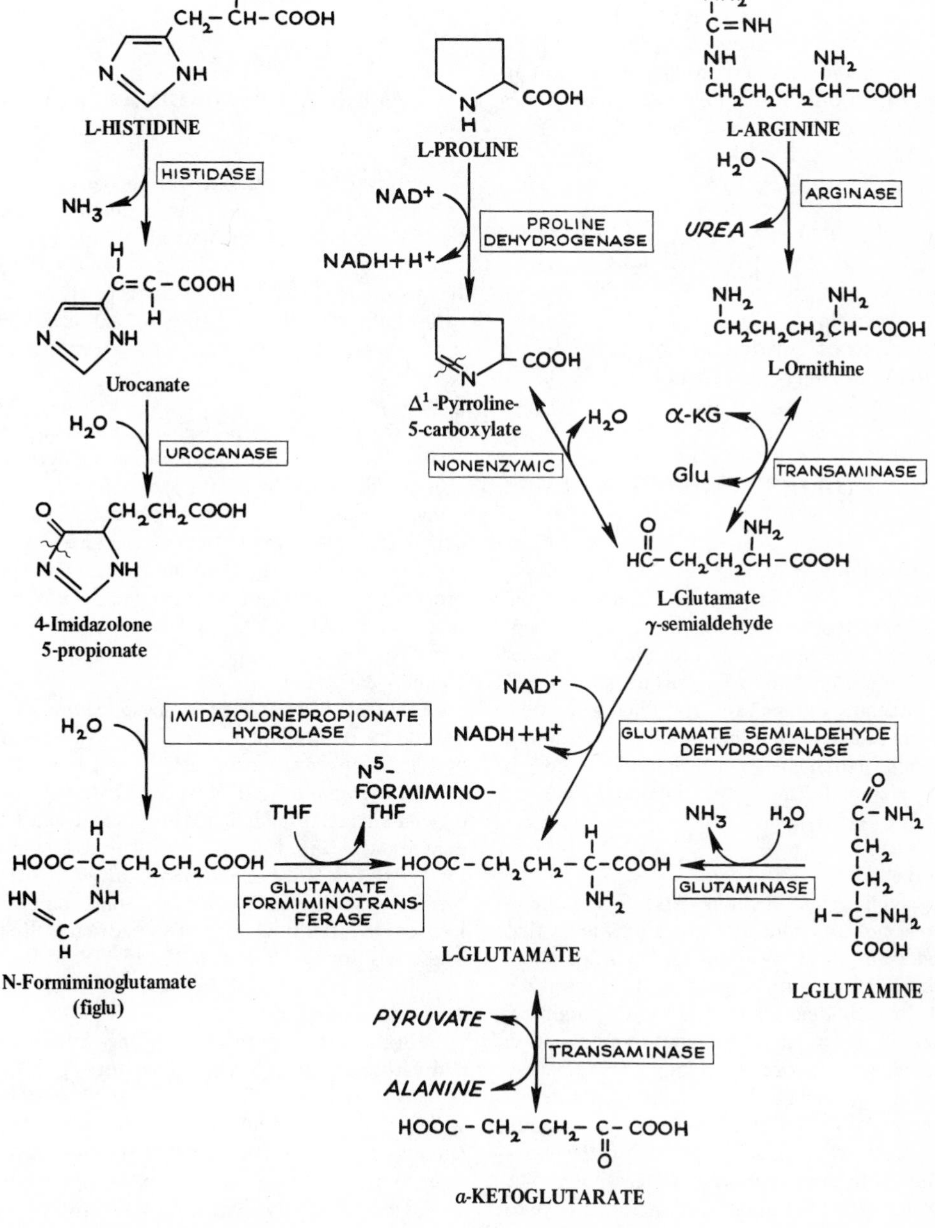

FIG 15–15. Catabolism of L-histidine, L-proline and L-arginine to α-ketoglutarate. (α-KG = α-ketoglutarate; Glu = glutamate; THF = tetrahydrofolic acid.)

isotopically labeled amino acids in the decade 1940–1950, supported the concept of the interconvertibility of fat, carbohydrate, and protein carbons and established that each amino acid is convertible either to carbohydrate (13 amino acids), fat (one amino acid), or both (5 amino acids) (Table 15–2). Although the knowledge of intermediary metabolism available at that time was incomplete and a detailed explanation of these interconversions was not possible, it was established that they indeed do occur.

How they occur is outlined in Fig 15–13. It will be noted that the carbon skeletons of all 13 "glycogenic" amino acids eventually form oxaloacetate, which is convertible to glycogen via phosphoenolpyruvate in the reactions of gluconeogenesis (Fig 13–1).

Individual amino acids are grouped for discussion on the basis of the principal amphibolic intermediates formed as end products of their catabolism.

AMINO ACIDS FORMING OXALOACETATE

Aspartate & Asparagine

All 4 carbons of aspartate and of asparagine are converted to oxaloacetate as shown in Fig 15–14).

AMINO ACIDS FORMING α-KETOGLUTARATE

Glutamate, Glutamine, & Proline

All 5 carbon atoms of glutamate, glutamine and proline are converted to oxaloacetate. Catabolism of glutamate and glutamine proceeds in a manner analogous to the conversion of aspartate and asparagine to oxaloacetate. Proline catabolism involves oxidation to a form of dehydroproline which on addition of water forms glutamate γ-semialdehyde. This is oxidized to glutamate and transaminated to α-ketoglutarate (Fig 15–15).

Arginine & Histidine

While arginine and histidine also form α-ketoglutarate, one carbon and 2 nitrogen atoms must first be removed from these 6-carbon amino acids. With arginine, this requires but a single step: the hydrolytic removal of the guanidino group catalyzed by arginase. The product, ornithine, then undergoes transamination of the δ-amino group, forming glutamate γ-semialdehyde, which is converted to α-ketoglutarate as described above for proline (Fig 15–15).

In the case of histidine, removal of the extra carbon and nitrogens requires 4 reactions. The product, glutamate, by transamination forms α-ketoglutarate. Deamination of histidine produces urocanate, so named because it was first detected as a histidine catabolite in the urine of dogs. The conversion of urocanate to 4-imidazolone-5-propionate, catalyzed by urocanase, involves both addition of H_2O and an internal oxidation-reduction. Although 4-imidazolone-5-propionate may undergo additional fates, conversion to α-ketoglutarate involves hydrolysis to N-formiminoglutamate followed by transfer of the formimino group on the α-carbon to tetrahydrofolate, forming N^5-formiminotetrahydrofolate (see p 110). In patients suffering from folic acid deficiency, this last reaction is partially or totally blocked and N-formiminoglutamate (figlu) is excreted in the urine. This forms the basis for a test for folic acid deficiency in which N-formiminoglutamate is detected in the urine following a large dose of histidine (see p 111).

AMINO ACIDS FORMING PYRUVATE

Conversion of the carbon skeletons of alanine, cysteine, cystine, glycine, threonine, and serine to pyruvate is shown diagramatically in Fig 15–16. Pyruvate may then be converted to acetyl-Co A (Fig 13–7). In this process, the 2 carbon atoms of glycine and all 3 carbon atoms of alanine, cysteine, and serine–but only 2 of the carbon atoms of threonine–form pyruvate.

Alanine & Serine

Two pyridoxal phosphate-dependent reactions suffice to convert alanine or serine to pyruvate: For alanine, the reaction is transamination (Fig 15–17). Conversion of serine to pyruvate, catalyzed by serine dehydratase, involves both addition and loss of water and, in addition, loss of ammonia. This probably proceeds via formation of an imino acid intermediate as shown in Fig 15–17.

Cysteine & Cystine

Cystine is converted to cysteine by an NADH-dependent oxidoreductase (Fig 15–18). Conversion of cysteine to pyruvate may then occur in any of 3 ways: (1) Via cysteine desulfhydrase, a pyridoxal phosphate-dependent reaction similar to that catalyzed by serine dehydratase (Fig 15–17). (2) By transamination and loss of H_2S. (3) By oxidation of the sulfhydryl group forming cysteine sulfinic acid, transamination, and loss of the terminal carbon's oxidized sulfur atom. These reactions are shown in Fig 15–19.

Threonine & Glycine

Both carbon atoms of glycine–but only the α- and carboxyl atoms of threonine–form pyruvate. The β- and γ-carbons of threonine form acetaldehyde, which is directly convertible to acetyl-Co A. The α- and carboxyl carbons are converted first to glycine and then to serine. Serine is then converted to pyruvate as shown in Fig 15–17. The net result of these reactions is that both carbon atoms of glycine and all 4 carbon atoms of threonine eventually form acetyl-Co A.

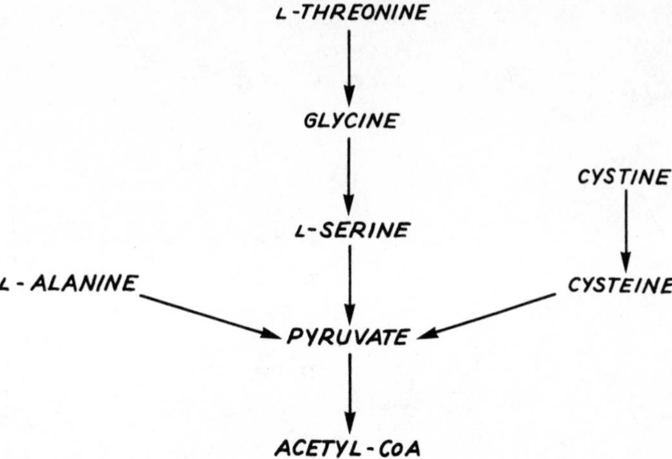

FIG 15–16. Diagramatic representation of the conversion of the carbon skeletons of alanine, cystine, cysteine, threonine, glycine, and serine to pyruvate and to acetyl-Co A.

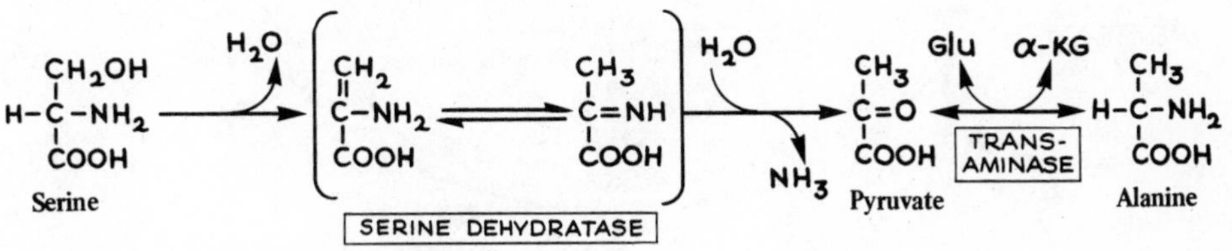

H_2O H_2O Glu α-KG

CH_2OH / $H-C-NH_2$ / $COOH$ → [CH_2 / $C-NH_2$ / $COOH$ ⇌ CH_3 / $C=NH$ / $COOH$] → CH_3 / $C=O$ / $COOH$ ⇌ CH_3 / $H-C-NH_2$ / $COOH$

NH_3

Serine Pyruvate Alanine

SERINE DEHYDRATASE TRANS-AMINASE

FIG 15–17. Conversion of alanine and serine to pyruvate. Both the alanine transaminase and serine dehydratase reactions require pyridoxal phosphate (PLP) as a coenzyme. The bracketed intermediates in the serine dehydratase reaction are hypothetical. This reaction may be thought of as proceeding via elimination of H_2O from serine forming an unsaturated amino acid. This rearranges to an α-imino acid which is spontaneously hydrolyzed to pyruvate plus ammonia. There is thus no net gain or loss of water during the serine dehydratase reaction. (Glu = glutamate, α-KG = α-ketoglutarate.)

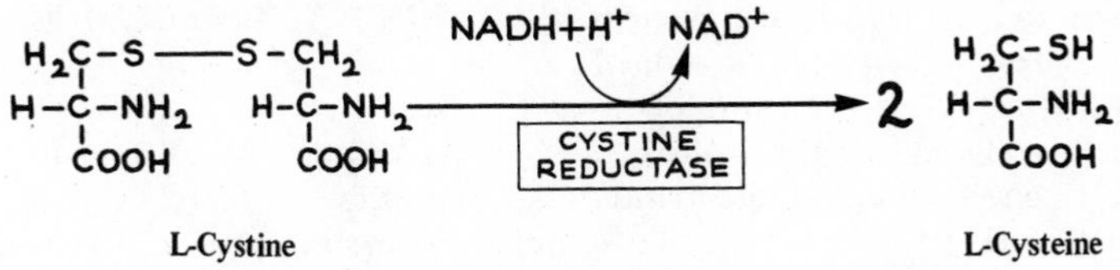

NADH+H$^+$ NAD$^+$

H_2C-S——$S-CH_2$ / $H-C-NH_2$ $H-C-NH_2$ / $COOH$ $COOH$ → 2 H_2C-SH / $H-C-NH_2$ / $COOH$

CYSTINE REDUCTASE

L-Cystine L-Cysteine

FIG 15–18. The cystine reductase reaction.

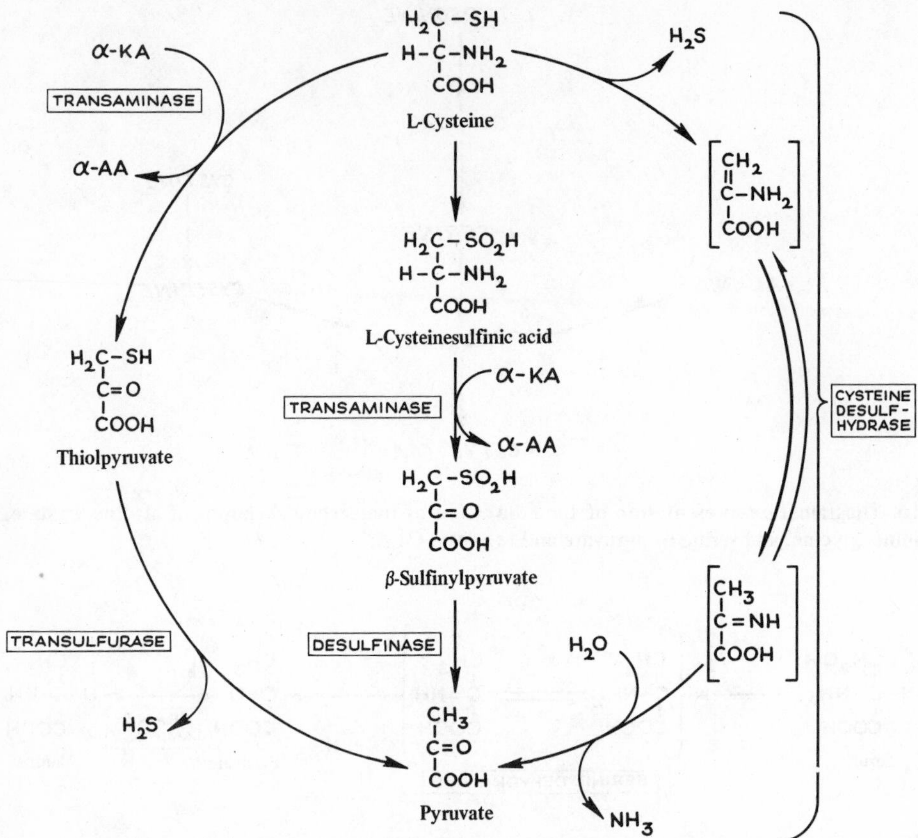

FIG 15–19. Conversion of cysteine to pyruvate. Bracketed intermediates in the cysteine disulfhydrase reaction are hypothetical. The reaction may be thought of as proceeding in a manner analogous to that catalyzed by serine dehydratase (Fig 15–17) except that the initial step involves loss of H_2S rather than H_2O. (a-AA = a-amino acid; a-KA = a-keto acid.)

Threonine is cleaved to acetaldehyde and glycine by **threonine aldolase.** Glycine then accepts a one-carbon moiety from $N^{5,10}$-methylene tetrahydrofolate (see p 110) in a reaction catalyzed by serine hydroxymethylase. Serine is then converted to pyruvate by the serine dehydratase reaction (Fig 15–17). Both pyruvate and acetaldehyde then form acetyl-Co A (Fig 15–20).

Hydroxyproline

Three of the 5 carbons of 4-hydroxy-L-proline are converted to pyruvate by the reactions shown in Fig 15–21. The remaining 2 form glyoxylate. A mitochondrial dehydrogenase catalyzes conversion of hydroxyproline to L-Δ^1-pyrroline-3-hydroxy-5-carboxylate. This is in nonenzymic equilibrium with γ-hydroxy-L-glutamate-γ-semialdehyde, which is formed by addition of 1 mol of water. The semialdehyde is oxidized to the corresponding carboxylic acid, erythro-γ-hydroxyglutamate, and transaminated to a-keto-γ-hydroxyglutarate. An aldol type cleavage then forms glyoxylate plus pyruvate.

AMINO ACIDS FORMING ACETYL-CO A

As shown above, all amino acids forming pyruvate are convertible to acetyl-Co A. In addition to alanine, cysteine, cystine, glycine, serine, and threonine, which form pyruvate prior to acetyl-Co A, 5 amino acids form acetyl-Co A directly without first forming pyruvate. These include the aromatic amino acids, phenylalanine, tyrosine, and tryptophan; the basic amino acid, lysine; and the neutral, branched-chain amino acid, leucine.

Conversion of Phenylalanine to Tyrosine

Conversion of phenylalanine to tyrosine was inferred from nutritional experiments which showed that sufficiently high levels of dietary phenylalanine can replace the requirement for dietary tyrosine. The reaction is not reversible, so that tyrosine cannot replace the nutritional requirement for phenylalanine. Direct evidence was later provided when it was shown that intact rats converted ^{2}H-DL-phenylalanine to

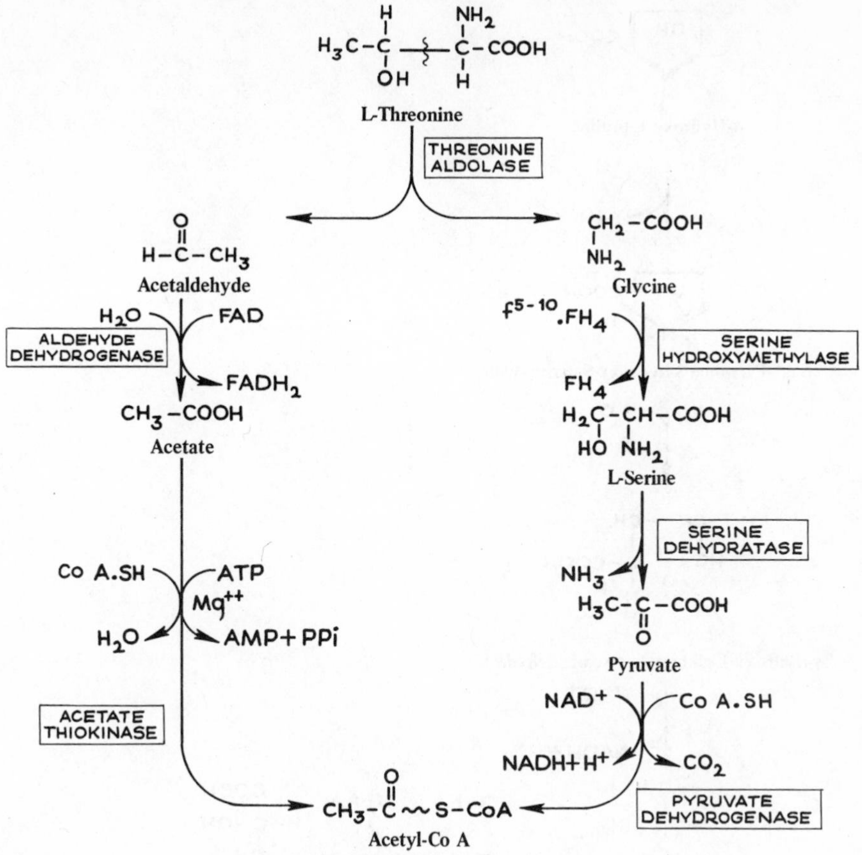

L-Threonine
THREONINE ALDOLASE
Acetaldehyde
Glycine
ALDEHYDE DEHYDROGENASE
H_2O FAD $FADH_2$
Acetate
$f^{5-10}\cdot FH_4$ FH_4
SERINE HYDROXYMETHYLASE
L-Serine
SERINE DEHYDRATASE
NH_3
Co A.SH ATP Mg^{++} H_2O AMP+ PPi
Pyruvate
ACETATE THIOKINASE
NAD^+ Co A.SH $NADH+H^+$ CO_2
PYRUVATE DEHYDROGENASE
Acetyl-Co A

FIG 15–20. Conversion of threonine and glycine to serine, pyruvate, and acetyl-Co A. For details of the pyruvate decarboxylase reaction, see p 106. ($f^{5-10}\cdot FH_4$ = formyl [5–10] tetrahydrofolic acid.)

^{2}H-L-tyrosine. Conversion of phenylalanine to tyrosine is catalyzed by the **phenylalanine hydroxylase** complex, a mixed function oxygenase present in mammalian liver but absent from other tissues. The overall reaction involves incorporation of one atom of molecular oxygen into the para position of phenylalanine while the other atom is reduced, forming water (Fig 15–22). The reducing power, supplied ultimately by NADPH, is immediately provided in the form of tetrahydrobiopterin, a pteridine resembling that in folic acid (see p 109 and Fig 15–23).

Conversion of Tyrosine to Acetoacetate & Fumarate

Five enzymatically catalyzed reactions are involved in conversion of tyrosine to fumarate and to acetoacetate: (1) transamination to *p*-hydroxyphenylpyruvate; (2) simultaneous oxidation and migration of the 3-carbon side chain and decarboxylation, forming homogentisate; (3) oxidation of homogentisate to maleylacetoacetate; (4) isomerization of maleylacetoacetate to fumarylacetoacetate; and (5) hydrolysis of fumarylacetoacetate to fumarate and acetoacetate. Acetoacetate may then undergo thiolytic cleavage to acetate plus acetyl-Co A. These reactions are shown in Figs 15–24 and 15–25.

The intermediates in tyrosine metabolism were discovered in part as a result of studies of the human genetic disease, alkaptonuria. Patients with alkaptonuria excrete substantial quantities of homogentisate in their urine, and much useful information was obtained by feeding suspected precursors of homogentisate to patients suffering from this disease. Early difficulties arising from the instability of several of the intermediates were resolved by the discovery that α-ketoglutarate and ascorbate were required for tyrosine oxidation by liver extracts. Subsequently, each of the individual enzymic steps was studied in detail.

Transamination of Tyrosine

Transamination of tyrosine to *p*-hydroxyphenylpyruvate is catalyzed by **tyrosine-α-ketoglutarate transaminase**, an inducible enzyme of mammalian liver tissue.

Oxidation of *p*-Hydroxyphenylpyruvate to Homogentisate

Although the reaction shown in Fig 15–24 appears to involve hydroxylation of *p*-hydroxyphenylpyruvate in the ortho position accompanied by oxidative loss of the carboxyl carbon, the reaction also

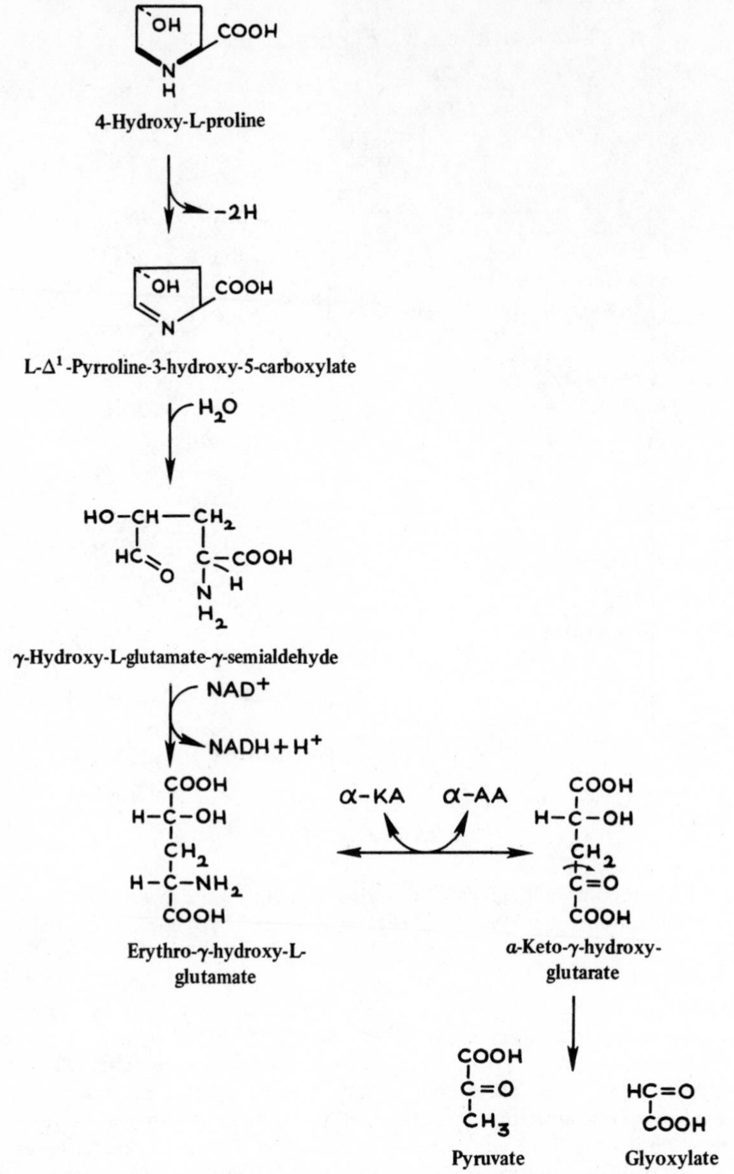

4-Hydroxy-L-proline

$-2H$

L-Δ^1-Pyrroline-3-hydroxy-5-carboxylate

H_2O

γ-Hydroxy-L-glutamate-γ-semialdehyde

NAD^+

$NADH + H^+$

α-KA α-AA

Erythro-γ-hydroxy-L-glutamate

α-Keto-γ-hydroxy-glutarate

Pyruvate

Glyoxylate

FIG 15–21 Intermediates in L-hydroxyproline catabolism in mammalian tissues. (α-KA = α-keto acid; α-AA = α-amino acid).

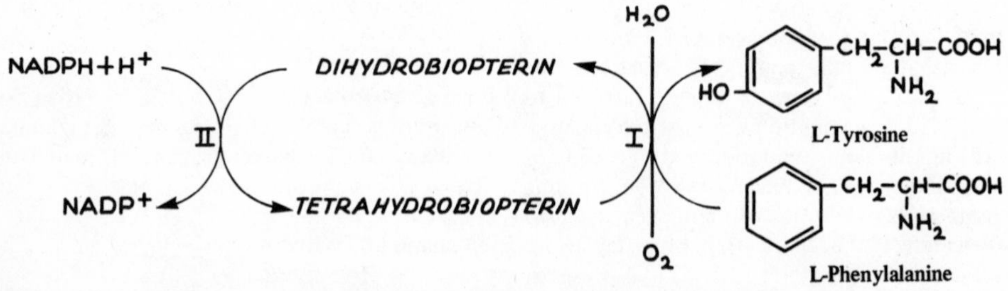

$NADPH + H^+$

$NADP^+$

II

DIHYDROBIOPTERIN

TETRAHYDROBIOPTERIN

I

H_2O

O_2

L-Tyrosine

L-Phenylalanine

FIG 15–22. The phenylalanine hydroxylase reaction. Two distinct enzymic activities are involved. Activity II catalyzes reduction of dihydrobiopterin by NADPH, and activity I the reduction of oxygen to water and conversion of phenylalanine to tyrosine.

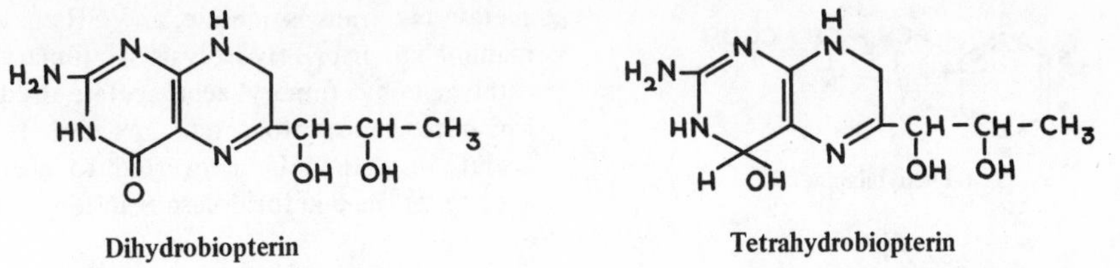

Dihydrobiopterin

Tetrahydrobiopterin

FIG 15–23. Structures of dihydrobiopterin and of tetrahydrobiopterin.

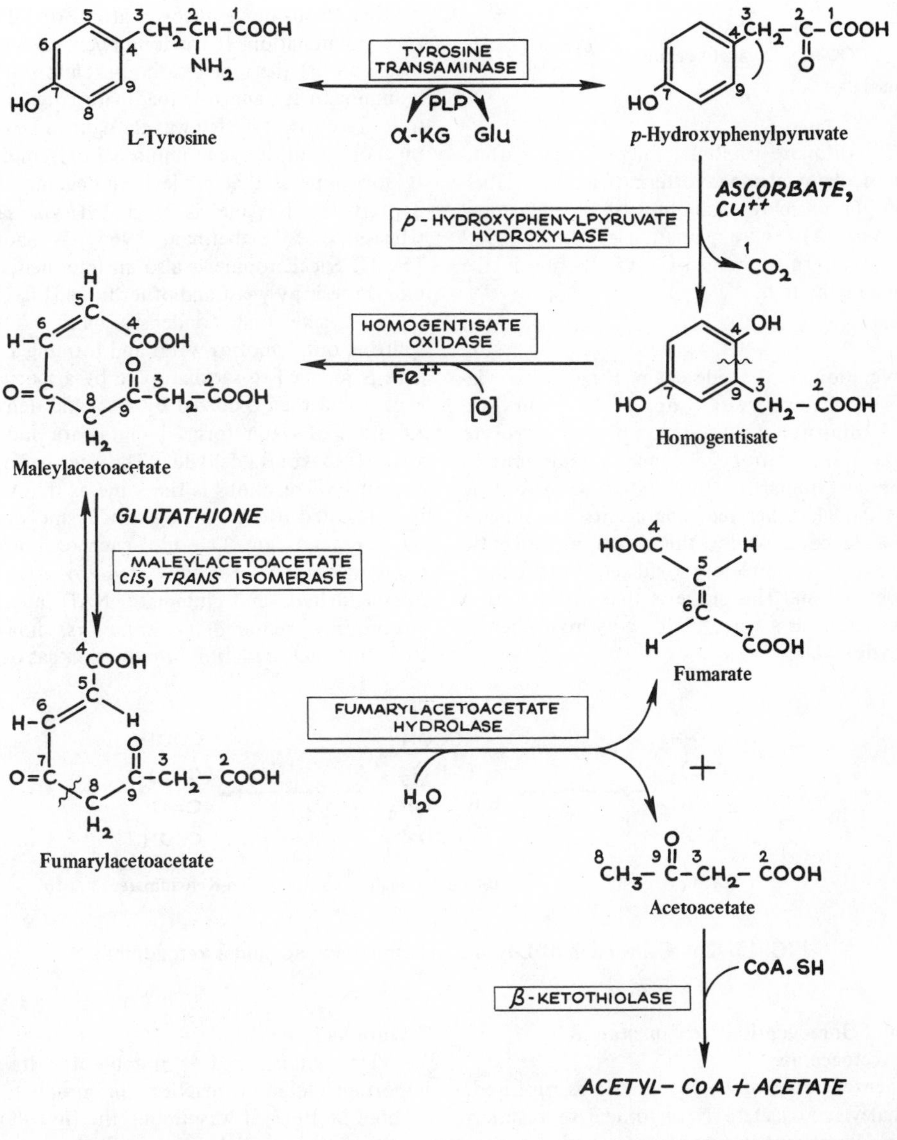

L-Tyrosine

TYROSINE TRANSAMINASE

PLP

α-KG Glu

p-Hydroxyphenylpyruvate

p-HYDROXYPHENYLPYRUVATE HYDROXYLASE

ASCORBATE, Cu^{++}

CO_2

Homogentisate

HOMOGENTISATE OXIDASE

Fe^{++}

[O]

Maleylacetoacetate

GLUTATHIONE

MALEYLACETOACETATE *CIS, TRANS* ISOMERASE

Fumarylacetoacetate

FUMARYLACETOACETATE HYDROLASE

H_2O

Fumarate

+

Acetoacetate

β-KETOTHIOLASE

CoA.SH

ACETYL– CoA + ACETATE

FIG 15–24. Intermediates in conversion of tyrosine to acetoacetate and fumarate. With the exception of β-ketothiolase (Fig 14–3), the reactions are discussed in the text. Certain of the carbon atoms of the intermediates are numbered to assist the reader in observing the ultimate fate of each carbon atom (see also Fig 15–25). (α-KG = α-ketoglutarate; Glu = glutamate; PLP = pyridoxal [B_6] phosphate.)

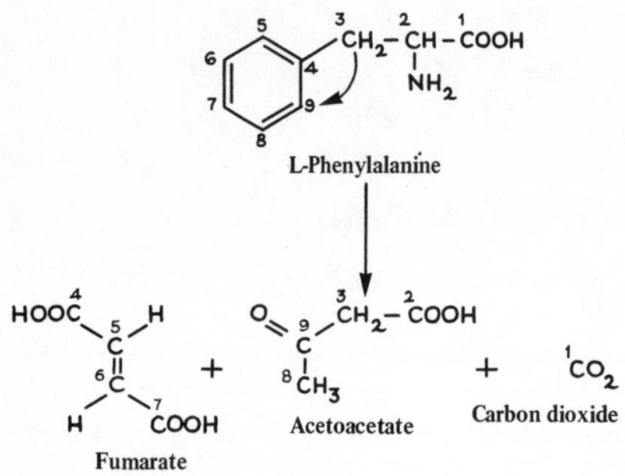

L-Phenylalanine

Fumarate + Acetoacetate + Carbon dioxide

FIG 15–25. Ultimate catabolic fate of each carbon atom of phenylalanine. Pattern of isotopic labeling in the ultimate catabolites of phenylalanine (and tyrosine). The explanation of the observed labeling pattern is given in Fig 15–24 and in the accompanying text.

involves migration of the side chain. Ring hydroxylation and side chain migration appear to occur in a concerted manner. ***p*-Hydroxyphenylpyruvate hydroxylase** appears to be a copper metalloprotein with properties similar to those of **tyrosinase** (Fig 15–59). Although other reducing agents can replace ascorbate as a cofactor for this reaction, scorbutic patients excrete incompletely oxidized products of tyrosine metabolism. This suggests that one function of ascorbic acid is as a cofactor for *p*-hydroxyphenylpyruvate hydroxylase.

acetate cis, trans isomerase, an –SH enzyme present in mammalian liver. Hydrolysis of fumarylacetoacetate, catalyzed by **fumaryl-acetoacetate hydrolase**, forms fumarate and acetoacetate. As noted above, acetoacetate can then be converted to acetyl-Co A plus acetate by the β-ketothiolase reaction.

Lysine

Lysine provides an exception to the general observation that the first step in catabolism of an amino acid is removal of its α-amino group by transamination. Neither the α- nor ϵ-nitrogen atoms of L-lysine undergo transamination. It has long been known that mammals convert the intact carbon skeleton of L-lysine to α-aminoadipate and α-ketoadipate (Fig 15–26). Until quite recently, L-lysine was thought to be degraded via pipecolic acid, a cyclic imino acid resembling proline. It now appears that while liver degrades D-lysine via pipecolate, L-lysine is degraded via saccharopine (Higashino & Lieberman, 1965) as shown in Fig 15–27. Saccharopine is also an intermediate in lysine biosynthesis by yeast and other fungi (Fig 15–44).

L-Lysine first condenses with α-ketoglutarate, splitting out 1 mol of water and forming a Schiff base. This is reduced to saccharopine by a specific dehydrogenase and then oxidized by a second dehydrogenase. Addition of water forms L-glutamate and L-α-aminoadipate-δ-semialdehyde. The net effect of this sequence of reactions is the same as that which would have resulted if the ϵ-nitrogen of lysine were removed by transamination. One mol each of L-lysine and of α-ketoglutarate are converted to α-aminoadipate-δ-semialdehyde and glutamate. NAD^+ and NADH are specifically required as cofactors, however, even though no net oxidation or reduction has occurred.

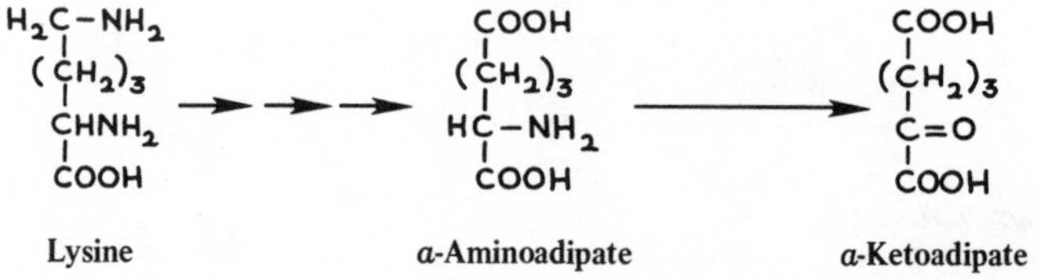

Lysine → → → α-Aminoadipate → α-Ketoadipate

FIG 15–26. Conversion of L-lysine to α-aminoadipate and α-ketoadipate.

Conversion of Homogentisate to Fumarate & Acetoacetate

The benzene ring of homogentisate is ruptured, forming maleylacetoacetate in an oxidative reaction catalyzed by **homogentisate oxidase**, an iron metalloprotein of mammalian liver. The reaction is inhibited by α-, α'-dipyridyl, a chelating agent that strongly binds iron. Treatment with α-, α'-dipyridyl has been used to induce alkaptonuria in experimental animals.

Conversion of maleylacetoacetate to fumarylacetoacetate, which involves cis to trans isomerization about the double bond, is catalyzed by **maleylaceto-**

Tryptophan

This amino acid is notable for its variety of important metabolic reactions and products. Originally isolated in 1901, it was among the first shown to be a nutritionally essential amino acid. Neurospora mutants and the bacterium Pseudomonas have proved invaluable aids in unraveling the details of tryptophan metabolism. Isolation of certain tryptophan metabolites from urine has also contributed valuable information.

Although a large portion of the isotope of administered ^{14}C-L-tryptophan is retained by the tissue proteins, a considerable fraction appears in the urine in

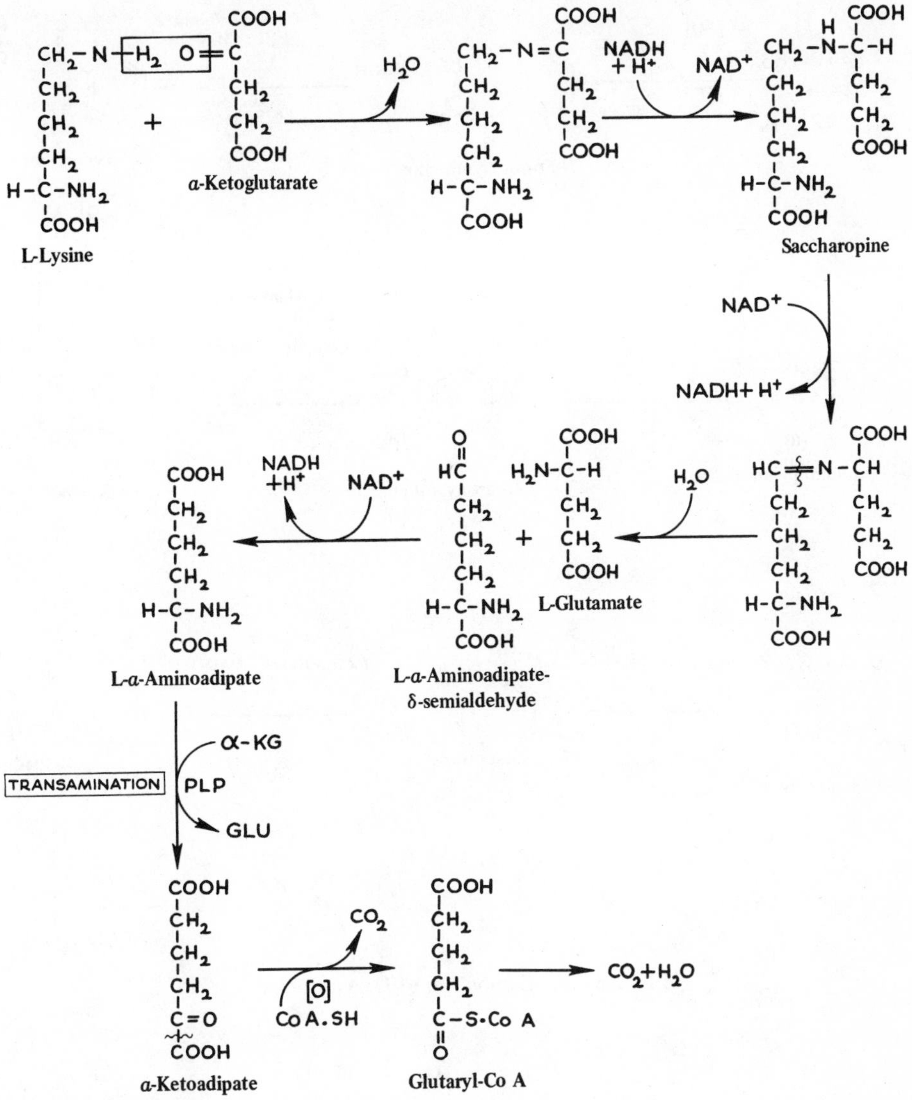

FIG 15–27. Catabolism of L-lysine. (α-KG = α-ketoglutarate; Glu = glutamate; PLP = pyridoxal phosphate.)

the form of various catabolites. The carbon atoms both of the side chain and of the aromatic ring may be completely degraded to amphibolic intermediates. This proceeds via what is known as the **kynurenine-anthranilate pathway** (Fig 15–28). The pathway is important not only for tryptophan degradation but also for conversion of tryptophan to niacin.

Tryptophan oxygenase (perhaps better known as **tryptophan pyrrolase**) catalyzes the cleavage of the indole ring with incorporation of 2 atoms of molecular oxygen, forming N-formylkynurenine. The oxygenase enzyme is an iron porphyrin metalloprotein which occurs in the liver of mammals, amphibians, birds, and insects and which has been obtained in highly purified form from liver and from Pseudomonas. Four forms of hepatic tryptophan pyrrolase have been described: the active holoenzyme, the apoenzyme, a third form that is combined with hematin, and a fourth that requires prolonged incubation or addition of small quantities of cell particles in order to be activated by hematin. Tryptophan pyrrolase is an inducible enzyme in liver. The chief inducing agents appear to be adrenal corticosteroids (eg, hydrocortisone) and tryptophan itself. De novo synthesis of tryptophan pyrrolase has been demonstrated by immunologic technics. Induction is blocked by administration of puromycin or dactinomycin. Tryptophan serves also to stabilize the enzyme toward proteolytic degradation. A considerable portion of newly synthesized enzyme is in a latent form that requires activation.

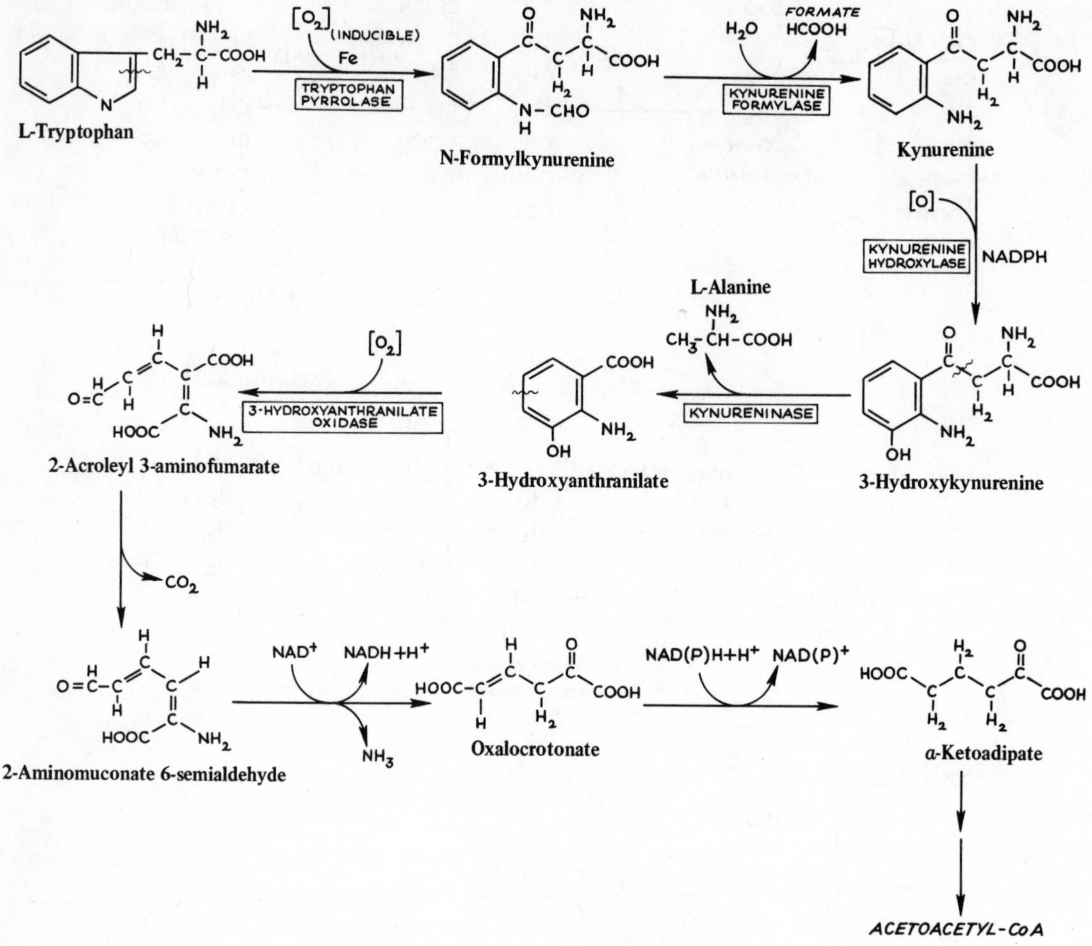

L-Tryptophan

$[O_2]$ (INDUCIBLE) Fe

TRYPTOPHAN PYRROLASE

N-Formylkynurenine

H_2O FORMATE HCOOH

KYNURENINE FORMYLASE

Kynurenine

[O] NADPH

KYNURENINE HYDROXYLASE

3-Hydroxykynurenine

L-Alanine

KYNURENINASE

3-Hydroxyanthranilate

$[O_2]$

3-HYDROXYANTHRANILATE OXIDASE

2-Acroleyl 3-aminofumarate

CO_2

2-Aminomuconate 6-semialdehyde

NAD^+ $NADH+H^+$ NH_3

Oxalocrotonate

$NAD(P)H+H^+$ $NAD(P)^+$

α-Ketoadipate

ACETOACETYL-CoA

FIG 15–28. Catabolism of L-tryptophan.

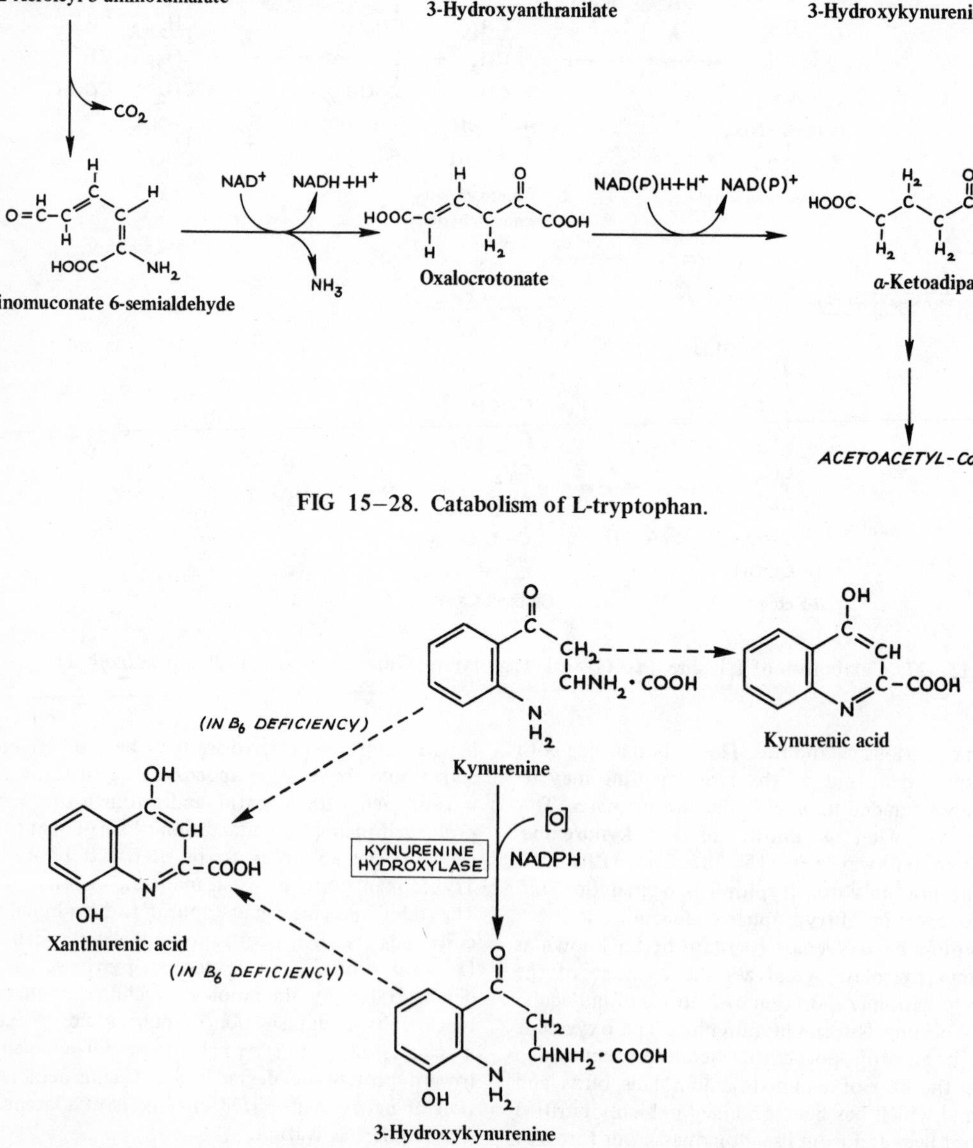

Kynurenine

Kynurenic acid

(IN B_6 DEFICIENCY)

Xanthurenic acid

KYNURENINE HYDROXYLASE

[O] NADPH

(IN B_6 DEFICIENCY)

3-Hydroxykynurenine

FIG 15–29. Conversion of kynurenine and hydroxykynurenine to xanthurenic acid in vitamin B_6 deficiency.

Hydrocortisone-induced activation of tryptophan pyrrolase consists of 2 steps: (1) conjugation of the apoenzyme with hematin, forming oxidized holoenzyme; and (2) reduction of the oxidized holoenzyme. The first step requires the presence of L-tryptophan or an analogue (eg, ω-methyltryptophan). The second step is promoted by L-tryptophan and by ascorbate.

It has been observed that tryptophan analogues that promoted reaction 2 were inducers of the enzyme. Tryptophan analogues thus appear to induce higher levels of tryptophan pyrrolase in liver, both by promoting synthesis of new protein and by stabilizing existing enzyme against degradation. Tryptophan pyrrolase is subject to feedback inhibition by a variety of nicotinic acid derivatives, including NADPH.

Hydrolytic removal of the formyl group of N-formylkynurenine is catalyzed by **kynurenine formylase** of mammalian liver. When hydrolysis is performed in the presence of $H_2{}^{18}O_2$, one equivalent of ^{18}O is incorporated into the formate formed. The enzyme is not specific for N-formylkynurenine and will catalyze similar reactions with a variety of arylformylamines.

The reaction catalyzed by kynurenine formylase produces **kynurenine** (Fig 15–28). This may be deaminated by transamination of the amino group of the side chain to ketoglutarate. The resulting keto derivative, *o*-aminobenzoyl pyruvate, loses water and then undergoes spontaneous ring closure to form **kynurenic acid.** This compound is actually a byproduct of kynurenine; it is not formed in the main pathway of tryptophan breakdown shown in Fig 15–28.

The further metabolism of kynurenine involves its conversion to hydroxykynurenine, which in turn is converted to 3-hydroxyanthranilate. The hydroxylation occurs with molecular oxygen in an NADPH-catalyzed reaction similar to that for the hydroxylation of phenylalanine to tyrosine previously described above.

The reaction by which kynurenine and hydroxykynurenine are converted to hydroxyanthranilate is catalyzed by the enzyme **kynureninase**, which requires vitamin B_6 (pyridoxal phosphate) as coenzyme. A deficiency of vitamin B_6 results in some degree of failure to catabolize these kynurenine derivatives, which thus reach various extrahepatic tissues where they are converted to **xanthurenic acid** (Fig 15–29). This abnormal metabolite has been identified in the urine of humans, monkeys, and rats when dietary intakes of vitamin B_6 were inadequate. The feeding of excess tryptophan can be used to induce excretion of xanthurenic acid if vitamin B_6 deficiency exists. The kidney is one organ which has been shown to produce xanthurenic acid derivatives from kynurenine.

AMINO ACIDS FORMING SUCCINYL-CO A

Whereas succinyl-Co A represents the amphibolic end product for the catabolism of methionine, isoleucine, and valine, only a portion of their carbon skeletons are in fact converted (Fig 15–30). Four-fifths of the carbons of valine–but only 3/5 of those of methionine and only ½ of those of isoleucine–contribute to the formation of succinyl-Co A. The carboxyl carbons of all 3 amino acids form CO_2, whereas the terminal 2 carbons of isoleucine form acetyl-Co A. In addition, the S-methyl group of methionine is removed as such.

The reactions leading from propionyl-Co A through methylmalonyl-Co A to succinyl-Co A–already discussed in connection with the catabolism of propionate and of fatty acids containing an odd number of carbon atoms (Fig 13–14)–will not be further discussed here. What follows relates only to the conversion of methionine and isoleucine to propionyl-Co A and of valine to methylmalonyl-Co A.

Methionine

The intermediates formed during conversion of methionine to propionyl-Co A are shown in Fig 15–31. L-Methionine first condenses with ATP, forming S-adenosylmethionine ("active methionine"; Fig 15–32). The now activated S-methyl group is then

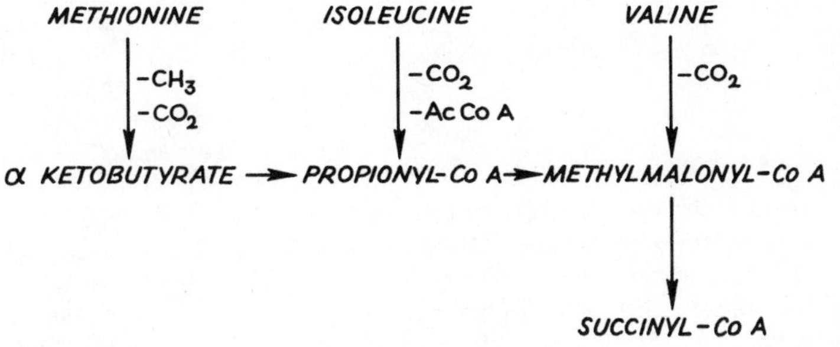

METHIONINE
$-CH_3$
$-CO_2$
ISOLEUCINE
$-CO_2$
$-$AcCo A
VALINE
$-CO_2$
α KETOBUTYRATE → PROPIONYL-Co A → METHYLMALONYL-Co A
SUCCINYL-Co A

FIG 15–30. Overall catabolism of methionine, isoleucine, and valine. Conversion to succinyl-Co A. (AcCo A = acetyl-Co A.)

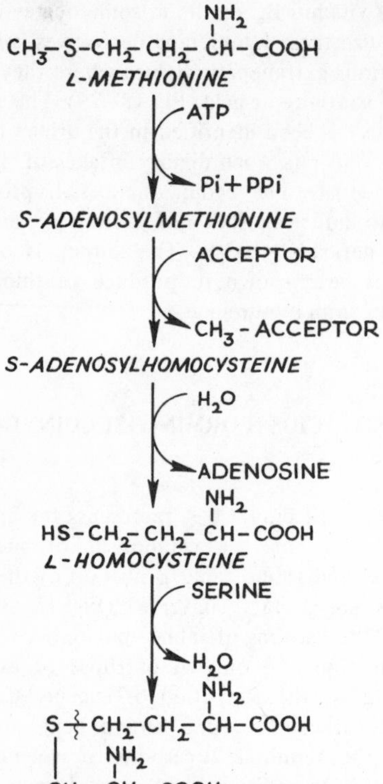

$CH_3-S-CH_2-CH_2-CH(NH_2)-COOH$
L-METHIONINE
ATP → Pi + PPi
S-ADENOSYLMETHIONINE
ACCEPTOR → CH_3-ACCEPTOR
S-ADENOSYLHOMOCYSTEINE
H_2O → ADENOSINE
$HS-CH_2-CH_2-CH(NH_2)-COOH$
L-HOMOCYSTEINE
SERINE → H_2O
$S-CH_2-CH_2-CH(NH_2)-COOH$ / $CH_2-CH(NH_2)-COOH$
CYSTATHIONINE
H_2O → CYSTEINE
$HO-CH_2-CH_2-CH(NH_2)-COOH$
L-HOMOSERINE
→ NH_3
$CH_3-CH_2-C(=O)-COOH$
α-KETOBUTYRATE
Co A·SH, NAD^+ → CO_2, NADH + H^+
$CH_3-CH_2-C(=O)-S-Co\ A$
PROPIONYL - Co A

FIG 15–31. Intermediates in conversion of methionine to propionyl-Co A.

transferred to any of a wide variety of acceptor compounds.* After removal of the methyl group, S-adenosylhomocysteine is formed. Hydrolysis of the S to C bond yields L-homocysteine plus adenosine. Homocysteine then condenses with a molecule of serine, forming the amino acid cystathionine. Hydrolytic cleavage of cystathionine forms L-homoserine plus cysteine, so that the net effect is the conversion of homocysteine to homoserine and of serine to cysteine. These 2 reactions are therefore also involved in biosynthesis of cysteine from serine (see p 328). Homoserine is then converted to α-ketobutyrate in a reaction catalyzed by homoserine deaminase (Fig 15–33). Conversion of α-ketobutyrate to propionyl-Co A then occurs in the usual manner for oxidative decarboxylation of α-keto acids to form acyl-Co A derivatives (eg, pyruvate, α-ketoglutarate).

*Compounds whose methyl groups are derived from S-adenosylmethionine include betaines, choline, creatine, epinephrine, melatonin, sarcosine, various N-methylated amino acids, and various alkaloids of plant origin.

Leucine, Valine, & Isoleucine

As might be suspected from their structural similarities, the catabolism of L-leucine, L-valine, and L-isoleucine initially involves the same reactions. Ultimately, this common pathway diverges, and each amino acid follows its own unique pathway to amphibolic intermediates (Fig 15–34). The nature of these amphibolic end products (β-hydroxy-β-methylglutaryl-Co A, succinyl-Co A, and acetyl-Co A) determines whether each amino acid is glycogenic (valine), ketogenic (leucine), or both (isoleucine) (Table 15–2). Many of the reactions involved are closely analogous to reactions of straight and branched chain fatty acid catabolism. The structures of intermediates in leucine, valine, and isoleucine catabolism are given in Figs 15–35, 15–36, and 15–37. Because of the similarities noted above (Fig 15–34), it is convenient to discuss the initial reactions in catabolism of all 3 amino acids together. In what follows, reaction numbers correspond to the numbered reactions of Figs 15–34, 15–35, 15–36, and 15–37.

A. Transamination: Reversible transamination (reaction 1, Figs 15–34, 15–35, 15–36, and 15–37) of all 3 branched L-α-amino acids in mammalian tissues probably is due to catalysis by a single transaminase. The reversibility of this reaction accounts for the ability of the corresponding α-keto acids to replace a dietary requirement for the L-α-amino acids.

B. Oxidative Decarboxylation to Acyl-Co A Thioesters: These reactions (reaction 2, Figs 15–34 through 15–37) closely resemble the analogous oxidations of pyruvate to CO_2 and acetyl-Co A and of α-ketoglutarate to CO_2 and succinyl-Co A. Indirect evidence suggests the presence in mammals of at least 2 oxidative decarboxylases specific for only one or 2 α-keto acids. A partially purified mammalian decarboxylase is known which catalyzes oxidative decarboxylation of α-ketoisocaproate (from leucine) and of α-keto-β-methylvalerate (from isoleucine) but not of α-ketoisovalerate (from valine). In man, the available evidence suggests a single oxidative decarboxylase for all 3 α-keto acids. In **maple syrup urine disease**, a rare genetic defect in infants, a metabolic block due to a nonfunctional oxidative decarboxylase prevents further catabolism of all 3 α-keto acids (Fig 15–34). These acids accumulate in the blood and urine, imparting to urine the characteristic odor for which the defect is named. The accumulation of all 3 α-keto acids suggests a single oxidative decarboxylase. Menkes &

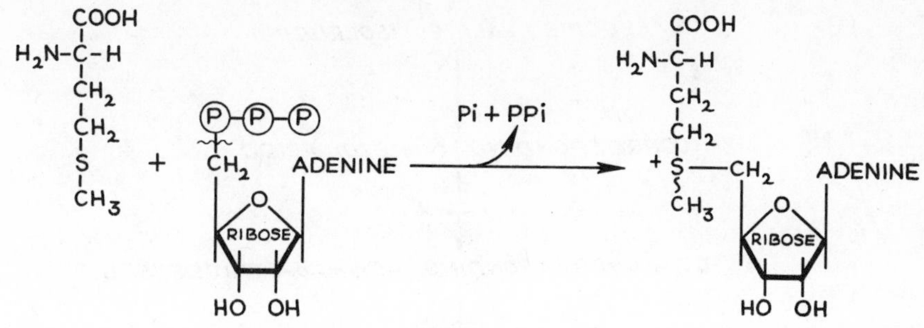

FIG 15–32. Formation of S-adenosylmethionine.

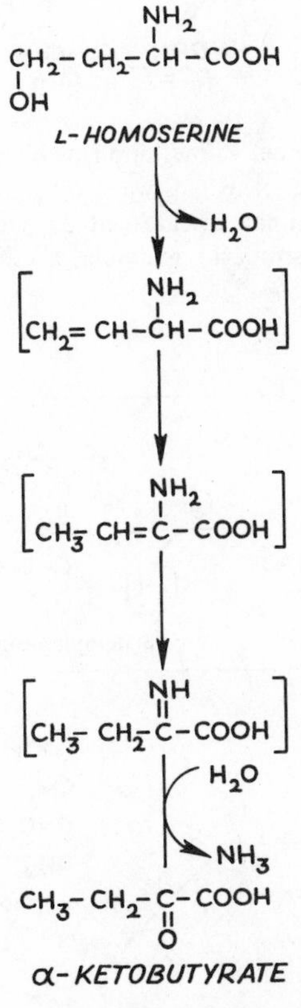

FIG 15–33. Conversion of L-homoserine to α-ketobutyrate, catalyzed by homoserine deaminase.

others (1954) described 4 cases occurring in one family in which the disease was associated with severe functional impairment of the CNS. Death occurred in all cases at an early age. Another patient described by MacKenzie & Woolf (1959) first showed symptoms at 4 months of age consisting of jerking movements of the legs with occasional episodes of respiratory distress and cyanosis. During these episodes, the EEG indicated a severe generalized abnormality with multifocal discharges typical of metabolic disorders accompanied by seizures. During these acute episodes of the disease, the urinary excretion of the keto acids of the branched-chain amino acids was much increased and the urine emitted a characteristic odor similar to that of maple syrup.

C. Dehydrogenation to α,β-Unsaturated Acyl-Co A Thioesters: This reaction is analogous to dehydrogenation of straight chain acyl-Co A thioesters in fatty acid catabolism. It is not known whether a single enzyme catalyzes dehydrogenation of all 3 branched acyl-Co A thioesters; indirect evidence suggests that at least 2 enzymes are required. This evidence derives from chemical observations in **isovaleric acidemia,** wherein, following the ingestion of protein-rich foods, there occurs an increase in blood of isovalerate. An increase in other branched α-keto acids does not occur. Isovalerate is formed by deacylation of isovaleryl-Co A, the substrate for the above dehydrogenase. Its formation suggests accumulation of isovaleryl-Co A, possibly due to a defective isovaleryl-Co A dehydrogenase. If a single dehydrogenase served to dehydrogenate all 3 branched acyl-Co A thioesters, accumulation of isobutyrate (from valine) and α-methylbutyrate (from isoleucine) would be anticipated following a protein-rich meal.

Reactions Specific to Leucine Catabolism (See Fig 15–35.)

Reaction 4L: Carboxylation of β-methylcrotonyl-Co A. A key observation leading to explanation of the ketogenic action of leucine (Table 15–2) was the discovery that 1 mol of CO_2 was "fixed" (ie, covalently bound) for every mol of isopropyl groups (from the terminal isopropyl group of leucine) converted to acetoacetate. This CO_2 fixation (reaction 4L, Fig 15–35) requires biotinyl-CO_2 formed from enzyme-bound biotin and CO_2 at the expense of ATP. Both in bacteria and in mammalian liver, this reaction forms β-methylglutaconyl-Co A as a free intermediate.

Reaction 5L: Hydration of β-methylglutaconyl-Co A. Very little is known about this reaction except that the product is β-hydroxy-β-methylglutaryl-Co A (HMG-Co A), a precursor not only of ketone bodies (reaction 6L, Fig 15–35) but also of mevalonate, and hence of cholesterol and other polyisoprenoids.

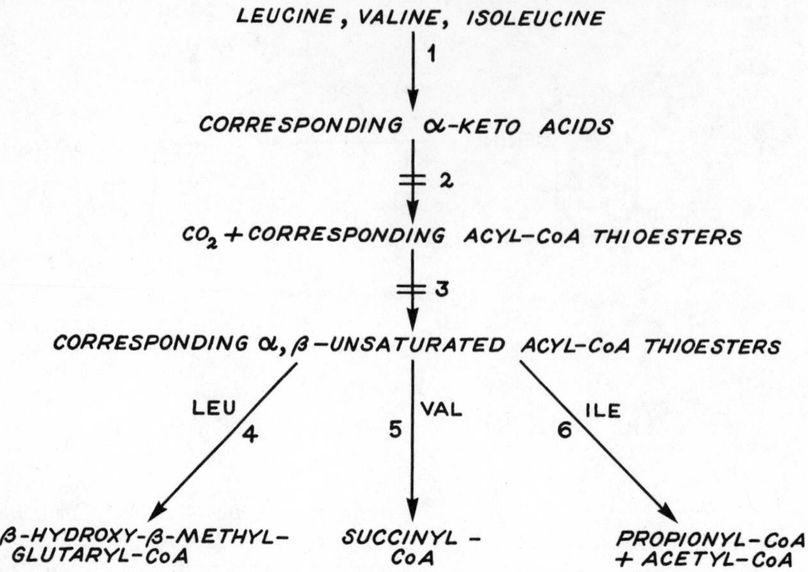

FIG 15–34. Overall catabolism of the branched chain amino acids, leucine, valine, and isoleucine in mammals. The first 3 reactions are common to all 3 amino acids; thereafter, the pathways diverge. Double lines intersecting arrows mark the sites of metabolic blocks in 2 rare human diseases: at 2, maple syrup urine disease, a defect in catabolism of all 3 amino acids; and at 3, isovaleric acidemia, a defect of leucine catabolism.

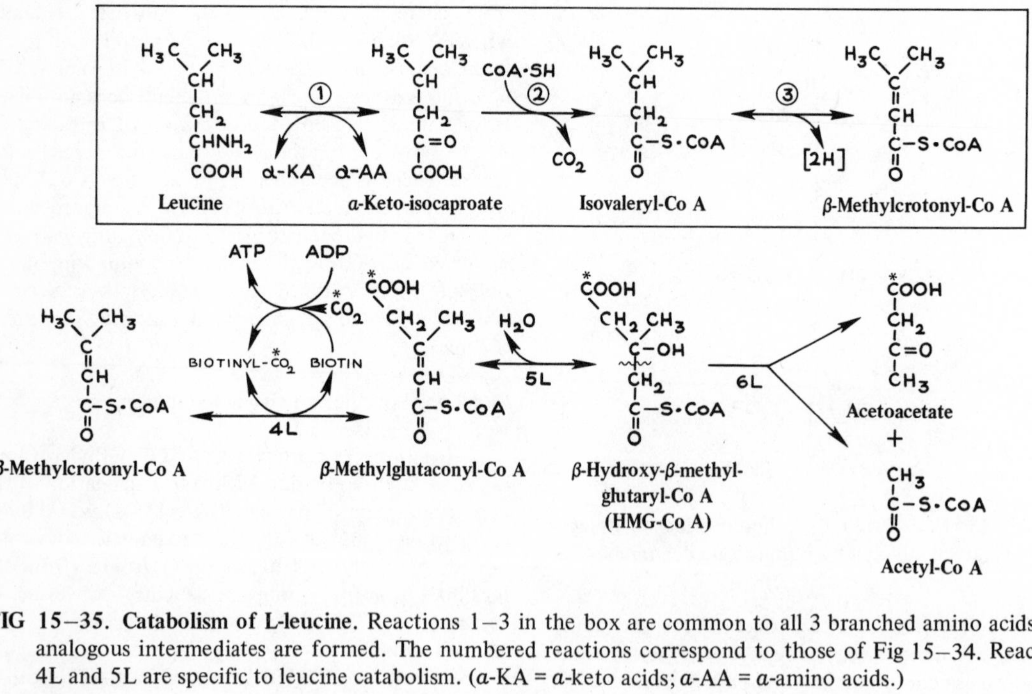

Leucine ① (α-KA, α-AA) α-Keto-isocaproate ② (CoA·SH, CO_2) Isovaleryl-Co A ③ ([2H]) β-Methylcrotonyl-Co A

β-Methylcrotonyl-Co A 4L (ATP, ADP, BIOTINYL-CO_2, BIOTIN, CO_2) β-Methylglutaconyl-Co A 5L (H_2O) β-Hydroxy-β-methyl-glutaryl-Co A (HMG-Co A) 6L Acetoacetate + Acetyl-Co A

FIG 15–35. Catabolism of L-leucine. Reactions 1–3 in the box are common to all 3 branched amino acids, and analogous intermediates are formed. The numbered reactions correspond to those of Fig 15–34. Reactions 4L and 5L are specific to leucine catabolism. (α-KA = α-keto acids; α-AA = α-amino acids.)

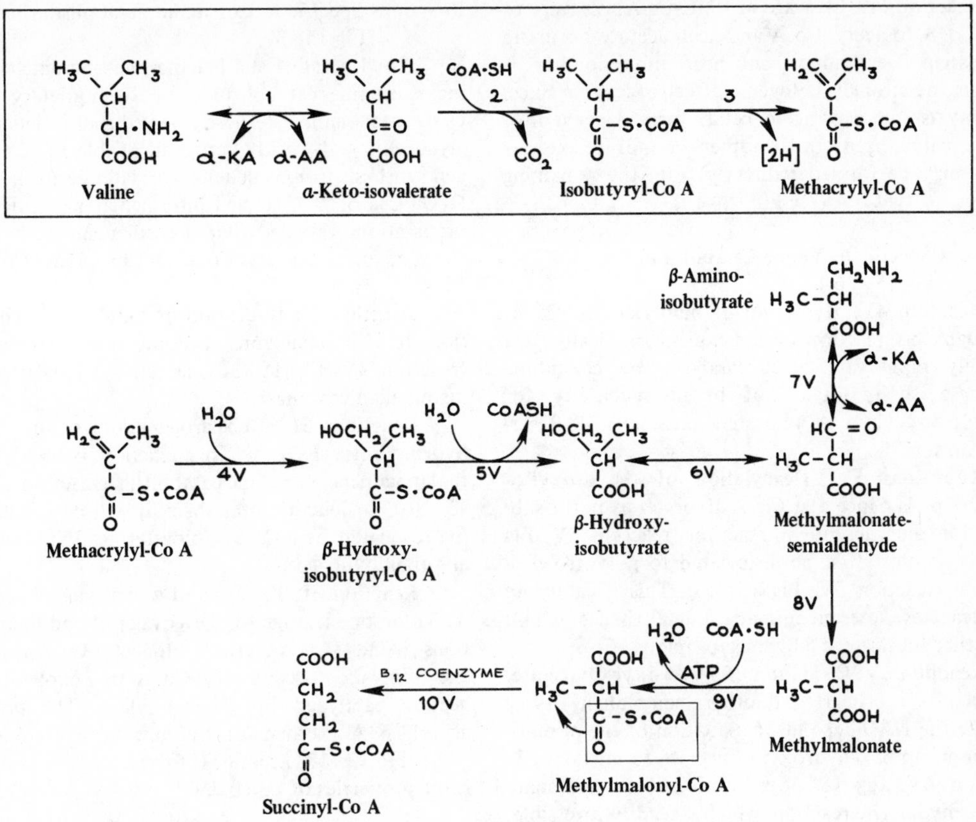

FIG 15–36. Catabolism of valine. Reactions 1–3 in the box are common to all 3 branched amino acids, and analogous intermediates are formed. The numbered reactions correspond to those of Fig 15–34. Reactions 4V through 10V are specific to valine catabolism.

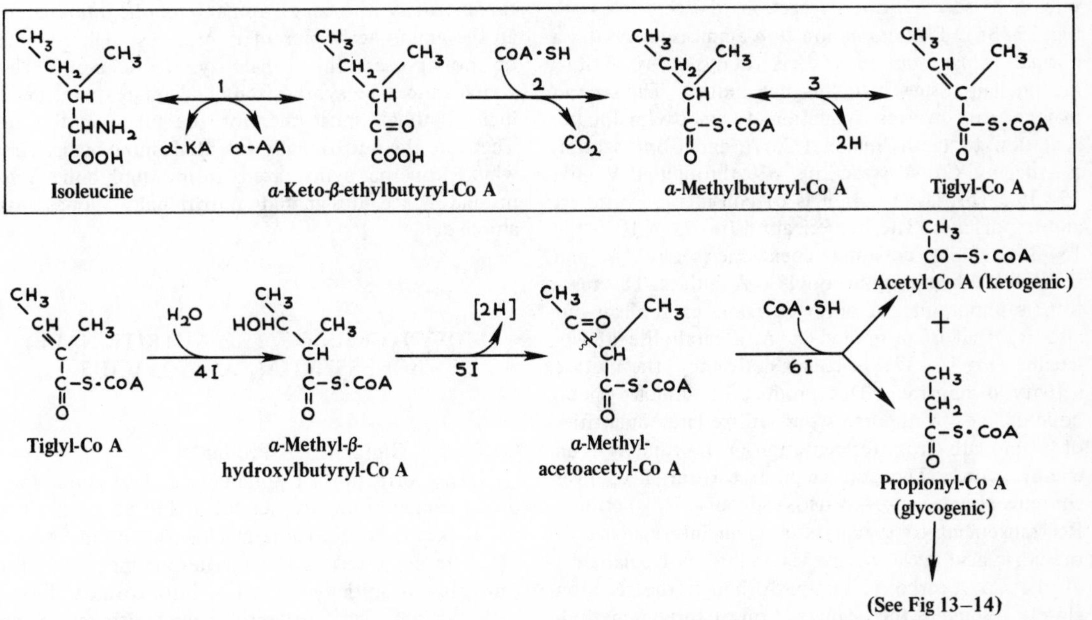

FIG 15–37. Catabolism of L-isoleucine. Reactions 1–3 in the box are common to all 3 branched amino acids, and analogous intermediates are formed. The numbered reactions correspond to those of Fig 15–34. Reactions 4I, 5I, and 6I are specific to isoleucine catabolism.

Reaction 6L: Cleavage of HMG-Co A. Cleavage of HMG-Co A to acetyl-Co A and acetoacetate occurs in mammalian liver, kidney, and heart mitochondria. It explains the strongly ketogenic effect of leucine since, not only is 1 mol of acetoacetate formed per mol of leucine catabolized, but another ½ mol of ketone bodies may be formed indirectly from the remaining product, acetyl-Co A (see Fig 14–16).

Reactions Specific to Valine Catabolism (See Fig 15–36.)

Reaction 4V: Hydration of methylacrylyl-Co A. Although this reaction occurs nonenzymatically at a relatively rapid rate, it is catalyzed by crystalline crotonase, a hydrolyase of broad specificity for L-β-hydroxyacyl-Co A thioesters possessing 4–9 carbon atoms.

Reactions 5V: Deacylation of β-hydroxyisobutyryl-Co A. Since the Co A thioester is not a substrate for the subsequent reaction (reaction 6V, Fig 15–36), it must first be deacylated to β-hydroxyisobutyrate (reaction 5V, Fig 15–36). This is catalyzed by a deacylase, present in many animal tissues, whose only other substrate is β-hydroxypropionyl-Co A.

Reaction 6V: Oxidation of β-hydroxyisobutyrate. Extracts of pig heart and other mammalian tissues catalyze the NAD-dependent oxidation of the primary alcohol group of β-hydroxyisobutyrate to an aldehyde (reaction 6V, Fig 15–36), forming methylmalonate semialdehyde. The reaction, which is readily reversible, is catalyzed by a purified, substrate-specific oxidoreductase from hog kidney.

Reaction 7V: Fate of methylmalonate semialdehyde. Two fates are possible for methylmalonate semialdehyde in mammalian tissues: transamination to β-aminoisobutyrate (reaction 7V, Fig 15–36) and conversion to succinyl-Co A (reactions 8V through 10V, Fig 15–36). Transamination to α-aminoisobutyrate, a normal urinary amino acid, is catalyzed by various mammalian tissues including hog kidney. The second major fate involves oxidation to methylmalonate, acylation to methylmalonyl-Co A, and isomerization to succinyl-Co A (reactions 8V through 10V, Fig 15–36). This last reaction is of considerable interest and importance. The isomerization (reaction 10V, Fig 15–36) requires cobamide coenzyme (see p 114) and is catalyzed by methylmalonyl-Co A mutase. This reaction is important not only for valine catabolism but also for that of propionyl-Co A, a catabolite of isoleucine (Fig 15–37). In cobalt deficiency, the mutase activity is impaired. This produces a "dietary metabolic defect" in ruminants that utilize large quantities of propionate (from fermentation in the rumen) as an energy source. The purified mutase from sheep liver contains about 2 mols of deoxyadenosyl-B_{12} per mol. Rearrangement to succinyl-Co A, an intermediate of the citric acid cycle, occurs via an intramolecular shift of the Co A-carboxyl group. Although the reaction closely resembles the isomerization of **threo**-β-methylaspartate to glutamate, the reaction mechanisms appear to differ in significant details.

Reactions Specific to Isoleucine Catabolism (See Fig 15–37.)

As with valine and leucine, the first data concerning isoleucine catabolism came from dietary studies using intact animals. These revealed that isoleucine was glycogenic and weakly ketogenic (Table 15–2). Glycogen synthesis from isoleucine was later confirmed using D_2O. Use of ^{14}C-labeled intermediates and liver slice preparations revealed that the isoleucine skeleton was cleaved, forming acetyl-Co A and propionyl-Co A (Fig 15–37).

Reaction 4I: Hydration of tiglyl-Co A. This reaction, like the analogous reaction in valine catabolism (reaction 4V, Fig 15–36), is catalyzed by crystalline mammalian crotonase.

Reaction 5I: Dehydrogenation of α-methyl-β-hydroxybutyryl-Co A. This reaction is analogous to that occurring in valine catabolism (reaction 5V, Fig 15–36). In leucine catabolism, it will be recalled, the hydroxylated acyl-Co A thioester is first deacylated and then oxidized.

Reaction 6I: Thiolysis of α-methylacetoacetyl-Co A. Thiolytic cleavage of the covalent bond linking carbons 2 and 3 of α-methylacetoacetyl-Co A resembles the thiolysis of acetoacetyl-Co A to 2 mols of acetoacetate catalyzed by β-ketothiolase. The products, acetyl-Co A (ketogenic) and propionyl-Co A (glycogenic; Fig 13–12) account for the ketogenic and glycogenic properties of isoleucine.

BIOSYNTHESIS OF AMINO ACIDS

Most procaryotic and many eurcaryotic cells are capable of synthesizing from amphibolic intermediates all the amino acids present in proteins; higher animals do not possess this capability. Those amino acids which cannot be synthesized in adequate quantities by higher animals must therefore be taken in the diet. These are the **nutritionally essential amino acids.** Those which can be synthesized from amphibolic intermediates are designated **nutritionally nonessential amino acids.**

BIOSYNTHESIS OF THE NUTRITIONALLY NONESSENTIAL AMINO ACIDS

Glutamate, Glutamine, & Proline

The formation of glutamate and of glutamine is of fundamental importance for amino acid biosynthesis. It occurs by the same mechanisms in all forms of life. The net effect is the fixation of inorganic nitrogen, present initially as NH_4^+, into covalent linkage with carbon. This covalently bound nitrogen is then transferred in the course of the biosynthesis of a wide variety of important biochemical products, including, but not limited to, other amino acids.

Reductive amination of α-ketoglutarate, catalyzed by L-glutamate dehydrogenase (Fig 15–7), forms L-glutamate. Formation of glutamine is achieved via the reaction catalyzed by glutamine synthetase (Fig 15–9). Proline is biosynthesized from glutamate by reversal of the catabolic sequence shown in Fig 15–15.

Hydroxyproline

Both 3- and 4-hydroxy-L-proline occur in mammalian tissues. Little else is known about 3-hydroxyproline other than its occurrence in rat tail tendon and in the antibiotic, telomycin. A great deal more is known about 4-hydroxyproline, which, like hydroxylysine, appears to be confined to collagen. Free dietary 4-hydroxyproline, like hydroxylysine, is not utilized by growing rats. Rather, it is synthesized directly from a combined form of proline. The substrate for proline hydroxylation appears to be a prolyl-containing polypeptide with a molecular weight of about 15,000 daltons (Juva & Prockop, 1966). The oxygen of hydroxyproline is derived from air rather than water. Hydroxylation, which requires enzymes of the microsomal fraction, presumably involves an oxygenase reaction. Both 5-ketoproline and 3,4-dihydroproline can be ruled out as intermediates on the basis of tritium-labeling data. Collagen synthesis, which occurs by mechanisms comparable to those utilized for other proteins, has been achieved in cell-free systems derived from chick embryo.

Both 4-hydroxyproline and 4-ketoproline occur in the polypeptide actinomycin antibiotics (see p 68). In contrast to mammalian systems, free hydroxyproline is incorporated directly into the actinomycin polypeptide elaborated by *Streptomyces antibioticus.*

Alanine, Aspartate & Asparagine

Alanine is formed by transamination of pyruvate and aspartate by transamination of oxaloacetate (Fig 15–3). Asparagine is formed from aspartate plus ammonia in a reaction catalyzed by asparagine synthetase. This reaction is analogous to that catalyzed by glutamine synthetase (Fig 15–9).

Tyrosine

Tyrosine is formed from phenylalanine by the reaction catalyzed by phenylalanine hydroxylase (Fig 15–22). Thus, whereas phenylalanine is a nutritionally essential amino acid, tyrosine is not provided the diet contains adequate quantities of phenylalanine.

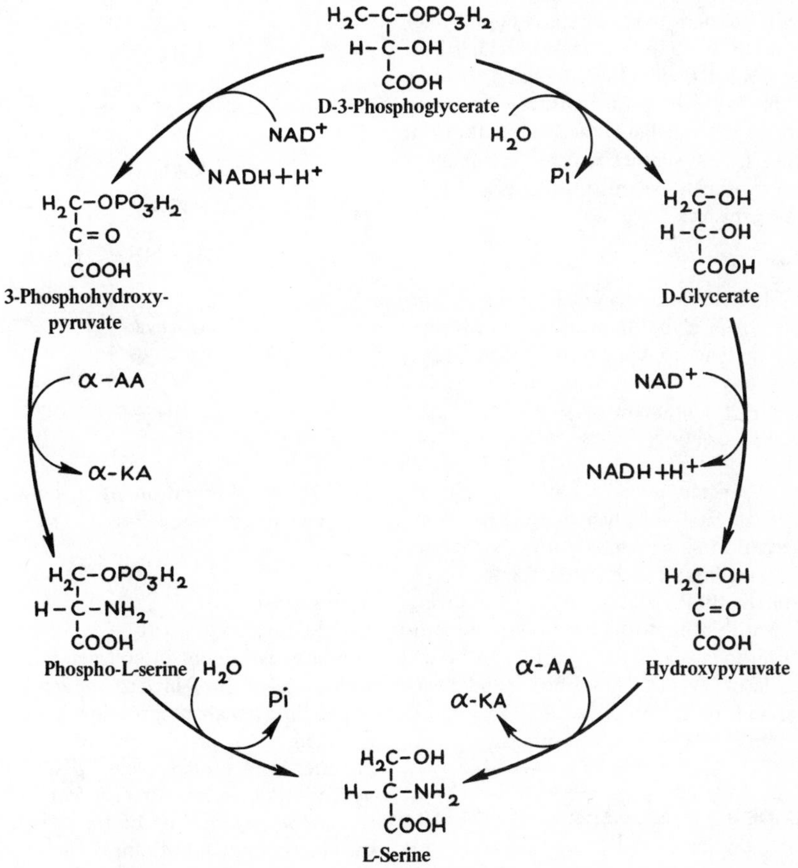

D-3-Phosphoglycerate
NAD⁺ / NADH+H⁺
H₂O / Pi
3-Phosphohydroxypyruvate
D-Glycerate
α-AA / α-KA
NAD⁺ / NADH+H⁺
Phospho-L-serine
Hydroxypyruvate
H₂O / Pi
α-AA / α-KA
L-Serine

FIG 15–38. Serine biosynthesis via phosphorylated and nonphosphorylated intermediates. (α-AA = α-amino acids; α-KA = α-keto acids.)

Cysteine

Cysteine, while not itself nutritionally essential, is formed from methionine (essential) and serine (nonessential) as shown in Fig 15–31. L-Methionine is converted to S-adenosylmethionine, to S-adenosylhomocysteine, and then to L-homocysteine. This condenses with L-serine, which provides the 3-carbon skeleton of cysteine, forming cystathionine. Hydrolysis of cystathionine then forms cysteine plus homoserine. Note that while the sulfur of cysteine derives from methionine by transulfuration, the carbon skeleton is provided by serine.

Serine

Two pathways for serine biosynthesis have been shown to exist in mammalian tissues. In both cases the carbon skeleton of serine is provided by D-3-phosphoglycerate, an intermediate in glycolysis (Fig 13–3). The 2 pathways differ with respect to the nature of the intermediates involved. One pathway uses nonphosphorylated intermediate and the other involves phosphorylated intermediates (Fig 15–38).

Synthesis via phosphorylated intermediates involves oxidation of 3-phosphoglycerate to phosphohydroxypyruvate, transamination to phosphoserine, and, finally, hydrolytic removal of the phosphate catalyzed by a phosphatase. For synthesis via nonphosphorylated intermediates, phosphoglycerate is dephosphorylated to glycerate by a phosphatase, oxidized to hydroxypyruvate, and, finally, transaminated to form L-serine. It is probable that the pathway involving phosphorylated intermediates accounts for the majority of the serine synthesized by mammalian tissues. This also appears to be true for plants and for a variety of microorganisms.

Glycine

Synthesis of glycine in mammalian tissues can occur in 2 ways: from glyoxylic acid or from serine. The cytosol of liver tissue contains active glycine transaminases that catalyze synthesis of glycine from glyoxylate and either glutamate or pyruvate. Unlike most transaminase reactions, these strongly favor glycine synthesis.

Mammalian liver mitochondria also contain an enzyme system that catalyzes glycine synthesis from serine by the reaction shown in Fig 15–39. Serine, ammonium ions, bicarbonate, tetrahydrofolate (FH_4), pyridoxal phosphate (PLP), and a source of reducing power (NADH or dithiothreitol) are all essential components. The bicarbonate and the β-carbon of serine form one molecule of glycine. The carboxyl and α-carbon atoms of serine form the other.

$$CO_2 + NH_3 + \text{SERINE} + 2H \underset{PLP}{\overset{FH_4}{\rightleftharpoons}} 2\,\text{GLYCINE} + H_2O$$

FIG 15–39. Synthesis of glycine by carboxylation of serine. (FH_4 = tetrahydrofolic acid; PLP = pyridoxal phosphate.)

A third route for glycine synthesis has been detected in clostridia, but it does not occur in mammalian tissues. L-Threonine or L-allothreonine (Fig 3–1) undergoes an aldol-type cleavage, forming glycine plus acetaldehyde. The presence of this pathway may reflect the absence in clostridia of the enzymes of the phosphorylated pathway for serine biosynthesis.

Glycine may also be formed from choline, as shown in Fig 15–40.

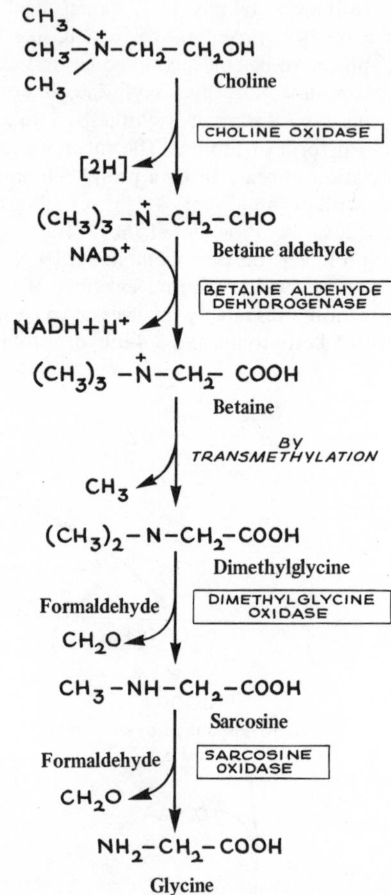

FIG 15–40. Formation of glycine from choline by way of betaine.

Hydroxylysine

5-Hydroxylysine (α,ϵ-diamino-δ-hydroxycaproate) is present in collagen and collagen products such as gelatin or isinglass but is probably absent from other mammalian proteins. Small quantities are reported to occur in wool, in trypsin, and as a phosphatide in *Mycobacterium phlei.* In rats, collagen hydroxylysine arises directly from dietary lysine. This was shown by feeding or injecting ^{14}C-lysine, which was incorporated both into the lysine and hydroxylysine of collagen. Significantly, ^{14}C-hydroxylysine is not incorporated into collagen when fed or injected.

These observations suggest that conversion of lysine to hydroxylysine occurs after incorporation of lysine into a polypeptide or protein—as is the case with proline, forming hydroxyproline. The mechanism of hydroxylysine synthesis has been extensively studied in developing chick embryo. Although the exact reactions are still unknown, several possibilities may be ruled out. During conversion of lysine to hydroxylysine, ^{18}O from $^{18}O_2$ (but not from $H_2{}^{18}O$), and tritium from $^{3}H_2O$ are incorporated into hydroxylysine. Furthermore, the conversion of 4,5-tetratritiolysine to hydroxylysine occurs with loss of a single atom of tritium, from C_5, the carbon atom ultimately bearing the OH group. The tritium retention data appear to exclude all mechanisms involving formation either of 5-ketonic or 4,5-unsaturated intermediates (Popenoe & others, 1965) (Fig 15–41).

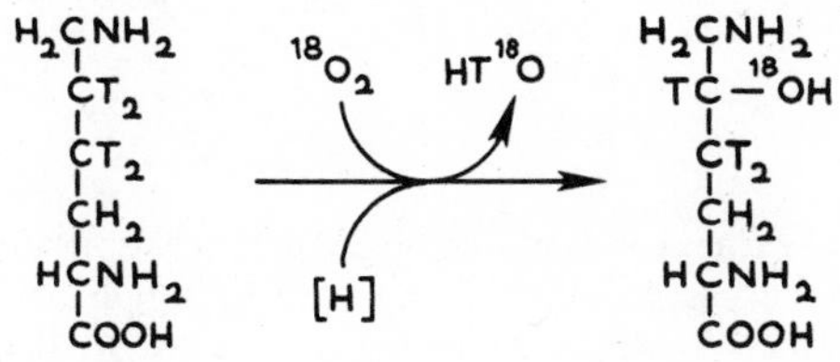

$$\begin{array}{c} H_2CNH_2 \\ | \\ CT_2 \\ | \\ CT_2 \\ | \\ CH_2 \\ | \\ HCNH_2 \\ | \\ COOH \end{array} \xrightarrow[{[H]}]{^{18}O_2 \quad HT^{18}O} \begin{array}{c} H_2CNH_2 \\ | \;\; {}^{18} \\ TC-OH \\ | \\ CT_2 \\ | \\ CH_2 \\ | \\ HCNH_2 \\ | \\ COOH \end{array}$$

Lysine **5-Hydroxylysine**

FIG 15–41. **Lysine hydroxylation.** 4,5-Tetratritiolysine is converted to hydroxylysine with loss of a single atom of tritium (T) and with incorporation of ^{18}O from $^{18}O_2$. The mechanism may involve a hydroxylase, with NADPH as the reductant.

BIOSYNTHESIS OF THE NUTRITIONALLY ESSENTIAL AMINO ACIDS

Arginine

Arginine is properly considered a nutritionally essential amino acid for man. It can be synthesized by rats, but not in quantities sufficient to permit normal growth. Microorganisms biosynthesize arginine from glutamate, utilizing N-acetylated intermediates (Fig 15–42). One intermediate in this pathway, N-acetylglutamate-γ-semialdehyde, is also a precursor of proline in bacteria. In man and other animals, however, glutamate forms proline, as shown in Fig 15–15.

Once ornithine has been formed, the reactions leading to arginine are those of the urea cycle (Fig 15–11). These enzymes are present not only in bacteria but also in abundance in mammalian liver. Mammals are, however, incapable of synthesizing ornithine by the reactions of Fig 15–42. Some ornithine is formed in mammalian liver by reversal of the reactions of ornithine catabolism (Fig 15–15). This accounts for the status of arginine as only a partially essential amino acid in man and other animals.

Leucine, Valine, & Isoleucine

While leucine, valine, and isoleucine are all nutritionally essential amino acids for man and other higher animals, mammalian tissues do contain transaminases that reversibly catalyze interconversion of all 3 amino acids with their corresponding α-keto acids (Figs 15–34 through 15–37). This explains the ability of the appropriate keto acids to replace their amino acids in the diet. Although D-leucine is utilized to some

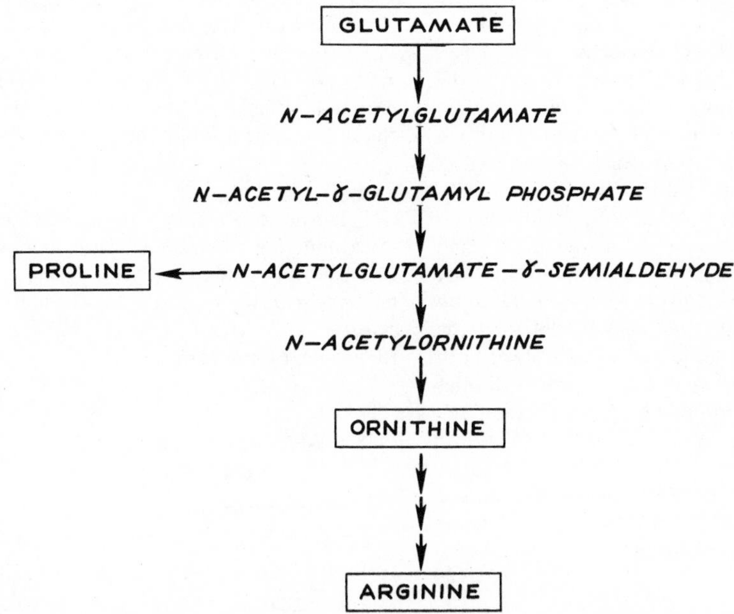

FIG 15–42. **Acylated intermediates in proline and arginine biosynthesis from glutamate by bacteria.** Synthesis of these nutritionally essential amino acids does not occur in mammalian tissues.

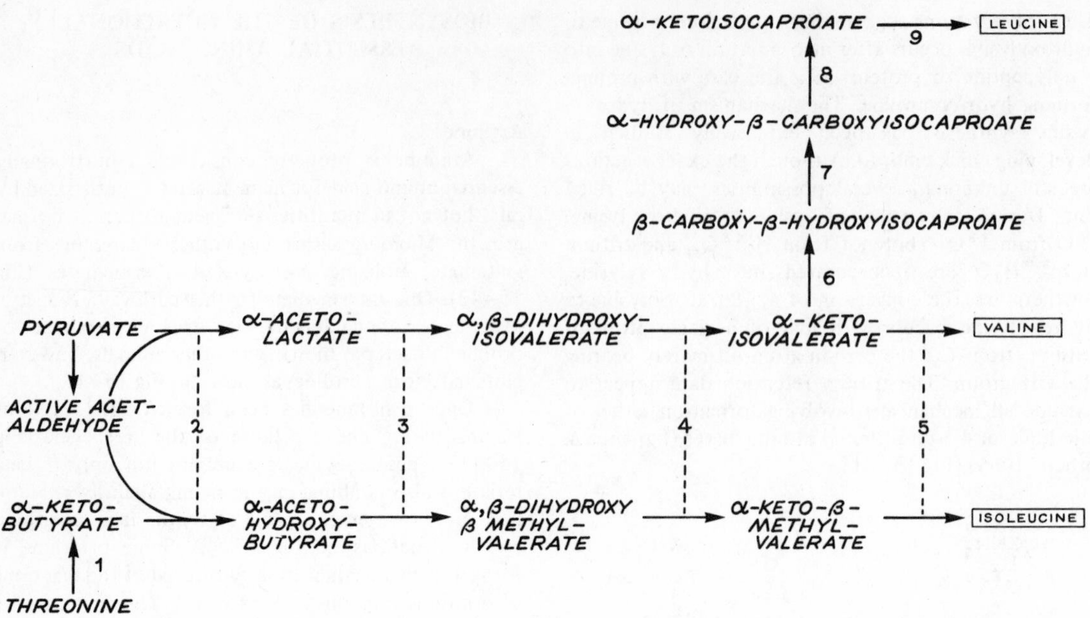

FIG 15–43. Intermediates in leucine, valine and isoleucine biosynthesis in bacteria. Since these reactions do not occur in man or other animals, these 3 amino acids are nutritionally essential and must be supplied in the diet. Reactions 2 through 5 appear to be catalyzed by single enzymes functional for synthesis of all 3 amino acids. Considerable information is available about feedback regulation of the enzymes of this pathway. (Redrawn from Freundlich & others, 1962.)

extent by chicks and rats, the rate of deamination of the D-isomers to the α-keto acids is too slow to support growth.

The intermediates in the biosynthesis of the 3 branched-chain amino acids are shown in Fig 15–43. The initial amphibolic starting materials are pyruvate (valine, leucine) or its next higher homologue α-ketobutyrate (isoleucine). These condense with active acetaldehyde derived from pyruvate. The details of the regulation of these pathways have been studied in considerable detail by Umbarger (1969) and others in a variety of organisms. While the exact mechanisms of organization and regulation differ from organism to organism, the biosynthesis of these amino acids is tightly controlled both at the level of the genome and at the level of enzymic activity. Reaction 6, the first reaction unique to leucine biosynthesis, is catalyzed by an allosteric enzyme that is feedback-inhibited by the end product, leucine. Addition of leucine alters both the molecular architecture and the substrate affinities of the salmonella **α-isopropylmalate synthetase**, the catalyst for reaction 6. In neurospora, a eucaryotic organism, leucine represses the synthesis of α-isopropylmalate synthetase. Both feedback inhibition and repression thus are involved in regulation of leucine biosynthesis. When bacteria are grown on media containing all 3 amino acids in adequate quantities, syntheses of the enzymes catalyzing all 9 reactions of branched-chain amino acid synthesis are repressed. Repression of enzymes 2 through 4 is "multivalent," ie, repression only occurs in the presence of **all** 3 amino acids. In addition, *Escherichia coli* contains a pyruvate oxidase whose function is closely linked to production of active acetaldehyde destined for valine synthesis. Growth of *E coli* on media containing growth-limiting quantities of valine then derepresses the synthesis of this biosynthetic pyruvate oxidase.

Lysine

Lysine appears not to participate in reversible transamination or deamination reactions in mammals. D-Lysine thus cannot replace L-lysine in the diet. Certain ε-substituted lysine derivatives such as ε-methyl or ε-acetyl lysine (see below) can replace dietary lysine for growth of young rats.

CH_3
CH_2-NH
$(CH_2)_3$
$HC-NH_2$
$COOH$

ε-N-Methyl lysine

CH_3
$C=O$
CH_2-NH
$(CH_2)_3$
$HC-NH_2$
$COOH$

ε-N-Acetyl lysine

Mammalian tissue preparations readily demethylate and deacylate the above compounds to lysine, thus accounting for their nutritional effect. α-Substituted lysine derivatives are not utilized, presumably owing to the absence from mammalian tissues of enzymes capable of catalyzing removal of the α-substituents.

Lysine biosynthesis in yeast (Fig 15–44) and in bacteria (Fig 15–45) occurs by distinct pathways whose intermediates have little in common. The yeast pathway (Broquist & Trupin, 1966) involves reactions and intermediates analogous to those of the citric acid cycle (Fig 13–7), whereas that of bacteria involves cyclic and N-acylated intermediates. Neither pathway is present in man or other animals, for whom lysine is thus a nutritionally essential amino acid. The absence of sufficient lysine in many cereal proteins is a prime reason for their failure to support maximal growth when fed as the sole source of dietary protein. The recent development of "high lysine corn" (Fig 3–12) thus holds rich promise for improving man's diet in areas where cereals constitute a primary source of dietary protein.

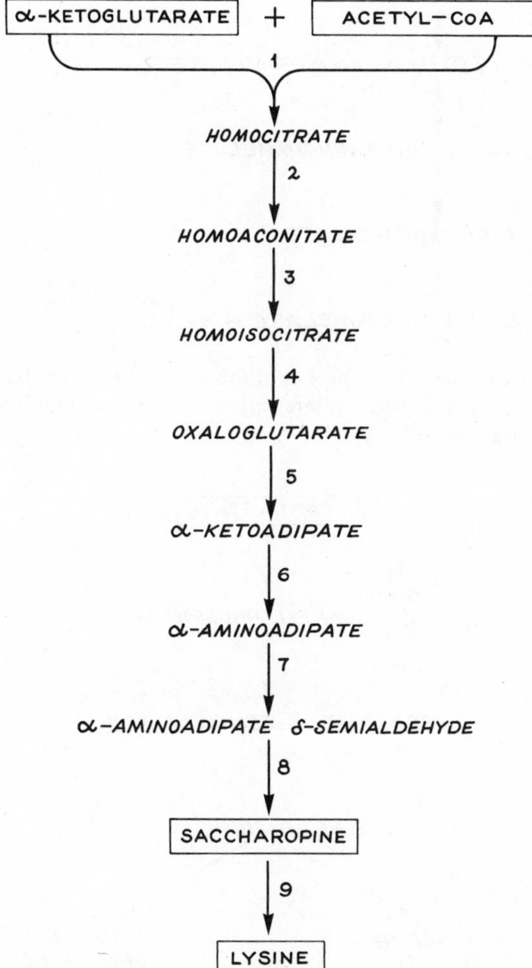

FIG 15–44. Lysine biosynthesis by yeast. Synthesis of this nutritionally essential amino acid does not occur in mammals. Reactions 1 through 5 are analogous to reactions of the citric acid cycle (Fig 13–7), with all intermediates having one more carbon atom each. Reaction 6 is analogous to the L-glutamate dehydrogenase reaction. Saccharopine (Fig 15–27) has recently been implicated in lysine catabolism by man and other mammals.

Histidine

This amino acid, like arginine, is nutritionally semi-essential. Adult human beings and adult rats have been maintained in nitrogen balance for short periods in the absence of histidine. The growing animal does, however, require histidine in the diet. If studies were to be carried on for longer periods, it is probable that a requirement for histidine in adult human subjects would also be elicited.

An outline of the intermediates involved in histidine biosynthesis in microorganisms is shown in Fig 15–46. Biosynthesis starts with 5-phosphoribosyl-1-pyrophosphate (PRPP) (see p 352), which condenses with ATP, forming N′-(5-phosphoribosyl)-ATP. This reaction thus closely resembles the initial reaction of purine biosynthesis. The catalyst for this reaction, PRPP-ATP phosphorylase, is a feedback-inhibited by histidine, the end product of the biosynthetic pathway.

Methionine & Threonine

The carbon skeletons of methionine and threonine both are formed from homoserine. Homoserine in turn is derived from aspartate by a sequence of reactions that do not occur in mammalian tissues (Fig 15–47). Bacterial aspartate kinase, the enzyme catalyzing the first reaction of threonine biosynthesis, is feedback-inhibited by threonine. The interconversion of homoserine and methionine have already been discussed (Fig 15–31). Conversion of threonine to homoserine (not shown) involves formation of homoserine phosphate.

Phenylalanine, Tryptophan, & Tyrosine

The conversion of phosphoenolpyruvate, an intermediate in glycolysis (Fig 13–3) and of erythrose-4-phosphate, an intermediate in the pentose-phosphate pathway (Fig 13–8), to phenylalanine and tryptophan is outlined in Fig 14–48. Chorismate, a key intermediate, is the precursor not only of the 3 aromatic amino acids but also of the quinone ring of coenzyme Q (Fig 9–4). These reactions do not occur in mammalian tissues.

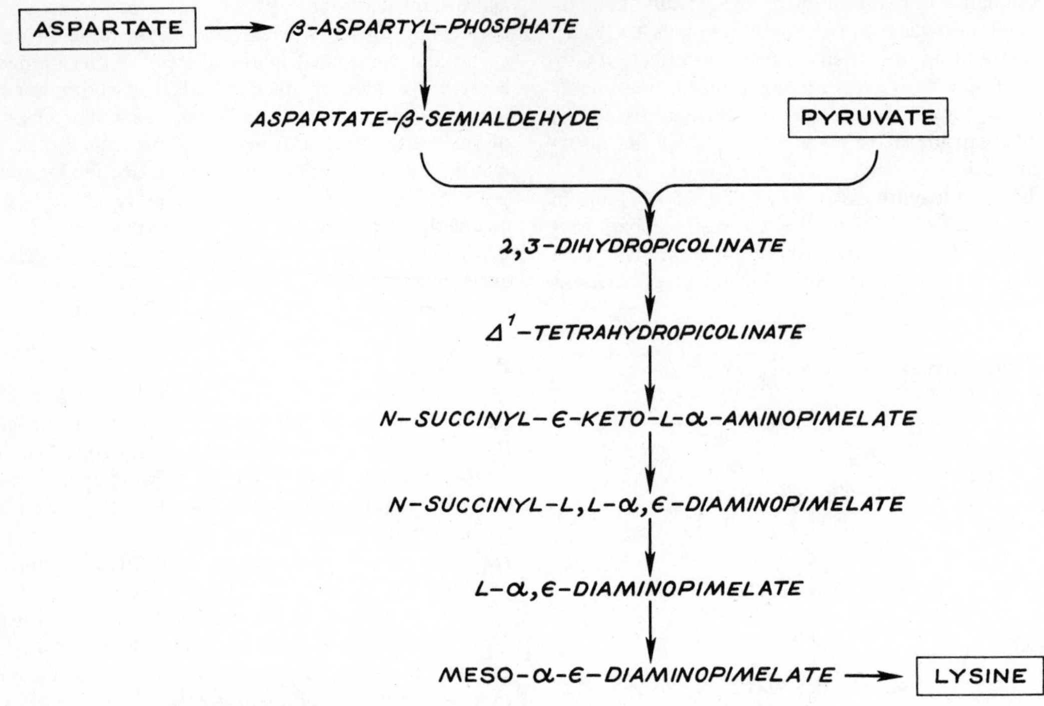

FIG 15–45. Lysine biosynthesis from aspartate and pyruvate by bacteria. Synthesis of this nutritionally essential amino acid does not occur in mammals. Intermediates in this pathway are utilized not only for lysine biosynthesis but also for cell wall synthesis and for spore germination.

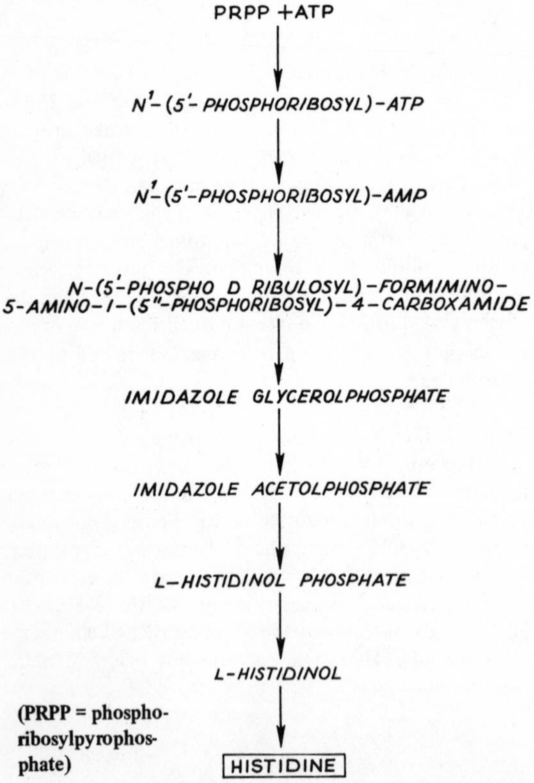

PRPP +ATP

↓

N1-(5'-PHOSPHORIBOSYL)-ATP

↓

N1-(5'-PHOSPHORIBOSYL)-AMP

↓

N-(5'-PHOSPHO D RIBULOSYL)-FORMIMINO-
5-AMINO-1-(5"-PHOSPHORIBOSYL)-4-CARBOXAMIDE

↓

IMIDAZOLE GLYCEROLPHOSPHATE

↓

IMIDAZOLE ACETOLPHOSPHATE

↓

L-HISTIDINOL PHOSPHATE

↓

L-HISTIDINOL

↓

HISTIDINE

(PRPP = phosphoribosylpyrophosphate)

FIG 15–46. Intermediates in histidine biosynthesis in microorganisms.

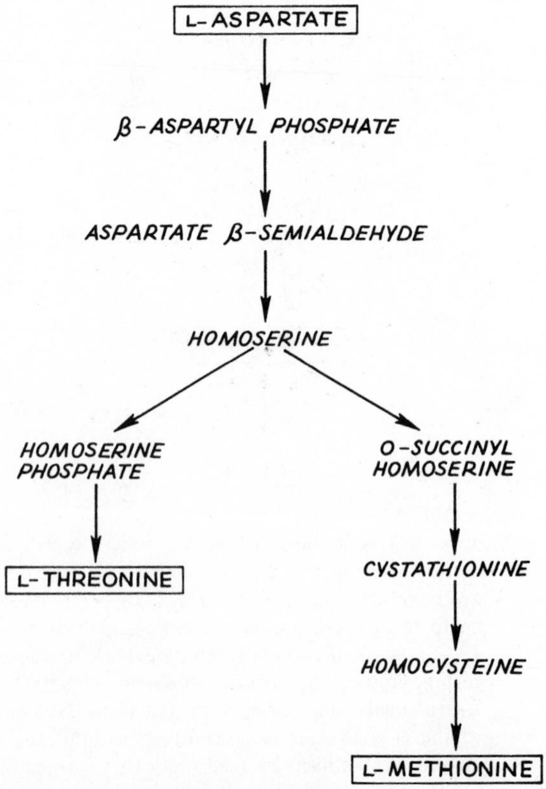

L-ASPARTATE

↓

β-ASPARTYL PHOSPHATE

↓

ASPARTATE β-SEMIALDEHYDE

↓

HOMOSERINE

↙ ↘

HOMOSERINE PHOSPHATE → L-THREONINE

O-SUCCINYL HOMOSERINE → CYSTATHIONINE → HOMOCYSTEINE → L-METHIONINE

FIG 15–47. Intermediates in biosynthesis of threonine and methionine in bacteria.

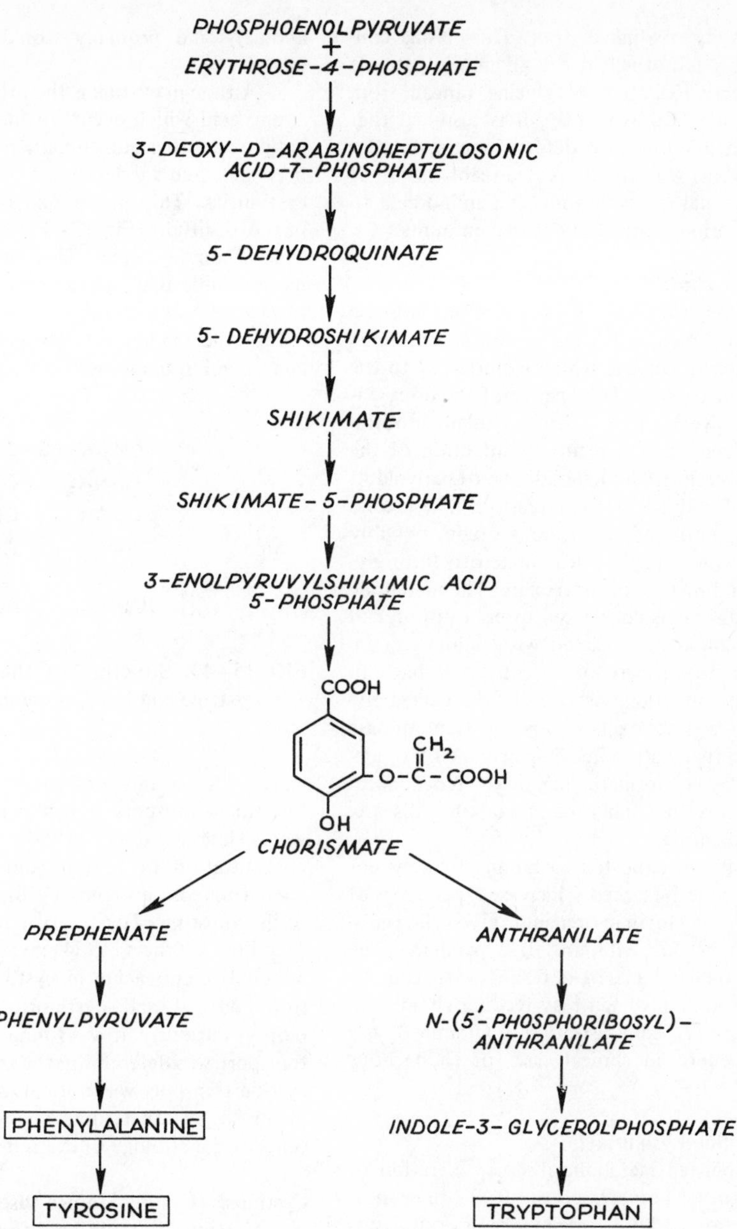

FIG 15–48. Intermediates in phenylalanine, tyrosine, and tryptophan biosynthesis in microorganisms.

METABOLIC DEFECTS IN AMINO ACID METABOLISM

This section discusses certain metabolic disorders of amino acid metabolism that occur in man. Although most are rare, several have historically played key roles in the elucidation of pathways of amino acid catabolism. These examples provide, in addition, illustrations of alternate pathways of amino acid catabolism which may become of importance only when a metabolic defect is present. Also included are several examples of nonmetabolic disorders, principally defects in renal reabsorptive mechanisms, which in some cases mimic true metabolic disorders.

Glycinuria

DeVries & others (1957) have reported a syndrome characterized by excess urinary excretion of glycine in association with a tendency to formation of oxalate renal stones, although the amount of oxalate excreted in the urine was normal. The disease must be extremely rare, having been described so far only in one family, in which it appears to be inherited as a dominant, possibly sex-linked, trait. The plasma content of glycine was normal in the glycinuric patients while the urinary excretion of glycine ranged from 600–1000 mg/day. Consequently, it is assumed that glycinuria is attributable to a defect in renal tubular transport of glycine whereby decreased reabsorption of glycine by the renal tubule permits the amino acid to escape into the urine in greatly increased amounts.

Primary Hyperoxaluria

Primary hyperoxaluria is a metabolic disease characterized biochemically by a continuous high urinary excretion of oxalate which is unrelated to the dietary intake of oxalate. The history of the disease is that of progressive bilateral calcium oxalate urolithiasis, nephrocalcinosis, and recurrent infection of the urinary tract. Death occurs in childhood or early adult life from renal failure or hypertension. The excess oxalate is apparently of endogenous origin, possibly from glycine, which may be deaminated to form glyoxylate, itself a direct source of oxalate. The metabolic defect in this disease is considered to be a disorder of glyoxylate metabolism associated with failure to convert glyoxylate to formate or to convert it back to glycine by transamination. As a result, the excess glyoxylate is oxidized to oxalate. Glycine transaminase deficiency, together with some impairment of oxidation of glyoxylate to formate, may be the biochemical explanation for the inherited metabolic disease, primary hyperoxaluria.

As might be expected, vitamin B_6-deficient animals (rats) excrete markedly increased quantities of oxalate because the glutamic or alanine glyoxylic transaminase reactions are vitamin B_6-dependent. The excretion of oxalate in B_6-deficient rats can be enhanced by feeding glycine or by feeding vitamin B_6 antagonists. However, administration of vitamin B_6 has not been of benefit in clinical cases of endogenous hyperoxaluria.

Cystinuria (Cystine-Lysinuria)

In this inherited metabolic disease, excretion of cystine in the urine is increased to 20–30 times normal. The excretion of lysine, arginine, and ornithine is also markedly increased. Cystinuria is considered to be due to a renal transport defect. The greatly increased excretion of lysine, arginine, and ornithine as well as cystine in the urine of cystinuric patients suggests that there exists in these individuals a defect in the renal reabsorptive mechanisms for these 4 amino acids. It is possible that a single reabsorptive site is involved. Thus, as far as renal mechanisms are concerned, cystinuria is not an uncomplicated defect which affects only cystine; the term "cystinuria" is therefore actually a misnomer, so that cystine-lysinuria may now be the preferred descriptive term for this disease.

Because cystine is a relatively insoluble amino acid, in cystinuric patients it may precipitate in the kidney tubules and form cystine calculi. This may be a major complication of the disease; were it not for this possibility, cystinuria would be an entirely benign anomaly and probably would escape recognition in many cases.

Although cystine is the principal sulfur-containing amino acid which occurs in the urine of the cystinuric patient, another sulfur-containing amino acid has also been detected in significant quantities in the urine of cystinurics. This amino acid has been identified as a mixed disulfide (Fig 15–49) composed of L-cysteine and L-homocysteine. This compound is somewhat more soluble than cystine; to the extent that it may be formed at the expense of cystine, it would therefore reduce the tendency to formation of cystine crystals and calculi in the urine.

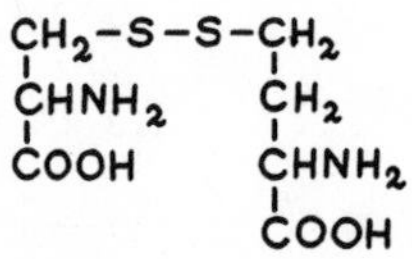

```
CH2—S—S—CH2
|          |
CHNH2      CH2
|          |
COOH       CHNH2
           |
           COOH

(Cysteine)  (Homocysteine)
```

FIG 15–49. Structure of the "mixed" disulfide of cysteine and homocysteine.

Evidence now indicates that there may be an intestinal transport defect for these amino acids as well. Thier & others (1964) have found a failure in concentration of cystine and lysine in cells of the jejunal mucosa obtained by biopsy of the jejunal area of the intestine of cystinuric patients.

Fox & others (1964) investigated transport of the affected amino acids in cystinuria into kidney slices from normal and cystinuric patients obtained by biopsy. Curiously, it was found that lysine and arginine transport was defective in the cystinuric tissue but that cystine transport was normal. All of the above experiments suggest that some revision of the present concepts of the etiology of cystinuria may be required.

Cystinosis (Cystine Storage Disease)

Cystinuria should be differentiated from cystinosis. In the latter disease, which is also inherited, cystine crystals are deposited in many tissues and organs (particularly the reticuloendothelial system) throughout the body. It is usually accompanied by a generalized amino-aciduria in which all amino acids are considerably increased in the urine (Harper & others, 1952). Various other renal functions are also seriously impaired, and these patients usually die at an early age with all of the manifestations of acute renal failure. On the other hand, except for the likelihood of the formation of cystine calculi, cystinuria is compatible with a normal existence.

Homocystinuria

In a metabolic screening program to detect abnormal excretion of amino acids among a group of mentally defective children, Carson & Neill (1962,

1963) discovered abnormally large quantities of homocystine in the urine of 2 sisters. Plasma homocystine as well as methionine levels were also elevated. The clinical findings in these 2 cases included, in addition to mental retardation, bilateral posterior dislocation of the lenses, fair complexion with blue eyes, and fine hair. One girl had a palpable liver which was demonstrated on biopsy to have undergone fatty changes.

A decrease in the activity of hepatic cystathionine synthetase has been shown to be responsible for homocystinuria. For example, Mudd & others (1964) found no cystathionine synthetase activity whatever in a specimen of tissue removed by needle biopsy from the liver of a homocystinuric patient. Finkelstein & others (1964) studied the synthetase activity of the livers of both parents of a homocystinuric patient and found the activity of this enzyme to be but 40% of that of unrelated control patients. These findings demonstrate that the metabolic defect is inherited and suggest that the parents, although clinically normal, nevertheless represent metabolically the heterozygous state.

It seems reasonable to suggest that providing extra cysteine in the diets of homocystinuric patients may be of benefit, particularly when protein intake is low, such as during infancy. This is because a lack of cystathionine synthetase impairs the pathway for the transfer of methionine sulfur to cystine. Since methionine is the major dietary source of sulfur for cystine formation, the metabolic error in homocystinuria should result in a shortage of cystine for incorporation into protein, which would be expected to have widespread effects on protein structure—particularly in the young, growing individual. It is of interest, however, that the daily excretion of homocystine (50–100 mg) in the homocystinuric patients who have so far been studied is far less than would be expected if all of the methionine sulfur is metabolized by the homocysteine pathway. Although methionine itself is also excreted in increased amounts in these patients, there is still evidence that some methionine may be metabolized by an as yet unknown alternate pathway.

Argininemia

A considerable increase in levels of arginine both in the serum and in the CSF was discovered in 2 sisters ages 18 months and 5 years who were referred to a pediatric clinic because of spastic diplegia, epileptic seizures, and severe mental retardation (Terheggen & others, 1969). A study of the arginase content of red blood cells obtained from these patients revealed that the activity of this enzyme was very low. The arginase activity in the red cells of the parents was also lower than normal—as might be expected if the defect occurred in the heterozygous state in both parents. Blood ammonia levels were high in the patients, and the urine amino acid pattern resembled that of lysine-cystinuria. The feeding of a low-protein diet resulted in lowering of the blood ammonia levels and disappearance of the urine lysine-cystinuria pattern, but the high serum arginine levels persisted.

Citrullinuria

In citrullinuria, increased levels of citrulline may be found in the blood and urine as well as in the CSF. Although the exact cause of the metabolic defect in citrullinuria is not known, it is hypothesized that there is impairment in the capacity of the urea cycle to form adequate amounts of urea when the nitrogen load is excessive, resulting in accumulation of urea cycle intermediates such as citrulline and ammonia.

Argininosuccinic Acidemia

Arginosuccinic acidemia is a rare disorder characterized by high levels of argininosuccinic acid in the blood and urine. Although the site of the biochemical defect is not known, the disorder is presumed to be the result of inadequate activity of the enzyme catalyzing the conversion of argininosuccinate to arginine and fumarate (argininosuccinase, Fig 15–11).

Phenylketonuria, Tyrosinosis, & Alkaptonuria

Absence of one or another of the various enzymes of phenylalanine or tyrosine catabolism results in any one of 3 rare heritable metabolic diseases. Metabolites prior to the metabolic block accumulate in the blood and are converted to "alternate catabolites" which accumulate in the blood and spill over into the urine.

A. Phenylketonuria: The genetic defect in phenylketonuria (phenylpyruvic oligophrenia) is absence of a functional component I of phenylalanine hydroxylase (Fig 15–22). The patient is thus unable to convert phenylalanine to tyrosine, so that phenylalanine and its "alternate catabolites" accumulate in the blood and urine (Table 15–3). Among infants and children exhibiting this metabolic defect, retarded mental development occurs for as yet unknown reasons. In the

TABLE 15–3. Metabolites of phenylalanine accumulating in the plasma and urine of phenylketonuric patients.

Metabolite	Plasma (mg/100 ml)		Urine (mg/100 ml)	
	Normal	Phenylketonuric	Normal	Phenylketonuric
Phenylalanine	1–2	15–63	30	300–1000
Phenylpyruvate		0.3–1.8		300–2000
Phenyllactate				290–550
Phenylacetate				Increased
Phenylacetylglutamine			200–300	2400

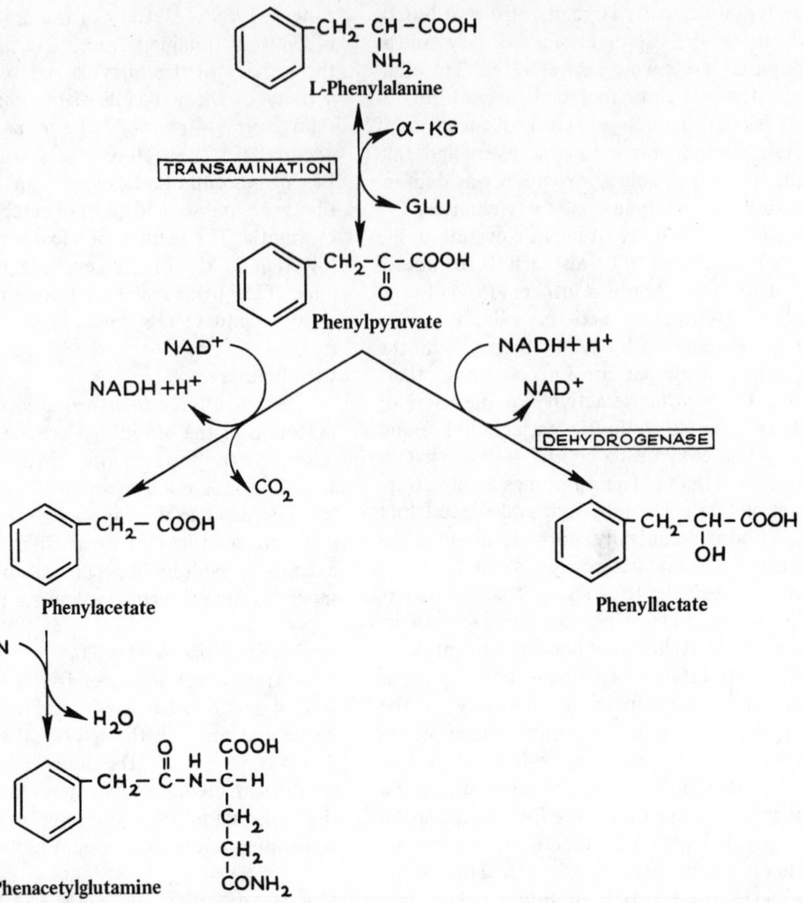

FIG 15–50. Alternate pathways of phenylalanine catabolism of particular importance in phenylketonuria. The reactions shown also occur in the liver tissue of normal individuals but are of minor significance if a functional phenylalanine hydroxylase is present. (α-KG = α-ketoglutarate; Glu = glutamate; Gln = glutamine.)

absence of a normal catabolic pathway for phenylalanine, several reactions of otherwise minor quantitative importance in normal liver assume a major catabolic role. In phenylketonurics, phenylpyruvate, phenyllactate, phenylacetate, and its glutamine conjugate phenyacetylglutamine are formed and occur in the blood and urine (Fig 15–50). Although phenylpyruvate which occurs in the urine of most phenylketonuric patients can be detected by a simple biochemical spot test, a definitive diagnosis requires determination of elevated plasma phenylalanine levels.

The mental performance of phenylketonuric children can be improved if they are maintained on a diet containing very low levels of phenylalanine. Clinical improvement is accompanied by a return to the normal range of blood phenylalanine levels and a reduced excretion of "alternate catabolites." Detection of the disease as early in infancy as possible is important if dietary treatment is to yield favorable results on mental development.

Plasma phenylalanine may be measured by an automated micro method that requires as little as 20 μl of blood per determination. It is important to note, however, that abnormally high blood phenylalanine levels may not occur in phenylketonuric infants until the third or fourth day of life. Furthermore, false positive tests may occur in premature infants due to delayed maturation of the enzymes required for phenylalanine catabolism. A useful but less reliable screening test depends on detecting the elevated urinary levels of *p*-hydroxyphenylpyruvate with ferric chloride.

It would be expected that the administration of phenylalanine to a phenylketonuric subject would result in prolonged elevation of the level of this amino acid in the blood, ie, diminished tolerance to phenylalanine. However, it has been found that an abnormally low tolerance to injected phenylalanine and a high fasting level of phenylalanine are also characteristic of the parents of the phenylketonuric individual. Evidently the recessive gene responsible for phenylketonuria can be detected biochemically in the phenotypically normal parents.

B. Tyrosinosis: The metabolic defect in tyrosinosis appears to be lack of a functional *p*-hydroxyphenylpyruvate oxidase (Fig 15–24). Although *p*-hydroxyphenylpyruvate is excreted in considerable amounts, the ability to oxidize homogentisate is normal.

C. Alkaptonuria: The metabolic defect in alkaptonuria is the essentially complete lack of homogentisate oxidase activity (Fig 15–24). Homogentisate therefore accumulates and appears in the urine of alkaptonurics. The urine darkens on exposure to air due to formation of oxidative products of homogentisate. The cartilage may also darken in alkaptonurics, presumably as a result of the deposition of oxidation products derived from homogentisate. This condition is known as **ochronosis.**

Hartnup Disease

Hartnup disease is a hereditary abnormality in the metabolism of tryptophan, characterized by a pellagra-like skin rash, intermittent cerebellar ataxia, and mental deterioration. The urine of patients with Hartnup disease contains greatly increased amounts of indoleacetic acid (α-N[indole-3-acetyl]glutamine), as well as tryptophan.

The indole acids of human urine have been studied by paper chromatography. A total of 38 different indole acids were chromatographed. The most strikingly "abnormal" patterns of indole acid excretion were found in the urine of severely mentally retarded patients and in urine from the mentally ill. The significance of these findings has been questioned insofar as the causes of mental disease were concerned, particularly in view of the fact that the urinary excretion patterns tended to revert to normal after administration of broad spectrum antibotics.

Imidazole Aminoaciduria

Three families in which there were 5 patients with cerebromacular degeneration have been studied by Bessman & Baldwin (1962). The patients and some members of their immediate families were found to have a generalized imidazole aminoaciduria. The patients excreted large amounts of carnosine and anserine as well as histidine and 1-methyl histidine. In normal urine, the excretion of carnosine and of anserine is 2–3 mg/day and 5–7 mg/day, respectively; in these patients, 20–100 mg/day were excreted. The patients also had a greatly increased urinary content of histidine and of 1-methylhistidine. The parents and unaffected siblings had urinary biochemical abnormalities similar to those of the patients but without the symptoms of neurologic and retinal disease (cerebral degeneration and blindness). The imidazoluria appears to be genetically transmitted as a dominant trait and the cerebromacular degeneration as a recessive trait. The fact that the 2 traits have been found in 3 unrelated families would suggest that both traits are manifestations of the same gene. The disease seems to resemble biochemically the findings in Hartnup disease (see above) in that both are characterized by defects in transport–one for the imidazoles and the other (Hartnup disease) for the indoles.

Histidinemia

Histidinemia is another inherited disorder of histidine metabolism (Auerbach & others, 1962). In addition to increased levels of histidine in the blood and urine, there is also increased excretion of imidazole pyruvic acid (which in a color test with ferric chloride may be mistaken for phenylpyruvic acid, so that a mistaken diagnosis of phenylketonuria could be made). Speech development may be retarded. The metabolic block in histidinemia is considered to be inadequate activity of liver histidase, which would impair conversion of histidine to urocanic acid. The alternative route of histidine metabolism, which involves transamination to form imidazole pyruvic acid, would then be favored and the excess imidazole pyruvic acid would be excreted in the urine. Imidazole acetic acid and imidazole lactic acid, the reduction product of imidazole pyruvic acid, have also been detected in the urine of histidinemic patients.

The quantity of histidine found in normal urine is relatively large. For this reason it may be more readily detected than most other amino acids. It has been reported that a conspicuous increase in histidine excretion is a characteristic finding in normal pregnancy but does not occur in toxemic states associated with pregnancy. The normally increased excretion of histidine during pregnancy apparently does not result from a metabolic defect in the metabolism of histidine. The phenomenon may be explained largely on the basis of the changes in renal function which are characteristic of normal pregnancy as well as the pregnancy toxemias. Furthermore, the alterations in amino acid excretion during pregnancy are not confined to histidine.

CONVERSION OF AMINO ACIDS TO SPECIALIZED PRODUCTS

This section considers the conversion of the carbon skeletons of the amino acids, the amino acids themselves, or portions of their structures to products of biochemical interest. Since most of these products are not themselves amino acids, the discussion merges at several points with metabolic pathways discussed elsewhere in this book.

Glycine

A. Synthesis of Heme: The α-carbon and nitrogen atoms of glycine are used in the synthesis of the porphyrin moiety of hemoglobin (see p 73). The nitrogen in each pyrrole ring is derived from the glycine nitrogen and an adjoining carbon from the α-carbon of glycine. The α-carbon is also the source of the methylene bridge atoms linking the pyrrole rings together as a tetrapyrrole. For every 4 glycine nitrogen atoms utilized, 8 α-carbon atoms enter the porphyrin molecule. The relationship of glycine to heme synthesis and to the citric acid cycle is summarized in Fig 15–51 as the **"succinate-glycine cycle."** Succinyl-Co A condenses on the α-carbon atom of glycine to form α-amino-β-ketoadipic acid. It is at this point that the metabolism of glycine is linked to the citric acid cycle, which provides succinyl-Co A. α-Amino-β-ketoadipic acid is then converted by loss of CO_2 to δ-aminolevulinic acid. This compound serves as a common precursor for porphyrin synthesis as well as, after deamination, a carrier molecule for the introduction of the ureido carbons (2 and 8) into the purine ring (Fig 16–1). Succinate and ketoglutarate, which may return to the citric acid cycle, are also formed.

B. Synthesis of Purines: The entire glycine molecule is utilized to form positions 4, 5, and 7 of the purine skeleton as shown in Fig 16–1.

The δ-carbon atom of δ-aminolevulinic acid (Fig 6–1) is derived from the α-carbon of glycine. This δ-carbon was found to be incorporated into positions 2 and 8 (the ureido carbons) of the purine nucleus to an even greater extent than α-carbons of glycine or β-carbons of serine. This suggests that δ-aminolevulinic acid is actually the carrier molecule for the transfer of this carbon atom to purines (Fig 15–51).

C. A Constituent of Glutathione: The tripeptide glutathione is a compound of glutamic acid, cysteine, and glycine. The nitrogen in glutathione is not available for transamination. Note that the peptide bond is with the γ-COOH of glutamate (Fig 15–52).

D. Conjugation With Glycine: Glycine conjugates with cholic acid, forming the bile acid glycocholic acid. With benzoic acid, it forms hippuric acid. This reaction (Fig 15–53) is used as a test of liver function (see p 368).

E. Synthesis of Creatine: The sarcosine (N-methyl glycine) component of creatine (Fig 15–60) is derived from glycine.

Alanine

Except as a constituent of proteins, alanine is not known to have any other specific function, but together with glycine it makes up a considerable fraction of the amino nitrogen in human plasma. Both D- and L-alanine appear to be utilized by the tissues, but at differing rates.

Alanine is a major component of the cell walls of bacteria. In a number of species, part of the alanine is present as the D-isomer–39–50% in *Streptococcus faecalis;* 67% in *Staphylococcus aureus.*

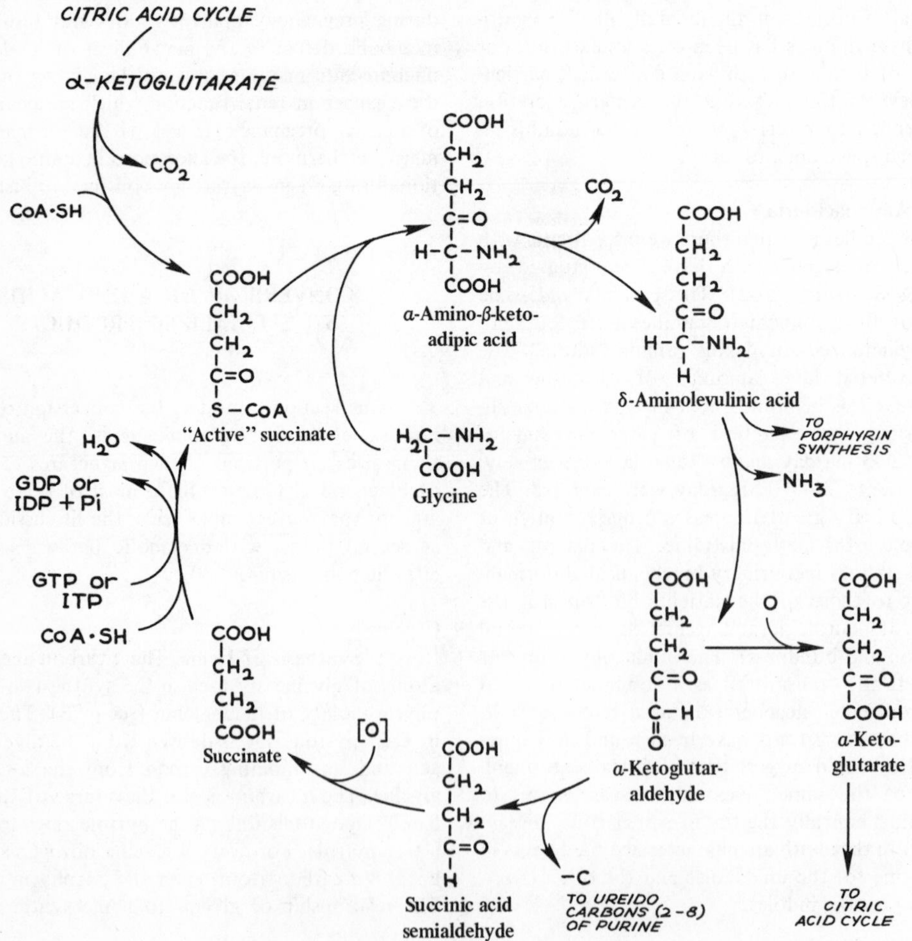

FIG 15–51. The succinate-glycine cycle.

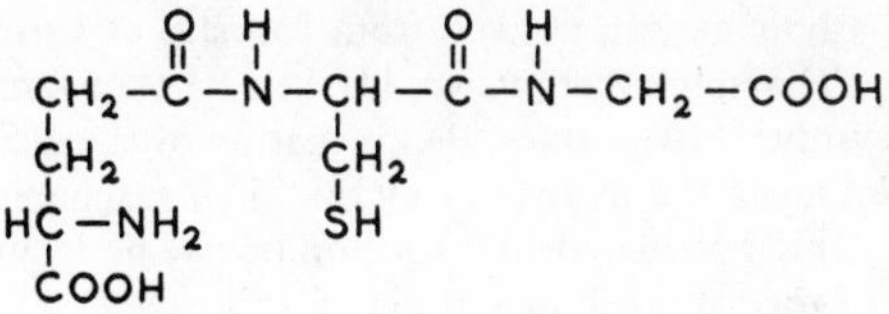

$$CH_2-\overset{O}{\overset{\|}{C}}-\overset{H}{\overset{|}{N}}-CH-\overset{O}{\overset{\|}{C}}-\overset{H}{\overset{|}{N}}-CH_2-COOH$$

CH_2 / $HC-NH_2$ / COOH; CH_2 / SH

FIG 15–52. Glutathione (reduced form).

Serine is involved directly in the synthesis of sphingol and therefore in the formation of sphingomyelins of brain. The details of this reaction are discussed on p 274.

Serine participates in purine and pyrimidine synthesis. The β-carbon is a source of the methyl groups of thymine (and of choline) and of the carbon in positions 2 and 8 of the purine nucleus.

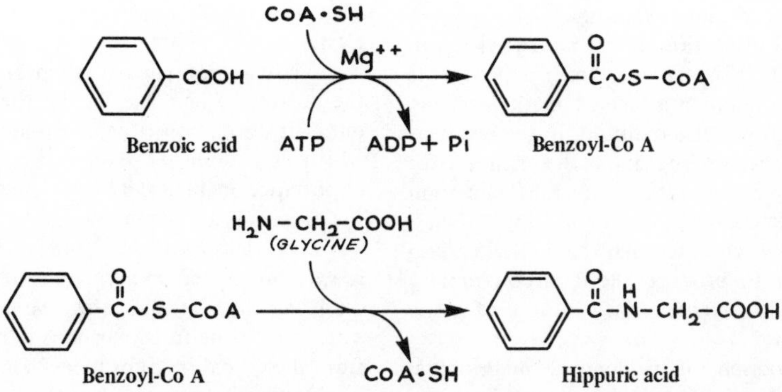

Benzoic acid (C_6H_5–COOH) + CoA·SH + ATP $\xrightarrow{Mg^{++}}$ Benzoyl-Co A (C_6H_5–CO∼S–CoA) + ADP + Pi

Benzoyl-Co A (C_6H_5–CO∼S–CoA) + H_2N-CH_2-COOH (GLYCINE) → Hippuric acid (C_6H_5–CO–NH–CH_2–COOH) + CoA·SH

FIG 15–53. Formation of hippuric acid.

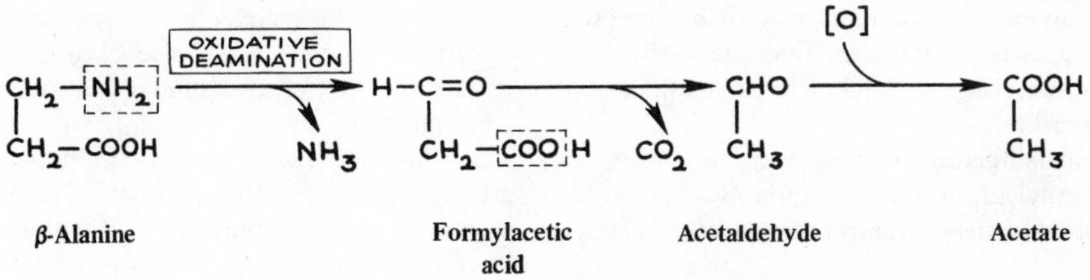

β-Alanine ($CH_2(NH_2)-CH_2-COOH$) → (OXIDATIVE DEAMINATION, − NH_3) → Formylacetic acid ($H-C{=}O$ / CH_2-COOH) → (− CO_2) → Acetaldehyde (CHO / CH_3) → ([O]) → Acetate (COOH / CH_3)

FIG 15–54. Catabolism of β-alanine in the rat.

β-Alanine is a constituent of pantothenic acid and an end product in the catabolism of certain pyrimidines (cytosine and uracil; Fig 16–7). Studies on the catabolism of β-alanine in the rat indicate that this amino acid is degraded to acetate as shown in Fig 15–54.

Serine

Much of the serine in phosphoproteins appears to be present in the form of O-phosphoserine.

A cephalin fraction containing serine is present in the brain. The production of this compound, phosphatidyl serine (see p 19), may be another lipotropic function of serine. Decarboxylation of the serine moiety of phosphatidyl serine would convert it to an ethanolamine moiety. Thus, phosphatidyl serine would be changed to a cephalin. It is possible that this is indeed the pathway of synthesis of cephalins and, after methylation, of lecithins.

Threonine

This essential amino acid does not participate in transamination reactions. The D-isomer is not utilized by the body, probably because of the nonconvertibility of the keto acid to the amino acid. A specific function for threonine other than as a constituent of body proteins has not been discovered. It may be related to the utilization of fat in the liver (see p 285). Threonine, like serine, may occur in proteins as O-phosphothreonine.

Methionine

This amino acid readily donates its terminal methyl group for methylation of various compounds (see p 285). This role of methionine in methylation reactions is an important function of this amino acid, because methionine is the principal methyl donor in the body. The methyl group may be transferred to other compounds for the synthesis of choline or of creatine, for example, or for use in detoxication processes, such as the methylation of pyridine derivatives such as nicotinic acid.

In the transmethylation reactions which utilize methionine as a methyl donor, it is first required that methionine be "activated." This requires adenosine triphosphate (ATP) and a methionine-activating enzyme of liver (Fig 15–32).

The S-methyl bond is "high-energy"; this is a reason for the lability of the methyl group acting as a source of methyl for transmethylations.

In addition to utilization of the methyl group in the intact form, there is evidence from experiments with methionine containing a labeled methyl carbon that the methyl group is also oxidized. In the rat, ¼ of the labeled methyl carbon appears in the expired CO_2 during the first day and about ½ of the labeled carbon is excreted in the urine, feces, or respiratory CO_2 in 2 days. It has already been noted that the methyl carbon may also be used to produce the 1-carbon moiety which conjugates with glycine in the synthesis of serine (Fig 15–12).

The demethylation of S-adenosyl methionine, either for transmethylation or oxidation of the methyl group, produces S-adenosyl homocysteine (Fig 15–31).

Homocysteine, together with a source of labile methyl (eg, betaine, or choline by way of betaine; see p 285), can be used to replace methionine in the diet. This observation suggests that the demethylation process is reversible.

The animal organism has considerable ability to synthesize methyl groups, and vitamin B_{12} and folic acid are involved in the synthesis of these labile methyl groups.

There is evidence that homocysteine is involved in the utilization of the 1-carbon (formate) moiety (see also Fig 7–8). This may occur through the formation of an intermediatary compound of homocysteine with a 1-carbon moiety derived from formate or formaldehyde. The single carbon could then be transferred to the synthesis of purines, the formation of the β-carbon of serine or the methyl of methionine. A scheme outlining this postulated role for homocysteine in utilization of formate is shown in Fig 15–55.

Methionine may undergo oxidative deamination to form the corresponding keto acid. This reaction is reversible, and inversion of the D- form is thus possible.

Cysteine

Cysteine is a particularly prominent amino acid in the protein of hair, hoofs, and the keratin of the skin, but it is also a constituent of many other proteins, in which it establishes S–S bonds which are of great importance in maintaining the secondary structure of the protein (see Chapter 3).

The D-isomer of cysteine is not utilized for growth in animal experiments. It is, however slowly oxidized since it increases the urinary sulfate. The sulfur occurring in the urine is derived almost entirely from the oxidation of cysteine. The sulfur contained in methionine is transferred to the formation of cysteine; consequently, methionine does not directly contribute to the sulfate pool of the body.

Many enzymes depend on a free –SH group for maintenance of their activity. The importance of the –SH groups on Co A has also been noted (see p 105). In the case of enzymes, mild oxidation, converting the –SH group to the S–S linkage, will inactivate them. They may be reactivated by reduction of the S–S group, as with glutathione. Heavy metals like mercury or arsenic also combine with –SH linkages and cause inactivation of enzyme systems (see pp 146–148).

The functions of glutathione are attributable to its cysteine content. In addition to the function of cysteine/cystine in glutathione synthesis, this amino acid is important in conjugation with aromatic halogens to form mercapturic acids.

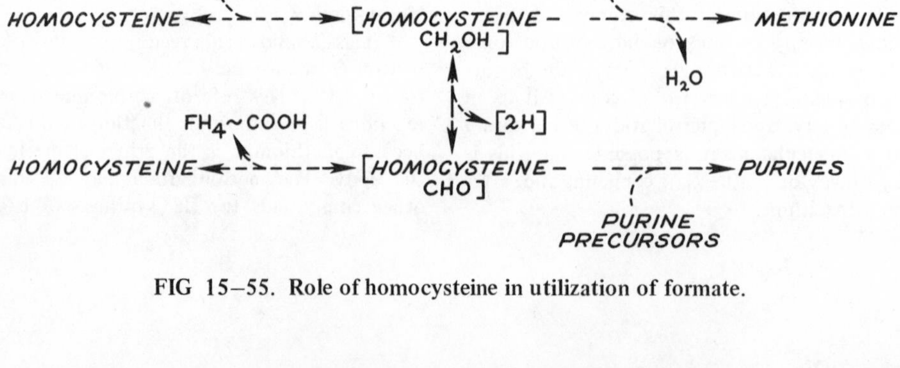

FIG 15–55. Role of homocysteine in utilization of formate.

Taurine, the cholic acid conjugate in bile which forms the bile acid, **taurocholic acid**, is also derived from cysteine.

The oxidation of cysteine to cystine proceeds readily. In fact, this conversion of 2 –SH radicals to S–S is probably an important oxidation-reduction system in the body.

$$2\begin{cases}CH_2SH\\ CHNH_2\\ COOH\end{cases} \underset{[2H]}{\rightleftharpoons} \begin{matrix}CH_2-S-S-CH_2\\ CHNH_2 \quad\quad CHNH_2\\ COOH \quad\quad COOH\end{matrix}$$

Cysteine — Cystine

Arginine

Arginine serves as a formamidine donor for creatine synthesis in primates (Fig 15–56) and for streptomycin synthesis in streptomyces (Fig 15–57). Other fates of interest include conversion to putrescine (see p 221), agmatine, spermine, and spermidine by enteric bacteria (Tabor & Tabor, 1969), and synthesis of arginine phosphate (functionally analogous to creatine phosphate) in invertebrate muscle.

Although a functional role for arginine-derived polyamines in bacteria has not been discovered, bacteria typically convert large quantities of arginine to spermine and spermidine.

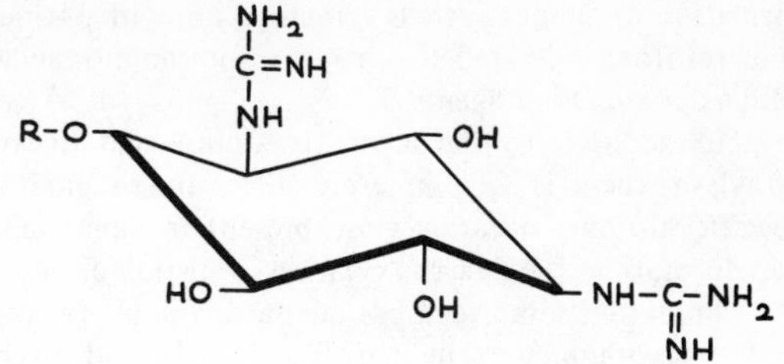

FIG 15–57. The streptidine portion of the streptomycins. In streptomyces, arginine serves as the donor of both guanido groups.

Histidine

Histamine is derived from histidine by decarboxylation. This reaction is catalyzed in mammalian tissues by an enzyme designated **aromatic L-amino acid decarboxylase**. The enzyme will also catalyze the decarboxylation of 3,4-dihydroxyphenylalanine (DOPA; Fig 15–59), 5-hydroxytryptophan (Fig 15–58), phenylalanine, tyrosine, and tryptophan. The amino acid decarboxylase is present in kidney and in other tissues such as brain and liver. It is inhibited by α-methyl amino acids, which also inhibit amino acid decarboxylation in vivo and thus have clinical application as antihypertensive agents by preventing the

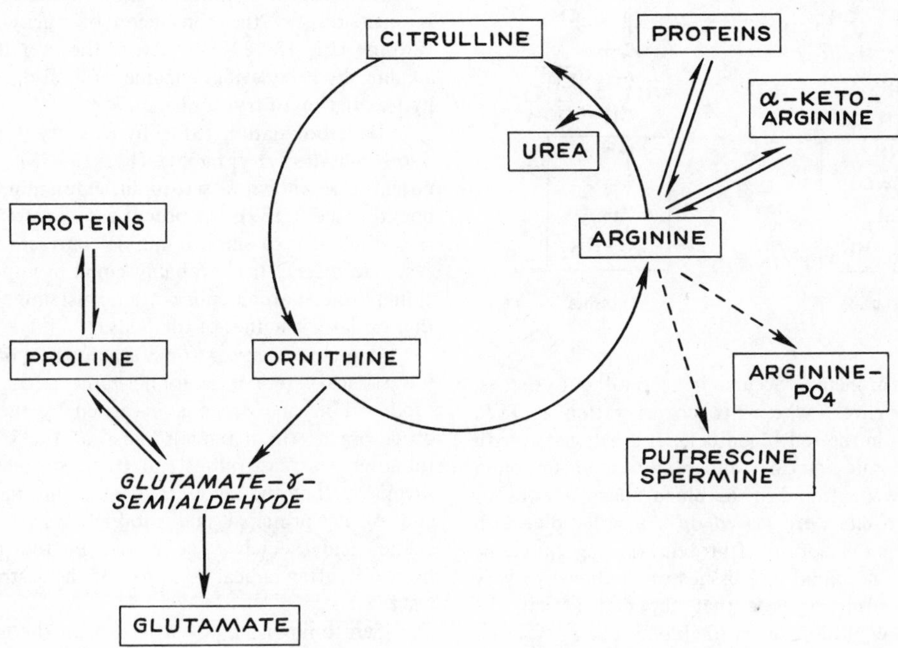

FIG 15–56. Arginine, ornithine, and proline metabolism. Reactions with solid arrows all occur in mammalian tissues. Putrescine and spermine synthesis occurs in *Escherichia coli*, a normal enteric bacterium. Arginine phosphate occurs in invertebrate muscle where it functions as a phosphagen analogous to creatine phosphate in mammalian tissues (see Fig 15–59).

formation of amines such as tyramine, norepinephrine, and serotonin derived from aromatic amino acids which act as pressor agents.

In addition to the aromatic amino acid decarboxylase, there is a completely different enzyme, a specific **histidine decarboxylase** present in most cells, which catalyzes the decarboxylation of histidine.

Three histidine compounds are found in the body: **ergothioneine**, in red blood cells and liver; **carnosine**, a dipeptide of histidine and α-alanine; and **anserine**, 1-methylcarnosine. The latter 2 compounds occur in muscle. The functions of these histidine compounds are not known.

Carnosine can replace histidine in the diet. When injected into animals, it has a circulatory depressant action similar to but not as potent as that of histamine, which in large doses may cause vascular collapse.

Ergothioneine

Carnosine

Anserine

Ergothioneine occurs in several rat tissues, particularly liver, where its concentration is even higher than in the red blood cells. It is associated with the cytoplasmic fraction, apparently in an unbound form. None was found in the blood plasma, testes, or brain. When rats were placed on a purified diet with casein as the sole source of protein, the ergothioneine content of the blood and tissues was reduced to very low levels. This suggests that the diet affects the amounts of ergothioneine in the blood.

Using radioactive sulfur (^{35}S), Melville & others (1955, 1956) could find no evidence in the pig or the rat that synthesis of ergothioneine occurs. In a search for the dietary precursors which are therefore presumed to serve as the sole source of ergothioneine, it was found that both corn and oats contained this compound.

When rabbits were placed on vitamin E deficient diets, 1-methylhistidine appeared in the urine in easily detectable amounts about 1 week after the deficient diet had first been given; and the excretion of this compound increased progressively until it became the major amino acid in the urine. The methylhistidinuria could usually be detected a few days earlier than the creatinuria which is also characteristic of this deficiency state in rabbits, and it preceded by 1–2 weeks the appearance of the physical symptoms of muscular dystrophy.

The presence of 1-methylhistidine in human urine has been reported. This probably is derived from anserine. Larger amounts were found in the urine after the ingestion of rabbit muscle, which is particularly high in anserine content. 3-Methylhistidine has been identified in human urine in amounts of about 50 mg/day. The origin of this compound, an isomer of 1-methylhistidine (a component of anserine), is not known. There is no evidence that it occurs in muscle as a constituent of a peptide similar to anserine. It is of interest that 3-methylhistidine is unusually low in the urine of patients with Wilson's disease (see Chapter 19).

Tryptophan

A. Serotonin: A secondary pathway for the metabolism of tryptophan involves its hydroxylation to 5-hydroxytryptophan (Fig 15–58). The hydroxylation step is probably carried out by a system similar to that involved in the formation of hydroxykynurenine. The oxidation of tryptophan to the hydroxy derivative is analogous to the conversion of phenylalanine to tyrosine (Fig 15–22). In fact, in the liver the phenylalanine hydroxylating enzyme will also accomplish hydroxylation of tryptophan.

Decarboxylation of 5-hydroxytryptophan produces **5-hydroxytryptamine** (Fig 15–58). This compound, also known as **serotonin**, enteramin, or thrombocytin (see p 189), is a potent vasoconstrictor as well as a stimulator of smooth muscle contraction. In these systemic effects it is probably equal in importance to epinephrine, norepinephrine, and histamine as one of the regulatory amines of the body.

It appears that serotonin is actually synthesized in the tissues where it is found rather than that it is produced in one organ and carried by the blood to other organs. In mammals most of the serotonin is found in the gastrointestinal tract, so that it is not surprising that the amounts of serotonin in the blood and of its principal end product, 5-hydroxyindoleacetic acid, which is excreted in the urine, fall markedly after radical resection of the gastrointestinal tract.

Serotonin has a potent effect on the metabolism of the brain. As is the case with other tissues, the compound must be produced within the brain itself from precursors which gain access to the brain from the blood because serotonin itself does not pass the blood-brain barrier to any significant degree. While the functions of serotonin in the brain are not yet entirely clear, it seems reasonable to assume that an excess of serotonin brings about stimulation of cerebral activity and that a deficiency produces a depressant effect.

Most of the serotonin is metabolized by oxidative deamination to form 5-hydroxyindoleacetic acid (5-HIAA). The enzyme which catalyzes this reaction is a **monoamine oxidase (MAO)**. A number of inhibitors of this enzyme have been found. Among them is iproniazid (Marsilid). It is hypothesized that the psychic stimulation which follows the administration of this drug is attributable to its ability to prolong the stimulating action of serotonin through inhibition of monoamine oxidase.

There is evidence that serotonin when first produced in the brain exists in a bound form which is not susceptible to the action of monoamine oxidase. The depressant drugs such as reserpine may effect a rapid release of the bound serotonin, thus subjecting it to rapid destruction by monoamine oxidase. The resultant depletion of serotonin would then bring about the calming effect which follows administration of reserpine.

The preparation and properties of a 5-hydroxytryptophan decarboxylase which forms serotonin from hydroxytryptophan were described by Clark & others (1954). The enzyme was said to be highly specific and to be present in the kidney (hog and guinea pig) as well as in the liver and stomach. However, it should also be recalled that the widely distributed aromatic L-amino acid decarboxylase described by Lovenberg (1962) will also catalyze the decarboxylation of 5-hydroxytryptophan, so that a specific enzyme may not be required for this important reaction.

Although the blood platelets contain a considerable amount of serotonin (see p 189), the lack of a decarboxylase in the platelets suggests that the serotonin is not manufactured but merely concentrated there.

The further metabolism of serotonin by deamination and oxidation results in the production of 5-hydroxyindoleacetic acid (5-HIAA), and this end product is excreted in the urine. In normal human urine, 2–8 mg of 5-HIAA are excreted per day, which indicates that the 5-hydroxyindole route is a significant pathway for the metabolism of tryptophan. Other metabolites of serotonin have been identified in the urine of patients with carcinoid. These include 5-hydroxyindoleaceturic acid (the glycine conjugate of 5-hydroxyindoleacetic acid), N-acetylserotonin, conjugated with glucuronic acid, some unchanged serotonin, and very small amounts of oxidation products of the nature of indican.

Greatly increased production of serotonin occurs in malignant **carcinoid** (argentaffinoma), a disease characterized by the widespread development of serotonin-producing tumor cells in the argentaffin tissue throughout the abdominal cavity. Patients exhibit cutaneous vasomotor episodes (flushing) and occasionally a cyanotic appearance. There may also be a chronic diarrhea. These symptoms are attributed to the effects of serotonin on the smooth muscle of the blood vessels and digestive tract. In over ½ of the patients observed, there is also respiratory distress with bronchospasm. Cardiac involvement may occur late in the disease. The serotonin content in the blood of carcinoid patients—all of it in the platelets—is 0.5–2.7 μg/ml (normal is 0.1–0.3 μg/ml). The most useful biochemical indication of increased production of serotonin, such as may occur in metastasizing carcinoid tumors, is the measurement of the urinary hydroxyindoleacetic acid. In the carcinoid patient, excretion of 5-HIAA has been reported as 76–580 mg in 24 hours (normal is 2–8 mg). Several assay methods for 5-HIAA have been described. For diagnostic purposes, it is recommended that a quantitative measurement of the 5-hydroxyindoleacetic excretion be made on a urine specimen collected over a 24-hour period. Random specimens of urine are useful for qualitative screening tests, but confirmation of the diagnosis should be made only on the 24-hour specimen.

From a biochemical point of view, carcinoid has been considered to be an abnormality in tryptophan metabolism in which a much greater proportion of tryptophan than normal is metabolized by way of the hydroxyindole pathway. One percent of tryptophan is normally converted to serotonin, but in the carcinoid patient as much as 60% may follow this pathway. This metabolic diversion markedly reduces the production of nicotinic acid from tryptophan; consequently, symptoms of pellagra as well as negative nitrogen balance may occur.

B. Melatonin: Melatonin (N-acetyl-5-methoxyserotonin) is a hormone derived from the pineal body and peripheral nerves of man, monkey, and bovine species. The hormone lightens the color of the melanocytes in the skin of the frog and blocks the action of the melanocyte-stimulating hormone (MSH; see p 465). It also blocks the action of adrenocorticotropic hormone.

The pathway for the biosynthesis and metabolism of melatonin is shown in Fig 15–58. It will be noted that the compound is derived from serotonin by N-acetylation in which acetyl-Co A serves as acetate donor, followed by methylation of the 5-hydroxy group in which S-adenoxyl methionine ("active" methionine; Fig 15–32) serves as methyl ($\sim CH_3$) donor. The reaction of methylation of the hydroxy group is localized in pineal body tissue. In addition to methylation of N-acetylserotonin, direct methylation of serotonin as well as of 5-hydroxyindoleacetic acid, the serotonin metabolite, also occurs.

The amines, serotonin as well as 5-methoxytryptamine, are metabolized to the corresponding acids through monoamine oxidase. Circulating melatonin itself is taken up by all tissues, including the brain. However, after administration of radioactively labeled melatonin to mice, it was noted that it was rapidly metabolized and only a small portion was bound and retained. The major catabolic pathway for degradation of melatonin is hydroxylation (in the liver) at position 6 followed by conjugation primarily with sulfate (70%) and with glucuronic acid (6%). A portion is also converted to nonindolic reacting compounds, which suggests that the indole ring has been opened.

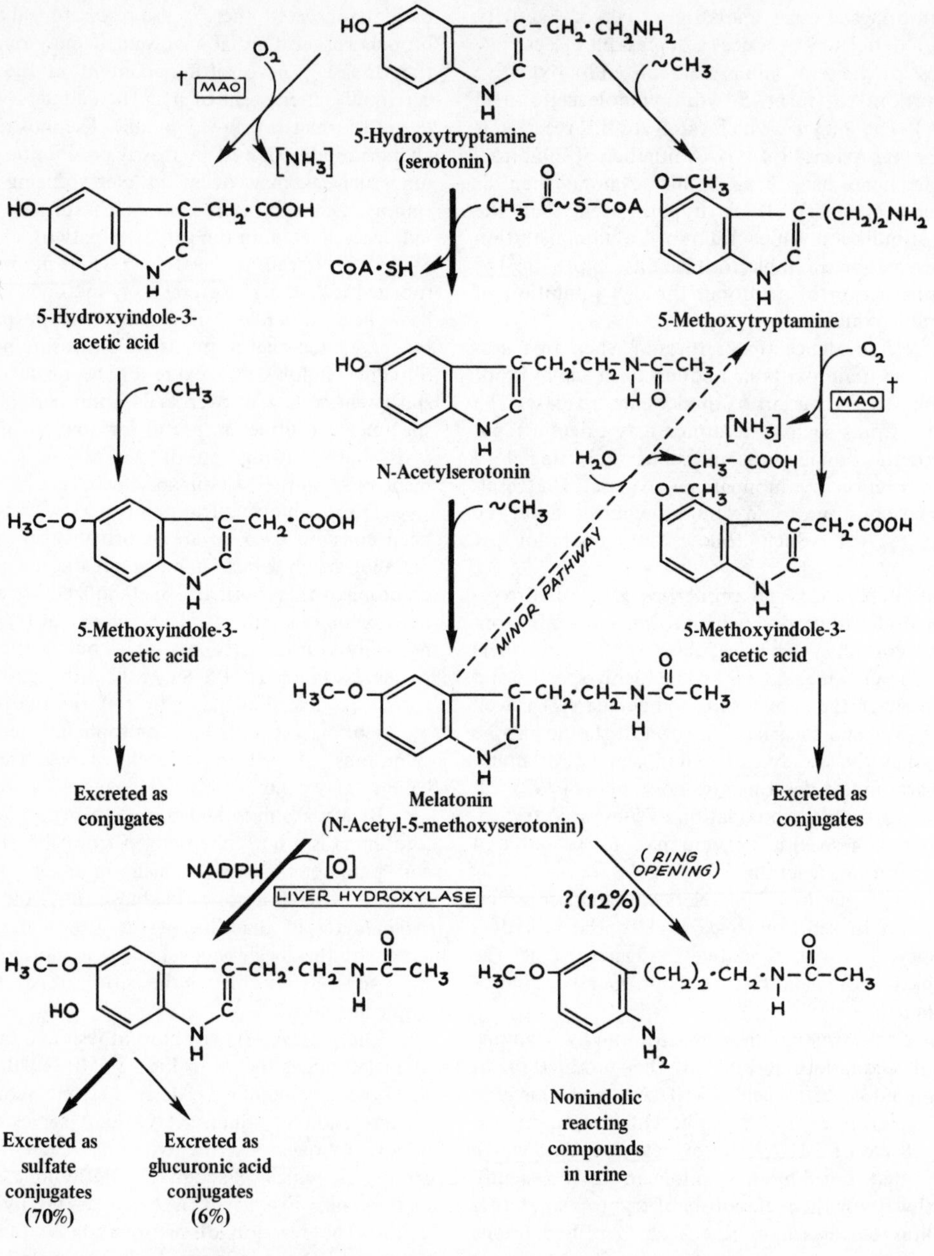

5-Hydroxytryptamine (serotonin)

5-Hydroxyindole-3-acetic acid

5-Methoxytryptamine

N-Acetylserotonin

5-Methoxyindole-3-acetic acid

5-Methoxyindole-3-acetic acid

Excreted as conjugates

Melatonin (N-Acetyl-5-methoxyserotonin)

Excreted as conjugates

NADPH [O] LIVER HYDROXYLASE

(RING OPENING) ? (12%)

MINOR PATHWAY

Excreted as sulfate conjugates (70%)

Excreted as glucuronic acid conjugates (6%)

Nonindolic reacting compounds in urine

FIG 15–58. Biosynthesis and metabolism of melatonin. (Principal pathways indicated by heavy arrows. [NH_3] = by transamination; MAO = monoamine oxidase.)

C. Conversion of Tryptophan to Niacin: Although the pathway for the breakdown of tryptophan shown in Fig 15–28 appears to be the major one, other pathways may also be utilized at the point of the metabolism of hydroxyanthranilate. One such alternate pathway results in formation of nicotinic acid.

In many animals, the conversion of tryptophan to nicotinic acid makes a supply of the vitamin in the diet unnecessary. In the rat, rabbit, dog, and pig, tryptophan can completely replace the vitamin in the diet; in man and other animals, tryptophan increases the urinary excretion of nicotinic acid derivatives (eg, N-methylnicotinamide). In vitamin B_6 deficiency, it has been noted that the synthesis of pyridine nucleotides (NAD and NADP) in the tissues may be impaired. This is a result of the inadequate conversion of tryptophan to nicotinic acid for nucleotide synthesis; if an adequate supplement of nicotinic acid is supplied, nucleotide synthesis proceeds normally even in the presence of the vitamin B_6 deficiency.

It is likely that in many diets tryptophan normally provides a considerable amount of the nicotinic acid requirement. In man, approximately 60 mg of tryptophan produces 1 mg of nicotinic acid. Nutritional deficiency states such as pellagra must therefore be considered combined protein (tryptophan) as well as vitamin (nicotinic acid) deficiencies.

D. Indole Derivatives in Urine: As shown in Fig 15–58, tryptophan may be converted to a number of indole derivatives. The end products of these conversions which appear in the urine are principally 5-hydroxyindoleacetic acid, the major end product of the hydroxy tryptophan-to-serotonin pathway, and indole-3-acetic acid, produced by decarboxylation and oxidation of indole pyruvic acid, the keto acid of tryptophan. The daily excretion of indole-3-acetic acid in man is generally in the range of 5–18 mg, but it may rise to as high as 200 mg/day in certain pathologic states. As might be expected, the excretion of this compound is markedly increased by tryptophan loading.

Mammalian kidney and liver as well as bacteria obtained from human feces have been shown to decarboxylate tryptophan to tryptamine, which can then be oxidized to indole-3-acetic acid. The formation of tryptamine and of indoleacetic acid in animals has been studied by Weissbach & others (1959).

As noted in Fig 15–50, patients with phenylketonuria have been found to excrete increased quantities of indoleacetic acid (and indolelactic acid, probably derived by reduction of indolepyruvic acid) as well as traces of many other indole acids.

Phenylalanine & Tyrosine

Melanin, the pigment of the skin and hair, is derived from tyrosine by way of 3,4-dihydroxyphenylalanine (DOPA) and its oxidation product, 3,4-dioxyphenylalanine (dopaquinone), which progresses to further melanin precursors (Fig 15–59).

The phenols which occur in the blood and urine are derived from tyrosine. In the urine, the phenols are largely conjugated with sulfate (see p 388), and this comprises a portion of the so-called ethereal sulfate fraction of the total urinary sulfur. Tyrosine itself is also excreted in the urine, not only in the free state but also as a sulfate in which the sulfate moiety is conjugated through the para-hydroxy group. The excretion of tyrosine-O-sulfate averaged 28 mg/day in the 5 adult males studied by Tallan & others (1955). This accounted for about ½ of the bound tyrosine and 3–8% of the ethereal sulfate sulfur in the urine. The enzymatic sulfuration of tyrosine derivatives has been studied by Segal & Mologne (1959). It was concluded that sulfuration of tyrosine in vivo may actually occur on an N-terminal tyrosine within a peptide because free tyrosine does not act as a sulfate acceptor in a liver sulfate transfer system.

As shown in Fig 15–59, tyrosine is a direct precursor of epinephrine and norepinephrine. Tyrosine is also the direct precursor of the thyroid hormones which are iodinated tyrosine compounds.

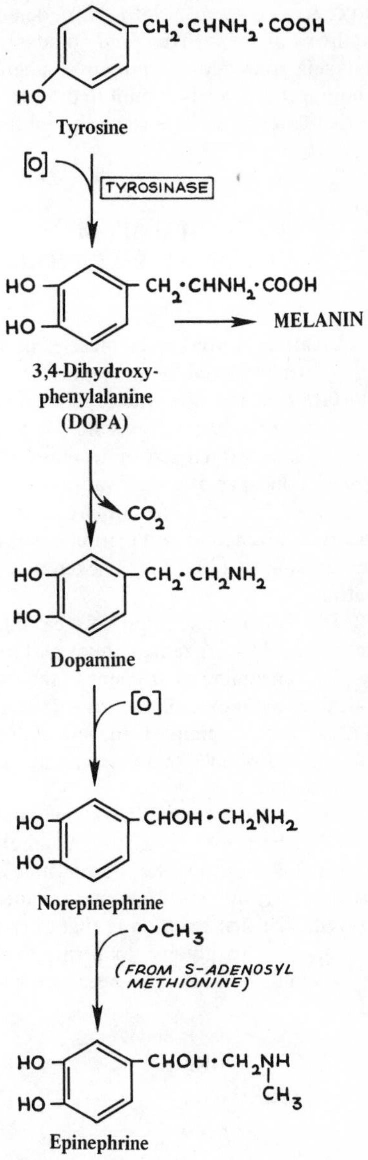

HO — $CH_2 \cdot CHNH_2 \cdot COOH$
Tyrosine
[O] → TYROSINASE
HO, HO — $CH_2 \cdot CHNH_2 \cdot COOH$ ⟶ MELANIN
3,4-Dihydroxyphenylalanine (DOPA)
→ CO_2
HO, HO — $CH_2 \cdot CH_2NH_2$
Dopamine
[O]
HO, HO — $CHOH \cdot CH_2NH_2$
Norepinephrine
$\sim CH_3$ (FROM S-ADENOSYL METHIONINE)
HO, HO — $CHOH \cdot CH_2NH(CH_3)$
Epinephrine

FIG 15–59. Conversion of tyrosine to epinephrine and norepinephrine.

Ascorbic acid and folic acid are both involved in tyrosine metabolism. Both vitamins prevent the defect in tyrosine oxidation observed in guinea pigs maintained on diets deficient in these substances. Alkaptonuria is observed not only in scorbutic guinea pigs but also in premature infants deprived of vitamin C. When the vitamin is supplied, the alkaptonuria promptly disappears. Vitamin C is not effective, however, in alkaptonuria of genetic origin. A direct association of ascorbic acid with tyrosine oxidation at the level of *p*-hydroxyphenylpyruvic acid as well as in the oxidation of homogentisic acid has been demonstrated.

Knox & Goswami (1960) have demonstrated that *p*-hydroxyphenylpyruvic acid oxidase, the enzyme catalyzing conversion of *p*-hydroxyphenylpyruvic acid to homogentisic acid, is inhibited by its own substrate but that ascorbic acid prevents this inhibition.

METABOLISM OF CREATINE & CREATININE

Creatine is present in muscle, brain, and blood, both phosphorylated as phosphocreatine and in the free state (Fig 15–60). Traces of creatine are also normally present in urine. Creatinine is the anhydride of creatine. It is formed largely in muscle by the irreversible and nonenzymatic removal of water from creatine phosphate. The free creatinine occurs in both blood and urine. Formation of creatinine is apparently a preliminary step required for the excretion of most of the creatine.

The 24-hour excretion of creatinine in the urine of a given subject is remarkably constant from day to day. The creatinine coefficient is the 24-hour urinary creatinine expressed in terms of body size. When expressed in this manner, the creatinine excretion of different individuals of the same age and sex is also quite constant.

The origin of creatine shown in the reactions below has been established by metabolic studies and confirmed by isotope technics. Three amino acids—glycine, arginine, and methionine—are directly involved. The first reaction is that of transamidination from arginine to glycine to form guanidoacetic acid (glycocyamine). This has been shown by in vitro experiments to occur in the kidney but not in the liver or in heart muscle. The synthesis of creatine is completed by the methylation of glycocyamine in the liver. In this reaction, "active" methionine is the methyl donor. Other methyl donors, such as betaine or choline after oxidation to betaine, may also serve indirectly by producing methionine through the methylation of homocysteine (Fig 15–31). The methylation of glycocyamine is not reversible. Neither creatine nor creatinine can methylate homocysteine to methionine. ATP and oxygen are required in the methylation of creatine.

The enzymatic mechanisms for the methylation of glycocyamine to form creatine have been studied by Cantoni & Vignos (1954). They were found to be similar to those required for the formation of N-methylnicotinamide. The first step is the formation of active methionine (S-adenosyl methionine; Fig 15–32), which requires ATP, magnesium ions, and glutathione (GSH), and a methionine-activating enzyme. The second step involves the methylation of guanidoacetic acid by active methionine, a reaction which is catalyzed by a soluble enzyme, **guanidoacetate methylferase**, found in cell-free extracts of guinea pig, rabbit, beef, and pig liver. Glutathione (GSH) or other reducing substances are required for the optimal activity of this enzyme; there is as yet no evidence for the need of metal ions or other cofactors.

Until recently, the only site of the transamidinating enzyme in mammals was thought to be the kidney. Evidence has now been obtained to show that bilaterally nephrectomized rats can still synthesize creatine. This is interpreted as proof for the existence of an extrarenal site or sites of transamidination in this animal.

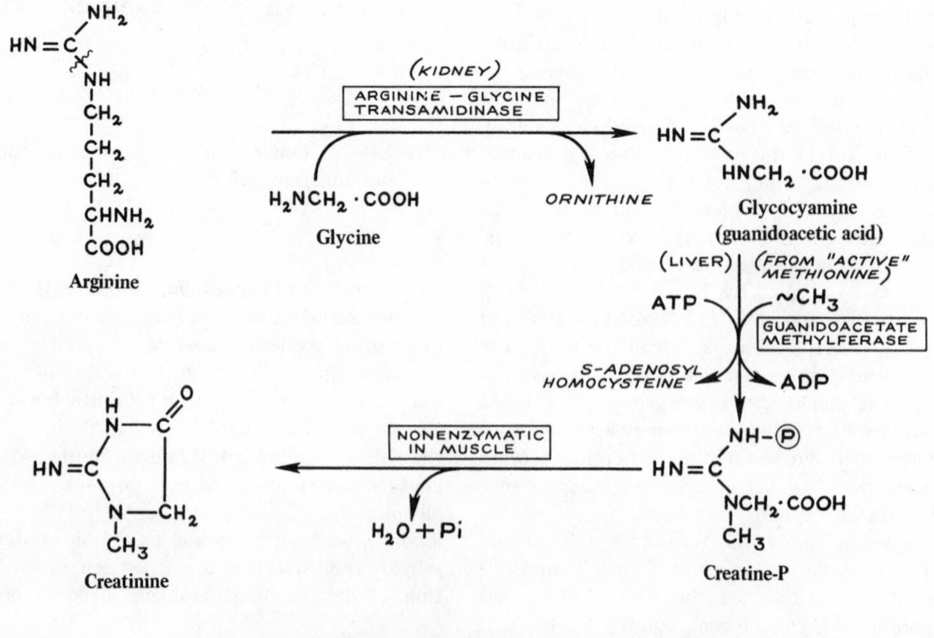

Arginine
(KIDNEY)
ARGININE – GLYCINE TRANSAMIDINASE
$H_2NCH_2 \cdot COOH$
Glycine
ORNITHINE
$HN{=}C(NH_2)HNCH_2 \cdot COOH$
Glycocyamine (guanidoacetic acid)
(LIVER)
(FROM "ACTIVE" METHIONINE) $\sim CH_3$
ATP
GUANIDOACETATE METHYLFERASE
S-ADENOSYL HOMOCYSTEINE
ADP
Creatine-P
NONENZYMATIC IN MUSCLE
$H_2O + Pi$
Creatinine

FIG 15–60. Biosynthesis of creatine and creatinine.

An enzyme preparation has been isolated from pancreatic tissue of beef as well as from the dog, which can catalyze the synthesis of creatine from glycocyamine and S-adenosyl methionine. It has also been found that the pancreas (in contrast to the liver) can synthesize glycocyamine. These observations suggest that the pancreas may play a unique role in the synthesis of creatine within the body of mammals.

The phenomenon of chemical feedback was discussed in Chapter 8. It is well illustrated by the effect of dietary creatine on creatine biosynthesis. In rats fed a complete diet containing 3% creatine, the transamidinase activity of the kidney was markedly lower than that of control animals (Walker, 1960). Gerber & others (1962) have studied the rate at which creatine is synthesized from glycocyamine in the isolated perfused rat liver (1) in the presence of creatine precursors at levels normally present in the blood, and (2) in the presence of high concentrations of precursors (methionine and glycocyamine) added to the perfusion fluid. When the concentration of glycocyamine in the blood fell from 0.4 to 0.2 mg/100 ml, the rate of creatine synthesis decreased, whereas increasing glycocyamine to as high as 20 mg/100 ml together with additional methionine did not alter creatine synthesis significantly. Dietary creatine or a high blood creatine had no effect on the rate of synthesis of creatine in the liver. However, the fact that hepatic synthesis of creatine is related to the blood glycocyamine levels—and that this compound is produced in the kidney—suggests that the rate of creatine biosynthesis is actually dependent on kidney transamidinase activity, and, as noted above, the activity of this enzyme is in fact affected by creatine, apparently as a feedback mechanism.

Hyperthyroidism is one of the diseases which is characterized by disturbances in creatine metabolism. Consequently, it is of interest that hyperthyroidism is also associated with reduced kidney transamidinase activity (Fitch, 1960). It may be that the effect of hyperthyroidism on kidney transamidinase is actually mediated by the elevated levels of blood creatine which occur in this disease, the creatine acting as described above to produce enzyme repression.

Some aspects of the biochemistry of creatine have been reviewed by Ennor & Morrison (1958).

γ-Aminobutyrate

Decarboxylation of glutamate produces γ-aminobutyrate (Fig 15–61). An enzyme that catalyzes its formation from glutamate by alpha decarboxylation is found in the tissues of the CNS, principally in the gray matter. This enzyme requires pyridoxal phosphate (PLP) as a coenzyme.

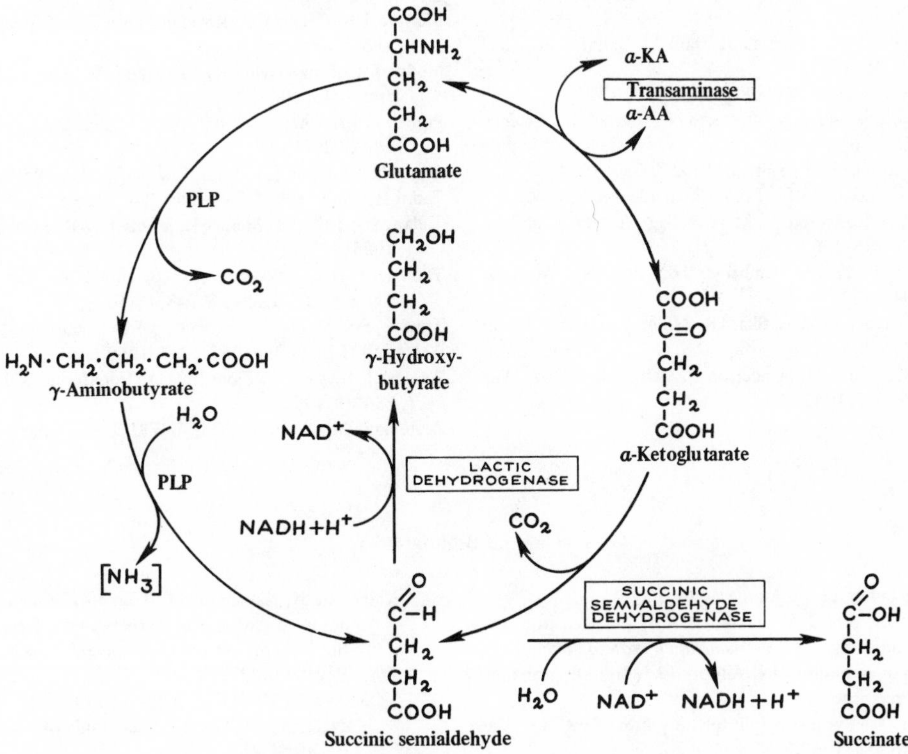

FIG 15–61. Metabolism of γ-aminobutyric acid. (α-KA = α-keto acids; α-AA = α-amino acids; PLP = pyridoxal phosphate.)

γ-Aminobutyrate is now known to serve as a normal regulator of neuronal activity, being active as an inhibitor when studied in various reflex preparations. It is further metabolized by deamination to succinic semialdehyde. The deamination is accomplished by a pyridoxal-dependent enzyme and the ammonia removed is transaminated to ketoglutarate, thus forming more glutamate.

Succinic semialdehyde may then be oxidized to succinate or reduced to γ-hydroxybutyrate, which has been found to exist in the brain in significant amounts. According to Fishbein & Bessman (1964), reduction of succinic semialdehyde by homogenates of brain can be accomplished by an enzyme in the soluble protein fraction which was indistinguishable from lactic dehydrogenase.

When succinic semialdehyde is oxidized to succinic acid, there is completed what amounts to a "bypass" around the citric acid cycle in the brain in the sense that ketoglutarate, rather than going directly to succinate (as in the citric acid cycle), is transaminated to glutamate and thence by decarboxylation to γ-aminobutyrate, which is the source of succinic semialdehyde to form succinate.

γ-Aminobutyrate has also been detected in the kidneys of humans, indicating that this compound is not a unique constituent of the CNS in man (Zachman, 1966).

• • •

References

Auerbach VH, DiGeorge AM, Baldridge RC, Tourtellote CD, Brigham MP: J Pediat 60:487, 1962.

Bessman SP, Baldwin R: Science 135:789, 1962.

Broquist HP, Trupin JS: Ann Rev Biochem 35:231, 1966.

Buniatian HC, Davtian MA: J Neurochem 13:743, 1966.

Cantoni G, Vignos PJ Jr: J Biol Chem 209:647, 1965.

Carson NAJ, Neil DW: Arch Dis Child 37:505, 1962; 38:425, 1963.

Clark CT, Weissbach H, Udenfriend S: J Biol Chem 210:139, 1954.

DeVries A, Kochwa S, Lazebnik J, Frank M, Djaldetti M: Am J Med 23:408, 1957.

Ennor AH, Morrison JF: Physiol Rev 38:631, 1958.

Finkelstein JD, Mudd SH, Irreverre F, Laster L: Science 146:785, 1964.

Fishbein WN, Bessman SP: J Biol Chem 239:357, 1964.

Fitch CD, Hsu C, Dinning JS: J Biol Chem 235:2362, 1960.

Fox M, Thier S, Rosenberg L, Kiser W, Segal S: New England J Med 270:556, 1964.

Freundlich M, Burns RO, Umbarger HE: Proc Nat Acad Sc 48:1804, 1962.

Gerber GB, Gerber G, Koszalka TR, Miller LL: J Biol Chem 237:2246, 1962.

Harper HA, Grossman M, Henderson P, Steinbach H: Am J Dis Child 84:327, 1952.

Harper HA, Hutchin ME, Kimmel JR: Proc Soc Exper Biol Med 80:768, 1952.

Higashino K, Lieberman I: Biochem Biophys Acta 111:346, 1965.

Juva K, Prockop DJ: J Biol Chem 241:4419, 1966.

Knox WE, Goswami MND: J Biol Chem 235:2662, 1960.

Lovenberg W, Weissbach H, Udenfriend S: J Biol Chem 237:89, 1962.

Melville DB, Eich S: J Biol Chem 218:647, 1956.

Melville DB, Otken CC, Kovalenko V: J Biol Chem 216:325, 1955.

Mudd SH, Finkelstein JD, Irreverre F, Laster L: Science 143:1443, 1964.

Popenoe EA, Aronson RB, Van Slyke, DD: J Biol Chem 240:3089, 1965.

Segal HL, Mologne LA: J Biol Chem 234:909, 1959.

Tabor H, Tabor CW: J Biol Chem 244:2286 and 6383, 1969.

Tallan HH, Bella ST, Stein WH, Moore S: J Biol Chem 217:703, 1955.

Terheggen HG, Schwenk A, Lowenthal A, Van Sande M, Colombo JB: Lancet 2:748, 1969.

Thier S, Fox M, Segal S, Rosenberg LE: Science 143:482, 1964.

Walker JB: J Biol Chem 235:2357, 1960.

Weissbach H, King W, Sjoerdsma A, Udenfriend S: J Biol Chem 234:81, 1959.

Zachmann M, Tocci P, Nyhan WL: J Biol Chem 241:1355, 1966.

General Bibliography

Nitrogen Catabolism of the Amino Acids

Allison JB, Bird JWC: Elimination of nitrogen from the body. Page 483 in: *Mammalian Protein Metabolism.* Vol 1. Munro HN, Allison JB (editors). Academic Press, 1964.

Fasella P: Pyridoxal phosphate. Ann Rev Biochem 36:185, 1967.

Hardy RWF, Burns RC: Biological nitrogen fixation. Ann Rev Biochem 37:331, 1968.

Katanuma N, Katsunuma T, Tomino I, Matsuda Y: Regulation of glutaminase activity and differentiation of the isozyme during development. Advances Enzym Regulat 6:227, 1968.

Katanuma N, Okada M, Nishi Y: Regulation of the urea cycle and TCA cycle by ammonia. Advances Enzym Regulat 4:317, 1966.

Meister A: The specificity of glutamine synthetase and its relationship to substrate conformation at the active site. Advances Enzymol 31:183, 1968.

Sallach HJ, Fahien LA: Nitrogen metabolism of amino acids. Page 1 in: *Metabolic Pathways.* Vol 3. Greenberg DM (editor). Academic Press, 1969.

Snell EE, Braunstein AE, Severin ES, Torchinsky Yu M: *Pyridoxal Catalysis: Enzymes and Model Systems.* Interscience, 1968.

Wurtman RJ: Time-dependent variations in amino acid metabolism: Mechanism of the tyrosine transaminase rhythm in rat liver. Advances Enzym Regulat 7:57, 1968.

Conversion of the Carbon Skeletons of the Common Amino Acids to Amphibolic Intermediates

Greenberg DM, Rodwell VW: Carbon catabolism of amino acids. Pages 95 and 191 in: *Metabolic Pathways.* Vol 3. Greenberg DM (editor). Academic Press, 1969.

Hayaishi O: Enzymic hydroxylation. Ann Rev Biochem 38:21, 1969.

Krebs HA: The metabolic fate of the amino acids. Page 125 in: *Mammalian Protein Metabolism.* Vol. 1. Munro HN, Allison JB (editors). Academic Press, 1964.

Meister A: *Biochemistry of the Amino Acids,* 2nd ed. Vol 2. Academic Press, 1965.

Biosynthesis of Amino Acids

Greenberg DM, Rodwell VW: Biosynthesis of amino acids and related compounds. Pages 237 and 317 in: *Metabolic Pathways.* Vol 3. Academic Press, 1969.

Holden JT (editor): *Amino Acid Pools.* Elsevier, 1962.

Meister A: *Biochemistry of the Amino Acids,* 2nd ed. Vol 2. Academic Press, 1965.

Stadtman ER: Allosteric regulation of enzyme activity. Advances Enzymol 28:41, 1966.

Truffa-Bachi P, Cohen GN: Some aspects of amino acid biosynthesis in microorganisms. Ann Rev Biochem 37:79, 1968.

Umbarger HE: Regulation of amino acid metabolism. Ann Rev Biochem 38:323, 1969.

Regulation of Amino Acid Metabolism

Cohen GN: The aspartokinases and homoserine dehydrogenases of *Escherichia coli.* Curr Topics Cellular Regulat 1:183, 1969.

Feigelson P: Studies on the allosteric regulation of tryptophan oxygenase; structure and function. Advances Enzym Regulat 7:119, 1968.

Kenney FT: Mechanism of hormonal control of rat liver tyrosine transaminase. Advances Enzym Regulat 1:137, 1963.

Knox WE: The regulation of tryptophan pyrrolase activity by tryptophan. Advances Enzym Regulat 4:287, 1966.

Knox WE, Greengard O: The regulation of some enzymes of nitrogen metabolism: An introduction to enzyme physiology. Advances Enzym Regulat 3:247, 1965.

Schimke RT, On the roles of synthesis and degradation in regulation of enzyme levels in mammalian tissues. Curr Topics Cellular Regulat 1:77, 1969.

Schimke RT, Doyle D: Control of enzyme levels in animal tissues. Ann Rev Biochem 39:929, 1970.

Umbarger HE: Regulation of the biosynthesis of the branched-chain amino acids. Curr Topics Cellular Regulat 1:57, 1969.

Wood WA: Allosteric L-threonine dehydrases of microorganisms. Curr Topics Cellular Regulat 1:161, 1969.

Metabolic Defects in Amino Acid Metabolism

Aebi HE: Inborn errors of Metabolism. Ann Rev Biochem 36:271, 1967.

Holt LE Jr, Snyderman SE: Anomalies of amino acid metabolism. Page 321 in: *Mammalian Protein Metabolism.* Vol 2. Munro HN, Allison JB (editors). Academic Press, 1964.

Hsia DY-Y: Inborn errors of metabolism. Page 301 in: *Diseases of Metabolism,* 5th ed. Duncan GB (editor). Saunders, 1964.

Larner J: Inborn errors of metabolism. Ann Rev Biochem 31:569, 1962.

Stanbury JB, Wyngaarden JB, Fredrickson DS (editors): *The Metabolic Basis of Inherited Disease,* 2nd ed. McGraw-Hill, 1966.

16 . . .
Metabolism of Purines & Pyrimidines

The chemistry of the nucleic acids derived from nucleoproteins, as well as of the purine and pyrimidines contained in the nucleic acids, has been described in Chapter 4. A summary of these derivatives is given in Table 16–1. In this chapter, the metabolism of these compounds will be discussed.

Digestion

The pancreatic juice contains enzymes (nucleases) which degrade nucleic acids into nucleotides. These include **ribonuclease** and **deoxyribonuclease**; each acts on the type of nucleic acid for which it is specific. In the intestinal juices there are enzymes which supplement the action of the pancreatic nucleases in producing mononucleotides from nucleic acids. These intestinal enzymes are called **polynucleotidases** or **phosphodiesterases**. Certain intestinal phosphatases (specifically termed **mononucleotidases**) may then remove phosphate from the mononucleotides to produce nucleosides.

The final step in this degradation is an attack on the nucleosides by the action of the **nucleosidases** of the intestinal secretions. One nucleosidase reacts with purine nucleosides liberating adenine and guanine, and another enzyme causes the breakdown of the pyrimidine nucleosides liberating uracil, cytosine, or thymine. The action of these nucleosidases is peculiar; the process actually involves a transfer of phosphate so that the end products are the free purine or pyrimidine and a phosphorylated pentose.

RIBOSE-PURINE (NUCLEOSIDE) + PHOSPHATE ⇌ [NUCLEOSIDASE (NUCLEOSIDE PHOSPHORYLASE)] RIBOSE-PHOSPHATE + PURINE

Studies of digestion and absorption of nucleotides using isolated surviving segments of rat or hamster small intestine have been performed by Wilson & Wilson (1958, 1962). Rapid hydrolysis of nucleotides to nucleosides occurred, followed by a much slower conversion to the free base by the **nucleoside phosphorylase** enzyme. In fact, a mixture of nucleoside and the free base was absorbed by the intestinal preparation in vitro. The intestinal mucosa is also capable of metabolizing the purine ribonucleotides to nucleosides as well as free bases. Furthermore, the bases themselves may be further metabolized—eg, conversion of guanine to xanthine and thence to uric acid; or adenosine to inosine, hypoxanthine, and uric acid.

Fate of the Absorbed Products

A. Free Purines and Pyrimidines: The metabolic fate of purine and pyrimidine derivatives has been studied by the oral or parenteral administration of isotopically labeled (^{15}N) compounds. When labeled guanine was fed to rats or pigeons, the tagged nitrogen appeared in large amounts in the urinary allantoin of the rat or the uric acid of the pigeon but did not appear to any appreciable extent in the tissue nucleoproteins. This indicates that ingested guanine is largely catabolized after absorption. A similar fate was indicated for uracil, cytosine, and thymine after the oral administration of the free purines or pyrimidines to rats. On the other hand, the oral administration of labeled adenine resulted in its incorporation into the tissue nucleoprotein and some of the labeled nitrogen was also found in guanine in the body. These experiments suggest that dietary adenine is a precursor of guanine. However, all of the guanine is probably not derived from adenine. After the feeding of labeled glycine, more of the isotope appeared in the guanine than in the adenine, indicating that all of the guanine had not been produced through adenine. In summary, it appears from these experiments that, with the exception of adenine, none of the free purines or pyrimidines of the diet serves as a direct precursor of the purines or pyrimidines of the tissue nucleic acids. It is also of interest that the naturally produced breakdown products of the nucleic acids in the tissues are not reutilized to any significant extent; this is demonstrated by the fact that ^{14}C labeled adenine incorporated into nucleoprotein of one test animal does not appear in that of another connected in a parabiotic experiment.

B. Nucleosides and Nucleotides: Somewhat different results were obtained when the purines or pyrimidines were administered not as the free bases but as

TABLE 16–1. The purine and pyrimidine bases and their related nucleosides and nucleotides.

Base	Nucleoside (Base + Sugar)	Nucleotide (Base + Sugar + Phosphoric Acid)	Source
Purines			
Adenine (6-aminopurine)	Adenosine	Adenylic acid	From RNA and DNA
Guanine (2-amino-6-oxypurine)	Guanosine	Guanylic acid	From RNA and DNA
Hypoxanthine (6-oxypurine)	Inosine (hypoxanthine riboside), hypoxanthine deoxyriboside	Inosinic acid (hypoxanthine ribotide), hypoxanthine deoxyribotide	From adenine by oxidative deamination
Xanthine (2,6-dioxypurine)	Xanthine riboside (or deoxyriboside)	Xanthine ribotide (or deoxyribotide)	From guanine by oxidative deamination
Pyrimidines			
Cytosine (2-oxy-6-amino-pyrimidine)	Cytidine	Cytidylic acid	From RNA and DNA
Thymine (2,6-dioxy-5-methylpyrimidine)	Thymidine	Thymidylic acid	From DNA
Uracil (2,6-dioxypyrimidine)	Uridine	Uridylic acid	From RNA
Uracil	Pseudouridine (5-ribosyl linkage; see pp 51 and 55)	Pseudouridylic acid	From RNA

nucleosides or nucleotides. Thus the subcutaneous injection of the labeled pyrimidine nucleoside, cytidine, into rats resulted in the incorporation of this compound, as well as the labeled nucleoside, uridine, into the ribonucleic acids (RNA), and, to a smaller extent, into the cytosine and thymine of the deoxyribonucleic acids (DNA). There was even some incorporation of the label into the purines of the DNA, although not those of the RNA. This is noteworthy since it indicates that pyrimidine nitrogen has been transferred to purines. The injection of uridine resulted in considerably less incorporation of the ^{15}N label, whereas injected thymidine remained unaltered. Apparently demethylation of thymidine did not take place since no labeled cytosine was detected. The fact that unchanged thymidine, when injected, can be incorporated into DNA has proved to be a valuable technic for labeling newly produced DNA in a great variety of biologic materials both in vivo and in vitro. For this purpose "tritiated thymidine," ie, thymidine containing tritium (^3H), the radioactive isotope of hydrogen, is used.

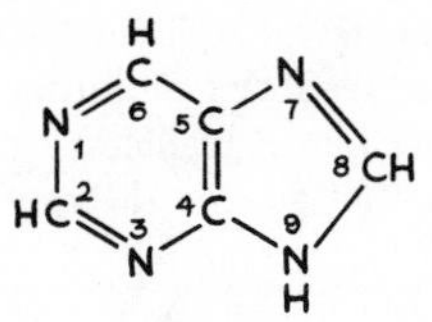

Purine 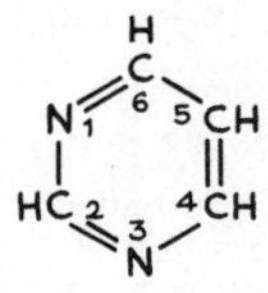 **Pyrimidine**

The utilization of purine nucleotides when given by mouth to rats was, however, not as effective as that of free adenine. Intraperitoneal injection of these nucleotides resulted in a somewhat improved utilization for the synthesis of tissue nucleic acids, but it was still not equivalent to that observed after the administration of adenine itself.

BIOSYNTHESIS OF PURINES & PYRIMIDINES

Preformed purines or pyrimidines are not required in the diet since the animal organism is able to synthesize these compounds. In fact, from the evidence cited above, it is likely that the tissue nucleic acids are very largely derived from synthetic or endogenous sources rather than from preformed, exogenous sources of the diet. The reactions by which these compounds are synthesized have been studied in both plant and animal tissues. Such studies have yielded considerable evidence concerning possible pathways for the synthesis of the purine and pyrimidine derivatives, although it must be remembered that the evidence is derived from various experimental preparations, notably microorganisms, and so does not necessarily represent the established pathways in any one species.

Biosynthesis of Purines

Information on the sources of the various atoms of the purine nucleus (see Fig 16–1) has been obtained by tracer studies in birds, rats, and man. The amino acid glycine is utilized in the intact form to form the carbons in positions 4 and 5 while its alpha-nitrogen forms position 7. The nitrogen at position 1 is derived from the amino nitrogen of aspartic acid; those at positions 3 and 9, from the amide nitrogen of glutamine. The carbon atom in position 6 is derived from respiratory CO_2 while the carbons in positions 2 and 8 come from a one-carbon compound such as formate or from

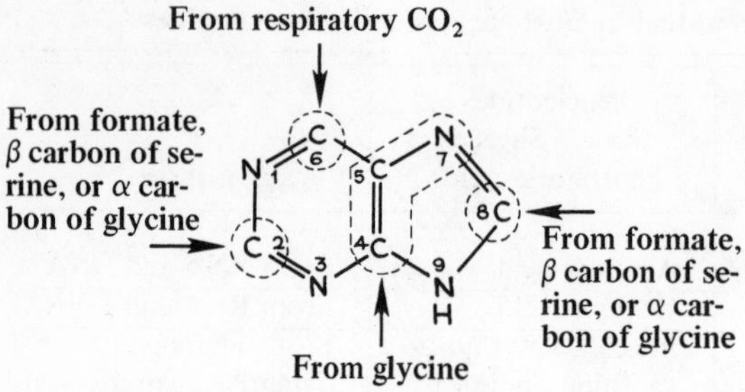

FIG 16–1. Sources of carbon and nitrogen in the purine nucleus.

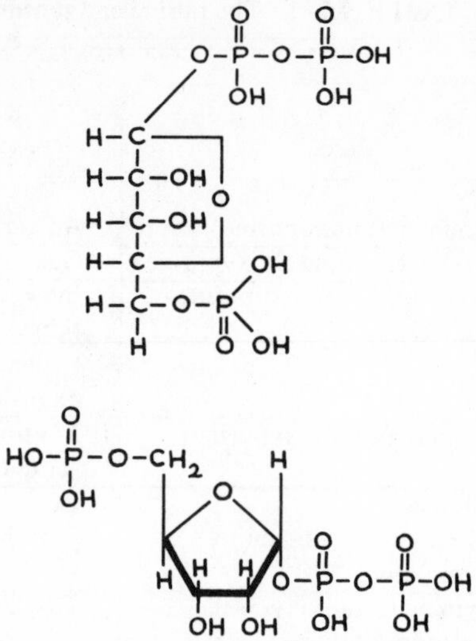

FIG 16–2. 5-Phosphoribosyl-1 pyrophosphate (PRPP).

the beta carbon of serine, which is itself derived from a one-carbon (formate) moiety, when glycine is converted to serine (Fig 7–8). The delta carbon of δ-aminolevulinic acid, which is itself derived from the alpha carbon of glycine, may actually serve as a carrier molecule for the transfer of the alpha carbons of glycine to the purine ring. A similar role has been proposed for homocysteine.

Tetrahydrofolic acid (FH_4) derivatives, acting as formyl carriers, are required for the incorporation of the one-carbon units into positions 2 and 8 (Fig 7–8).

The biosynthetic pathway for the synthesis of purines is shown in Fig 16–4 (Buchanan, 1957). The initial step includes the formation of a nucleotide structure with glycine.* For this purpose, 5 phosphoribosylpyrophosphate (PRPP; see Fig 16–2) serves as phosphate donor. PRPP is frequently used in nucleotide synthesis, eg, in the formation of NAD. This sugar-phosphate is formed in the liver according to the following reaction (Carter, 1956):

$$\text{Ribose-5-phosphate} + \text{ATP} \xrightarrow{Mg^{++}} \underset{\text{(Adenylic acid)}}{\text{Adenosine-5-P}} + \underset{\text{(PRPP)}}{\text{5-Phosphoribosyl-1-pyrophosphate}}$$

In deoxyribonucleotides, the sugar is deoxyribose rather than ribose. Consequently, it is required that ribonucleotides, synthesized as described above and below, must later be convertible to deoxyribonucleotides for incorporation into DNA. This is accomplished by conversion of ribose to deoxyribose, after incorporation of ribose in nucleotide structure, without cleavage of the glycosidic bonds. The conversion is catalyzed by enzymes (**ribonucleotide reductases**) which have been detected in bacterial, avian, and mammalian tissues. The vitamin B_{12}-containing coenzyme, DBC (see p 115), functions as an essential hydrogen-transferring agent in the cobamide-dependent ribonucleotide reductase reaction. The mechanism of action of the cobamide coenzyme in the reductase reaction is discussed by Abeles & Beck (1967).

In *E coli,* the preferred substrates for the reductases are ribonucleotide diphosphates such as cytidylates, uridylates, and the purine ribonucleotide diphosphates. It is likely that control mechanisms operating on these reductase enzymes play an important role in regulation of growth, since it is known that the reductases are subject to feedback inhibition (see p 154) as well as to repression (see pp 69 and 153).

PRPP then reacts (reaction 1, Fig 16–3) with glutamine in a reaction catalyzed by the enzyme **PRPP-amidotransferase** to form 5-phosphoribosylamine, which (reaction 2) reacts in turn with glycine, resulting in the production of glycinamide-ribosyl-phosphate (GAR). This compound is the source of positions 4, 5, 7, and 9 of the purine nucleus. The enzyme catalyzing reaction 2 is designated **GAR-kinosynthetase.**

PRPP-amidotransferase activity is inhibited by ATP and ADP and, less strongly, by GMP, GDP, and IMP. This is an example of **"feedback inhibition"** (see p 154), the important biochemical regulatory mechanism by which an end product adjusts the rate of its own production by inhibiting the activity of an enzyme which is required very early in the biosynthetic pathway involved (see p 356 for other examples in pyrimidine biosynthesis). Caskey & others (1964) have reported on the enzymology of feedback inhibition of the PRPP-amidotransferase by purine ribonucleotides. They postulate that there are at least 2 distinct feedback regulatory sites (see also p 154) on the enzyme protein since 2 types of inhibitors (6-aminopurine and

*There is considerable evidence that the ribosides or ribotides are biologically much more active than the free bases. This has previously been noted in connection with the administration of pyrimidines. None is utilized for the synthesis of nucleic acids if fed as the free base. In the synthesis of purines and pyrimidines all of the intermediate compounds are also first converted into ribotides or deoxyribotides.

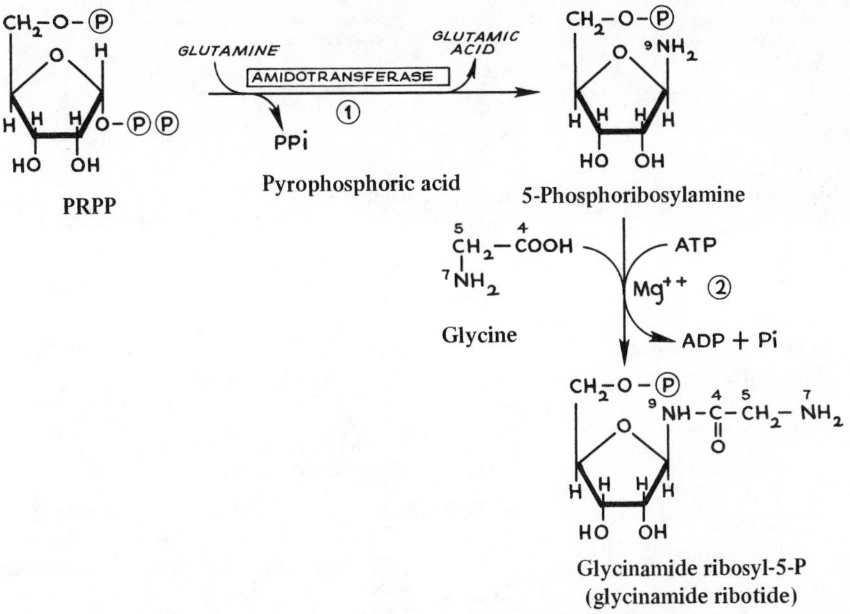

PRPP

GLUTAMINE — AMIDOTRANSFERASE (1) — GLUTAMIC ACID

PPi

Pyrophosphoric acid

5-Phosphoribosylamine

Glycine

ATP, Mg^{++} (2), ADP + Pi

Glycinamide ribosyl-5-P (glycinamide ribotide)

FIG 16–3. Biosynthesis of glycinamide ribosyl-5-phosphate (GAR).

6-hydroxypurine ribonucleotide inhibitors) may act simultaneously upon the enzyme molecule.

Glycinamide-ribosyl-phosphate is then formylated in reaction 3, Fig 16–4, which requires FH_4 and the enzyme **transformylase**, to transfer the one-carbon moiety which will become position 8 of the purine nucleus. In reaction 4, with glutamine as the amino donor, amination occurs at carbon 4 of the formylated glycinamide. The added nitrogen will be position 3 in the purine. Ring closure (reaction 5) forms an amino imidazole which progresses (reaction 6) to 5-amino-4-imidazole-N-succinyl carboxamide ribotide. This is formed by addition of a carbamyl group to the precursor compound. The source of the carbon is respiratory CO_2, and the source of the nitrogen is the amino nitrogen of aspartic acid. The utilization of CO_2 as in other reactions of CO_2 fixation, apparently requires biotin; in fact, the precursor substance, amino-imidazole-ribosyl phosphate, has been found to accumulate in biotin-deficient animals. Next (reaction 7), fumaric acid is split off from the amino-imidazole-succinyl carboxamide and 5-amino-4-imidazole carboxamide ribotide remains. This latter compound is then formylated (reaction 8) to form 5-formamido-4-imidazole carboxamide ribotide. The formyl carbon is transferred from a tetrahydrofolic acid derivative catalyzed by a transformylase. This newly added carbon, which, like carbon 8 of the purine nucleus, is derived from the one-carbon pool, will become carbon 2 of the purine nucleus. Ring closure now occurs (reaction 9), and the first purine to be synthesized is formed.

It is clear from the above scheme that hypoxanthine nucleotide (inosinic acid) is the first purine to be synthesized and that adenine (reactions 10 and 11) and guanine nucleotides (reactions 12 and 13) are then derived from the hypoxanthine or xanthine nucleotides, respectively, by amination.

The amination of inosinic acid is accomplished by the intermediate formation of a compound in which aspartic acid is attached to inosinic acid to form succinoadenine nucleotide (adenylosuccinic acid). The reaction is similar to that of a preceding reaction (reaction 7) in which the nitrogen at position 1 of the purine nucleus was added from aspartic acid. The splitting off of fumaric acid from succinoadenine nucleotide then produces the final product, adenine mononucleotide or adenylic acid (adenosine monophosphate).

The biosynthesis of guanosine nucleotide has been studied by Lagerkvist (1958). Using preparations of pigeon liver, he found that xanthine nucleotide is first formed from inosine nucleotide (reaction 12). Guanosine nucleotide is then produced by amination of xanthosine nucleotide at position 2, using the amide nitrogen of glutamine as nitrogen donor (reaction 13).

Several antimetabolites are effective at various points in purine biosynthesis. Azaserine (O-diazoacetyl-L-serine; see p 30) is an antagonist to the action of glutamine; it blocks reactions 1, 4, and 13, but its action is exerted mainly at reaction 4 (Fig 16–4). Deoxynorleucine (DON, 6-diazo-5-oxo-L-norleucine) is also an antagonist to glutamine. It blocks reaction 4 (Fig 16–4) in purine biosynthesis. Folic acid antagonists (eg, amethopterin) block reaction 8. Purinethol (6-mercaptopurine) is presumed to block reaction 11. A more recently reported inhibitor of purine biosynthesis is hadacidin (N-formylhydroxyaminoacetic acid; $OHC.N[OH].CH_2.COOH$) (Shigeura & Gordon, 1962).

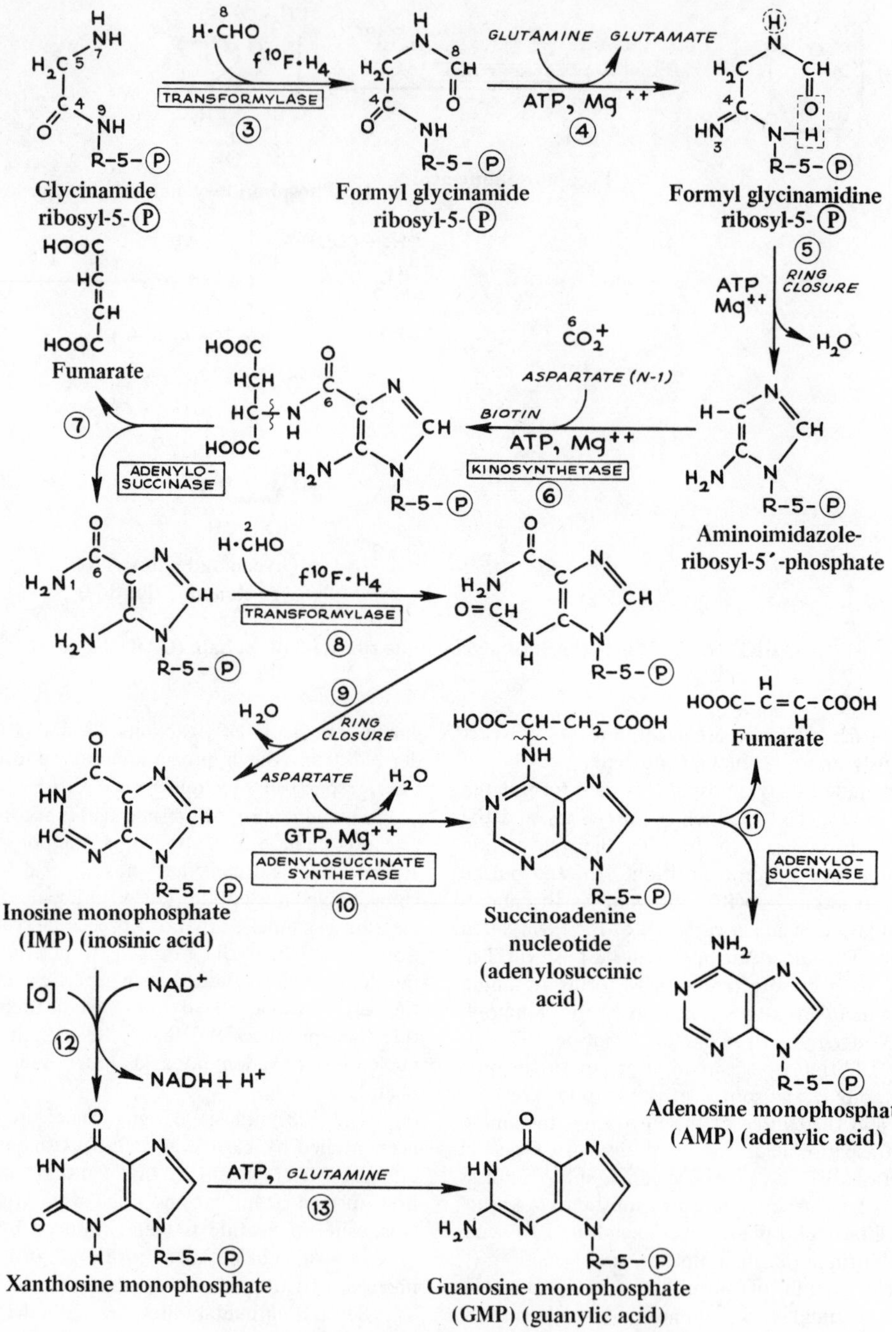

FIG 16–4. The biosynthetic pathway for purines.

This compound markedly suppresses biosynthesis of adenylic or of deoxyadenylic acid. The site of inhibition is probably at reaction 10 because the enzyme adenylosuccinate synthetase was found to be inhibited by hadacidin and its effects reversed by aspartic acid. Unfortunately, hadacidin is a relatively weak agent when tested for antitumor activity probably because of the presence in the intact system of considerable quantities of preformed adenine.

It was noted above that the first step in purine biosynthesis involves formation of a nucleotide structure in a reaction between 5-phosphoribosyl-1-pyrophosphate (PRPP) and glycine. All subsequent compounds leading to formation of purines are therefore formed as nucleotides. However, purine nucleotides can also be formed directly from free purines or purine nucleosides. The pathways which provide for this are useful to permit reutilization of free purines or

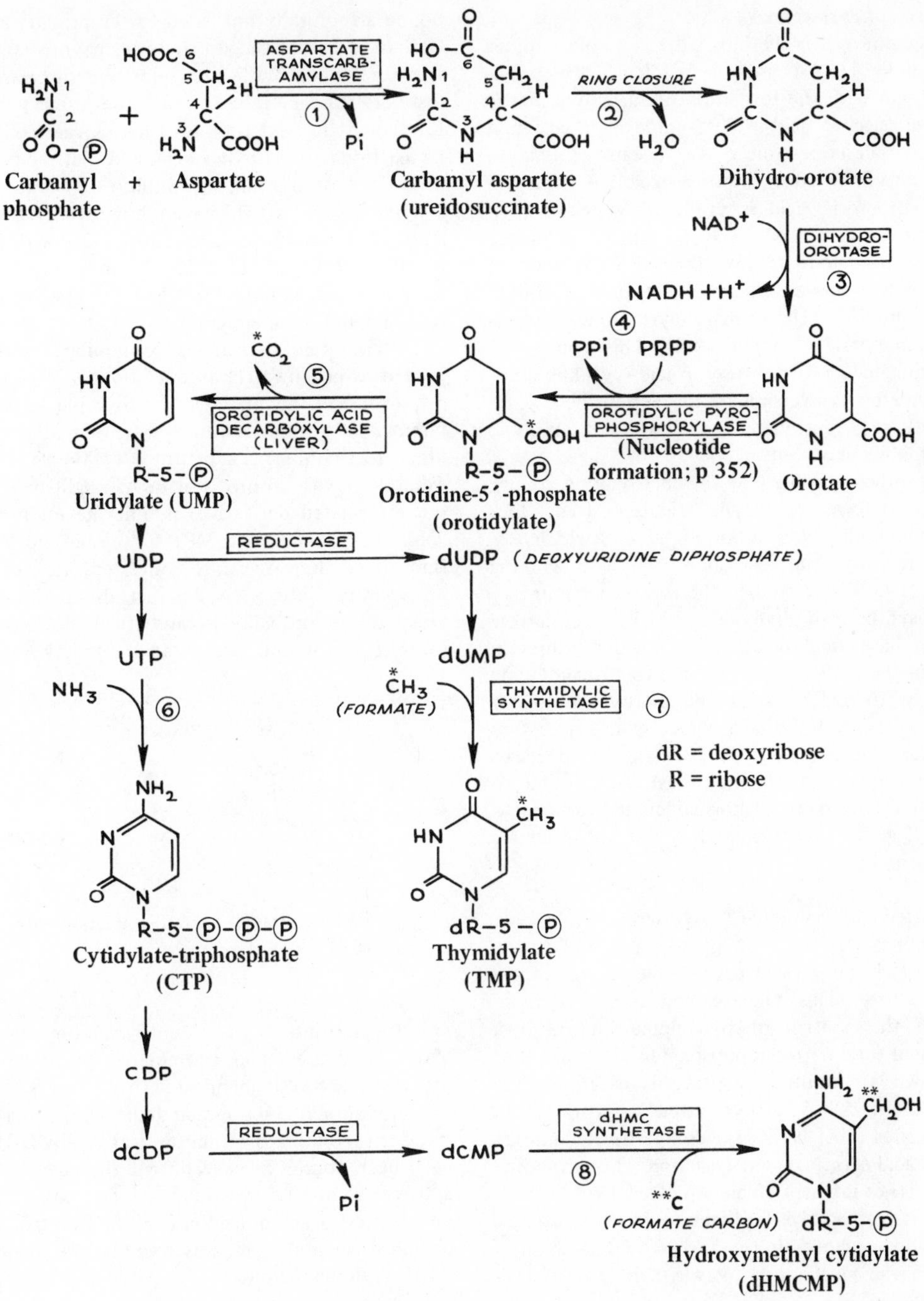

FIG 16–5. The biosynthetic pathway for pyrimidines.

purine nucleosides formed in the course of nucleic acid breakdown in the tissues or possibly absorbed from the intestinal tract. Thus adenine can be converted to adenylic acid in a reaction with PRPP catalyzed by an enzyme in liver, **adenine phosphoribosyl transferase.** Guanine can be converted to guanylic acid and hypoxanthine to inosinic acid, the respective nucleotides, by reacting with PRPP in reactions catalyzed by a liver enzyme, **hypoxanthine-guanine phosphoribosyl transferase.** Purine nucleosides can be formed from free purines by reacting with ribose-1-phosphate in reactions catalyzed by a nucleoside phosphorylase (see p 350). All of the above pathways should be considered as secondary pathways of nucleotide or nucleoside formation. There is no evidence that free purines are synthesized in the tissues. Rather, the purine nucleotides are synthesized from simpler structures as detailed in Figs 16–3 and 16–4, beginning with nucleotide structures formed with PRPP.

Biosynthesis of Pyrimidines

The pathway for the biosynthesis of pyrimidines is shown in Fig 16–5 (Carter, 1956). It may be considered to begin with the formation of carbamylaspartic acid (ureidosuccinic acid). (The synthesis of carbamyl phosphate is described on p 308 because this compound is also involved in the urea cycle.) By ring closure, dihydroorotic acid is produced which is then oxidized to orotic acid. The nucleotide structure orotidine-5′-phosphate is next formed by a reaction with PRPP (see above). Decarboxylation of orotidine phosphate produces the primary pyrimidine, uridine-5′-phosphate (UMP; uridylic acid). This compound may be aminated to form cytidylic acid (cytidine phosphate) or, after conversion to the deoxyribose compound (dUMP), methylated to produce thymidylic acid (thymidine phosphate).

The carbon required in the methylation of uridylic acid to form thymidylic acid is derived either from formate, the beta-carbon of serine, which is a major source, or the alpha-carbon of glycine. Dinning & Young (1959) have studied the role of vitamin B_{12} in the biosynthesis of thymine by chick bone marrow cells. They have concluded that vitamin B_{12} is directly involved in the reduction of formate to formaldehyde preparatory to synthesis of the methyl group for thymine synthesis. Folic acid, which is also necessary in thymine synthesis (as well as in other aspects of purine and pyrimidine biosynthesis), was found to increase the conversion of formaldehyde carbons to the methyl group of thymine; B_{12} did not influence this reaction.

Factors Affecting Pyrimidine Biosynthesis

Reference was made on pp 154 and 352 to the principle of chemical feedback as a factor influencing enzyme activity. This phenomenon is well demonstrated in the control of pyrimidine biosynthesis. Uridylic acid (uridine monophosphate, UMP) has thus been shown to inhibit aspartate transcarbamylase (reaction 1, Fig 16–5, in pyrimidine biosynthesis) in Ehrlich ascites cells. UMP also competitively inhibits orotidylic acid decarboxylase (reaction 5 in pyrimidine biosynthesis) of rat liver (Blair & Potter, 1961). These are examples of negative feedback in which a product of a reaction sequence—in this instance UMP—exerts a control on the mechanism by which it is itself produced.

Gerhart & Pardee (1962) investigated the activity of aspartate transcarbamylase in *Escherichia coli* and found that the activity of this first enzyme in pyrimidine biosynthesis is closely controlled by an end product of the biosynthetic pathway, probably a cytosine derivative, cytidine triphosphate (CTP). This inhibitory action by a nucleotide end product serves as a very important control of the whole biosynthetic pathway. The specific affinity of an enzyme for its substrate is a well established concept. In this case the enzyme is specific to carbamyl phosphate and aspartate. However, there must also be a device whereby the same enzyme protein can be sensitive to another compound structurally unrelated to the primary substrate. Gerhart & Pardee suggest that the enzyme has a second site (the "feedback" site, allosteric site, see p 154) distinct from its active site for the primary substrate, for which the end product (in this case, CTP) has a high affinity. It is further suggested that the bound end product may accomplish inhibition by deforming the enzyme so that it will have a low affinity for its substrate. The observations described above with reference to aspartate transcarbamylase feedback inhibition may serve as a general model for the method of action of all feedback-inhibited enzymes.

Pyrimidine derivatives containing fluorine have proved to be powerful antimetabolites. An example is 5-fluorouracil, an analogue of thymine in which the number 5 position (which on methylation converts uracil to thymine) is substituted by fluorine. Another example is 5-fluoroorotic acid. As might be supposed, the fluorinated derivatives inhibit the methylation of deoxyuridylic acid (dUMP) to thymidylic acid. The fluorinated derivatives are converted into nucleotides, incorporated into RNA, but act also to inhibit RNA biosynthesis, probably because they block incorporation of uracil and orotic acid into RNA (Harbers, 1959).

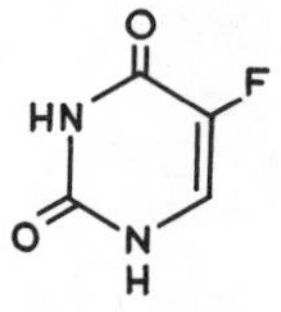

5-Fluorouracil 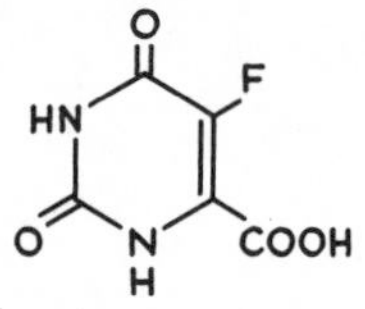 **5-Fluoroorotate**

In addition to the fluoropyrimidines mentioned above, iodinated and brominated pyrimidines have proved to be useful chemical antiviral agents. Examples are 5-iodouracil, 5-iodouracil deoxyribose (IUDR), and the corresponding 5-bromo derivative (BUDR), which have been found to have potent therapeutic antiviral activity against herpes simplex infections of the cornea. Indeed, the iodouridines are the first true antiviral chemical agents to be discovered. The structure of IUDR is shown below.

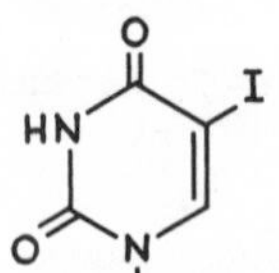

(dR = 2-deoxyribose)

5-Iodouracil deoxyribose (IUDR)

In mammalian cell systems iodouracil and iodouridine have been shown to inhibit utilization of thymidine, formate, and orotic acid for the biosynthesis of the precursors of the thymine of DNA. However, they do not inhibit the utilization of orotic acid for the biosynthesis of cytosine in DNA or RNA (Prusoff, 1960).

Kaufman & Heidelberger (1964) have reported on the therapeutic antiviral activity of the fluorinated pyrimidine, 5-trifluoromethyl-2′-deoxyuridine (F_3TDR). This compound is incorporated into the DNA of human cells grown in culture and confers on such cells increased radiosensitivity. It is claimed that F_3TDR is more potent than IUDR in the treatment of herpes simplex infection of the cornea when tested in rabbits. Strains of the virus both sensitive and resistant to IUDR are equally sensitive to F_3TDR. The new compound is also active against vaccinia virus, so that it may prove effective in the treatment of corneal infection with this virus. The fluorinated pyrimidine as the free base rather than as the nucleoside has no demonstrable therapeutic effect.

(dR = 2-deoxyribose)

Trifluorothymidine (F_3TDR)

Orotic Aciduria

Among the inherited metabolic diseases is a syndrome characterized biochemically by increased excretion of orotic acid and so termed orotic aciduria. In the first case to be described (Huguley, 1959), that of a very young infant born of a consanguineous marriage, there was also a severe anemia observed at age 5 months, characterized by hypochromic erythrocytes and megaloblastic marrow, unrelieved by treatment with iron, pyridoxine, vitamin B_{12}, folic acid, or ascorbic acid. The biochemical defect in orotic aciduria is believed to be attributable to decreased activity of 2 enzymes, orotidylic pyrophosphorylase and orotidylic acid decarboxylase (reactions 4 and 5, Fig 16–5). In support of this hypothesis is the observation that the antimetabolite 6-azauridine, which acts as an antagonist to orotidylic acid decarboxylase, will produce transient orotic aciduria in normal individuals. When uridine and cytidine (or the corresponding nucleotides) are given, the excess excretion of orotic acid is abolished, the pyrimidine nucleosides or their nucleotides acting as feedback inhibitors of the early stages of orotic acid synthesis (see pp 154, 352, and 356). Additional evidence to support the above-designated enzyme defects in orotic aciduria is the finding by Smith & others (1961) of reduced levels of both the pyrophosphorylase and the decarboxylase in the erythrocytes of both parents and 2 surviving siblings of the original case, although none of these cases showed anemia, any morphologic abnormality of the red cells, nor detectable excretion of orotic acid.

The above-described patient with orotic aciduria was treated with a mixture obtained from yeast extract of cytidylic and uridylic acids. As noted above, this mixture of nucleotides effected a marked reduction in excretion of orotic acid as well as an excellent reticulocyte response and a rise in hemoglobin to normal levels. The erythrocytes became normal in appearance, and megaloblasts disappeared from the marrow. The child also gained weight for the first time in 18 months, and developmental retardation (walking, talking, etc) was reversed. Supplying the end products of pyrimidine synthesis evidently permitted normal development of the erythrocytes as well as alleviating other signs of the defect.

Synthesis of Pentoses

Ribose, the pentose required for the synthesis of nucleosides and nucleotides, is readily manufactured from glucose. The HMP shunt, an alternate pathway of glycolysis (see p 241), would provide a method for the synthesis of this sugar. It is probable that pentose is also formed from the combination of 3-carbon and 2-carbon intermediates and by decarboxylation of uronic acids as well (Fig 13–15).

CATABOLISM OF PURINES & PYRIMIDINES

Various tissues, particularly in the liver, contain enzyme systems similar to those described in the intestine for the digestion of nucleic acids.

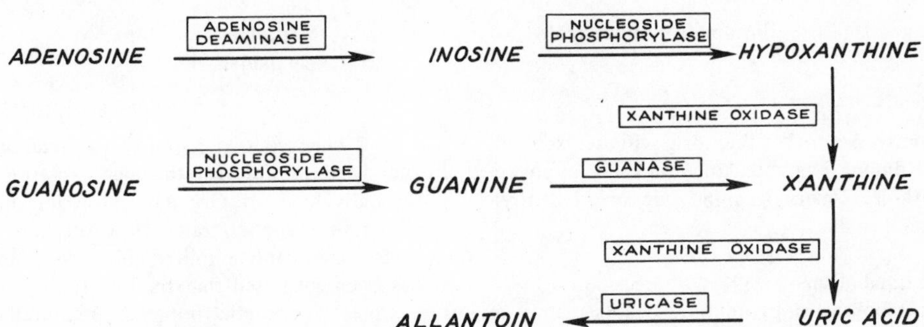

FIG 16.–6. Metabolism of purine nucleosides to uric acid and allantoin.

Purine Catabolism

The further metabolism in the tissues of the purine nucleosides may proceed according to the scheme shown in Fig 16–6.

Adenase is an enzyme corresponding in its action to guanase. These enzymes deaminate the free amino purines. Adenase is present in only very low concentrations in animal tissues, whereas an adenosine deaminase as well as the other enzymes listed below are more generally distributed. It is therefore likely that adenine is deaminated while still in the nucleoside form, as shown below.

Xanthine oxidase, the enzyme which oxidizes hypoxanthine and xanthine, has been isolated from liver and from milk. Recent studies on the activity of this enzyme indicate that the enzyme complex contains riboflavin as a prosthetic group and that trace amounts of iron and molybdenum are also part of the enzyme molecule. Both of these minerals are required in the diet to provide for the deposition and maintenance of normal levels of xanthine oxidase in rat liver and intestine.

Abnormalities of xanthine metabolism have been reported in several patients who excreted only minute quantities of uric acid, although the amounts of xanthine excreted were about equal to the amount of uric acid expected in a normal individual. The disease is considered to be an inherited metabolic defect caused by extremely low xanthine oxidase activity. The low levels of blood uric acid are believed to be due to the rapid glomerular filtration of xanthine together with a failure of reabsorption by the renal tubules, so that little or no xanthine remains to form uric acid. In the xanthinuric patient, xanthine stones may be formed in the kidney. The xanthine stone is not opaque to x-ray; hence this condition might be suspected in any patient presenting symptoms of urinary calculi without x-ray evidence of their presence.

Several purine derivatives in addition to uric acid have been identified in the urine of normal subjects. On a purine-free diet, a total of more than 30 mg/day of purine derivatives other than uric acid are excreted. In order of diminishing abundance they are: hypoxanthine, xanthine, 7-methylguanine, adenine, 1-methylguanine, and guanine.

Pyrimidine Catabolism

The catabolism of pyrimidines occurs mainly in the liver. The release of the ureido carbon (carbon 2) of the pyrimidine nucleus as respiratory CO_2 represents a major pathway for the catabolism of uracil. Diets rich in thymine or in deoxyribonucleic acid (DNA) have been found to evoke in rats increased excretion of **β-aminoisobutyric acid.** Dihydrothymine is even more effective as a precursor of this amino acid than is thymine. The feeding of ribonucleic acid (RNA), which contains no thymine, does not induce the excretion of the amino acid. The hypothetical pathways for the breakdown of the pyrimidines, based on the fragmentary evidence mentioned, are shown in Fig 16–7. β-Alanine and β-aminoisobutyric acids, according to this scheme, are the major end products of cytosine-uracil or thymine catabolism, respectively.

Armstrong & others (1963) have studied the excretion of β-aminoisobutyric acid (BAIB) in man. As indicated below, thymine is the precursor of BAIB both in laboratory animals as well as in humans. The excretion of BAIB is increased in leukemia as well as after the body has been subjected to x-ray radiation. This is undoubtedly a reflection of increased destruction of cells. A familial incidence of abnormally high excretion of BAIB has also been observed in otherwise normal individuals. The genetic trait for this appears to be traceable to a recessive gene since high excretors result only when the trait is homozygous. It is of interest that approximately 25% of persons tested who were of Chinese or Japanese ancestry consistently excreted large amounts of BAIB. Berry (1960) has provided a summary of the various conditions, in addition to increased tissue destruction and the familial high excretors, in which high levels of BAIB are excreted. Although little is known about the mechanisms whereby BAIB is degraded in man, an enzyme has been identified in pig kidney which catalyzes the reversible transamination reaction, shown below, whereby BAIB is converted to methylmalonic semialdehyde. This compound is also a component of the catabolic pathway for the degradation of valine. Consequently, BAIB might be metabolized by conversion to methylmalonic semialdehyde and thence to propionic acid which in turn proceeds to succinate.

$$H_2N-CH_2-\underset{CH_3}{\overset{H}{C}}-COOH \xrightarrow[\text{TRANSAMINASE}]{[NH_3]} H-\overset{O}{C}-\underset{CH_3}{\overset{H}{C}}-COOH$$

β-Aminoisobutyric acid → **Methylmalonic semialdehyde**

It is believed that inability to metabolize BAIB further is responsible for the high excretion noted in those individuals of Oriental extraction having the responsible genetic trait. However, even normal humans possess only a limited ability to destroy BAIB. It has been suggested that the occurrence of BAIB in the urines of genetic nonexcretors, under circumstances where there is no evidence for increased destruction of tissues, may reflect an impairment of liver function.

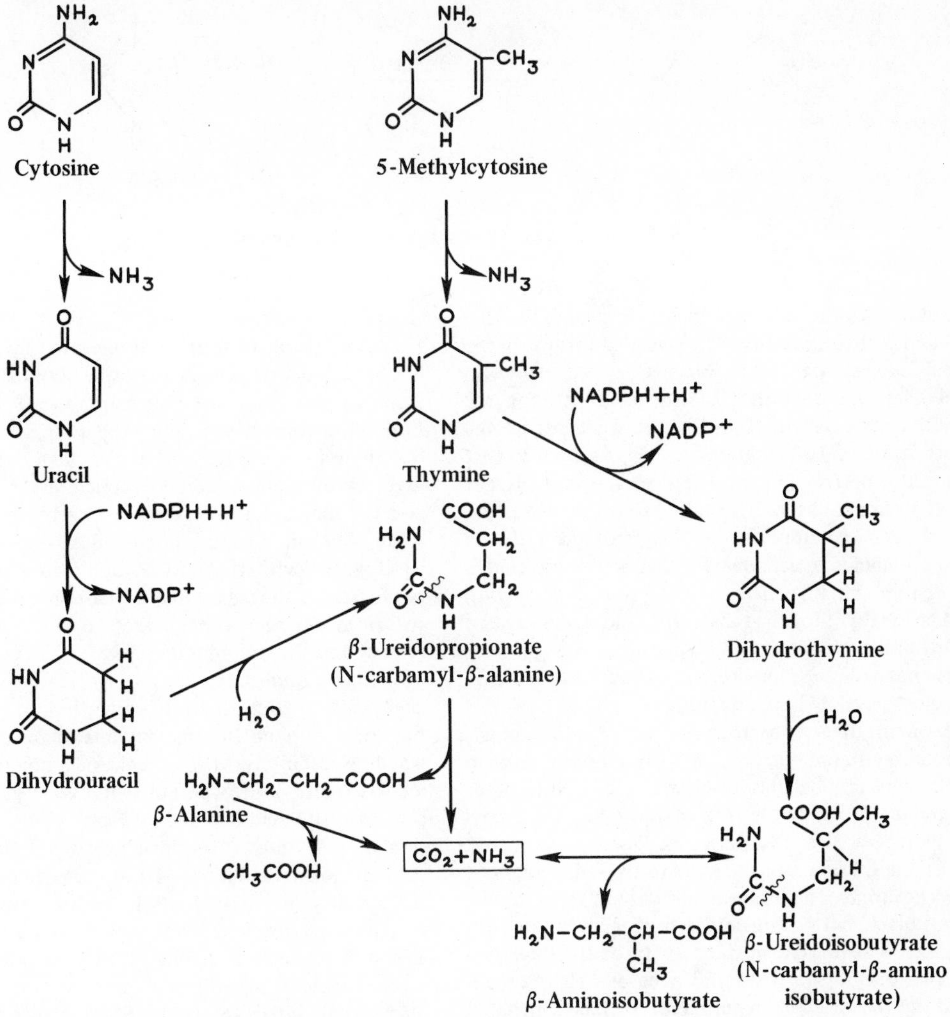

FIG 16–7. Catabolism of pyrimidines.

Uric Acid Metabolism (Wyngaarden, 1957, 1966)

In man, the end product of purine metabolism is mainly uric acid, but in subprimate mammals the uric acid is further oxidized to **allantoin** by the action of **uricase**. Allantoin is, therefore, the principal end product of purine metabolism in such animals.* (See Fig 16–8.)

In birds and reptiles uric acid is synthesized, and this corresponds in its function to urea in man since it is the principal end product of nitrogen metabolism. Such animals are said to be **uricotelic**. Man is **ureotelic**.

*In the Dalmatian dog more uric acid is excreted in comparison to body size than in man, although the plasma levels are only 10–20% as high. This is not due to a deficiency of uricase but is apparently a renal phenomenon caused by a failure in reabsorption of uric acid by the kidney tubules. In fact, there is some evidence for tubular secretion as well as filtration of uric acid in this animal.

The metabolism of uric acid in man has been studied by the use of isotopically labeled uric acid (^{15}N in carbons 1 and 3). Single doses of the labeled uric acid were injected intravenously into a normal human subject and into patients suffering from gout, a disease which is characterized by a disturbance in purine metabolism. The dilution of the injected labeled compound was used to calculate the quantity of uric acid which is present in the body water. This quantity, the so-called "miscible pool," contained in the normal subjects an average of 1131 mg of uric acid. The plasma was found to be considerably higher in uric acid than other portions of the body water. In gouty subjects, the miscible pool was much larger: for example, 4742 mg in a mild case having a serum uric acid of 6.9 mg/100 ml and no symptoms of the disease; up to 31,000 mg in a patient with severe symptoms.

From the rate at which ^{15}N declined in the uric acid, it was possible to estimate the "turnover" of this

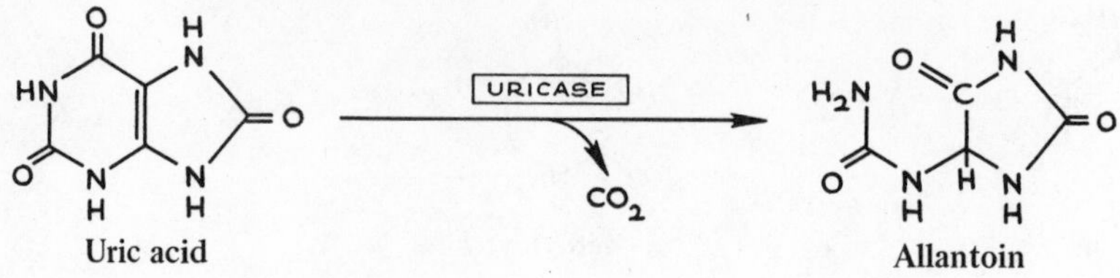

FIG 16–8. Conversion of uric acid to allantoin.

compound, ie, the rate at which uric acid is synthesized and lost to the body. The uric acid formed in the normal subject was 500–580 mg/day. It was also noted that the quantities of uric acid entering the miscible pool exceeded those which were lost by the urinary route by 100–250 mg/day. This suggested that about 20% of the uric acid lost to the body is not excreted as such but is chemically broken down, a conclusion which is supported by the finding of ^{15}N in urea and ammonia. It is known that some uric acid is excreted in the bile. This would then be subject to degradation by intestinal bacteria, and the liberated ^{15}N would be absorbed and reappear in various nitrogenous metabolites. More recent experiments indicate that the breakdown of uric acid in man is, however, independent of the bacteria of the intestinal tract. When administered intravenously to a normal human subject, 18% of labeled uric acid was degraded to other nitrogenous products and 6% of the label was recovered in the feces over a period of 2 weeks. However, 78% of the uric acid injected was recovered unchanged from the urine. The experiment was repeated in the same subject, but sulfonamides were given by mouth to induce a reduction in the activity of the intestinal bacteria. The results were the same as had been obtained in the original experiment. It was concluded that some uricolysis does occur in normal human beings and that the intestinal flora does not make a major contribution to the process.

Excretion of Uric Acid

Uric acid in the plasma is filtered by the glomeruli but is later partially reabsorbed by the renal tubules. Glycine is believed to compete with uric acid for tubular reabsorption. Certain uricosuric drugs block reabsorption of uric acid, eg, salicylates, cinchophen, neocinchophen, caronamide, and probenecid (Benemid). The urinary excretion of uric acid by human subjects can also be increased by the administration of hormones of the adrenal cortex (the 11-oxysteroids) as well as by corticotropin (ACTH).

There is now good evidence for tubular secretion of uric acid by the kidney. Therefore, uric acid may be considered to be cleared from the plasma both by glomerular filtration and by tubular secretion.

Uric acid is very slightly soluble, so that in acid urines it tends to precipitate on standing. This factor may also increase the tendency to form renal calculi in the gouty patient. The urates, alkali salts of uric acid, are much more soluble.

Gout

Gout is characterized by elevated levels of urate in blood and uric acid in the urine. Deposits of urates (tophi) in the joints are also common, so that the disease in this form is actually a type of arthritis.

It has been noted above that the miscible pool of urate in the gouty patient is much larger than in the normal subject. This excess is largely stored in the tophi. The urate in the body fluids is in some degree equilibrated with that in the tophi, since the miscible pool can be decreased by the administration of uricosuric drugs even though these changes may not be reflected in the serum urate levels.

The administration of ^{15}N labeled glycine has been used to study the rate of synthesis of uric acid in the gouty patient. In some patients there were findings which were interpreted as evidence not only for an increased rate of incorporation of dietary glycine nitrogen into uric acid but also of a cumulative incorporation of ^{15}N into uric acid which was 3 times greater than in normal subjects. It was suggested, because of the rapidity and extent of the isotope incorporation, that in some gouty subjects there may be a metabolic alteration whereby glycine nitrogen can enter the purine nucleus of uric acid more promptly than in normal individuals (Stetten, 1954). A diet rich in protein would be expected to increase the formation of uric acid, and it has in fact been found that uric acid synthesis from glycine is accelerated both in normal and in gouty individuals when the protein content of the diet is increased.

Elevated blood levels of uric acid may occur in other diseases in which there is abnormally great turnover of nucleic acids. Examples are polycythemia, myeloid metaplasia, chronic leukemia, and other hematopoietic diseases. Occasionally in such cases there may be attacks of gout. This has also been observed in some patients maintained for long periods on a total fast, undertaken to accomplish weight reduction. Such cases are referred to as secondary metabolic gout. This form of gout is thought to be due to increased catabolism of nucleic acids and a consequent flooding of the organism with the products of nucleic acid breakdown. The etiology of secondary metabolic gout is therefore different from that of primary metabolic gout discussed above, which is presumably due to an inherited metabolic error.

An elevation in the levels of urate in the blood (hyperuricemia) which is not secondary to increased destruction of nucleic acids is the most consistent biochemical criterion of primary metabolic gout. Hyperuricemia may be the only manifestation of the genetic trait for gout, occurring in as many as 25% of a group of asymptomatic individuals in families of gouty patients. In men with the genetic trait for gout, hyperuricemia occurring after puberty may be the first sign of the disease. There may never be any further evidence of the abnormality, or gouty arthritis may develop in middle age. However, women carrying the trait usually do not develop hyperuricemia until after the menopause, and clinical gout, if it occurs at all, does not appear until some years later.

As indicated above, the majority of cases of primary metabolic gout are considered to be due to an inherited metabolic defect in purine metabolism leading to an excessive rate of conversion of glycine to uric acid. The mechanisms for elimination of uric acid are normal, but are inadequate to cope with the increased load. However, there is now increasing evidence that in a smaller number of cases the primary defect may be in the kidney, where there is faulty enzymatic transport (secretion) of urates by the renal tubules. In consequence, there is a lowering of the rate of excretion of urate even though other functions of the kidneys appear to be normal. This type of "primary renal gout" is to be distinguished from secondary renal gout as may occur in a patient with progressive renal failure incident to glomerulonephritis or some other destructive process. Indeed, renal impairment with a decline in glomerular filtration is a common complication of gout in its later stages. In all of these situations, the accumulation of uric acid occurs along with that of many other catabolites which are normally removed by glomerular filtration.

The search is continuing for specific metabolic defects which may cause gout. As indicated above, in some patients, excessive formation of purines seems to be the cause of the excess production of uric acid. More recently, it has been reported that some patients with inherited primary gout have reduced amounts of certain enzymes involved in purine metabolism. The first of these is phosphoribosyl pyrophosphate (PRPP) amidotransferase (reaction 1, Fig 16–3). This enzyme is the key regulator of the rate of purine biosynthesis, involved as it is in feedback regulation of the pathway as described on p 353. It might therefore be suggested that a deficiency of activity of this enzyme would impair normal regulation of rate of production of purines. A second enzyme, which has been found in less than normal amounts in some patients with gout, is hypoxanthine-guanine phosphoribosyl transferase, mentioned on p 355 as involved in catalyzing the formation of nucleotides from free purine bases. Reduced activity of this enzyme would restrict formation of nucleotides, which is essential to the incorporation of purines into nucleic acids for use by the tissues. Under these circumstances, the free purine bases would then be broken down to yield uric acid.

It has also been observed that the urate-binding capacity of plasma from many patients with primary gout is much lower than normal. This reduced ability to bind urate may explain the tendency of the gouty patient to precipitate crystals of sodium urate from the body fluids into the affected tissues. Sodium urate crystals have indeed been recovered from synovial fluid obtained from the joints of acute gouty patients. It is believed that the pain and inflammation that occur during acute attacks of gout is attributable to the deposition of these crystals. Previously it has been stated that uric acid itself is not the offending substance in the causation of acute gouty arthritis. This statement is based upon the fact that acute gout cannot be produced in normal or gouty subjects by the administration of uric acid orally, intravenously, or subcutaneously when injected around the joints. However, Seegmiller has found that, following injection of large, needle-shaped crystals of sodium urate into the unaffected joints of gouty patients, inflammation occurred in almost all cases. The injection of small, amorphous particles of uric acid, on the other hand, produced in these patients little or no inflammation. It was concluded that the urate crystals bring about inflammatory reactions because of their size and shape rather than by a direct chemical reaction.

For the treatment of gout, various uricosuric drugs may be used. These drugs lower the levels of urate in the blood by increasing the excretion (diminishing renal tubular reabsorption) of uric acid. Recently, an extremely useful drug has been introduced for the treatment of gout. This drug (allopurinol, Zyloprim) is an inhibitor of the enzyme xanthine oxidase (see Purine Catabolism, above), which is involved in the conversion of hypoxanthine to xanthine and the later conversion of xanthine to urate. Allopurinol is structurally similar to hypoxanthine, so that it acts as an antimetabolite to hypoxanthine and xanthine, blocking the receptor sites for the natural substrates on the xanthine oxidase enzyme molecule. An important advantage of allopurinol over the uricosuric drugs is that it will be effective in patients with impaired renal function in whom filtration of uric acid is much re-

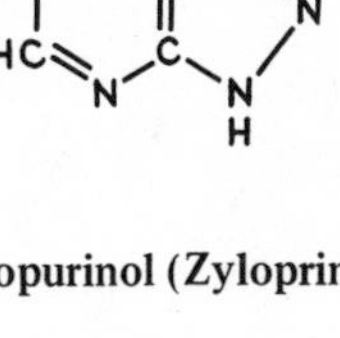

Allopurinol (Zyloprim)

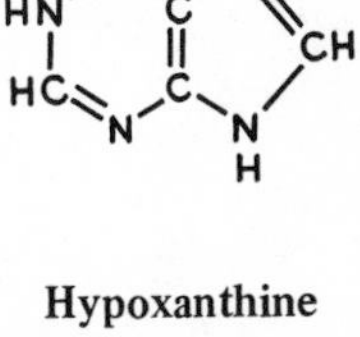

Hypoxanthine

duced. In this situation, the uricosuric drugs are of limited usefulness since they can be effective as tubular blocking agents only after uric acid has been filtered and hence is available for excretion. Furthermore, allopurinol, by reducing the formation of uric acid, decreases the possibility of formation of uric acid stones in the kidney. As expected, allopurinol does lead to an increase in levels of hypoxanthine and xanthine in blood and urine; but these compounds are much more soluble than uric acid, so that the complications associated with high uric acid levels in blood, tissues, and urine are avoided.

In view of the recent information on the etiology of the various forms of gout, as summarized above, it may now be seen why dietary sources of purines contribute very little to the causation of the disease. It follows that the use of purine-free diets cannot be expected to contribute to any significant degree in the management of the disease.

• • •

References

Abeles RH, Beck WS: J Biol Chem 242:3589, 1967.

Armstrong MD, Yates K, Kakimoto Y, Taniguchi K, Kappe T: J Biol Chem 238:1447, 1963.

Berry HK: Metabolism 9:373, 1960.

Blair DGR, Potter VR: J Biol Chem 236:2503, 1961.

Buchanan JM: Texas Rep Biol Med 15:148, 1957.

Carter CE: Ann Rev Biochem 25:123, 1956.

Caskey CT, Ashton DM, Wyngaarden JB: J Biol Chem 239:2570, 1964.

Dinning JS, Young RS: J Biol Chem 234:3241, 1959.

Gerhart JC, Pardee AB: J Biol Chem 237:891, 1962.

Harbers E, Chaudhuri NK, Heidelberger C: J Biol Chem 234:1255, 1959.

Huguley CM Jr, Bain JA, Rivers S, Scoggins R: Blood 14:615, 1959.

Kaufman HE, Heidelberger C: Science 145:585, 1964.

Lagerkvist U: J Biol Chem 233:138, 143, 1958.

Prusoff WH: Cancer Res 20:92, 1960.

Shigeura HT, Gordon CN: J Biol Chem 237:1932, 1962.

Smith LH Jr, Sullivan M, Huguley CM Jr: J Clin Invest 40:656, 1961.

Stetten D Jr: Geriatrics 9:163, 1954.

Wilson DW, Wilson HC: J Biol Chem 237:1643, 1962.

Wilson TH, Wilson DW: J Biol Chem 233:2544, 1958.

Wyngaarden JB: Metabolism 6:245, 1957.

Wyngaarden JB: In: *Metabolic Basis of Inherited Disease,* 2nd ed. Stanbury JB & others (editors). McGraw-Hill, 1966.

17 . . .

The Functions & Tests of the Liver

The liver is the largest and, from a metabolic standpoint, the most complex internal organ in the body. Metabolic disturbances in hepatic disease are therefore quite characteristic and may serve as diagnostic aids. However, because the liver performs so many diverse metabolic functions a great many tests of its functions have been devised, some of which are not clinically practicable. Furthermore, these tests differ widely in sensitivity in various pathologic processes. This is particularly notable in assessing the extent of liver damage, since the less sensitive tests may give normal results even when only about 15% of the liver parenchyma is functioning.

Anatomic Considerations

The basic structure of the liver is the lobule (Fig 17–1), consisting of cords of cells extending outward from the portal triad, which contains the intralobular bile duct and the final small branches of the circulatory vessels: portal vein, hepatic artery, and lymphatics. A system of capillaries and open spaces (sinusoids) containing blood spreads from the lobules to surround individual liver cords and to conduct blood to the central hepatic vein, by which blood leaves the liver. Every lobule is well supplied with a capillary network originating from the portal vein. In this manner, each hepatic cell is provided with an adequate amount of blood. Furthermore, in each lobule there are 5 afferent veins to each efferent vein. Such an arrangement slows down blood flow through the liver and so facilitates the exchange of materials between the blood and the liver tissue.

The liver is supplied with arterial blood by the hepatic arteries. Within the liver, the arterial and portal venous circulation anastomose. In general, about 70% of the blood supply to the liver is delivered by the portal veins, and only 30% by the hepatic arteries.

The Functions of the Liver

The functions of the liver may be classified in 5 major groups:

A. Circulatory Functions: Transfer of blood from portal to systemic circulation; activity of its reticuloendothelial system (Kupffer cells) in immune mechanisms; and blood storage (regulation of blood volume).

B. Excretory Functions: Bile formation and excretion of bile into the intestine; secretion in the bile of products emanating from the liver parenchymal

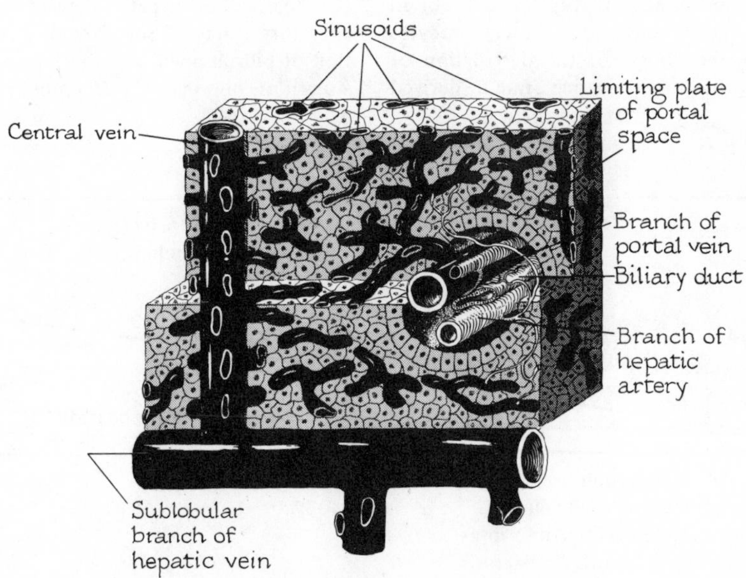

FIG 17–1. The liver lobule.

cells, eg, bilirubin conjugates, cholesterol, cholic acid as bile salts; and excretion of substances withdrawn from the blood by hepatic activity, eg, heavy metals, dyes such as Bromsulphalein, and alkaline phosphatase.

C. Metabolic Functions: Carbohydrate, protein, lipid, mineral, and vitamin metabolism; and heat production.

D. Protective Functions and Detoxification: Kupffer cell activity in removing foreign bodies from the blood (phagocytosis); detoxification by conjugation, methylation, oxidation, and reduction; and removal of ammonia from blood, particularly that absorbed from the intestine by way of the portal vein.

E. Hematologic Functions (Hematopoiesis and Coagulation): Blood formation in the embryo and, in some abnormal states, in the adult; production of fibrinogen, prothrombin, and heparin; and erythrocyte destruction.

PHYSIOLOGIC & CHEMICAL BASIS FOR TESTS OF LIVER FUNCTION

Many functions of the liver have been discussed in other portions of this book. Certain other functions and their applications in tests of liver function will now be described.

Bile Pigment Metabolism

The 2 principal bile pigments are bilirubin and biliverdin. Bilirubin is the chief pigment in the bile of carnivora, including man. Biliverdin is present in only small amounts in human bile, although it is the principal pigment of avian bile.

Bile pigments originate in the reticuloendothelial cells of the liver, including the Kupffer cells, or in other reticuloendothelial cells where the erythrocytes are destroyed. In the course of the destruction of erythrocytes, the protoporphyrin ring of heme derived from hemoglobin is opened to form the bile pigment, biliverdin (see Chapter 6).

Normally, 0.1–1.5 mg of bilirubin loosely associated with protein (mainly albumin) is present in 100 ml of human serum. It has been estimated that 1 g of hemoglobin yields 35 mg of bilirubin.

Bilirubin produced in the reticuloendothelial tissue from the catabolism of heme pigments is carried to the liver, where it is conjugated with glucuronic acid. The bilirubin glucuronide is much more soluble in an aqueous medium than is the free (unconjugated) bilirubin. For this reason, the bilirubin conjugate is readily excreted into the intestine with the bile. Indeed, the formation of the conjugated bilirubin within the liver seems to be a necessary preliminary to its excretion in the bile (Fig 17–2).

Within the intestine, bilirubin ($C_{33}H_{36}O_6N_4$) is successively reduced ultimately to stercobilinogen ($C_{33}H_{48}O_6N_4$), also called L-urobilinogen, the reductions being accomplished by the metabolic activity of the intestinal bacteria. A portion of the urobilinogen is then absorbed from the intestine into the blood. Some of this is excreted in the urine (1–4 mg/day); the remainder is re-excreted in the bile. The unabsorbed urobilinogen is excreted in the stool as fecal urobilinogen (40–280 mg/day). On exposure to air, urobilinogen is oxidized to urobilin. This is the cause of the darkening of the stools on exposure to air.

Jaundice

When bile pigment in the blood is excessive, it escapes into the tissues, which then become yellow. This condition is known as jaundice or icterus.

Jaundice may be due to the production of more bile pigment than the normal liver can excrete, or it may result from the failure of a damaged liver to excrete the bilirubin produced in normal amounts. In the absence of hepatic damage, obstruction of the excretory ducts of the liver by preventing the excretion of bilirubin will also cause jaundice. In all of these situations, bile pigment accumulates in the blood; and

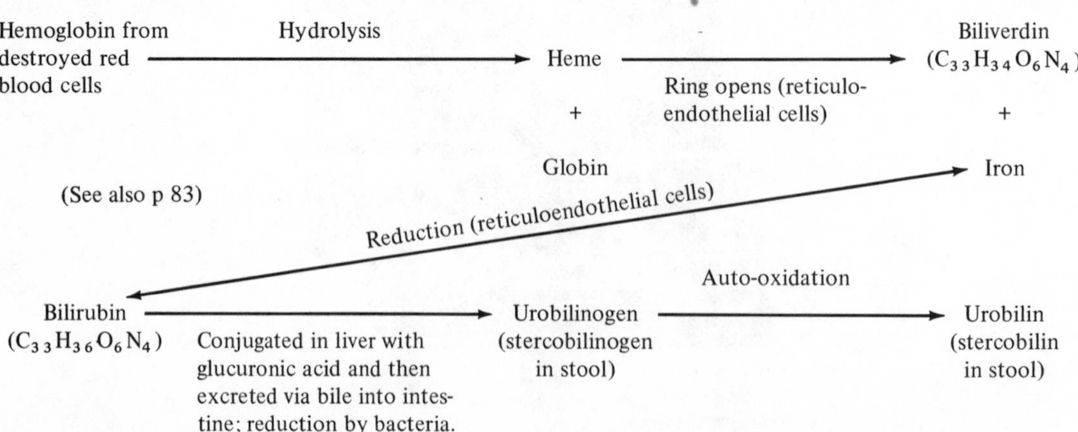

FIG 17–2. Formation of bile pigments.

when it reaches a certain concentration, it diffuses into the tissues. Jaundice is frequently due to a combination of factors.

According to its mode of production, jaundice is sometimes subdivided into 3 main groups: hemolytic, hepatic, or obstructive.

A. Hemolytic Jaundice: Any condition which increases erythrocyte destruction also increases the formation of bile pigment. If erythrocytes are destroyed faster than their products, including bilirubin, can be excreted by the liver, the concentration of bilirubin in the serum rises above normal; hemolytic jaundice is the result.

B. Hepatic Jaundice: This type of jaundice is caused by liver dysfunction resulting from damage to the parenchymal cells. Examples are the jaundice caused by various liver poisons (chloroform, phosphorus, arsphenamine, carbon tetrachloride), toxins, hepatitis virus, engorgement of hepatic vessels in cardiac failure, and cirrhosis.

C. Obstructive (Regurgitation) Jaundice: This condition results from blockage of the hepatic or common bile ducts. The bile pigment is believed to pass from the blood into the liver cells as usual; however, failing to be excreted by the bile capillaries, it is absorbed into the hepatic veins and lymphatics.

Constitutional Nonhemolytic Hyperbilirubinemia

In addition to the occurrence of elevated levels of bilirubin in the serum as a result of the various pathologic states described above, hyperbilirubinemia has been detected in individuals who are otherwise free of any symptoms of hepatic disease or of hemolysis as a cause of hyperbilirubinemia. The abnormality is familial and is therefore believed to be genetically transmitted (Gilbert, 1902; Baroody, 1956). An impairment of excretion of bilirubin has been suggested as the cause of this syndrome, and the results of the bilirubin tolerance test (see below) in these subjects support this suggestion. The increased serum bilirubin is entirely of the indirect-reacting type (ie, unconjugated bilirubin; see below). This could be the result of excessive red blood cell destruction, ie, a hemolytic jaundice; but, as noted above, the metabolism of the red cells is not abnormal. An alternate explanation is that hyperbilirubinemic subjects have an impaired ability to conjugate bilirubin and thus to excrete it in the bile. This latter explanation has proved to be correct, the defect having been found to reside in an impairment in the mechanism for conjugation of bilirubin due to a deficiency of the hepatic enzyme, **glucuronyl transferase**, which catalyzes the transfer of the glucuronide moiety from uridine diphosphoglucuronic acid (UDPGluc) to bilirubin (Arias, 1957). Because the bile of an individual with constitutional nonhemolytic hyperbilirubinemia does contain some direct-reacting bilirubin, it must be assumed that the conjugative defect is not complete. However, there are also reports of rare instances of what appears to be a complete absence of conjugation of bilirubin. These individuals develop a very intense jaundice (serum bilirubin as high as 80 mg/100 ml). In one such case the bile contained no bilirubin ("white" bile), and there was also complete inability to form glucuronide conjugates with substances such as menthol and tetrahydrocortisone that are normally excreted in the urine as glucuronides. This defect resembles that which has also been found in a mutant strain of rats wherein there seems to be a complete absence of the mechanism for detoxification by conjugation with glucuronic acid (Carbone, 1957). It is possible that the mild disease in humans (so-called Gilbert's disease) represents the "trait" or heterozygous form of inheritance of the genetic defect and that the severe form represents the homozygous inheritance of the defect.

The occurrence of some degree of jaundice in the newborn (icterus neonatorum) is not infrequent. Ordinarily, bilirubin levels in the serum rise on the first day of life to reach a maximum on the third or fourth day, after which a rapid decline to normal occurs. The maximum concentration of bilirubin in the serum rarely exceeds 10 mg/100 ml except in premature infants, in whom this so-called physiologic icterus may be more intense and more prolonged than in normal term infants. The phenomenon of icterus neonatorum is considered to be simply a reflection of immaturity ascribable to temporary inadequacy of the hepatic system for conjugating bilirubin. If, however, hemolysis (as in Rh incompatibility) occurs during this period of decreased ability to excrete bilirubin, it is likely that very high levels of bilirubin will be found in the serum. As a result bilirubin may accumulate in the tissues, producing a generalized jaundice. In the brain, the localized deep pigmentation of basal ganglia which occurs is termed "kernicterus." Such "brain jaundice" is associated with objective signs of disturbances in nervous system function, and permanent damage to the CNS will occur if death does not supervene.

Prolonged hyperbilirubinemia in the neonatal period has also been reported by Newman & Gross (1963) in a series of cases which appeared to be associated with breast feeding of the affected infants. Only the unconjugated ("indirect") bilirubin was elevated (as high as 8–19 mg/100 ml on the third to fifth days of life). All of the infants were in otherwise normal health, and there was no evidence of blood group incompatibility or a hemolytic process to account for the elevated indirect bilirubin. Although the bilirubin gradually declined during the first month of life, the rate of decline was much slower than normal unless breast feeding was discontinued in favor of feeding with cow's milk, when the elevated bilirubin levels promptly declined to normal. It was concluded that the hyperbilirubinemia in these infants was in some way related to breast feeding. Earlier it had been shown that serum from pregnant women and newborn infants exerted an inhibitory effect on bilirubin conjugation in rat liver slices and that the inhibitory substance in the serum was pregnanediol. This steroid hormone in quantities as high as 1 mg/day has been identified in the breast milk of mothers whose infants exhibited hyperbilirubinemia. In cow's milk the hormone cannot be detected. It should be emphasized that these effects of pregnanediol in breast milk were observed only in the very young infant at a time when the bilirubin conjugating system may not yet be fully developed.

TESTS BASED ON SECRETORY & EXCRETORY FUNCTIONS

(See Table 17–1.)

Tests of liver function based on the secretory and excretory functions of the liver and on bile pigment metabolism are of major importance.

Estimation of Serum Bilirubin; Van den Bergh Test

A method of quantitatively assaying the bilirubin content of the serum was first devised by Van den Bergh (1913) by application of Ehrlich's test for bilirubin in urine (1883). The Ehrlich reaction is based on the coupling of diazotized sulfanilic acid (Ehrlich's diazo reagent) and bilirubin to produce a reddish-purple azo compound. In the original procedure as described by Ehrlich, alcohol was used to provide a solution in which both bilirubin and the diazo reagent were soluble. Van den Bergh (1916) inadvertently omitted the alcohol on an occasion when assay of bile pigment in human bile was being attempted. To his surprise, normal development of the color occurred "directly." This form of bilirubin which would react without the addition of alcohol was thus termed "direct-reacting." It was then found that this same direct reaction would also occur in serum from cases of jaundice due to obstruction. However, it was still necessary to add alcohol to detect bilirubin in normal serum or that which was present in excess in serum from cases of hemolytic jaundice where no evidence of obstruction was to be found. To that form of bilirubin which could be measured only after the addition of alcohol, the term "indirect-reacting" was applied.

In the years intervening since Van den Bergh first described the 2 types of bilirubin, a number of theories have been proposed in an attempt to explain the chemical and clinical significance of his observations. It has now been demonstrated that the indirect bilirubin is "free" (unconjugated) bilirubin en route to the liver from the reticuloendothelial tissues where the bilirubin was originally produced by the breakdown of heme porphyrins. Since this bilirubin is not water-soluble, it requires extraction into alcohol to initiate coupling with the diazo reagent. In the liver, the free bilirubin becomes conjugated with glucuronic acid, and the conjugate, bilirubin glucuronide, can then be excreted into the bile. Furthermore, conjugated bilirubin, being water-soluble, can react directly with the diazo reagent so that the "direct bilirubin" of Van den Bergh is actually a bilirubin conjugate (bilirubin glucuronide). In jaundice due to obstruction (regurgitation jaundice), the conjugated bilirubin may return to the blood, which accounts for the presence of direct as well as indirect bilirubin in such cases. In hemolytic jaundice there is no obstruction and the increase in bilirubin is confined to the indirect type, since the cause of the bilirubinemia is increased production and not an abnormality in hepatic conjugation or excretion.

When bilirubin appears in the urine, it is almost entirely of the direct type. It is thus probable that bilirubin is excreted into the urine only when conjugated and so made water-soluble. It also follows that bilirubinuria will be expected to occur only when there is an increase in the direct bilirubin content of the serum.

"Direct-reacting" bilirubin is a diglucuronide of bilirubin in which the 2 glucuronyl groups transferred from "active" glucuronide (UDPGluc) by the catalytic action of glucuronyl transferase are attached through ester linkages to the propionic acid carboxyl groups of bilirubin to form an acyl glucuronide. In the conjugation reaction, the monoglucuronide is formed first; this is followed by attachment of a second glucuronyl group to form the diglucuronide. Some monoglucuronide is normally found in the bile, but in hepatic disease, because of impairment of the conjugating mechanism, greater quantities of the monoglucuronide may be formed. Because the monoglucuronide does not react as rapidly with the Van den Bergh reagent as does the diglucuronide, it is possible that the so-called "delayed direct-reacting" bilirubin may reflect the presence of the monoglucuronide.

The determination of serum bilirubin is frequently carried out by the colorimetric method of Malloy & Evelyn as modified by Ducci & Watson (1945). In this method, the color developed within 1 minute after addition of diluted serum (usually 1:10 with distilled water) to the diazo reagent is a measure of the "direct" bilirubin. This is therefore referred to as the "1-minute bilirubin." Another sample of diluted serum is then added to a mixture of the diazo reagent and methyl alcohol and the color developed after 30 minutes is measured. This "30-minute bilirubin" represents the total bilirubin content of the serum, ie, both conjugated (direct) and unconjugated (indirect). The indirect bilirubin concentration is then obtained simply as the difference between the 1-minute and the 30-minute readings. In summary:

Total serum bilirubin = 30-minute bilirubin
Direct (conjugated) bilirubin = 1-minute bilirubin
Indirect (unconjugated) bilirubin = 30 minute minus 1-minute bilirubin

The normal concentrations of these various fractions of the serum bilirubin are given by Watson as follows:

	(mg/100 ml)		
	Total	Direct	Indirect
Mean	0.62 ± 0.25	0.11 ± 0.05	0.51 ± 0.20
Upper limit of normal	1.50	0.25	1.25

A quantitative measurement of the total serum bilirubin is of considerable value in the detection of latent jaundice, which is represented by concentrations between 1.5–2 mg/100 ml. Hyperbilirubinemia may be noted not only in diseases of the liver or biliary tract but also in disease states involving hemolysis such as occurs in infectious diseases, pernicious anemia, or hemorrhage.

The progress of a case of manifest jaundice may be followed by repeated serum bilirubin determinations. A rising concentration is an unfavorable sign; a progressive decline in serum bilirubin signifies improvement in the course of liver disease or of biliary obstructions.

The Icterus Index (Meulengracht Test)

In this test, the intensity of the yellow color of the serum is compared with a 1:10,000 solution of potassium dichromate as a standard. One ml of serum is diluted to 10 ml with 5% sodium citrate solution, and the density of the color of the diluted serum is measured in a photoelectric colorimeter at a wavelength of 420 nm. The colorimetric density of the standard has an icterus index of 10; the ratio of the colorimetric density of the unknown to that of the standard multiplied by 10 equals the icterus index. Normal values for the icterus index are 4–6 units. In latent jaundice (increased bilirubin in the blood without clinical signs of jaundice), the index is between 6 and 15. Above this value symptoms of icterus are usually noted. Since the yellow color is chiefly due to bilirubin, an icterus index of 5 corresponds to about 0.1-0.2 mg of bilirubin/100 ml of serum. Certain other pigments which resemble the yellow color due to bilirubin (eg, the presence of carotene in the blood following the ingestion of carrots) lead to apparent high icterus indices which are, of course, not due to increased blood bilirubin; carrots should not be eaten on the day before the test. Blood should be drawn before breakfast in order to obtain a clear serum, and hemolysis must be carefully avoided.

This test is no longer frequently used because a direct determination of serum bilirubin is more accurate and specific and, with modern analytic methods, almost as simple to perform.

Fecal Urobilinogen & Urobilin

Fecal urobilinogen usually varies directly with the rate of breakdown of red blood cells. It is therefore increased in hemolytic jaundice provided the quantity of hemoglobin available is normal. A decrease in fecal urobilinogen occurs in obstruction of the biliary tract or in extreme cases of diseases affecting the hepatic parenchyma. It is unusual to find a complete absence of fecal urobilinogen. When it does occur, a malignant obstructive disease is strongly suggested. The normal quantity of urobilinogen excreted in the feces per day is from 50–250 mg. Fecal urobilinogen is determined by conversion of urobilin into urobilinogen and determination of urobilinogen by methods similar to those used in the urine analyses (see below).

Test for Bilirubin in the Urine

The presence of bilirubin in the urine suggests that the direct bilirubin concentration of the blood is elevated due to hepatic parenchymatous or duct disease. Bilirubin in the urine may be detected even before clinical levels of jaundice are noted. Usually only direct-reacting bilirubin is found in the urine. The finding of bilirubin in the urine will therefore accompany a direct Van den Bergh reaction in blood (see above).

Bilirubin may be detected in the urine by the Harrison test, the Gmelin test, or the Huppert-Cole test. This first test is the most sensitive.

A. Harrison Test: Mix 5 ml of urine and 5 ml of a 10% barium chloride solution in a test tube. Collect the precipitate on filter paper and spread it to dry on another filter paper. When dry, add 1 or 2 drops of Fouchet's reagent (25 g trichloroacetic acid and 0.9 g ferric chloride in 150 ml water) to the precipitate. A green color appears in the presence of bilirubin. Disregard other colors. Strips of filter paper impregnated with barium chloride may also be used. The strip of barium chloride paper is moistened with urine, and a drop of Fouchet's reagent is then added to the wet area.

B. Gmelin Test: Two or 3 ml of urine are carefully layered over about 5 ml of concentrated nitric acid in a test tube. If bilirubin is present in the urine, rings of various colors–green, blue, violet, red, and reddish-yellow–will appear at the zone of contact.

C. Huppert-Cole Test: Five ml of a suspension of calcium hydroxide in water are added to 10 ml of urine; the tube is well shaken and the mixture filtered. The bile pigment is removed with the calcium hydroxide. The residue on the filter is dissolved with about 10 drops of concentrated hydrochloric acid; the pigment set free is dissolved by adding 10 ml of alcohol to the filter; and the alcoholic solution of the bile pigments is then caught in a clean tube. The filtrate in the test tube is warmed on a water bath; the development of a green color indicates the presence of bilirubin.

Urine Urobilinogen

Normally, there are mere traces of urobilinogen in the urine (average, 0.64 mg; maximum normal, 4 mg [in 24 hours]). In complete obstruction of the bile duct, no urobilinogen is found in the urine since bilirubin is unable to get to the intestine to form it. In this case, the presence of bilirubin in the urine without urobilinogen suggests obstructive jaundice, either intrahepatic or posthepatic. In hemolytic jaundice, the increased production of bilirubin leads to increased production of urobilinogen, which appears in the urine in large amounts. Bilirubin is not usually found in the urine in hemolytic jaundice, so that the combination of increased urobilinogen and absence of bilirubin is suggestive of hemolytic jaundice. Increased blood destruction from any cause (eg, pernicious anemia) will, of course, also bring about an increase in urine urobilinogen. Furthermore, infection of the biliary passages may increase the urobilinogen in the absence of any reduction in liver function because of the reducing activity of the infecting bacteria.

Urine urobilinogen may also be increased in damage to the hepatic parenchyma because of inability of the liver to re-excrete into the stool by way of the bile the urobilinogen absorbed from the intestine.

The **Wallace-Diamond test** is used for detection of urobilinogen in the urine. One ml of the aldehyde reagent of Ehrlich (a solution of *p*-dimethylaminobenzaldehyde acidified with hydrochloride) is added to 10 ml of undiluted urine and allowed to stand 1–3 minutes. The quantity of urobilinogen in the urine is estimated by noting the rapidity and intensity of color development. If the color remains a light red, which is normal, further testing with diluted urine is not required. When urobilinogen is present in larger concentrations than normal, additional tests with various dilutions from 1:10–1:200 or higher must be carried out. The highest dilution which shows a faint pink discoloration is sought, and the results are expressed in terms of this dilution. The appearance of color in dilutions up to 1:20 is considered normal, whereas persistence of color in higher dilutions is indicative of abnormal concentrations of urine urobilinogen.

For quantitative measurement of urinary urobilinogen the method of Schwartz & others (1944) as improved by Balikov (1955) may be used.

Excretion Tests

A. The Bromsulphalein Excretion Test (BSP): In this test the ability of the liver to remove a dye from the blood is determined, and this is considered to be indicative of its efficiency in removing other substances from the blood which are normally excreted in the bile. In normal persons, a constant proportion (10–15% of the dye present in the blood stream) is removed per minute. In hepatic insufficiency, BSP removal is impaired by cellular failure or decreased blood flow; in fact, the test gives more useful information than does the concentration of serum bilirubin if impairment of function is still marginal.

Removal of BSP by the liver has now been found to involve conjugation of the dye as a mercaptide with the cysteine component of glutathione (Grodsky, 1961). The reaction of conjugation of BSP with glutathione is rate limiting, and thus it exerts a controlling influence on the rate of removal of the dye by the liver (Philp, 1961).

The BSP test is carried out by the intravenous injection of 5 mg of the dye/kg body weight, withdrawal of a blood sample 30 or 45 minutes later, and estimation of the concentration of the dye in the plasma.

The normal retention of this dye is less than 10% at 30 minutes and less than 6% at 45 minutes. The BSP excretion test is a useful index of liver damage, particularly when the damage is diffuse and extensive, as in portal or biliary cirrhosis. The test is of no value if obstruction of the biliary tree exists.

The rate of removal of substances other than BSP has also been studied in liver disease.

B. Rose Bengal Dye Test: Ten ml of a 1% solution of the dye are injected intravenously. Normally, 50% or more of the injected dye disappears within 8 minutes.

C. Bilirubin Tolerance Test: One mg/kg body weight of bilirubin is injected intravenously. If more than 5% of the injected bilirubin is retained after 4 hours, the excretory (or bilirubin conjugating?) function of the liver is considered abnormal.

The bilirubin excretion test has been recommended by some authorities as a better test of excretory function of the liver than the dye tests mentioned above because bilirubin is a physiologic substance whereas these dyes are foreign substances. However, the test is not used extensively because of the high cost of bilirubin and the difficulty of measuring small differences in blood bilirubin by the analytical technics in common use.

The 3 substances listed above, with the exception of BSP, are excreted almost entirely by the liver. No significant amounts are taken up by the reticuloendothelial cells.

Plasma Alkaline Phosphatase (See p 161.)

The normal values for plasma alkaline phosphatase are 2–4.5 Bodansky units/100 ml of plasma for adults, and 3.5–11 for children. Since alkaline phosphatase is normally excreted by the liver, these values are increased in obstructive jaundice. In a purely hemolytic jaundice there is no rise. Unfortunately, various other factors may affect phosphatase activity so that the results of this test must be correlated with clinical findings and with other tests.

DETOXIFICATION TESTS

Hippuric Acid Test

The best known and most frequently used test of the protective functions of the liver is based on a conjugation reaction: the detoxication of benzoic acid with glycine to form hippuric acid, which is excreted in the urine. In a sense, the test also measures a metabolic function of the liver since the rate of formation of hippuric acid depends also on the concentration and amount of glycine available.

A. Technic: The test may be carried out by either of 2 methods. (1) Intravenous injection of 1.77 g of sodium benzoate in 20 ml of water over a 5- to 10-minute period with collection of all of the urine secreted during the hour after injection. (2) Oral administration of 6 g of sodium benzoate with a 4-hour collection of urine.

B. Normal Values: The hippuric acid in the urine is then isolated, hydrolyzed, and the benzoic acid produced by hydrolysis measured by titration. Using the intravenous test, more than 0.7 g of hippuric acid should be excreted in 1 hour by normal persons; in the oral test, excretion of 3 g in the 4-hour period is normal.

C. Interpretation of Results: This test is a valuable method of determining the presence or absence of intrinsic liver disease as distinguished from extrahepatic involvement as in obstruction. It is in no way affected by jaundice and may therefore be used to distinguish between intrahepatic and extrahepatic jaundice. It may also be used to determine liver function and reserve in jaundiced patients when biliary operation is contemplated. However, there is difficulty in the interpretation of the hippuric acid test when the liver damage is secondary to extrahepatic obstruction. Therefore it is useful in distinguishing between intrahepatic and extrahepatic jaundice only early in the course of obstruction.

The output of hippuric acid is low in intrahepatic disease such as hepatitis or cirrhosis. Results are normal in cholecystitis, cholelithiasis, and biliary obstruction from stones in the common duct if the condition is of short duration. Later in the disease, liver damage secondary to the obstruction or inflammatory process leads to reduced hippuric acid output.

The hippuric acid test is valueless if renal function is impaired, and in the presence of normal renal function a minimum output of 100 ml of urine is necessary for a valid result.

TESTS BASED ON METABOLIC FUNCTIONS

Carbohydrate Metabolism

The most important tests of this type are based on tolerance to various sugars since the liver is involved in removal of these sugars by glycogenesis or in the conversion of other sugars to glucose in addition to glycogenesis.

A. Glucose Tolerance: (See Chapter 13.) This test may be used to evaluate hepatic function in the absence of other abnormalities in glucose metabolism.

B. Galactose Tolerance: This test is used primarily to detect liver cell injury. It is applicable in the presence of jaundice. The normal liver is able to convert galactose into glucose; but this function is impaired in intrahepatic disease, and the amount of galactose in the blood and urine is excessive. Since the test measures an intrinsic hepatic function, it may be used to distinguish obstructive and nonobstructive jaundice, although in the former case, after prolonged obstruction, secondary involvement of the liver leads to abnormality in the galactose tolerance.

1. In the intravenous test, 0.5 g of galactose/kg body weight is given after a 12-hour fast. Blood galactose is then measured at various intervals. Normally, none is found in the 75-minute sample; but in intrahepatic jaundice, the 75-minute blood value is greater than 20 mg/100 ml. In obstructive jaundice without liver cell damage, galactose is present at 75 minutes; but it is less than 20 mg/100 ml.

2. In the oral test, 30 g of galactose in 500 ml of water are taken by mouth after a 12-hour fast. Normally, or in obstructive jaundice, 3 g or less of galactose are excreted in the urine within 3–5 hours; and the blood sugar returns to normal in 1 hour. In intrahepatic jaundice, the excretion amounts to 4–5 g or more during the first 5 hours.

The intravenous galactose tolerance test is a good liver function test, whereas the oral test is a poor one since in addition to liver function other factors, such as rate of absorption of the sugar from the intestine, affect the outcome of the test.

C. Glycogen Storage (Epinephrine Tolerance): The response to epinephrine as evidenced by elevation of the blood sugar is a manifestation of hepatic glycogenolysis. However, this is directly influenced by hepatic glycogen stores. This test is therefore designed to measure the glycogen storage capacity of the liver.

The subject is placed on a high-carbohydrate diet for 3 days before the test. After an overnight fast, the blood sugar is determined and 0.01 ml of a 1:1000 solution of epinephrine/kg body weight is injected. The blood sugar is then determined at 15-minute intervals up to 1 hour. Normally, in the course of an hour, the rise in blood sugar over the fasting level exceeds 40 mg/100 ml. In hepatic disease, the rise is less.

This test may also be used for diagnosis of glycogen storage disease (Von Gierke's disease).

Protein Metabolism

A. Hypoalbuminemia: The importance of the liver in the manufacture of albumin and other blood proteins has been noted on p 197. In acute and chronic liver disease, such as cirrhosis, there is a general tendency to hypoproteinemia which is manifested mainly in the albumin fraction. Parenchymal liver damage may also elicit a rise in globulin, mainly gamma globulins, although the $alpha_2$ and beta globulins are notably changed in patients with relapsing hepatitis, between the 14th and 30th days.

The severity of hypoalbuminemia in chronic liver disease is of diagnostic importance and may serve as a criterion of the degree of damage. Furthermore, a low serum albumin which fails to increase during treatment is usually a poor prognostic sign.

B. Prothrombin Time and Response to Vitamin K: Measurement of plasma prothrombin (see p 189) and of the response to vitamin K is a useful test of liver function. Prothrombin is measured in a system which relates the amount of time required for clotting of plasma to the prothrombin levels of the plasma. The test is therefore referred to as **"prothrombin time."** Normal levels of prothrombin in control sera give prothrombin times of about 15 sec. Where there is impairment of liver function, plasma prothrombin times may be increased from 22 to as much as 150 sec when marked reduction of liver function has occurred. The persistence of a prothrombin deficit (as evidenced by prolonged prothrombin time) or the occurrence of only a slight increase after administration of vitamin K is evidence of severe liver damage. In conducting this liver function test, vitamin K must be given parenterally in order to ensure absorption. The test must be properly interpreted since many other diseases as well as drugs may reduce the prothrombin content of the plasma. However, the measurement of prothrombin before and after the administration of vitamin K, although not very sensitive, is considered a valuable test in the diagnosis of liver disease, particularly in the distinction between early obstructive and hepatic jaundice. It is also a valuable prognostic sign. It should be noted that the test cannot be used unless the prothrombin time is prolonged by at least 25% of normal before administration of vitamin K; furthermore, a decrease in prothrombin time by at least 20% must occur after vitamin K has been given if the response is to be considered diagnostically significant. The test dose of vitamin K should be no more than 4 mg. Larger doses may produce a false-positive response by mass action of a large, unphysiologic dose of the vitamin.

C. Amino Acid Tolerance: Tests based on the rates of disappearance from the blood of intravenously or orally administered amino acids may also measure a protein metabolic function of the liver. Methionine and tyrosine have been so used. In liver disease the rates of removal of these amino acids are retarded.

D. Tests of Altered Protein Fraction Production: In liver disease there is evidence that altered protein fractions are produced, and these can be detected in the blood. A group of tests to detect these proteins has been devised for clinical use.

1. The cephalin-cholesterol flocculation (Hanger) test is said to be positive if (1) gamma globulin is increased, (2) albumin of the plasma is decreased below the concentrations which can inhibit the reaction, or (3) the inhibiting ability of plasma albumin is decreased. The test is conducted by mixing 4 ml of saline with 1 ml of the cephalin-cholesterol emulsion and adding to this mixture 0.2 ml of serum. This is then kept in the dark at room temperature and read at 24 and 48 hours.

In normal serum the emulsion remains stable, and flocculation does not occur. With pathologic serum, varying degrees of both precipitation and flocculation occur. The test is reported in terms of plus signs, 4+ indicating complete precipitation and flocculation. Equivocal (± to 2+) results are considered negative. It is not a quantitative test but is intended to aid in the diagnosis of an active pathologic process in the liver such as in acute hepatitis. Thus, it is possible that the cephalin-cholesterol flocculation test may be negative and the results of the hippuric acid test abnormal, eg, in a case of currently inactive chronic liver disease with permanent liver damage. It is also valuable in infectious hepatic disease in measuring the progress of the infection. Its value in distinguishing between intrahepatic and surgical obstructive jaundice is questionable.

2. The thymol turbidity test–0.1 ml of serum is added to 6 ml of a thymol solution and allowed to stand for 30 minutes. Turbidity is then read in a colorimeter against a barium sulfate standard. Normal is 0–4 units.

Sera with high beta and gamma globulin fractions give positive tests. The thymol turbidity test correlates well with the findings of the cephalin-cholesterol test, although it is more sensitive. It measures only an acute process in the liver, but the degree of turbidity is not proportional to the severity of the disease. Positive results are obtained in the thymol turbidity test for a longer period in the course of the disease than in the case of the cephalin-cholesterol flocculation test. A negative thymol turbidity test in the presence of jaundice is very useful for distinguishing between hepatic and extrahepatic jaundice.

The mechanisms operating in the thymol turbidity test are not identical with those of cephalin-cholesterol flocculation. The thymol test requires lipids, which are not necessary in the cephalin-cholesterol test; and, conversely, gamma globulin, so important in cephalin flocculation, is not necessarily involved in thymol turbidity. The flocculate in the thymol test is a complex lipothymoprotein. The thymol seems to decrease the dispersion and solubility of the lipids; and the protein is mainly beta globulin, although some gamma globulin is also precipitated.

Thymol flocculation may occur in a thymol turbidity test some time after maximal turbidity has developed. The occurrence of thymol flocculation may also be used as a diagnostic test in liver disease. Flocculation greater than 1+ in 18 hours is considered abnormal. Early in hepatitis, thymol flocculation may be abnormal before turbidity.

3. The zinc turbidity test–When serum with an abnormally high content of gamma globulin is diluted with a solution containing a small amount of zinc sulfate, a turbid precipitate forms; the amount of precipitation is proportional to the concentration of gamma globulin. In conducting this test, 0.05 ml of serum is mixed with 3 ml of the buffered zinc reagent and the mixture allowed to stand for 30 minutes. The tubes are then shaken and the turbidity is read in a spectrophotometer at 650 nm against barium sulfate standards similar to those used in the thymol turbidity test. Normal values are 4–12 units, with an average normal of 8 units.

The reaction in the zinc turbidity test is specific for gamma globulin and not for liver disease in itself; thus it may reflect only antibody formation. The principal advantage of the zinc turbidity test over the cephalin-cholesterol flocculation or thymol turbidity tests is that a single alteration in the serum is measured, ie, an elevation in the gamma globulin fraction. The test is useful, when serial determinations are made, in following the course of infectious hepatitis.

Naganna & others (1962) have described the chemical findings on a patient in whom high globulin levels were present in the serum but both zinc and thymol turbidity reactions were normal. It was suggested that a gamma globulin fraction was present which interfered with the expected turbidity reactions by acting as a "protective protein." The occurrence of such a protein may explain the observation that, in some cases of myeloma with high serum globulin levels, normal turbidity and flocculation reactions may nonetheless occur.

4. Turbidimetric measurement of gamma globulin–A direct turbidimetric measurement of gamma globulin by precipitation with ammonium sulfate-sodium chloride solutions has been described (De La Huerga, 1950).

Lipid Metabolism

A. Cholesterol-Cholesterol Ester Ratio: The liver plays an active part in the metabolism of cholesterol, including its synthesis, esterification, oxidation, and excretion. Normal total blood cholesterol ranges between 150–250 mg/100 ml, and about 60–70% of this is esterified.

In obstructive jaundice an increase in total blood cholesterol is common, but the ester fraction is also raised so that the percentage esterified does not change. In parenchymatous liver disease there is either no rise or even a decrease in total cholesterol, and the ester fraction is always definitely reduced. The degree of reduction roughly parallels the degree of liver damage.

B. Phenol Turbidity: A measure of the total lipids in the serum can be obtained by a turbidimetric reaction with a phenol reagent. The method was described by Kunkel & others (1948).

Other aspects of lipid metabolism in liver disease are discussed in Chapter 14. They are not used in chemical tests of liver function, however.

Enzymes in Liver Disease

The activity of a number of enzymes in the blood has been studied in liver disease. The increase in serum alkaline phosphatase which may occur in obstructive hepatic disease (see above) is considered to be due to the fact that this enzyme is normally excreted by the liver. However, alterations in the serum concentrations of a number of other enzymes may reflect other aspects of hepatic dysfunction. An example is the appearance in the serum of certain enzymes which are liberated from the liver as a result of the tissue breakdown which occurs in hepatocellular disease. Thus, in acute hepatitis during its clinical peak, high levels of serum lactic dehydrogenase (LDH) have occasionally been found (Wróblewski, Science, 1956). The content of this enzyme is also reported to be elevated in the serum of patients with obstructive jaundice or with metastatic disease in the liver (Hsieh, 1956).

TABLE 17–1. Relative value of various liver function tests in jaundice and disease.

Physiologic Basis for Test	Test		Acute "Hepatic" Disease: Pre-icteric Phase	Acute "Hepatic" Disease: Icteric Phase: Obstructive	Icteric Phase: Intrahepatic	Icteric Phase: Hemolytic	Acute "Hepatic" Disease: Convalescent Phase	Latent or Subclinical Hepatic Disease	Chronic Hepatic Disease	Normal Range of Values
Bile pigment metabolism	Urine bilirubin	Presence	A (incr)	B	B	–	B (decr)	–	–	None
		Absence	–	–	–	A	–	–	–	
	Urine urobilinogen	Increase	A (incr)	–	A	A	B (decr)	B	B	1–4 mg/24 hours
		Absence	–	B	–	–	–	–	–	
	Fecal urobilinogen	Increase	–	–	–	A	B (decr)	–	–	50–250 mg/24 hours
		Absence	–	A	–	–	B (incr)	–	–	
	Serum bilirubin (icterus index)		A (incr)	–	–	–	A (decr)	B	B	0.05–0.50 mg/100 ml 3–8 II units
Enzyme activity	Plasma alkaline phosphatase		–	A (incr)	–	–	–	–	–	2–4.5 units (Bodansky)
	Serum transaminase (SGOT†)		–	A	A	±	±	±	B	8–40 units
Cholesterol metabolism	Plasma cholesterol	Esterified	–	–	B (decr)	–	–	–	–	60–75% of total
		Total	–	B (incr)	–	–	–	–	–	100–250 mg/100 ml‡
Protein synthesis	Plasma prothrombin	Before vitamin K	–	–	–	–	–	–	–	90–100%
		After vitamin K	–	A (response)	A§	–	–	–	B§	15% incr in 48 hours
	Serum proteins	Albumin	–	–	–	–	–	B	B	3.4–6.5 g %
		Globulin	–	–	–	–	–	B	B	2.0–3.5 g %
		Total	–	–	–	–	–	–	B	5.7–8.2 g %
	Serum thymol turbidity		–	–	–	–	B	B	A	0–4 units
	Cephalin flocculation		B	–	B	–	–	B	A	0–1 plus
Dye excretion	Bromsulphalein (5 mg/kg)		–	–	–	–	B	A	A	< 10% (30 minutes)
			–	–	–	–	B	A	A	< 6% (45 minutes)

Legend: A = excellent; B = good; [–] = limited or no value.

*Modified, with permission, from Chatton & others: *Handbook of Medical Treatment*, 12th ed. Lange, 1970.

†SGOT = Serum glutamic-oxaloacetic transaminase.

‡Varies widely with different laboratories. Serial determination important.

§No response.

Glutamic-pyruvic transaminase (GPT) as compared to glutamic-oxaloacetic transaminase (GOT) activity is relatively greater in liver than in other tissues. It is not surprising, therefore, that measurement of serum GPT has been found to be useful in the diagnosis and study of acute hepatic disease (Wróblewski, Proc, 1956). GPT is not significantly altered by acute cardiac necrosis, as is GOT.

Aldolase (Bruns & Puls, 1954; Cook, 1954) and phosphohexose isomerase (Bruns & Hinsberg, 1954; Bruns & Jacob, 1954) are both markedly increased in the serum of patients with acute hepatitis. No increase is found in cirrhosis, latent hepatitis, or biliary obstruction.

The liver is a major if not the only source of amylase found in the serum under normal physiologic conditions. Consequently, it is not surprising that the serum amylase levels may be low in liver disease. The activity of cholinesterase may also be lowered when disease involves the hepatic parenchyma (Mann, 1952).

The activity in the serum of the NADP-dependent enzyme, isocitric dehydrogenase (ICD), is increased in the early stages of hepatitis of viral origin. There is a lesser elevation of the activity of this enzyme in serum in some malignancies with metastases to the liver. The ICD levels are well within the normal range in cirrhosis of the liver and in extrahepatic obstructive jaundice. There is also no change in the activity of this enzyme in a number of other diseases, including myocardial infarction.

An enzyme acting on alpha-hydroxybutyric acid (α-hydroxybutyric acid dehydrogenase) has been identified in the serum and studied as a diagnostic aid in liver disease (Elliott, 1961). Elevated levels of this enzyme were observed both in acute hepatitis and in myocardial infarction. An estimate of the ratio of the activities of serum lactic dehydrogenase to the hydroxybutyric acid dehydrogenase is said to be useful. Values over 1:5 are strongly suggestive of acute liver disease.

A number of other enzymes occurring in the serum have been studied as possible aids to diagnosis of liver disease. These include ornithine-carbamyl transferase (Fig 15–7), which is increased in several diseases affecting the liver; and leucine aminopeptidase, which appears to behave like alkaline phosphatase in relation to liver dysfunction, increased activity in the serum resulting from interference with excretion of bile regardless of the cause. High values of leucine aminopeptidase also occur frequently in diseases of the pancreas with or without an accompanying jaundice.

Excretion of Porphyrins

Coproporphyrin excretion in the urine of patients with liver disease often rises markedly (Watson, 1949, 1951). In viral hepatitis the increase is mainly in type I coproporphyrin; in "alcoholic" cirrhosis, it is mainly type III. Uroporphyrin and porphobilinogen may also be detected in the urine of patients with liver disease (Watson, 1954).

Vitamin Metabolism

The liver converts carotene to vitamin A and stores both this vitamin and vitamin D. Since the absorption of all of the fat-soluble vitamins is dependent upon a supply of bile to the intestine, some abnormalities in fat-soluble vitamin metabolism accompany obstructive jaundice. The concentration of vitamin A in the blood is reduced, but it rises after a test dose of the vitamin. It is possible that the lowered serum calcium observed in obstructive jaundice is traceable to a vitamin D deficiency occasioned by the absence of bile in the intestine. Failure to absorb vitamin K in obstructive jaundice produces a prothrombin deficiency. It must also be noted that a prothrombin deficiency may result from hepatic insufficiency even though vitamin K is present in adequate amounts. This is further discussed on p 369.

Iron Metabolism

The concentration of iron in the serum is significantly altered by liver disease. In hepatitis, 2- or 3-fold increases may occur; in cirrhosis, serum iron tends to be lower than normal, whereas in biliary obstruction normal values for iron are the rule.

Hepatic disease histologically suggestive of cirrhosis is a feature of hemochromatosis. The serum iron in these cases is unusually high and the unsaturated iron-binding capacity extremely low. There is, therefore, a high percentage of saturation (see p 407). For example, 74–90% saturation occurs in hemochromatosis as compared to 14–69% in cirrhosis and 28–58% in normal subjects (Gitlow, 1952). These changes in serum iron-binding capacity are not confined to hemochromatosis, so that histologic studies by liver biopsy are also required to confirm the diagnosis.

• • •

References

Arias IM, London IM: Science 126:563, 1957.

Balikov B: Clin Chem 1:264, 1955.

Baroody WG, Shugart RT: Am J Med 20:314, 1956.

Bruns F, Hinsberg K: Biochem Z 325:532, 1954.

Bruns F, Jacob W: Klin Wschr 32:1041, 1954.

Bruns F, Puls W: Klin Wschr 32:656, 1954.

Carbone JV, Grodsky GM: Proc Soc Exper Biol Med 94:461, 1957.

Cook JL, Dounce AL: Proc Soc Exper Biol Med 87:349, 1954.

De La Huerga J, Popper H: J Lab Clin Med 35:459, 1950.

Ducci H, Watson CJ: J Lab Clin Med 30:293, 1945.

Ehrlich P: Centralb.f.klin.Med. 45:721, 1883.

Elliott BA, Wilkinson JH: Lancet 1:698, 1961.

Gilbert A, Lereboullet P: Gas.hebd.d.sc.med.de Bordeaux 49:889, 1902.

Gitlow SE, Beyers MR, Colmore JP: J Lab Clin Med 40:541, 1952.

Grodsky GM, Carbone JV, Fanska R: Proc Soc Exp Biol Med 106:526, 1961.

Hsieh KM, Blumenthal HT: Proc Soc Exper Biol Med 91:626, 1956.

Kunkel HG, Ahrens EH, Eisenmenger WJ: Gastroenterology 11:499, 1948.

Maclagan NF: P 105 in: *Biochemical Disorders in Human Disease,* 2nd ed. Thompson RHS, King EJ (editors). Academic Press, 1964.

Mann JD, Mandel WI, Eichmann PL, Knowlton MA, Sborov VM: J Lab Clin Med 39:543, 1952.

Naganna B, Rama Rao B, Venkaiah KR, Lakshmana Rao P: J Clin Path 15:73, 1962.

Newman AJ, Gross S: Pediatrics 32:995, 1963.

Philp JR, Grodsky GM, Carbone JV: Am J Physiol 200:545, 1961.

Schwartz S, Sborov Y, Watson CJ: Am J Clin Path 14:598, 1944.

Sterkel RL, Spencer JA, Wolfson SK Jr, Williams-Ashman HG: J Lab Clin Med 52:176, 1958.

Van den Bergh AAH, Müller P: Biochem Z 77:90, 1916.

Van den Bergh AAH, Snapper I: Deutsch Arch Klin Med 110:540, 1913.

Watson CJ, Hawkinson V, Capps RB, Rappaport AM: J Clin Invest 28:621, 1949.

Watson CJ, Lowry P, Collins S, Graham A, Ziegler NR: Trans A Am Physicians 67:242, 1954.

Watson CJ, Sutherland D, Hawkinson V: J Lab Clin Med 37:8, 1951.

Wróblewski F, La Due JS: Proc Soc Exper Biol Med 91:569, 1956.

Wróblewski F, Ruegsegger P, La Due JS: Science 123:1122, 1956.

Bibliography

Henry RJ: *Clinical Chemistry. Principles and Technics.* Hoeber, 1964.

Lathe GH: *The Chemical Pathology of Animal Pigments.* P 34 in: Biochemical Society Symposium No. 12. Cambridge Univ Press, 1954.

Reinhold JG: Chemical evaluation of the functions of the liver. Clin Chem 1:351, 1955.

18 . . .

The Kidney & the Urine

The extracellular fluid constitutes the internal environment of the cells of the body. It is in this medium that the cells carry out their vital activities. Since changes in extracellular fluid necessarily are reflected in changes in the fluid within the cells and thus also in cell functions, it is essential to the normal function of the cells that this fluid be maintained relatively constant in composition.

This internal environment (the "milieu intérieur" of Claude Bernard) is regulated mainly by 2 pairs of organs: the lungs, which control the concentrations of oxygen and CO_2; and the kidneys, which maintain optimal chemical composition of the body fluids. Thus, the kidney is an organ which does not merely remove metabolic wastes but actually performs highly important homeostatic functions. It also has a considerable metabolic capacity.

Role of the Kidney in Homeostasis

The regulation of the internal environment by the kidneys is a composite of 4 processes: (1) Filtration of the blood plasma by the glomeruli. (2) Selective reabsorption by the tubules of materials required in maintaining the internal environment. (3) Secretion by the tubules of certain substances from the blood into the tubular lumen for addition to the urine. (4) Exchange of hydrogen ions and production of ammonia for conservation of base.

Urine is formed as the result of these 4 processes. The anatomic unit which carries out these functions is called a nephron, of which there are about 1 million in each kidney. The urine is collected by the collecting tubules and carried to the renal pelvis; from the renal pelvis it is carried by way of the ureter to the bladder.

Anatomy of the Nephron

Urine formation begins as the blood enters the glomeruli, which are tufts or networks of arteriolar capillaries. Each glomerulus is surrounded by Bowman's capsule, a double-walled epithelial sac (like a rounded funnel) which leads to the tubule (Mueller, 1955).

The tubule includes Bowman's capsule, with which it begins, and the following components: a proximal convolution, the descending limb of the loop of Henle, the loop of Henle itself, an ascending limb, and a distal convolution. The distal convolution joins a collecting tubule or duct which carries the urine to the remainder of the renal drainage system. The anatomy of the nephron is illustrated in Fig 18–1.

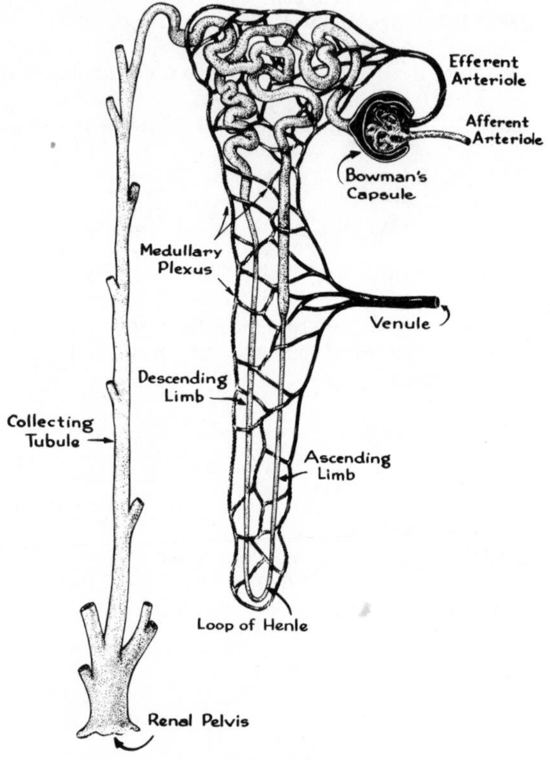

FIG 18–1. Anatomy of the nephron. (Reproduced, with permission, from Merck Sharp & Dohme: *Seminar.* Vol 9, No. 3, 1947.)

FORMATION OF URINE

Filtration

The first step in urine formation is filtration of the blood. A large volume of blood, approximately 1 liter/minute (or 25% of the entire cardiac output at rest), flows through the kidneys. Thus in 4–5 minutes a volume of blood equal to the total blood volume passes through the renal circulation. This is made possible by a very extensive circulatory system in these organs. By the same token, the kidneys are particularly susceptible to damage by diffuse vascular disease.

The energy for filtration is derived from the hydrostatic pressure of the blood. About 70% of the mean pressure in the aorta is actually exerted on the glomerular capillaries–about 75 mm Hg. Thus, these capillaries are able to sustain a considerably higher pressure than that of the other capillaries of the body, in which the pressure is only 25–30 mm Hg. The osmotic pressure of the plasma proteins (about 30 mm Hg) opposes this outflow pressure exerted by the aortic blood. The interstitial pressure on the capillaries themselves, together with the resistance to flow in the tubular system, also contributes about 20 mm Hg. This controls the pressure supplying the energy for filtration to a net of 25 mm Hg. The control of filtration through the glomeruli is illustrated in Fig 18–2. It is apparent that the quantity of filtrate produced by the glomerulus is governed by the net filtration pressure and by the amount of blood flowing through the kidneys. Normally, these factors are maintained relatively constant by compensatory adjustments.

A fall in blood pressure too severe for compensation would of course reduce renal filtration. For example, a decline in the aortic systolic pressure to 70 mm Hg results in a pressure of about 50 mm Hg in the glomerular capillaries. This would reduce the net glomerular hydrostatic pressure–after correction for resistance to flow within the system–to zero, and filtration would cease. Urine would not be formed (anuria) until the blood pressure was restored.

Other factors which affect filtration include obstruction of the arterial pathway of the glomerulus; increased interstitial pressure as may be caused by an inflammatory process; and increased resistance to flow in the tubular systems of the excretory system, such as obstruction in the collecting tubules, ureters, or urethra.

The glomerular membrane may also be so injured by disease that it fails to function as a filter for the blood. Ultimately the capillary may be completely occluded and thus removed from the active circulation. During the progress of such a disease, blood cells and plasma protein will leak through the injured capillary and will be excreted in the urine. Such a pathologic process is illustrated by the syndrome of glomerulonephritis.

Glomerular Filtration Rate (GFR)

In the normal adult, 1 liter of blood is filtered each minute by the 2 million nephrons of both kidneys. At a net filtration pressure of 25 mm Hg, 120 ml of glomerular filtrate are formed at Bowman's capsule. The **glomerular filtration rate (GFR)** in adults is therefore about 120 ml/minute. Chemically, glomerular filtrate is an essentially protein-free extracellular fluid or a protein- and cell-free filtrate of whole blood.

The Action of the Tubule

The composition of the urine is quite different from that of glomerular filtrate. There is also a vast difference in the volume of fluid formed at the glomerulus each minute and the amount which arrives during the same period at the collecting tubule. The glomeruli act only as a filter; the composition of the glomerular filtrate is thus determined solely by permeability of the capillary membrane to the constituents of the blood. As a result, the glomerular filtrate contains many substances necessary for normal metabolism,

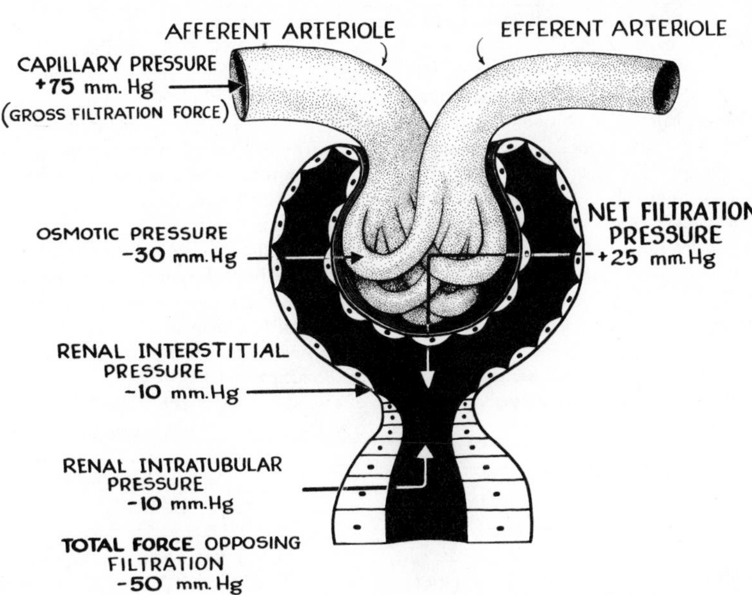

FIG 18–2. Factors affecting the net filtration pressure. (Reproduced, with permission, from Merck Sharp & Dohme: *Seminar.* Vol 9, No. 3, 1947.)

such as water, glucose, amino acids, and chlorides, as well as substances to be rejected, such as urea, creatinine, and uric acid. Furthermore, under various conditions, greater or lesser amounts of essential substances are retained in accordance with the need to maintain constancy in the internal environment.

This highly selective function of the kidney is the task of the tubule. By reabsorption and secretion it modifies the glomerular filtrate and thus produces the urine.

Threshold Substances

Certain substances are reabsorbed almost completely by the tubule when their concentrations in the plasma are within the normal range but appear in the urine (ie, are not completely reabsorbed) when their normal plasma levels are exceeded. These are spoken of as threshold substances. A substance reabsorbed only slightly or not at all is a low-threshold substance (eg, creatinine, urea, and uric acid). Those materials necessary to the body which are reabsorbed very efficiently are high-threshold substances (eg, amino acids and glucose). The reabsorption of glucose is typical of this mechanism of the renal tubule.

Reabsorption of Glucose

At an arterial plasma level of 100 mg/100 ml and a glomerular filtration rate of 120 ml/minute, 120 mg of glucose are delivered into the glomerular filtrate each minute. Normally, all of this glucose is reabsorbed into the blood in the proximal convoluted tubule.

The capacity of the glucose transporting system is limited. If the arterial plasma level of glucose rises, say to 200 mg/100 ml, and the glomerular filtration rate remains the same, twice as much glucose (240 mg/minute) is presented for reabsorption as before. All of this additional glucose will be reabsorbed until the full capacity of the tubular transfer system is reached; the excess, which is filtered but cannot be reabsorbed, will remain in the tubular fluid and pass on into the urine. This excess glucose will also carry water with it, and this results in the characteristic diuresis of glycosuria.

The maximum rate at which glucose can be reabsorbed has been determined to be about 350 mg/minute. This is designated the Tm_G, or tubular maximum for glucose. If the glomerular filtration rate decreases, less glucose is presented per minute to the reabsorbing cells. This would permit the concentration of glucose in the blood to rise above those levels at which glucose would normally spill into the urine, without glycosuria actually being observed. The reabsorptive capacity of the tubule for glucose probably does not change under normal conditions, although it may be above normal in hyperthyroidism and in diabetes mellitus.

Reabsorption of Water

Under normal circumstances, the osmotic pressure of the plasma varies only slightly despite wide variations in the intake of fluid and of solutes. The normal osmolarity of the plasma (285–295 mOsm/liter of water*) is largely attributable to its content of inorganic salts (electrolytes). It is maintained by the kidney by varying the volume flow as well as the osmolar concentration of the urine. For example, an excess of water, which would tend to dilute the plasma and thus reduce its osmotic pressure, produces a renal response which results in excretion of an increased volume of urine with an osmolarity less than that of plasma, ie, less than 285 micro-osmols (μOsm)/ml of urine water. By thus excreting water in excess of solute, the kidney defends the osmolarity of the

*The osmotic activity of solutions of substances which ionize is higher than that of substances which do not. Thus, a solution of sodium chloride which is 100% ionized would have an osmotic pressure twice that calculated from its concentration because it dissociates into 2 particles (ions)/mol. A solution of glucose, on the other hand, which does not ionize, would have an osmotic pressure equal only to that calculated from its molar concentration since it contains in solution only one particle/mol. These relationships between molar concentration and osmotic activity are conveniently expressed by the use of the term osmol. The osmolarity of a solution of a given compound is determined by multiplying the molar concentration by the number of particles/mol obtained by ionization.

$$\text{Osmolarity} = \text{Molarity} \times \begin{array}{l}\text{Number of particles}\\ \text{per mol resulting}\\ \text{from ionization}\end{array}$$

The concentrations of ionizable substance in physiologic fluids are so weak that complete ionization is assumed to occur. It is therefore possible to convert molar concentrations directly to osmolarity by reference to the number of ions/mol produced. It is also more convenient to express these dilute concentrations in terms of millimols and milliosmols, ie, 1/1000 mol or 1/1000 osmol.

Example:

NaCl ($Na^+ + Cl^-$) at a concentration of 70 mM/liter = 140 mOsm/liter

Na_2HPO_4 ($2Na^+ + HPO_4^{=}$) at a concentration of 1.3 mM/liter = 3.9 mOsm/liter

The osmotic activity of physiologic fluids such as plasma and urine is due to the combined osmotic activity of a number of substances which are dissolved in them. The osmolarity of such fluids is therefore most easily determined by measurement of the freezing point depression rather than by attempting to calculate it from the concentration and degree of ionization of each constituent in the mixture. Electrically operated instruments for measurement of osmotic pressure by freezing point depression (osmometers) have been devised to measure very small changes in temperature. A depression in the freezing point of a solution of 0.00186° C below that of water (taken as 0° C) is equivalent to 1 mOsm of osmotic activity/liter.

$$1 \text{ mOsm} = \Delta 0.00186^\circ \text{ C}$$

Example: The freezing point of a sample of human blood plasma was found to be −0.59° C; the osmolarity would then be 317 mOsm/liter (0.59/0.00186).

plasma. On the other hand, an increase in plasma osmolarity would be corrected by the excretion of a "concentrated" urine with an osmolarity higher than that of plasma, which indicates that, relative to the plasma, solute is being excreted in excess of water. The capacity of the kidney to control water loss from the body is illustrated by the fact that, although as much as 100 ml/minute of glomerular filtrate may normally be formed, as little as 1 ml of urine may remain by the time the fluid has passed down the collecting tubule. Under conditions of water deprivation an even greater renal conservation of water may be effected.

The modification of the chemical composition of the glomerular filtrate, as pointed out above, involves primarily the active, selective reabsorption of solutes. Water is reabsorbed secondarily in accordance with the anatomic and functional nature of the various parts of the nephron as described below. According to the most recent evidence, water is able to diffuse passively out of the proximal convoluted tubule, the descending limb of the loop of Henle, and the distal convoluted tubule, as well as out of the collecting tubule. Movement of water is accomplished by osmotic gradients provided by the prior reabsorption of solutes, including inorganic electrolytes and osmotically active organic compounds such as urea, glucose, and uric acid. It is believed that the proximal tubule and the descending limb of the loop of Henle are freely permeable to water at all times but that the distal convoluted tubule and the collecting tubule become freely permeable to water only as a result of the action of the antidiuretic hormone (ADH) of the posterior pituitary gland. The ability of the kidney to alter water loss in response to ADH-induced changes in permeability to water in the distal convoluted tubule and in the collecting tubule is the basis for the excretion of urines of appropriate volume and osmolarity under varying conditions of plasma osmolarity, as mentioned above.

Approximately 80% of the volume of the glomerular filtrate is ordinarily reabsorbed in the proximal convoluted tubule. Thus, the total volume of glomerular filtrate from the kidneys, which normally forms at a rate of about 100 ml/minute, is reduced to a flow of about 20 ml/minute of a fluid still isosmotic with the original glomerular filtrate. This is accomplished by the passive reabsorption of water, as a solvent for the actively reabsorbed solutes such as sodium chloride and glucose. Water is absorbed at such a rate as to maintain isosmotic conditions in this portion of the nephron as well as in the surrounding renal cortex. This component of water reabsorption is termed **obligatory reabsorption** to indicate that it occurs secondary to solute reabsorption without regard for the water requirements of the body, being controlled instead by the necessity to maintain an isosmotic environment in this area of the kidney.

It is possible, of course, that the glomerular filtrate and the proximal tubular fluid may retain such a high concentration of solutes that osmotic factors would operate in the opposite direction, thus causing an increased loss of water from the body. This is referred to as an **osmotic diuresis.** It is exemplified by the excess water loss which accompanies a glycosuria. The diuresis takes place because failure to reabsorb all of the filtered glucose increases the solute content of the tubular filtrate. Those drugs which impair the tubular reabsorption of sodium chloride, such as the mercurials or the thiazides, accomplish a diuretic effect in a similar manner—in this case by increasing the content of inorganic salts in the tubular filtrate.

According to Wirz (1956), the tubular filtrate becomes progressively more concentrated (hypertonic or hyperosmotic) as it passes down the descending limb of the loop of Henle. The renal medulla and papilla which surround the loop are also said to be hyperosmotic to the plasma, in contrast to the renal cortex and the proximal tubule which, as noted above, are isosmotic. Apparently, water without solute is being lost from the descending limb into the surrounding hypertonic environment. However, in the ascending limb, active loss of sodium chloride now takes place without water (to which this portion of the nephron is impermeable), so that the tubular filtrate up to the distal convoluted tubule now becomes hyposmotic. It is at this point that the action of ADH will be exerted. If a dilute urine is to be excreted, little or no ADH activity will occur. The distal tubule and the collecting tubule become relatively impermeable to water, and the hypotonic urine delivered to them from the ascending limb of the loop of Henle is excreted as such. In contrast, when conservation of water is required, ADH activity is increased and water is then allowed to diffuse freely from the distal and collecting tubules. The urine thus becomes more concentrated (hyperosmotic) in the collecting tubule. Diffusion of water is also facilitated here by the presence of a hypertonic environment in the surrounding renal medulla.

The ADH-controlled water reabsorption in the distal and collecting tubules is termed **facultative reabsorption** to indicate that it occurs independently of active solute transport and in accordance with the needs of the body for water.

The various components of water reabsorption by the nephron are represented diagrammatically in Fig 18–3.

The hypertonicity of the renal medulla and papilla as proposed by Wirz (1953) is said to be due to deposition of sodium chloride absorbed from the ascending limb of the loop of Henle into the surrounding tissue. As described above, the urine is concentrated in one limb of the loop (the descending limb) and diluted in the other (the ascending limb). This process of concentration and dilution of urine in opposing segments of a loop, where the flow of urine in one is opposite in direction to that of the other, is a feature of the so-called countercurrent theory of Wirz. An advantage of this arrangement is that the concentration of NaCl in the ascending limb at the beginning is about the same as that of the environment. Consequently, sodium salts need not be transported out of the tubule against a precipitous concentration gradient which could only be accomplished by the expenditure of considerable energy.

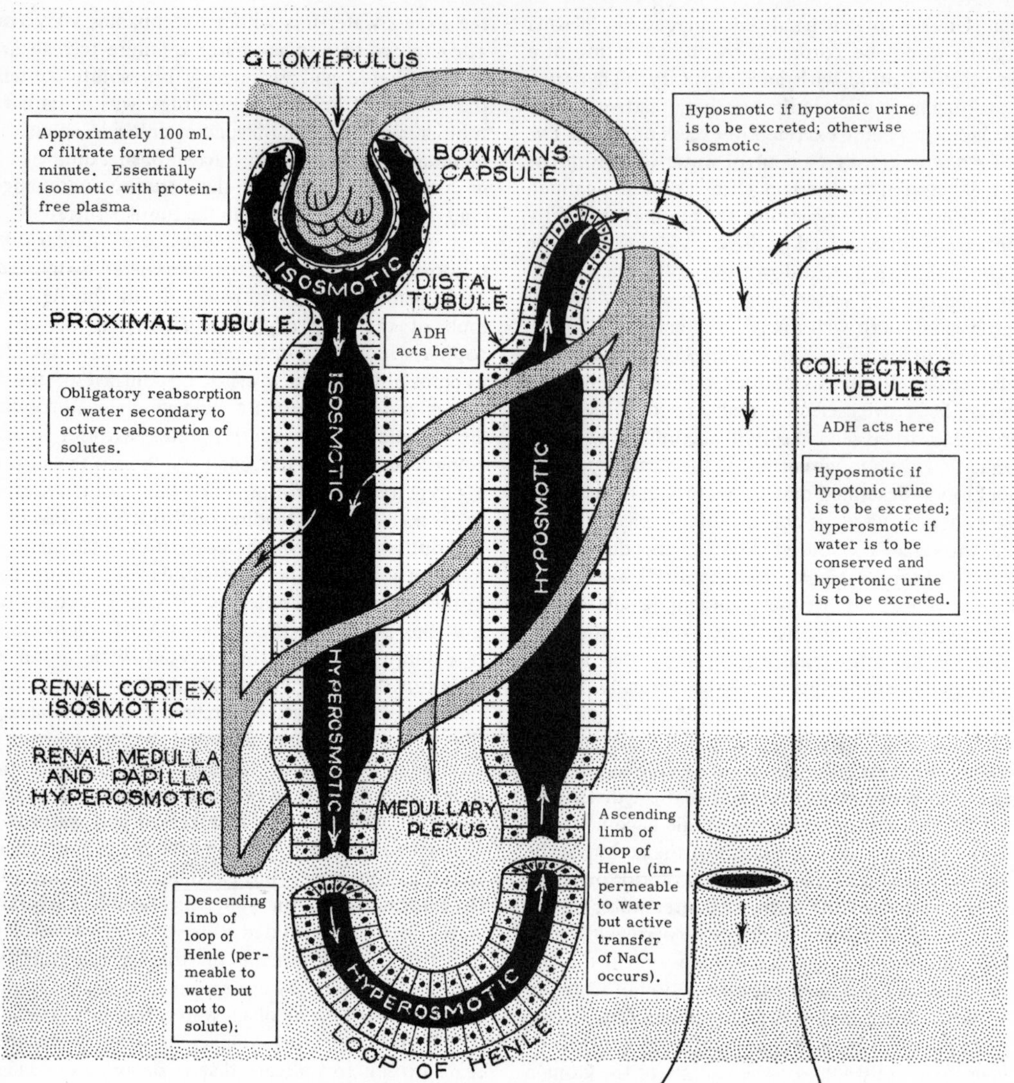

FIG 18–3. Summary of normal physiology of the nephron.

Mechanism of Central Control of ADH Secretion

According to Verney (1958), the mechanism for facultative reabsorption of water which operates through pituitary ADH is controlled by the activity of osmoreceptors located in the anterior hypothalamic region of the brain. When the blood is diluted, as by the ingestion of large amounts of water, the osmoreceptors detect the resultant decrease in the osmolarity of the blood brought to that region via the internal carotid artery. As a result, the osmoreceptors transmit, over nervous connections to the posterior pituitary, impulses which produce inhibition of pituitary secretion of antidiuretic hormone. The resultant suppression of ADH then permits excretion of more water in the distal and collecting tubules by impairment of facultative reabsorption, and a large volume of dilute urine results. In contrast, after deprivation of water, the blood becomes more hypertonic and the osmoreceptors then act to stimulate ADH secretion, which returns more water to the blood in an effort to compensate for the hypertonicity. Under these circumstances, a concentrated urine of small volume results. The tubular reaction to ADH is one of the most sensitive of the homeostatic mechanisms of the kidney. For this reason, determination of maximal concentrating power of the tubule (see p 385) is one of the most useful clinical tests of renal function.

A number of drugs, including ingested alcohol, act to suppress ADH secretion and thus to increase urine flow. Certain stresses such as that of surgery or severe trauma as well as some drugs used in anesthesia all cause excessive production of ADH. In the immediate postoperative period these effects on ADH secretion contribute to excessive retention of water by the kidney and thus to oliguria. Failure to recognize this has led to serious overhydration of patients in an effort to correct the oliguria which, under the circumstances, is actually a normal physiologic response.

Defects in Reabsorption by the Renal Tubules

The mechanisms for selective reabsorption in the kidney tubules may exhibit various defects, which are frequently inherited. Such tubular defects may be confined to a failure in the reabsorption of a single substance, such as phosphates or glucose, or a combination of several reabsorptive defects may be found. The presence of glucose in the urine when the level of glucose in the blood is normal is indicative of a tubular defect in the reabsorption of glucose. Such a reduction in the "renal threshold" for glucose is called renal diabetes or renal glycosuria.

Those individuals who have a decreased capacity for the reabsorption of phosphates exhibit changes in the metabolism of bone as a result of excessive losses of phosphorus from the body. Evidence for such changes can be found in the blood, where the phosphorus is low and alkaline phosphatase activity is high. Such cases may be designated vitamin D-resistant rickets in children, or idiopathic osteomalacia (Milkman's syndrome) in adults. These diseases are "resistant" to vitamin D only when the vitamin is given in ordinary doses. Large doses of the vitamin (50,000 units/day) will bring about a decrease in urinary excretion of phosphates and healing of the bones.

The existence of hyperphosphaturia of renal tubular origin may be detected by measurement of the serum phosphate level and of the amounts of phosphate excreted in a 24-hour urine sample. These data are then used to calculate the phosphate clearance (see p 383). In normal individuals, the clearance of phosphate does not exceed 12% of the glomerular filtration rate. It should be emphasized that the metabolic effects of a renal tubular defect may be masked by a decline in glomerular filtration. Thus a marked reduction in filtration would compensate for a decreased tubular reabsorptive capacity for phosphate, so that the serum levels would rise and healing of the bones could then occur. The urinary excretion of phosphate is therefore unreliable as a diagnostic tool in the presence of reduced glomerular filtration.

A third tubular defect which often occurs in combination with those mentioned above is characterized by excessive excretion of amino acids. In normal subjects, the content of amino acids in the urine is very low; but in individuals having this renal tubular deficiency a generalized aminoaciduria occurs even though the levels of amino acids in the blood are normal.

The occurrence of all 3 tubular defects was first described in children by Fanconi and termed by him "hypophosphatemic-glycosuric rickets." It is now referred to as the De Toni-Fanconi syndrome. Similar cases have since been reported in adults. A peculiar feature of the disease in children is the accumulation of cystine crystals in the reticuloendothelial system and other tissues, including the cornea. Such cystine storage (cystinosis) without any of the other features of the Fanconi syndrome has also been found in an otherwise normal individual (Cogan, 1957). This suggests that the genetic defect responsible for cystine storage is not directly related to that involved in the causation of the renal tubular abnormalities of the Fanconi syndrome, as might otherwise have been inferred from the almost inevitable coexistence of cystinosis and Fanconi disease in children.

In severe cases of the De Toni-Fanconi syndrome there is also a failure to reabsorb water and potassium properly as well as to acidify the urine. Such patients, if untreated, suffer from a metabolic acidosis and dehydration as well as a severe potassium deficiency, and this combination of tubular defects is usually fatal at an early age.

As was pointed out above in connection with tubular hyperphosphaturia, a decline in the glomerular filtration rate may obscure the aminoaciduria which would otherwise occur in the De Toni-Fanconi syndrome. For this reason it is important to measure the GFR (eg, creatinine clearance; see p 384) in order to interpret correctly the extent of aminoaciduria in cases where this tubular defect is suspected.

Another group of renal tubular defects, which, unlike those described above, is usually acquired rather than inherited, is characterized by increased urinary losses of calcium, phosphate, and potassium and a failure to acidify the urine. There may also be a lack of ammonia formation by the kidneys (nephrocalcinosis, hyperchloremic nephrocalcinosis, Butler-Albright syndrome). The primary defect in these cases is the inability to form an acid urine. A chronic acidosis results, leading to increased urinary excretion of alkali cations (sodium, potassium, and calcium), which are used to neutralize the excess acid. The losses of calcium and phosphorus in the urine produce changes in the bones (osteomalacia). Correction of the acidosis by administration of an alkaline salt (sodium bicarbonate) produces healing of the bones.

In contrast to the generalized aminoaciduria of the De Toni-Fanconi syndrome, a limited excretion of only a few amino acids without evidence of any other renal tubular defect may also occur. An example is the excessive excretion of cystine, lysine, arginine, and ornithine in "cystinuria."

Tubular Secretion

It is believed that some creatinine is secreted by the tubules when the blood levels rise above normal. There may be minimal tubular secretion of uric acid. Potassium is also secreted from the blood by the renal tubules. The exchange of hydrogen ions described below might also be considered a secretory function of the tubules. These are the only normal constituents of the urine thus added to the tubular filtrate. It is also possible that additional secretory activity may be resorted to in glomerulonephritis when filtration is impaired. However, many foreign substances are readily secreted by the tubules. An example is the dye, phenol red, which is secreted by the proximal tubule utilizing a carrier system similar to that for glucose but acting, of course, in reverse. The same carrier system is mainly responsible for the secretion of iodopyracet (Diodrast; 3,5-diiodo-4-pyridone-N-acetic acid which is loosely combined with ethanolamine), iodohippurate (Hippuran; *o*-iodohippuric acid), penicillin, and para-aminohippurate (PAH).

Investigations of some of the energy sources for renal tubular transport indicate that energy from oxidation of succinate and other citric acid cycle oxidations is directly involved in tubular secretory processes. Various inhibitors of the succinic oxidase system are known to alter the secretion of PAH and of the dye phenolsulfonphthalein (PSP).

The maximal rate at which the tubule can accomplish secretion is determined by the capacity of these carrier systems. This is measured and expressed as the **maximal tubular secretory capacity**, or **secretory Tm**. For PAH this is about 80 mg/minute in normal persons. This value is useful in measuring the amount of functioning tubular tissue since it represents the total effect of all the tubules working together. Iodopyracet, which is opaque to x-ray, is used for roentgenologic examinations of the urinary tract. It may also be used like PAH to measure tubular activity. The normal secretory Tm for iodopyracet is about 50 mg/minute.

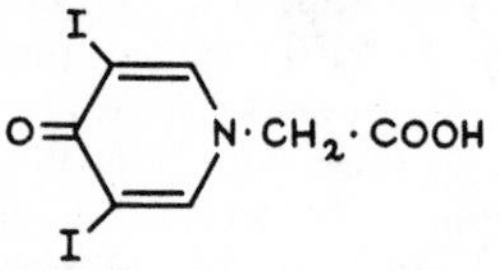

Iodopyracet (Diodrast)

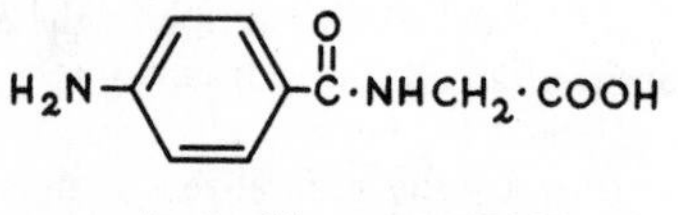

***p*-Aminohippurate (PAH)**

THE ROLE OF THE KIDNEY IN ACID-BASE BALANCE

The kidney plays a very important role in regulating the electrolyte content of body fluids and thus in maintaining pH of the body.

Reabsorption of Electrolytes

Sodium, the principal cation, and chloride and bicarbonate, the principal anions of the extracellular fluid and of the glomerular filtrate, are selectively reabsorbed, mainly in the proximal tubule. In general, chloride reabsorption and excretion roughly parallel that of sodium.

Potassium is also present in small but important quantities in the extracellular fluid and therefore in the glomerular filtrate. When a large excess of potassium is excreted the clearance of potassium may exceed the glomerular filtration rate. This indicates that potassium is excreted not only by filtration but also by tubular secretion. However, at normal rates of excretion, virtually all of the potassium which is filtered is later reabsorbed in the proximal tubule. That which appears in the urine is therefore added to the urine by tubular secretion in the distal tubule. The secretion of potassium by the renal tubule is closely associated with hydrogen ion exchange and with acid-base equilibrium. This will be discussed further on p 383.

The fact that potassium can be excreted by tubular secretion as well as by filtration explains why serum potassium is usually normal or only slightly elevated in chronic renal failure when reduced filtration produces a marked rise in urea and other substances which depend entirely on filtration for their excretion. Tubular secretion of potassium does not usually fail until late in the course of chronic renal disease.

The adrenal corticoids, particularly aldosterone (see p 442), have an important effect on the handling of sodium and potassium by the renal tubules. These hormones favor the reabsorption of sodium and the excretion of potassium. In adrenocortical insufficiencies, such as Addison's disease, there is, therefore, excess loss of sodium, and potassium tends to be retained. Conversely, hyperactivity of the adrenal cortex or the administration of corticoid hormones produces excess reabsorption of sodium and urinary loss of potassium. Hyperactivity of the adrenal may be caused by the presence of adrenal tumors (see Chapter 20). An adrenal tumor (aldosteroma) which is characterized by the secretion of large amounts of aldosterone (aldosteronism) leads to marked increases in sodium retention and to potassium loss. Thus, whereas the clearance of potassium in normal individuals is about 5–10% of the GFR, in aldosteronism it may be as much as 40%.

Acute loss of sodium, as may occur in Addison's disease or in chronic renal disease characterized by inadequate tubular reabsorption of sodium (salt-losing nephritis), leads to severe dehydration, a decline in plasma volume, and shock.

Acid-Base Regulatory Mechanisms

(Pitts, 1950; and Gilman, 1953.)

The kidney also affects acid-base equilibrium by providing for the elimination of nonvolatile acids, such as lactic acid, the ketone bodies, sulfuric acid produced in the metabolism of protein, and phosphoric acid produced in the metabolism of phospholipids. These acids buffered with cations (principally sodium) are first removed by glomerular filtration. The cation is then recovered in the renal tubule by reabsorption in exchange for hydrogen ions which are secreted.

Mobilization of hydrogen ions for tubular secretion is accomplished by ionization of carbonic acid, which is itself formed from metabolic CO_2 and water. In the proximal tubule, the exchange of hydrogen ions proceeds first against sodium bicarbonate. The process as it occurs in the proximal tubule is shown in Fig

18–4. The formation of carbonic acid is catalyzed by **carbonic anhydrase**. The result of these reactions is not only to provide for the complete reabsorption of all of the sodium bicarbonate filtered from the plasma but also to effect a reduction in the hydrogen ion load of the plasma with little change in the pH of the urine. It will be noted that the bicarbonate moiety filtered is not that which is reabsorbed into the blood and that the CO_2 resulting from the decomposition of the carbonic acid formed in the tubule may diffuse back into the tubule cells for re-utilization in the hydrogen ion secretory system.

At normal levels of bicarbonate in the plasma, about half of that filtered at the glomerulus is reabsorbed in the proximal tubule and the remainder in the distal tubule. After all of the bicarbonate has been reabsorbed, hydrogen ion secretion then proceeds against Na_2HPO_4. The exchange of a sodium ion for the secreted hydrogen ion changes Na_2HPO_4 to NaH_2PO_4, with a consequent increase in the acidity of the urine and a decrease in the urinary pH. The secretion of hydrogen ions in the distal tubule is shown in Fig 18–5.

A third mechanism for the elimination of hydrogen ions and the conservation of cation is the production of ammonia by the distal renal tubule cells. The ammonia is obtained by deamidation of amino acids. Deamidation of glutamine by glutaminase serves as the principal source of the urinary ammonia. The ammonia formed within the renal tubule cell may react directly with hydrogen ions so that ammonium ions rather than hydrogen ions are secreted; or ammonia may diffuse into the tubular filtrate and there form ammonium ion as shown. Such a mechanism operating against sodium chloride is illustrated in Fig 18–6.

The lower the pH of the urine, ie, the greater the concentration of hydrogen ions, the faster will ammonia diffuse into the urine. Thus ammonia production is greatly increased in metabolic acidosis and negligible in alkalosis. It has also been proposed that the activity of renal glutaminase is enhanced by an acidosis (Rector, 1955). The ammonia mechanism is a valuable device for the conservation of fixed base. Under normal conditions 30–50 mEq of hydrogen ions are eliminated per day by combination with ammonia and about 10–30 mEq as titratable acid, ie, buffered with phosphate.

Influence of Carbon Dioxide Tension on Bicarbonate Reabsorption

An increase in the CO_2 tension of the body fluids will accelerate the formation of carbonic acid and thus of hydrogen ions for secretion by the renal tubule cells. This would be expected to facilitate the reabsorption of bicarbonate. Decreased CO_2 tensions would act in reverse, ie, cause a reduction in the reabsorption of bicarbonate. This response of the renal tubule to the CO_2 tension of the body fluids provides an explanation for the renal response to states of respiratory acidosis or alkalosis (Fig 11–4). In respiratory acidosis, compensation is achieved by an increase in the bicarbonate levels of the blood in an attempt to restore the normal 1:20 ratio of carbonic acid to bicarbonate. The increased bicarbonate may be obtained by the response of the kidney to the high CO_2 tension which prevails in respiratory acidosis. The situation is reversed in respiratory alkalosis. Here compensation is achieved by elimination of bicarbonate, and the lowered CO_2 tension which prevails in this condition provokes a renal response which reduces the rate of reabsorption of bicarbonate until the normal ratio of carbonic acid to bicarbonate is restored.

Action of a Carbonic Anhydrase Inhibitor on Excretion of Bicarbonate

The enzyme carbonic anhydrase, which catalyzes the important reaction in the renal tubules by which hydrogen ions are produced for secretion, may be inhibited by sulfonamide derivatives. The most potent inhibitor yet discovered is acetazolamide (Diamox, acetylamino-1,3,4-thiadiazole-5-sulfonamide). When this drug is administered, the urine becomes alkaline, and increased amounts of sodium bicarbonate appear in the urine. There is also a reduction in the titratable acidity and in the ammonia of the urine together with an increase in potassium excretion (see below). All of these effects of acetazolamide may be explained by its action on hydrogen ion secretion. The drug is used clinically to induce a loss of sodium and water in patients who will benefit from such a regimen (eg, patients with congestive heart failure or hypertensive disease).

Failure of Acid-Base Regulation by the Kidneys in Disease

Metabolic acidosis is a characteristic complication of renal disease. It is most frequently caused by a decrease in the glomerular filtration rate, which causes retention of fixed acid catabolites such as phosphates and sulfates, accompanied by a rise in nonprotein nitrogen in the blood. There may also be a failure of the renal tubular mechanisms for secretion of hydrogen ions and for the manufacture of ammonia. The decrease in glomerular filtration can be expected to contribute to acidosis because the retained acidic anions, such as phosphates and sulfates, are buffered by cations which cannot be recaptured and returned to the blood in exchange for hydrogen ions by the tubule until these buffered compounds have been delivered to the tubular filtrate. If, even after filtration, tubular secretion of hydrogen ions and concomitant reabsorption of sodium are also defective, there will necessarily occur a considerable loss of cation into the urine. Under these circumstances reduction in sodium content of the plasma may become so severe that dehydration secondary to the electrolyte depletion will result. The consequent fall in extracellular fluid volume is a factor in reducing blood flow to the kidney, which further impairs its function.

The so-called "renal tubular acidosis" mentioned above is relatively rare. This defect is confined to failure of the renal tubular mechanisms for secretion of

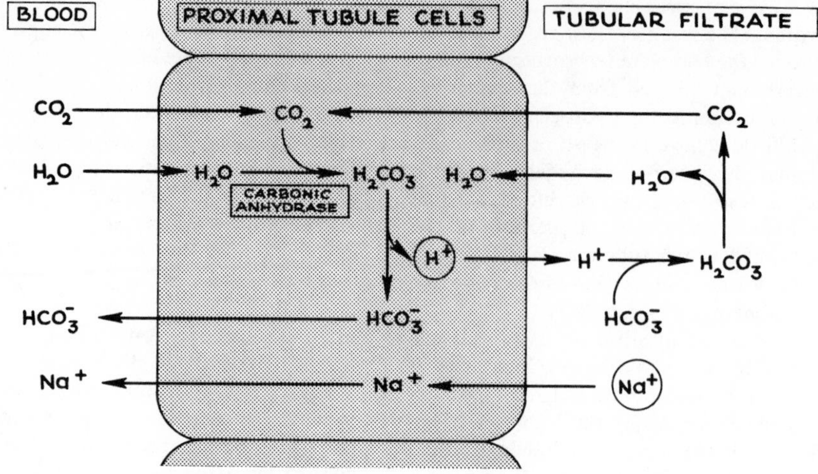

FIG 18–4. Mobilization of hydrogen ions in proximal tubule.

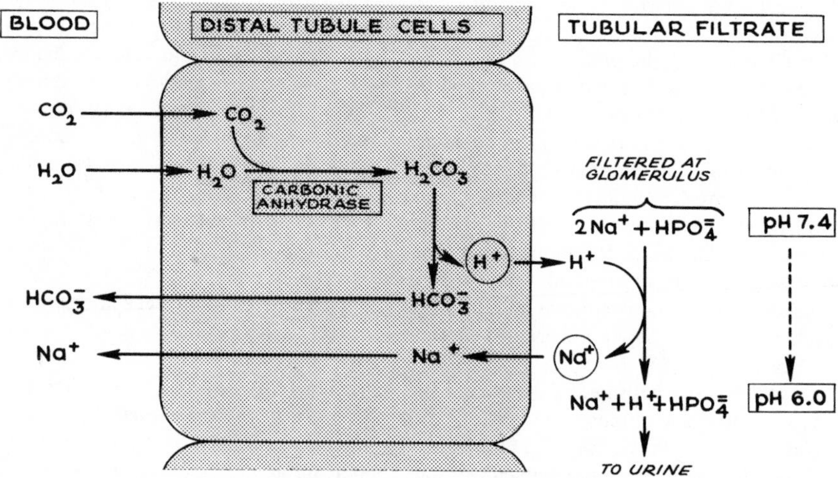

FIG 18–5. Secretion of hydrogen ions in distal tubule.

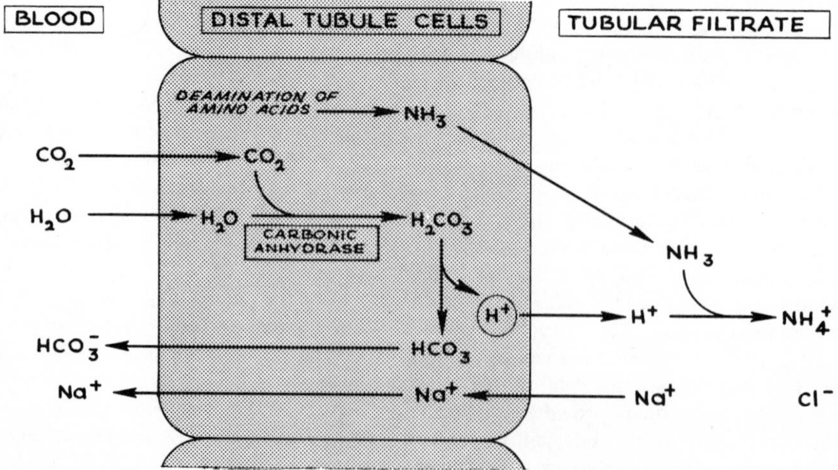

FIG 18–6. Production of ammonia in distal tubule.

hydrogen ions and for manufacture of ammonia. It may be differentiated from the classical acidosis of renal insufficiency, described above, by the fact that the glomerular filtration rate (eg, creatinine clearance) is normal and there is therefore no elevation in the plasma of urea, phosphate, or sulfate. (In fact, phosphate may actually be low.) The pH of the urine is close to neutral (6.0–7.0). Sodium depletion does not occur, probably because calcium and potassium are substituted as sources of cation to buffer acids. However, such substitution results in depletion of calcium from bone (osteomalacia) and increased calcium in urine (hypercalciuria), with a resultant tendency to formation of renal calculi. Furthermore, depletion of potassium may be so severe as to produce paralysis of muscle. All of the metabolic abnormalities resulting from a renal tubular acidosis will be corrected when sufficient quantities of sodium bicarbonate are administered to maintain the plasma bicarbonate at normal levels.

Relation of Potassium Excretion to Acid-Base Equilibrium

The administration of potassium salts in excess produces a decline in hydrogen ion concentration within the cells and an increase (acidosis) in the extracellular fluid accompanied by the excretion of an alkaline urine. Conversely, potassium depletion is associated with the development of an alkalosis in the extracellular fluid and an increase in hydrogen ion concentration within the cell (intracellular acidosis) followed by the excretion of a highly acid urine despite the high bicarbonate content of the plasma (paradoxic aciduria). These latter circumstances may occur in patients treated with cortisone or corticotropin (ACTH) or in those with the hypercorticism of Cushing's syndrome. It is also a common occurrence in postoperative patients maintained largely on potassium-free fluids in whom depletion of potassium may result because of continued excretion of this cation in the urine and in the gastrointestinal fluids. Although the alkalosis in these surgical cases is usually accompanied by depletion of chloride (hypochloremic alkalosis), correction cannot be attained by the administration of sodium chloride alone but only by the administration of potassium salts as well. When adequate repletion of potassium has been accomplished, a fall in serum bicarbonate and a rise in serum chloride together with elevation of the urine pH to normal levels will then occur.

As noted above, potassium is secreted by the distal tubule. Furthermore, the same mechanisms which provide for the secretion of hydrogen ions in exchange for sodium are similarly utilized for the secretion of potassium, also in exchange for sodium. The capacity of this system is limited. Ordinarily it is represented by the sum of the secretion of hydrogen and potassium ions. When intracellular potassium levels are low and intracellular hydrogen ions are elevated, as occurs in potassium-deficient states, more hydrogen ions can be secreted by the distal tubule cells. Under these circumstances bicarbonate reabsorption proceeds together with excess hydrogen ion secretion, and a highly acid urine is formed.

The effect of acid-base balance (in this instance, alkalosis) on the excretion of potassium in the urine can be further demonstrated by administration of a quantity of sodium bicarbonate sufficient to lower hydrogen ion concentration both intracellularly and extracellularly. Within the renal tubular cells, the decline in hydrogen ion concentration permits an increase in secretion of potassium because of the normal competition for secretion which exists between hydrogen and potassium ions in the distal tubule. Marked enhancement of potassium excretion will thus be produced in response to the alkalosis following bicarbonate administration. In fact, in man, the excretion of potassium is so closely related to alkalosis that simple hyperventilation (producing respiratory alkalosis) will raise the rate of excretion of potassium from 87–266 μEq/minute.

The ability of acetazolamide (Diamox) to increase excretion of potassium into the urine is attributable to its inhibiting effect on carbonic anhydrase activity, which thus brings about a reduction in hydrogen ion concentration within the renal tubule cell. Consequently, secretion of potassium can be increased.

Summary of the Mechanisms for Sodium Reabsorption

The relative importance of each of the mechanisms for the conservation of sodium by the kidney may be assessed by a consideration of the following data (Mudge, 1956).

A. Sodium Reabsorbed With Fixed Anion (Mostly Chloride): 12,000 μEq/minute.

B. Na^+-H^+ Exchange:

1. $NaHCO_3$–3200 μEq/minute.
2. NH_4–20 μEq/minute.
3. Titratable acid (decrease in urine pH)–30 μEq/minute.
4. Free acid (1 ml urine at pH 5.0)–0.01 μEq/minute.

C. Na^+-K^+ Exchange (K^+ Excreted): 50 μEq/minute.

TESTS OF RENAL FUNCTION

Clearance

As a means of expressing quantitatively the rate of excretion of a given substance by the kidney, its "clearance" is frequently measured. This is a volume of blood or plasma which contains the amount of the substance which is excreted in the urine in 1 minute. Alternatively, the clearance of a substance may be defined as that volume of blood or plasma cleared of the amount of the substance found in 1 minute's excretion of urine. The calculation of clearance can be illustrated by measurement of the clearance of inulin.

A. Inulin Clearance: The polysaccharide inulin is filtered at the glomerulus but neither secreted nor reabsorbed by the tubule. The clearance of inulin is therefore a measure of glomerular filtration rate (GFR). **Mannitol** can also be used for the same purpose. These clearances, like many other physiologic phenomena, vary with body size. They are therefore expressed on the basis of a given size, eg, normal inulin clearance (GFR) is 120 ml/1.73 sq m body surface area. To facilitate interpretation, the results of an actual clearance study are usually calculated ("corrected clearance") on the basis of ml/1.73 sq m.

In measuring inulin clearance it is desirable to maintain a constant plasma level of the test substance during the period of urine collections. Simultaneous measurement of the plasma inulin level and the quantity excreted in a given time supplies the data necessary to calculate the clearance according to the following formula:

$$C_{in} = \frac{U \times V}{B}$$

Where C_{in} = Clearance of inulin (ml/minute)
U = Urinary inulin (mg/100 ml)
B = Blood inulin (mg/100 ml)
V = Volume of urine (ml/minute)

B. Endogenous Creatinine Clearance: At normal levels of creatinine in the blood, this metabolite is filtered at the glomerulus but not secreted nor reabsorbed by the tubule. Consequently, its clearance may also be measured to obtain the GFR. This is a convenient clinical method for estimation of the GFR since it does not require the intravenous administration of a test substance, as is the case with an exogenous clearance study using inulin. Normal values for creatinine clearance are 95–105 ml/minute.

C. Urea Clearance: This test is used clinically to appraise renal function. The subject is given 2 glasses of water to ensure adequate urine flow. He then voids, and the urine is discarded. One hour later, the patient voids a specimen of urine and a blood sample is taken. One hour later, a second urine specimen is voided. The blood and urine specimens are analyzed for urea.

In calculating the results, the volume of urine excreted per minute is noted. If it is less than 2 ml/minute, the test is not valid. The formula above is used to calculate the urea clearance substituting urea for inulin.

The normal value for urea clearance is 75 ml/minute. This means that the maximal amount of urea removed in 1 minute by the kidney is equivalent to that normally contained in 75 ml of blood. It will be noted that urea clearance is less than that of inulin, which indicates that some of the filtered urea is subsequently passively reabsorbed by the tubules.

The clearance of many dyes or other substances like PAH which are secreted as well as filtered may therefore exceed that of inulin. It is customary to report urea clearances in percent of normal (75 ml/minute). The normal range is 75–120%.

With severe renal damage, the clearance may fall to 5% or less. It may fall to 50% or lower in early acute nephritis, and to 60% or lower in early glomerulonephritis. Impaired urea clearance is due principally to a reduction of glomerular filtration.

Measurement of Renal Plasma Flow (RPF)

Para-aminohippurate (PAH) is filtered at the glomeruli and secreted by the tubules. At low blood concentrations (2 mg or less/100 ml of plasma), PAH is removed completely during a single circulation of the blood through the kidneys. Thus the amount of PAH in the urine becomes a measure of the volume of plasma cleared of PAH in a unit of time. In other words, PAH clearance at low blood levels measures **renal plasma flow**. This is about 574 ml/minute for a surface area of 1.73 sq m.

Filtration Fraction (FF)

The filtration fraction, ie, the fraction of plasma passing through the kidney which is filtered at the glomerulus, is obtained by dividing the inulin clearance by the PAH clearance (GFR/RPF = FF). For a GFR of 125 and RPF of 574, the FF would then be 125/574 = 0.217 (21.7%). The filtration fraction tends to be normal in early essential hypertension, but as the disease progresses the decrease in renal plasma flow is greater than the decrease in glomerular filtration. This produces an increase in the FF. In the malignant phase of hypertension, these changes are much greater; consequently the filtration fraction rises considerably. The reverse situation prevails in glomerulonephritis. In all stages of this disease, a decrease in the filtration fraction is characteristic because of the much greater decline in glomerular filtration than in renal plasma flow.

Measurement of Tubular Secretory Mass

This is accomplished by measuring the Tm for PAH, ie, the maximal secretory capacity of the tubule for PAH. In this case, PAH must be raised to relatively high levels in the blood, eg, 50 mg/100 ml. At these levels, the tubular secretory carriers are working at maximal capacity. By correcting the urine PAH for that filtered as calculated from the glomerular filtration rate, the quantity of PAH secreted is obtained. The normal maximum is about 80 mg/minute/1.73 sq m. Iodopyracet (Diodrast) clearance may be similarly used to measure tubular excretion. The Tm for PAH can be used to gauge the extent of tubular damage in renal disease because, as tubule cells cease to function or are destroyed, excretion of PAH is proportionately diminished.

Phenolsulfonphthalein Test (PSP)

Dyes are widely used for excretion tests. An example is phenolsulfonphthalein (PSP, phenol red). The test is conducted by measuring the rate of excretion of the dye following intramuscular or intravenous administration. The intravenous test is the more valid since it eliminates the uncertainties of absorption which exist in the intramuscular test. Urine specimens may be collected at 15, 30, 60, and 120 minutes after the injection of the dye. If the 15-minute urine contains 25% or more of the injected PSP, the test is normal. Forty to 60% of the dye is normally excreted in the first hour and 20–25% in the second. The most useful information is obtained from the original 15-minute specimen since by the end of 2 hours the amount of dye excreted, although originally delayed, may now appear normal. The dye is readily excreted by the tubules, and therefore the result is not abnormal until impairment of renal function is extreme.

Concentration Tests

Impairment of the capacity of the tubule to perform osmotic work is an early feature of renal disease. The determination of the specific gravity of the urine after a period of water deprivation becomes, therefore, a valuable and sensitive indicator of renal function. If the kidneys do absolutely no work, a fluid is excreted with a specific gravity the same as that of the glomerular filtrate, 1.010. As has been pointed out, any deviation from this specific gravity, ie, dilution or concentration, requires osmotic work by the renal tubule.

In the **Addis test**, fluids are withheld for 24 hours (from 8:00 a.m. on one day to 8:00 a.m. on the next). This must never be done in cases of obviously impaired renal function or in hot weather. Other contraindications to the test include diabetes with polyuria, and adrenal insufficiency. The urine excreted up to 8:00 p.m. of the first day is discarded, but that excreted from 8:00 p.m. of the first day to 8:00 a.m. of the next day is collected and its specific gravity determined. Normally. this urine should have a specific gravity of more than 1.025 (up to 1.034). If the concentrating power of the kidney is such that after this period of water deprivation the specific gravity of the urine is still less than 1.025, renal damage is indicated (except during pregnancy, receding edema, or on diets inadequate in protein or salt).

The **Mosenthal test** is somewhat less rigorous than the Addis test. The patient is not restricted with respect to fluid intake. The bladder should be emptied at 8:00 a.m. on the first day and the urine discarded. Urine collections are made at 2-hour intervals from 8:00 a.m. to 8:00 p.m., and all urine excreted during the 12-hour period from 8:00 p.m. to 8:00 a.m. is collected as one specimen. The specific gravity of each 2-hour specimen and the volume and specific gravity of the 12-hour specimen are noted. Normally, the specific gravity of one or more 2-hour specimens should be 1.018 or more, with a difference of not less than 0.009 between the highest and lowest readings. The volume should be less than 725 ml for the 12-hour night specimen, with a specific gravity of 1.018 or above.

COMPOSITION OF URINE

Characteristics of Urine (See Table 18–1.)

A. Volume: In the normal adult, 600–2500 ml of urine are formed daily. The quantity normally depends on the water intake, the external temperature, the diet, and the mental and physical state. Urine volume is less in summer or in warm climates, for it is more or less inversely related to the extent of perspiration. Nitrogenous end products and coffee, tea, and alcoholic beverages have a diuretic effect. About ½ as much urine is formed during sleep as during activity.

B. Specific Gravity: This normally ranges from 1.003 to 1.030, varying according to concentration of solutes in the urine. The figures in the second and third decimal places, multiplied by 2.66 (Long's coefficient), give roughly the total solids in the urine in g/liter; 50 g of solids in 1200 ml are an average normal for the day.

C. Reaction: The urine is normally acid, with a pH of about 6.0 (range: 4.7–8.0). Ordinarily, over 250 ml of 0.1 N acid, 25 mEq H ion (titratable acidity), are excreted daily. When the protein intake is high, the urine is acid because excess phosphate and sulfate are produced in the catabolism of protein. Acidity is also increased in acidosis and in fevers.

The urine becomes alkaline on standing because of conversion of urea to ammonia. It may also be alkaline in alkalosis such as after excessive vomiting, at least in the early stages (see also p 214).

D. Color: Normal urine is pale yellow or amber. The color varies with the quantity and concentration of urine voided. The chief pigment is urochrome, but small quantities of urobilin and hematoporphyrin are also present.

In fever, because of concentration, the urine may be dark yellow or brownish. In liver disease, bile pigments may color the urine green, brown, or deep yellow. Blood or hemoglobin give the urine a smoky to red color. Methemoglobin and homogentisic acid color it dark brown. Drugs may color the urine. For example, methylene blue gives the urine a green appearance; and cascara and some other cathartics give it a brown color.

The urine is usually transparent, but in alkaline urine a turbidity may develop by precipitation of calcium phosphate. Strongly acid urine precipitates uric acid salts, which have a pink color.

E. Odor: Fresh urine is normally aromatic but the odor may be modified by substances in the diet such as asparagus (methyl mercaptan odor?). In ketosis, the odor of excreted acetone may be detected.

Normal Constituents of Urine (See Table 18–2.)

Urea constitutes about ½ (25 g) of the urine solids. Sodium chloride constitutes about ¼ (9–16 g).

A. Urea: This is the principal end product of protein metabolism in mammals. Its excretion is directly related to the protein intake. Normally it comprises 80–90% of the total urinary nitrogen; but on a low-

TABLE 18–1. Composition of normal urine.*

Specific gravity: 1.003–1.030
Reaction (pH): 4.7–8.0 (avg 6.0)
Volume: Normal range: 600–2500 ml/24 hours (avg 1200 ml). Night/day ratio of volume: 1:2–1:4 if 8:00 a.m. and 8:00 p.m. are the divisions. Night urine usually does not exceed 500–700 ml and usually has a specific gravity of more than 1.018.
Titratable acidity of 1000 ml (depending on pH): 250–700 ml of 0.1 N NaOH for acid urine.
Total solids: 30–70 g/liter (avg 50 g). Long's coefficient to estimate total solids per liter: multiply last 2 figures of specific gravity by 2.66.

Inorganic Constituents (per 24 hours):

Chlorides (as NaCl)	10 (9–16) gm on usual diet	Sulfur (total) (as SO_3)	2 (0.7–3.5) g
Sodium (varies with intake)	4 g on usual diet	Calcium	0.2 (0.1–0.2) g
		Magnesium	0.15 (0.05–0.2) g
Phosphorus	2.2 (2–2.5) g	Iodine	50–250 μg
Potassium (varies with intake)	2 g	Arsenic	50 μg or less
		Lead	50 μg or less

Organic Constituents (per 24 hours):

		Nitrogen Equivalent
Nitrogenous (total)	25–35 g	10–14 g
Urea (half of total urine solids; varies with diet)	25–30 g	10–12 g
Creatinine	1.4 (1–1.8) g	0.5 g
Ammonia	0.7 (0.3–1) g	0.4 g
Uric acid	0.7 (0.5–0.8) g	0.2 g
Undetermined N (amino acid, etc)		0.5 g
Protein, as such ("albumin")	0–0.2 g	
Creatine	60–150 mg (increased in liver or muscle diseases or thyrotoxicosis)	

Other Organic Constituents (per 24 hours):

Hippuric acid 0.1–1 g	Oxalic acid 15–20 mg	Indican 4–20 mg	Coproporphyrins 60–280 μg
Purine bases 10 mg	Ketone bodies 3–15 mg	Allantoin 30 mg	Phenols (total) 0.2–0.5 g

Sugar: 50% of people have 2–3 mg/100 ml after a heavy meal. A diabetic can lose up to 100 g/day.

Ascorbic Acid: Adults excrete 15–50 mg/24 hours; in scurvy, less than 15 mg/24 hours.

*Modified from Krupp & others: *Physician's Handbook*, 16th ed. Lange, 1970.

protein diet this is less because certain other nitrogenous constituents tend to remain relatively unaffected by diet.

Urea excretion is increased whenever protein catabolism is increased, as in fever, diabetes, or excess adrenocortical activity. In the last stages of fatal liver disease, decreased urea production may lead to decreased excretion. There is also a decrease in urine urea in acidosis since some of the nitrogen which would have been converted to urea is diverted to ammonia formation. The urea does not, however, give rise to the ammonia directly.

B. Ammonia: Normally there is very little ammonia in freshly voided urine. Its formation by the kidney in acidosis has been described on p 381. In acidosis of renal origin, this mechanism may fail. Therefore, such acidosis is accompanied by a low concentration of ammonia in the urine. On the other hand, the ketosis and resultant acidosis of uncontrolled diabetes mellitus, in which renal function is unimpaired, will cause a high ammonia output in the urine.

C. Creatinine and Creatine: Creatinine is the product of the breakdown of creatine (see p 346). In a given subject it is excreted in relatively constant amounts regardless of diet. The creatinine coefficient is the ratio between the amounts of creatinine excreted in 24 hours to the body weight in kg. It is usually 20–26 mg/kg/day in normal men and 14–22 mg/kg/day in normal women. Because this rate is so constant in a given individual, the creatinine coefficient may serve as a reliable index of the adequacy of a 24-hour urine collection. The excretion of creatinine is decreased in many pathologic states.

Creatine is present in the urine of children and, in much smaller amounts, in the urine of adults as well. In men the creatine excretion is about 6% of the total creatinine output (probably 60–150 mg/day). In women creatinuria is much more variable (usually 2-2½ times that of normal men), although in about 1/5 of

TABLE 18–2. Variations in some urinary constituents with different protein levels in the diet.*†

	Usual Protein Intake		Protein-Rich Diet		Protein-Poor Diet	
	g	%N	g	%N	g	%N
Total urinary nitrogen	13.20		23.28		4.20	
Protein represented by above N	82.50		145.50		26.25	
Urea nitrogen	11.36	86.1	20.45	87.9	2.90	69.0
Ammonia nitrogen	0.40	3.0	0.82	3.5	0.17	4.0
Creatinine nitrogen	0.61	4.6	0.64	2.7	0.60	14.3
Uric acid nitrogen	0.21	1.6	0.30	1.3	0.11	2.6
Undetermined nitrogen	0.62	4.7	1.07	4.6	0.52	12.4
Titratable acidity (ml 0.1 N)	284.0 ml		655.0 ml		160.0 ml	
Volume of urine	1260.0 ml		1550.0 ml		960.0 ml	
Total sulfur (as SO_3)	2.65 g		3.55 g		0.86 g	
Inorganic sulfate (as SO_3)	2.16		2.82		0.64	
Ethereal sulfate (as SO_3)	0.18		0.36		0.11	
Neutral sulfate (as SO_3)	0.31		0.37		0.11	
Total inorganic phosphate (as P_2O_5)	2.59		4.07		1.06	
Chloride (as NaCl)	12.10		15.10		9.80	

*Reprinted, with permission, from Bodansky: *Introduction to Physiological Chemistry*, 4th ed. Wiley, 1938.
†A balance between intake of protein nitrogen and the excretion of nitrogen (nitrogen balance) is presumed to exist in these experiments.

the normal women studied the creatinuria did not exceed that found in men. In pregnancy, creatine excretion is increased. Creatinuria is also found in pathologic states such as starvation, impaired carbohydrate metabolism, hyperthyroidism, and certain myopathies and infections. Excretion of creatine is decreased in hypothyroidism.

Creatinine is measured colorimetrically by adding alkaline picrate to the urine. In the presence of creatinine, the mixture develops an amber color (Jaffé reaction). The color is read against a creatinine standard similarly treated with alkaline picrate solution.

Creatine, when heated in acid solution, is converted to creatinine, which can be measured as described. The difference in the creatinine content of the urine before and after boiling with acid gives the creatine content.

D. Uric Acid: This is the most important end product of the oxidation of purines in the body. It is derived not only from dietary nucleoprotein but also from the breakdown of cellular nucleoprotein in the body.

Uric acid is very slightly soluble in water but forms soluble salts with alkali. For this reason it precipitates readily from acid urine on standing.

The output of uric acid is increased in leukemia, severe liver disease, and in various stages of gout.

The blue color which uric acid gives in the presence of arsenophosphotungstic acid sodium cyanide is the basis of the Folin colorimetric test. This is not a specific reaction, however. Salicylates raise the color value due to the excretion of gentisic acid and other similar metabolites. This may raise the apparent excretion of uric acid as much as 25% in 24 hours after a large dose of acetylsalicylic acid.

The specificity of the analysis for uric acid may be increased by treatment of the sample with uricase, the enzyme (from hog kidney) which causes the conversion of uric acid to allantoin. The decline in apparent uric acid concentration after uricase treatment is taken as a measure of the true uric acid content of the sample.

E. Amino Acids: In adults, only about 150–200 mg of amino acid nitrogen are excreted in the urine in 24 hours. The full-term infant at birth excretes about 3 mg amino acid nitrogen per pound body weight; this excretion declines gradually up to the age of 6 months, when it reaches a value of 1 mg/lb that is maintained throughout childhood. Premature infants may excrete as much as 10 times as much amino acid nitrogen as the full-term infant.

The reason such very small amounts of amino acids are lost into the urine is that the renal thresholds for these substances are quite high. However, all of the naturally occurring amino acids have been found in the urine, some in relatively large quantities when compared to the trace quantities characteristic of most. It is also of interest that a high percentage of some excreted amino acids is in combined forms and can be liberated by acid hydrolysis. Diet alters the pattern of amino acid excretion to a slight extent.

The amount of free amino acids found in the urine of normal male human subjects, after a 12-hour fast, is shown in Table 18–3.

In terminal liver disease and in certain types of poisoning (chloroform, carbon tetrachloride), the quantity of amino acids excreted is increased. This "overflow" type of aminoaciduria is to be distinguished from renal aminoaciduria due to an inherited tubular defect in reabsorption. In "cystinuria" a considerable increase in excretion of only 4 amino

TABLE 18–3. Free amino acids in urine.* (Normal male subjects after a 12-hour fast.)

	Range (mg/hour)	Mean
Glycine	3.0–36.0	11.7 ± 9.45
Histidine	2.7–23.0	7.7 ± 1.33
Glutamine	1.6– 4.6	2.9 ± 0.96
Cystine	0.4– 4.3	2.3 ± 1.22
Tryptophan, tyrosine, serine, proline, threonine, phenylalanine, lysine, arginine, alanine	1.0– 3.4 (for any of these amino acids)	
Leucine, isoleucine, valine, aspartic acid, glutamic acid, methionine	Less than 1 (for any of these amino acids)	

*Reproduced, with permission, from Harper, Hutchin, Kimmell: Proc Soc Exper Biol Med 80:768, 1952.

acids occurs: arginine, cystine, lysine, and ornithine. The amounts of all other amino acids excreted remain normal.

F. Allantoin: This is derived from partial oxidation of uric acid. There are very small quantities in human urine, but in other mammals allantoin is the principal end product of purine metabolism, replacing uric acid.

G. Chlorides: These are mainly excreted as sodium chloride. Because most of the chlorides are of dietary origin, the output varies considerably with the intake. The excretion of chlorides is reduced in certain stages of nephritis and in fevers.

H. Sulfates: The urine sulfur is derived mainly from protein because of the presence of the sulfur-containing amino acids, methionine and cystine, in the protein molecule. Its output therefore varies with the protein intake. The total urine sulfur is usually partitioned into 3 forms. It is customary to express all sulfur concentrations in the urine as SO_3.

1. Inorganic (sulfate) sulfur–This is the completely oxidized sulfur precipitated from urine when barium chloride is added. It is roughly proportional to the ingested protein with a ratio of 5:1 between urine nitrogen and inorganic sulfate (expressed as SO_3). Together with the total urinary nitrogen, this fraction of urine sulfur is an index of protein catabolism.

2. Ethereal sulfur (conjugated sulfates)–This fraction (about 10% of the total sulfur) includes the organic combinations of sulfur excreted in the urine. Examples are the phenol and cresol sulfuric acids, indoxyl and skatoxyl sulfuric acids, and other sulfur conjugates formed in detoxication.

The ethereal sulfate fraction is in part derived from protein metabolism; but in indican and some of the phenols, putrefactive activity in the intestine is also represented.

After hydrolysis with hot hydrochloric acid, the ethereal sulfates may be precipitated with barium chloride.

3. Neutral sulfur–This fraction is the sulfur which is incompletely oxidized, such as that which is contained in cystine, taurine, thiocyanate, or sulfides. It does not vary with the diet to the same extent that the other fractions do.

Neutral sulfur is determined as the difference between the total sulfur and the sum of the inorganic and ethereal sulfur.

I. Phosphates: The urine phosphates are combinations of sodium and potassium phosphate (the alkaline phosphates) as well as of calcium and magnesium (so-called earthy phosphates). These latter forms are precipitated in alkaline urines.

The diet, particularly the protein content, influences phosphate excretion. Some is also derived from cellular breakdown.

In certain bone diseases, such as osteomalacia and renal tubular rickets, the output of phosphorus in the urine is increased. In hyperparathyroidism the excretion of phosphorus is also markedly increased. A decrease is sometimes noted in infectious diseases, in hypoparathyroidism, and renal disease.

J. Oxalates: Ordinarily the amount of oxalate in the urine is low, but in a recently discovered inherited metabolic disease (primary hyperoxaluria) relatively large quantities of oxalate may be continuously excreted.

K. Minerals: Sodium, potassium, calcium, and magnesium–the 4 cations of the extracellular fluid–are present in the urine. The sodium content varies considerably with intake and physiologic requirements. Urine potassium rises when the intake is increased or in the presence of excessive tissue catabolism, in which case it is derived from intracellular materials. The excretion of potassium is also affected by acid-base equilibrium, most notably by alkalosis, which inevitably increases potassium excretion (see p 383). Sodium and potassium excretions are also controlled by the activity of the adrenal cortex.

Most of the calcium and magnesium is excreted by the intestine; the content of these elements in the urine is, therefore, relatively low. However, this will vary in certain pathologic states, particularly those involving bone metabolism.

L. Vitamins, hormones, and enzymes can be detected in small quantities in normal urine. The urinary content of these substances is often of diagnostic importance (see pp 449–451).

Abnormal Constituents of the Urine

A. Proteins: Proteinuria (albuminuria) is the presence of albumin and globulin in the urine in abnormal concentrations. Normally not more than 30–200 mg of protein are excreted daily in the urine.

1. Physiologic proteinuria, in which less than 0.5% protein is present, may occur after severe exercise, after a high-protein meal, or as a result of some temporary impairment in renal circulation when a

person stands erect (orthostatic or postural proteinuria).

In 30–35% of the cases, pregnancy is accompanied by proteinuria.

2. Pathologic proteinurias are sometimes classified as prerenal, when the primary causes are factors operating before the kidney is reached, although the kidney may also be involved; renal, when the lesion is intrinsic to the kidney; and postrenal, when the proteinuriá is due to inflammation in the lower urinary tract. In glomerulonephritis proteinuria is marked during the degenerative phase; the lowest excretion of albumin is during the latent phase and may increase terminally. In nephrosis a marked proteinuria occurs, accompanied by edema and low concentrations of serum albumin. Nephrosclerosis, a vascular form of renal disease, is related to arterial hypertension. The proteinuria observed in this disease increases with the increasing severity of the renal lesion. The loss of protein in nephrosclerosis is generally less than that in glomerulonephritis. Proteinuria is also observed in poisoning of the renal tubules by heavy metals like mercury, arsenic, or bismuth unless the poisoning is severe enough to cause anuria.

3. Albumin may be detected by heating the urine, preferably after centrifuging to remove the sediment, then adding a little dilute acetic acid. A white cloud or precipitate which persists after addition of the acid indicates that protein is present. In quantitative measurement of urine protein, the protein is precipitated with trichloroacetic acid and then separated for analysis, either colorimetrically (biuret) or by Kjeldahl analysis.

4. Bence Jones proteins (see also p 197)–These peculiar proteins are globulins which occur in the urine most commonly in multiple myeloma and rarely in leukemia, Hodgkin's disease, and lymphosarcoma. They may be identified in the urine by their ability to precipitate when the urine is warmed to 50–60° C and to redissolve almost completely at 100° C. The precipitate reforms on cooling.

B. Glucose: Normally not more than 1 g of sugar is excreted per day. Glycosuria is indicated when more than this quantity is found. The various causes of glycosuria have been discussed. Transient glycosuria may be noted after emotional stress, such as an exciting athletic contest. Fifteen percent of cases of glycosuria are not due to diabetes. Usually, however, glycosuria suggests diabetes; this must be confirmed by blood studies to eliminate the possibility of renal glycosuria.

A simple test for the presence of glucose in the urine is often performed by using paper test strips contining the peroxidase enzyme. It has been reported that peroxidase is inhibited by homogentisic acid (found in alkaptonuric urine), bilirubin glucuronide (as occurs in urine of jaundiced patients), ascorbic acid, and epinephrine. In the presence of these inhibitory compounds, the enzyme paper strips may therefore give false-negative results when used to test for glucose in the urine.

C. Other Sugars:

1. **Fructosuria** is a rare anomaly in which the metabolism of fructose but not that of other carbohydrates is disturbed.

2. **Galactosuria and lactosuria** may occur occasionally in infants and in the mother during pregnancy, lactation, and the weaning period. In congenital galactosemia, the inherited disease which is characterized by impaired ability to convert galactose to glucose (see p 259), the blood levels of galactose are much elevated and galactose spills over into the urine.

3. **Pentosuria** may occur transiently after ingestion of foods containing large quantities of pentoses, such as plums, cherries, grapes, and prunes. Congenital pentosuria is a benign genetic defect characterized by inability to metabolize L-xylulose, a constituent of the uronic acid pathway (Fig 13–15).

All of the above sugars reduce Benedict's solution. When it is suspected that sugars other than glucose are present, it has been customary to perform a fermentation test with baker's yeast. If all of the reducing action is removed by the yeast, this suggests that only glucose is present. However, more specific tests are preferred. The recent introduction of a specific analytic test for glucose by the use of the enzyme glucose oxidase is one such test. A comparison of the apparent glucose content of the urine, as determined by total reducing action with the absolute glucose content (as determined by glucose oxidase), would indicate more definitely whether sugars other than glucose were present. If so, these other sugars can be identified readily by paper chromatography or in some cases by preparation of specific osazones.

D. Ketone Bodies: Normally, only 3–15 mg of ketones are excreted in a day. The quantity is increased in starvation, impaired carbohydrate metabolism (eg, diabetes), pregnancy, ether anesthesia, and some types of alkalosis. In many animals, excess fat metabolism will also induce a ketonuria. The acidosis accompanying ketosis will cause increased ammonia excretion as the result of the body's effort to conserve fixed base.

E. Bilirubin: The presence of bilirubin in the urine and its relationship to jaundice are discussed in Chapter 17.

F. Blood: In addition to its occurrence in nephritis, blood in the urine (hematuria) may be the result of a lesion in the kidney or urinary tract (eg, after trauma to the urinary tract). However, free hemoglobin (hemoglobinuria) may also be found in the urine after rapid hemolysis, eg, in blackwater fever (a complication of malaria) or after severe burns.

G. Porphyrins: (See Chapter 6.) The urine coproporphyrin excretion of the normal human adult is 60–280 μg/day. The occurrence of uroporphyrins as well as increased amounts of coproporphyrins in the urine is termed porphyria.

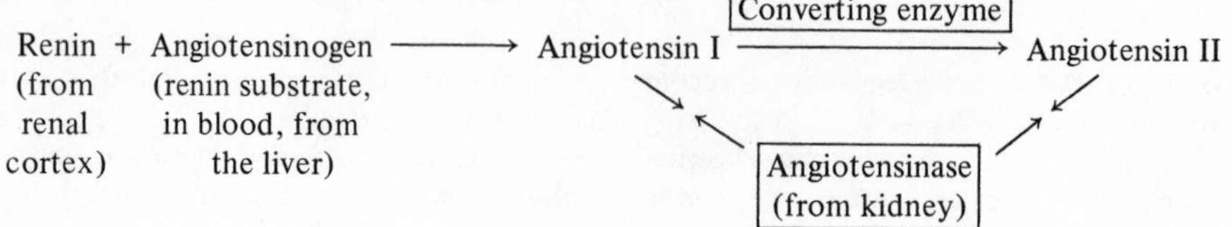

FIG 18–7. Reactions of the renal pressor system.

THE RENAL PRESSOR SYSTEM

In addition to excretory functions, the kidneys have what resembles an endocrine function in elaborating a chemical substance which is added to the blood and subsequently contributes to a reaction affecting tissues throughout the body. This function of the kidneys has been demonstrated by the results of a decrease in the blood supply to these organs, produced experimentally by compressing the renal artery (Goldblatt) or by inducing contraction around the kidneys. Perinephritis, which results from contraction around the kidneys, has also been effected experimentally by enclosure of the kidneys in silk cloth (Page). Clinically, it is observed as a result of contraction around the kidney of a scar caused by injury due to a foreign body. Occlusion of the renal artery and arterioles by atherosclerotic changes may also compromise the renal blood supply. In all of these situations, ie, reduction of the blood supply to the kidneys (renal ischemia) or the induction of perinephritis, a persistent hypertensive state develops.

The chemical mechanism of the renal hypertension described above is as follows: The ischemic renal cortex forms a proteolytic enzyme called **renin** and secretes it into the blood by way of the renal vein. In the blood, renin acts upon its specific substrate, an $alpha_2$ globulin normally present in blood plasma and formed in the liver. This globulin is termed renin substrate or, more recently, **angiotensinogen.** The action of the enzyme, renin, is to split off from angiotensinogen a polypeptide containing 10 amino acids (a decapeptide) which is called **angiotensin I.** Then another enzyme in the blood, the converting enzyme, acts on angiotensin I to split off 2 amino acids and thus to form an octapeptide, **angiotensin II** (Skeggs, 1954). The decapeptide, angiotensin I, is only slightly active. Angiotensin II is the active material, having a pressor activity about 200 times that of norepinephrine. Angiotensin increases the force of the heartbeat and constricts the arterioles, and this often results in diminished renal blood flow even though peripheral blood flow may remain unchanged. In addition to raising blood pressure, angiotensin also brings about contraction of smooth muscle (myotropic effect).

Normal kidneys, and other tissues to a lesser extent, contain proteolytic enzymes called **angiotensinases** which are capable of destroying angiotensin.

The reactions of the renal pressor system may be summarized as shown in Fig 18–7.

The amino acid sequence of the angiotensins is shown in Fig 18–8. It will be noted that conversion of angiotensin I to angiotensin II involves removal of the 2 C-terminal amino acids, leucine and histidine. It is probable that the secondary form of angiotensin involves a helical structure which may be maintained by hydrogen bonds. This is indicated by the fact that treatment of angiotensin with 10% solutions of urea, which is known to rupture hydrogen bonds, brings about a 50% decrease in myotropic activity of the pressor compound.

Laragh (1960) has shown that intravenous infusion of angiotensin inevitably produces an increase in the rate of secretion of aldosterone, the electrolyte-regulating corticoid of the adrenal cortex. This suggests that aldosterone secretion may be affected by a renal-adrenal endocrine system in which angiotensin is acting

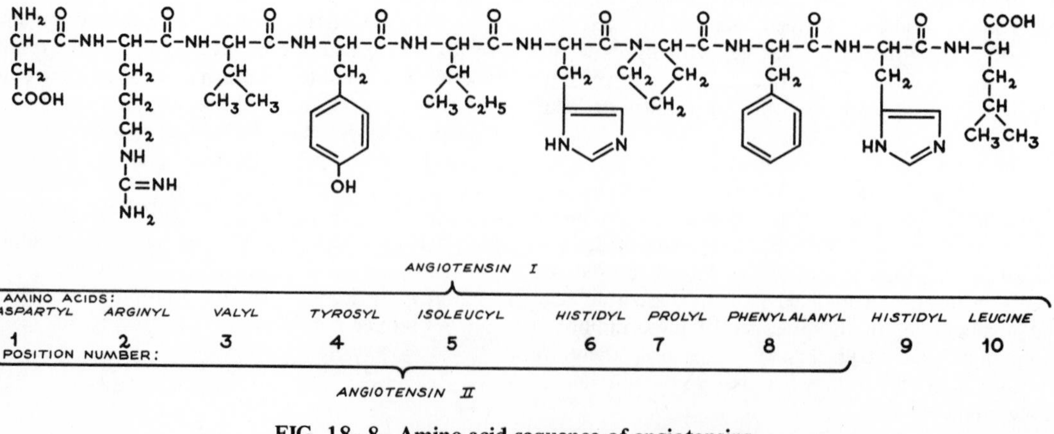

ANGIOTENSIN I

AMINO ACIDS:	ASPARTYL	ARGINYL	VALYL	TYROSYL	ISOLEUCYL	HISTIDYL	PROLYL	PHENYLALANYL	HISTIDYL	LEUCINE
POSITION NUMBER:	1	2	3	4	5	6	7	8	9	10

ANGIOTENSIN II

FIG 18–8. Amino acid sequence of angiotensins.

as a tropic hormone in a manner similar to that of pituitary ACTH as it relates to production of the adrenocorticoids other than aldosterone. Aldosterone, by causing retention of sodium and water, would tend to improve the circulation to the kidney and thus effect suppression of renin production.

It is still uncertain whether the renin-angiotensin system plays any role in normal blood pressure regulation. When renal circulation is impaired, however, the pressor system is certainly more active than in the presence of normal renal circulation, and there is now increasing clinical evidence that some forms of hypertension may be greatly benefited if not cured by methods successful in restoring adequate circulation to the kidneys. In malignant hypertension, aldosterone hypersecretion is a much more consistent finding than in any other type. This suggests that malignant hypertension is a clinical counterpart to the Goldblatt experimental hypertension which was produced by compression of the renal artery in animals but which may occur clinically by the formation of many small occlusions to the circulation of both kidneys. The finding of increased aldosterone in the urine of the hypertensive patient should aid in the differential diagnosis of hypertension as a means of evaluating whether a renal pressor substance is involved in the etiology of the disease.

The discovery of the renal pressor system has clarified the long recognized association between renal and cardiovascular disease, but increased amounts of renin have not been found consistently in many types of chronic hypertension. It remains for further study to delineate the true role of the renin-angiotensin system in clinical hypertension.

• • •

References

Cogan DG, Kuwubara T, Kinoshita J, Sheehan L, Merola L: JAMA 164:394, 1957.

Gilman A, Brazeau P: Am J Med 15:765, 1953.

Laragh JH, Angers M, Kelly WG, Lieberman S: JAMA 174:234, 1960.

Mudge GH: Am J Med 20:448, 1956.

Mueller CB, Mason AD Jr, Stout DG: Am J Med 18:267, 1955.

Pitts RF: Am J Med 9:356, 1950.

Rector FC Jr, Seldin DW, Copenhaver JH: J Clin Invest 34:20, 1955.

Skeggs LT, Marsh WH, Kahn JR, Shumway NP: J Exp Med 99:275, 1954.

Verney EB: Surg Gynec Obst 106:441, 1958.

Wirz H: Helv physiol pharmacol acta 11:20, 1953.

Wirz H: Helv physiol pharmacol acta 14:353, 1956.

Bibliography

Milne MD: Diseases of the kidney and genito-urinary tract. In: *Biochemical Disorders in Human Disease,* 2nd ed. Thompson RHS, King EJ (editors). Academic Press, 1964.

Smith H: *The Kidney.* Oxford Univ Press, 1951.

Smith H: *Principles of Renal Physiology.* Oxford Univ Press, 1956.

Stanbury JB, Wyngaarden JB, Fredrickson DS (editors): *The Metabolic Basis of Inherited Disease,* 2nd ed. McGraw-Hill, 1966.

19 . . .
Water & Mineral Metabolism

WATER METABOLISM

Body Water & Its Distribution

The total body water is equal to 45–60% of the body weight; the average is 55% for adult men and 50% for adult women. It has been assumed that the total body water is distributed throughout 2 main compartments: the extracellular compartment (which was subdivided into plasma and interstitial fluid) and the intracellular compartment. More recently it has become clear that although the concept of a single intracellular component of the body water is still useful, the extracellular component must now be recognized as more heterogeneous and should be subdivided into 4 main subdivisions: (1) plasma, (2) interstitial and lymph fluid, (3) dense connective tissue, cartilage, and bone, and (4) transcellular fluids (Edelman, 1959).

Plasma volume, measured as described below, comprises in general the fluid within the heart and blood vessels. The interstitial and lymph fluid may be considered to represent an approximation of the actual fluid environment outside the cells. Dense connective tissue, cartilage, and bone, because of differences in structure and relative avascularity, do not exchange fluid or electrolyte readily with the remainder of the body water. For this reason these tissues must be classified as a distinct subdivision of the extracellular water. Finally, the extracellular water must include also a variety of extracellular fluid collections formed by the transport or secretory activity of the cells. Examples are the fluids found in the salivary glands, pancreas, liver and biliary tree, thyroid gland, gonads, skin, mucous membranes of the respiratory and gastrointestinal tracts, and the kidneys, as well as the fluids in spaces within the eye, the cerebrospinal fluid, and that within the lumen of the gastrointestinal tract. The distribution of body water is shown in Table 19–1.

TABLE 19–1. Distribution of body water in "average" normal young adult male.*

Source	ml/kg of Body Weight		% of Total Body Water	
Intracellular	330		55.0	
Extracellular	270		45.0	
Plasma		45		7.5
Interstitial-lymph		120		20.0
Dense connective tissue and cartilage		45		7.5
Inaccessible bone		45		7.5
Transcellular		15		2.5
Total body water	600		100.0	

*Edelman & Leibman: Am J Med 27:256, 1959.

Measurement of Distribution of Body Water

Total body water has been determined in animals by desiccation. More recently the distribution of heavy water, deuterium oxide (D_2O), or of tritium oxide has been used in the living animal and in human subjects as a method of measuring total body water. Antipyrine is also used.

The value for body water given above was obtained by measuring the volume of distribution of D_2O. However, in all studies of the proportion of the body which is water, considerable variation is to be expected when different subjects are compared even by the same analytic method. This is due mainly to variation in the amount of fat in the body. The higher the fat content of the subject, the smaller the percentage of water that subject will contain in his body. If a correction for the fat content of the subject is made, the total body water in various subjects is relatively constant (60–70% of body weight) when expressed as percentage of the "lean body mass," ie, the sum of the fat-free tissue.

The composition of an adult human body was determined by direct chemical analysis (Forbes, 1953). The whole body contained 19.44% ether-extractable material (lipid), 55.13% moisture, 18.62% protein, and 5.43% ash, including 1.907% calcium and 0.925% phosphorus. When these data were recalculated to the fat-free basis, the moisture content was then 69.38%, which is in agreement with data obtained by indirect methods as described above.

Specific gravity of the body may also serve as a basis for the calculation of total body water. The body is considered as a mixture of fat, which is of relatively low density, and fat-free tissue, which is of relatively high density. By measuring the specific gravity of the body (weighing the subject in air and under water), it is possible to calculate the proportion of the body which is fat tissue and that which is fat-free tissue. This technic has been used to arrive at an estimate of the lean body mass, described above.

A. Plasma Volume: Plasma volume may be measured by the Evans blue dye (T-1824) technic. In this procedure a carefully measured quantity of the dye is injected intravenously. After a lapse of time to allow for mixing, a blood sample is withdrawn and the concentration of the dye in the plasma is determined colorimetrically. The normal figures for plasma volume thus determined are 47–50 ml/kg body weight.

Other methods of plasma (or blood) volume measurements are based on the intravenous injection of radiophosphorus-labeled red cells (^{32}P) or radioiodine-labeled human serum albumin. These substances distribute themselves in the blood stream and after a mixing period of 10 or more minutes their volume of distribution may be calculated from their concentration in an aliquot of blood or plasma.

B. Extracellular Fluid Volume: The volume of the extracellular fluid would be measured by the dilution of a substance which does not penetrate into the cells and which is distributed rapidly and evenly in all of the plasma as well as the remainder of the extracellular fluid. No such ideal substance has yet been found. However, the volume of distribution of certain saccharides such as mannitol or inulin has been found to give a reasonably accurate measurement of the volume of the interstitial and lymph fluid, although it is believed that about 25% of the extracellular phase of dense connective tissue and cartilage also equilibrates rapidly with the saccharides, so that the measured volume must be corrected by this amount.

The volume of the remaining components of the extracellular fluid shown in Table 19–1 has been estimated by calculations from direct chemical analyses of representative samples of the individual tissues.

C. Intracellular Water Volume: Intracellular water volume is calculated simply as the difference between the total body water and the extracellular water.

The Availability of Water (Water Intake)

The 2 main sources of water (about 2500 ml/day) are as follows:

A. Preformed Water: Liquids imbibed as such, 1200 ml; water in foods, 1000 ml.

B. Water of Oxidation (300 ml): The water of oxidation (sometimes termed "metabolic water") is derived from the combustion of foodstuffs in the body. The oxidation of 100 g of fat yields 107 g of water; oxidation of 100 g of carbohydrate, 55 g of water; oxidation of 100 g of protein, 41 g of water.

Losses of Water (See Table 19–2.)

Water is lost from the body by 4 routes: the skin, as sensible and insensible perspiration; the lungs, as water vapor in the expired air; the kidneys, as urine; and the intestines, in the feces. It is customary to refer to the sum of the dermal loss (exclusive of visible perspiration) and the pulmonary loss as the insensible losses.

Additional Water Losses in Disease

In kidney disease in which concentrating ability is limited, renal water loss may be twice as high as that listed in the table. Insensible losses may rise much higher than normal as a consequence of operations, in fever, or in the physically debilitated. When subjected to high environmental temperatures, patients will also sustain extremely high extrarenal water losses, as much as 2000–5000 ml in some instances. Water losses from the intestine may be considerable, particularly in diarrhea and vomiting.

Factors Which Influence the Distribution of Body Water

Water is retained in the body in a rather constant amount, but its distribution is continuously subject to change. Osmotic forces are the principal factors which control the location and the amount of fluid in the various compartments of the body. These osmotic forces are maintained by the solutes, the substances dissolved in the body water.

A. Solutes in the Body: The solutes in the body fluids are important not only in directing fluid distribution but also in maintaining acid-base balance (see Chapter 11). In a consideration of the various substances in solution in the fluids of the body and of their effect on water retention and distribution, it is convenient to divide them into 3 categories:

TABLE 19–2. Daily water losses and water allowances for normal individuals who are not working or sweating.*

	Losses				Allowances	
Size	**Urine (ml)**	**Stool (ml)**	**Insensible (ml)**	**Total (ml)**	**ml/person**	**ml/kg**
Infant (2–10 kg)	200–500	25–40	75–300 (1.3 ml/kg/hr)	300–840	330–1000	165–100
Child (10–40 kg)	500–800	40–100	300–600	840–1500	1000–1800	100–45
Adolescent or adult (60 kg)	800–1000	100	600–1000 (0.5 ml/kg/hr)	1500–2100	1800–2500	45–30

*Butler & Talbot: New England J Med 231:585, 1944.

1. Organic compounds of small molecular size (glucose, urea, amino acids, etc)–Since these substances diffuse relatively freely across cell membranes, they are not important in the distribution of water; if present in large quantities, however, they aid in retaining water and thus do influence total body water.

2. Organic substances of large molecular size (mainly the proteins)–The importance of the plasma proteins in the exchange of fluid between the circulating blood and the interstitial fluid has been discussed (see Chapter 10). The effect of the protein fraction of the plasma and tissues is mainly on the transfer of fluid from one compartment to another, not on the total body water.

3. The inorganic electrolytes–(Table 19–3.) Because of the relatively large quantities of these materials in the body, the inorganic electrolytes are by far the most important, both in the distribution and in the retention of body water.

B. Measurements of Solutes: In describing chemical reactivity, particularly acid-base balance, all reacting ions must be expressed in identical concentration units. Since one chemical equivalent of any substance is exactly equal in chemical reactivity to one equivalent of any other, this can be accomplished by converting the concentrations of each into equivalents/liter. Because of the small quantities involved in body fluids, the milliequivalent (mEq, 1/1000 Eq) is preferred. Furthermore, when changes occur in the chemistry of the body fluid, there are usually compensatory shifts of one ion to make up for losses of another. For example, excessive losses of chloride over sodium in vomiting from the stomach result in a chloride deficit in the extracellular fluid. This is promptly compensated by an increase in bicarbonate to accompany the sodium left uncovered by the chloride loss. These changes can be readily understood and calculated when all reactants are expressed in the same units.

1. Conversion of electrolyte concentrations to mEq–For conversion of mg/100 ml to mEq/liter, (1) express the concentration on a per liter basis, ie, multiply the number of mg (per 100 ml) by 10 to determine the number of mg/liter, and (2) divide the mg/liter by the appropriate mEq weight given in Table 19–4. The mEq weight of an element is the millimolecular weight divided by the valence.

TABLE 19–3. Plasma electrolyte concentrations (from Gamble).

Cations (+)	mEq/liter	Anions (-)	mEq/liter
Na^+	142	HCO_3^-	27
K^+	5	Cl^-	103
Ca^{++}	5	$HPO_4^=$	2
Mg^{++}	3	$SO_4^=$	1
		Organic acids	6
		Protein	16
Totals	155		155

TABLE 19–4. Milliequivalent weights.

Na	23	Cl	35.5
K	39	Cl (as NaCl)	58.5
Ca	20	HPO_4 (as P)	17.2*
Mg	12	SO_4 (as S)	16

*The inorganic phosphorus in the serum exists as a buffer mixture in which approximately 80% is in the form of $HPO_4^=$ and 20% as $H_2PO_4^-$. For this reason the mEq weight is usually calculated by dividing the atomic weight of phosphorus by 1.8. Thus the mEq weight for phosphorus in the serum is taken as 31/1.8, or 17.2. To avoid the problem presented by the 2 valences of the serum phosphorus, some laboratories prefer to express the phosphorus as millimols (mM) rather than mEq. One mM of phosphorus is 31 mg. To convert mg of phosphorus to mM, divide mg/liter by 31, eg, serum phosphorus = 3.1 mg/100 ml = 31/31 = 1 mM/liter.

Examples: Plasma sodium: 322 mg/100 ml. Multiply the mg by 10 (to express on a per liter basis). Then divide by mEq weight of sodium, 23.

322 X 10 = 3220 mg/liter;
divided by 23 = 140 mEq/liter

Chloride (reported as NaCl): 603 mg/100 ml.

603 X 10 = 6030 mg/liter;
divided by 58.5 = 103 mEq/liter

Calcium: 10 mg/100 ml.

10 X 10 = 100 mg/liter;
divided by 20 = 5 mEq/liter

2. Conversion of bicarbonate to milliequivalents–The bicarbonate of the plasma is measured by conversion to CO_2 and reported in volumes percent (Vol %); to convert to mEq of bicarbonate per liter, divide CO_2 combining power, expressed as Vol %, by 2.3.*

3. Conversion of organic acids and proteins to mEq–The organic acids and the proteins in the anion column of plasma are calculated from their combining power with base. The base equivalence of protein, in

*The conversion of the CO_2 combining power to mEq of bicarbonate is based on the following facts. One mol of a gas occupies 22.4 liters (at 0° C and 760 mm Hg), and therefore 1 mM occupies 22.4 ml or, what is the same thing, each 22.4 ml of gas is equivalent to 1 mM; 600 ml of CO_2/liter (a normal total blood CO_2) thus equals 600/22.4 = 26.8 mM total CO_2/liter (1 mM of CO_2 is the same as 1 mEq of CO_2).

The total CO_2 as determined in the blood includes carbonic acid, free CO_2, and bicarbonate. The bicarbonate fraction alone can be calculated by assuming that a 1:20 ratio exists between carbonic acid and bicarbonate. Under these conditions, the plasma bicarbonate fraction is derived by dividing the total CO_2 (as CO_2 combining power in Vol %) by 2.3.

mEq/liter, is obtained by multiplying the g of total protein per 100 ml by 2.43.

C. Electrolyte Composition of Blood Plasma and of Intracellular Fluid: (See Fig 19–1.)

1. Composition of interstitial fluid–The electrolyte composition of the interstitial fluid is similar to that of the plasma except that chloride largely replaces protein in the anion column.

2. Composition of intracellular fluids–The intracellular fluid differs in electrolyte composition from that of the plasma in that potassium rather than sodium is the principal cation and, largely due to the presence of phosphorylated organic compounds, phosphate rather than chloride is the principal anion. The intracellular chloride content is variable in accordance with the metabolic circumstances. The amount of protein within the cell is also considerably larger than that in its extracellular environment.

The intracellular concentration of sodium is higher than had previously been assumed. Deane & Smith (1952) have reported that the normal average intracellular sodium is 37 mEq/liter of intracellular water. Furthermore, it is now clear that sodium may replace potassium within the cell when sodium salts are administered to potassium-deficient subjects. The adrenocortical steroids and ACTH also influence the concentration of sodium and potassium within the cell. Under the influence of these hormones, intracellular sodium may be increased.

D. Importance of Sodium and Potassium in Water Metabolism: Both from the standpoint of osmotic forces (directing the movement of water from one compartment to another in the body) and in controlling the total hydration of the body, sodium and potassium are the most important elements in the body fluids. As has been pointed out, in the normal individual sodium is largely confined to the extracellular space and potassium to the intracellular space.

1. Sodium–As Gamble has so well expressed it, sodium is the "backbone" of the extracellular fluid in that it, more than any other element, determines the quantity of extracellular fluid to be retained. This is the reason that sodium intake is restricted in order to control overhydration in various pathologic states.

2. Potassium–Under certain conditions, potassium leaves the cells. Important examples are found in prolonged gastrointestinal losses due to vomiting, diarrhea, or prolonged gastric suction. Replacement of lost electrolytes with only sodium salts leads to migration of sodium into cells to replace the potassium deficit. This produces profound alterations in cellular metabolism such as persistent alkalosis even after apparently adequate salt and water therapy. It is preventable by the prompt and concomitant replacement of potassium deficits as well as sodium deficits.

3. Electrolyte influence on water shifts–Since water is freely diffusible across the cell barrier, its movement is determined by the changes in concentration of the osmotically effective electrolytes (principally sodium and potassium) on either side. Changes in extracellular electrolyte concentration are most commonly the basis for these shifts of water.

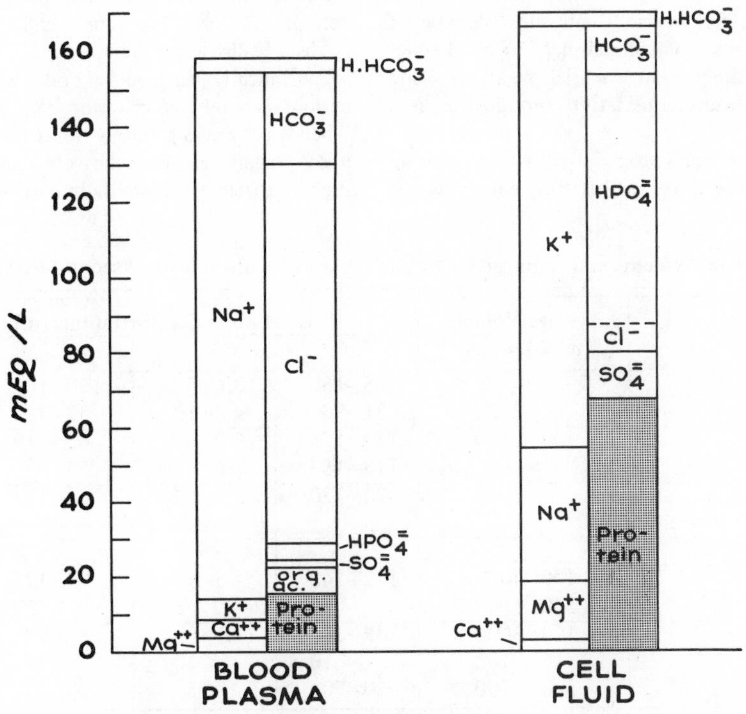

FIG 19–1. Electrolyte composition of blood plasma and intracellular fluid. (Modified from Gamble.)

Dehydration

This term should not imply only changes in water balance. Almost always there must also be accompanying changes in electrolytes.

A. Water Loss or Restriction Causing Dehydration: When the supply of water is restricted for any reason, or when the losses are excessive, the rate of water loss exceeds the rate of electrolyte loss. The extracellular fluid becomes concentrated and hypertonic to the cells. Water then shifts from the cells to the extracellular space to compensate.

The symptoms of this intracellular dehydration are severe thirst, nausea and vomiting, a hot and dry body, a dry tongue, loss of coordination, and a concentrated urine of small volume. Intracellular dehydration is corrected by giving water by mouth, or dextrose and water parenterally, until symptoms are alleviated and the urine volume is restored.

B. Electrolyte Deficit: A relative deficit of electrolytes may occur when an excess of water is ingested. This condition of overhydration is commonly observed when large amounts of electrolyte-free solutions are administered to patients. More frequently, however, water and electrolytes are both lost, and replacement with only water leads to a deficiency of electrolytes in the presence of normal or excess total body water. The deficiency of sodium in the extracellular fluid is mainly responsible for the resulting hypotonicity of this fluid compartment. Some water passes into the cells, which are hypertonic to the extracellular fluid, causing the so-called intracellular edema. There follows a diminution in extracellular fluid volume which is very damaging. The resulting decrease in blood volume is conducive to a fall in blood pressure, slowing of circulation, and consequent impairment in renal function. Since the kidney is an essential aid in restoring the normal equilibrium, this latter complication is a serious one.

The patient becomes progressively weaker, but he does not complain of thirst and his urine volume is not notably changed. There is, however, reliable evidence of this type of dehydration in the elevated hematocrit or plasma total protein and the lowered sodium and chloride concentration in the plasma.

C. Correction of Dehydration: Because of the high content of electrolytes in the gastrointestinal secretions, loss of fluid from the gastrointestinal tract will readily produce serious fluid and electrolyte deficits if prompt and accurate replacement of the losses does not take place. In Table 19–5 the volume and composition of gastrointestinal fluids and of sweat are shown. Loss of chloride in excess of sodium will be expected when fluid is withdrawn from the upper gastrointestinal tract, as may occur in high intestinal obstruction, pyloric stenosis, gastric vomiting, or in continuous gastric suction. Ordinarily, sodium chloride solutions may be given parenterally to repair the losses since, in the presence of adequate kidney function, a proper adjustment of the electrolyte imbalance will occur. The importance of simultaneous replacement of potassium must also be recalled (see p 402).

Fluid and electrolyte losses originating from the intestinal tract (as in prolonged diarrhea, pancreatic or biliary fistulas, etc) are characterized by the removal of a fluid high in sodium and bicarbonate. This leads to a relative chloride excess and a bicarbonate deficit. This condition might best be repaired initially by the intravenous administration of a mixture of 2/3 isotonic saline solution and 1/3 sodium lactate solution (1/6 molar).

Dehydration is frequently a complication of gastrointestinal tract disturbances, but it is not confined to these conditions. Other disorders in which dehydration is a problem include diabetes mellitus, Addison's disease, uremia, extensive burns, and shock.

Clinically, change in body weight during short periods is a reliable criterion of changes in hydration. When a patient is properly nourished and hydrated, his body weight remains relatively constant, with only a slight variation. Rapid daily gain in weight indicates

TABLE 19–5. Volume and composition of blood plasma, gastrointestinal secretions, and sweat.*

Fluid	Average Volume (ml/24 hours)	Electrolyte Concentrations (mEq/liter)			
		Na^+	K^+	Cl^-	HCO_3^-
Blood plasma		135–150	3.6–5.5	100–105	24.6–28.8
Gastric juice	2500	31–90	4.3–12	52–124	0
Bile	700–1000	134–156	3.9–6.3	83–110	38
Pancreatic juice	> 1000	113–153	2.6–7.4	54–95	110
Small bowel (Miller-Abbott suction)	3000	72–120	3.5–6.8	69–127	30
Ileostomy					
Recent	100–4000	112–142	4.5–14	93–122	30
Adapted	100–500	50	3	20	15–30
Cecostomy	100–3000	48–116	11.1–28.3	35–70	15
Feces	100	< 10	< 10	< 15	< 15
Sweat	500–4000	30–70	0–5	30–70	0

*Lockwood & Randall: Bull New York Acad Med 25:228, 1949; and Randall: S Clin North America 32:3, 1952.

overhydration. Loss of 8–12% in body weight represents a significant degree of dehydration if it is due to loss of fluids.

MINERAL METABOLISM

Although the mineral elements constitute a relatively small amount of the total body tissues, they are essential to many vital processes. The function of individual minerals has been mentioned at various points in this book; for example, blood calcium and its role in neuromuscular irritability and in the clotting of blood, the effect of various ions on activation of enzymes, and the activities of electrolytes in acid-base regulation.

The balance of ions in the tissues is often of importance. For example, normal ossification demands a proper ratio of calcium to phosphorus; the normal ratio between potassium and calcium in the extracellular fluid must be maintained to ensure normal activity of the muscle.

Certain mineral elements, principally sodium and potassium, are the major factors in osmotic control of water metabolism, as described in the previous section of this chapter. Other minerals are an integral part of important physiologic compounds such as iodine in thyroxine, iron in hemoglobin, zinc in insulin, cobalt in vitamin B_{12}, sulfur in thiamine, biotin, coenzyme A, and lipoic acid.

The animal body requires 7 principal mineral elements: calcium, magnesium, sodium, potassium, phosphorus, sulfur, and chlorine. These minerals constitute 60–80% of all the inorganic material in the body. At least 7 other minerals are utilized in trace quantities: iron, copper, iodine, manganese, cobalt, zinc, and molybdenum. Several other elements are present in the tissues, but their functions, if any, are not clearly defined. These include fluorine, aluminum, boron, selenium, cadmium, and chromium.

CALCIUM

Functions

Calcium is present in the body in larger amounts than any other cation. Almost all of it is in the bones and teeth. The very small quantity not in the skeletal structures is in the body fluids and is in part ionized. Ionized calcium is of great importance in blood coagulation, in the function of the heart, muscles, and nerves, and in the permeability of membranes.

Sources

Dietary sources of calcium include milk, cheese, egg yolk, beans, lentils, nuts, figs, cabbage, turnip greens, cauliflower, and asparagus.

Requirements (See also Table 21–3.)

Men and women after 18 years of age: 800 mg daily.

During second and third trimesters of pregnancy and during lactation: 1.2–1.3 g daily.

Infants under 1 year: 400–600 mg daily.

Children 1–18 years: 0.7–1.4 g daily.

To supply additional calcium, the carbonate, lactate, or gluconate salts as well as dicalcium phosphate may be administered.

The requirements for calcium listed above are thought to be excessive by some nutritional authorities. There are also reports (Burnett, 1949) of the occurrence of metastatic calcifications associated with high intakes of calcium and of alkali in connection with dietary supplements prescribed for patients with peptic ulcer (the milk-alkali syndrome). A high intake of calcium in the presence of a high intake of vitamin D such as may occur in children is also a potential source of hypercalcemia and possibly of widespread excessive calcification. These considerations may indicate a need for revision of the presently recommended allowances for calcium, particularly in the case of adults other than pregnant and lactating women.

Absorption

The ability of different individuals to utilize the calcium in foods varies considerably. On a high-protein diet, 15% of the dietary calcium is absorbed; on a low-protein diet, 5%. Phytic acid in cereal grains interferes with calcium absorption by forming insoluble calcium phytate in the intestine. Oxalates in foods (eg, spinach) may have a similar effect. Other intestinal factors which influence absorption of calcium include:

A. pH: The more alkaline the intestinal contents, the less soluble the calcium salts. An increase in acidophilic flora (eg, the lactobacilli) is recommended to lower the pH, which favors calcium absorption.

B. Phosphate: If the Ca:P ratio is high, much $Ca_3(PO_4)_2$ will be formed and absorption diminished.

C. Presence of Free Fatty Acids: When fat absorption is impaired, much free fatty acid is present. These free fatty acids react with free calcium to form insoluble calcium soaps.

D. Vitamin D: Vitamin D promotes the absorption of calcium from the intestine.

Distribution

The calcium other than that in the bones and teeth is distributed as shown in Table 19–6.

Metabolism

The blood cells contain very little calcium. Most of the blood calcium is therefore in the plasma, where it exists in 3 fractions: ionized (so-called diffusible calcium), protein-bound (nondiffusible), and a small

TABLE 19–6. Distribution of calcium in body fluids or tissues.

Fluid or Tissue	mg/100 ml or 100 g	mEq/liter
Serum	9.0–11	5
CSF	4.5–5	2
Muscle	70	
Nerve	15	

amount complexed probably as the citrate. All of these forms of calcium in the serum are in equilibrium with one another. In the usual determination of calcium, all 3 fractions are measured together.

An ultracentrifugal method for separation of protein-bound and free (ionized) calcium which requires only 5 ml of serum for a determination in duplicate has been devised by Loken & others (1960). According to these authors, at a pH of 7.35 and a temperature of 37° C, the normal range of free calcium in the serum is 49.7–57.8% of the total calcium (mean, 53.1 ± 2.6%).

A decrease in the ionized fraction of serum calcium causes tetany. This may be due to an increase in the pH of the blood (alkalotic tetany; gastric tetany) or to lack of calcium because of poor absorption from the intestine, decreased dietary intake, increased renal excretion as in nephritis, or parathyroid deficiency. Increased retention of phosphorus, as in renal tubular disease, also predisposes to low serum calcium levels.

The Ca:P ratio is important in ossification. In the serum, the product of Ca X P (in mg/100 ml) is normally in children about 50. In rickets, this product may be below 30.

A relatively small quantity of the calcium lost from the body is excreted in the urine. In man, approximately 10 g of calcium are filtered in a 24-hour period by the renal glomeruli, but only about 200 mg appear in the urine. The maximal renal tubular reabsorptive capacity for calcium (Tm_{Ca}) is about 4.99± 0.21 mg/minute.

Most (70–90%) of the calcium eliminated from the body is excreted in the feces. This actually represents almost entirely unabsorbed dietary calcium. The amount of calcium reexcreted into the intestine is very small.

Disease States

A. Relationship of Parathyroids: Calcium metabolism is profoundly influenced by the parathyroids.

1. In hyperparathyroidism caused by hyperactive, hyperplastic, or adenomatous parathyroid glands, the following signs are noted: hypercalcemia (serum calcium 12–22 mg/100 ml), decrease in serum phosphate, decreased renal tubular reabsorption of phosphate, increased phosphatase activity, rise in urinary calcium and phosphorus from bone decalcification, and dehydration and hemoconcentration. These signs are due to increased renal losses of phosphorus, causing a decrease in serum phosphate which elicits an increase in calcium to maintain the Ca:P product. The extra calcium and phosphorus is lost from soft tissues and from bone by increased osteoclastic (bone-destroying) activity.

In many cases of hyperparathyroidism, the total serum calcium may not be elevated sufficiently to permit diagnosis of this disease with certainty. However, Lloyd & Rose (1958) have found that in a group of patients with hyperparathyroidism due to parathyroid adenoma, ionized calcium ranged from 6.1–9.5 mg/100 ml (normal, 5.9–6.5 mg/100 ml), whereas the majority had levels of protein-bound calcium within normal limits or only slightly above normal. After surgical removal of the diseased parathyroid glands, there was a drop in total plasma calcium, again most marked in the ionized fraction. It has therefore been suggested that determination of the ionized fraction of the serum calcium may enhance the utility of the serum calcium for the diagnosis of hyperparathyroidism, particularly in patients who have normal total calcium levels.

2. In hypoparathyroidism, such as occurs after operative removal of the parathyroid glands, the concentration of the serum calcium may drop below 7 mg/100 ml. There is a concomitant increase in serum phosphate and a decrease in urinary phosphates. The urinary calcium is extremely low as well.

B. Rickets: This disease is characterized by faulty calcification of bones due to a low vitamin D content of the body, a deficiency of calcium and phosphorus in the diet, or a combination of both. Usually the serum phosphate concentration is low or normal, except in renal disease (in which it may be elevated), and the serum calcium remains normal or may be lowered. There is an increase in fecal phosphate and calcium because of poor absorption of these elements, accompanied by a decrease in urine phosphate and calcium. An increase in alkaline phosphatase activity is also characteristic of rickets.

C. Renal Rickets: This condition is caused by renal tubular defect (usually inherited) which interferes with reabsorption of phosphorus. The disease is not relieved by vitamin D in ordinary dosages.

D. Decrease of Serum Calcium: In severe renal disease the serum calcium may decrease, in part because of increased losses in the urine but mainly because of increase in serum phosphorus, which causes a compensatory decrease in serum calcium.

PHOSPHORUS

Functions

Phosphorus is found in every cell of the body, but most of it (about 80% of the total) is combined with calcium in the bones and teeth. About 10% is in combination with proteins, lipids, and carbohydrates, and in other compounds in blood and muscle. The remaining 10% is widely distributed in various chemical compounds. The great importance of the phosphate ester in energy transfer is discussed in Chapter 9.

Requirements & Sources (See also Table 21–3.)

A ratio of calcium to phosphorus of 1:1 is recommended except in infancy, when the ratio starts at 2:1 based on the ratio found in human milk. For older infants, the recommended intake of phosphorus is raised to about 80% of the calcium allowance to provide a ratio similar to that of cows' milk. Since their distribution in foods is very similar, an adequate intake of calcium generally ensures an adequate intake of phosphorus also.

Distribution

The distribution of phosphorus in the body is shown in Table 19–7.

Metabolism

The metabolism of phosphorus is in large part related to that of calcium, as described heretofore. The Ca:P ratio in the diet affects the absorption and excretion of these elements. If either element is given in excess, excretion of the other is increased. The optimal ratio is 1:1 when the intake of vitamin D is adequate.

An increase in carbohydrate metabolism, such as during absorption of carbohydrate, is accompanied by a temporary decrease in serum phosphate. A similar decrease may occur during absorption of some fats. In diabetes mellitus, there is a lower concentration of organic phosphorus but a higher concentration of inorganic phosphorus in the serum.

In rickets of the common low-phosphate variety, serum phosphate values may go as low as 1–2 mg/100 ml (0.64–1.3 mEq/liter).

Phosphate retention is a prominent cause of the acidosis in severe renal disease, and the resultant elevated serum phosphorus also contributes to the lowered serum calcium. Blood phosphorus levels are also high in hypoparathyroidism. A relationship of phosphorus metabolism to growth hormone is possibly indicated by the fact that growing children usually have high blood phosphorus levels and that in acromegaly an elevation of the blood phosphorus also occurs.

TABLE 19–7. Distribution of phosphorus in body fluids or tissues.

Fluid or Tissue	mg/100 ml or 100 g	mM/liter
Blood	40	
Serum (inorganic)		
Children	4–7	1.3–2.3
Adults	3–4.5	0.9–1.5
Muscle	170–250	
Nerve	360	
Bones and teeth	22,000	

Blood phosphorus levels are low in hyperparathyroidism and in sprue and celiac disease. A low blood phosphorus together with an elevated alkaline phosphatase is also a characteristic finding in patients with an inherited or acquired renal tubular defect in the reabsorption of phosphate. Such cases include so-called vitamin D-resistant rickets, Milkman's syndrome, and the De Toni-Fanconi syndrome (see also p 379). The greatly increased excretion of phosphate in the urine of these patients distinguishes them from those in which a deficiency of vitamin D is the cause of the low blood phosphorus and the accompanying defects in calcification of bone.

MAGNESIUM

Functions & Distribution

The body contains about 21 g of magnesium. Seventy percent is combined with calcium and phosphorus in the complex salts of bone; the remainder is in the soft tissues and body fluids. Magnesium is one of the principal cations of soft tissue. Whole blood contains 2–4 mg/100 ml (1.7–3.4 mEq/liter). The serum contains less than 1/2 that in the blood cells (1.94 mEq/liter). This is in contrast to calcium, almost all of which is in the serum. Cerebrospinal fluid is reported to contain about 3 mg/100 ml (2.40 mEq/liter). The magnesium content of muscle is about 21 mg/100 g. Here it probably functions in carbohydrate metabolism as an activator for many of the enzymes of the glycolytic systems.

A comprehensive listing of the magnesium content of foods is found in the reports of McCance & Widdowson (1960). Derivatives of cocoa, various nuts, soybeans, and some seafoods are relatively rich in magnesium (100 to as high as 400 mg/100 g). Whole grains and raw dried beans and peas may contain 100–200 mg magnesium per 100 g.

Requirements (See also Table 21–3.)

The most recent recommendation for magnesium in the diet is 350 mg/day for adult men and 300 mg/day for adult women.

In the diet of Oriental peoples there is a preponderance of foods high in magnesium, so that their intake of this mineral may approximate 6–10 mg/kg/day. The diet of Western peoples, however, provides less than 5 mg/kg/day for most adults (an average of 250–300 mg/day). It is claimed that several dietary constituents may interfere with retention or increase the requirement of magnesium. Examples are calcium, protein, and vitamin D. Alcohol is said to increase magnesium loss from the body. Since the Western diet may be high in calcium, protein, and vitamin D and because alcohol is also more commonly ingested by Western peoples, some authorities have suggested that the optimal daily intake of magnesium under these circumstances may be as high as 7–10 mg/kg/day.

As noted below, a subacute or chronic deficiency of magnesium is not readily detectable. Speculation therefore seems justified on the possible role of long-term magnesium deficiencies as a factor in the causation of chronic disease of those systems (cardiovascular, renal, and neuromuscular) particularly susceptible to magnesium deficiency.

Metabolism

The metabolism of magnesium is similar to that of calcium and phosphorus. Absorption and excretion of magnesium from the intestine has been measured in human subjects with the aid of isotopic ^{28}Mg (Graham, 1960). On an average diet (20 mEq Mg/day), 44.3% of the ingested radioisotope was absorbed. On a low Mg diet (1.9 mEq/day), 75.8% was absorbed; whereas on a high Mg intake (47 mEq/day), only 23.7% was absorbed. The rate and duration of absorption of the ingested magnesium indicated that most was absorbed from the small intestine and little or none from the colon. Absorption of magnesium from the intestine did not appear to be related to the status of magnesium stores in the body. In the first 48 hours after administration of radioactive magnesium, about 10% of the amount absorbed was excreted in the urine. Thus renal conservation of body magnesium appears to be excellent; in fact, the average urinary magnesium content is only about 6–20 mEq/liter. Aldosterone (see p 442) increases the renal clearance of magnesium as it does also the excretion of potassium.

An antagonism between magnesium and calcium has been noted in certain experiments. The intravenous injection of magnesium in a quantity sufficient to raise the magnesium ion concentration in the serum to about 20 mg/100 ml (normal, 2.4 mg/100 ml) results in immediate and profound anesthesia together with paralysis of voluntary muscles. The intravenous injection of a corresponding amount of calcium results in an instantaneous reversal of this effect. It is suggested that these 2 cations are exerting differing effects on cell permeability. In the case of magnesium, there is about 10 times as much of this element in the cells as in the extracellular fluid. For example, in plasma there is an average of 2.4 mg/100 ml; in muscle cells, 23 mg/100 g. This differential distribution between plasma and muscle cells is not observed with calcium, but it is particularly prominent in the case of sodium and potassium as well as magnesium. Apparently magnesium and potassium are normally concentrated within the cell and sodium without. An alteration in this relationship is followed by profound physiologic changes. In this connection it is of interest, as already noted, that both magnesium and potassium excretion by the kidney are increased to the same extent by aldosterone.

In rats on a very low magnesium diet (0.18 mg/100 g of food), vasodilation and hyperemia, hyperirritability, cardiac arrhythmia, and convulsions developed which were subsequently fatal. The tetany which developed when the diet was low in magnesium was probably due to the low magnesium content of the serum since the calcium levels remained normal.

It is difficult to produce a serious depletion of magnesium experimentally in man (Fitzgerald, 1956). However, Flink & others (1957) have described a clinical syndrome characterized by muscle tremor, twitching, and more bizarre movements, occasionally with convulsions and often with delirium which is attributed to magnesium deficiency. The patients studied included a large group with chronic alcoholism and tremulousness and a few postoperative patients as well as those with pyloric obstruction and hypochloremic alkalosis. In all of these patients, however, the levels of magnesium in the serum were only moderately reduced. A decrease in the concentration of serum magnesium has also been noted in clinical hyperparathyroidism. It is possible that prolonged hyperparathyroidism could deplete the body stores of magnesium. After surgical correction of hyperparathyroidism, the development of tetany which is refractory to the administration of large amounts of calcium may indicate the need for magnesium.

As indicated above, serum magnesium levels may not correlate well with the intracellular concentration of this ion. Therefore, in an effort to assess the intracellular magnesium concentration, the content of this ion in the red blood cells has been measured. In healthy adults, the mean erythrocyte Mg was found to be 5.29 ± 0.34 mEq/liter when the mean plasma level was 1.80 ± 0.13 mEq/liter (Smith, 1959). In 12 patients with delirium tremens who were disoriented and confused, many with a pronounced tremor, some with hallucinations, the mean erythrocyte magnesium concentration was 3.9 ± 0.75 mEq/liter, the plasma concentration, 1.5 ± 0.28 mEq/liter. Following treatment with magnesium sulfate given intramuscularly, the erythrocyte and plasma magnesium levels rose to normal.

In renal failure, magnesium tends to rise in the serum. Indeed, some have wondered whether this might not contribute to somnolence and weakness characteristic of the uremic state. Thus it is of interest that 14 patients with uremia and associated depression of the CNS had mean erythrocyte magnesium levels of 8.84 ± 1.71 mEq/liter (upper limit of normal, 6.0 mEq/liter), and plasma levels of 3.17 ± 1.30 mEq/liter. Smith & Hammarsten (1959) concluded that elevated serum or plasma levels are reliable evidence of a total body excess of magnesium. On the other hand, a deficit of magnesium may not be readily apparent from measurement of the serum level; in this instance, erythrocyte levels may be more informative.

SODIUM

Functions

This element is the major component of the cations of the extracellular fluid. It is largely associated with chloride and bicarbonate in regulation of acid-base equilibrium. The other important function of sodium is the maintenance of the osmotic pressure of body fluid and thus protection of the body against excessive fluid loss. It also functions in the preservation of normal irritability of muscle and the permeability of the cells.

Requirements & Sources

Daily requirements of 5 to as much as 15 g of sodium chloride have been recommended for adults by various authorities. These requirements were established from observations on urinary losses in subjects who were not on controlled low intakes of sodium chloride, and much of the salt in the urine therefore represented merely the excretion of the excess intake. Dahl (1958) has appraised the need for sodium chloride under conditions of controlled intakes. In his experiments, adults maintained on daily intakes of only 100–150 mg sodium lost a total of less than 25 mg sodium per day, which probably represents the minimum losses in the sweat. Dahl estimates the normal obligatory (irreducible) daily losses of sodium as follows: urine, 5–35 mg; stool, 10–125 mg; skin (not sweating), 25 mg; total, 40–185 mg.

The most variable source of salt loss is the sweat, but even this route of salt loss can be minimized during prolonged exposure to high temperatures if a period of a few days is allowed for adaptation. It is concluded that a maximum sodium chloride intake of about 5 g/day may be recommended for adults without a family history of hypertension. This is about 1/2 the daily amount which is ordinarily voluntarily consumed. Furthermore, an intake of 5 g of sodium chloride per day is 10 times the amount at which adequate sodium chloride balance can apparently be maintained. For persons with a family history of hypertension, Dahl recommends a diet containing no more than 1 g of sodium chloride per day.

The main source of sodium is the sodium chloride used in cooking and seasoning; ingested foods contain additional sodium.* It is estimated that about 10 g of sodium chloride (4 g of sodium) is thus ingested each day. In addition to salted foods, the content of sodium is high in bread, cheese, clams, oysters, crackers, wheat germ, and whole grains; relatively high in sodium are such foods as carrots, cauliflower, celery, eggs, legumes, milk, nuts, spinach, turnips, oatmeal, prunes, and radishes.

About 95% of the sodium which leaves the body is excreted in the urine. Sodium is readily absorbed, so that the feces contain very little except in diarrhea, when much of the sodium excreted into the intestine in the course of digestion escapes reabsorption.

Distribution

About 1/3 of the total sodium content of the body is present in the inorganic portion of the skeleton. However, most of the sodium is found in the extracellular fluids of the body, as shown in Table 19–8.

TABLE 19–8. Distribution of sodium in body fluids or tissues.

Fluid or Tissue	mg/100 ml or 100 g	mEq/liter
Whole blood	160	70
Plasma	330	143
Cells	85	37
Muscle tissue	60–160	
Nerve tissue	312	

Metabolism

The metabolism of sodium is influenced by the adrenocortical steroids. In adrenocortical insufficiency, a decrease of serum sodium and an increase in sodium excretion occur.

In chronic renal disease, particularly when acidosis coexists, sodium depletion may occur due to poor tubular reabsorption of sodium as well as to loss of sodium in the buffering of acids.

Unless the individual is well adapted to a high environmental temperature, extreme sweating may cause the loss of considerable sodium in the sweat; muscular cramps of the extremities and abdomen, headaches, nausea, and diarrhea may develop.

The levels of sodium as measured in the serum may not reflect accurately the total body sodium. Thus a low concentration of serum sodium (hyponatremia) may develop if patients are given large quantities of salt-free fluids. This obviously is not an indication of actual depletion of body sodium but rather the effect of overhydration. A similar situation prevails in edematous states such as cirrhosis or congestive heart failure, wherein low serum sodium is frequently observed although the total body sodium may actually be excessive.

However, in those clinical situations where depletion of sodium occurs (such as after excessive losses of gastrointestinal fluids or in renal disease accompanied by some degree of salt wasting), the low serum sodium which is found truly indicates depletion of total body sodium. In such hyponatremic states there will also be loss of water, which will be evident by rapid weight loss. Observations of changes in weight are of value in differentiating hyponatremic states due to dilution and overhydration, in which case a weight gain will be noted, from those in which true sodium depletion has occurred, in which case a weight loss due to dehydration will be the rule.

Increased serum sodium (hypernatremia) is rare. It may occur as a result of rapid administration of sodium salts, or may be due to hyperactivity of the adrenal cortex, as in Cushing's disease. After the administration of corticotropin (ACTH), cortisone, or deoxycorticosterone, as well as some of the sex hor-

*A detailed compilation of the sodium and potassium content of foods and of water obtained from the drinking supply of many cities has been prepared by Bills & others (1949).

mones, a rise in serum sodium concentration may also occur unless the concomitant retention of water acts to mask the sodium retention. The most common cause of hypernatremia, however, is rapid loss of water, such as in the dehydration associated with diabetes insipidus. Occasionally hypernatremia may also follow excessive sweating. This is so because sweat is a hypotonic salt solution; consequently, loss of this fluid engenders a loss of water at a rate exceeding loss of salt insofar as the ratio of water to salt in the body fluids is concerned.

Addison's disease, which is characterized by an increased sodium loss because of adrenocortical insufficiency, is ameliorated during pregnancy presumably because of the production of steroid hormones, which cause sodium retention. It has also been shown that the placenta elaborates hormones with sodium-retaining effects, and it is believed that these hormonal substances are responsible for the sodium and water retention, accompanied by rapid gains in weight, commonly observed in certain stages of pregnancy.

Meneely & Ball (1958) have made a study in rats of the effects of chronic ingestion of large amounts of sodium chloride on an otherwise standardized diet. Their test diets contained varying amounts of sodium chloride ranging from 2.8–9.8%. Among the animals eating a diet with 7% sodium chloride or more there occurred a syndrome resembling nephrosis, characterized by the sudden onset of massive edema and by hypertension, anemia, pronounced lipemia, severe hypoproteinemia, and azotemia. All of the affected animals died, and at autopsy showed evidence of severe arteriolar disease. Significant hypertension was uniformly observed at all levels of sodium chloride (from 2.8–9.8%), and there was a tendency for the degree of elevation in blood pressure to parallel the amount of salt in the diet. At the higher levels of salt intake, there was also a significant decrease in the survival time of the experimental animals. However, the addition of potassium chloride to the high sodium chloride diets produced a striking increase in the survival times on the various diets, although a moderating effect of potassium on the blood pressure was observed only on the high levels of sodium chloride intake.

POTASSIUM

Functions

Potassium is the principal cation of the intracellular fluid; but it is also a very important constituent of the extracellular fluid because it influences muscle activity, notably cardiac muscle. Within the cells it functions, like sodium in the extracellular fluid, by influencing acid-base balance and osmotic pressure, including water retention.

Requirements & Sources

The normal intake of potassium in food is about 4 g/day (see p 401, footnote). It is so widely distributed that a deficiency is unlikely except in the pathologic states discussed below.

A high content of potassium is found in the following foods (300–600 mg per serving): veal, chicken, beef liver, beef, pork; dried apricots, dried peaches, bananas, the juices of oranges, tangerines, and pineapples; yams, winter squash, broccoli, potatoes, and Brussels sprouts. There are other foods high in potassium but also high in sodium. Since in many situations the need for high potassium intake parallels that for a low sodium intake, these foods are of less value as sources of potassium.

Distribution

The predominantly intracellular distribution of potassium is illustrated in Table 19–9.

TABLE 19–9. Distribution of potassium in body fluids or tissues.

Fluid or Tissue	mg/100 ml or 100 g	mEq/liter
Whole blood	200	50
Plasma	20	5
Cells	440	112
Muscle tissue	250–400	
Nerve tissue	530	

Metabolism

Variations in extracellular potassium influence the activity of striated muscles so that paralysis of skeletal muscle and abnormalities in conduction and activity of cardiac muscle occur. Although potassium is excreted into the intestine in the digestive fluids (Table 19–5), much of this is later reabsorbed. The kidney is the principal organ of excretion for potassium. Not only does the kidney filter potassium in the glomeruli but it is also secreted by the tubules. The excretion of potassium is markedly influenced by changes in acid-base balance as well as by the activity of the adrenal cortex. (The renal mechanisms for potassium excretion are discussed on p 379). The capacity of the kidney to excrete potassium is so great that hyperkalemia will not occur, even after the ingestion or intravenous injection at a moderate rate of relatively large quantities of potassium, if kidney function is unimpaired. This, however, is not the case when urine production is inadequate. It is important to emphasize that potassium should **not** be given intravenously until circulatory collapse, dehydration, and renal insufficiency have been corrected.

A. Elevated Serum Potassium (Hyperkalemia):

1. Etiology—Toxic elevation of serum potassium is confined for the most part to patients with renal failure, advanced dehydration, or shock. A high serum potassium, accompanied by a high intracellular potassium, also occurs characteristically in adrenal insufficiency (Addison's disease). This elevated serum potassium is corrected by the administration of desoxycorticosterone acetate (Doca). Hyperkalemia may also occur if potassium is administered intravenously at an excessive rate.

2. Symptoms—The symptoms of hyperkalemia are chiefly cardiac and CNS depression; they are related to the elevated plasma potassium, not to increases in intracellular levels. The heart signs include bradycardia and poor heart sounds, followed by peripheral vascular collapse and, ultimately, cardiac arrest. ECG changes are characteristic and include elevated T waves, widening of the QRS complex, progressive lengthening of the P–R interval, and then disappearance of the P wave. Other symptoms commonly associated with elevated extracellular potassium include mental confusion; weakness, numbness, and tingling of the extremities; weakness of respiratory muscles; and a flaccid paralysis of the extremities.

B. Low Serum Potassium (Hypokalemia):

1. Etiology—Potassium deficiency is likely to develop in any illness, particularly in postoperative states when intravenous administration of solutions which do not contain potassium is prolonged. Potassium deficits are likewise to be expected in chronic wasting diseases with malnutrition, prolonged negative nitrogen balance, gastrointestinal losses (including those incurred in all types of diarrheas and gastrointestinal fistulas, and in continuous suction), and in metabolic alkalosis. In most of these cases intracellular potassium is transferred to the extracellular fluid, and this potassium is quickly removed by the kidney. Because adrenocortical hormones, particularly aldosterone, increase the excretion of potassium, overactivity of the adrenal cortex (Cushing's syndrome or primary aldosteronism), or injection of excessive quantities of corticosteroids or corticotropin (ACTH) may induce a deficit.

The excretion of potassium in the urine is increased by the activity of certain diuretic agents, particularly acetazolamide (Diamox) and chlorothiazide (Diuril). It is therefore recommended that potassium supplementation be provided when these drugs are used for more than a few days.

A prolonged deficiency of potassium may produce severe damage to the kidney (Kark, 1958). This may be associated secondarily with the development of chronic pyelonephritis. There is evidence that the initial damage to the kidney in potassium-depleted animals affects particularly the mitochondria in the collecting tubule.

During heart failure, the potassium content of the myocardium becomes depleted; with recovery, intracellular repletion of potassium occurs. However, intracellular deficits of potassium increase the sensitivity of the myocardium to digitalis intoxication and to arrhythmias. This fact is of importance in patients who have been fully digitalized and are then given diuretic agents which may produce potassium depletion. Administration of potassium may prevent or relieve such manifestations of digitalis toxicity.

Potassium deficits often become apparent only when water and sodium have been replenished in an attempt to correct dehydration and acidosis or alkalosis. Darrow states that changes in acid-base balance involve alterations in both intracellular and extracellular fluids and that the normal reaction of the blood cannot be maintained without a suitable relation between the body contents of sodium, potassium, chloride, and water.

When 1 g of glycogen is stored, 0.36 mM of potassium is simultaneously retained. In treatment of diabetic coma with insulin and glucose, glycogenesis is rapid and potassium is quickly withdrawn from the extracellular fluid. The resultant hypokalemia may be fatal.

Familial periodic paralysis is a rare disease in which potassium is rapidly transferred into cells, lowering extracellular concentration.

2. Symptoms—The symptoms of low serum potassium concentrations include muscle weakness, irritability, and paralysis; tachycardia and dilatation of the heart with gallop rhythm are also noted. Changes in the ECG record are also a prominent feature of hypokalemia, including first a flattened T wave; later, inverted T waves with sagging ST segment and A-V block; and finally cardiac arrest.

It is important to point out that a potassium deficit may not be reflected in lowered (less than 3.5 mEq/liter) extracellular fluid concentrations until late in the process. This is confirmed by the finding of low intracellular potassium concentrations in muscle biopsy when serum potassium is normal. Thus the serum potassium is not an accurate indicator of the true status of potassium balance.

3. Treatment—In parenteral repair of a potassium deficit, a solution containing 25 mEq of potassium (KCl, 1.8 g) per liter may be safely given intravenously after adequate urine flow has been established. A daily maintenance dose of at least 50 mEq of potassium (KCl, 3.6 g) intravenously is probably necessary for most patients, with additional amounts to cover excessive losses, as from gastrointestinal drainage, up to 150 mEq of potassium per day. When these larger doses are required, as much as 50 mEq of potassium may be added to a liter of intravenous solution, although in this concentration a slower rate of injection is required (2½–3 hours). The potassium salts may be added to saline solutions or to dextrose solutions. Some prefer to add also magnesium and calcium in order to provide a better ionic balance, suggesting 10 mEq each of calcium and magnesium for each 25 mEq of potassium. The following formula contains these 3 cations in that proportion:

KCl	1.8 g (25 mEq K)
$MgCl_2$	0.5 g (10 mEq Mg)
$CaCl_2$	0.6 g (10 mEq Ca)

Whenever possible, the correction of a potassium deficit by the oral route is preferred. For adults, 4–12 g of KCl (as 1–2% solution) per day in divided doses is recommended.

In muscle, the proportion of potassium to nitrogen is 3 mM to each g. Storage of nitrogen as muscle protein therefore demands additional potassium. It has been suggested that a loss of 5 kg of muscle protein requires 600 mEq of potassium together with the protein nitrogen necessary for its replacement. For this reason, the administration of potassium along with parenterally-administered amino acids has been recommended. Frost & Smith (1953) recommend that 5 mEq of potassium be given for each g of amino acid nitrogen to provide for optimal nitrogen retention.

CHLORINE

Function

As a component of sodium chloride, the element chlorine (as chloride ion) is essential in water balance and osmotic pressure regulation as well as in acid-base equilibrium. In this latter function, chloride plays a special role in the blood by the action of the chloride shift (Fig 11–3). In gastric juice, chloride is also of special importance in the production of hydrochloric acid (Fig 12–2).

Requirement & Metabolism

In the diet, the chloride occurs almost entirely as sodium chloride, and therefore the intake of chloride is satisfactory as long as sodium intake is adequate. In general, both the intake and output of this element are, in fact, inseparable from those of sodium. On low-salt diets, both the chloride and sodium in the urine drop to low levels.

Abnormalities of sodium metabolism are generally accompanied by abnormalities in chloride metabolism. When losses of sodium are excessive, as in diarrhea, profuse sweating, and certain endocrine disturbances, chloride deficit is likewise observed.

In loss of gastric juice by vomiting or in pyloric or duodenal obstruction, there is a loss of chloride in excess of sodium. This leads to a decrease in plasma chloride, with a compensatory increase in bicarbonate and a resultant hypochloremic alkalosis (see Chapter 11). In Cushing's disease or after the administration of an excess of corticotropin (ACTH) or cortisone, hypokalemia with an accompanying hypochloremic alkalosis may also be observed. Some chloride is lost in diarrhea because the reabsorption of chloride in the intestinal secretions is impaired.

TABLE 19–10. Distribution of chloride in body fluids or tissues.

Fluid or Tissue	mg/100 ml or 100 g	mEq/liter
Whole blood	250	70
Plasma or serum	365	103
Cells	190	53
CSF	440	124
Muscle tissue	40	
Nerve tissue	171	

Distribution in the Body

The chloride concentration in CSF is higher than that in other body fluids, including gastrointestinal secretions (Table 19–10).

SULFUR

Function

Sulfur is present in all cells of the body, primarily in the cell protein, where it is an important element of protein structure (see p 41). The metabolism of sulfur and of nitrogen thus tend to be associated.

The importance of sulfur-containing compounds in detoxication mechanisms and of the SH group in tissue respiration has been noted in Chapter 15. A high-energy sulfur bond similar to that of phosphate plays an important role in metabolism (see Chapter 7).

Sources & Metabolism

The main (if not the only) sources of sulfur for the body are the 2 sulfur-containing amino acids, cysteine and methionine. Elemental sulfur or sulfate sulfur is not known to be utilized. Organic sulfur is mainly oxidized to sulfate and excreted as inorganic or ethereal sulfate. The various forms of urinary sulfur are described on p 388.

Utilization of sulfate in organic combination requires "activation" as described on p 53.

Distribution in the Body

In addition to cystine and methionine, other organic compounds of sulfur are heparin, glutathione, insulin, thiamine, biotin, coenzyme A, lipoic acid, ergothioneine, taurocholic acid, the sulfocyanides; sulfur conjugates, like phenol esters and indoxyl sulfate; and the chondroitin sulfuric acid in cartilage, tendon, and bone matrix. Small amounts of inorganic sulfates, with sodium and potassium, are present in blood and other tissues.

Keratin, the protein of hair, hoofs, etc, is rich in sulfur-containing amino acids. The sulfur (methionine and cystine) requirement of hairy animals, such as the rat and the dog, is higher than that of human beings, possibly because of their additional hair.

IRON

Function & Distribution

The role of iron in the body is almost exclusively confined to the processes of cellular respiration. Iron is a component of hemoglobin, myoglobin, and cytochrome, as well as the enzymes catalase and peroxidase. In all of these compounds the iron is a component of a porphyrin. The remainder of the iron in the body is almost entirely protein-bound; these forms include the storage and transport forms of the mineral. The approximate distribution of iron-containing compounds in the normal adult human subject is shown in Table 19–11.

Requirements & Sources

The need for iron in the human diet varies greatly at different ages and under different circumstances. During growth, pregnancy, and lactation, when the demand for hemoglobin formation is increased, additional iron is needed in the diet. However, in other adults, iron deficiency is very unlikely unless some loss of blood occurs. An exception is the possibility of iron deficiency as a result of malabsorption from the gastrointestinal tract such as may occur after surgical resection of portions of the stomach and the small intestine, or in patients with diseases characterized by malabsorption. In women, the average loss of blood during a menstrual period is 35–70 ml, which represents a loss of 16–32 mg of iron. This amount of iron is easily obtained from the diet. For women with excessive menstrual blood loss and a resultant chronic iron-deficiency anemia, a supplement of 100 mg of iron per day (as ferrous sulfate) is sufficient to produce a maximal response toward correction of the anemia. However, in the healthy adult male, or in healthy women after the menopause, the dietary requirement for iron is almost negligible.

Traces of copper are required for utilization of iron in hemoglobin formation. The recommended daily amounts of iron and copper currently suggested by nutritional authorities are as follows: (See also Table 21–3.)

A. Infants: 6–15 mg.

B. Children: 1–3 years of age, 15 mg; 3–10 years of age, 10 mg.

C. Older Children and Adults: *Males:* 10–12 years of age, 10 mg; 12–18 years of age, 18 mg; after 18 years of age, 10 mg. *Females:* 10–55 years of age, and during pregnancy or lactation, 18 mg. After 55 years of age, 10 mg.

The recommended allowance for males of 10 mg/day is readily obtained in the average diet in the USA, which provides about 6 mg/1000 Cal. However, the recommended allowance for females (18 mg/day), based on 2000 Cal/day, is difficult to obtain from dietary sources without further iron fortification of foods.

The best dietary sources of iron are "organ meats": liver, heart, kidney, and spleen. Other good sources are egg yolk, whole wheat, fish, oysters, clams, nuts, dates, figs, beans, asparagus, spinach, molasses, and oatmeal.

Absorption From the Stomach & Intestine

A peculiar and possibly unique feature of the metabolism of iron is that it occurs in what is virtually a closed system. Under normal conditions very little dietary iron is absorbed, and the amounts excreted in the urine are minimal. Because there is no way to excrete excess iron, its absorption from the intestine must be controlled if it is not to accumulate in the tissues in toxic amounts. In the ordinary diet, 10–20 mg of iron are taken each day, but less than 10% of this is absorbed. The results of a study with labeled iron (^{59}Fe) illustrate this fact. After the administration of a test dose of the isotopic iron, only 5% was found in the blood, 87% was eliminated in the feces, and 8% remained unaccounted for. Infants and children absorb a higher percentage of iron from foods than do adults. Iron-deficient children absorb twice as much from foods as normal children; therefore, iron deficiency in infants can usually be attributed to dietary inadequacy.

In the male, excretion from the body is less than 1 mg/day; in the female during the childbearing years, 1.5–2.0 mg/day.

TABLE 19–11. Distribution of iron-containing compounds in the normal human adult.

	Total in Body (g)	Iron Content (g)	Percent of Total Iron in Body
Iron porphyrins (heme compounds)			
Hemoglobin	900	3.0	60–70
Myoglobin	40	0.13	3–5
Heme enzymes			
Cytochrome c	0.8	0.004	0.1
Catalase	5.0	0.004	0.1
Other cytochromes	. . .	. . .	. . .
Peroxidase	. . .	. . .	. . .
Nonporphyrin iron compounds			
Siderophilin (transferrin)	10	0.004	0.1
Ferritin	2–4	0.4–0.8	15.0
Hemosiderin	. . .	. . .	. . .
Total available iron stores		1.2–1.5	
Total iron		4.0–5.0	

Most of the iron in foods occurs in the ferric (Fe^{+++}) state either as ferric hydroxide or as ferric organic compounds. In an acid medium, these compounds are broken down into free ferric ions or loosely bound organic iron. The gastric hydrochloric acid as well as the organic acids of the foods are both important for this purpose. Reducing substances in foods, SH groups (eg, cysteine), or ascorbic acid convert ferric iron to the reduced (ferrous) state. In this form, iron is more soluble and should therefore be more readily absorbed.

Absorption of iron occurs mainly in the stomach and duodenum. Impaired absorption of iron is therefore observed in patients who have had subtotal or total removal of the stomach or in patients who have sustained surgical removal of a considerable amount of the small bowel. Iron absorption is also diminished in various malabsorption syndromes, such as steatorrhea.

In iron deficiency anemias the absorption of iron may be increased to 2–10 times normal. It is also increased in pernicious anemia and in hypoplastic anemia.

A diet high in phosphate causes a decrease in the absorption of iron since compounds of iron and phosphate are insoluble; conversely, a diet very low in phosphates markedly increases iron absorption. Phytic acid (found in cereals) and oxalates also interfere with the absorption of iron.

Not all of the iron in foods is available to the body. An ordinary chemical determination for total iron of foodstuffs is thus not an accurate measure of nutritionally available iron. This can be determined by measuring the amount that will react with the *a,a*-dipyridyl reagent.

There appears to be some control of the absorption of iron by the intestine itself, although the exact mechanism of this "control" is not certain. At one time the so-called "mucosal block" theory was considered to account for the intestinal control of iron absorption. According to this theory, the amount of iron-binding protein, **apoferritin**, in the mucosal cells was the controlling factor. Ferrous iron, once within the mucosal cells, is oxidized to the ferric state and then combined with apoferritin to form the iron-containing protein, **ferritin**, which contains 23% of iron by weight. It was believed that the binding capacity of apoferritin for iron limited the further absorption of iron; when saturated with iron, no further storage of iron in the intestine could occur in the form of ferritin. Normally, the apoferritin content of the mucosal cells is low, but there is evidence for a rapid increase in its formation by the intestinal mucosal cells as a result of the administration of iron. Furthermore, the previous administration of iron hinders the subsequent absorption of iron for 12–24 hours.

There is now evidence for the presence of a ferritin-independent active transport system for iron, and it appears that variation in the activity of this system regulates iron absorption (Dowdle, 1960). Experimental support for the existence of nonferritin control of iron absorption has been provided by studies in mice with an X-linked inherited defect in iron absorption (Edwards & Bannerman, 1970).

Transport in the Plasma

Iron released from storage as ferritin is reduced again to ferrous iron and leaves the intestine to enter the plasma. There is a possibility that diminished oxygen tension in the blood may increase the activity of the intestinal mucosal cells transferring iron to the plasma. In the presence of CO_2 the iron from the plasma forms a complex with a metal-binding beta globulin known as **transferrin** or **siderophilin** which occurs in Cohn fraction IV-7. Transferrin is a glycoprotein containing hexose, hexosamine, sialic acid, and, possibly, fucose. Jamieson (1965) has investigated the glycopeptide structure of human transferrin. The analytical data indicate that transferrin contains 4 mols of sialic acid, 8 mols of N-acetylglucosamine, 4 mols of galactose, and 8 mols of mannose per molecular weight of glycoprotein (90,000 daltons). The carbohydrate is arranged as 2 branched chains, identical in composition, terminating in sialic acid residues and linked to the protein moiety by means of asparaginyl linkages. There may also be other linkages, particularly O-glycosidic linkages to serine residues in the protein.

As indicated above, the iron entering the blood stream from the intestine is mostly in the ferrous (Fe^{++}) state. In the plasma, Fe^{++} is rapidly oxidized to the ferric (Fe^{+++}) state and is then incorporated into the specific iron-binding protein, transferrin. This protein can bind 2 atoms of Fe^{+++} per molecule of protein to form a red ferric-protein complex. If ferrous iron is added to the protein (apotransferrin), oxygen is required for the formation of the red complex and the rate of formation of the color depends on the rate of oxidation of Fe^{++} to Fe^{+++} by molecular oxygen.

Under normal circumstances almost all of the iron bound to transferrin is taken up rapidly by the bone marrow. It appears that only the reticulocytes are capable of utilizing the Fe^{+++} bound to transferrin, although both the reticulocytes and the mature erythrocyte can take up unbound ferric iron. Thus transferrin in some manner diverts the plasma iron into those cells which are actively making hemoglobin.

Ceruloplasmin is described below as a copper-binding protein occurring in the plasma. However, Osaki & others (1966) have found that ceruloplasmin exerts a catalytic activity in plasma to convert Fe^{++} to Fe^{+++} and thus promote the rate of incorporation of iron into transferrin. Indeed these authors propose to classify ceruloplasmin as a ferro-O_2 oxidoreductase and suggest that it can be redesignated serum ferroxidase.

The normal content of protein-bound iron (BI) in the plasma of males is 120–140 μg/100 ml; in females, 90–120 μg/100 ml. However, the total iron-binding capacity (TIBC) is about the same in both sexes: 300–360 μg/100 ml. This indicates that normally only about 30–40% of the iron-binding capacity of the serum is utilized for iron transport and that the iron-free siderophilin, ie, the unsaturated iron-binding capacity (UIBC), is therefore about 60–70% of the total.

In iron deficiency anemias the plasma-bound iron is low, whereas the total iron-binding capacity tends to rise, resulting in an unsaturated iron-binding capacity which is higher than normal. In hepatic disease, both the bound iron and the total iron-binding capacity of the plasma may be low, so that the percentage of the total iron-binding capacity which is unsaturated is not significantly altered from normal.

The amount of bound iron in the plasma is reported to exhibit a diurnal variation which can be as much as 60 μg/100 ml over a 24-hour period. The lowest values were found 2 hours following retirement for sleep; the highest values were found 5–7 hours later.

The failure of the kidney to excrete iron is probably due to the presence of iron in the plasma as a protein-bound compound which is not filtrable by the glomerulus. By the same token, losses of iron into the urine may occur in proteinuria. In nephrosis, for example, as much as 1.5 mg of iron per day may be excreted with protein in the urine.

Metabolism

The storage form of iron, ferritin, is found not only in the intestine but also in liver (about 700 mg), spleen, and bone marrow. If iron is administered parenterally in amounts which exceed the capacity of the body to store it as ferritin, it accumulates in the liver as microscopically visible **hemosiderin**, a form of colloidal iron oxide in association with protein. The iron content of hemosiderin is 35% by weight.

The level of iron in the plasma is the result of a dynamic equilibrium; the factors which influence it include the rate of breakdown of hemoglobin, uptake by the bone marrow in connection with red blood cell synthesis, removal and storage by the tissues, absorption from the gastrointestinal tract, and the rate of formation and decomposition of siderophilin (transferrin). Studies of the turnover of iron, using isotopic ^{59}Fe, indicate that about 27 mg are utilized each day, 75% of this for the formation of hemoglobin. About 20 mg are obtained from the breakdown of red blood cells, a very small amount from newly absorbed iron, and the remainder from the iron stores. Normally there is a rather slow exchange of iron between the plasma and the storage iron; in fact, following an acute hemorrhage in a normal individual, the level of iron in the plasma may remain low for weeks, a further indication that mobilization of iron from the storage depots is a slow process.

Iron deficiency anemias are of the hypochromic microcytic type. In experiments with rats made iron-deficient, it was found that cytochrome c levels were reduced even in the absence of anemia. This suggests that some of the symptoms in anemia may be due to increased activity of intracellular enzymes rather than to low levels of hemoglobin.

A deficiency may result from inadequate intake (eg, a high cereal diet, low in meat) or inadequate absorption (eg, gastrointestinal disturbances such as diarrhea, achlorhydria, steatorrhea, or intestinal disease, after surgical removal of the stomach, or after extensive intestinal resection) as well as from excessive loss of blood. If absorption is adequate, the daily addition of ferrous sulfate to the diet will successfully treat the iron deficiency type of anemia. A preparation of iron (iron dextran) for intramuscular injection in patients who cannot tolerate or absorb orally administered iron has been used. Caution must be exercised when iron is given parenterally because of the possibility of oversaturation of the tissues with resultant production of hemosiderosis.

Studies with isotopic iron have been used to determine the rate of red blood cell production. ^{59}Fe is given intravenously in tracer doses and the rate of disappearance of the label is measured. Normally ½ of the radioactivity disappears exponentially from the circulating blood in 90 minutes. In hemolytic anemias, where there is hyperplasia of the erythroid tissue, and in polycythemia vera, ½ of the activity disappears in 11–30 minutes. In aplastic anemia the opposite situation prevails; the disappearance time is prolonged to as long as 250 minutes. The reappearance of the label in newly formed blood cells is then noted. In an iron deficiency type of anemia, the uptake of iron in the erythrocytes is accelerated; in aplastic anemia, it is diminished.

Because of the absence of an excretory pathway for iron, excess amounts may accumulate in the tissues. This is observed in patients with aplastic or hemolytic anemia who have received many blood transfusions over a period of years. The existence in some individuals of an excessive capacity for the absorption of iron from the intestine has been detected by studies with radioactive iron. Such individuals absorb 20–45% of an administered dose of the labeled iron; a normal subject absorbs 1.5–6.5%. The anomaly in iron absorption may be inherited. In such patients, a very large excess (as much as 40–50 g) of iron accumulates in the tissues after many years. This hemosiderosis may be accompanied by a bronzed pigmentation of the skin, **hemochromatosis**, and, presumably because of the toxic effect of the unbound iron in the tissues, there may be liver damage with signs of cirrhosis, diabetes, and a pancreatic fibrosis. The condition is sometimes referred to as bronze diabetes. It is of interest that the unsaturated iron-binding capacity of the serum of the patient with hemochromatosis is very low. Thus, whereas the iron-binding proteins of the serum in a normal individual are only about 30% saturated, in patients with hemochromatosis they are about 90% saturated. This is doubtless due to the excess absorption of iron from the intestine.

An acquired siderosis of dietary origin has been reported to occur with great frequency among the Bantu peoples of Africa. This condition has been termed "Bantu siderosis" and is believed to be caused by the fact that the natives consume a diet which is very high in corn and thus low in phosphorus and that their foods are cooked in iron pots. The combination of a low-phosphate diet and a high intake of iron enhances absorption of iron sufficiently to produce siderosis with accompanying organ damage as described above. It is of further interest that iron deficiency anemias, common among pregnant women in other areas of the world, are virtually unknown among the Bantu.

COPPER

Functions & Distribution

The functions of this essential element are not well understood. Copper is a constituent of certain enzymes or is essential in their activity; these include cytochrome, cytochrome oxidase, catalase, tyrosinase, monoamine oxidase, and ascorbic acid oxidase as well as uricase, which contains 550 μg of copper per g of enzyme protein. Along with iron, copper is necessary for the synthesis of hemoglobin.

A copper-containing protein, **erythrocuprein** (Stansell & Deutsch, 1965), has been found in the red blood cells. In contrast to the blue-colored cuproprotein of the plasma, ceruloplasmin, described below, erythrocuprein is a nearly colorless protein containing 2 atoms of copper per mol and with a molecular weight of about 33,000 daltons. The normal adult human red blood cell contains about 30–36 mg of erythrocuprein per 100 ml packed cells. The copper content of such cells is 93–114 μg/100 ml. Since erythrocuprein contains at least 3.2 μg copper per mg protein, this protein accounts for most if not all of the copper in the red cell (Markowitz, 1959).

Cerebrocuprein is a copper protein isolated from human brain by Porter & Folch (1957). It differs from erythrocuprein (and from ceruloplasmin) in that the copper in cerebrocuprein reacts directly with diethyldithiocarbamate (see below).

Other functions of copper include a postulated role in bone formation and in maintenance of myelin within the nervous system. Hemocyanin is a copper-protein complex in the blood of certain invertebrates, where it functions like hemoglobin as an oxygen carrier.

The adult human body contains 100–150 mg of copper; about 64 mg are found in the muscles, 23 mg in the bones, and 18 mg in the liver, which contains a higher concentration of copper than any of the other organs studied. It is of interest that the concentration of copper in the fetal liver is 5–10 times higher than that in the liver of an adult. Both the blood cells and serum contain copper; but the copper content of the red blood cell is constant, while that of the serum is highly variable, averaging about 90 μg/100 ml.

The serum copper is present in 2 distinct fractions. The so-called "direct-reacting" copper is that fraction which reacts directly with diethyldithiocarbamate, the reagent used to determine copper colorimetrically (Gubler, 1952). This copper is loosely bound to albumin and may represent copper in transport. Relatively little of the serum copper is present in this form. Most of the copper in the serum (96%) is bound to an alpha globulin contained in the Cohn fraction IV-1. Consequently, in determination of serum copper it is necessary first to treat the serum with hydrochloric acid to free the copper from the globulin-bound form so that it can react with the color reagent.

A copper-binding protein of the serum or plasma is called **ceruloplasmin.** It has a molecular weight of about 151,000 daltons and contains 0.34% copper, or about 8 atoms of copper per mol. Normal plasma contains about 30 mg of this protein per 100 ml. The absorption peak for the protein is at 610 nm. According to Holmberg & Laurell (1948) this copper-containing protein acts in vitro as an enzyme—a polyphenol oxidase. However, as noted on p 406, it may also function in oxidation of plasma iron to permit formation of transferrin-bound iron.

Metabolism

Experiments have been carried out with labeled copper (^{64}Cu). The copper was found largely associated with the albumin fraction of the plasma immediately after its ingestion; 24 hours later, most of it was in the globulin fraction, associated with ceruloplasmin. Since the copper in the plasma is largely bound to protein, it is not readily excreted in the urine. Apparently most of it is lost through the intestine.

Experimental animals on a copper-deficient diet lose weight and die; the severe hypochromic microcytic anemia which they exhibit is not the cause of death since an iron-deficiency anemia of equal proportions is not fatal. This suggests that copper has a role in the body in addition to its function in the metabolism of red cells. This additional role of copper may be related to the activity of oxidation-reduction enzymes of the tissues, such as the cytochrome system. A relation between copper and iron metabolism has been detected. In the presence of a deficiency of copper, the movement of iron from the tissues to the plasma is decreased and hypoferremia results. Copper favors the absorption of iron from the gastrointestinal tract.

A bone disorder associated with a deficiency of copper in the diet of young dogs has been described (Baxter, 1953). The bones of these animals were characterized by abnormally thin cortices, deficient trabeculae, and wide epiphyses. Fractures and deformities occurred in many of the animals. Anemia was present, and the hair turned gray. The disorder did not occur in any of the control animals, and was relieved by the administration of copper.

Wilson's disease (hepatolenticular degeneration) is associated with abnormalities in the metabolism of copper. In this disease the liver and the lenticular nucleus of the brain contain abnormally large amounts of copper, and there is excessive urinary excretion of copper and low levels of copper and of ceruloplasmin in the plasma. A generalized aminoaciduria also occurs in this disease.

According to Bearn & Kunkel (1954) the absorption of copper from the intestine is considerably increased in the patient with Wilson's disease. As a result copper accumulates in the tissues and appears in the urine. If the deposition of copper in the liver becomes excessive, cirrhosis may develop. Accumulation of copper in the kidney may give rise to renal tubular damage which leads to increased urinary excretion of amino acids and peptides and, occasionally, glucose as well.

Others (Earl, 1954) feel that Wilson's disease is characterized by a failure to synthesize ceruloplasmin at a normal rate (although administration of ceruloplasmin does not ameliorate the disease) or a defect at the stage of incorporation of copper into the copper-binding globulin, or both. As a result there is present in these patients a quantity of "unattached" copper which is free to combine in an abnormal manner, such as with the proteins of the brain or liver. The excessive excretion of copper in the urine is also explained as a result of the presence of abnormal amounts of unbound copper. In the normal individual, copper is bound to ceruloplasmin within 24 hours of its ingestion; in the patient with Wilson's disease, copper is still associated with the albumin fraction at this time. As a result the so-called "direct-reacting" fraction of the serum copper (see above) is not reduced in these patients; in fact, it may actually be increased. Thus the total serum copper may appear normal or only slightly reduced.

There have been some suggestions that inadequate excretion of copper via the intestine may be a factor in the genesis of Wilson's disease.

Hypercupremia occurs in a variety of circumstances. It does not seem to have any diagnostic significance.

Requirements

The human requirement for copper has been studied by balance experiments. A daily allowance of 2.5 mg has been suggested for adults; infants and children require about 0.05 mg/kg body weight. This is easily supplied in average diets, which contain 2.5–5 mg of copper.

A nutritional deficiency of copper has never been positively demonstrated in man, although it has been suspected in cases of sprue or in nephrosis. However, there are reports (Gitlin, 1958) of a syndrome in infants which is characterized by low levels of serum copper and iron and by edema and a hypochromic microcytic anemia. Therapy with iron easily cures the disease; and "spontaneous" cures are also reported.

IODINE

Function

The only known function of iodine is in the thyroid mechanism. Its metabolism is discussed under that subject (see Chapter 20).

Requirements & Source (See also Table 21–3.)

The requirement for iodine is 5 μg/100 Cal or, for adults, about 100–150 μg/day. The use of iodized salt regularly will provide more than this.

The need for iodine is increased in adolescence and in pregnancy. Thyroid hypertrophy occurs if iodine deficiency is prolonged.

MANGANESE

Ordinary diets yield about 4 mg/day of manganese. This amount is in excess of that suggested as required by man. Animal experiments have indicated that this element is essential, but there is no evidence of a manganese deficiency in man. Perosis (slipped tendon disease) in chickens and other fowl may be produced by a manganese deficiency.

The total body content of manganese is about 10 mg. The kidney and the liver are the chief storage organs for manganese. In blood, values of 4–20 μg/100 ml have been reported. Most of the manganese is excreted in the intestine by way of the bile (determined by experiments with isotopic manganese). Very little manganese is found in the urine.

The functions of manganese are not known. In vitro, manganese activates several enzymes: blood and bone phosphatases, yeast, intestine and liver phosphatases, arginase, carboxylase, and cholinesterase.

When isotopic manganese (^{56}Mn) was injected intraperitoneally, a correlation was found between the mitochondrial content of a given organ and its ability to concentrate manganese (Maynard, 1955). Fractionation studies of liver cells confirmed the assumption that the mitochondria are the principal intracellular sites of manganese uptake. Since the mitochondria are also the structures with which the majority of the intracellular respiratory enzyme systems are associated, a role of manganese as a coenzyme for these enzymes is strongly suggested.

COBALT

This element is a constituent of vitamin B_{12}. Cobalt affects blood formation. A nutritional anemia in cattle and sheep living in cobalt-poor soil areas has been successfully treated with cobalt. Microorganisms in the rumen of these animals utilize cobalt to synthesize vitamin B_{12}. Consequently a deficiency of cobalt results in limitation of the vitamin B_{12} supply which undoubtedly is a factor in causation of the anemia.

There are reports of the favorable use of cobalt in the anemias of children.

An excess of red cells (polycythemia) has been produced in rats when the element is fed or injected. Cobaltous chloride administration to human subjects also produced an increase in red blood cells.

Isotopic cobalt is quickly eliminated almost completely via the kidneys.

ZINC

When experimental animals (rats) are maintained on a diet which is very low in zinc, impaired growth and poor development of their coats are noted.

Zinc is a structural and functional component of the enzyme carboxypeptidase, and this element participates directly in the catalytic action of the enzyme (Vallee, 1955). Zinc is also closely associated with several other enzymes, including carbonic anhydrase and alcohol dehydrogenase, which contains one atom of zinc per mol of enzyme protein.

Insulin is known to contain zinc and it is therefore of interest to note that the pancreas of diabetics contains only about ½ the normal amount of zinc.

The zinc content of leukocytes from the blood of normal human subjects is reported to be $3.2 \pm 3 \times 10^{-10}$ μg/million cells. However, the zinc content of white blood cells in human leukemia patients is reduced to 10% of the normal amount (Hoch, 1952). Temporary therapeutic amelioration of the leukemic process is accompanied by a rise in the zinc content of these cells to normal. Because there is no carbonic anhydrase in white blood cells, the function of the intracellular zinc in these cells is not known. The metal is firmly bound to protein within the cell; in fact, it is found in a constant ratio to the protein: 82–117 μg of zinc per g of protein.

Although it is believed that zinc is an essential element, little is known concerning the human requirement. As a result of balance studies, a daily intake of 0.3 mg/kg has been recommended. This amount is easily obtained in the diet since the zinc content of natural foods approximates that of iron.

Evidence for a conditioned deficiency of zinc in patients with so-called postalcoholic cirrhosis has been presented by Vallee & others (1959). Patients with Laennec's cirrhosis had a mean concentration of zinc of 66 ± 19 μg/100 ml serum (normal, 120 ± 19 μg/100 ml). There was also increased urinary excretion of zinc in the cirrhotic patients. When studied at autopsy, the livers of patients who had died of cirrhosis were found to have significantly less zinc than liver tissues from patients who had died of nonhepatic disease. It was suggested that the zinc deficiency in the cirrhotic patients was due to secondary factors related to the underlying disease which act to render a normal intake of zinc inadequate. Because of the importance of zinc in the activity of certain enzymes, as mentioned above, it may be supposed that a zinc deficiency would cause significant biochemical changes in the metabolism of those substrates affected by the enzyme concerned. One such substrate is alcohol, whose role in the etiology and exacerbation of cirrhosis has not been fully explained. Further studies of zinc metabolism in liver disease may therefore contribute to our understanding of this important clinical entity.

FLUORINE

This element is found in certain tissues of the body, particularly in bones and teeth.

Fluoride is a poison for some enzyme systems. Specifically, it inhibits the conversion of glyceric acid to pyruvic acid by enolase in anaerobic glycolysis.

In very small amounts, it seems to improve tooth development, but a slight excess causes mottling of the enamel of the tooth. Mild mottling of the teeth occurs in less than 2% of the children living in areas where the fluorine content of the water is between 0.6 and 1.2 ppm. Severity and incidence of mottling increase when the fluorine in the water exceeds these amounts. However, at levels above 1 ppm, the incidence of dental caries in children is much lower. The topical application of fluorine to the teeth during the developmental stage—or the addition of 1 ppm fluoride to drinking water in areas where it is normally low—is now utilized in an effort to reduce the incidence of dental caries.

Studies with animals on low-fluorine diets have not provided evidence of its essentiality. In large quantities, it is definitely toxic.

ALUMINUM

While this element is widely distributed in plant and animal tissues, there is still no evidence that it is essential. Its physiologic role, if any, remains obscure. Rats subsisting on a diet which supplied as little as 1 μg/day showed no abnormalities. Large amounts fed to rats produced rickets by interfering with the absorption of phosphates.

The daily intake of aluminum in the human diet ranges from less than 10 mg to over 100 mg. In addition to the very small amounts naturally present in foods and that derived from cooking utensils, aluminum may be added to the diet as sodium aluminum sulfate in baking powder and as alum, sometimes added to foods to preserve firmness. However, absorption of aluminum from the intestine is very poor. From measurements of urinary excretion it is estimated that only about 100 μg of aluminum are absorbed per day, most of the ingested aluminum being excreted in the feces even on a reasonably high intake. The total amount of aluminum in the body is about 50–150 mg.

BORON

Boron is essential for the growth of plants, and traces are found in animal tissues. The growth of rats maintained on diets very low in boron is not impaired.

MOLYBDENUM

In studies of the activity of the flavoprotein enzyme, xanthine oxidase, it was found that traces of molybdenum are required for the deposition and maintenance of normal levels of this enzyme in the intestine and liver of the rat. Highly purified preparations of the enzyme as obtained from milk contain molybdenum in a form which indicates that it is actually a part of the enzyme molecule. Liver aldehyde oxidase, a flavoprotein which catalyzes the oxidation of aldehydes, also contains molybdenum.

SELENIUM

Selenium is usually thought of as an element with pronounced toxic properties (Moxon, 1943). When present in the diet in concentrations above 5–15 ppm, selenium is indeed highly toxic to animals as has been demonstrated in particular with ruminants maintained in pastures where the soil is rich in selenium. However, below about 3 ppm, selenium improves growth in sheep and is effective in combating several diseases in sheep as well as other animals, as mentioned below.

In very small quantities, selenium may prove to be an essential factor in tissue respiration. This suggestion emanates from the work of Schwartz and coworkers on the identity of "factor 3," which is protective against a hepatic necrosis produced in the rat by dietary means (see pp 93 and 285). The preventive factor is an organic compound containing selenium as the active ingredient. It is extremely active, as evidenced by the finding that 4 parts of selenite selenium per 100 million parts of diet were found to be protective against dietary liver necrosis. Factor 3 not only prevents dietary liver necrosis in rats but also multiple necrosis in mice, muscular dystrophy and heart necrosis in minks, exudative diathesis in chicks and turkeys, stiff lamb disease and ill thrift in sheep, white muscle disease in calves, and liver dystrophy and muscle degeneration in pigs (Schwartz, 1960). In some of these fatal diseases, factor 3 seemed to act synergistically with vitamin E and with cystine. It is now known that cystine was effective only because it was contaminated with traces of factor 3-active selenium. Although vitamin E enhances the action of factor 3-selenium, there is evidence that selenium does not spare vitamin E nor does it substitute for it but that the selenium compound must be considered an essential dietary constituent in its own right.

CADMIUM

Cadmium resembles zinc in its chemical characteristics. It is therefore not surprising that metalloproteins containing cadmium—analogous to those containing zinc (see above)—might exist in the tissues. Such a protein has been isolated from equine renal cortex (Kagi & Vallee, 1960). It contains 2.9% cadmium, 0.6% zinc, and 4.1% sulfur. Because of its high metal and sulfur content, it has been termed **metallothionein.** Prior to the isolation of this protein, cadmium had not been found as part of a macromolecule in any natural materials, although the existence of cadmium in the tissues in trace amounts has been reported.

The concentration of cadmium in the renal cortex apparently exceeds that in all other organs. In metallothionein, the high sulfur content is probably due to a large number of cysteine residues which provide SH groups for binding the cadmium and zinc. Although the biologic function of metallothionein has not yet been elucidated, its existence does suggest a role for cadmium in a biologic system.

CHROMIUM

Chromium occurs as a trace element in the tissues and can be detected in dairy products, in meats, and in fish. Some studies have suggested that trivalent chromium may act together with insulin in promoting utilization of glucose, a conclusion drawn from the observation that impairment of glucose tolerance occurs in animals maintained on low-chromium diets and that the diminished glucose tolerance can be reversed by supplementation of the diet with chromium. A severely restricted intake of chromium in the diet of rats and mice has been shown to impair growth and survival of these experimental animals. Supplementation with 5 ppm of chromium improved growth and survival of the animals.

• • •

References

Baxter JH, Van Wyk JJ: Bull Johns Hopkins Hosp 93(1):25, 1953.

Bearn AG, Kunkel HG: J Clin Invest 33:400, 1954.

Bills CE, McDonald FG, Niedermeier W, Schwartz MC: J Am Dietet A 25:304, 1949.

Burnett CH, Commons RR, Albright F, Howard JE: New England J Med 240:787, 1949.

Dahl LK: New England J Med 258:1152, 1205, 1958.

Deane N, Smith HW: J Clin Invest 31:197, 1952.

Dowdle EB, Schachter D, Schenker H: Am J Physiol 198:609, 1960.

Earl CJ, Moulton MJ, Selverstone B: Am J Med 17:205, 1954.

Edelman IS, Leibman J: Am J Med 27:256, 1959.

Edwards JA, Bannerman RM: J Clin Invest 49:1869, 1970.

Fitzgerald MG, Fourman P: Clin Sc 15:635, 1956.

Flink EB, McCollister R, Prasad AS, Melby JC, Doe RP: Ann Int Med 47:956, 1957.

Forbes RM, Cooper AR, Mitchell HH: J Biol Chem 203:359, 1953.

Frost PM, Smith JL: Metabolism 2:529, 1953.

Gitlin D, Janeway CA: Pediatrics 21:1034, 1958.

Graham LA, Caesar JJ, Burgen ASV: Metabolism 9:646, 1960.

Gubler CJ, Lahey ME, Ashenbrucker H, Cartwright GE, Wintrobe MM: J Biol Chem 196:209, 1952.

Hoch FL, Vallee BL: J Biol Chem 195:531, 1952.

Holmberg CG, Laurell CB: Acta chem scandinav 2:550, 1948.

Jamieson GA: J Biol Chem 240:2914, 1965.

Kagi JHR, Vallee BL: J Biol Chem 235:3460, 1960.

Kark RM: Am J Med 25:698, 1958.

Lloyd HM, Rose GA: Lancet 2:1258, 1958.

Loken HF, Havel RJ, Gordan GS, Whittington SL: J Biol Chem 235:3654, 1960.

Markowitz H, Cartwright GE, Wintrobe MM: J Biol Chem 234:40, 1959.

Maynard LS, Cotzias GC: J Biol Chem 214:489, 1955.

Meneely GR, Ball COT: Am J Med 25:713, 1958.

Moxon AL, Rhian M: Physiol Rev 23:305, 1943.

Osaki S, Johnson DA, Frieden E: J Biol Chem 241:2746, 1966.

Porter H, Folch J: J Neurochem 1:260, 1957.

Schwartz K: Nutr Rev 18:193, 1960.

Smith WO, Hammarsten JF: Am J M Sc 237:413, 1959.

Stansell MJ, Deutsch HF: J Biol Chem 240:4306, 1965.

Vallee BL, Neurath H: J Biol Chem 217:253, 1955.

Vallee BL, Wacker WAC, Bartholomay AF, Hoch FL: Ann Int Med 50:1077, 1959.

Bibliography

Bland JH: *Clinical Metabolism of Body Water and Electrolytes.* Saunders, 1963.

Elkinton JR, Danowski TS: *The Body Fluids.* Williams & Wilkins, 1955.

Gamble JL: *Chemical Anatomy, Physiology and Pathology of Extracellular Fluid,* 6th ed. Harvard Univ Press, 1954.

Goldberger E: *A Primer of Water, Electrolyte and Acid-Base Syndromes,* 3rd ed. Lea & Febiger, 1965.

Hardy JD: *Fluid Therapy,* 2nd ed. Lea & Febiger, 1962.

Maxwell MH, Kleeman CR (editors): *Clinical Disorders of Fluid and Electrolyte Metabolism.* Blakiston-McGraw, 1962.

McCance RA, Widdowson EM: *The Composition of Food.* Great Britain Med Res Council Special Report Series. Her Majesty's Stationery Office, London, 1960.

Recommended Dietary Allowances. National Academy of Sciences-National Research Council Publication 1694, Washington, 1968.

Thompson RHS, King EJ (editors): *Biochemical Disorders in Human Disease,* 2nd ed. Academic Press, 1964. [See chapter by Passmore & Draper.]

Weisberg HF: *Water, Electrolyte and Acid-Base Balance,* 2nd ed. Williams & Wilkins, 1962.

20...

The Chemistry & Functions of the Hormones

With Gerold M. Grodsky, PhD*

Most glands of the body deliver their secretions by means of ducts. These are the exocrine glands. Other glands manufacture chemical substances which they secrete into the blood stream for transmission to various "target" tissues. These are the endocrine or ductless glands. Their secretions, the hormones, catalyze and control diverse metabolic processes. Despite their varying actions and different specificities, depending on the target organ, the hormones have several characteristics in common. They act as body catalysts, resembling enzymes in some aspects since they are required only in very small amounts and are not used during their catalytic action. They differ from enzymes in the following ways:

(1) They are produced in an organ other than that in which they ultimately perform their action.

(2) They are secreted into the blood prior to use. Thus, circulating levels can give some indication of endocrine gland activity and target organ exposure. Because of the small amounts of hormones required, blood levels can be extremely low. For example, circulating levels of protein hormones range from 10^{-10} to 10^{-12} mols, whereas the circulating levels of thyroid and steroid hormones are from 10^{-6} to 10^{-9} mols.

(3) Structurally, they are not always proteins, the known hormones include proteins of molecular weights 30,000 daltons or less, small polypeptides, single amino acids, and steroids.

The action of a hormone at a target organ is regulated by 4 factors: (1) rate of synthesis and secretion of the stored hormone from the endocrine gland of origin; (2) in some cases, specific transport systems in the plasma; (3) hormone specific receptors in target cell membranes which differ from tissue to tissue; and (4) ultimate degradation of the hormone, usually by the liver or kidneys. Variation in any of these factors can result in a rapid change in the amount or activity of a hormone at a given tissue site.

It is characteristic of the endocrine system that a state of balance is normally maintained among the various glands. This is particularly notable in the relationship of the anterior pituitary tropic principles to the various target glands which they affect. Characteristically, elevated hormone levels result in both direct and indirect feedback inhibition of their production by the originating gland.

General Mechanism of Action of Hormones

Although the exact site of action of any hormone is still not established, 5 general sites have been proposed:

A. Induction of Enzyme Synthesis at the Nuclear Level: Many hormones, particularly the steroids, may act to stimulate RNA production in the target cell nucleus and thereby increase the synthesis of a specific enzyme or group of enzymes catalyzing a specific metabolic pathway. Some hormones, when isotopically labeled, are found to be localized in the nucleus. Other evidence of nuclear action is the frequent demonstration of an increase in RNA synthesis as measured by incorporation of labeled precursors, such as orotic acid or glycine, into the nuclear RNA fraction. The steroid hormones increase the synthesis of total RNA, including messenger, transfer, and ribosomal RNA, by the nucleus. Since the various messenger RNAs synthesized by the nucleus cannot be identified with the specific enzymes whose synthesis they determine, it is not as yet possible to relate unequivocally a change in total RNA with the rise or fall of the activity of certain enzymes. Thus, a hormone which does not effect a net increase in RNA synthesis could still act by this mechanism by increasing the RNA for a specific enzyme while decreasing RNA synthesis for others. Evidence for nuclear activity of hormones is also indicated by demonstration of changes in chromosomal appearance such as the puffing phenomenon observed after administration of the hormone ecdysone (Beerman & Clever, 1964). Finally, the increase of activity of an enzyme after hormone administration can often be blocked by the administration of inhibitors of RNA synthesis such as dactinomycin, indicating that the hormonal action on enzyme activity was mediated by an effect on RNA synthesis. Current evidence, though not conclusive, indicates that hormones acting on nuclear regulating of RNA and protein synthesis do so by influencing the protein regulators of repression and derepression (see p 69); a direct chemical reaction of the hormones with the polynucleotides of either DNA or RNA is less probable. Hormone action leading to a change in the rate of RNA and enzyme synthesis with a consequent effect on cellular metabolism is comparatively slow. This type of action may therefore require hours or even days of exposure to the hormone before the effect may be detectable.

*Professor of Biochemistry and Research Biochemist, University of California, San Francisco.

B. Stimulation of Enzyme Synthesis at the Ribosomal Level: Activity is at the level of translation of information on the messenger RNA to the sites or mechanism involved in the ribosomal production of the enzyme protein. Ribosomes taken from a growth hormone–treated animal, for example, have a modified capacity to synthesize protein in the presence of normal messenger RNA (Korner, 1965).

C. Direct Activation at the Enzyme Level: Although the direct effect of a hormone on a pure enzyme is difficult to demonstrate, treatment of the intact animal or of isolated tissue with some hormones results in a change of enzyme activity not related to de novo synthesis. These hormonal effects are usually extremely rapid and, in rare instances, have been demonstrated in broken cell preparations (eg, the activation of the phosphorylase system by glucagon or epinephrine).

D. Hormonal Action at the Membrane Level: Many hormones seem specifically involved in the transport of a variety of substances across cell membranes, including carbohydrates, amino acids, cations, and nucleotides. In general, these hormones specifically bind to cell membranes. They cause rapid metabolic changes in the tissue but have little effect on metabolic activity of broken cell preparations.

E. Hormonal Action as It Relates to the Level of Cyclic AMP: Cyclic AMP (cyclic 3′,5′-AMP) is a nucleotide which plays a unique role in the action of many hormones. Its level may be increased or decreased by hormonal action; the effect varies, depending on the tissue. Thus, glucagon may cause large increases of cyclic AMP in the liver but comparatively small increases in muscle. In contrast, epinephrine produces a greater increase of cyclic AMP in muscle than in liver. Insulin can decrease hepatic cyclic AMP in opposition

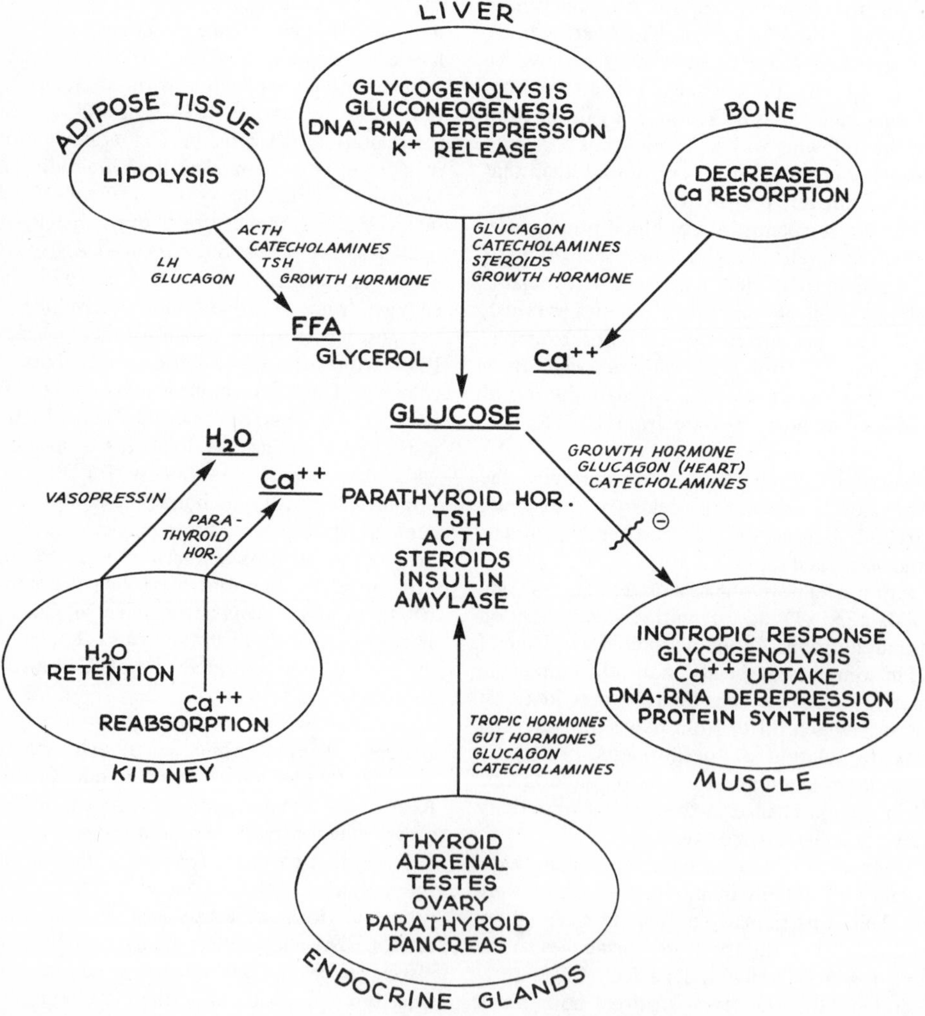

FIG 20–1. Tissue mechanisms increased by cyclic AMP and the hormones which generate it. Insulin and prostaglandin decrease cyclic AMP and reverse the mechanisms in most tissues.

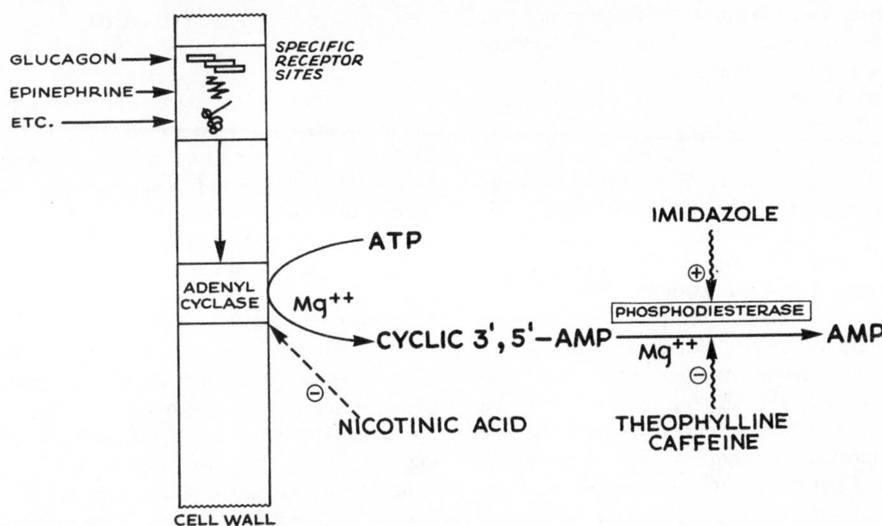

FIG 20–2. Factors involved in the production and degradation of cyclic 3′,5′-AMP.

to the increase caused by glucagon. Direct action of a hormone on purified adenyl cyclase (the enzyme responsible for the synthesis of cyclic AMP from ATP) has been difficult to demonstrate. The hormones probably act at specific receptor sites in the different cell membranes which in turn activate the cyclase (Fig 20–2).

Most of the varied effects of cyclic AMP appear to reflect its general ability to activate a large variety of kinase enzymes (Kuo & Greengard, 1969). Thus, cyclic AMP activation of phosphorylase is the result of a specific activation of the enzyme, phosphorylase kinase, which results ultimately in the conversion of inactive dephosphophosphorylase to active phosphorylase (see p 440). In adipose tissue, cyclic AMP may activate lipolysis by a similar stimulation of protein kinase which causes increased lipase activity. Cyclic AMP can also increase the activity of protein kinases which phosphorylate nuclear histones and possibly other nuclear proteins. Thus, changing levels of cyclic AMP may influence the function of protein repressors in the nucleus and explain how some hormones regulate RNA and enzyme synthesis. The hydrolysis of cyclic AMP results in the liberation of 1.6 Cal/mol more energy than the hydrolysis of a high energy bond from ATP; this has led to the yet unconfirmed suggestion that cyclic AMP may activate kinase enzymes by adenylation.

Tissue levels of cyclic AMP can be influenced not only by hormones but also by nicotinic acid, imidazole, and methyl xanthines acting on its synthesis and degradation (Fig 20–2). However, the effects of these agents can vary with concentration or type of tissue and may not always relate to cyclic AMP metabolism.

Adenyl cyclase is localized in the cell membrane; hormonal action on cyclic AMP levels and membrane phenomena may therefore be related.

Though originally discovered as an endocrinologic phenomenon, cyclic AMP is now recognized as an ubiquitous nucleotide which may be as important as ATP or AMP in controlling enzymatic reactions. A rise in cyclic AMP is usually associated with beta-adrenergic reactions (see p 440), and levels can be influenced by most metabolic changes caused by stress or diet.

All of the 5 mechanisms described above may be involved in the action of a given hormone and may vary in significance when the action of a hormone is studied in different tissues. For example, insulin has a major and rapid effect on membrane transport in adipose and muscle tissue, but its action is slower in liver, where it may function at the nuclear or translation level. Finally, all 5 mechanisms are intimately related. An effect on transport would thus permit the entrance of substances which could act as enzyme activators or, at the nuclear level, as repressors or derepressors of RNA synthesis. Similarly, a direct effect on one enzyme system could modify the availability of substrates or products for other pathways or for activation or inactivation at the nuclear or membrane level. At present, therefore, because of the interdependency of these mechanisms, the primary action of a given hormone is not easily established.

Assay of Hormones

A. Biologic Assays: Biologic assays, in which an aspect of hormonal activity is measured in vitro or in vivo, remain important since they measure levels of functional activity. However, these assays often lack precision and sensitivity and are usually not specific.

B. Chemical Assays: These are often used in conjunction with isotope dilution and employ classical isolation and purification technics, including gas and column chromatography, electrophoresis, differential solvent extraction, and isotope dilution. They provide a measure of the absolute quantity of a given hormone but can be burdensome and, in the case of protein hormones, are generally not applicable.

C. **Radiodisplacement Chemical Assays:** These have been widely adopted in recent years for both protein and nonprotein hormones. They are based on the competition for a specific binding protein of radiolabeled hormones with unlabeled hormone (Fig 20–3). For protein hormones, the binding protein is usually a specific antibody; for steroids, specific serum transport proteins are used. Unlabeled hormone, present either as a standard or an unknown, competitively displaces the labeled hormone, resulting in an increase in radioactivity in the unbound fraction. Although many methods are available, they differ primarily in the technics used to separate the bound and free hormone fractions. These include hydrodynamic flow, electrophoresis, preferential salt precipitation, precipitation of the bound complex with anti-gamma globulin serum, and adsorption of the free hormone to charcoal or to cellulose. Radiodisplacement methods are more sensitive than most bioassays since they permit detection of hormones in concentrations less than 1 ng/ml. The radiodisplacement methods are highly specific and relatively convenient. One defect of these assays is that they can measure fragments of precursors of the hormone which retain some binding activity but are biologically inert. When possible, both chemical and biologic assays should be performed on identical samples.

THE THYROID

The thyroid gland consists of 2 lobes, one on each side of the trachea, with a connecting portion making the entire gland more or less H-shaped in appearance. In the adult, the gland weighs about 25–30 g. Although there is some evidence of extrathyroidal production of thyroid-like hormones, the thyroid gland is the primary source of their production.

Function

The principal function of thyroid hormone is that of a catalyst—of the nature of a "spark"—for the oxidative reactions and regulation of metabolic rates in the body. The tissues of hypothyroid animals exhibit a low rate of oxygen consumption; conversely, those of hyperthyroid animals take up oxygen at an accelerated rate. In the absence of adequate quantities of the hormone, bodily processes decrease as manifested by a slow pulse, lowered systolic blood pressure, decreased mental and physical vigor, and often, although not always, obesity. Cholesterol levels in the blood are increased, but lipolysis and the liberation of fatty acids from the tissues decrease. In hyperthyroid states, the reverse occurs. Some of the noteworthy symptoms are increased oxygen consumption and pulse rate, increased irritability and nervousness, and, usually, loss of weight. Because of increased metabolic activity, there is increased requirements for vitamins.

Traces of radiolabeled hormone (●) incubated with excess antisera or specific binding protein (O). Little radioactivity remains in free fraction.

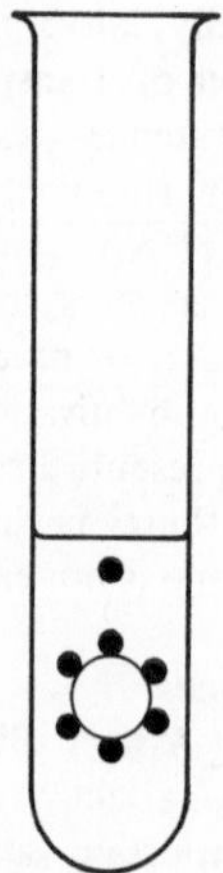

Addition of unlabeled hormone (sample or standard) (○) increased radioactivity in supernatant.

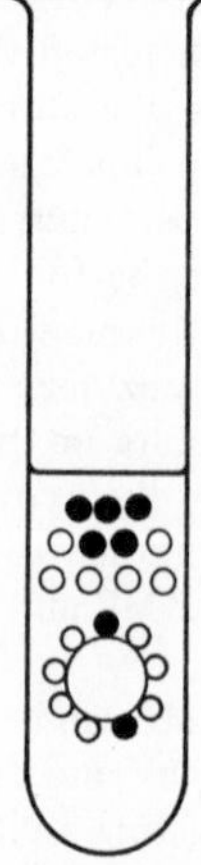

FIG 20–3. Principle of immunochemical assay for measuring hormone in biologic fluids. The amount of radioactivity in the supernatant is a direct function of the amount of hormone in the specimen.

Although a number of effects of thyroid hormone on specific metabolic reactions have been demonstrated, a unifying concept of the mechanism by which it produces acceleration of metabolism is not yet apparent. This is in part due to the different effects noted when the hormone is studied at physiologic levels or at unphysiologic high doses.

In moderate concentrations the thyroid hormone has an anabolic effect, causing an increase in RNA and protein synthesis, an action which precedes increased basal metabolic rate. Facilitation of protein synthesis occurs not only by increasing RNA synthesis at the nuclear level but also by increasing translation of the message contained in messenger RNA at the ribosome where protein synthesis occurs (Sokoloff & others, 1963). It may also increase protein synthesis by stimulating growth hormone production (see p 460). When injected into animals, thyroxine appears to stimulate most of the oxidative enzyme systems that have been investigated. In high concentrations, an opposite effect occurs: negative nitrogen balance is observed, and protein synthesis is depressed. Carbohydrate and lipid turnover is increased, and calcium is mobilized from bone.

An effect of thyroid hormone to uncouple oxidative phosphorylation and increase swelling in the mitochondria has been described. Such an action results in the production of heat rather than the storage of energy as ATP. However, these effects are observed only with very high concentrations of the hormone and may not reflect its effect in the small physiologic doses characteristic of the intact organism. Other uncoupling agents such as antimycin A or dinitrophenol also increase body heat but fail to duplicate the other biologic actions of thyroid hormone.

Intestinal absorption of glucose is increased by thyroid hormone. This rapid absorption may be a factor in the abnormal glucose tolerance often observed in hyperthyroidism. Thyroid hormone normally has a long latent period, requiring many hours or days before an effect is noted. Consequently, effects may occur early in the period after administration of the hormone which, at present, are not detectable.

Chemistry & Normal Physiology

The inorganic iodine in the body is largely taken up by the thyroid in connection with the synthesis of thyroid hormone. Of a total of 50 mg of iodine in the body, about 10–15 mg are in thyroid.

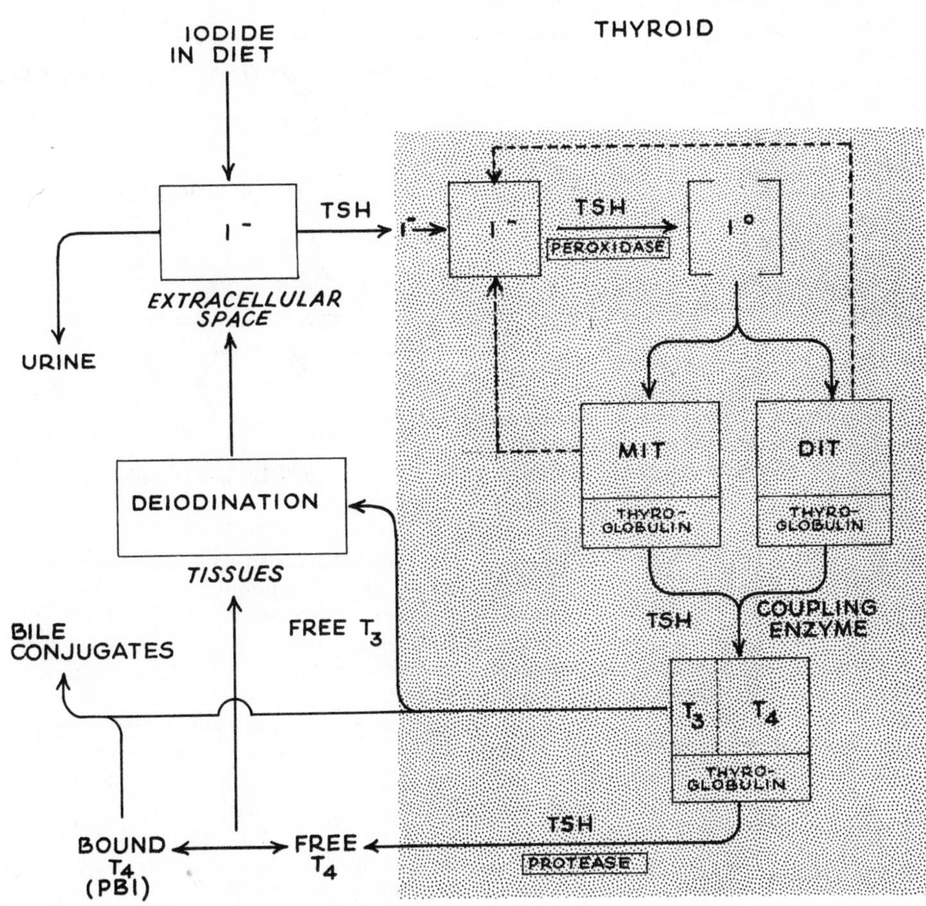

FIG 20–4. Metabolism of iodine and thyroid hormones.

The normal daily intake of iodide is 100–200 μg. This iodide, absorbed mainly from the small intestine, is transported in the plasma in loose attachment to proteins. Small amounts of iodide are secreted by the salivary glands, stomach, and small intestine, and traces occur in milk. About 2/3 of the ingested iodide are excreted by the kidney; the remaining 1/3 is taken up by the thyroid gland. Thyroid-stimulating hormone (TSH, thyrotropic hormone) of the pituitary (see p 462) stimulates iodide uptake by the gland. Thyroid activity appears to be related to the total amount of iodine in the gland. For example, in the normal gland, an iodine content of 2 mg/g of dried tissue is found, whereas in hyperthyroidism the content may fall to 0.25 mg/g. It may be that a fall in the iodine content of the gland stimulates the production of more TSH, which therefore prolongs and exacerbates the hyperthyroid state.

Iodine uptake is against a gradient, ratios of thyroid gland to plasma iodide varying from 10:1–100:1. Uptake is energy-dependent and can be inhibited by cyanide or dinitrophenol. Despite this high concentration effect, free iodide represents only about 1% of the total iodide in the thyroid. As shown in Fig 20–4, intracellular iodide exists in 2 pools. One pool is freely exchangeable with blood iodide; the other represents iodide arising from deiodination of unused iodotyrosines (Fig 20–4). Thiocyanate or perchlorate competes with iodide for the uptake mechanism and causes the rapid discharge of the exchangeable iodide from the thyroid gland.

Within the thyroid, iodide is oxidized to an active form of "iodine," a reaction which is catalyzed by a particulate-bound peroxidase which requires hydrogen peroxide as a source of oxygen. Oxidation of iodide thus involves both the production of peroxide and oxidation of the iodide by the peroxidase enzyme, though neither of these mechanisms has been elucidated, and the peroxidase has yet to be isolated. However, it has been established that the peroxidase (molecular weight ~ 62,000 daltons) contains heme as a prosthetic group. TSH is also active in stimulating this reaction. The receptors of the activated iodine are the tyrosine residues in the macroglobulin protein, thyroglobulin. This 19 S protein has a molecular weight of 660,000 daltons, each molecule containing 115 tyrosine residues. The molecule is composed of a polymer of dissimilar subunits (Alexander, 1968). Iodination of the tyrosines in thyroglobulin occurs first in position 3 of the aromatic nucleus and then at position 5, forming monoiodotyrosine and diiodotyrosine, respectively. Normally, the 2 are present in approximately equal concentration, but with iodine deficiency more monoiodotyrosine is formed. It is not likely that significant amounts of free tyrosine are iodinated and then incorporated into thyroglobulin since thyroidal transfer-RNA will not accept mono- or diiodotyrosine and since organification of iodine is not inhibited by concentrations of puromycin which block tyrosine incorporation into thyroglobulin.

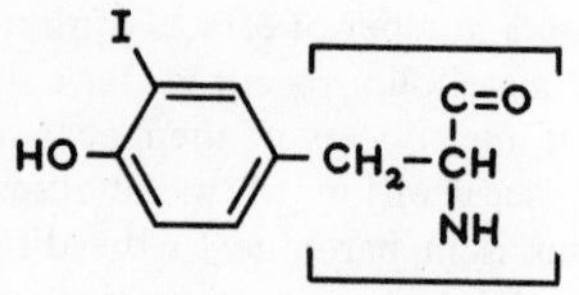

I
HO–[ring]–CH_2–CH(–C=O)(–NH)

3-Monoiodotyrosine (MIT) (shown in peptide linkage in thyroglobulin)

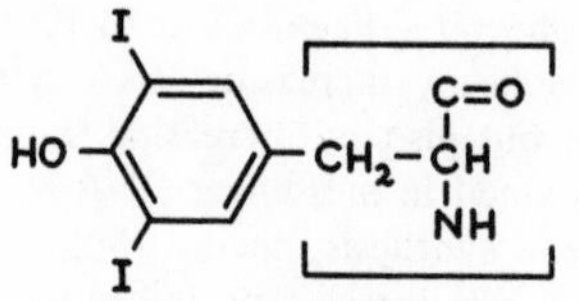

I, I
HO–[ring]–CH_2–CH(–C=O)(–NH)

3,5-Diiodotyrosine (DIT) (in peptide linkage)

It is assumed that coupling of 2 molecules of diiodotyrosine (DIT) then occurs to form tetraiodothyronine, or thyroxine (T_4). Coupling of monoiodotyrosine (MIT) with DIT also occurs to form triiodothyronine (TIT; T_3).

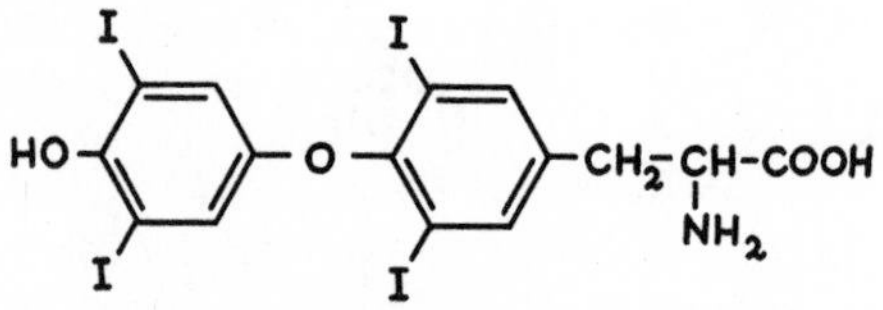

HO–[ring(I,I)]–O–[ring(I,I)]–CH_2–CH(NH_2)–COOH

Thyroxine (T_4)

HO–[ring(I)]–O–[ring(I,I)]–CH_2–CH(NH_2)–COOH

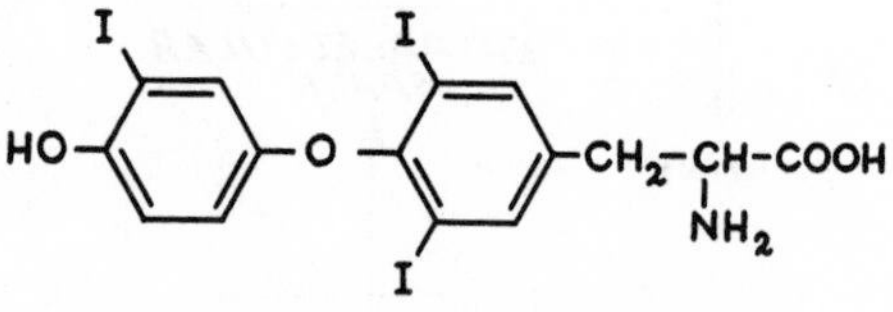

3,5,3′-Triiodothyronine (T_3)

The extensive structural changes in thyroglobulin to be expected following the combination of internal molecules of tyrosine do not occur. It is possible that the aromatic ring of a free iodotyrosine is transferred to the aromatic ring of an iodotyrosine located in thyroglobulin. A possible active form of free iodotyrosine is 3,5-diiodo-4-hydroxyphenylpyruvic acid, whose iodinated ring was shown to be incorporated into thyroxine when it was incubated with thyroglobulin.

Only 8–10% of the total tyrosine-bound iodine in the thyroid is in the form of thyroxine; most is found as the diiodotyrosine or monoiodotyrosine.

Thyroglobulin is broken down by the action of proteolytic enzymes from lysosomes or possibly vesicle membranes, and thyroxine and triiodothyronine are then released from the gland. The hydrolysis of thyroglobulin also liberates MIT and DIT. If these iodinated amino acids were lost from the gland, considerable amounts of iodide would be biologically unavailable for the synthesis of active hormone. However, particulate thyroidal deiodinase enzymes rapidly remove the iodide and permit its reutilization for the synthesis of new hormone (Fig 20–4). Approximately 1/3 of the total iodide in the thyroid is recycled in this manner. Both pituitary TSH and exposure to a cold environment stimulate thyroglobulin breakdown and the release of active hormone.

Besides thyroglobulin, some albumin-like and hormonally inactive 4 S iodoproteins can appear in the serum and contribute to the measured protein-bound iodine (PBI).

Within the plasma, thyroxine is transported almost entirely in association with 2 proteins, the so-called thyroxine-binding proteins, which act as specific carrier agents for the hormone. A glycoprotein (molecular weight 50,000 daltons) which migrates electrophoretically in a region between the $alpha_1$ and $alpha_2$ globulins is designated **thyroxine-binding globulin (TBG).** Another protein, **thyroxine-binding prealbumin (TBPA),** is detectable electrophoretically just ahead of the albumin fraction. When large amounts of thyroxine are present and the binding capacities of these specific carrier proteins are exceeded, thyroxine is bound to serum albumin. The comparative binding affinities for thyroxine and triiodothyronine are shown in Table 20–1. Approximately 0.05% of the circulating thyroxine is in the free, unbound state. This "free" thyroxine is the metabolically active hormone in the plasma.

TABLE 20–1. Binding of thyroid hormones to serum proteins.

Protein	Relative Binding	Affinity	Capacity
Thyroxine-binding globulin (TBG)	$T_4 > T_3$; tetrac = 0	High	Low
Thyroxine-binding prealbumin (TBPA)	Tetrac > T_4; T_3 = 0	Moderate	High
Serum albumin (SA)	Same for all	Low	Very high

Thyroxine can be dissociated from its binding proteins by competing anions such as diphenylhydantoin, salicylates, or dinitrophenol.

In a normal subject only about 1/3 of the maximum binding capacity is utilized; in a hyperthyroid patient, this may increase to ½ or more. In certain circumstances, TBG levels may become abnormal. An increase occurs in pregnancy as well as after the administration of estrogens. In nephrosis and after treatment with androgenic or anabolic steroids, decreased levels of TBG may occur. Generally, those circumstances which cause a decrease in TBG result in a reciprocal rise in TBPA. However, thyroxine is less tightly bound to TBPA and may quickly exchange to the "free" hormonal fraction. TBG should therefore be considered the stable reservoir for both T_3 and T_4.

The total circulating thyroid hormone is measured as the so-called **protein-bound iodine (PBI).** In euthyroid individuals, its concentration is 4–8 μg/100 ml.

When TBG levels are normal, the measured amounts of PBI correlate well with the expected metabolic results. Thus, in hyperthyroidism, PBI is elevated, and with a normal TBG level the amounts of TBG left unsaturated (as measured by the T_3 test, to be described below) would be low; free hormone is increased. Conversely, in hypothyroidism, when TBG is normal, the PBI would be lower than normal and the fraction of TBG remaining unsaturated would be high.

PBI is normally elevated in pregnancy, but the concomitant rise in TBG which was mentioned above causes a great amount to be bound. In consequence, "free" hormone, which is actually responsible for the metabolic end results of the thyroid hormone, remains normal and the patient is therefore euthyroid. Thus, an individual may be euthyroid even though the PBI is elevated, provided that there is a compensatory increase in TBG.

Unfortunately, the PBI procedure also measures nonhormonal iodine and iodotyrosines, which can limit its usefulness. If iodine has been given by mouth or if iodine-containing compounds (eg, Diodrast) have been used therapeutically or diagnostically, the PBI may remain high for some time. By extraction of the serum with n-butanol and subsequent washing of the extracts with an alkaline solution, it is possible to remove inorganic iodine and iodotyrosines. This modification is the basis for the **"butanol-extractable iodine" (BEI)** method for determination of the true hormonal iodine. In normal serum, the BEI level averages about 4–8 μg/100 ml, or 80% of that obtained by simple measurement of PBI without prior extraction. The BEI measures principally T_4 and T_3 and is not affected in those instances where the serum contains increased amounts of iodine due to administration of Lugol's solution or of potassium iodide. However, whereas inorganic iodides are not detected in the BEI test, organically combined iodine arising from the use of radiopaque dyes in x-ray contrast media will still interfere. In an effort to improve the procedure for measuring hormonal iodine when organic iodide contamination may be present, a method utilizing column chromatography for removal of contaminants has been proposed (West & others, 1965).

An indirect measure of thyroxine-binding protein (TBG) can be obtained by the use of a test based on the in vitro uptake of radioiodide (^{131}I)-labeled T_3 by adsorbents such as charcoal, resin, or the red blood cells. This T_3 test depends upon the observation that, when labeled T_3 is added to whole blood, the T_3 is taken up and bound by the carrier proteins. The more T_3 taken up by the serum carrier proteins, the less will be available for uptake by the adsorbent, and vice versa. Consequently, the amount of T_3 bound by the adsorbent is an inverse measure of the degree of saturation of thyroxine-binding protein in the serum. As noted above, the thyroxine-binding capacity is more saturated than normal in hyperthyroid patients. As a result, the residual serum-binding capacity for T_3 is decreased and adsorbent uptake is correspondingly increased. In hypothyroid patients, the reverse circumstances would prevail and uptake would be less than normal.

Total serum thyroxine can now be measured by a radiodisplacement method using TBG and labeled T_4 (Murphy, 1969).

The accumulation of inorganic iodide and its conversion to T_3 and thyroxine in the thyroid are completed over about a 48-hour period, but the labeled protein-bound iodine does not appear in the plasma for several days after the original administration of the isotope. About 80% of the organic iodine stored by the thyroid gland is thyroxine (T_4), and 20% is probably triiodothyronine (T_3). Because of its poor binding to serum proteins, "free" plasma levels of T_3 are, however, relatively high (1.5 ng/100 ml), approximately 25–35% of the levels of "free" thyroxine. T_3 is not only almost twice as active as T_4; its onset of action is also more rapid. This may be because it is loosely bound by the serum proteins and is cleared from the blood and penetrates into the cells more rapidly than T_4. T_3 is also more rapidly degraded in the body than T_4. T_4 is cleared slowly from plasma, with a half-time of 6–7 days.

Although the exact chemical mechanism of action of T_3 and T_4 is not known, iodine is required at the 3,5 position, suggesting that binding of T_3 and T_4 to tissue receptors involves this portion of the molecule.

Both T_4 and T_3 are metabolized in the peripheral tissues by deamination and decarboxylation to tetraiodothyroacetic acid (**tetrac**) or triiodothyroacetic acid (**triac**). These metabolites are about ¼ as active on a weight basis as their hormonal precursors, although their onset of action, according to some observers, is much more rapid. Both tetrac and triac have been used as agents to decrease serum cholesterol levels with a minimum of the less desired thyroxine actions.

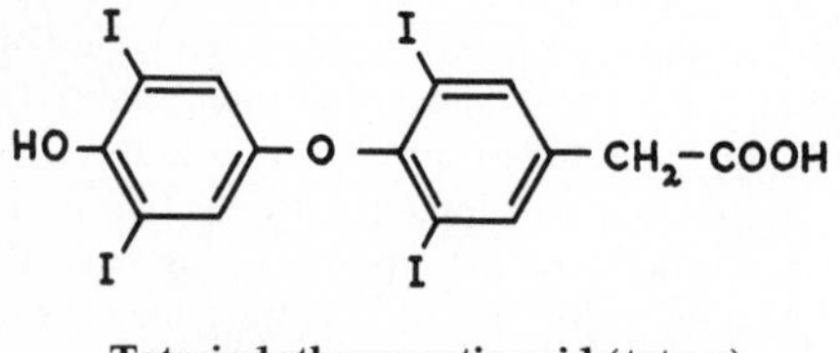

Tetraiodothyroacetic acid (tetrac)

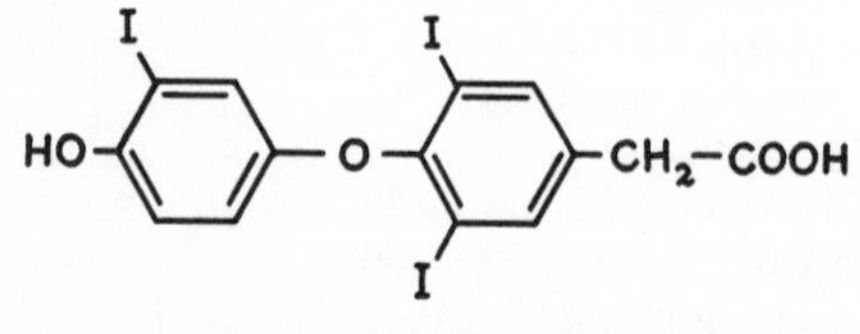

Triiodothyroacetic acid (triac)

Deiodination may also occur in the peripheral tissues, the liberated iodide being excreted in the urine. In the liver, thyroid hormone is rapidly conjugated with glucuronic acid and, to a lesser extent, with sulfate, and the inactive conjugates are excreted into the bile. Part of the conjugated thyroxine may be reabsorbed and transported to the kidney, where it may be deiodinated or excreted as intact conjugate.

Control of Thyroid Release

Thyroid-stimulating hormone (TSH) of the pituitary gland has a general activating effect on uptake and oxidation of iodine as well as on synthesis and secretion of thyroxine by the thyroid gland. The broad effect of TSH on these various metabolic steps may be due to its ability to increase the vascularity of the gland. Exposure to cold causes a release of thyroid hormone, but this is probably mediated through TSH release by the pituitary. Experimental lesions made in the hypothalamus can result in the stimulation of thyroid release, indicating the presence of a humoral factor in the hypothalamus which can activate the pituitary to secrete TSH. This factor (**thyroglobulin releasing factor, TRF**) has been identified in the sheep as a tripeptide.

Thyroxine is a feedback inhibitor of its own secretion. This inhibitory effect occurs at the pituitary by inhibition of TSH secretion; it may be due to a direct action of thyroxine which decreases the pituitary sensitivity to TRF. Except during cold exposure, TRF secretion is fairly constant. Thus, most regulation occurs by interaction in the pituitary between thyroxine, TRF, and TRF-destroying enzymes. There is some evidence (Kajihata & Kendall, 1967) that thyroxine can (to a lesser extent) also inhibit TRF secretion at the hypothalamic level.

The controlling effect of thyroxine on TSH release in the pituitary is blocked by substances inhibiting protein synthesis whereas the effect of TRF is not. Thus, the 2 agents operate by different mechanisms.

Iodine Metabolism

Since iodine is essential to the production of thyroid hormone and since this is the only known function of iodine in the body, iodine metabolism and thyroid function are closely related.

Abnormalities of Thyroid Function

A. Hypothyroid States: A deficiency of thyroid hormone produces a number of clinical states depending upon the degree of the deficiency and the age at which it occurs.

It has recently become clear that hypothyroidism may result from an inherited inability to synthesize thyroid hormone. In such cases marked physical and mental retardation may result unless the condition is recognized and treated (with thyroid hormone) within the first years of life.

Though comparatively rare, congenital defects in iodine uptake, organification, coupling, deiodination, and hormone secretion have all been described.

1. **Cretinism** results from the incomplete development or congenital absence of the thyroid gland.

2. **Childhood hypothyroidism (juvenile myxedema)** appears later in life than cretinism. It is generally less severe, and some of the typical cretinoid symptoms are absent. The most important signs of juvenile myxedema are stunting because of lack of bone growth, cessation of mental development, and, in some cases, changes in the skin as noted in cretinism.

3. **Myxedema** is caused by hypothyroidism in the adult; as such, it is the adult analogue of cretinism. The BMR and body temperature are lowered, and the patient complains of undue sensitivity to cold. Other characteristic findings include puffiness of the face and extremities, thickening and drying of the skin, falling hair, and, in some patients, obesity. Anemia and slowing of physical and mental reactions are also present.

4. **Hashimoto's disease** is a form of hypothyroidism in which all aspects of thyroid function may be impaired. This has now been established as an auto-immune disease in which the thyroid has been subjected to attack by cellular antibodies.

B. Hyperthyroid States and Goiter: Enlargement of the thyroid is called goiter. With the exception of malignant goiter and inflammatory diseases of the thyroid, as in thyroiditis, goiter involves either (1) simple enlargement of the gland (without hyperactivity) or (2) enlargement of the gland with associated hyperthyroidism. Hyperthyroid states caused by excessive secretion of TSH from the pituitary are rare.

1. **Simple (endemic or colloid) goiter** is a deficiency disease caused by an inadequate supply of iodine in the diet. Although treatment with iodine or with sodium iodide should be adequate for cases of simple goiter, thyroid hormone (L-thyroxine or desiccated thyroid) itself is more commonly used since there may be undetected defects in the pathway for synthesis of thyroxine, such as those, previously mentioned above, attributable to an inherited thyroid metabolic defect.

Simple goiter is common where the soil and water are low in iodine (eg, Great Lakes area). The use of iodized salt has done much to reduce the incidence of simple goiter.

2. **Toxic goiter (hyperthyroidism)** differs from simple goiter in that enlargement of the gland is accompanied by the secretion of excessive amounts of thyroid hormone, ie, hyperthyroidism together with enlargement (goiter). The term "toxic" does not refer to the secretion of the gland but to the toxic symptoms incident to the hyperthyroidism.

The most common form of hyperthyroidism is exophthalmic goiter (Graves' disease, Basedow's disease). The enlargement of the gland may be diffuse or nodular. In the latter case the terms **nodular toxic goiter** and **toxic adenoma** (Plummer's disease) are sometimes used, but the symptoms are the same since hyperthyroidism is, of course, common to both.

Hyperthyroidism occasionally occurs without goiter.

The most important signs and symptoms of hyperthyroidism, in addition to the goiter, are nervousness, easy fatigability, loss of weight in spite of adequate dietary intake, increased body temperature with excessive sweating, and an increase in the heart rate. A characteristic protrusion of the eyeballs (exophthalmos) usually accompanies hyperthyroidism, but this may be absent.

The causes of nodular toxic goiter and of diffuse goiter (Graves' disease) are probably different. In the serum of patients with Graves' disease there occurs a thyroid-stimulating protein factor which is immunologically different from thyroid-stimulating hormone (TSH) of the pituitary. The site of origin of this factor is not yet known, but it is apparently not the pituitary. When the factor is given intravenously it disappears from the circulation more slowly than does TSH. Qualitatively, it duplicates most actions of TSH, but its maximal thyroid-stimulating effect occurs many hours after that of TSH. For these reasons, it has been designated **long-acting thyroid stimulator (LATS).**

Current evidence indicates that LATS is an antibody developed as an auto-immune phenomenon against thyroid protein. Solomon & Beale (1967) have produced hyperthyroidism by immunizing animals with thyroid microsomes. LATS activity is neutralized by anti-gamma globulin antibodies but not by antibody to TSH. Many of the phenomena of Graves' disease, particularly the exophthalmos, may be related to this auto-immune reaction. In addition, transplacental transfer of LATS antibody from mother to fetus may be responsible for neonatal thyrotoxicosis. (Reviewed by Kriss, 1968.)

Laboratory Diagnosis of Thyroid Abnormalities

A. Hypothyroid States: A basal metabolic rate (BMR) below −30% suggests hypothyroidism, although this is not necessarily diagnostic of hypothyroidism alone; a PBI of less than 4 μg/100 ml and a ^{131}I uptake of less than 20% are also of diagnostic value. Serum cholesterol may be elevated. Urinary excretion of neutral 17-ketosteroids is usually reduced in myxedema.

B. Hyperthyroid States: Laboratory signs in hyperthyroidism include elevated BMR, blood hormonal iodine (PBI) exceeding 8 μg/100 ml, and increased pertechnetate or ^{131}I uptake (> 40%) measured over the thyroid gland. Lowered serum cholesterol (see p 298) and elevated serum and urinary creatinine may occasionally be observed.

Treatment of Hypothyroidism & Hyperthyroidism

A. Hypothyroid States: "Endemic cretinism," which is due to lack of iodine in the geographical area involved, is best treated prophylactically, eg, by administration of iodine to pregnant women or by the use of iodized salt. In most patients with actual occurrence of simple goiter, treatment with thyroid (L-thyroxine or thyroid extract) is preferred.

Desiccated thyroid or thyroxine is useful in cretinism, but it must be given in early infancy. It will maintain normal growth and physical development, but mental development usually remains inadequate, probably because of irreversible changes which occur during intrauterine life.

Thyroid is also used in the treatment of childhood hypothyroidism and of myxedema. When rapid response is necessary, sodium liothyronine (Cytomel) may be used.

B. Hyperthyroid States and Goiter:

1. **Subtotal thyroidectomy,** after preliminary treatment with iodine and propylthiouracil, causes permanent ablation but can often result in hypothyroidism.

2. **Radioactive iodine** (^{131}I) therapy consists of concentrating radioiodine solution in the gland and permitting the beta and gamma emissions of the solution to irradiate the tissue in the same way as x-rays, thus decreasing thyroid activity. Radioiodine is usually not effective in the treatment of malignant tumors since most such tumors do not concentrate the isotope. It remains the method of choice, however, for the treatment of hyperthyroidism in elderly patients when the glands are of normal size or only slightly enlarged. As with surgery, permanent hypothyroidism can occur.

3. **Iodine (Lugol's solution)** sometimes benefits hyperthyroid patients, at least briefly. The iodine depresses the activity of the gland, probably by decreasing TSH output. However, after a time the gland seems to "escape" from its influence since iodine no longer retards the production of TSH.

4. **Antithyroid drugs**–Certain compounds act as antithyroid agents, inhibiting the production of thyroxine by preventing the gland from incorporating inorganic iodide into the organic form. Examples of these antithyroid drugs are the goitrogens, thiouracil, propylthiouracil, methylthiouracil, carbimazole, thiourea, and methimazole (Tapazole, 1-methyl-2-mercaptoimidazole). Thiouracil is relatively toxic; the other compounds less so.

The antithyroid compounds have an almost immediate effect since they act during the early stages after iodide uptake by the gland when the iodine is still in the inorganic form. Under their influence the thyroid gland does not manufacture the hormone; the protein-bound iodine gradually falls, and the symptoms of hyperthyroidism gradually decrease.

Studies of thyroid glands which have been "blocked" with thiouracil show that iodide is trapped in the gland but not incorporated into the organic form. In the normal thyroid which has been blocked with thiouracil, the accumulation of iodide is equal to the amount in about 500 ml of plasma. However, in a thyrotoxic patient in which excess pituitary thyroid-stimulating hormone (TSH) activity accelerates the activity of the gland, the iodide accumulation within the blocked gland may be equivalent to that contained in as much as 35 liters of plasma, a gradient of iodide in the gland as compared to plasma of 500:1.

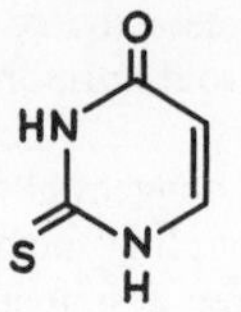

Thiouracil 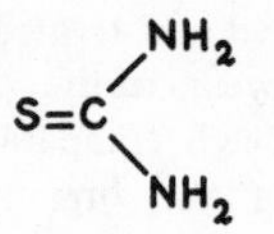 **Thiourea**

These agents may inhibit thyroid synthesis by preferentially combining with the active iodine and thereby preventing its incorporation into the tyrosine moieties.

A direct action to decrease maturation of thyroglobulin to its final polymeric form may also occur.

Iodide uptake can be prevented by agents which compete for the uptake mechanism. These include thiocyanate and perchlorate, the latter having less toxic side effects.

Thyrocalcitonin (Calcitonin)

Though now believed to be a hormone elucidated primarily by the thyroid gland, thyrocalcitonin's metabolic actions and historical interest relate closely with parathyroid hormone. It is therefore discussed in the following section.

THE PARATHYROIDS

The parathyroid glands are 4 small glands so closely associated with the thyroid that they remained unrecognized for some time and were often removed during thyroidectomy. In man, the parathyroids are reddish or yellowish-brown egg-shaped bodies 2–5 mm wide, 3–8 mm long, and 0.5–2 mm thick. The 4 glands together weigh about 0.05–0.3 g.

Function

The primary function of the parathyroid glands, mediated by their secretion of parathyroid hormone, is to maintain the concentration of ionized calcium in the plasma within the narrow range characteristic of this electrolyte despite wide variations in calcium intake, excretion, and deposition in bone. In addition to its effect on plasma ionized calcium, parathyroid hormone controls renal excretion of calcium and of phosphate.

Parathyroid hormone exerts a profound effect on the metabolism of calcium and phosphorus. There may be other metabolic effects which have not yet been clarified. The metabolism of calcium and phosphorus was described in Chapter 19.

Chemistry

The first stable crude parathyroid extracts were obtained by extracting bovine glands with hot dilute hydrochloric acid. Aurbach (1959) developed a method of extracting the active principle of the parathyroid glands using aqueous solutions of phenol.

The structure of bovine parathyroid hormone has been determined. It is a linear polypeptide consisting of 84 amino acids. The amino acid sequence of the polypeptide is shown below (Fig 20–5). Potts & others (1971) have reported the synthesis of a peptide consisting of the first 34 amino acid residues of the naturally occurring hormone, ie, the amino acids from the N-terminal (alanine) to the 34th amino acid (phenylalanine). The specific biologic effects of this synthetic 34-amino acid peptide on bone and kidney are qualitatively identical to those of the natural hormone. It therefore appears that the essential requirements for the physiologic actions of the peptide hormone on both skeletal and renal tissues are contained within the 34 N-terminal amino acids.

There is evidence that parathyroid hormone secreted in vivo is degraded rapidly in the circulation to smaller fragments. Thus, the hormone appears to be quickly metabolized, the half-life in the circulation of cows being about 18 minutes. The potency of the synthetic peptide (34 amino acids) relative to that of the naturally secreted 84-amino acid compound is greater in vitro than in vivo, suggesting that the portion of the peptide from amino acid No. 35 (valine) to the C-terminal may serve to protect the native peptide in the circulation from extremely rapid destruction.

Parathyroid extracts have been assayed biologically by their ability to increase the blood calcium in dogs or in rats after subcutaneous injection. In dogs, 1 unit of parathyroid activity is 0.01 of the amount necessary to raise the serum calcium by 1 mg/100 ml within 16–18 hours after injection. Other changes which may be observed in bioassays are a rapidly induced rise in excretion of urinary phosphate and of 3′,5′-cyclic AMP. Bioassay in vitro is based on activation of renal adenyl cyclase and increase in cyclic AMP in fetal rat skull tissue. Immunoassay (see p 460) for parathyroid hormone has also been developed for the purpose of detecting levels of the hormone in the circulation.

Physiology

The administration of parathyroid hormone (1) raises the serum calcium and lowers the serum phosphorus; (2) increases the elimination of calcium and phosphorus in the urine; (3) removes calcium from the bone, particularly if the dietary intake of calcium is inadequate; and (4) increases serum alkaline phosphatase if changes in bone have been produced.

The action of parathyroid hormone to increase bone resorption results in the release not only of calcium but also of collagenase, lysosomal enzymes, and hydroxyproline. In the kidney, it affects renal tubular reabsorption of calcium and reabsorption or secretion of phosphate. There is also evidence that

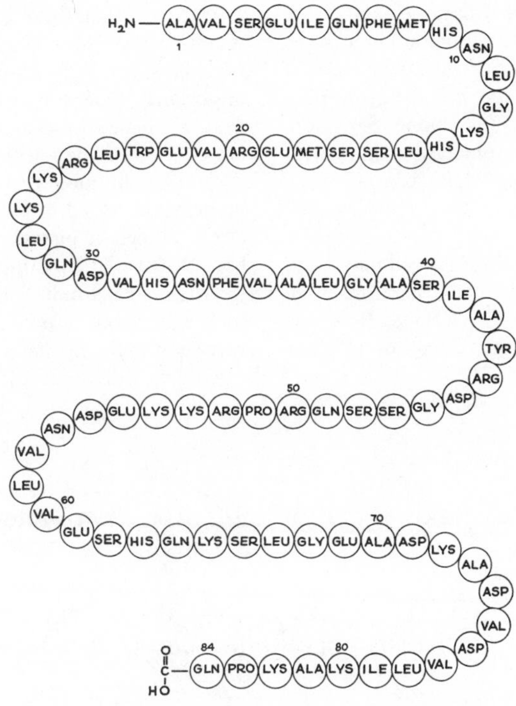

FIG 20–5. Structure of parathyroid hormone: A linear polypeptide of 84 amino acids. The physiologic activities of the peptide on both skeletal and renal tissues are contained within the 34 amino acids counting from the amino terminal end of the molecule.

parathyroid hormone increases the rate of absorption of calcium from the intestine, an effect which is, however, comparatively minor.

The actions of the parathyroid hormone on bone and kidney are independent processes, as indicated by the fact that the hormone effectively mobilizes bone calcium in nephrectomized animals as well as from bone tissue incubated in vitro. The mechanism by which parathyroid hormone stimulates bone resorption is not known. The hormone increases the amounts of both lactic and citric acid in the tissues and both of these acids could act to make bone soluble. However, the amount of acid produced appears insufficient to explain the degree of bone resorption observed.

Rasmussen & others (1964) have suggested that parathyroid hormone may act to stimulate protein synthesis in the osteoclasts, which, in turn, effect resorption of bone. This idea is supported by the observation that inhibition of RNA synthesis (and thus, indirectly, protein synthesis) by dactinomycin blocks the activity of the hormone in vivo. However, the rapid acute effect of parathyroid hormone on bone resorption is not inhibited by dactinomycin, indicating that at least part of the activity is independent of RNA and possibly of protein synthesis as well.

Sallis & others (1963) have demonstrated that parathyroid hormone can stimulate the transport of calcium and phosphate across the membranes of isolated mitochondria. Although these studies suggest that the hormone may act to influence ion transport in bone, the quoted experiments employed large concentrations of parathyroid hormone and were performed with mitochondria of liver or kidney—the former, at least, not being an important target organ of the hormone. Cyclic AMP may prove to be involved in the action of parathyroid hormone on bone resorption. Imidazole, which increases phosphodiesterase activity and thereby decreases cyclic AMP, causes hypocalcemia (Wells, 1968). Cyclic AMP itself causes increased resorption of calcium.

Parathyroid hormone requires also vitamin D to bring about its effects on bone. It is therefore comparatively inactive in subjects with rickets. However, the effect of the hormone on renal excretion of phosphate is not dependent on vitamin D.

Histologic evidence indicates that vitamin D and the hormone act by different mechanisms. The action of vitamin D is primarily on calcium ion transport systems both in the intestine and in bone.

Parathyroid hormone may affect organs other than bones and kidneys; indeed, patients with hyperparathyroidism may present without prominent renal or bone symptoms but with involvement of the central nervous system, the gastrointestinal tract, or the peripheral vascular system. Repeated doses of parathyroid extract may cause severe symptoms such as oliguria or anuria, anorexia, gastrointestinal hemorrhage, nausea and vomiting, and finally loss of consciousness and death. These symptoms are probably due to water and electrolyte depletion as well as to other changes in the body which are not clearly understood.

Control of Release of Parathyroid Hormone

In contrast to many protein hormones, parathyroid hormone is not stored in the gland; no storage granules are present. It is thus synthesized and secreted continuously.

Secretion of parathyroid hormone is subject to control by a negative feedback mechanism relating to the levels of ionized calcium in the plasma (Aurbach & Potts, 1967). These investigators have shown that parathyroid hormone concentrations are decreased abruptly by administration of calcium ion and rise when circulating ionized calcium is lowered by the administration of the chelating agent ethylenediaminetetraacetate (EDTA). Calcium loss associated with uremia also results in an increase in circulating parathyroid hormone. Although calcium appears to be an important homeostatic regulator of parathyroid secretion, a change in the amount of phosphate has no effect on hormone release.

Other hormones in addition to parathyroid hormone are involved in the regulation of plasma calcium levels. A specific plasma calcium lowering hormone has been postulated by Copp & others (1962) and termed by them **"calcitonin."** Evidence for the existence of such a hormone was obtained by thyroid-parathyroid perfusion experiments in the dog. Although Copp indicated that calcitonin was derived from the parathyroid

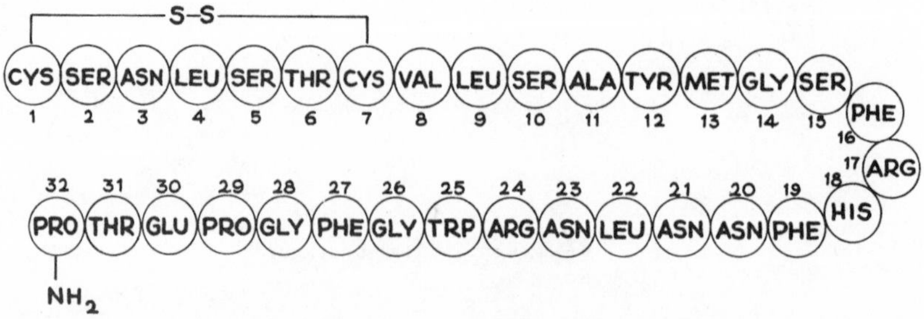

S–S (1–7)
CYS 1, SER 2, ASN 3, LEU 4, SER 5, THR 6, CYS 7, VAL 8, LEU 9, SER 10, ALA 11, TYR 12, MET 13, GLY 14, SER 15, PHE 16, ARG 17, HIS 18, PHE 19, ASN 20, ASN 21, LEU 22, ASN 23, ARG 24, TRP 25, GLY 26, PHE 27, GLY 28, PRO 29, GLU 30, THR 31, PRO 32
NH_2

Structure of porcine thyrocalcitonin.

glands, others—while confirming the existence of such a calcium lowering hormone—have indicated that it is probably derived from the thyroid gland. The designation **"thyrocalcitonin"** has been assigned to the calcium lowering hormone arising specifically from the thyroid gland. Its release is stimulated by high calcium ion levels.

Thyrocalcitonin is directly effective on bone, where it results in metabolic effects opposite to those of parathyroid hormone. Thyrocalcitonin inhibits bone resorption in vitro (Friedman & Raisz, 1965); studies with isotopic Ca^{++} indicate the hormone blocks release of bone calcium rather than facilitating Ca^{++} uptake. It acts on circulating calcium levels faster than parathyroid hormone, but the effects in general are quantitatively less as well as less prolonged. Thyrocalcitonin may therefore be specifically involved in maintaining the constancy of calcium ion in the plasma only when minor changes are involved.

In most species including man (Potts, and others, 1969) thyrocalcitonin is a peptide of molecular weight 3600 daltons (32 amino acids, see above). The fish hormone is particularly potent in man.

The clearance of thyrocalcitonin is 4–12 minutes, about twice the rate for parathyroid hormone.

Abnormalities of Parathyroid Function

A. Hypoparathyroidism: Usually this is the result of surgical removal of the parathyroids, although idiopathic hypoparathyroidism, of unknown etiology, has been reported. The symptoms of hypoparathyroidism include muscular weakness, tetany, and irritability. X-ray examination of the skull may reveal calcifications of the basal ganglia, and the bones may be denser than normal. Early cataracts may be detected by slit lamp examination of the eyes. If hypoparathyroidism begins early in childhood, there may be stunting of growth, defective tooth development, and mental retardation.

B. Hyperparathyroidism: An increase in parathyroid hormone production is usually due to a tumor of the gland (parathyroid adenoma). Decalcification of the bones, which results from hyperparathyroid activity, causes pain and deformities in the bones, including cystic lesions as well as spontaneous fractures. Anorexia, nausea, constipation, intractable peptic ulcer, polydipsia, and polyuria are other symptoms. Recurrent pancreatitis occurs in some patients. Deposits of calcium may form in the soft tissues, and renal stones may also occur. In fact, hyperparathyroidism should always be suspected in patients with chronic renal lithiasis. Deficiencies of magnesium may also result from long-continued hyperparathyroidism. Some extra-parathyroid tumors associated with hyperparathyroidism contain active material immunologically indistinguishable from parathyroid hormone.

Enlargement of the parathyroid glands (secondary hyperparathyroidism), probably as a result of increased serum phosphate levels, is often a feature of chronic renal disease.

Laboratory Diagnosis of Parathyroid Abnormalities

A. Hypoparathyroidism: Hypoparathyroidism is usually caused by accidental damage during thyroid surgery; it may also be caused by an auto-immune reaction. Serum calcium is low (less than 5–6 mg/100 ml), serum phosphate elevated (4–6 mg/100 ml or even higher), serum alkaline phosphatase normal, urinary calcium low to absent, and urinary phosphate low (tubular reabsorption of phosphate above 95%) in the absence of renal failure. Serum magnesium levels are reduced (1.5–1.8 mg/100 ml). Hydroxyproline levels are also reduced. A decrease in the normally low levels of circulating hormone (1 ng/ml) is difficult to detect.

B. Hyperparathyroidism: Serum calcium is high, serum phosphate low; serum alkaline phosphatase is usually normal but may be elevated. Urine calcium is increased as well as urine phosphate (tubular reabsorption of phosphate decreased). In chronic renal disease, with secondary hyperactivity of the parathyroid, serum calcium is low as a consequence of the high serum phosphate characteristic of renal failure. Urinary calcium and phosphate are both low, and there is resistance to the action of vitamin D in the presence of uremia and acidosis. As a result of associated bone abnormalities, serum alkaline phosphatase is often elevated. Radioimmunologically detectable parathyroid hormone is also elevated.

Treatment

A. Hypoparathyroidism: In the presence of tetany, prompt intravenous or intramuscular injection of calcium salts (calcium gluconate) and intramuscular injection of parathyroid hormone are required.

Parathyroid hormone is effective for only a short time, presumably because of the formation of an anti-hormone antibody. Dihydrotachysterol (Hytakerol) is also used in the maintenance management of hypoparathyroidism in conjunction with a high-calcium, low-phosphorus diet supplemented with oral calcium salts. Dihydrotachysterol is formed in the course of irradiation of ergosterol to produce vitamin D_2. Its action, similar to that of vitamin D, is to increase the absorption of calcium from the intestine. It also increases phosphate excretion by the kidney. Vitamin D_2 (calciferol) may be used in place of dihydrotachysterol and may, in fact, be preferred. However, the danger of toxicity due to accumulation of vitamin D after prolonged therapy must not be overlooked.

It is now apparent that therapy with vitamin D is likely to produce hypercalciuria because of the ability of the vitamin to decrease renal tubular reabsorption of calcium. The hypercalciuria occurs only in patients with true hypoparathyroidism and not in those with so-called **pseudohypoparathyroidism**, in which the disease is caused by a failure of the renal tubules to respond to the action of parathyroid hormone rather than to a deficiency of the hormone itself. Treatment must be guided by the results of analyses of the serum calcium, which should be maintained within the normal range.

B. Hyperparathyroidism: Hyperparathyroidism is usually associated with an adenoma of the parathyroid gland. Tumors arising from lung and kidney can also secrete parathyroid hormone-like material and produce clinical states similar to primary hyperparathyroidism.

Recent studies suggest that circulating parathyroid hormone in some hyperparathyroid states is immunologically different from the normal hormone.

THE PANCREAS

The endocrine function of the pancreas is localized in the islets of Langerhans, epithelial cells that are dispersed throughout the entire organ. Two hormones which affect carbohydrate metabolism are produced by the islet tissue: insulin by the beta cells, and glucagon by the alpha cells. A delta cell has also been described, but its function is not known.

INSULIN

Insulin plays an important role in general metabolism, causing increased carbohydrate metabolism, glycogen storage, fatty acid synthesis, amino acid uptake, and protein synthesis. It is thus an important anabolic hormone which acts on a variety of tissues including liver, fat, and muscle.

Chemistry

Insulin is a protein hormone which has been isolated from the pancreas and prepared in crystalline form. Crystallization of insulin requires traces of zinc, which may be a constituent of stored insulin since normal pancreatic tissue is relatively rich in this element.

Digestion of insulin protein with proteolytic enzymes inactivates the hormone, and for this reason it cannot be given orally. The structure of human insulin is shown in Fig 20–6. In all animals from which insulin has been obtained, the molecule consists of 2 chains connected by disulfide bridges. A third intradisulfide bridge also occurs on the A chain. Breaking the disulfide bonds with alkali or reducing agents inactivates insulin.

The structures of a number of insulins obtained from various animal sources have been elucidated (Smith, 1966). That of pork pancreas is the most similar to human insulin. The 2 insulins differ only in the terminal amino acid (number 30) of the B chain, which is alanine in porcine insulin and threonine in human insulin. This terminal amino acid is easily removed with carboxypeptidase, and the resulting altered molecule retains its biologic activity. Thus it is possible to convert porcine insulin into a compound with an active primary structure identical to human insulin save for the fact that the altered insulin is shorter by one amino acid in the B chain.

Insulins from the pig, whale, and dog are structurally identical. Those from the sheep, horse, and cow differ from porcine insulin only in 3 amino acids under the disulfide bridge in the A chain. Other species may differ in as much as 29 out of the 51 amino acids. Two insulins have been isolated from a single rat pancreas,

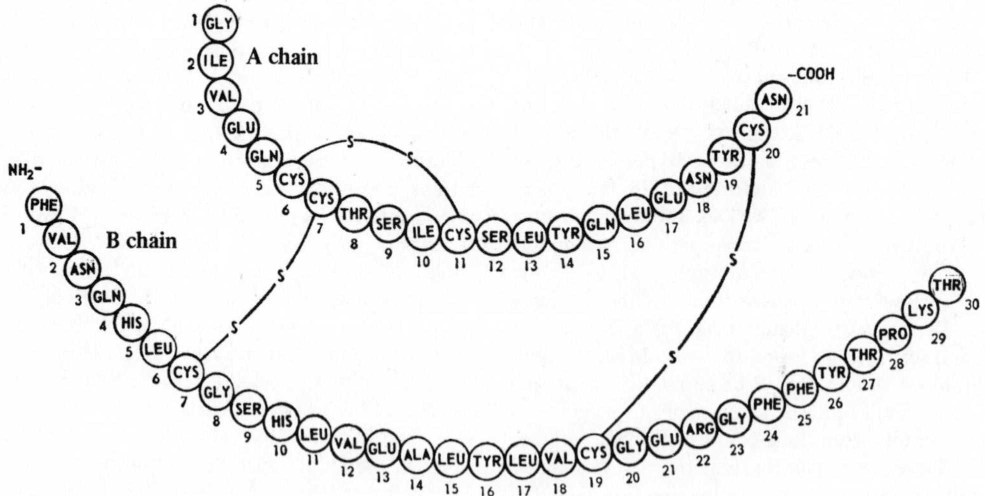

FIG 20–6. Structure of human insulin. (Nicol & Smith, 1960.)

differing by a single amino acid (lysine or methionine) in the A chain. The pancreas of certain fish contains more than one insulin, differences being found in both the A and B chains.

Despite the wide variation in primary amino acid structure, the biologic activity per unit weight is remarkably constant for all insulins.

Although the minimum calculated molecular weight of insulin is 5734 daltons, it can exist in different polymeric forms depending on the temperature, concentration, and pH. Most estimates of molecular weight by physical measurements range from 12,000–48,000 daltons. Molecular weight is also dependent upon the zinc content of the molecule, increasing to 300,000 daltons with elevated zinc concentrations. The molecular weight of insulin under normal physiologic conditions is not certain. Whether insulin can exist in more than one polymeric form under different physiologic conditions is also not established.

The secondary and tertiary structures of bovine insulin have been determined by x-ray crystallography (Adams & others, 1969). These studies indicate that the A chain portion of the molecule is the more exposed, including, the 6–11 disulfide bridge, possibly involved in hormonal activity (see below). The B chain is in the internal portion of the molecule; noncovalent binding between B chains is responsible for the formation of the insulin dimer and probably higher polymers. The crystallized hexamer of insulin forms around zinc molecules which are held in place by imidazoles (on histidines) in the B chain.

Attempts have been made to determine the core of biologic activity in insulin by the controlled modification of side chains and by partial degradation of the molecule with proteolytic enzymes. Amino groups at the terminal amino acids or the epsilon amino groups of the lysine residues are not required since their acetylation causes no loss in biologic activity. Similarly, the removal of the amide from the terminal asparagine on the A chain (Fig 20–6) has no effect, though removal of the aspartic acid itself causes complete loss of activity. Half of the theoretical number of carboxyl groups can be esterified without loss, but further esterification results in an inactive product. Removal of the terminal octapeptide on the B chain inactivates insulin, although removal of the terminal carboxyamino acid of the B chain and some amino acids in the amino terminal portion of the molecule does not affect activity. The hydroxyl groups of serine and threonine can be modified by sulfation with little loss of potency, while destruction of the histidines by photo-oxidation results in inactivation. Although most side chains of insulin can be modified without interfering with activity, reaction at the same sites with molecules large enough to produce steric hindrance can cause inactivation.

Most of the modifications which result in inactivation of insulin cause large changes in the secondary and tertiary structure as demonstrated by altered sedimentation rate and optical rotary dispersion. Thus, inhibition of insulin by an agent acting at a specific site may result from the gross nonspecific changes induced and may not indicate the site to be an essential part of the core activity.

Iodination of tyrosine residues (usually those of the exposed A chain) up to one atom per mol has little effect on the biologic activity of insulin, but increasing iodination causes progressive inactivation. Therefore, in the preparation of radioiodine tagged insulin (used as a biologic and immunologic tracer), iodination must be restricted to less than one atom per mol.

Altered insulins are also used clinically. Sulfated bovine insulin, porcine insulin (intact or dalanated* at amino acid 30 of the B chain), and fish insulins are used in resistant diabetics because of their reduced antigenicity and cross-reactivity with circulating antibody. A highly purified insulin **(monocomponent insulin)** with greatly reduced immunogenicity has recently been prepared. This insulin may ultimately prove preferable for routine clinical use.

Other modifications reduce the absorption of insulin from the injected sites, thus prolonging the action of the hormone. These preparations have the occasional disadvantage of being more immunogenic than crystalline insulin. **Protamine zinc insulin** is a combination of insulin with protamine which is absorbed more slowly than ordinary insulin; one injection of protamine zinc insulin may lower the blood glucose for more than 24 hours, whereas 2 or 3 injections of regular (crystalline) insulin might be required for the same effect.

Globin insulin, another combination of insulin with a protein (in this case globin) has an effect somewhere between those of regular and protamine insulin (a 12- to 15-hour duration of action).

Ultralente insulin is a slow-acting insulin prepared by controlled crystallization in the presence of high concentrations of zinc and acetate in order to produce large crystals which are therefore slowly absorbed. **Lente insulin** is a 7:3 mixture of ultralente and regular insulin which has a duration of effect between the two.

Assay of Insulin

Insulin preparations are standardized by measuring their effect on the blood glucose of rabbits. One unit of insulin is the amount required to reduce the blood glucose level of a normal 2 kg rabbit after a 24-hour fast from 120–45 mg/100 ml. The international standard contains 24 units/mg recrystallized insulin. Most preparations of crystalline insulin average 25 units/mg. Commercial insulin is usually sold in 2 concentrations–U40 and U80, the numbers referring to units/ml.

Measuring the hypoglycemic effect of insulin in the intact animal as described above is a satisfactory method for assay of the potency of purified preparations of the hormone. However, it is much too insensi-

*Dalanated insulin is now the accepted term for insulin with alanine removed (dealaninated) at position 30 of the B chain.

tive and nonspecific for measurement of the small amounts of insulin which may be present in the circulating blood. Consequently, several other more sensitive assay technics have been developed. In the rat diaphragm assay, the effects on glucose metabolism in the rat diaphragm incubated in vitro are observed after addition of plasma or of known amounts of insulin. Parameters of glucose metabolism measured variably include glucose uptake, $^{14}CO_2$ production from ^{14}C glucose, and glycogen deposition, the latter being the most specific for insulin action. The method measures all of the activity attributable to insulin, referred to as "insulin-like activity" (ILA). It is of interest that, with use of the rat diaphragm assay, the apparent insulin content of the plasma progressively increases as the plasma is more and more diluted. This phenomenon has been explained by Vallance-Owen (1960) as being due to the presence of an inhibitor substance, associated with the serum albumin (synalbumin), which becomes a less effective inhibitor as plasma is diluted. Synalbumin has been suggested as a genetic marker for diabetics. However, the chemical nature of synalbumin—or, indeed, its actual existence—is still unsettled.

Another in vitro bioassay for insulin-like activity (ILA) is based upon the effect of insulin or plasma on glucose metabolism in the epididymal fat pad of the rat. Minimum sensitivity of the method and the effects of plasma dilution are similar to those observed with the rat diaphragm, though it may be less specific for the insulin molecule. ILA measured with the fat pad may be affected by the presence of other hormones which affect glucose metabolism (eg, growth hormone or epinephrine). Furthermore, the activity of plasma is not impaired completely by insulin antisera and there is some persistence of activity after pancreatectomy. It has also been noted that the mean ILA measured with the fat pad is higher than when measured with the rat diaphragm. This may relate to the existence in the serum of a "bound" form of an insulin-like substance which the fat pad is capable of utilizing but which the diaphragm cannot (Antoniades & Gundersen, 1961). At pH 9.8, or in the presence of heparin, or after incubation with an extract of fat, the ILA factor can be released and is then active on diaphragm. Others have defined that insulin-like activity of plasma which cannot be blocked by insulin antibody as "nonsuppressible" or "atypical" insulin and that portion of the activity which is inhibited by antibody as "suppressible" or "typical" insulin. Levels of suppressible or typical insulin rise during glucose stimulation, are depressed after pancreatectomy and, in general, seem to reflect levels of circulating insulin. Levels of nonsuppressible or atypical insulin tend to be unchanged. At present it is unclear whether "bound," "nonsuppressible," or "atypical" insulins actually represent protein-bound insulin or insulin in any form. However, they may be of physiologic significance as noninsulin humoral factors facilitating carbohydrate metabolism. Whatever their role, they are not present in sufficient quantities to prevent ketosis in animals whose insulin supply is removed by pancreatectomy.

Radioimmunochemical assays are more sensitive than the previously described bioassays since they permit detection of insulin in concentrations less than 1 μU/ml. The insulin content of serum in the fasting state is reported to be about 25 μU/ml when measured immunochemically. In recent years this method in one form or another has gained ascendancy over the less specific and less reproducible bio-assays. The principles of the method have been extended to the measurement of a variety of other protein hormones. One defect of the radioimmunoassay is the fact that it may measure fragments or precursors of insulin in plasma. Those substances may retain some immunologic activity but would be biologically inert. (See section below.)

Biosynthesis of Insulin*

In the beta cells of the pancreas, insulin is synthesized as is any other protein by the ribosomes of the endoplasmic reticulum. Initially, it was suggested that the 2 chains were synthesized independently and subsequently combined by formation of connecting disulfide bonds. Recently, however, a precursor of insulin **(proinsulin)**, has been isolated and purified after gel filtration of pancreatic extracts or of commercial crystalline insulin (Steiner & Cunningham, 1967).

The structure of porcine proinsulin is shown in Fig 20–7. The molecule consists of a single polypeptide chain which begins with the normal B chain sequence at its amino terminus but contains a linking polypeptide of 33 amino acids which connects the carboxy terminus of the B chain to the amino terminus of the A chain amino acid sequence. The molecular weight of porcine proinsulin is 9082 daltons, about 50% greater than that of insulin. The connecting link is about the same size in proinsulins from other species, but it varies in specific amino acid content. The molecule is almost biologically inactive, but it can cross-react with antisera prepared against insulin. Proinsulin, after reduction to its open chain structure, is readily reconverted in high yield to the proper disulfide configuration with mild oxidation. Since the yields are higher than normally obtained with free A or B chains, the connecting link appears important in providing the proper alignment of the molecule for correct disulfide synthesis. Incubation of proinsulin with trypsin removes the linking peptide and liberates a completely biologically active product. Cleavage is not exact, however, and the final product, after trypsinization, is not insulin but dalanated insulin. Therefore, the biologic mechanism by which the activation of proinsulin to insulin occurs requires proteolysis but may employ enzymes other than or in conjunction with trypsin itself. The conversion of proinsulin to insulin occurs in the granule package, not in the endoplasmic reticulum, where proinsulin is synthesized. It may be that lysosomal proteolytic enzymes perform the conversion.

During biologic proteolysis, the 2 basic amino acids at either end of the connecting peptide (C-peptide) are removed (Arg 31, 32, and Lys 62, Arg

*For a detailed review, see Grodsky & Forsham (1966).

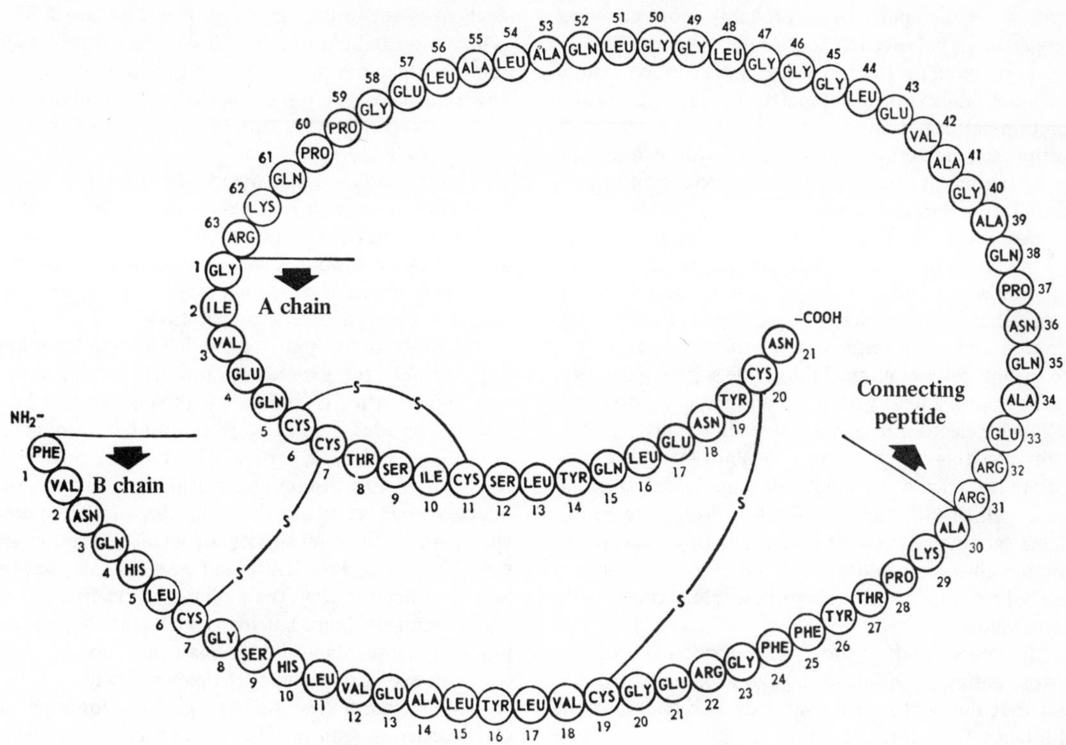

FIG 20–7. Structure of porcine proinsulin. (Reproduced, with permission, from Chance & others: Science 161:166, 1968.)

63). The free biologically inactive C-peptide is retained in the granule and is ultimately secreted in equal molar ratio with the insulin.

Proinsulin can be measured by special specific radioimmunoassays or after separation from insulin by "molecular sieving" chromatography. Normally, proinsulin represents only a small proportion of the insulin stored in the pancreas or found in the plasma. (Proinsulin in the plasma has been referred to as "big insulin," by Roth & others, 1968.) Plasma proinsulin is not elevated in human diabetics or in normals after glucose stimulation, but it may be the predominant circulating form in some subjects with islet cell tumors.

After conversion, the insulin inside the vesicle package condenses and forms the typical beta granules enclosed by membranous sacs. Current evidence suggests that stored insulin may exist in different "compartments" which differ in their sensitivity to stimulating agents (Grodsky & others, 1969).

Insulin Secretion

Approximately 50 units of insulin per day are required; this is about 1/5 of the amount stored in the human pancreas.

The secretion process for stored insulin after stimulation by glucose or tolbutamide has been visualized by electron microscopy (Lacy, 1964). It is similar to the secretion of other proteins stored as granules, including those found in pituitary and pancreatic acinar cells. During secretion, the granules move to the plasma membrane of the cell, where the granule surface membrane fuses with the cell membrane. The fused membranes then rupture, and the granular contents are liberated into the pericapillary space. This process is called **emeiocytosis.** Lacy & others (1969) observed microtubules to which some granules are attached and suggested that they facilitate or direct this granular movement. Agents which destroy microtubules (vincristine, deuterium oxide) inhibit insulin secretion.

Since secretion of insulin is the resultant of a variety of phenomena, agents may influence secretion at different levels. Those stimulating release from labile storage compartments cause insulin secretion in seconds; those acting on insulinogenesis or provision of insulin to the labile compartments may require 15–60 minutes. Agents acting on more than one phenomenon can produce multiphasic patterns of insulin release (Fig 20–8).

Glucose directly stimulates insulin release within 30–60 seconds, both in vivo and in vitro. It acts directly on the mechanism controlling the release of stored insulin, and the effect is not inhibited by complete blocking of insulinogenesis with dinitrophenol or puromycin (Grodsky & others, 1963). In addition, glucose also stimulates insulin synthesis, an effect which requires a longer period and probably more chemical energy than that needed for insulin release. Glucose may also stimulate the provision of preformed insulin to a more labile storage compartment. Thus, an insulin secretion pattern during constant glucose stimulation is multiphasic, reflecting the release of stored hormone during the initial phase and insulin production or provision during the later phase.

An intermediate produced in the course of glucose metabolism, rather than glucose itself, may be responsible for stimulation of insulin secretion. In vitro, other sugars which are readily metabolized—eg, mannose and, to a lesser extent, fructose—can stimulate insulin release; nonmetabolizable sugars, galactose, L-arabinose, 2-deoxyglucose, and xylose do not. Although galactose may cause a rise in circulating insulin in vivo, this may be due to the ability of this carbohydrate to release an unknown hepatic or intestinal factor which, in turn, facilitates insulin secretion (Goetz & others, 1964). Stimulation of insulin secretion by glucose is blocked by inhibitors of glucose metabolism such as mannoheptulose, glucosamine, and 2-deoxyglucose.

Evidence that xylitol, an intermediate in the pentose pathway, stimulates insulin release in vitro suggests that the pentose pathway may also be important (Montague & Taylor, 1968).

In contrast to the "metabolite" theory, Matchinsky & others (1970) failed to observe any immediate increase of intracellular metabolites during glucose stimulation; this suggested that glucose itself may act directly on a receptor site at the membrane level to stimulate the initial release phase.

Nucleotide metabolism also plays a significant role since insulin secretion can be stimulated by cyclic AMP and hormones or agents such as theophylline which increase intracellular cyclic AMP (Fig 20–2).

Although inhibition of glucose metabolism will inhibit the glucose stimulatory effect, the pancreatic mechanism for insulin secretion is not impaired. Thus, tolbutamide and citrate can still stimulate insulin secretion when glucose metabolism is blocked. In man, this may explain why tolbutamide can act in mild diabetics in whom hyperglycemia alone is not an effective stimulus for insulin secretion.

Fatty acids, such as octanoate, stimulate the pancreas, but the role of the higher fatty acids in physiologic concentrations is not established. The influence of ketones on insulin secretion also is not clear, variably being reported as stimulatory, inactive, or inhibitory in in vitro pancreatic preparations.

Amino acids, particularly leucine and arginine, can stimulate the pancreas to produce insulin both in vivo and in vitro. In children with spontaneous hypoglycemic episodes and in subjects with functioning islet cell tumors, leucine is particularly effective in causing a rise in circulating insulin. Pretreatment of subjects with hypoglycemic agents such as insulin or tolbutamide will increase their sensitivity to leucine. The metabolites of leucine, isovalerate and acetoacetate, have no effect, indicating that the amino acid itself is the primary stimulant. Since leucine can decrease hepatic output of glucose, its action in man may not be exclusively, or even primarily, at the pancreatic level.

Many hormones, such as growth hormone and glucocorticoids, can produce an increase in circulating insulin when administered to the intact animal but are inactive in vitro. Since these agents also cause hyperglycemia, it is possible that they act on the pancreas primarily by way of increasing stimulation by glucose rather than by a direct effect. ACTH and TSH in vivo or in vitro result in a slow release of insulin. Glucagon in vivo in man or in vitro stimulates insulin release under conditions where its hyperglycemic effect is minimized. It may act primarily in the pancreas to increase cyclic AMP, which in turn facilitates the action of glucose and other agents which stimulate insulin secretion. Thus, glucagon is effective only in

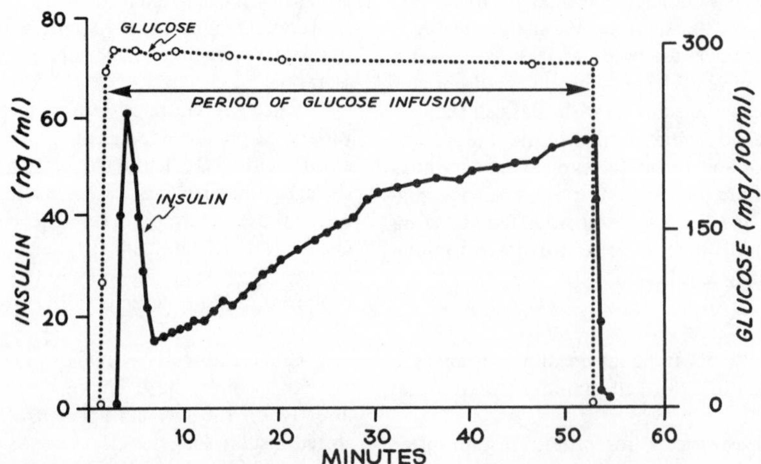

FIG 20–8. Multiphasic response of the in vitro perfused pancreas during constant stimulation with glucose. (Modified from Grodsky & others, 1969.)

the presence of a primary stimulant such as amino acids or glucose (or its precursor, glycogen) (Malaisse & others, 1967). The in vitro effects of glucagon on insulin release usually require high concentrations. At physiologic levels in vivo, the major effect of glucagon is probably exerted on the liver.

Epinephrine, both in vivo and in vitro, is a potent and highly effective inhibitor of insulin secretion regardless of the blood glucose concentration. Thus, under extreme stress, epinephrine not only provides glucose to the circulation by glycogenolysis but preferentially preserves it for utilization by the brain since it simultaneously depresses insulin release. At the same time, it supplies fatty acids mobilized from adipose tissue to provide the major fuel for the exercising muscle.

Insulin secretion is stimulated by agents increasing beta-adrenergic action and is inhibited by agents stimulating the alpha-adrenergic systems. Epinephrine is both a beta- and alpha-adrenergic stimulator. Presumably, the alpha stimulation in the islet tissue is responsible for its inhibitory effect. When the alpha-adrenergic action is blocked with phentolamine, the beta stimulating action persists; under these circumstances, epinephrine actually increases insulin release (Porte & others, 1966).

Zinc is usually found associated with insulin in the beta cells and the amount of zinc in those cells declines after administration of glucose. Alloxan also causes a parallel decline in zinc and insulin stores. These observations indicate a role of zinc in insulin storage or release. However, this element is not present in the beta cells of all species. Insulin secretion is intimately controlled by certain other cations. In vitro, calcium is an absolute requirement for insulin secretion regardless of glucose concentration. Its uptake by the beta cells may be a necessary first step prior to release. Magnesium is inhibitory. A decrease in the sodium pump or an increase of potassium ion causes an immediate secretion of insulin from the pancreas in vitro. In certain hypertensive diabetics with hypokalemia and impaired insulin secretion, replacement of potassium has improved the pancreatic sensitivity to normal stimulation.

Insulin release can also be indirectly influenced by the central nervous system. Lesions of the ventral medial nucleus or stimulation of the vagus increases insulin release. Finally, the sensitivity of pancreas to the above stimuli may vary with the developmental state. For example, pancreas of fish, amphibians, ruminants, and the mammalian fetus are remarkably insensitive to glucose though they may respond normally to amino acids or other stimulants.

Hypoglycemic Agents

There are several hypoglycemic drugs, effective when taken by mouth, that are useful for the control of diabetes. One class of these drugs, the sulfonylureas, are not hypoglycemic in alloxanized diabetic animals, in animals or patients after pancreatectomy, or in juvenile diabetics whose pancreas contains little or no insulin. They are useful in the treatment of diabetics of the "adult-onset" type who have retained some pancreatic function. The orally effective sulfonamide agents most extensively used are tolbutamide (Orinase), chlorpropamide (Diabinese), and tolazamide (Tolinase), the last 2 being long acting (Fig 20–9).

Although tolbutamide may have some peripheral effects on glucose metabolism, its primary action appears to be on the pancreas. In vitro, tolbutamide causes the prompt release of insulin from pancreas even under conditions where the glucose stimulation of insulin secretion has been blocked by mannoheptulose or glucosamine. Tolbutamide may therefore act on insulin secretion by a different mechanism than that of glucose, which may explain its effectiveness in maturity-onset diabetes or in patients with islet cell tumors in whom the pancreas does not respond normally to glucose. In contrast to glucose, tolbutamide does not acutely stimulate insulin synthesis. With long-term treatment, it may have, however, a betatropic effect.

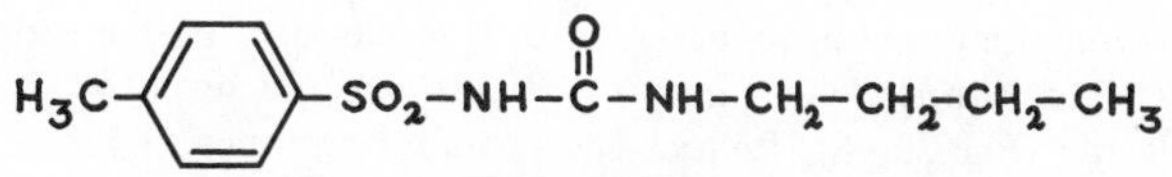

$$H_3C-C_6H_4-SO_2-NH-\overset{O}{\overset{\|}{C}}-NH-CH_2-CH_2-CH_2-CH_3$$

Tolbutamide

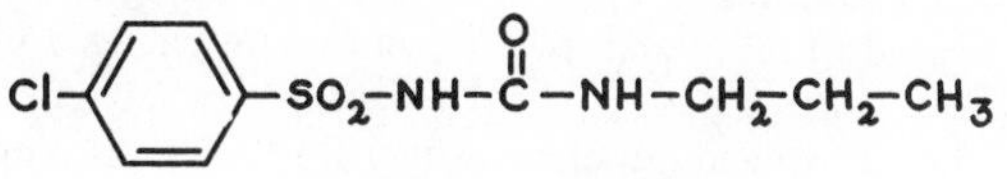

$$Cl-C_6H_4-SO_2-NH-\overset{O}{\overset{\|}{C}}-NH-CH_2-CH_2-CH_3$$

Chlorpropamide

$$C_6H_5-CH_2-CH_2-NH-\overset{NH}{\overset{\|}{C}}-NH-\overset{NH}{\overset{\|}{C}}-NH_2$$

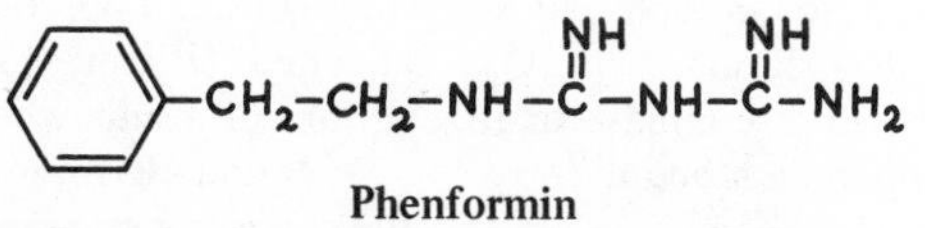

Phenformin

FIG 20–9. Orally effective hypoglycemic agents.

Phenethyl biguanide (phenformin, DBI) and chemically related drugs also produce hypoglycemia. The biguanides are active in severe diabetes as well as alloxanized or pancreatectomized animals. In contrast to the sulfonylureas, the action of these hypoglycemic agents is exerted on the peripheral tissues. In muscle, glucose uptake is increased, partly as a result of an acceleration of glycolysis due to an uncoupling action on oxidative phosphorylation. In addition, the biguanides may produce hypoglycemia by decreasing gluconeogenesis in the liver and by decreasing intestinal glucose absorption.

An increase in serum lactate levels may often result from biguanide treatment, presumably as a result of its uncoupling action in muscle.

Intestinal Factors

Oral glucose tolerance tests cause greater insulin secretion than a comparable intravenous glucose tolerance test, even though blood glucose levels are usually higher in the latter case. Recent evidence indicates that glucose administered orally stimulates the release of intestinal factors which in turn act on the pancreas. Gastrin, pancreozymin, secretin, and a glucagon-like substance are 4 such substances which are found in the intestine that can directly stimulate insulin secretion both in vivo and in vitro.

Transport of Insulin

In connection with observations on the apparent insulin content of the serum when tested in vitro by use of the epididymal fat pad, it was suggested that a bound form of insulin existed in plasma which was active on adipose tissue but not on diaphragm. Such a bound form of insulin may represent a specific insulin-transporting protein. However, the existence of "bound insulin" remains highly controversial. Binding of crystalline insulin cannot be demonstrated in normal or diabetic serums by ultracentrifugal technics.

Metabolism of Insulin

Insulin is degraded primarily in liver and kidney by the enzyme **glutathione insulin transhydrogenase** (Tomizawa, 1962). This enzyme brings about reductive cleavage of the S–S bonds which connect the A and B chains of the insulin molecule (Fig 20–6). Reduced glutathione, acting as a coenzyme for the transhydrogenase, donates the H atoms for the reduction and is itself thus converted to oxidized glutathione. After insulin has been reductively cleaved, the A and B chains are further degraded by proteolysis.

Not only the liver but also the kidney and muscle possess active insulin-degrading systems. In plasma, however, insulin destruction proceeds only at a very slow rate. When insulin is bound to antibody, it is much less sensitive to enzymatic degradation.

The insulin inactivating systems are rapid-acting; the half-life of circulating insulin in man is about 7–15 minutes.

Physiology: Mode of Action of Insulin

Despite the fact that insulin has for many years been available as a comparative pure protein hormone, its primary site of action (if indeed there is a single primary site) is still virtually unknown. Insulin acts in such a variety of ways that it is difficult to establish whether a given effect is a primary or a secondary one. In addition, observations made in vivo can be misleading since a change in the metabolism of one tissue may occur as a result of the ability of insulin to influence the provision of metabolic substrates or inhibitors from a completely different tissue. In one or another tissue, insulin exhibits all of the activities ascribed to hormones, including transport at the membrane site, RNA synthesis at the nuclear site, translation at the ribosome for protein synthesis, and an influence on tissue levels of cyclic AMP. Insulin is active in skeletal and heart muscle, adipose tissue, liver, the lens of the eye, and possibly, leukocytes. It is comparatively inactive in renal tissue, red blood cells, and the gastrointestinal tract. The major metabolic actions of insulin are centered in the muscle, adipose tissue, and liver.

Muscle & Adipose Tissue

A primary effect of insulin in muscle and adipose tissue is to facilitate transport of a variety of substances. These include glucose and related monosaccharides, amino acids, potassium ion, nucleosides, and inorganic phosphate. These effects are not secondary to glucose metabolism since they can be demonstrated in in vitro systems when glucose is not present. Insulin does not have to enter the cell to activate transport; Cuatrecasas (1970) found that insulin bound covalently to large inert particles was fully active on the much smaller fat cells.

Studies by Edelman & Schwartz (1966) indicate that insulin incubated with muscle tissue is rapidly concentrated in the sarcolemma; little, if any, reaches the cell nucleus. Although the exact nature of the sarcolemma is not known, it may possess a tubular component which can be activated by insulin, thereby permitting selective transport to occur.

Insulin is firmly bound to the membrane fractions of muscle and adipose tissue. This binding is to a highly specific receptor site, since other proteins and protein hormones of similar size do not compete. The amount of membrane-bound insulin quantitatively parallels its biologic activity in the tissue, which suggests that binding is a requisite of hormone activity. Insulin is not changed in this process, as evidenced by the fact that, when reextracted from the membranes to which it is bound, it remains fully potent and immunologically fully reactive. Further evidence that insulin acts at the membrane level is the recent observation that certain lipolytic (Rodbell, 1966) or proteolytic enzymes (Rieser, 1966) can duplicate many of the actions of insulin when incubated with intact cells.

In muscle or adipose tissue, uptake of glucose by the cell is the rate-limiting step for all subsequent intracellular glucose metabolism. The ability of insulin to facilitate transport thus leads to an increase in all pathways of glucose metabolism, including glycogen deposition, stimulation of the HMP shunt resulting in increased production of NADPH, increased glycolysis, increased oxidation (reflected by an increase in oxygen uptake and CO_2 production), and increased fatty acid synthesis. In adipose tissue, insulin increases lipid synthesis by providing acetyl-Co A and NADPH required for fatty acid synthesis, as well as the glycerol moiety (glycerophosphate) for triglyceride synthesis.

The action of insulin on carbohydrate transport does not require subsequent intracellular metabolism of the sugar. Insulin will increase transport and facilitate an increase in intracellular concentration of nonmetabolizable sugars such as L-arabinose and xylose, as well as galactose, which is metabolized only in the liver. The hormone promotes the entry into the cells of those sugars possessing the same configuration as D-glucose at carbons 1, 2, and 3. Fructose does not require insulin for transport into the cells, possibly because of the ketone group at position 2. Morgan & others (1964) have described a theoretical carrier transport system for glucose in muscle which may involve a transient combination of the sugar with a component of the membrane transport system. Polymers of glucose or complexes of glucose with amino acids or peptides have been observed to result from insulin action. These have been postulated to be the chemical components of the carrier system. Intracellular transport of glucose is enhanced by anoxia or uncoupling agents such as dinitrophenol. This indicates that exclusion of glucose from muscle or adipose tissue may require energy. Glucose uptake after insulin administration can be demonstrated within a few minutes. Furthermore, it occurs in the presence of dactinomycin. Therefore, the insulin-mediated transport system does not require enzyme induction or the synthesis of RNA.

As previously stated, insulin increases the uptake of amino acids into muscle in the absence of glucose. This effect is not secondary to stimulation of protein synthesis since the effect can be observed when all protein synthesis is blocked with puromycin. Insulin also increases the uptake of nonmetabolizable amino acids such as alpha-aminoisobutyrate.

In addition to its effect on transport in muscle and adipose tissue, insulin may also activate certain enzymes. Activation of hexokinase by insulin has been indicated by indirect studies but a direct effect on a purified preparation of hexokinase is difficult to verify, particularly when physiologic concentrations of insulin are employed. Bessman (1966) has suggested that insulin may act to translocate hexokinase to a site at the mitochondrial level where the enzyme may serve to increase respiration by producing ADP. This highly speculative hypothesis has received little experimental support as yet. Incubation of muscle tissue with insulin increases glycogen synthetase activity. Until this effect is demonstrated on a partially purified enzyme, there remains the possibility that the change observed was the result of insulin action at a completely different site.

An effect of insulin in connection with oxidative phosphorylation has been proposed. Hall & others (1960) showed a definite defect in oxidative phosphorylation in diabetic mitochondria which were sensitive to addition of insulin both in vivo and in vitro. These authors also studied the effect of insulin in vitro on diabetic mammalian muscle; they found that the respiratory rate of such muscle is below normal and that insulin restores the rate to normal.

In adipose tissue insulin sharply depresses the liberation of fatty acids induced by the action of epinephrine or glucagon. Part of this effect of insulin may be its role in glycolysis, which produces glycerophosphate from glucose and thus facilitates the deposition of the fatty acids as triglyceride. However, insulin will also depress fatty acid release in the absence of glucose, indicating that the hormone may act specifically on lipolysis. Since liberation of fatty acids from adipose tissue is stimulated by cyclic AMP, insulin may decrease fatty acid liberation by reducing tissue levels of cyclic AMP. The effect may still be a membrane action of the hormone since adenyl cyclase, the enzyme responsible for the synthesis of cyclic AMP, is membrane bound. The reduction by insulin of fatty acid liberation from adipose tissue is extremely important since circulating fatty acid levels are responsible for many effects on intracellular metabolic events, both in muscle and liver. Randle & others (1963) have suggested that the release of fatty acids in various pathologic states may contribute to impairment of glucose metabolism by blocking glycolysis at several steps in the pathway. Indeed, many of the effects on liver noted after insulin administration in vivo may be the result of secondary changes induced by reduction in circulating free fatty acids.

Although insulin may directly increase the intracellular transport of free amino acids, it may have additional action directly on protein synthesis. The hormone facilitates the incorporation of labeled intracellular amino acids into protein. A direct effect of insulin on protein synthesis in muscle or adipose tissue is probably independent of the synthesis of RNA since it can be demonstrated when RNA synthesis is inhibited with dactinomycin. Studies by Martin & Wool (1968) indicate that insulin may act at the ribosomal level to increase the capacity of this organelle to translate information from messenger RNA to the protein-synthesizing machinery. In the diabetic animal, the polysomes become disaggregated; insulin, in vivo, restores them to the normal aggregated form. This may not be a direct action of the hormone since it cannot be demonstrated in vitro.

Liver

Unlike muscle and adipose tissue, where a barrier for the transport of glucose exists, there is no barrier in liver. In this organ, extracellular and intracellular concentrations of glucose are approximately equal. Although a primary action of insulin on the hepatic cell membrane seems relatively unimportant, insulin

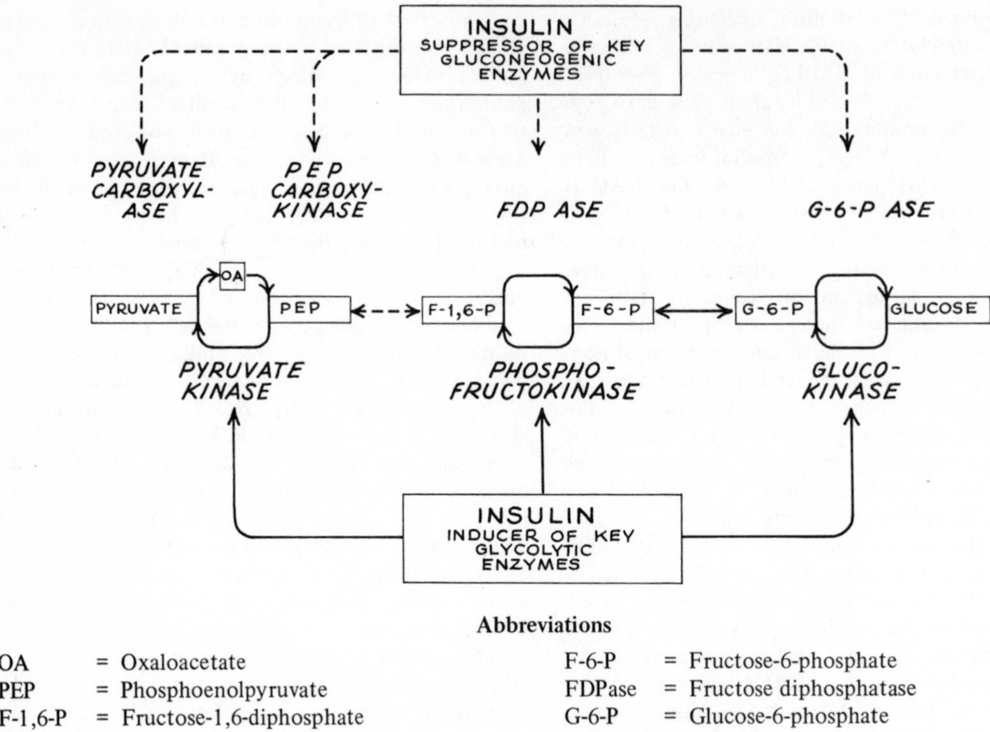

Abbreviations

OA	= Oxaloacetate	F-6-P	= Fructose-6-phosphate
PEP	= Phosphoenolpyruvate	FDPase	= Fructose diphosphatase
F-1,6-P	= Fructose-1,6-diphosphate	G-6-P	= Glucose-6-phosphate

FIG 20–10. Suppressor and inducer function of insulin on key liver enzymes.

still affects several aspects of hepatic metabolism. In vivo this can be explained, in part, as effects secondary to a decrease in the amounts of amino acids, potassium ion, glucose, and fatty acids presented to the liver. In addition, insulin acts directly; the following actions are demonstrable on the isolated perfused liver: decreased glucose output, decreased urea production, decreased cyclic AMP, increased potassium uptake, and increased phosphate uptake.

Weber & others (1966) have postulated a hepatic role for insulin on induction of specific enzymes involved in glycolysis and the simultaneous inhibition of specific gluconeogenic enzymes (Fig 20–10). It is suggested that insulin may act on a genetic locus in the nucleus which contains the functional genetic unit (genome) for a group of specific enzymes. Thus, insulin stimulates glycolysis by effecting a simultaneous increase in synthesis of the enzymes glucokinase, phosphofructokinase, and pyruvate kinase. Simultaneously, insulin represses the enzymes controlling gluconeogenesis: pyruvate carboxylase, phosphoenolpyruvate carboxykinase, fructose-1,6-diphosphatase, and glucose-6-phosphatase. These changes in hepatic enzyme activity can be inhibited by blocking RNA and protein synthesis with dactinomycin and puromycin. Enzymes which are relatively unimportant in the control of either gluconeogenesis or glycolysis are not affected by insulin.

The effects of insulin on specific enzyme synthesis in the liver are observed in vivo and may not imply direct action of the hormone at the nuclear level. Also the relative pattern of the enzyme induction is grossly influenced by diet (Friedland & others, 1966). Therefore, changes of intracellular metabolites or enzyme activities arising from the direct actions of the hormone on glucose output, potassium uptake, etc or from the effect of "signals" in the form of metabolites from the peripheral tissues could be responsible. Glucose itself may not be one of the signals, since glycolytic enzyme levels are not increased during the hyperglycemia of diabetes.

The Diabetic State Caused by Insulin Deprivation (Fig 20–11.)

In diabetes, hyperglycemia occurs as a result of impaired transport and uptake of glucose into muscle and adipose tissue. Repression of the key glycolytic enzymes and derepression of gluconeogenic enzymes promotes gluconeogenesis in the liver, which further contributes to hyperglycemia. Transport and uptake of amino acids in peripheral tissues is also depressed, causing an elevated circulating level of amino acids which provide fuel for gluconeogenesis in the liver. The amino acid breakdown in the liver results in increased production of urea nitrogen.

Because of the decreased production of ATP—and possibly because of a direct requirement for insulin—protein synthesis is decreased in all tissues. A decrease in acetyl-Co A, ATP, NADPH, and glycerophosphate in all tissues results in decreased fatty acid and lipid synthesis. Stored lipids are hydrolyzed by increased lipolysis and the liberated fatty acids may then inter-

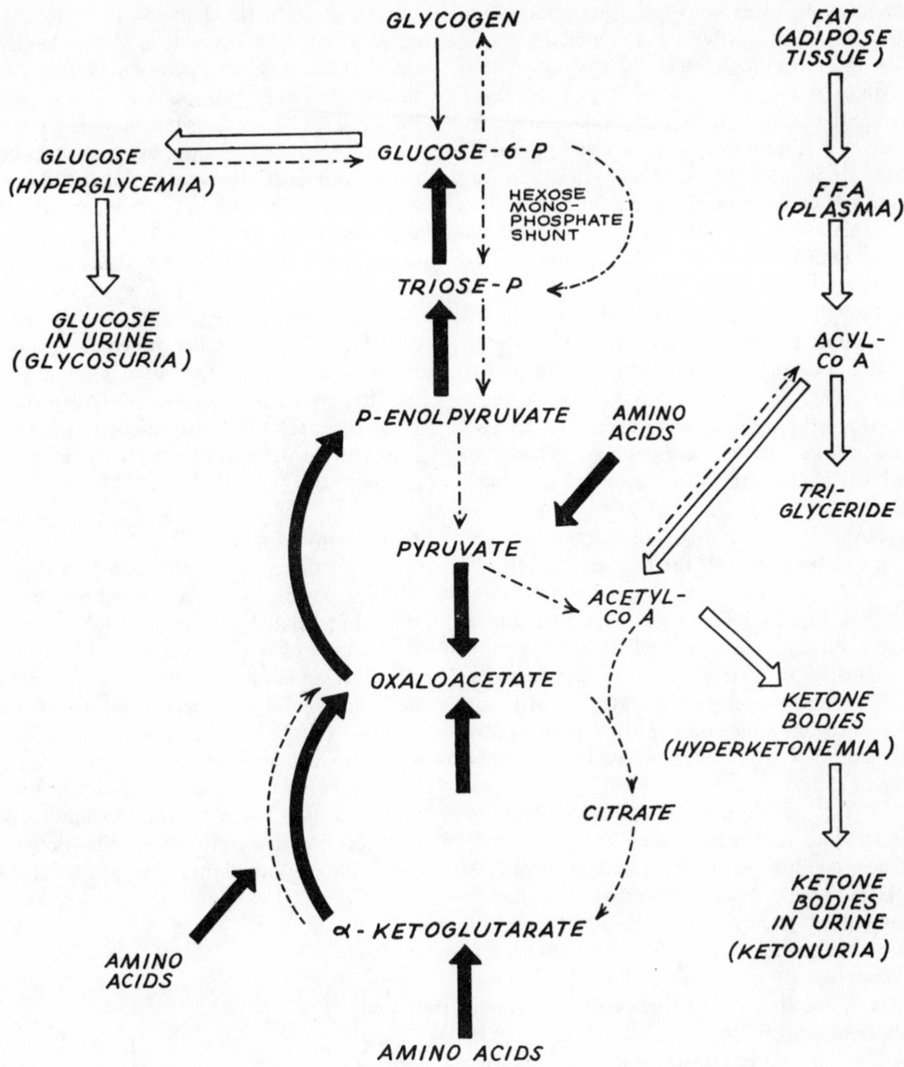

FIG 20–11. Abnormal metabolism in the liver during uncontrolled diabetes.

fere at several steps of carbohydrate phosphorylation in muscle, further contributing to hyperglycemia. Fatty acids reaching the liver in high concentration inhibit further fatty acid synthesis by a feedback inhibition at the acetyl-Co A carboxylase step. Increased acetyl-Co A from fatty acids activates pyruvate carboxylase, stimulating the gluconeogenic pathway required for the conversion of the amino acid carbon skeletons to glucose. Fatty acids also stimulate gluconeogenesis by entering the citric acid cycle and increasing production of citrate, an established inhibitor of glycolysis (at phosphofructokinase). Eventually the fatty acids inhibit the citric acid cycle at the level of citrate synthetase and possibly pyruvate dehydrogenase. The acetyl-Co A which no longer can enter either the citric acid pathway or be used for fatty acid synthesis is shunted to the synthesis of ketones or cholesterol. Glycogen synthesis is depressed as a result of decreased glycogen synthetase activity, by activation of phosphorylase through the action of epinephrine or glucagon, and by the increased ADP:ATP ratio.

In general, insulin deficiency results in an abnormal balance in inter-hormone control. Thus the insulin-deficient animal is in a state of hormonal imbalance favoring the action of corticosteroids, growth hormone, and glucagon, which add to the stimulation of gluconeogenesis, lipolysis, and decreased intracellular metabolism of glucose.

Metabolic Abnormalities in Diabetes

Diabetes mellitus in man is characterized by hyperglycemia and glycosuria. Secondary disturbances in the metabolism of protein and fat lead to acidosis and to dehydration because of the water required to excrete the excess glucose in the urine. Diabetes mellitus in man is due to an insufficiency of insulin relative to the requirements by the tissues. This is supported by the fact that pancreatectomy or destruction of the islet tissue by specific drugs (alloxan) or infection produces typical diabetes.

The juvenile diabetic has little detectable circulating insulin and his pancreas fails to respond to a glucose load. On the other hand, the maturity-onset diabetic may show a delayed response to glucose but because of the continued elevated glucose levels he may ultimately secrete more insulin for a given glucose load than a normal individual. Usually, however, continued impaired release is indicated since plasma glucose:insulin ratios are much higher than normal in these subjects. Excessive insulin release after a glucose load occurs in obese individuals who are not diabetic (Karam & others, 1963) or who have only mild abnormalities in glucose tolerance. It has been suggested that this hyperinsulinism may be attributable both to a peripheral insensitivity to insulin and to a hypersensitivity of the pancreatic islet cells to glucose. Most maturity-onset diabetics are also obese; it may be therefore, that hyperinsulinism, when observed in the maturity-onset diabetic, is more closely related to obesity than to the diabetes. In nonobese prediabetic subjects (ie, persons with a strong family history of diabetes but no abnormalities in glucose tolerance), the insulin response to glucose is normal or slightly impaired. Subjects whose diabetes is secondary to acromegaly, Cushing's disease, or pheochromocytoma may suffer primarily from a peripheral defect in carbohydrate metabolism produced by growth hormone, corticosteroids, or catecholamines, and, in an effort to compensate, produce increased amounts of insulin for a given glucose load.

Liver and kidney are involved in the degradation of insulin. Thus, in renal or hepatic disease, there is an apparent increase in the potency of administered insulin (hyperinsulinemia) and a decrease in the requirement for insulin. This has also been observed in certain instances in diabetics with associated kidney or liver disease.

Experimental Diabetes

Experimental diabetes may be produced by total pancreatectomy or by a single injection of alloxan, a substance related to the pyrimidines; or with streptozoticin, an N-nitroso derivative of glucosamine. Such chemical ablation of insulin production is a simpler method of producing permanent diabetes than surgical removal of the pancreas, although it is not equally effective in all animals.

Diabetes may also be produced by injection of diazoxide (see below), a sulfonamide derivative which inhibits insulin secretion (an inhibition which can be overcome by tolbutamide but not by glucose or glucagon).

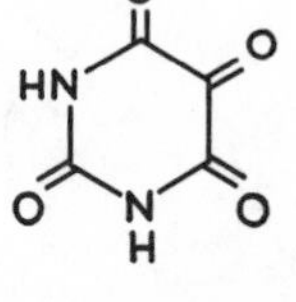

Alloxan

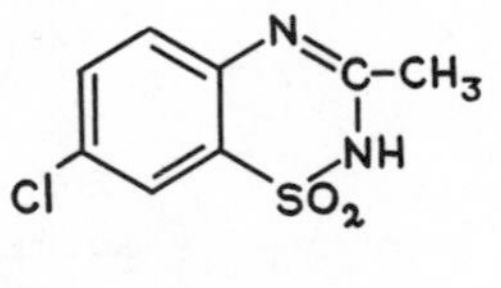

Diazoxide

Since a specific acute deficiency of insulin may be produced by the injection of large amounts of antibodies to insulin, this may also be considered another method for the experimental production of diabetes.

Phlorhizin diabetes, a syndrome produced by injection of the drug phlorhizin, is actually a renal diabetes in which glycosuria is produced by failure of reabsorption of glucose by the renal tubules rather than by virtue of any endocrine abnormality.

Treatment of Hypo- & Hyperinsulinism

A. Hypoinsulinism (Juvenile Diabetes): Juvenile diabetes is usually treated with insulin and dietary regimens; a perfect maintenance treatment is not yet possible. In the maturity-onset diabetic having only a mild degree of diabetes, the orally effective hypoglycemic agents may be used.

B. Hyperinsulinism: Hyperinsulinism results from excessive production of insulin or overdosage of the hormone. A pancreatic tumor affecting islet tissue is a frequent cause of hyperinsulinism; however, other cases of hyperinsulinism are not associated with a demonstrable tumor but are characterized by idiopathic hypoglycemia associated with a marked leucine sensitivity. Hyperinsulinism is characterized by nervousness, weakness, depression, cold sweats, anxiety, confusion, delirium, and, in some cases, convulsions and final collapse. Blood sugar levels are considerably reduced. The administration of glucose by any route will alleviate the condition. If the hyperinsulinism is due to hyperplasia or tumor of the islet tissue, surgical removal is necessary to correct the condition; however, cortisone, glucagon, and, recently, diazoxide, or streptozoticin have been used successfully in treating hyperinsulinism.

Antibodies to Insulin

The repeated injection of insulin results in the production of low levels of an antibody to insulin in all subjects, whether diabetic or not, after 2 or 3 months of treatment. On occasion high concentrations of the antibody are found in subjects who have clinically demonstrated resistance to insulin (Berson, 1958). By the use of differential immunologic technics (Grodsky, 1965), it was found that antibodies formed against exogenously administered insulin were able to bind and thus to inactivate insulin formed endogenously. Antibodies induced in animals by exogenous insulin can produce lesions in the pancreas, islet cell destruction, and, occasionally, severe diabetes. Antibody-bound insulin is not available to the cells and is only slowly degraded; thus much of the insulin, whether administered or secreted, is actually wasted. Occasionally, it is released, and after an acute episode of insulin resistance which has been treated with massive doses of insulin the released insulin may cause repeated bouts of hypoglycemia.

GLUCAGON

Glucagon is now recognized as an important hormone involved in the rapid mobilization of hepatic glucose and, to a lesser extent, fatty acids from adipose tissue. Glucagon has been purified and crystallized from pancreatic extracts and shown to be a polypeptide with a molecular weight of 3485 daltons. It contains 15 different amino acids with a total of 29 amino acid residues, arranged in a straight chain. The sequence of the amino acids in the chain has also been determined, and the molecule has been completely synthesized. Histidine is the N-terminal amino acid and threonine is the C-terminal amino acid. In contrast to insulin, glucagon contains no cystine, proline, or isoleucine but does contain considerable amounts of methionine and tryptophan. Further, it can be crystallized in the absence of zinc or other metals. The structure of the glucagon polypeptide is shown below.

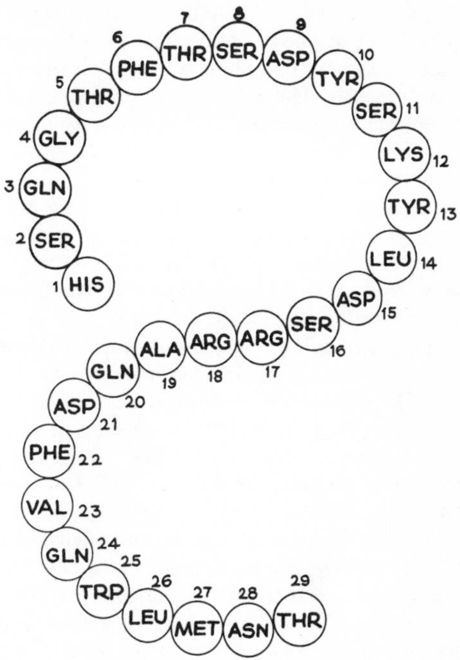

Glucagon polypeptide

Glucagon originates in the alpha cells of the pancreas, and a glucagon-like hormone (GLI) has also been demonstrated in the gastric and duodenal mucosa. GLI is immunologically similar though not identical to the pancreatic hormone. Furthermore, it is less active than pancreatic glucagon in stimulating adenyl cyclase and therefore cannot duplicate many of the actions of the pancreatic hormone. Of great interest is the structural similarity between glucagon and another intestinal hormone, secretin. Of the 27 amino acids in secretin, 14 are identical in position to amino acid residues in glucagon.

Levels of glucagon in the plasma can be measured by radioimmunoassay. Circulating glucagon increases slowly with prolonged starvation, an effect opposite to that seen for insulin. The intravenous injection of glucose reportedly causes reversal of the elevation. In contrast, oral glucose produces an immediate rise in glucagon; this paradoxical situation is explained by the fact that oral glucose causes a release of intestinal "glucagon." Glucagon directly stimulates insulin release; therefore, the mobilization of intestinal glucagon may explain, at least in part, why oral glucose loads stimulate insulin release more effectively than glucose administered intravenously. Some amino acids, in particular arginine, cause a rapid secretion of glucagon from the pancreas. The overall effect of arginine in vivo is not always predictable since this amino acid stimulates hormones with antagonistic action such as insulin and growth hormone as well as glucagon.

The adenyl cyclase receptor sites in the liver are particularly sensitive to glucagon. Within minutes after the presentation of glucagon to the liver, cyclic AMP levels increase. The cyclic AMP in turn activates the enzyme dephosphophosphorylase kinase, and an increase of hepatic phosphorylase results. The activation of phosphorylase results in rapid glycogenolysis and hepatic output of glucose. In the dog, increased phosphorylase activity reaches its peak 10–15 minutes after administration of glucagon, further indicating that enzyme activation rather than the more delayed process of enzyme synthesis is involved. A stimulating effect of glucagon on gluconeogenesis has been observed in the isolated perfused liver (Exton & others, 1966). Glucagon stimulated the conversion of lactic acid and amino acids to glucose. In the same system, glucagon did not effect the conversion of fructose to glucose, indicating that the site of action was below the entry of fructose into the gluconeogenetic pathway. It has been suggested that this enhanced gluconeogenesis is mediated by a cyclic AMP activation of hepatic lipase which produces a fatty acid activation of the gluconeogenic process. The increased hepatic cyclic AMP after glucagon has been shown to increase protein kinases which catalyze nuclear histone phosphorylation (Langan, 1969). Thus, glucagon with time may effect the repression or derepression of hepatic enzyme synthesis.

Glucagon also increases potassium release from the liver, an action which may or may not be related to its glycogenolytic activity.

In adipose tissue, and possibly liver, glucagon increases the breakdown of lipids to fatty acids and glycerol. In general, glucagon and epinephrine act similarly to increase cyclic AMP synthesis and glycogen and lipid breakdown. However, glucagon is proportionately more active in liver whereas epinephrine is more active in adipose tissue and skeletal muscle.

Glucagon may contribute to the etiology of the diabetic state in man. It is elevated in severe diabetes (with ketoacidosis) and rises to abnormally high levels when milder diabetics are stimulated with arginine (Unger & others, 1970).

Totally depancreatized patients do not require as much insulin as do some diabetics with an intact pancreas. This may be explained by the loss of the anti-insulin action of glucagon which would accompany pancreatectomy. Hyperglucagonemia has been reported in a patient with an alpha cell tumor.

Little is known concerning the manner in which glucagon is metabolized. However, an enzyme capable of degrading glucagon has been identified in beef liver (Kakiuchi & Tomizawa, 1964). The action of the enzyme is exerted at the N-terminal position of the glucagon polypeptide (see above), where it removes the first 2 amino acids by hydrolysis of the peptide bond between serine and glutamine.

Crystalline glucagon polypeptide is now commercially available in the form of glucagon hydrochloride. It may be given by the intramuscular, subcutaneous, or intravenous route for treatment of hypoglycemic reactions, which may occur after overdosage with insulin used for the control of diabetes. Response is usually seen within 5–20 minutes after doses of 0.5–1 mg of glucagon. It is also used as a diagnostic test for glycogen storage disease.

THE ADRENALS

THE ADRENAL MEDULLA

Function

The adrenal medulla is a derivative of the sympathetic portion of the autonomic nervous system. Despite its diverse physiologic functions, it is not essential to life.

The hormones of the adrenal medulla are epinephrine and norepinephrine. Epinephrine (adrenaline) in general duplicates the effect of sympathetic stimulation of an organ. Epinephrine is necessary to provide a rapid physiologic response to emergencies such as cold, fatigue, shock, etc. In this sense, it mobilizes what has been termed the "fight or flight" mechanism, a cooperative effort of the adrenal medulla and the sympathetic nervous system.

In addition to bringing about effects similar to those which follow stimulation of the sympathetic nervous system, both epinephrine and the closely related hormone norepinephrine induce metabolic effects, including glycogenolysis in the liver and skeletal muscle, and an increase in circulating free fatty acid levels as a result of stimulation of lipolysis in adipose tissue.

Chemistry

The adrenal medulla contains a high concentration of epinephrine, 1–3 mg/g of tissue. The chemical structure of epinephrine is shown in Fig 20–12. It is a catecholamine derivative of phenylalanine and tyrosine. The initial step in conversion of tyrosine to dihydroxyphenylalanine (DOPA) requires the enzyme tyrosine hydroxylase. Inhibition of this enzyme (eg, with *α*N-methyl-*p*-tyrosine) is used to block adrenergic activity in pheochromocytoma (see p 441).

Naturally occurring **epinephrine** is the L-isomer, the D-form being only 1/15 as active. L-Epinephrine produced synthetically is identical in activity with the natural product.

Although in man and in the dog epinephrine represents 80% of the catecholamine activity in the adrenal medulla, the closely related hormonal compound **norepinephrine (arterenol)** is also present. Norepinephrine is found principally in the sympathetic nerves, occurring both as a result of synthesis in the nervous tissue and uptake by this tissue from the circulation. Norepinephrine has been isolated from several

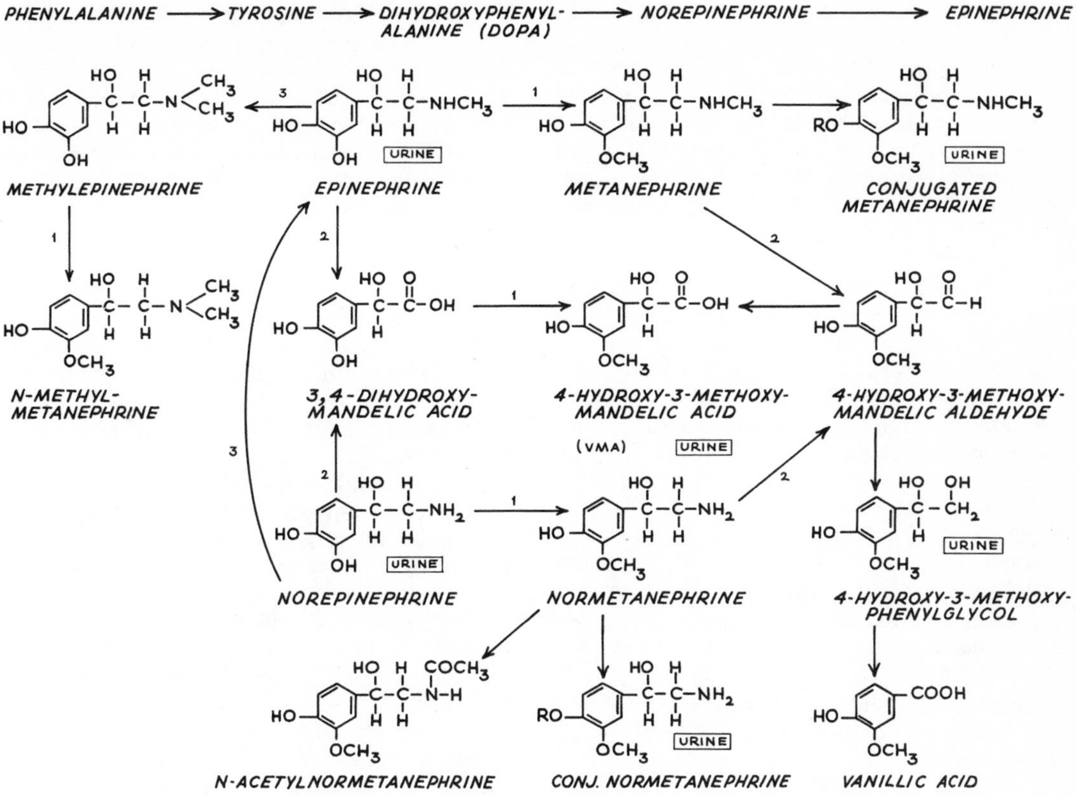

PHENYLALANINE → TYROSINE → DIHYDROXYPHENYLALANINE (DOPA) → NOREPINEPHRINE → EPINEPHRINE

METHYLEPINEPHRINE · EPINEPHRINE (URINE) · METANEPHRINE · CONJUGATED METANEPHRINE (URINE)

N-METHYL-METANEPHRINE · 3,4-DIHYDROXY-MANDELIC ACID · 4-HYDROXY-3-METHOXY-MANDELIC ACID (VMA) (URINE) · 4-HYDROXY-3-METHOXY-MANDELIC ALDEHYDE

NOREPINEPHRINE (URINE) · NORMETANEPHRINE · 4-HYDROXY-3-METHOXY-PHENYLGLYCOL (URINE)

N-ACETYLNORMETANEPHRINE · CONJ. NORMETANEPHRINE (URINE) · VANILLIC ACID

FIG 20–12. Pathways for the metabolism of norepinephrine and epinephrine. 1, Catechol-O-methyltransferase; 2, monoamine oxidase; 3, phenylethanolamine N-methyltransferase. (Redrawn and reproduced, with permission, from Axelrod J: Recent Progr Hormone Res 21:598, 1965.)

commercial lots of natural epinephrine. From the chemical structures of epinephrine and norepinephrine as shown in Fig 20–12, it can be seen that epinephrine differs from norepinephrine only in that it is methylated on the primary amino group; S-adenosylmethionine is the methyl donor.

Both epinephrine and norepinephrine occur in the urine in free as well as conjugated forms. For a 24-hour period, the excretion of epinephrine is approximately 10–15 μg, whereas that of norepinephrine is somewhat higher, 30–50 μg. This reflects the higher concentrations of norepinephrine in the plasma (0.3 μg/liter as compared to 0.06 μg/liter for epinephrine).

When radiotagged epinephrine is injected into animals, only about 5% is excreted in the urine unchanged, most of the hormone being metabolized in the tissues by a series of methylations of the phenolic groups or oxidations on the amine side chains (Fig 20–12). The main enzymes involved are monoamine oxidase (MAO) for the oxidation reactions and catechol-O-methyltransferase (COMT) for catalysis of the methylations. MAO is a mitochondrial enzyme (actually a series of isoenzymes) with a broad specificity capable of catalyzing the oxidation of side chains on a large variety of catechols. Inhibition of MAO does not cause a rise in epinephrine or norepinephrine, indicating that there is another important route of metabolism. COMT serves in this capacity by rapidly catalyzing the inactivation of the catecholamines by methylation of the hydroxyl group at the 3 positions. This Mg^{++}-dependent enzyme is located in the cytosol. The enzyme is capable of methoxylating a variety of catecholamine intermediates with utilization of S-adenosylmethionine as the source of the methyl groups. COMT requires an active SH group and can therefore be inhibited by *p*-chloromercuric benzoate and iodoacetic acid. Pyrogallol and other catecholamines can act as competitive inhibitors of epinephrine metabolism by serving as substrate for this enzyme. Although MAO and COMT are found in most tissues, their activity is particularly high in the liver where most of the degradation of the catecholamines takes place.

The first step in the metabolism of the catecholamines can be either methoxylation or oxidation of the side chain, the preferred step varying with circumstances that have not yet been well established. The norepinephrine component which is tightly bound to tissue is initially metabolized by the mitochondrial MAO, whereas the less tightly bound component is initially methoxylated by the methyltransferase (COMT). Since both enzymes usually react with the metabolic products, the final compounds appearing in the urine are often the same regardless of which of the 2 reactions occurred first.

One of the principal metabolites of epinephrine and of norepinephrine which occurs in the urine is **4-hydroxy-3-methoxy-mandelic acid.** This substance has also been called **vanilmandelic acid (VMA).** Other metabolites occurring in the urine in significant quantities are 3-methoxyepinephrine (metanephrine) and 4-hydroxy-3-methoxy-phenylglycol. Minor excretion products are the catechols (dihydroxy compounds) which correspond to the methylated compounds mentioned above: 3,4-dihydroxymandelic acid and 3,4-dihydroxyphenylglycol, as well as vanillic acid.

Small quantities of acetylated derivatives are also found. The typical distribution of urinary products found after injection of radiotagged epinephrine are shown below (LaBrosse & others, 1961):

Unchanged epinephrine	6%
Metanephrine	40%
VMA	41%
4-Hydroxy-3-methoxy-phenylglycol	7%
3,4-Dihydroxymandelic acid	2%
Miscellaneous	4%

The urinary products are excreted mostly as conjugates with sulfate or glucuronide, sulfate being the preferred conjugation moiety in man.

Injected catecholamines are rapidly concentrated in the heart, spleen, lung, and adrenal gland. Only small amounts are concentrated in the brain. Presumably, the large amounts of catecholamines in brain tissue are synthesized from dopa (Fig 20–12) in situ. According to Axelrod (1965), catecholamines concentrated by the tissues may be strongly bound, and in the bound form they are biologically unavailable. Cocaine, reserpine, chlorpromazine, guanethidine, and adrenergic blocking agents inhibit the fixation of the catecholamines by tissues, resulting in increased biologic availability. However, this is probably not the sole action and may not even be the major action of these drugs. Most norepinephrine is specifically bound at the sympathetic nerve endings. The bound catecholamines, whether occurring in the tissue as a result of synthesis in situ or uptake from the plasma, are stoichiometrically associated with ATP.

Many agents structurally similar to epinephrine and norepinephrine, though with less biologic activity, can be stored in the tissue sites normally reserved for the active hormones. The agents, known as "false neurotransmitters," prevent either the synthesis or storage of hormones and are released during normal sympathetic stimulation in their place. Thus, these agents or their precursors can be used clinically to reduce release of active hormone and therefore serve as hypotensive agents. "False transmitters" such as β-hydroxytyramine, α-methylnorepinephrine, and metaraminol (Aramine) are produced by administration of tyramine, methyldopa, α-methyltyrosine, or Aramine itself. Initially, these agents often cause increased circulating hormones by preventing hormone binding in tissues.

Normal Physiology of Epinephrine & Norepinephrine

A. Action on Cardiovascular System: Epinephrine causes vasodilation of the arterioles of the skeletal muscles and vasoconstriction of the arterioles of the skin, mucous membranes, and splanchnic viscera. It is also effective as a stimulant of heart action, increasing the irritability and the rate and strength of contraction of cardiac muscle and increasing cardiac output.

Norepinephrine has less effect on cardiac output than epinephrine, although it has an excitatory effect on most areas of the cardiovascular system. Norepinephrine exerts an overall vasoconstrictor effect, whereas epinephrine exerts, in general, an overall vasodilator effect, with exceptions as noted above. Both hormones lead to an elevation of blood pressure, more marked in the case of norepinephrine, as a result of their action on the heart and blood vessels. Norepinephrine has found important application in the treatment of hypotensive shock other than that caused by hemorrhage. Its marked action on the arterioles without producing tachycardia makes it particularly useful for this purpose.

B. Action on Smooth Muscle of the Viscera: Epinephrine causes relaxation of the smooth muscles of the stomach, intestine, bronchioles, and urinary bladder, together with contraction of the sphincters in the case of the stomach and bladder. Other smooth muscles may be contracted. The relaxing effect of epinephrine on bronchiolar smooth muscle makes this hormone particularly valuable in the treatment of asthmatic attacks.

C. Metabolic Effects: In the liver, epinephrine stimulates the breakdown of glycogen, an action that contributes to the ability of this hormone to elevate the blood glucose. The primary hepatic effect of epinephrine may be to increase cyclic AMP by activating the enzyme, adenyl cyclase. Cyclic AMP in turn activates dephosphophosphorylase kinase, which catalyzes the conversion of the inactive form of phosphorylase to the phosphorylated active form. This effect of epinephrine on increasing hepatic cyclic AMP is similar to that of glucagon. However, measurements of the cyclic AMP levels after epinephrine or glucagon indicate that glucagon is by far the more active hormone in liver tissue.

In muscle, epinephrine also causes the breakdown of glycogen by increasing cyclic AMP, and in this tissue it is more active than glucagon. In exercising muscle, this can result in increased lactate secretion into the plasma. Increases in cyclic AMP after administration of epinephrine can be determined in the isolated heart within 2–4 seconds; the effect of epinephrine on cardiac output (inotropic effect) is seen shortly afterward, whereas the activation of phosphorylase is not detectable for 45 seconds (Sutherland & others, 1965). Thus, the inotropic effect of epinephrine is not the result of an initial action on glycogenolysis, though both actions may result from the increased cyclic AMP. In vivo, epinephrine action can result in an increase in heart glycogen. This is probably secondary to the action of the hormone on adipose tissue to increase circulating fatty acids which are rapidly utilized in the heart as fuel. Although total glucose uptake may be decreased, that glucose entering the heart is preferentially shunted to glycogen. In vivo, the lactic acid from muscle and the fatty acids released from adipose tissue are taken up by the liver and metabolized through the citric acid cycle. Eventually the fatty acids activate the reversal of the glycolytic pathway. This permits lactate to be converted to glucose or glycogen. Therefore, despite the specific effect of epinephrine on glycogen breakdown in liver, liver glycogen can often be increased when the drug is administered in vivo.

In adipose tissue epinephrine has a marked effect on lipolysis resulting in the rapid release of both fatty acids and glycerol. These fatty acids serve as fuel in the muscle and can activate gluconeogenesis in the liver. The breakdown of fat in adipose tissue is usually accompanied by a compensating increase in glucose uptake and the synthesis of glycerophosphate. Epinephrine has a direct inhibitory action on insulin release in the pancreas. It therefore serves as an emergency hormone by (1) rapidly providing fatty acids, which are the primary fuel for muscle action; (2) mobilizing glucose, both by increasing glycogenolysis and gluconeogenesis in the liver and by decreasing glucose uptake in the muscle; and (3) decreasing insulin, thereby preventing the glucose from being taken up by peripheral tissues and preserving it for the central nervous system.

As mentioned earlier, epinephrine can stimulate both beta- and alpha-adrenergic receptors in tissue. Norepinephrine in small doses acts primarily, though not exclusively, on alpha receptors. Historically, adrenergic responses were classified according to the comparative effectiveness of a series of related catechols on vascular, cardiac, and pulmonary physiologic responses. Thus, typical norepinephrine-stimulated responses such as vascular venous constriction were denoted as alpha-adrenergic. Typical epinephrine effects, such as increased heart rate and atrial contractibility, were a beta phenomenon. Fatty acid mobilization and glycogenolysis are beta-type metabolic reactions since they can be duplicated with beta-adrenergic stimulating drugs, eg, isoproterenol. It now appears that beta effects are those associated with an increase in cyclic AMP, whereas alpha effects are associated with cyclic AMP depression. Since epinephrine can stimulate both adrenergic responses, its effect in a tissue depends on the quantity or relative sensitivity of the alpha and beta receptors. Thus, in the pancreas, the alpha-adrenergic response to epinephrine predominates, cyclic AMP decreases, and insulin release is inhibited. However, in the presence of an alpha-adrenergic blocker such as phentolamine (Regitine), the beta effect predominates and epinephrine causes increased cyclic AMP and increased insulin release.

D. Other Effects: After injection of epinephrine, skeletal muscle becomes fatigued at a slower rate than usual; the rate and depth of respiration are increased; and the metabolic rate is accelerated with an increase in the respiratory quotient.

Abnormal Physiology

The adrenal medulla is not often involved in disease. No clinical state directly attributable to a deficiency of the adrenal medulla is known. However, certain tumors of the medullary tissue (chromaffin cell tumors, **"pheochromocytoma"**)) produce a condition which simulates hyperactivity of the adrenal medulla. The symptoms of these tumors include intermittent hypertension which may progress to permanent hypertension and lead eventually to death from complications such as coronary insufficiency, ventricular fibrillation, and pulmonary edema.

Laboratory tests for adrenal medullary hyperactivity, as in suspected pheochromocytoma, include chemical analyses for catecholamines in the blood and urine–in particular, VMA (see above), the principal urinary metabolite of epinephrine and norepinephrine. Excretion of VMA in the urine is normally 0.7–6.8 mg/24 hours. Phentolamine (Regitine) is a specific antagonist to norepinephrine. In the presence of sustained hypertension due to pheochromocytoma (pressures exceeding 170/110 mm Hg), the rapid intravenous injection of 5 mg of phentolamine should produce a sustained fall of at least 35/25 mm Hg within 2–5 minutes. This test is another diagnostic aid in testing for pheochromocytoma.

It has often been noted that the norepinephrine content of adrenal medullary tumors is much higher than that of epinephrine. It is believed that the attacks of hypertension produced by these tumors are attributable to their high norepinephrine content.

THE ADRENAL CORTEX*

The outer portion of the adrenal gland, the adrenal cortex, is essential to life. Its embryologic origin is quite different from that of the adrenal medulla. The adrenal cortex originates from the mesodermal tissue of the nephrotome area; in a 20 mm human embryo the chromaffin cells, which develop into medullary tissue, migrate into the developing cortical tissue so that eventually the complex organ is formed.

An extract of the adrenal cortex, **cortin**, contains a number of potent hormones all of which are steroid derivatives having the characteristic cyclopentanoperhydrophenanthrene nucleus. As will be noted later, the hormones of the gonads are also steroid hormones not remarkably different from those of the adrenal cortex. The similarity of embryologic origin of the adrenal cortex and of the gonads is of interest in connection with the close relationship of the chemistry of their respective hormones.

*In cooperation with Tawfik ElAttar, PhD, Professor of Biochemistry, School of Dentistry, University of Missouri–Kansas City.

General Function

The steroid hormones of the adrenal cortex fall into 3 general classes, each with characteristic functions:

(1) The **glucocorticoids**, which primarily affect metabolism of protein, carbohydrate, and lipids.

(2) The **mineralocorticoids**, which primarily affect the transport of electrolytes and the distribution of water in tissues.

(3) The **androgens** or **estrogens**, which primarily affect secondary sex characteristics in their specific target organs.

Individual steroids usually have activities which are predominantly in one of the above categories but which may overlap into one or both of the others.

General Chemistry

All steroid hormones have a cyclopentanoperhydrophenanthrene ring system as their chemical nucleus. This 4-membered ring and its conventional numbering system is illustrated in the structure of cholesterol in Fig 20–13. Most naturally occurring steroids contain alcohol side chains and are therefore usually referred to as sterols.

A variety of stereoisomeric forms of the steroids are possible: (1) the A and B rings may be joined either in a trans or cis configuration. Estrogens are not capable of this form of isomerism since their A ring is aromatic; (2) hydrogens or other groups may be attached to the rings with an orientation either above (β-) or below (α-) the plane of the ring. The β-orientation is conventionally assigned to groups in the same plane as the C_{19} methyl group and are diagramatically represented by solid lines. The opposite α-groups are normally represented by dashed lines. In natural steroids both the chains attached at C_{17} and various substitutions at C_{11} are in the β-configuration. Some general terms of steroid nomenclature are given in Table 20–2.

TABLE 20–2. Nomenclature of steroids.

Prefix	Suffix	Chemical Significance
allo-		Trans (as opposed to cis) configuration of the A and B rings.
epi-		Configuration different from parent compound at a single carbon atom.
	-ane	Saturated carbon atom.
	-ene	A single double bond in ring structure.
hydroxy-, dihydroxy-, etc	-ol, -diol, etc	Alcohols.
oxo-	-one, -dione	Ketones.
dehydro-		Conversion of –C–OH to –C=O by loss of 2 hydrogen atoms.
dihydro-		Addition of 2 hydrogen atoms.
cis-		Arrangement of 2 groups in same plane.
trans-		Arrangement of 2 groups in opposing planes.
α-		A group trans to the 19-methyl.
β-		A group cis to the 19-methyl.
nor-		One less carbon in a side chain as compared to parent molecules. (*Example:* 19-Nor signifies that the methyl group constituting carbon 19 of a steroid is deleted.)

About 50 steroids have been isolated from the adrenal gland, but only 8 are known to possess physiologic activity. These include compounds A, B, E (cortisone), F (hydrocortisone, cortisol), 17-hydroxycorticosterone, aldosterone, and the 2 androgens, androstenedione (androst-4-ene-3,17-dione) and dehydroepiandrosterone (DHEA) (Figs 20–13 and 20–14). Cortisol is the major free circulating adrenocortical hormone in human plasma. The normal level of cortisol in plasma is about 12 μg/100 ml. The other steroid hormones are present in human plasma in relatively small concentrations.

Biosynthesis of Adrenal Hormones (See Figs 20–13 and 20–14.)

Acetate is the primary precursor for the synthesis of all steroids. The pathway involves the initial synthesis of cholesterol, which, after a series of side chain cleavages and oxidations, is converted to Δ5-pregnenolone. Pregnenolone is the "pivotal" steroid from which all the other steroid hormones are produced. There is evidence that pregnenolone (or progesterone) can be synthesized from acetate by a pathway other than through cholesterol, possibly from 24-dehydrocholesterol. However, in normal tissue this path is relatively minor. The adrenal cortex contains relatively large quantities of cholesterol, mostly as cholesterol esters which are derived both from synthesis and from extra-adrenal sources.

Pregnenolone is converted to progesterone by a dehydrogenase or to 17-hydroxypregnenolone by a specific 17-hydroxylase. As shown in Figs 20–13 and 20–14, those 2 steroids are converted to a variety of active hormones by specific oxygenases and dehydrogenases which require molecular oxygen and NADPH. The result of these combined enzymatic reactions is the addition of hydroxyl or keto groups at the C_{11}, C_{17}, or C_{21} positions.

In general C-21 hydroxylation is necessary for both glucocorticoid and mineralocorticoid activities. Those steroids with an additional –OH or C=O at the C_{11} position or an –OH at C_{17} have greater glucocorticoid and lesser mineralocorticoid action. Examples of the glucocorticosteroids are corticosterone (Kendall's compound B), 11-dehydrocorticosterone (Kendall's compound A), cortisol (Kendall's compound E), and hydrocortisone or cortisol (Kendall's compound F). The 2 most important glucocorticoids are cortisol and corticosterone. Cortisol predominates in man and the fish, whereas corticosterone is the more important hormone in rodents.

The most potent mineralocorticoid is **aldosterone.** It has been detected in extracts of the adrenal cortex

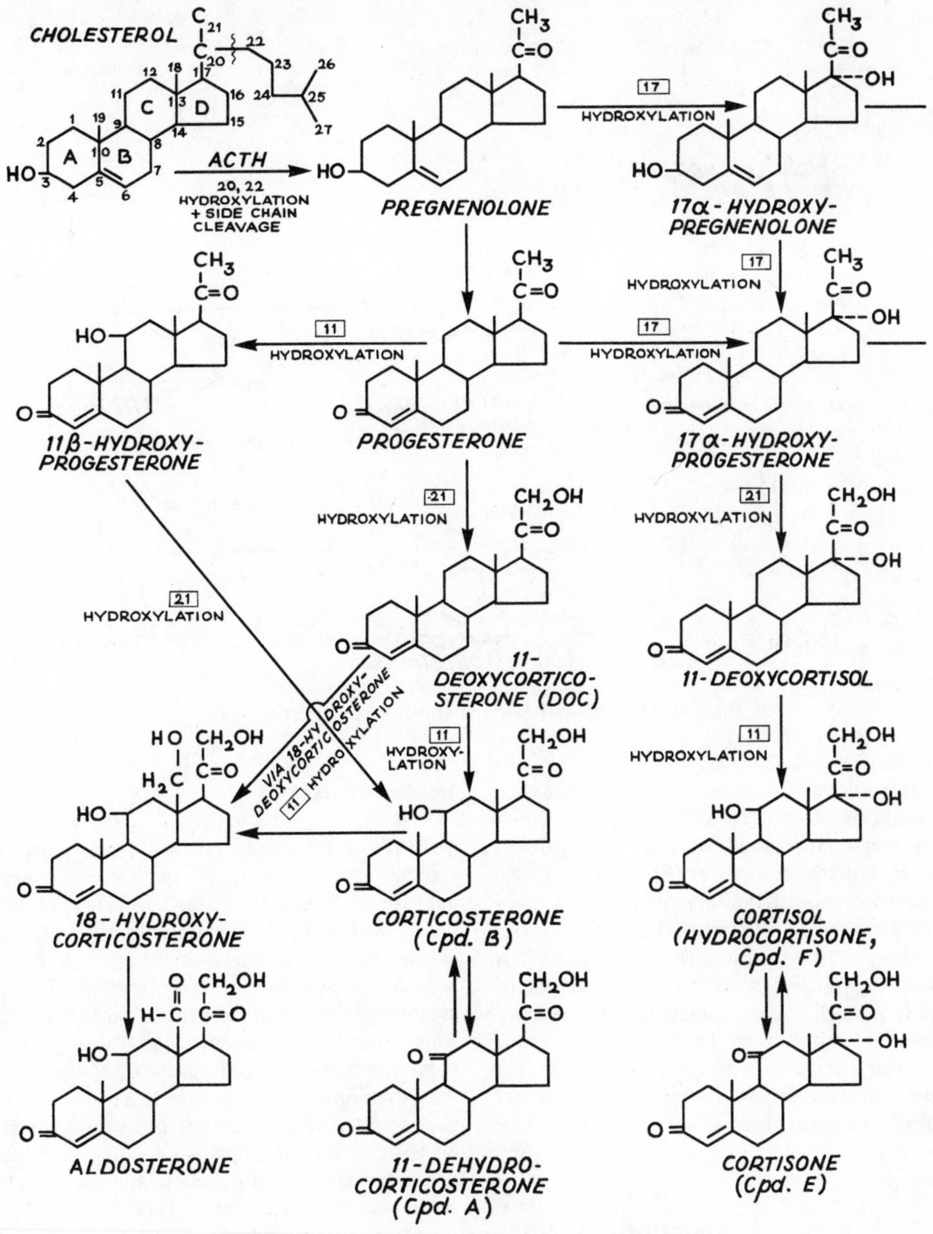

FIG 20–13. Biosynthesis of adrenal corticosteroids.

and in the blood of the adrenal vein. Its major pathway of synthesis requires a unique 18-hydroxylation (Fig 20–13). Although most hydroxylases involved in adrenal steroid synthesis are found throughout the gland, the 18-hydroxylase activity is restricted to the glomerular layer below the capsule. Thus, aldosterone synthesis is limited to this area.

The structure of aldosterone is shown in Fig 20–15 in the aldehyde form and in the hemiacetal form. It is believed that the hormone exists in solution in the hemiacetal form. It will be noted that aldosterone has the same structure as corticosterone except that the methyl group at position 18 is replaced by an aldehyde group. Deoxycorticosterone appears to be the precursor in the adrenal of both aldosterone and corticosterone.

11-Deoxycorticosterone (DOC) is only 1/25 as potent as aldosterone. However, the fact that it can be prepared synthetically (as the acetate, DOCA) and that aldosterone is not yet available for therapeutic use make the former compound important in the treatment of Addison's disease. DOCA may be administered sublingually, since absorption from the buccal mucosa occurs.

The major adrenal androgen, dehydroepiandrosterone, is produced by side chain cleavage of 17-

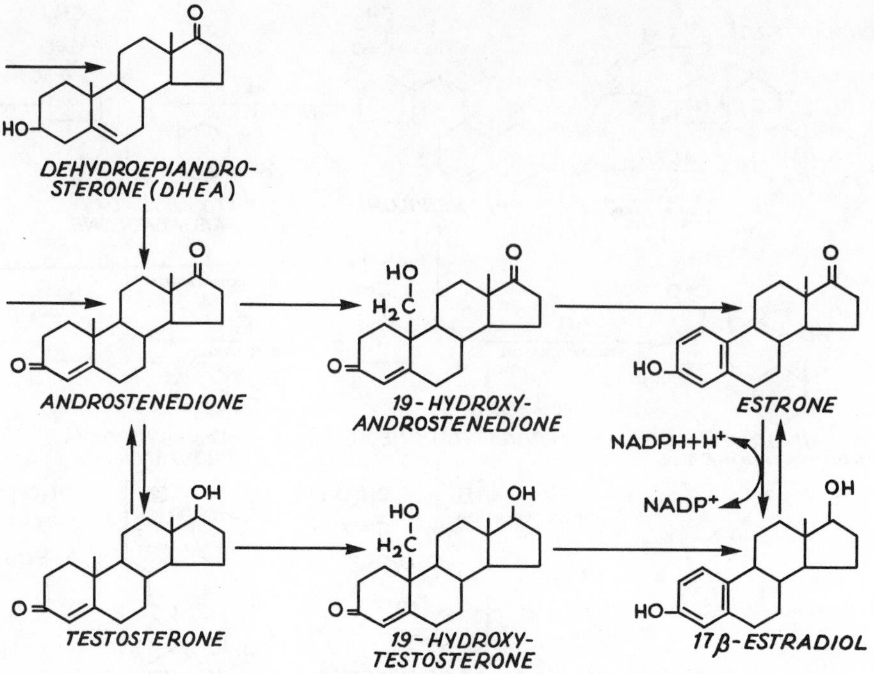

DEHYDROEPIANDROSTERONE (DHEA)

ANDROSTENEDIONE

19-HYDROXYANDROSTENEDIONE

ESTRONE

NADPH+H$^+$

NADP$^+$

TESTOSTERONE

19-HYDROXYTESTOSTERONE

17β-ESTRADIOL

FIG 20–14. Biosynthesis of androgens and estrogens.

hydroxypregnenolone. The smaller amounts of adrenal estrogens can arise from testosterone produced either from dehydroepiandrosterone or from 17-hydroxyprogesterone. Sulfate conjugates of some of the steroids, most notably of the androgen, DHEA, have been detected in the adrenal gland and in adrenal secretions. Conversion of pregnenolone sulfate to DHEA sulfate has been reported to occur without the loss of the sulfate. Although sulfate conjugation is generally associated with inactivation mechanisms in the liver for drugs and other hormones, these results indicate that sulfate conjugates are involved in some pathways of biosynthesis of steroid hormones.

18-Aldehyde form

11-Hemiacetal form

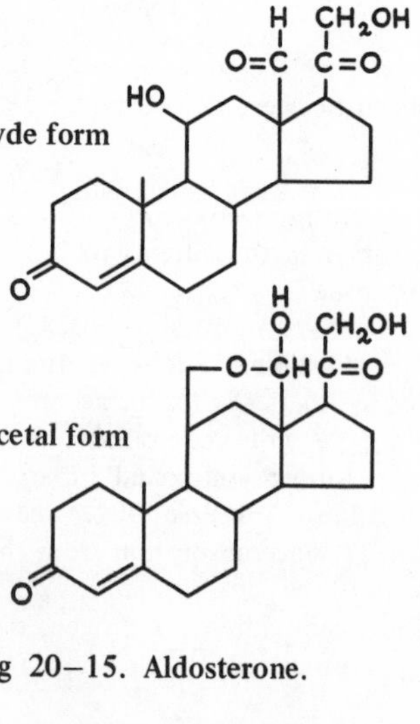

Fig 20–15. Aldosterone.

Metabolic Functions

A. The Glucocorticoids: The glucocorticoids, notably cortisone, cortisol, and corticosterone, produce the following effects (note in many cases antagonism to the effects of insulin): (1) Elevation of blood glucose. (2) Decrease in carbohydrate oxidation. (3) Increase in glycogen synthesis. (4) Decrease in hepatic lipogenesis from carbohydrate. (5) Increase in gluconeogenesis and a protein anti-anabolic effect which increases protein catabolism. This includes reduction of the osteoid matrix of bone, with a resulting osteoporosis and excessive loss of calcium from the body. (6) A sparing action on carbohydrate by an increase in fat breakdown, including mobilization of depot fat as free fatty acids to the liver. (7) In severe circumstances, ketones may also be increased.

The specific mechanism by which the glucocorticoids act is still unknown. In vivo, the hyperglycemia, particularly during later periods of treatment, is a result of increased gluconeogenesis in the liver and decreased glucose uptake in peripheral tissues. Though the primary source of the glucose moiety in the process of gluconeogenesis is usually considered to be amino acids, the amount of glucose produced cannot be entirely accounted for by amino acid breakdown. It is possible that lactate and glycerol derived from muscle and adipose tissue, respectively (the latter a product of the increased lipolysis), can also serve as sources of carbon for hepatic glucose synthesis. In liver, steroids not only increase amino acid conversion to glucose but also conversion of CO_2 to glucose, suggesting that they may act on CO_2 fixation, particularly at the level of pyruvate carboxylase, a key enzyme involved in gluconeogenesis (see below). Conversion of fructose or glycerol to glucose is not specifically increased in vitro, thus supporting the concept of an action at some stage lower than the entry of these metabolites into the gluconeogenic pathway. The increase in glucose, glycogen, and protein synthesis observed in the liver indicates an important action of the steroids on increased metabolic availability of amino acids. However, steroids have little effect on the concentration gradient of amino acids across cell membranes. In muscle, steroids impair both glucose uptake and the subsequent production of CO_2, as well as protein synthesis. Inhibition of glycolysis occurs, as does a reduction in glucose transport. In adipose tissue the lipolytic action of the corticosteroids causes a breakdown of triglyceride with consequent release of fatty acids and glycerol. Glucose oxidation and fatty acid synthesis are both depressed.

The influence of the glucocorticoids on metabolism has been identified with the effects of the hormones on the synthesis of specific enzymes. These effects are mediated through RNA synthesis which in the liver is increased by steroids. Most increases in enzyme activity after steroid administration can be blocked by puromycin and often by dactinomycin, suggesting that de novo synthesis of enzyme protein has occurred. Isotopic and radioautographic studies show that the steroids are rapidly concentrated in the nuclei of their target cells; at this level, they may influence the activity of the polypeptide repressors controlling the synthesis of RNA.

In particular, the steroids increase the amount of enzymes involved in amino acid metabolism such as alanine-α-ketoglutarate and tyrosine transaminases as well as tryptophan pyrrolase. The key enzymes in the regulation of gluconeogenesis (pyruvate carboxylase, phosphoenolpyruvate carboxykinase, fructose-1,6-diphosphatase, and glucose-6-phosphatase) are also increased, possibly by the derepression of a functional genetic unit in the nucleus which controls their synthesis (Weber & others, 1966). This seems to be a comparatively specialized action of the steroids since many other hepatic enzymes are not increased.

Although induction of the synthesis of specific enzymes is an important and quantitatively major action, it may not represent the initial function of the steroids. Hyperglycemia is often noted before induction can be detected. Some initial action of the steroids could therefore subsequently produce the metabolic "signals" responsible for the RNA and enzyme synthesis which occurs later.

Other effects of the glucocorticoids can be extremely important:

1. Anti-inflammatory effects–At high concentrations, glucocorticoids decrease cellular protective reactions and in particular retard the migration of leukocytes into traumatized areas. Thus, cortisol is an anti-inflammatory agent and is used in this capacity in the so-called collagen diseases such as rheumatoid arthritis.

2. Immunosuppressive effects–Cortisol also decreases immune responses involved in infections, allergic states, and anaphylactic shock. The glucocorticoids are used for this purpose to repress antibody formation in connection with organ transplant procedures. Within 30 minutes after injection of cortisol, there is a decreased synthesis of DNA and DNA-dependent RNA polymerase in lymphocytes (Hofert, 1968). Thus the primary action of steroids to depress immune reactions may be at the sites of nucleic acid and protein synthesis in immunologically active cells.

3. Exocrine secretory effects–Chronic treatment with glucocorticoids causes increased secretion of hydrochloric acid and pepsinogen by the stomach and trypsinogen by the pancreas; this can enhance the formation of gastrointestinal ulcers.

B. The Mineralocorticoids: With the exception of the androgens, all of the active corticosteroids increase the reabsorption of sodium and chloride by the renal tubules and decrease their excretion by the sweat glands, salivary glands, and the gastrointestinal tract. However, there is a considerable difference in the extent of this action. Cortisol has the least sodium-retaining action, whereas aldosterone is extremely potent, being at least 1000 times as effective as cortisol and about 35 times as effective as 11-deoxycorticosterone (DOC) in electrolyte regulation. Accompanying the retention of sodium, potassium excretion is increased by exchange of intracellular potassium with extracellular sodium.

As a result of the shifts in electrolyte distribution there are accompanying changes in the volume of the fluid compartments within the body. For example, an increase in the extracellular fluid volume of as much as 20% may occur after the administration of large doses of cortisone as sodium is mobilized from tissue and transferred to the extracellular fluid. There will also be an increase in the volume of the circulating blood and in the urinary output. After removal of the adrenal glands, these effects on water and salt metabolism are quickly reversed. The loss of water and of sodium can result in acute dehydration and death. Aldosterone also increases the renal clearance of magnesium to a degree which parallels its effect on potassium excretion.

Although the exact site of action of aldosterone is not clear, it may act primarily at the nuclear level. In vitro, aldosterone is concentrated in the nuclear fraction of the cell before its effects on sodium transport are noted. It also causes a rapid increase in RNA synthesis and therefore indirectly influences the synthesis of enzymes or other proteins. The fact that the hormone is ineffective in tissue where RNA synthesis or protein synthesis are inhibited further indicates its primary role in mechanisms controlling protein synthesis at the nuclear level. The comparatively long period of exposure of tissue, both in vitro and in vivo, which is required before physiologic effects are noted is also consistent with this mechanism of action.

Although aldosterone is at least 25 times as potent as DOC in its sodium-retaining effects, it is only about 5 times as potent in increasing the excretion of potassium. When used in excess, deoxycorticosterone leads to overhydration within the cells and a consequent water intoxication; this effect is much less prominent with aldosterone. Although aldosterone resembles deoxycorticosterone in many of its metabolic effects, it also possesses other physiologic properties which distinguish it. These include increasing the deposition of glycogen in the liver, decreasing the circulating eosinophils, and maintaining resistance to stress, such as exposure to low temperatures (so-called cold stress test). In maintaining the life of the adrenalectomized animal, aldosterone is more potent than any other known steroid.

C. Sex Hormones (C-19 Corticosteroids): The primary adrenal androgens are dehydroepiandrosterone (DHEA) and androstenedione (Fig 20–14). Testosterone can be detected in certain adrenal tumors.

The adrenal origin of some sex hormones accounts for the fact that the urine of castrates still contains androgen derivatives. These adrenocorticosteroids of the androgenic type cause retention of nitrogen (a protein anabolic effect), phosphorus, potassium, sodium, and chloride. If present in excessive amounts, they also lead to masculinization in the female.

D. Other Effects (Anti-Inflammatory Effects): The adrenocortical hormones exert many other effects which cannot at present be explained on the basis of their known metabolic activities. A wide variety of disease states have been modified by treatment with corticosteroids or by injection of corticotropin (ACTH), which increases the production of corticosteroids by the intact adrenal. The so-called collagen diseases, such as rheumatoid arthritis, are outstanding examples. These hormones also decrease immune responses involved in infection and allergic states.

Synthetic Analogues of Natural Steroids

During recent years, adrenal hormones have been synthesized which are in many instances more potent than the naturally-occurring hormones and often more specific in their action (Fig 20–16).

The introduction of halogen atoms (eg, fluorine) in the 9*α* position of cortisone, cortisol, or corticosterone has produced compounds that are much more potent than the parent compounds. However, the increase in salt-retaining activity which occurs as a result of 9*α* halogenation is relatively greater than that of anti-inflammatory or metabolic activities. For this reason the clinical usefulness of these derivatives is limited. On the other hand, the introduction of a double bond between carbon atoms 1 and 2 results in the production of cortisone and cortisol analogues which in therapeutically useful doses are relatively inert as far as salt-retaining properties are concerned, although they retain the anti-inflammatory activity of the natural steroids. The cortisone analogue is known as prednisone; the cortisol analogue is prednisolone.

Synthetic steroids which contain a methyl group at position 2 have also been prepared. In steroids which have a hydroxy group on position 11 (eg, cortisol, 9*α*-fluorocortisol, or 11*β*-hydroxyprogesterone), the addition of the 2-methyl group markedly enhances the sodium-retaining and potassium-losing activity of the hormone. The most active derivative of this type which has been prepared is 2-methyl-9*α*-fluorocortisol. It is 3 times as potent as aldosterone.

A synthetic analogue of prednisolone, having a similar but more potent anti-inflammatory action, is dexamethasone (9*α*-fluoro-16*α*-methylprednisolone). It is about 30 times more potent than cortisol.

Regulation of Steroid Secretion

The synthesis and secretion of adrenal steroids is controlled by adrenocorticotropin (ACTH) from the pituitary. The secretion of ACTH, in turn, is regulated by corticotropin-releasing factor (CRF) which is released from the hypothalamus during stress.

After stimulation of the gland, there is a rapid decline in the concentration of cholesterol within the adrenal. This and other evidence indicates that ACTH has its effect at some step involving conversion of cholesterol to pregnenolone.

9α-Fluorocortisone

Prednisone

Dexamethasone (9α-fluoro-16α-methyl-prednisolone)

Prednisolone

FIG 20–16. Synthetic steroids.

Most current evidence indicates that ACTH acts to increase the "desmolase" step, a series of oxidative cleavages of the cholesterol side chain employing NADPH as cofactor. The ultimate products of these reactions are the C-21 steroids, 20α,22ξ-dihydroxycholesterol and 17α,20α-dihydroxycholesterol. These compounds are converted directly to pregnenolone or 17α-pregnenolone by loss of an isocaproic aldehyde moiety from their side chains. Since ACTH stimulates synthesis of the substrate for all steroid hormone synthesis, it does not preferentially stimulate synthesis of a particular class of steroids.

The role of the large amounts of ascorbic acid found in the adrenal cortex is not known. It may act to provide reducing equivalents for the NADPH-dependent hydroxylations required for steroid synthesis mentioned below. Ascorbic acid is not synthesized in the adrenal but is concentrated there from extra-adrenal sources. ACTH reduces its uptake into the gland. The measurement of cholesterol or ascorbic acid depletion in the adrenal glands of hypophysectomized animals after the injection of ACTH was an early method of assay for the tropic hormone.

Stimulation of steroid synthesis and release by ACTH may be mediated through cyclic AMP since the level of this substance is increased by the tropic hormone within minutes in adrenal slices. Cyclic AMP itself can directly duplicate ACTH action. Since this agent stimulates activation of the enzyme phosphorylase in adrenal tissue (cyclic AMP is also involved in activation of phosphorylase in liver), Haynes (1958) suggested that it acts by increasing the breakdown of glycogen to glucose-1-phosphate, which in turn gives rise to glucose-6-phosphate. The metabolism of glucose-6-phosphate through the direct oxidative pathway would then produce NADPH necessary for corticosteroid synthesis. The more recent discovery that cyclic AMP can act at many metabolic sites other than on phosphorylase indicates that this hypothesis for the action of cyclic AMP in the adrenal may be an oversimplification.

Stimulation of steroid synthesis is usually associated with alterations in structure of adrenal mitochondrial membrane and with Ca^{++} dependency. The ultimate effect of ACTH and cyclic AMP, therefore, may involve changes in ionic flux across adrenal cell membranes. As in most reactions stimulated by cyclic AMP, ATP is inhibitory.

The secretion of ACTH is under feedback control by circulating steroids; in man, cortisol is the most important regulator. Thus, when cortisol levels decrease, there is a concomitant rise in ACTH. For example, since ACTH nonspecifically stimulates all adrenal steroids, a defect in cortisol production will foster overproduction of androgens, resulting in various forms of adrenogenital syndrome. Pregnenolone is a feedback inhibitor of steroidogenesis, possibly by some unspecified "allosteric effect."

Unlike the other corticosteroids, the production of aldosterone by the adrenal is relatively uninfluenced by ACTH. However, aldosterone production in normal subjects is markedly increased by deprivation of sodium. Conversely, the administration of sodium decreases aldosterone excretion. Normal subjects secrete about 50–150 μg aldosterone per day; but, after a period on a diet low in salt, as much as 300–500 μg/day may be secreted.

Aldosterone production is accompanied by variations in body weight, presumably because of changes in the extracellular fluid volume of the body. It has therefore been suggested (Bartter, 1956) that aldosterone secretion is mainly controlled by changes in extracellular fluid volume and is independent of total body sodium or of sodium concentration. Recent experiments have shown that a specific **aldosterone-stimulating hormone (ASH)** is also an immediate effector of aldosterone production by the adrenal cortex. Reference was made (see p 390) to the fact that infusion of angiotensin produces an increase in the rate of secretion of aldosterone. This fact suggests that the kidney, by means of the renin-angiotensin system (renal pressor system; see p 391), is an important organ controlling aldosterone secretion.

Renin is believed to be secreted by the so-called **juxtaglomerular cells** of the kidney. These cells are located in the walls of the renal afferent arterioles, and it may well be that the "volume receptors" mentioned above are located here. It is assumed that a decreased stretch of the receptors acts as a stimulus to juxtaglomerular cell secretion of renin. Such a stimulus would be brought about by decreased renal arterial pressure and renal blood flow, as might occur in any situation which causes a reduction in intravascular volume, including a deficiency of aldosterone. Renin secreted by the stimulated juxtaglomerular cells converts hypertensinogen to angiotensin I; in plasma, angiotensin I is converted to angiotensin II, which acts directly on the aldosterone-producing cells of the zona glomerulosa of the adrenal cortex. At very high concentrations, renin may act in a similar way to increase nonspecifically the production of the other corticosteroids.

A relation of aldosterone secretion to potassium has also been established. When potassium is withdrawn from the diet, aldosterone secretion is lowered; administration of potassium increases aldosterone production, which thus aids in removal of the excess potassium from the body because of the ability of aldosterone to increase renal excretion of this cation. It is believed that potassium has an effect on aldosterone secretion which is independent of that mediated by changes in fluid volume.

Secretion of aldosterone is increased in several diseases such as cirrhosis, nephrosis, and some types of cardiac failure. There is also an increase in malignant (accelerated) hypertension but not in the benign form of hypertension. The cause of this increase is not known, but the result is enhancement of retention of sodium and water, which further aggravates the edema characteristic of certain of these diseases. In nephritis, the renal loss of sodium and water may directly increase aldosterone production, which then leads to greater losses of potassium and hydrogen ion.

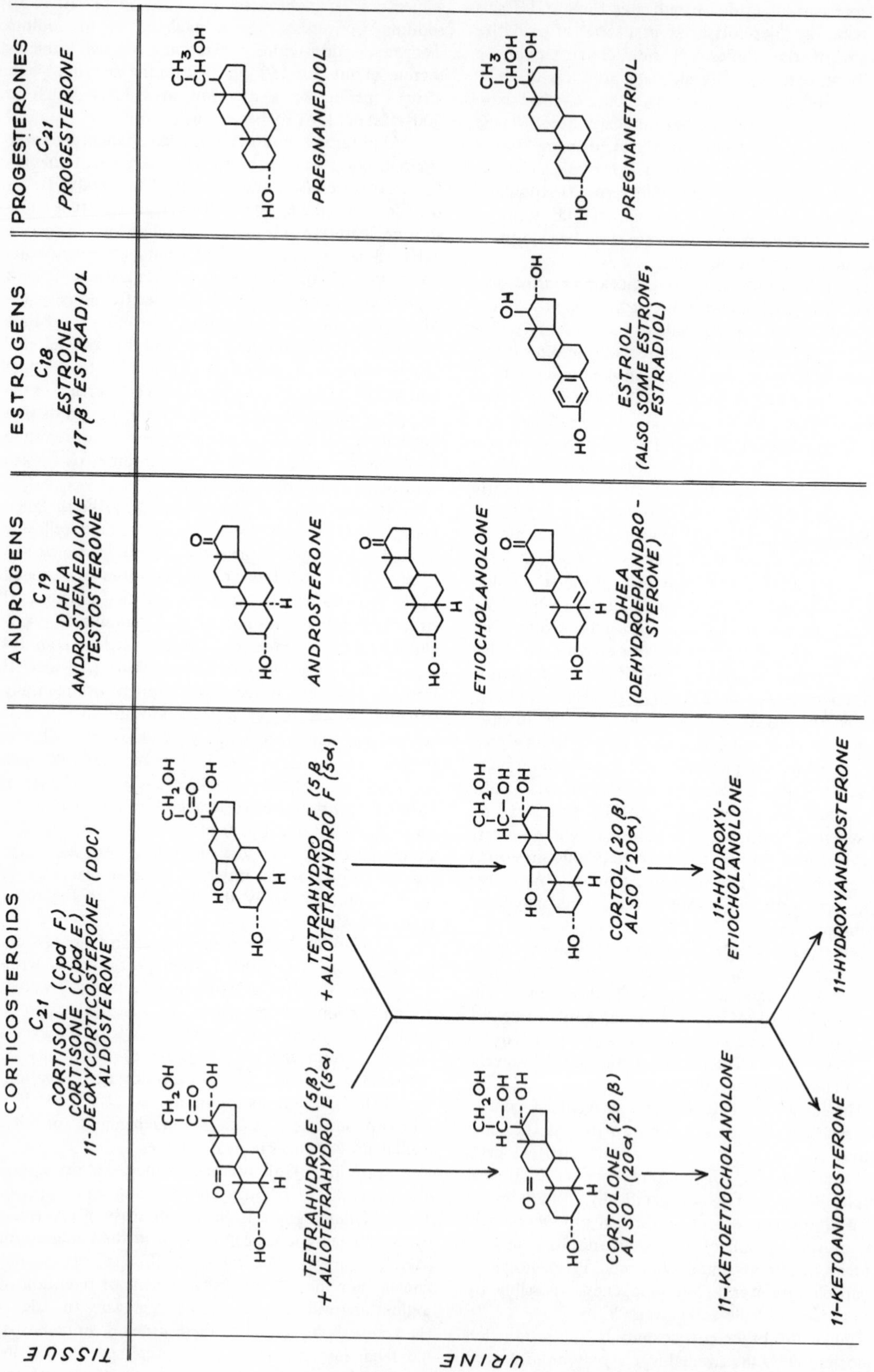

	CORTICOSTEROIDS	ANDROGENS	ESTROGENS	PROGESTERONES
TISSUE	C_{21} CORTISOL (Cpd F) CORTISONE (Cpd E) 11-DEOXYCORTICOSTERONE (DOC) ALDOSTERONE	C_{19} DHEA ANDROSTENEDIONE TESTOSTERONE	C_{18} ESTRONE 17-β-ESTRADIOL	C_{21} PROGESTERONE
URINE	TETRAHYDRO E (5β) + ALLOTETRAHYDRO E (5α) TETRAHYDRO F (5β) + ALLOTETRAHYDRO F (5α) CORTOLONE (20β) ALSO (20α) CORTOL (20β) ALSO (20α) 11-KETOETIOCHOLANOLONE 11-HYDROXY-ETIOCHOLANOLONE 11-KETOANDROSTERONE 11-HYDROXYANDROSTERONE	ANDROSTERONE ETIOCHOLANOLONE DHEA (DEHYDROEPIANDRO-STERONE)	ESTRIOL (ALSO SOME ESTRONE, ESTRADIOL)	PREGNANEDIOL PREGNANETRIOL

FIG 20–17. Primary excretion products of steroids. (As conjugates with glucuronic acid.)

Compounds which block the action of aldosterone on sodium retention may be of value as diuretic agents in treatment of the edema which occurs in those disease states characterized by excess aldosterone production. Such an aldosterone blocking agent which is in clinical use is spironolactone (Aldactone), a compound belonging to a group of steroid spirolactones, all of which are effective diuretics.

Transport & Metabolism of Adrenal Steroids

About ½ of the hydroxycorticosteroids are transported in the blood bound loosely to the serum proteins. Glucocorticoids are bound specifically to an α-globulin **(corticosteroid-binding globulin, CBG, transcortin).** CBG can be increased by estrogen. The bound hormone is essentially inactive, so estrogens can decrease the effectiveness of glucocorticoids by decreasing the amount of free hormone available to the tissues. Progesterone, on the other hand, is one of the few steroids with a high affinity for this binding protein and can cause displacement of cortisol to the free, active fraction. Some reversible binding to albumin may occur at high concentrations of circulating steroids.

In the resting state, the plasma contains 5–15 μg/100 ml of cortisol, the major corticosteroid of the blood. In the human subject the level of cortisol in the plasma is highest in the early morning and lowest during the night hours. The content of aldosterone in plasma is very low (about 0.01 μg/100 ml). During a 24-hour period, the normal adult human secretes about 5–30 mg of cortisol, 1–6 mg of corticosterone, and 30–75 μg of aldosterone.

The disappearance from the body of steroidal compounds such as cortisol is normally very rapid. ^{14}C-labeled cortisol injected intravenously has a half-life of about 4 hours. Within 48 hours, 93% of the injected dose has disappeared from the body: 70% by way of the urine, 20% by the stool, and the remainder presumably through the skin. The steroid nucleus is eliminated in the intact form; no significant breakdown to CO_2 and water occurs.

In the course of their passage through the liver, the corticosteroids are inactivated by ring reduction catalyzed by NADPH-requiring hydrogenases and by reduction of the 3-ketone group by NADH or NADPH, requiring reversible dehydrogenases (Fig 20–17). The resulting tetrahydro derivatives are in turn conjugated, mainly with glucuronic acid. Other tissues (eg, connective tissue; El Attar, 1970) do not inactivate steroids since they lack the necessary hydrogenases and dehydrogenases.

Both free and conjugated corticosteroids are excreted into the intestine by way of the bile and, in part, reabsorbed from the intestine by the enterohepatic circulation. Excretion of free and, in particular, conjugated corticosteroids by the kidney takes place, although there is some tubular reabsorption.

Tait & Horton (1964) have reported that large amounts of aldosterone are produced from androstenedione in the liver. Aldosterone thus produced in the liver may, however, have little peripheral biologic activity since it could be inactivated by reduction and conjugation before leaving the liver. Aldosterone synthesized in the liver would, however, contribute to the conjugates of this hormone measured in the urine.

The inactivating effect of the liver on corticosteroids may decline during periods of prolonged malnutrition (protein and B vitamins) or in liver disease. Decreased excretion may also occur in renal insufficiency. Under any of these circumstances the levels of corticosteroids in the blood are markedly elevated. It has been suggested that in chronic liver disease (cirrhosis) as well as in congestive heart failure the liver does not completely inactivate salt-retaining steroids. This would lead to excessive salt retention, which may be an important cause of edema and ascites found in certain stages of these diseases.

Because of the effect of the liver on their metabolism, the dosage of many corticosteroids, when administered by mouth, must be increased, since after absorption from the gastrointestinal tract the corticosteroids are carried by the portal circulation to the liver, where a portion of the dose may be inactivated.

The blood also contains androgens which are carried both free and in conjugated form in association with the serum proteins. These consist mostly of DHEA, which is present in plasma at a level of about 50 μg/100 ml. About ½ as much is androsterone, the male hormone found in both males and females. The conversion of small amounts of cortisol and cortisone to 11-hydroxy androgens is believed to occur in the liver. In certain forms of liver disease blood levels of 17-hydroxycorticoids are elevated whereas androgens are reduced. Most of the androgens are excreted into the urine as 17-ketosteroids (Fig 10–17).

Laboratory Studies of Adrenocortical Function

A. Urinary 17-Hydroxycorticosteroids: Determination of urinary 17-hydroxycorticosteroids during periods of rest or stimulation with corticotropin (ACTH) is used to assess adrenocortical function. The normal 24-hour excretion of these corticoids (expressed as cortisol) in adult males is 10 ± 4 mg; in females, 7 ± 3 mg. Most of the excreted corticoids are in the conjugated form with glucuronic acid, so that a preliminary hydrolysis with glucuronidase is necessary. After an 8-hour intravenous infusion with 20 units of ACTH, an increase in corticoid excretion of as much as 300% may be seen in normal subjects. In patients with adrenal insufficiency, the levels of 17-hydroxycorticosteroids in the urine will be low and, after administration of ACTH, no increase occurs. If the adrenal insufficiency is secondary to a lack of ACTH, basal steroid levels may be low but some stimulation by ACTH will occur.

An elevation of the 24-hour excretion of 17-hydroxycorticosteroids or abnormally increased secretion after stimulation with ACTH strongly suggests that the patient has hyperactivity of the adrenal cortex, such as Cushing's syndrome (unless the patient is under stress), or has liver disease with hepatic insufficiency.

A useful diagnostic test is based upon the use of the enzyme inhibitor metyrapone (Metopirone). This drug inhibits adrenal 11β-hydroxylase, the enzyme which catalyzes the addition of the hydroxy group at position 11 of the steroid nucleus in the biosynthetic pathway for corticosteroid formation (Figs 20–13 and 20–14). Thus, a transient deficiency of cortisol is produced which normally results in stimulation of secretion of ACTH from the pituitary (see discussion of braking action of cortisol, p 447), followed in turn by a resultant increase in 11-deoxycortisol secretion. The test may also be used to assess adrenocortical function. Failure of corticoids to rise after Metopirone infusion favors the diagnosis of adrenal neoplasm rather than hyperplasia.

Colorimetric assay of adrenocorticoid excretion is usually based on the **Porter-Silber** color reaction using phenylhydrazine and sulfuric acid. This reaction is positive with 17,21-dihydroxy-20-ketosteroids such as cortisone, cortisol, and 11-deoxycortisol, as well as their reduced (tetrahydro) derivatives. Although the reaction is not specific to these steroids, preliminary extraction and chromatographic separation improves the sensitivity and specificity of the assay.

B. Urinary 17-Ketosteroids: The 17-ketosteroids may be subdivided into 3 general types: acidic, phenolic, and neutral (Fig 10–18). The acidic compounds are derived from the bile acids; the phenolic, from the estrogens; the neutral reflect mainly the excretion of metabolic derivatives from the endocrine secretions of the gonads and the adrenal cortex. The principal components of this so-called "neutral 17-ketosteroid" fraction of the urinary steroids are androsterone and its isomers, etiocholanolone and isoandrosterone, as well as DHEA (Fig 20–18). The neutral 17-ketosteroids may be further subdivided into alpha and beta fractions, alpha and beta referring to the stereochemical configuration at carbon 3 of the steroid molecule. The beta fractions are those which are precipitated by digitonin; the alpha fractions are not. Since the 17-ketosteroids are present as conjugates, they are first hydrolyzed with concentrated hydrochloric acid or sulfuric acid or the enzyme β-glucuronidase. The ketosteroids are estimated colorimetrically by means of the **Zimmermann reaction**, which involves coupling of the reactive 17-keto group of these compounds with **metadinitrobenzene** to form a blue complex. The standard in the estimation of the urinary 17-ketosteroids is crystalline dehydroepiandrosterone (DHEA). In some cases, the neutral fraction obtained as described above may be treated with **Girard's reagent T**, which separates it into ketonic and nonketonic components. The nonketonic fraction comprises only 10–15% of the original neutral fraction. The remaining 85–90% is the ketonic fraction; it is separable into alpha and beta components by digitonin as described above. A flow chart of these operations is shown in Fig 20–18.

The urinary neutral 17-ketosteroids are a reflection of the androgenic function of the subject. In the female, 17-ketosteroids are produced entirely by the adrenal cortex; in the male, by the adrenal cortex and testes. The testes contribute 1/3 of the neutral urinary 17-ketosteroids. DHEA is probably derived mainly from the adrenal. It is found in the urine of both normal men and women, and it is greatly increased in some cases of hyperadrenocorticism, particularly those with excess production of androgens. While most androgens increase the excretion of neutral 17-ketosteroids, some injected forms (eg, methyltestosterone) do not.

Normal values for neutral 17-ketosteroids in a 24-hour urine sample are as follows:

Adult males: 10–20 mg (avg, 15 mg)
Adult females: 5–15 mg (avg, 10 mg)
Ratio alpha to beta ketonic fractions, 9:1

Only a small portion of the 17-ketosteroids excreted is contributed by the breakdown of 17-hydroxycorticoids. Therefore, 17-ketosteroids do not represent a direct measurement of adrenocortical activity as do 17-hydroxycorticoids. Only a slight elevation occurs in Cushing's syndrome, acromegaly, simple hirsutism, pregnancy, and ovarian hyperthecosis. However, the simultaneous determination of 17-hydroxycorticoid excretion and 17-ketosteroid excretion is of value in differentiating Cushing's syndrome from the adrenogenital syndrome. Whereas in Cushing's syndrome the 17-ketosteroid excretion barely exceeds normal except in the presence of a carcinoma (when very high values usually are observed), 17-ketosteroid values are elevated in adrenogenital syndrome. In adrenocortical carcinoma, bilateral hyperplasia of the cortex, and testicular tumors (Leydig cell tumors), urinary 17-ketosteroid excretion is also often increased. During the administration of corticotropin, increases in urinary 17-ketosteroids may be noted if the adrenal is competent to produce steroid hormones. Cortisol may effect a fall in 17-ketosteroid excretion because of inhibition of ACTH production. Excretion of 17-ketosteroids is low in Addison's disease, pituitary dwarfism, Simmonds' disease, and occasionally in anorexia nervosa and myxedema. Slight reduction in normal values for urinary 17-ketosteroids may be noted in eunuchoidism or in castrates, as well as in diabetes mellitus, after the climacteric, and during any chronic debilitating disease. The levels normally decline somewhat in old age.

C. Urinary Aldosterone: Measurement of aldosterone in the urine is of great value in the diagnosis of primary or secondary hyperaldosteronism. At present, quantitative determination of this hormone requires chromatographic separation and quantitative determination by double isotope dilution (see below).

D. Specific Methods for Assay of Steroids: A great variety of methods are being devised for the assay of specific steroids. These newer methods are more precise than the older technics, which usually depended on generalized characteristics of groups of steroids. Gas-liquid chromatography is particularly valuable for separation and quantitation of steroids

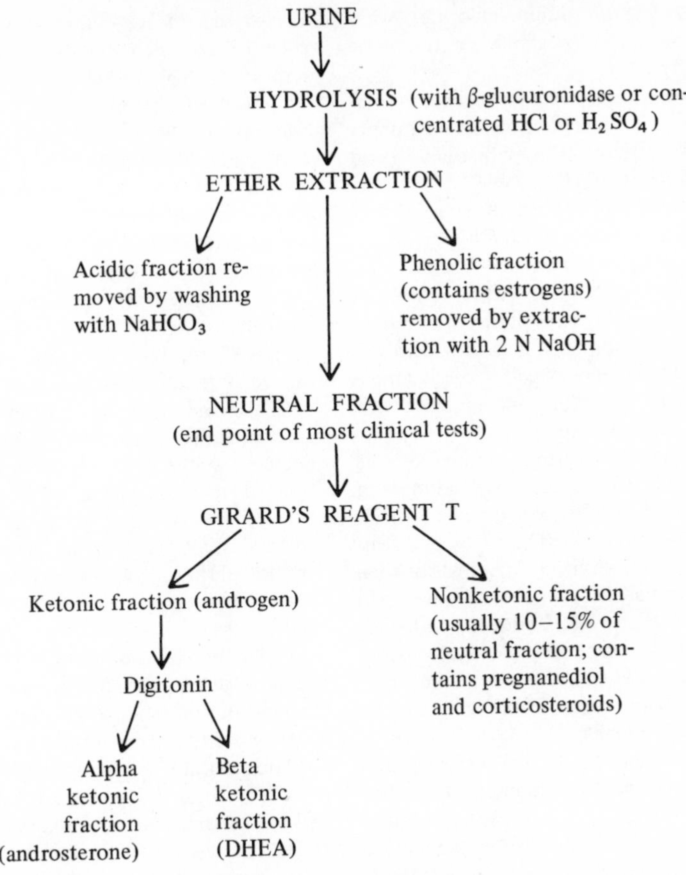

FIG 20–18. Flow chart for detection of urinary ketosteroids.

(including androgens and estrogens) after partial purification by column or paper chromatography.

The double isotope dilution technic uses 2 different isotopic forms (usually tritium and ^{14}C) of the steroid to be measured. One form is injected into the patient; the second is added in vitro to the sample taken for analysis. The steroid is rigorously purified by a series of chromatographic procedures, and the final chemical recovery is calculated from the recovery of the isotopic form added in vitro. The specific activity of the recovered injected steroid is used to calculate secretion rate. (For discussion of this method see Eik-Nes & Hall, 1965.)

Abnormal Physiology

A. Hypoadrenocorticism:

1. Adrenalectomy—With adrenalectomy, sodium is lost as a result of the failure of the renal tubules to reabsorb sodium ion. Sodium loss is accompanied by retention of potassium and losses of chloride and bicarbonate. There is associated loss of water from the blood and tissue spaces, resulting in severe dehydration and hemoconcentration. These effects on salt and water metabolism are promptly reversed by adrenocortical extract when adequate water and minerals are made available. In the adrenalectomized rat, merely providing sodium chloride suffices to keep the animal alive indefinitely.

2. Addison's disease—In humans, degeneration of the adrenal cortex, often due to a tuberculous process, results in Addison's disease. Depending on the extent of adrenocortical hypofunction, the effects are similar to those observed in the adrenalectomized animal. They include decreased 17-hydroxycorticoid and aldosterone excretion, excessive loss of sodium chloride in the urine, elevated levels of potassium in the serum, low blood pressure, muscular weakness, gastrointestinal disturbances, low body temperature, hypoglycemia, and a progressive brownish pigmentation which increases over a period of months. The pigmentation is caused by the MSH activity inherent in the structure of ACTH which is present in increased amounts as a result of the deficiency of cortisol.

The symptoms of Addison's disease, with the exception of hypoglycemia and pigmentation, are mainly due to the lack of salt- and water-regulating hormones, particularly aldosterone. Addisonian patients may be maintained on deoxycorticosterone acetate (DOCA), although most are now maintained on small amounts of cortisone acetate together with a daily intake of 10–15 g of sodium chloride. Addisonian patients withstand any shock or traumatic experience very poorly.

B. Hyperadrenocorticism: Adrenocortical hyperfunction may be caused by benign or malignant tumors of the cortex or by adrenocortical hyperplasia initiated by increased production of ACTH.

The continuous administration of steroid hormones or ACTH may also induce signs of hyperadrenocorticism. These include (1) hyperglycemia and glycosuria (diabetogenic effect); (2) retention of sodium and water, followed by edema, increased blood volume, and hypertension; (3) negative nitrogen balance (protein anti-anabolic effect and gluconeogenesis); (4) potassium depletion and hypokalemic alkalosis; and (5) hirsutism and acne.

Hyperplasia as well as certain tumors of the adrenals cause the production of increased amounts of androgenic (C-19) steroids. The congenital form is almost always due to hyperplasia. The resulting disturbance is termed **congenital virilizing hyperplasia** when it is present at birth and **adrenogenital syndrome** when it occurs in the postnatal period. Under the influence of excess androgens, the female assumes male secondary sex characteristics. When it occurs in the male, there is excessive masculinization. Feminizing adrenal tumors may rarely occur in males.

There are 3 varieties of the adrenogenital syndrome. The first and most common is the uncomplicated form which in either sex apparently causes little or no disturbance in salt and water metabolism in the child. Under these circumstances the metabolic defect seems to lie in the virtual absence of C-21 hydroxylase. Consequently, cortisol and cortisone are not produced in normal amounts; aldosterone production is not impaired. The urine of these patients contains large amounts of 17-ketosteroids and of pregnanetriol, but cortisol is very low in both blood and urine. It is assumed that cortisol is necessary to maintain a proper balance between the pituitary and the adrenal. In the absence of this hormone, excess ACTH production occurs because of a lack of the "braking effect" which cortisol normally exerts. The resulting hyperactivity of the adrenal leads to the production of increased amounts of androgenic steroids of the C-19 type. There is also increased production of pregnanetriol, which reflects the shunting of the intermediates of the C-21 pathway which cannot be converted to cortisol (and cortisone). Treatment of these patients with cortisol restores pituitary-adrenal balance. Androgenic steroid production by the adrenal is then reduced, as evidenced by a fall in the excretion of 17-ketosteroids to normal.

A variant of this adrenogenital syndrome is a more severe salt-losing form. Death may occur, as in addisonian crisis, shortly after birth, if the disease is not promptly recognized. In this case, both 17- and 21-hydroxylation are inhibited, resulting in decreased production of both glucocorticoids and the mineralocorticoids.

A second and rather rare variety of the adrenogenital syndrome is the hypertensive form. It is characterized by high blood pressure and vastly increased excretion of compound S (11-deoxycortisol), which may not be identified in the ordinary assays for urine corticosteroids and hence is measured as a corticosteroid of the 11-oxy type. The characteristic hypertension is apparently due to the production of deoxycorticosterone. The defect may lie in a failure of hydroxylation at C-11; but since there is no defect in C-21 hydroxylation, pregnanetriol excretion is not particularly high.

A third, comparatively rare, variety is characterized by extensive adrenal insufficiency, sodium loss, collapse, and early death. Deficiency of both glucocorticoids and mineralocorticoids is pronounced. The defect is at the $\Delta 5$-isomerase level, resulting in inhibition of the conversion of pregnenolone to progesterone. Thus the glucocorticoids and mineralocorticoids, all of which are derived from progesterone, are not produced, whereas androgens, all of which are derived from pregnenolone, are increased (Figs 20–13 and 20–14).

The urinary excretion of 17-hydroxycorticoids is elevated in adrenocortical adenoma. However, the increase in excretion is greater in carcinoma, and it is not elevated after an intravenous ACTH test as it is in adrenocortical adenoma.

The adrenal suppression test may also be used as an aid to diagnosis of abnormal adrenocortical function. The test is based upon the observation that giving fluorohydrocortisone in low dosage will suppress ACTH secretion and thereby suppress the activity of normal adrenals. Hyperplastic glands are not suppressed unless the drug is given in large doses, when the activity of hyperplastic adrenals is suppressed but not that of glands in which the hyperactivity is due to adenoma or carcinoma.

Cushing's Disease & Syndrome

This disease was first described by Cushing in 1932 and was attributed at that time to pituitary hyperactivity due to a basophilic tumor (adenoma) of the pituitary. The prevailing opinion is that the disease is indeed probably of pituitary origin except in those cases wherein a unilateral tumor of the adrenal is the direct cause of the hyperadrenocorticism. To distinguish between the 2 etiologic forms of the disease (pituitary and adrenal), the term **Cushing's disease** has been restricted to those cases which are of pituitary origin, ie, pituitary basophilism. **Cushing's syndrome** denotes adrenocortical hyperfunction directly involving the adrenal gland.

Both cortisol and ACTH, when administered for prolonged periods, induce the Cushingoid state, which is reversible when these agents are discontinued.

Cushing's syndrome or Cushing's disease is characterized by a rapidly increasing adiposity of the face, neck, and trunk (buffalo fat distribution), a tendency to purpura, purplish striae on the abdomen, hirsutism, sexual dystrophy, hypertension, and osteoporosis.

Hyperglycemia (insulin resistance) is associated with protein catabolism and negative nitrogen balance. There is also a tendency to depletion of potassium and to alkalosis.

Aldosteronism

"Primary aldosteronism" (Conn, 1955) results from tumors (aldosteronomas) of the adrenals in which the hyperactivity of the adrenal cortex is apparently confined to excess production of aldosterone. The primary metabolic defect may be an inability of the adrenals to perform 17-hydroxylations, thereby shunting progesterone to aldosterone (Fig 20–13). Patients with these tumors exhibit intermittent tetany, paresthesias, periodic muscle weakness, alkalosis, persistent low serum potassium levels, hypertension, and polyuria and polydipsia but no edema. The significant metabolic findings are hypokalemic alkalosis, hyperaldosteronuria, and hyposthenuria which is unresponsive to antidiuretic hormone (ADH). After sodium restriction, the tendency to alkalosis and hypokalemia is eliminated and sodium disappears from the urine. Measurement of aldosterone excretion in the urine, particularly after sodium loading, is also helpful in establishing the diagnosis.

A consistently low level of potassium in the serum is a characteristic finding in primary hyperaldosteronism. If the administration of the aldosterone antagonist, spironolactone (Aldactone) restores serum potassium to normal levels in a patient, the presence of this disease may be strongly suspected.

The lack of edema and the presence of polyuria in primary hyperaldosteronism may be explained by failure of the renal tubule to respond to ADH because of a potassium depletion induced by the excess urinary potassium losses which the tumor engenders.

Conditions associated with excessive sodium and water loss such as congestive heart failure, cirrhosis, and nephrosis may produce **secondary aldosteronism** and a resultant edema.

THE SEX HORMONES

The testes and ovaries, in addition to their function of providing spermatozoa or ova, manufacture steroid hormones which control secondary sex characteristics, the reproductive cycle, and the growth and development of the accessory reproductive organs, excluding the ovary and testis themselves. The sex hormones also exert potent protein anabolic effects.

Most of the regulation of hormone production in the testes and ovaries is controlled by tropic hormones from the pituitary.

THE MALE HORMONES

A number of androgenic hormones (C-19 steroids) have been isolated either from the testes or the urine. Their physiologic activity may be tested by the effects on the seminal vesicles and prostate of their administration to castrated animals. In the capon, restoration of the growth of the comb and wattles, the secondary sex characteristics of the rooster, follows the administration of androgenic hormones.

Androsterone is used as the international standard of androgen activity; 1 IU = 0.1 mg of androsterone. Although this hormone does not occur in the testes, it is excreted in the urine of males (Fig 20–17) as a metabolite of testosterone.

The principal male hormone, **testosterone,** is synthesized by the interstitial (Leydig) cells of the testes from cholesterol through pregnenolone, progesterone, and hydroxyprogesterone, which is then converted to the 19-ketosteroid, androstenedione, the immediate precursor of testosterone. Alternatively, the pathway through hydroxypregnenolone and dehydroepiandrosterone (DHEA) can be used to produce androstenedione. A direct conversion of DHEA to testosterone has been established in which androstenedione is bypassed; DHEA in this pathway is initially reduced to its 17-hydroxy derivatives, which are then converted to testosterone. These reactions are shown in Figs 20–13 and 20–14 since they are also a part of the biosynthetic pathway in the adrenal for the formation of the androgenic (C-19) steroids. It will also be noted that pregnenolone is a common precursor of the adrenocortical hormones and testosterone as well as of progesterone.

Circulating DHEA sulfate from the adrenal can be converted in the testes to free DHEA by a sulfatase and thus provide an additional source of testosterone precursor in this tissue.

In addition to testosterone, androstenedione and DHEA (androgens also produced by the adrenal) are synthesized in the testes although in amounts and with a total androgenic potency far less than that of testosterone.

The androgens are in part transported by binding to a specific plasma protein. This protein increases in pregnancy or estrogen therapy, which results in a reduction of effective "free" androgenic action. About 2/3 of the testosterone circulating in the plasma are bound to protein.

Recent evidence indicates that testosterone is converted in target tissues to the more potent dihydrotestosterone, which may be the active intracellular androgen (Bruchovsky & Wilson, 1968). In general, the testes and the adrenals have similar qualitative capacities to synthesize androgens. Since the testis lacks 11-hydroxylase activity, however, only the adrenal can synthesize the glucocorticoids and mineralocorticoids (Figs 20–13 and 20–14).

In the normal male, 4–12 mg of testosterone are secreted per day. Direct measurement of testosterone in plasma by isotope dilution or radiodisplacement assay indicates that about 0.6 μg/100 ml is present in the normal male and about 0.1 μg/100 ml in the normal female. The small amount of testosterone in female plasma results mainly from peripheral conversion of androstenedione to testosterone by the ovary. Dehydroepiandrosterone (DHEA) is secreted in greater amounts than testosterone in normal men (15–50 mg/24 hours).

Testicular function is controlled by both the pituitary FSH and LH; increased testosterone levels cause feedback inhibition of LH secretion.

The principal metabolites of testosterone are androsterone and etiocholanolone, the major 17-ketosteroids in the urine (Fig 20–17). In addition, small amounts of DHEA (as the sulfate) are excreted. The principal pathway of degradation of testosterone involves oxidation in the liver to androstenedione and subsequent saturation of the double bond in ring A and reduction of the keto groups (Figs 20–14 and 20–17). Some 11-oxy or 11-hydroxy derivatives of androsterone and androstenedione are produced in the liver from adrenal cortisol and cortisone.

Testosterone promotes the growth and function of the epididymis, vas deferens, prostate, seminal vesicles, and penis. It is of value as replacement therapy in eunuchoidism. Its metabolic effect as a protein anabolic steroid exceeds that of any other naturally occurring steroid. It also contributes to the muscular and skeletal growth that accompanies puberty; this function is doubtless a necessary concomitant to its androgenic functions.

In vivo, androgens promote protein synthesis in male accessory glands by causing increased RNA and RNA polymerase in the nucleus and increased aminoacyl transferase at the ribosomal level. As a possible result of these actions, androgens increase the activity of glycolytic enzymes, of hexokinase, and of phosphofructose kinase. The androgens also act at the mitochondrial level to increase respiration rate, number of mitochondria, and synthesis of mitochondrial membranes.

The protein anabolic effect of testosterone (nitrogen-retaining effect) is certainly as important as its androgenic effects, if not more so. In many clinical situations where promotion of protein anabolism is required, testosterone has proved quite effective, but the accompanying androgenic effects, particularly at the high dosages required, are often undesirable. Consequently, efforts have been directed toward the production of synthetic steroids which, while retaining the protein anabolic action of testosterone, are relatively free of androgenicity. Although a number of such compounds have been introduced, their effectiveness has not yet been completely evaluated. Some synthetic androgens include fluoxymesterone and 2α-methyldehydrotestosterone.

Excretion of 17-ketosteroids as their sulfates and glucuronides is in part a reflection of testicular hormone production. The testis contributes about 1/3 of the urinary neutral 17-ketosteroid. In normal children up to the eighth year of life, there is a gradual increase in 17-ketosteroid excretion up to 2.5 mg/day; an increase up to 9 mg occurs between the eighth year and puberty in both boys and girls; after puberty, the sex differences in ketosteroid excretion which were noted on p 450 occur. In eunuchs, ketosteroids may be normal or lowered, whereas in testicular tumors with associated hyperactivity of the Leydig cells ketosteroid excretion may be considerably increased.

THE FEMALE HORMONES

Two main types of female hormones are secreted by the ovary: the **follicular** or **estrogenic** hormones produced by the cells of the developing graafian follicle; and the **progestational** hormones derived from the corpus luteum that is formed in the ovary from the ruptured follicle.

The Follicular Hormones

The estrogenic (follicular) hormones are C-18 steroids, differing from androgens in lacking C-19 substitution and the methyl group at C_{10}. In contrast to all other natural steroids, ring A is aromatic (Fig 20–14).

The principal estrogenic hormone in the circulation is estradiol, which is in metabolic equilibrium with estrone.

The international standard for estrogen activity is estrone; 1 International Unit (IU) = 0.1 mg estrone. Comparable biologic activity of the estrogens to increase uterine weight in rodents varies with the mode of administration: estradiol > estrone > estriol when given subcutaneously; estriol > estradiol > estrone after oral administration.

Estriol (Fig 20–17) is the principal estrogen found in the urine of pregnant women and in the placenta. It is produced by hydroxylation of estrone at C_{16} and reduction of the ketone group at C_{17}.

The androgens, testosterone and androstenedione, are precursors for the synthesis of the estrogens in testes, ovaries, adrenals, and placenta (Fig 20–14). The conversion from testosterone involves 3 enzyme-catalyzed steps which require oxygen and NADPH: (1) 19-hydroxylation to 19-hydroxytestosterone or 19-hydroxyandrostenedione; (2) 19-oxidation to the keto derivatives; and (3) aldehyde lyolysis to remove the C_{19} keto group and cause aromatization of the A ring. **Metyrapone** inhibits estrogen synthesis by blocking 19-hydroxylation. As with the other steroids, estrogens (primarily estriol) can be found in urine either as a conjugate with sulfate or as a glucuronide (Fig 20–17).

Physiologic Effects of Estrogenic Hormones

In the lower animals, the estrogenic hormone induce estrus, a series of changes in the female reproductive system associated with ovulation. These changes may be detected by the histologic appearance of the vaginal smear. The **Allen-Doisy test**, used for the detection of estrogenic activity, is based on the ability of a compound to produce estrus in ovariectomized, sexually mature rats.

In women, the follicular hormones prepare the uterine mucosa for the later action of the progestational hormones. The changes in the uterus include proliferative growth of the lining of the endometrium, deepening of uterine glands, and increased vascularity; changes in the epithelium of the fallopian tubes and of the vagina also occur. All of these changes begin immediately after menstrual bleeding has ceased.

The estrogens also suppress the production of the pituitary hormone (follicle-stimulating hormone, FSH) which initially started the development of the follicle. In contrast, they appear to stimulate pituitary LH; peak levels of estrogens precede the peak level of LH by 1–2 days. Estrogens are effective in maintenance of female secondary sex characteristics, acting antagonistically to testosterone.

Mechanism of Action

A number of experiments have indicated that the activities of several enzyme systems in the uterus as well as the placenta are stimulated by the prior administration of estradiol to the intact rat. The estrogens are bound to specific cytosol receptor proteins in the cells which transport them into the nucleus where they act to increase RNA synthesis. It is possible that estrogens may act on target tissues at the level of RNA synthesizing enzymes; in ovariectomized rats, RNA polymerase activity in the nuclear fraction of uterine homogenates (but not liver homogenates) is increased by 2 hours after the administration of estrogen.

Estrogens may also act as cofactors in a transhydrogenation reaction (Fig 20–14) in which H ions and electrons are transferred from reduced NADP to NAD:

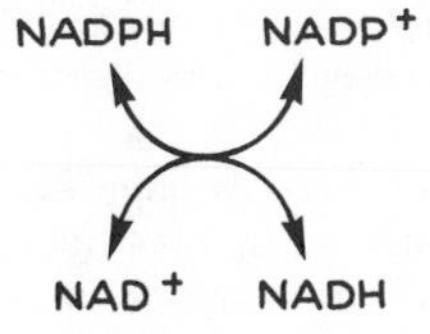

NADPH → NADP$^+$
NAD$^+$ → NADH

The estrogen-dependent transhydrogenase which catalyzes the transfer of hydrogen from NADPH to NAD may bring about an increased rate of biologically useful energy in 2 ways: (1) Considering that the concentration of NADP in the cells is ordinarily very low, increased oxidation of NADPH would maintain availability of NADP for the activity of those dehydrogenases which required the oxidized form of this cofactor. (2) Because the direct oxidation of NADP yields little if any high-energy phosphate, a method of deriving high-energy phosphate indirectly would be made available by transfer of hydrogen to NAD, thus forming NADH which could then be oxidized through the respiratory chain. As a result of an increased rate of production of biologically useful energy, as described above, stimulation of growth might be brought about in those tissues affected by estrogens. This mechanism may in part explain the ability of estradiol to stimulate the rate of oxygen consumption, the conversion of acetate to CO_2, and the incorporation of acetate into protein, lipid, and the adenine and guanine of the total nucleic acid fraction of target tissues. A soluble 17β-hydroxysteroid dehydrogenase from human placenta has been isolated and purified. This enzyme catalyzes the interconversion of estrone and estradiol-17β. According to Jarabak & others (1962), the dehydrogenase reacts with both NAD and NADP and is responsible for most, if not all, of the estradiol-17β-mediated transfer of hydrogen between pyridine nucleotides which occurs in soluble extracts of human placenta.

Synthetic Estrogens

A number of synthetic estrogens have been produced; the following 2 are clinically valuable:

A. Ethinyl estradiol is a synthetic estrogen which when given orally is 50 times as effective as water-soluble estrogenic preparations or 30 times that of estradiol benzoate injected intramuscularly.

B. Diethylstilbestrol is an example of a group of para-hydroxyphenyl derivatives which, while not steroidal in structure, nonetheless exert potent estrogenic effects. However, as shown in the formula below, it is possible that ring closure may occur in the body to form a structure resembling the steroid nucleus.

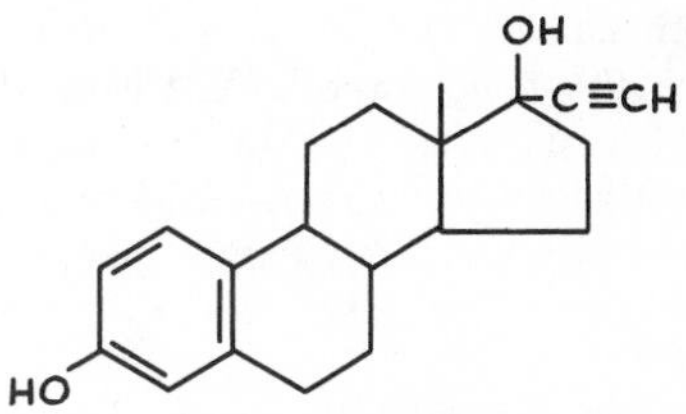

OH, C≡CH, HO

Ethinyl estradiol

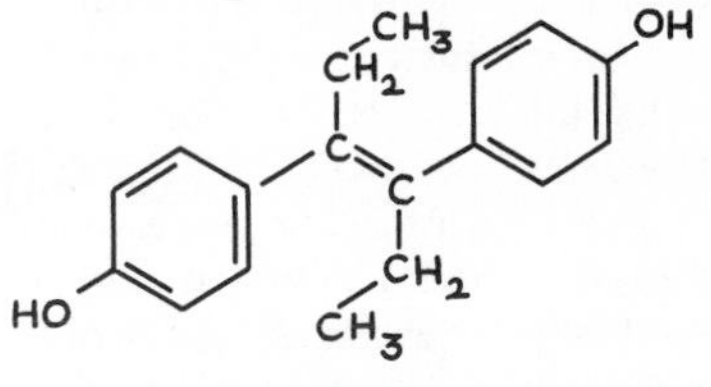

CH_3, CH_2, C, C, CH_2, CH_3, OH, HO

Diethylstilbestrol

The Progestational Hormones (Luteal Hormones)

Progesterone (Fig 20–13) is the hormone of the corpus luteum, the structure which develops from the ruptured follicle. It is formed also by the placenta, which secretes progesterone during the latter part of pregnancy. Progesterone is also formed in the adrenal cortex as a precursor of both C-19 and C-21 corticosteroids (Fig 20–13). In all of the above tissues, progesterone is synthesized from its immediate precursor, pregnenolone, by a combined dehydrogenase and isomerase reaction. The steroid analogue, cyanotrimethylandrostenolone, can inhibit this conversion.

Functions of progesterone. This hormone appears after ovulation and causes extensive development of the endometrium, preparing the uterus for the reception of the embryo and for its nutrition. The hormone also suppresses estrus, ovulation, and the production of pituitary luteinizing hormone (LH), which originally stimulated corpus luteum formation. Progesterone antagonizes the action of estrogens in various tissues, including the cervical mucus, vaginal epithelium, and fallopian tubes. Progesterone also stimulates the mammary glands. When pregnancy occurs, the corpus luteum is maintained and menstruation and ovulation are suspended. The concentration of progesterone decreases near term. The Corner and Hisaw test for progesterone is based on the increase of secretory action caused by the hormone on the uterine endometrium.

If fertilization does not occur, the follicular and progestational hormones suddenly decrease on about the 28th day of the cycle; the new cycle then begins with menstrual bleeding and sloughing of the uterine wall (Fig 20–19).

The metabolic fate of progesterone has been studied by the injection of ^{14}C-labeled hormone. About 75% of injected progesterone (or its metabolites) is transported to the intestine by way of the bile and eliminated in the feces. Large amounts occur in the urine only if the biliary route of excretion is blocked.

Other progestational hormones. In addition to progesterone, the corpus luteum may also produce a second hormone which has been termed **relaxin** because of its ability to bring about relaxation of the symphysis pubis of the guinea pig or of the mouse. Extracts of corpora lutea from the sow contain the active relaxing principle, and the blood of pregnant females of a number of species (including man) does also. Relaxin also occurs in the placenta. It is active only when injected into an animal in normal or artificially induced estrus. Relaxin appears chemically to be a polypeptide with a molecular weight of about 9000 daltons. It is inactivated by treatment with proteolytic enzymes or by reagents which convert (by reduction) disulfide linkages to sulfhydryl groups.

The chief excretory product of progesterone is **pregnanediol** (Fig 20–17). It is present in quantities of 1–10 mg/day (as the glucuronide) during the latter ½ of the menstrual cycle. Its presence in the urine signifies that the endometrium is progestational rather than follicular.

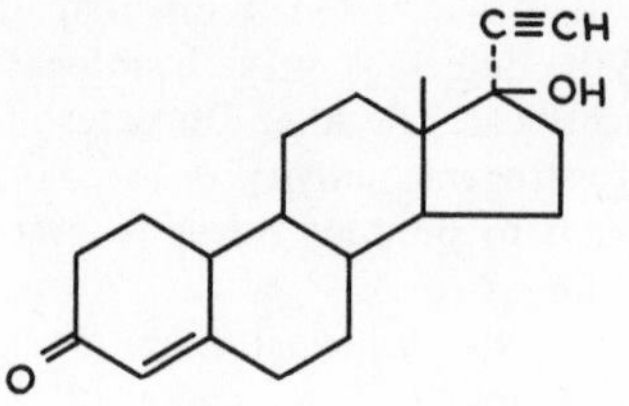

Norethindrone (Norlutin; 17α-ethinyl-19-nortestosterone)

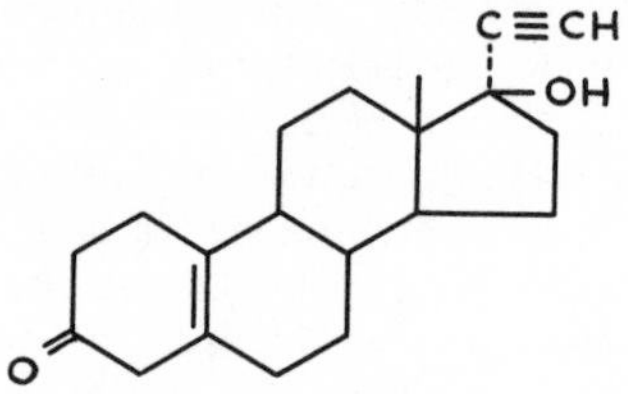

Norethynodrel (Enovid*; 17α-ethinyl-17-hydroxy-5(10)-estren-3-one)

Pregnenolone also has some progestational activity. It may be used therapeutically to produce a secretory endometrium or to inhibit uterine motility in threatened abortion.

Orally effective progestational agents. Progesterone is relatively ineffective when taken by mouth. In recent years, several synthetically produced progestational agents have been devised which are much more effective biologically than progesterone when taken orally. Two such compounds are shown above. Because, like progesterone, they have the ability to suppress ovulation, they have found application in association with estrogens as oral contraceptives.

Chorionic Gonadotropin (Anterior Pituitary-Like Hormones, APL); Pregnancy Tests

The gonadotropic hormone found in urine during pregnancy is used as the basis for a test for pregnancy. For many years, the routine pregnancy test required maintenance of animal colonies. Injection of urine of a pregnant woman into immature female mice or rats caused rapid changes in the ovaries which could be easily seen as hemorrhagic spots and yellowish corpora lutea. This was the **Aschheim-Zondek test**. For the **Friedman test**, a virgin rabbit was used. The test urine was injected into an ear vein, and the ovaries were examined 24 hours later for ruptured or hemorrhagic follicles. The male frog (*Rana pipiens*), the female toad

*Enovid is a mixture containing, in addition to norethynodrel, a small portion of mestranol (ethinyl estradiol 3-methyl ether).

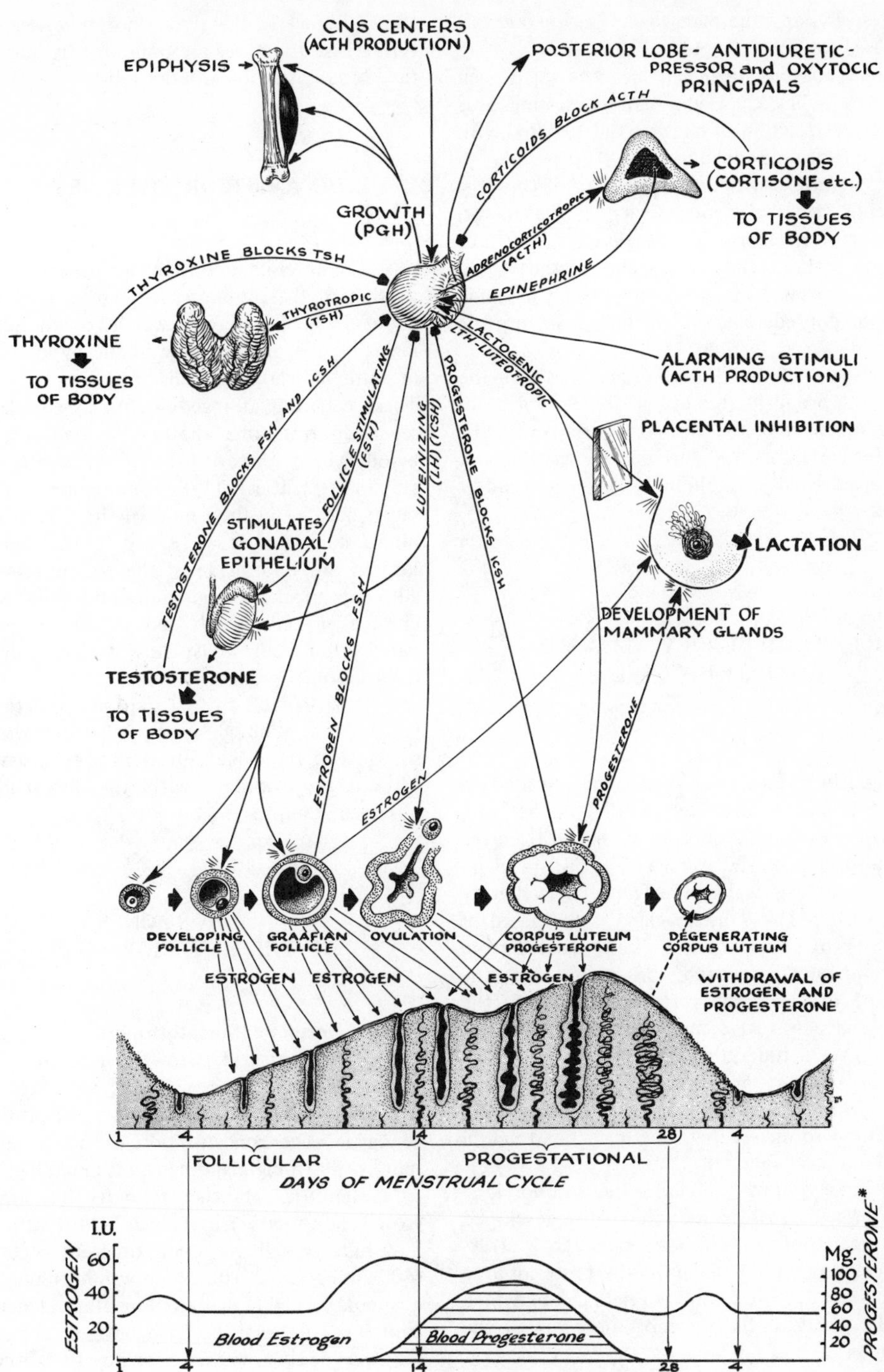

*Progesterone as determined from the urinary pregnanediol.

FIG 20–19. Relationships of the pituitary hormones to the target glands and tissues.

(*Xenopus laevis*), and the male toad (*Bufo arenarum*) were also used for pregnancy tests.

An immunoassay pregnancy test has since been developed which has superseded biologic testing. It is based upon precipitation of latex particles coated with antibodies to chorionic gonadotropin.

Chorionic gonadotropin is a product of the very early placenta. It is also produced when there is abnormal proliferation of chorionic epithelial tissue such as hydatidiform mole or chorionepithelioma; this would give rise to false results in pregnancy tests. Chorionic gonadotropin is used to induce ovulation in infertile women.

The effects of the female hormones on the menstrual cycle are illustrated in Fig 20–19.

For a summary of the interrelationships of FSH, LH, and female hormones during menstruation, see Catt references listed in the bibliography at the end of this chapter.

THE PITUITARY GLAND (HYPOPHYSIS)

The human pituitary is a reddish-gray oval structure, about 10 mm in diameter, located in the brain just behind the optic chiasm as an extension from the floor of the thalamus. The average weight of the gland in the male is 0.5–0.6 g; in the female it is slightly larger, 0.6–0.7 g. The pituitary gland is composed of different types of tissue embryologically derived from 2 sources: a neural component, originating from the infundibulum (a downgrowth from the floor of the thalamus); and a buccal component which develops upward from the ectoderm of the primitive oral cavity (stomodeum) to meet and surround the infundibular rudiment. The terms **adenohypophysis** and **neurohypophysis** are used to differentiate the buccal and neural components, respectively.

The adenohypophysis includes the anterior lobe and the intermediate or middle lobe of the developed endocrine organ, both of which are glandular in structure. The neurohypophysis includes the posterior lobe of the gland and the infundibular or neural stalk which attaches the gland to the floor of the brain at the hypothalamus.

Complete removal of the pituitary (hypophysectomy) in young animals (eg, rats) causes a cessation of growth and a failure in maturation of the sex glands. Removal of the pituitary in the adult animal is followed by atrophy of the sex glands and organs, involution of the thyroid, parathyroids, and adrenal cortex, and a depression of their functions. In addition, there are alterations in protein, fat, and carbohydrate metabolism. A notable characteristic of a hypophysectomized animal or of the human patient suffering from pituitary insufficiency is abnormal sensitivity to insulin and resistance to the glycogenolytic effect of epinephrine. These functions are undoubtedly due to a lack of the hormones of the anterior lobe.

THE ANTERIOR PITUITARY GLAND

The anterior lobe is the largest and most essential portion of the pituitary. In man, this lobe comprises about 70% of the total weight of the gland. Histologically, the anterior lobe of the hypophysis consists of epithelial glandular cells of varying sizes and shapes. These cells are arranged in broad, circular columns, separated from one another by sinusoids containing blood. Three types of cells are differentiated by their staining qualities: (1) Chromophobe cells (neutrophils), located in the center of the columns of epithelium, stain poorly; (2) and (3) Chromaffin cells, located on the outside of the columns and therefore adjacent to the blood sinusoids, stain well. Those chromaffin cells which take the acid dyes are designated eosinophilic cells; those which stain with basic dyes, basophilic cells.

It is difficult to assign specific functions to each type of cell, although attempts have been made to do so. Most of the evidence is derived from association of physiologic disorders with the histologic changes observed at autopsy.

HORMONES OF THE ANTERIOR PITUITARY

Growth Hormone (Somatotropin)

A. Chemistry: Growth hormone (GH), also known as somatotropin (STH), was first isolated in sufficient quantity for study from the pituitary glands of cattle. The concentration of bovine growth hormone is 5–15 mg/gland, which is much higher than the μg/g quantities of other pituitary hormones. Bovine growth hormone is a branched, Y-type structure with a molecular weight of approximately 46,000 daltons (400 amino acids). Human growth hormone (HGH) has a simple straight chain polypeptide structure which contains 2 internal disulfide bridges. The amino acid sequence of HGH was described by Li & others (1969). Li & Yamashiro (1970) have reported the synthesis of a polypeptide possessing growth-promoting and lactogenic activities. Niall (1971) has indicated that the amino acid sequence of HGH reported by Li is not entirely correct and has proposed certain changes which result in the formation of a polypeptide containing 190 amino acids. It is expected that the precise structure of HGH will soon be reported.

Several years ago, on the basis of physiologic and immunologic studies, a close structural similarity between HGH and human placental lactogen (HPL)–

also designated human chorionic somatomammotropin (HCS)—was suggested. Studies of the amino acid sequences of these hormones confirmed that there were many similarities in structure. Similarities in structure of many areas of HGH and of sheep prolactin have also been demonstrated. This probably explains the lactogenic activity of preparations of growth hormone.

Growth hormone from cattle is effective in stimulating growth in rats but not in man. The monkey and human preparations are active in both man and rat. It has been hypothesized that activity resides in a portion of the molecule since partial chymotrypsin hydrolysis of cattle hormone does not abolish its activity. The "active" fragment of the molecule which is essential to activity may be common to all 3 hormones. It is possible that the rat can degrade the cattle hormone to the active form, whereas man cannot.

B. Function: Growth hormone has a variety of effects on different tissues, including muscle, adipose tissue, and liver. In general, the hormone acts slowly, requiring from 1–2 hours to several days before its biologic effects are detectable. This slow action, plus its stimulatory effect on RNA synthesis (see below), suggest that part of the activity of the hormone involves protein or enzyme synthesis. Support for this has been recently obtained by the demonstration of an increase in kidney transamidinase and hepatic sulfatase activities in hypophysectomized rats following the administration of growth hormone. Growth hormone may also act at the membrane level to facilitate transport, particularly of the amino acids.

1. **Protein synthesis**—Growth hormone stimulates over-all protein synthesis in the intact animal, resulting in a pronounced increase in nitrogen retention with an associated retention (actually increased renal tubular reabsorption) of phosphorus. Blood amino acids and urea are decreased. Thus, growth hormone in this regard acts synergistically with insulin. In muscle, growth hormone can stimulate protein synthesis by increasing the transport of amino acids into the cells, an action that can be demonstrated in vitro with rat diaphragm muscle. This effect is not inhibited by puromycin. Therefore, the facilitation of amino acid transport by the hormone is not mediated by a direct action on protein synthesis. In addition, growth hormone facilitates protein synthesis in muscle tissue by a mechanism independent of its ability to provide amino acids. Thus, Kostyo & Schmidt (1962) were able to demonstrate increased protein synthesis in vitro even when amino acid transport was blocked. In the intact animal, growth hormone administration results in an increase in DNA and RNA synthesis. It remains unclear whether increased RNA synthesis is a result of direct action of the hormone or a secondary result of an increase in metabolic signal substance, possibly amino acids themselves. The hormone increases collagen synthesis. Since this protein is rich in hydroxyproline, an increase in turnover of collagen after growth hormone is reflected by an increase in urinary hydroxyproline and hydroxyproline peptides. The measurement of urinary hydroxyproline therefore can be used to assess growth hormone activity in the intact animal.

2. **Fat metabolism**—Growth hormone is mildly lipolytic when incubated in vitro with adipose tissue, promoting release of FFA and glycerol. In vivo, 30–60 minutes later, administration of growth hormone is followed by increase in circulating free fatty acids and increased oxidation of fatty acids in the liver. Under conditions of insulin deficiency (eg, diabetes), increase ketogenesis may occur. This increase of fatty acid mobilization may provide the chemical energy required for the action of the hormone in stimulating protein synthesis, although the 2 phenomena may in fact not be directly related.

3. **Carbohydrate metabolism**—In muscle growth hormone antagonizes the effects of insulin. Impairment of glycolysis may occur at several steps, as well as inhibition of transport of glucose. Whether this latter effect is a direct effect on transport or a result of the inhibition of glycolysis has not yet been established. The mobilization of fatty acids from triglyceride stores may also contribute to the inhibition of glycolysis in the muscle. In liver there is an increase in liver glycogen, probably arising from activation of gluconeogenesis from amino acids. Hyperglycemia after growth hormone administration is a combined result of decreased peripheral utilization of glucose and increased hepatic production via gluconeogenesis. These combined hyperglycemic actions cause impaired glucose tolerance which can be detected within 2 hours after administration of growth hormone. Prolonged administration of growth hormone results in an enhanced release of insulin from the pancreas during glucose stimulation. This effect may be secondary to the peripheral diabetogenic action of growth hormone to increase circulating pancreatic stimulants such as glucose, fatty acids, and ketones or it may be caused by a direct but slow action on the pancreas.

4. **Ion or mineral metabolism**—Growth hormone increases intestinal absorption of calcium as well as its excretion. Since growth hormone stimulates the growth of the long bones at the epiphyses, as well as the growth of soft tissue, increased calcium retention results for the most part from the increased metabolic activity of the bone. In addition to calcium, sodium, potassium, magnesium, phosphate, and chloride are also retained. Serum phosphate levels are usually elevated in the acromegalic and are often measured as an indication of the degree of excess growth hormone "activity" in the patient. Growth hormone also increases a "sulfation factor" responsible for the incorporation of sulfate into cartilage.

5. **Prolactin properties**—In those species without a separate prolactin molecule, growth hormone has many prolactin properties such as stimulation of the mammary glands, lactogenesis, and stimulation of the pigeon crop sac.

6. Hypophysectomy in young animals results in reduction of growth and decreased replication of muscle and liver cells. Injection of pituitary extracts or of purified growth hormone restores the rate of growth

to nearly normal. Similarly, patients with specific growth hormone deficiency and retarded growth respond to human growth hormone. The injection of growth hormone into normal animals causes an extensive increase in growth, or gigantism.

C. Control of Secretion: The daily production rate of growth hormone in man is about 500 μg/day. With the advent of immunoassay technics, it has become possible to measure amounts of circulating intact growth hormone. In normal males, growth hormone ranges from 0–5 ng/ml of plasma; in females, values may occasionally be considerably higher, though most fall in the same range; the occasional high levels in females may be related to the fact that estrogens enhance the growth hormone released in response to various stimuli. Progesterones, in contrast, are inhibitory. In acromegalics, levels range from 15–80 or more ng/ml. Growth hormone is usually elevated in the newborn, decreasing to adult levels by 4 years of age (Glick & others, 1965).

Contrary to previous belief, the plasma growth hormone level in the adult is not stable; depending on the nature of a stimulus, it may change as much as tenfold within a few minutes. Plasma growth hormone concentration is increased following stress (pain, cold, surgical stress, severe insulin hypoglycemia, and apprehension) and exercise, the response to exercise being greater in females than in males. The rapid rise after exercises necessitates careful control to ensure that the subject is resting before blood samples are taken for growth hormone measurements. The increased growth hormone after stress may be caused by increased ACTH and catecholamines; infusion of either of these agents stimulates growth hormone secretion. Factors decreasing glucose availability to the hypothalamic regulating centers (see below) also stimulate release. This can be accomplished (1) by fasting, (2) by hypoglycemia associated with an insulin tolerance test, or (3) by administration of an agent such as 2-deoxyglucose, which inhibits the normal glycolysis of glucose, making it unavailable to the regulating centers even though circulating blood sugar becomes elevated. Since 2-deoxyglucose causes the prompt release of growth hormone while at the same time producing an elevated blood sugar, it is apparent that regulation in the hypothalamus is dependent on the normal metabolism of glucose, not on the circulating level of glucose as such.

Stimulation is also increased by protein meals and by amino acids, particularly arginine. This provides a regulatory system whereby increases in amino acids result in secretion of growth hormone which itself facilitates uptake of amino acids into protein. It will be recalled that arginine also facilitates the secretion of insulin, which, like growth hormone, is required for protein synthesis. Curiously, growth hormone is elevated in malnutrition with kwashiorkor. This may be due to abnormal glucose metabolism and decreased glucose availability at the control sites.

Both glucose and the glucocorticoids inhibit growth hormone release. In the acromegalic there is loss of the normal control mechanisms for growth hormone release. This is reflected in an inability to suppress plasma growth hormone values by administration of glucose. In subjects with mild acromegaly or those who have been treated and in whom moderate hormonal overactivity persists, levels of growth hormone may overlap with those occasionally high normal values seen in normal females. In these cases, inability to suppress hormone production with glucose is taken as further evidence that acromegaly exists. Conversely, growth hormone deficiency may be documented in suspected hypopituitary patients by demonstration of inadequate responses to insulin-induced hypoglycemia or arginine infusion.

Much of the control of growth hormone secretion occurs at the level of the hypothalamus. Section of the neural stalk completely obliterates control by hypoglycemia and exercise. The control centers appear to be located in the anterior and central eminence (Abrams & others, 1966). A specific **growth hormone releasing factor (GHRF, SRF)** has been extracted from the hypothalamus. The substance is an acetylated peptide containing 11 amino acids. GHRF levels in the hypothalamus are reduced by starvation or corticosteroids and increased by thyroxine. The effect of thyroxine may explain its capacity to stimulate synthesis and storage of pituitary growth hormone as well as increased pituitary growth. Drug-induced inhibition and stimulation of growth hormone release also appear to be mediated through the hypothalamic releasing factor.

In contrast to insulin release, growth hormone secretion is enhanced by alpha-adrenergic agents, although it is not established whether this regulation is at the pituitary or hypothalamic level.

Growth hormone deficiency in children results in low levels of growth hormone in the serum and dwarfism. However, dwarfism may also occur as a result of (1) the possible production of immunologically measurable but inactive hormone, or (2) tissue insensitivity to normal amounts of hormone. Peptides of variable metabolic activity which are not necessarily detectable by immunoassay have been prepared from growth hormone. Whether these peptides are released from the pituitary or indeed play any physiologic role is not yet established.

The Pituitary Tropins

The most characteristic function of the anterior pituitary is the elaboration of hormones which influence the activities of other endocrine glands, principally those involving reproduction or stress. Such hormones are called **tropic** hormones. They are carried by the blood to other target glands and aid in maintaining these glands and stimulating production of their own hormones. For this reason, atrophy and decline in the function of many endocrine glands occur in pituitary hypofunction or after hypophysectomy.

With the possible exception of prolactin and MSH, the tropic hormones are under the positive control of peptide factors from the hypothalamus

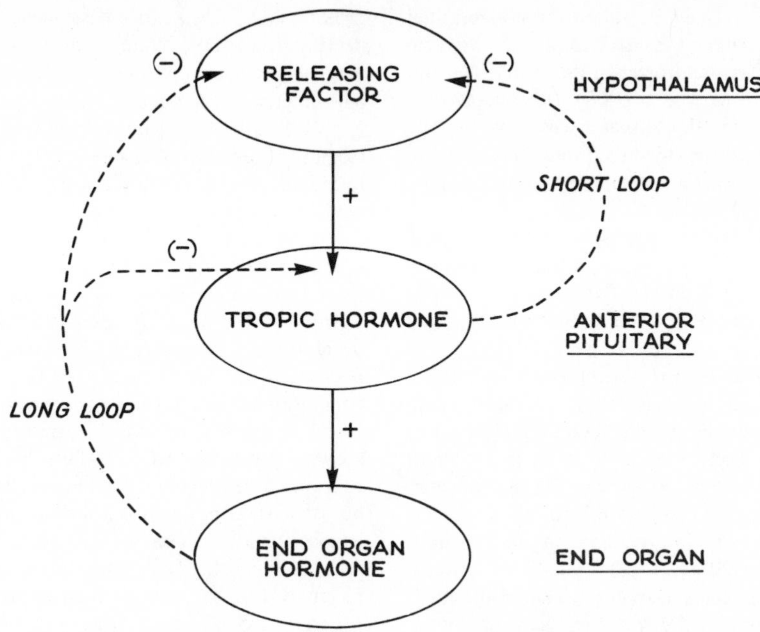

FIG 20–20. General feedback control of endocrine systems involving the hypothalamus, anterior pituitary, and end organ.

(releasing factors or hypothalamic neurohormones). The production and release of the neurohormones, in turn, are sensitive to neural and metabolic stimuli and can be inhibited by their respective tropic hormones ("short loop feedback") (Fig 20–20).

In addition, the tropic hormones are usually subject to feedback inhibition at the pituitary or hypothalamic level by the hormone product of the final target gland. Thus, hydrocortisone, sex steroids, and thyroxine inhibit the release of their respective tropic hormones.

Lactogenic Hormone (Prolactin, Mammotropin, Luteotropic Hormone, LTH)

Pituitary LTH is a protein hormone with a molecular weight of approximately 23,000 daltons which, like growth hormone, is produced by the pituitary acidophil cells. It may be that lactogenic hormone and growth hormone share a common structure since they can cross-react immunologically. LTH has been demonstrated in birds, ruminants, and rats; however, it is still unclear if LTH and growth hormone are distinct pituitary hormones in man.

In animals, LTH activates the corpus luteum and stimulates continued progesterone production by the developed corpus luteum. It also stimulates enlargement of crop gland and formation of "crop milk" in pigeons. In man, the sheep hormone acts as an anabolic agent, mimicking the effects of growth hormone, but it is less active. Although pituitary LTH is biologically and immunochemically similar to chorionic gonadotropin, its exact role is not known. It increases during pregnancy and may stimulate mammary development and growth hormone-like metabolic changes. Its rapid decrease after parturition, however, suggests that it may have little to do with mammalian lactation. LTH can be inhibited by a hypothalamic factor, **prolactin inhibiting factor (PIF)**.

A. The Gonadotropins: These tropic substances influence the function and maturation of the testis and ovary. They are glycoproteins with molecular weights of about 25,000 (FSH) and 40,000 (LH) daltons. The gonadotropins contain 2 protein subunits. The α subunit is almost identical in each of the gonadotropins. The carbohydrate content of the gonadotropins consists of sialic acid, hexose, and hexosamine.

1. Follicle-stimulating hormone (FSH)—This hormone promotes follicular growth, prepares the follicle for the action of LH, and enhances the release of estrogen induced by LH. In the male it stimulates seminal tubule and testicular growth and plays an important role in the early stages of spermatogenesis. FSH plasma concentrations increase through puberty from the low levels of infancy. In the female there is marked cycling of levels with peaks of the order of tenfold or more over basal levels being reached at or slightly before the time of ovulation. The secretion of FSH is mediated by a hypothalamic releasing factor **(follicle stimulating hormone releasing factor, FRF)**, and is inhibited by the administration of estrogens, testosterone, and possibly FSH itself. Under certain conditions, testosterone may increase the pituitary content of FSH. The half-time of disappearance of FSH is about 10 minutes.

2. Luteinizing hormone (LH, ICSH) in the female stimulates final maturation of the graafian follicle, ovulation, and the development of the corpora lutea. Both estrogen and progesterone secretion are stimulated.

In the ovary, LH can stimulate the nongerminal elements, which contain the interstitial cells, to produce the androgens, androstenedione, DHEA, and testosterone (Savard & others, 1965). In subjects with polycystic ovaries (Stein-Leventhal disease), part of the observed masculinization (hirsutism) may result from overactivity of the ovarian stroma to produce these androgens.

In the male, LH stimulates testosterone production by the testis, which in turn maintains spermatogenesis and provides for the development of accessory sex organs such as the vas deferens, prostate, and seminal vesicles.

The actual site of LH action on progesterone synthesis in corpora lutea is unknown, but experimentally it is possible to show a specific activation of the conversion of acetate to squalene (the precursor for cholesterol synthesis). However, LH also accelerates the conversion of cholesterol to 20α-hydroxycholesterol, a necessary intermediate in the synthesis of progesterone (Fig 20–21).

LH activates phosphorylase. The subsequent increased glycogen breakdown produces glucose-6-phosphate, which, by entering the HMP shunt pathway, would cause increased production of NADPH, which is required for steroid synthesis. However, although increased NADPH can facilitate conversion of cholesterol to progesterone in corpora lutea, it cannot duplicate the entire action of LH.

LH may act to increase cyclic AMP production. It has been shown that cyclic AMP levels are elevated after addition of LH to incubating corpora lutea and that added cyclic AMP duplicates the stimulatory action of LH both at the site of acetate incorporation into cholesterol and at the site of cholesterol incorporation into progesterone (Fig 20–21). The effects of cyclic AMP and luteinizing hormone are reportedly blocked by puromycin, indicating that their action may require protein synthesis, but the rapid rise in cyclic AMP after hormone administration indicates that more direct modifications at enzyme or membrane sites also occur.

The secretion of LH is mediated by a specific **hypothalamic releasing factor (HRF)**. For a discussion of hypothalamic releasing factors, see Harris & others, 1966.

The plasma concentration and pituitary content of LH increase through puberty. In the female there is cycling of plasma LH levels with midcycle (ovulatory) peaks many times the basal level. There is some evidence that the cycling observed in the adult female may also occur to a lesser extent before adulthood, particularly in pubescence.

3. Pituitary-like hormones from the placenta—Gonadotropic hormones not of pituitary origin are also found (cf, the pregnancy hormone, chorionic gonadotropin; see p 458).

a. Human chorionic gonadotropin (HCG)—This is a protein with immunologic similarity to pituitary LH. It is derived from the syncytiotrophoblast and is elevated in the plasma of pregnant females. The action of chorionic gonadotropin is similar to that of LH. Because of its long half-life, HCG may remain in the circulation several days after parturition.

b. Human placental lactogen (human chorionic growth prolactin, HCGP)—This hormone has many physicochemical and immunologic similarities to human GH, although it is a slightly larger molecule and is more acidic. There is also some similarity in the sequence of the first few amino acids from the N terminal. For the complete structure of sheep prolactin, see Li & others (1970). This hormone has lactogenic and luteotropic activity. It also has some metabolic effects which are qualitatively similar to those of GH, including inhibition of glucose uptake, stimulation of free fatty acid and glycerol release, enhancement of nitrogen and calcium retention (despite increased urinary calcium excretion), reduction in the urinary excretion of phosphorus and potassium, and an increase in the turnover of hydroxyproline, as reflected by increased urinary excretion of that amino acid.

c. TSH—Recent evidence indicates that the placenta may also be the source of a TSH-like substance.

B. Thyrotropic Hormone; Thyroid-Stimulating Hormone (TSH): Pure preparations of TSH are not yet available. The hormone is a mucoprotein with a molecular weight of 25,000 daltons. Injection of TSH will bring about all of the symptoms of hyperthyroidism. It increases thyroid growth and general metabolic activity, including glucose oxidation, oxygen consumption, and synthesis of phospholipids and RNA. Within minutes, TSH rapidly increases each phase of thyroxine metabolism, including iodine uptake, organification, and, finally, the breakdown of thyroglobulin with the concomitant release of thyroid hormone. It has long been suggested that TSH activates thyroidal adenyl cyclase on the cell surface and that

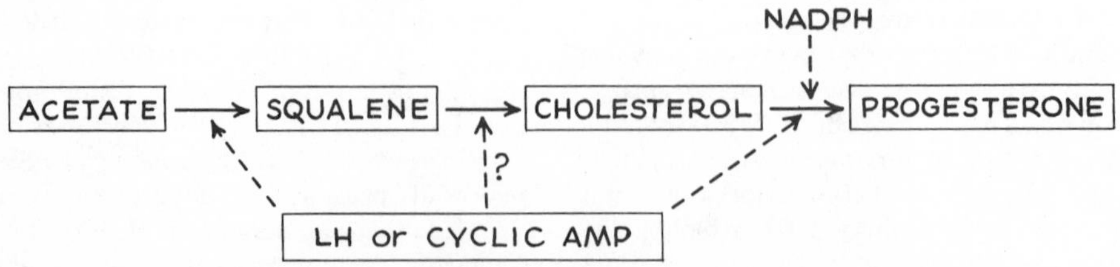

FIG 20–21. Effect of LH (ICSH) on progesterone synthesis in corpora lutea.

the resulting increase in cellular cyclic AMP increases the availability of the enzymes necessary for the various thyroidal functions (Pastan, 1969). It is not known if the major action of cyclic AMP is to rapidly activate these enzymes or to increase their synthesis at the translation or transcription levels. It may be that it has both actions. Greenspan & Hargadine (1965) found that TSH 15 minutes after injection into intact animals could be localized in the nuclei of the thyroid cells. These observations suggest that TSH may act at the nuclear level and, ultimately, influence protein synthesis. It is unlikely, however, that such a nuclear action could explain all of the rapid effects of the hormone.

TSH is used clinically to differentiate primary hypothyroidism (myxedema) from that due to pituitary insufficiency (secondary).

Control of TSH release. The reciprocal relationship between the target gland and pituitary thyrotropin is demonstrated by a reduction in TSH after thyroxine administration. The thyroxine effect is primarily at the pituitary level and is mediated through a **TSH releasing factor (TRF)** from the hypothalamus.

TRF from the sheep has been identified as the tripeptide, Glu-His-Pro-(NH_3) by Nair & others, 1970. It is present in very small quantities (approximately 1 ng per sheep pituitary). TRF has been synthesized; it does not appear to be species specific.

Plasma levels of TSH. Earlier bioassay technics in which the various parameters of TSH activity–such as iodine uptake into the thyroid or release of organically bound iodine–were measured did not provide reliable or reproducible estimates of plasma levels of this hormone. Blood levels determined by radioimmunoassay range from 0–3 ng/ml. Levels are low in subjects with hyperthyroidism and are increased to levels of 7–156 ng/ml in myxedema. In simple goiter, TSH levels are elevated only if hypothyroidism is still present.

C. Adrenocorticotropic Hormone (ACTH, Corticotropin): Hypophysectomy causes atrophy of the adrenal cortex. This can be prevented by injections of pituitary ACTH. Pituitary hyperfunction is accompanied by adrenocortical hypertrophy, presumably due to excess ACTH production by the pituitary.

ACTH is a straight chain polypeptide with a molecular weight of 4500 daltons containing 39 amino acids. Its amino acid sequence has been determined; only the first 23 amino acids (from the N-terminal end of the chain) are required for activity. The sequence of these 23 amino acids in the peptide chain was the same in all animals tested as well as in humans, whereas the remaining biologically inactive 16 amino acid chain varies according to the animal source.

1. Physiologic effects–ACTH not only increases the synthesis of corticosteroids by the adrenal but also stimulates their release from the gland. It also increases total protein synthesis as indicated by increased incorporation of ^{14}C-labeled acetate into adrenal tissue proteins and increased adrenal RNA. Thus, ACTH produces both a tropic effect on steroid production and a trophic effect on adrenal tissue. In general, the tropic effects can be observed between 1–3 hours after administration of hormone, whereas the trophic actions, including stimulation of RNA synthesis, are much slower. Adrenal target organ specificity is apparent since other hormones such as TSH, growth hormone, and gonadotropins will not stimulate adrenal RNA synthesis, although they have this action on their own respective target tissues.

ACTH affects steroid hormone synthesis in the adrenal at an early stage in the conversion of cholesterol to pregnenolone, the primary precursor for the synthesis of all adrenal steroids (Figs 20–13 and 20–14). Thus, ACTH stimulation results in an increase in mineralocorticoids, glucocorticoids, and androgens. As noted on p 447, ACTH has only a mild effect on the output of aldosterone. The administration of ACTH to normal human beings, therefore, causes the following effects: (1) increased excretion of nitrogen, potassium and phosphorus; (2) retention of sodium, chloride, and secondary retention of water; (3) elevation of fasting blood sugar and a diabetic glucose tolerance curve; (4) increase in circulating free fatty acids; (5) increased excretion of uric acid; (6) increased androgenicity (in extreme cases); and (7) decline in circulating eosinophils and lymphocytes and elevation of polymorphonuclear leukocytes.

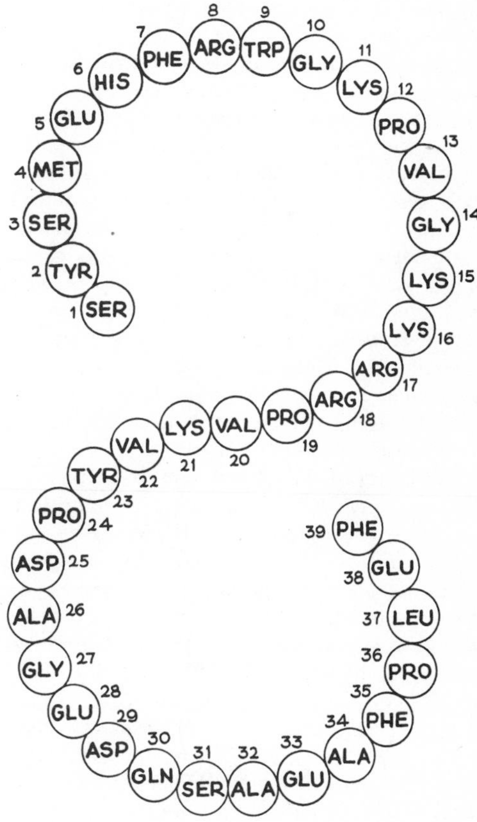

Human ACTH

The mechanism by which ACTH performs its tropic action remains unclear. However, its action appears to be primarily at the cell membrane level; ACTH diazotized to large cellulose fibers was still active in steroid biosynthesis when the large complex was suspended with adrenal cell cultures (Schimmer & others, 1968). Furthermore, the hormone is active only on preparations where the cell membrane is intact.

It was originally believed that ACTH had a primary effect on increasing phosphorylase activity. This resulted in glycogen breakdown and increased glucose-6-phosphate for entry into the hexose monophosphate shunt, resulting in increased production of NADPH. In turn, NADPH is required for several steps in the synthesis of steroids, particularly in the hydroxylation reactions. This hypothesis no longer appears valid since, even after maximum stimulation with ACTH, the addition of an NADPH generating system produces an additive effect. Furthermore, the action of the hormone on steroid secretion occurs before a detectable increase in phosphorylase activity. Finally, although ACTH increases adrenal glucose metabolism, there is no preferential diversion of glucose metabolites into the shunt pathway.

The primary action of ACTH may involve cyclic AMP, since this compound is increased in the adrenal gland by ACTH and, if added directly, will in itself stimulate steroidogenesis. The effect of ACTH on the adrenal can be inhibited by blocking protein synthesis with puromycin, whereas the blocking of RNA synthesis by dactinomycin has little inhibitory effect. This action is consistent with a cyclic AMP mediated effect on protein kinases in the nucleus and indicates that ACTH may have a delayed effect on synthesis of adrenal enzymes. Perhaps because of its ability to activate adenyl cyclase and increase intracellular levels of cyclic AMP, ACTH can increase lipolysis in adipose tissue and stimulate insulin secretion from the pancreas. However, the primary membrane receptor is the adrenal cell. Thus, these extra-adrenal effects are small, require large concentration of the hormone, and are comparatively unimportant under normal physiologic conditions.

2. Control of ACTH secretion–As with most of the other tropic hormones, ACTH is controlled by corticotropin releasing factors (CRF) found in the hypothalamus. (For discussion, see Vernikos-Danellis, 1965.) Three factors have been designated as α_1-, α_2-, and β-CRF, respectively. β-CRF may be vasopressin or a peptide similar in structure. (The similarity of vasopressin and one form of CRF may explain why vasopressin is an effective stimulator of ACTH production.) α_1- and α_2-CRF are similar in structure to α-MSH. Activation of the hypothalamic centers through the cerebral cortex by nonspecific stresses such as cold, pyrogens, insulin hypoglycemia, epinephrine, estrogens, trauma, or by psychic reactions results in increased production of ACTH, leading to increased adrenal cortical activity and protective compensation against the stress. High levels of ACTH may inhibit its further synthesis by a "short loop" inhibitor of CRF production in the hypothalamus (Fig 20–20).

A reciprocal relationship between corticosteroid production and ACTH secretion is well established; exogenous cortisol causes feedback inhibition of ACTH release at the hypothalamic level. Androgens and progesterone may have an inhibitory action at the same site.

3. Circulating ACTH–ACTH is present in low concentration in normal plasma (0.1–2 mU/100 ml). Its activity can be measured by bioassay or radioimmunologic assay (Yalow & others, 1964). The pituitary stores 5–10 mg of hormone and releases only a small amount daily. Pituitary capacity to secrete ACTH is tested by the administration of metyrapone. This agent inhibits 11-β-hydroxylase, resulting in decreased cortisol production and stimulation of ACTH production by hypothalamic CRF. The increased ACTH stimulates secretion of total steroids except those which require 11-hydroxylation, resulting in an increase in urinary-17-hydroxycorticoids. (The advantages and disadvantages of this test are discussed by Kaplan, 1963.)

Abnormalities of Pituitary Function

A. Hyperpituitarism:

1. Excess production of growth hormone (eosinophilic adenoma)–Gigantism results from hyperactivity of the gland during childhood or adolescence, ie, before closure of the epiphyses. The long bones increase in length so that the patient reaches an unusual height. There are also associated metabolic changes attributed to a generalized pituitary hyperfunction.

Acromegaly results from hyperactivity that begins after epiphysial closure has been completed and growth has ceased. The patient exhibits characteristic facial changes (growth and protrusion of the jaw, enlargement of the nose), growth and enlargement of the hands, feet, and viscera, and thickening of the skin.

2. Excess production of ACTH (basophilic adenoma) produces Cushing's disease.

B. Hypopituitarisms: Hypopituitarism may occur as a result of certain types of pituitary tumors, or after hemorrhage (especially postpartum), infarct, or atrophy of the gland.

1. Dwarfism is a result of hypoactivity of the gland, sometimes caused by chromophobe tumors or craniopharyngioma. In either case, the underactivity is due to pressure of the tumor on the remainder of the gland. If the tumor begins early in life, dwarfism will result; if it occurs later, there will be cessation of growth and metabolic abnormalities similar to those observed after total hypophysectomy.

2. Pituitary myxedema, due to lack of TSH, produces symptoms similar to those described for primary hypothyroidism.

3. Panhypopituitarism refers to deficiency of function of the hypophysis which involves all of the hormonal functions of the gland. This can result from destruction of the gland because of hemorrhage or infarct.

Milder forms of long-standing panhypopituitarism may result from the pressure of a tumor on the pituitary gland. The symptoms in these cases are similar to those of the early stages of acute panhypopituitarism, including a tendency to hypoglycemia with sensitivity to insulin.

The excretion of neutral 17-ketosteroids in the urine is much reduced in panhypopituitarism. The degree of reduction depends upon the severity of the disease.

Hypopituitarism cannot be treated with extracts of the hypophysis, since no adequate preparation is available. Speific replacement therapy for the target glands involved, such as the thyroid, adrenal cortex, and gonads, is therefore usually resorted to. The recent availability of human growth hormone has made possible the treatment of a limited number of cases of pituitary dwarfism with strikingly favorable results.

THE MIDDLE LOBE OF THE PITUITARY

The middle lobe of the pituitary secretes a hormone, **intermedin,** which was first detected by its effect on the pigment cells in the skin of lower vertebrates. This hormone apparently also increases the deposition of melanin by the melanocytes of the human skin. In this role it is referred to as melanocyte-stimulating hormone (MSH). Both hydrocortisone and cortisone inhibit the secretion of MSH; this is similar to their action on ACTH. Epinephrine and, even more strongly, norepinephrine inhibit the action of MSH. When production of the corticosteroids is inadequate, as in Addison's disease, MSH is secreted, the synthesis of melanin is increased, and there is an accompanying brown pigmentation of the skin. In totally adrenalectomized patients, as much as 30–50 mg of orally administered cortisone must be supplied each day to prevent excess deposition of pigment in the skin. In patients suffering from panhypopituitarism (see above), in which case there is a lack of MSH as well as corticosteroids, pigmentation does not occur.

Chemistry

Two peptides (α-MSH and β-MSH) have been isolated from extracts of hog pituitaries. In α-MSH the amino acid sequence is that of the first 13 amino acids of ACTH (as shown on p 463) except that the N-terminal amino acid, serine, is acetylated and the C-terminal, valine, is in the amide form. β-MSH is a peptide containing 18 amino acids and has a minimum molecular weight of 2177 daltons. The amino acid sequence in β-MSH is as follows (Geschwind, 1956):

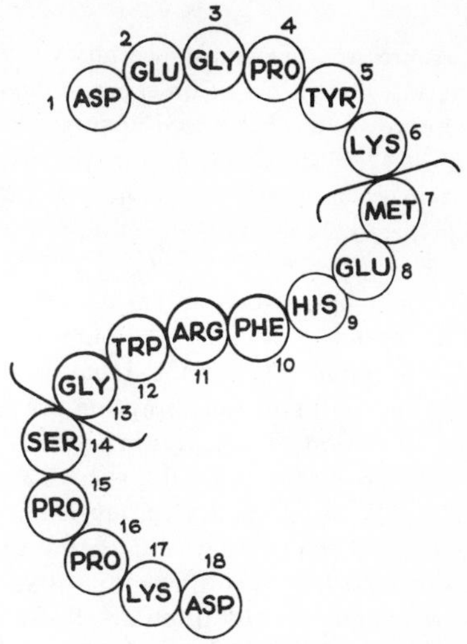

1 ASP – 2 GLU – 3 GLY – 4 PRO – 5 TYR – 6 LYS – 7 MET – 8 GLU – 9 HIS – 10 PHE – 11 ARG – 12 TRP – 13 GLY – 14 SER – 15 PRO – 16 PRO – 17 LYS – 18 ASP

Amino acid sequence of β-MSH

The amino acid sequence between the brackets is identical with that between amino acids 4 and 10 of ACTH. This is of interest in view of the fact that ACTH has small but definite melanocyte-stimulating activity (about 1% that of MSH). This may be due to its content of the so-called intermedin sequence of amino acids, as described above. Although α-MSH has some corticotropic activity, this is not the case with β-MSH.

Melatonin (N-acetyl-5-methoxy serotonin) is a hormone which lightens the color of the melanocytes of the skin of the frog and blocks the action of MSH as well as of ACTH. The chemistry and metabolism of melatonin are described in Chapter 15.

THE POSTERIOR LOBE OF THE PITUITARY

Extracts of the posterior pituitary contain at least 2 active substances: a pressor-antidiuretic principle, **vasopressin** (Pitressin); and an oxytocic principle, **oxytocin** (Pitocin). Both are produced primarily, however, in the neurosecretory neurons, specifically the supraoptic neurons of the paraventricular nuclei of the hypothalamus. The hormones are stored in the pituitary in association with 2 proteins, **neurophysin I and II,** with molecular weights of 19,000 and 21,000 daltons, respectively (Hollenberg & Hope, 1968). Each neurophysin can bind either hormone. When secreted, the hormones are transported in association with serum proteins, preferentially to kidney, mammary gland, and liver. The half-life in plasma is extremely short, varying from 1–5 minutes depending on the chemical structure of the hormone and the particular species. Part of the vasopressin concentrated by the kidney is excreted in the urine, but most is degraded.

Function

A. Vasopressin: This substance raises blood pressure by its vasopressor effect on the peripheral blood vessels. Vasopressin has been used in surgical shock as an adjuvant in elevating blood pressure. It may also be used occasionally in obstetrics in the case of delayed postpartum hemorrhage and, at delivery, for uterine inertia.

Vasopressin also exerts an antidiuretic effect as the so-called posterior pituitary antidiuretic hormone, ADH. The hormone affects the renal tubules and provides for the facultative reabsorption of water. The mechanism of action of vasopressin in the kidney is unknown. It binds firmly to renal tissue, an action that can be inhibited by sulfhydryl group (SH) blocking agents. Such observations have led to the suggestion (Fong & others, 1960) that the action of vasopressin involves the opening of the disulfide (S–S) ring (Fig 20–22), which is followed by a combination of the now available SH groups with other SH groups on membranes. Although such a binding can occur, there is as yet little support that this is a required step in the action of the hormone. Cyclic AMP can duplicate many of the actions of vasopressin in kidney tissue, indicating that the hormone may act to increase levels of this substance.

ADH may be increased by a variety of stimulators of neural activity. Emotional and physical stress, electrical stimulation, acetylcholine, nicotine, and morphine increase ADH secretion as does dehydration or increased blood osmolality experimentally induced by injection of hypertonic saline. In general, these agents appear to increase synthesis of the hormones since their action does not cause a depletion of stored hormone. Furthermore, these stimulations are usually associated with an increase in RNA synthesis in the neuron, indicating an increased protein synthetic activity. Epinephrine and factors increasing blood volume are effective inhibitors of ADH secretion. Alcohol also inhibits ADH secretion.

In the absence of ADH, **diabetes insipidus** occurs. This is characterized by extreme diuresis—up to 30 liters of urine per day. The disease may be controlled by subcutaneous administration of posterior pituitary extract or even by nasal instillation of the extract.

B. Oxytocin: (Fig 20–22.) This substance is the principal uterus-contracting hormone of the posterior pituitary. It is employed in obstetrics when induction of uterine contraction is desired. It also causes contracting of the smooth muscles in the mammary gland, resulting in milk excretion. Oxytocin levels are increased by suckling.

Chemistry

Du Vigneaud and his collaborators (1953) determined the structure and accomplished the synthesis of the posterior pituitary hormones. The structure of oxytocin is shown below. It is a cyclic polypeptide containing 8 different amino acids and has a molecular weight of about 1000 daltons. Note that 5 amino acids, including cystine, are arranged to form a cyclic disulfide (S–S) structure; the other 3 amino acids are attached as a side chain to the carboxyl group of ½ of the cystine.

The structure of vasopressin is quite similar to that of oxytocin. The differences are in 2 amino acids: (1) isoleucine of oxytocin is replaced in vasopressin by phenylalanine; and (2) leucine of oxytocin is replaced in vasopressin obtained from hog pituitary by lysine **(lysine vasopressin)** and in that from beef as well as many other animals, by arginine **(arginine vasopressin)**.

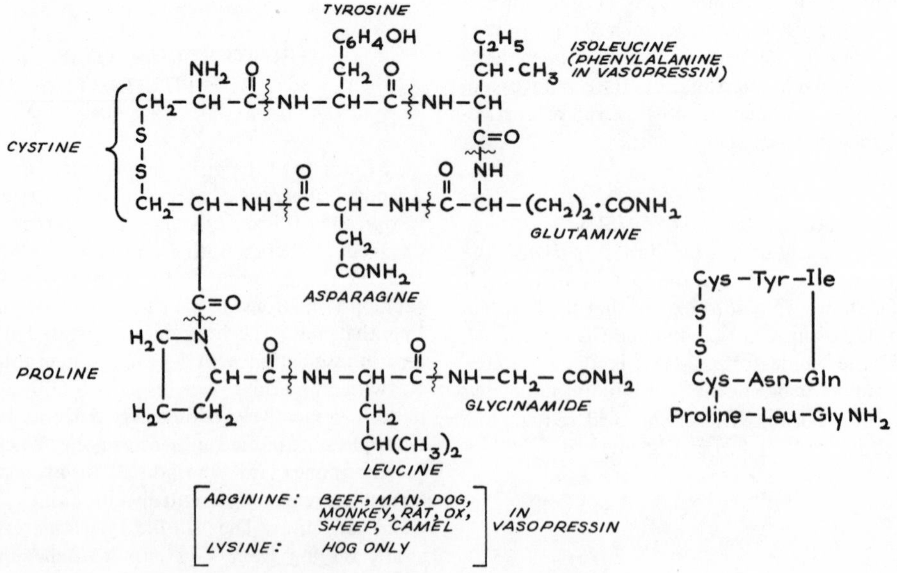

TYROSINE
C_6H_4OH
C_2H_5
ISOLEUCINE (PHENYLALANINE IN VASOPRESSIN)
CYSTINE
GLUTAMINE
ASPARAGINE
PROLINE
GLYCINAMIDE
LEUCINE

Cys–Tyr–Ile
S
S
Cys–Asn–Gln
Proline–Leu–Gly NH_2

ARGININE:	BEEF, MAN, DOG, MONKEY, RAT, OX, SHEEP, CAMEL	IN VASOPRESSIN
LYSINE:	HOG ONLY	

FIG 20–22. Structure of oxytocin.

Vasopressin has been synthesized and made available for clinical use.

The functional chemical groups in oxytocin—besides the necessity for the peptide linkages—include the primary amino group of cystine; the phenolic hydroxyl group of tyrosine; the 3 carboxamide groups of asparagine, glutamine, and glycinamide; and the disulfide (S—S) linkage. By removal of certain of these groups, analogues of oxytocin have been produced. Examples are deoxy oxytocin, which lacks the phenolic group of tyrosine; and desamino oxytocin, which lacks the free primary amino group of the terminal cysteine residue. Desamino oxytocin has 5 times the antidiuretic activity of oxytocin itself. Various compounds which represent modifications of the naturally occurring oxytocin and vasopressin have been synthesized. Their activities are often significantly different from the parent hormones.

For details of the comparative activities of a vast assortment of synthetic analogues and homologues, see Berde & Boissonnas (1966).

• • •

References

Abrams RL, Parker ML, Blanko S, Reichlin S, Daughaday WH: Endocrinology 78:605, 1966.

Adams MJ, Blundell TL, Dodson EJ, Dodson GG, Vijayan M, & others: Nature 224:491, 1969.

Alexander NM: Endocrinology 82:925, 1968.

Antoniades HN, Gundersen K: Endocrinology 68:36, 1961.

Aurbach GD: J Biol Chem 234:3179, 1959.

Aurbach GD, Potts JI: Am J Med 24:1, 1967.

Axelrod J: Recent Progr Hormone Res 21:597, 1965.

Bartter FC: Metabolism 5:369, 1956.

Beerman W, Clever U: Sc Am 210:50, April, 1964.

Berde B, Boissonnas RA: The Pituitary Gland 3:624, 1966.

Berson SA, Yalow RS: Am J Med 24:155, 1958.

Bessman S: Am J Med 40:740, 1966.

Bruchovsky N, Wilson JD: J Biol Chem 243:2012, 1968.

Chance RE, Ellis RM, Bromer WW: Science 161:166, 1968.

Conn JW: J Lab Clin Med 45:3, 1955.

Copp DH, Cameron EC, Cheney BA, Davidson EGF, Henze KG: Endocrinology 70:638, 1962.

Cuatrecasas P: J Biol Chem 245:3059, 1970.

DuVigneaud V, Ressler C, Trippett S: J Biol Chem 205:949, 1953.

Edelman PM, Schwartz IL: Am J Med 40:695, 1966.

Eik-Nes K, Hall PF: Vitamins Hormones 23:153, 1965.

El Attar TMA: J Clin Endocrinol 31:334, 1970.

Exton JH, Jefferson LS, Butcher RW, Park CR: Am J Med 40:709, 1966.

Fong CTO, Silver L, Christman DR, Schwartz IL: Proc Nat Acad Sc 46:1273, 1960.

Freedland RA, Cunliffe TL, Zinkl JG: J Biol Chem 241:5448, 1966.

Friedman J, Raisz LG: Science 150:1465, 1965.

Geschwind II, Li CH, Barnafi L: J Am Chem Soc 78:4494, 1956.

Glick SM, Roth J, Yalow RS, Berson SA: Recent Progr Hormone Res 21:241, 1965.

Goetz FC, Maney J, Greenberg BZ: Excerpta Medica 74:135, 1964.

Greenspan FS, Hargadine JR: J Cell Biol 26:177, 1965.

Grodsky GM, Bennett LL: Proc Soc Exper Biol Med 114:769, 1963.

Grodsky GM: Diabetes 14:396, 1965.

Grodsky GM, Forsham PH: Ann Rev Physiol 28:347, 1966.

Grodsky GM, Curry D, Landahl H, Bennett L: Acta Diabet Latina 6 (Suppl 1):554, 1969.

Hall JC, Sordahl LA, Stefko PL: J Biol Chem 235:1536, 1960.

Harris GW, Reed M, Fawcett CP: Brit M Bull 22:266, 1966.

Haynes RC Jr: J Biol Chem 233:1220, 1958.

Hofert JF, White A: Endocrinology 82:767, 1968.

Hollenberg MD, Hope DB: Biochem J 106:557, 1968.

Jarabak J, Adams JA, Williams-Ashman HG, Talalay P: J Biol Chem 237:345, 1962.

Kajihata A, Kendall JW: Clin Res 15:123, 1967.

Kakiuchi S, Tomizawa HH: J Biol Chem 239:2160, 1964.

Kaplan NM: J Clin Endocrinol 23:945, 1963.

Karam JH, Grodsky GM, Forsham PH: Diabetes 12:197, 1963.

Korner A: Recent Progr Hormone Res 21:205, 1965.

Kostyo JL, Schmidt JE: Endocrinology 70:381, 1962.

Kriss JP: Advances Metab Dis 3:209, 1968.

Kuo JF, Greengard P: Proc Nat Acad Sc 64:1349, 1969.

LaBrosse EH, Axelrod J, Kopin IJ, Kety SS: J Clin Invest 40:253, 1961.

Lacy P: Ciba Foundation Colloq 15:75, 1964.

Lacy PE, Howell SL, Young DA, Fink CJ: Nature 219:1177, 1968.

Langan TA: Proc Nat Acad Sc 64:1276, 1969.

Li CH, Dixon JS, Knud TL, Schmidt D, Pankov YA: Arch Biochem 141:705, 1970.

Li CH, Dixon JS, Liu WK: Arch Biochem 130:70, 1969.

Li CH, Liu WK, Dixon JS: J Am Chem Soc 88:2050, 1966.

Li CH, Yamashiro D: J Am Chem Soc 92:26, 1970.

Li Ch, Yamashiro D: J Am Chem Soc 92:7608, 1970.

Malaisse WJ, Malaisse-Lagae F, Mayhew D: J Clin Invest 46:1724, 1967.

Martin TE, Wool IG: Proc Nat Acad Sc 60:569, 1968.

Matschinsky FM, Ellerman J, Kotter-Brajtbrug J, Krzanowski J, Fertel R, Landgraef R: Diabetes 19:365, 1970.

Montague W, Taylor KW: Biochem J 109:333, 1968.

Morgan HE, Regen DM, Park CR: J Biol Chem 239:369, 1964.

Murphy BEP: Recent Progr Hormone Res 00:563, 1969.

Nair RMG, Barrett JF, Bowers CY, Schally HV: Biochemistry 9:1103, 1970.

Niall HD: Nature New Biology 230:90, 1971.

Nicol DSHW, Smith LF: Nature 187:483, 1960.

Pastan I: Page 577 in: *Progress in Endocrinology.* Gaul C (editor). Amsterdam, 1969.

Porte D, Graber A, Kuzuya T, Williams R: J Clin Invest 45:228, 1966.

Potts JT Jr, Tregear GW, Keutmann HT, Niall HD, Saver R, Deftos LJ, Dawson BF, Hogan ML, Aurbach GD: Proc Nat Acad Sc 68:63, 1971.

Randle PJ, Hales CN, Sarland PB, Newsholme EA: Lancet 1:785, 1963.

Rasmussen H, Arnaud C, Hawker C: Science 144:1019, 1964.
Rieser P: Am J Med 40:759, 1966.
Rodbell J: J Biol Chem 241:140, 1966.
Roth V, Gorden P, Paston I: Proc Nat Acad Sc 61:138, 1968.
Sallis JD, DeLuca HF, Rasmussen H: J Biol Chem 238:1098, 1963.
Savard K, Marsh JM, Rice BF: Recent Progr Hormone Res 21:285, 1965.
Shimmer BP, Veda K, Sato GH: Biochem Biophys Res Commun 32:806, 1968.
Smith LF: Am J Med 40:662, 1966.
Sokoloff L, Kaufman S, Campbell PL, Francis CM, Gilboin HV: J Biol Chem 238:1432, 1963.
Solomon DH, Beale GN: Clin Res 15:127, 1967.
Steiner DF, Cunningham D, Spigelman L, Aten B: Science 157:697, 1967.
Sutherland EW, Øye I, Butcher RW: Recent Progr Hormone Res 21:623, 1965.
Sutherland EW, Robison GA, Butcher RW: Circulation 37:279, 1968.
Tait JF, Horton R: Steroids 4:365, 1964.
Tomizawa HH: J Biol Chem 237:428; 3393, 1962.
Unger RH, Arquilar-Parada E, Müller WA, Eisentraut AM: J Clin Invest 49:837, 1970.
Vallance-Owen J: Brit M Bull 16:214, 1960.
Vernikos-Danellis J: Vitamins Hormones 23:97, 1965.
Weber G, Singhal RL, Stamm NG, Lea MA, Fisher EA: Advances Enzym Regulat 4:59, 1966.
Wells H, Lloyd W: Endocrinology 83:521, 1968. West CD, Chavré V, Wolfe M: J Clin Endocrinol 25:1189, 1965.
Yalow RS, Glick S, Roth J, Berson SA: J Clin Endocrinol 24:1219, 1964.

Bibliography

Catt KJ: Hormones in general. Lancet 1:763, 1970.
Catt KJ: Pituitary function. Lancet 1:827, 1970.
Catt KJ: Growth hormone. Lancet 1:933, 1970.
Catt KJ: Reproductive endocrinology. Lancet 1:1097, 1970.
Catt KJ: Adrenal cortex. Lancet 1:1275, 1970.
Catt KJ: Hormonal control of calcium homeostasis. Lancet 2:255, 1970.
Recent studies on the hypothalamus. Brit M Bull 22, No. 3, 1966.
Werner SC, Nauman JA: The thyroid. Ann Rev Physiol 30:213, 1968.
Williams R (editor): *Textbook of Endocrinology*, 4th ed. Saunders, 1968.

21...

Calorimetry: Elements of Nutrition

Food is the source of the fuel which is converted by the metabolic processes of the body into the energy for vital activities. Calorimetry deals with the measurement of the energy requirements of the body under various physiologic conditions and of the fuel values of foods which supply this energy.

The Unit of Heat: Large Calorie or Kilocalorie (C)

Vital energy and the fuel value of foods are most conveniently measured in calories. A calorie is the amount of heat required to raise the temperature of 1 g of water 1° C (from 15° to 16° C). This is a very small quantity of heat. In nutrition it is therefore more common to use the large calorie (written with a capital "C") or kilocalorie, 1000 times the small calorie. All references to caloric values will henceforth be to the kilocalorie.

Measurement of the Fuel Value of Foods

The combustion of a foodstuff in the presence of oxygen results in the production of heat. The amount of heat thus produced can be measured in a bomb calorimeter. By this technic, the caloric value of a foodstuff can be determined.

The caloric content of the 3 principal foodstuffs, determined by burning in a bomb calorimeter, is given in Table 21–1.

These are average figures, since variations occur within each class, eg, monosaccharides do not have exactly the same caloric content as polysaccharides.

When utilized in the body, carbohydrate and fat are completely oxidized to CO_2 and water, as they are in the bomb calorimeter also. Proteins, however, are not burned completely since the major end product of protein metabolism, urea, still contains some energy which is not available to the body. For this reason, the energy value of protein in the body (4.1 Calories/g) is less than that obtained in the bomb calorimeter.

It is customary to round off the energy value of foods as utilized in the body to the figures given in Table 21–1. These figures also correct for the efficiency of the digestion of foods.

TABLE 21–1. Fuel values of foods.

	Kilocalories/g	
	In Bomb Calorimeter	In the Body*
Carbohydrate	4.1	4
Fat	9.4	9
Protein	5.6	4

*Figures are expressed in round numbers.

Control of Body Heat

The heat generated by the body in the course of the metabolism of foodstuffs maintains the body temperature. Warm-blooded animals, such as birds and mammals, have heat-regulating mechanisms which either increase heat production or radiate or otherwise dissipate excess heat, depending on the temperature of their external environment. When the external temperature rises above the normal body temperature, 98.6° F (37° C), evaporation from the surface of the body becomes the only mechanism available for cooling the body.

Animal Calorimetry

Since all of the energy produced in the body is ultimately dissipated as heat, measurement of the vital heat production of an animal is a way to estimate its energy expenditure. There are 2 methods of accomplishing this.

A. Direct Calorimetry: In the direct calorimeter, the subject is placed in an insulated chamber; his heat production is measured directly by recording the total amount of heat transferred to a weighed quantity of water circulating through the calorimeter. The oxygen intake, the CO_2 output, and the nitrogen excretion in the urine and feces are also measured during the entire period of observation. These data are used as described below.

B. Indirect Calorimetry: Direct calorimetry is attended by considerable technical difficulties. By measuring gas exchange and determining the respiratory quotient, energy metabolism studies are considerably simplified and thus rendered applicable to field studies and to clinical analysis.

Respiratory Quotients (RQ) of Foodstuffs

The respiratory quotient is the ratio of the volume of carbon dioxide eliminated to the volume of oxygen utilized in oxidation.

A. Carbohydrates: The complete oxidation of glucose, for example, may be represented as follows:

$$C_6H_{12}O_6 + \boxed{6O_2} \longrightarrow \boxed{6CO_2} + 6H_2O$$

The RQ for carbohydrate is therefore:

$$\frac{CO_2}{O_2} = \frac{6}{6} \text{ or } 1$$

B. Fats have a lower RQ because the oxygen content of their molecule in relation to the carbon content is quite low. Consequently they require more oxygen from the outside. The oxidation of tristearin will be used to exemplify the RQ for fat.

$$2C_{57}H_{110}O_6 + \boxed{163O_2} \longrightarrow \boxed{114CO_2} + 110H_2O$$

$$\frac{CO_2}{O_2} = \frac{114}{163} = 0.70$$

C. Proteins: The oxidation of proteins cannot be so readily expressed because their chemical structure is not known. By indirect methods the RQ of proteins has been calculated to be about 0.8.

D. RQ of Mixed Diets Under Varying Conditions: In mixed diets containing varying proportions of protein, fat, and carbohydrate, the RQ is about 0.85. As the proportion of carbohydrate metabolized is increased, the RQ approaches closer to 1. When carbohydrate metabolism is impaired, as in diabetes, the RQ is lowered. Therapy with insulin is followed by an elevation in the RQ. A high carbohydrate intake, as used in fattening animals, will result in an RQ exceeding 1. This rise is caused by the conversion of much of the carbohydrate, an oxygen-rich substance, to fat, an oxygen-poor substance; a relatively small amount of oxygen is required from the outside, and the ratio of CO_2 eliminated to the oxygen taken in (the RQ) will be considerably elevated.

A reversal of the above process, ie, the conversion of fat to carbohydrate, would lower the RQ below 0.7. This has been reported but has not been generally confirmed.

Performance of Indirect Calorimetry

The use of the indirect method for calculating the total energy output and the proportions of various foodstuffs being burned may be illustrated by the following example:

The subject utilized oxygen at a rate of 414.6 liters/day and eliminated 353.3 liters of CO_2 in the same period. The urinary nitrogen for the day was 12.8 g.

Because of the incomplete metabolism of protein, the gas exchange is corrected for the amount of protein metabolized; a nonprotein RQ (nonprotein portion of the total RQ) is thus obtained.

One g of urinary nitrogen represents the combustion of an amount of protein which would require 5.92 liters of oxygen and would eliminate 4.75 liters of CO_2.

A. Calculate the Nonprotein RQ:

1. Multiply the amount of urinary nitrogen by the number of liters of oxygen required to oxidize that amount of protein represented by 1 g of urinary nitrogen.

$$12.8 \text{ g} \times 5.92 = 75.8 \text{ liters}$$

2. Multiply the amount of urinary nitrogen by the number of liters of CO_2 which result from this oxidation.

$$12.8 \text{ g} \times 4.75 = 60.8 \text{ liters of } CO_2$$

Thus, 75.8 liters of the total oxygen intake was used to oxidize protein, and 60.8 liters of the CO_2 eliminated was the product of this oxidation. The remainder was used for the oxidation of carbohydrates and fats. Therefore, to determine the nonprotein RQ, subtract these values for protein from the totals for the day.

$$\text{Oxygen: } 414.6 - 75.8 = 338.8 \text{ liters}$$
$$CO_2\text{: } 353.3 - 60.8 = 292.5 \text{ liters}$$

$$\frac{CO_2}{O_2} = \frac{292.5}{338.8} = 0.86 \text{ (nonprotein RQ)}$$

B. Convert the Nonprotein RQ to Grams of Carbohydrate and Fat Metabolized: Reliable tables have been worked out which give the proportions of carbohydrate and of fat metabolized at various RQs. According to the tables of Zuntz and Shunberg (as modified by Lusk and later by McLendon), when the nonprotein RQ is 0.86, 0.622 g of carbohydrate and 0.249 g of fat are metabolized per liter of oxygen used. Therefore, to determine the total quantities of carbohydrate and fat used, multiply these figures by the number of liters of oxygen (derived from the nonprotein RQ) consumed during the combustion of carbohydrate and fat.

$$338.8 \times 0.622 \text{ g} = 210.7 \text{ g carbohydrate}$$
$$338.8 \times 0.249 \text{ g} = 84.4 \text{ g fat}$$

C. Determine the Amount of Protein Metabolized: Each gram of urinary nitrogen represents the oxidation of 6.25 g of protein. Therefore, to determine the amount of protein metabolized, multiply the total urinary nitrogen by 6.25.

$$12.8 \text{ g} \times 6.25 = 80 \text{ g of protein}$$

D. Calculate the Total Heat Production: Multiply the quantity in grams of each foodstuff oxidized by the caloric value of that food to obtain the heat production due to its combustion. The sum of these caloric values equals the total heat production of the diet.

Carbohydrate:	210.7 g × 4 C =	842.8 C
Fat:	84.4 g × 9 C =	759.6 C
Protein:	80.0 g × 4 C =	320.0 C

Total heat production = 1922.4 C

BASAL METABOLISM

The total heat production or energy expenditure of the body is the sum of that required merely to maintain life (basal metabolism), together with such additional energy as may be expended for any additional activities. The lowest level of energy production consonant with life is the **basal metabolic rate**, or BMR.

Conditions Necessary for Measurement of the Basal Metabolic Rate

1. A post-absorptive state; patient should have had nothing by mouth for the past 12 hours.
2. Mental and physical relaxation immediately preceding the test; usually ½ hour of bed rest is used, although ideally the patient should not arise from bed after a more prolonged rest.
3. Recumbent position during the test.
4. Patient awake.
5. Environmental temperature of between 20° and 25° C.

Factors Influencing Basal Metabolism

A. Surface Area: The basal metabolic rates of different individuals, when expressed in terms of surface area (sq m), are remarkably constant. In general, however, smaller individuals have a higher rate of metabolism per unit of surface area than larger individuals.

B. Age: In the newborn the rate is low; it rises to maximum at age 5, after which the rate begins to decline, continuing into old age. There is, however, a relative rise just before puberty. Examples of influence of age on BMR: At age 6, the normal BMR is between 50 and 53 C/sq m/hour; at age 21 it is between 36 and 41 C/sq m/hour.

C. Sex: Women normally have a lower BMR than men. The BMR of females declines between the ages of 5 and 17 more rapidly than that of males.

D. Climate: The BMR is lower in warm climates.

E. Racial Variations: When the BMR of different racial groups is compared, certain variations are noted. For example, the BMRs of Oriental female students living in the USA average 10% below the standard BMR for American women of the same age; the BMRs of adult Chinese are equal to or below the lower limit of normal for Occidentals; high values (33% above normal) have been reported for Eskimos living in the region of Baffin Bay.

F. State of Nutrition: In starvation and undernourishment, the BMR is lowered.

G. Disease: Infectious and febrile diseases raise the metabolic rate, usually in proportion to the elevation of the temperature. Diseases which are characterized by increased activity of cells also increase heat production because of this increased cellular activity. Thus, the metabolic rate may increase in such diseases as leukemia, polycythemia, some types of anemia, cardiac failure, hypertension, and dyspnea—all of which involve increased cellular activity. Perforation of an eardrum causes falsely high readings.

H. Effects of Hormones: The hormones also affect metabolism. Thyroxine is the most important of the hormones in this respect, and the principal use of calorimetry in clinical practice is in the diagnosis of thyroid disease. The rate is lowered in hypothyroidism and increased in hyperthyroidism. Changes in the BMR are also noted in pituitary disease.

Other than thyroxine, the only hormone which has a direct effect on the rate of heat production is epinephrine, although the effect of epinephrine is rapid in onset and brief in duration. Tumors of the adrenal (pheochromocytoma) cause an elevation in the BMR. In adrenal insufficiency (Addison's disease), the basal metabolism is subnormal, whereas adrenal tumors and Cushing's disease may produce a slight increase in the metabolic rate.

Measurement of Basal Metabolism

In clinical practice, the BMR can be estimated with sufficient accuracy merely by measuring the oxygen consumption of the patient for 2 6-minute periods under basal conditions. This is corrected to standard conditions of temperature and barometric pressure. The average oxygen consumption for the 2 periods is multiplied by 10 to convert it to an hourly basis and then multiplied by 4.825 C, the heat production represented by each liter of oxygen consumed. This gives the heat production of the patient in C per hour. This is corrected to C per square meter body surface per hour by dividing the C per hour by the patient's surface area. A simple formula for calculating the surface area is as follows:

$$\frac{\text{Circumference of}}{\text{midthigh (in cm)} \times 2} \times \frac{\text{Height}}{\text{(in cm)}} = \frac{\text{Surface area}}{\text{(in sq cm)}}$$

The classical formula is that of Du Bois, as follows:

Du Bois' Surface Area Formula

$$A = H^{0.725} \times W^{0.425} \times 71.84$$

where A = surface area in sq cm,
H = height in cm,
and W = weight in kg

(Surface area in sq cm divided by 10,000 = surface area in sq m)

In practice, a nomogram which relates height and weight to surface area is used. It is based on the Du Bois formula.

Calculation of Basal Metabolic Rate

The normal BMR for an individual of the patient's age and sex is obtained from standard tables. The patient's actual rate is expressed as a plus or minus percentage of the normal. ***Example:*** A male, age 35 years, 170 cm in height and 70 kg in weight, consumed an average of 1.2 liters of oxygen (corrected to normal temperature and pressure: 0° C, 760 mm Hg) in a 6-minute period.

1.2 × 10 = 12 liters of oxygen/hour
12 × 4.825 = 58 C/hour
Surface area = 1.8 sq m (from Du Bois' formula)
BMR = 58 C/1.8 = 32 C/sq m/hour

The normal BMR for this patient, by reference to the Du Bois standards, is 39.5 C/sq m/hour. His BMR, which is below normal, is then reported as:

$$\frac{39.5 - 32}{39.5} \times 100 = 18.98\% \text{ or } \textbf{minus 18.98}$$

A BMR between −15 and +20% is considered normal. In hyperthyroidism, the BMR may exceed +50 to +75%. The BMR may be −30 to −60% in hypothyroidism.

MEASUREMENT OF ENERGY REQUIREMENTS

The metabolic rate increases with activity. Maximal increases occur during exercise—as much as 600–800% over basal. The energy requirement for the day will therefore vary considerably in accordance with the amount of physical activity. It is sometimes important to know how much energy a given subject uses in performing various tasks in order to recommend an appropriate caloric intake. The gas exchange methods of indirect calorimetry, discussed above, are used for this purpose. The **Douglas bag,** which may be carried on the back of the subject like a knapsack, is filled with a measured quantity of oxygen; after the task is completed, the oxygen used and the CO_2 eliminated during the study are measured. These data, together with the urinary nitrogen, are then used to calculate the energy expenditure. (Consolazio & others, 1951.)

Such studies indicate, for example, that for a 70 kg man, 65 C/hour are expended in sleeping, 100 C/hour sitting at rest, 200 C/hour walking slowly, 570 C/hour running, and as much as 1100 C/hour walking up a flight of stairs.

SPECIFIC DYNAMIC ACTION (SDA)

The specific dynamic action of a foodstuff is the extra heat production, over and above the caloric value of a given amount of food, which is produced when this food is used by the body. For example, when an amount of protein which contains 100 C (25 g) is metabolized, the heat production in the body is not 100 C but 130 C. This extra 30 C is the product of the specific dynamic action of the protein. In the body, a 100-C portion of fat produces 113 C, and a 100-C portion of carbohydrate produces 105 C. The origin of this extra heat is not clear, but it is attributable to the activity of the tissues which are metabolizing these foodstuffs.

The specific dynamic action of each foodstuff, as given above, is obtained when each foodstuff is fed separately; but when these foods are taken in a mixed diet, the dynamic effect of the whole diet cannot be predicted by merely adding the individual effects of each foodstuff in accordance with its contribution to the diet, eg, 30% of the caloric value of protein, 13% of the caloric value of fat, and 5% of the caloric value of carbohydrate. According to Forbes, the dynamic effects of beef muscle protein, glucose, and lard, when each was fed separately, were 32, 20, and 16%, respectively, of their caloric content. But when the glucose and protein were combined, the dynamic effect was 12.5% less than predicted from the sum of their individual effects; combining lard, glucose, and protein produced 22% less dynamic effect than predicted; a glucose-lard combination was 35% less, and a protein-lard mixture 54% less than calculated from the individual percentages of each foodstuff.

These observations are of importance because they indicate that the high specific dynamic action of protein can be reduced depending on the quantities of other foodstuffs in the diet. Forbes's data show that fat (lard) has a greater influence on specific dynamic action than does any other nutrient, ie, fat decreases the SDA more than any other nutrient.

In calculating the total energy requirement for the day, it is customary to add 10% to the total required C to provide energy for the SDA, ie, expense of utilization of the foods consumed. This figure may be too large, particularly if the fat content of the diet is high.

THE ELEMENTS OF NUTRITION

The Components of an Adequate Diet

There are 6 major components of the diet. **Carbohydrate, fat,** and **protein** yield energy, provide for growth, and maintain tissue subjected to wear and tear. **Vitamins, minerals,** and **water,** although they do not yield energy, are essential parts of the chemical mechanisms for the utilization of energy and for the synthesis of various necessary metabolites such as hormones and enzymes. The minerals are also incorporated into the structure of the tissue and, in solution, play an important role in acid-base equilibrium.

The Energy Aspect of the Diet

Energy for physiologic processes is provided by the combustion of carbohydrate, fat, and protein. The daily energy requirement or the daily caloric need is the sum of the basal energy demands plus that required for the additional work of the day. During periods of growth, pregnancy, or convalescence, extra calories must be provided.

While all 3 major nutrients yield energy to the body, carbohydrate and, to a lesser extent, fat are physiologically the most economical sources. Protein serves primarily to provide for tissue growth and repair; but if the caloric intake from other foods is inadequate, it is burned for energy.

The caloric requirements of persons of varying sizes and under various physiologic conditions are tabulated in Table 21–3.

Obesity is almost always the result of excess consumption of calories. Treatment is therefore directed at reducing the caloric intake from fat and carbohydrate but maintaining the protein, vitamin, and mineral intake at normal levels. An adequate diet containing not more than 800–1000 C should be maintained and normal or increased amounts of energy expended until the proper weight is reached.

"Protein-Sparing Action" of Carbohydrates & Fats

Carbohydrate and fat "spare" protein and thus make it available for anabolic purposes. This is particularly important in the nutrition of patients, especially those being fed parenterally, when it is difficult or even impossible for them to take in enough calories. If the caloric intake is inadequate, giving proteins orally or amino acids intravenously is a relatively inefficient way of supplying energy because the primary function of proteins is tissue synthesis and repair and not energy production.

Distribution of Calories in the Diet

In a well balanced diet, 10–15% of the total calories is usually derived from protein, 55–70% from carbohydrate, and 20–30% from fat. These proportions vary under different physiologic conditions or in various environmental temperatures. For example, the need for calories is increased by the need to retain a constant body temperature; in extreme cold, the caloric intake of the diet must therefore increase and this requirement is usually met by increasing the fat content of the diet.

A. The Carbohydrate Intake: Carbohydrate is the first and most efficient source of energy for vital processes. Cereal grains, potatoes, and rice, the staple foods of most countries, are the principal sources of carbohydrate in the diet. Usually 50% or more of the daily caloric intake is supplied by carbohydrate; this is equivalent to 250–500 g/day in the average diet, but it varies within wide limits. Carbohydrate intake is often the principal variable in gain or loss of weight. A minimum of 5 g of carbohydrate/100 C of the total diet is required to prevent ketosis.

B. The Fat Intake: Because of its high fuel value, fat is an important component of the diet. Furthermore, the palatability of foods is generally increased by their content of fat. As a form of energy storage in the body, fat has more than twice the value of protein or carbohydrate.

The human requirement for fat is not precisely known. An important aspect of the contribution of fats to nutrition may be their content of the so-called "essential fatty acids," linoleic and arachidonic acids. These unsaturated fatty acids have been shown to be essential in the diet of experimental animals, and although they seem to exert important effects on lipid metabolism in man, a deficiency of these substances has not yet been unequivocally demonstrated in human subjects. However, it is now clear from studies on the metabolic effects of the polyunsaturated fatty acids that the requirement for fats in the diet must be considered from a qualitative as well as from a quantitative standpoint.

Up to a certain point, the isocaloric replacement of carbohydrate or protein by fat results in better growth in animals. This may be due to the effect of fat in reducing the SDA of the ration, thus improving the caloric efficiency of the diet.

At ordinary levels of intake, 20–25% of the calories of the diet should probably be derived from fats (for a 3000-C intake, 66–83 g of fat), including 1% of the total calories from "essential" fatty acids. At higher caloric levels, 30–35% will probably come from fats.

C. The Protein Requirement: A minimal amount of protein is indispensable in the diet to provide for the replacement of tissue protein, which constantly undergoes destruction and resynthesis. This is often spoken of as the **wear and tear quota.** The protein requirement is considerably increased, however, by the demands of growth, increased metabolism (as in infection with fever), in burns, and after trauma. The recommended intake of protein for individuals of various age groups is shown in Table 21–3. As already noted, these presume that the caloric demand is adequately supplied by other foods so that the ingested protein is available for tissue growth and repair.

The requirement for protein in the diet is, however, not only quantitative; there is also an important qualitative aspect since the metabolism of protein is inextricably connected with that of its constituent amino acids. Certain amino acids are called indispensable in the diet in the sense that they must be obtained preformed and cannot be synthesized by the animal organism. The remainder, the so-called dispensable amino acids, are also required by the organism since they are found in the protein of the tissues, but they can apparently be synthesized, presumably from alpha-keto acids, by amination. The list of indispensable or "essential" amino acids varies with the animal species tested (eg, the chick requires glycine). It may also vary with the physiologic state of the animal. For some amino acids (eg, arginine) the rate of synthesis may be too slow to supply fully the needs of the animal. Histidine in the diet is also necessary to maintain growth during childhood. Such amino acids are said to be **relatively** indispensable. The nutritive value of a protein is now known to be dependent on its content of essential amino acids. Examples of incomplete proteins are gelatin, which lacks tryptophan; and zein of corn, which is low in both tryptophan and lysine. Such incomplete proteins are unable to support growth if given as the sole source of protein in the diet.

The experiments of Rose have supplied for normal adult human subjects the requirements for the 8 amino acids which are essential to the maintenance of nitrogen balance (Table 21–2). The diet must also furnish sufficient nitrogen for the synthesis of other amino acids. Normal adult male subjects were maintained in nitrogen balance only when all 8 amino acids listed below were simultaneously present in their diets. The elimination of any one promptly produced a negative nitrogen balance. Although tyrosine and cystine are not listed among the essential amino acids, tyrosine can spare 70–75% of the phenylalanine requirement and cystine can spare 80–90% of the methionine requirement in humans. This is undoubtedly because the normal metabolism of phenylalanine and methionine involves their conversion, in whole or in part, to tyrosine or cystine, respectively.

Some of the unnatural (D-) forms of the amino acids have been found to fulfill in whole or in part the requirement for a given amino acid. Thus DL-methionine was as effective as L-methionine in supplying the requirement for this amino acid. Significant amounts of D-phenylalanine are utilized by the human organism, perhaps as much as 0.5 g/day. On the other hand, D-valine, D-isoleucine, and D-threonine individually do not exert a measurable effect upon nitrogen balance when the corresponding L-form is absent from the diet.

In summarizing his observations on the amino acid requirements of man, Rose has stated that when the diet furnishes the 8 "essential" amino acids at their "recommended" levels of intake (Table 21–2) and extra nitrogen is provided as glycine to provide a total daily intake of only 3–5 g, nitrogen balance could be maintained in his subjects. Expressed as protein (N × 6.25), this quantity of nitrogen represents about 22 g of protein, an amount far below the recommended daily intake for adults. This finding emphasizes the importance of a consideration of the individual amino acid intake in assessing protein requirements rather than whole proteins of varying essential amino acid content.

TABLE 21–2. Amino acids essential to maintenance of nitrogen balance in human subjects.

	Minimum Daily Requirement (g)	Recommended Daily Intake* (g)
L-Tryptophan	†0.25	0.5
L-Phenylalanine	1.10	2.2
L-Lysine	0.80	1.6
L-Threonine	0.50	1.0
L-Valine	0.80	1.6
L-Methionine	1.10	2.2
L-Leucine	1.10	2.2
L-Isoleucine	0.70	1.4

*Note that the recommended daily intake is twice the minimum daily requirement.

†The tryptophan requirement may vary with the niacin intake (see p 99).

The caloric intake necessary for nitrogen balance in human subjects receiving amino acids (eg, casein hydrolysate) is higher than for subjects receiving whole protein (eg, whole casein, by mouth). For example, with whole protein, nitrogen balance could be attained with 35 C/kg, whereas as much as 53–60 C/kg were required when amino acids were used as the sole source of nitrogen.

Proteins differ in "biologic value" depending on their content of essential amino acids. The proteins of eggs, dairy products, kidney, and liver have high biologic values because they contain all of the essential amino acids. Good quality proteins, which are somewhat less efficient in supplying amino acids, include shellfish, soybeans, peanuts, potatoes, and the muscle tissue of meats, poultry, and fish. Fair proteins are those of cereals and most root vegetables. The proteins of most nuts and legumes are of poor biologic value. It is important to point out, however, that 2 or more proteins in themselves only "poor" or "fair" in quality may have a "good" biologic value when taken together because they may complement one another in supplying the necessary amino acids.

It is of interest to recall that proteins, in addition to their other important functions, also constitute the most important sources of nitrogen, sulfur, and phosphorus for the body.

Amino acid requirements for the human infant are, in general, higher than for the adult when expressed on the basis of body weight. It is also important to note that histidine is to be considered an essential amino acid for requirements of growth in the young infant.

Many amino acids have specific functions in metabolism in addition to their general role as constituents of the tissue proteins. Examples are cited in Chapter 15. They include the role of methionine as a methyl donor, cystine as a source of SH groups, the dicarboxylic acids in transamination, tryptophan as a precursor of niacin, arginine and the urea cycle, etc.

Vitamins

The chemistry and physiologic functions of the vitamins are discussed in Chapter 7. Normal individuals on an adequate diet can secure all of the required vitamins from natural foods; no supplementation with vitamin concentrates is necessary. In disease states in which digestion and assimilation are impaired or the normal requirement for the vitamins is increased, they must of course be supplied in appropriate quantities from other sources.

Many of the vitamins are destroyed by improper cooking. Some of the water-soluble vitamins, for example, are partially lost in the cooking water. Overcooking of meats also contributes to vitamin loss. Vitamin C is particularly labile both in cooking and storage. In fact, one can hardly depend on an adequate vitamin C intake unless a certain quantity of fresh fruits and vegetables is taken each day.

The refinement of cereal grains is attended by a loss of B vitamins. Enrichment of these products with thiamine, riboflavin, niacin, and iron is now used to restore these nutrients. Other foods are often improved by the addition of vitamins, eg, the addition of vitamin D to milk and of vitamin A to oleomargarine.

Minerals

The minerals, while forming only a small portion of the total body weight, are nonetheless of great importance in the vital economy. Their functions and the requirements for each as far as now known are discussed in Chapter 19. Fruits, vegetables, and cereals are the principal sources of the mineral elements in the diet. Certain foods are particularly outstanding for their contribution of particular minerals, eg, milk products, which are depended on to supply the majority of the calcium and phosphorus in the diet.

Water

Water is not a food, but since it is ordinarily consumed in the diet it is included as one of its components. The water requirements and the functions of water in the body are discussed in Chapter 19.

RECOMMENDED DIETARY ALLOWANCES

The Food and Nutrition Board of the National Academy of Sciences, National Research Council, has collected the best available data on the quantities of various nutrients required by normal persons of varying body sizes and ages and in different physiologic states. The first recommendations derived from such data were published in 1943. At intervals of approximately 5 years since that time, revisions of the recommendations have been issued. In the 1968 edition, the seventh revision, recommendations for vitamins B_6, B_{12}, folacin, and vitamin E as well as for phosphorus, iodine, and magnesium have been added to the recommendations previously made for calories, protein, calcium, iron, thiamine, riboflavin, niacin, ascorbic acid, vitamin A, and vitamin D.

Unless the purposes of the Recommended Dietary Allowances are kept in mind, it might be concluded that the allowances are too high. The recommendations are designed to set levels suitable for maintenance of good nutrition in the majority of healthy individuals in the USA. Thus, they are proposed as a means of fixing goals in planning food supplies and interpreting food intakes of groups of people. It follows that the allowances may be higher than required for some individuals. Indeed, the allowances should not be interpreted as specific requirements for individuals unless it is understood that the stated amounts of each nutrient may in fact be too high for a given person. Therefore, the diet of an individual may not necessarily be regarded as deficient if it does not equal in every respect the recommendations specified.

Sebrell (1968), in commenting on the new allowances, points out that they are intended for use in the USA under current conditions of living. Therefore, they take into account climate, economic status, distribution of population, and various other factors that make them particularly suitable for the USA. While they may serve as a reference for use in other countries, many variables such as population, food supply, climate, body size, and energy expenditure make it desirable that each country develop recommendations specific to the needs of its own people.

In order to simplify the concept of an adequate diet for normal individuals, foodstuffs have been arranged into 4 groups, each of which makes a major contribution to the diet. It is recommended that some food from each group be taken daily, as specified below.

(1) Milk group: For children, 3 or more 8 oz glasses (smaller servings for some children under 9 years of age); for teenagers, 4 or more glasses; for adults, 2 or more glasses. Cheese, ice cream, and other milk products can supply part of the milk recommended. 1 oz American cheese = ¾ glass milk; ½ cup creamed cottage cheese = 1/3 glass milk; ½ cup ice cream = ¼ glass milk.

(2) Meat group: Two or more servings of meat, fish, poultry, eggs, or cheese. Dry beans, peas, and nuts are alternates, although they cannot substitute entirely for meat, fish, poultry, or eggs.

(3) Vegetables and fruits: Four or more servings, including some dark green or yellow vegetables, citrus fruits, or tomatoes. Some uncooked vegetables should be included.

(4) Breads and cereals: Four or more servings of enriched or whole grain breads and cereals.

Parenteral Nutrition

In the nutrition of patients unable to take food by mouth, all nutrients must be given by a parenteral (usually intravenous) route. However, it is difficult to supply parenterally the requirements of complete nutrition. Dextrose solutions are used to supply calories, but the quantity of dextrose which can be administered is limited. In most individuals it is not desirable to give dextrose intravenously at a rate exceeding 0.5 g/kg body weight/hour. At faster rates, there is likely to be considerable glycosuria and accompanying diuresis. One hundred g of dextrose is required to prevent ketosis and to spare protein. Additional amounts of dextrose improve this sparing action somewhat, but, considering the maximal rate of administration, it is inconvenient to administer dextrose in significantly larger quantities.

A major defect in parenteral nutrition is caloric inadequacy. This could be minimized by the provision

TABLE 21–3. Recommended daily dietary allowances, revised 1968.*[1]

								Fat-Soluble Vitamins			Water-Soluble Vitamins							Minerals				
	Age[2] Years	Weight (kg)	Weight (lb)	Height (cm)	Height (in)	Kcal	Protein (g)	Vitamin A Activity (IU)	Vitamin D Activity (IU)	Vitamin E Activity (IU)	Ascorbic Acid (mg)	Folacin[3] (mg)	Niacin (mg eq[4])	Riboflavin (mg)	Thiamine (mg)	Vitamin B_6 (mg)	Vitamin B_{12} (μg)	Calcium (g)	Phosphorus (g)	Iodine (μg)	Iron (mg)	Magnesium (mg)
Infants	0–1/6	4	9	55	22	kg × 120	kg × 2.2[5]	1500	400	5	35	0.05	5	0.4	0.2	0.2	1.0	0.4	0.2	25	6	40
	1/6–1/2	7	15	63	25	kg × 110	kg × 2.0[5]	1500	400	5	35	0.05	7	0.5	0.4	0.3	1.5	0.5	0.4	40	10	60
	1/2–1	9	20	72	28	kg × 100	kg × 1.8[5]	1500	400	5	35	0.1	8	0.6	0.5	0.4	2.0	0.6	0.5	45	15	70
Children	1–2	12	26	81	32	1100	25	2000	400	10	40	0.1	8	0.6	0.6	0.5	2.0	0.7	0.7	55	15	100
	2–3	14	31	91	36	1250	25	2000	400	10	40	0.2	8	0.7	0.6	0.6	2.5	0.8	0.8	60	15	150
	3–4	16	35	100	39	1400	30	2500	400	10	40	0.2	9	0.8	0.7	0.7	3	0.8	0.8	70	10	200
	4–6	19	42	110	43	1600	30	2500	400	10	40	0.2	11	0.9	0.8	0.9	4	0.8	0.8	80	10	200
	6–8	23	51	121	48	2000	35	3500	400	15	40	0.2	13	1.1	1.0	1.0	4	0.9	0.9	100	10	250
	8–10	28	62	131	52	2200	40	3500	400	15	40	0.3	15	1.2	1.1	1.2	5	1.0	1.0	110	10	250
Males	10–12	35	77	140	55	2500	45	4500	400	20	40	0.4	17	1.3	1.3	1.4	5	1.2	1.2	125	10	300
	12–14	43	95	151	59	2700	50	5000	400	20	45	0.4	18	1.4	1.4	1.6	5	1.4	1.4	135	18	350
	14–18	59	130	170	67	3000	60	5000	400	25	55	0.4	20	1.5	1.5	1.8	5	1.4	1.4	150	18	400
	18–22	67	147	175	69	2800	60	5000	400	30	60	0.4	18	1.6	1.4	2.0	5	0.8	0.8	140	10	400
	22–35	70	154	175	69	2800	65	5000	. . .	30	60	0.4	18	1.7	1.4	2.0	5	0.8	0.8	140	10	350
	35–55	70	154	173	68	2600	65	5000	. . .	30	60	0.4	17	1.7	1.3	2.0	5	0.8	0.8	125	10	350
	55–75+	70	154	171	67	2400	65	5000	. . .	30	60	0.4	14	1.7	1.2	2.0	6	0.8	0.8	110	10	350
Females	10–12	35	77	142	56	2250	50	4500	400	20	40	0.4	15	1.3	1.1	1.4	5	1.2	1.2	110	18	300
	12–14	44	97	154	61	2300	50	5000	400	20	45	0.4	15	1.4	1.2	1.6	5	1.3	1.3	115	18	350
	14–16	52	114	157	62	2400	55	5000	400	25	50	0.4	16	1.4	1.2	1.8	5	1.3	1.3	120	18	350
	16–18	54	119	160	63	2300	55	5000	400	25	50	0.4	15	1.5	1.2	2.0	5	1.3	1.3	115	18	350
	18–22	58	128	163	64	2000	55	5000	400	25	55	0.4	13	1.5	1.0	2.0	5	0.8	0.8	100	18	350
	22–35	58	128	163	64	2000	55	5000	. . .	25	55	0.4	13	1.5	1.0	2.0	5	0.8	0.8	100	18	300
	35–55	58	128	160	63	1850	55	5000	. . .	25	55	0.4	13	1.5	1.0	2.0	5	0.8	0.8	90	18	300
	55–75+	58	128	157	62	1700	55	5000	. . .	25	55	0.4	13	1.5	1.0	2.0	6	0.8	0.8	80	10	300
Pregnancy						+200	65	6000	400	30	60	0.8	15	1.8	+0.1	2.5	8	+0.4	+0.4	125	18	450
Lactation						+1000	75	8000	400	30	60	0.5	20	2.0	+0.5	2.5	6	+0.5	+0.5	150	18	450

*Reproduced from Publication 1694, Food and Nutrition Board, National Academy of Sciences, National Research Council, 1968.

[1] The allowance levels are intended to cover individual variations among most normal persons as they live in the USA under usual environmental stresses. The recommended allowances can be attained with a variety of common foods, providing other nutrients for which human requirements have been less well defined.

[2] Entries on lines for age range 22–35 years represent the reference man and woman at age 22. All other entries represent allowances for the midpoint of the specified age range.

[3] The folacin allowances refer to dietary sources as determined by *Lactobacillus casei* assay. Pure forms of folacin may be effective in doses less than ¼ of the recommended daily allowance.

[4] Niacin equivalents include dietary sources of the vitamin itself plus 1 mg equivalent for each 60 mg of dietary tryptophan.

[5] Assumes protein equivalent to human milk. For proteins not 100% utilized, factors should be increased proportionately.

of fat because of its high caloric content. An emulsion of fat (cottonseed oil) suitable for intravenous use has been given in many cases where prolonged intravenous feeding was necessary. The added calories from fat also improve the nutritive efficiency of intravenously administered amino acids which otherwise are largely catabolized to supply needed calories.

The protein requirements of sick and injured persons are quite variable, but about 100 g a day of protein is considered the minimal desired intake. This may be supplied intravenously by the use of amino acid preparations, usually made by hydrolyzing proteins so that they are not antigenic. Plasma, including concentrated solutions of albumin, and blood may also contribute to the protein nutrition of the body, but these substances are not used primarily for this purpose.

The requirements for water and sodium chloride have been discussed in Chapter 19. Here also there are considerable variations depending upon the clinical situation.

A. Short-term Parenteral Nutrition: Assuming no unusual losses, a patient can be satisfactorily maintained for a few days by the use of the following regimen every 24 hours:

1. Sodium chloride solution–500 ml of isotonic (0.9%) sodium chloride solution.

2. Dextrose solution–2000 to 2500 ml of dextrose, 5 or 10%, in water. This should not be given faster than 0.5 g/kg/hour.

B. In More Prolonged Parenteral Nutrition:

1. Dextrose solution with amino acids–If the patient must be deprived of any other dietary source of protein for more than a few days some of the dextrose solutions (as above) should contain 5% amino acids. The rate of utilization of dextrose by the tissues will also be increased by the added amino acids.

2. Potassium salts–After the third or fourth day of parenteral nutrition, potassium must be added to the regimen. The quantities necessary and the precautions in administering it parenterally are discussed on p 403.

3. Vitamins are not required during short illnesses unless the patient was previously undernourished. In a prolonged disability they may be added to the parenteral regimen.* The most important vitamins and the quantities recommended to be given each day are: thiamine, 5–10 mg; riboflavin, 5–10 mg; niacin, 50–100 mg; calcium pantothenate, 10–20 mg; pyridoxine, 1–2 mg; vitamin B_{12}, 3–4 μg; and ascorbic acid, 100–250 mg.

• • •

References

Consolazio CF, Johnson RE, Marek E: *Metabolic Methods.* Mosby, 1951.

National Academy of Sciences-National Research Council: *Recommended Daily Dietary Allowances,* rev ed. Publication No. 1694. NAS-NRC, 1968.

Sebrell WH Jr: Nutr Rev 26:355, 1968.

General Bibliography

Albritton EC (editor): *Standard Values in Nutrition and Metabolism.* Saunders, 1954.

Block RJ, Weiss KW: *Amino Acid Handbook.* Thomas, 1956.

Brock JF: *Recent Advances in Human Nutrition.* Little, Brown, 1961.

Burton BT (editor): *Heinz Handbook of Nutrition,* 2nd ed. McGraw-Hill, 1965.

Cooper LF, Barber EM, Mitchell HS: *Nutrition in Health and Disease.* Lippincott, 1960.

Crampton EW, Lloyd LE: *Fundamentals of Nutrition.* Freeman, 1960.

Davidson S, Passmore R: *Human Nutrition and Dietetics,* 4th ed. Williams & Wilkins, 1969.

Goldsmith G: *Nutritional Diagnosis.* Thomas, 1959.

Pollack H, Halpern SL: *Therapeutic Nutrition.* Publication 234. National Research Council, 1952.

Sebrell WH Jr, Harris RS: *The Vitamins.* 3 vols. Academic Press, 1954.

Wohl MG, Goodhart RS (editors): *Modern Nutrition in Health and Disease,* 4th ed. Lea & Febiger, 1968.

*Vitamins added to fluids given intravenously may be lost in considerable quantity because of excretion in the urine. Oral administration or intramuscular injection are therefore the preferred routes for most efficient utilization of vitamin supplements.

22...

The Chemistry of the Tissues

The tissues of the body are usually divided into 5 categories: vascular (including the blood), epithelial, connective, muscle, and nerve tissue. The chemistry of blood has been discussed in Chapter 10.

EPITHELIAL TISSUE

Epithelial tissue covers the surface of the body and lines hollow organs such as those of the respiratory, digestive, and urinary tracts.

Keratin

A major constituent of the epidermal portion of the skin and of other epidermal derivatives, such as hair, horn, hoof, feathers, and nails, is the albuminoid protein, keratin. This protein is notable for its great insolubility and resistance to attack by proteolytic enzymes of the stomach and intestine, and for its high content of the sulfur-containing amino acid, cystine.

The composition of human hair differs in accordance with its color and with the race, sex, age, and genetic origin of the individual. The amino acid, cystine, accounts for about 20% of the amino acid content of the protein in human hair; this very high cystine content differentiates human hair from all other types of hair.

The keratin of wool can be rendered more soluble and digestible by grinding it in a ball mill. Such finely ground keratin, when supplemented with tryptophan, histidine, and methionine, can serve as a source of dietary protein in animals.

Melanin

The color of the skin is due to a variety of pigments, of which melanin, a tyrosine derivative, is the most important. It is said that racial differences in skin, hair, and eye color are due entirely to the amount of melanin present and that there is no qualitative difference in skin pigmentation of different races.

CONNECTIVE TISSUE

Connective tissue comprises all of that which supports or binds the other tissues of the body. In the widest sense it includes the bones and teeth as well as cartilage and fibrous tissue.

White Fibrous Tissue

The Achilles tendon of the ox is generally used as a characteristic example of white fibrous tissue. It contains about 63% water and 37% solids, of which only 0.5% is inorganic matter. The 3 distinct proteins identified in the tendon are:

Collagen	31.6%
Elastin (see p 479)	1.6%
Tendomucoid	0.5%

The remainder of the organic material consists of fatty substances (1%) and extractives (about 0.98%).

A. Collagen: An albuminoid, the protein **collagen,** is the principal solid substance in white fibrous connective tissue. Collagen is difficult to dissolve and somewhat resistant to chemical attack, although not to the same extent as keratin. The amino acid distribution of collagen is, however, quite different from that of keratin; glycine replaces cystine as the principal amino acid, and in fact accounts for as much as 1/3 of all the amino acids present. Proline and hydroxyproline constitute another 1/3. In addition to the most common form of hydroxyproline (2-hydroxyproline), small amounts (0.26% in cattle Achilles tendon) of 3-hydroxyproline have also been found in collagen (Ogle, 1962). Collagen can be slowly digested by pepsin and hydrochloric acid; it can be digested by trypsin only after pepsin treatment or at temperatures over 40° C.

An important property of collagen is its conversion to **gelatin** by boiling with water or acid. This seems to involve only a physical change, since there is no chemical evidence of hydrolysis. Gelatin contains no tryptophan and very small amounts of tyrosine and cystine. It differs from collagen and keratin in being easily soluble and digestible. It may therefore be used as a source of protein in the diet, but only in a supplementary capacity because of its amino acid deficiencies.

The collagen of bone, skin, cartilage, and ligaments differs in chemical composition from that of white fibrous tissue.

B. Tendomucoid: This is a glycoprotein similar to salivary mucin. There are similar mucoids in bone (osseomucoid) and in cartilage (chondromucoid). The chemistry of these glycoproteins is discussed below.

Yellow Elastic Tissue

The nuchal ligaments exemplify yellow elastic tissue. This tissue is composed of about 40% solid matter, of which 31.7% is **elastin**, 7.2% is collagen, and about 0.5% is mucoid.

The albuminoid, elastin, the characteristic protein of this type of connective tissue, is insoluble in water but digestible by enzymes. It is not converted to gelatin by boiling. The sulfur content of elastin is low. Ninety percent of the amino acid content of the protein is accounted for by only 5 acids: leucine, isoleucine, glycine, proline, and valine.

Cartilage

The 3 principal solids in the organic matrix of cartilage are **chondromucoid, chondroalbumoid,** and **collagen** (see above).

A. Chondromucoid: When chondromucoid is degraded it yields a protein and a prosthetic group belonging to that group of polysaccharides known as mucopolysaccharides, which are a part of the structural elements of many tissues throughout the body. The mucopolysaccharides may be defined from a chemical standpoint as heteropolysaccharides composed of equal amounts of an N-acetyl hexosamine and a uronic acid alternately linked through a glycosidic bond. Such a structural disaccharide unit is illustrated in Fig 22–1. It consists of glucuronic acid linked to N-acetylgalactosamine sulfate. This is the fundamental recurring unit of chondroitin sulfate, the mucopolysaccharide prosthetic group of chondromucoid. The mucopolysaccharides such as chondroitin sulfate are actually relatively large molecules, having molecular weights of about 50,000 daltons. Because their complete structure is not known, only the fundamental disaccharide unit can be shown.

Four chondroitin sulfates have been isolated. The best known is chondroitin sulfate C, which is present in cartilage and tendon. The fundamental disaccharide unit which it contains is shown below. The biosynthesis of D-glucosamine has been described on p 261 and that of D-glucuronic acid has been described on p 256. N-Acetylgalactosamine with a sulfate attached at carbon 6 is the other component of the chondroitin C disaccharide. The biosynthesis of this compound has been studied by Glaser (1959). Glaser has identified in rat liver an enzyme capable of converting N-acetylglucosamine to N-acetylgalactosamine. The enzyme, which is an epimerase, acts upon uridine diphosphate-N-acetylglucosamine to catalyze its conversion to uridine diphosphate-N-acetylgalactosamine. This epimerase reaction is obviously quite similar to that described on p 259 for the interconversion of galactose and glucose.

Chondroitin sulfate A, present in cartilage, adult bone, and corneal tissue, differs from chondroitin sulfate C only in that the sulfate moiety is attached at carbon 4.

Chondroitin sulfate B, present in skin, tendon, and heart valves, has L-iduronic acid in place of the glucuronic acid of chondroitin sulfates C or A; but like A, the sulfate moiety is attached at carbon 4. Chondroitin sulfate B is sometimes designated beta-heparin because it has weak anticoagulant properties.

L-Iduronic acid, which occurs in chondroitin sulfate B, is an epimer of D-glucuronic acid at carbon 5; that is, it may be derived from D-glucuronic acid merely by reversing the position of the H and OH groups around carbon 5. The direct metabolic origin of L-iduronic acid from D-glucose has been demonstrated by Roden & Dorfman (1958); more recently, Jacobson & Davidson (1962) have elucidated the further inter-

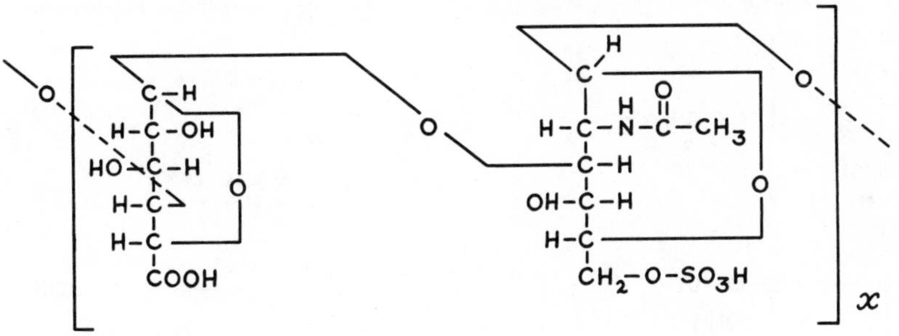

FIG 22–1. The fundamental disaccharide unit of chondroitin sulfate C (β-1,3-glucuronido-N-acetylgalactosamine-6-sulfate).

mediates in the biosynthetic pathway. As shown in Fig 13–15, uridine diphosphoglucuronic acid ("active glucuronic acid") is formed in the liver from uridine diphosphoglucose through the catalytic action of uridine diphosphoglucose dehydrogenase and NAD. This dehydrogenase has also been identified in extracts of rabbit skin and has been shown to be capable of producing uridine diphospho-D-glucuronic acid from uridine diphospho-D-glucose. Also present in the rabbit skin extracts is a **uridine diphospho-D-glucuronic acid-5-epimerase**, which brings about the conversion of the glucuronic acid derivative to the corresponding derivative of L-iduronic acid. The reactions shown in Fig 22–2 seem to account for the origin of L-iduronic acid in the skin, where it is required for synthesis of chondroitin sulfate B.

The fourth chondroitin (chondroitin sulfate D) has been isolated from the cartilage of the shark (Suzuki, 1960). It resembles chondroitin sulfate C except that it has a second sulfate attached probably at carbon 2 or 3 of the uronic acid moiety. (A similar disulfate has also been isolated from a preparation of chondroitin B.)

B. Chondroalbumoid is somewhat similar to elastin and keratin. In contrast to keratin, however, the sulfur content of chondroalbumoid is low and it is soluble in gastric juice.

Other Mucopolysaccharides

Other examples of mucopolysaccharides are heparin, mucoitin sulfuric acid, and hyaluronic acid.

A. Mucoitin sulfuric acid of the saliva contains a glucosamine which is probably similar to that in the chitin of the exoskeleton of lower animals.

B. Hyaluronic acid is a mucopolysaccharide in which the fundamental recurring disaccharide units consist of glucuronic acid linked to N-acetylglucosamine. Its composition is therefore similar to chondroitin, the essential difference between the 2 structures being the presence of galactosamine in chondroitin instead of glucosamine, as is the case with hyaluronic acid. Hyaluronic acid is a component of the capsules of certain strains of pneumococci, streptococci, and certain other organisms, as well as of vitreous humor, synovial fluid, and umbilical cord (Wharton's jelly). The hyaluronic acid of the tissues acts as a lubricant in the joints, as a jelly-like cementing substance, and as a means of holding water in the interstitial spaces.

An enzyme, **hyaluronidase**, is present in certain tissues, notably testicular tissue and spleen, as well as in several types of pneumococci and the hemolytic streptococci. An enzyme similar to testicular hyaluronidase has been detected also in rat and guinea pig liver, lung, and kidney, and in human kidney, urine, plasma, and synovial fluid and tissue.

Hyaluronidase, by destroying tissue hyaluronic acid, reduces viscosity and thus permits greater spreading of materials in the tissue spaces. Hyaluronidase is therefore sometimes designated the **"spreading factor."** Its activity may be measured by the extent of spread of injected India ink as indicator. The invasive power

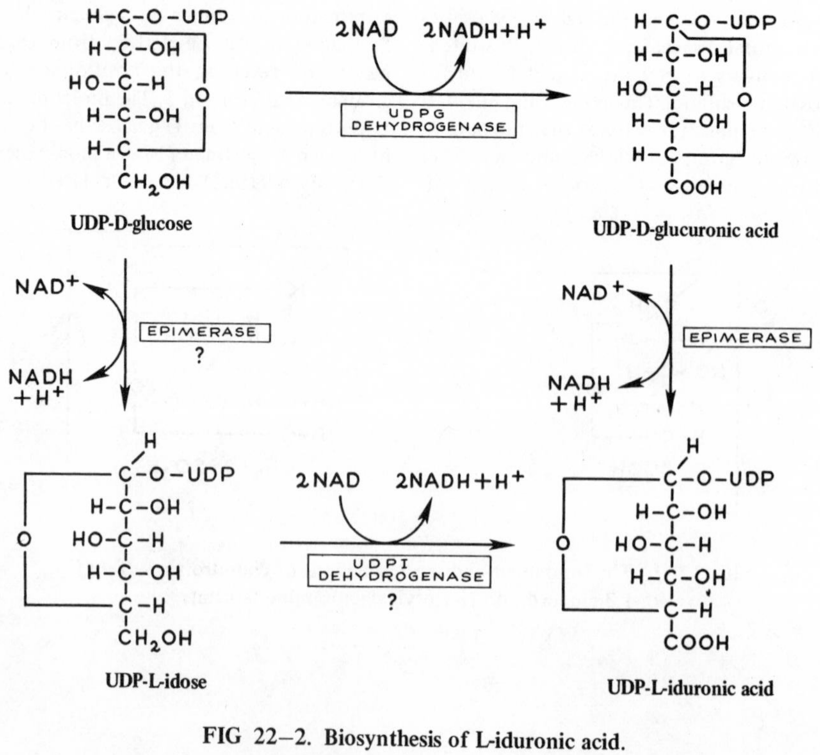

UDP-D-glucose → (2NAD → 2NADH+H⁺; UDPG DEHYDROGENASE) → UDP-D-glucuronic acid

UDP-D-glucose → (NAD⁺ / NADH + H⁺; EPIMERASE ?) → UDP-L-idose

UDP-D-glucuronic acid → (NAD⁺ / NADH + H⁺; EPIMERASE) → UDP-L-iduronic acid

UDP-L-idose → (2NAD → 2NADH+H⁺; UDPI DEHYDROGENASE ?) → UDP-L-iduronic acid

FIG 22–2. Biosynthesis of L-iduronic acid.

of some pathogenic organisms may be increased because they secrete hyaluronidase. In the testicular secretions, it may dissolve the viscid substances surrounding the ova to permit penetration of the sperm cell. Hyaluronidase is used clinically to increase the efficiency of absorption of solutions administered by clysis.

C. Glycoproteins: The carbohydrates of the glycoproteins probably have mucopolysaccharides as prosthetic groups. Such proteins are not coagulated by heat nor are they digested by the enzymes of the gastrointestinal tract. They therefore act as protective substances for the mucosal tissue.

Bone

A. Chemistry: The water content of bone varies from 14–44%. From 30–35% of the fat-free dry material is organic. In some cases as much as 25% is fat. The organic material in bone, the bone matrix, is similar to that of cartilage in that it contains collagen, which can be converted to gelatin (ossein gelatin). There are also a glycoprotein named **osseomucoid** and an **osseo-albumoid.** The presence of citrate (about 1%) in bone has been reported recently.

The inorganic material of bone consists mainly of phosphate and carbonate salts of calcium. There are also small amounts of magnesium, hydroxide, fluoride, and sulfate. A study of the x-ray pattern of the bone salts indicates a similarity to the naturally occurring mineral hydroxyapatite. The formula of hydroxyapatite is said to be

$$Ca(OH)_2 \cdot 3\ Ca(PO_4)_2$$

or

$$Ca_{10}(OH)_2(PO_4)_6$$

It is believed that the crystal lattices of bone are similar to the lattices of these apatite minerals but that elements may be substituted in bone without disturbing the structure. For example, calcium and phosphorus atoms may be replaced by carbon; magnesium, sodium, and potassium may replace calcium; and fluorine may replace hydroxide. This probably accounts for the alterations in bone composition which occur with increasing age, in rickets, as a result of dietary factors, or subsequent to changes in acid-base equilibrium (eg, the acidosis of chronic renal disease).

B. Metabolism: Like all of the tissues of the body, the constituents of bone are constantly in exchange with those of the plasma. Demineralization of bone occurs when the intake of minerals necessary for bone formation is inadequate or when their loss is excessive.

The calcium and phosphorus content of the diet (see Chapter 19) is obviously an important factor in ossification. Vitamin D (see Chapter 7) raises the level of blood phosphate and calcium, and this may in turn raise the calcium-phosphorus product to the point where calcium phosphate is precipitated in the bone. There is some evidence that the vitamin not only acts to promote better absorption of the minerals but also acts locally in the bone. In rickets, osteitis deformans, and other bone disorders the blood alkaline phosphatase rises, possibly in an effort to supply more phosphate. Treatment with vitamin D is accompanied by a reduction in phosphatase.

The influence of endocrines on calcification has been described in Chapter 20. The parathyroids, thyroid, anterior pituitary, adrenal, and sex glands are all important in this respect. They act either at the site of calcification or by altering absorption or excretion of calcium and phosphorus.

Ossification supposedly involves precipitation of bone salts in the matrix by means of a physicochemical equilibrium involving Ca^{++}, $HPO_4^{=}$, and $PO_4^{\equiv}$. The enzyme alkaline phosphatase, which liberates phosphate from organic phosphate esters, may produce the inorganic phosphate; this phosphate then reacts with the calcium to form insoluble calcium phosphate. Phosphatase is not found in the matrix but only in the osteoblasts of the growing bone.

The deposition of bone salt is not entirely explainable as due simply to physicochemical laws governing the solubility of the inorganic components of bone. The deposition of the bone salt in the presence of concentrations of inorganic phosphate and calcium similar to those of normal plasma may require the expenditure of energy from associated metabolic systems. In cartilage, calcification in vitro can be blocked with iodoacetate, fluoride, or cyanide, substances which are known to inhibit various enzymes involved in glycolysis. Furthermore, the deposition of bone salts in calcifying cartilage is preceded by swelling of the cartilage cells due to intracellular deposition of glycogen. Just prior to or simultaneously with the appearance of bone salt in the matrix of the cartilage, the stores of glycogen seem to disappear, which suggests that the breakdown of glycogen is necessary for the calcification of cartilage. All of the enzymes and intermediate compounds involved in glycolysis have been identified in calcifying cartilage, and, as noted above, enzyme inhibitors interfere with calcification in vitro.

It must be remembered that there are 2 important components in bone: the matrix, which is rich in proteins, and the mineral or inorganic component. Demineralization may result from effects on either. Steroids aid in the maintenance of osteoblasts and matrix; thus, in the absence of these hormones, osteoporosis may occur even though alkaline phosphatase is normal and, presumably, a favorable mineral environment exists. Changes in the concentration of alkaline phosphatase reflect activity of the osteoblasts which are stimulated by stress on a weakened skeleton (as may be found in rickets).

Teeth

Enamel, dentin, and cementum of the teeth are all calcified tissues containing both organic and inorganic matter. In the center of the tooth is the **pulp,** a soft, uncalcified organic mass containing also the blood vessels and nerves.

A. Composition and Structure: The average composition of human enamel and of dentin is shown in Table 22–1.

TABLE 22–1. Average composition of human enamel and of dentin.*

	(% Dry Weight)	
	Enamel	Dentin
Calcium	35.8	26.5
Magnesium	0.27	0.79
Sodium	0.25	0.19
Potassium	0.05	0.07
Phosphorus	17.4	12.7
CO_2 (from carbonate)	2.97	3.06
Chlorine	0.3	0.0
Fluorine	0.0112	0.0204
Iron	0.0218	0.0072
Organic matter	1.0	25.0

*From Hawk, Oser, Summerson: *Practical Physiological Chemistry,* 12th ed. Blakiston, 1947.

According to x-ray studies, the inorganic matter in the enamel and dentin of the teeth is arranged similarly to that in bone; it consists mainly of hydroxyapatite salts.

Keratin is the principal organic constituent of the enamel. There are also small amounts of cholesterol and phospholipid. In the dentin, collagen and elastin occur together with a glycoprotein and the lipids of the enamel.

Collagen is a major organic constituent in the cementum. Both dentin and, to a lesser extent, enamel contain citrate.

B. Metabolism of Teeth: Studies with radioactive isotopes (radiophosphorus) indicate that the enamel and especially the dentin undergo constant turnover; this is slow in adult teeth. The diet must contain adequate calcium and phosphorus and also vitamins A, C, and D to ensure proper calcification. However, when the diet is low in calcium and phosphorus, the demineralization of bone exceeds that of the teeth, which may actually calcify during the restricted period but at a slower than normal rate. This and other data suggest that the mineral metabolism of teeth and that of bones is not necessarily parallel.

The relation of fluoride to teeth has been discussed in Chapter 19.

MUSCLE TISSUE

There are 3 types of muscle tissue in the body: striated (voluntary) or skeletal muscle, nonstriated (involuntary) or smooth muscle, and cardiac muscle.

The chemical constitution of the skeletal muscle has been most completely studied: 75% is water, 20% is protein, and the remaining 5% is composed of inorganic material, certain organic "extractives," and carbohydrate (glycogen and its derivatives).

The Proteins in Muscle

The muscle fibrils are mainly protein. These fibrillar proteins are characterized by their elasticity or contractile power. **Myosin** is the most abundant protein in muscle, and, except for **actin**, in the complex known as **actomyosin**, it is probably the only other protein in the fibril—other muscle proteins being considered either constituents of the sarcoplasm, the material which imbeds the myofibrils, or at least extracellular in origin. Myosin is a globulin, soluble in dilute salt solutions (eg, alkaline 0.6 M KCl) and insoluble in water.

Myosin is composed of subunits about 80 nm in length with molecular weights of approximately 200,000 daltons. Artificial cleavage of a myosin subunit with proteolytic enzymes produces **light meromyosins** and **heavy meromyosins.** The latter subunit retains the adenosine triphosphatase activity associated with myosin.

Actin is a globulin, molecular weight 60,000 daltons, which binds one ATP per mol. If actin is prepared by extraction with dilute salt solutions, the protein can be obtained in the form of G- (globular) actin. When the ionic strength of a solution of G-actin is increased, and in the presence of Mg^{++}, G-actin polymerizes to F- (fibrous) actin. During polymerization of G-actin to F-actin, ATP is hydrolyzed to ADP and Pi is released. Depolymerization to G-actin occurs when ATP is added to replace ADP.

Actomyosin is a protein complex of 3 myosins with one actin molecule. It can be obtained by extraction of muscle tissue with water followed by prolonged extraction of the residue with alkaline 0.6 M KCl. This latter extract is highly viscous and exhibits birefringence of flow as a result of the presence of actomyosin.

Globulin X is a protein which remains after myosin has been removed from a saline extract of muscle. It is probably a protein of the sarcoplasm. It coagulates at 50° C; its molecular weight is about 160,000 daltons.

Myogen is a term used to describe a solution of proteins extractable from muscle with cold water. It consists largely of enzymes involved in glycolysis.

Myoglobin is a conjugated protein, often called muscle hemoglobin. It is indeed chemically similar to hemoglobin (ie, a hemoprotein), and it may function as an oxygen carrier. The molecular weight of myoglobin is about ¼ that of hemoglobin. In crush injuries myoglobin is liberated and may appear in the urine since it is filterable by the glomerulus. It may also precipitate in the renal tubules, obstructing them and resulting in anuria—a feature of the **crush syndrome.** The precipitation of hemoglobin in the tubule is said to cause an etiologically similar anuria after the hemolytic episodes of a transfusion reaction.

Muscular Contraction

The contractile unit of muscle, the **myofibril,** is composed of the 2 principal proteins of muscle tissue mentioned above, myosin and actin. Each protein is believed to be involved in the contractile process. Examination of the myofibril by light microscopy shows the myofibril to consist of a series of light bands (I bands) alternating with dark bands (A bands) along its length. With the greater magnification afforded by the electron microscope (Huxley, 1956), it appears that each myofibril is built up of 2 types of longitudinal filaments. One type, confined to the A band and thought to be composed of myosin, appears to consist of filaments about 16 nm in diameter arranged in cross-section as a hexagonal array. The other filaments are considered to be of actin and to lie in the I band but to extend also into the A band. These latter filaments are smaller than those of myosin (about 6 nm in diameter), and in the A band they are arranged around the myosin filaments as a secondary hexagonal array. Thus, each actin filament lies symmetrically between 3 myosin filaments. On contraction, the A bands remain the same length but the I bands disappear. Huxley (1956) has suggested that on contraction the individual filaments of myosin and actin remain unchanged in length but that a change in length of the muscle, ie, contraction, is brought about by the sliding of the 2 arrays of filaments into, or out of, each other. A diagrammatic representation of these events is shown in Fig 22–3.

Chemical studies have shown that myosin and actin can combine in solution to form the protein complex, **actomyosin.** This protein forms threads which actually "contract" in vitro on the addition of ATP. Simultaneously, ATP is hydrolyzed by the myosin because of the high adenosine triphosphatase activity associated with myosin. The action of ATP is to cause a dissociation of the actomyosin into its individual constituents, actin and myosin, but how these properties of the proteins of muscle can be translated into terms which account for the contraction process in muscle remains a matter for further investigation.

The immediate source of the energy required for muscular contraction is derived from the hydrolysis of ATP to ADP. The metabolic processes associated with production of ATP in oxidative phosphorylation provide ATP but not at a sufficient rate to sustain muscle during bursts of intense activity. Consequently, a store of high-energy phosphate is present in muscle in the form of **phosphocreatine** which, by acting as a source of high-energy phosphate for the prompt resynthesis of ATP, serves to maintain adequate amounts of ATP, the source of energy for muscular contraction. In the resting state, mammalian muscle contains 4–6 times as much phosphocreatine as ATP. The transfer of high-energy phosphate from creatine phosphate to ADP (the Lohmann reaction) is catalyzed by the enzyme, **creatine kinase**. The reaction is reversible, so that resynthesis of creatine phosphate can take place when ATP later becomes available, as during the recovery period following a period of muscular contraction. Transfer of phosphate from ATP to creatine to form creatine phosphate is catalyzed by the enzyme, **ATP-creatine transphosphorylase** (Kuby, 1954). These relationships between ATP and creatine phosphate are represented in Fig 22–4.

A further source of ATP in muscle is attributable to the presence of another enzyme, **myokinase** (adenylate kinase), which catalyzes the transfer of a high-energy phosphate from one molecule of ADP to another to form ATP and adenosine monophosphate (AMP).

$$2\text{ADP} \xrightleftharpoons{\text{MYOKINASE}} \text{AMP} + \text{ATP}$$

Resynthesis of ATP and phosphocreatine may be blocked by poisoning an isolated muscle with iodoacetate, which prevents glycolysis. The muscle may contract for a while, but contraction ceases when all reserves of ATP and phosphocreatine have been used. This experiment demonstrates that the energy required for regeneration of ATP under such conditions is derived mainly from glycolysis. However, it is probable that in vivo other fuels such as free fatty acids (FFA)

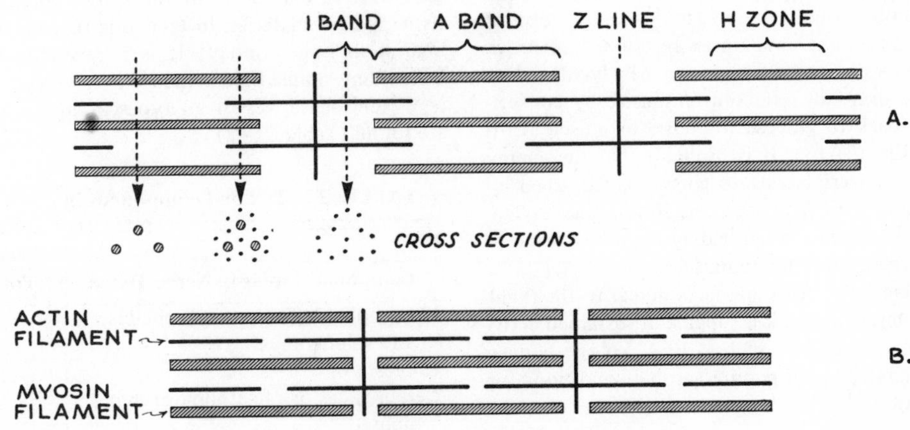

FIG 22–3. Arrangement of filaments in striated muscle.
***A.* Extended. *B.* Contracted.**

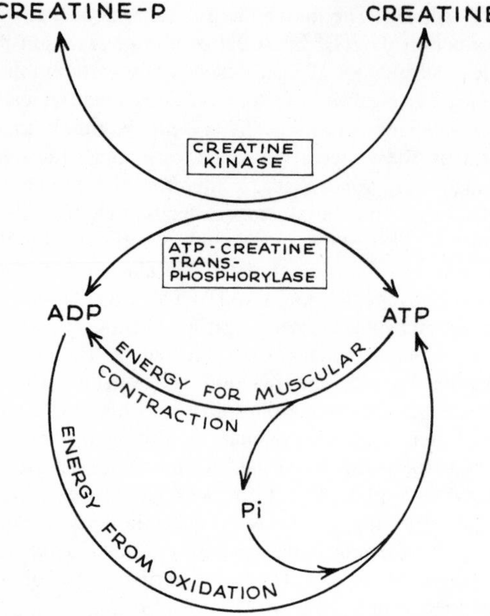

FIG 22–4. Formation and breakdown of creatine phosphate and the relationship of these events to ATP in muscular contraction.

and ketone bodies are used as well as glucose to supply contracting muscle with ATP. Fat must be the ultimate source of energy for long periods of muscular exertion, such as in migratory birds, whose total carbohydrate stores are quite inadequate to serve as the sole source of energy and whose fat stores become depleted during migration. The flight muscles of birds ("red meat") are particularly well developed for aerobic oxidation of fuels such as pyruvate, FFA, and ketone bodies, having a well developed vasculature to increase oxygenation and a high content of enzymes of the respiratory chain as well as of cytochromes and myoglobin.

Thus, skeletal muscle is adapted to its function of providing for a very rapid output of energy by means of various mechanisms which allow it to produce ATP under anaerobic conditions, viz, the decomposition of phosphocreatine, the myokinase reaction, and provision of glycogen stores which can be glycolyzed to lactate. The liver aids oxidation in muscle by converting lactate back to glucose for reuse in muscle (Cori cycle; see Fig 13–13). It is probable that the energy required to convert lactate to glucose is furnished by the oxidation of FFA in the liver. In this way, oxidation in the liver could be providing indirectly some of the energy for contraction in muscles.

Biochemically, heart muscle is similar to the flight muscles of birds in that it is capable of sustained activity and can utilize fuels such as FFA, ketone bodies, and even lactate, which require aerobic conditions for their oxidation.

Inorganic Constituents of Muscle

The cations of muscle (potassium, sodium, magnesium, and calcium) are the same as in extracellular fluids except that, in muscle, potassium predominates. The anions include phosphate, chloride, and small amounts of sulfate. The apparent high inorganic phosphate content may actually be an artefact produced by the breakdown of the organic phosphates of ATP and phosphocreatine in the course of the analysis.

Intracellular potassium plays an important role in muscle metabolism. When glycogen is deposited in muscle and when protein is being synthesized, a considerable amount of potassium is also incorporated into the tissue. Muscle weakness is a cardinal sign of potassium deficiency. The calcium and magnesium of muscle appear to function as activators or inhibitors of intramuscular enzyme systems.

NERVE TISSUE

The tissues of the brain, spinal cord, the cranial and spinal nerves and their ganglia and plexuses, and those of the autonomic nervous system contain a considerable quantity of water. The gray matter, which represents a concentration of nerve cell bodies, always contains more water than the white matter, where the nerve fibers are found. In the adult brain, where gray and white matter are mixed, the water content averages 78%; in the cord the water content is slightly less, about 75%.

The solids of nerve tissue consist mainly of protein and lipids. There are also smaller amounts of organic extractives and of inorganic salts.

The Proteins of Nerve Tissue

These constitute 38–40% of the total solids. They include various globulins, nucleoprotein, and a characteristic albuminoid called **neurokeratin.**

The Lipids of Nerve Tissue

Over ½ (51–54%) of the solid content of nerve tissue is lipid material. In fact, this tissue is one of the highest in lipid content. It is noteworthy that very little if any simple lipid is present.

Representatives of all types of lipid compounds are found (Table 22–2).

TABLE 22–2. Lipid compounds in nerve tissue.

Compound Lipids in Nerve Tissue	Percent of Total Solids
Phospholipids (lecithins, cephalins, and sphingomyelins)	28%
Cholesterol	10%
Cerebrosides or galactolipids (glycolipids)	7%
Sulfur-containing lipids, aminolipids, etc	9%
	54%

The chemistry of these substances is discussed in Chapter 2.

The rate at which the lipids of the brain are exchanged is relatively slow in comparison to that in an active organ such as the liver. Tracer studies with deuterium indicate that while 50% of the liver fats may be exchanged in 24 hours, only 20% of the brain fat is replaced in 7 days.

Inorganic Salts

These substances in nervous tissue are components of the 1% ash produced as the result of combustion. The principal inorganic salts are potassium phosphate and chloride, with smaller amounts of sodium and other alkaline elements. The potassium of the nerve is thought to be important in the electrical nature of the nerve impulse, which depends on depolarization and repolarization at the membrane boundary of the nerve fiber.

Metabolism of Brain & Nerve

When a nerve is stimulated to conduct an impulse, a small but measurable amount of heat is produced. The heat is produced in 2 stages, as it is in working muscle. This suggests that the rapidly released initial heat represents the energy involved in transmission of the impulse, and that the delayed or recovery heat (which may continue for 30–45 minutes) is related to restoration of the energy mechanisms. Similarly, the nerve may conduct impulses and develop heat under anaerobic conditions as, for example, in an atmosphere of nitrogen; but recovery depends on the admission of oxygen, as indicated by the extra consumption of readmitted oxygen.

The respiratory quotient of metabolizing nerve is very close to 1, which suggests that the nerve is utilizing carbohydrate almost exclusively. Recent studies with rat brain mitochondria (Tanaka & Abood, 1963) indicate that, contrary to earlier assumptions, brain mitochondria are not capable of oxidizing glucose. Glycolytic enzymes are apparently contained only in soluble cytoplasmic material of some cellular fragments present as contaminants of the original mitochondrial preparations, which thus appear to possess glycolytic activity. Mitochondria do, however, appear to possess significant quantities of the hexokinase activity identified in preparations from rat brain.

The metabolism of carbohydrate in nerve tissue seems to be similar to that of muscle, since lactic and pyruvic acids appear under anaerobic conditions. These end products disappear very slowly; oxygen does not accelerate the process.

The synthesis of glycogen by brain tissue has been shown to take place by way of uridine diphosphate glucose as was described for liver and muscle (see p 233). The glycogen stores of brain and nerve are very small; hence a minute-to-minute supply of blood glucose is particularly important to the nervous system. This may be the major reason for the prominence of nervous symptoms in hypoglycemia. In contrast to muscle extracts, brain extracts act more readily on glucose than on glycogen.

Glutamic acid seems to be the only amino acid metabolized by brain tissue. However, this amino acid is of considerable importance in brain metabolism. It serves as a precursor of γ-aminobutyric acid (see below and p 347) and is a major acceptor of ammonia produced either in the metabolism of the brain or delivered to the brain when the arterial blood ammonia is elevated (see p 306). In this latter reaction, glutamic acid accepts 1 mol of ammonia and is thus converted to glutamine. Although it has been shown that the brain can form urea (Sporn & others, 1959), the formation of urea does not play a significant role in removal of ammonia in the brain. This is accomplished almost entirely by reactions involving the formation of glutamic acid by amination of ketoglutaric acid as well as by the formation of glutamine.

When the levels of ammonia in the brain are elevated, usually as a result of increased ammonia in the blood, the supply of glutamic acid available from the blood may be insufficient to form the additional amounts of glutamine required to detoxify the ammonia in the brain. Under these circumstances glutamic acid is synthesized in the brain by amination of the ketoglutaric acid produced in the citric acid cycle within the brain itself. However, continuous utilization of ketoglutaric acid for this purpose would rapidly deplete the citric acid cycle of its intermediates unless a method of replenishing the cycle were available. Repletion is accomplished by CO_2 fixation, involving pyruvate (as shown in Fig 13–7), to form oxaloacetic acid, which enters the citric acid cycle and proceeds to the formation of ketoglutarate. The reaction in brain is precisely analogous to that which occurs in the liver.

Fixation of CO_2 in isolated retinal tissue was demonstrated by Crane & Ball (1951). Studies of CO_2 fixation in brain have been described by Berl & others (1962). These authors have found that CO_2 fixation into amino acids occurs to a significant degree in the cerebral cortex. The highest specific activity among metabolites in the brain was found in aspartate, which would be expected if the initial reaction involved formation of malate and then oxaloacetate, this latter compound forming aspartate by transamination. However, after infusion of ammonia, it was apparent that oxaloacetate was now being used for the synthesis of ketoglutarate, glutamate, and glutamine at a faster rate than it was being converted to aspartate. This suggests that ammonia causes channeling of oxaloacetate to the formation of glutamine.

The formation of γ-aminobutyrate in CNS tissue from glutamate has been discussed on p 347. The significance of γ-aminobutyrate as an important regulatory factor in neuronal activity has also been mentioned.

The synthesis of long-chain fatty acids by enzyme preparations from rat brain tissue has been demonstrated by Brady (1960). The pathway of synthesis was that of the extramitochondrial system, in which malonyl coenzyme A is a required intermediate (as described on p 267).

From the examples cited above and the results of other recent investigations of brain metabolism, it is becoming apparent that cerebral tissue possesses all of the enzymatic activities necessary to support the major metabolic pathways which are found in other organs of the body. This has been referred to as the "autonomy of cerebral metabolism."

The ability of the brain to fix CO_2 introduces interesting speculations with respect to the influence of CO_2 on the operations of the citric acid cycle in the brain. Because the metabolism of glucose provides virtually the sole source of energy for brain metabolism, if CO_2 tension did exert a controlling influence on the citric acid cycle an additional explanation for the effects of CO_2 on the brain might be forthcoming.

Chemical Mediators of Nerve Activity

At the junction of the nerve fiber and the effector organ, such as the muscle which is stimulated (the **myoneural junction**), a chemical substance is elaborated by the action of the nerve impulse. This substance actually brings about the activity of the effector. Such a chemical substance is called a **chemical mediator** of the nerve impulse or a **neurohormone.**

A. The Neurohormones:

1. **Acetylcholine**—In parasympathetic and in voluntary nerves to the skeletal muscles, the chemical mediator is acetylcholine.

2. **Sympathin**—Stimulation of the sympathetic nerves produces sympathin, with an effect opposite to that of acetylcholine. Sympathin exerts 2 effects—one excitatory, referred to as sympathin E; the other inhibitory, referred to as sympathin I. The action of sympathin suggests that it may be identical with epinephrine, and, in fact, sympathetic stimulation does cause liberation of epinephrine from the adrenal medulla.

B. Chemistry of Acetylcholine:

1. **Breakdown of acetylcholine; action of acetylcholine esterase (cholinesterase)**—Acetylcholine is readily hydrolyzed to choline and acetic acid by the action of the enzyme acetylcholine esterase, found not only at the nerve endings but also within the nerve fiber. The reaction is shown in Fig 22–5.

The action of acetylcholine in the body is controlled by the inactivating effect of acetylcholine esterase (designated ACh-esterase by Nachmansohn to distinguish it from a pseudocholinesterase found in the serum, which hydrolyzes other esters).

2. **Resynthesis of acetylcholine**—The breakdown of acetylcholine is apparently an exergonic reaction since energy is required for its resynthesis. Active acetate (Co A-acetate) serves as acetate donor for the acetylation of choline. The enzyme, **choline acetylase,** which is activated by potassium and magnesium ions, catalyzes the transfer of acetyl from Co A-acetate to choline. The regeneration of ATP from ADP is accomplished by phosphocreatine, which, in turn, is resynthesized with the aid of the energy derived from glycolysis. All of these reactions are apparently similar to those in the muscle.

3. **Anticholinesterases**—Inhibition of acetylcholine esterase with resultant prolongation of parasympathetic activity is effected by physostigmine (eserine). The action is reversible.

Neostigmine (Prostigmin) is an alkaloid which is thought to function also as an inhibitor of cholinesterase and thus to prolong acetylcholine or parasympathetic action. It has been used in the treatment of myasthenia gravis, a chronic progressive muscular weakness with atrophy.

A synthetic compound, diisopropylfluorophosphate (DFP), also inhibits the esterase activity but in an irreversible manner.

This compound appears to be the most powerful and specific enzyme inhibitor yet discovered. It inhibits acetylcholine esterase when present in molar concentrations as low as 1×10^{-10} M. A mechanism for detoxifying DFP exists in the body in the action of an enzyme capable of bringing about the hydrolysis of the compound to fluoride and diisopropylphosphoric acid. This enzyme, diisopropylfluorophosphatase, has been identified in the kidney. The enzyme is activated by Mn^{++} or Co^{++} and specific co-factors such as imidazole and pyridine derivatives (eg, proline or hydroxyproline).

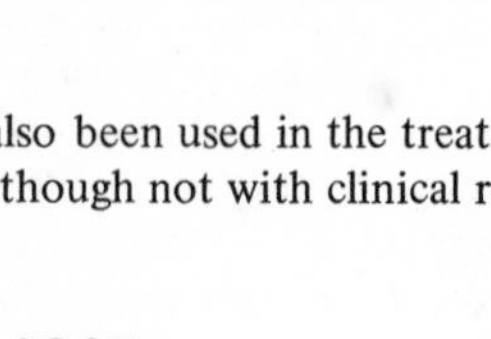

$$(CH_3)_2CH{-}O\diagdown\;\;\;\diagup O$$
$$P$$
$$(CH_3)_2CH{-}O\diagup\;\;\;\diagdown F$$

Diisopropylfluorophosphate (DFP)

DFP has also been used in the treatment of myasthenia gravis, although not with clinical results equal to

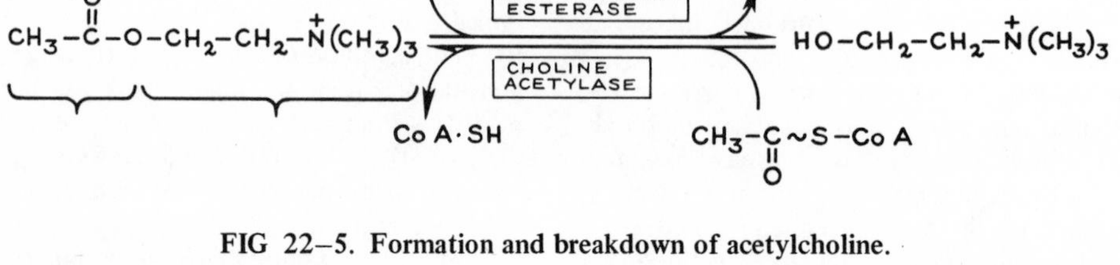

$$CH_3{-}\overset{O}{\overset{\|}{C}}{-}O{-}CH_2{-}CH_2{-}\overset{+}{N}(CH_3)_3 \;\underset{\text{CHOLINE ACETYLASE}}{\overset{\text{ACETYLCHOLIN-ESTERASE}}{\rightleftharpoons}}\; HO{-}CH_2{-}CH_2{-}\overset{+}{N}(CH_3)_3$$

(H_2O in, CH_3COOH out with esterase; $CH_3{-}\underset{O}{\underset{\|}{C}}{\sim}S{-}Co\,A$ in, $Co\,A{\cdot}SH$ out with choline acetylase)

FIG 22–5. Formation and breakdown of acetylcholine.

those obtainable with Prostigmin. It is a dangerous drug to use since the toxic dose is too close to the effective dose.

A number of **anticholinesterases** similar in their action to DFP have been investigated. They serve as the active principle of insecticides. The so-called "nerve gases" proposed for gas warfare, as well as many insecticides (eg, parathion), are also anticholinesterases. These insecticides may produce toxic effects in individuals exposed to high concentrations when these are used as plant sprays.

Atropine is used as an antidote to the toxic effects of DFP and other anticholinesterases, together with a stimulant such as nikethamide (Coramine), for treatment of respiratory paralysis.

• • •

References

Berl S, Takagaki G, Clarke DD, Waelsch H: J Biol Chem 237:2570, 1962.

Brady RO: J Biol Chem 235:3099, 1960.

Crane RK, Ball EG: J Biol Chem 189:269, 1951.

Glaser L: J Biol Chem 234:2801, 1959.

Huxley HE: Endeavour 15:177, 1956.

Jacobson B, Davidson EA: J Biol Chem 237:638, 1962.

Kuby SA, Noda L, Lardy HA: J Biol Chem 209:191, 1954.

Ogle JD, Arlinghaus RB, Logan MA: J Biol Chem 237:3667, 1962.

Roden L, Dorfman A: J Biol Chem 233:1030, 1958.

Sporn MB, Dingman W, Defalco A, Davies RK: Nature 183:1520, 1959.

Suzuki S: J Biol Chem 235:3580, 1960.

Tanaka R, Abood LG: J Neurochem 10:571, 1963.

General Bibliography

Elliott KAC, Page IH, Quastel JH (editors): *Neurochemistry: The Chemistry of Brain and Nerve,* 2nd ed. Thomas, 1962.

McLean FC, Urist MR: *Bone: An Introduction to the Physiology of Skeletal Tissue,* 2nd ed. Univ of Chicago Press, 1961.

Oser BL: *Hawk's Physiological Chemistry,* 14th ed. Blakiston, 1965.

Sourkes TL: *Biochemistry of Mental Disease.* Hoeber, 1962.

Waelsch H (editor): *Biochemistry of the Developing Nervous System.* Academic Press, 1955.

West ES & others: *Textbook of Biochemistry,* 4th ed. Macmillan, 1966.

White A, Handler P, Smith EL: *Principles of Biochemistry,* 4th ed. McGraw-Hill, 1968.

Appendix

General & Physical Chemistry

Physiologic chemistry is the science which is concerned with the chemical reactions associated with biologic processes. It has developed as a natural outgrowth of general chemistry, which deals with the reactions of all elements, radicals, and compounds, organic and inorganic, living and nonliving. Physiologic chemistry incorporates many of the working principles of physical chemistry, ie, the theoretical and physical properties inherent in the chemical elements. Although physiologic chemistry deals with biologic processes, continuing effort is being made to define these processes in mathematical, physical, or chemical terms. Such an effort is exemplified by the area of molecular biology.

Certain physicochemical concepts which have immediate application in the field of physiologic chemistry will be discussed in the following sections.

ELECTROLYTIC DISSOCIATION

Ions

When an electric current is passed through an aqueous solution of a compound in an electrolytic cell, some elements collect at the anode, or positive pole; others at the cathode, or negative pole. This indicates that these elemental particles bear electric charges. Such charged particles are termed **ions.** In accordance with the physical principle that unlike charges attract each other and like charges repel, those ions which have migrated to the anode (anions) are electronegative, and those to the cathode (cations) are electropositive.

Electrolytes & Nonelectrolytes

Compounds composed of elements which may be electrolytically dissociated into ions are called **electrolytes.** Many other compounds, particularly organic compounds, cannot be so dissociated and are thus designated **nonelectrolytes.**

ACIDS & BASES

Definitions

A. **Acids** are compounds of electronegative elements or groups plus ionizable hydrogen (H^+).

B. **Bases** are compounds of electropositive elements or radicals plus ionizable hydroxyl groups (OH^-).

Standard Solutions

A. A **molar solution** of an acid or base is one which contains in each liter an amount in grams of the acid or base equal to the molecular weight of that compound.

B. A **normal solution** is exactly the same as a molar solution if only one hydrogen ion or hydroxyl ion is available. If more than one H^+ or OH^- ion is available, a normal solution is one which contains in grams per liter an amount of acid or base equal to the molecular weight (the gram-molecular weight) **divided by** the number of such H^+ or OH^- ions present.

Examples: 1-Molar acetic acid (CH_3COOH) contains 60.032 g/liter. 1-Normal acetic acid is the same since only one H^+ is available.

1-Molar sulfuric acid (H_2SO_4) contains 98.016 g/liter. But 1-normal sulfuric acid contains 98.016 g divided by 2 (2 H ions are available), or 49.08 g/liter. It follows that a 1-molar solution of sulfuric acid is 2-normal.

Degree of Acidity

This is expressed as follows:

A. **Actual Acidity**: The concentration of H **ions** in a solution (this is usually spoken of as the pH of the solution; see p 490). It must be remembered, however, that all of the H atoms available for ionization may not actually be in ionic form and so may not be chemically reactive.

B. **Titratable Acidity**: The concentration of H **ions** in a solution plus those available for ionization although not actually ionized at the time. This is determined by titration against a base, and the result is a measure of the **total H^+** concentration, both actually and potentially ionized.

Total acidity is a term sometimes used to express the amount of free acid in a solution in addition to that in combination with organic compounds (eg, proteins) and that present in acid salts. Actually, it is synonymous with titratable acidity.

C. Free Acidity: The amount of acid, not combined with other substances, which is present in a solution.

DISSOCIATION OF WATER

Dissociation

Acids, bases, and salts in solution dissociate into ions of the elements or radicals of which they are composed, although such dissociation does not occur to the same extent in each case.

Examples: Sodium chloride and sodium acetate dissociate almost completely as follows:

$$NaC_2H_3O_2 \longrightarrow Na^+ + C_2H_3O_2^-$$

$$NaCl \longrightarrow Na^+ + Cl^-$$

Weak electrolytes such as acetic acid and ammonium hydroxide dissociate to a considerably lesser degree as follows:

$$CH_3COOH \rightleftarrows H^+ + CH_3COO^-$$

$$NH_4OH \rightleftarrows NH_4^+ + OH^-$$

Water itself dissociates, although to a very slight degree, in accordance with the following equilibrium:

(1) $$HOH \rightleftarrows H^+ + OH^-$$

Law of Mass Action: The relationship between the dissociated and undissociated molecules is a constant which may be expressed as:

(2a) $$\frac{[H^+][OH^-]}{[HOH]} = K$$

or

(2b) $$[H^+][OH^-] = K[HOH]$$

where the bracketed values refer to concentrations of the ions involved. This dissociation constant, K, for water is approximately 1×10^{-14} at room temperature. At the same time the amount of dissociated water is so small in comparison to the total amount present that the undissociated water (HOH) may be considered constant, with a value of unity. Equation (2b) may then be written as:

(3) $$[H^+][OH^-] = K = 1 \times 10^{-14}$$

However, in pure water,

$$[H^+] = [OH^-]$$

Therefore,

$$[H^+]^2 = 1 \times 10^{-14}$$

and

$$[H^+] = 1 \times 10^{-7} \text{ mols/liter}$$

Effects of Acids & Bases on Dissociation of Water

From the definition of acids and bases given above, it is evident that the addition of either acid or base to water will affect the ratio between the concentrations of H^+ and OH^-. Since the product of the 2 is a constant (see equation [3] above), the addition of one ion in excess requires a lowering of the concentration of the other. It should also be apparent from the above equation that in any aqueous solution of an acid there are still some hydroxyl ions present, and vice versa.

Examples: Give the OH^- concentration of a solution of acid with an H^+ concentration of 10^{-5} mols/liter.

$$10^{-5}[OH^-] = 10^{-14}$$

$$[OH^-] = 10^{-9} \text{ mols/liter}$$

If base is added to water to make $[OH^-]$ of 1/10 mols/liter, what is the $[H^+]$?

$$[H^+] \times 1 \times 10^{-1} = 10^{-14}$$

$$[H^+] = 10^{-13}$$

Methods of Expressing Concentration

The concentration of a solution may be expressed by one of several methods (normality, mols/liter, etc) as follows.

Examples: What is the $[H^+]$ of a 10-thousandth-normal hydrochloric acid solution, assuming complete dissociation?

Molar equivalent: 0.0001 mols/liter

By powers of ten: 1×10^{-4} mols/liter

By logarithms: $\bar{4}.0000$ ($\log 1 \times 10^{-4}$)

Express the $[H^+]$ of blood by various means, if it is known to be 0.0000000501 mols/liter.

By powers of ten: 5.01×10^{-8}

By logarithms: $\log 5.01 \times 10^{-8}$

$= \bar{8}.700$

THE CONCEPT OF pH

Hydrogen ion concentrations in the physiologic range are usually considerably less than one. In order to express such concentrations in a convenient form, Sørensen introduced the concept of pH, in which whole numbers and decimal fractions thereof are used. All vital activities are affected by H ion concentration, and this convenient system of expressing it has become universally accepted in physiologic chemistry.

pH may be defined either as (a) the negative of the logarithm (base 10) of the H ion concentration or, what is equivalent, (b) the logarithm of the reciprocal of the H ion concentration, ie:

$$\text{(a)}\quad pH = -\log[H^+]$$

or

$$\text{(b)}\quad pH = \log\frac{1}{[H^+]}$$

Note that the H ion concentration must be ascertained before the pH can be calculated. For strong electrolytes this may be substantially the same as the total concentration, if complete ionization is assumed; but for weak electrolytes the H ion concentration must be obtained by calculation from the ionization constant, as illustrated in subsequent paragraphs.

Examples: What is the pH of a solution with an H ion concentration of 3.2×10^{-4}?

$$[H^+] = 3.2 \times 10^{-4}$$

$$pH = \log\frac{1}{[H^+]} = \log\frac{1}{3.2 \times 10^{-4}}$$

$$= \log\frac{1}{10^{.505} \times 10^{-4}} = \log\frac{1}{10^{-3.495}}$$

$$= 3.495$$

Answer: pH = 3.495 (or 3.50).

In the above example one proceeds as follows: (1) Express H ion concentration by a power of 10. (2) Find the log of the coefficient, subtract this from the exponent, and then change the sign.

$$[H^+] = 3.2 \times 10^{-4}$$

$$pH = -4 - \log 3.2^* = -3.50 = +3.50$$

To determine the H ion concentration from the pH, the order of calculation is reversed.

Example: Give the H ion concentration of a solution at pH 4.72.

$$pH\ 4.72 = \log 10^{4.72} = \log\frac{1}{10^{-4.72}}$$

$$= \log\frac{1}{10^{.28} \times 10^{-5}} = \log\frac{1}{1.9 \times 10^{-5}}$$

Answer: $[H^+] = 1.9 \times 10^{-5}$.

*Log 3.2 = 0.50.

Variations in pH & Interpretation

It will be noted that the **greater** the pH is, the **lower** will be the acidity, and that one unit of pH represents a difference of 10 times in H ion concentration; eg, pH 5.0 = 0.00001 mols/liter $[H^+]$; pH 4.0 = 0.0001 mols/liter $[H^+]$.

From what has already been said concerning the ionization of water, it is apparent that when a solution is neutral (ie, H and OH ions in equal proportions), it has a pH of 7.0.

$$[H^+][OH^-] = K_W = 1 \times 10^{-14}\ (pH + pOH = 14.0)$$

$$[H^+]^2 = 1 \times 10^{-14}$$

$$[H^+] = 1 \times 10^{-7}$$

$$pH = 7.0$$

As the H ions increase, the pH will decrease (solution becomes more acid); and conversely, as the H ions decrease, the pH will increase (solution becomes more alkaline). Therefore solutions whose pH values lie between 0 and 7.0 are acid solutions and those whose pH values lie between 7.0 and 14.0 are alkaline solutions.

pOH

This is the value obtained by subtracting the pH from 14.0. For example, a solution of pH 3.0 has a pOH of 11.0; a solution of pH 10.0 has a pOH of 4.0. This expression is based on OH ion concentration. It is useful in calculating the pH of an alkaline solution.

Example: What is the pH of a 0.0004-normal solution of sodium hydroxide?

$$\text{OH ion concentration} = 4 \times 10^{-4}$$

$$pOH = 4.000 - 0.602 = 3.398$$

$$pH = 14.0 - pOH$$

$$= 14.0 - 3.398 = 10.6$$

Answer: pH = 10.6.

Determination of pH

In order to arrive at the pH of a solution, one must first obtain the H **ion** concentration. In the case of a strong electrolyte this is obtained from the original concentration by assuming complete ionization. For weak electrolytes, with which physiologic chemistry is more often concerned, this must be calculated by the use of the law of mass action:

$$\frac{[H^+][A^-]}{[HA]} = Ka$$

where H = cation, A = anion, and HA = the undissociated acid.

This states that the ratio of the product of the concentration of the ionized components to that of the un-ionized is equal to a constant (Ka), the dissociation (ionization) constant for that acid (see Table 2, p 494). The following examples illustrate the use of this expression in order to calculate the pH of a weak acid, such as acetic acid, or of a weak base, such as ammonium hydroxide.

Examples: What is the pH of a 0.1-M solution of acetic acid (HAc)? (Ka, HAc = 1.86×10^{-5}.)

$$HAc \rightleftharpoons H^+ + Ac^-$$

$$\frac{[H^+][Ac^-]}{[HAc]} = Ka = 1.86 \times 10^{-5}$$

Since $[H^+] = [Ac^-]$, we may represent the numerator as $[H^+] \times [H^+]$ or $[H^+]^2$. Therefore

$$\frac{[H^+]^2}{0.1} = 1.86 \times 10^{-5}$$

Actually, the denominator should be $0.1 - H^+$, the original concentration (0.1-M) less that in ionized form, to give the un-ionized fraction. To avoid a quadratic expression, one may assume that the ionized fraction is so small that the original concentration of the solution, uncorrected for the ionized fraction, will suffice.

$$[H^+]^2 = 1.86 \times 10^{-6}$$

$$[H^+] = \sqrt{1.86} \times 10^{-3}$$

$$[H^+] = 1.36 \times 10^{-3}$$

and $$pH = 3.00 - 0.13 = 2.87$$

What is the pH of a solution of NH_4OH, 0.1-N? (Kb, $NH_4OH = 2 \times 10^{-5}$.)

$$NH_4OH \rightleftharpoons NH_4^+ + OH^-$$

$$\frac{[NH_4^+][OH^-]}{[NH_4OH]} = K_b = 2 \times 10^{-5}$$

$$\frac{[OH^-]^2}{0.1} = 2 \times 10^{-5}$$

$$[OH^-] = 1.42 \times 10^{-3}$$

$$pOH = 2.85$$

$$pH = 14.0 - 2.85 = 11.15$$

STANDARDIZATION OF ACIDS & BASES

Solutions for Standardization

To prepare solutions of known acidity, a primary standard of reference is used. This may be a chemically pure, dry substance (such as potassium acid phthalate), which can be weighed accurately and dissolved in a known volume of solution. The concentration of a solution of alkali is then obtained by titration against this primary standard. The solution of known alkaline strength may now be used to ascertain the strength of the acid solution. The following equation is useful in calculation of concentrations (normalities) from comparison with a solution of known strength:

$$N_1 V_1 = N_2 V_2$$

$$N_1 = \frac{N_2 V_2}{V_1}$$

where V_1 and N_1 refer to the volume and concentration (normality) of the unknown, and V_2 and N_2 refer to the corresponding known values.

Indicators

An **indicator** is used to determine the **end point** of an acid-base reaction. The indicator is a very weak organic acid or base which undergoes a change of structure and a consequent change of color in the presence of certain concentrations of hydrogen and hydroxyl ions. The color change actually takes place over a range of concentration, as shown in Table 1.

Two factors must be considered in making titrations for the standardization of acids and bases: (1) the choice of the strength of the standard, which should usually be of approximately the same normality as that of the unknown; and (2) the end point of the reaction. If solutions of equivalent concentrations of a strong base and a strong acid are placed together, both will have approximately equal dissociation constants; the pH at the end point will be near 7.0. If, however, the acid is stronger than the base, the former will dissociate more completely than the latter, the $[H^+]$ will rise, and the pH will be less than 7.0; the solution will be acid in spite of the fact that equivalent concentrations of the acid and base have been used. Similarly, when a weak acid and strong base are together, the reaction at the end point will be basic. It is therefore necessary to choose an indicator which will undergo a change of color at the actual end point of the reaction for the particular system under study.

TABLE 1. Some common indicators.

Indicator	pH Range	Color Change
Töpfer's reagent	2.9–4.0	Red to yellow
Congo red	3.0–5.0	Blue to red
Methyl orange	3.1–4.4	Orange red to yellow
Bromcresol green	4.0–5.6	Yellow to blue
Methyl red	4.2–6.3	Red to yellow
Litmus	4.5–8.3	Red to blue
Alizarin red	5.0–6.8	Yellow to red
Bromcresol purple	5.4–7.0	Yellow to purple
Phenol red	6.6–8.2	Yellow to red
Phenolphthalein	8.3–10.0	Colorless to red

In gastric analyses, free acidity is determined by the use of an indicator which shows a color change at a low pH (2.9) (Töpfer's reagent, dimethylaminoazobenzene). Phenolphthalein is then added and the titration continued to its end point to determine the total acid. The latter fraction measures the free acid plus the acid bound in acid-salt and protein combination.

The determination of pH measures the effective or actual acidity of a solution, as it indicates the amount of H^+ present in the solution at the time, which is often the most important consideration from a physiologic standpoint. Indicators may be used for pH measurement (Table 1). Test samples are treated with various indicators which are sensitive to different pH values; they are then compared with standards held at a definite pH by means of a buffer (see below) in order to determine the exact pH. Electrometric methods of determination of pH are much better since they are very sensitive, more accurate, and can be applied to colored solutions. Hence, they find considerable application in biologic work. The hydrogen electrode is the standard. In actual practice, the glass electrode together with a calomel electrode are used in most pH meters.

BUFFER SOLUTIONS

Buffers, which are usually mixtures of weak acids with their salts of strong bases, possess the ability (known as buffer action) to resist change in hydrogen ion concentration. The buffer capacity is a measure of this ability to resist change in pH and is expressed as the rate of change in pH when acid or alkali is added. Buffer action is very important in physiologic chemistry because it is part of the homeostatic mechanism whereby the neutrality of the body is regulated within the relatively narrow limits of pH compatible with the life of most cells.

Principles of Buffers

To illustrate the action of buffers we may take the usual condition where a weak acid and its salt with a strong base are together in solution (eg, acetic acid and sodium acetate). The conditions of the buffer mixture may be shown as:

$$Na^+ + Ac^- + HAc \longleftrightarrow Na^+ + H^+ + 2Ac^- \text{ (+ some HAc)}$$

If base (OH ions) is added to this system, it will react with the acid to form the salt:

$$OH^- + HAc \longrightarrow Ac^- + H_2O$$

which salt is less alkaline than the original base. The addition of acid (hydrogen ions) will cause another reaction:

$$H^+ + Ac^- \longrightarrow HAc$$

to form a relatively undissociated acid and thus no free hydrogen ions. In either case the change in hydrogen ion concentration, and therefore the pH change, is relatively smaller than would be the case if the buffer were not present. The buffer acts almost as if it were "absorbing" the added free hydrogen or hydroxyl ions.

Use of Buffers

A. Standard Solutions: Buffers are used for making standard solutions in which it is desired to maintain a constant pH, as in colorimetric determination of the pH of unknown solutions (see previous section).

B. Maintenance of Hydrogen Ion Concentration: Buffers are also used to maintain a given hydrogen ion concentration, which may be necessary for optimal activity of a reaction. This function of buffers is most important in the action of enzymes, both in vivo and in vitro, and buffer action is, therefore, of great importance in all physiologic systems.

The pH of a buffer solution may be determined by the **Henderson-Hasselbalch equation:**

$$pH = pKa + \log \frac{\text{salt}}{\text{acid}}$$

where pH is log $1/H^+$ and pKa is log 1/Ka (Ka is dissociation constant of the acid in question).

The equation may also be used to determine the relative normal concentrations of the salt and acid to be used in reaching a desired pH. For practical purposes, 2 systems of buffers are in common use, those of Clark & Lubs or those of Sørensen. Details are to be found in any handbook of chemistry. The tables are arranged so that by mixing solutions as directed, buffers of any desired pH over the range from 1.0–14.0 may be prepared.

Example: Calculate the pH of a solution prepared by mixing 25 ml of acetic acid with 10 ml of sodium hydroxide, both solutions 0.1-normal.

When these 2 solutions are mixed, the sodium hydroxide will neutralize 10 ml of the acetic acid to form 10 ml of the salt, sodium acetate. Fifteen ml of unneutralized acid will remain. This produces a buffer mixture in which the ratio of salt to acid is 10:15. The Ka for acetic acid is 1.86×10^{-5}, and therefore the pKa (log 1/Ka) is 4.73.

According to the Henderson-Hasselbalch equation,

$$pH = 4.73 + \log \frac{10}{15}$$

$$pH = 4.73 + \log 0.66$$

$$pH = 4.73 + \bar{1}.8195$$

Answer: pH = 4.55.

It will be seen by examination of the equation that if the ratio of salt to acid is 1:1, ie, at the exact halfway point in the neutralization of an acid by a base, the pH of the resultant mixture would equal the pKa (log 1:1, log 1 = zero). Under these circumstances the **buffer capacity** (see above) is at its maximum, for here it possesses maximal ability to react either as a base or an acid.

For dissociation constants of common organic acids, see Table 2.

TABLE 2. Dissociation constants of common organic acids.

Acid	K	pKa
Acetic acid	1.86×10^{-5}	4.73
Carbonic acid	3.0×10^{-7}	6.52
Second hydrogen	4.5×10^{-11}	10.35
Phosphoric acid	8.0×10^{-3}	2.10
Second hydrogen	7.5×10^{-8}	7.13
Third hydrogen	5.0×10^{-13}	12.30
Lactic acid	1.5×10^{-4}	3.82

THE SOLUBILITY OF GASES IN AQUEOUS SOLUTIONS

The normal environments of all life processes in both animals and plants are water and air. Water is the solvent in which the chemical reactions of the cell occur, and in order to gain access to the cell the gases of the air must first be dissolved in this aqueous medium. Certain physicochemical laws governing the solubility of gases in water or in other aqueous solutions such as the body fluids will, therefore, be discussed at this point.

The amount of a gas dissolved in water varies with (1) the nature of the gas, (2) the temperature, and (3) the pressure.

Henry's Law

For a given gas, Henry's law applies: The temperature remaining constant, the **amount** (ie, weight) of gas absorbed or dissolved by a given volume of liquid is proportionate to the pressure of the gas.

Absorption Coefficient

It is convenient to express the extent to which a gas is absorbed by a liquid as the absorption coefficient (α), ie, the volume of a gas (reduced to normal temperature and pressure [NTP, 273° absolute, 760 mm]) which will be dissolved at a pressure of 1 atmosphere by 1 volume of the liquid.

Partial Pressure

When dealing with a mixture of gases, such as air, the pressure under which each constituent is dissolved is called its partial pressure (p). The partial pressure of each gas can be calculated from its percentage (by volume) of the total.

Example: Air is 79% nitrogen, 20.98% oxygen, and 0.04% CO_2. If the total pressure 760 mm, after correction for the vapor pressure of water (17.5 mm at 20° C) it becomes 742.5 mm, and the partial pressure of each gas is then as follows:

$$N: 0.79 \times 742.5\text{ mm} = 586.6\text{ mm}$$
$$O_2: 0.2096 \times 742.5\text{ mm} = 155.6\text{ mm}$$
$$CO_2: 0.0004 \times 742.5\text{ mm} = 0.297\text{ mm}$$

Calculation of Amount of Gas in Solution

To calculate the amount, V (volume), at normal temperature and pressure of each gas dissolved:

$$V = \frac{\alpha \cdot v \cdot p}{760}$$

where α = the absorption coefficient for the gas, v = the volume of the liquid, and p = the partial pressure of the gas.

Example: At an external pressure of 760 mm, how much (volume at NTP) oxygen, nitrogen, and CO_2 will be absorbed by 100 ml water at 20° C?

After correction for vapor pressure of water, the external pressure becomes 742.5 mm.

The absorption coefficients (in water at 20° C) are as follows:

$$O_2: 0.0310$$
$$N: 0.0164$$
$$CO_2: 0.878$$

O_2 dissolved:

$$V = \frac{0.0310 \times 100 \times 155.6}{760} = 0.635\text{ ml}$$

N dissolved:

$$V = \frac{0.0164 \times 100 \times 586.6}{760} = 1.266\text{ ml}$$

CO_2 dissolved:

$$V = \frac{0.878 \times 100 \times 0.297}{760} = 0.034\text{ ml}$$

Total gases dissolved: = 1.935 ml

Solubility of Gases in Solutions of Electrolytes & Nonelectrolytes

In the example above, the total, 1.935 ml, is the amount of air which would be dissolved in 100 ml of water under those conditions. The solubility of a gas, however, is lower in solutions of salts or nonelectrolytes than in water; the solubility decreases as the concentration of the solute increases.

These considerations are important because, with respect to their absorption of gases, blood and body fluids act like aqueous solutions of various salts. For example the a values in blood plasma at 38° C are:

O_2: 0.024
N: 0.012
CO_2: 0.510

The amounts of each gas dissolved in 100 ml of plasma at 760 mm Hg may be calculated as follows: After correction for vapor pressure of water in the alveolar air (47 mm Hg at 37° C), the pressure of the gas mixture becomes 713 mm Hg. The partial pressure of each gas is then (alveolar air oxygen, 15%; nitrogen, 80%; CO_2, 5%):

O_2: 0.15 × 713 mm Hg = 107 mm Hg
N: 0.80 × 713 mm Hg = 570 mm Hg
CO_2: 0.05 × 713 mm Hg = 36 mm Hg

O_2 dissolved:

$$V = \frac{0.024 \times 100 \times 107}{760} = 0.338 \text{ ml}$$

N dissolved:

$$V = \frac{0.012 \times 100 \times 570}{760} = 0.900 \text{ ml}$$

CO_2 dissolved:

$$V = \frac{0.510 \times 100 \times 36}{760} = 2.416 \text{ ml}$$

Oxygen and, to a certain extent, CO_2 are, however, absorbed by **whole blood** to a much greater degree than one would predict from Henry's law. This is due to the ability of hemoglobin to accept reversibly these gases (see p 208).

Varying Solubility of Gases With Varying External Pressure

The relationship between external pressure and the solubility of gases in blood and tissue fluids becomes important when the external pressure varies significantly from normal.

A. Low External Pressures: At high altitudes, although the composition of the air is substantially the same as at sea level, there is a relative lack of each gas because of the diminished barometric pressure. This lack particularly affects the supply of oxygen to the blood. At an elevation where the barometric pressure is ½ that at sea level, the partial pressure of oxygen and, consequently, the amount dissolved in the plasma would be reduced to a similar extent.

B. High External Pressures: The various gases of the air are dissolved in the body fluids to an increased degree when the external pressure is higher than normal. With return to a normal environment, the gases leave the blood and tissue fluids. This may cause harmful effects in the remote areas of the body where gaseous equilibrium cannot be accomplished. Nitrogen is the most serious offender because of its high concentration in the air. Deep sea divers or other workers subject to high external pressures may suffer from the effects of this rapid evolution of gaseous nitrogen if returned to lower pressures too rapidly. These effects are called decompression sickness ("bends").

OSMOTIC PRESSURE

Because of the nature of the cell membrane, osmotic pressure plays a very important role in the exchange of water and dissolved materials between the cell and its extracellular environment.

Diffusion

When a soluble substance such as sodium chloride is added to water, it quickly dissolves and distributes itself equally throughout the liquid. This process of diffusion is produced by the constant movement of the ions of the salt when in solution.

Osmosis

If, however, a semipermeable membrane is interposed between a solution and its pure solvent (eg, water and a solution of NaCl), or between 2 solutions differing in concentrations of solute, the water molecules will diffuse through the membrane more rapidly than the salt molecules. If the salt solution is enclosed within the membrane, water will diffuse across the membrane into the salt solution (Fig 1).

Osmotic Pressure

This movement of water (osmosis) results in the building up of a definite pressure (osmotic pressure) within the membrane-enclosed area. The amount of excess pressure which must be imposed on a solution in order to prevent the passage of a solvent into it through a semipermeable membrane is the osmotic pressure.

Determination of Osmotic Pressure

A. Direct Measurement of Osmotic Pressure: Osmotic pressure can be measured directly with an osmometer. The simplest form of an osmometer is a thistle tube, the opening of which is covered with a semipermeable membrane, containing a solution (of sucrose, for example). When the tube is inverted and immersed in water, the solvent (water) passes by osmosis into the thistle tube. The osmotic pressure which develops within the tube causes the liquid to rise in the stem of the tube. When the pressure is high enough to prevent further inflow of water, a state of equilibrium is reached. The hydrostatic pressure of the column of solution is the measure of the osmotic pressure (Fig 2).

B. Indirect Measurement of Osmotic Pressure: The direct determination of osmotic pressure by means

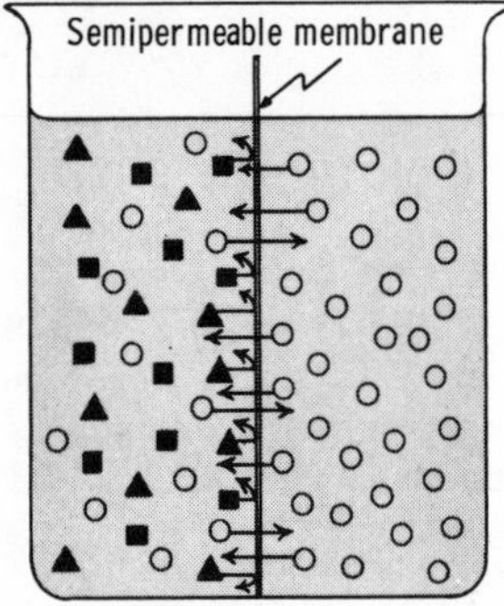

FIG 1. Osmosis.

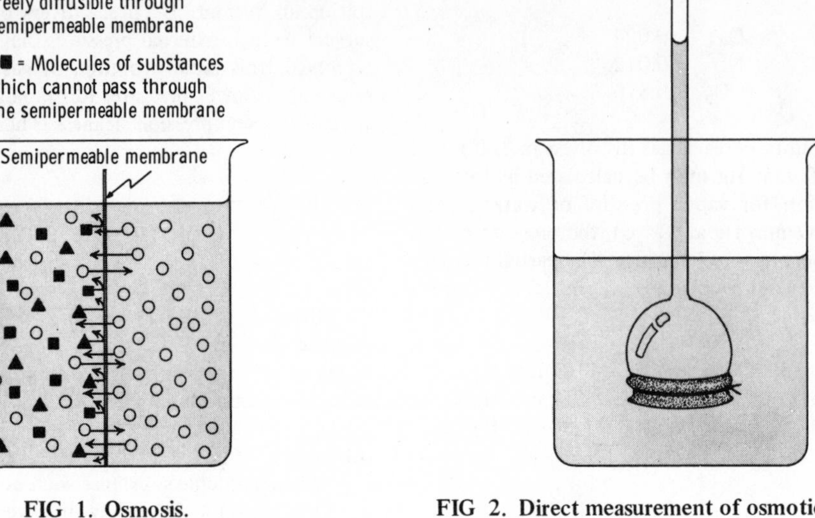

FIG 2. Direct measurement of osmotic pressure.

of osmometers has been mentioned, but osmotic pressures are more conveniently determined by indirect methods. For physiologic fluids, the depression of the freezing point, or the delta value (Δ), is used. According to Blagden's law, for a given solvent (eg, water) the depression of the freezing point is proportionate to the concentration of the solute. For aqueous solutions a 1-M solution of a substance which does not ionize freezes at −1.86° C. Such a solution has an osmotic pressure of 22.4 atmospheres (1 atmosphere = a pressure of 760 mm Hg); therefore, a depression of 1° C corresponds to an osmotic pressure of 12.04 atmospheres or 9150 mm Hg (12.04 × 760) at 0° C.

The freezing point of mammalian blood plasma is about −0.53° C, corresponding to an osmotic pressure at **body temperature** of about 7.2 atmospheres. This pressure is maintained remarkably constant, largely by the regulatory action of the kidneys. Solutions which have an osmotic pressure equal to that of the body fluids are said to be iso-osmotic or isotonic, while those which have lower or higher osmotic pressures are hypotonic or hypertonic, respectively.

Relationship of Osmotic Pressure to the Gas Laws

The osmotic pressure of a solution is equal to the pressure which the dissolved substance would exercise in the gaseous state if it occupied a volume equal to the volume of the solution. For this reason, the gas laws may also be applied to osmotic pressure (at least in dilute solutions).

Avogadro's law applied to gases states that at 0° C 1 g mol of a gas at a pressure of 760 mm of mercury occupies a volume of 22,400 ml. This law also holds for osmotic pressure. In this case it may be expressed as follows: At a temperature of 0° C the osmotic pressure of a solution is 760 mm if it contains 1 g mol (of solute) in 22,400 ml.

This relationship between osmotic pressure and the concentration of a solution serves as a basis for the calculation of osmotic pressure, as will be discussed in the next section. In a mixture, each solute will produce an osmotic pressure equal to that which it would produce according to its molar concentration if it were the only solute present. The total osmotic pressure is equal to the sum of the partial pressures. This property is of physiologic importance since the body fluids are actually mixtures of osmotically active substances. The total osmolar concentration of these fluids controls the quantity and distribution of the body water, as described in Chapter 19.

Influence of Ionization & Molecular Size on Osmotic Pressure

Osmotic pressure is a "colligative" property of solutions, which means that it is affected by the number of discrete particles in solution. Therefore, in the case of substances which ionize, the osmotic pressure which would be expected from concentration will actually be higher because of the presence of an increased number of particles per mol over those present when a substance which ionizes slightly or not at all is examined. A similar effect of molecular size on osmotic pressure is exemplified by a consideration of solutions containing organic molecules, such as proteins. In the blood plasma, for example, serum albumin, although constituting only about 60% of the total proteins, is responsible for 80% of their osmotic pressure. The globulin fraction, which consists of proteins of considerably higher molecular weight (and, therefore, of fewer particles per unit weight) is for this reason less effective than the albumins in the maintenance of plasma osmotic pressure.

Example: A 0.2-M solution of cane sugar (which does not ionize) will depress the freezing point 0.372° C, while NaCl of the same concentration will be observed to depress the freezing point 0.693° C. The osmotic pressure of the 0.2-M NaCl is not the 4.48 atmospheres expected by virtue of its molar concentration; due to ionization, it is 0.693 × 12.04, or 8.34 atmospheres. This 0.2-M solution (1.17%) is slightly hypertonic to the blood of mammals.

Isotonic Coefficient

To calculate the osmotic pressure of solutions which ionize, it is necessary to consider the degree of ionization and the number of ions per mol. From these data, the isotonic coefficient (i) may be calculated. The theoretical osmotic pressure (calculated simply from concentration), after multiplication by the isotonic coefficient, is then corrected to the actual osmotic pressure. In general, if a salt molecule yields n ions and the degree of ionization is α, then i, the isotonic coefficient, is calculated as follows:

$$i = 1 + \alpha(n - 1)$$

Example: A molar solution of NaCl is about 86.3% ionized; $\alpha = 0.863$ and $n = 2$ because NaCl ionizes into 2 ions:

$$NaCl \longleftrightarrow Na^+ + Cl^-$$

Therefore

$$i = 1 + 0.863(2 - 1) = 1.863$$

The theoretical osmotic pressure of a molar solution is 22.4 atmospheres. For NaCl, the actual osmotic pressure would be 22.4 × 1.863, or 41.73 atmospheres at 0° C.

Osmols & Milliosmols

The osmotic activity of solutions of substances which ionize is higher than that of substances which do not. Thus, in the example given above, the osmotic pressure of a molar solution of NaCl is 1.863 times greater than predicted from its concentration. This is because NaCl molecules in solution dissociate into smaller particles or ions. A molar solution of glucose, which does not ionize, would have an osmotic pressure equal to that calculated from its concentration, ie, 22.4 atmospheres. These relationships between molar concentration and osmotic activity are conveniently expressed by the use of the term **osmol** (Osm). The osmolarity of a solution of a given compound is determined by multiplying the molar concentration by the number of particles per mol obtained by ionization.

Osmolarity = Molarity × number of particles per mol resulting from ionization

Example: NaCl, if it is 100% ionized, dissociates into 2 particles per mol, one Na ion and one Cl ion:

$$NaCl \longleftrightarrow Na^+ + Cl^-$$

Under these conditions (100% ionization), the osmolar concentration would be twice the molar concentration. However, in the previous example it was noted that a molar solution of NaCl is only 86.3% ionized. Thus the osmolar concentration of a molar solution of NaCl would be 1.863 Osm.

For solutions of substances which do not ionize, molar and osmolar concentrations are equivalent.

The concentrations of ionizable substances in physiologic fluids are so weak that complete ionization is assumed to occur. It is therefore possible to convert molar concentrations directly to osmolarity by reference to the number of ions produced per mol. It is also more convenient to express these dilute concentrations in terms of mM and mOsm, ie, 1/1000 mol or 1/1000 osmol.

Examples: NaCl ($Na^+ + Cl^-$) at a concentration of 70 mM/liter is equivalent to 140 mOsm/liter.

Na_2HPO_4 ($2\ Na^+ + HPO_4^{=}$) at a concentration of 1.3 mM/liter is equivalent to 3.9 mOsm/liter.

The osmotic activity of physiologic fluids such as plasma or urine is due to the combined osmotic activity of a number of substances which are dissolved in them. The osmolarity of such fluids is therefore conveniently determined by measurement of freezing point depression rather than by calculation from the concentration and ionization of each constituent in the mixture. Electrical instruments for measurement of osmotic pressure by freezing point depression (osmometers) have been devised to measure very small changes in temperature. A depression in the freezing point of a solution of 0.00186° C below that of water (taken as 0° C) is equivalent to 1 mOsm of osmotic activity per liter, ie, 1 mOsm is equivalent to Δ0.00186° C.

If the freezing point of human blood plasma is found to be −0.53° C, the osmolar concentration would be reported as 285 mOsm/liter (0.53/0.00186).

Influence of Temperature on Osmotic Pressure

The osmotic pressure increases directly with absolute temperature, just as does the pressure of a gas. Therefore, in order to express osmotic pressures at temperatures other than 0° C, corrections for temperature are made as follows:

$$\frac{P_1}{P_2} = \frac{T_1}{T_2}$$

(P_1 and P_2 are the original and corrected osmotic pressures; T_1 and T_2 the corresponding absolute temperatures.)

DONNAN EQUILIBRIUM

When 2 compartments are separated by a semipermeable membrane, water and dissolved particles of crystalloidal size will pass freely through the membrane but larger particles of colloidal dimensions will be held back. At equilibrium it will be noted that the diffusible substances will be unequally distributed on either side of the membrane because of the influence of the nondiffusible component. The Donnan theory is concerned with the distribution of electrolytes in systems separated by membranes which are impermeable to certain of the components. The resultant unequal distribution of ions is referred to as a **Donnan effect.** It may be illustrated as follows: [Assuming a semipermeable membrane separates a solution of NaCl from a solution of a protein (NaR) and the membrane is permeable to Na^+ and Cl^- but not to R^- (or to undissociated NaR), a shift of electrolytes will occur as shown in the second diagram below.]

Na^+	Na^+
R^-	Cl^-

A. Before shift of ions

Na^+	Na^+
R^-	
Cl^-	Cl^-

B. After shift of ions

The distribution of the various components of the system at the start is shown at A. When equilibrium has been reached, their distribution is as shown in B. From thermodynamic considerations as well as from actual measurement it will be found that, at equilibrium, the product of the concentrations of the diffusible ions (Na^+ and Cl^-) on one side of the membrane equals the product of the concentrations of the same ions on the other side; ie:

$$[Na^+] \times [Cl^-] = [Na^+] \times [Cl^-]$$

(Left side in B) (Right side in B)

Therefore, if the products of the concentration of the diffusible ions are to be equal, the concentration of Na ions on the left (in B) must be greater than on the right; while that of Cl ions would be greater on the right than on the left.

However, to maintain electrical neutrality, the cation concentrations would have to be equal to those of the anions; ie, in B above, Na^+ would equal the sum of R^- and Cl^- on the left side of the membrane, and Na^+ would equal Cl^- on the right. If such equal concentrations are not present, electrical neutrality is not maintained.

It is obvious that the Donnan effect is of universal applicability in biologic systems since the cell membranes act to separate various substances into diffusible and nondiffusible components, just as described above. The resultant differences in concentration of diffusible ions cause differences in electrical potential across the membrane; this may explain many bioelectric phenomena. Furthermore, the Donnan effect may also be involved in absorption and secretion and in the maintenance of differential concentrations between the various compartments of the body. The dynamic character of cellular activity makes it difficult to predict exactly the relationships at any one time, since metabolic processes (eg, protein anabolism and catabolism) change the concentration of nondiffusible components and thus alter the Donnan equilibrium.

• • •

Organic Chemistry (A Brief Review)

By Victor Rodwell, PhD*

The Elemental Basis of Life

All cells, regardless of origin (animal, plant, or microbial), contain the same elements in approximately the same proportions (Table 1).

TABLE 1. The elemental composition of living cells.

Element	Composition by Weight (%)	Element	Composition by Weight (%)
O	65		
C	18	Cu, Zn	0.70
H	10	Se, Mo	
N	3	F, Cl, I	
Ca	1.5	Mn, Co, Fe	
P	1.0		
K	0.35	Li, Sr	Traces†
S	0.25	Al, Si, Pb	
Na	0.15	V, As	
Mg	0.05	Br	
Total	99.30		

†Variable occurrence in cells. No known function in most cases.

Only 19 of the more than 100 known elements are essential to life. Six nonmetals (O, C, H, N, P, and S), which contribute almost 98% of the total mass of cells, provide the structural elements of protoplasm. From them the functional components of cells (walls, membranes, genes, enzymes, etc) are formed. These 6 elements all occur in the first 3 periods of the periodic table (Table 2).

The relative abundance of these 6 elements in the seas, crust, and atmosphere of earth does not by itself explain their utilization for life. Aluminum is more abundant, but performs no known function essential to life. By contrast, the intrinsic chemical properties of these 6 elements suggest their unique suitability as building blocks for life. Desirable features for structural elements apparently are as follows: (1) Small atomic radius. (2) The versatility conferred by the ability to form 1-, 2-, 3-, and 4-electron bonds. (3) The ability to form multiple bonds.

Small atoms form the tightest, most stable bonds—a distinct advantage for structural elements. H, O, N, and C are the **smallest atoms capable of forming 1-, 2-, 3-, and 4-electron bonds**, respectively. Utilization of all possible types of electron bonds permits maximum versatility in molecular design. So also does the ability to form multiple bonds, a property confined almost entirely to P, S, and the elements of period 2. Advantages of C- versus Si- based life include: (1) Greater chemical stability of C–C versus Si–Si bonds. (2) The ability of C, but not of Si, to form multiple bonds (eg, the oxides of C are diffusible, monatomic gases, whereas the oxide of Si is a viscous polymer). (3) The stability of C–C bonds, but not of Si–Si bonds, to rupture by nucleophilic reagents* such as O_2, H_2O, or NH_3.

Similar factors uniquely qualify P and S for utilization in energy transfer reactions. Energy transfer is facilitated by bonds susceptible to nucleophilic attack† (eg, nucleophilic attack of the 6-OH of glucose on the terminal P–P bond of ATP, forming ADP plus glucose-6-phosphate). P and S resemble Si in that P–P or S–S bonds, like Si–Si bonds, are susceptible to nucleophilic rupture by virtue of their unoccupied third orbitals. However, unlike Si, P and S form multiple bonds (more versatile), a consequence of their smaller atomic diameters. Most energy transfer reactions in biochemistry may be visualized as resulting from attack of a nucleophil (N) on the unoccupied third orbital of a phosphorus atom:

O
⁻O–P–O–R
⁻O N̈

TABLE 2. The structural elements of protoplasm.

Period	Group I	II	III	IV	V	VI	VII	VIII
1	**H**							He
2	Li	Be	B	**C**	**N**	**O**	F	Ne
3	Na	Mg	Al	Si	**P**	**S**	Cl	Ar

*Electron-rich elements or compounds.

†Attack of an electron-rich center upon an electron-deficient center.

*Associate Professor of Biochemistry, Purdue University, Lafayette, Indiana.

The Covalent Bond

The region in space where an electron is most likely to be found is termed an **orbital.** The sizes and shapes of different orbitals may be thought of as determining the spatial arrangements of atoms in molecules. The most fundamental of the "rules" that describe the electronic configurations of **atoms** is the **Pauli exclusion principle: only 2 electrons can occupy any given orbital, and these must have opposite spins.** Electrons of like spin tend to get as far away from each other as possible. Electrons in **molecules** occupy orbitals in accordance with similar rules.

To form a covalent bond, 2 atoms must be positioned so that an orbital of one overlaps an orbital of the other. Each orbital must contain a single electron, and these must have opposite spins. The 2 atomic orbitals merge, forming a single **bond orbital** containing both electrons. Since this new arrangement contains less energy (ie, is more stable) than that of the isolated atoms, **energy is evolved when bonds are formed.** The amount of energy (per mol) given off when a bond is formed is called the **bond dissociation energy.** For a given pair of atoms, the greater the overlapping of atomic orbitals, the stronger the bond.

The carbon atom (atomic number = nuclear charge = 6) has 6 electrons, 2 of which are unpaired and occupy separate $2p$ orbitals:

C 1s 2s 2p

Although this suggests that C should form 2 bond orbitals with H, 4 bonds are formed, giving CH_4. Since bond formation is an exergonic (stabilizing) process, as many bonds as possible tend to be formed. This occurs even if the resulting bond orbitals bear little resemblance to the original atomic orbitals.

To produce a tetravalent C atom, mentally "promote" one of the $2s$ electrons to the empty p orbital:

C 1s 2s 2p

While this representation suggests C should form 3 bonds of one type (using the p orbitals) and a fourth of another type (using the s orbital), the 4 bonds of methane are known to be equivalent. The **molecular** orbitals have a mixed or hybridized character and are termed sp^3 orbitals since they are considered to arise from mixing of one s and 3 p orbitals:

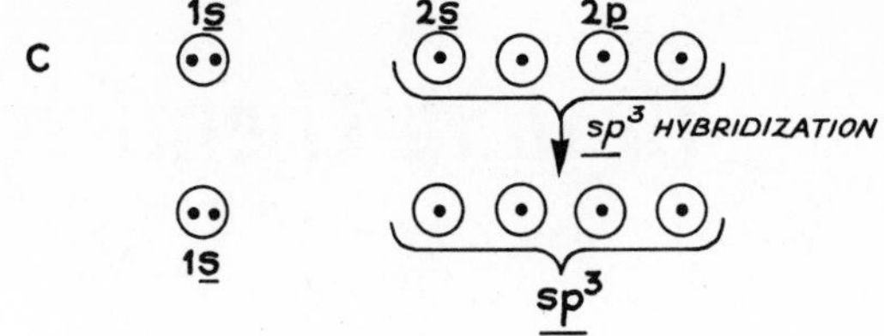

sp^3 Orbitals have the following shape:

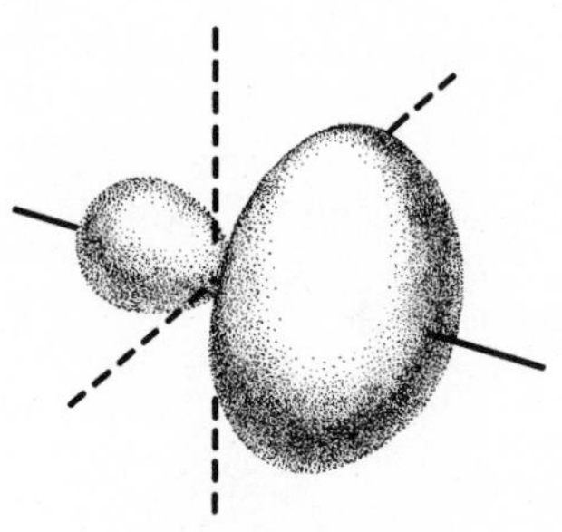

We shall neglect the back lobe and represent the front lobe as a sphere:

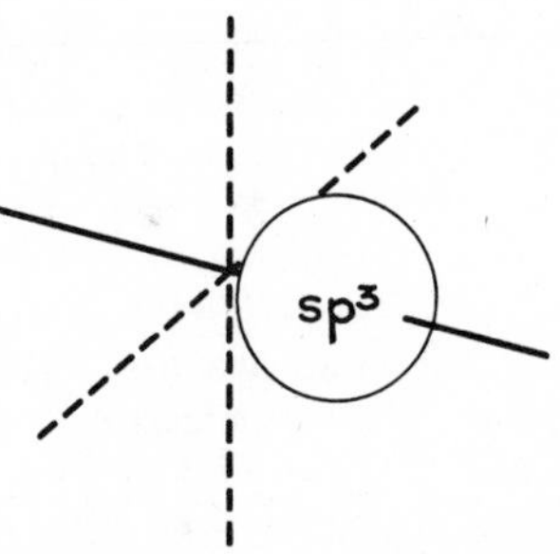

Concentrating atomic orbitals in the direction of a bond permits greater overlapping and strengthens the bond. The most favored hybrid orbital is therefore much more strongly directed than either s or p orbitals, and the 4 orbitals are exactly equivalent. Most important, these hybrid orbitals are directed toward the corners of a regular **tetrahedron.** This permits them to be as far away from each other as possible (recall Pauli exclusion principle).

Bond Angle

For maximum overlapping of the sp^3 orbitals of C with the s orbitals of hydrogen, the 4 H nuclei must be along the axes of the sp^3 orbitals and at the corners of a tetrahedron. The angle between any 2 C–H bonds must therefore be the **tetrahedral angle 109.5°**:

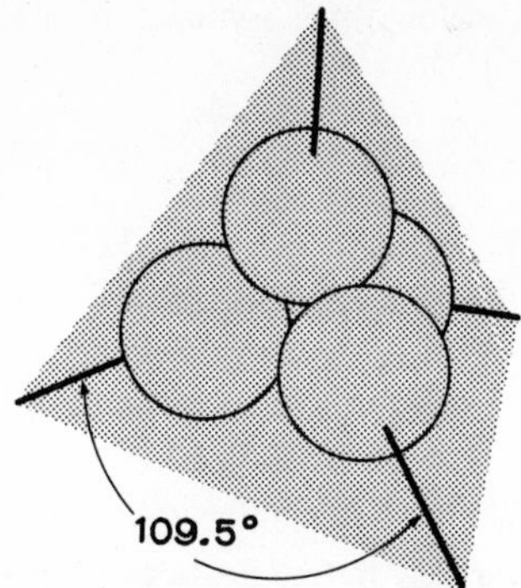

Methane has been shown experimentally to conform to this model. Each C–H bond has exactly the same length (.109 nm) and dissociation energy (102 kcal/mol), and the angle between any pair of bonds is 109.5°. **Characteristic bond lengths, bond energies, and bond angles thus are associated with covalent bonds.** Unlike the ionic bond, which is equally strong in all directions, **the covalent bond has directional character.** Thus, the chemistry of the covalent bond is much concerned with molecular size and shape. Three kinds of C atom are encountered: **tetrahedral** (*sp^3* hybridized), **trigonal** (*sp^2* hybridized), and **digonal** (*sp* hybridized).

In ammonia (NH_3), nitrogen (atomic number = 7) has a valence state similar to that described for carbon: 4 *sp^3* orbitals directed to the corners of a tetrahedron.

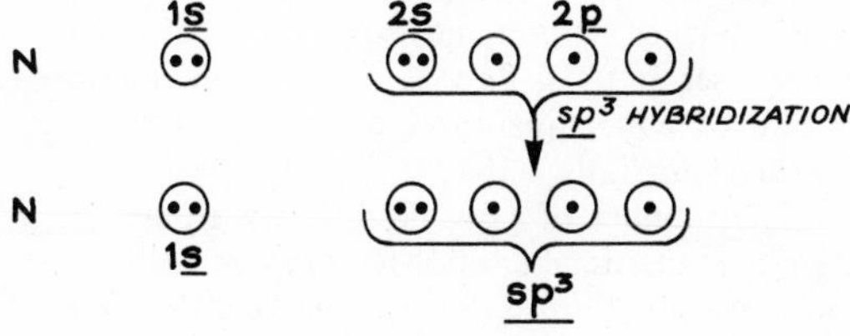

Each of the unpaired electrons of N occupying one of the *sp^3* orbitals can pair with that of a H atom, giving NH_3. The fourth *sp^3* orbital contains an unshared electron pair. The unshared electron pair appears to occupy more space and to compress the bond angles slightly to 107°. It is a region of high electron density and confers on NH_3 its basic properties (attracts protons).

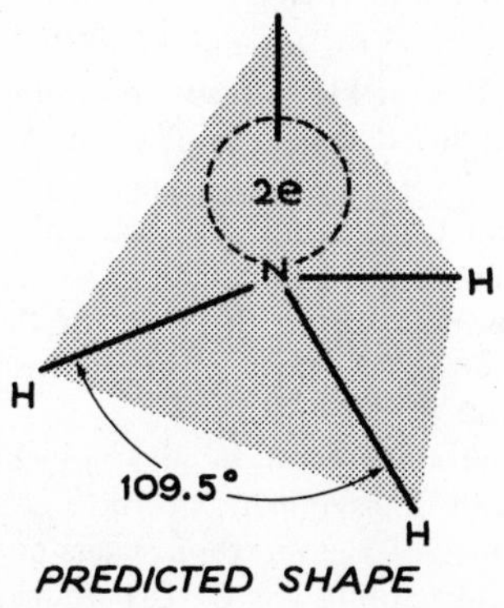

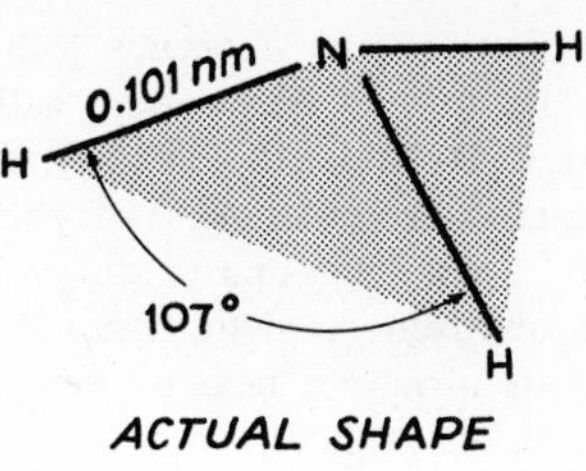

In H_2O, the O (atomic number = 8) has only 2 unpaired electrons and hence bonds to only 2 hydrogens.

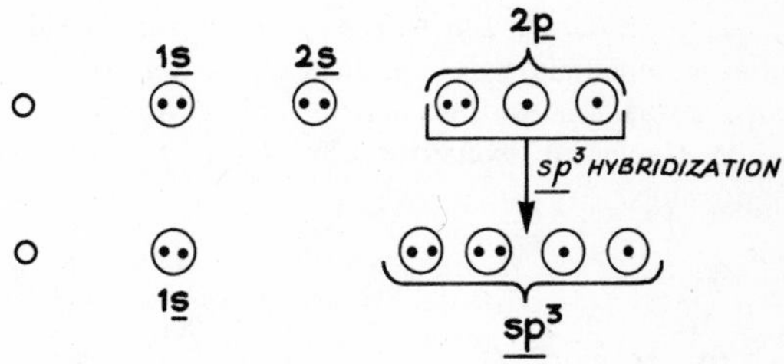

Water also is tetrahedral. The 2 hydrogens occupy 2 corners of the tetrahedron and the 2 unshared electron pairs the remaining corners. The bond angle (105°) is even smaller than that in NH_3.

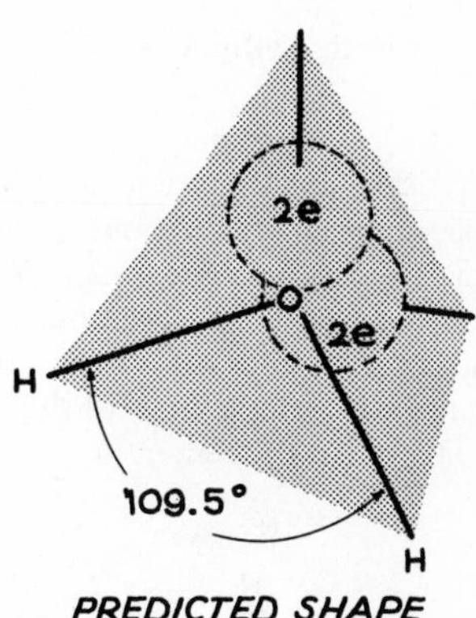

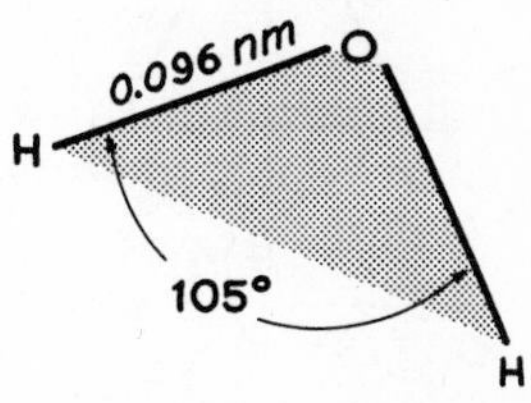

ACTUAL SHAPE

Hetero atoms (Greek *heteros* = "other") such as O, N, and S also form covalent bonds with carbon, eg, in ethylamine, $C_2H_5NH_2$, ethyl alcohol, C_2H_5OH, and ethyl mercaptan (thioethanol), C_2H_5SH. As noted above, hetero atoms have one or more pairs of electrons not involved in covalent bonding. Since these unshared electrons have a negative field, **compounds with hetero atoms** attract protons, ie, they **act as bases.***

$$C_2H_5NH_2 + H^+ \rightleftharpoons C_2H_5\overset{+}{N}H_3$$

Isomerism

Isomers (Greek *isos* = same, *meros* = part) are different compounds with identical elemental compositions. Several classes of isomerism are known.

A. Geometric Isomerism: For C_3H_6O, 3 isomers are possible.

$CH_3CH_2CH_2OH$

n-Propyl alcohol

$(CH_3)_2CHOH$

iso-Propyl alcohol

$CH_3-O-CH_2CH_3$

Methyl ethyl ether

Chemical, physical, and physiologic properties of geometric isomers are totally different.

B. Optical Isomerism: (See p 2.) Organic compounds have a 3-dimensional structure. In methane, CH_4, the 4 hydrogen atoms are at the vertices of an imaginary equilateral tetrahedron (4-sided pyramid) with the carbon atom at the center.

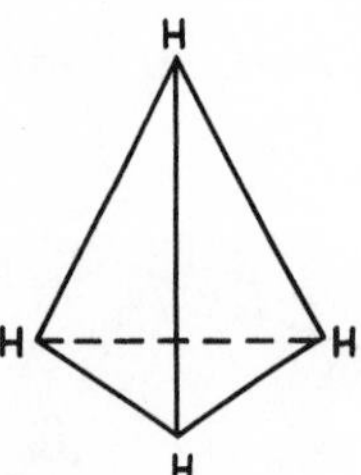

A carbon atom to which 4 different atoms or groups of atoms are attached is known as an asymmetric carbon atom (see p 1). For example, in the formula for alanine, shown below, the asymmetric (alpha) carbon atom is starred(*).

$$CH_3-\overset{H}{\underset{NH_2}{C^*}}-COOH$$

Alanine

Many carbohydrates, peptides, steroids, nucleic acids, etc contain 2 or more asymmetric C atoms. A thorough understanding of the stereochemistry of systems with more than one asymmetric center is therefore essential. Certain relationships are readily visualized using ball-and-stick atomic models. A compound having asymmetric carbon atoms exhibits **optical isomerism.** Thus, lactic acid has 2 nonequivalent optical isomers, one being the mirror image or **enantiomer** of the other (Fig 1).

The reader may convince himself that these structures are indeed different by changing the positions of either enantiomer by rotation about any axis and attempting to superimpose one structure on the other.

Although enantiomers of a given compound have the same chemical properties, certain of their physical and essentially all of their physiologic properties are different. Enantiomers rotate plane-polarized light to an equal extent but in opposite directions (see p 2). Since enzymes act on only one of a pair of enantiomers (see p 127), only ½ of a **racemic mixture** (a mixture of equal quantities of both enantiomers) is thus physiologically active.

The number of possible different isomers is 2^n, where n = the number of different asymmetric carbon atoms. An aldotetrose, for example, contains 2 asymmetric carbon atoms; hence, there are $2^2 = 4$ optical isomers (Fig 2).

To represent 3-dimensional molecules in 2 dimensions, **projection formulas,** introduced by Emil Fisher, are used. The molecule is placed with the asymmetric carbon in the plane of the projection. The groups at the top and bottom project **behind** the plane of projection. Those to the right and left project equally **above** the plane of projection. The molecule is then projected in the form of a cross (Fig 3).

Unfortunately, the orientation of the tetrahedron differs from that of Fig 1. **Fisher projection formulas may never be lifted from the plane of the paper and turned over.** Since the vertical bonds are really **below** the projection plane while the horizontal bonds are **above** it, **it also is not permissible to rotate the Fisher projection formula within the plane of the paper by either a 90° or a 270° angle, although it is permissible to rotate it 180°.**

A special representation and nomenclature for molecules with 2 asymmetric carbon atoms derives from the names of the 4-carbon sugars erythrose and threose. If 2 like groups (eg, 2 –OH groups) are on the

*This definition of bases (see p 489) may be enlarged to include all compounds which can accept protons.

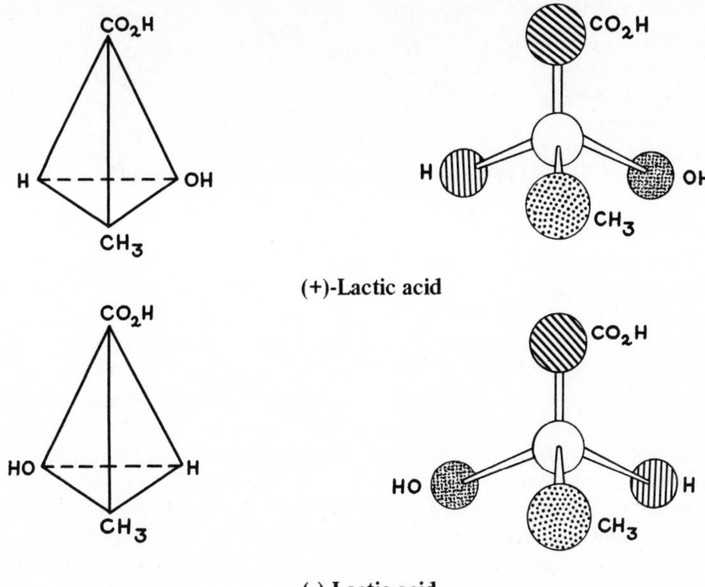

FIG 1. Tetrahedral and ball-and-stick model representation of lactic acid enantiomers.

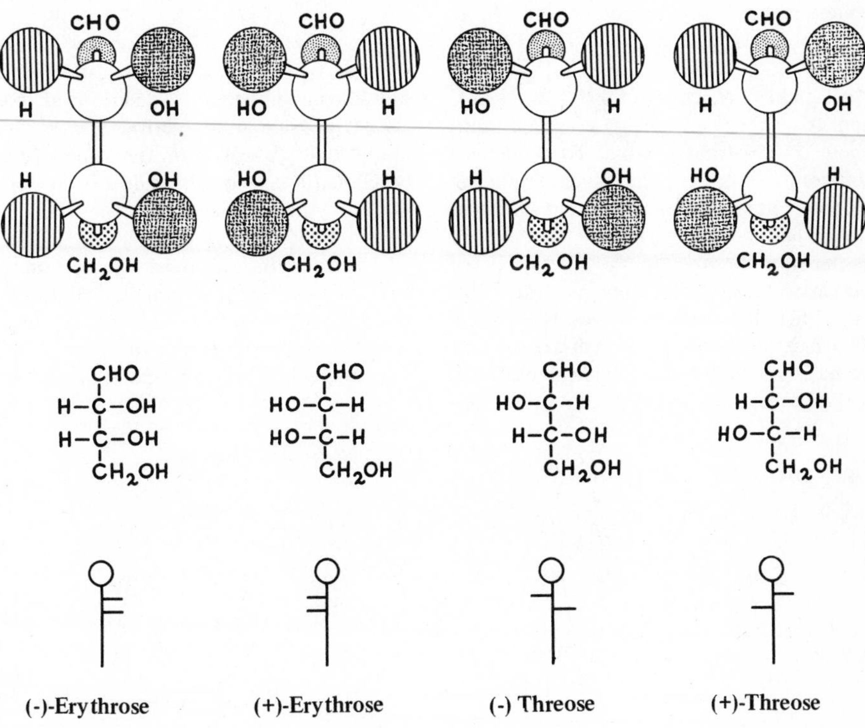

FIG 2. The aldotetroses. *Top:* Ball-and-stick models. *Middle:* Fisher projection formulas. *Bottom:* Abbreviated projection formulas.

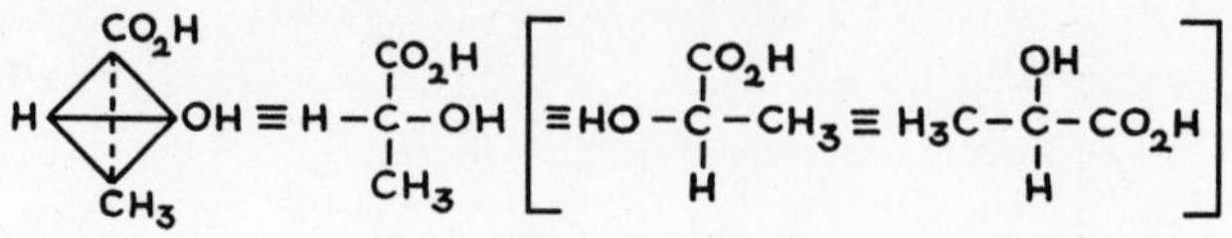

$$\text{H}\!-\!\overset{\text{CO}_2\text{H}}{\underset{\text{CH}_3}{\text{C}}}\!-\!\text{OH} \equiv \text{H}-\overset{\text{CO}_2\text{H}}{\underset{\text{CH}_3}{\text{C}}}-\text{OH} \left[\equiv \text{HO}-\overset{\text{CO}_2\text{H}}{\underset{\text{H}}{\text{C}}}-\text{CH}_3 \equiv \text{H}_3\text{C}-\overset{\text{OH}}{\underset{\text{H}}{\text{C}}}-\text{CO}_2\text{H} \right]$$

FIG 3. Fisher projection formula of (-)-lactic acid.

same side, the isomer is called the **"erythro"** form; if on the opposite side, the **"threo"** isomer. Fisher projection formulas inadequately represent one feature of these molecules. Look at the models from which these formulas are derived. The upper part of Fig 2 represents molecules in the **"eclipsed"** form in which the groups attached to C_2 and C_3 approach each other as closely as possible. The real shape of the molecule more closely approximates an arrangement with C_2 and C_3 rotated with respect to each other by an angle of 60°, so that their substituents are **staggered** with respect to each other and are as far apart as possible. One way to represent "staggered" formulas is to use **"sawhorse"** representations (Fig 4).

A second representation is the **Newman projection formula** (Fig 5). The molecule is viewed front-to-back along the bond joining the asymmetric carbon atoms. These 2 atoms, which thus exactly eclipse each other, are represented as 2 superimposed circles (only one is shown). The bonds and groups attached to the asymmetric C atoms are projected in a vertical plane and appear as "spokes" at angles of 120° for each C atom. The spokes on the rear atom are offset 60° with respect to those on the front C atom. To distinguish the 2 sets of bonds, those for the front carbon are drawn to the center of the circle and those for the rear carbon only to its periphery (Fig 5).

It is desirable to be able to shift between the Fisher projection formulas most often used in books and articles to either the sawhorse or Newman projection formulas, which most accurately illustrate the true shape of the molecule and hence are most useful in understanding its chemical and biologic properties. One way is to build a model* corresponding to the Fisher projection formula, stagger the atoms, and draw the sawhorse or Newman formulas. Fig 6 shows how to interconvert these formulas without models. The Fisher projection formula is converted to an "eclipsed sawhorse" or Newman projection which then is rotated 180° about the C_2-C_3 bond, producing a staggered sawhorse or Newman projection.

C. Cis-Trans Isomerism: (See p 16.) This occurs in compounds with double bonds. Since the double bond is rigid, the atoms attached to it are not free to rotate as about a single bond. Thus the structures

H−C−COOH ‖ H−C−COOH	H−C−COOH ‖ HOOC−C−H
Maleic acid (cis)	**Fumaric acid (trans)**

are not equivalent and have different chemical and physiologic properties. Fumaric acid, but not maleic acid, is physiologically active. The *cis* isomer has the 2 more "bulky" groups on the same side of the double bond (Latin *cis* = on this side). If they are on opposite sides of the double bond, the **trans** isomer (Latin *trans* = across) is produced.

Introduction of **trans** double bonds in an otherwise saturated hydrocarbon chain deforms the shape of the molecule relatively little. A **cis** double bond, by contrast, entirely changes its shape. The reader can thus appreciate why cis and trans isomers of a com-

H OH H₃C CH₃ H₂N H — **Erythro (one enantiomer)**

HO H H₃C CH₃ H₂N H — **Threo (one enantiomer)**

FIG 4. Sawhorse representations of the erythro and threo isomers of 3-amino-2-butanol. The **threo** and **erythro** refer to the relative positions of –OH and NH_2 groups. Note that there are 3 ways to stagger C_2 with respect to C_3. That shown represents a structure with the bulky CH_3 groups oriented as far away from each other as possible.

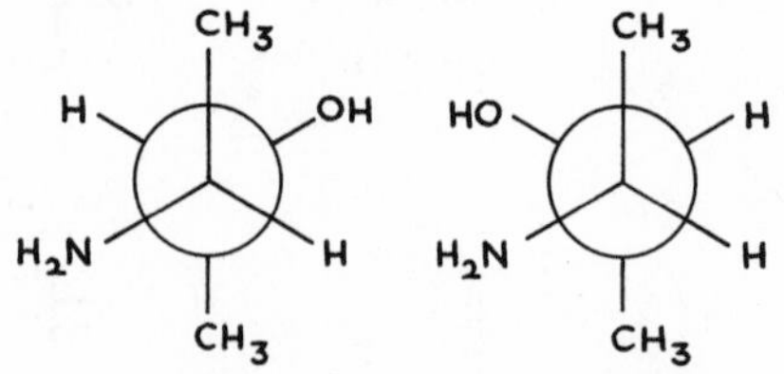

CH_3, H, OH, H_2N, H, CH_3 — **Erythro (one enantiomer)**

CH_3, HO, H, H_2N, H, CH_3 — **Threo (one enantiomer)**

FIG 5. Newman projection formulas for the erythro and threo isomers of 3-amino-2-butanol.

*The student is urged to purchase an inexpensive set of models ($10 or less). These will prove invaluable in studying the chemistry of sugars, amino acids, and steroids in particular.

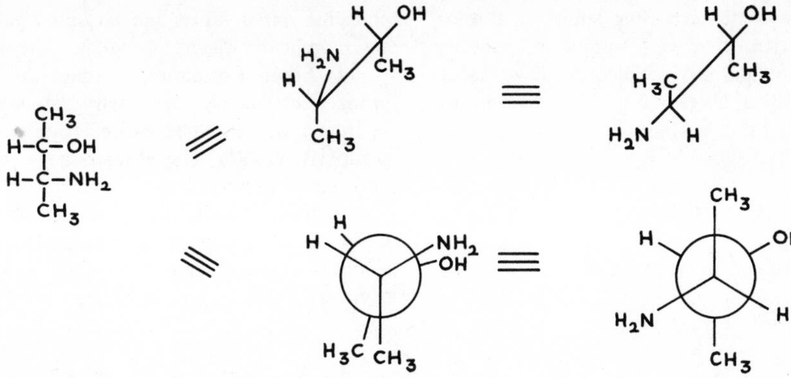

(Erythro) "Eclipsed" "Staggered"

FIG 6. Transformation from Fisher to sawhorse or Newman formula.

pound are not interchangeable in cells. Membranes composed of trans and cis isomers would have entirely different shapes. Enzymes acting on one isomer might be expected to be entirely inert with the other.

Again, the usual formulas fail to represent the actual shape of the molecules. Portions of the hydrocarbon backbone of a saturated fatty acid such as stearic acid ($CH_3[CH_2]_{16}COOH$) and of the cis and trans isomers of an 18-carbon unsaturated fatty acid ($CH_3[CH_2]_7CH{=}CH[CH_2]_7COOH$) are represented in Fig 7.

Alcohols

Alcohols have both a polar (hydroxy, –OH) and a nonpolar (alkyl) character. They are thus best regarded both as **hydroxylated hydrocarbons** and as **alkyl derivatives of water.** Although alcohols with up to 3 carbon atoms are infinitely soluble in water, water

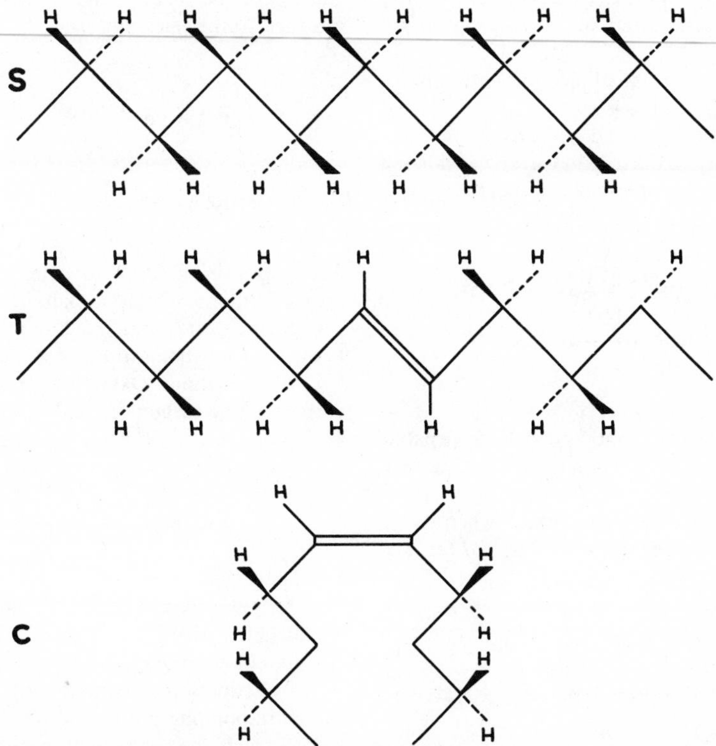

FIG 7. Representation of portions of the hydrocarbon backbones of a saturated fatty acid *(S)*, an unsaturated fatty acid with a single cis double bond *(C)*, and one with a single trans double bond *(T)*. Bonds drawn as solid lines are in the plane of the paper. Bonds drawn as dotted lines project behind, and those drawn ◄ project in front of the plane of the paper.

solubility decreases with increasing length of the carbon chain, ie, with increasing nonpolar character. Primary, secondary, and tertiary alcohols have respectively one, 2, and 3 alkyl groups attached to the carbon atom bearing the –OH group.

$$CH_3CH_2CH_2CH_2-OH$$

Primary butyl alcohol

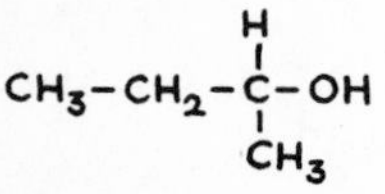

$$CH_3-CH_2-\underset{CH_3}{\overset{H}{C}}-OH$$

Secondary butyl alcohol

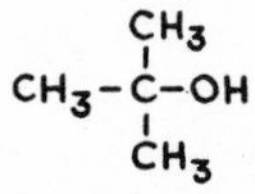

$$CH_3-\underset{CH_3}{\overset{CH_3}{C}}-OH$$

Tertiary butyl alcohol

Both monohydric (one –OH group) and polyhydric (more than one –OH group) alcohols are of physiologic significance. The sugars (see Chapter 1) are derivatives of polyhydric alcohols. Cyclic or ring-containing alcohols such as the sterols (see pp 17, 18, and 442) or inositol (see p 19) are polyhydric alcohols which also occur in nature. Their highly polar character makes polyhydric alcohols far more water-soluble than corresponding monohydric alcohols with equivalent numbers of carbon atoms. Thus, even polyhydric alcohols with 6 or more carbon atoms (eg, sugars) are highly water-soluble.

Some chemical reactions of alcohols with physiologic analogies include:

A. Oxidation: Primary and secondary (but not tertiary) alcohols are oxidized by strong oxidizing agents to carboxylic acids or ketones, respectively:

PRIMARY:

$$R-CH_2OH \xrightarrow{[O]} R-COOH$$

SECONDARY:

$$R_2-CHOH \xrightarrow{[O]} R_2C=O$$

Tertiary alcohols, which cannot be dehydrogenated without rupture of a C–C bond, are not readily oxidized.

B. Esterification: An ester is formed when water is split out between a primary, secondary, or tertiary alcohol and an acid.

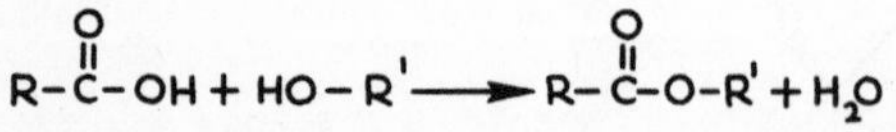

$$R-\overset{O}{\overset{\|}{C}}-OH + HO-R' \longrightarrow R-\overset{O}{\overset{\|}{C}}-O-R' + H_2O$$

Many lipids (see p 14) contain carboxylic ester linkages. The acid may be an organic acid, as shown above, or an inorganic acid. Thus, the esters of H_3PO_4 (see phosphorylated sugars and phospholipids) and H_2SO_4 are of great significance in biochemistry.

C. Ether Formation: Ethers are derivatives of primary, secondary, or tertiary alcohols in which the hydrogen of the –OH group is replaced by an alkyl group (R–O–R′). The ether linkage is comparatively uncommon in living tissues.

Sulfur, which is in the same group of the periodic table as oxygen, forms similar compounds. Thioalcohols (mercaptans), thioesters, and thioethers all occur in nature.

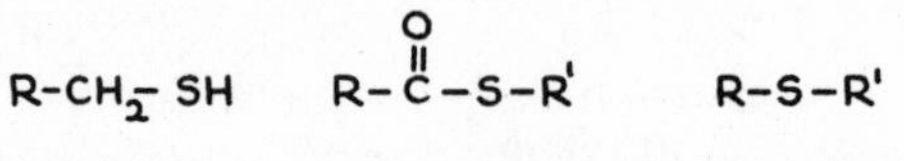

$$R-CH_2-SH$$

Thioalcohol

$$R-\overset{O}{\overset{\|}{C}}-S-R'$$

Thioester

$$R-S-R'$$

Thioether

In addition, the disulfides,

$$R-S-S-R'$$

which have no oxygen counterpart in biology, play an important role in protein structure (see p 41).

Aldehydes & Ketones

Aldehydes and ketones possess the strongly reducing carbonyl group >C=O (see pp 7 and 11). Aldehydes have one and ketones 2 alkyl groups attached to the carbon bearing the carbonyl group:

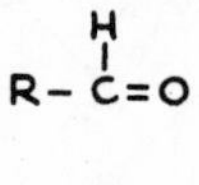

$$R-\overset{H}{\overset{|}{C}}=O$$

Aldehyde

$$\begin{matrix} R \\ & C=O \\ R' \end{matrix}$$

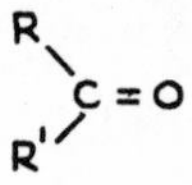

Ketone

The sugars, in addition to being polyhydric alcohols, are also either aldehydes or ketones.

Some chemical reactions of aldehydes and ketones of biochemical interest include the following:

A. Oxidation: Oxidation of an aldehyde to the corresponding carboxylic acid

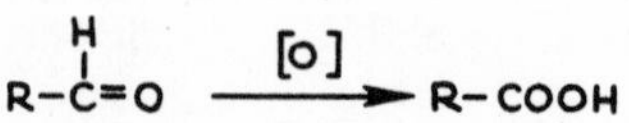

$$R-\overset{H}{\overset{|}{C}}=O \xrightarrow{[O]} R-COOH$$

Ketones are not readily oxidized since, like tertiary alcohols, they cannot lose hydrogen without rupture of a C–C bond.

B. Reduction: Reduction of an aldehyde yields the corresponding primary alcohol, and reduction of a ketone yields the corresponding secondary alcohol.

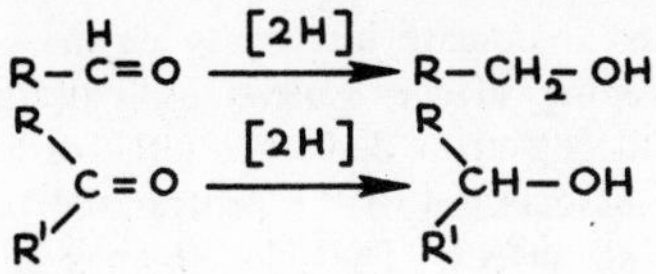

$$R{-}\overset{H}{C}{=}O \xrightarrow{[2H]} R{-}CH_2{-}OH$$

$$\begin{matrix}R\\ \quad\diagdown\\ \quad C{=}O\\ \quad\diagup\\ R'\end{matrix} \xrightarrow{[2H]} \begin{matrix}R\\ \quad\diagdown\\ \quad CH{-}OH\\ \quad\diagup\\ R'\end{matrix}$$

C. Hemiacetal and Acetal Formation: Under acidic conditions, aldehydes can combine with one or 2 of the hydroxyl groups of an alcohol, forming, respectively, a hemiacetal or an acetal:

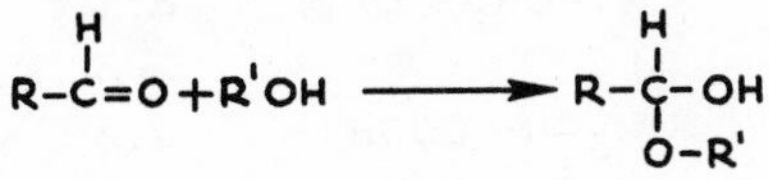

$$R{-}\overset{H}{C}{=}O + R'OH \longrightarrow R{-}\overset{H}{\underset{O-R'}{C}}{-}OH$$

Hemiacetal

$$R{-}\overset{H}{C}{=}O + 2R'OH \longrightarrow R{-}\overset{H}{\underset{OR'}{C}}{-}OR' + H_2O$$

An acetal

The carbonyl and alcohol functions may be part of the same molecule. For example, the aldose (aldehyde) sugars exist in solution primarily as internal hemiacetals.

Aldehydes may also form thiohemiacetals and thioacetals with thioalcohols:

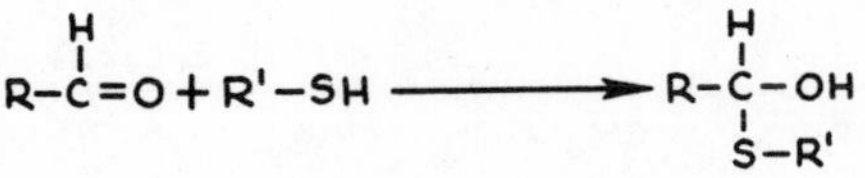

$$R{-}\overset{H}{C}{=}O + R'{-}SH \longrightarrow R{-}\overset{H}{\underset{S-R'}{C}}{-}OH$$

A thiohemiacetal

Thiohemiacetals are involved as enzyme-bound intermediates in the enzymic oxidation of aldehydes to acids.

D. Aldol Condensation: Under alkaline conditions, aldehydes and, to a lesser extent, ketones undergo condensation between their carbonyl and their α-carbon atoms to form aldols or β-hydroxy aldehydes or ketones.

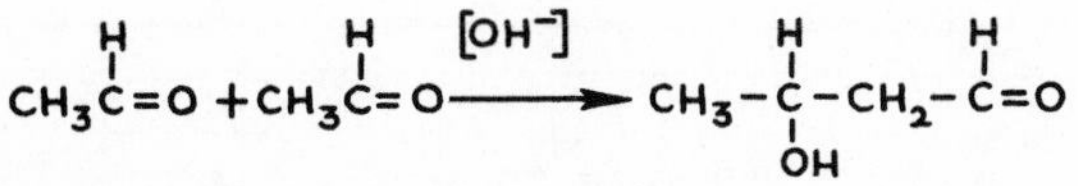

$$CH_3\overset{H}{C}{=}O + CH_3\overset{H}{C}{=}O \xrightarrow{[OH^-]} CH_3{-}\overset{H}{\underset{OH}{C}}{-}CH_2{-}\overset{H}{C}{=}O$$

The β-hydroxy acids derived from these are of great importance in fatty acid metabolism.

Carboxylic Acids

Carboxylic acids have both a carbonyl (>C=O) and a hydroxyl group on the same carbon atom. They are typical weak acids and only partially dissociate in water to form a hydrogen ion (H^+) and a **carboxylate anion** ($R.COO^-$) with the negative charge shared equally by the 2 oxygen atoms.

Some reactions of carboxylic acids of physiologic interest include the following:

A. Reduction: Complete reduction yields the corresponding primary alcohol.

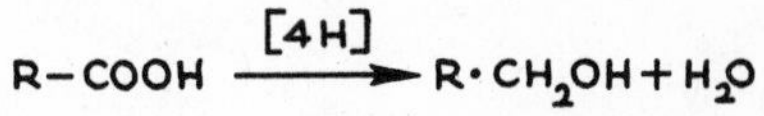

$$R{-}COOH \xrightarrow{[4H]} R{\cdot}CH_2OH + H_2O$$

B. Ester and Thioester Formation: See alcohols, p 505.

C. Acid Anhydride Formation: A molecule of water is split out between the carboxyl groups of 2 acid molecules.

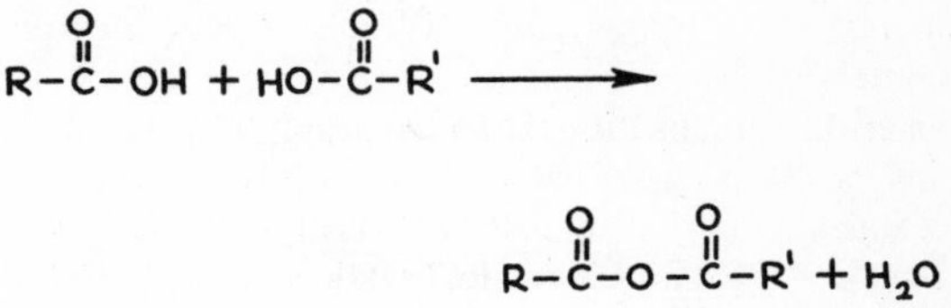

$$R{-}\overset{O}{\overset{\|}{C}}{-}OH + HO{-}\overset{O}{\overset{\|}{C}}{-}R' \longrightarrow R{-}\overset{O}{\overset{\|}{C}}{-}O{-}\overset{O}{\overset{\|}{C}}{-}R' + H_2O$$

When both acid molecules are the same, a **symmetric anhydride** is produced. Different molecules yield **mixed anhydrides.** Anhydrides found in nature include those of phosphoric acid (in ATP) and the mixed anhydrides formed from phosphoric acid and a carboxylic acid, eg:

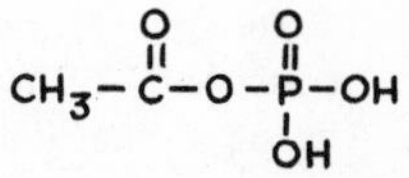

$$CH_3{-}\overset{O}{\overset{\|}{C}}{-}O{-}\overset{O}{\overset{\|}{\underset{OH}{P}}}{-}OH$$

Acetyl phosphate

D. Salt Formation: Carboxylic acids react stoichiometrically (equivalent for equivalent) with bases to form salts which are 100% dissociated in solution.

E. Amide Formation: Splitting out a molecule of water between a carboxylic acid and an amine forms an amide:

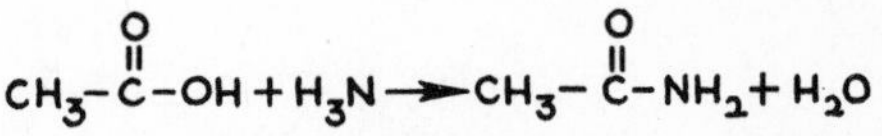

$$CH_3{-}\overset{O}{\overset{\|}{C}}{-}OH + H_3N \longrightarrow CH_3{-}\overset{O}{\overset{\|}{C}}{-}NH_2 + H_2O$$

Acetamide

Particularly important amides are **peptides** (see p 42), formed from the amino group of one amino acid and the carboxyl group of another.

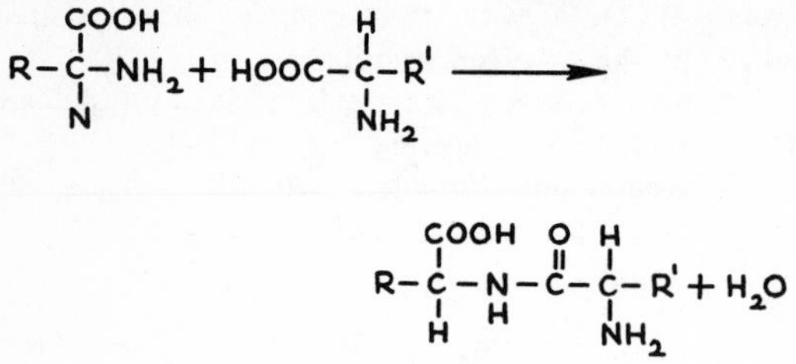

$$R-\underset{N}{\overset{COOH}{C}}-NH_2 + HOOC-\underset{NH_2}{\overset{H}{C}}-R' \longrightarrow$$

$$R-\underset{H}{\overset{COOH}{C}}-\underset{H}{N}-\overset{O}{\overset{\|}{C}}-\underset{NH_2}{\overset{H}{C}}-R' + H_2O$$

Peptide bond

The acid strengths of carboxylic acids are expressed in terms of pK_a values (see also p 493). For the dissociation

$$R-COOH \longrightarrow R-COO^- + H^+$$

we may write the dissociation constant

$$K_a = \frac{[RCOO^-]\,[H^+]}{[RCOOH]}$$

The pK_a value is defined as $-\log K_a$. Weak acids have high pK_a values and strong acids low pK_a values. For example:

	K_a	pK_a
Acetic acid	1.86×10^{-5}	4.73
Chloroacetic acid	1.40×10^{-3}	2.86
Trichloroacetic acid	1.30×10^{-1}	0.89

pK_a values may be determined by adding one equivalent of alkali to 2 equivalents of acid and measuring the resulting pH. The pK_a value is thus the pH of a half-neutralized solution.

Inductive Effects of Neighboring Groups on Acid Strength

The electrons of a covalent bond between 2 dissimilar atoms tend to associate with the more electronegative (electron-attracting) atom. The result is a dipole:

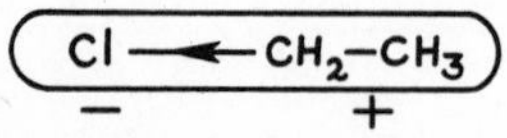

$$\underset{-}{Cl} \longleftarrow \underset{+}{CH_2-CH_3}$$

The arrow $\longleftarrow$ represents the direction of electron "drift." Factors which increase the electron density on the carboxyl group from which the positively charged proton must dissociate hinder its leaving and have an **acid weakening effect.** Conversely, anything which decreases the electron density on the carbonyl group will assist dissociation of the proton and have an **acid strengthening effect.** The closer an electronegative atom is to the carboxyl group, the more pronounced the acid strengthening effect. These effects are readily seen with the strongly electronegative atom chlorine:

	pKa
CH_3CH_2COOH	4.9
$\underset{Cl}{CH_2}-CH_2-COOH$	4.1
$CH_3\underset{Cl}{CH}-COOH$	2.8

Alkyl groups supply electrons, but in a less dramatic manner:

	pKa
CH_3COOH	4.7
CH_3-CH_2COOH	4.9
$(CH_3)_3C-COOH$	5.0

Charged groups may either supply or withdraw electrons:

		pKa FOR CARBOXYL
ACETIC ACID	CH_3COOH	4.7
GLYCINE	$\underset{NH_3^+}{CH_2}-COOH$	2.3
GLUTAMIC ACID	$HOOC-\underset{NH_3^+}{CH_2}-CH_2-\underset{COOH}{CH_2}$	2.2

The second carboxyl dissociation of glutamic acid ($pK_a = 4.2$) is intermediate in acid strength between that of glycine and acetic acid since the molecule has both + and − charged groups.

The carbonyl group and hydroxyl groups also exert inductive effects and are acid strengthening:

		pKa
PROPIONIC ACID	CH_3CH_2COOH	4.9
LACTIC ACID	$CH_3CHOHCOOH$	2.9
PYRUVIC ACID	$CH_3\underset{O}{\underset{\|}{C}}-COOH$	2.7

Amines

Amines, which are alkyl derivatives of ammonia, are usually gases or fairly volatile liquids with odors similar to ammonia, but more "fish-like." Primary, secondary, and tertiary amines are formed by replacement of one, 2, or 3 of the hydrogens of ammonia, respectively.

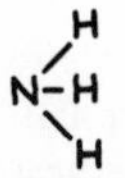

NH_3 (H–N(–H)–H) — **Ammonia**

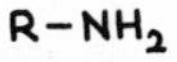

$R-NH_2$ — **Primary amine**

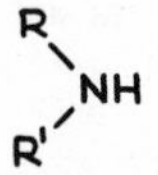

$R(R')NH$ — **Secondary amine**

$R(R')(R'')N$ — **Tertiary amine**

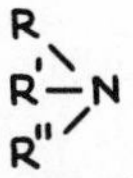

Ammonia in solution exists in both charged and uncharged forms:

$$NH_3 + H^+ \rightleftharpoons NH_4^+$$

Ammonia **Ammonium ion**

The amines behave in an entirely analogous way:

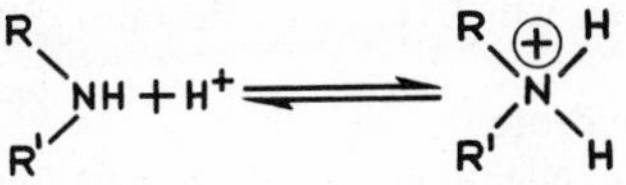

$$R(R')NH + H^+ \rightleftharpoons R(R')N^{\oplus}H_2$$

An amine **An alkylammonium ion**

The uncharged forms are bases, ie, proton acceptors, whereas the charged forms are acids, ie, proton donors. The relative strengths of various amines may be expressed by the pK_a values for the dissociation:

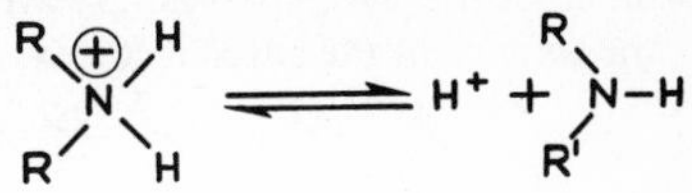

$$R(R')N^{\oplus}H_2 \rightleftharpoons H^+ + R(R')N-H$$

Some prefer to use pK_b values for amines. Conversion of pK_b to pK_a is accomplished from the relationship:

$$pK_a = 14 - pK_b$$

The pK_a values show that the aliphatic amines are weaker acids, or stronger bases, than ammonia. They also show that **at pH 7.4 essentially all of an aliphatic amine is in the charged form.** In body fluids, therefore, these amines are associated with an anion such as Cl^-.

The aromatic amines such as aniline and nitrogen atoms of cyclic amines such as pyridine or purines and pyrimidines are, by contrast, moderately strong acids. Aromatic amines, therefore, exist for the most part in the dissociated or uncharged form at pH 7.4. Their acidity is attributable to their aromatic "electron sink" which reduces the negative charge on the nitrogen and facilitates dissociation of a proton.

The reaction of amines with acids to form amides was mentioned on p 507. Amines and amine derivatives are involved in many important reactions of amino acids, lipids, and nucleic acids. Many drugs and other pharmacologically active compounds are amines.

TABLE 3. Acid dissociation constants of amines.*

	Acid Form	pKa
Ammonia	NH_4^+	9.26
Methylamine	$CH_3NH_3^+$	10.64
Dimethylamine	$(CH_3)_2NH_2^+$	10.72
Trimethylamine	$(CH_3)_3NH^+$	9.74
Aniline	$C_6H_5NH_3^+$	4.58
Pyridine	$C_5H_5NH^+$	5.23

*From Weast RC (editor): *Handbook of Chemistry & Physics,* 46th ed. Chemical Rubber Publishing Co., 1965–1966.

• • •

ABBREVIATIONS & ALTERNATIVE TERMINOLOGY FREQUENTLY USED IN PHYSIOLOGICAL CHEMISTRY

ACTH (adrenocorticotropic hormone, corticotropin): A hormone of the anterior pituitary which influences the activity of the adrenal cortex.

Acyl-Co A: An acyl derivative of Co A (eg, butyryl-Co A)

ADH (antidiuretic hormone): Vasopressin; a hormone of the posterior pituitary which influences reabsorption of water by the renal tubule.

ADP (adenosine diphosphate): A nucleotide which participates in high-energy phosphate transfer.

AHG (antihemophilic globulin): A thromboplastic factor a deficiency of which is the cause of "classical" hemophilia (hemophilia A).

AMP (adenosine monophosphate): A nucleotide which participates in high-energy phosphate transfer.

2′-AMP, 3′-AMP (5′-AMP), etc: the 2′-, 3′-, (and 5′- when needed for contrast) monophosphates of adenosine. Similar conventions in nomenclature are used in designations of other purine or pyrimidine nucleotides. The number refers to the linkage position of the phosphate, the prime mark indicating that the phosphate is attached at the numbered position on the sugar moiety of the nucleotide, not at a position on the purine (or pyrimidine) base.

APL (anterior pituitary-like hormone, chorionic gonadotropin): A hormone which is excreted in the urine in pregnancy.

ATP (adenosine triphosphate): A nucleotide which is an important source of high-energy phosphate.

ATPase (adenosine triphosphatase): An enzyme which catalyzes the dephosphorylation of ATP.

BEI: Butanol-extractable iodine.

BMR: Basal metabolic rate.

BSA: Bovine serum albumin.

BSP (Bromsulphalein, sulfobromophthalein): A dye used in tests of liver function.

CMP (cytidine monophosphate): 5′-Phosphoribosyl cytosine.

Co A.SH (free [uncombined] coenzyme A): A pantothenic acid-containing nucleotide which functions in the metabolism of fatty acids, ketone bodies, acetate, and amino acids.

Co A.S.C(=O).CH$_3$ (acetyl-Co A, "active acetate"): The form in which acetate is "activated" by combination with coenzyme A for participation in various reactions.

Co I (coenzyme I): Now called NAD (q.v.).

Co II (coenzyme II): Now called NADP (q.v.).

Co III (coenzyme III): A nicotinamide-containing nucleotide which functions in oxidation of cysteine.

Compound
- **A** (11-dehydrocorticosterone):
- **B** (corticosterone):
- **E** (cortisone):
- **F** (hydrocortisone, cortisol):
- **S** (11-deoxycortisol):

} Adreno-cortical steroids

dAMP (deoxy adenosine monophosphate): A nucleotide (adenine-deoxyribose-phosphate) in which the sugar is deoxyribose, as indicated by the prefix d. Where no prefix is used, the sugar in the nucleotide is assumed to be ribose. A similar convention in nomenclature is used for other nucleotides, eg, dGMP (deoxy guanosine monophosphate), dCMP (deoxy cytidine monophosphate), etc.

DCA (DOCA, deoxycorticosterone acetate): A salt-retaining steroid.

DIT: Diiodotyrosine.

DNA (deoxyribonucleic acid): The characteristic nucleic acid of the nucleus.

DOPA: Dioxyphenylalanine or dihydroxyphenylalanine.

DPN: See NAD.

EDTA: Ethylenediaminetetraacetate, a chelating compound used to complex divalent metals.

EFA (essential fatty acids): Polyunsaturated fatty acids essential for nutrition.

FAD (flavin adenine dinucleotide): A riboflavin-containing nucleotide which participates as a coenzyme in oxidation-reduction reactions.

FADH$_2$: Reduced form of FAD.

FFA: Unesterified, free fatty acid (also called NEFA, UFA).

FH$_4$: Tetrahydrofolic acid.

Figlu (formiminoglutamic acid): A metabolite of histidine which can be identified in the urine of folic acid-deficient individuals.

FMN (flavin mononucleotide): A riboflavin-containing cofactor in cellular oxidation-reducing systems.

FSH (follicle-stimulating hormone): A gonadotropic hormone of the anterior pituitary.

GFR: Glomerular filtration rate.

GMP (guanosine monophosphate): 5′-Phosphoribosyl guanine.

GOT: See SGOT.

GPT: See SGPT.

GSH (glutathione, reduced):
GS-SG (glutathione, oxidized):
} A tripeptide, γ-glutamyl-cysteinyl-glycine.

GTP: Guanosine triphosphate.

HMP shunt (hexose monophosphate shunt): An alternate pathway of carbohydrate metabolism.

IDP (inosine diphosphate): A hypoxanthine-containing nucleotide which participates in high-energy phosphate transfer.

IMP (inosine monophosphate): 5′-Phosphoribosyl hypoxanthine.

ITP (inosine triphosphate): A deamination product of ATP. Functions similarly to ATP as a source of high-energy phosphate.

IU: International unit(s), as in expressing content of vitamins in foods.

LDH (lactic acid dehydrogenase): An enzyme whose activity may be measured in serum for diagnosis of certain acute diseases, eg, acute myocardial infarction.

LH (luteinizing hormone): A gonadotropic hormone of the anterior pituitary.

LTH (luteotropic hormone): A hormone of the anterior pituitary possibly identical with lactogenic hormone.

MAO: Monoamine oxidase.

MIT: Monoiodotyrosine.

mRNA: Messenger RNA.

MSH (melanocyte-stimulating hormone): A hormone of the middle lobe of the pituitary which increases deposition of melanin by the melanocytes.

NAD (nicotinamide adenine dinucleotide): Formerly termed DPN (diphosphopyridine nucleotide). A nicotinamide-containing nucleotide which functions in electron and hydrogen transfer in oxidation-reduction reactions.

NADP (nicotinamide adenine dinucleotide phosphate): Formerly termed TPN (triphosphopyridine nucleotide). A nucleotide with functions and structure similar to those of NAD.

NEFA (nonesterified fatty acids): The major form of circulating lipid used for energy.

NMN: Nicotinamide mononucleotide.

NPN: Nonprotein nitrogen.

OAA: Oxaloacetic acid.

OD: Optical density.

PABA (para-aminobenzoic acid): A factor among the B vitamins.

PBI (protein-bound iodine): Thyroid hormone iodine of the plasma.

PGH (pituitary growth hormone): The growth hormone of the anterior pituitary.

P_i: Inorganic (ortho) phosphate.

P.Pi: Pyrophosphate.

ΨMP (pseudouridine monophosphate): 5′-Phosphoribosylpseudouracil.

PSP (phenolsulfonphthalein): A dye used for measurement of renal tubular function.

PTA: Plasma thromboplastin antecedent (clotting factor XI).

PTC (plasma thromboplastin component): A thromboplastic factor a deficiency of which is the cause of hemophilia B.

RNA (ribonucleic acid): The characteristic nucleic acid of cytoplasm involved in protein synthesis.

rRNA: Ribosomal RNA.

SGOT (serum glutamic oxaloacetic transaminase): An enzyme often measured in serum for diagnosis of acute myocardial infarction or acute liver disease.

SGPT (serum glutamic-pyruvic transaminase): An enzyme which may be measured in serum for diagnosis of certain types of liver disease.

SPCA: Serum prothrombin conversion accelerator.

sRNA: Soluble RNA (same as tRNA [transfer RNA]).

T_3, T_4 (triiodothyronine, tetraiodothyronine [thyroxine]): Hormones of the thyroid gland.

TBG: Thyroxine-binding globulin.

TMP (thymidine monophosphate): 5′-Phosphoribosyl thymine.

TPN: See NADP.

TPP (thiamine pyrophosphate): The thiamine-containing coenzyme which is a cofactor in decarboxylation.

Tris (tris[hydroxymethyl]aminomethane): A buffer frequently used in biologic preparations for in vitro studies, as with enzymes.

tRNA (transfer RNA): Same as sRNA.

TSH (thyroid-stimulating hormone): A hormone of the anterior pituitary which influences the activity of the thyroid.

UDPG (uridine diphosphoglucose): An intermediary in glycogen synthesis.

UDPGal: Uridine diphosphogalactose.

UDPGluc: Uridine diphosphoglucuronic acid.

UFA: Unesterified free fatty acids.

UMP (uridine monophosphate): 5′-Phosphoribosyl uracil.

UTP: Uridine triphosphate.

VMA: Vanilmandelic acid (3-methoxy-4-hydroxymandelic acid).

Index